COAL PRODUCTION AND PROCESSING TECHNOLOGY

T0172713

COAL PRODUCTION AND PROCESSING TECHNOLOGY

edited by
M.R. Riazi
Rajender Gupta

CRC Press
Taylor & Francis Group
Boca Raton London New York

CRC Press is an imprint of the
Taylor & Francis Group, an **informa** business

CRC Press
Taylor & Francis Group
6000 Broken Sound Parkway NW, Suite 300
Boca Raton, FL 33487-2742

First issued in paperback 2021

© 2016 by Taylor & Francis Group, LLC
CRC Press is an imprint of Taylor & Francis Group, an Informa business

No claim to original U.S. Government works

Version Date: 20150923

ISBN 13: 978-0-367-78330-3 (pbk)
ISBN 13: 978-1-4822-5217-0 (hbk)

This book contains information obtained from authentic and highly regarded sources. Reasonable efforts have been made to publish reliable data and information, but the author and publisher cannot assume responsibility for the validity of all materials or the consequences of their use. The authors and publishers have attempted to trace the copyright holders of all material reproduced in this publication and apologize to copyright holders if permission to publish in this form has not been obtained. If any copyright material has not been acknowledged please write and let us know so we may rectify in any future reprint.

Except as permitted under U.S. Copyright Law, no part of this book may be reprinted, reproduced, transmitted, or utilized in any form by any electronic, mechanical, or other means, now known or hereafter invented, including photocopying, microfilming, and recording, or in any information storage or retrieval system, without written permission from the publishers.

For permission to photocopy or use material electronically from this work, please access www.copyright.com (http://www.copyright.com/) or contact the Copyright Clearance Center, Inc. (CCC), 222 Rosewood Drive, Danvers, MA 01923, 978-750-8400. CCC is a not-for-profit organization that provides licenses and registration for a variety of users. For organizations that have been granted a photocopy license by the CCC, a separate system of payment has been arranged.

Trademark Notice: Product or corporate names may be trademarks or registered trademarks, and are used only for identification and explanation without intent to infringe.

Library of Congress Cataloging-in-Publication Data

Coal production and processing technology / edited by M.R. Riazi and Rajender Gupta.
pages cm
Includes bibliographical references and index.
ISBN 978-1-4822-5217-0 (hardcover : alk. paper)
1. Coal mines and mining. 2. Coal--Analysis. 3. Coal-handling. 4. Coal gasification. 5. Coal liquification. I. Riazi, M. R., editor. II. Gupta, Rajender, editor.

TN800.C635 2016
553.2'4--dc23 2015031034

Visit the Taylor & Francis Web site at
http://www.taylorandfrancis.com

and the CRC Press Web site at
http://www.crcpress.com

Dedication

To Shiva, Touraj, and Nazly

and

Rekha, Richa, and Raveen

Contents

SECTION III Coal Conversion Processes

SECTION IV Global Issues and Trends

Preface

Coal has been used as the most abundant and widely distributed source of energy for centuries and presently provides about 30% of the world's energy, second after oil. Coal alone provides 40% of current global electricity needs. The cost of electricity produced from coal is less than that produced from any other energy source. A significant proportion of coal is used for metallurgical processes. According to the International Energy Agency, in the twenty-first century, coal has been the fastest growing energy source; between 2000 and 2012 coal consumption increased by 60%, an average of 4% per year. For example, Japan announced the building of seven new coal-fired power plants in early 2015 to make up for the loss of power generation from nuclear plants. The main producers of coal are China, the United States, India, Australia, Indonesia, Russia, and South Africa, while the main importers of coal are China, Japan, South Korea, India, and Taiwan. While China produces 46% of the world's coal, it consumes more than 49% of global coal production.

Coal is a combustible solid organic rock, which is mainly composed of carbon, along with other elements such as hydrogen, oxygen, sulfur, and nitrogen. However, there are always some ash-forming minerals, such as quartz, clays, pyrites, and carbonates consisting of silicon, aluminum, iron, calcium, magnesium, sodium, potassium, sulfur, and phosphorus, responsible for almost all the operational and environmental issues in coal utilization. There are also some toxic trace elements, such as mercury, selenium, cromium, and antimony as well as some valuable elements such as germanium, gallium, uranium, molybdenum, and beryllium, which may also be present as traces in a coal sample.

Excessive use of coal has led to the utilization of low grade coals with higher mineral, sulfur, and moisture content, resulting in higher emissions of SO_x, NO_x, and trace elements apart from increased operational problems related to corrosion, erosion, and ash deposition. The biggest challenge related to coal has been greenhouse gas emissions, either as coalbed methane or as carbon dioxide in flue gas from combustion. Coal bed methane is currently a source of power. Carbon capture and storage is the most promising technique to reduce carbon dioxide emission from coal combustion; research in this area is growing.

It was due to the above-mentioned importance of coal as a source of energy and associated issues in its utilization that led us to preparing this book. The present book is the result of more than two years of collective effort with contributions from more than 40 scientists and experts around the globe. The book has 23 chapters and is divided into 4 sections: Section I: Coal Reserves and Characterization of Coal, Section II: Coal Mining and Production, Section III: Coal Conversion Processes, and Section IV: Global Issues and Trends. Section II includes a chapter on coal bed methane, whereas Section IV includes a chapter on carbon capture and storage, a chapter on evaluation of coal companies, and a chapter on future trends in coal technologies. Although we attempted to cover all aspects of coal production and processing, this proved to be a difficult task, in which some topics have received less attention than they deserve. Readers may find some minor overlaps between the chapters; each chapter focuses on a unique topic specific to that chapter.

We are grateful to a number of experts who helped us in reviewing some chapters with their constructive comments and suggestions. Among them we are particularly indebted to Peter Cain, T. K. Das, Bulent Erdem, John R. Grace, Cigdem Keles, Allan Kolkler, Paul M. Mathias, Sachin K. Sharma, Luis Silva, Robert Sutherland and Deepak Pudasainee. We are also thankful to Taylor & Francis Group for publication of this book and in particular to Allison Shatkin, the managing editor, who was helpful in every stage of this project; her initial contacts and encouragement were instrumental in taking such a major task. We appreciate the work of Ashley Weinstein, project coordinator at CRC Press, who reviewed and edited this manuscript.

M. R. Riazi
Kuwait University

Rajender Gupta
University of Alberta

Editors

M. R. Riazi, PhD, (www.RiaziM.com) is currently a professor and the chairman of chemical engineering at Kuwait University. He earned MSc and PhD degrees at Pennsylvania State University, where he served as an assistant professor of chemical engineering. He has also been a visiting scholar/faculty at Illinois (Chicago), Texas (Austin), McGill (Montreal), Waterloo (Canada), Wright State (Ohio), Norwegian University of Science and Technology (Trondheim), and various universities in the Middle East. He has written 140 publications, including 5 books and a solutions manual, as well as about 100 conference presentations, mainly in the fields of petroleum and chemical technology. He has been an invited speaker and consultant at more than 50 major oil companies and research centers worldwide and has conducted about 70 workshops and short courses in more than 40 countries. He is the founding editor and editor-in-chief of IJOGCT (London, UK) and an editor of the *Journal of Petroleum Science and Engineering*. He has been awarded distinguished researcher and teaching awards at several universities, including an award from the former Amir of Kuwait. He has also received a diploma of honor from the National (American) Petroleum Engineering Society and is a fellow of the American Institute of Chemical Engineers.

Rajender Gupta, PhD, is a professor of chemical and materials engineering at the University of Alberta. He graduated in 1972 from IIT Kharagpur and earned a PhD at the University of Newcastle. He has been researching in the general area of clean coal technologies for well over 35 years. He has published more than 250 technical and research papers, including about 100 in peer-reviewed international journals. He is a member on the editorial board of a number of international journals.

Dr. Gupta, before his current tenure at the University of Alberta, led several research projects at the Co-operative Research Centre for Coal in Sustainable Development at the University of Newcastle, Australia, on advanced coal characterization, mineral-to-ash transformation, and the performance prediction of coal-fired boilers.

Dr. Gupta has been leading coal research at the Canadian Centre of Clean Coal Carbon and Mineral Processing at the University of Alberta. His current research interests include upgrading of low-grade coals by reducing minerals and moisture, coal to liquid fuels, coal/biomass/petcoke/asphaltene utilization, gasification and hot gas clean up, oxy-firing, post-combustion capture, and underground coal gasification.

Contributors

C. Okay Aksoy
Department of Mining Engineering
Engineering Faculty
Dokuz Eylül University
Izmir, Turkey

Faisal S. AlHumaidan
Petroleum Research Center
Kuwait Institute for Scientific Research (KISR)
Safat, Kuwait

Tianhang Bai
School of Mechanical and Mining Engineering
University of Queensland
Brisbane, Australia

Bimala P. Baruah
Coal Chemistry Division
CSIR-North East Institute of Science and Technology
Jorhat, India

Hakan Basarir
Mining Engineering Department
Middle East Technical University
Ankara, Turkey

T. Gouri Charan
Central Institute of Mining and Fuel Research
CSIR, Dhanbad, India

Zhongwei Chen
School of Mechanical and Mining Engineering
University of Queensland
Brisbane, Australia

Arno de Klerk
Department of Chemical and Materials Engineering
University of Alberta
Edmonton, Canada

Nuray Demirel
Mining Engineering Department
Middle East Technical University
Ankara, Turkey

Ali Elkamel
Department of Chemical Engineering
University of Waterloo
Waterloo, Canada

Xing Fan
Key Laboratory of Coal Processing and Efficient Utilization
Ministry of Education
China University of Mining and Technology
Jiangsu, China

Bill Gunter
G BACH Enterprises Inc.
Edmonton, Canada

Rajender Gupta
Department of Chemical and Materials Engineering
University of Alberta
Edmonton, Canada

Mamoru Kaiho
National Institute of Advanced Industrial Science
Tokyo, Japan

Mark J. Kaiser
Center for Energy Studies
Louisiana State University
Baton Rouge, Louisiana

Celal Karpuz
Mining Engineering Department
Middle East Technical University
Ankara, Turkey

M. P. Ketris
Komi Science Center
Ural Division of the Russian Academy of Science
Moscow, Russia

Yoichi Kodera
National Institute of Advanced Industrial Science and Technology (AIST)
Tokyo, Japan

Halil Köse
Department of Mining Engineering
Engineering Faculty
Dokuz Eylül University
Izmir, Turkey

Zhan-Ku Li
Key Laboratory of Coal Processing and Efficient Utilization
Ministry of Education
China University of Mining and Technology
Jiangsu, China

Hisao Makino
Energy Engineering Research Laboratory
Central Research Institute of Electric Power Industry
Yokosuka, Japan

Zhejun Pan
Earth Science and Resource Engineering
CSIRO, Kensington, Australia

Vinay Prasad
Department of Chemical and Materials Engineering
University of Alberta
Edmonton, Canada

Deepak Pudasainee
Department of Chemical and Materials Engineering
University of Alberta
Edmonton, Canada

Mohan S. Rana
Petroleum Research Center
Kuwait Institute for Scientific Research (KISR)
Safat, Kuwait

L. Reijnders
IBED
University of Amsterdam
Amsterdam, The Netherlands

Ananya Saikia
Coal Chemistry Division
CSIR-North East Institute of Science and Technology
Jorhat, India

Binoy K. Saikia
Coal Chemistry Division
CSIR-North East Institute of Science and Technology
Jorhat, India

Vinod K. Saxena
Department of Chemical Engineering
and
Department of Fuel and Mineral Engineering
Indian School of Mines Dhanbad
Dhanbad, India

Gouthami Senthamaraikkannan
Department of Chemical and Materials Engineering
University of Alberta
Edmonton, Canada

Zarook Shareefdeen
Department of Chemical Engineering
American University of Sharjah
Sharjah, United Arab Emirates

Kenji Tanno
Energy Engineering Research Laboratory
Central Research Institute of Electric Power Industry
Yokosuka, Japan

Hari P. Tiwari
Research and Development Division
Coal and Coke Making Research Group
Tata Steel
Jamshedpur, India

Xian-Yong Wei
Key Laboratory of Coal Processing and Efficient Utilization
Ministry of Education
China University of Mining and Technology
Jiangsu, China

Raymond Yeung
Department of Chemical Engineering
University of Waterloo
Waterloo, Canada

Ya. E. Yudovich
Komi Science Center
Ural Division of the Russian Academy of Sciences
Moscow, Russia

Zhi-Min Zong
Key Laboratory of Coal Processing and Efficient Utilization
Ministry of Education
China University of Mining and Technology
Jiangsu, China

Section I

Coal Reserves and Their Characteristics

1 Nature and Chemistry of Coal and Its Products

Binoy K. Saikia, Ananya Saikia, and Bimala P. Baruah

CONTENTS

Abstract: Coal, a non-renewable resource, is far more plentiful than oil or gas with considerable amount of coal reserves worldwide. Coal and coal products play an important role in fulfilling the energy needs of the society. Basic knowledge about coal, its origin, and its formation will help in its future applications in power generation, metal processing, and chemical production. The greatest coal-forming period in geological history is during the Carboniferous era, while large deposits during the Permian, with lesser but still significant in the Triassic and Jurassic periods, and minor in Cretaceous. The formation of coal has been explained either by in situ or by drift origin of vegetable matter under different conditions. The continuing effects of temperature and pressure over millions of years progressively increased the maturity of coal formation, that is, from peat to anthracite. The process of coalification starts with the biochemical process of peatification/humification followed by geochemical process. The types of coal have been defined based on lithotypes, coal ranks, microlithotypes, and maceral contents. In this chapter, the chemical nature of coal and its products are summarized. It includes discussion on coal formation, types and nature of coals, coal ranks and classification, micro-constituents of coal, chemical composition of coal, heteroatoms in coal, coal chemical structure including molecular structure, analytical methods for coal characterization, rock types and mineral matters in coal, coal sampling (in situ and ex situ), coal analysis, elemental analysis of coal (C, H, N, O, Cl, Hg, etc.), analytical methods for coal analysis, coal products, chemical compositions of coal liquids, and other coal-derived products. The maceral composition, heteroatoms, mineralogy, and the presence of trace elements and rare earth elements in coals found in the different parts of the world are also reported. Emphasis is also given on X-ray diffraction technique for the structural characterization on coals, evaluation of the carbon stacking structure, estimating the size of the aromatic lamellae and the average distance between lamellae, and the mean bond distance. The contents of this chapter will be basically of help to the beginners and the coal fraternity in developing and designing advanced coal conversion technologies. Moreover, this chapter will serve as a basis for going through the subsequent chapters in the book.

1.1 INTRODUCTION

Coal is one of the abundant fossil fuels in most parts of the world. The largest coal reserves of the world are found in the United States (237,295 mt), Russia (157,010 mt), China (114,500 mt), Australia (76,400 mt), and India (60,600 mt) as on 2011 (World Energy Council 2013). Coal now becomes globally important in generating electricity, which could remain so for the next 200 years. It is a heterogeneous combustible sedimentary rock, which is mainly composed of carbon, hydrogen, and oxygen. It also contains other elements such as nitrogen, sulfur, and traces of mineral matter. This is the most plentiful resource of conventional energy, and it forms the backbone of the modern industrial civilization. Coal is also a source material for producing a wide range of chemicals, fertilizers, and liquid fuels. It currently provides 29.9% of the global primary energy needs and generates 41% of the world's electricity. Thus, coal is considered as the first source for electricity generation after oil and is considered to be the second source of primary energy in the world. However, the efficiency and sustainability of the processes utilized for processing of this resource will depend on an adequate, as well as advance, knowledge on the nature and chemistry of the coals (Saikia and Baruah 2008; Saikia et al. 2007a, b, c, 2009a, b, c). It is to be mentioned that coal appears black and rather homogenous at first sight; however, it consists of a heterogeneous type of chemistry.

1.2 ORIGIN AND FORMATION OF COAL

Coal originates from the prehistoric accumulation of vegetation in swamps and peat bogs followed by an aggregation of heterogeneous materials composed of organic materials with some amount of moisture and inorganic mineral matters. Two main theories have been proposed to explain the origin and formation of coal. The first theory, which supposes growth-in-place of vegetable material, is called the *autochthonous theory*. It stated that the plants that compose the coal were accumulated in large freshwater swamps or peat bogs during many thousands of years. The second theory suggests that coal strata accumulated from plants, which had been rapidly transported and deposited under flood conditions afterward covered with sediments such as sands or clays. It claims transportation of vegetable debris and is called as the *allochthonous theory*.

In general, the formation of the coal can be attributed to the accumulation and preservation of plant remains in some areas of the earth where favorable climate prevails for the luxuriant growth of flora (see Figure 1.1). Different coals undergo different degrees of chemical and physical alteration of the parent organic debris under different temperature and pressure conditions. Peat is the precursor of coal, and is formed in the first phases. The formation of peat is controlled by several factors, including: (1) the evolutionary development of plant life and (2) the physical conditions of the area. Warm and moist climates are thought to produce broad bands of bright coal. Cooler and temperate climate, on the other hand, are thought to produce dirty coal with relatively little brightness.

The whole process of coal formation from such type of plant materials is a very complex and a long-term process, which involves physical, chemical, and biochemical reactions. Thus, the nature and chemistry of coal depend upon the properties of different constituents present in the mass of plants and vegetables, the nature and extent of the changes on which these

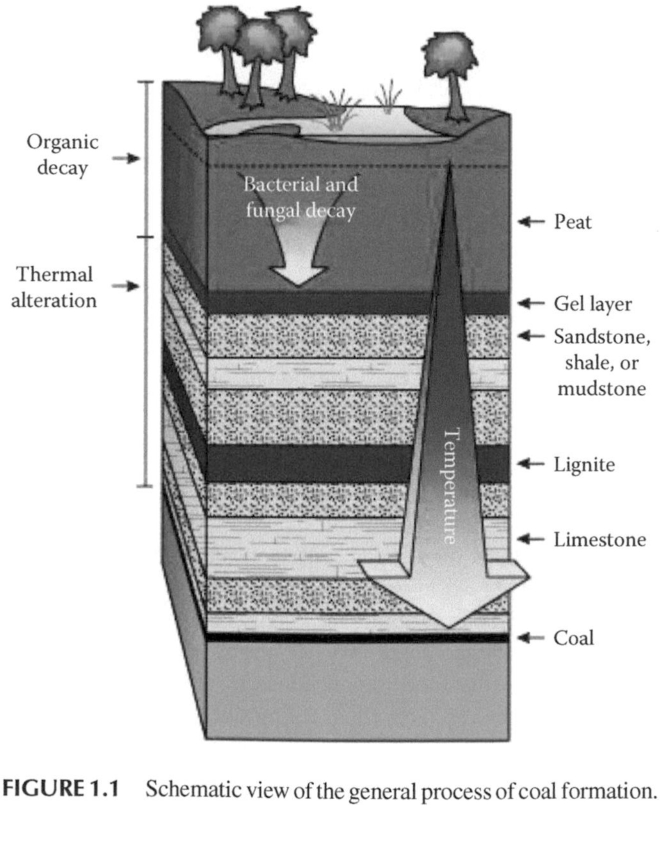

FIGURE 1.1 Schematic view of the general process of coal formation.

constituents have undergone since their deposition, as well as the nature and quantity of the inorganic matter present. This process is commonly called as *coalification*. The general sequence of coalification is from lignite to sub-bituminous to bituminous to anthracite. The process of coalification starts with the biochemical process of peatification of the dead plant materials. It is continued initially biochemically and then as a very complex geochemical process when peatification has ceased and the peat has been covered with other sediments. Depending upon the degree of coalification, different types of coal are formed. The main factors that are decisive for the coalification are primarily time, pressure, and temperature.

In the biochemical stage, the plant materials are decomposed and degraded under atmospheric oxidation by fungi and aerobic bacteria, which are further transformed into coal micro-constituents under the main controlling factors such as hydrogen-ion (pH) concentration and redox potential (Eh) of the medium (Chandra et al. 2000). With increased burial, bacteriological activity ceases and is considered absent at depths greater than 10 m (Stach et al. 1982). The quality of the coal produced also depends on the nature and extent of the decomposition and degradation of the plant material. The thickness of the coal seams can be determined from the rate of accumulation and subsidence of the coal-forming basins. Slower subsidence and greater accumulation of plant materials leads to greater thickness of coal seams. The coal formation is also related to the tectonic disturbances, which in turn help in the transformation by bringing sufficient overburden pressure and temperature gradient required for the coal generation. The geochemical or metamorphic stage of the coalification process involves progressive changes through increase in carbon content and decrease in the hydrogen and oxygen contents, resulting in a loss of volatiles within the coal. This, together with continued water loss and compaction, results in the reduction of the coal volume. There is reduction in moisture content and increase in methane–to–carbon dioxide ratio with the increase of coal rank. These changes in the physical and chemical properties of the coal are in reality the changes to the inherent coal constituents during coalification. The optically homogeneous discrete organic material in coal called as *macerals* are produced during peatification and coalification, while the inorganic materials consist primarily of mineral matter, chiefly clay minerals, quartz, carbonates, sulfides, and sulfates; many other matters in very small quantities were accumulated as well (Ting 1982). The total bulk of inorganic constituents present in coal ranges from a few percent to more than 50%. If the inorganic constituent is >50 wt%, it is classified as *carbonaceous shale*. Schopf (1956) has classified aggregates containing organic material amounting to >50 wt% and >70 vol.% as coal (Ting 1982). In a coalification process, the degradation of peat occurs through complex chemical reactions and through the activity of microorganisms. As peat formation and coalification proceeds, living biomass, which have more or less open structures, are broken down, and new compounds, primarily aromatic and hydro-aromatic, are produced. These compounds are connected by cross-linked oxygen, sulfur, and other functional groups such as methylene. However, the initial chemical changes of vegetable matters in the coalification process have been debated since the beginning of this century, with a cellulose theory and a lignin theory (Ghosh and Prelas 2009). A number of researchers concluded that the cellulose in plants was the main path toward the ultimate formation of the coal. Both the theories have been reviewed and their applicability to various deposits of the world had been discussed. However, it was concluded that a single theory could not be applied to explain these deposits (Ghosh and Prelas 2009).

Time is another important factor, which plays a significant role in coalification. It has been suggested that the process from accumulation to coal formation requires at least one million years (Chandra et al. 2000). According to the geologists, sufficient coal deposits were formed in the Upper Palaeozoic era ranging in age from 358.9 $\pm$ 0.4 to 252.2 $\pm$ 0.5 million years, particularly in the Carboniferous and Permian periods. The three major episodes of coal accumulation were established within the geological column from the Carboniferous to Quartenary periods (Figure 1.2). The first episode took place during the Late Carboniferous to Early Permian periods, where coals formed the bulk of the black coal reserves of the world. The coals formed during these periods are of high rank undergoing significant structural changes. These coal deposits extend across the Northern Hemisphere from Canada and the United States, through Europe and the Commonwealth of Independent States to the Far East. In the Southern Hemisphere, these Carboniferous–Permian coals of Gondwanaland are preserved in South America, Africa, the Indian subcontinent, Southeast Asia, Australasia, and Antarctica. The second episode occurred during the Jurassic–Cretaceous period and is present in Canada, the United States, China and the Commonwealth of Independent States (Russia included). The third major episode occurred during the Paleogene–Neogene periods, where coals formed are mainly lignite to anthracitic types. These coals form the bulk of the world's brown coal reserves. Paleogene–Neogene coals are also found worldwide, and are the focus of current exploration and production as the traditional Carboniferous coalfields become depleted or geologically too difficult to mine (Thomas 2013). The coal deposits found in Turkey are found to be of Oligo–Miocene in age in accordance with its fossil findings (Tozsin 2014).

1.3 TYPES OF COALS

Based on the macroscopically recognizable components termed as *lithotypes* (Stach et al. 1982), coals are divided into two main groups: sapropelic and humic coals. Sapropelic coals are derived from variety of microscopic plant debris such as resins, waxes, or fats under sub-oxic to anoxic conditions and have a homogeneous appearance. These coals are characteristically fine-grained, homogeneous, dark in color and display a marked conchoidal fracture (Thomas 2013). These deposits are usually lenticular in shape, local in extent, and occur at the top of a coal bed (Parks 1952). They may occur in association with humic coals or as individual coal layers. Humic coals are composed of a diversified mixture of macroscopic plant debris such as wood, bark, and leaves, deposited under oxic to sub-oxic conditions and they typically have a banded

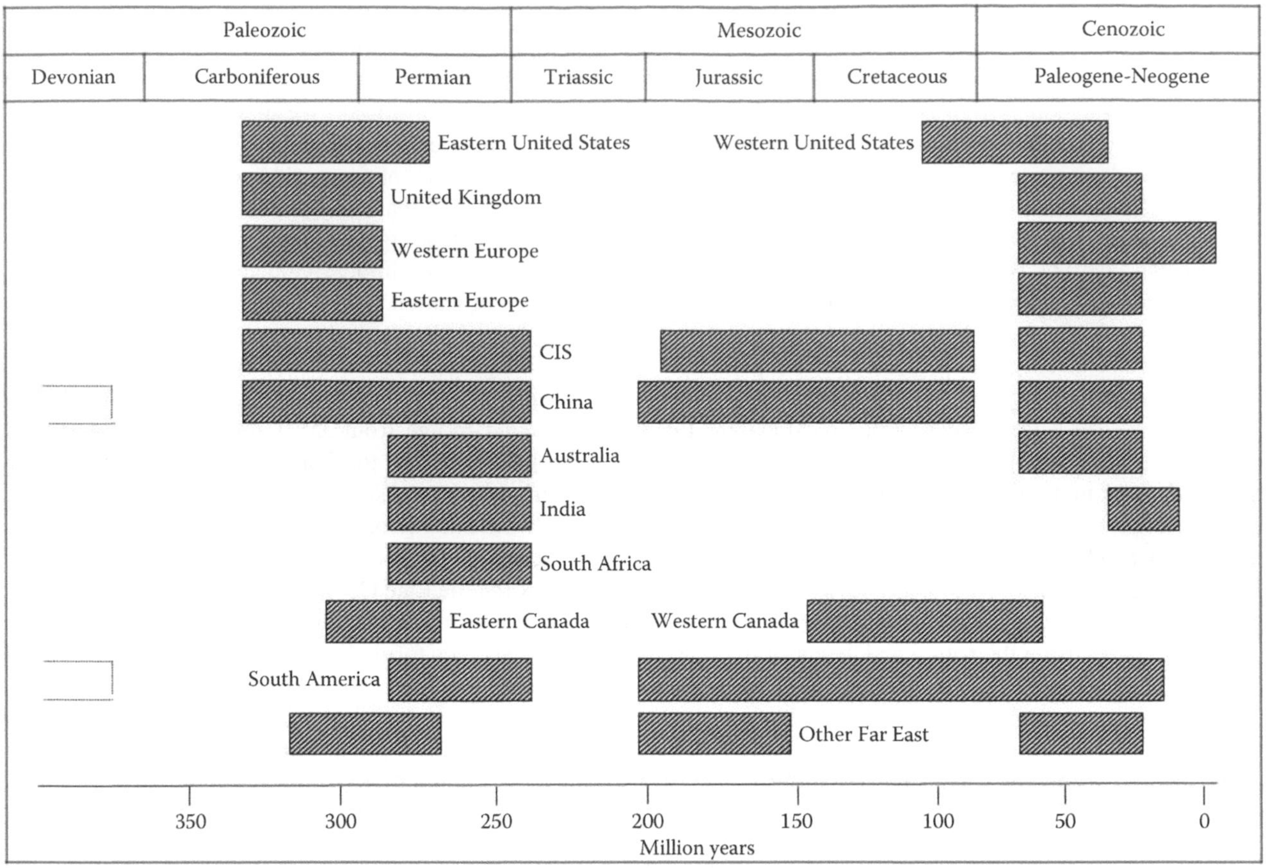

FIGURE 1.2 Geological age distribution of the world's black coal and lignite deposits (CIS: Commonwealth of Independent States).

appearance. The hydrogen content of sapropelic coals is more than in humic coals.

There are two types of sapropelic called as *boghead* (torbanite) and cannel. Boghead coal is an algal coal (>5% algae) and a variety of fine-grained black oil shale, usually occurring as lenticular masses. It may also be associated with the deposits of Permian coals. In boghead coal, the content of the inorganic matter is less than about 25% (Temperley 1936). On the other hand, cannel coal (about 5% or no algae) is black, dull, and homogeneous and breaks with a conchoidal fracture. It is composed largely of microspores and organic mud lay down under water, such as in a shallow lake (Thomas 2013). The cannel coal grades into boghead coal by the prolific occurrence of algal bodies, especially when the concentration of the microspores exceeds the algal matter (Temperley 1936). However, the transitional or intermediate forms such as cannel–boghead or boghead–cannel are also recognized (Thomas 2013).

The humic coals are composed of megascopic (macroscopic) components known as *ingredients* (Stopes 1919) or *lithotypes* (Stach et al. 1982), which includes vitrain, clarain, durain, and fusain. The description of these lithotypes has been developed as follows:

1. Vitrain is intense black, glassy, with uniform texture, having vitreous luster, occurring as thin bands and breaking with conchoidal fracture. It is found mostly in humic coals.
2. Durain is grey to black with a dull luster and breaks with an irregular and rough surface. It can occur as extensive layers within a coal seam.
3. Fusain is black, soft, friable, and easily disintegrates into a black fibrous powder. It occurs in coals as lenses, usually several millimeters thick, often concentrating in discrete layers. In most coals, fusain is a minor lithotype composed of the microlithotype fusite.
4. Clarain is bright with a silky luster between vitrain and durain and is a thinly banded lithotype formed by alternate laminations of bright (vitrain) and dull bands (durain and fusain). It does not break with conchoidal fracture. Vitrain and clarain together are classified as *brights* or *bright bands*.

1.4 COAL RANKS AND CLASSIFICATION

1.4.1 Coal Ranks

Due to variation in its chemical composition and properties, coal has been classified by variety of methodologies. The coalification process has been an important bearing on its physical and chemical properties. Prior to the nineteenth century, coal was classified according to appearance, for example, bright coal, black coal, and brown coal. The best common method of classification of coal is by its *rank*. The rank of the coal is generally determined by the degree of diagenesis and coalification that the coal undergoes during burial and tectonic effects and is assessed in terms of moisture and carbon content, volatile matter, and vitrinite reflectivity. The ranks of coals, from most to least carbon, are lignite, sub-bituminous, bituminous, and anthracite (Thomas 2013). Increasing rank is also accompanied by a rise in the carbon and energy contents and a decrease in the moisture content of the coal. According to Tissot and Welte (1984), the process of coalification occurs in three stages: diagenesis, catagenesis, and metagenesis. In the diagenetic stage, coals formed include peat, lignite (A and B), and sub-bituminous (C and B), which are formed by compaction, loading, and chemical transformation due to bacterial activity and temperature. The lignite and bituminous coals were further divided into A, B, C, and D levels under *American Society for Testing and Materials* (ASTM) classification based on decreasing heating value on moisture mineral matter free basis. The coals formed within catagenetic stage are sub-bituminous A and high-, medium-, low-volatile bituminous, which are formed through the effects of heat, pressure, and time, whereas at the metagenetic stage, semi-anthracite, and anthracite are formed by prolonged action of temperature and pressure.

The basic rank progression of coals is summarized in the Table 1.1. The extent of peatification and coalification can be examined by using the Van Krevelen diagram (Figure 1.3). The loss of oxygen-containing functional groups during diagenesis can be clearly seen in the decreasing Oxygen/ Carbon (O/C) ratios of the humic coal band, which is primarily attributable to the evolution of CO_2 and H_2O, while the Hydrogen/ Carbon (H/C) changes very little (Killops and Killops 2013). With increasing burial of the peat, chemical and physical changes occur. Due to compaction, the moisture content decreases and carbon content increases shown by decreasing atomic O/C ratios (Figure 1.3). And the carbon content continues to increase with increasing rank, while at the same time, the oxygen and hydrogen contents decrease.

1.4.2 Classification of Coal

Due to variation in chemical composition and properties, coal is classified by variety of methodologies. The coalification process has an important bearing on its physical and chemical properties. A number of classification systems have been developed (Hawe-Grant 1993). On the basis of utilization in different fields, classification may be divided into two types: scientific and commercial. The scientific classifications are Regnault Grüner-Bone, Fraser, Seyler, ASTM, and National Coal Board classifications. The commercial classification comprises classification by grade.

1.4.2.1 Scientific Classifications

1.4.2.1.1 Regnault Grüner-Bone Classification

The Regnault Grüner-Bone classification is based on the elemental composition of coal (C, H, O, N, and S), volatile matter contents, and caking nature of coals (Chandra et al. 2000).

1.4.2.1.2 Fraser Classification

Fraser classified coal on the basis of fuel ratio (fixed carbon/ volatile matter) and divided into four groups: bituminous, semi-bituminous, semi-anthracite, and anthracite (Chandra et al. 2000).

TABLE 1.1
Characteristics of the Different Ranks of Coals

Rank (From Lowest to Highest)	Properties
Lignite	1. The lowest rank of coal that has been transformed into a brown-black coal from peat
	2. Contains recognizable plant structures
	3. Excessive moisture (25%–50%)
	4. Heating value less than 8300 BTU/lb on a mineral matter free basis
	5. Carbon content between 60% and 70% on a dry ash-free basis
Sub-bituminous	1. Denser and harder than lignite
	2. Lower moisture content (12%–25%)
	3. With higher carbon content (71%–77%) on dry ash-free basis
	4. Heating value between 8300 and 13000 BTU/lb on mineral matter free basis
	5. On the basis of heating value, it is subdivided into sub-bituminous A, sub-bituminous B, and sub-bituminous C ranks
Bituminous	1. Alternate bright and dull bandings
	2. Harder and denser than sub-bituminous coal
	3. Carbon content between 77% and 87% on dry ash-free basis
	4. Heating value much higher than lignite or sub-bituminous coal
	5. On the basis of volatile content, bituminous coals are subdivided into low volatile bituminous, medium volatile bituminous, and high volatile bituminous
Anthracite	1. Anthracite is the highest rank of coal
	2. Carbon content over 87% on dry ash-free basis
	3. Anthracite coal generally has the highest heating value per ton on a mineral matter free basis
	4. It is often subdivided into semi-anthracite, anthracite, and meta-anthracite on the basis of carbon content

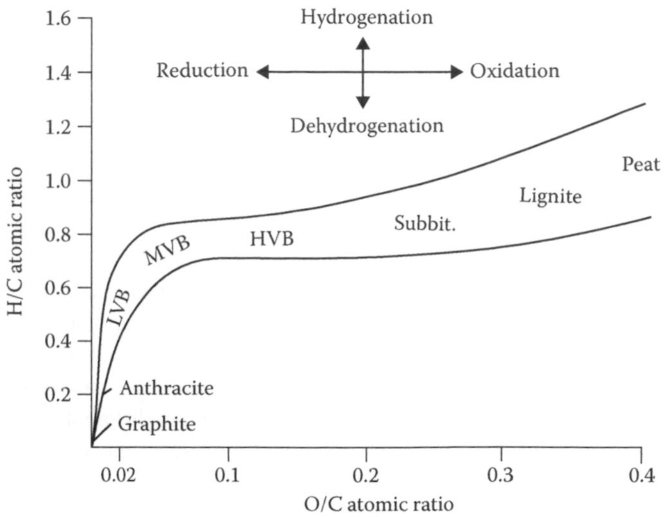

FIGURE 1.3 Van Krevelen diagram illustrating the range of H/C and O/C atomic ratios for various coalification stages. (Data from Van Krevelen, D.W., *Coal: Typology-Chemistry-Physics-Constitution.* Elsevier Science, Amsterdam, The Netherlands, 514, 1961; Langenberg, W. et al., *Coal Geology and Its Application to Coal Bed Methane Reservoirs. Lecture Notes for Short Course,* Information Series No. 109, Alberta Research Council, Alberta Geological Survey, Edmonton, 9, 1990.)

1.4.2.1.3 Seyler Classification

The Seyler classification is based on the carbon and hydrogen content of coals determined on dry mineral matter basis. In the scientific category, the Seyler chart method has considerable value. It gives the range and interrelationship of the properties of coal including parameters such as moisture content and swelling indices (Seyler 1938).

1.4.2.1.4 ASTM Classification

The ASTM classification system adopted in 1938 as a standard means of specification is used in the United States and in many other parts of the world, and is designated as D388 in ASTM standard. In this classification, the higher rank coals are specified by fixing carbon >69 wt%, or from volatile matter <31 wt%, on a dry mineral free basis. Lower rank coals are classified by calorific value (on moisture and mineral matter free basis).

1.4.2.1.5 National Coal Board Classification

This classification of coal proposed in 1946 by the UK Department of Scientific and Industrial Research. There are two parameters: the quantity of volatile matter determined on dry mineral matter free basis and the Gray-King coke-type assay, a measure of coking as designed in the British Standards. The Gray-King coke-type assay is used as a primary means of classification for lower rank coals having less than 10 wt% ash content.

1.4.2.1.6 Classification Based on Source of Genesi

This type of classification depends upon the character or the original vegetable matter. The different types are humic and sapropelic coals.

1.4.2.1.7 Classification by Rank

Classification by rank is made on the degree of the coalification, which gives an information about its carbon content with anthracite having the highest carbon content followed by bituminous, sub-bituminous, and lignite coal (in descending order). However, peat is not regarded as coal. Coalification or metamorphosis of coal results in changing of the physical and chemical properties of coal in response to temperature and time. With extreme metamorphism, anthracite can change to graphite (Chandra et al. 2000).

1.4.2.2 Commercial Classification

1.4.2.2.1 Classification by Grade

This classification is made on the basis of the proportion of impurities contained in coal. The extraneous matter forms most of the noncombustible impurities of coal. The classification of commercial grades of coals in India was evolved based on the ash contents, moisture, and calorific values (Chandra et al. 2000).

1.5 MICRO-CONSTITUENTS OF COAL

Coal is primarily composed of two parts: organic and inorganic. The micro-components and microstructures present in the organic part are called *macerals*, which are considered to be the building blocks of coal. There are three maceral groups known as *inertinite*, *liptinite* (formerly exinite), and *vitrinite* as shown in Table 1.2, along with their subgroup and photomicrographs. Inertinite is derived from the partial carbonization of the coal-forming materials by fire or intensive degradation by microorganisms. Vitrinite is derived from woody tissues of plants, which are chemically composed of the polymers, cellulose, and lignin. Vitrinite is the most abundant maceral in coal; however, most coals from Gondwanaland and some coals from western Canada are vitrinite poor. Exinite, also known as *liptinite*, is derived from waxy and resinous parts of plants such as spores, cuticles, and similar materials. This group is very sensitive to advanced coalification and it begins to disappear in medium-volatile-rank coals and are absent in coals of low-volatile rank. Vitrinite is the most important maceral in coals and is generally found to be the most abundant maceral in coals. However, inertinite and more rarely liptinite is the most abundant maceral group in some coals of Palaeozoic or Mesozoic age. The chemical composition of the three macerals is different. Liptinite macerals are more enriched in aliphatic hydrogen than vitrinite. The aliphatic CH and CH_2 carbon decrease in the order liptinite > vitrinite > inertinite in coals. Thus, the liptinites are distinguishable by a higher aliphatic (i.e., paraffin) fraction. Vitrinites contain more oxygen than the other macerals at any given rank level. In low-rank coals, the carbon atoms present in vitrinite are divided between aliphatic and aromatic bonding, but hydrogen occurs predominantly as aliphatic grouping. The aromaticity of the maceral groups increases with increasing rank. At low rank, in reflected white light, liptinite appears dark, vitrinite appears as a mesostasis of medium grey appearance, and inertinite has a higher reflectance than the other maceral groups. The reflectance of vitrinite and inertinite converge within the semi-anthracite to anthracite range.

However, the bireflectance of vitrinite is much greater than that of inertinite except in the case of some contact-altered vitrinites and these two maceral groups can still be distinguished in plane-polarized light or with partially crossed polars (ICCP 2011). It is seen from the laboratory studies of coal macerals carried worldwide that the coals in the United States and the United Kingdom are enriched in vitrinite and exinite, whereas coals from Southern Hemisphere, India, China, and those originating from Gondwanaland have inertinite as their dominant maceral (Sanyal 1997). The tertiary Indian coals are usually characterized by high vitrinite content (80% average), with the nonvitrinite fraction being predominantly inertinite,

TABLE 1.2
Maceral Groups, Subgroups, and Photomicrographs of Macerals of Coals

Maceral Group	Maceral Subgroup	Maceral	Photomicrographs
Vitrinite	Telovitrinite	Textinite Texto-ulminite Ulminite Collotelinite	
	Detrovitrinite	Attrinite Densinite Vitrodetrinite Collodetrinite Corpohuminite	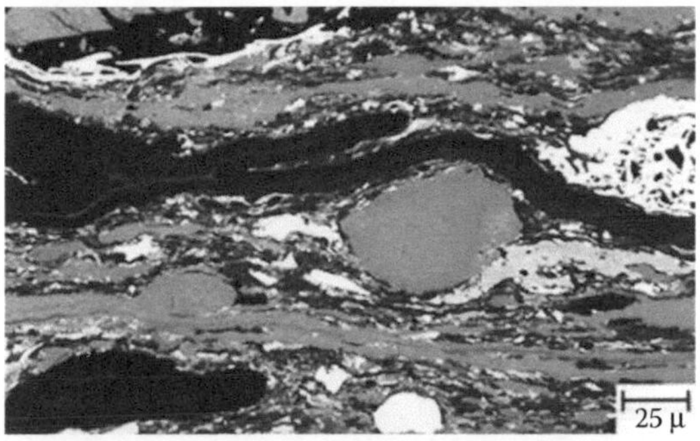
	Gelovitrinite	Corpogelinite Gelinite	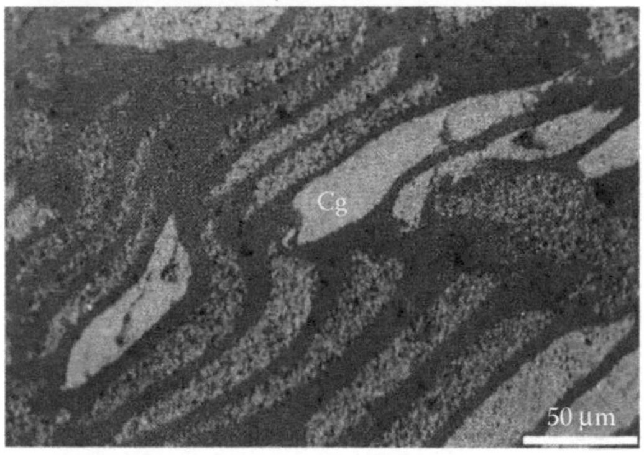
Inertinite	Macerals with plant cell structures	Fusinite Semifusinite Funginite	

(Continued)

TABLE 1.2 (*Continued*)
Maceral Groups, Subgroups, and Photomicrographs of Macerals of Coals

Maceral Group	Maceral Subgroup	Maceral	Photomicrographs
	Macerals lacking plant cell structures	Secretinite Macrinite Micrinite	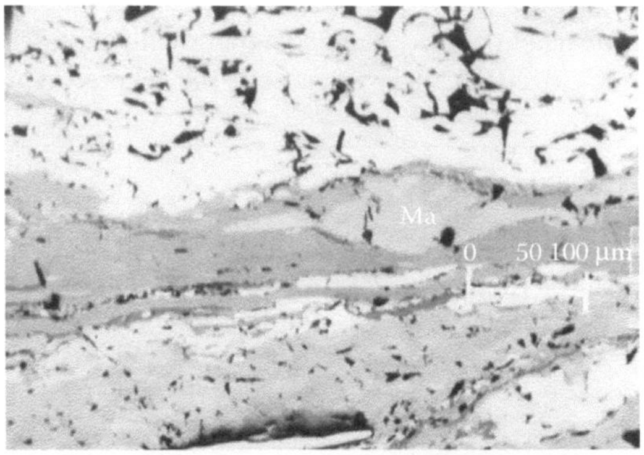
	Fragmental inertinite	Inertodetrinite	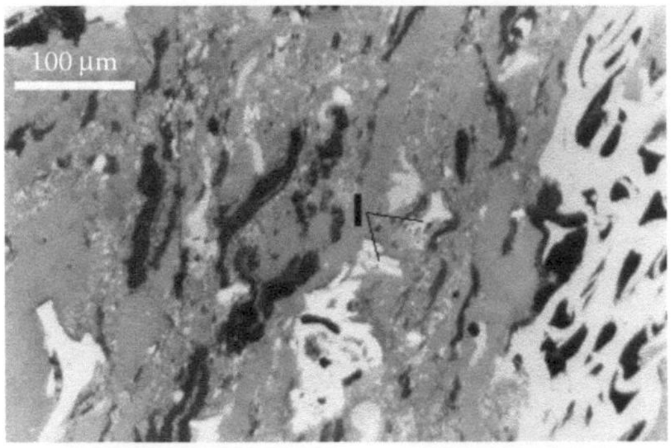
Liptinite		Sporinite Cutinite Suberinite Resinite Fluorinite Liptodetrinite Alginite Bituminite	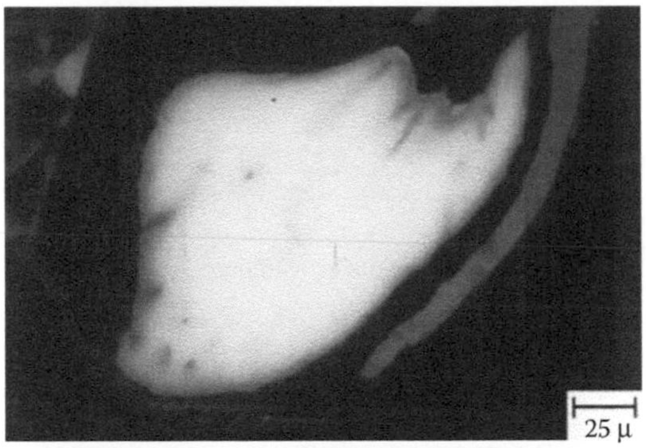
	Strictly a bitumen rather than a maceral	Exsudatinite	

Source: ICCP, Organic petrology, macerals, microlithotypes, lithotypes, minerals, rank. Chapter 2. *Training Course on Dispersed Organic Matter*, Department of Geoscience, Portugal, 4–71, 2011.

which are low in proportions. The liptinite content is usually found to be less than 20% (Sharma et al. 2012). The Çardak coal deposits in south-western Turkey are characterized by the dominance of vitrinite, mainly huminite group (up to 92 vol.%, on mmf basis), followed by liptinite (up to 14 vol.%), and inertinite (up to 10 vol.%) (Oskay et al. 2013).

1.5.1 Microlithotypes

The term *microlithotype* is for the designation of rock types within coal that are at a microscopic scale. The definition is based on maceral percentages in coals. Thus, layers with >0.05 mm in thickness and consisting of more than 95% vitrinite are termed as *vitrite*. Bimaceral (clarite, durite, and vitrinertite) and trimaceral (duroclarite and clarodurite) microlithotypes are derived from lithotypes rather than macerals. In Palaeozoic coals, a wide range of microlithotypes, such as vitrite, clarite, durite, and inertite, can be present. Inertinite is present as durite in Carboniferous coals, whereas it is present as inertite in Gondwana coals. The microlithotypes present in significant proportions in tertiary coals are vitrite and clarite (liptinite greater than 5% and inertinite less than 5%). Use of the microlithotype terms assists by indicating the ratio of telovitrinite (preferentially associated with vitrite) to detrovitrinite (preferentially associated with clarite) and the amount of liptinite present. However, these distinctions can commonly be made more directly by using maceral analyses (ICCP 2011). The classification of microlithotypes is shown in Table 1.3.

1.6 CHEMICAL COMPOSITION AND HETEROATOMS IN COAL

1.6.1 Chemical Composition

The principal chemical constituents of coal are carbon, hydrogen, and oxygen with fewer amounts of nitrogen and sulfur. The chemical formula of coal may be approximated as $C_{135}H_{96}O_9NS$ (Raghuvanshi et al. 2006). With the increase of

TABLE 1.3
Microlithotypes of Coal

Maceral Composition (Mineral Free)	Microlithotype	Maceral Group Composition	Microlithotype Group
Monomaceral		Vitrinite > 95%	Vitrite
Collinite > 95%			
Telinite > 95%			
Vitrodetrinite > 95%			
Sporinite > 95%	Sporite	Exinite(liptinite) > 95%	Liptite
Cutinite > 95%	Algite		
Resinite > 95%			
Alginite > 95%			
Liptodetrinite > 95%			
Semifusinite > 95%	Semifusinite	Inertinite > 95%	Inertite
Fusinite > 95%	Fusinite		
Sclerotinite > 95%	Inertodetrite		
Inertodetrinite > 95%	Macroite		
Macrinite > 95%			
Bimaceral	Sporoclarite	Vitrinite + exinite(liptinite) > 95%	Clarite
Vitrinite + sporinite > 95%	Cuticoclarite		
Vitrinite + cutinite > 95%			
Vitrinite + resinite > 95%			
Vitrinite + liptodetrinite > 95%			
Vitrinite + macrinite > 95%		Vitrinite + inertinite > 95%	Vitrinertite
Vitrinite + semifusinite > 95%			
Vitrinite + fusinite > 95%			
Vitrinite + sclerotinite > 95%			
Vitrinite + liptodetrinite > 95%			
Inertinite + sporinite > 95%	Sporodurite	Inertinite + exinite(liptinite) > 95%	Durite
Inertinite + cutinite > 95%			
Inertinite + resinite > 95%			
Inertinite + liptodetrinite > 95%			
Trimaceral	Duroclarite	Vitrinite > inertinite, exinite(liptinite)	Trimacerite
Vitrinite + inertinite + exinite > 5%	Vitrinertoliptite	Exinite > inertinite, vitrinite	
	Clarodurite	Inertinite > vitrinite, exinite(liptinite)	

Sources: ICCP, *International Handbook of Coal Petrology*, CNRS, Paris, France, 160, 1963; ICCP, *Supplement to Second Edition of the Handbook*, CNRS, Paris, France, 1971; ICCP, *International Handbook of Coal Petrology*, CNRS, Paris, France, 1975.

maturity of coal from lignite to anthracite, the carbon content increases, while the oxygen content decreases. The result is that the calorific value of coal increases with rank. Apart from these main constituents of coal, the presence of inorganic elements in coals from various coalfields of the world is also reported (Chandra et al. 2000; Saikia et al. 2009a, b, c). They include Ag, Al, As, Au, B, Ba, Be, C, Ca, Cd, Cl,Co, Cr, Cs, Cu, F, Fe, Ga, Ge,He, Hf, Hg, I, Br, In, K, Na, Li, Mg, Mn, Mo, Nb, Ni, P, Pb, Pt, Rh, Pd, Ra, Rb, Re, S, Sb, Cs, Se, Si, Sn, Sr, Ti, Th, Te, Ta, U, V, W, Y, La, Pr, Nd, Sm, Eu, Gd, Tb, Ho, Dy, Er, Yb, Lu, Zn, and Zr. Some of the elements present in coals are also identified as hazardous air pollutants in the 1990 Clean Air Act Amendments, United States. However, the amount of these inorganic elements in different coals widely varies with their different geochemical aspects. The determination of these elements in coals have been conducted by using X-ray photoelectron spectroscopy, neutron activation analysis, isotope dilution mass spectroscopy, spark source mass spectrometry, atomic absorption spectrometry, and other analytical techniques (Chandra et al. 2000; Ward 2002).

The inorganic matter present in coal may be divided into two groups of chemical elements: first group as the main or ash-forming elements [Si, Al, Fe, Ca, Mg, Na, K, Ti, (S, P)] and the second as the trace or rare elements, which usually increase after coal combustion and utilization. The trace elements include Cl, F, Hg, As, Se, and Cr along with rare elements such as Ge, Ga, U, Mo, Be, Sc, and rare earth elements. Some trace elements prove to be valuable (silver, zinc, and germanium), while some have the potential to be hazardous (cadmium and selenium) if their concentration is more than trace amounts (Yudovich and Ketris 2005). Significant concentration of trace elements such as Ge, Se, Cd, Cl, Co, As, Cr, Hg, Mn, Gd, Hf, Li, In, Nb, Ni, Sr, Ba, Th, and rare earth elements are found in high-sulfur coals of Russia (Yudovich and Ketris 2002). The trace elements such as As, B, Pb, Se, Cu, V, Ba, Ge, Mn, and Sn (Baruah et al. 2003) are found in coals of Assam, India. The coals in the United Kingdom have concentrations of As, which is significantly higher than worldwide average, while elements such as Zn are lower (Zandi and Russell 2004).

1.6.2 Heteroatoms

The heteroatoms in coal is a very important aspect of coal structure and is extremely significant in coal conversion processes and design of catalysts for these processes (Speight 2013). The preliminary study on the presence or absence of functional groups of coal is performed by Fourier infrared spectroscopic analysis of coals. But, the recent advanced level studies and examination of the functional entities in coal have brought to light some interesting features the occurrence and distribution of the heteroatoms (Gupta 2006; Speight 2013). The assessment of the macromolecular structure of the nitrogen, sulfur, and oxygen heteroatom species in coal and coal-derived products is a complex and yet challenging analytical problem in coal chemistry. Oxygen occurs predominately as phenolic or etheric groups, with fewer amounts of carboxylic acids or esters and some carbonyls. Nitrogen occurs predominantly as pyridine or pyrrolic-type rings in coal. Sulfur has similar chemistry to oxygen, and it also creates environmental problems during coal utilization.

1.6.2.1 Oxygen

Out of the three heteroatoms found in coal, oxygen has received the most important attention, and it is divided into four categories: (1) carboxyl, (2) carbonyl, (3) hydroxyl, and (4) ether, out of which only the first three categories are amenable to quantitative analysis. The ether bond in coal is mainly is of diaryl (Ar—O—Ar) and benzyl-aryl ($C_6H_5CH_2$—O—Ar) types, although methoxyl groups (CH_3O—) and dialkyl ethers (R—O—R′) are found in low-rank coals (Attar and Hedrickson 1982). The amount of oxygen in carboxylic groups decrease continually with the increase in coal rank until about 80% carbon, where the only oxygen groups present are phenolic and carbonyl groups. Aliphatic hydroxyl groups have been reported in brown coals. Thus, for coals of rank higher than lignites, oxygen can be accounted for as carbonyls, hydroxyl, or carboxyl with minor amount of heteronuclear oxygen, although there is now growing evidence for various ethers in coals. A great deal of information regarding the location of oxygen in coal structure has been derived from infrared spectroscopic investigations. The high-sulfur coals from northeast region of India were extensively studied by using Fourier transform infrared (FT-IR) spectroscopy and found to contain long aliphatic chains with parallel orientations (Baruah and Khare 2010; Saikia et al. 2007a, b, c, 2009a, b, c). The oxygen-containing functional groups found in these coals include phenols, alcohols, ethers, carboxylic acid, and carbonyls (Saikia et al. 2007a, b, c). However, the infrared spectroscopic data tend to suffer from their inability to be fully quantitative, and can be used to illustrate trends throughout the progression from low-rank coals to high-rank coals (see Table 1.4). A series of coals of varying rank, from brown coal to anthracite, of the Ordos Basin, China, were studied by FT-IR and the oxygen-containing groups found in these coals are aromatic hydrocarbons, aliphatic carbons, aromatic carbons, and carboxyl (Yao 2011).

The oxygen content of sub-bituminous coal was also discussed by Ruberto et al (1977) in his solvation studies, where it was concluded that a significant portion of the oxygen occurs in saturated ether functional groups α or β to the aromatic moieties or as furan systems. Hayatsu et al. (1978a, b) found significant quantities of phthalan and xanthone in lignite; xanthone and dibenzofuran in bituminous coal, and dibenzofuran in anthracite coals (Davidson 1982). Phenolic acids were also isolated from coal, which show a relationship between lignins and coals. From the available works in the literature, it can be also stated that the oxygen content of coal generally decreases with rank.

1.6.2.2 Nitrogen

The nitrogen functionality in coal is the representative of plant and animal proteins, nitrogen-rich bacteria, and plant alkaloids. Nitrogen concentrations have been reported to be in the range of 0.5%–2.0% (Meyers 2012). It has also been reported that coals commonly contain 1%–2% nitrogen, bituminous coals typically contain 1.5%–1.75% nitrogen, and anthracites generally contain <1% nitrogen. Most of the nitrogen in coal-tar

TABLE 1.4
General Assignments of Some Important FT-IR Stretching Frequencies for Functional Groups in Few Northeast Indian Coals

Stretching Frequencies cm^{-1}					
Coal Sample-1	Coal Sample-2	Coal Sample-3	Coal Sample-4	Coal Sample-5	Assignments
3752	3996		3698	3697	O–H str.
	3651		3655	3653	O–H str.
	3620		3621	3621	O–H str.
3412	3396	3411	3400	3389	O–H str. NH str.
3240	3280	3240	3200	3242	O–H str., NH str.
3020	3160	3020	2953	2926	CH_3 str., CH_2 str., al. CH str.
2953	3120	2919	2921	2920	CH_3 str., CH_2 str., al. CH str.
2920	2920	2850	2848	2851	CH_3 str., CH_2 str., al. CH str.
2849	2851				CH_3 str., CH_2 str., al. CH str.
1610	1607	1682	1610	1607	ar. C=C str., C=O str.
1558		1611			ar. C=C str., C=O str.
1436	1438	1433	1432	1436	CH_3 assym. def., CH_2 scissor. ar C=C str.
1372	1400	1376	1401	1400	Aliph CH_3, CH_2 str.
1320	1376	1300	1375	1165	CH_3 symm. def.
1163	1300	1202	1165	1090	CH_3. def., C–O str.
1091	1167	1165	1092	1031	C–O str., Car–O–Car str.
1031	1090	1093	1032	1008	C–O str., Car–O–Car str., Q
1008	1051	1032	1008	938	C–O str. (alcohol), C–O–C str., Cal–O–Cal str.
910	1008	1007	939	913	Car–O–Car str.
960	939	866	913	798	Aromatic band
800	913	810	798	778	Aromatic rings
780	798	748	779	752	Substituted benzene rings with 2 neighbors H
744	779	695	750		Hydrogen atoms in meso-position (anthracene) angular condensed ring system/α-substituted benzene rings, monosubstituted benzene rings, H, Q
692	754	628	695		Condensed ring system
627	629	599	535		H-atom in condensed ring system
533	537	513	470		H-atom in condensed ring system
471	471	472			Quartz
420	431		430		Pyrite

and in liquefaction products is found as pyridines, condensed pyridines, pyrroles, nitriles, quinolines, indoles, and carbazoles. Thus, it is generally believed that most of the nitrogen in coal is incorporated in heterocyclic rings with small amounts of nitrogen-containing side chains (Davidson 1982).

1.6.2.3 Sulfur

Sulfur is of most important concern due to its environmental impacts during coal utilization. Hence, there is a considerable amount of works relating to the sulfur chemistry in coal (Baruah et al. 2006; Saikia et al. 2014a, b, c, d). Sulfur has a similar chemistry to oxygen, but less is known about organic sulfur in coal compared with organic oxygen. Heterocyclic sulfur compounds have been identified in coals but very few studies have concentrated on the organic sulfur functional groups in coal. However, Attar and Dupuis (1979) have reported data on these functionalities in different coals. Table 1.5 shows the distribution of organic sulfur groups in five coals, which shows that the content of thiols (—SH) is substantially larger in lignites and high-volatile bituminous coals than in low-volatile bituminous coals. The fraction of aliphatic sulfides (R—S—R) remains approximately constant at 18%–25%. The data indicate that larger fractions of the organic sulfur are present as thiophenic sulfur in higher ranked coals than in lower ranked ones. The data on —SH, R—S—R, and thiophenic sulfur are explained by suggesting that the coalification process causes the organic to change from —SH through R—S—R to thiophenes in condensation reactions (Speight 2013).

1.6.2.3.1 High-Sulfur Coals

High-sulfur coals are found throughout the world with exception to Australia, where most of the coals are low in sulfur content (Stephen 1986). However, the sulfur content of coals is seen to be quite variable. The low sulfur content in coal becomes an important consideration during coal utilization in thermal plants. It is present in coal in forms such as pyritic sulfur, sulfate, and organic sulfur. Organic sulfur is chemically bonded to the carbon atoms in the coal structure.

TABLE 1.5
Distribution of Organic Sulfur Groups in Five Coals

Coal	Organic S (wt%)	Organic S Accounted (%)	Thiolic	Thiophenolic	Aliphatic Sulfide	Aryl Sulfide	Thiophenes
Illinois	3.2	44	7	15	18	2	58
Kentucky	1.43	46.5	18	6	17	4	55
Martinka	0.60	81	10	25	25	8.5	21.5
Westland	1.48	97.5	30	30	25.5	–	14.5
Texas lignite	0.80	99.7	6.5	21	17	24	31.5

Source: Attar, A. and Dupuis, F., *Am. Chem. Soc.*, 24(1), 166, 1979.

TABLE 1.6
Distribution of Sulfur Functional Groups in Assam Coal (wt%)

Coal	Mercaptan	Disulfide/Thiol	Thio-Ether	Thiophene	Pyrite	Sulfate	Thioketone
Ledo (India)	0.30	1.70	0.50	0.90	0.76	0.35	1.59

Pyritic sulfur occurs in coal as grains of the mineral pyrite (FeS_2). The amount of pyritic sulfur in coal is highly variable, depending on the geologic conditions. Sulfate occurs generally as iron or calcium sulfates. Moreover, the occurrence of another type of sulfur, termed as *secondary sulfur*, containing Fe-S moieties associated with coal organic matter in high-sulfur Assam coals was reported elsewhere (Baruah 1984, 1992). The sulfur phases are reported to be rich in ^{34}S isotopes for coals in the Pannonian basin, indicating the marine bacterial sulfate reduction during their formation (Hámor-Vidó and Hámor 2007). The vitrinites in coals show extreme anomalies in reflectance due to the high proportions of organic sulfur (Ward et al. 2007). Turner and Richardson (2004) examined the geological factors influencing the regional, stratigraphic, and between- and within-seam variations in sulfur content of coals in the Westphalian A and B Coal Measures in the Northumberland Coalfield of Northeast England. The relationships between sulfur abundance in coal seams and depositional environments of coals from the United States, China, the United Kingdom, Germany, Hungary, Turkey, Indonesia, and Brazil were reviewed and it was concluded that the variation of sulfur in coals is closely related to the depositional environments of coal seams (Chou 2012). The coals from northeastern region of India have high sulfur contents (Baruah et al. 2006). They generally contain 2%–8% sulfur, where 75%–90% is organically bound, while the rest is in inorganic form, namely sulfate and pyritic sulfur. The high organic sulfur coals are also found in the Upper Permian marine carbonate successions (Heshan Formation) in the Heshan Coalfield, central Guangxi, southern China (Longyi et al. 2003). These Late Permian low-volatile bituminous coals of Heshan Coalfield of southern China are termed as *super high-organic sulfur* (5.13%–10.82%) (Dai et al. 2013). The organic sulfur present in these coals is generally aliphatic or aromatic thiols, aliphatic or mixed sulfides, aliphatic or aromatic disulfides, heterocyclic compounds of the thiophenic types, and so on. Kumar and Srivastava (2013) have concluded that the sulfur present in Assam coal (India) is in the forms of mercaptan, disulfide/thiol, thioether, thiophene, pyrite, sulfate, and thioketone (Table 1.6).

1.7 CHEMICAL STRUCTURE OF COAL

The concept of coal structure is difficult to define, as the macromolecules of coal are not composed of repeating mono-organic units (Gorbaty and Larsen 1982). The structures had been built on the basis that they were characterized by values of the above parameters (such as elemental composition, carbon and hydrogen aromaticities, number of rings, and size of aromatic cluster), as well as the total organic matter of the coal or its major petrographic components. Thus, the macromolecular character of coal (quasi-polymeric chains and/or polymeric sheets depending on the particular type of coal) and their nonpolymeric component of small organic molecules are embedded in a polymeric matrix. Coal from different geological time periods and different geographic areas may differ in composition and physicochemical properties. The structural information, which had been thought to be generally valid for coal, obtained after different investigations may, in fact, only be so for coals of a particular region or geological age. The macromolecular skeletal structural models for coal were well described in the literature (Gorbaty and Larsen 1982). Two such model structures of bituminous coal are shown in Figures 1.4 and 1.5. The investigation of the chemical structures of coal has led to comprehensive and well-defined results on the basis of development of spectroscopic methods. These are FT-IR, ultraviolet-visible spectroscopy, X-ray structural analysis, solid-state nuclear magnetic resonance (NMR), and so on. X-ray diffraction studies provide useful information about the internal arrangement of atoms in

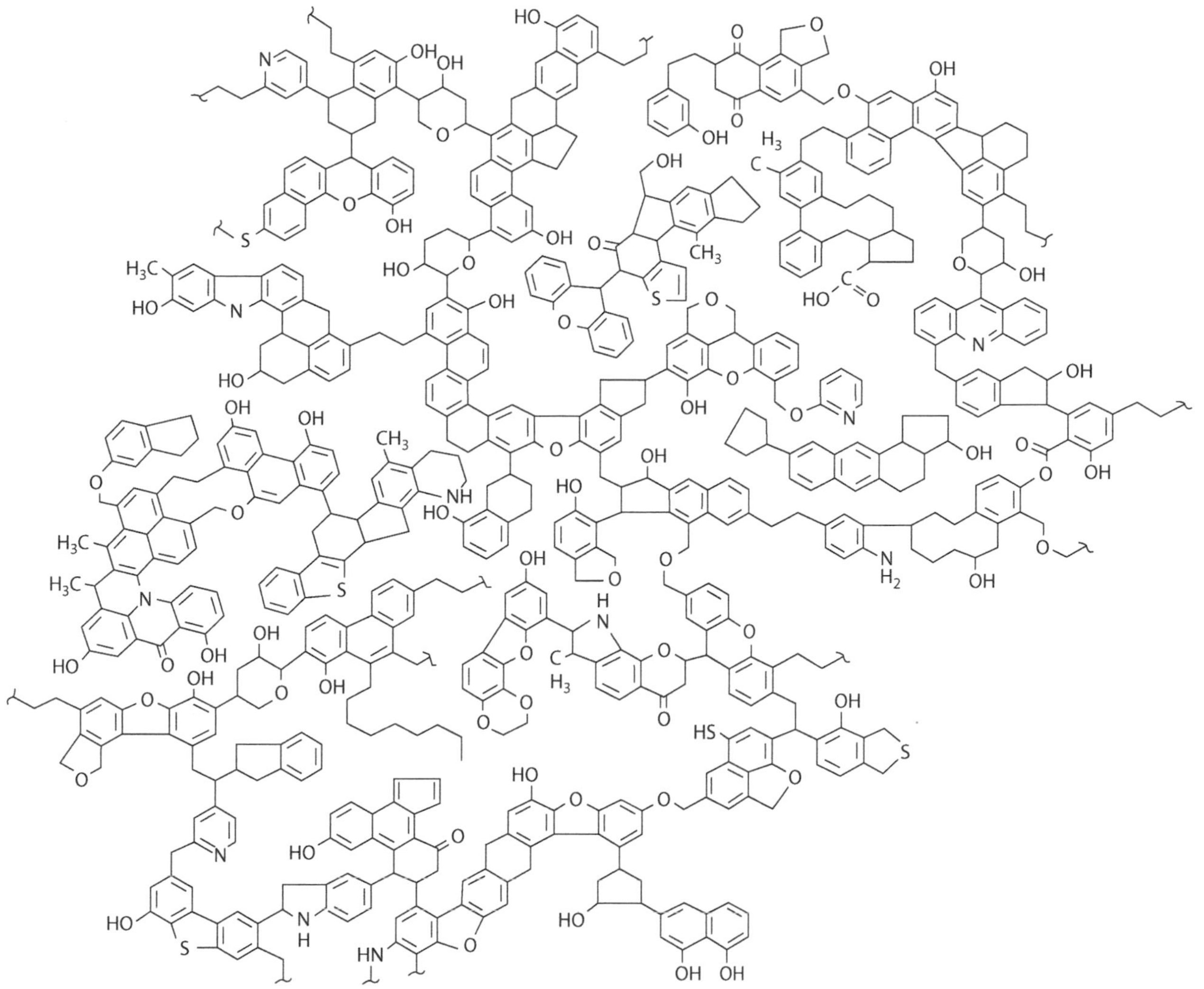

FIGURE 1.4 Model of coal structure provided by Wiser. (From Wiser, W.H., Conversion of bituminous coal 'to liquids and gases: Chemistry and representative processes, *Magnetic Resonance: Introduction, Advanced Topics and Applications to Fossil Energy*. L. Petrakis, and J.P. Fraissard (Eds.), Reidel, Dordrecht, The Netherlands, 325, 1984.)

coal (Saikia et al. 2007a, b, 2009a, b, c). This arrangement in coal is a vital factor, which affects many physical properties in relation with coal utilization.

Modern conceptions of the chemical structure of coals is well described by various researchers in numerous publications (Alvarez et al. 2012a, b, 2013; Barsky et al. 2009; Castro-Marcano and Mathews 2011; Castro-Marcano et al. 2012; Duane et al. 1982; Margriet et al. 1992; Mathews and Chaffee, 2012; Mathews and Sharma 2012; Mathews et al. 2011, 2014; Tselev et al. 2014). Summarizing the experimental results of coal structure allows considering the *molecule* of this substance as a set of various individual structures, mainly of aromatic and aliphatic nature. They are connected with each other by the forces of different nature, value, and kind of the bond (covalent, donor-acceptor or hydrogen bonds, polyconjugatings owing to electrons delocalization, van der Waals forces, etc.). It is also reported that the organic coal substance is a totality of packs of condensed aromatic nuclei with side nonaromatic groups, including oxygen, nitrogen, sulfur, and other hetero-atoms, which chemically bond the adjoined packs into three-dimensional polymer (Barsky et al. 2009). Such approach suggests that during coalification, the carbon-oxygen bonds are changed for the carbon-carbon ones in the side radicals.

1.7.1 Random Layers (Turbostratic Structure) in Chemical Structure of Coal

X-ray diffraction technique has been widely applied for the structure characterization of carbon materials and has given useful information. It may be applied for evaluating the carbon-stacking structure of coals. Different X-ray structural parameters can be determined from different parts of the scattering curve. The large variety of organic and inorganic materials involved in the formation of coal makes them highly heterogeneous, both in physical and chemical structures. One aspect of heterogeneity in coals is thus related to their physical structure, which is composed of a macromolecular cross-linked network and a molecular compound. X-ray diffraction

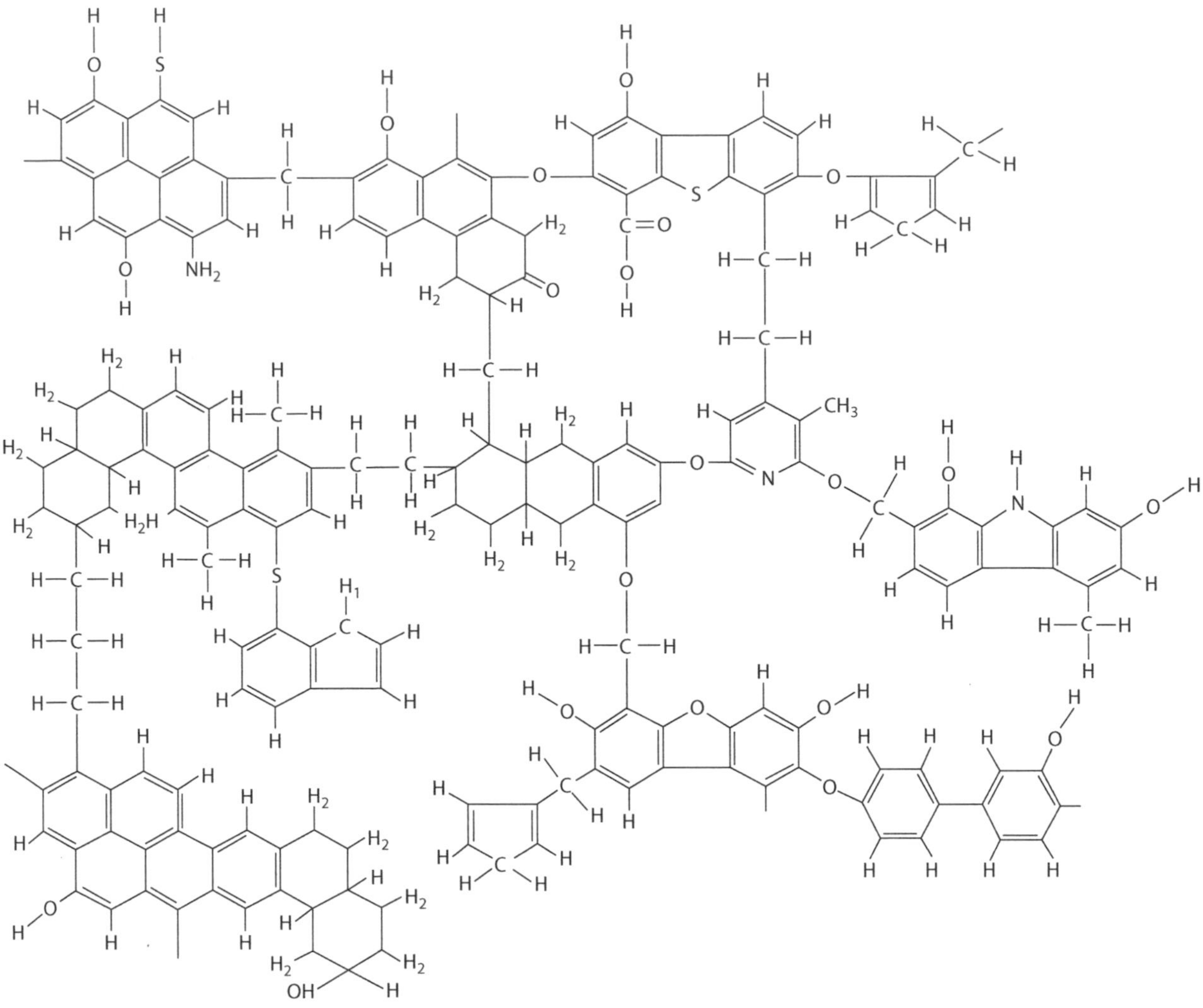

FIGURE 1.5 Model of coal structure provided by Shinn. (From Shinn, J.H., *Fuel*, 63, 1187, 1984.)

has been one of the few orthodox methods that have been used to estimate the size of the aromatic lamellae in coal, the average distance between lamellae and the mean bond distance. According to the visual inspection of the reduced intensity profile of the Assam (India) coal, the observation that can be made is that it contains a short-range graphite-like structure, that is, crystalline carbon, which is extremely small, giving rise to three diffuse peaks (Boruah et al. 2008). These crystalline carbons have an intermediate structure between graphitic and amorphous state, so-called turbostratic structure or random layer lattice structure (Biscoe and Warren 1942). This means that coal contains stacked aromatic layers, which are strongly parallel and equidistant, but each having a completely random orientation in plane and about the layer normal. It is reported to understand the short-range structural features, to determine the relationship(s) between the aryl/alkyl carbon ratio, and the size of the average polycyclic aromatic unit in coal from Makum coalfield, Assam, India. An X-ray scattering analysis of the average polycyclic aromatic unit in coal indicates that the aromatic fraction in this coal is 74%, with the aliphatic fraction correspondingly estimated to be 26% (Boruah et al. 2008) (Table 1.7).

1.8 ROCK TYPES AND MINERAL MATTERS IN COAL

1.8.1 Rock Types

Physically distinguishable bands that are observed in lignite (brown soft coal) or coal (hard coal) are known as *rock types* or *lithotypes*. According to the International Committee for Coal Petrology (ICCP) (1993), the lithotype group of soft brown coal is further divided as lithotype based on structure as given in Table 1.8. These groups are designated based on the occurrence of components such as xylite (coalified woody material), groundmass (fine detrital humic material), and mineral matter. Thus, the lithotype groups recognized by ICCP classification are as follows (see also Table 1.8):

1. *Matrix coal* consists of a fine detrital humic groundmass and is homogeneous in appearance with yellow to dark brown in color. Plant fragments may also be embedded in the groundmass and these coals show some stratification. Matrix coals are common in Paleogene–Neogene soft brown coals (Thomas 2013).

TABLE 1.7
X-Ray Structural Parameters of an Assam Coal

Coal	D-Value for (002) Plane, Å	L_{002} (Lc) Å	No of Layers	No of Atoms Per Layer	L_{10} (La) Å	γ-Band Å
Ledo	3.42	7.58	2	8	4.86	4.42

Source: Boruah, R.K. et al., *J. Appl. Crystallogr.*, 41, 27–30, 2008.

TABLE 1.8
Lithotype Classification for Soft Brown Coals

Lithotype Group (Constituent Elements)	Lithotype (Structure)	Lithotype Variety (Color; Gelification)
Matrix coal	Stratified coal	Brown (weakly gelified) coal
		Black (gelified) coal
	Unstratified coal	Yellow (ungelified) coal
		Brown (weakly gelified) coal
		Black (gelified) coal
Xylite-rich coal		
Charcoal-rich coal		
Mineral-rich coal		

Sources: Taylor, G.H. et al., *Organic Petrology*, Gebruder Borntraeger, Berlin, Germany, 704, 1998; ICCP, *International Handbook of Coal Petrography*, 3rd Supplement to 2nd Edition. University of Newcastle up on Tyne, London, 1993.

2. *Xylite-rich coal* includes coals with more than 10% of xylite (woody fibrous tissue). The groundmass is detrital and may or may not be stratified. It is the dominant lithotype and found in all brown coals. Its characteristics are thought to be the decomposition of trees and shrubs in the peat-forming mire.
3. *Charcoal-rich coal* contains more than 10% charcoal. The coal can be weakly or strongly stratified, occurring as lenses and occasionally with more persistent layers. The coal is brownish-black and has a coke-like appearance. It is considered to be the product of burned forest swamps. If such coal is stratified, it is indicative of water or wind transported residues in an open-swamp environment.
4. *Mineral-rich coal* includes all kinds of mineralization of the different brown coal lithotype groups and should be visible to the naked eye. The inorganic materials present typically include quartz, clay, carbonates, sulfides, and other minerals (Thomas 2013). The hard coals may be classified into two coal types at macroscopic scale as the humic or banded coals and the sapropelic or nonbanded coals, whose descriptions have already been discussed in Section 1.3.

1.8.2 Mineral Matters in Coal

Mineral matters of coal are the important structural components being studied recently. The determination of major and trace elements in coals is of primary importance and represents a significant challenge for assessments and management of their environmental issues of coal processing. Mineral matter refers to all forms of inorganic materials present in coal and therefore, includes various metals and anions as well as mineral phases. It also refers to the elements that are bonded in various ways to the organic (C, H, O, N, and S) components in coal. The distribution and concentrations of different metals and mineral matter content in coal significantly vary for different coalfields. The term *mineral* refers only to the discrete mineral phases. Minerals in coal occur as discrete grains or flakes in one of five physical modes: (1) disseminated, as tiny inclusions within macerals; (2) layers or partings (also lenses), where fine-grained minerals predominate; (3) nodules, including lenticular or spherical concretions; (4) fissures (cleat and fracture fillings and also small void fillings); (5) rock fragments megascopic masses of rock replacements of coal as a result of faulting, slumping, or related structures (Harvey and Ruch 1986). The most common minerals and mineral groups in coal seams are quartz, clays, feldspars, sulfides, and carbonates. Of the clay minerals, kaolinite is quite common (Gluskoter 1967; Rao and Gluskoter 1973), as is illite–smectite. Siderite, calcite, and dolomite are commonly observed carbonate minerals in coal. Pyrite is the most commonly found sulfide mineral in coal (Gluskoter et al. 1977). Clays (primarily illite and kaolinite) and quartz are common clastic minerals in coal, and in some cases, can account for nearly all of the mineral matter present (Stach et al. 1982). Syngenetic quartz is also present in coal seams. Quartz dissolution has been reported from mire-type environments under reducing conditions (Bennett et al. 1988). Syngenetic quartz and clay may appear as cell and pore infillings (Ward 2002), or as coatings or overgrowths on clastic mineral grains.

The majority of iron present in coal mineral matter is in the form of syngenetic and epigenetic iron disulfides. The most common sulfide mineral in coal is found to be pyrite. The pyrite in coal generally has syngenetic or early epigenetic origins. However, organically bound sulfur is also common in coals. Iron is also present in the form of siderite, which is usually present in coal as small (1–2 mm) nodules, often associated with vitrinite and clay layers (Stach et al. 1982). Siderite is thought to form syngenetically or during early diagenesis. The sulfides make up a significant portion of the epigenetic mineral fraction in coal. Pyrite is also found associated with other sulfides or oxides, in quartz veins in coals as well as in coal beds.

Carbonates are found as epigenetic minerals in the form of veins or cleat fillings in coal (Stach et al. 1982). Common epigenetic carbonates include calcite, ankerite, siderite, and dolomite. The authigenic calcium is thought to be derived from groundwater (Cecil et al. 1978). Thus, there is more possibility of calcium carbonate minerals to be found in coals produced from topogenous mires than in coals produced by ombrogenous mires (Cecil et al. 1985; McCabe 1993).

Phosphate minerals may contain significant amounts of the rare earth elements present in coal (Willett et al. 2000). Apatite is typically the most common of the phosphate minerals occurring in coals. Other reported forms of phosphate minerals include aluminophosphates (Crowley et al. 1993; Finkelman and Stanton 1978; Rao and Walsh 1997, 1999; Ward et al. 1996). Phosphate minerals are present as cell and pore fillings, in concentrated layers within coal units, and aluminophosphates are known to be associated with tonsteins (Bohor and Triplehorn 1993; Hill 1988; Schatzel and Stewart 2012).

The mineral matters present in coal are generally determined through the X-ray diffraction investigation. Some of the principal minerals present in coal are given in Table 1.9. The coals in the United States are enriched with clay minerals, pyrite, quartz, calcite, marcasite, sphalerite, feldspar, hematite, siderite, ankerite, and so on (Harvey and Ruch 1986).

TABLE 1.9
Principal Minerals Found in Coal

Silicates	
Quartz	SiO_2
Chalcedony	SiO_2
Kaolinite	
Ilite	$Al_2Si_2O_5(OH)_4$
Smectite	$K_{1.5}Al_4(Si_{6.5}Al_{1.5})O_{20}(OH)_4$
Chlorite	$Na_{0.35}(Al_{1.67}Mg_{0.33})Si_4O_{10}(OH)_2$
Interstratified clay	$(MgFeAl)_6(AlSi)_4O1_0(OH)_8$
mineral	$KAlSi_3O_8$
Feldspar	$NaAlSi_3O_8$
Toumaline	$CaAl_2Si_2O_8$
Analcime	$Na(MgFeMn)_3Al_6B_3Si_6O_{27}(OH)_4$
Clinoptilolite	$NaAlSi_2O_6.H_2O$
Heulandite	$(NaK)_6(SiAl)_{36}O_{72}.20H_2O$
	$CaAl_2Si_7O_{18}.6H_2O$

Sulfides	
Pyrite	FeS_2
Marcasite	FeS_2
Pyrrhotite	$Fe_{(1-x)}S_2$
Sphalertite	ZnS
Galena	PbS
Stibnite	SbS
Milerite	NiS

Phosphates	
Apatite	$Ca_5F(PO_4)_3$
Crandalite	$CaAl_3(PO_4)_2(OH)_5.H_2O$
Gorceixite	$BaAl_3(PO_4)_2(OH)_5.H_2O$
Goyazite	$SrAl_3(PO_4)_2(OH)_5.H_2O$
Monazite	$(Ce,La,Th,Nd)PO_4$
Xenotime	$(Y,Er)PO_4$

Carbonates	
Calcite	$CaCO_3$
Argonite	$CaCO_3$
Dolomite	$CaMg(CO_3)_2$
Ankerite	$(Fe,Ca,Mg)CO_3$
Siderite	$FeCO_3$
Dawsonite	$NaAlCO_3(OH)_2$
Strontianite	$SrCO_3$
Witherite	$BaCO_3$
Alsonite	$CaBa(CO_3)_2$

Sulfates	
Gypsum	$CaSO_4.2H_2O$
Bassanite	$CaSO_4.1/2H_2O$
Anhydrite	$CaSO_4$
Barite	$BaSO_4$
Coquimbite	$Fe_2(SO_4)_3.9H_2O$
Rozentite	$FeSO_4.4H_2O$
Szomolnkite	$FeSO_4.H_2O$
Natrojaarosite	$NaFe_3(SO_4)_2(OH)_6$
Thenardite	Na_2SO_4
Glauberite	$Na_2Ca(SO_4)_2$
Hexahydrite	$MgSO_4.6H_2O$
Tschermigite	$NH_4(SO_4)_2.12H_2O$

Others	
Anatase	TiO_2
Rutile	TiO_2
Boehmite	Al.O.OH
Geothite	$Fe(OH)_3$
Crocoite	$PbCrO_4$
Chromite	$(Fe,Mg)Cr_2O_4$
Clausthalite	PbSe
Zircon	$ZrSiO_4$

Source: Ward, C.R., *Int. J. Coal Geol.*, 50, 135–168, 2002.

It has been reported that in northeast Indian high-sulfur coals, the major minerals (>5%) are identified in the crystalline matter of coal, which include quartz, kaolin, illite, feldspar, calcite, pyrite, and gypsum (Baruah 2008, 2009). The other minerals commonly present in minor (1%–5%) and accessory (<1%) amounts in coal.

1.9 COAL SAMPLING (IN SITU AND EX SITU) AND COAL ANALYSIS

1.9.1 Coal Sampling

Samples are the representative fractions of a body of material that are required for testing and analysis in order to assess the nature and composition of the parent body. They are collected by approved methods (e.g., ASTM and BIS) to protect them from contamination and chemical changes. The variability of coal makes coal sampling a difficult task. Sampling of coal is required to determine the suitability of coal for further investigation, for mine development program and for the quality study of the coal mined in opencast and underground mines. Sampling is done in two methods: in situ and ex situ sampling.

In situ coal samples are taken from surface exposures, exposed coal seams in opencast, underground workings, and from drill cores and cuttings. Ex situ samples are taken from run of mine coal streams, coal transport containers, and coal stockpiles. Sampling may have to be undertaken in widely differing conditions, particularly those of climate and topography. It is important to avoid weathered coal sections, coals contaminated by extraneous clay or other such materials, coals containing a bias of mineralization, and coals in close contact with major faults and igneous intrusions (Thomas 2013).

1.9.1.1 In Situ Sampling

One of the best sampling methods for in-seam coals is the channel sampling method. In this method, the coal is normally sampled perpendicular to the bedding. A channel of uniform cross section is cut manually into the coal seam, and all the coal within the cut section is collected on a plastic sheet placed at the base of the channel. Most channels are around 1.0 m across and samples should not be less than 15 kg m^{-1} of coal thickness (Thomas 2013). The sampling of the full seam actually provides overall quality of the seam including the mineral matters within the coal (Speight 2013).

In the pillar sampling method, a large block of undisturbed coal is usually sampled from some specific areas of potential or known problems in underground coal mining, where the sampling scheme is similar to that of the channel sampling. Core sampling is an integral part of coal exploration and mine development, which has the advantage of producing nonweathered coal including the coal seam floor and roof, and unlike channel samples, core samples preserve the lithological sequence within the coal seam. The coal samples, cylindrical in nature, preserve the lithological sequence within the coal seam. However, the cleaning of the core samples has to be done if drilling fluids have been used, and then also lithologically logged to study the coal thickness and to adjust the core recovery (Speight 2013; Thomas 2013). Bulk sampling may also be collected from a site already channel sampled, loaded into drums, numbered and shipped to the selected test center. This type of sampling checks the swelling properties of coal and is used to rank coal as high-pressure coal and low-pressure coal (Speight 2013; Thomas 2013).

1.9.1.2 Ex Situ Sampling

Ex-situ coal sampling is carried out on moving streams of coal, from rail wagons, trucks, barges, grabs, or conveyors unloading ships, from the holds of ships and from coal stockpiles. The various practices used in collecting ex situ samples and the mathematical analysis for representative of the samples, that is, quality control, is reported by Laurila and Corriveau (1995). Increments of samples are taken by using four methods:

1. Systematic sampling, where increments are spaced evenly in time or in position over the unit.
2. Random sampling, where increments are spaced at random but prerequisite numbers are taken.
3. Stratified random sampling, where the unit is divided by time or quantity into a number of equal strata and one or more increments are taken at random from each (Thomas 2013).
4. Coning and quartering, which involve pouring the sample so that it takes on a conical shape, and then flattening it out into a cake. The cake is then divided into quarters; the two quarters that sit opposite one another are discarded, while the other two are combined and constitute the reduced sample. The same process is continued until an appropriate sample size remains.

1.9.2 Coal Analysis

The analysis of coal involves specific analytical methods to measure the particular physical and chemical properties of coals. These methods are used primarily to determine the suitability of coal for coking, power generation, or for iron ore smelting in the manufacture of steel. The coal analyses are done in three categories: proximate analysis, ultimate analysis, and miscellaneous analysis. Proximate analysis is the determination, by prescribed methods, of the contents of moisture, volatile matter, ash, and fixed carbon (by difference) content. On the other hand, the ultimate analysis of coal and coke involves the determination of carbon, hydrogen, total sulfur, nitrogen, and ash content of the coal, and the estimation of oxygen content by difference. Miscellaneous analysis is a collective category for various types of physical and chemical tests for coal that are commonly requested by coal producers and buyers. This category includes tests such as the determination of calorific value, analysis of the forms of sulfur, carbon, chlorine analysis, major and minor elements in ash analysis, trace element analysis, carbon dioxide analysis, determinations of free-swelling index, grindability, plastic properties of coal, and ash fusibility.

1.9.2.1 Proximate Analysis

The standard test method for proximate analysis (ASTM D-3172) covers the methods of analysis associated with the proximate analysis of coal and coke.

1. *Moisture*: Moisture in coal may occur in four possible forms. It can be surface, hydroscopic, decomposition, and mineral moisture. The surface moisture results from water held on the surface of coal particles or macerals. Hydroscopic moisture is caused by water held by capillary action within the microfractures of coal. Decomposition moisture is due to water held within the coal's decomposed organic compounds. Mineral moisture is a result of water, which comprises part of the crystal structure of hydrous silicates, especially clay minerals. Thus, the moisture is usually determined as total moisture, calculated as the loss of weight between the untreated (1 g coal) and analyzed samples. Moisture can be determined by different methods, including heating coal with toluene, drying in a minimum free-space oven at 150°C in nitrogen atmosphere, and drying in air at 100°C to 105°C. The first two methods can be used for low-rank coals. The third method is for high-rank coals (for low-rank coals, oxidation may take place). The third method can also be used for determination of the inherent moisture. However, the analysis should be run in vacuum.
2. *Volatile matter*: Volatile matter of coal is determined by using standard test methods, that is, ASTM D-3175; ISO 562, which is related to the percentage of volatile products excluding the moisture vapor released during the heating of coal or coke under rigidly controlled conditions (Speight 2013).
3. *Ash content*: The residue remained after the combustion of coal under standard conditions (ASTM D-3174; ISO 1171) is called as *ash content*, which is composed primarily of oxides and sulfates. The ash is formed due to the chemical changes taking place in the mineral matter during the ashing process. The quantity of ash can be more than, equal to, or less than the quantity of mineral matter in coal, depending on the nature of the mineral matter and the chemical changes that take place during the ashing process.
4. *Fixed Carbon*: Fixed carbon is the difference of these three values summed and subtracted from 100. In low-volatile materials, such as coke and anthracite coal, the fixed-carbon value equates approximately to the elemental carbon content of the sample.

The ranges of proximate and ultimate analyses of various ranks of coals are as shown in the Table 1.10.

1.9.2.2 Ultimate Analysis

It is already stated that the ultimate analysis of coal involves determination of the carbon and hydrogen, as well as sulfur, nitrogen, and oxygen (usually estimated by difference). Trace elements that occur in coal are often included as a part of the ultimate analysis. Thus, the standard method for the ultimate analysis of coal and coke (ASTM D-3176-89) includes the determination of elemental carbon, hydrogen, sulfur, and nitrogen, together with the ash in the material as a whole. Oxygen is usually calculated by difference. The test methods recommended for elemental analysis also include the determination of carbon and hydrogen (ASTM D-3178), nitrogen (ASTM D-3179), and sulfur (ASTM D-3177-02; ISO 334; ISO 351), with associated determination of moisture (ASTM D-3173) and ash (ASTM D-3174) to convert the data to a moisture-ash-free basis.

The carbon determination also includes carbon present as organic carbon in the coal substance and any carbon present as mineral carbonate. The hydrogen determination includes hydrogen present in the organic materials as well as hydrogen in all of the water associated with the coal. In the absence of evidence to the contrary, all of the nitrogen is assumed to occur within the organic matrix of coal. Oxygen occurs in both the organic and inorganic portions of coal. In the organic portion, oxygen is present in phenol groups,

TABLE 1.10
Composition and Property Ranges for Various Ranks of Coal (wt%)

	Anthracite	Bituminous	Sub-Bituminous	Lignite
Moisture (%)	3–6	2–15	10–25	25–45
Volatile matter (%)	2–12	15–45	28–45	24–32
Ash (%)	4–15	4–15	3–10	3–15
Fixed Carbon (%)	75–85	50–70	30–57	25–30
Carbon (%)	75–85	65–80	55–70	35–45
Hydrogen (%)	1.5–3.5	4.5–6	5.5–6.5	6–7.5
Nitrogen (%)	0.5–1	0.5–2.5	0.8–1.5	0.6–1.0
Oxygen (%)	5.5–9	4.5–10	15–30	38–48
Sulfur (%)	0.5–2.5	0.5–6	0.3–1.5	0.3–2.5

Source: Speight, J.G., *The Chemistry and Technology of Coal*, 3rd Edition, CRC Press, Taylor & Francis Group, Boca Raton, FL,2013.

carboxyl, methoxyl, and carbonyl. The inorganic materials in coal that contain oxygen are the various forms of moisture, silicates, carbonates, oxides, and sulfates.

On the other hand, sulfur occurs in three forms in coal: (1) organic sulfur; (2) inorganic sulfur, that is, iron sulfides, pyrite, and marcasite (FeS); and (3) sulfate sulfur (e.g., Na_2SO_4 and $CaSO_4$). The elemental sulfur has also been reported to be present in coal (Baruah 1994). The reaction of elemental sulfur with coal aromatics account for the formation of organosulfur compounds in coals (Narayan et al. 1989).

The determination of chlorine content in coal is also important for its further utilization. It occurs predominantly as sodium, potassium, and calcium chlorides, with magnesium and iron chlorides present in some coals. There are two standard methods of determining chlorine in coal (ASTM D-2361 and ASTM D-4208). Method of determination of chlorine in coal generally includes combusting the coal sample with or without Eschka mixture, in an oxygen bomb and heating the mixture in an oxidizing atmosphere. Eschka mixture is a combination of two parts by weight of magnesium oxide and one part of anhydrous sodium carbonate.

The presence of mercury in coal is identified as a very dangerous environmental contaminant and its emission is an environmental concern. The test for total mercury (ASTM D-3684-01; ISO 15237) involves combusting a weighed sample in an oxygen bomb with dilute nitric acid absorbing the mercury vapors. The ultimate analyses also include determination of carbon dioxide (ASTM D-1756-02; ASTM D-6316), arsenic, and selenium (ASTM D-4606) (Speight 2013).

1.9.2.3 Thermochemistry of Coal

The thermochemistry of coal is important in determining the applicability of coal to a variety of conversion processes such as combustion, carbonization, gasification, and liquefaction. The thermal properties of coal include energy of activation, calorific value, ash fusibility, caking index, and swelling index. Generally, thermogravimetric analysis (TGA) is used to study the thermal properties of coals.

The calorific value is an important parameter for assessment of coal quality, which signifies the heat produced by the combustion of a unit quantity of coal in a bomb calorimeter, with oxygen and under a specified set of conditions (standard methods: ASTM D-121; ASTM D-2015; ASTM D-3286; ISO 1928). The calorific value is usually expressed as the gross calorific value or the higher heating value and the net calorific value or lower calorific value. The energy content of the coal can also be expressed as the *useful* heating value, which is an expression derived from the ash and moisture contents for noncaking coals through the formula

$$\text{Useful heating value (kcal/kg)} = 8900 - 138 \times \left[\text{ash content (wt\%)} + \text{moisture content (wt\%)}\right]$$

The differences between the gross calorific value and the net calorific value are given by

$$\text{Net calorific value (BTU/lb)} = \text{gross calorific value} - \left[(1030 \times \%\ \text{total hydrogen} \times 9)/100\right]$$

Calorific value of coal can also be determined by means of various formulas, the most popular of which are as follows:

1. Dulong formula:

$$\text{Calorific value} = 144.4(\%\text{C}) + 610.2(\%\text{H}) - 65.9(\%\text{O}) - 0.39(\%\text{O})^2$$

2. Dulong–Berthelot formula:

$$\text{Calorific value} = 81370 + 345 \left\{\left[\%\text{H} - (\%\text{O} + \%\text{N} - 1)\right]/8\right\} + 22.2(\%\text{S})$$

where:

% C, % H, % N, % O, and % S are the respective carbon, hydrogen, nitrogen, oxygen, and organic sulfur contents of the coal (all of which are calculated to a dry, ash-free basis). In both the cases, the values calculated are in close agreement with the experimental calorific values (Speight 2013)

The fusion characteristics of the coal ash either in a reducing or in an oxidizing atmosphere is provided by the ash fusibility test method (ASTM D-1857). Thus, it gives an approximation of the temperatures at which the ash remaining after the combustion of coal will sinter, melt, and flow. Sintering is the process by which the solid ash particles weld together without melting. The temperature points are measured by observation of the behavior of triangular pyramids (cones) produced from coal ash when heated at a specified rate in a controlled atmosphere (Speight 2013). The critical temperature points are as follows:

1. *Initial deformation temperature*: The temperature at which the first rounding or bending of the apex of the cone occurs. Cone is prepared from the coal ash.
2. *Softening temperature*: The temperature at which the cone has fused down to a spherical lump, in which the height is equal to the width of the base.
3. *Hemispherical temperature*: The temperature at which the cone has fused down to a hemispherical lump, at which point the height is one-half the width of the base.
4. *Fluid temperature*: The temperature at which the fused mass has spread out in a nearly flat layer.

The caking index of coal gives a measure of the caking properties, while swelling index indicates the degree of swelling that a sample of coal will undergo during carbonization. Both the parameters are important in the coal carbonization industries

(Speight 2013). In the caking index test, the maximum whole-number ratio of sand to coal in a 25 g mixture, on carbonization under standard conditions, produces a carbonized mass capable of supporting a 500 g weight and yielding less than 5% of loose, unbound material. The powdered coal, when heated in the absence of air coalesces into coherent mass. Then, this coherent mass swells and resolidifies into coke. The amount of swelling helps in the evaluation of the coking behavior of coal for industrial production of coke (Speight 2013).

1.10 ANALYTICAL METHODS FOR COAL ANALYSIS

The chemistry of coal has been receiving much attention among coal chemists because of their importance in chemical reactivity during various utilizations. However, detailed chemical characterization has been found to be extremely difficult and therefore, research on coal chemistry is still a challenging task and continues to be pursued intensively (Saikia 2009). There are comprehensive and well-defined results on the chemistry of coal on the basis of advance analytical methods. Thus, the coal can be analyzed through conventional and advanced bulk analytical techniques. The conventional techniques include proximate and ultimate analyses, ash fusion temperature, and petrographic analysis, which have already been discussed in Sections 1.9.2.1 through 1.9.2.3. The petrographic analysis can be done through various characterization techniques such as reflectance microscopy, NMR techniques, FT-IR spectroscopy, and X-ray techniques to study the organic matters in coal.

The advanced analytical techniques include differential thermal analysis (DTA), differential scanning calorimetry (DSC), thermomechanical analysis (TMA), chemical fractionation, X-ray fluorescence, X-ray diffraction, Mössbauer spectroscopy, TGA, Carbon-13 (^{13}C) and Proton (1H) Nuclear Magnetic Resonance (NMR) spectroscopy, FT-IR spectroscopy, chromatographic techniques, and electron microscopy (scanning electron microscopy/transmission electron microscopy) (Gupta 2006). However, there are other advanced level multi analytical methods to the characterization of minerals associated with coals and to diagnose their potential risk (Silva et al. 2012).

TGA, DSC, and DTA techniques are useful tools for the study of combustion, pyrolysis behavior, and kinetics of coals. The TGA/DTA/DSC reveals the different types of reactions during coal combustion, formation of solid, liquid and gaseous products, and records the mass loss of the sample with time and temperature. Shi et al (2014), Gomez et al. (2014), and Huang et al. (1995) studied the different reactions during pyrolysis of coals and determined the coal rank through thermogravimetric analysis. DTA and DSC are employed to track the reaction type (e.g., endothermic or exothermic), calculate the kinetics and heat flow rates, and delineate Thermogravimetry (TG)/derivative thermogravimetry (DTG) combustion curves (Sis 2007). Different researchers have worked on pyrolysis, combustion characteristics of coals and coal blends, and the effect of coal particle size through study of TGA-DTG and DSC curves (Chang et al. 2009; Mustafa et al. 1998; Sahu et al. 2010; Saikia et al. 2009a, b, c; Xiangguo et al. 2006).

Thermomechanical analysis is used to acquire knowledge on the fusibility, melting, and sintering behavior of coal particles, as well as information such as deposit strength and its influence on heat transfer. Both Bryant et al. (2000) and Gupta et al. (1999) had characterized the thermal properties of coal ash by using thermomechanical analysis.

The chemical fractionation of coal is also one of the techniques to determine the trace element's chemistry in coal. It provides species-specific information using three successive selective extractions of elements based on solubility that reflects their association in coal. There are three types of successive extractions: (a) by using water for removal of water-soluble salts containing elements such as sodium; (b) by using ammonium acetate to remove elements such as sodium, calcium, and magnesium that are ion exchangeable; and (c) by using hydrochloric acid or other to remove acid-soluble species such as alkaline earth sulfates and carbonates. The residual material typically consists of silicates, oxides, and sulfides.

The FT-IR spectroscopy is the most important tool for assessment of the functional groups present in coal structure and chars. It characterizes the inorganic and organic matter present in coal (Saikia et al. 2007a, b, c). FT-IR has been extensively used in the identification of the chemical structure of coals (Baruah and Khare 2010; Saikia et al. 2009a, c). Saikia et al. (2007a, b, c) used the difference FT-IR spectroscopy for evaluation of changes in coal structure during solvent extraction.

The X-ray fluorescence analysis is one of the most powerful techniques used for determination of trace elements, sulfur content, ash content, and yield of liquefied coal and also for analysis of effluents from coal conversion processes. Inductively coupled plasma and atomic absorption spectrometry are also used for determining the major and trace element contents of ash and coal and inductively coupled plasma can detect the hazardous trace elements such as As, Cd, and Pd in low parts per billion concentration range (Whateley 2002).

On the other hand, X-ray diffraction reveals the structural information of coal and chars and the changes in the crystalline structure of coal during conversion processes. X-ray diffraction can be used to identify the changes in the crystallite size for samples produced at different temperatures in a wide temperature range (~900°C–1500°C). Saikia et al. (2009b) reported the presence of graphene layers in Assam coal using X-ray diffraction technique. There are extensive studies on coal chemical structure by using X-ray diffraction methods (Saikia 2010; Saikia et al. 2009a, b, c; Van Krevelen 1993; Wertz 1998).

Electron probe microanalysis describes the chemical composition of the coal sample, while scanning electron microscopy describes the particle size, swelling of the particles during conversion, and the structure of coal/char/ash. The scanning electron microscopy and transmission electron microscopy identify the maceral composition, as well as the mineral matters in coal and explore the relationships between coal minerals and certain maceral types (Gluskoter and Lindahl 1973; Saikia et al. 2009c; Wall et al. 1998). Computer-controlled scanning electron microscopy (CCSEM) is an advanced technique that provides information on mineral matters in coal including mineral types, their size distribution, and relationship between included and excluded minerals (Gupta et al. 1997; Saikia and Ninomiya 2011). It also provides a more robust analysis of coal mineral matters and their association, and it would be invaluable for resolving many complex questions in clean coal technology (Saikia and Ninomiya 2011).

Mössbauer spectroscopy provides information on the iron-bearing minerals in coal, which plays a significant importance in the study of slag formation in a coal combustion system. The speciation of iron and sulfur in coals has been extensively studied by Mössbauer spectroscopy and is well established (Waanders et al. 2003; Waanders and Bunt 2006). Due to the presence of iron in a large percentage of mineral matter appearing in coal, Mössbauer spectroscopy is a useful, and to a certain degree is a unique, analytical tool in the identification of iron-bearing minerals (Long and Stevens 1986; Montano 1981). Huggins and Huffman (1979) also studied the coal through the application of Mössbauer spectroscopy.

^{1}H NMR spectroscopy enables the functional groups present in coal to be identified and indicates the low and high molecular weight fractions of coals (Saikia et al. 2013). On the other hand, the solid-state ^{13}C NMR spectroscopy describes the average carbon skeletal structure of coal. The cross-polarization, magic angle spinning, and dipolar dephasing techniques permit direct measurement of the number and diversity of aromatic and aliphatic regions of the coal matrix (Hambly 1998). The ^{13}C NMR spectroscopy also determine the number of carbons per cluster, number of attachments per cluster, the number of bridges and loops, the ratio of bridge to total attachments, the average aromatic cluster molecular weight, and the average side chain molecular weight (Gupta 2006).

Pyrolysis-gas chromatography-mass spectrometry (Py-GC-MS) is one of the latest developments in coal chemistry, which studies the molecular components or structural units of the polymeric organic solids. The combined use of solid-state NMR and Py-GC-MS provides information on the average structure and specific molecular components along with insights into the major and minor changes in coal structures during coal liquefaction processes (Song et al. 1991).

1.11 COAL PRODUCTS

Coal-based products play an increasingly important role in fulfilling the needs of energy, chemicals, and other gaseous products (Levine et al. 1982). The coal-based products can be classified as coal combustion products, coal-derived products, and other products.

1.11.1 Coal Combustion Products

Coal combustion products are produced primarily from the combustion of coal, which include fly ash, bottom ash, boiler slag, flue gas, fluidized bed combustion ash, cenospheres, and scrubber residues. The size, shape, and chemical composition of these material determine their beneficial reuse in different purposes; for example, coal fly ash could be used as building materials. Moreover, these products when used beneficially can generate environmental, economic, and performance benefits (improved strength, durability, and workability of materials like concrete). However, coal fly ash also imposes health hazards to communities through exposure (Silva et al. 2012; Oliveira et al. 2014).

1.11.2 Coal-Derived Products

Coal-derived products are associated with coal gasification, liquefaction, and carbonization. In coal gasification process, coal is typically exposed to steam and controlled amounts of air or oxygen under high temperatures and pressures. The coal structure breaks apart, initiating chemical reactions that typically produce a mixture of carbon monoxide, hydrogen, and other gaseous compounds. The syngas (H_2 v CO) produced can be used for power generation or manufacture of hydrogen, synthetic natural gas, or liquid fuel. The different coal-derived products from coal gasification are summarized in Figure 1.6.

Coal liquefaction is the process of conversion of coal into liquid fuels such as petroleum oil. The liquefaction processes are known as *direct* and *indirect liquefaction processes* (Figure 1.7). In direct liquefaction process, the pulverized coal is directly converted to liquid products by adding hydrogen in the presence of suitable catalysts at high pressure and temperature. Hydrogen is added to increase H/C ratio in the process. On the other hand, the Fischer–Tropsch process, known as *indirect liquefaction* of coal, is a process where the coal structure is completely broken down into synthesis gas (H_2 and CO) by gasification with steam and oxygen. The CO and H_2 molecules in the syngas are then combined catalytically to produce either hydrocarbon fuels such as synthetic gasoline or synthetic diesel, or oxygenated fuels (Williams and Larson 2003). The direct and indirect liquefaction processes along with their products are summarized in the Figure 1.7.

Carbonization of coal is a promising process to produce various commercial coal-based products. It is the destructive distillation of organic compounds in the absence of air and the production of carbon as coke, liquid such as fuel oil, and tar and gaseous products. This process is also used by steel industries to produce metallurgical coke for use in iron-making blast furnaces and other smelting processes. The gaseous by-product referred to as *coke oven gas* or *coal gas* is also formed along with ammonia, water, and sulfur compounds. Coke oven gas is a valuable heating fuel, which could also be used for generation of electricity. Other carbonization by-products are usually refined within the coke plant to produce commodity chemicals (Kaegi et al. 2000).

1.11.3 Other Products

Coal tar produced from the carbonization of coal is a source of other valuable chemical products. The needle coke is produced from coal tar that also appears during coke production. It has excellent physical properties, it has low coefficient of thermal expansion and low electric resistance, along with less spalling and less breakage. Thus, several commercially

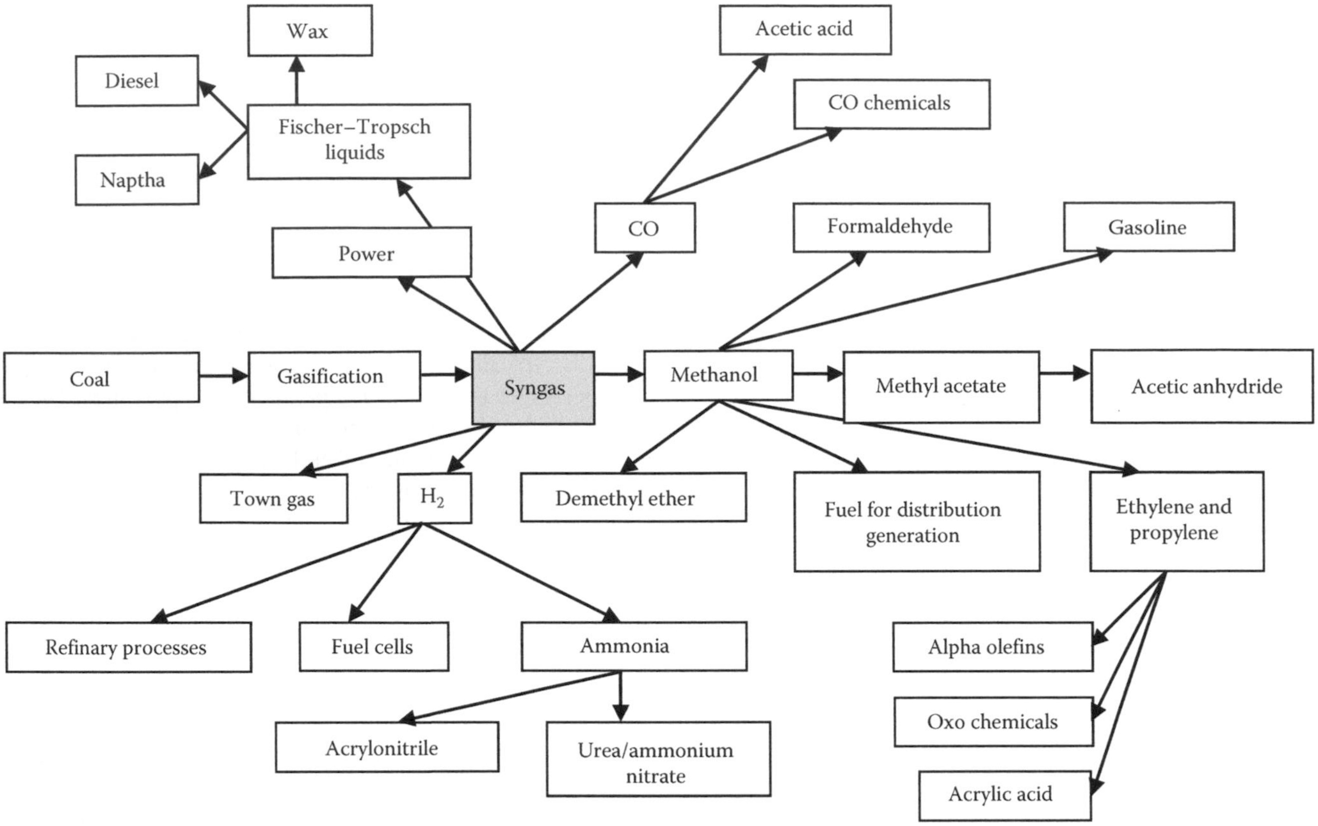

FIGURE 1.6 Summary of the products obtained from gasification process.

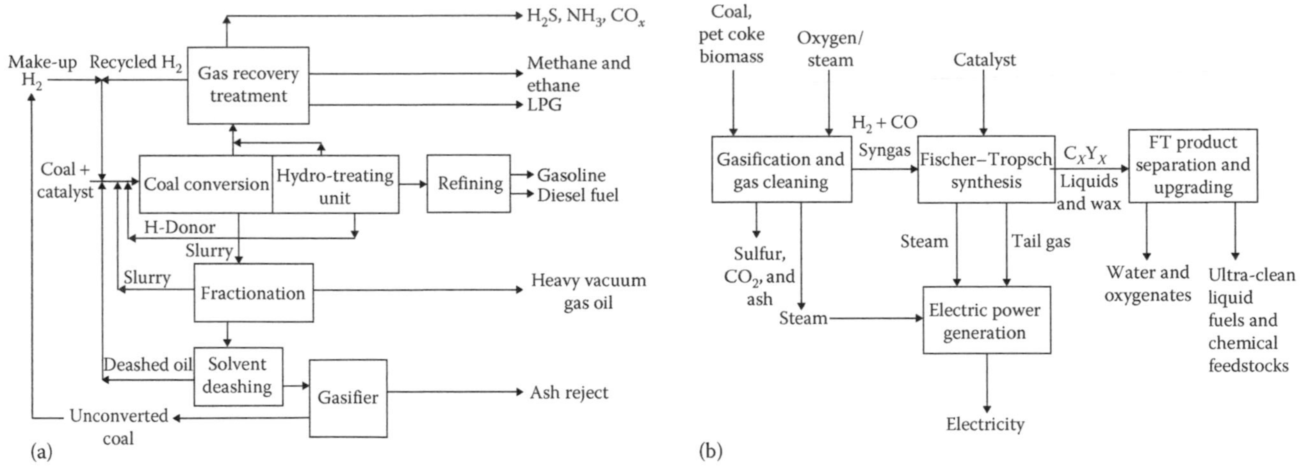

FIGURE 1.7 Products obtained through (a) direct and (b) indirect liquefaction of coals.

valuable chemical products can be produced from the by-products of coal. Refined coal tar is used in the manufacture of chemicals, such as creosote oil, naphthalene, phenol, and benzene. Ammonia gas recovered from coke ovens is used to manufacture ammonia salts, nitric acid, and agricultural fertilizers. Thousands of different products have coal or coal by-products as components: soap, aspirins, solvents, dyes, plastics, and fibers such as rayon and nylon. Coal is also an essential ingredient in the production of specialist products such as activated carbon, carbon fiber, carbon foam, carbon nanotube, and fullerene. There is other value-added product such as humic acid, which could also be chemically produced. Das et al. (2013) reported the isolation of coal-based humic acid from the inferior coal and coal wastes from Mongchen coalfield, Nagaland (India). The humic acid has its utilization in industrial sectors particularly in the soil conditioning and complexation of heavy metals in environmentally polluted sites.

REFERENCES

Alvarez, Y.E., Moreno, B.M., Klein, M.T., Watson, J.K., Castro-Marcano, F., and Mathews, J.P. 2013. A novel simplification approach for large-scale structural models of coal: 3D molecules to 2D lattices. 3. Reactive lattice simulations. *Energy & Fuels* 27: 2915–2922.

Alvarez, Y.E., Watson, J.C.K., and Mathews, J.P. 2012a. A novel simplification approach for large-scale structural models of coal: 3D molecules to 2D lattices. 1. Lattice creation. *Energy & Fuels* 26: 4938–4945.

Alvarez, Y.E., Watson, J.K., Pou, J.O., and Mathews, J.P. 2012b. A novel simplification approach for large-scale structural models of coal: 3D molecules to 2D lattices. 2. Visualization capabilities. *Energy & Fuels* 26: 4946–4952.

ASTM D121. 2007. Standard Terminology of Coal and Coke, ASTM International, West Conshohocken, PA.

ASTM D1756-02. 2007. Standard Test Method for Determination as Carbon Dioxide of Carbonate Carbon in Coal, ASTM International, West Conshohocken, PA.

ASTM D-1857. 2003. Standard Test Method for Fusibility of Coal and Coke Ash, ASTM International, West Conshohocken, PA.

ASTM D-2015. 1997. Standard Test Method for Gross Calorific Value of Coal and Coke by the Adiabatic Bomb Calorimeter, ASTM International, West Conshohocken, PA.

ASTM D-2361. 1995. Standard Test Method for Chlorine in Coal, ASTM International, West Conshohocken, PA.

ASTM D-3172. 2007. Standard Practice for Proximate Analysis of Coal and Coke, ASTM International, West Conshohocken, PA.

ASTM D-3173. 2002. Standard Test Method for Moisture in the Analysis Sample of Coal and Coke, ASTM International, West Conshohocken, PA.

ASTM D-3174. 2011. Standard Test Method for Ash in the Analysis Sample of Coal and Coke from Coal, ASTM International, West Conshohocken, PA.

ASTM D-3175. 2002. Standard Test Method for Volatile Matter in the Analysis Sample of Coal and Coke, ASTM International, West Conshohocken, PA.

ASTM D3176-89. 2002. Standard Practice for Ultimate Analysis of Coal and Coke, ASTM International, West Conshohocken, PA.

ASTM D3177-02. 2007. Standard Test Methods for Total Sulfur in the Analysis Sample of Coal and Coke, ASTM International, West Conshohocken, PA.

ASTM D-3178. 1997. Standard Test Method for Carbon and Hydrogen in the Analysis Sample of Coal and Coke, ASTM International, West Conshohocken, PA.

ASTM D-3179. 2002. Standard Test Methods for Nitrogen in the Analysis Sample of Coal and Coke, ASTM International, West Conshohocken, PA.

ASTM D-3286. 1996. Standard Test Method for Gross Calorific Value of Coal and Coke by the Isoperibol Bomb Calorimeter, ASTM International, West Conshohocken, PA.

ASTM D3684-01. 2006. Standard Test Method for Total Mercury in Coal by the Oxygen Bomb Combustion/Atomic Absorption Method, ASTM International, West Conshohocken, PA.

ASTM D-4208. 2002. Total Chlorine in Coal by the Oxygen Bomb Combustion/ion Selective Electrode Method, ASTM International, West Conshohocken, PA.

ASTM D-4606. 2003. Standard Test Method for Determination of Arsenic and Selenium in Coal by the Hydride Generation/Atomic Absorption Method, ASTM International, West Conshohocken, PA.

ASTM D-6316. 2009. Standard Test Method for Determination of Total, Combustible and Carbonate Carbon in Solid Residues from Coal and Coke, ASTM International, West Conshohocken, PA.

Attar, A. and Dupuis, F. 1979. Preprints, division of fuel chemistry. *American Chemical Society* 24(1): 166.

Attar, A. and Hedrickson, G.G. 1982. In *Coal structure*, R.A. Meyers (Ed.), Academic Press, New York.

Barsky, V., Vlasov, G., and Rudnitsky, A. 2009. Composition and structure of coal organic mass. Analytical review. *Chemistry & Chemical Technology* 3: 4.

Baruah, B.P. 2008. *Environmental Studies Around Makum Coalfield, Margherita.* Thesis, Dibrugarh University, Assam, India, 118.

Baruah, B.P. 2009. *Environmental Studies Around Makum Coalfield, Assam, India.* Lambert Academic Publishing AG & Co. KG, Germany, 16–18.

Baruah, M.K. 1984. Infra-red identification of Fe-S bond in Assam coal. *Current Science* 53: 1242–1243.

Baruah, M.K. 1992. The chemical structure of secondary sulfur in Assam coal. *Fuel Processing Technology* 31(2): 115–126.

Baruah, M.K. 1994. Is elemental sulfur responsible for high-sulfur coal, *Fuel Processing Technology*, 40(1), 97–100.

Baruah, B.P. and Khare, P. 2010. Mobility of trace and potentially harmful elements in the environment from high sulfur Indian coal mines. *Applied Geochemistry* 25: 1621–1631.

Baruah, M.K., Kotoky, P., and Borah, G.C. 2003. Distribution and nature of organic/mineral bound elements in Assam coals, India. *Fuel* 82: 1783–1791.

Baruah, B.P., Saikia, B.K., Kotoky, P., and Rao, P.G. 2006. Aqueous leaching on high sulfur sub-bituminous coals, in Assam, India. *Energy Fuels* 20: 1550–1555.

Bennett, P.C., Melcer, M.E., Seigel, D.I., and Hassett, J.P. 1988. The dissolution of quartz in dilute aqueous solutions of organic acids at 25°C. *Geochimica et Cosmochimica Acta* 52: 1521–1530.

Biscoe, J. and Warren, B. E. 1942. An X-ray study of carbon black. *Journal of Applied Physics* 13: 364–371.

Bohor, B.F. and Triplehorn, D.M. 1993. Tonsteins: Altered volcanic ash layers in coalbearing sequences. *Geological Society of America Special Paper* 285: 44.

Boruah, R.K., Saikia, B.K., Baruah, B.P., and Dey, N.C. 2008. X-ray scattering study of the average poly-cyclic aromatic unit in Ledo coal. *Journal of Applied Crystallography* 41: 27–30.

Bryant, G.W., Browning, G.J., Gupta, S.K., Lucas, J.A., Gupta, R.P., and Wall, T.F. 2000. Thermomechanical analysis of coal ash: The influence of the material for the sample assembly. *Energy & Fuels* 14(2): 326–335.

Castro-Marcano, F., Lobodin, V.V., Rodgers, R.P., McKenna, A.M., Marshall, A.G., and Mathews, J.P. 2012. A molecular model for the Illinois no. 6 Argonne Premium coal: Moving towards capturing the continuum structure. *Fuel* 95: 35–49.

Castro-Marcano, F. and Mathews, J.P. 2011. Constitution of Illinois no. 6 Argonne Premium coal: A review. *Energy & Fuels* 25(3): 845–853.

Castro-Marcano, F., Winans, R.E., Chupas, P., Chapman, K., Calo, J.M., Watson, J.K., and Mathews, J.P. 2012. Fine structure evaluation of the pair distribution function with molecular models of the Argonne Premium coals. *Energy & Fuels* 26: 4336–4345.

Cecil, C.B., Stanton, R.W., Allshouse, S.D., and Finkelman, R.B. 1978. Geologic controls on mineral matter in the Upper Freeport coal bed. *Proceedings: Symposium on Coal Cleaning to Energy and Environmental Coals*, vol. 1, U.S. Environmental Protection Agency, E.P.A. 60017-79-0998a, 1: 110–125.

Cecil, C.B., Stanton, R.W., Neuzil, S.G. Dulong, F.T., Ruppert, L.F., and Pierce, B.S. 1985. Paleoclimate controls on late Paleozoic sedimentation and peat formation in the Central Appalachian Basin, USA. *International Journal of Coal Geology* 5: 195–230.

Chandra, D., Singh, R.M., and Singh, M.P. 2000. *Text Book of Coal*, 1st Edition, Tara Book Agency, Kamachha, Varanasi.

Chang, L., Yan-min, Z., and Ming-gao, Y. 2009. Research on low-temperature oxidation and pyrolysis of coal by thermal analysis experiment. *Procedia Earth and Planetary Science* 1(1): 718–723.

Chou, C.L. 2012. Sulfur in coals: A review of geochemistry and origins. *International Journal of Coal Geology* 100: 1–13.

Crowley, S.S., Ruppert, L.F., Belkin, H.E., Stanton, R.W., and Moore, T.A. 1993. Factors affecting the geochemistry of a thick, subbituminous coal bed in the Powder River Basin: Volcanic, detrital, and peat forming processes. *Organic Geochemistry* 20: 843–853.

Dai, S., Zhang, W., Seredin, V.V., Ward, C.R., Hower, J.C., Song, W., Wang, X. et al. 2013. Factors controlling geochemical and mineralogical compositions of coals preserved within marine carbonate successions: A case study from the Heshan Coalfield, southern China. *International Journal of Coal Geology* 109–110: 77–100.

Das, T., Saikia, B.K., and Baruah, B.P. 2013. Feasibility studies for isolation of humic acid from coal of Mongchen coalfiled, Nagaland. *Journal of Indian Chemical Society* 90: 2007–2014.

Davidson, R.B. 1982. Molecular structure of coal. In *Coal Science*, M.L. Gorbaty, J.W. Larsen, and I. Wender (Eds.), Vol. 1, Academic Press, New York, 84–155.

Duane, G.L., Schlosberg, R.H., and Silbernagel, B.G. 1982. Understanding the chemistry and physics of coal structure (A review). *Proceedings of the National Academy of Sciences of the United States of America* 79: 3365–3370.

Ferguson, A., Ho, A., Graham, B., Kennedy, E., Stockenhuber, M., Friggieri, J., and Mahoney, M. 2013. Characterization of coal tars by gel permeation chromatography and NMR analysis.

Finkelman, R.B. and Stanton, R.W. 1978. Identification and significance of accessory minerals from a bituminous coal. *Fuel* 57: 763–768.

Ghosh, T.K. and Prelas, M.A. 2009. *Energy Resources and Systems: Volume 1: Fundamentals and Non-Renewable Resources*, Springer Science & Business Media, 778.

Gluskoter, H.J. 1967. Clay minerals in Illinois coals. *Fuel* 44: 285–291.

Gluskoter, H.J. and Lindahl, P.C. 1973. Cadmium: Mode of occurrence in Illinois coal. *Science* 181: 264–266.

Gluskoter, H.J., Ruch, R.R., Miller, W.G., Cahill, R.A., Dreher, G.B., and Kuhn, J.K. 1977. Trace elements in coal: Occurrence and distribution. *Illinois State Geological Survey Circular* 499: 154.

Gomez, A., Silbermann, R., and Mahinpey, N.A. 2014. Comprehensive experimental procedure for CO_2 coal gasification: Is there really a maximum reaction rate? *Applied Energy* 124: 73–81.

Gorbaty, M.L. and Larsen, J.W. 1982. *Coal Science*, vol. 1, Academic Press, New York, 23.

Gupta, R. 2006. Advanced coal characterization: A review. *Australia Symposium on Advanced Coal Utilization Technology*, China.

Gupta, S.K., Gupta, R.P., Bryant, G.W., Juniper, L., and Wall, T.F. 1999. Thermomechanical analysis and alternative ash fusibility temperatures. In *Impact of Mineral Impurities in Solid Fuel Combustion*, R.P. Gupta, T.F., Wall, and L. Baxter (Eds.), Kluwer Academic/Plenum Press, New York, 155–169.

Gupta, R.P., Kennedy, E., Wall, T.F., and Masson, M. 1997. Computer controlled electron microscopy (CCSEM) bureau service. *Proceedings of the 1st Annual Conference of Participants, CRC for Black Coal Utilisation*, Brisbane, Australia.

Hambly, E.M. 1998. *The Chemical Structure of Coal Tar and Char during Devolatilization*. A M.S. Thesis Presented to the Department of Chemical Engineering, Brigham Young University.

Hamor-Vido, M. and Hámor, T. 2007. Sulphur and carbon isotopic composition of power supply coals in the Pannonian Basin, Hungary. *International Journal of Coal Geology* 71(4): 425–447.

Harvey, R.D. and Ruch, R.R. 1986. Mineral matter in illinois and other U.S. coals, In *Mineral Matter and Ash in Coal*, K.S. Vorres (ed.), Washington, D.C., ACS Symposium Series 301, Chapter 2, 10–40.

Hawe-Grant, M. 1993. Encyclopedia of Chemical Technology, 4th Edition, vol. 6, John Wiley & Sons, New York, 423–447.

Hayatsu, R., Winans, R.E., Scott, R.G., Moore, L.P., and Studier, M.H. 1978a. Trapped organic compounds and aromatic units in coals. *Fuel* 57(9): 541–548.

Hayatsu, R., Winans, R.E., Scott, R.G., Moore, L.P., and Studier, M.H. 1978b. Lignin-like polymers in coals. *Nature (London)*, 275, 116–118.

Hill, P.A. 1988. Tonsteins of Hat Creek, British Columbia: A preliminary study. *International Journal of Coal Geology* 10: 155–175.

Huang, H., Wang, K., Klein, M.T., and Calkins, W.H. 1995. Determination of coal rank by Thermogravimetric Analysis. *Fuel and Energy Abstracts* 37: 170.

Huggins, F.E. and Huffman, G.P. 1979. Mössbauer analysis of iron-containing phases in coal, coke and ash. In *Analytical Methods for Coal and Coal Products*, C. Karr (Ed.), vol. 3. Academic Press, New York, 372–422.

ICCP (International Committee for Coal Petrology). 1963. *International Handbook of Coal Petrology*, CNRS, Paris, France, 160.

ICCP (International Committee for Coal Petrology). 1971. *Supplement to Second Edition of the Handbook*, CNRS, Paris, France.

ICCP (International Committee for Coal Petrology). 1975. *International Handbook of Coal Petrology*, CNRS, Paris, France.

ICCP (International Committee for Coal Petrology). 1993. *International Handbook of Coal Petrography*, 3rd Supplement to 2nd Edition. University of Newcastle upon Tyne, London.

ICCP (International Committee for Coal Petrology). 2011. Organic petrology, macerals, microlithotypes, lithotypes, minerals, rank. Chapter 2. *Training Course on Dispersed Organic Matter*, Department of Geoscience, Portugal, 4–71.

ISO 334. 1992. Solid mineral fuels—Determination of total sulfur-Eschka method, International Standards Organization, Geneva, Switzerland.

ISO 351. 1996. Solid mineral fuels—Determination of total sulfur-High temperature combustion method, International Standards Organization, Geneva, Switzerland.

ISO 562. 2010. Hard coal and coke—Determination of volatile matter, International Standards Organization, Geneva, Switzerland.

ISO 1171. 2010. Solid mineral fuels—Determination of ash, International Standards Organization, Geneva, Switzerland.

ISO 1928. 2009. Solid mineral fuels—Determination of gross calorific value by the bomb calorimetric method and calculation of net calorific value, International Standards Organization, Geneva, Switzerland.

ISO 15237. 2003. Solid mineral fuels—Determination of total mercury content of coal, International Standards Organization, Geneva, Switzerland.

Kaegi, D., Addes, V., Valia, H., and Grant, M. 2000. Coal conversion processes, carbonization. *Kirk-Othmer Encyclopedia of Chemical Technology*, John Wiley & Sons, New York, 27.

Killops, S.D. and Killops, V.J. 2013. *Introduction to Organic Geochemistry*, 2nd Edition, Wiley-Blackwell, UK.

Kumar, A. and Srivastava, S.K. 2013. Disposition pattern of sulphur functional groups in high sulphur ledo coals of Assam. *Journal of Applied Chemistry* 4(3): 1–8.

Langenberg, W., Kalkreuth, W., Levine, J., Strobl, R., Demchuk, T., Hoffman, G., and Jerzykiewicz, T. 1990. *Coal Geology and Its Application to Coal Bed Methane Reservoirs. Lecture Notes for Short Course*, Information Series No. 109, Alberta Research Council, Alberta Geological Survey, Edmonton, 9.

Laurila, M.J. and Corriveau, M.P. 1995. *The Sampling of Coal*, Intertec Publishing, Chicago, IL.

Levine, D.G., Schlosberg, R.H., and Silbernage, B.G. 1982. Understanding the chemistry and physics of coal structure (A review). *Proceedings of the National Academy of Sciences of the United States of America* 79: 3365–3370.

Long, G.J. and Stevens, J.G. 1986. *Industrial Applications of the Mössbauer Effect*. Plenum Press, New York, 796.

Longyi, S., Peter, J.T., Rod, G., Shifeng, D., Shengsheng, L., Yaofa, J., and Pengfei, Z. 2003. Petrology and geochemistry of the high-sulphur coals from the Upper Permian carbonate coal measures in the Heshan Coalfield, southern China. *International Journal of Coal Geology* 1040: 1–26.

Margriet, N., Leeuw, J.W.D., and Crelling, J.C. 1992. Chemical structure of bituminous coal and its constituting maceral fractions as revealed by flash Pyrolysis. *Energy & Fuels* 6(2): 125–136.

Mathews, J.P. and Chaffee, A. 2012. The molecular representations of coal—A review. *Fuel* 96: 1–14.

Mathews, J.P., Krishnamoorthy, V., Louw, E., Tchapda, A.H.N., Castro-Marcano, F., Karri, V., Alexis, D.A., and Mitchell, G.D. 2014. A review of correlations of coal properties with elemental composition. *Fuel Processing Technology* 121: 104–113.

Mathews, J.P. and Sharma, A. 2012. The structural alignment of coals and the analogous case of Argonne Upper Freeport. *Fuel* 95: 19–24.

Mathews, J.P., Van Duin, A., and Chaffee, A. 2011. The utility of coal molecular models. *Fuel Processing Technology* 92(4): 718–728.

McCabe, P.J., 1993. Sequence stratigraphy of coal bearing strata. *AAPG short course*, AAPG New Orleans, LA, 81.

Meyers, R. 2012. *Coal Structure*, Academic Press, New York.

Montano, P.A. 1981. Application of Mossbauer spectroscopy to coal characterisation and utilisation. In *Mössbauer Spectroscopy and Its Chemical Applications. Advances in Chemistry*, J.G. Stevens and G.K. Shenoy (Eds.), vol. 194, American Chemical Society, Washington, DC, 135–175.

Mustafa, V.K., Esber, O., Ozgen, K., and Cahit, H. 1998. Effect of particle size on coal pyrolysis. *Journal of Analytical and Applied Pyrolysis* 45(2): 103–110.

Narayan, R., Kullerud, G., and Wood, K.V. 1989. A new perspective on the nature of organic sulfur in coal. *Abstracts of Papers National Meeting*, American Chemical Society, Washington, D.C, 16.

Oliveira, M.L.S., Marostega, F., Taffarel, S.R., Saikia, B.K., Waanders, F.B., DaBoit, K., Baruah, B.P., and Silva, L.F.O. 2014. Nano-mineralogical investigation of coal and fly ashes from coal-based captive power plant (India): An introduction of occupational health hazards. *Science of the Total Environment* 468–469: 1128–1137.

Oskay, R.G., Karayiğit, A.İ, and Christanis, K. 2013. Coal-petrography and mineralogical studies of the Cardak coal deposit (SW Turkey). *Proceedings of the 65th Annual Meeting of the ICCP, Sosnowiec, Poland*, 103.

Parks, B.C. 1952. Origin, Petrography and classification of coal. In *Chemistry of Coal Utilization*, H.H. Lowry (Ed.), John Wiley & Sons New York, Supplementary, 1, 1–34.

Raghuvanshi, S.P., Chandra, A., and Raghav, A.K. 2006. Carbon dioxide emissions from coal based power generation in India. *Energy Conversion and Management* 47: 427–441.

Rao, P.D. and Gluskoter, H.J., 1973. Occurrence and distribution of minerals in Illinois coal. *Illinois State Geological Survey, Circular* 476: 56.

Rao, P.D. and Walsh, D.E. 1997. Nature and distribution of phosphorous minerals in Cook Inlet coals, Alaska. *International Journal of Coal Geology* 33: 19–42.

Rao, P.D. and Walsh, D.E. 1999. Influence of environments of coal deposition on phosphorus accumulation in a high latitude, northern Alaska coal seam. *International Journal of Coal Geology* 38: 261–284.

Ruberto, R.G., Cronauer, D.C., Jewell, D.M., and Seshadri, K.S. 1977. Structural aspects of sub-bituminous coal deduced from solvation studies. 2. Hydrophenanthrene solvents, *Fuel* 56(1): 25–32.

Sahu, S.G., Sarkar, P., and Chakraborty, N. 2010. Thermogravimetric assessment of combustion characteristics of blends of a coal with different biomass chars. *Fuel Processing Technology* 91(3): 369–378.

Saikia, B.K. 2009. Scanning electron microscopy of Assam coals, India. *Journal of the Geological Society of India* 74: 749–752.

Saikia, B.K. 2010. Inference on carbon atom arrangement in the turbostatic grapheme layers in Tikak coal (India) by X-ray pair distribution function analysis. *International Journal of Oil, Gas and Coal Technology* 3(4): 362–373.

Saikia, B.K. and Baruah, R.K. 2008. Structural studies of some Indian coals by using X-ray diffraction techniques. *Journal of X-Ray Science and Technology* 16(2): 89–94.

Saikia, B.K., Boruah, R.K., and Gogoi, P.K. 2007a. XRD and FT-IR investigations of sub-bituminous Assam coals. *Bulletin of Materials Science* 30(4): 421–426.

Saikia, B.K., Boruah, R.K., Gogoi, P.K. 2007b. FT-IR and XRD analysis of coal from Makum coalfield of Assam. *Journal of Earth System Science* 116(6): 575–579.

Saikia, B.K., Sahu O.P., and Boruah, R.K. 2007c. FT-IR spectroscopic investigation of high sulphur Assam coals and their solvent-extracts. *Journal of Geological Society of India* 70: 917–922.

Saikia, B.K., Boruah, R.K., and Gogoi, P.K. 2009a. X-ray (RDF) and FT-IR analysis of high sulphur Assam coals. *Journal of the Energy Institute* 82(2): 106–108.

Saikia, B.K., Boruah, R.K., and Gogoi, P.K. 2009b. X-ray diffraction analysis on graphene layers of Assam coal. *Journal of Chemical Sciences* 121(1): 1–4

Saikia, B.K., Boruah, R.K., Gogoi, P.K., and Baruah, B.P. 2009c. A thermal investigation on coals from Assam (India). *Fuel Processing Technology* 90: 196–203.

Saikia, B.K., Dutta, A.M., and Baruah, B.P. 2014a. Feasibility studies of de-sulfurization and de-ashing of low grade medium to high sulfur coals by low energy ultrasonication. *Fuel* 123: 12–18.

Saikia, B.K., Dutta, A.M., Saikia, L., Ahmed, S., and Baruah, B.P. 2014b. Ultrasonic assisted cleaning of high sulphur Indian coals in water and mixed alkali. *Fuel Processing Technology* 123: 107–113.

Saikia, B.K., Khound, K., and Baruah, B.P. 2014c. Extractive de-sulfurization and de-ashing of high sulfur coals by oxidation with ionic liquids. *Energy Conversion and Management* 81: 298–305.

Saikia, B.K. and Ninomiya, Y. 2011. An evaluation of heterogeneous mineral matters in Assam (India) coals by CCSEM. *Fuel Processing Technology* 92: 1068–1077.

Saikia, B.K., Sahu, O.P., and Boruah, R.K. 2007c. FT-IR spectroscopic investigation of high sulphur Assam coals and their solvent-extracts. *Journal of Geological Society of India* 70: 917–922.

Saikia, B.K., Sharma, A., Khound, K., and Baruah, B.P. 2013. Solid state 13C-NMR spectroscopy of some oligocene coals of Assam and Nagaland. *Journal Geological Society of India* 82: 295–298.

Saikia, B.K., Ward, C.R., Oliveira, M.L.S., Hower, J.C., Baruah, B.P., Braga, M., and Silva, L.F. 2014d. Geochemistry and nano-mineralogy of two medium-sulfur northeast Indian coals. *International Journal of Coal Geology* 121: 26–34.

Sanyal, A. 1997. Role of Macerals- An underappreciated coal quality parameter for unburned carbon characterization and control. *Proceedings on Third Annual Conference on Unburned Carbon on Utility Fly Ash*, U.S. Department of Energy, Federal Energy Technology Center, Pennsylvania, 63–66.

Schatzel, S.J. and Stewart, B.W. 2012. A provenance study of mineral matter in coal from Appalachian Basin coal mining regions and implications regarding the respirable health of underground coal workers: A geochemical and Nd isotope investigation. *International Journal of Coal Geology* 94: 123–136.

Schopf, J.M. 1956. A definition of coal. *Economic Geology* 51: 521–527.

Seyler, C.A. 1938. Petrology and the classification of coal. *Proceedings of the South Wales Institute of Engineering* 53: 254–327.

Sharma, A., Saikia, B.K., and Baruah, B.P. 2012. Maceral contents of tertiary Indian coals and their relationship with calorific values. *International Journal of Innovative Research & Development* 1(7): 196–203.

Shi, L., Liu, Q., Guo, X., He, W., and Liu, Z. 2014. Pyrolysis of coal in TGA: Extent of volatile condensation in crucible. *Fuel Processing Technology* 121: 91–95.

Shinn, J.H. 1984. From coal to single-stage and two-stage products: A reactive model of coal structure. *Fuel* 63: 1187.

Silva, L.F.O., Sampaio, C.H., Guedes, A. et al. 2012. Multianalytical approaches to the characterisation of minerals associated with coals and the diagnosis of their potential risk by using combined instrumental microspectroscopic techniques and thermodynamic speciation. *Fuel* 94: 52–63.

Sis, H. 2007. Evaluation of combustion characteristics of different size Elbistan Lignite by using TG/DTG and DTA. *Journal of Thermal Analysis and Calorimetry* 88(3): 863–870.

Song, C., Schobert, H.H., and Hatcher, P.G. 1991. Solid-state CPMAS 13C NMR and Pyrolysis-GC-MS studies of coal structure and liquefaction reactions. *Proceedings of the International Conference on Coal Science*, University of Newcastle-Upon-Tyne, UK,, 664.

Speight, J.G. 2013. *The Chemistry and Technology of Coal*, 3rd Edition, CRC Press, Taylor & Francis Group, Boca Raton, FL.

Stach, E., Mackowsky, M-Th., Teichmuller, M., Taylor, G.H., Chandra, D., and Teichmuller, R. 1982. *Stach's Textbook of Coal Petrology.* Gebruder Borntraeger, Berlin, Germany.

Stephen, N. 1986. Opening Address. *Fuel* 65(12): 1627–1628.

Stopes, M.C. 1919. On the four visible ingredients in banded bituminous coals. *Proceedings of the Royal Society* 90B: 470–487.

Taylor, G.H., Teichmuller, M., Davis, A., Diessel, C.F.K., Littke, R., and Robert, P. 1998. *Organic Petrology*, Gebruder Borntraeger, Berlin, Germany, 704.

Temperley, B.N. 1936. *Botryococcus* and algal coal, pt.2: The boghead controversy and the morphology of boghead algae. *Trans. Royal. Soc. Edinb.* 58: 855–858.

Thomas, L. 2013. *Coal Geology*, 2nd Edition, John Wiley and Sons, Chichester, UK.

Ting, F.T.C. 1982. Coal Macerals. *Coal Structure*, R.A. Meyers (Ed.), Academic Press, New York, 7–49.

Tissot, B.P. and Welte, D.H. 1984. *Petroleum Formation and Occurrence*, 2nd Edition. Springer-Verlag, Berlin, Germany, 699.

Tozsin, G. 2014. Hazardous elements in soil and coal from the Oltu coal mine district, Turkey. *International Journal of Coal Geology* 131: 1–6.

Tselev, A., Ivanov, I.N., Lavrik, N.V., Belianinov, A., Jesse, S., Mathews, J.P., Mitchell, G.D., and Kalinin, S.V. 2014. Mapping internal structure of coal by confocal micro-Raman spectroscopy and scanning microwave microscopy. *Fuel* 126(15): 32–37.

Turner, B.R. and Richardson, D. 2004. Geological controls on the sulphur content of coal seams in the Northumberland Coalfield, Northeast England. *International Journal of Coal Geology* 60(2–4): 169–196.

Van Krevelen, D.W. 1961. Coal: Typology-Chemistry-Physics-Constitution. Elsevier Science, Amsterdam, the Netherlands, 514.

Van Krevelen, D.W. 1993. *Coal-Typology: Physics, Chemistry and Constitution*, Elsevier, Amsterdam, New York.

Waanders, F.B. and Bunt, J.R. 2006. Transformation of the Fe-mineral associations in coal during gasification. *Hyperfine Interactions* 171: 287–292.

Waanders, F.B., Vinken, E., Mans, A., and Mulaba-Bafubiandi, A.F. 2003. Iron minerals in coal, weathered coal and coal ash—SEM and Mössbauer results. *Hyperfine Interactions* 148–149: 21–29.

Wall, T.F., Gupta, R.P., Bryant, G.W., and Wall, T.F. 1998. The effect of potassium on the fusibility of coal ashes with high silica and alumina levels. *International Conference on Ash Behaviour Control in Energy Conversion Systems*, Japan,127–134.

Ward, C.R. 2002. Analysis and significance of mineral matter in coal seams. *International Journal of Coal Geology* 50:135–168.

Ward, C.R., Corcoran, J.F., Saxby, J.D., and Read, H.W. 1996. Occurrence of phosphorus minerals in Australian coal seams. *International Journal of Coal Geology* 30: 185–210.

Ward, C.R., Li, Z., and Gurba, L.W. 2007. Variations in elemental composition of macerals with vitrinite reflectance and organic sulphur in the Greta Coal Measures, New South Wales, Australia. *International Journal of Coal Geology* 69(3): 205–219.

Wertz, D.L. 1998. X-ray scattering analysis of the average polycyclic aromatic unit in Argonne Premium Coal 401. *Fuel* 77: 43.

Whateley, M.K.G. 2002. Measuring, understanding and visualizing coal characteristics-innovations in coal geology for the 21st century. *International Journal of Coal Geology* 50: 303–315.

Willett, J.C., Finkelman, R.B., Mroczkowski, S., Palmer, C.A., and Kolker, A. 2000. Semiquantitative determination of the modes of occurrence of elements in coal: Results from an international round robin project. In *Modes of Occurrence of Trace Elements in Coal*, Davidson, R.M. (Ed.), IEA Coal Research, London, UK. CD-ROM.

Williams, R.H. and Larson, E.D. 2003. A comparison of direct and indirect liquefaction technologies for making fluid fuels from coal. *Energy for Sustainable Energy* 4: 103–129.

Wiser, W.H. 1984. Conversion of bituminous coal 'to liquids and gases: Chemistry and representative processes. In *Magnetic Resonance: Introduction, Advanced Topics and Applications to Fossil Energy*. L. Petrakis, and J.P. Fraissard (Eds.), Reidel, Dordrecht, the Netherlands, 325.

World Energy Council. 2013. *World Energy Resources: 2013 Survey: Summary*, London, 11.

Xiang-Guo, L., Bao-Guo, M., Li, X., Zhen-Wu, H., and Xin-Gang, W. 2006. Thermogravimetric analysis of the co-combustion of the blends with high ash coal and waste tyres. *Thermochimica Acta* 441(1): 79–83.

Yao, S., Zhang, K., Jiao, K., and Hu, W. 2011. Evolution of coal structures: FTIR analyses of experimental simulations and naturally matured coals in the Ordos Basin, China. *Energy Exploration & Exploitation* 29(1): 1–19.

Yudovich, Ya.E. and Ketris, M.P. 2002. *Inorganic Matter in Coal*, Ural Division, Russian Academy of Science, Ekaterinburg, Russia, 422 pp., (in Russian).

Yudovich, Ya.E. and Ketris, M.P. 2005. *Toxic Trace Elements in Fossil Coal*, Ural Division, Russian Academy of Science, Ekaterinburg, Russia (in Russian): 655.

Zandi, M. and Russell, N.V. 2004. Fate of Ca, Mg and trace elements in re-fired flyash. *Proccedings of the 5th European Conference on Coal Research and Its Applications*, Edinburgh.

2 Statistical Data on Worldwide Coal Reserves, Production, Consumption, and Future Demand

Mohan S. Rana and Faisal S. AlHumaidan

CONTENTS

Abstract: Coal is a solid fossil fuel that is primarily composed of carbon with variable quantities of other elements such as hydrogen, oxygen, sulfur, and nitrogen. Coal has formed over the course of many millions of years from dead organic material that was exposed to elevated temperature and pressure. This chapter is based on a worldwide analysis of geological inventory of coal resources, reserves, production, and consumption. The chapter also takes into consideration the development of coal commodity markets with respect to global and regional export, import, and consumption of fossil fuels. Global statistics indicate that coal provides around 30% of international primary energy needs and generates about 41% of the world's electricity. The international coal market has recently reshaped mainly due to large energy demand in China and India, both of which are major coal producers and consumers. The coal production trend is expecting some variation due to changes related to *shale gas revolution* and environmental restrictions, which have made coal a less competitive option, particularly in the energy sector. In the United States, coal demand dropped by 4% in 2011 and 11% in 2012, with many power plants using coal shutdown or converted to natural gas. Despite that, a significant drop in coal domestic demand in the United States was partially compensated by increasing the coal export to China and other countries in the Asian market. These changes, although not directly linked, have had a huge impact on the pricing of coal in global market. The total coal production worldwide reached a record level of 7831 Mt (million tons) in 2012, increasing by 2.9% when compared to the previous year. Currently, coal trails oil as the second most important fossil fuel, with primary consumption energy share of about 30%.

The environmental impact of the coal industry will continue to play a decisive role in the demand for coal worldwide, where the governments are expected to take actions to restrict greenhouse gas emissions and to control issues related to land use and waste management. An important factor in the future of coal will be the introduction of clean coal technologies that can increase efficiency and reduce gas emission. This will require significant efforts and realistic options for future research and development.

2.1 WORLDWIDE ENERGY RESOURCES AND ROLE OF COAL

The conventional (i.e., fossil fuels) and unconventional (i.e., nuclear, hydraulic, and renewable) energy resources around the world are still undergoing continuous change. However, the unconventional ones are still limited at a global scale; thus, changes are slow and barely visible. When the global energy markets are discussed, two main aspects are always considered: the energy source and its impact on environment. The selection of energy source depends on the geographical location, economic situation, availability of resources, and security of supply (political considerations). The environmental concerns related to gas emissions and global warming are forcing the world to move away from fossil fuels (McCarthy and Copeland 2012; Venkatesh et al. 2012; Rana et al. 2013). Therefore, the price of an energy source today is not only based on its energy density or heating value but also based on its environmental impact.

Energy consumption is an authentic human development index that measures a country's achievements in basic aspects of human development (i.e., longevity, literacy, and living standards). Considering the importance of energy to human development, various sources of energy were attempted to explore the possibility of using their calorific values. Generally speaking, fossil fuels (coal, natural gas, and oil) have been widely used in power generation for many decades mainly due to their relatively low cost and high-quality attributes. Hence, fossil fuels today account for 80% of global primary energy consumption (BP 2014a; BP 2014c). Although the global demand of energy continued to increase, the relative weight of various energy sources keep slightly shifting because of economic recession or political instabilities. Apart from other sources of energy, fossil fuels are likely to continue their domination and grow in the future, as illustrated in Figure 2.1. Within fossil fuels, coal is clearly the fastest growing source of energy, where its demand has increased to 38% between 2005 and 2013 and is expected to further increase to 73% by 2030 (BGR 2013). Thus, coal is not only among the world's fastest growing sources of energy but also considered as the principal fuel for generating electricity in many parts of the world, satisfying almost 41% of electricity worldwide as shown in Figure 2.2. In addition, coal is the most abundant among fossil fuels and their large reserves make it a possible candidate to meet increasing energy demand of the global community. However, an increase in coal consumption has a major shortcoming related to global warming, where coal is currently responsible for 43% of the overall CO_2 emission as shown in Figure 2.2 (IEA 2012a).

Apart from their importance as energy sources, coal and other fossil fuels remain one of the biggest dilemma from an environmental point of view (Kharecha et al. 2010; Singh 2011; Bloch et al. 2012; IEA 2012; Ruhl et al. 2012; EIA 2013a, 2014a, b; Govindaraju et al. 2013). However, satisfying energy requirements, particularly for industrial activities, without fossil fuels in the near future is almost impossible (Mohr et al. 2009, 2013; IEA 2013c; BP 2014a). Thus, the environmental impacts, especially coming from carbon dioxide (CO_2) emissions, should not be overlooked. CO_2 emission coefficients for various types of fuel are reported in Figure 2.3.

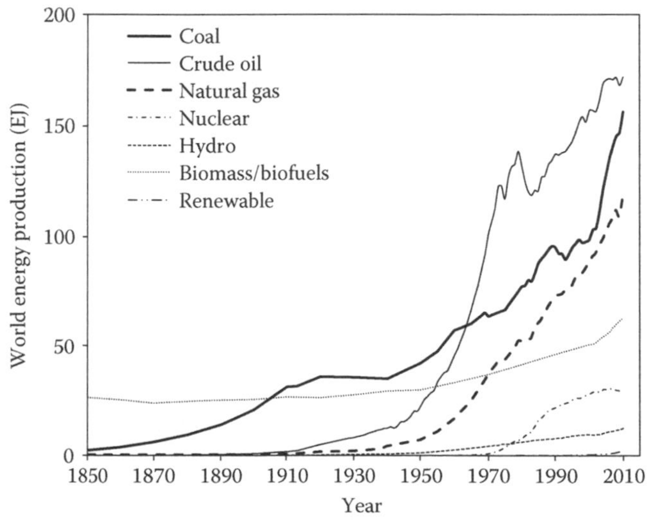

FIGURE 2.1 Worldwide energy production as a function of fuel and time (1850–2010). (Data from IEA, *World Energy Outlook 2013*, IEA, Paris, France, 2013a.)

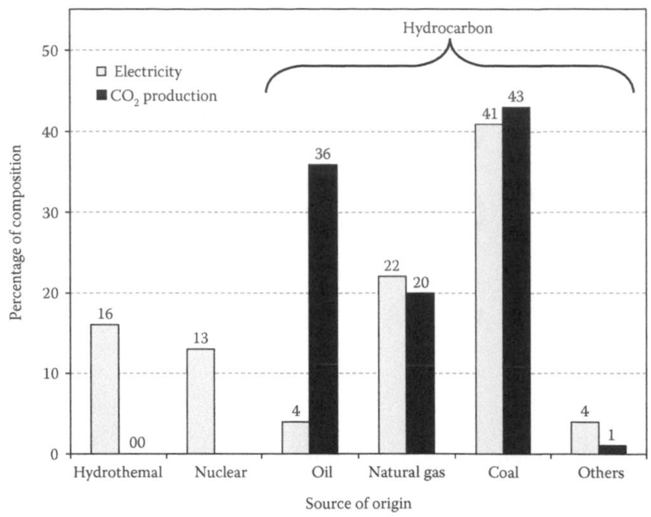

FIGURE 2.2 Worldwide percentage composition of electricity and CO_2 production as a function of source. (Data from IEA, *Coal Medium-Term Market Trends and Projections to 2017*, http://www.worldcoal.org/bin/.../global_coal_market_price (01_06_2009).pdf, 2012a; IEA, *Technology Roadmap: High Energy Efficiency, Low Emissions Coal-Fired Power Generation*, IEA, Paris, France, 2012b.)

The worldwide energy demand is likely to increase at a rate of 10% to 15% every 10 years. This will subsequently raise the CO_2 emissions to around 50% by 2030, when compared to the current level of emissions (IEA 2013d; EIA 2013a). Developed or industrial countries such as North America, Europe, and the Organization for Economic Co-operation and Development (OECD) Pacific countries (Australia, New Zealand, Japan, and South Korea) contribute to the increase in emissions by 70% when compared to the rest of the world. More than 60% of these emissions come from power generation and other industrial sectors. The OECD Pacific countries have high per capita energy consumption in common, and have emission reduction obligations under the Kyoto protocol except South Korea (Jinke et al. 2008).

In order to satisfy the energy requirement, the use of fossil fuels is projected to represent 88% of world's energy consumption at the end of 2030 (BP 2005, 2013, 2014a, b). Therefore, the demands for oil, natural gas, and coal are expected to increase by 1.6%, 3%, and 2.5%, respectively, every year (EIA 2013b). From 2000 to 2012, global coal consumption grew from 4,762 to 7,697 Mt. This represents a 60% growth or a 5% average growth per year (IEA 2012b). This substantial growth in coal demand was mainly attributed to the rapidly growing economies in China and India.

Despite the worldwide economic recession in recent years, both China and India have been among the fastest growing economies for the last two decades. From 1990 to 2010,

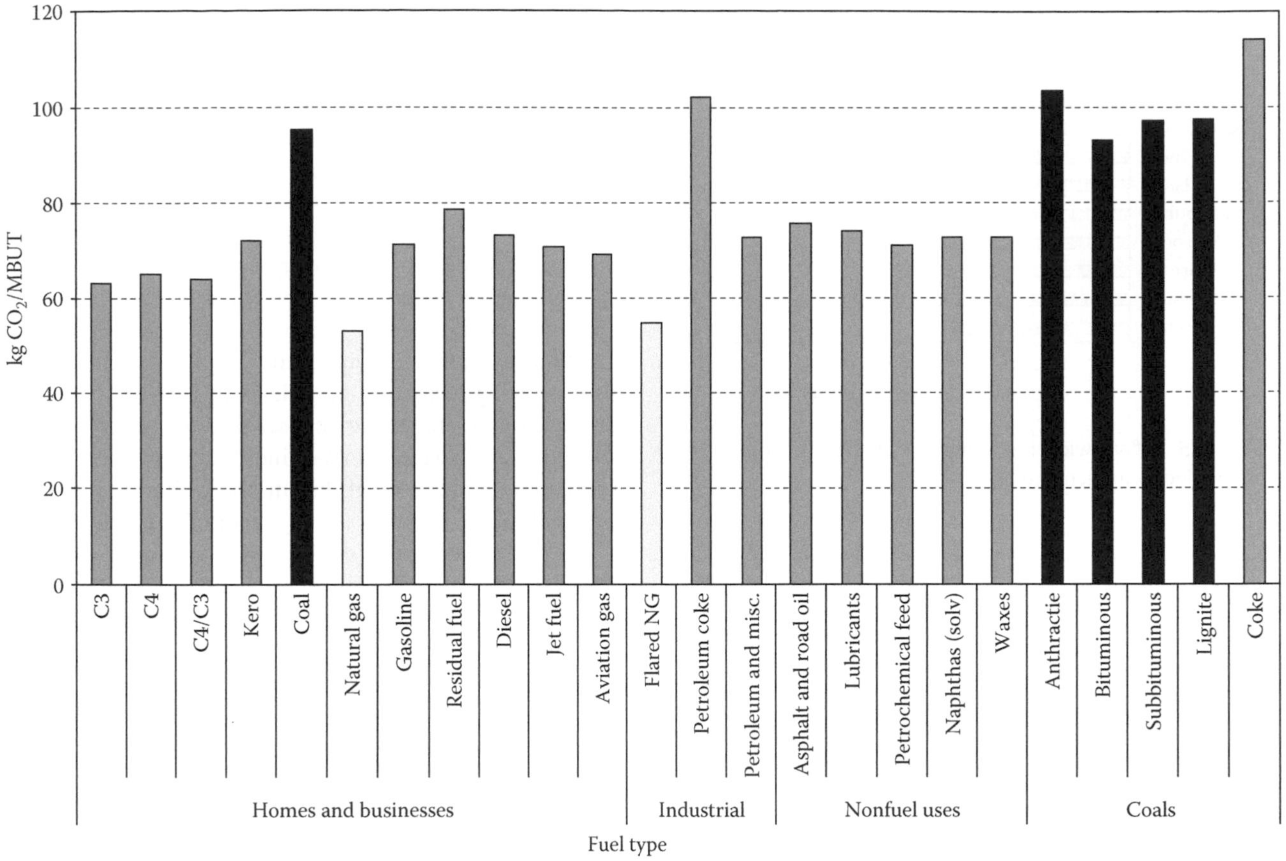

FIGURE 2.3 Carbon dioxide emissions coefficients by fuel. (From EIA, *Carbon Dioxide Emissions Coefficients by Fuel*, http://www.eia.gov/environment/emissions/co2_vol_mass.cfm, February 14, 2013a.)

China's economy grew by an average of 10.4% per year, while that of India's grew by 6.4% per year (Li et al. 2008, 2012; Pfeifer 2013). In both countries, coal played an important role in generating energy and sustaining development. Although the economic growth in both countries slowed in 2012, where the gross domestic product (GDP) declined 7.2% and 5.5% for China and India respectively, their demand and consumption of energy will continue to increase. In the near future, world's growth in energy consumption is expected to rise by an average of 3.6% per year. That growth rate is expected to be around 4.7% for non-OECD regions and 2.1% for OECD regions as shown in Figure 2.4 (EIA 2013b).

The global annual energy demand in 2013 enhanced the economic growth about 2.8%, the energy consumption by 2.1% (10.7 GToe), along with 2.0% (26.1 $GtCO_2$) increase in CO_2 emissions (Hook 2010; Patzek and Croft 2010; IEA 2012a). This demand in energy is anticipated to further increase in the coming years and will eventually reflect on fossil fuel demand. The recent forecast indicates that between 2012 and 2018, the expected yearly average rate of growth in coal demand is around 2.3%. This rate is relatively lower than the previously mentioned average growth rate between the years 2000 and 2012, which was 4%. Despite that, coal will still be one of the main driving forces in the economic growth of developing countries, mainly China and India. The importance of coal to human prosperity and development has made it a truly global industry, where it is commercially mined in more than 50 countries. Figure 2.5 illustrates the top countries in the world that heavily rely on coal for electricity generation while Table 2.1 illustrates the approximate values of coal energy sources and equivalent conversion factors and their calorific values.

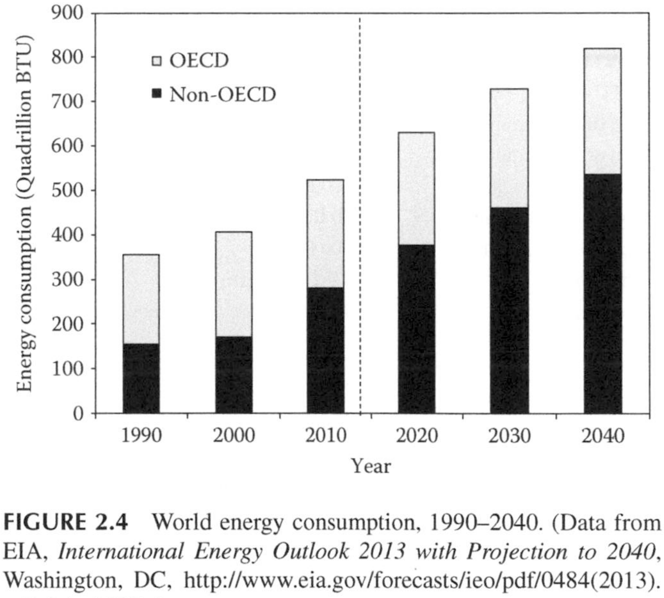

FIGURE 2.4 World energy consumption, 1990–2040. (Data from EIA, *International Energy Outlook 2013 with Projection to 2040*, Washington, DC, http://www.eia.gov/forecasts/ieo/pdf/0484(2013).pdf, July 2013b.)

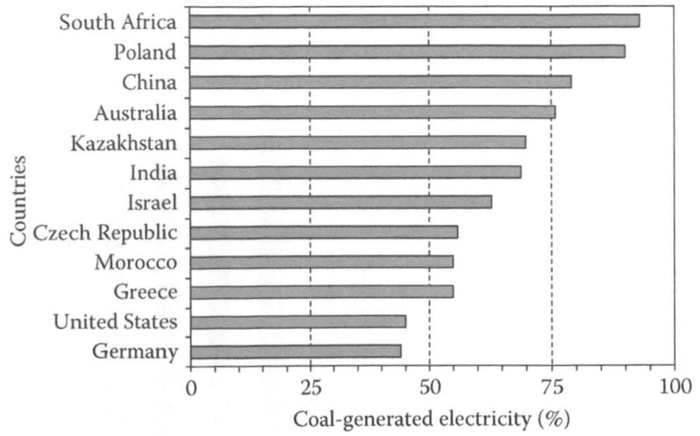

FIGURE 2.5 Worldwide top countries that use coal as a major source for electricity generation.

TABLE 2.1
Approximate Values of Coal Energy Sources and Conversion Factors

Energy Equivalent Conversions or Calorific Values	
1 ton of coal equivalent =	0.7 ton of oil equivalent
1 exajoules [XJ] =	34120842.37 ton of coal equivalent (TCE)
1 kilojoules =	1.0E−15 exajoules
1 exajoule [EJ] =	1018 J = 278 × 109 TWh = 34.1 × 106 t TCE
1 cal =	0.003967 BTU = 4.184 J
1 BTU =	1.0550559E-15 XJ
1 TCE =	0.70 toe = 770.7 m^3 natural gas = 29.3 × 109 J
1 kcal =	0.0041868 megajoule (MJ)
1 short ton =	25.18 MBTU
1 short ton =	26.57 gigajoule (GJ) = 2896.13 joule
1 MBTU =	106 BTU (IT) = 0.111836 MJ
1 BTU (IT) =	1055.06 joule = 0.001055 MJ
1 ton of coal (equiv) =	29.3 GJ = 27.8 MBTU
1 ton of coal (equiv) =	26.6 GJ = 25.2 MBTU
1 lb coal (anthracite) =	12,700 BTU
1 lb coal (sub-bituminous) =	8,800 BTU
1 lb coal (bituminous) =	11,500 BTU
1 million BTU =	0.94782 giga (10^9) joules
1 million BTU =	27.778 metric ton of coal equivalent (TCE)
Mass Equivalent Conversions	
1 short ton =	1.12 long ton
1 long ton =	1.016 metric ton
1 short ton =	0.9072 metric ton
1 ton =	0.9842 UK long ton
1 metric ton = ton =	1000 kg = 2204.6 lb = 1.1023 short ton
1 ton =	2204.62 pounds (lb)
1 ton =	1.1023 US short ton

2.2 COAL TYPES AND CLASSIFICATIONS

Coal is basically the hydrocarbon material that transforms from one phase to another under certain conditions during long period of time through a process known as *coalification*. Coal usually has a high content of carbon; thus, its physical state is solid and black in nature. Due to the presence of hydrocarbon, coal is widely used as a solid fossil fuel that generates electricity and heat. The heat content of coal depends upon the amount of fixed carbon, which relies on the origin of coal, the aging process, and the formation conditions. The transformation of dead biotic material and vegetal matter to coal passes through different stages, from which various types of coals take shape. The coal formation starts with the peat, the coal precursor, which successively transforms under suitable conditions to lignite, sub-bituminous, bituminous, anthracite, and finally graphite. Temperature plays a very important role in converting coal from one phase to another. For example, an increase in temperature causes peat to convert to lignite, a soft, low-rank coal. A further increase in temperature transforms the lignite into sub-bituminous, bituminous, and then into anthracite, which is the highest rank of coal.

As previously stated, coalification is defined as the degree of change that coal undergoes as it matures from peat to anthracite. The degree of coalification determines the coal's physical and chemical properties. Based on these properties, different types of coal are classified today into mainly two categories: *soft or low rank* and *hard or high rank*. Coal ranking primarily reflects the degree of progressive alteration in the transformation from lignite to anthracite; therefore, the primary two ranks of coal can be further divided into different classes (Parr 1928; Krishnan 1940; Wood et al. 1983). Low-ranking coal represents about 47% of global coal reserve, while high-ranking coal represents 53%. Low-rank coals, which include lignite (17%) and sub-bituminous (30%), are typically soft and friable with a monotonous look (World Coal 2015). The high-rank coals, on the other hand, are generally shiny hard black solid, and they include bituminous (52%) and anthracite (1%). High-rank coals have more carbon and, therefore, higher heat value than the low-rank coals. The high-rank coals are also characterized by lower moisture content when compared to the low-rank coals. Anthracite is on top of the coal rank and correspondingly, has the highest carbon and energy content and the lowest level of moisture. Bituminous, which is the most abundant high-rank coal, is normally divided into thermal and metallurgical coals. Thermal coal is also known as *steam coal* and is frequently used in power generation and cement manufacturing. The metallurgical coal, on the other hand, is sometimes named *coking coal*, and it is industrially utilized in iron and steel manufacturing. The United States is an example of a country that produces different types of coal, where U.S. coal production consists of 7.8% lignite, 44% sub-bituminous, 48% bituminous and 0.2 anthracite.

Table 2.2 illustrates the properties of the different types of coal (Krishnan 1940; Wood et al. 1983), including their calorific value. Anthracite is almost hydrogen-free coal with the highest energy content of +30 MJ/kg. Bituminous coal contains small amounts of hydrogen and water and its energy content ranges between 18 and 29 MJ/kg (Krishnan 1940; EIA 2013c, e, 2014b). Sub-bituminous and lignite, in contrast, have comparatively lower heating values of 8 to 25 MJ/kg and 5 to 14 MJ/kg, respectively. The heating values of the coal types

TABLE 2.2
Coal Classifications and Properties

Type of Coal	Fixed Carbon (%)	Volatile Matter (%)	Moisture (%)	Calorific Values (MJ/kg)	Physical State/Uses
Low-Rank Coal, 47%					
Peat	<50	10–12	60–85	6.69	Low energy level Subject to bacterial and fungal action
Brown coal	–	–	40–60	8.0	Water squeeze out
Lignite (17%)	65–78	40–50	35–50	5–14	Hard and massive, largely power generation
Sub-bituminous (30%)	–	–	20–30	8–25	Hard and brittle
Hard Coal, 53%					
Bituminous (52%)	60–80	–	41	18–30	Soft and shiny
High volatile bituminous	–	>31	32	32.2	–
Medium volatile bituminous	–	22–31	40	35.2	Power generation, cement manufactureing, and industrial uses
Low volatile bituminous	–	8–14	30	35.6	–
Semi-anthracite	86–92	8–15	15	36.0	–
Anthracite (1%)	92–98	5–10	<12	30.0	Very shiny domestic/industrial smokeless fuel
Meta-anthracite	≥98	<2	–	–	Power generation, cement manufacturer, and industrial uses
Thermal (coking coal/steam coal)	43–48	20–36	5–12	25–28	Manufacturing of iron and steel
Metallurgic coal (coking coal)	88–93	20–25	8–12	31–33	Demand by steel producers

suggest that 1 kg of anthracite has the same heating value of 2–6 kg of lignite. Although the different types of coal are classified into two categories as stated previously, these types sometimes overlap and the energy content of the specified coal is not always apparent and can vary within a broad range. In addition to calorific values, the quality of coal also depends on its composition, which is a typically conventional composition of organic compounds that is made of fine and coarse aggregates, carbon, hetero-elements, and other chemicals as shown in Figure 2.6.

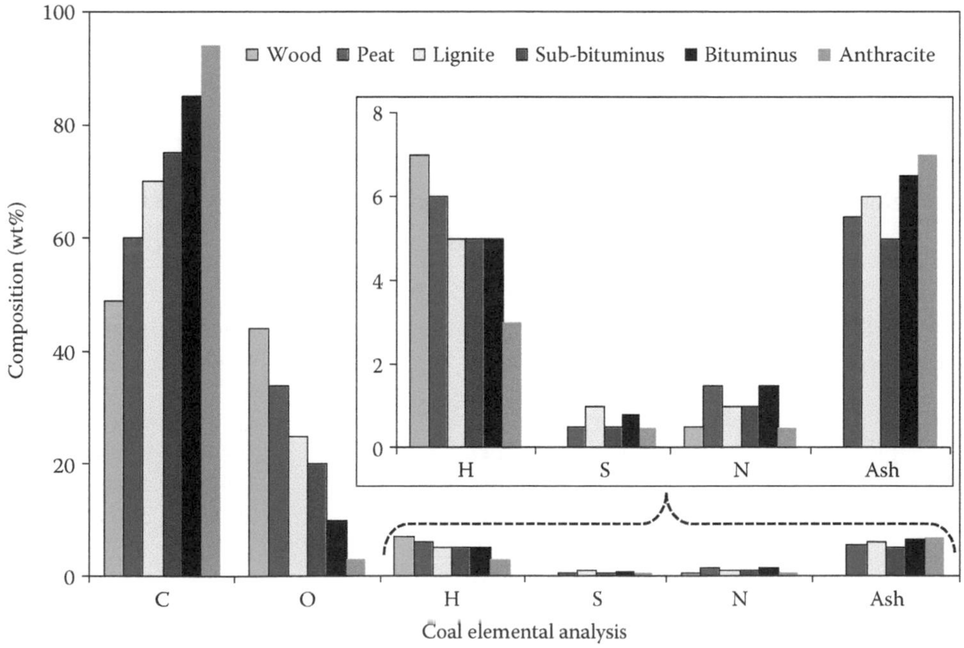

FIGURE 2.6 Average of coal elemental composition.

In the high-rank coal (i.e., bituminous and anthracite), carbon contents tend to increase, while oxygen and hydrogen contents tend to decrease, which consequently lower the oxygen/carbon (O/C) and hydrogen/carbon (H/C) ratios, respectively. The amounts of sulfur and nitrogen in coal, in contrast, do not show any clear tendency against the variation of coal sources or types. Although the value of a coal primarily depends on its rank (high- or low-rank coal), one should not lose sight of the fact that other characteristics have an impact on its utilization, such as the content of trace elements (S and N), ash content, moisture, and volatile hydrocarbon. Metallurgical coal is primarily used in steel industries, while steam coal has application to generate electricity and heat to produce steam used in industrial processes.

2.3 WORLD COAL RESERVES AND RESOURCES

Coal is one of the oldest known fossil fuels used. The United Kingdom issued the first analysis of coal reserves in 1871, followed by the U.S. reserves analysis of Pennsylvania anthracite fields in 1926 (Harnsberger, 1919; Avrtitt 1975; Grubert 2012; BP 2013; Milici et al. 2013). The first comprehensive survey at the world level was The Coal Resources of the World, produced in 1913 for the 12th World Geological Congress (Avrtitt 1975; Wood et al. 1983). After World War I, surveys were continued by the World Power Conference, which is now called the *World Energy Council*. The World Energy Council published its latest (23rd) survey in 2013.

Based on the definition, coal resources correspond to the estimated amounts of coal that are believed to be physically contained in a certain place, at a given time. The extraction of this coal might not be technically viable or economically feasible, but is geologically possible. Coal reserves, on the other hand, refer to the estimated amounts of coal that can be feasibly exploited using existing technologies. The developments taking place in coal extraction technology has a significant impact on coal reserves. Recent report shows that the global coal reserve has increased by more than 100%, particularly in China and India (Lester and Steinfeld 2007; EIA 2013b; IEA 2013b; Pfeifer 2013; BP 2014b). Although there are several predictions for worldwide coal reserve by major institutes such as British Petroleum (BP), International Energy Agency, Energy Information Administration, and Bundesanstalt für Geowissenschaften und Rohstoffe (BGR), inconsistent data are still found in the literature due to the difficulty associated with accurately estimating the amount of coal in underground (Fettweis 1976; BGR 2009; BP 2013; EIA 2013c; Rutledge 2013).

However, considering the rapidly increasing trend in coal reserves and the continuous development in mining industry, one might suggest that the actual coal reserve is much higher than the one currently reported. The reserves and resources of hard coal and lignite are adequate to cover the foreseeable demand for many decades from a geological point of view. Coal boasts the largest potential of all nonrenewable energy resources (BGR 2013) with a share of around 56% of reserves and 89% of resources as shown in Table 2.3.

Coal is uniformly distributed throughout the world as depicted in Figure 2.7. Proven coal reserves are present in 70 countries (Fettweis 1976). Countries with top proven coal reserves are shown in the Figure 2.8. A comprehensive regional coal distribution by type is shown in Figure 2.9. The figures show that around 28% of global coal reserves are located in North America, mainly in the United States. More than 34% is present in Europe and Eurasia, dominated by Russia followed by Kazakhstan and Ukraine, and approximately 31% is located in Asia Pacific, mainly China, India, and Australia.

In 2013, the EIA has estimated that the United States has about 481 billion tons of coal reserve, which is larger than the remaining natural gas and oil resources (EIA 2013d). Africa and the Middle East's share of coal reserves vary between 4% and 8%, while South America holds 2% of world reserves (BP 2014b). As previously stated, these statistical data are not absolute and there are some discrepancies in the literature about them. The discrepancy in data can be attributed to

TABLE 2.3
Worldwide Regional Distribution of Reserves and Resources of Non-Renewable Coal Fuels, 2012

	Hard Coal (Mt)			Lignite Coal (Mt)		
Regions	**Reserves**	**Resources**	**Percent Use**	**Reserves**	**Resources**	**Percent Use**
Europe	19,452	470,429	4.1	70,134	325,038	21.6
Commonwealth of Independant States	130,362	2,839,068	4.6	93,065	1,278,553	7.3
Africa	36,210	81,438	44.5	66	402	16.4
Middle East	1,203	40,000	3.0	na	na	na
Australasia	342,917	6,891,042	5.0	81,884	1,054,481	7.8
North America	229,914	6,645,013	3.5	32,912	1,486,144	2.2
Latin America	8,943	26,491	33.8	5,073	20,118	25.2
Antarctica	0	150,000	0.0			
Rest of the world	768,999	17,143,481	4.5	283,134	4,164,736	6.8

Source: BGR (Bundesanstalt für Geowissenschaften und Rohstoffe/German Federal Institute for Geosciences and Natural Resources), *Energy Study 2013, Reserves, Resources and Availability of Energy Resources*, BGR, Hannover, Germany, December 2013.

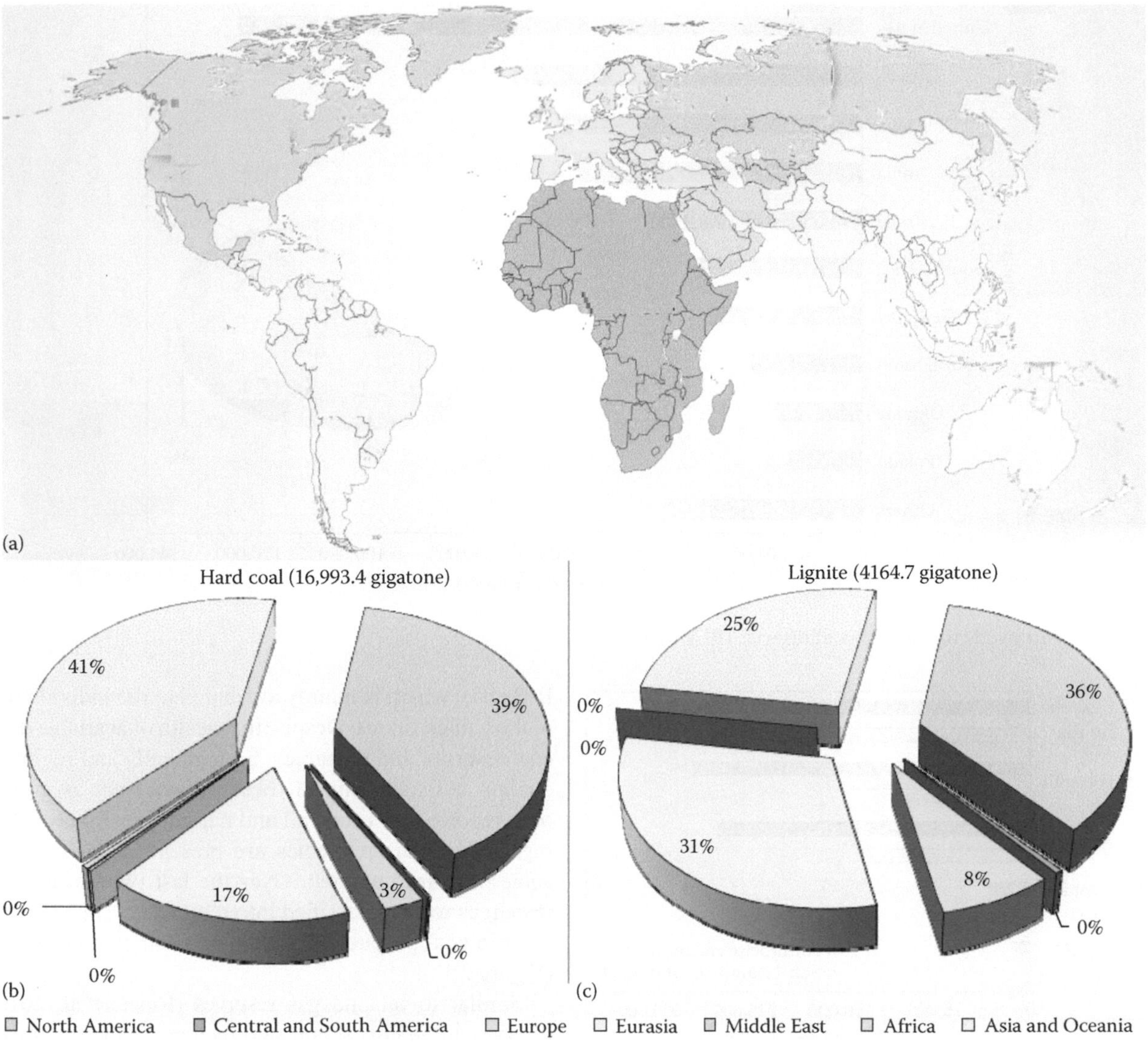

FIGURE 2.7 Worldwide distribution (a) of hard (b) and lignite (c) coals.

various factors such as the difficulty associated with accurately estimating the amount of coal in underground, the variation in update to statistics data, and the inconsistency in region's definitions as in the case of Eurasia and Asia Pacific. In addition, one should also take into consideration the overlap between the coal types, where, for example, sub-bituminous coal is frequently referred to as *black lignite*. Thus, the low-rank coal (sub-bituminous and lignite) is sometimes called *lignite*.

The top 10 countries that have the largest hard and lignite coal resources and reserves are shown in Figures 2.10 and 2.11, respectively. On a regional basis, many European countries reduced their hard coal reserves by 25% to 35% over the past few decades. For example, the European lignite reserves have considerably reduced, particularly for Germany, which was the world's largest lignite producer until 1997. Similarly, Poland has reduced its hard coal reserves by 50% since 1997. On the other hand, the Indian and Australian hard coal reserves have substantially increased since 1990 (EIA 2014b).

2.3.1 Global Coal Scenario

Coal plays a unique role in securing the world's energy demand and its resources are more widely distributed in the world compared to oil and natural gas. The role of coal as a principle fuel is expected to increase in the near future because it is less expensive and more available than other fossil fuels (Wolde-Rufael 2010). According to previous studies, coal is expected to be a primary source of energy in the coming 150–170 years, whereas oil and natural gas, roles are expected to end in about 50–75 years (Campbell and Laherrere 1998; BP 2005). However, in the current scenario, the world is not running out of fossil fuel for at least next several decades (Hook and Aleklett 2010; Riazi et al. 2013). Clean fossil fuel burning may be an issue that humanity does face but their (gas, oil, and coal) abundance is not limited to the twenty-first century. The life expectancy of coal reserves is normally determined by the ratio between the proven reserve and the production rate (R/P ratio). North America has the highest

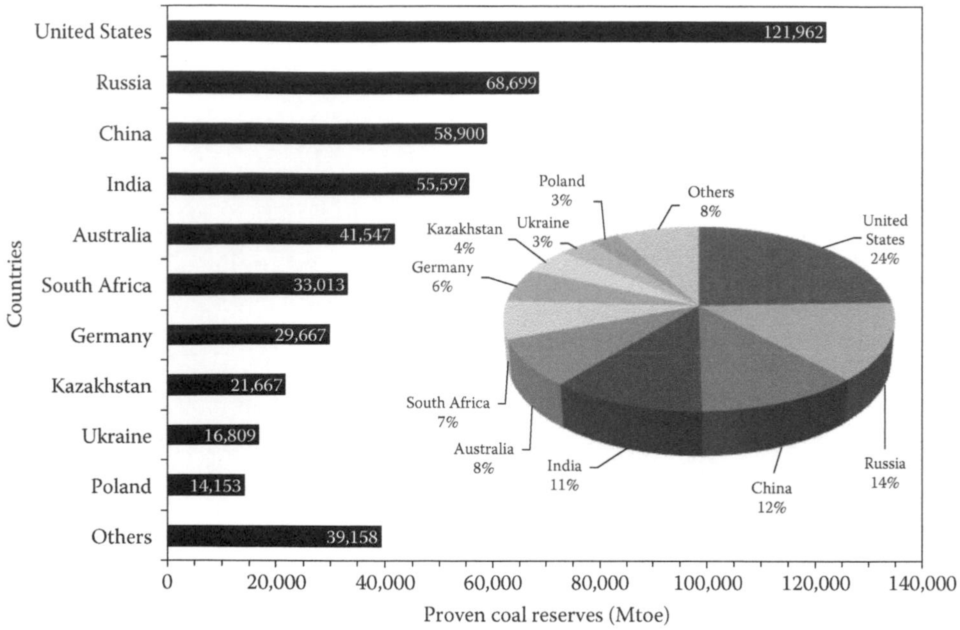

FIGURE 2.8 Worldwide top proven coal reserve. (EIA, 2013.)

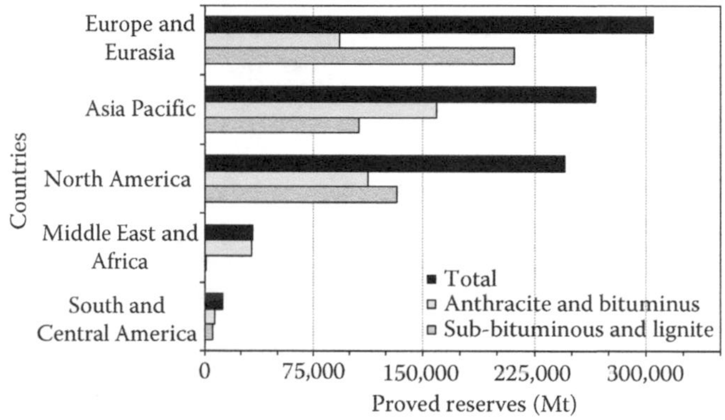

FIGURE 2.9 Worldwide regional distribution of reserves of various coals. (Data from EIA, *International Energy Statistics*, http://www.eia.gov/cfapps/ipdbproject/IEDIndex3.cfm?tid=1&pid=7&aid=1, 2014b.)

R/P ratio, which is mainly attributed to the individual reserve of the United States. Despite the wealth of available data about coal reserves and resources, both globally and regionally, the quality and reliability of these data are not as good as the ones reported for crude oil and natural gas. Practically speaking, obvious discrepancies are present in the literature and some data are outdated. Over the last two decades, limited resources were reclassified into reserves, despite the enhancement in coal production technology and coal production rate (Figure 2.12).

Similar to oil and gas reserves (Riazi et al. 2013), coal reserves are mainly dominated by a group of countries known as the *Big Six*. These countries are the United States, Russia, China, India, Australia, and South Africa in descending order of reserves (Hook et al. 2010). The United States holds around 25% of global reserves and is considered to be the second

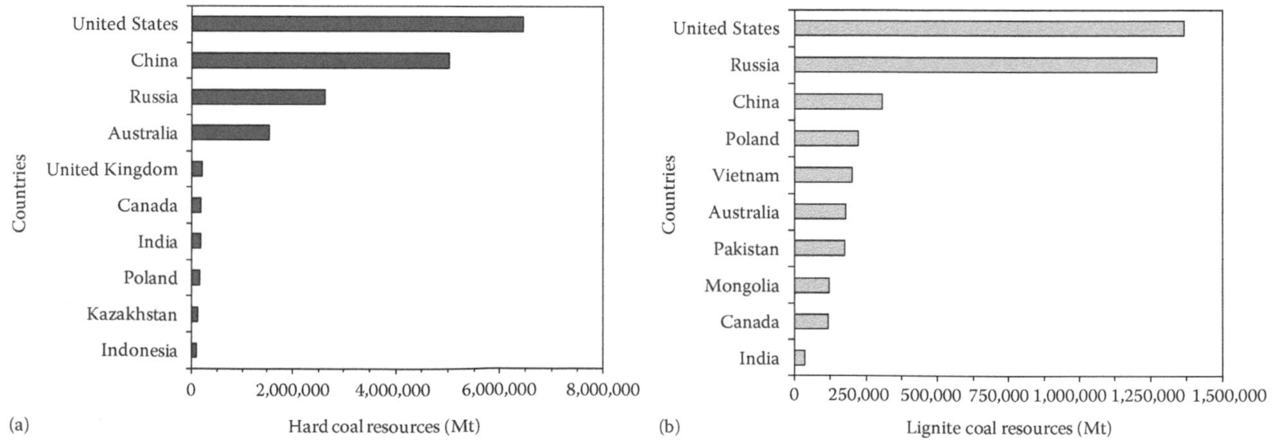

FIGURE 2.10 Top 10 countries, resources for hard coal (a) and lignite coal (b). (Data from BGR (Bundesanstalt für Geowissenschaften und Rohstoffe/German Federal Institute for Geosciences and Natural Resources), *Energy Study 2013, Reserves, Resources and Availability of Energy Resources*, BGR, Hannover, Germany, 2013.)

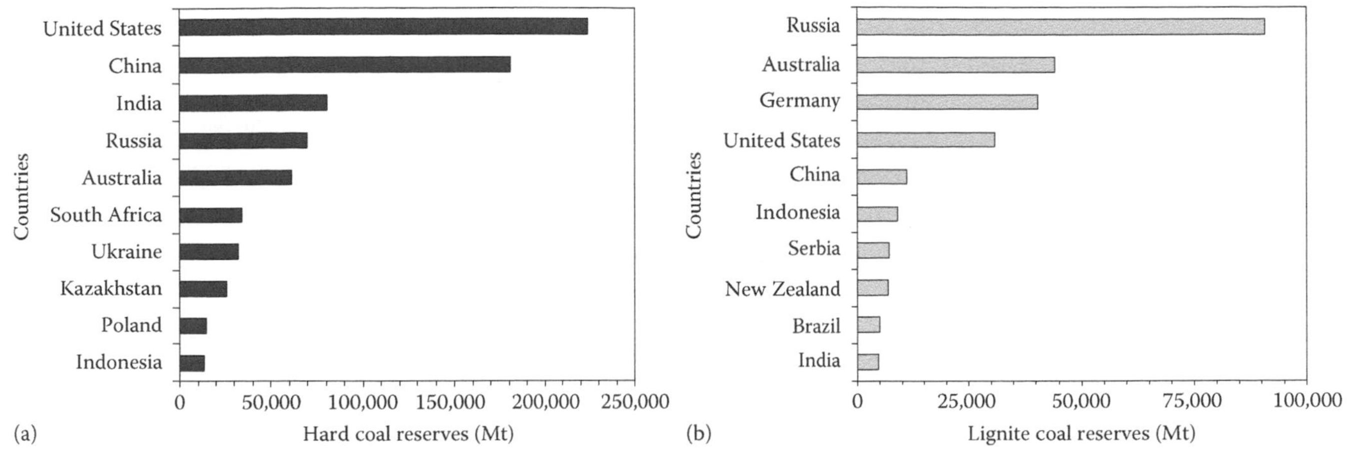

FIGURE 2.11 Top 10 countries, reserves for hard coal (a) and lignite coal (b). (Data from BGR (Bundesanstalt für Geowissenschaften und Rohstoffe/German Federal Institute for Geosciences and Natural Resources), *Energy Study 2013, Reserves, Resources and Availability of Energy Resources*, BGR, Hannover, Germany, 2013.)

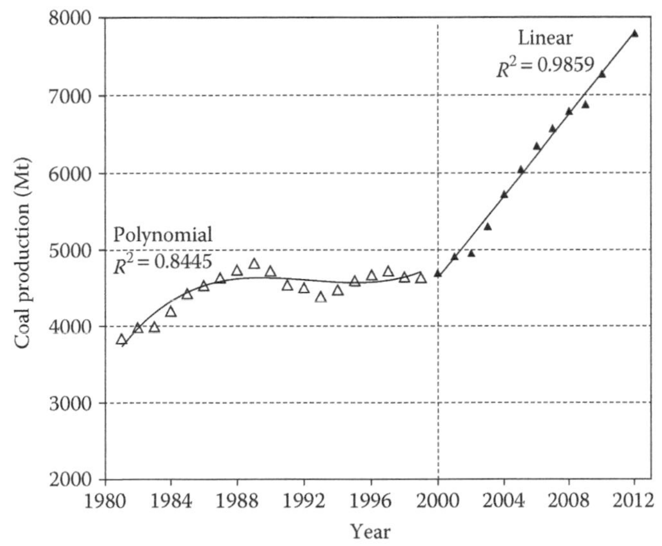

FIGURE 2.12 Worldwide coal production between 1980 and 2012. (Data from BP, *Statistical Review of World Energy June 2014*, http://www.bp.com/statisticalreview, 2014b.)

largest producer of coal after China. On the other hand, China is the largest producer of coal, despite the fact that it possesses only 12% of the global reserves. The United States and China together control more than 90% of world's hard coal reserves (Michieka and Fletcher 2012; BP 2013), in addition to their substantial resources of lignite (BGR 2009; BP 2009). The Big Six group currently produces around 80% of the world's coal, with a share of approximately 50% from China. Russia is another major producer that has around 14% of the world's total coal reserve (BP 2009). The BP report indicates that China and India, which have around 12% and 11%, respectively, of world reserves, are of particular interest for the coal industry as they need more inexpensive energy to cope with their rapidly growing economies (BP 2009). Subsequently, in the order, Australia and South Africa have around 8% and 7%, respectively, of the world's coal reserves, and both countries are considered to be significant exporters. The previously illustrated statistical data on coal reserves and resources were collected from BGR Energy, United States Geological Survey, EIA, the BP Statistical Review of World Energy, IEA, and several other sources as reported in the references.

2.4 COAL PRODUCTION

Coal mining has been an important economic activity throughout human history and it continues to be so. Large-scale coal mining and production begun in the United Kingdom more than 200 years ago during the European Industrial Revolution, and gradually spread to other parts of Europe and North America (Hook and Aleklett 2009; Hook et al. 2010). Tower Colliery, in South Wales, is the world's oldest continuously operating coal mine, in operation for about 203 years (1805–2008). Just like other fossil fuels, coal is also created over millions of years in the form of solid hydrocarbon, whose production has depleted or is in decline with time (Thomas 2002; Rodriguez and Arias 2008; Bardi and Lavacchi 2009; Hook and Aleklett 2009). Today, coal continues to be an important energy source, particularly for electricity generation, because of its relatively low cost and abundance when compared to the other fossil fuels.

Although coal reserves are widespread, the largest reserves are limited to a number of countries, notably the United States, China, Russia, India, Australia, and South Africa. The coal production in the United States reached an all-time high of 1175 Mt in 2008, which declined by 1100 Mt from 2009 to 2011 (IEA 2013a). On the other hand, world coal production has increased by 60% since 2002 (Humphries and Sherlock 2013). The cumulative and regional coal production values over the last three decades are shown in Figures 2.12 and 2.13, respectively. Since the beginning of the twenty-first century, coal production has increased severalfold to fulfill the growing demand for energy in the world. Most of the growth in coal consumption came from Asia, particularly China, while coal consumption growth in the OECD region dawdled due

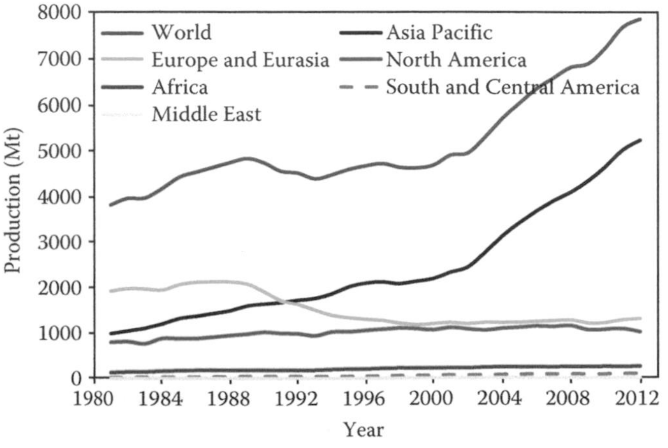

FIGURE 2.13 Regional coal production and recent trends. (Data from EIA, *International Energy Statistics*, http://www.eia.gov/cfapps/ipdbproject/IEDIndex3.cfm?tid=1&pid=7&aid=1, 2014b.)

TABLE 2.4
Variation in Coal Production

Region	Production Variation (%/Year) 2012–2013	2000–2013
Europe	−1.7	−1.2
European Union	−0.6	−1.4
Commonwealth of Independant States	1.4	2.8
North America	2.5	1.2
Latin America	0.0	1.7
Asia	2.1	4.9
Pacific	6.9	2.5
Africa	−4.4	1.5
Middle East	0.0	2.7
Rest of the world	1.1	2.4

to the weak demand except from Australia. The variation in coal consumption and demand influences coal price, as shown in Figure 2.14. Table 2.4 illustrates coal production since the start of the twenty-first century, where obvious variation among various regions of the world is observed. The strict environmental legislations in Europe and the United States forced coal production to consistently decrease over the last 12 years (Hook et al. 2010; Patzek and Croft 2010).

This was encountered by a robust increase in coal production in Asia to fulfill the growing demand for energy by China and India (Singh 2011; EIA 2013d; Bloch et al. 2012; Govindaraju and Tang 2013; Lin and Ouyang 2014). In future, global coal production is likely to increase (mainly in non-OECD countries) by about 516 Mtoe, reaching around 3917 Mtoe by 2035 (EIA 2013b). For OECD Europe, however, coal production is expected to drop almost to 6%–7%, particularly in Poland and Germany, where the governments and industries have adopted a plan to phase out hard coal production by 2018 (IEA 2013b).

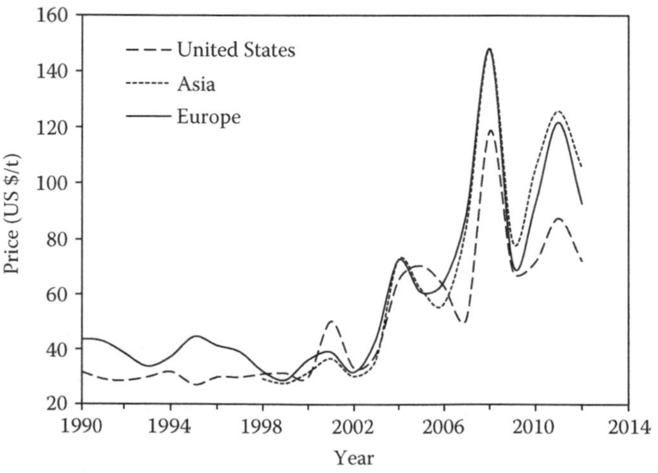

FIGURE 2.14 Variation in coal price for different regions in the world between 1990 and 2014. (Data from BP, *Statistical Review of World Energy June 2014*, http://www.bp.com/statisticalreview, 2014b.)

Similarly, coal production in the United States is anticipated to decrease around 9%–10%, where the coal energy will be compensated by alternative sources such as natural gas.

The top coal-producing countries in the world are shown in Figure 2.15, along with the enhancements in their production capacities from 2003 to 2011. China is the dominant producer of coal, with 3470 Mt of total coal production, followed by the United States, and then India. Currently, China alone produces around 44% of global coal (EIA 2013b). The domination of China in coal production is also shown in Figure 2.16. The significant increase in coal production in China is mainly driven by the demand for energy, which forces China to even import 5% of its total consumption. India is the second largest non-OECD coal producer and the third largest in the world. Between 1980 and 2011, the average growth in India's production of coal was around 5%–6% per year. Power sector consumes 70% of India's coal, while the rest goes to the iron and/or steel industry. In addition to China and India, Indonesia is also rapidly increasing their production rate and has recently become a major producer and exporter of coal in the world (BGR 2013; BP 2013; Cornot-Gandolphe 2013; EIA 2013b). Colombia and Ukraine, although not among the top producers of coal, are working toward increasing their coal production.

This discussion clearly shows how diversified coal production is in the world. The fastest and the most dominant region in coal production is Asia, where the world's largest production share goes to China (51%), followed by the United States (12%), and then India (8.2%). Figure 2.17 illustrates the top 10 countries in the world that produce hard and lignite coals.

2.5 COAL CONSUMPTION

As previously indicated, coal is considered a principal fuel for electricity generation today. As with the other commodity markets, the development of the coal market is mainly governed by supply and demand, therefore, production and consumption. A comparison between the top 10 countries in terms of coal production and consumption is shown in Figure 2.18,

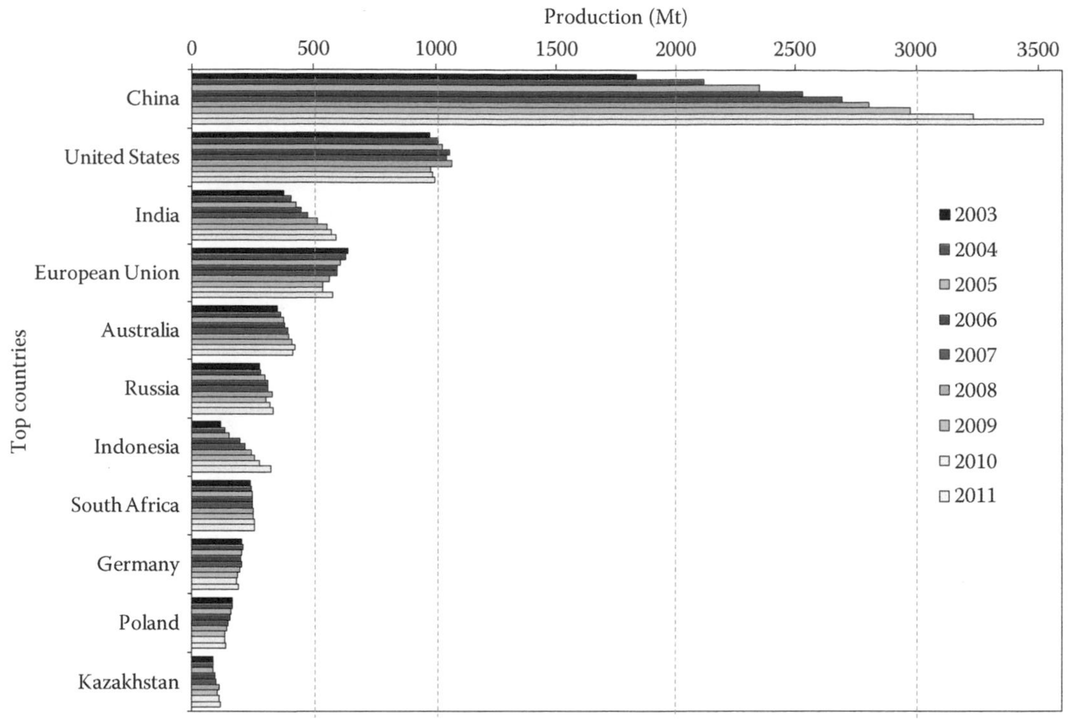

FIGURE 2.15 Top countries that produce coal and their production since 2003.

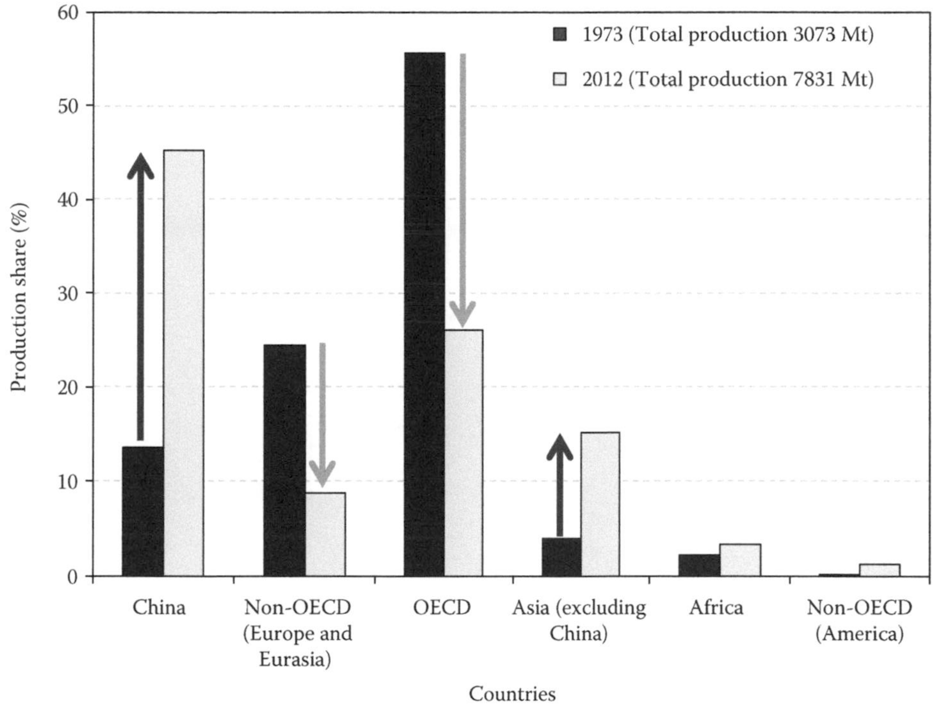

FIGURE 2.16 Regional shares of coal production, comparison between 1973 and 2012. (Data from EIA, *International Energy Outlook 2013 with Projection to 2040*, Washington, D.C., http://www.eia.gov/forecasts/ieo/pdf/0484(2013).pdf, July 2013b.)

while Figure 2.19 shows the regional coal consumption over the last five decades. Similar to coal production, the consumption is mainly dominated by Asia, more specifically China and India. China's global share of coal consumption is 47%, followed by the United States (14%), and then India (9%) (BP 2014c). Thus, approximately 70% of global coal consumption belongs to these three countries. In recent years, the share of coal consumption has shifted from Europe and the former Soviet Union to Asia (Hook and Aleklett 2010). Figure 2.19 shows that Europe and the former Soviet Union

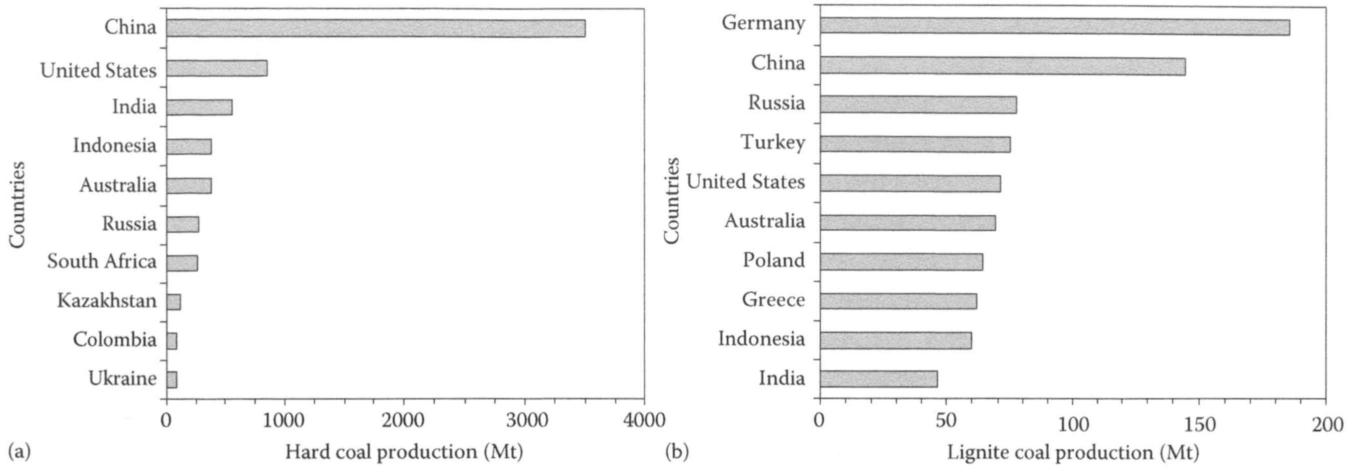

FIGURE 2.17 Top 10 countries producing hard coal (a) and lignite coal (b). (Data from BGR (Bundesanstalt für Geowissenschaften und Rohstoffe/German Federal Institute for Geosciences and Natural Resources), *Energy Study 2013, Reserves, Resources and Availability of Energy Resources*, BGR, Hannover, Germany, December 2013.)

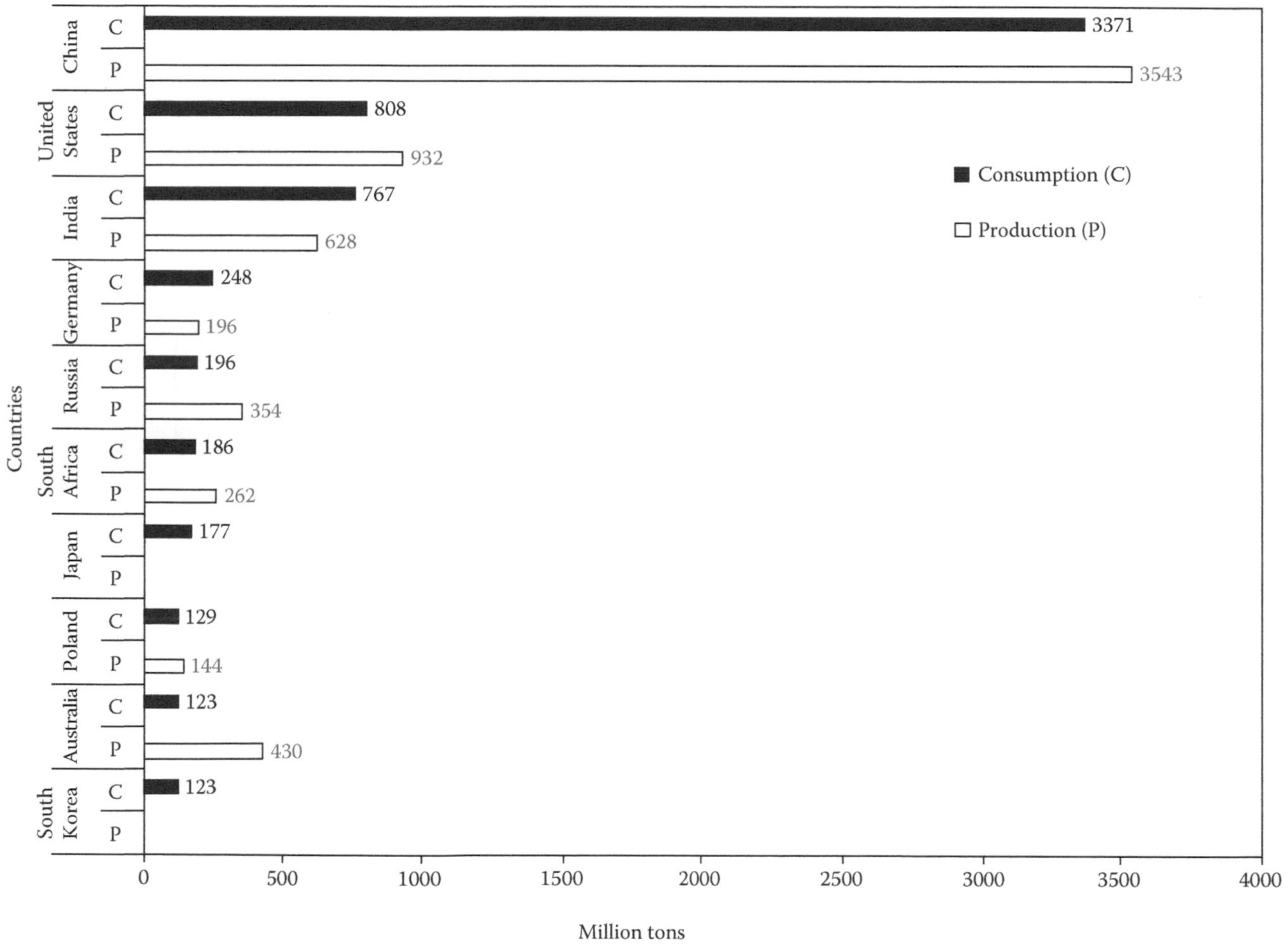

FIGURE 2.18 A comparison between top 10 countries for production and consumption of coal.

(Europe and Eurasia) are the main two regions of where there was a significant decline in coal consumption between 1980 and 2010, where the consumption falls between 32% and 42%, respectively. Conversely, Asia's share of global coal consumption increased from 24% to 63% during this period. From a long-term perspective, the growth rate in coal consumption is expected to slow down gradually due to strict environmental regulations and an improvement in the feasibility of alternative cleaner energy sources. For example, natural gas becomes more economically competitive as a result of shale

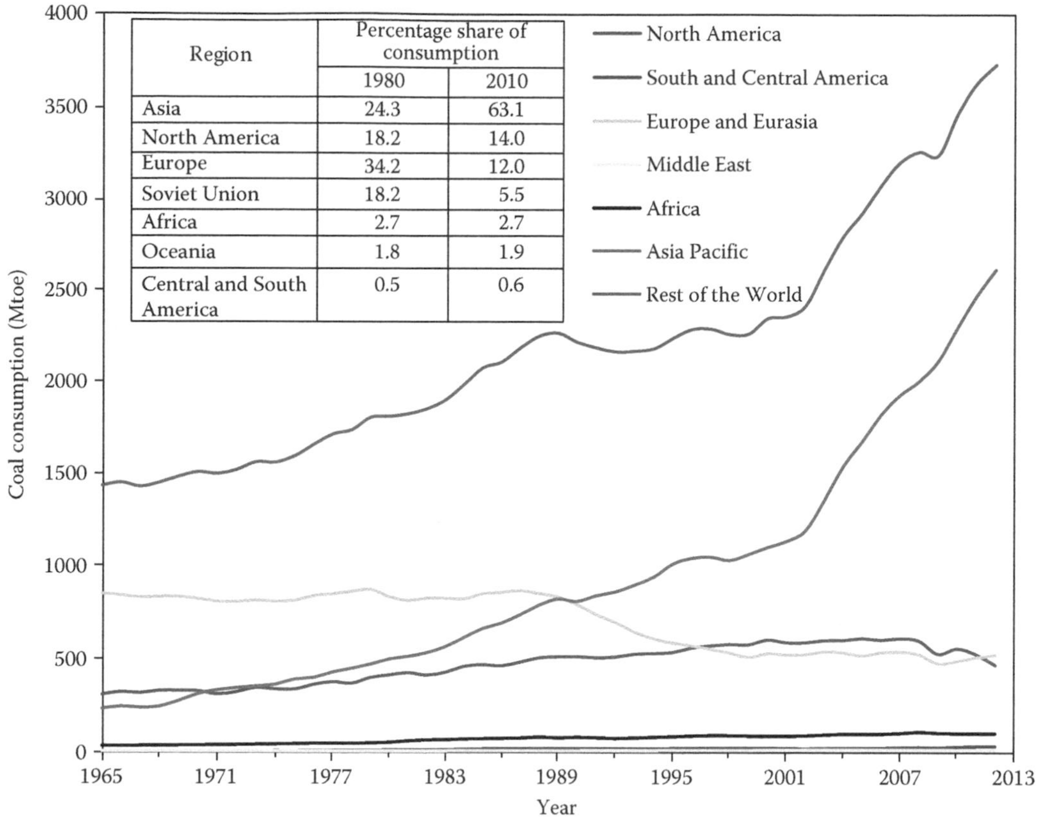

Region	Percentage share of consumption	
	1980	2010
Asia	24.3	63.1
North America	18.2	14.0
Europe	34.2	12.0
Soviet Union	18.2	5.5
Africa	2.7	2.7
Oceania	1.8	1.9
Central and South America	0.5	0.6

FIGURE 2.19 Regional coal consumption and its percent variation with time. (Data from EIA, *International Energy Statistics*, http://www.eia.gov/cfapps/ipdbproject/IEDIndex3.cfm?tid=1&pid=7&aid=1, 2014b.)

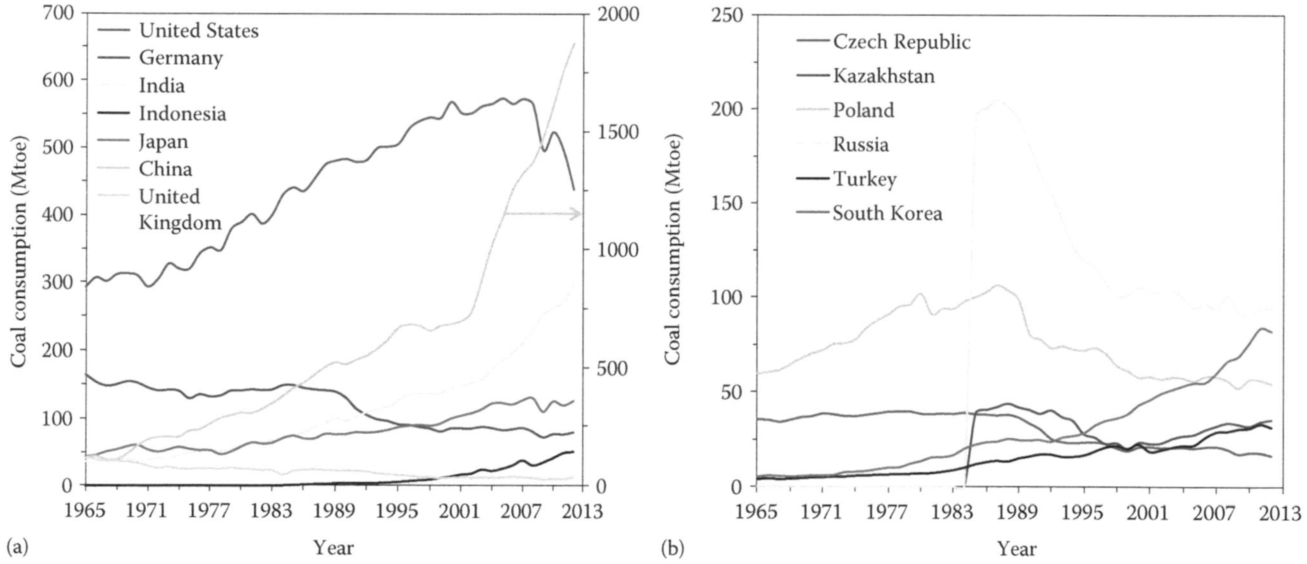

FIGURE 2.20 Variation in coal consumption for various countries (a and b). (Data from EIA, *International Energy Statistics*, http://www.eia.gov/cfapps/ipdbproject/IEDIndex3.cfm?tid=1&pid=7&aid=1, 2014b.)

gas development. The impact of environmental legislation and natural gas utilization is notable in the consumption of coal in the United States and Europe, where an obvious decline can be observed. The variation in coal consumption for different countries is illustrated in Figures 2.20a and 2.20b. Countries with emerging and developing economies, which are in the process of industrialization, usually exemplify an increasing trend in coal consumption, as in the case of China, India, Indonesia, Kazakhstan, Turkey, and South Korea. In contrast, developed countries, with postindustrial economies such as the United States, the United Kingdom, and Germany are showing the opposite trend.

There are mainly three sectors that consume coal in the world. Nearly 4% of coal is consumed by the households and the commercial sector, whereas the industrial sector consumes 34%. The remaining 62% is mainly consumed by electricity-producing units (BGR 2013). Coal consumption can be also classified based on the type of coals, where the consumption is proportional to the calorific value. Figure 2.21 shows the top 10 consumers of hard and lignite coals. Hard coal is more widely traded worldwide because it has a relatively higher energy content that can compensate for the transportation cost. Lignite coal, on the other hand, is more locally consumed as in the case of Germany, which boosted domestic production to 185.4 Mt (York 2007).

Future forecast of coal consumption suggests an increase of 50% by 2030, with developing countries responsible for 97% of that increase, primarily to meet electricity demand. China and India are the most populous developing countries and their influence on global coal consumption and CO_2 emissions are expected to increase in the future (Lu and Ma 2004; Garg and Shukla 2009; Parikh and Parikh 2011; Rout et al. 2011; Singh 2011). Recent studies suggest that China's coal consumption is expected to progressively increase until 2035, after which the demand is likely to decline, mainly motivated by efficiency improvements and more stringent environmental legislations. India, on the other hand, is expected to overtake the United States and become the second consumer of coal by 2025. Coal demand in India is expected to continue growing, even after 2035, as the country's industrialization continues. Variation in coal consumption, along with energy demand and CO_2 emission, for OECD and non-OECD countries are illustrated in Figure 2.22. OECD's share of global coal consumption is slightly declining, while the share of non-OECD countries is rapidly escalating (Jinke et al. 2008; Li et al. 2008; Apergis and Payne 2010a, b; Li and Leung 2012; EIA 2013b). Nowadays, the non-OECD countries, particularly the emerging Asian markets, where coal consumption and CO_2 emission levels are at highest levels in the world, are under pressure to decrease their greenhouse gas emissions, including CO_2 and methane (CH_4)

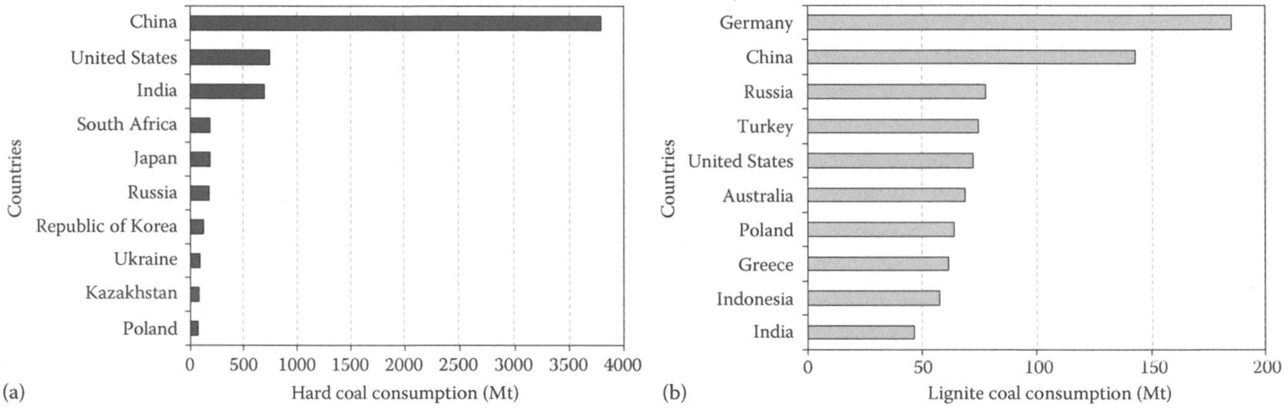

FIGURE 2.21 Top 10 countries consuming hard coal (a) and lignite coal (b). (Data from BGR (Bundesanstalt für Geowissenschaften und Rohstoffe/German Federal Institute for Geosciences and Natural Resources), *Energy Study 2013, Reserves, Resources and Availability of Energy Resources*, BGR, Hannover, Germany, 2013.)

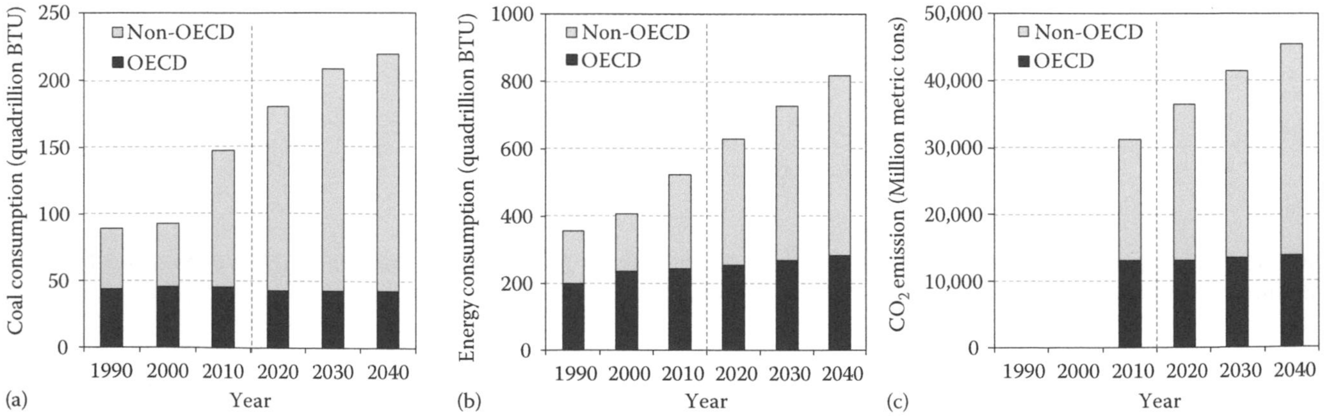

FIGURE 2.22 Variation in coal consumption (a) and energy consumption (b) in OECD and Non-OECD countries along with its effect on CO2 emission (c). (Data from EIA, *International Energy Outlook 2013 with Projection to 2040*, Washington, D.C., http://www.eia.gov/forecasts/ieo/pdf/0484(2013).pdf, 2013b.)

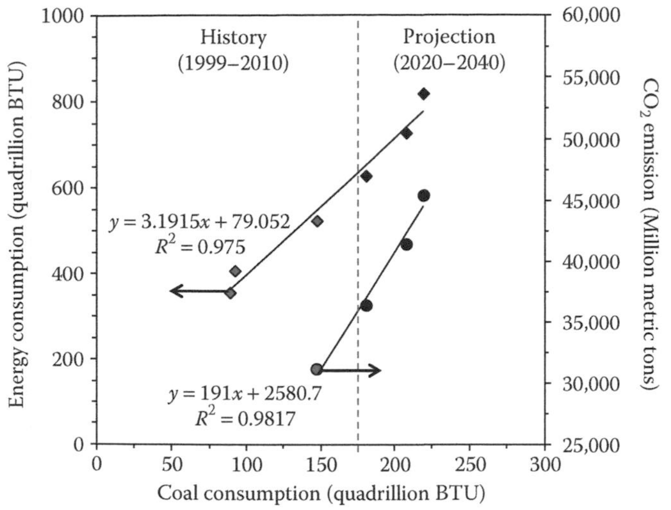

FIGURE 2.23 A relationship between coal consumption, energy consumption, and CO_2 emissions.

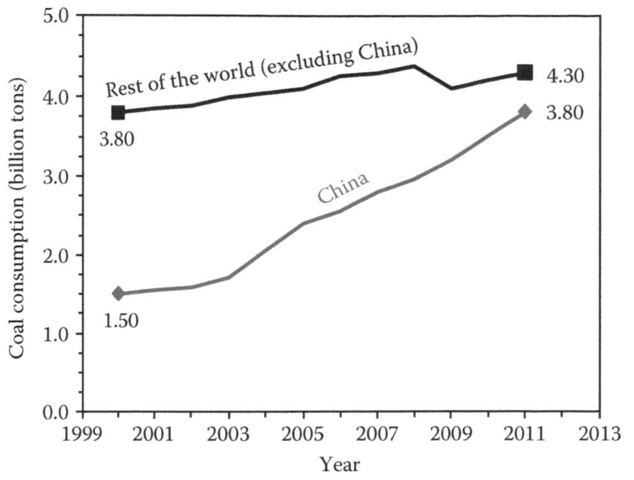

FIGURE 2.24 Coal consumption in China and rest of the world. (EIA, May 2014.)

because of their link to climate change (IEA 2013b, c, d). The correlations between coal consumption with energy and CO_2 emissions are given in Figure 2.23.

Coal consumption in China increased significantly since 2000 to account for almost 47% of global coal consumption today. Figure 2.24 illustrates and compares the rate of increase in coal consumption in China and the rest of the world. Developing countries such as China and India have large electricity requirements and their consumption of coal is mandatory for their economic growth. China's GDP grew 7.7% in 2012, following an average GDP growth rate of 10% per year from 2000 to 2011. Nevertheless, on a per capita basis, electricity consumption in Southeast Asia is still relatively low when compared to the OECD countries. Figure 2.25 shows the annual percentage rate of change for coal demand in major coal-consuming countries of the world. The relationship between coal consumption and GDP was extensively covered in the literature for other countries, in addition to China, such as the United States, India, European Union, Japan, and South Korea (Yang 2000; Sari and Soytas 2004; Lee and Chang 2005; Yoo 2006; Hu and Lin 2008; Apergis and Payne 2010b; Apergis et al. 2010). Coal consumption in the Middle East accounts less than 1% of the total primary energy, which is

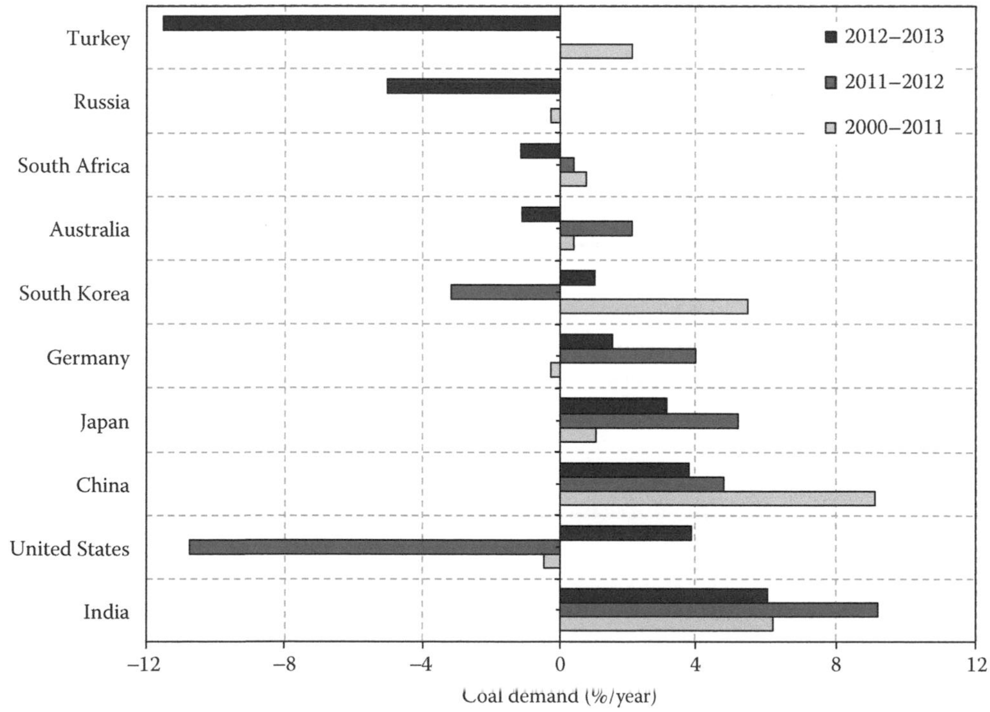

FIGURE 2.25 Annual percentage rate of change for coal demand in major coal-consuming countries of the world.

mainly used in few countries (Iran, Lebanon, Israel, and Syria) in the region (EIA 2013b). The conclusions from these studies are mixed and the findings are often conflicting because of a variation in chosen time span, geographic region, data structure, and econometric methodology. The uncertainty associated with the long-term projections is attributed to the fact that many statistics are incomplete or unreliable. Thus, some specialists believe that the sufficiency of coal supply for the coming two centuries, as often asserted, is still a speculation.

2.6 GLOBAL COAL MARKET TRADE

As previously indicated, coal is a global industry, which is commercially produced in about 50 countries and consumed in over 70 countries (IEA 2013c). Today, coal has a wide variety of sources with a large number of suppliers who can efficiently transport it to the world market using ships and rails. Despite its activity and competitiveness, the international coal trade in 2011 was around 1142 Mt, which only accounts for about 16% of the total coal produced. The relatively small percentage of coal trade can be mainly attributed to the fact that the vast majority of coal is mainly consumed in the producing countries. In addition, coal transportation cost is high and it accounts for a large percentage of the total selling price.

The international coal trade market is mainly divided into two regional markets: the Atlantic market and the Pacific market. The Atlantic market consists of importing countries from Western Europe such as the United Kingdom, Germany, and Spain, while the Pacific market is made up of developing and OECD importers such as Japan, Korea, China, and India. As all other markets, the supply (production) and demand (consumption) explains how the market operates (BGR 2013; EIA 2013b). For example, the current surplus in high-quality hard coal has forced the low calorific value coals to be sold at a discounted price. Coal consumption can be also driven by other factors such as export and/or import policies and strict environmental legislations as in the case of European countries. Figure 2.26 summarizes the coal market over the last 42 years, while Figures 2.27 and 2.28, respectively, illustrate the top 10 countries exporting and importing coal. The figures clearly indicate that Indonesia and Australia are the leading coal exporters, while China and Japan are the leading importers. In the last decade, coal trade has experienced a significant shift, such as Indonesia posting remarkable increase in export, displacing Australia as the world's leading exporter in 2011 (EIA 2013b). Coal production and exportation is rapidly growing in South Africa and it is mainly driven by the domestic electricity demand and the international demand from countries such as China and India. Since the beginning of this century, coal consumption has stagnated in Europe and North America. Since 2012, coal export from the United States to various countries in Europe declined, which reflects both the lower demand for steam coal in the European Union and increased coal supply from Australia and Indonesia.

In recent years, China has been controlling most of the international coal trade (Pfeifer, 2013). China was a major coal exporter till 2008 but their heavy consumption of coal made the country a major importer, as it overtook Japan and became the top importer in 2011. Despite the recent economic recession, China continued their rising trend and reached a record

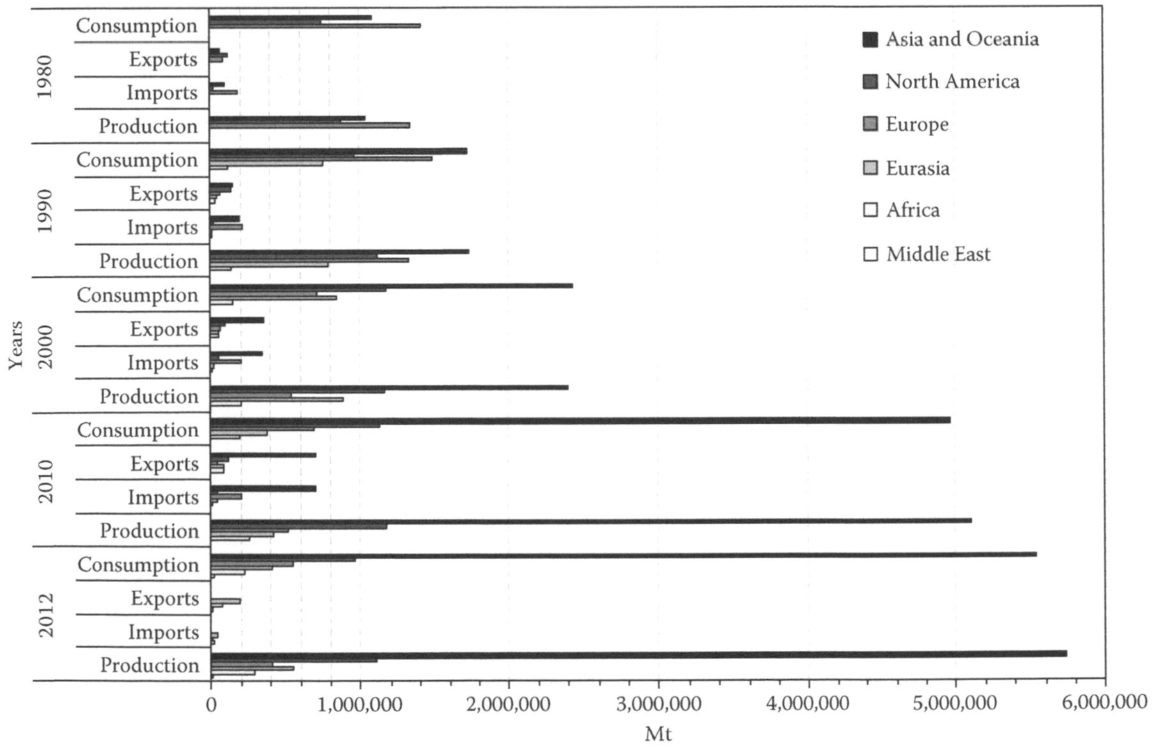

FIGURE 2.26 Worldwide trend in coal production, consumption, import, and export.

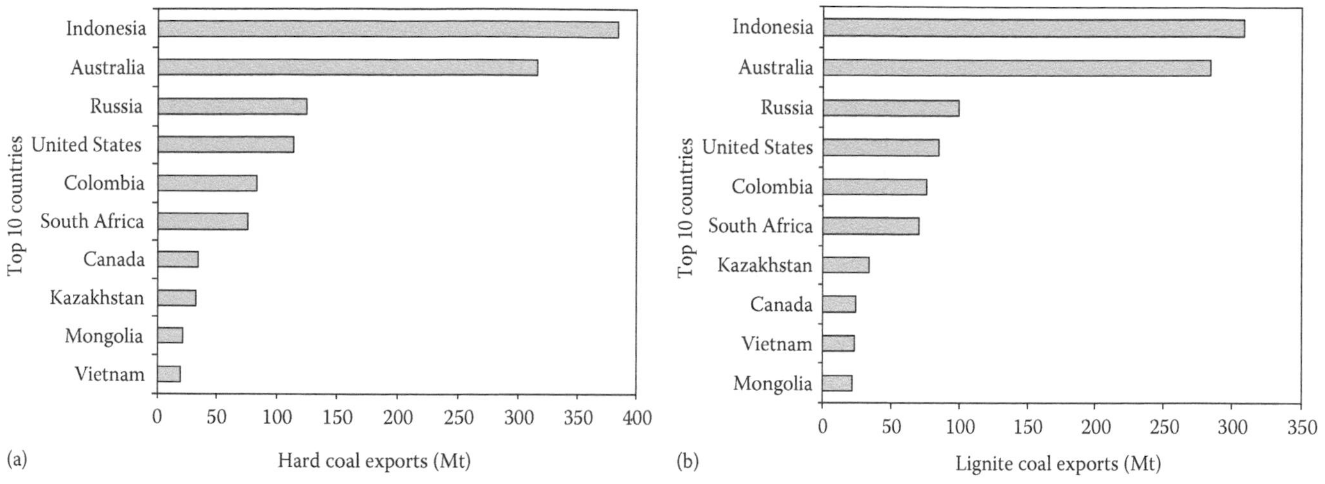

FIGURE 2.27 Top 10 countries exporting coal (includes steam and coking): hard coal (a) and lignite coal (b). (Data from BGR (Bundesanstalt für Geowissenschaften und Rohstoffe/German Federal Institute for Geosciences and Natural Resources), *Energy Study 2013, Reserves, Resources and Availability of Energy Resources*, BGR, Hannover, Germany, December 2013.)

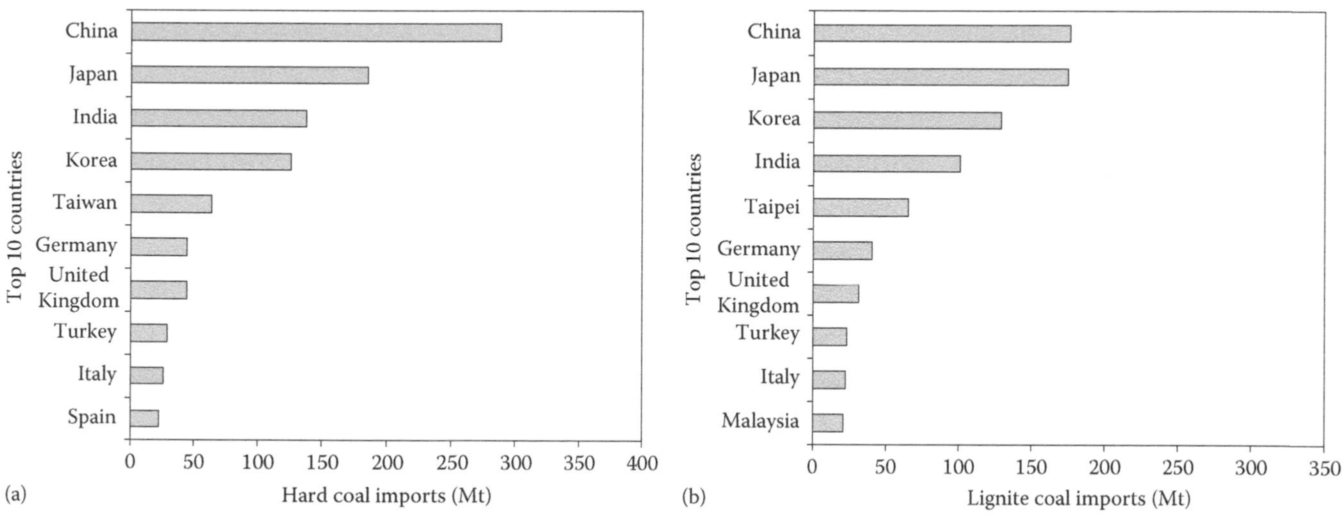

FIGURE 2.28 Top 10 countries importing coal (includes steam and coking): hard coal (a) and lignite coal (b). (Data from BGR (Bundesanstalt für Geowissenschaften und Rohstoffe/German Federal Institute for Geosciences and Natural Resources), *Energy Study 2013, Reserves, Resources and Availability of Energy Resources*, BGR, Hannover, Germany, December 2013.)

level of importing 289 Mt of coal in 2011, which represented around 25% of global coal trade. China relies on coal to supply approximately 69% of its energy needs, which happens to be the largest in the world. Along with coal usage, China overtook the United States as the world's biggest carbon emitter in 2007 and became the world's largest energy consumer in 2010.

2.7 FUTURE PROSPECTIVE OF COAL AND ENVIRONMENTAL ISSUES

Coal, as compared to other fossil fuels, is abundant, relatively less expensive, and is more distributed worldwide (IEA 2012a). The demand for coal is rapidly growing with a substantial enhancement in energy requirement. In 2012, coal became the second-most important source of energy in the world, and its role is expected to increase in the developing countries of Asia, considering the fact that Asian coal import accounts for 70% of global hard coal trading volume and 50% of world total consumption of coal belongs to China (Lester and Steinfeld 2007; EIA 2013d). Coal demand worldwide has increased to 38% between 2005 and 2014 and is anticipated to reach 73% by 2030. Such an increase in coal consumption has a major environmental shortcoming, which is related to greenhouse gas emissions. Environmental policies have already forced many developed countries in Europe and North America to substantially cut down their consumption of coal. For example, coal demand in the United States dropped 4% in 2011 followed by 11% in 2012. The increase in the interest in shale gas might also impact the future of coal. For example, shale gas has recently become an important source of natural gas

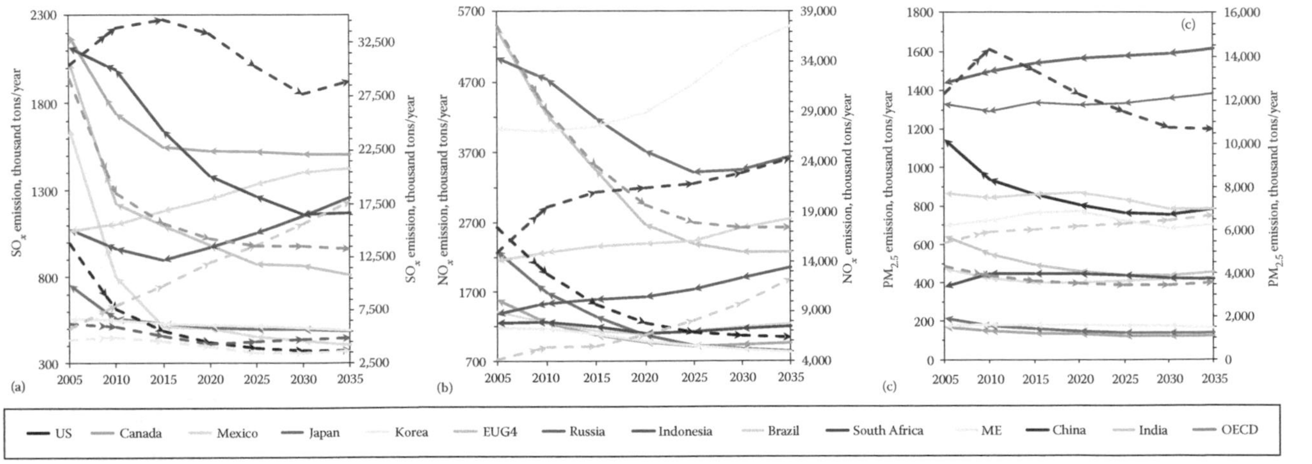

FIGURE 2.29 Emissions of SO_x (a), NO_x (b), and $PM_{2.5}$ (c) by country group in the New Policies Scenario, thousand tons/year. (Data from Cofala, J. et al., *Atmos. Environ.*, 41/38, 8486–8499, 2007; Cofala, J. et al., *Emissions of Air Pollutants for the World Energy Outlook 2011 Energy Scenarios*, International Energy Agency, Paris, France under Contract for Services between IEA and IIASA Contract No. 11-121, September 2011.)

in the United States and this interest is rapidly spreading to the rest of the world. In 2000, shale gas contributed to only 1% of USA natural gas production whereas in 2010, the figure increased to 20% (BP 2013).

The climatic changes in recent years are frequently associated with large industrialization and huge demand for energy, which mainly comes from fossil fuels. Currently, fossil fuels account for 80% of energy demand, of which 34% comes from crude oil, 25% from coal, and 21% from natural gas. Burning fossil fuels has a negative impact on the environment including large amount of smog, acid rain, toxic air pollution, and global warming. Coal, in particular, has a large amount of sulfur, nitrogen, and other impurities, which usually generate pollutants after combustion. The examples of pollutants are sulfur oxides (SO_x), nitrogen oxides (NO_x), carbon dioxide (CO_2), particulate materials (PM), and heavy metals such as mercury. The introduction of clean coal technology to mitigate these environmental impacts will definitely endorse the future of coal. Examples of existing clean coal technologies are chemically washing minerals and impurities for coal, gasification, improved technology for treating flue gases to remove pollutants, carbon capture and storage, and dewatering lower rank coal to improve calorific values. Recent statistics from U.S. Environmental Protection Agency indicate that the advancements in clean coal technologies notably regulated emission per unit of energy produced.

Despite its environmental impact, coal is likely to remain as an important source of energy in any feasible future energy scenario, particularly in developing countries such as China and India. A major challenge faced by many countries today is to come up with applicable strategies that can secure energy need and also reduce greenhouse gas emissions even with the utilization of coal (Cofala et al. 2007, 2011; Bell et al. 2011). Overcoming this challenge requires the introduction of highly effective clean coal technologies in mainly three areas: (1) reduction or elimination of pollutant emissions such as sulfur and nitrogen oxides; (2) reduction of CO_2 emissions, and (3) improvement of conversion technologies to increase efficiency. Figure 2.29 illustrates the gas emissions for major developed and developing countries. In *World Energy Outlook*, the New Policies Scenario takes account of broad policy commitments and plans that have been announced by countries to reduce greenhouse gas emissions and to phase out fossil-energy subsidies, even if the measures to implement these commitments and plans have not yet been identified or announced. In this scenario, the emissions of air pollutants for $PM_{2.5}$, SO_x, and NO_x are expected to decrease to nearly 5%, 16%, and 11%, respectively, by 2035.

2.8 CONCLUDING REMARKS

The demand for fossil fuels is expect to grow in the near future to satisfy the rapid increase in energy requirement. By 2030, fossil fuels will represent 88% of the world's energy consumption, where the demands of oil, coal, and natural gas are expected to annually increase by 1.6%, 2.5%, and 3%, respectively. Thus, coal is among the fastest growing source of energy and its role as a principal fuel is expected to increase in the near future, where the demand for coal increased by 38% between 2005 and 2013, and might reach 73% by 2030. Between 2000 and 2012, the world's consumption of coal increased from 4762 to 7697 Mt, which represents a 60% growth. This substantial growth was mainly driven by the rapidly growing economies of China and India.

Despite the wealth of available data about coal reserves and resources, the quality and the reliability of statistical data is not as good as the ones reported for crude oil and natural gas. Obvious discrepancies and disputes are there in the literature and some data are outdated. Although coal is more evenly distributed in the world when compared to the other fossil fuels, a limited number of countries control the vast majority of the world's coal resources. About 76% of global coal reserves belongs to the Big Six countries that include the United States (24%), Russia (14%), China (12%), India (11%), Australia (8%), and South Africa (7%). From a coal production aspect, China is the leading producer (47%) followed by the United States (13%), India (7.2%), Australia (4.9%), and Indonesia (4.6%). In terms of consumption, recent statistics indicate that China alone consumes 45%–50% of worldwide production followed by the United States (10.5%), India (8.8%), Russia (3.2%), and Germany (3.2%). Thus, approximately 70% of coal global consumption belongs to China, the United States, and India. Although the growth in coal consumption is gradually slowing down in developed countries due to environmental restrictions and feasibility of cleaner sources (i.e., natural gas), the increasing trend in coal consumption is expected to continue in countries with developing economies such as China, India, and South Korea. Coal, today, is a global industry with an active and vibrant world market. Yet, the international coal trade accounts for less than 20% of total coal production. This is mainly due to the fact that coal is normally consumed in the producing countries because of high transportation costs. Indonesia and Australia are the world's leading coal exporters, while China and Japan are the leading importers.

Climate changes and other environmental problems are frequently associated with fossil fuel and coal in particular. Problems associated with coal production and consumption include acid mine drainage, sediment and waste of mining, greenhouse gases, and other pollutants such as CO_2, sulfur dioxide, nitrogen oxides, particulate matters, and heavy metals like mercury. The introduction of clean coal technologies to mitigate these negative environmental impacts is mandatory for the future of coal.

REFERENCES

Apergis, N., D. Loomis, and J. E. Payne, 2010. Are fluctuations in coal consumption transitory or permanent? Evidence from a panel of US states, *Applied Energy* 87: 2424–2426.

Apergis, N., and J. P. Payne, 2010a. Coal consumption and economic growth: Evidence from a panel of OECD countries, *Energy Policy* 38: 1353–1359.

Apergis, N., and J. E. Payne, 2010b. The causal dynamics between coal consumption and growth: Evidence from emerging market economies, *Applied Energy* 87: 1972–1977.

Avrtitt, P. 1975. Coal resources of the United States, Jan. 1, 1974, *US Geological Survey Bulletin*, Vol. 1412, p. 131.

Bardi, U., and A. Lavacchi, 2009. A simple interpretation of Hubbert's model of resource exploitation, *Energies* 2: 646–661.

Bell, D. A., B. F. Towler, and M. Fan, 2011. *Coal Gasification and Its Application*, Elsevier, Oxford.

BGR, 2009. German Federal Institute of Geology and Natural Resources (BGR). *Reserves, Resources and Availability of Energy Resources*. http://www.bgr.bund.de/.

BGR, December 2013. (Bundesanstalt für Geowissenschaften und Rohstoffe/German Federal Institute for Geosciences and Natural Resources), *Energy Study 2013, Reserves, Resources and Availability of Energy Resources*, BGR, Hannover, Germany.

Bloch, H., S. Rafiq, and R. Salim, 2012. Coal consumption, CO_2 emission and economic growth in China: Empirical evidence and policy responses, *Energy Economics* 34: 518–528.

BP, 2005. *BP Statistical Review of World Energy*, BP, London, June 2013.

BP, 2009. *Statistical Review of World Energy*, http://www.bp.com/statisticalreview.

BP, June 2013. *Statistical Review of World Energy*, http://www.bp.com/content/dam/bp/pdf/statisticalreview/statistical_review_of_world_energy_2013..pdf.

BP, 2014a. *BP Energy Outlook 2035*. January 2014. *Technical Report*.

BP, 2014b. *Statistical Review of World Energy June 2014*, http://www.bp.com/statisticalreview.

BP, 2014c. *Statistical Review of World Energy, The Future of Coal in China, India, Australia, the US, EU, and UK*.

Campbell, C. J., and J. H. Laherrere, March 1998. *The End of Cheap Oil*. Scientific American, Nature Publishing Group, 78–83.

Cofala, J., M. Amann, Z. Klimont, K. Kupiainen, and L. Hoglund-Isaksson, 2007. Scenarios of global anthropogenic emissions of air pollutants and methane until 2030, *Atmospheric Environment* 41/38: 8486–8499.

Cofala, J., J. C. H. Borken-Kleefeld, Z. Klimont, P. Rafaj, R. Sander, W. Schöpp, and M. Amann, September 2011. *Emissions of Air Pollutants for the World Energy Outlook 2011 Energy Scenarios*, International Energy Agency, Paris, France under Contract for Services between IEA and IIASA Contract No. 11-121.

Cornot-Gandolphe, S. 2013. *Global Coal Trade from Tightness to Oversupply*, Institut français des relations internationales, http://www.ifri.org/?page=contribution-detail&id=7570.

EIA, February 14, 2013a. *Carbon Dioxide Emissions Coefficients by Fuel*, http://www.eia.gov/environment/emissions/co2_vol_mass.cfm.

EIA, July 2013b. *International Energy Outlook 2013 with Projection to 2040*, Washington, D.C., http://www.eia.gov/forecasts/ieo/pdf/0484(2013).pdf.

EIA, 2013c. *Executive Summary*, http://www.iea.org/Textbase/npsum/MTCoalMR2013SUM.pdf.

EIA, September 2013d. *South East Asia: Energy Outlook*, http://www.iea.org/publications/freepublications/publication/southeastasiaenergyoutlook_weo2013specialreport.pdf.

EIA, December 16, 2013e. *U.S. Coal Reserves*, http://www.eia.gov/coal/reserves/.

EIA, 2014a. *Short-Term Energy Outlook*, EIA, Washington, D.C.

EIA, 2014b. *International Energy Statistics*, http://www.eia.gov/cfapps/ipdbproject/IEDIndex3.cfm?tid=1&pid=7&aid=1.

Fettweis, G., 1976. *World Coal Resources: Method of Assessment and Result, Technology & Engineering*, Elsevier Scientific Publication, The Netherlands, 1–416.

Garg, A., and P. R. Shukla, 2009. Coal and energy security for India: Role of carbon dioxide (CO_2) capture and storage (CCS), *Energy* 34: 1032–1041.

Govindaraju, V. G. R. C., and C. F. Tang, 2013. The dynamic links between CO_2 emissions, economic growth and coal consumption in China and India. *Applied Energy* 104: 310–318.

Grubert, E. 2012. Reserve reporting in the United States coal industry, *Energy Policy* 44: 174–184.

Harnsberger, T. K. 2012. *The Geology and Coal Resources of the Coal-Bearing Portion of Tazewell County, Virginia*, Virginia Geological Survey, Issues 19–21, Nabu Press, Las Vegas, NV.

Hook, M., and K. Aleklett, 2009. Historical trends in American coal production and a possible future outlook, *International Journal of Coal Geology* 78: 201–216.

Hook, M., and K. Aleklett, 2010. A review on coal to liquid fuels and its coal consumption, *International Journal of Energy Research* 34(10): 848–864.

Hook, M., W. Zittel, J. Schindler, and K. Aleklett, 2010. Global coal production outlooks based on a logistic model, *Fuel* 89: 3546–3558.

Hu, J. L. and C. H. Lin, 2008. Disaggregated energy consumption and GDP in Taiwan: A threshold co-integration analysis, *Energy Economics* 30: 2342–2358.

Humphries, M. and M. F. Sherlock, March 14, 2013. US and world coal production federal taxes, and incentives, *CRS Report for Congress R43011*, http://www.crs.gov.

IEA, 2012a. *Coal Medium-Term Market Trends and Projections to 2017*, http://www.worldcoal.org/bin/.../global_coal_market_price(01_06_2009).pdf.

IEA, 2012b. *Technology Roadmap: High Energy Efficiency, Low Emissions Coal-Fired Power Generation*, IEA, Paris, France, 2012.

IEA, 2013a. *World Energy Outlook 2013*, IEA, Paris, France.

IEA, 2013b. *Energy Balances of Non-OECD Countries*, IEA, Paris, France.

IEA, 2013c. *Energy Balances of OECD Countries*, IEA, Paris, France.

IEA, 2013d. *CO_2 Emissions from Fuel Combustion*, IEA, Paris, France.

Jinke, L., S. Hualing, and G. Dianming, 2008. Causality relationship between coal consumption and GDP: Difference of major OECD and non-OECD countries, *Applied Energy* 85(6): 421–429.

Kharecha, P. A., C. F. Kutscher, J. E. Hansen, and E. Mazria, 2010. Options for near-term phase out of CO_2 emissions from coal use in the United States, *Environmental Science & Technology*, 44(11): 4050–4062.

Krishnan, M. S., 1940. Classification of coal, *Geological Survey of India* VI(3): 549–559.

Lee, C. C., and C. P. Chang, 2005. Structural breaks, energy consumption and economic growth revisited: Evidence from Taiwan. *Energy Economics* 27: 857–872.

Lester, R. K. and E. Steinfeld, 2007. Chapter 5: Coal consumption in China and India, In *The Future of Coal: Options for a Carbon-Constrained World*, an interdisciplinary MIT study, in MIT-IPC-07-001, Cambridge, MA.

Li, R., and G. C. K. Leung, 2012. Coal consumption and economic growth in China, *Energy Policy* 40: 438–443.

Li, J., H. Song, and D. Geng, 2008. Causality relationship between coal consumption and GDP: Difference of major OECD and non-OECD countries. *Applied Energy* 85: 421–429.

Lin, B., and X. Ouyang, 2014. Energy demand in China: Comparison of characteristics between the US and China in rapid urbanization stage, *Energy Conversion and Management* 79: 128–139.

Lu, W., and Y. Ma, 2004. Image of energy consumption of well off society in China. *Energy Converse Manage* 45: 1357–1367.

McCarthy, J. E., and C. Copeland, 2012. *EPA's Regulation of Coal-Fired Power: Is a "Train Wreck" Coming?*, CRS Report for Congress R41914, August 8, 2011, http://www.crs.gov.

Michieka, N. M., and J. J. Fletcher, 2012. An investigation of the role of China's urban population on coal consumption, *Energy Policy* 48: 668–676.

Milici, R. C., R. M. Flores, and G. D. Stricker, 2013. Coal resources, reserves and peak coal production in the United States, *International Journal of Coal Geology* 113: 109–115.

Mohr, S. H., and G. M. Evans, 2009. Forecasting coal production until 2100. *Fuel* 88: 2059–2067.

Mohr, S., G. Mudd, L. Mason, T. Prior, and D. Giurco, 2013. *Coal: Production Trends, Sustainability Issues and Future Prospects*. Prepared for CSIRO Minerals Down Under Flagship, Institute for Sustainable Futures, UTS, Sydney, and Monash University.

Parikh, J., and K. Parikh, 2011. India's energy needs and low carbon options. *Energy* 36: 3650–3658.

Parr, S. W., 1928. *The Classification of Coal*, Experiment Station Bulletin No. 180, University of Illinois, Urbana, IL.

Patzek, T. W., and G. D. Croft, 2010. A global coal production forecast with multi-Hubbert cycle analysis, *Energy* 35: 3109–3122.

Pfeifer, S. October 14, 2013. Energy: Demand for coal will be driven by China and India, http://www.ft.com/intl/cms/s/0/a2325138-252e-11e3-b349-00144feab7de.html#axzz3J2i3Vunv.

Rana, M. S., J. Ancheyta, M. Riazi, and M. Marafi, 2013. Chapter-34, Future direction in petroleum and natural gas refining. In *American Society for Testing and Materials (ASTM) Manual Series MNL 58*, edited by M. R, Riazi et al., 769–800, ASTM International, West Conshohocken, PA.

Riazi, M. R., M. S. Rana, and J. L. Pena Diez, 2013. Chapter-3, Worldwide statistical data on proved reserves, production and refining capacities of crude oil and natural gas. In *American Society for Testing and Materials (ASTM) Manual Series MNL 58*, edited by M. R, Riazi et al., 33–78, ASTM International, West Conshohocken, PA.

Rodriguez, X. A., and C. Arias, 2008. The effects of resource depletion on coal mining productivity, *Energy Economics* 30: 397–408.

Rout, U. K., A. Vob, A. Singh, U. Fahl, M. Blesl, and B. P. O. Gallachoir, 2011. Energy and emissions forecast of China over a long-time horizon, *Energy* 36: 1–11.

Ruhl, C., P. Appleby, J. Fennema, A. Naumov, and M. E. Schaffer, 2012. Economic development and the demand for energy: A historical perspective on the next 20 years. *Energy Policy* 50: 109–116.

Rutledge, D. 2013. Compilation of coal production and reserve data (Excel sheets). http://rutledge.caltech.edu/.

Sari, R., and U. Soytas, 2004. Disaggregate energy consumption, employment, and income in Turkey, *Energy Economics* 26: 335–344.

Singh, K. 2011. India's emissions in a climate constrained world, *Energy Policy* 39: 3476–3482.

Thomas, L. 2002. *Coal Geology*. 1st Ed. John Wiley & Sons, Chichester.

Venkatesh, A., P. Jaramillo, W. M. Griffin, and H. S. Matthews, 2012. Implications of near-term coal power plant retirement for SO_2 and NO_x and life cycle GHG emissions, *Environmental Science Technology* 46(18): 9838–9845.

Wolde-Rufael, Y., 2010. Coal consumption and economic growth revisited, *Applied Energy* 87(1): 160–167.

Wood, G. H., T. M. Kehn, M. D. Carter, and W. C. Culbertson, 1983. *Coal Resource Classification System of the U.S. Geological Survey*, U.S. Geological Survey, circular 891, p. 65.

World Coal Institute, 30th Anniversary, 1985-2015, Coal facts, 2015. http://www.worldcoal.org/pages/content/index.asp?PageID=188.

Yang, H. Y., 2000. Coal consumption and economic growth in Taiwan. Energy Sources Part A: Recovery, *Utilization, and Environmental Effects* 22: 109–115.

Yoo, S. H., 2006. Causal relationship between coal consumption and economic growth in Korea, *Applied Energy* 83: 1181–1189.

York, R., 2007. Demographic trends and energy consumption in European Union Nations, 1960–2025, *Social Science Research* 36(3): 855–872.

3 Geochemistry of Coal
Occurrences and Environmental Impacts of Trace Elements

Ya. E. Yudovich and M. P. Ketris[*]

CONTENTS

[*] The authors have worked in the field of coal geochemistry for about 50 years. Therefore, this chapter is found in the 13 Russian monographs by them, each containing an extensive bibliography (see Further Reading). Also, in 2003–2009, some of their outlines were published in English (refer to Ketris and Yudovich 2009; Yudovich 2003a, b; Yudovich and Ketris 2005a, b, c; 2006a, b). All these allows us to use *only selected references* in the chapter text—more often, Russian monograph-2002 (Yudovich and Ketris 2002), and only several other references.

Abstract: This chapter gives the full presentation about contents, modes of occurrence, and possible importance of the trace elements in coal. The main topics of the inorganic matter of coal (IOM) are studied based on modern analytical data. The following topics are discussed in detail: mineralogical regularities, with special attention to micro- and nanominerals in coal; modal and virtual classes of IOM; IOM composition influenced by the ash yield, maceral content, and vertical column of the coal seam; IOM composition influenced by the peat-forming facies, weathering crusts (WCs), metal-bearing (aureole) waters, the magmatism, and volcanism; IOM composition influenced by general factors, such as coal-bearing basin geotectonical type, proximity of the source rocks, and their geochemical peculiarities, coal metamorphism, and thermal epigenesis. Thus, IOM geochemistry and mineralogy are discussed at three levels: (1) coal seam, (2) coal fields, and (3) coal-bearing basins. Besides, the world average trace element contents in coal (called *coal Clarke values*) are calculated and discussed, with special attention to toxic (such as Hg, As, and Se) and valuable (such as Ge, Sc, and rare-earth elements (REE)) elements.

3.1 INTRODUCTION: COAL—ITS ORGANIC AND INORGANIC MATTER

The following text uses some basic notions and terminology, and abbreviations that are broadly applicable in the field of geochemistry.

3.1.1 Some Terms and Abbreviations

The combustible part of coal (in which enter C, O, N, and a part of S) is named as *organic*, and all the rest is named as *inorganic*, in spite of the fact that this list includes the same chemical elements: C (in carbonates), N and O (in silicates and the other minerals), and S (in sulfides and sulfates). There is no logic in such division, but this is an old geological tradition.

As is accepted in organic geochemistry, the organic matter of coal is marked as *coal OM* or simply OM. Accordingly, the abbreviation IOM is used for inorganic matter of coal. If the average ash content of coal produced is taken at 15%, it follows that annually, from the coal bearing strata no less than 250–280 mT coal IOM is extracted—this specific fossil, which is somewhat *useful*, is somewhat *bad* or *harmful*.

It should be noted that *inorganic* cannot be viewed as a synonym of *mineral*. As *mineral*, we shall name only such inorganic matter that is present in coal in the form of *minerals* (the natural crystalline matter with distant order in structure) and *mineraloids* (the natural noncrystalline matter with some near order in structure). This is because in coals (particularly in lignites), *there is inorganic matter, which is not mineral in the specified sense*! This question is one of the central ones in coal geochemistry (Yudovich and Ketris 2005d, p. 7).

3.1.2 Chemical Elements of IOM

A chemical composition of IOM may be divided into two groups of chemical elements. The first group is the main or ash-forming elements: Si, Al, Fe, Ca, Mg, Na, K, and (S, P). These are the same elements that form 99% of the earth crust's rocks (rock-forming). In coal, such elements make up 99% of all IOM. The second group contains the minor or trace elements, which usually make up no more than 1% of the whole IOM. In English language studies, they are referred to as *minor*, or *rare*, or *trace elements (TE)*, and in German studies—as *Spurenelemente*. For brevity, such elements can be referred to as TE or rare elements (RE). The latter term is more ingenious, but is broadly used in monographic literature—for instance, Yudovich and Ketris (2002).

There are valuable rare metals amongst TE, such as Ge, Ga, U, Mo, Be, Sc, REE. On the other hand, amongst TE are present such elements, as Cl, F, Hg, As, Se, Cr and others. Together with sulphur, nitrogen and phosphorus compounds, they sharply reduce the quality of cast iron and steel, and form (by the use of coke in metallurgy) ash sediments on different parts of boilers, powerfully corrode the walls of cauldrons and pipes, and poison the air, water, and vegetation near coal-using enterprises (power stations, coke plants, and others).

Therefore, coal should not be considered only as an energy resource; it is a complex useful fossil and at the same time a complex *bad (harmful) fossil*, the utilization of which is associated with a significant negative impact on the environment (Kizilshtein 2002; Kolker and Finkelman 1999; Ruch et al. 1974; Seredin and Danilcheva 2000a; Yudovich and Ketris 2005d).

At present, a dozen methods of complex utilization of organic and mineral coal matter are known, helping reduce to a minimum waste production. Besides these, in some cases, the cost of rare metals contained in coal can be more than the cost of the coal itself. Such *metalliferous coal* can be considered as rare-metal bearing ore, and their organic matter can be considered as a by-product (Seredin 1994, 1995, 1996, 1997, 2010; Seredin and Danilcheva 2000b; Seredin et al. 1997, 1999a, b; Seredin and Shpirt 1995; Yudovich and Ketris 2005d, p. 15–18; Yudovich and Ketris 2006c).

3.2 SIX GENETIC CLASSES OF IOM

In spite of there being a broad range of the mineral and non-mineral forms of IOM, usually the ash form (or whole coal, and its sink-float fractions) is analyzed, in which all these forms are mixed. Therefore, a given figure of any chemical element contained in coal always presents a certain *total sum of its different modes of occurrence contributions*.

The tens of thousands of such figures make up the subject of coal geochemistry (Abernethy and Gibson 1963; Arbuzov et al. 1999, 2003; Bouska 1981; Breger 1958; Briggs 1934; Denson 1959; Finkelman 1993, 1988, 1981; Gluskoter 1975; Gluskoter et al. 1977; Goldschmidt 1950; Hawley 1955; Headlee and Hunter 1953, 1955; Ketris and Yudovich 2009; Leutwein and Roesler 1956; Otte 1953; Ren et al. 1999; Swaine 1990; Valcovic 1983; Yudovich 1978; Yudovich and Ketris 2001, 2004; 2005a, b, c; 2006a, b, c; Yudovich et al. 1985). Therefore, for the interpretation of such data, the need arises for the creation of some conceptual models of IOM (marked as total *coal ash*) (Yudovich 1978; Yudovich and Ketris 2002, p. 105–155).

These models operate with certain virtual (imaginary) essences—the *genetic classes* (= *ash* classes) of the IOM. Such classes are present only in the consciousness of the researcher, and only sometimes and/or somewhat can *assume a material form*, being coinciding with modal modes of occurrence. Since in scientific practice the repeated mixing of modal and virtual objects occurs, the need to identify the concepts of virtual essences in coal geochemistry became obvious. The most popular of these models is the model of the genetic classes of the IOM (*coal ashes—A*), introduced by the authors in 1966 and further advanced in a set of later publications (Yudovich and Ketris 2002). Altogether, there are six such classes (Figure 3.1):

(1) *Biogenic ash* of the coal-forming plants (A_{bio}); (2) *sedimentary (chemogenic) ash*, which is formed in some peat bogs (A_{chem}); (3) *sorption ash*, formed by means of sorption of the dissolved species on the OM of peat or lignite, or on their mineral matter (A_{sorb}); (4) *concretion ash* (A_{conc}), formed in diagenesis or from the A_{sorb}, or by itself; (5) *clastogenic ash* (A_{clast}), presented by terrigenic or volcanogenic particles, which have been brought into the peat bog by water or wind; (6) *infiltration ash* (A_{inf}), presented by epigenic (cold-water or hydrothermal) mineralization into the coal cracks.

By coal cleaning, virtual classes $A_{bio} + A_{sorb}$ and very small particles of $A_{conc} + A_{chem}$ remain in light fractions, since they are not exposed under usual crushing. This way, they form an *inherent ash*. The genetic classes $A_{clast} + A_{inf}$ and larger particles of the A_{conc} are accumulated in heavy factions of coal, presenting an *extraneous ash*.

There are several indirect methods that allow us to estimate the chemical composition of the genetic classes.

The composition of *biogenic ash* can be estimated by analysis of coal-forming plants or their modern analogues.

A composition of *clastogenic ash* may be indirectly evaluated by means of analyses of host rocks, and directly by analysis of the clastogenic minerals in coal (Gluskoter 1975; Mackowsky 1968; Mraw et al. 1983; Parks 1952; Parzentny 1995; Renton 1982; Sprunk and O'Donnell 1942; Stach et al. 1982; Yudovich 1978; Yudovich and Ketris 2002). The composition of the *concretion and infiltration ashes*, which has a

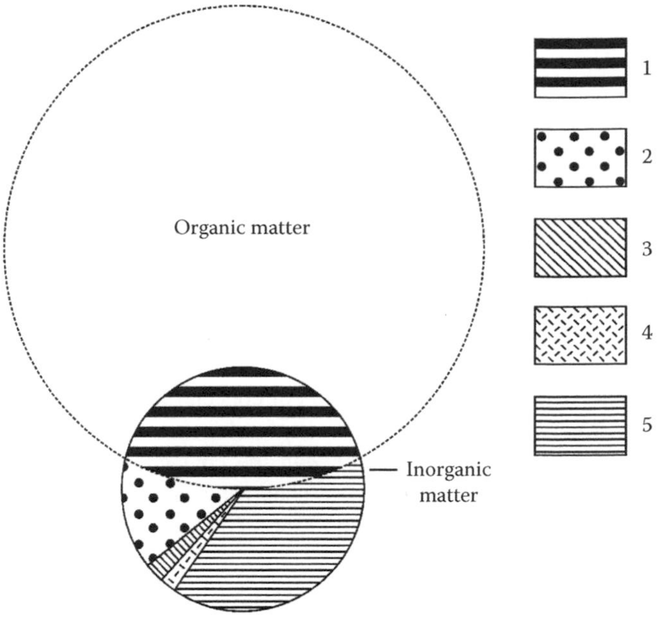

FIGURE 3.1 Virtual structure of the coal IOM—genetic classes of *coal ash*. The above segment (inside of combustible OM, composed of C_{org}, H_{org}, O_{org}, N_{org}, and S_{org}) is *inherent ash* by coal cleaning. The lower segment (outside of combustible OM) is *extraneous ash* by coal cleaning. *Inherent ash*: 1—A_{sorb} (sorption ash) + A_{bio} (biogenic plant ash). *Extraneous ash*: 2—A_{conc} (concretion ash); 3—A_{chem} (chemogenic ash); 4—A_{inf} (infiltration ash); 5—A_{clast} (clastogenic ash).

mineral form, is defined by the composition of the corresponding authigenic minerals (kaolinite, carbonates, and sulfides).

In the past few years, the methods of quantitative mineral analysis of lower temperature ash (LTA) have become well known and practiced. Such LTA is produced in rarefied oxygen plasma, which has been subjected to radio irradiation (Gluskoter 1965).

The most important concept of the all coal geochemistry is the genetic class 3, *a sorption ash* introduced into coal geochemistry by Uzbek geochemists (Yudovich 1978, p. 42–54; Yudovich and Ketris 2002, p. 116–121). Further development of this concept has allowed to explain satisfactorily the *ash paradox* (Yudovich 1978, p. 51; Yudovich and Ketris 2002, p. 183–198), and a number of other empirical regularities in coal geochemistry. A composition of the sorption ash (always in sum with biogenic ash) can be determined by the analysis of low-ash peats, ultra-low-ash coal fractions, vitrain or xylene coal ingredients, and directly by microprobe analysis of coals, and model experiments with peat, lignites, and humic acids, which demonstrate the ability of the peat or coal organic matter to strong sorption of TE from solution.

All of the above-listed genetic classes of IOM can be easily combined in two larger groups: aquagenic, *A*1(A_{bio}, A_{chem}, A_{sorb}, and A_{inf}), and clastogenic, *A*2(A_{ter} and A_{volc} clastics):

$$IOM = A1\left(A_{bio} + A_{chem} + A_{sorb} + A_{inf}\right) + A2\left(A_{ter} + A_{volc}\right)$$

*A*1 is authigenic, sedimentary-diagenetic, and somewhat epigenetic, and the group *A*2 is allothigenic and sedimentary. At the same time, *A*1 contains both mineral and non-mineral matter, whereas *A*2 contains only mineral or mineraloid matter (Yudovich and Ketris 2002, p. 31).

The introduction of the abovementioned virtual concepts (the genetic classes) turned out to be the key to leaving the dead end of the paradoxical empirical dependencies, where a geochemist studying plots in the coordinates of *element in coal–ash content* and *element in ash–ash content* and trying to present coal as simple bi-component model of *ash-free coal OM + mineral matter* would inadvertently be stuck. Only having rejected this model and handling the concepts of concentration of elements in virtual *ash* and of correlations of *ash species* between *ash-carrier* and *ash-concentrator*, can the coal geochemist fully interpret all the paradoxes of coal geochemistry (Yudovich 1978, p. 96–111; Yudovich and Ketris 2002, p. 183–198).

3.3 CONCEPT OF COALPHILE (TYPOMORPH) ELEMENTS

The concept of coalphile, or typomorph, elements is the *main dogma* of coal inorganic geochemistry. This is because the specific chemical TE for coal, same as for biolithic rock, are only the ones for which coal OM serves as a geochemical barrier. On this barrier, dissolved chemical elements are bound as simple salt-like humates (the metal substitutes a proton in the functional group of the OM), or as complex chelates (the metal co-ordinates two or more functional groups of OM) (Gluskoter 1975; Yudovich 1978; Zubovic 1966). One more type of the Me (metal) bonds with OM appears in hard coals—notably the bond Me-C, which characterizes the so-called metal-organic compositions (Seredin et al. 1999b; Yudovich 1978; Yudovich and Ketris 2002; Zubovic 1966; Zubovic et al. 1961a). The elements that have an enhanced affinity to coal OM were named *typomorph elements* many years ago (Yudovich 1978), or to be precise as organic-phile (organic affinity) elements.

However, coal (or peat) organic matter, which realizes its barrier function by capturing TE from solution, is not the only type of geochemical barrier found in coals. TE can also be concentrated in authigenic minerals of coal: sulfides, carbonates, silicates, and phosphates, or may form their own minerals (for instance, PbSe), or isomorph into minerals (for instance, Sr in phosphate), or, finally, be sorbed on mineral matter (for instance, Ga or Ti in kaolinite). In all, any addition of the authigenic genetic fractions of TE to clastogenic IOM enriches the coal ash—as opposed to host rocks. Those elements that show such enrichment are now referred to as the *coalphile*[*] elements.

[*] The term *typomorph elements of coal* was introduced by us in 1968 and has since then received wide usage in scientific literature, though the akwardness of the term was clear to us at that time (after all *morphine* is a form). Unfortunately, we followed the example of A. E. Fersman, the authoritative Russian geochemist who had identified *typomorph* as chemical elements that were typical for *some certain geochemical process*. Indeed, the correct use of the term *typomorphism* and *typomorph* is possible only in mineralogy.

Coalphile elements in coal can be revealed by several methods:

1. By comparing the composition of the coal ash with the composition of host rocks. The more a given element is concentrated in the ash, the more will be its coalphile affinity.
2. By comparing the average contents of the elements in the world coal ash (*ash-basis Clarkes*) with the Clarkes of the sedimentary rocks. The derived figure is the *ash concentration Clarke* (CC) (Yudovich et al. 1985). The higher the CC figure, the higher the average coalphile affinity of the given element.
3. By comparing similar coals, which have different ash yield: the coalphile elements enrich the ashes of coals with less ash yield.
4. By comparing in the same way, but for sink-float factions of the same coal: coalphile elements enrich the ashes of light low-ash fractions.

By using all of these methods, it is possible to range the TE in coal based on the degree of their coalphile affinity (the high-coalphile, moderate-coalphile, low-coalphile, etc.).

The result of this ranging is a set, which usually has at its head Ge, Be, Mo, W, Ga, Sc, and REE, and has in its tail Sr, Ba, and Mn (Yudovich 1978; Yudovich and Ketris 2002; Gluskoter et al. 1977). The most complex-forming elements are the most coalphile ones, and at the same time, are less mobile in supergene waters; and the most mobile elements have the least coalphile affinity. The *transport function* of humic acids (HA) and fulvic acids (FA) of such very mobile elements is stronger than their *barrier function* (Yudovich et al. 1985). The concept of the *geochemical functions* of the living and dead organic matter of the biosphere was introduced by V. I. Vernadsky and turned out to be extremely fruitful for the geochemistry of coal and black shales (Yudovich and Ketris 1997; 2002).

It is important to bear in mind that general set of coalphile affinity for all coals can be built only in a rough approximation. Each specific coal deposit will be found to have its own set of coalphile affinity. For example, in the Lower Cretaceous lignites of the south Lena basin (in Siberia), Sr is the high-coalphile element, and Ge is the non-coalphile element. For all world coals (taken as a whole), titanium is the non-coalphile (or low-coalphile) element, whereas in the Visean coals of the South Timan, in which terrigenic ash is presented as the by-product of the WC, titanium is the coalphile element. In the majority of Bulgarian coal deposits, nickel is more coalphile than cobalt (i.e., $CC_{Ni} > CC_{Co}$), but in the Cretaceous and Paleogene coals of the Russian Far East, the picture is reversed: $CC_{Co} > CC_{Ni}$ (Kitaev 1989).

3.4 TE MODES OF OCCURRENCE IN COAL

One of the basic concepts of geochemistry is V. I. Vernadsky's concept of the *chemical element mode of occurrence in the nature* (in rocks, minerals, living matter, etc.). In reference to geological objects, the mode of occurrence is a totality of the concrete chemical compounds (including an elementary form) of the given chemical element, in which it actually exists in given rock or mineral. At the end of the twentieth century, with the introduction of electronic microscopic and microprobe mineralogy methods in geochemistry, which allow for identification of particles ranging in size from 0.01 up to 1.0 μm (Finkelman 1981, 1986, 1988, 1993; Seredin 1994, 1995, 1997; Seredin and Danilcheva 2000a, b; Seredin et al. 1999a; Seredin et al. 1997; Seredin and Shpirt 1995), an amazing variety of the forms of chemical element sites was found. Such a variety could not even have been conceived between the 1950s and the 1960s, when geochemistry of rare elements was rapidly expanding postwar, influenced by the strategic competition between the USSR and the United States.

3.4.1 Importance of the Exact Knowledge about TE Modes of Occurrence

Contamination of the environment, brought about by coal burning or gasification, is much more influenced by the toxic TE mode of occurrence, than by its gross contents in coal. For instance, if we have high-sulfur coal with significant contents of S, Se, As, and Hg, and all of these are concentrated mainly in pyrite, and particles of the pyrite are exposed by coal grinding (i.e., they are freed from coal OM) under standard grinding (pulverizing) of coal, then these toxic admixtures can be effectively removed by standard coal cleaning before coal combustion or gasification. Conversely, even with comparatively low contents of toxic Se, As, and Hg, which form compounds like *metal-organic* and submicroscopic mineral phases of sulfide and selenides, these elements will not be removed by standard (i.e., economically justified) coal cleaning and could present a significant threat for the environment during the process of coal combustion (Kizilshtein 2002; Kolker and Finkelman 1999; Yudovich and Ketris 2005d).

3.4.2 Methods of Determining TE Mode Occurrences in Coal

These methods can be categorized into direct and indirect (Finkelman et al. 1990). Before the 1980s, indirect methods were predominant, but rapid development of coal micromineralogy is slowly bringing about their replacement by direct methods.

The direct methods include (1) scanning electronic microscopy; (2) microprobe analysis; (3) X-ray analysis of coal or LTA-ash; (4) modern fine methods of X-ray spectroscopy, named as proton induced X-ray emission (PIXE), X-ray absorption fine structure (XAFS), near edge X-ray absorption fine structure (XANE), synchrotron radiation X ray fluorescence (SXRF), and others (Huggins 2002). Methods use for X-ray analysis, specify X-ray diffraction (XRD), X-ray fluorescence (XRF), or both. For spectroscopic methods, define the methods listed. For bulk analysis, add ICP-MS and for microanalysis, add laser ablation ICP-MS.

The indirect methods include (1) ashing and analysis of the ash; the microprobe analysis is generally limited to major elements due to detection limits; (2) selective leaching and analysis of the corresponding extracts; (3) sink-float coal cleaning and analysis of the fractions produced; (4) study of correlations between ash yield and TE contents both in coal and in ash; (5) Humic acids may be extracted from the peat and lignite. These (in lab extracted) HA are very strong sorbents of the elements dissolved. Such modeling allows to judge with greater confidence the possible presence of E_{org}—*organic bound* forms of TE in coal.

All these methods have their own merits and inevitable limits. The direct methods allow us to realistically determine the modal mineral phases, though a microprobe analysis is usually limited to only some elements, while not covering the set of others.

The indirect methods are simpler and more available, but exactly because of their indirect nature, the interpretation of data obtained can be associated with serious mistakes. This is because of the heterogeneity of coal IOM—presence of several virtual *ashes* in coal. For instance, an interpretation of the broadly occurring positive correlation—*contents of the element in coal–ash yield*—does not always mean that a given element is present in mineral form. Such a correlation can be shown by the presence of the most typical rare element in coal—germanium, which forms bonds of *Ge–OM* type and occurs in coal no less than 90% as Ge_{org}. As a result, all efforts to find any amount of mineral germanium in coal were to no avail.

Today, after Finkelman et al.'s (1990) offered methods of stepped selective leaching of coal, and after a certain modification of this procedure by Palmer et al. (1993), it has received broad international usage as a simple, low-cost (although indirect) method of determining the TE mode of occurrence in coal. The procedure of the stepped selective leaching (demineralization of coal) is as follows.

5 g of coal is ground up to −60 mesh (i.e., passed through a sieve with 0.25 mm apertures), and loaded into a cone-shaped polypropylene 50 ml tube, where it is subjected under ambient temperature to stepped leaching by 35 ml 1N CH_3COONH_4, 3N HCl, and 48% HF. The tubes fit in the shaker, where they are shaken for 18 h on each step of leaching. Aliquots from the remaining coal are selected for analysis by methods instrumental neutron activation analysis, inductively coupled plasma (atomization), and mass spectra analysis (determination of the concentration) (ICP-MS), and others; extracts are also analyzed by atomic-absorption analysis. After this, the remaining coal is carried in flat-bottomed Erlenmeyer flasks, where it is leached by 2N HNO_3.

It is considered that the ion-exchanging and water-soluble forms pass to solution by ammonium acetate processing, HCl processing dissolves the carbonates, monosulfides and oxides; HF dissolves the silicates, and HNO_3 dissolves pyrite. The elements that remain in coal after all these processing are interpreted in two ways: either as organic or as *micromineral* form—not exposed under the given coal grinding and labeled as *shielded minerals*. It should be noted that the latter makes interpretation of the organic form contribution very ambivalent, especially for element-sulfophiles. There are also other particularities of the interpretation, which should be taken into account when using this method, which is being introduced in coal-mining countries, with a tenacity particular to Americans.*

3.4.3 Nine Possible Element Mode of Occurrence in Coal

At present, at least nine forms of elements sites in coal are known; these forms can be categorized into three distinct groups: mineral, mineraloid, and non-mineral forms.

3.4.3.1 Mineral Forms

1. Plain minerals (zircon and elementary tantalum)
2. Isomorph admixtures in mineral (Y in zircon)
3. Non-isomorph admixtures in mineral (Ti in pyrite); V. V. Seredin has very intensely studied the microminerals in coals, and has opened many native minerals among such forms
4. Sorbed admixtures in mineral (Au in pyrite).

3.4.3.2 Mineraloid Forms

1. Plain mineraloid (for instance, carbon in vitrain)
2. Isomorph admixtures in mineraloids (Co in semi-amorphous Mn oxides) (Yudovich and Ketris 2002).

3.4.3.3 Organo-Mineral Forms

Organo-mineral forms, for instance, include Ge in vitrain of lignites. The type C forms are typical not only for coal but also for black shales and bitumens (Yudovich and Ketris 1997). Such forms are of a special variety, being centaur E–OM, that is, a compound of the element (E) with OM. In this type of organo-mineral forms, the forms can be distinguished into several varieties: (1) sorbed, (2) salt-like (a substitution of the carboxyl group by metal), (3) chelate (the coordination by metal of two functional groups, creating *chela-like* structure), (4) metal–organic compound (entering the element in carbon skeleton with shaping E–C bond), and (5) fullerene (holding of the E-atom in cavities of fullerene molecules).

3.4.3.4 Fluid Form

1. In liquid phase, NaCl in clay—as a component of free-bound interstitial water; Cl^- sorbed on surfaces of coal OM, but existing in a firmly bound liquid phase of the pores (Yudovich and Ketris 2002; Huggins 2002).
2. In a gaseous phase, Ar occluded in mica.

All the forms listed are *modal*—in the sense that they actually exist as measurable (or, at least, observable in experiments) material objects. Here, we give a brief characterization of the most important of the above listed forms.

Macromineral form: These forms have been known for a long time in coal namely terrigenic, volcanic clastics (syngenetic), and authigenic (diagenetic and epigenetic) minerals: quartz, clay minerals (kaolinite, hydromicas, mixed-layered), carbonates, sulfides (as well as sulfates in oxidized coals), and heavy accessory minerals, which can be observed in a usual optical microscope, that is, having the individual size of no less than 0.01 mm (Francis 1961; Gluskoter 1975; Mackowsky 1968; Parks 1952; Renton 1982; Sprunk and O'Donnell 1942; Stach et al. 1982; Yudovich and Ketris 2002). Macrominerals make up three virtual genetic classes of the ash: clastogenic (A_{clast}), concretion (A_{conc}), and infiltration (A_{inf}) ones.

Micromineral form: The development of modern methods of scanning electronic microscopy and microprobe analysis during the 1970s and the 1980s (Dutcher et al. 1963; Finkelman 1981) has brought about the discovery of multiple *micromineral* forms of trace elements—as particles of micrometer sizes (from the first microns up to hundredth fractions of microns!).† It was found that such microminerals are closely associated with coal OM and are not exposed under standard crushing of coal. Consequently, under sink-float cleaning of coal, all such minerals pass into light *organic* fractions. These remarkable findings have generated (by some amount of researchers) the wrong idea that *organic bound forms* in general do not exist (or, in a lighter wording, that estimations of the contribution of this form, received earlier by indirect methods are invalid). This idea is wrong because it ignores the geological history of coal, in the course of which the initial TE modes of occurrences in peat have been strongly changed in the process of lithogenesis, that is, on the way peat → lignite → hard coal → anthracite.

Non-mineral sorbed form (which is capable of ion exchange): Clay minerals, coal OM, as well as

* Querol et al. (2001) indicate in their article two more complicating occasions besides uncertainties of the interpretation of the remaining insoluble faction.
1. The procedure is meant only for mineral associations (the pyrite, calcite, and clay minerals), typical for the majority of coals. But, if atypical minerals (such as the sulfides, Pb and Zn, phosphates, and ferrous carbonates) are present, the interpretation of extracts becomes doubtful, since PbS and ZnS can dissolve in HCl, whereas ankerite and siderite are poorly soluble, but will dissolve later during acid processing. As for phosphates, they can be dissolved during different stages of leaching.
2. An interpretation of the ammonium-acetic extraction as ion-exchanging does not in itself carry information on concrete sites of elements, because an element can be sorbed on coal OM, clay material, zeolites, and other sorbents. For example, Na, after passing into ammonium extract, can be bonded with all listed phases; besides this, Ca-carbonates may also in part pass into this extract.

† Technical and instrumentation backwardness in the USSR and Russia is an old and well-recognized problem; V. I. Vernadsky wrote about it with alarm during World War II. Still, the best laboratory technology comes to Russia from the West. In this regard, more astonishing is the fact that one of the pioneers of the coal micromineralogy was the Russian coal geologist G. S. Kalmykov, who did a remarkable study of micromineral forms of the pyrite in coals as early as in 1959 (Yudovich and Ketris, 2002, p. 91).

sulfides (the presence in such a form a part of mercury and gold is theoretically possible) can serve as sorbents. It is considered that this fraction can be determined by processing coal by 1N ammonium acetate, NH_4CH_3COO (Finkelman et al. 1990). Under an electronic microscope, such forms are invisible. The given form is a component of two virtual genetic classes of ash: A_{sorb} and A_{inf}.

Mineral salt-like water-soluble form: Most often, in such a form in coal, chlorine is present (with an equivalent amount of sodium). In order to determine the water-soluble fraction, coal is processed by hot distilled water.

Non-mineral form, bonded with coal OM: These can be salt-like compounds of the humate type, as well as complex compounds of the chelate type. The most known among such compounds are water insoluble Ca-humates (from ash-forming elements), and Ge-chelates (from TE). It is considered that such *organic bound* fractions can be determined via processing of coal by a solution 1:3 HCl (Finkelman et al. 1990).

3.5 GEOLOGICAL FACTORS CONTROLLING THE GEOCHEMISTRY OF IOM IN COAL

Obviously, coal is a complex geochemical system since its geochemical characteristics are controlled by a set of geological factors—both relatively independent and interrelated ones (Lindhal and Finkelman 1986; Yudovich and Ketris 2002, p. 175–182).

3.5.1 Geochemical Processes in Coals

The accumulation of TE in coal depends on internal and external factors of the coal (peat) beds. The external factors define the intensity of the TE intake into the bed, and the internal factors—the conditions of TE capture in the coal OM. If these processes run during the stage of peat accumulation and diagenesis, they are qualified as syngenetic, and if they run during the catagenesis or late hypergenesis stage, they are qualified as epigenetic (Table 3.1).

3.5.2 Hierarchy of the Geological Factors in Coal Geochemistry

In turn, internal factors of the TE distribution in coal (the ash yield, the petrographic composition, the degree of coal metamorphism, the structure and thickness of the coal bed, and the presence and intensity of the epigenetic mineralization) are themselves derivatives of some *factors of the higher rank*. As a result, the following regularities were found:

The thickness of coal bed closely correlates with the rate of sinking and the tectonic pulsation mode. *The structure of* the bed may be changed (a) in lateral direction (due to change of peat-forming facies) and (b) in vertical direction (due to change of the climate or tectonic conditions during peat accumulation.

Petrographic composition of coals is defined by both the botanical nature of the coal-forming plants and the conditions of their decomposition in the original peat bog, but the latter is again a result of facial, climatic, and geotectonic factors operated during peat accumulation.

The ash yield of coal is closely related to the landscape of the original peat bog and its type, and the composition of clastogenic ash is defined either by the composition of the source rocks in the framing of coal basin, or by the composition of the volcanic clastics.

The rank of coal (i.e., the processes of strictly metamorphism of coal, the catagenesis degree of the coal-bearing strata) is predetermined by the geotectonic type of the coal basin. For the foreland formations that were submerged to the depths of 5–7 km, the intensive manifestation of catagenesis is

TABLE 3.1
Typification of the TE Accumulation Processes in Coal

Syngenetic Processes		Epigenetic Processes	
Accumulation Factors		Accumulation Factors	
Extraneous	**Internal**	**Extraneous**	**Internal**
TE input into peat bog	Favorable conditions for TE fixation on peat	TE input into coal bed (more rarely—into buried peat layer)	Favorable conditions for TE fixation on coal (buried peat)
With hydrotherms (Ge, B, TR, Mo, Sb, and As)	Marginal and partly near-bottom facies of the peat bogs (Ge and W)	With hydrotherms (Hg, Pb, Zn, Cu, Cu, Sn, Sb, As, Cd, Ag, Ba, and U)	Marginal zones of coal beds (more often near-roof ones) (Ge, P, and TR)
With volcanic exhalations (B and Ga)	Enhanced ash yield of peat (a majority of TE)	With connate waters (B, Be, Mo, U, and Cl)	Zones of tectonic fissuring in coal (Ge and Se)
With surface (including marine) and groundwaters (S, Ge, Cu, U)	An availability in peat of woody fragments–precursors of future lignites, xylens, vitrains (Ge, Ga, W, and Be)	With groundwaters (Cu, Pb, Zn, Mo, and U)	An availability of specific sorbents, such as fusains in non-weathered coals (Cu, Pb, Ba, Sr, and Mn), and regenerated humic acids in the oxidized coals (Cu, Pb, Zn, and U)
With terrigenic or volcanogenic particulates (Ti, Sc, TR, Be, Zr, and Th)	Sapropel or gyttia facies of the peat bogs (V, Ni, Mo, and Cr)		

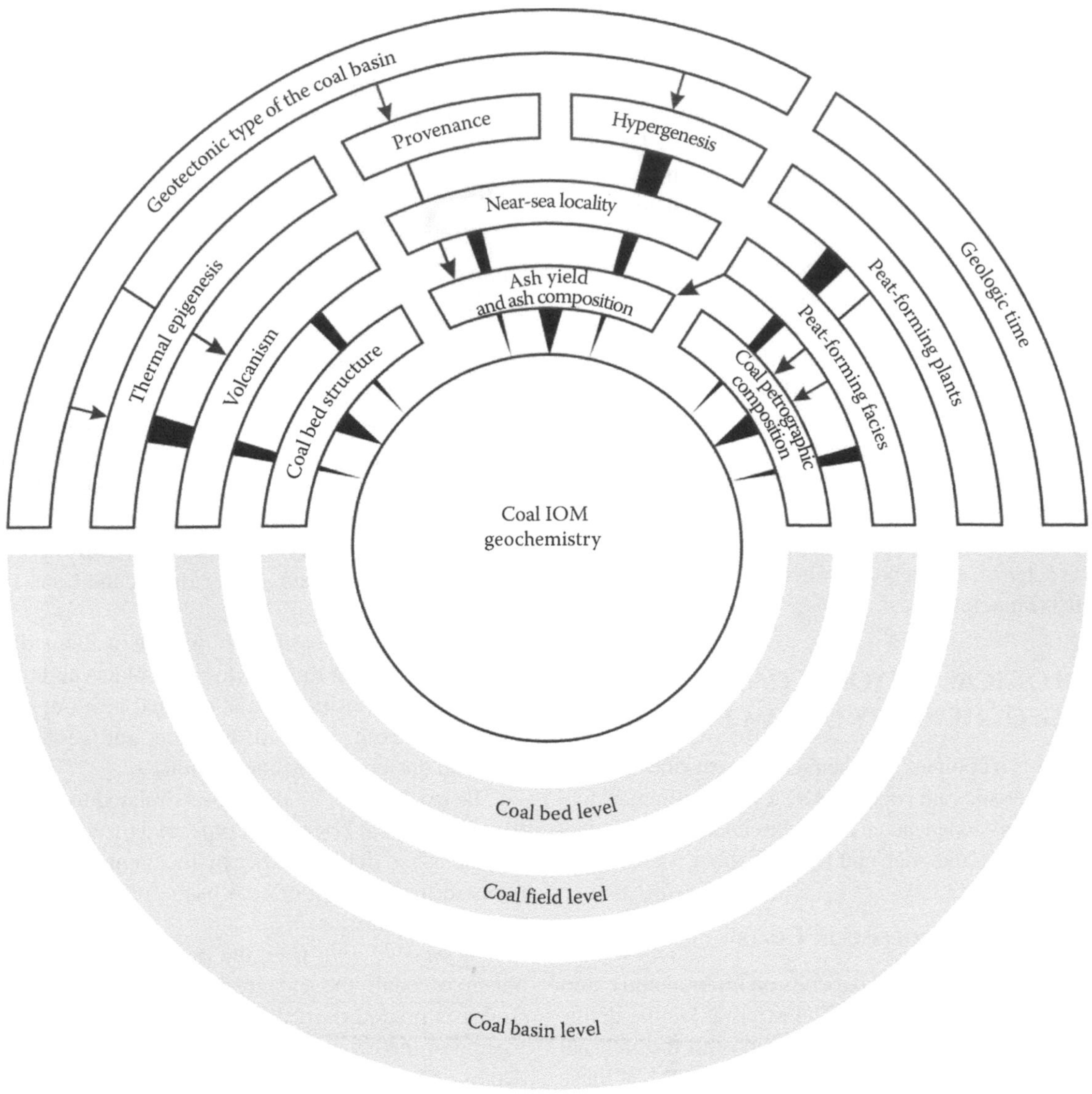

FIGURE 3.2 Factor scheme of the coal geochemistry.

typical; often, it is accompanied by hydrothermal mineralization, while coals may be metamorphosed up to the anthracitic rank. In the coal-bearing strata of the ancient stable platforms, catagenesis processes were much weaker and coal may be brown (lignitic); for the coal-bearing strata of the young mobile platforms, typical phenomena are tectonic inversions, which lead to strong hypergene transformations of original coals. Typical examples are carbon basins: hard coals of the Kuznetsk basin with many volcanogenic tonsteins, and brown coals of the Near-Moscow basin without tonsteins.

It is possible to select three hierarchical levels of the geological factors in the coal geochemistry (Figure 3.2).

Level 1—the coal layer: The majority of empirical regularities revealed in the coal geochemistry correspond mainly to this, the lowest, level. In terms of the Table 3.1, the processes of TE accumulation are controlled, in general, by the *internal* factors.

Level 2—the coal deposit (coal field): Here, in play, the external landscape-facies factors also enter, determining the change of the coal-bearing strata laterally and in its vertical section. On the level of a single coal bed, such factors might be unnoticeable (being fixed).

Level 3—the coal basin: Four most general external factors move to the forefront, defining *the type of coal-bearing formation*: paleogeographical, geotectonical, paleoclimatic, and provenance. On the level of a single coal deposit, such factors might be unnoticeable (being fixed).

3.5.3 Factors Operating on the Coal Bed Level

Among such factors are the ash yield, the petrographic composition of coal, and the position of the coal sample in the vertical profile of the coal bed.

3.5.3.1 Factor of the Ash Yield of Coal

This problem was always in the highlight since the ash yield is the most important characteristic of the coal quality. The analysis of the problem has shown that though TE contents in coal are related to the ash yield in a regular way, these regularities are complex—precisely in consequence of polygene nature of the ash. Therefore, in the 1960s, in the coal geochemistry, the *paradox of ash yield* has appeared: TE contents *in coal* grew (or stayed nearly constant) with the increase of the ash yield, but decreased clearly *in the ash* of coal. This paradox has revealed a disadvantage of the primitive bi-component model of coal in the form of *ash-free organic matter (as the carrier and concentrator of TE) + ash (as a matter sterile on TE).* This model was replaced by a three-component model in the form of *sorption ash (as the carrier of TE) + clastogenic ash (containing certain amount of TE, but acting as a diluent of TE content in the total ash + ash-free organic matter (as matter sterile on TE)* (Yudovich and Ketris 2002, p. 183–198).

These considerations are enough to construct simple mathematical models, permitting satisfactory interpretation of a large variety of empirical plots of the types *Ash yield—the contents of TE in coal* and *Ash yield—the contents of TE in ash*. Since the total ash of coal (A_{tot}) is composed of six genetic classes (a_1, a_2, … a_6), the concrete chemical element can be in all genetic classes. Hence, the content of the concrete TE in coal (C_0) must be defined by two factors:

1. Contents of TE in genetic classes of the ash (s_1, s_2, … s_6)
2. Contribution of each genetic class to coal ($a_i \times c_i$).

As regards to TE contents in the ash of coal (C_A), it will be defined additionally by the third factor:

3. By the contribution of a given genetic class in the total ash (a_i/A_{tot}).

All above mentioned can be formalized in the following manner:

$$A_{tot} = a_1 + a_2 + a_3 + a_4 + a_5 + a_6$$

$$C_0 = c_1a_1 + c_2a_2 + c_3a_3 + c_4a_4 + c_5a_5 + c_6a_6$$

$$C_A = c_1\left(\frac{a_1}{A_{tot}}\right) + c_2\left(\frac{a_2}{A_{tot}}\right) + c_3\left(\frac{a_3}{A_{tot}}\right) + c_4\left(\frac{a_4}{A_{tot}}\right) + c_5\left(\frac{a_5}{A_{tot}}\right) + c_6\left(\frac{a_6}{A_{tot}}\right)$$

where:

C_A is a content of TE in the total ash

Virtual ash classes a_1, a_2, a_3, a_4, a_5, and a_6 correspond accordingly to the following types of ash: biogene (A_{bio}), sorption (A_{sorb}), chemogene (A_{chem}), concretion (A_{concr}), clastogene (A_{clast}), and infiltration (A_{inf}).

Then, we shall simplify the six-component ash model as follows: we shall exclude completely the epigenic (infiltration) ash (as not being directly related to the coal-forming process), and shall unite kindred classes A_{bio}, A_{sorb}, A_{chem}, and A_{concr} in a single class. Then, the model of the total ash will change into the bi-component model, consisting of only authigenic (A_1) and allothigenic (A_2) components:

$$A_{tot} = A_1 + A_2$$

where:

$$A_1 = A_{bio} + A_{sorb} + A_{chem} + A_{conc}$$

and

$$A_2 = A_{clast}$$

For a description of possible relationships, concepts of *concentrator, carrier*, and *diluent* are very helpful used long ago in geochemistry. The *ash concentrator* is such genetic class, in which the content of a given TE is maximal. The *ash-carrier* is such genetic class, which contributes the most part into gross content of the given TE in coal. The *ash-diluent* is such genetic class, in which the content of the given TE is minimal; therefore, it *dilutes* the gross TE content in the total ash of coal. Thereby, the first two terms make a sense with reference *to coal*, whereas the third one makes sense with reference only *to the ash* of coal. It is obvious that, *in a special case, the ash-concentrator may also act as the ash-carrier, but in general, it is not necessary.*

Example 3.1: Strong Coalphile Element—Germanium

Let A_{tot} = 20%, A_1:A_2 = 1:4, c_1 = 10 ppm, c_2 = 1 ppm.

We shall calculate the contributions of the ashes and Ge contents in the ash and in the coal.

The contribution A_1 = 10 ppm (0.20 × 20%) = 0.40 ppm

The contribution A_2 = 1 ppm (0.80 × 20%) = 0.16 ppm

Ge content in the coal C_0 = 0.56 ppm

Ge content in the ash C_A = 0.56 ppm/0.20 = 2.8 ppm

Thereby, the carrier of Ge in coal is the ash A_1 (by convention, the sorption ash); it contributes into the gross Ge content, the part consisting of 71%. In the total ash, the clastogene ash is an ash-diluent, since it is by the order of magnitude impoverished in Ge, compared to the sorption ash. In this example, the ash-concentrator is simultaneously the ash-carrier.

Example 3.2: Weak Coalphile Element—Titanium

Let A_{tot} = 20%, A_1:A_2 = 1:4, c_1 = 2000 ppm, c_2 = 1000 ppm.

We shall calculate the contributions of the ashes and Ti contents in the ash and in coal.

The contribution A_1 = 2000 ppm (0.20 × 20%) = 80 ppm

The contribution A_2 = 1000 ppm (0.80 × 20%) = 160 ppm

Ti content in the coal C_0 = 240 ppm

Ti content in the ash C_A = 240 ppm/0.20 = 1200 ppm

Thereby, the carrier of titanium in coal is the clastogene ash A_2—it contributes into gross Ti content, the main share, consisting of 67%. However, in the total ash, the clastogene ash appears as a diluent of the sorption ash, since the former is impoverished by half in titanium content. In this example, the ash-concentrator is not the ash-carrier!

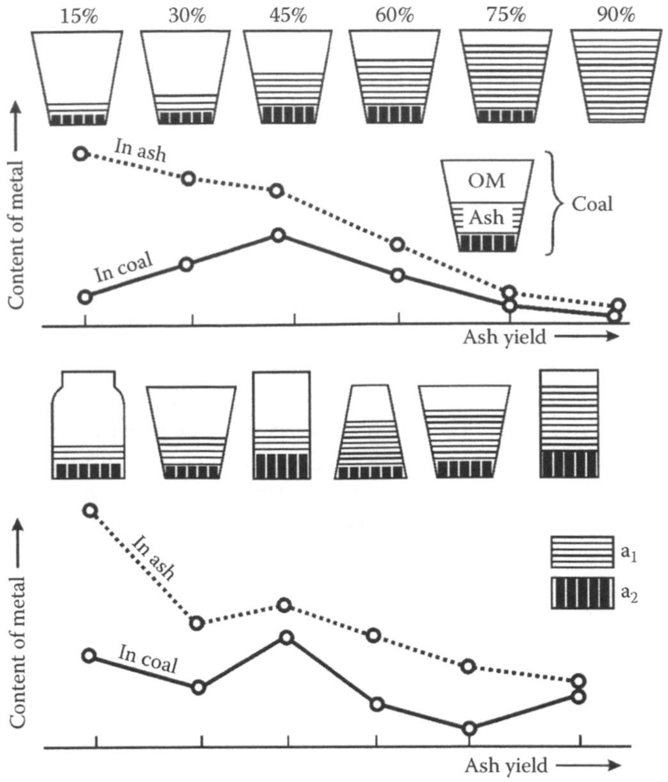

FIGURE 3.3 Schematic explanation of the *ash paradox*. As the total ash (A_{tot}) grows up to a certain limit (in given example—up to average value near 45%), the content of the ash-carrier a_1 *in coal* increases, and accordingly increases the a content of a metal (c). Thereby, function $C_0 = f(A_{tot})$ goes through its maximum. This is *the sorption optimum* of the organophile elements. Since the growth of the A_{tot} is accompanied by a much faster increase of the ash-diluent a_2, the content of the metal in ash falls: function $C_A = f(A_{tot})$ looks as monotonically (or nearly monotonically) decreasing. Such picture exists only for homogeneous statistical coal samples of single-type coals (the upper part of the Figure 3.3). But if statistical coal samples are non-homogeneous (or the value a_1 is a complex function of the value a_2), the plots $C_0 = f(A_{tot})$ and $C_A = f(A_{tot})$ became more complex (the lower part of the Figure 3.3). (Data from Yudovich, Ya. E., *Geochemistry of Fossil Coals (Inorganic Components)*, Nauka (in Russian), Leningrad, Russia, 262 pp, 1978.)

These simple considerations help to interpret the ash paradox completely.

Evidently, if with growing A_{tot}, *any ash-carrier is increased*, then TE contents in coal—C_0, will also increase. But if the contents of the ash-carrier decrease with the growth of the A_{tot}, then the contents C_0 will decrease as well. The increment of C_0 will be linear if the ash-carrier increases linearly with A_{tot}, and the change of C_0 will be nonlinear otherwise. For instance, in the scheme of Figure 3.3 (at the top), it is shown how the contents of the sorption ash-carrier a_1 grow with growing A_{tot} up to the value A_{tot} = 45%. It means, that in an ash interval of 15%–45%, the contents (c_1) of all coalphile elements related to the sorption ash (for instance, Ge) will grow.

Meantime, as applied *to the ash*, a character of resulting plots will be different: it will depend exclusively on the dynamics of the ratio *ash-concentrator:ash-diluent*. If with growing A_{tot} the content of the ash-diluent grows faster than the content of the ash-concentrator (the most usual case), then in coordinates *Ash yield–TE content in ash* a negative correlation will be observed (Figure 3.3, at the top). If, however, the a_1 value is not independent of the a_2 value, and is some function of it, or when a given statistical coal sampling is not homogeneous, the pattern of the change of C_A with the A_{tot} will complicate considerably (Figure 3.3).

3.5.3.2 Factor of the Petrographic Composition of Coal

As is well known, coals consist of three groups of authigenic organic microcomponents (the macerals), in a way *coal minerals*: gelified (the vitrinite group), fusain and mixtinite (the inertinite group), and liptinite and alginite (the exinite group). The first group is the product of anaerobic gelification of primary plant tissues of coal-forming vegetation; the second group is the product of their (plant tissues) partial oxidation, and the third group consists of bituminous components of plants (the cuticle, spores, and others). The fourth petrographic group in coal is formed from allotigenic and authigenic *mineral admixtures*: quartz, clay minerals, carbonates, sulfides, and various heavy accessory minerals (including microminerals).

At that, the amount and composition of mineral admixtures are not completely independent—they are correlated in a certain manner with the amount of coal organic macerals. The analysis of very contradictory data has allowed to reveal several general empirical regularities.

1. The majority of coalphile elements enter vitrinite group. Among them, the vitrain by itself (bright ingredient—structured gelified remains of wood) is a specific concentrator of Ge and some other metals, which very clearly manifests itself in brown (lignitic) coals, and less clearly in bituminous coals (called *Ratynsky's regularity*—refer to Yudovich and Ketris (2002, p. 203). The cause of vitrain's special affinity to Ge is its primary enrichment in lignin, since in the course of lignin transformation, humic species were generated containing ortho-diphenol and ortho-oxyquinol ligands, capable of strong co-ordination with germanium (Yudovich 2003a; Yudovich and Ketris 2002, p. 248–249).
2. The fusain (fibrous, porous, and carbonized structured ingredient, resembling charcoal) usually carries the lowest TE concentrations.
3. Lipoid bituminous components of coal usually are very poor in TE content, but sometimes may contain V and Ni.
4. The majority of sulfophile elements are present as authigene sulfide minerals (or microminerals), and only in a small percentage—as organomineral species (bonded with OM).
5. In brown (lignitic) coals, ash-forming elements, such as Ca and Mg, and, from trace elements, Mn, Sr, and Ba, are usually associated with gelified macerals, and in bituminous coals with carbonates (besides, Ba quite often forms a sulfate—the barite).

6. The elements of hydrolyzates—ash-forming Al and trace ones Ti, Ga, Sc, Sn, and some others—are often associated with authigenic kaolinite (may be also with zeolites, but these minerals in coals are hardly studied).
7. Ash-forming phosphorus, and from TE—Sr, Th, and REE—are bound with authigenic micromineral phosphates such as apatite, monazite, and others more often than with organic macerals.
8. The rest of the TE in coal basically enter into the composition of terrigenic mineral admixtures, most often clay minerals.

As I. V. Kitaev demonstrated with an example of the Russian Far East coals, the petrographic composition of coals has a strong impact on the pattern of the plots *Ash yield–content of the element in coal (or in ash).* It was found that for the one (in the ash of bright coal, is associated negatively with ash yield) and the same element (in the ash of dull coal, is associated positively with ash yield), but in different coal lithotypes (bright, semibright, semidull, and dull), these plots differed greatly. This is explained by different contributions of the forms TE_{org} and TE_{min} (Kitaev 1989). We think it would be fair to call this empirical regularity as the *Kitaev's regularity.*

Often enough, there are exceptions from these general regularities. As a rule, such exceptions have important genetic sense, for instance, the enrichment in TE-sulfophiles of fusains. In general, the above-listed regularities are typical for the syngenetic peat accumulation and diagenesis, whereas either one or another exception usually characterizes the epigenetic processes (the catagenic or hypergenic stages of coal-bearing strata).

The character of the TE distribution (and, in particular, of toxic TE) between organic macerals and the mineral admixtures in many respects defines the behavior of toxic TE in processes of coal combustion. Therefore, organic-bonded trace elements pass into gaseous phase and fly ash, whereas inorganic ones pass into bottom ash and slag. Important exception is Hg: regardless of Hg mode of occurrence, the most part of Hg in coal goes to the gaseous phase by coal combustion.

3.5.3.3 Factor of the Coal Sample Position in the Bed Column

Coal beds have marginal zones of enrichment in TE near the bottom, floor, and partings. In such zones, the TE contents may exceed the TE contents in central part of the layer by 1–2 orders of magnitude. The thickness of such *marginal zones* is on the average 0.10 m with usual fluctuation between 0.05 and 0.20 m. Thin layers are always richer in TE than thick ones, due to the greater contribution of the marginal zones to gross TE contents. It is most likely that the enrichment in the marginal zones (at least in the near-bottom ones) was generated during peat formation, but the main development occurs during diagenesis.

It is suggested (though there is not enough evidence) that a relationship must exist between the lithology type of the marginal host rocks and particularities in the TE enrichment of marginal zones.

In the 1970s, we had advanced a mathematical model of the post-depositional diffusive TE enrichment in the marginal zones, together with Vorkuta's geophysicist I. V. Ryasanov. According to Ryasanov's calculation, the diffusion was apparently a two-stage process. In the first (fast) stage, a stock of a metal was created in the fine film on the coal/rock interface. Later, in the second (slow) stage, the stock was redistributed throughout the marginal zone by means of a solid-phase self-diffusion mechanism. The advancement of the theory can be experimentally checked via sampling of marginal zones by very short benches (several centimeters) and by comparison of the measured TE profiles with the predicted ones.

The review of a great body of data (Yudovich 1978, 2003b) has shown that the phenomenon of the marginal enrichment in germanium (as well as in many other TE) may manifest itself differently, depending, first, on the coal bed structure, and, second, on the host rocks lithology.

Model 1: The coal bed is uniform in ash yield and petrographic composition, with the same host lithologies at the bottom and the roof. This model is the simplest and serves as a starting point for all others. The phenomenon of marginal enrichment is not complicated by anything; the thicknesses of marginal zones are nearly equal. The germanium content in coal is approximately proportional to logarithm of the distance between the coal sample and the contact with host rocks.

Model 2: It is the same as model 1, but the composition of the host rocks in the bed roof and bottom is not the same (for instance, sand in the roof and clay in the bottom). In this case, one can expect the asymmetry in the marginal enrichments.

Model 3: It is the same as model 1, but the bed contains clay parting, near which additional (intra-bed) marginal zones appear. Usually, the thicker the parting, the thicker will be the zones; as a rule, they are non-symmetrical, more distinct above the parting than below it.

Model 4: It is the same as model 1, but the bed is non-homogenous in its petrographic composition: there is a band of bright low-ash coal with enhanced Ge content. It complicates the germanium profile with a *noise*, unrelated to the position of the sample in the bed column.

Model 5: It is the same as model 1, but marginal zones are themselves nonhomogenous. For instance, in the near-roof zone, there are alternated bands of coal with different ash yield and various petrographic compositions, which lead to complication in the germanium profile.

Model 6: It is a combination of all models in the same coal bed, which brings about the most complex germanium profile (Figure 3.4).

Besides, some trace elements (for instance, B, Sr, Ba, P, and Mn) show a tendency of *carrying-out* from the coal bed during diagenesis. If this process has gone far enough, then

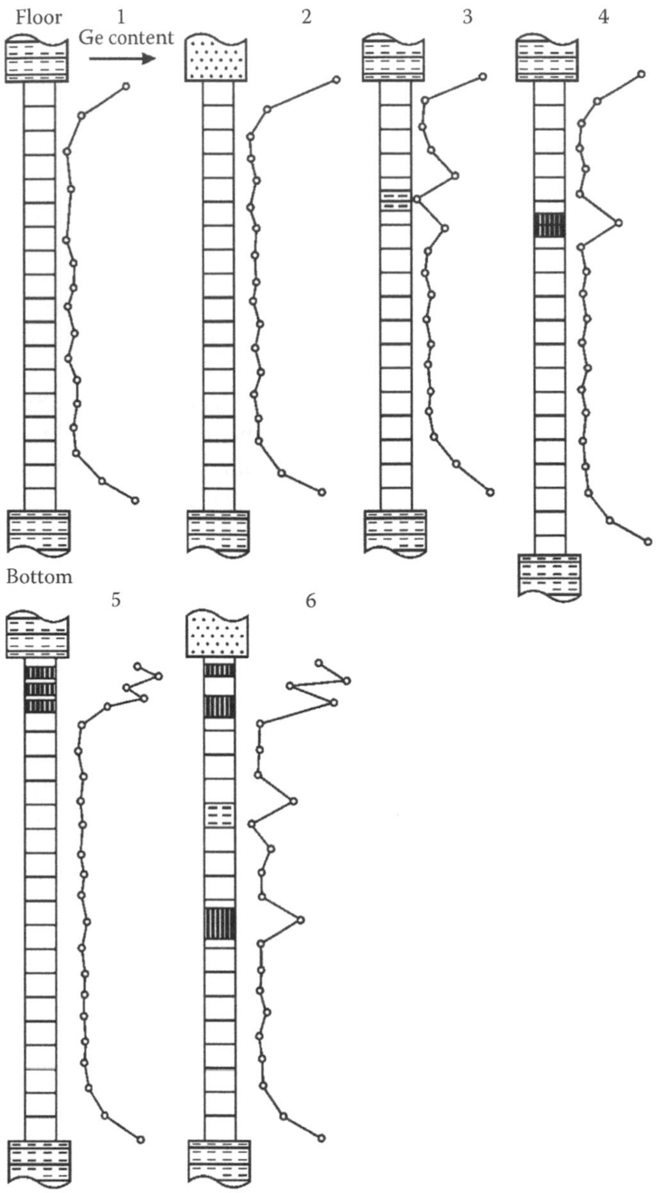

FIGURE 3.4 The phenomenon of marginal enrichment of coal beds in germanium (*Zilbermints Low*). A coal bed is shown, sampled by equal benches.

marginal parts of the bed will not be enriched, but on the contrary impoverished in the listed elements.

The phenomenon of marginal enrichment of coal beds in valuable TE can have significant economic importance—for organization of selective extraction of such enriched coal at the coal mining. This method of mining avoids *diluting* of the metalliferous coal benches by poor coals taken from central parts of the coal layer.

3.5.4 Factors That Affect the Level of Coal Deposits (Coal Fields)

It is possible to refer to factors such as the paleo-environment particularities of coal (a proximity of the original peat bogs to sea, the landscape position of the bog), presence of the WCs and ores in the source area, and local influence of magmatism and volcanism.

3.5.4.1 Peat-Forming Facies

It is well known that the concept of *facies* can be interpreted in different ways in geology. In the given context, as *facies*, we mean either *hydrofacies* (the water masses with different hydrochemical characteristics) or *landscape* facies.

Many geologists have tried (most often by means of geochemical methods) to determine the influence of seawater during the peat formation. To this end, ash-forming elements (Na, Mg, Ti, Ca, and S), as well as TE, including minor (Rb, Sr, Ti, and Mn) and rare ones (B, halogens, and some others) were used. The detailed analysis of the facial factor is given in the monograph (Yudovich and Ketris 2002, p. 256–271). It is of note that some geologists refer to the formation of the so-called saline coals as *marine-influenced*. Such opinions were voiced, in particular, by Professor L. Ya. Kizilstein, A. V. Ivanova, and other researchers (Yudovich and Ketris 2002, p. 82–86).

For example, a comparison of two Canadian high-moor peats has shown (Shotyk et al. 1990) that under similar average ash yield (2.4% and 1.5%), the peat bog Barrington (New Scotland), located on the coast of Atlantic, contains 350 ± 40 ppm Na against 100 ± 20 ppm in the intracontinental peat bog Lazer (Ontario). Mg is similarly partitioned: 1500 ± 390 ppm against 440 ± 150 ppm. It is quite obvious that this difference is created by periodic washing of the coast peat bog by seawater spray. As a result of this washing, the Na^+ and Mg^{2+} in the acid mileu of the lower (eutrophic) Barrington peat layers were intensively leached, but in the more alkaline Lazer peat bog mileu (in the layers of the same stratigraphic position) such leaching does not occur. Apart from this, freshwater clays with kaolinite are appreciably richer in titanium than the brackish ones without kaolinite (Bustin and Lowe 1987). This difference is the same for sedge peats (sphagnum peat is depleted in titanium). The content of titanium *in ash* of the freshwater sedge peat is three times as much when compared with its underclay. This means that such clay (even though it does consist entirely of peat ash) cannot provide the contents of titanium in peat. Consequently, there existed a mechanism of the Ti accumulation in peat: (1) either in the form of Ti_{org}, (2) or in the form of some mineral carrier of Ti, different from those existing in terrigenic ash. A mineral carrier such as this may be, for instance, TiO_2 in authigenic kaolinite.

In other words, we must assume that the same authigenic mineral (kaolinite) in host sediment at the bottom of the peat bog (as of a coal bed precursor) and in the peat bog itself have different chemical compositions; in particular, the second one is enriched with titanium. This important conclusion can be widened with respect to some other elements of hydrolyzates (such as Ga, Th, Zr, Hf, and light REE).

The modern subtropical peat bogs present a model of some ancient coal basins (Yudovich and Ketris 2002, p. 258). In Kolkhida's (modern-day Georgia) lowland peat bogs near the Black Sea coast, there are partings consisting of grey-blue clay. Near such partings, the pH of peat is increased from normal acid values of 5.3–6.36 up to near-neutral values of 7.10–7.85. This is accompanied by an ash yield increment of 1.4%–2%, and also by increased release of volatile matter under pyrolysis (obviously as a result of the increase of the peat decomposition degree), and by a typical increase in the contents of ash-forming elements Ca, Mg, and Na—accordingly in 1.5, 2.5, and 1.5 times. The Fe content, on the other hand, falls to 20%. All these changes are related to the ingression of seawater into peat bogs.

Upper Permian coals of southern China present a unique example of the coal forming facies, very similar to a modern mangrove swamp (Shao et al. 1998). They are situated on a carbonate bottom (and have a rather broad set of facies of the tidal carbonate platform), and even directly alternate with carbonates. At the bottom and at the roof of the coal beds lie laminated clay-algal sediments. As a result of such an unusual situation, these coals are anomalously enriched in Na_2O (up to a weight of 4.36%), MgO (up to 2.6%), and CaO (up to 4%) and contain not only dolomite and calcite but also shells of sea fauna!

Among ash-forming elements, the most informative indicator of sea influence is sulfur, and among TE, such an indicator is boron. It is of no doubt that boron is accumulated in the sea-influenced coals, as it is well known that seawater is enriched in boron by 2 orders of magnitude when compared to freshwater. This difference must be inherited not only by clays but also by coals. The most thorough studies of the use of boron as a facies indicator concern the coals of the Australia, New Zealand (Kear and Ross 1961; Swaine 1971), Russia (Yudovich and Ketris 2002, p. 261–264, 2005d, p. 171–192), and Canada (Goodarzi and Swaine 1994).

Thus, in Australia, Carboniferous coals of the New South Wales containing <40 ppm B were considered as freshwater, 40–120 ppm—as brackish, and >120 ppm—as influenced by sea. Permian coals of the Bowen and Sidney basins, which contain less than 20 ppm B (in some instances, less than 10 ppm) were considered as freshwater ones, whereas the undoubtedly *sea-influenced coals* of the *Theodor* bed contain 15–150, on the average 80 ppm B. The comparison of the average B contents of coals similar in ash yield and rank of two U.S. provinces also provides a meaningful example. The Interior Province coals are three times richer in boron (30 $\Rightarrow$ 100 ppm) than the Appalachian ones. Such a difference is caused by marine facies of the Interior Province coal. Such an explanation corresponds to Ca contents: 0.12 $\Rightarrow$ 1.2%. Since syngenetic Ca accumulation is caused by enhanced pH value in paleo-peat bogs, the conclusion about more seaward facies of the Midcontinental Carboniferous coals formation gets indirect proof. In the monograph by F. Goodarzi and D. Swaine, dedicated to boron in coal, the following gradations of the boron contents (in ppm) related to specified coal-forming environments are accepted (Goodarzi and Swaine 1994):

> Less than 50—freshwater, 50–110—mildly brackish, and more than 110—brackish.

For instance, clear freshwater lignites of the Canadian West (super-thick Hat Creek seam in British Columbia) contain less than 50 ppm B, whereas Cretaceous brackish Alberta coals contain as much as 150 and 675 ppm B. Herewith, the consistency of boron contents on the coal bed area (on a distance from 6 up to 14 km) is considered as evidence of the consistent facies during peat formation.

In the opinion of E. Ya. Podelko, who studied the Mesozoic coal-bearing Trans-Baikal deeps, contents of boron in coals (as in argillites) can serve as a paleosalinity indicator. For the Jurassic lignites of the Kansk-Achinsk basin, there is clear facial difference (Yudovich and Ketris 2002, p. 262):

7.8–36 ppm B—lake-moor coals
24–53 ppm B—fluvial–near sea coals
87–148 ppm B—sea coast coals

Similar data were found for the young brown (lignitic) coals of the East Germany, Bulgaria, and Greece (Yudovich and Ketris 2002, p. 133):

As a salinity indicator, it is possible to use the boron content not only in coal but also in coal illite (obtained by LTA procedure). Thus, illites from the Illinois Herrin bed contain 75–200 ppm B. Herewith, the minimal boron contents were found in illites from the splitted bed zone, near the delta *channels* filled by sandstone. On the contrary, the highest boron contents were found in the most seaward, solid bed zone. It is of note that B contents in illite are in a weak inverse correlation with the illite contribution to clay fraction. *In general, a seawater influence on the buried peat beds can be presumed only when any other explanations of the enhanced boron content in coal appear to be doubtful.*

For instance, Oligocene coals of the Hessen (Germany) have high contents of B: on average, more than 100 ppm, but on the Helmschtedt deposit (coal bed *Treue*), 420 ppm (and more than 4600 ppm on the ash basis). The leading role of the B_{org} form in these coals is evidenced by a relatively low ash yield and accumulations of boron in redeposited debris of xylites and coals (up to 1400 ppm). Oelschlegel's (1964) interpretation is ambivalent: the B accumulation was either from coal-forming plants growing on the saline soils, or from seawater. But the former is doubtful, since an arid environment does not correspond well with the formation of thick peat beds; consequently, a facial factor appears to be more realistic.

Since seawater is a source of both boron and sulfur in coal, a positive correlation between these elements could appear. This looks as paradoxical, since usually such a correlation indicates a sulfide form of an element. But this is not the case, although such a correlation is actually found in some Canadian coals, and very clearly appears in bituminous coals of the East-Pennine basin (England), where boron closely correlates with pyrite iron (Cavender and Spears 1995).

During the formation of the young Meso- and Cenozoic coals, both known peat-formation types—low-land (typical for Paleozoic) and high-moor—had already existed. As is well known, modern high-moor peat deposits, supplied by atmospheric precipitations, sharply differ from the low-land ones by their very low ash yield (1%–5%), low pH values (2.8–3.6), a small degree of the decomposition of vegetable material (the abundant structured components), and high contents of antiseptics—sphagnum mosses and coniferous, as a result of this, by low contents in ash of CaO, P_2O_5, Fe_2O_3, and high SiO_2. The corresponding features of the low-land peats are directly opposite. The similar regularities must also characterize trace elements: under otherwise equal conditions, coals formed from high-moor peats must be much more depleted in TE than coals generated from the low-land peats.

In particular, we have already given attention to *a xylene paradox*, which appears in many Bulgarian lignites, studied by Greta Eskenazi. In spite of the reliably established *Ratynsky's regularity*—Ge enrichments in vitrains, some Bulgarian xylenes (which, as is known, are the precursors of vitrain)—strangely depleted in Ge and other TE. Something similar was noted for xylites from the East Germany and Poland Neogene coals. As a rule, they were depleted in TE by comparison with hosted attrite mass (the exception was Ge, and sometimes 2–3 other elements). The cause of such a picture may be a formation of such xylenes in very acidic environments of the high moor. In these conditions (1) pH is too low for optimum sorption of the majority of TE from peat water and (2) these waters themselves, supplied by ultra-fresh atmospheric precipitation, contain miserable TE concentrations.

Valuable information about paleo-peat bog types are given by the manganese contents, since powerful dispersion of the Mn contents in coal, caused by variable contribution of the Mn_{org} form, reflects, supposedly, greatly different facies on the peat-formation area (Yudovich and Ketris 2002, p. 268). It is obvious that Mn contents in peat depend strongly on peat pH and Eh. In oligotrophic peat (which may be the precursor for low-ash coal), the conditions for the bonding of Mn by the peat did not exist, and opposite, intensive leaching of the primary-vegetable manganese occurred. On the contrary, at large inputs of the terrigenic particles (especially those enriched in dark-colored minerals, the carriers of Mn) to running-water peat bogs, an alkalization of peat water occurred, and conditions for Mn_{org} formation were created. However, the following diagenetic processes of carbonate and pyrite formation could bring about a destruction of the weak-bonding Mn_{org} compounds, and to the redistribution of Mn in carbonatic (siderite, ankerite, and calcite) and sulfidic (pyrite) forms.

For instance, Mn accumulations (up to 0.1%–0.3% Mn on dry peat basis) in some peats of the South Urals and Trans-Urals were noted in those places where vivianite is present. Since a presence of the Fe phosphate or inheriting Ca phosphate is possible also in coal, it may be interesting to check the reality of such mechanism of the Mn fixation. An indication of its reality could be the positive correlation of the triad Mn-Fe-Ca in coals.

The peat-bog sediments form a single whole row with lake ones, and often pass into them laterally (Yudovich and Ketris 2002, p. 269–271). The lake organic-rich sediments, genetically connected to peats, are *gyttia* and *sapropel*, differing by the oxygen regime of sedimentation: gyttia facies are disaerobic (O_2 from 0.2 to 2.0 mg/L; Yudovich and Ketris 2011, p. 740), whereas sapropel facies are anaerobic. As compared to moor peat, in the composition of organic matter of such sediments, the contribution of an aquagenic plant component is increased—specifically the algae and plankton. Accordingly, the contribution of terrigenic high vegetation fragments is depleted.

If peat is a precursor of coal, then sapropels on land and sea are precursors of the *black shales*, gyttia is a precursor of the intermediate type between typical coal and typical black shale; the products of gyttia facies are humic-sapropelic coals.

As it is well known (Yudovich and Ketris 1997), for bituminous (aquagenic) OM of black shales, an enhanced content of N and accumulation of TE, such as Mo, Re, Se, Cr, Ni, Co, V, and Cu, are comparatively typical; for typical marine black shales also I, Br, and P. Besides, in the water of anaerobic sapropel facies, Mn sometimes is strongly concentrated, and may be later deposited in carbonate form—by aeration of the stagnant deeps. The deposition of Mn-carbonates was noted not only in typically marine facies (the deeps of the Baltic and Black seas) but also in sapropel lakes, for instance, in Belorussia.

A contribution of the enriched in P, zooplankton in the black shale OM predestines the possibility of accumulation of phosphate in them. Biogenic P in acid environment carries out from the buried peat. It meets Ca-barrier in host rocks, and may record as Ca-phosphate.

Finally, *anaerobic* (gley) hydrofacies may pass into *hydrosulfuric*, which brings about intensive sulfurization of organic matter, its strong pyritization, and accompanying enrichment in many TE-sulfophiles, such as Mo, Ni, Co, and Cu.

As a result, all this leads to lacustrine-swamp coals (humites-sapropelites) receiving certain geochemical features, which distinguishes them from typical humolites (Yudovich and Ketris 2002, p. 269–271).

3.5.4.2 Influence of the WCs

The paragenesis of coal-forming peats with synchronous WCs is important, but is not yet an enough-elaborated problem of coal geochemistry. There is some parallelism between processes of the peat and WC formation, since both are controlled by similar landscape-tectonics and climatic factors. The detailed analysis of the problem is given in the monograph by Yudovich and Ketris (2002, p. 271–277).

As it is well known, for the WC formation, it is necessary that the territory is in relatively stable condition, since fast uplift prevents development of WCs, and quick descent buries the substratum under sediments. The relief must be leveled, otherwise erosion quickly destroys the WC. For the kaolinite WC formation, warm or moderate humid climate is enough. The formation of lateritic (bauxite-bearing) WCs requires an alternation of short dry seasons (for decomposition of substratum) and humid ones (for a washout of mobile soluble compounds) in subtropical or tropical zone. But for peat accumulation, humid climate (which can be temperate) is also necessary, as well as still even relief, with a trend to constant submersion (otherwise peat formation stops), and high level of the groundwater—an essential condition for swamping.

Obviously, realization of these prerequisites can bring about the paragenesis of coal and WC. In addition, in the sections of many bauxite-bearing platform strata, lacustrine-swamp sediments occupy the top of the column, that is, bauxite formation was followed by peat formation. According to Golowkinsky-Walter rule, such vertical sequence corresponds also to lateral sequence: coal facies are developed in the paleo-deeps, whereas WCs are developed on the neighboring paleo-hills, which are gradually submerged and swamped. Such paragenesis is more probable in the intra-cratonic coal basins than in typically paralic ones, a characteristic for the Paleozoic.

Theoretically, two variants of the correlation of synchronous WCs with peat bogs could be suggested.

1. In conditions of tectonic rest, WCs are formed on the catchment area. Tectonic rise cuts off these WCs, and their friable products are redeposited in relief depression, which leads to swamping. Acid peat water actively reacts with substratum of the peat beds, leaching from them coalphile elements. The latter passes to a solution and is bonded by peat OM and/or diagenetic sulfides. Obviously, in this variant, geochemical specifics of coals will be more distinct in the near-bottom part of the coal layer: high contents of Al_2O_3, TiO_2, as well as other element-hydrolyzates with outstanding complex-forming characteristics could be expected. Due to the aluminum accumulation in sorption ash (Al_{sorb}), the Al_2O_3/SiO_2 module in the total ash can considerably exceed the kaolinite norm (0.85), and a microprobe study may show the presence of free aluminum minerals, such as diaspore.
2. In tectonically active areas, a peat formation can occur in mid-mountain deeps, submersion of which is conjugated with a rising of the surrounding catchments, when it is slow enough for development of WCs. This variant is most favorable, since at first, dissolved products will leach from WCs (for instance, B, Ba, Ge, Sr, and Ca), will come in the peat bog, working as a collector, and later will also hydrolyzate products of the WC destruction (for instance, Be, Ga, Sc, TR, and Th). One is inclined to think that lower and upper bands of the coal layer will be selectively enriched in listed TE.

Other variants of the paragenesis of coals and WCs are also possible. However, in all cases of paragenesis of coal and WC formations one should expect some geochemical specifics of coals. Main feature of such coals is significant accumulations of the ash-forming and rare element-hydrolyzates, often along with accumulation of forming element-sulfophiles. For instance, such coals were described by the authors as the Mississippian lignites of the South Timan, where strong geochemical anomalies of Al, Ti, Be, Sc, REE, Nb, Zn, Cu, As, Cd, Bi, and Ag, along with evident deficit of Ba, Sr, Ge, and Mn, were noted. Higher sulfur content of these coals can be explained by feeding of ancient peat bogs by hard sulfate water due to proximity to arid zone (Yudovich and Ketris 2002, p. 273). Similar particularities of the geochemistry have even-aged coals of the Near-Moscow basin. However, unlike the Timanic ones, these coals are more enriched in Sc and in elements derived from the acid or alkaline substratum—Ga and Zr (Yudovich 1978, p. 202).

3.5.4.3 Influence of Aureole Waters

Sharp geochemical anomalies in coals, exceeding significantly the limits of the geochemical background, require high TE contents in solutions; that is, water, enriched in dissolved elements, brings them into coal-forming peat bogs. Such concentrations may occur in the halo of the ore deposits. For instance, in one of the bogs on the Yukon peninsula (Canada), waters discharge, draining the Zn-Pb sulfide mineralization in the Silurian black cherts (Shotyk 1988). In comparison with a hydrogeochemical background waters, the aureole water is more alkaline, with pH, corresponding to carbonate-bicarbonate buffer (8.17–8.53 against 5.14–8.21 in background water) and more sulfatic (SO_4^{2-} 10.2–158.6 mg/L against 2.3–21.5 mg/L), and have Zn contents of 2 orders of magnitude higher: 15–1357 mcg/L against 3–29 mcg/L. As a result, the *ore* peat was formed here, in which up to weight 60% is accounted by zinc minerals, presented by carbonates and sulfides (wurtzite), which are gradually transform to water-contained Zn silicate hemimorphite $Zn_4[Si_2O_7](OH)_2$. According to the results of three different analyses, the peat contains: Zn up to weight 22.3%, Pb up to 0.6%, Cd up to 2225 ppm, Cu up to 357 ppm, Ni up to 613 ppm, Co up to 60 ppm, Ag up to 2.4 ppm, Mn up to 3910 ppm, and U up to 30 ppm. This example is far from singular: there are many references, illustrating different variants of the enrichment in TE (Cu, Au, Zn, Pb, As, Sb, Mo, and U) of modern peat bogs.

The facts of such sort were long ago used for actualistic (= uniformitarian) comparisons. For instance, the enrichment in Ge of some Ohio and Montana coals was connected to nearby zinc ores. It was assumed that Ge could get into coal-forming peats during Zn ores formation or at their weathering (Stadnichenko et al. 1953).

A similar idea is put forward in respect of Ge in the East Germany and Czech bituminous coals: it is assumed that a syngenetic enrichment in Ge can be a result of the erosion of ancient ores (pre-Carboniferous or Carboniferous). In the Upper Eocene lignites of the South Island in New Zealand, anomalous contents of Ge (up to 250 ppm) were noted, which is commonly associated with erosion of the Ge-bearing pegmatites during peat formation (Yudovich and Ketris 2002, p. 277).

3.5.4.4 Influence of Magmatism and Volcanism

These factors depend most strongly on the geotectonic type of the coal basin, which is in part associated with its geological age. The detailed analysis of this issue is given in the monograph by Yudovich and Ketris (2002, p. 286–298). The enrichment in TE of the Hessen's Neogene lignites due to basalt intrusion may serve as an example of the direct magmatic influence (Oelschlegel 1964). Basalt intrusion has impacted sands of coal bed bottom (in Zeche Stelberg), and sands became red color (hematitized). A heat from intrusion influences the accumulation of Pb, Zn, Cu, Ge, Ga, Be, Mo, Co, and Ni in coal with the source of the elements being presumably host rocks.

It is also assumed that during contact metamorphism, the mode of element occurrence in coal might be changed.

Among the Lower Permian coals of the Shangsi province (North China), strong variation of the Cl, Br, Hg, Zn, Pb, Se, and Sb contents was noted. This was associated with the influence of the Mesozoic magmatism during Yang-Shang orogenese. For instance, the anomalies of Cl, Br, Hg, Zn, and Pb in the Guiyao deposit are related to alkaline intrusions associated with hydrothermnal activity (Ren et al. 1999). In China, Lower Cretaceous coal beds are cross-cut by diabase dikes, with high Mn concentrations noted at the contacts (Querol et al. 1997).

The report appeared recently about discovery of the world's largest germanium-coal deposit Linkang, with Ge reserves of nearly 800 t, in China (Hu et al. 2000; Seredin 2006, 2010; Seredin and Finkelman 2008). The deposit is situated in south Yunnan province and is located in Neogene (?) intramountain trough. The Ge contents in lignites are numbered by the hundreds of ppm. The deposit has two pecularities: (1) there are granites in the basement of the trough, which contain Ge 2–3 times more than Clarke (world background) values; (2) coals are intercalated with hydrothermal-sedimentary cherts. It is assumed that Si-Ge-hydrothermal fluids (associated with the basement granites) were discharged into coal-forming peat bog.

Sometimes, an appearance of *contact metamorphism* is noted in coals described, for instance, in the East Germany and in Russia, in the Tunguska coal basin (near abundant trap intrusions). Unlike the slow regional coal metamorphism through the submergence of coal bearing measure, the contact metamorphism of coal occurs rapidly, under higher temperature, and is accompanied by intensive gas discharge. On the whole, this process resembles coal coking. See also the material on contact metamorphism of coal by Finkelman et al. (1998).

One of the indicators of the coal contact metamorphism can be *epigenetic calcite mineralization.* To illustrate, in 13 Tunguska Permian coal deposits with ash yield of 5%–33%, unusually high hard coal contents of CaO in ash was discovered: from 9.2% up to 37.3%. For comparison, in even-aged coals of the Pechora basin, the CaO contents in ash are only 4.2%–6.0%, in the Pennsylvanian Donets basin coals—2.0%–4.0%. Obviously, Ca accumulation is caused by epigenetic calcite mineralization in the contact halo of trap intrusions. As a result of such mineralization, the coal ash yield increased from 5% to 10% (Yudovich and Ketris 2002, p. 288).

During peatforming processes, (approximately at the same time), volcanic processes might occur. These processes might input into peat bogs pyroclastics volatile compounds (such as iodides, chlorides, and borates) and postvolcanic hydrothermal fluids. Pyroclastics was leached in acidic peat waters with the formation of tonsteins (Yudovich and Ketris 2002, p. 75–82); volatile products and the hydrotherms might be dissolved in peat waters.

According to the available data about thermal waters of the Czech, Caucasus, Kamchatka, and New Zealand, TE contents can exceed regional hydrochemical background by many orders of magnitude and make up the concentrations (mcg/L): Be—430, Ti—15,000, Ge—100, Rb—7700, Zr—150, Sb—900,000, and Cs—2900 (Yudovich and Ketris 2002, p. 289). A unique example of the volcanic influence on Jurassic peat-formation is the Elginsk coal deposit in the South Yakutian coal basin, described in a brilliant article by V. M. Zhelinsky and D. A. Mitronov (Yudovich and Ketris 2002, p. 290). Volcanism is considered as one of the important factors of the syngenetic formation of the *saline coals*, enriched in NaCl (Yudovich and Ketris 2002, p. 82–86).

A. E. Fersman was the first to link an accumulation of ore elements in coal with subsynchronous volcanism. It was this process that he associated himself in 1915 with an enrichment in Pb, Zn, Ag, and As of the Mississippian Borovichi coals. Later V. A. Zilbermints (in 1938, he was killed by Stalin's butchers) had used this idea for interpretation of Ge anomalies in Donets coals. In 1960s, Yu. E. Baranov has noted that enrichment of coals in TE rather clearly occurred in coal formations, which are spatially connected with subsynchronous volcanism, which is particularly characteristic for the Pacific belt. Such formations may belong to both stabilized folded zones (West Canada, Sakhalin Island, and Tohoku Island in Japan), and peripheries of young platforms (the western United States, Lower Silesian basin). V. R. Kler, who for many years supervised geochemical study of the USSR coals, in his generalizations also emphasized that coal deposits with enhanced Ge contents are localized to such coal-bearing formations that were associated with volcanic and hydrothermal processes (Yudovich and Ketris 2002, p. 289).

Clear enough evidence of the volcanic influence can be found both in modern peat bogs and in Meso-Cenozoic lignites. With respect to bituminous coals, such an influence is established with less confidence—only as a result of more or less reliable data interpretation. For instance, on the Kamchatka peninsula was described a unique phenomenon of modern formation of the Ge-bearing peat. It gives a pictorial model of formation of the so-called germanium-coal deposits. Here, water at 60°C, containing up to 28 mcg/L Ge (which is 500 times higher than Ge concentrations in seawater), discharge into peat bog and enrich the peat in Ge up to the concentration of 220 ppm (Yudovich and Ketris 2002, p. 290).

In nearby region of the Russian Primorye, lignitic beds are known with thicknesses up to 10 m and REE concentrations as much as 300–1000 ppm on coal and up to 1% on ash basis (Seredin 1995; Seredin and Finkelman 2008; Seredin and Shpirt 1995; Seredin et al. 1999b). One of the probable mechanisms of the REE accumulation is considered to be hydrothermal input of REE into paleo-peat bogs (or later, into lignitic beds), where formation of a $(REE)_{org}$ form occurred. Also, here, lignitic deposits are known, with extraordinarily high Ge contents: these are the above-mentioned *germanium-coal deposits* (Seredin 2006, 2010; Seredin et al. 1999b; Seredin and Finkelman 2008; Seredin and Shpirt 1995).

It was noticed as far back as in 1970s that areas of coals and coaly argillites mostly enriched in Ge (as well as in Be, Zr, Nb,

Mo, and W) are controlled by faults in the base of coal-bearing troughs, penetrating to some extent into their sedimentary cover, and having age from Lower Cretaceous up to Neogene. In some places, distinct hydrothermal alteration of host rocks is observed: silicification, carbonatization, chloritization, and leucoxenization. It was suggested that exhalative Ge (in the form of $GeCl_4$?) has been partly dissolved in meteoric waters and then migrated from volcanic areas to nearby peat bogs, localized in the relief deeps. Similar empirical regularities are noted in coals found in Japan, where also there are Ge-bearing deposits (Yudovich and Ketris 2002, p. 295).

The geochemistry of the Trans-Carpathian young lignites is considered to be undoubtedly influenced by the synchronous volcanism. Say, on Upper Pliocene Ilnitskoe deposit, a coal formation occurred on the territories, near ancient volcanoes. It seems that the majority of trace elements were accumulated here after peat layers' burial. Among recent publications, we can mention the studies of the Jurassic Yakutian coals, Cretaceous Canadian coals, and Turkey Miocene lignites (Yudovich and Ketris 2002, p. 290–296).

3.5.5 Factors Operating on the Coal Basin Level

3.5.5.1 Geotectonics

Though geotectonic type of the coal basin does define many characteristics of the coal-bearing strata, direct relationship of geotectonic basin type to coal geochemical particularities is tracked rather poorly. The cause is that at such a high hierarchical level, the whole of statistical *noise* from the lower-level factors is integrated. Thus, as was shown in the special analysis (Yudovich 1978, p. 207–208), attempts of some geologists to find a dependency of the Ge and other TE contents in coal on a tectonics type of coal basins appeared to be eventually hardly convincing.

3.5.5.2 Proximity of Source Area

A location of paleo-peat bogs relative to the source area depended both on tectonics type of coal basin, and on landscape environments during formation of peat. The majority of the researchers, who held to the ideas of syngenetic accumulations Ge in coal, attached very big importance to the paleogeography—a distance of coal beds from the borders of coal basin, and on local level—from the borders of coal fields.

Both Russian and western geologists attached much attention to paleogeographic factor. Most often, they indicated that in marginal parts of coal basins situated near the source areas, contents of many TE were higher in comparison to the ones in middle parts of basins. In case the terrigenous input was weakening during the coal strata deposition, a vertical zonality in TE distribution appears along with the lateral one. For instance, A. B. Travin pointed to TE accumulation in the lowest coal beds of the Kuznetsk basin. Such regularity is noted also for W in the Bulgar coals and for Be in the Hungarian Trans-Danube Eocene lignites (Yudovich and Ketris 2002, p. 300).

An explanation of paleogeographic regularitiy in TE distribution in coals is quite obvious: on the long ways of its migration, terrigenous material was leached and depleted in its TE resource and entered the peat bogs strongly impoverished in TE contents. This particularly concerns the dark-colored minerals, the least stable in the weathering environment. Besides, surface waters, as they are removed from watersheds, are rapidly depleted in contents of dissolved TE. In the monograph of Yudovich and Ketris (2002, p. 300–304), quite significant data are published, illustrating an influence of the proximity of the sources' areas on the TE accumulation in modern bogs (U) and in coals (Be, Sr, Ge, and others).

For instance, an accumulation of Be in coal on the peripheries of coal basins was first noted for the Donets basin; later, similar regularity was noted also for the coals found in the United States. To illustrate, in the North Great Plains Province, Paleocene and Eocene lignites, located nearer to the source area, are richer in Be, than distant ones (Zubovic et al. 1961b). Likewise, the Appalachian Pennsylvanian coals are enriched in beryllium from NW to SE, that is, toward the ancient source area (Stadnichenko et al. 1961). The West Virginian bituminous coals are located closer to the source area, and they are richer in Be, than coals of the northern part of the state (Headlee and Hunter 1955).

In some cases, geochemical anomalies of coal can be explained by geochemical specificity of the terrigenous ash. In the monograph of Yudovich and Ketris (2002, p. 304–314), extensive data are presented, proving this the most important empirical regularity in coal geochemistry. In particular, considerable interest presents the experience of undertaking in the USSR of broad comparison of coal geochemistry—with a provenance of the large blocks of the earth's crust, called *metal-bearing provinces*. Comparing average TE contents in commodity coals of the given metal-bearing province (normalized to the average over the USSR coals) and selecting the TE group, enriching coal in 2.5–3 times against average, N. V. Olkhovik came to the conclusion that "The data presented show the distinct dependency of the metal-bearing nature of coals on the general metal-genesis of the province" (Yudovich and Ketris 2002, p. 304–305). Though this statement has a very general and non-concrete character, it is more or less obvious that geochemical specialization of coal must correlate with the coal-bearing strata of the entire coal basins, that is, eventually must reflect the geochemical particularities of the terrigenous or volcanogenic source material.

Western geologists also came to similar conclusions. For example, J. Hawley, who studied the Pennsylvanian New Scotland (Canada) coals, relates its enrichment in Ba, Mn, Sr, As, and Pb to syngenetic erosion of Mn-bearing limestones and Pb-bearing ores (Hawley 1955). Among the coals found in China, Eocene coals of the Laoning province and Jurassic coals of the Beipyao province stand out due to higher contents of Cr, Ni, and Zn. It is believed that such enrichment is explained by vicinity of the coal formation area to the basaltic source rocks (Ren et al. 1999). Upper Eocene lignites of the Buller region of South Island, New Zealand, is a good example of provenance influence (Sim 1977). In spite of low ash yield, not exceeding 3%–5%, they are highly enriched in sulfophiles elements (in ppm): As (up to 550), Sb (up to 120),

Tl (up to 180), Ga (up to 60), Ge (up to 250), Sn (up to 150), Cd (up to 25), as well as Ag (up to 0.5). In the fly ash of the coke plant, the concentration of some volatile elements is even higher, reaching (ppm) for Pb—5000; As—10,000; Sb—450; Tl—250; Ga—10,000; Ge—600; and Ag—10. The cause of such accumulations is that while other New Zealand coal deposits during the peat formation had in their framing Paleozoic greywackes, here in the base and framing of coal basin were the Precambrian igneous and metamorphic rocks, containing Be-Tl-Ga-Sn mineralization! For instance, Be content in pegmatites makes up to 250 ppm, Tl—15 ppm.

As far long as the 1960s, the distinct relationship between source rocks composition and geochemical specificity of coals was found by Yu. A. Tkachev in his studies of Jurassic lignites of the intra-mountain troughs of the Soviet Middle Asia. If carbonate rocks dominated in a framing of troughs, coals were enriched in Ag and Cd, whereas the contents of *terrigenic* elements such as Ti, Zr, Cr, V, and Ga were very low; characteristic ratio Sr/Ba = 3. If mainly acid igneous rocks were in the source area, then Mo content in coals was increased, and Sr/Ba ratio was extremely low (~0.05). Besides, in Middle Asian coals, geochemical anomalies of *exotic* Tl were noted. There are many Tl-bearing sulfide ore deposits in the former Soviet Middle Asia. Yu. E. Baranov thought that this is the cause of the presence of thallium in the Middle Asian coals (Yudovich and Ketris 2002, p. 305).

As was shown in some recent studies, quite similar regularities are revealed for Greece young lignites. They were formed in multiple but small intra-mountain troughs, and have very clear geochemical specialization, reflecting the source rocks composition. For instance, Miocene lignites Moshopotamos (Peloponnes Peninsula) are strongly enriched in Cr and Ni due to vicinity of ultramafic Cr-Ni ores, whereas Miocene Serres lignites in NE Greece are enriched in U and Mo due to vicinity of felsic effusive rocks, enriched in these elements (Foscolos et al. 1989).

According to the Canadian geochemists, in Vancouver Cretaceous coal basin, within Comoks coal field, contents of Ni, Rb, and Ta in coal ash decrease, and contents of Sb, Zr, and heavy REE increase in the direction from South to North. It is supposed that during the peat accumulation, grano-diorite source rocks dominated on the North-West area of Comoks, whereas in the other localities of the coal basin, effusive one (tholeite basalts and alkaline basalt, in ratio approximately 1:1) dominated. As a result of grano-dioritic provenance, the Comoks coal ash as a whole is enriched in light REE, Ga, and Mo, in contrast with coal ash in two other areas (Van der Flier-Keller and Goodarzy 1991, p. 262–263).

The good example presents the Pliocene-Eocene lignites of Elhovo basin in Bulgaria (Eskenazy and Mincheva 1995). These lignites are enriched in many coalphile elements (Ge, V, Cr, Mo, Au, as well as W, Sr, As, Mn, Co, Ni, REE, and U). In the opinion of Greta Eskenazy, such enrichment can be explained by the provenance variety in the paleo-Tundzhi river system, which had riverbed close to the modern one. On its way, the river drained a variety of source rocks with different ore deposits, including granitoids with Cu- and Fe-bearing scarns; the complex multiphase Chernozem-Razdelninsk gabbro-diorite massif with titanomagnetite and sulfide mineralization; complex gabbro-diorite massif with dyke facies and Cu, Fe, and Ti-bearing scarns; the quartz porphyries, diabases, and andesites of the Sveti Ili highlands with Cu-deposit Prohorovo; and, finally, broadly developed in fundament of the coal-bearing strata—Upper Cretaceous volcanics with multiple manifestations of the Mo and Cu mineralization.

3.5.5.3 Coal Metamorphism and Thermal Epigenesis

At present, it is very difficult to separate the processes of strictly metamorphism of coal organic matter from the processes of thermal epigenesis. By this reason, separate consideration of some geochemical materials in Yudovich and Ketris (2002, p. 314–338) should be considered as only conditional methodical approach.

The term *coal metamorphism* (= coalification, carbonization) traditionally exists by itself and independently of the term *metamorphism of rocks*, but meanings of these terms do not coincide: coal metamorphism in lithology and petrology corresponds to pre-metamorphical rock changes—catagenesis (= late diagenesis in the West literature). The reason for such a discrepancy is that coal organic matter reacts to temperature and pressure increase much stronger than mineral material of the sedimentary rocks. That is why, time and again attempts were made to abandon the term *coal metamorphism* and bring it in accordance with lithogenesis terminology. According to Ya. F. Kanana's calculations (Yudovich and Ketris 2002, p. 317), temperatures of the coal conversions in the metamorphic row (from lignites up to anthracites) are within the range from 50°C up to 210°C. Notice that these estimations are substantially lower than the ones accepted previously. For instance, it was believed that formation of bituminous coals occurs is in the interval from 200°C to above 250°C.

At present, the coal organic geochemistry within metamorphic row is studied in detail. It is found that many coal characteristics change not linearly, but with extremum in the region of fat or coke coals. Herewith, the general trend is homogenization of the coal organic matter with development of metamorphism. It is very important that in the region of fat/coke coals transition (V^{daf} 30%–20%), porosity (and accordingly, sorption ability) of coal passes through deep minimum, falling drastically from 80–100 mg/g down to 20 mg/g. We think that such change must lead to a loss of a large variety of TE sorbed by coal (for instance, Cl and set of others). As to coal inorganic matter, it is well established, that it also changes during coal metamorphism. Obviously, the changes must in the first place affect A_{sorb}—closely related with coal organic matter, and to a lesser extent A_{clast}. In the course of A_{sorb} *mineralization*, that is its transformation from the form E_{org} into mineral form E_{min}, it may dissolve in water and be partly put out.

And really, a loss of peripheral functional groups and general trend to condensations of the aromatic carbon net of the coal OM *macromolecule* lead to destruction of organic-mineral bonds and to depletion of metals in coal. Thus, the coal metamorphism must be accompanied by carry-over of the coalphile elements. This process appears to be proved for Ca,

Mg, Ge, Sr, Ba, and B. For many other coalphile elements, such a process is very much possible. As far long as in 1970s, Soviet coal chemists have conducted a profound research of the *Ge bonding in bituminous coals*. It was shown that the part of Ge, which remains in coal during metamorphism enters into much stronger (metal-organic) bonding with coal OM, than on the lignitic stage. It is important that an output of the sorption and biogenic components from coal could be incomplete, but rather have a manner of the *mineralization*, that is, conversion of the sorption and biogenic ashes into mineral salts (calcite and others). Probably, the effect of the coal cleaning change in the metamorphic row can be connected with this process (Yudovich 1978; Yudovich and Ketris 2002, p. 321–322).

Thus, in the process of submersion of the coal-bearing strata, due to increase of temperature and pressure, coal OM transformation develops accompanied by deep changes of the IOM. Along with this, these processes can be accompanied by a hydrothermal mineralization of coals under influence of hot brines, appearing on deep horizons of the sedimentary basins. *So, alongside with safely proved output of the coalphile TE from coal in the course of metamorphism, exists also opposite process—epigenic mineralization of coal under influence of hydrothermal brines.*

As far long as 1968, we have noted that contrary to the widely accepted opinion at that time, bituminous coals of many basins are much richer in TE contents than lignites (brown coals). Later on, this idea was solidly proved by calculation of the world averages of TE contents (called *coal Clarkes*)—separately for bituminous and lignitic (brown) coals (Yudovich 1978). After we have calculated the coal Clarkes—as a result, the group of TE was revealed; the Clarke values in bituminous coals were higher than in lignitic (brown) ones (for instance, Pb, As, Ge, and a set of others). Additionally, the epigenetic hydrothermal mineralization is associated with the phenomenon of the *anomalous fusains*, which seemingly against established regularity of TE accumulation in gelified coals are richer in some TE than vitrains (see Section 3.5.3.2).

As it was noted above, an appearance in coal of the hydrothermal sulfide mineralization (Zn, Pb, Cu, Cd, Ag, and Sn) is characteristic of thick coal-bearing strata of the forelands, which have submerged to zones of high temperature and pressures. In these conditions, metalliferous hydrotherms were formed in depths. Mineralizations of the coal in Ruhr, Illinois, and Nikitovka (Donets Basin)—belong to the same type—*Mississippi Valley Type* of the Zn-sulfide mineralization. One of the consequences of coals exposure to hot brines, may be formation of the so-called saline coals (Shendrik et al. 1997; Yudovich and Ketris 2002, p. 82–86). Though this idea does not have many supporters in Russia, it is very popular in the West. There, it was shown by the examples of the North-East England and Illinois coals, the dependency of salinification of coals on vicinity of evaporites in sedimentary column, on the depth of coal setting down (and on coal rank as well!), and on the connate waters salinity.

For the Donets basin coals, the *distinct relationship between the coal rank and epigenetic ore zonality* was demonstrated (Yudovich and Ketris 2002, p. 326). The Pb-Zn-mineralization zones are associated with anthracites, Sb-mineralization—with hard coals, Hg—with coals of the gas, fat, coking ranks. Temperatures of the ore mineralization are in agreement with temperatures of coalification, which points clearly enough to genetic relationship of the hydrothermal ore mineralization and catagenesis of the coal-bearing strata. According to Ya. F. Kanana, the underlaying rocks may be a source of ore matter.

3.5.5.4 Influence of Hypergenesis

The final result of coal basin inversion is a displacement of coals to the near-surface hypergenese zone, where they are subjected to weathering in the zone of active water-exchange. This is accompanied by emergence of newly formed fulvic acids (capable of washing out the metals as dissolved complexes) and humic acids (HA) (capable, on the contrary, of extracting efficiently the metals from surface and groundwaters). The composition of newly formed (*regenerated*) HA in hypergene zone depends on coal rank. With increase of the rank, the contribution of carboxyl groups in such HA increases and contribution of hydroxyl ones decreases. Similar regularity is found also for even-ranking coals of different petrographic composition: fusinites (i.e., more carbonized ingredients), unlike vitrinites, produce during oxidation HA, completely deprived of hydroxyl groups. These intriguing regularities suggest that, during interaction of newly-formed HA with metals, a formation of dissolved salt-like humates is more probable, than formation of hard-soluble chelate complexes (Yudovich and Ketris 2002, p. 338–350).

At present, a hypergene alteration of bituminous coals is studied extensively. It is found that on early weathering stages, coals lose previously accumulated TE, but during later stages they could be additionally enriched in TE due to emergence of free HA. For example, presence of Ge in mine water of the Donets basin, carrying 0.12–0.38 mcg/L Ge is direct evidence of the Ge output in the oxidation zone of coal-bearing strata. In the Kizelovsk basin mine water, containing hundreds of mg/L free H_2SO_4 due to oxidation of pyrite, Ge contents of 1–3 mg/m^3 were found, that much exceeds a hydrochemical background. The calculations have shown that by mine pumping, nearly 200 kg of Ge is removed annually! In collective monograph by Moscow hydrogeologists Yudovich and Ketris (2002, p. 339), a great body of new data is presented from studies of mine (*by-passing*) pumping of the USSR coal basins. An accumulation of large group of complex-forming TE (including REE) in acid sulfuric acid mine water is demonstrated.

The oxidation of coal organic matter by the air oxygen could bring not only to loss of Ge, but also to secondary enrichment in Ge of oxidized coals due to production in them of secondary humic acids. According to Yu. P. Kostin (Yudovich and Ketris 2002, p. 339), bituminous coals just lose Ge during oxidation, while brown coals can be enriched in Ge, if they are localized on watershed or in higher parts of glasics. The TE input–output processes during hypergene change of coals are noted not only for Ge but also for other TE. There are supergene-altered coals that were studied in the United States (Arkansas and Oklahoma) and Russia

(Tungus, Minusinsk, Kansk-Achinsk, and Kuznets basins). In such coals, many TE accumulations were described (Branson 1967; Zubovic et al. 1967).

For instance, comparison of the contents of yttrium and lanthanum in four profiles of the Interior Province, United States, on weathered and nonweathered coals, has shown that the former were enriched in Y by 14.9 times, and La by 14.3 times (Zubovic et al. 1967). The accumulation by weathering of Y is noted also for the Maicuben coal deposit (Kazakhstan): 37 ppm (weakly oxidized coal) and 104 ppm (strongly oxidized coal). An accumulation of Sc (up to 100 ppm) is noted in oxidized lignites of the Tomsk area (Russia).

The upper 30 m of coals in Kazakhstan (near 7 billion tons) are strongly weathered. Such coals may serve as efficient fertilizers since they contain humic acids (HA) and trace elements (TE). For instance, on the Ekibastuz opencast mines alone, tens of million tons of weathered coals are removed for stripping annually. These coals contain as yet appreciable amount of organic matter (8%–10%) and high concentrations of TE (ppm): Cu (2000), Co (50), Pb (300), Zn (800), Mo (30), Ni (20), Mn (7000), and Cr (300). Agronomic tests have shown that adding these coals into soil, in amount of 2–3 metric centner/hectare, brings to average fertility increase of 15%–25% in conditions of the Alma-Ata area.

F. Goodarzi, who studied Canadian bituminous coals of the Fording mine, has noted the decrease of Al, Mn, as well as Lu, Sc, Th, and Sb contents—from the top to bottom of the coal beds. In his opinion, this points to leaching of the above-mentioned components by groundwater (Goodarzi 1988). In other Canadian coals, significant inverse correlation between Cl and O is found, indicating a partial loss of Cl_{org} during hypergene alteration of coal (Goodarzi 1995; Goodarzi and Cameron 1987). On the Bulgarian Pirin deposit, hypergene oxidation of coal resulted in partial loss of W (Eskenazy et al. 1982).

Of course, the most prominent example of influence of the hypergene alteration on coal geochemistry is regional uranium-bearing of lignites and young bituminous coals of the western United States (Denson 1959; Denson et al. 1959). It has long been established that the higher the coal bed is localized, the more it is oxidized, and the higher is uranium contents in coal. It is very important that all uranium-bearing coal beds have elevated ash yield—on an average 40%. We have emphasized genetic relationship of these two phenomena since an increase of ash yield of uranium-bearing lignites is also caused by hypergene alteration of coal: by means of Ca and Mg sorption with newly formed humic acids from hard groundwater in arid climatic zone (Yudovich and Ketris 2001).

3.6 AVERAGE TE CONTENTS IN THE WORLD COALS (COAL CLARKE VALUES)

In 1923, the famous Russian geochemist A. E. Fersman introduced the term *clark* (= Clarke, in English), in honor of the prominent American scientist, one of the founders of geochemistry, F. W. Clarke (who worked many years as chief chemist at U.S. Geological Survey), who first calculated average composition of various rocks, and later, the earth's crust.

Fersman defined Clarke as the average content of given chemical element in the Earth's crust and also in hydrosphere. This term very soon took root in the Russian literature, but up to now, is nearly unknown to western researchers. With time, however, a sense of the term was strongly extended: many other geochemical averages were named as *Clarkes*, for example, *Clarkes of granites*, *Clarkes of basalts*, *Clarkes of sedimentary rocks*, including *coal Clarkes*, that is, average trace element contents in the world coals. For example, coal Clarke of Ge is 3.0 ± 0.3 ppm (hard coals) and 2.0 ± 0.2 ppm (brown coals) (Ketris and Yudovich 2009).

3.6.1 Coal-Basis and Ash-Basis Clarkes

During 1970–1980, coal was mostly not directly analyzed for TE but through an analysis of coal ash. Standard coal ashing was performed at 750°C (in the United States and many western countries), or at 850°C (in the former USSR and now in Russia).* It is well known that some elements may be almost fully (Hg, I, and Br), or partly (Ge and Mo) volatilized by *high-temperature* ashing. TE loss may be minimized by *low-temperature* ashing (~130°C–150°C) by means of radiofrequency exposure on oxygen plasma (Gluskoter 1965). This excellent method is, however, time-consuming and for this reason is not widely acceptable. So, up to end of the 1980s, TE content in coal was obtained by recalculation from the content in ash (*ash-basis content*). Such recalculation may lead to underestimation of coal-basis figures, and, as a result, to underestimation of coal-basis Clarkes.

During the last few decades of the twentieth century, several *direct methods* of coal analysis were introduced in coal geochemistry, and first of all instrumental neutron activation analysis. Therefore, many directly obtained coal-basis figures appeared in the literature. Now, a new opportunity appears to recalculate coal-basis figures to ash-basis ones. It is of note, that instrumental neutron activation analysis is mentioned as a direct method of analysis, but its application is very limited because a reactor is required to apply this method, and for bulk analysis, it has largely been replaced by ICP-MS and equivalent methods.

If the analysis of coal ash was earlier simply *a technical tool* (because a direct analysis of coal was too hard and unreliable procedure), have we any need for the *ash Clarkes* today? Yes, we have such need: for the calculation of important geochemical values, *coal affinity indexes* (or *coalphile coefficients*), as we will discuss below.

3.6.2 *Coal-Affinity* (Coalphile) Indexes

Goldschmidt (1950) first calculated the *enrichment coefficients* of coal ash by comparison of element content in coal ash and earth's crust Clarke value. For example, the earth's

* Nonstandard ashing (550°C, with an air access) used in USSR and the United States for special intent—following analysis of ash for trace elements.

crust-As Clarke value was assumed to be 5 ppm, and As content in coal ash-rich ashes was determined as 500 ppm. So, enrichment coefficient was 500/5 = 100.

Later Yudovich (1978) used coal ash Clarke values (instead of *enriched ashes*) and Clarkes of sedimentary rocks for such calculation. These figures (enrichments coefficients) were named as *typomorph coefficients* and were widely cited in Russian and Bulgarian literature.* More recently, this poor term was substituted for *coalphile coefficient (index)*, or *coal affinity index* (Yudovich and Ketris 2002).

What does a coal affinity index mean? *It shows, how efficiently coal acted as geochemical barrier for trace elements, during all its geologic history.* The more coal concentrated trace elements from environment compared with sedimentary rocks, the greater would be the coal affinity index. A researcher could compare coal affinity indexes for different elements in given coal deposit, coal field, coal basin, or province; for the coals of different rank; and for the same coal field (basin, province) but for different elements. For example, As coal affinity index is 50 ppm/11 ppm = ~5 (Yudovich and Ketris 2005a, d), and Hg coal affinity index is 0.75 ppm/0.0 5 ppm = 15 (Yudovich and Ketris 2005 a, b, c, d). So, Hg is threefold more coalphile element than As.

3.6.3 Calculation Results

The calculation procedure is described earlier in detail (Ketris and Yudovich 2009), and the results of the calculations are shown in Table 3.2.

The last column in Table 3.2 represents coal affinity indexes, calculated using our coal Clarkes for hard and brown coals, and modern weighed Clarkes for sedimentary rocks (see Ketris and Yudovich 2009 for details). It is of note, the coal affinity indexes are calculated on the base of averaged ash Clarkes, that is, (element Clarke in hard coal ash + element Clarke in brown coal ash)/2. For example, lithium: (49 ppm+82 ppm)/2 = 66 ppm; 66 ppm/33 ppm = 2 (coal affinity index). In should be noted that some Clarkes in Table 3.2 are based on small statistical samples or/and not very reliable analyses. Such (preliminary) Clarkes could be strongly changed in future; corresponding figures are shown in the brackets.

3.6.4 Comparison of Brown and Hard Coal Clarkes

Such a comparison allows analysis of coal rank influence on coal geochemistry. As was mentioned above (see Section 3.5.5.3), coal *metamorphism* is the thermal epigenetic process, involving hot brines and fluids influencing coal beds. This process not only changed coal organic matter, but may greatly change the TE contents by means of their input or output (Yudovich 1978; Yudovich and Ketris 2002). However, this (obvious) issue is far from universal. It is known that, *in general*, hard coals are Paleozoic, and brown coals are Mesozoic and Cenozoic. This means that in some instances the geochemical differences *brown coals versus hard coals* may be *primary*, accounted for the large initial difference of Paleozoic and Mesozoic–Cenozoic coal-forming flora.

As seen from Table 3.2, brown coal is enriched in B, U, and Mn. For all three elements, one can suggest their output during thermal metamorphism of the coal organic matter. More elements are enriched in hard coals compared to brown ones. Weak enrichments include Co, Ge, V, Pb, and Se, and strong ones include Rb, Be, Zn, and Ni. Only for evident sulfophile elements (such as Pb, Se, and Zn), such enrichment could be caused by hydrothermal input related to coal metamorphism. However, for litho- and siderophile elements (such as Be, Cr, Co, and Ni) such explanation appears to be dubious. It is not excluded that primary coal-forming flora acted here as actual factor of difference (Yudovich 1978). Such (nontrivial) conclusion could highlight some problems dealing with the biosphere evolution.

3.6.5 Coal Affinity Indexes

By analyzing ranges of the coalphile indexes (the last column in Table 3.2), one can divide the elements in four groups:

1. *Non-coalphile elements* (coal affinity indexes are <1): I, Cl, Mn, Br, Rb, and Cs
2. *Weak or moderate coalphile elements* (coal affinity indexes range from 1 to 2): Ti, Zr, F, Cd, V, Ta, Cr, Y, Li, and P
3. *Coalphile elements* (coal affinity indexes range from 2 to 5): Ni, Hf, Sn, La, Co, Ba, Sc, Nb, Sr, Th, Ga, Cu, REE, Zn, W, Au, In, Pb, U, B, and Be
4. *Highly coalphile elements* (coal affinity indexes >5): Ag, Sb, Tl, As, Mo, Ge, Hg, Bi, and Se

The greater the coal affinity index, the greater will be the contribution of an *authigenic fraction* of the given trace element (represented by organic or micromineral forms), and the less is one of a *clastogenic fraction* (represented by macromineral forms [e.g., silicatic]).

Due to the new (weighed) Clarkes of sedimentary rocks (Grigoriev 2003—refer to Ketris and Yudovich 2009), the weighed Clarkes of halogens Cl, Br, and I in sedimentary rocks increased. This results in a corresponding sharp decrease of their coal affinity indexes: halogens *have transformed* from highly coalphile elements (Yudovich et al. 1985; Yudovich and Ketris 2006b) to non-coalphile ones.† Also, the coalphile indexes of Au, Cd, Y, V, U, and Cr have unexpectedly decreased, and, in turn, indices of Tl, Zn, and In have unexpectedly increased.

* See numerous papers published by Greta Eskenazy, Jordan Kortensky, Stanislav Vassilev, and some others.

† However, J.C. Hower believes (personal communication, December 2008) that halogens are coalphile on lithotype basis; vitrains have high Cl and Br where these elements are high. But Cl (or Br) is not universally high, so averaging takes in low-Cl coals.

TABLE 3.2
Coal Clarke Values for Trace Elements (ppm; median ± S.D)

Elements	Coals			Coal Ashes			Clarke of Sedimentary Rocks	CAI
	Brown	Hard	All	Brown	Hard	All		
Typical Cation-Forming Lithophile Elements								
Li	10 ± 1	14 ± 1	12	49 ± 4	82 ± 5	66	33	2.0
Rb	10 ± 0.9	18 ± 1	14	48 ± 5	110 ± 10	79	94	0.84
Cs	0.98 ± 0.10	1.1 ± 0.12	1.0	5.2 ± 05	8.0 ± 0.5	6.6	7.7	2.0
(Tl)	0.68 ± 0.07	0.58 ± 0.04	0.63	5.1 ± 0.5	4.6 ± 0.4	4.9	0.89	2.0
Sr	120 ± 10	100 ± 7	110	740 ± 70	730 ± 50	740	270	2.0
Ba	150 ± 20	150 ± 10	150	900 ± 70	980 ± 60	940	410	2.0
Cation- and Anion-Forming Elements with Stable Valency								
Be	1.2 ± 0.01	2.0 ± 0.1	1.6	6.0 ± 0.5	12 ± 1	9.4	1.9	4.9
Sc	4.1 ± 0.2	3.7 ± 0.2	3.9	23 ± 1	24 ± 1	23	9.6	2.4
Y	8.6 ± 0.04	8.2 ± 0.5	8.4	44 ± 3	57 ± 2	51	29	1.8
La	10 ± 0.5	11 ± 1	11	61 ± 3	76 ± 3	69	32	2.2
Ce	22 ± 1	23 ± 1	23	120 ± 10	140 ± 10	130	52	2.5
Pr	3.5 ± 0.3	3.4 ± 0.2	3.5	13 ± 2	26 ± 3	20	6.8	2.9
Nd	11 ± 1	12 ± 1	12	58 ± 5	75 ± 4	67	24	2.8
Sm	1.9 ± 0.1	2.2 ± 0.1	2.0	11 ± 1	14 ± 1	13	5.5	2.4
Eu	0.50 ± 0.02	0.43 ± 0.02	0.47	2.3 ± 0.2	2.6 ± 0.1	2.5	0.94	2.7
Gd	2.6 ± 0.2	2.7 ± 0.2	2.7	16 ± 1	16 ± 1	16	4.0	4.0
Tb	0.32 ± 0.03	0.31 ± 0.03	0.32	2.0 ± 0.1	2.1 ± 0.1	2.1	0.69	3.0
Dy	2.0 ± 0.1	2.1 ± 0.1	2.1	12 ± 1	15 ± 1	14	3.6	3.9
Ho	0.50 ± 0.05	0.57 ± 0.04	0.54	3.1 ± 0.3	4.8 ± 0.2	4.0	0.92	4.3
Er	0.85 ± 0.08	1.00 ± 0.07	0.93	4.6 ± 0.2	6.4 ± 0.3	5.5	1.7	3.2
Tm	0.31 ± 0.02	0.30 ± 0.02	0.31	1.8 ± 0.3	2.2 ± 0.1	2.0	0.78	2.6
Yb	1.0 ± 0.05	1.0 ± 0.06	1.0	5.5 ± 0.2	6.9 ± 0.3	6.2	2.0	3.1
Lu	0.19 ± 0.02	0.20 ± 0.01	0.20	1.1 ± 0.1	1.3 ± 0.1	1.2	0.44	2.7
Ga	5.5 ± 0.3	6.0 ± 0.2	5.8	29 ± 1	36 ± 1	33	12	28
Ge	2.0 ± 0.1	2.4 ± 0.2	2.2	11 ± 1	18 ± 1	15	14	11
Cation- and Anion-Forming Elements with Non-Stable Valency								
Ti	720 ± 40	890 ± 40	800	4000 ± 200	5300 ± 200	4650	3740	1.2
Zr	35 ± 2	36 ± 3	36	190 ± 10	230 ± 10	210	170	1.2
Hf	1.2 ± 0.1	1.2 ± 0.1	12	7.5 ± 0.4	9.0 ± 0.3	8.3	3.9	2.1
Th	3.3 ± 0.2	3.2 ± 0.1	3.36	19 ± 1	23 ± 1	21	7.7	2.7
Sn	0.79 ± 0.09	1.4 ± 0.1	1.1	4.7 ± 0.4	8.0 ± 0.4	6.4	2.9	2.2
V	22 ± 2	28 ± 0.1	25	140 ± 10	170 ± 10	155	91	1.7
Nb	3.3 ± 0.3	4.0 ± 0.4	3.7	18 ± 1	22 ± 1	20	7.6	2.6
Ta	0.26 ± 0.03	0.30 ± 0.02	0.28	1.4 ± 0.1	2.0 ± 0.1	1.7	1.0	1.7
Mo	2.2 ± 0.2	21 ± 0.1	2.2	15 ± 1	14 ± 1	14	1.5	9.3
W	1.2 ± 0.2	0.99 ± 0.11	1.1	6.0 ± 1.7	7.8 ± 0.6	6.9	2.0	3.5
U	2.9 ± 0.3	1.9 ± 0.1	2.4	16 ± 2	15 ± 1	16	3.4	4.7
Typical Anion-Forming Lithophile Elements								
B	56 ± 3	47 ± 3	52	410 ± 30	260 ± 20	335	72	4.7
P	200 ± 30	250 ± 10	230	1200 ± 100	1500 ± 100	1350	670	2.0
F	90 ± 7	82 ± 6	88	630 ± 50	580 ± 20	605	470	1.3
(Cl)	120 ± 20	340 ± 40	180	770 ± 120	2100 ± 300	1440	2700*	0.53
(Br)	4.4 ± 0.8	6.0 ± 0.8	5.2	32 ± 5	32 ± 9	32	44	0.73
(I)	2.3 ± 0.4	1.5 ± 0.3	1.9	13 ± 2	12.2 ± 5.40	12.6	1100	0.01

(Continued)

TABLE 3.2 (*Continued*)
Coal Clarke Values for Trace Elements (ppm; median ± S.D)

	Coals			Coal Ashes				
Elements	Brown	Hard	All	Brown	Hard	All	Clarke of Sedimentary Rocks	CAI
				Metals-Sulfophiles				
Cu	15 ± 1	16 ± 1	16	74 ± 4	110 ± 5	92	31	3.0
Ag, ppb	90 ± 20	100 ± 10	95	590 ± 90	630 ± 100	610	120	5.1
Au, ppb	3.0 ± 0.6	4.4 ± 1.4	3.7	20 ± 5	24 ± 10	22	6.0	3.7
Zn	18 ± 1	28 ± 2	2.3	110 ± 10	170 ± 10	140	43	3.3
Cd	0.24 ± 0.04	0.20 ± 0.04	0.22	1.1 ± 0.2	1.20 ± 0.30	1.2	0.80	1.5
Hg	0.10 ± 0.01	0.10 ± 0.01	0.10	0.6 ± 0.1	0.9 ± 0.1	0.75	0.068	11
(In, ppb)	21 ± 20	40 ± 20	31	110 ± 10	210 ± 18	160	43	3.7
Pb	6.6 ± 0.4	9.0 ± 0.7	7.8	38 ± 2	55 ± 6	47	12	3.9
(Bi, ppb)	840 ± 90	1100 ± 100	970	4300 ± 800	7500 ± 400	5900	260	23
				Non-Metals-Sulfophiles				
As	7.6 ± 1.3	9.0 ± 0.7	8.3±	48 ± 7	46 ± 5	47	7.6	6.2
Sb	0.84 ± 0.09	1.00 ± 0.09	0.92	5.0 ± 0.4	7.5 ± 0.6	6.3	1.2	5.3
Se	1.0 ± 0.15	1.6 ± 0.1	1.3	7.6 ± 0.6	10.0 ± 0.7	8.8	0.27	33
				Elements-Siderophiles				
Cr	15 ± 1	17 ± 1	16	82 ± 5	120 ± 5	100	58	1.7
Mn	100 ± 6	71 ± 5	86	550 ± 30	430 ± 30	490	830	0.59
Co	4.2 ± 0.3	6.0 ± 0.2	5.1	26 ± 1	37 ± 2	32	14	2.3
Ni	9.0 ± 0.9	17 ± 1	13	52 ± 5	100 ± 5	76	37	2.1
(Pd, ppb)	13 ± 6	1 ± 2	7.4	66 ± 27	7 ± 11	37		
(Ir, ppb)	20 ± 6	1 ± 0.3	2	13 ± 31	7 ± 3	10		
(Pt, ppb)	65 ± 18	5 ± 3	35	220 ± 40	38 ± 18	130		

Notes: Cl Clarke value in sedimentary rocks—according to Ronov et al. (1990)—see Ketris and Yudovich (2009) for details.
The brackets mean that calculated Clarke value (based on small statistical samples or not муке reliable analyses) is of minor adequacy (reliability). Such values could be strongly changed in future.
CAI, coal affinity indexes (*coalphile indexes*).
CAI = (average element concentration in coal ash)/(Clarke value in sedimentary rocks)

The extreme coalphile indexes of Bi and Se appear to be very doubtful. Such strange figures of Bi and Se could be accounted for errors in the Clarke values for sedimentary rocks (and not for coal Clarkes errors?). It is of note that these elements are often not analyzed due to analytical difficulties; see, for example, Yudovich and Ketris (2006b) for Se. This may be a factor.

3.7 GEOCHEMISTRY OF COAL INCLUSIONS IN SEDIMENTARY ROCKS—A GEOCHEMICAL PHENOMENON

In many coal-bearing basins, there are numerous coalified fragments of ancient plants (coal-precursors) enclosed in host rocks. Such fragments occur in isolated positions out the coal beds. In the Russian literature, these coal fragments are named *coal inclusions*. Coal inclusions are mostly the remains of stems, trunks, and branches, as well as the roots of trees.

In our monograph (Yudovich 1972), an outline on the geochemistry of such coal inclusions was given. Later, the English review was presented (Yudovich 2003a) including some new geochemical and micromineralogical data. The review has covered (1) definition and classification of fossil woods; (2) relations between coalification and mineralization of fossil woods; (3) some special topics dealing with different and zonal coalification degree of coal inclusions embedded different host rocks; (4) some historical data on geochemistry of coal inclusions; (5) basic empirical regularities in geochemistry, observed worldwide; (6) some data about chemical nature of humin substance—a precursor of lignite and vitrain matter in coalified wood; (7) results of calculation modeling the Ge enrichment in coal inclusions; (8) economic

TABLE 3.3
Median Trace Element Contents in the Ash of Coal Inclusions and Coals from the Beds

Element	Number of Random Samples	Number of Analyses	Contents of Elements, ppm (Median ± S.D.) in the Coal Ash: Coal Inclusions	Coal Beds (See Table 3.7.1)	Concentration Ratio (4:5)
1	2	3	4	5	6
Sr	7	1140	1200 ± 150	740	1.6
Ba	7	1140	1800 ± 340	940	1.9
Be	11	1150	26 ± 13	9.4	2.8
Sc	9	860	32 ± 8	23	1.4
Y	12	1170	160 ± 87	51	3.1
Yb	7	560	9 ± 8	6.2	1.5
La	5	610	150 ± 96	69	2.2
Ce	7	340	500 ± 130	130	3.8
Ga	10	1160	220 ± 82	33	6.7
Ge	27	1780	3100 ± 2200	15	207
Ti	12	1170	8600 ± 3200	4650	1.8
Zr	7	1110	380 ± 300	210	1.8
Sn	10	1150	28 ± 17	6.4	4.4
V	30	1680	640 ± 600	155	4.1
Nb	4	540	50 ± 23	20	2.5
Mo	10	1150	39 ± 24	14	2.8
B	5	1060	430 ± 200	335	1.3
Cu	11	1170	160 ± 110	92	1.7
Ag	11	560	5.6 ± 3.8	0.61	9.2
Zn	8	1000	310 ± 300	140	2.2
Pb	9	1140	51 ± 16	5.9	8.6
As	8	610	290 ± 240	47	6.2
Sb	5	80	31 ± 21	6.3	4.9
Cr	16	1160	1400 ± 950	100	14
Mn	7	1130	630 ± 280	490	1.3
Co	12	1170	110 ± 36	32	3.4
Ni	6	1180	290 ± 200	76	3.8

importance of Ge in coal inclusions; and (9) use of coal inclusion geochemistry for indication of some diagenetic and catagenetic processes, and as a tool for stratigraphic correlation.

The materials, outlined early in the monograph (Yudovich 1972) were added to some later work by Bulgarian (Greta Eskenazy) and Russian (Vladimir Seredin) geologists, performed with use of modern analytical methods. In has been shown that (1) coalified wood may contain very exotic micromineral phases, sometimes far unexpected; (2) apart from Ge, coalified wood may contain high concentrations of some other trace elements, which were earlier not detected because of analytical limitations (REE and As). These special peculiarities can be partly contributed by epigenetic hydrothermal processes.

As a summary, the review shows that coal inclusions are unique geochemical phenomenon, sharply different from even neighboring coal beds in trace element content (Table 3.3). Among the most extreme elements is germanium, its mean concentration in the ash of coal inclusions being up 220 times higher than in the ash of coal beds. The most important peculiarity of the coalified wood is the good preservation of the original lignin structures, which may effectively scavenge Ge from solutions, whereas peat-born coals (in beds) contained such structures in far fewer amounts. In addition, a reservoir of dissolved germanium in peat bog waters was a lower concentration than in sediments, which buried the coal inclusions. Finally, the peat bog acidic environment may act as an unfavorable factor.

Ge-enrichment can be completed in a time ranging from a few thousand years up to tens of million years. However, if the waters are enriched in Ge, the process can proceed more rapidly and would be completed even under the most unfavorable parameters (compared to the model condition). This implies that enrichment can take place during the early diagenesis stage. Such a scenario was supported by geologic considerations.

Some Canadian and Soviet works performed from 1950 to 1960 with some recent Russian studies show that Ge in coal inclusions can be of economic interest.

Geochemistry and mineralogy of coal inclusions are of great interest and need further detailed study.

SELECTED REFERENCES

Abernethy, R.F. and F.H. Gibson. 1963. *Rare Elements in Coal*. U.S. Bureau of Mines Information Circular 8163, 69 pp.

Arbuzov, S.I., Ershov, V.V., Pozeluev, A.A., and L.P. Rikchvanov. 1999. *Trace-Element Potential of the Kuznetsk Basin Coals*. Kemerovo, Russia. 248 pp. (in Russian).

Arbuzov, S.I., Ershov, V.V., Rikchvanov, L.P. et al. 2003. *Trace-Element Potential of the Minusinck Basin Coals*. Novosibirsk, Russia: SO RAN. 347 pp. (in Russian)

Bouska, V. 1981. *Geochemistry of Coal*. Amsterdam, Elsevier. 284 pp. (Coal Science and Technology 1.)

Branson, C.C. 1967. Trace elements in Oklahoma coals. *Oklah. Geol. Notes* 27: 150.

Breger, I.A. 1958. Geochemistry of coal. *Econ. Geol.* 53: 823–841.

Briggs, H. 1934. Metals in coal. *Colliery Eng.* 11: 303–304.

Bustin, R.M. and L.E. Lowe. 1987. Sulphur, low temperature ash and minor elements in humid-temperate peat of the Fraser River delta, British Columbia. *J. Geol. Soc. London* 144, pt. 3: 435–450.

Cavender, P.F. and D.A. Spears. 1995. Analysis of forms sulphur within coal, and minor and trace element associations with pyrite by ICP analysis of extraction solutions. In *Coal Science*, ed. J.A. Pajares and J.M.D. Tascon, 1653–1656. Amsterdam: Elsevier.

Denson, N.M. 1959. Uranium in coal in the western United States: Introduction. *U.S. Geol. Surv. Bull.* 1055-A: 1–10.

Denson, N.M., Bachman, G.O., and H.D. Zeller. 1959. Uranium-bearing lignite in northwestern South Dakota and adjacent states. *U.S. Geol. Surv. Bull.* 1055-B: 11–57.

Dutcher, R.R., White, E.W., and W. Spackman. 1963. Elemental ash distribution in coal components—Use of electron probe, In *Proceedings of the Ironmaking Conference, Iron and Steel Division Metallurgical Society of American Institute of Mining, Metallurgical, and Petroleum Engineers* 22: 463–483. London.

Eskenazy, G. and E. Mincheva. 1995. Geochemical characterization of the Elchov basin coals. Yearbook Sofia University. *Geol. Geogr. Fac.* 84: 65–84 (in Bulgarian).

Eskenazy, G., Petrov, P., and V. Simeonova. 1982. Wofram-bearing coals of Bulgaria. *Geol. Balcan* 12: 99–114 (in Bulgarian).

Finkelman, R.B. 1986. Characterisation of the inorganic constituents in coal. In *Fly Ash and Coal Conversion By-Products: Characterization, Utilization and Disposal II*, ed. G.J. McCarthy, F.P. Glasser, and D.M. Roy, 71–76. (Materials Research Society Symposia Proceedings, vol. 65.)

Finkelman, R.B., February 22–26, 1988. Coal geochemistry: Practical applications. In *Mineral Matter and Ash Deposition from Coal*, ed. R.W. Bryers and K.S. Vorres, 1–12. Santa-Barbara, CA. (Engineering Foundation Conference.)

Finkelman, R.B. 1981. Modes of occurrence of trace elements in coal. *U.S. Geol. Surv. Open-File Rep.*81–99, 322 pp.

Finkelman, R.B. 1993. Chapter 28: Trace and minor elements in coal. In *Organic Geochemistry*, ed. M.H. Engel and S.A. Macko, 593–607. New York: Plenum Press.

Finkelman, R.B., Palmer, C.A., Krasnow, M.R. et al. 1990. Combustion and leaching behaviour of elements in the Argonne Premium coal samples. *Energy & Fuels* 4: 755–767.

Finkelman, R.B., Bostick N.H., Dulong F.T. 1992. Influence of an igneous intrusion on the element distribution of a bituminous coal from Pitkin County, Colorado. *Ninth Annual Meeting of the Society for Organic Petrology, Program and Abstracts*, p. 112–114. University Park, PA.

Foscolos, A.E., Goodarzi, F., Koukouzas, C.N., and G. Hatziyannis. 1989. Reconnaissance study of mineral matter and trace elements in Greek lignites. *Chem. Geol.* 76: 107–130.

Francis, W. 1961. *Coal. Its Formation and Composition*. London: E. Arnold, 806 pp.

Gluskoter, H.J. 1965. Electronic low-temperature ashing of bituminous coal. *Fuel*. 44: 285–291.

Gluskoter, H.J. 1975. Mineral matter and trace elements in coal. In *Trace Elements in Fuel*, ed. S.P. Babu, 1–22. Washington, DC: American Chemical Society.

Gluskoter, H.J., Ruch, R.R., Miller, W.G., Cahill, R.A., Dreher, G.B., and J.K. Kuhn. 1977. *Trace Elements in Coal: Occurrence and Distribution*, 154 pp. (Geological Survey Circular 499.)

Goldschmidt, V.M. 1950. Occurrence of rare elements in coal ashes. (Lecture delivered to B.C.U.R.A. Staff.: May 8, 1943). In *Progress in Coal Science. Vol. 1*, ed. D.H. Bangham, 238–247. London: Butterworths Science Publication.

Goodarzi, F., 1988. Elemental distribution in coal seams at the Fording Coal Mine, British Columbia, Canada. *Chem. Geol.* 68: 129–154.

Goodarzi, F., 1995. The effects of weathering and natural heating on trace elements of coal. In *Environmental Aspects of Trace Elements in Coal.*, ed. F. Goodarzi and D. Swaine, 76–92. the Netherlands: Kluwer Academic Publication.

Goodarzi, F. and A.R. Cameron. 1987. Distribution of major, minor and trace elements in coals of the Kootenay group, Mount Allan, Alberta. *Can. Miner.* 25, pt. 3: 555–565.

Goodarzi, F. and D.J. Swaine. 1994. *Paleoenvironmental and Environmental Implications of the Boron Content of Coals*, 76 pp. Ottawa, Canada: Canada Communication Group Publications.

Hawley, J.E., 1955. Germanium content of some Nova Scotian coals. *Econ. Geol.* 50: 517–532.

Headlee, A.J.W., and R.G. Hunter. 1953. Elements in coal ash and their industrial significance. *Ind. Eng. Chem.* 45: 548–551.

Headlee, A.J.W. and R.G. Hunter. 1955. Characteristics of minable coals of West Virginia. Pt. 5. The inorganic elements in the coals. *West Va. Geol. Econ. Surv. Bull.* 13A: 36–122.

Hu, R.Z., Bi, X.W., Su, W.C. and H.W. Qi. August 6–17, 2000. A superlarge germanium deposit hosted in coal seams, China. In *Proceedings of the 31st International Geological Congress*. Rio de Janeiro, Brazil. CD-ROM.

Huggins, F.E. 2002. Overview of analytical methods for inorganic constituents in coal. *Int. J. Coal. Geol.* 50: 169–214.

Kear, D. and J.B. Ross. 1961. Boron in New Zealand coal ashes. *N. Z. J. Sci.* 4: 360–380.

Ketris, M.P. and Ya. E. Yudovich. 2009. Estimations of Clarkes for carbonaceous biolithes: World averages for trace element contents in black shales and coals. *Int. J. Coal. Geol.* 78: 135–148.

Kitaev, I.V. 1989. *Ash-forming and Minor Elements in Far East Coals*, 138 pp. Vladivostok, Russia: DVO AN SSSR (in Russian).

Kizilshtein, L.Ya. 2002. *Eco-Geochemistry of Trace Elements in Coals*, 296 pp. Rostov-on-Don, Russia: SKNC VS (in Russian).

Kolker, A. and R.B. Finkelman. 1999. Potentially hazardous elements in coal: Modes of occurrence and summary of concentration data for coal components. In *Toxic Elements in Coal.*, ed. B.K. Parekh, 133–157. (Coal Prep. J., Spec. Issue, vol.19, No 3–4.)

Leutwein, F. and H.J. Rösler. 1956. *Geochemische Untersuchungen an paläozoischen und mesozoischen kohlen Mittel- und Ostdeutschlands*, 196 SS. Berlin: Akademie Verlag.

Lindhal, P.C. and R.B. Finkelman. 1986. Factors influencing major, minor, and trace element variations in U.S. coals. In *Mineral Matter and Ash in Coal,* ed. K.S. Vorres, 61–69. Washington, DC: American Chemical Society.

Mackowsky, M.-Th. 1968. Mineral matter in coal. In *Coal and Coalbearing Strata*, ed. D.C. Murchison and T.S. Westoll, 309–321. Edinburgh, London: Oliver & Boyd.

Mraw, S.C., de Neufville, J.P., Freund, H. et al. 1983. The science of mineral matter in coal. In *Coal Science*, ed. M.L. Corbaty, J.W. Larsen, and I. Wender, 1–63. New York: Academic.

Oelschlegel, H.G. 1964. Geochemische Untersuchungen an nordwestdeutschen und nordhesslischen tertiären Braunkohle. *N. Jahrb. Min. Abh.* 101, H. 1: 67–96.

Otte, M.U. 1953. Spurenelemente in einigen deutschen Steinkohlen. *Chem. Erde* 16, H. 3: 237–294.

Palmer, C.A., Krasnow, M.R., Finkelman, R.B., and W.M. D'Angelo. 1993. An evaluation of leaching to determine modes of occurrence of selected toxic elements in coal. *J. Coal Qual.* 12: 135–141.

Parks, B.C. 1952. Mineral matter in coal. In *Proceedings of the 2nd Conference on the Origin and Constitution of Coal,* 272–299. Crystal Cliffs, Nova Scotia.

Parzentny, H.R. 1995. *The Influence of Inorganic Mineral Substances on Content of Certain Trace Elements in the Coal of the Upper Silesian Coalfield,* 90 pp. Katowice, Poland: Silesia University. (in Polish).

Querol, X., Alastuey, A., Lopez-Soler, A. et al. 1997. Geological controls on mineral matter and trace elements of coals from the Fuxin basin, Lianing Province, northeast China. *Int. J. Coal. Geol.* 34: 89–109.

Querol, X., Klika, Z., Weiss, Z. et al. 2001. Determination of element affinities by density fractionation of bulk coal samples. *Fuel* 80: 83–96.

Ren, D., Zhao, F., Wang, Y. and S. Yang. 1999. Distribution of minor and trace elements in Chinese coals. *Int. J. Coal Geol.* 40: 109–118.

Renton, J.J. 1982. Mineral matter in coal. In *Coal Structure*, ed. R.A. Meyers, 283–326. New York: Academic.

Ronov, A.B., Yaroshevsky, A.A., and A.A. Migdisov. 1990. *Chemical Composition of the Earth's Crust and Main Elements Geochemical Balance*, 182 pp. Moscow: Nauka (Science) (in Russian).

Ruch, R.R., Gluskoter, H.J., and N.F. Shimp. 1974. *Occurence and Distribution of Potentially Volatile Trace Elements in Coal: A Final Report*, 96 pp. Illinois State Geological Survey Environmental Geology Notes 72.

Seredin, V.V., August 12–18, 1994. Stratiform rare metal deposits of the Russian Far East cenozoic coal-bearing basins. In *Proceedings of the 9th Symposium of International Association on the Genesis of Ore deposits*, Abstract. Vol. 2, 596–597. Beijing, China.

Seredin, V.V., 1995. New types of REE and Au-PGE mineralization of the Russian coal-bearing depressions. In *Mineral Deposits*, ed. J. Pašava, B. Kribek and K. Zak, 799–801. Rotterdam, the Netherlands: Balkema.

Seredin, V.V. 1996. Rare earth element-bearing coals from the Russian Far East deposits. *Int. J. Coal. Geol.* 30: 101–129.

Seredin, V. September 7–12, 1997. Elemental metals in metalliferous coal-bearing strata. In *Proceedings of the 9th International Conference Coal Science.* Vol. 1, ed. A. Ziegler et al., 405–408. Essen, Germany: DGMK.

Seredin, V.V. 2006. Germanium deposits. In *Large and Superlarge Ore Deposits*, ed. N.P. Laverov and D.V. Rundkvist), 707–736. Moscow, Russia: IGEM RAS [in Russian].

Seredin, V.V. and Danilcheva, J.A., August 6–17, 2000a. Hazardous elements in metalliferous coal deposits of the Russian Far East. In *Proceedings of the 31st International Geological Congress.* Rio de Janeiro, Brazil. CD-ROM.

Seredin, V.V. and Danilcheva, J.A. August 6–17, 2000b. Mineral deposits associated with coals and black shales: A comparative analysis. In *Proceedings of the 31st International Geological Congress.* Rio de Janeiro, Brazil. CD-ROM.

Seredin, V.V., Danilcheva, J.A., Cherepovsky, V.F., and A.M. Vassianovich. 1999a. Ge-bearing coals of Russian Far East deposits: An example from the Luzanovka openmine. In *Prospects for Coal Science in 21st Century,* ed. B.Q. Li and Z.Y. Liu, 137–140. Shanxi, China: Science And Technology Press.

Seredin, V.V., Evstigneeva, T.L. and M.E. Generalov. 1997. Au-PGE mineralization in Cenozoic coal-bearing strata of the Pavlovka deposit, Russian Far East: Mineralogical evidence for a hydrothermal origin. In *Mineral Deposits: Research and Exploration*, ed. H. Papunen, 107–110. Rotterdam, the Netherlands: Balkema.

Seredin, V.V. and R.B. Finkelman. 2008. Metalliferous coals: A review of the main genetic and geochemical types. *Int. J. Coal Geol.* 76: 253–289.

Seredin, V.V. and M.Y. Shpirt. 1995. Metalliferous coals: a new potential source of valuable trace elements as by-products. In *Coal Science,* ed. J.A. Pajares and J.M.D. Tascón, 1349–1652. Amsterdam: Elsevier.

Seredin, V.V., Shpirt, M.Y., and A. Vassianovich. 1999b. REE contents and distribution in humic matter of REE-rich coals. In *Mineral Deposits: Processes to Processing,* ed. C.J. Stanley et al., 267–269. Rotterdam, the Netherlands: Balkema.

Shao, L., Zhang, P., Ren, D., and J. Lei. 1998. Late Permian coal-bearing carbonate successions in southern China: Coal accumulation on carbonate platforms. *Int. J. Coal. Geol.* 37: 235–256.

Shendrik, T.G., Simonova, V.V., and L.V. Pashchenko. September 7–12, 1997. The mineral matter of Ukrainian salty coals in connection with the environmental problem of their application. In *Proceedings of the 9th International. Conference on Coal Science.* Vol. 1, ed. A. Ziegler et al., 409–412. Essen, Germany: DGMK.

Shotyk, W., 1988. Review of the inorganic geochemistry of peats and peatland waters. *Earth-Sci.* Rev. 25: 95–176.

Shotyk, W., Nesbitt, H.W., and W.S. Fyfe. 1990. The behaviour of major and trace elements in complete vertical peat profiles from three Sphagnum bogs. *Int. J. Coal Geol.* 15: 163–190.

Sim, P.G. 1977. Concentration of some trace elements in New Zealand coals. In *Geochemistry.* 132–137. Wellington, New Zealand: DSIR. (New Zealand Department of Science Industry Research Bulletin 218.)

Sprunk, G.C. and H.J. O'Donnell. 1942. *Mineral Matter in Coal.* U.S. Bureau of Mines Technical Paper 648, 67 pp.

Stach, E., Mackowsky, M.-Th., Teichmüller, M. et al. 1982. *Coal Petrology,* 535 pp. Berlin, Germany: Bornträger.

Stadnichenko, T.M, Murata, K.J., Zubovic, P., and E.L. Hufschmidt. 1953. *Concentration of Germanium in the Ash of American Coals. A Progress Report.* U.S. Geological Survey Circular 272, 34 pp.

Stadnichenko, T., Zubovic, P., and N.B. Sheffey. 1961. Beryllium content of American coals. *U.S. Geol. Surv. Bull* 1084-K: 253–295.

Swaine, D.J. 1971. Boron in coals of the Bowen Basin as an environmental indicator. In *Proceedings of the Second Bowen Basin Symposium.* October, 1970, ed. A. Davis, 41–48. (Geological Survey of Queensland Report 62.)
Swaine, D.J. 1990. *Trace Elements in Coal*, 278 pp. London: Butterworths.
Valkovic, V., 1983. *Trace Elements in Coal.* Vol. 1, 210 pp. Vol. 2, 281 pp. Boca Raton, FL: Chemical Rubber Company.
Van der Flier-Keller, E. and F. Goodarzy. 1991. Geological controls on major and trace element contents of Cretaceous coals of Vancouver Island, Canada. *France Soc. Géol. Bull.* 162: 255–265.
Yudovich, Ya. E. 1972. *Geochemistry of Coal Inclusions in Sedimentary Rocks,* 84 pp. Leningrad, Russia: Nauka (in Russian).
Yudovich, Ya. E. 1978. *Geochemistry of Fossil Coals (Inorganic Components),* 262 pp. Leningrad, Russia: Nauka (in Russian).
Yudovich, Ya. E. 2003a. Coal inclusions in sedimentary rocks: A geochemical phenomenon. A review. *Int. J. Coal. Geol.* 56: 203–222.
Yudovich, Ya. E. 2003b. Notes on the marginal enrichment of Germanium in coal beds. *Int. J. Coal. Geol.* 56: 223–232.
Yudovich, Ya. E, and M.P. Ketris. 1997. *Geochemistry of Black Shales*, 212 pp. Syktyvkar, Russia: Prolog.
Yudovich, Ya. E. and M.P. Ketris. 2001. *Uranium in Coal*, 84 pp. Syktyvkar, Russia: Komi Science Center (in Russian).
Yudovich, Ya. E. and M.P. Ketris. 2002. *Inorganic Matter in Coal*, 422 pp. Ekaterinburg, Russia: Ural Division of the Russian Academy of Science (in Russian).
Yudovich, Ya. E. and M.P. Ketris. 2004. *Germanium in Coal*, 204 pp. Syktyvkar: Komi Sci. Center (in Russian).
Yudovich, Ya. E. and M.P. Ketris. 2005a. Arsenic in coal: A review. *Int. J. Coal. Geol.* 61: 141–196.
Yudovich, Ya. E. and M.P. Ketris. 2005b. Mercury in coal: A review. Pt. 1. *Int. J. Coal. Geol.* 62: 107–134.
Yudovich, Ya. E. and M.P. Ketris. 2005c. Mercury in coal: a review. Pt. 2. *Int. J. Coal. Geol.* 62: 135–165.
Yudovich, Ya. E. and M.P. Ketris. 2005d. *Toxic Trace Elements in Fossil Coal,* 655 pp. Ekaterinburg, Russia: Ural Division of the Russian Academy of Science. (in Russian).
Yudovich, Ya. E. and M.P. Ketris. 2006a. Chlorine in coal: A review. *Int. J. Coal. Geol.* 67: 127–144.
Yudovich, Ya. E. and M.P. Ketris. 2006b. Selenium in coal: a review. *Int. J. Coal. Geol.* 67: 112–126.
Yudovich, Ya. E. and M.P. Ketris. 2006c. *Valuable Trace Elements in Coal,* 538 pp. Ekaterinburg, Russia: Nauka (in Russian)
Yudovich, Ya. E., Ketris, M.P., and A.V. Merts. 1985. *Trace Elements in Fossil Coal*, 239 pp. Leningrad, Russia: Nauka (in Russian).
Zubovic, P. 1966. Physicochemical properties of certain minor elements as controlling factors in their distribution in coal. In *Coal Science*, ed. R.F. Gould, 211–231. Washington, DC: American Chemical Society.
Zubovic, P., Sheffey, N.B., and T.M. Stadnichenko. 1967. Distribution of minor elements in some coals in the western and southwestern regions of the interior coal province. *U.S. Geol. Surv. Bull.* 1117-D, 33 pp.
Zubovic, P., Stadnichenko, T.M., and N.B. Sheffey. 1961a. Chemical basis of minor-element associations in coal and other carbonaceous sediments. *U.S. Geol. Surv. Profess. Pap.* 424-D: D345–D348.
Zubovic, P., Stadnichenko, T.M., and N. Sheffey. 1961b. Geochemistry of minor elements in coals of the northern great plains coal province. *U.S. Geol. Surv. Bull.* 1117-A, 58 pp.

4 Coal Reservoir Characterization

Zhongwei Chen, Tianhang Bai, and Zhejun Pan

CONTENTS

Abstract: Characterizing coal reservoir is a critical part of gaining reservoir information for the development of a reservoir model. This chapter comprehensively reviews the techniques and/or methods used to characterize a coal reservoir with respect to the following three key parameters: coal porosity, coal permeability, and coalbed gas in place. The principle for each technique is explained with some specific examples; the strengths and limitations of each method are also provided in the context. It is expected that this chapter provides a useful introduction to coal reservoir characterization for all professionals working on this field, particularly the people in the industry.

4.1 INTRODUCTION

Coal reservoir characterization provides the key data for accurate estimation of gas resources, production system design and control, construction of reservoir simulation model, prediction of reservoir production performance, and for the optimization of its lifetime performance.

Specifically, coal reservoir characterization includes but not limited to: measurements of the basic properties of coal seam and inter-burden materials, such as coal rank, type, grade, depth, thickness, and structure; tests of in-situ coal seam gas content, composition, and reservoir temperature; laboratory measurements of gas sorption isotherms that, together with gas content, pore pressure, and reservoir temperature, are used to determine gas saturation and critical desorption pressure. The value of desorption time can also be gained in the lab via desorption test; and permeability, which is probably the most important parameter.

In this chapter, the above key properties needed for the characterization of a coal reservoir is discussed. A detailed discussion regarding coal porosity, permeability, and the determination of gas-in-place is included.

4.2 COAL POROSITY

Coal porosity is a fundamental property that makes coal a unique material; furthermore, the majority of methane is adsorbed in the pore system of the coal matrix. Therefore, the study of coal porosity is essential.

Coal's pore system is generally accepted as the combination of the pores that are in coal matrix and the cleats that separate matrix, both of which form the dual-porosity structure of coal. The aperture of a cleat is much larger than the size of a pore in the matrix. However, 1cm^3 of coal typically may hold a few square meters of internal surface area due to extraordinarily plentiful pores inside (Mares et al. 2009).

As with the pores in coal matrix, there are a number of methods to categorize such pores, the most frequently used classification method is to categorize them into three groups in terms of sizes: macro, meso, and micropores (Gan et al. 1972). For macropores, the diameters are >50 nm, whereas they are between 2 and 50 nm for mesopores, and <2 nm for micropores (Rouquerol et al. 1994). Coal porosity is often presented as the proportion of pore volume over bulk volume, or sometimes the void volume in a unit weight of coal, cm^3/g and thus, described by pore size or pore volume distribution.

In reservoir engineering, coal porosity is categorized as total porosity, effective porosity, and cleat porosity. The total porosity related to coalbed methane production (Jarzyna et al. 2013) denotes the porosity with the entire porous volume, including closed pores; effective porosity, in other words, the porosity with connectivity, refers to the porosity excluding closed pores, whereas cleat porosity represents the porosity that only involves the cleat system, and is often used in the development and evaluation of permeability models (Seidle 2011a).

A number of researchers indicated that coal porosity is primarily predominated by its composition and coal rank (Gan et al. 1972; Harris and Yust 1976). Vitrinite maceral mainly contains micropores, whereas inertinite maceral mostly encloses meso and macropores. Yao et al. (2011) concluded that the average pore size of high-rank coals is generally smaller than that of low-rank coals. In addition, Levine (1996) and Moore (2012) demonstrated that there is an opposite relationship between the proportions of macro and micropores against coal ranks, as shown in Figure 4.1. In general, micropores increase with increasing carbon content (i.e., coal rank), while macropores increase with decreasing carbon content. However, the carbon content ranging from 55% to 82% is observed to have contrary trends in total pore volume, which stands between the ranks of sub-bituminous and bituminous coals. Francis (1961) illustrated the general relationship between porosity and carbon content as shown in Figure 4.2; the line shows that carbon content around 88% (bituminous coal) has the lowest porosity of approximately 3%, and the shaded area represents the experimental data distributed along the solid porosity line.

A bunch of methods can be employed to measure and calculate coal porosity, such as helium porosity, traditional mercury intrusion porosimetry (MIP), nitrogen adsorption at 77 K, microfocus X-ray computerized tomography (μCT),

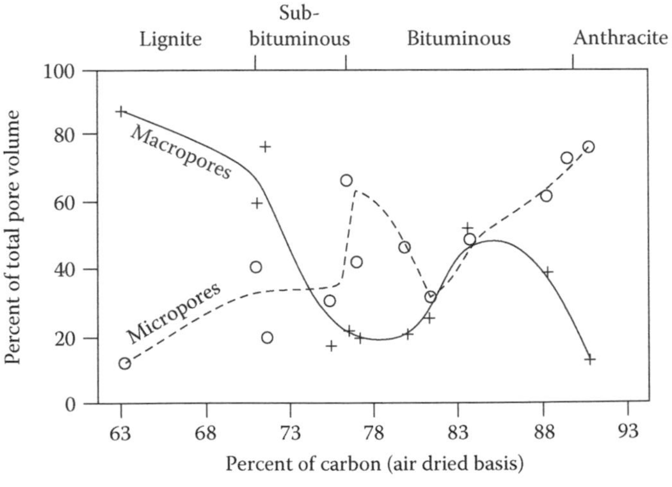

FIGURE 4.1 Relationship between percentages of total pore volume of macropores and micropores against coal ranks. (Adapted from Moore, T.A., *Int. J. Coal Geol.*, 101, 36–81, 2012.)

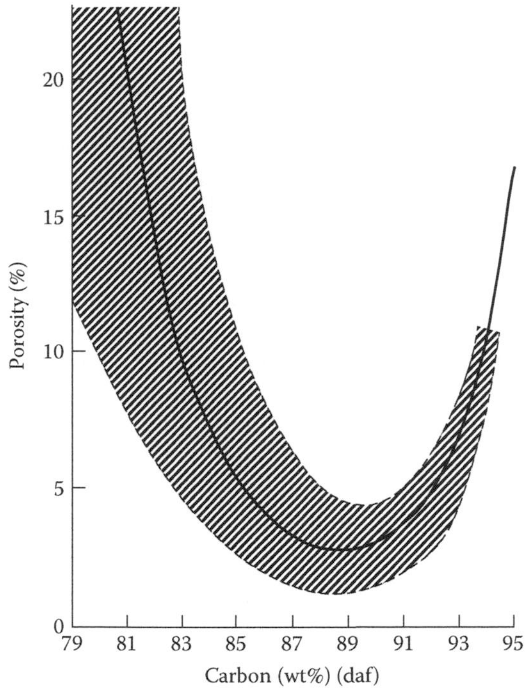

FIGURE 4.2 Relationship between porosity and carbon content (coal rank). (Adapted from Francis, W., *Coal: Its Formation and Composition*, 2nd ed. E. Arnold, London, 1961.)

scanning electron microscopy (SEM), and nuclear magnetic resonance (NMR) and well logging.

By the use of the formula below, coal porosity (ϕ) can be obtained by coal true density (ρ_t), g/cm^3 and apparent density (ρ_a), g/cm^3 (Speight 2005):

$$\phi = \frac{\rho_t - \rho_a}{\rho_t} \times 100\% \tag{4.1}$$

where the true density is described as the mass of coal solid per unit volume without pores inside, and is achieved by helium displacement, while the apparent density, as the mass of a unit coal solid with pores inside, is determined by mercury displacement. The mechanism of this displacement method is that helium is capable of penetrating into all coal pores without any chemical reactions, whereas mercury cannot enter the pores at all without applied external pressure.

4.2.1 Helium Porosity

This measurement is based on the assumptions that helium does not adsorb in the coal or the adsorption amount is negligible, and it can be regarded as an ideal gas. In this chapter, a direct method based on measuring the void volume of a coal sample simply using helium is described (Torsæter and Abtahi 2000). In this technique, the pore space of a coal sample is replaced by helium, the amount of which is detected to develop the porosity (Hedenblad 1997).

Boyle's law of ideal gas is the mechanism underlying the experiment to calculate the coal porosity (Ishaq et al. 2009; Rodrigues and Lemos de Sousa 1999; Rodrigues and Lemos de Sousa 2002):

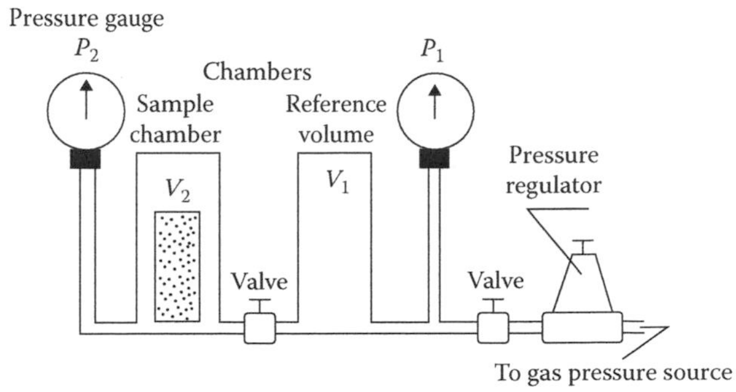

FIGURE 4.3 A schematic of a helium porosimeter. (Adapted from Torsæter, O. and Abtahi, M., *Experimental Reservoir Engineering Labortory Work Book*. Norwegian University of Science and Technology, Trondheim, Norway, 2000.)

$$P_1V_1 = P_2V_2 \tag{4.2}$$

One typical apparatus utilized for this type of measurements is a helium porosimeter, whose scheme is shown in Figure 4.3.

There are two chambers in a helium porosimeter: the *sample chamber* and the *reference chamber*. Tubing is fixed between these two chambers to establish the connection. The measurement is performed at a constant temperature, so that Equation 4.2 can be used to calculate coal porosity because the expansion of helium takes place isothermally. Readings can be recorded when the system reaches equilibrium, then the effective coal porosity (ϕ) can be computed using Equation 4.3:

$$\phi = \frac{V_b - V_g}{V_b} \tag{4.3}$$

where:
V_b is the bulk volume of a coal sample (core)
V_g is the volume of coal sample skeleton and closed pores

Since helium has the smallest molecular size available, it can penetrate the whole coal structure more thoroughly than any other gases. Moreover, the low molecular weight and high diffusivity features make helium capable to acquire the information of coal with low porosity. This measurement also allows repeatable experiments with high precision results (Kazimierz et al. 2004). However, there are presences of closed pores that stop helium from permeating (Kotlensky and Walker 1960). Thus, the coal porosity acquired by helium porosity measurement is the effective porosity.

In the following case study, one example of porosity measurement using high pressure helium was given and explained. This example is a little bit different from the porosity measurements using porosimeter, which is conducted at low pressure range, so that ideal gas law is applied. Instead, in the example below, the real gas law for helium was used to calculate the porosity.

4.2.1.1 Samples

A series of measurements of porosity were conducted on a core sample from the Jincheng basin, China. The coal sample was anthracite coal and cored to 49.86 cm in diameter and 102.25 cm in length. The sample was wrapped with a thin lead foil and then by a rubber sleeve before it was installed in the cell. The thin lead foil was to prevent gas diffusion from the core to the confining fluid at high sample pressures.

4.2.1.2 Experimental Settings

The schematic of the rig used to conduct these series of porosity measurements is shown in Figure 4.4. When testing porosity, gas is injected from the upstream injection pump to the sample. Then gas is allowed to flow through the sample from the upstream cylinder to the downstream cylinder. The gas volume change of the pump for a setting pressure and gas pressure variation in the sample are monitored simultaneously. The sample cell and other parts of the rig are in a temperature controlled cover to maintain constant temperature during the experiment. All measurements were conducted at 35°C.

The void volume, V_{void}, in the cell is determined by injecting known quantities of helium from a calibrated gas injection pump. Since helium is not significantly adsorbed, the void volume can be determined from the measured values of temperature, pressure, and the helium volume injected into the cell. In this calculation, gas density and mass change with respect to different gas pressures and temperatures that were used. The values of different parameters such as densities for helium were obtained from the NIST web book at http://webbook.nist.gov/chemistry/fluid/ (Pan et al. 2010). The void volume was then used to calculate porosity. This helium void

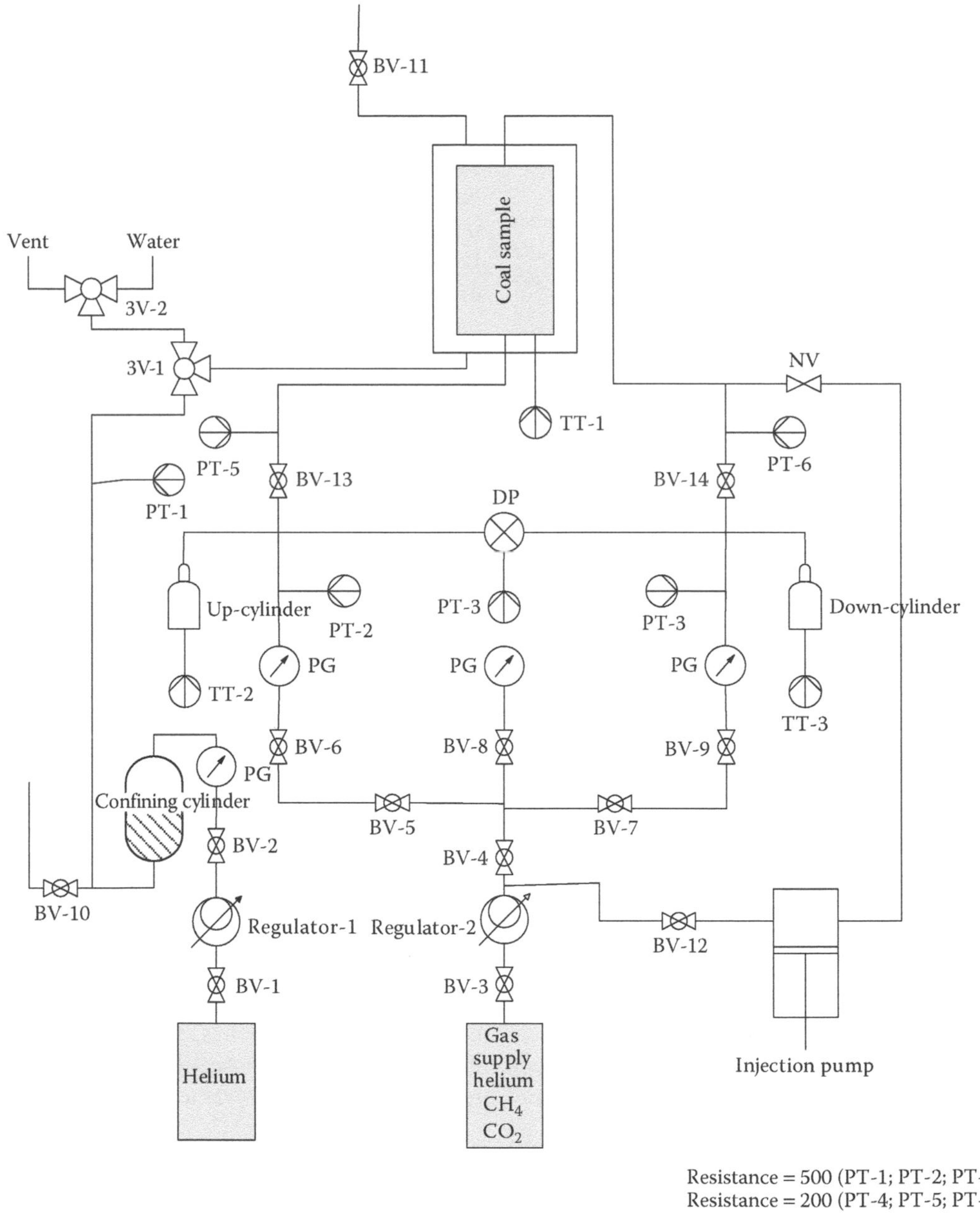

FIGURE 4.4 The schematic plot of the sorption rig. 3-V: three-way valve; BV: ball valve; DP: differential pressure; NV: needle valve; PG: pressure guage; PT: pressure transducer; TT: temperature transducer.

volume measurements were performed at different pressures to investigate the consistency of the calculated volume.

4.2.1.3 Results and Analysis

As porosity also changes with effective stress and low pressure, pycnometer helium measurement is not enough to reveal the full information of porosity; a series of helium porosity measurements were conducted with varying effective stresses. The confining stresses changed from 2.08 to 5.13 MPa, and the pore pressures were varied from 0.43 to 3.12 MPa. A typical porosity calculation with pump pressure being 4.1 MPa was listed in Table 4.1, and the whole series of porosity tests on this sample were summarized in Table 4.2.

TABLE 4.1
Illustration of Porosity Calculation

Pump system ($T = 35°C$)	Parameter	Pump pressure (MPa) = 4.1
	Initial volume (ml) (V_1)	203.81
	After injection (ml) (V_2)	193.19
	Gas density (mol/l) (D_1)	1.57
	Mole volume injected (× 10^{-3}mol) (V_3)	$(V_1 - V_2)/D_1 = 16.68$
Cell system ($T = 35°C$)	Tube volume (ml) (V_4)	10.53
	Initial sample gas pressure (MPa) (P_1)	1.10
	Gas density in sample (mol/l) at P_1 (D_2)	0.43
	Gas pressure after injection (MPa) (P_2)	2.80
	Gas density in sample (mol/l) at P_2 (D_3)	1.07
	Mole volume from previous step (ml) (V_5)	17.93
	Mole volume occupied by tube (× 10^{-3} mol) (V_6)	6.81
Void calculation	V_{void} (ml)	$(V_3 - V_6 + V_5 \times D_2)/D_3 = 15.07$
Total sample volume (ml) (V)		199.61
Porosity calculation (%)		7.55

TABLE 4.2
Summary of Porosity Values

Confining Pressure (MPa)	Pore Pressure (MPa)	Effective Stress (MPa)	Porosity (%)
2.12	1.02	1.09	7.48
3.34	2.69	0.65	8.06
3.61	3.17	0.44	7.84
3.84	2.79	1.04	7.56
4.10	2.80	1.30	7.55
4.12	3.11	1.01	8.10
4.81	2.80	2.01	7.53
5.13	3.12	2.01	8.05

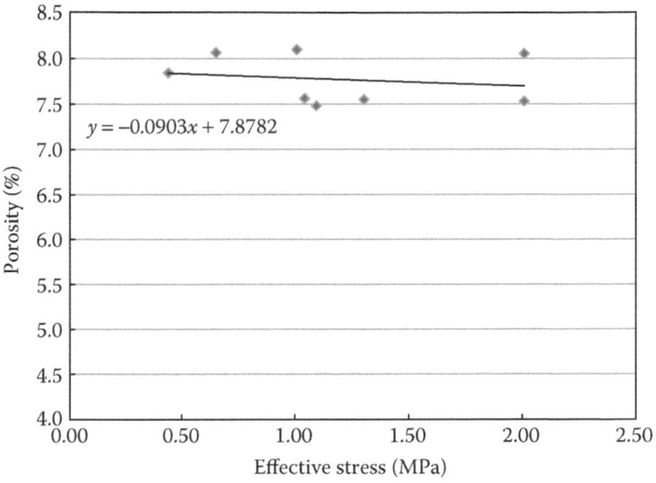

FIGURE 4.5 Relationship between effective stress and porosity.

To evaluate porosity variation against the effective stress, their relationship was plotted in Figure 4.5. This figure shows that very slight porosity variation was observed for this sample at these particular experimental conditions. By regressing the data, it is reasonable to consider the initial porosity of the sample to be 7.88%.

4.2.2 Mercury Intrusion Porosimetry

Referring to MIP, the pores in coal are assumed as tiny cylinders or capillaries, and the basic physical principle is that mercury is a kind of nonwetting fluid, so that capillary action does not enable mercury to permeate through pores unless an external pressure is applied. If a set of test specimens is placed in the mercury injection apparatus, or also known as *mercury porosimeter*, by the application of gradual injection of liquid mercury into an evacuated pore system under continuously increasing external pressures by small increments, and with the help of Equation 4.4, which was developed by Washburn (1921), MIP can depict the intruded mercury volume curve against pressure to analyze and estimate coal pore size distribution.

$$R_c = \frac{-2\sigma\cos\theta}{P_c} \quad (4.4)$$

where:
- P_c represents the absolute injection pressure, MPa
- R_c is the radius of the pore when mercury is injected, μm
- σ is the interfacial tension of coal and mercury, J/m^2
- θ states the contact angle between mercury and the pore surface

The porosity can be then acquired by transferring the volume-pressure curve to the porosity-pressure curve using software. The mercury pressure is raised up by small increments, and the corresponding injected mercury volumes are recorded periodically (Yao and Liu 2012). It can be seen from Equation 4.4 that the higher the external pressure is, the smaller size of a pore can be penetrated by mercury.

MIP also determines the effective coal porosity, whose results are usually intuitive and reliable (Chen and Li 2006); the testing scale of this porosimetry ranges from 2 nm to 100 μm in terms of the radii of the pores; thus, MIP can be suitable for measuring pore size distributions for the most of coal. However, there are a number of flaws regarding this method (Mahajan and Walker 1978; Yao and Liu 2012), such as the cylindrical geometry assumption, sample size limit, and the possible destruction to the pore structure during the mercury penetration when pressure is high.

Please note that MIP does not measure the internal size of a pore, it rather determines the largest entrance toward a specific pore (Giesche 2006), and due to the pore throat structure, the intruded mercury may not fill the whole space of a pore, which is called the *ink bottle* effect (Abell et al. 1999; Wardlaw and McKellar 1981), an underrated porosity or pore volume can be induced in common cases. Moreover, high pressure may either deform or destroy the original pore systems in coal samples (Suuberg et al. 1995), and can cut the number of macropores resulted from the shielding effects of smaller pores (Gane et al. 2004), both of which definitely exert adverse influence on experiment results. Therefore, pore compressibility and coal structure had better be taken into account when analyzing coal porosity and pore size distributions.

4.2.2.1 Samples

The coal samples were from Wangpo Coal Mine in Qinshui Basin, Shanxi, China, which lay in the No.3 Shanxi Formation, the Lower Permian Series. They were anthracite excavated from 700 m depth at about 35°C. The samples were crushed into the size as small as 0.3 mm to fit into the mercury porosimeter (Bo et al. 2006).

4.2.2.2 Experimental Settings

The experiment was conducted using a Micromeritics 9310 Micropore Analysis Instrument (Micromeritics Corporation, USA). In this case, θ was assumed to be 140°, and σ was set to 0.48 J/m², then Equation 4.4 was converted into Equation 4.5:

$$P_c = \frac{0.735}{R_c} \tag{4.5}$$

The measurement ran from a pressure of 0 up to 207 MPa by small increments, inferring that the pore throats with more than 3.5 nm radii can be penetrated by mercury.

4.2.2.3 Results and Analysis

The relationship between the volume of injected mercury and the external pressure is plotted in Figure 4.6.

Figure 4.6 shows that at the primary stage of intrusion the larger fractures contributed to the quick increase of mercury volume. As the penetrated pore size got smaller, it became increasingly difficult to force mercury into the porous space due to the increasing capillary pressure. The cumulative mercury volume peaked at about 0.028 cm³/g when the external pressure reached 207 MPa, which represented the total pore volume in the samples. The solid density of coal grain is 1.44 g/cm³, then void ratio of this sample can be calculated as

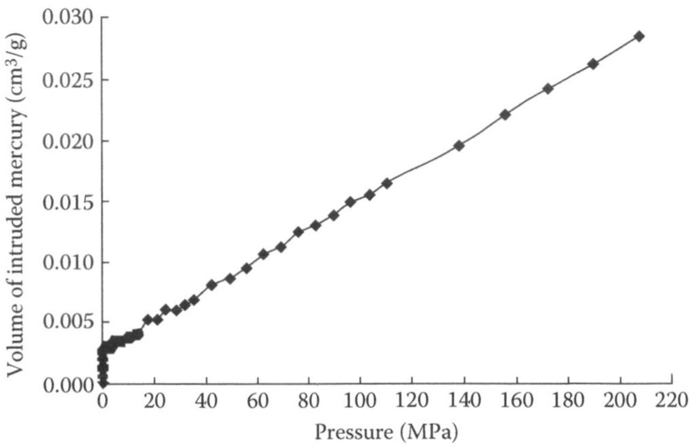

FIGURE 4.6 The relationship between mercury volume and the pressure. (Adapted from Bo, L. et al., Analysis to porosity of coal by mercury porosimetry, Sciencepaper Online, 2006.)

4.02%, indicating that the porosity of the sample was 3.87%, which was verified by other conventional measurements (Bo et al. 2006).

In order to estimate the pore size distribution, such functions had been established (Xing and Yan 2007):

$$D_v(r) = \frac{dV}{dr} \tag{4.6}$$

and

$$V = \int_{r_1}^{r_2} D_v(r)\,dr \tag{4.7}$$

where:

V is the intruded mercury volume, cm³/g
r is the radius of a pore

The pore size distribution curve can be plotted using Equation 4.6, while Equation 4.7 helps in calculating the total volume of intruded mercury between two certain pore radii. Figure 4.7 illustrates the pore size distribution of the coal sample.

The results also show that the changes in the mercury volume are small in the pore size ranging from 50 to 250 nm. However, the volume starts to increase dramatically when the pore size drops down to 50 nm and afterward. In general, the pore size is in inverse proportion to the corresponding total pore volume with the same pore size. For this sample, the pore volume with the sizes less than 50 nm contributes the majority to the total pore volume.

4.2.3 Nitrogen Adsorption

Nitrogen adsorption at 77 K is also a commonly used method for the determination of coal surface area and characterization of coal pore size distribution. The first attempt of the nitrogen adsorption at 77 K was conducted by Brunauer et al. (1938), and has been gradually accepted as a standard procedure

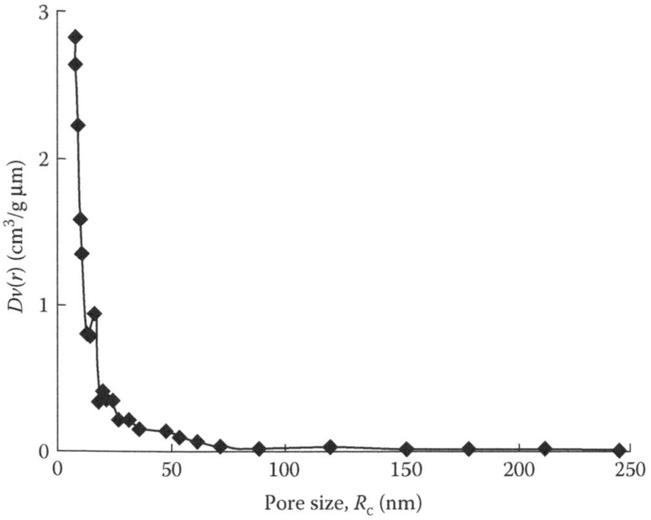

FIGURE 4.7 The differential function curve with pore size less than 250 nm. (Adapted from Bo, L. et al., Analysis to porosity of coal by mercury porosimetry, Sciencepaper Online, 2006.)

for the estimate of microstructure of porous media due to a number of advantages, including high chemical stability, less destruction compared to mercury intrusion method, and easy implementation compared to other visual methods such as SEM (Wang et al. 2014). Brunauer–Emmett–Teller (BET) and Barrett–Joyner–Halenda (BJH) models are the commonly used methods for adsorption/desorption isotherm (Wang et al. 2014). BET is employed to evaluate the surface area of a porous media, and BJH method is employed to calculate porosity pore size distribution using the Kevin model, but applies only to the range of mesopore and macropore size (Bansal and Goyal 2005). The illustration of nitrogen adsorption theory is displayed in Figure 4.8.

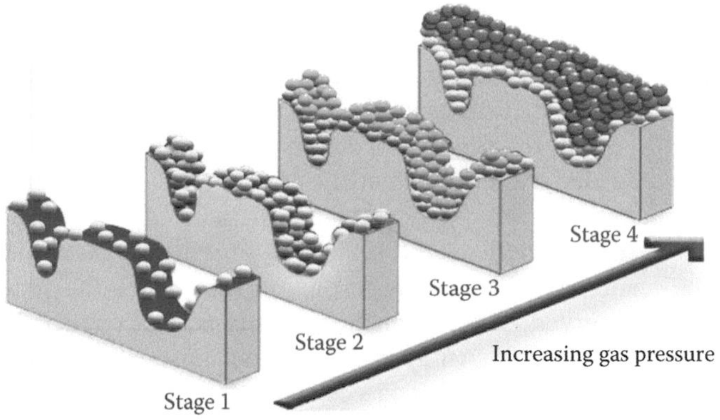

FIGURE 4.8 Illustration of nitrogen adsorption theory: Stage 1—the start of adsorption at low pressure stage; Stage 2—with increasing nitrogen pressure, adsorption molecules increase to form a monolayer; Stage 3—further increasing nitrogen pressure forms multilayer coverage; BET theory can be used to determine the surface area; Stage 4—a further increase in nitrogen pressure causes the saturation of all pores. BJH equation is applied to calculate porosity and pore size distribution. (From Bansal, R.C. and Goyal, M., *Activated Carbon Adsorption*, Taylor & Francis Group, Boca Raton, FL, 2005.)

However, the main drawback of nitrogen adsorption at 77 K is the diffusional problems of nitrogen molecule into the narrow coal pores at such low temperature, and consequently, the results obtained using nitrogen adsorption are often smaller than other methods such as carbon dioxide adsorption at 195 K (Bansal and Goyal 2005).

4.2.4 Microfocus X-Ray Computed Tomography

Computed tomography (CT) technique implements an X-ray detector and mathematical algorithms for tomographic imaging reconstruction, producing cross-sectional image slices of the inner structure of a sample, which is resulted from the measurement and analysis of penetrating multiple radiation beams through the sample in a same plane (Flannery et al. 1987). The beams generated by CT scanners can be manipulated from many different angles, which consequently form a series of projections. Through a back-projection algorithm in the scanner's computer, a cross-sectional image can be acquired. Such image slices can also be grouped, yielding a 3D image (Wellington and Vinegar 1987). Once a 3D image is obtained, any plane of the sample is available in the computer.

X-ray CT technology has also been used in detecting the petrophysical properties of rocks as a reservoir engineering tool (Wellington and Vinegar 1987). The attenuations of X-rays are distinct when the X-rays reach the pores and the solid coal matrix. In the light of this statement, CT applications in coal have been conducted in characterizing pores and fractures (Verhelst et al. 1996), as well as in assessing gas adsorption/desorption and migration in coal (Karacan and Mitchell 2003; Karacan and Okandan 2001). Microfocus X-ray CT (μCT), embraces the same principles as CT but has a better resolution, was used to determine coal characterizations (Van Geet and Swennen 2001). The superior resolution can be narrowed down to $10 \times 10 \times 10\ \mu m^3$ in 3D (Van Geet et al. 2000). Compared with CT technique, μCT uses a tiny focal spot of X-ray emission.

The μCT is a nondestructive method, allowing visualization of the inner microstructure of coal (Remeysen and Swennen 2006), and therefore, is suitable for estimating coal porosity. This technique can provide coal porosity at any position of a sample (Yao et al. 2009). Moreover, μCT makes it possible to measure coal porosity including those closed pores inside.

Nevertheless, the measured results may not be accurate due to the partial-volume effect, which is the loss of small objects or regions due to the limited resolution (Ketcham and Carlson 2001); moreover, the comparatively low resolution does not allow the micro- and mesopores to be detected (Yao et al. 2009), giving a far underestimated result of total porosity.

4.2.4.1 Sample

A coal block was acquired from an underground coal mine in Hongyang mining area, North China Basin. After the block was cut into a cylindrical core of about 25 mm in diameter, the μCT scan was performed to measure the coal porosity (Yao et al. 2009).

4.2.4.2 Experimental Settings

The µCT scan was conducted on a piece of equipment that was manufactured by BIR Corporation, Maple Grove, Minn, USA. The X-ray focal spot was 225 kV. The Toshiba 3D detector enabled the data obtained from the apparatus to be captured, analyzed, and digitized.

The sample core was placed perpendicular to the sample couch to get a set of imaging results by using various combinations of scanner settings, until one optimal setting was achieved as the result had the least pixel noise and image artifacts. Moreover, calibrations were made to reduce the interference of such defects (Yao et al. 2009).

4.2.4.3 Results and Discussion

The µCT scan created a series of 2D image slices, which recorded the entire X-ray attenuation coefficient data distributed all over the scanning plane to distinguish pores from mineral matters and coal matrix (Auzerais et al. 1996). In this experiment, 80 slices were scanned. In order to get the coal porosity, each image slice was described as 512 × 512 pixels, and each pixel was allocated with a gray value ranging from 0 to 255, which can be used as a guideline to identify pores. A Boolean calculation was done in Equation 4.8 for the gray image after scanning and defining a threshold T:

$$g(i,j)=\begin{cases}0\, f(i,j)\le T\\ 1\, f(i,j)>T\end{cases} \tag{4.8}$$

where:

$f(i,j)$ was the gray value of a pixel at coordinate (i,j)
$g(i,j)$ was the binary value of this pixel

The threshold T was determined by Mathematica™ software (Nakashima and Yamaguchi 2004). After such manipulation, the original gray image slice was transformed into a black-and-white one, where the black areas ($g = 0$) were identified as pores, and the white parts ($g = 1$) were coal solids, as shown in Figure 4.9.

Equation 4.9 provides the algorithm to calculate the porosity of a plane:

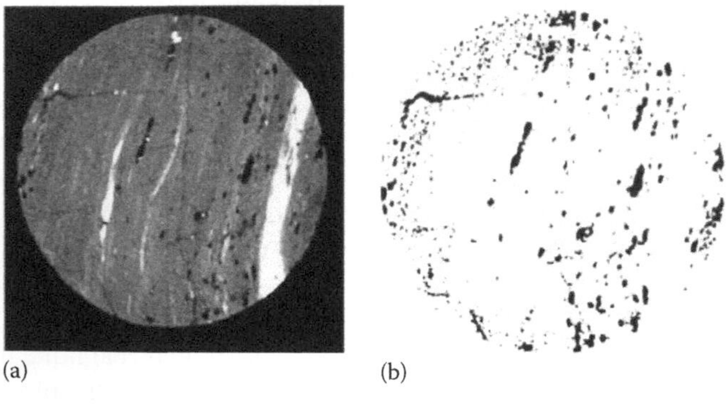

(a) (b)

FIGURE 4.9 The original image (a) and the image after binary manipulation (b). (Adapted from Yao, Y. et al., *Int. J. Coal Geol.*, 80, 113–123, 2009.)

$$\varphi_{ct}=\frac{N_p}{N_c}\times 100\% \tag{4.9}$$

where:

N_p is the number of pore pixels
N_c is the number of other components pixels

With the help of the Mathematica software, the porosity of the above image can be computed as 7.63%.

For this particular coal sample, the average total porosity (volume) of all 80 slice porosities was 10.29%, with the minimum and maximum slice porosities of 7.2% and 16.2%, respectively. The result was verified by other conventional estimating methods and NMR as well.

4.2.5 Small Angle X-Ray/Neutron Scattering

Small angle X-ray (SAXS) and small angle neutron scattering (SANS) can be used to detect the cross-sectional geometry of pore space of coal samples, including pore size distribution, as well as total porosity (Bale Harold et al. 1984; Cohaut et al. 2000; Prinz et al. 2004; Radlinski et al. 2004). Dealing with SAXS, when a beam of X-ray penetrates a coal sample, the X-ray diffracts from the original beam with very low angles (typically $\theta < 5°$) due to the microscopic heterogeneities presence within the sample, and the scattering intensity is associated with the microstructure of such sample (Radlinski and Radlinska 1999). Radically, it is the diverse densities of electrons that give rise to various scattering intensities (Schnablegger and Singh 2011). The contrasts of electron densities among varied components are distinguished by SAXS, which makes it possible to detect both open and closed pores in coal (Kalliat et al. 1981; Tricker et al. 1983). Radlinski et al. (2004) demonstrated that SAXS is capable of delivering coal structural information from about 1 nm to 20 µm. On the other hand, SANS works in the similar way to SAXS, but employs a beam of neutron instead of X-ray.

The pores in coal are seen as spheres with varied radii, regardless of their original shapes (Sakurovs et al. 2012). A typical graph from SAXS or SANS is given in Figure 4.10, showing the scattering intensity $I(Q)$ (cm^{-1}) as a function of scattering vector Q($Å^{-1}$), where the scattering intensity shows the strength of the scattering in units of cm^{-1}, and the scattering vector is the resolution with which the sample is observed. The higher the intensity is, the rougher the sample surface is. An approximate relationship between a pore radius (r, nm) and Q was established by Radlinski et al. (2000) as shown in Equation 4.10. Moreover, $I(Q)$ relates to the pore number density $f(r)$, which can be obtained by the standard spherical pore approximation procedure (Hinde 2004). The total coal porosity is determined as the sum of all pore volumes in the pore size distribution derived from $f(r)$ and r.

$$r=\frac{2.5}{Q} \tag{4.10}$$

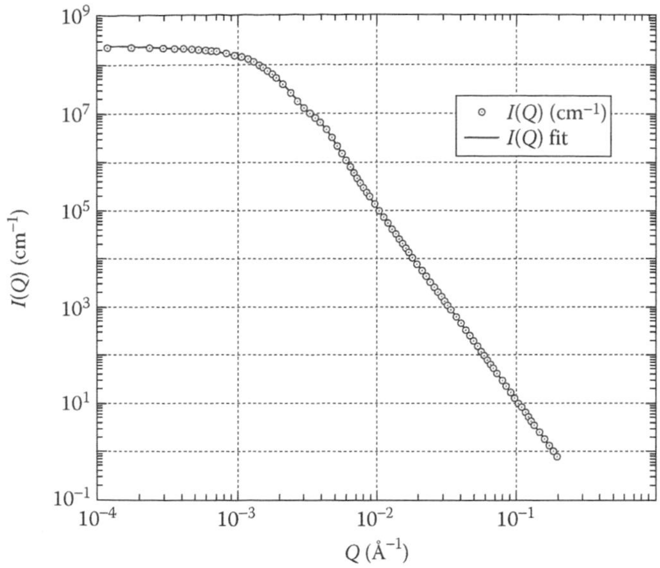

FIGURE 4.10 A typical graph showing intensity versus scattering vector. (Adapted from Hinde, A., *J. Appl. Crystallogr.*, 37, 1020–1024, 2004.)

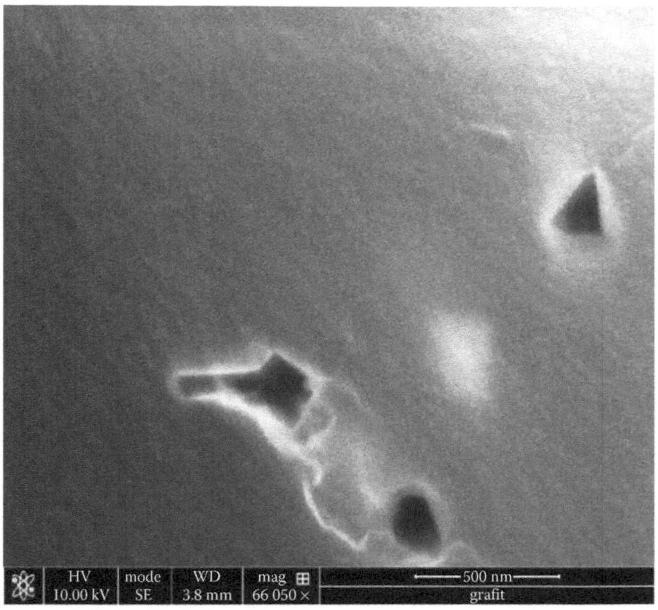

FIGURE 4.11 A scanning electron micrograph of the selected area of a coal sample. (Adapted from Miedzinska, D. et al., *Bull. Polish. Acad. Sci. Tech. Sci.*, 61, 499–505, 2013.)

SAXS and SANS are nonintrusive techniques that are cost-efficient, and do not require much sample preparation (Calo and Hall 2004; Dragsdorf 1956). Though they only give the cross-sectional information of a coal sample, the results derived from these methods can be regarded as representative of the whole sample if taken the average (Radlinski et al. 2004). In the light of these statements, the total porosity can be estimated.

4.2.6 Scanning Electron Microscopy

SEM can produce coal sample surface images with morphology information with a beam of electrons, which interact with atoms in the sample surface. The interaction generates different signals, like secondary electrons and backscattering electrons, which can be picked up by detectors that are sensitive to varied energies (Dunlap and Adaskaveg 1997), and analyzed by image processing software (Tang et al. 2005). The whole process makes the sample surface magnified to a certain extent to be visualized (Gupta 2007).

The similar interpreting technique as μCT is taken as a guideline to estimate coal porosity. That is to say that the sample surface image obtained by SEM is described as pixels assigned with gray values, and a threshold is set to discriminate between coal matrix and pores (Tang et al. 2005). The sample porosity is calculated as ratio of the number of pore pixels to the total number of pixels. Generally, pores are visible as black spots in scanning electron micrographs due to a low intensity of electron signals (Mancktelow et al. 1998). Figure 4.11 shows a selected area image of a coal sample surface taken by SEM, with black dots referring to pores.

Considering as a measurement for coal porosity, the sample preparation is simple and fast for SEM (Harpalani 1987). The resolution of SEM narrows down the scales of the investigated pores to 10 nm (Giffin et al. 2013), which results in the absence of micropores.

However, SEM only delivers the surface porosity information rather than the whole sample and thus, taking average may provide a representative result (Bonnie and Fens 1992). Therefore, to gain an accurate value of the total porosity, incorporating SEM with other methods would be a reasonable choice.

Together with the porosity estimate, SEM can recognize individual macerals at the same time, providing an important tool for coal characterization (Davis et al. 1986).

4.2.7 Nuclear Magnetic Resonance (Logging)

NMR only responds to the nuclei with an odd number of nucleons. Since a single hydrogen nucleus encloses one proton and no neutron, which results in nonzero spin (Rabi et al. 1938), it is prone to being influenced by external magnetic fields to trigger NMR phenomenon; therefore, hydrogen nuclei are the key elements to NMR imaging (Tang and Pan 2011). Dunn et al. (2002) argued that NMR measures the gross population of hydrogen protons in coal formation, including those of the water and gas in coal porosity systems, but excluding those in the solid matrix. That is to say that in general cases, NMR signal solely reveals the hydrogen nuclei in the pore space of coal. Because this measurement echoes hydrogen proton spins in the fluids only, the estimated porosity is considered to be mineralogy independent. Furthermore, Yao et al. (2010b) pointed out that an NMR measurement contains two major parts: aligning hydrogen magnetic moments by the employment of an external magnetic field and generating a dipole moment in the hydrogenous module of a coal mass. The amplitude of such dipole moment, whose time evolution can be divided into longitudinal (T_1) and transverse (T_2) relaxation time distributions, is determined by the number of hydrogen nuclei, and thus is suitable for estimating the pore volumes that filled with fluids (gas and water). According to Kleinberg et al. (1993a, b), T_2 relaxation is believed to be faster than T_1, and delivers similar distribution and the same information with T_1 (low-field NMR), and thus it is used to obtain coal porosity. The amplitudes of the spin echo decays of nuclei can be converted into T_2 distribution curves through multiexponential process of inversion (Coates et al. 1999).

The NMR coal porosity measurement is in a low magnetic ambience, so that a few paramagnetic substances in coal samples can hardly disturb the results (Yao et al. 2010a); the influences of those nuclei that are in solid status can be shielded as well (Kleinberg et al. 1993a).

For the coal NMR phenomena, there are three kinds of relaxation mechanisms: diffusion, bulk, and surface relaxations. Therefore, T_2 can be calculated from Equation 4.11 (Bloembergen et al. 1948; Carr and Purcell 1954; Cohen and Mendelson 1982):

$$\frac{1}{T_2} = \frac{1}{T_{2B}} + \frac{\gamma^2 G^2 D \tau^2}{3} + \rho_2 \frac{S}{V} \tag{4.11}$$

where:

T_{2B} is the bulk relaxation, which is related to the fluid properties

the second and last terms in the equation represent the reciprocals of diffusion and surface relaxations, respectively

Among these terms, the bulk and diffusion relaxations can be considered as negligible under low and constant magnetic conditions (below 150k gauss) and a short pulse spacing (smaller than 2 ms) (Kenyon 1992; Straley et al. 1997; Timur 1969). Then Equation 4.11 is converted into Equation 4.12, the function of the specific surface of pores (Kleinberg 1996):

$$\frac{1}{T_2} = \rho_2 \frac{S}{V} \tag{4.12}$$

where:

ρ_2 is the intensity of surface relaxation

S/V is the surface-to-volume ratio of the pore space in coal samples

As the S/V ratio increases with the loss of pore size, a smaller pore size indicates a shorter T_2 relaxation, and a larger pore size implies a longer T_2 relaxation.

Based on the coal porosity estimation model developed by Yao et al. (2010b), the coal porosity can be computed as the sum of movable fluid porosity and irreducible fluid porosity. Furthermore, the cumulative total porosity curve can be gained using the T_2 distribution curve.

NMR can be used as an efficient tool to characterize the total coal porosity, and it does not require sample destruction when compared with other conventional methods. It is an advanced and quantitative method in such application (Li et al. 2012). Another key feature of NMR is that it allows duplicable measurements, which makes it possible to calibrate the measured results (Cherry 1997; Murphy 1995). Furthermore, the NMR technology is applied to in-situ logging, exhibiting relatively high practical applicability in measuring coal porosity (Yao and Liu 2012; Yao et al. 2010b). Cheng and Ling (2008) even indicated that among various ways to obtain porosity, the NMR logging was the most direct and efficient, while the case study below was performed using laboratory NMR coal porosity measurement.

4.2.7.1 Sample

A cylindrical core plug (2.5 cm in diameter) was cut from the end of the coal sample, which was collected from Hongyang Coal Mine, North China Basin (Yao et al. 2010b).

4.2.7.2 Experimental Settings

Two kinds of NMR experiments were conducted using the core plug. One was under a 100% water-saturated condition (S_w), and the other was under an irreducible water condition (S_{ir}). The plug was first measured at S_w, and was then centrifuged at 1.4 MPa to reach the S_{ir} to be performed for the second measurement. A Rec Core 2500 instrument was employed in the NMR measurements, in which a uniform magnetic field of 1200 gauss and a resonance frequency of 2.38 MHz were generated. The parameters of the experiment were set as follows: the echo spacing was 0.6 ms; the waiting time was 5 s; the number of echoes was 2048; and the number of scans was 64. The T_2 distribution curve was calculated by a mathematical inversion process of the spin echo data as a function of time, which was aligned logarithmically from 0.1 ms to 10 s (Coates et al. 1999).

4.2.7.3 Results and Discussion

The irreducible porosity and the producible porosity can be obtained from Figure 4.12. The irreducible porosity and producible porosity were 1.21% and 4.67%, respectively, and thus, the total porosity was 5.88% for this sample. The porosity corresponded with the estimates from other measurements very well.

4.2.8 Well Logging

Well logging delivers in-situ coal petrologic information by using logging devices in boreholes (Wonik and Olea 2007). Several types of well logging are used to test the total coal porosity, including density logging and sonic logging. Since coal adsorbs neutrons, the neutron porosity logging is not directly applicable to coal.

4.2.8.1 Density Logging

A density log utilizes a gamma ray source to measure the bulk density of coal. The basis of this measurement is the Compton effect (Compton 1923), indicating that the gamma ray is partially scattered by the atoms in coal formation, and partially adsorbed, and the gamma ray count is inversely related to coal density (Wood et al. 1983). If the densities of

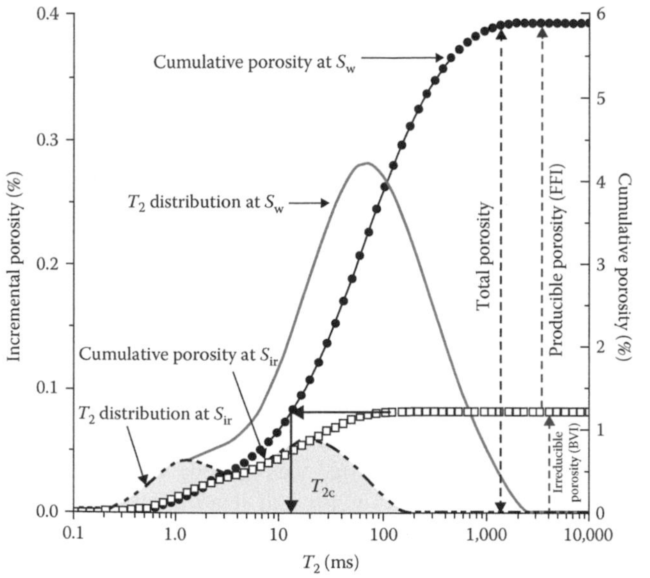

FIGURE 4.12 NMR measurement at S_w and S_{ir} conditions to determine porosity. (Adapted from Yao, Y. et al., *Fuel*, 89, 1371–1380, 2010b.)

coal matrix (ρ_{ma}) and pore fluids (ρ_f) are given, with the help of the density (ρ) acquired from the density log, coal porosity (ϕ) can be derived from Equation 4.13.

$$\phi = \frac{\rho_{ma} - \rho}{\rho_{ma} - \rho_f} \quad (4.13)$$

4.2.8.2 Sonic Logging

A sonic log (also called as an *acoustic log*) employs acoustic measurements to record interval transit time of a sound wave traveling through the coal formation (Δt) (Wonik and Olea 2007). An increase in coal porosity witnesses an analogous increase in the transit time of a sound wave. Similar to the density log, together with the transit times of coal matrix (Δt_{ma}) and pore fluids (Δt_f), coal porosity (ϕ) is obtained by Equation 4.14.

$$\phi = \frac{\Delta t - \Delta t_{ma}}{\Delta t_f - \Delta t_{ma}} \quad (4.14)$$

Figure 4.13 illustrates the density and acoustic logging images of a coal seam gas.

4.3 COAL PERMEABILITY

Coal permeability describes the transportability of water and gas to flow through interconnected pores and cleats of coal formations, and plays a key role in achieving economic methane flow and production rates (Guo and Cheng 2013). However, it is affected greatly by gas desorption and effective stress change due to gas depletion. On the one hand, as the reservoir pressure reduces below the desorption pressure, methane is released from coal matrix to fracture network, and consequently coal matrix shrink. As a result of the shrinkage, the cleats dilate and cleat permeability correspondingly improves. On the other hand, the gas depletion reduces reservoir pressure, and in turn increases effective stress, which closes cleat apertures and

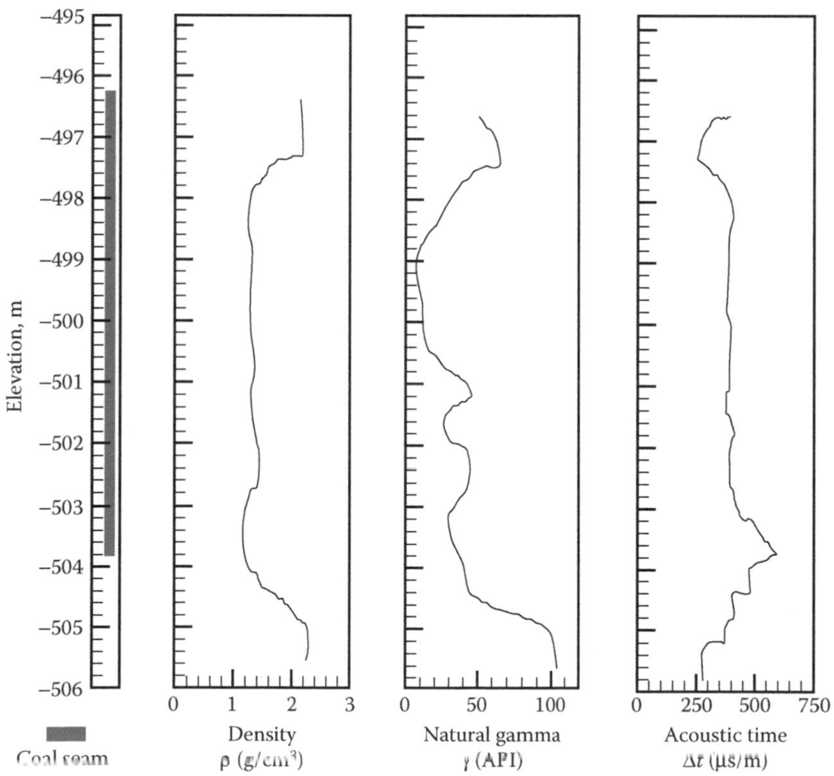

FIGURE 4.13 Density and acoustic logging images of a coal seam. (Adapted from Fu, X. et al., *Fuel,* 88, 2269–2277, 2009.)

reduces permeability. Whether the resultant dynamic permeability is greater or less than the original permeability value depends on the net results of these two competing mechanisms (Chen et al. 2011).

The heterogeneities of coalbeds also play an important role in the coal permeability distribution feature. It makes coal permeability vary not only from basin to basin but also within a certain seam, depending on coal rank, geological age, purity, in-situ stress, gas and water saturations, gas concentrations, seam depth, tectonics, and localized shear zone (Moore 2012; Seidle 2011a). Actually, the permeability variation within a basin over the course of depletion is usually greater than that between basins. The definition of coal permeability is segregated as the combination of absolute permeability and gas or water relative permeability, yet in reservoir engineering, coal permeability generally refers to the absolute permeability. In order to acquire the coal permeability, in-situ and laboratory approaches are employed (Cox et al. 1995).

4.3.1 Laboratory Measurements

4.3.1.1 Absolute Permeability

Absolute permeability is an inherent property of coal, rather than the fluids running through it; however, it does rely on in-situ stress and gas concentration. Millidarcy (mD) is the most commonly used unit for coal reservoir permeability. Moreover, absolute permeability can be further divided into matrix permeability and cleat permeability due to the dual-porosity system of coal. To be specific, matrix permeability is determined only by the pores within coal matrix, while cleat permeability is determined by the naturally developed fractures including face cleat and butt cleat, which are commonly found perpendicular to each other, as illustrated in Figure 4.14. Face cleats are formed first and throughgoing, and generally aligned in the direction of the maximum principal stress, while butt cleats end at intersections with face cleats. Face

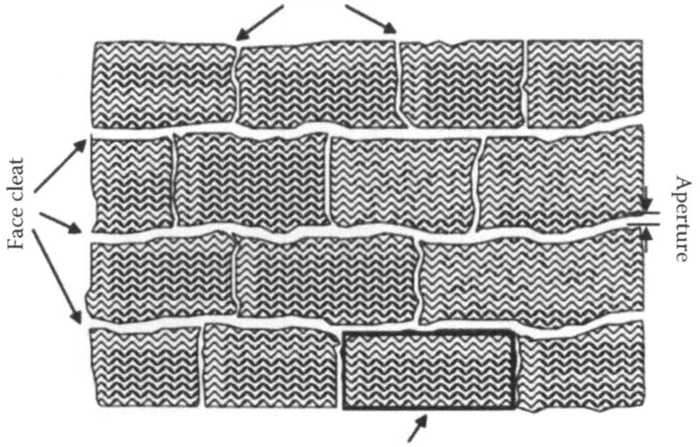

FIGURE 4.14 A schematic of coal structure plan view. (Adapted from Harpalani, S., *Influence of Microscopic Structure of Coal on Methane Migration*, pp. 113–118, 1999.)

cleats are usually dominant in contributing to permeability instead of butt cleats. Compared with matrix permeability, cleat permeability can be several orders of magnitude higher (often no less than 3 orders of magnitude). Therefore, cleat permeability is entitled as the single most important parameter governing fluids flow in a coal reservoir (Moore 2012). With regard to the acquisition of absolute permeability, both in-situ measurements and lab measurements can be employed to do so. In Sections 4.3.1.1.1 through 4.3.1.1.3, an example of measuring coal total permeability in the laboratorial condition is explained in detail.

4.3.1.1.1 *Sample*

One bituminous coal sample from the Bulli seam, southern Sydney basin Australia, was tested. The sample was 4.51 × 10.55 cm² in diameter and length. The sample was vacuumed for a few days inside the cell to remove the residual gas and to reach the desired temperature. Then the core was consolidated in the cell with a few load cycles before carrying out experiments to make sure results were repeatable (Pan et al. 2010).

4.3.1.1.2 *Experimental Settings*

A similar gas rig as shown in Figure 4.4 was used for this work. The pressure transient method was selected to conduct the gas flow experiments due to its shorter test durations compared to steady state measurements while maintaining the same accuracy. Permeability was measured after the equilibrium state was reached for changes of either confining stress or pore pressure. To measure the permeability value at a particular stress and pore pressure condition, the upstream cylinder pressure was first increased while the downstream cylinder pressure was reduced to achieve a differential pressure of around 80 kPa. Then gas was allowed to flow through the sample from the upstream cylinder to the downstream cylinder. Permeability can be calculated from the pressure decay curve using the following equation (Brace et al. 1968):

$$\frac{P_{\mathrm{u}} - P_{\mathrm{d}}}{P_{\mathrm{u,0}} - P_{\mathrm{d,0}}} = e^{-mt} \tag{4.15}$$

where:
- P_{u} and P_{d} are the pressures for the up and downstream cylinder, respectively
- $P_{\mathrm{u}} - P_{\mathrm{d}}$ is measured via the differential pressure transducer
- $P_{\mathrm{u,0}} - P_{\mathrm{d,0}}$ is the initial pressure difference between two cylinders
- t is the time used for individual permeability test
- m is defined in Equation 4.16:

$$m = \frac{k}{\mu C_{\mathrm{g}} L^2} V_{\mathrm{R}} \left(\frac{1}{V_{\mathrm{u}}} + \frac{1}{V_{\mathrm{d}}} \right) \tag{4.16}$$

where:
- k is the permeability
- μ and C_{g} are the gas viscosity and compressibility, respectively
- L is the sample length

TABLE 4.3
Values of Some Parameters

Parameters	Values
Sample length, L (m)	0.1055
V_R ($\times 10^{-6}$ m^3)	168.56
V_u ($\times 10^{-6}$ m^3)	580
V_d ($\times 10^{-6}$ m^3)	580
μ (Pa/s)	0.011913
C_g(1/P$_{abs}$)	1.10×10^{-6}

V_R is the sample volume
V_u and V_d are the volumes of the up and downstream cylinder, respectively

The specific value of each parameter for this case is listed in Table 4.3.

Once m is obtained from the pressure decay curve, permeability will be the only unknown in the equation, thus can be obtained directly.

4.3.1.1.3 Results and Discussion

Four permeability measurements were conducted using CH_4 at the temperature of 35°C, and a constant effective stress condition, defined as the difference between confining stress and gas pressure, was maintained. Figure 4.15 shows a typical pressure evolution during the measurements in both upstream and downstream cylinders, and the change of the pressure difference between them. The confining stress and gas pressure of the sample for this measurement were 2.82 and 0.82 MPa, respectively. The regression data shows that for this particular point m equals to 0.00327.

Substituting all the values of Table 4.3 into Equation 4.16 gives $k = 8.22 \times 10^{-16}$ m$^2 \approx 0.83$ mD. Following the same process, the change trend of coal permeability with respect to pore pressure increase can be obtained by plotting permeability values together, as shown in Figure 4.16.

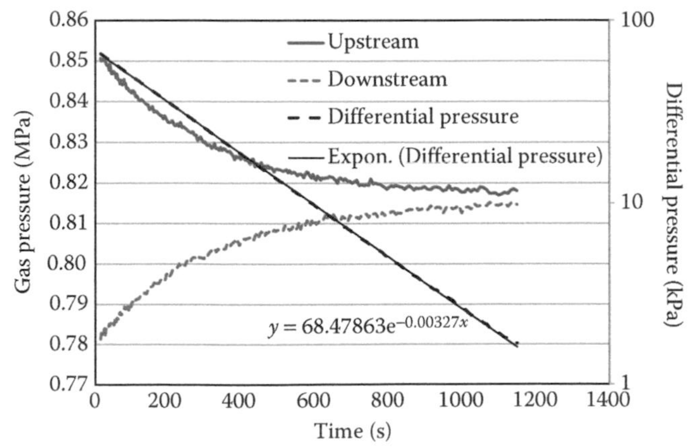

FIGURE 4.15 Pressure decay curve for permeability calculation. (Modified from Pan, Z. et al., *Int. J. Coal Geol.*, 82, 252–261, 2010.)

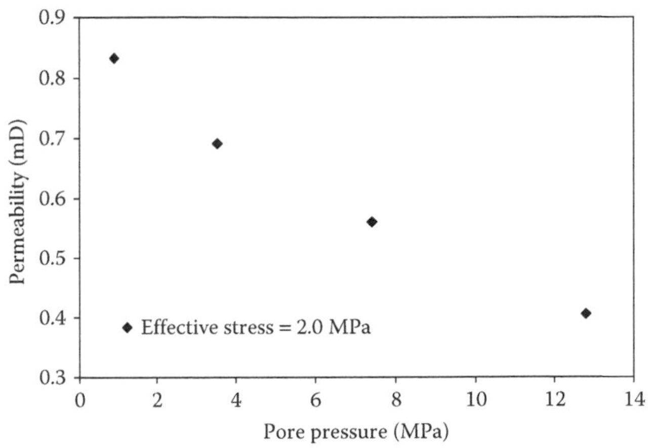

FIGURE 4.16 Permeability change versus pore pressure for a constant effective stress of 2.0 MPa. (Adapted from Pan, Z. et al., *Int. J. Coal Geol.*, 82, 252–261, 2010.)

4.3.1.2 Relative Permeability

Relative permeability of gas or water represents the ratio of the permeability of one phase (i.e., effective permeability of that phase) relative to the permeability where there is the presence of only that phase (i.e., absolute permeability). Thereby, relative permeability is often presented as a fraction scaling from 0 to 1. Absolute permeability values and gas or water saturations play a key role in determining relative permeability amount, and in turn significantly impact coalbed methane production performance.

Compared with absolute permeability in terms of measurements, relative permeability can be obtained from in-situ and laboratory measurements. They can be broken down into steady-state and unsteady-state approaches. As with the steady-state approach, the mixture of water and gas are injected simultaneously into a coal sample at constant flow rates until the expelled fluids have the same fraction as the injected ones, whereas unsteady-state approaches apply one fluid to displace the second fluid in a second-fluid-saturated coal sample, consuming much shorter time than that of steady-state methods (Shen et al. 2011).

4.3.2 Well Testing

Coal permeability can be determined through well testing from the pressure response to a change in gas or water flow rate. The theory underlying well tests is that when a well pressure change is made to the fluids that are in the equilibrium state, a down-hole pressure change is produced, and such change diffuses radially with time and finally reaches a new equilibrium (Chen and Lian 2003). The well tests used in coal reservoir is very similar to those in conventional oil and gas reservoirs, and requires only little or even no modification. These tests involve injection/falloff tests (IFTs) and drill stem tests (DSTs), which will be described in detail later on; In addition, drawdown and build-up tests, slug tests, tank tests, and interference tests are also commonly used in coal reservoir characterization. Aminian (2006) and Zhuang (2013) stated that well testing is the most effective way to gain coal reservoir permeability, and the results are the most representative among many available approaches.

Log–log curves are commonly produced through well testing, illustrating pressure change and pressure change rate as a function of time. A typical curve can be divided into several portions: the wellbore storage portion, the transition portion, the radial flow portion, and sometimes the boundary portion, which correspond to the wellbore, the combination of wellbore and coal reservoir, the coal reservoir and the boundary effect, respectively. Among these four portions, the radial flow portion deserves major concentration, as it delivers the key information of coal permeability. To deal with such portion, Horner's (1951) method is introduced to interpret the collected data from log–log curves into semilog ones, and subsequently coal permeability. In this method, a semilog curve with a half straight line of the bottom-hole pressure against Horner time is plotted, whose slope of the radial flow section is used to calculate the absolute permeability. The use of Horner's method will be explained in detail in the case study section below.

4.3.2.1 Injection/Falloff Tests

An IFT is performed by injecting water into a water-saturated coal seam and thus, the permeability resulting from this method is the absolute permeability. It monitors the instant pressure change that responds to the injection rate. The procedure of an IFT essentially follows four steps:

1. Setting down test devices and connecting them to surface equipment.
2. Opening down-hole test valve and wellhead controlling valve and injecting water with a constant rate.
3. Shutting in the well and stopping injection.
4. Monitoring pressure falloff curves and analyzing the data.

With respect to well shut in, the preferred method is bottom-hole shut in, because it can minimize the wellbore storage effect, which may influence the accuracy of the data (Zhuang 2012).

As already mentioned, the Horner's method is used, and the dimensionless Horner time is defined from Equation 4.17 (Earlougher 1977):

$$t_H = \frac{t_i + \Delta t}{\Delta t} \tag{4.17}$$

where:

t_i is injection time, in h
Δt is shut-in time, in h

The permeability is gained from Equation 4.18 (Earlougher 1977):

$$k = -\frac{162.6 q_w B_w \mu_w}{mh} \tag{4.18}$$

where:

q_w is the water injection rate, bpd
B_w is defined as the water formation volume factor, and equals to water volume in reservoir over its surface volume
μ_w refers to the water viscosity, cp
m is the straight line slope of the semi-log, psia/cycle
h is the effective thickness of the tested coal reservoir, ft

One thing that requires careful consideration when conducting IFT is that the injection pressure needs to be less than the coal seam breakdown pressure, otherwise destruction of the original coal microstructure may happen, and consequently the test accuracy is reduced.

IFT method avoids the two-phase flow, thus greatly simplifies data analysis process, and makes itself a widely used and efficient method to measure coal permeability (Zhao et al. 2010), including coal formations with very low permeability.

Seidle et al. (1991) conducted a pressure falloff test in a coal well in San Juan Basin to determine the coal permeability. The well was set in a 42 ft thick water-saturated coal seam. The water formation volume factor was 1.0, and the water viscosity was 0.5 cp. The water injection lasted for 12 h at a rate of 360 bpd prior to a well shut in for 28 h at the pressure of 1856 psia. The well test data are listed in Table 4.4.

According to Table 4.4, the semilog plot of Horner time is depicted as shown in Figure 4.17.

Since the radial flow portion, which is related to the coalbed formation, occurs at late time, in this case, the slope of the Horner time semilog curve was computed by using points from 29 to 46, as shown in Equation 4.19 (Seidle 2011b):

$$m = \frac{1348.97 - 1418.19}{\log(1.436) - \log(10.77)} = -79.10 \text{ psia/cycle} \tag{4.19}$$

With this figure, the permeability is derived from Equation 4.20:

$$k = -\frac{162.6 \times 360 \times 1.0 \times 0.5}{-79.10 \times 42} = 8.8 \text{ mD} \tag{4.20}$$

This number agreed well with the one of 8.2 mD reported by Seidle et al. (1991), which validated this well test method.

4.3.2.2 Drill Stem Tests

A DST is a procedure that isolates disturbing influences, and tests the permeability of a coal seam, and can be conducted in both open holes and cased holes, allowing formation fluids to migrate to the surface or to a chamber in the DST tool. The typical procedure of DSTs includes the following, corresponding to numbers 1 to 6 in Figure 4.18. Specifically, it includes (Assaad 2008) the following:

- Tester is lowered into hole: Increasing pressure of mud column.
- Packers are set: Gauge records slight pressure increase as packers expand.
- Tester valve opens and pressure drops: Formation fluids enter hole.
- Tester valve closes: Gauge records increase of pressure to shut-in pressure.
- Packers are released: Gauge records abrupt pressure increase under mud column.
- Tester is withdrawn: Gauge records decrease pressure of mud column.

TABLE 4.4
IFT Data of a Coal Well in San Juan Basin

Point No.	Δt, h	Pressure, psia	Horner Time, h	Δp, psia
1	0.00	1788.51	2958.85	67.49
2	0.01	1719.46	1462.99	136.54
3	0.01	1693.27	962.54	162.73
4	0.02	1669.46	723.46	186.54
5	0.02	1645.65	577.92	210.35
6	0.03	1633.75	480.23	222.25
7	0.03	1621.84	413.51	234.16
8	0.03	1614.70	359.10	241.30
9	0.04	1609.94	320.32	246.06
10	0.04	1605.18	286.92	250.82
11	0.05	1586.13	240.62	269.87
12	0.06	1576.60	206.23	279.40
13	0.07	1567.08	180.72	288.92
14	0.07	1562.32	161.88	293.68
15	0.08	1550.41	145.01	305.59
16	0.09	1543.27	132.78	312.73
17	0.10	1538.51	120.05	317.49
18	0.13	1523.83	96.62	332.17
19	0.15	1513.62	82.14	342.38
20	0.17	1507.32	69.81	348.68
21	0.20	1500.90	60.97	355.10
22	0.25	1484.26	49.08	371.74
23	0.30	1478.52	40.83	377.48
24	0.35	1470.32	35.46	385.68
25	0.45	1457.04	27.79	398.96
26	0.55	1450.83	22.99	405.17
27	0.69	1444.52	18.51	411.48
28	0.89	1431.58	14.51	424.42
29	1.23	1418.19	10.76	437.81
30	1.78	1408.61	7.75	447.39
31	2.06	1403.83	6.81	452.17
32	2.53	1397.57	5.75	458.43
33	3.04	1393.89	4.94	462.11
34	3.54	1390.20	4.39	465.80
35	4.03	1385.05	3.98	470.95
36	4.57	1382.84	3.63	473.16
37	5.09	1379.16	3.36	476.84
38	5.60	1367.58	3.14	488.42
39	6.07	1374.37	2.98	481.63
40	7.09	1371.79	2.69	484.21
41	8.08	1368.11	2.48	487.89
42	9.04	1366.27	2.33	489.73
43	12.02	1360.75	2.00	495.25
44	16.18	1356.33	1.74	499.67
45	20.10	1353.39	1.60	502.61
46	27.53	1348.97	1.44	507.03

Source: Seidle, J., Coal well pressure transient tests, in: *Fundamentals of Coalbed Methane Reservoir Engineering*, J. Seidle (ed.), PennWell, Tulsa, Oklahoma, pp. 185–215, 2011b.

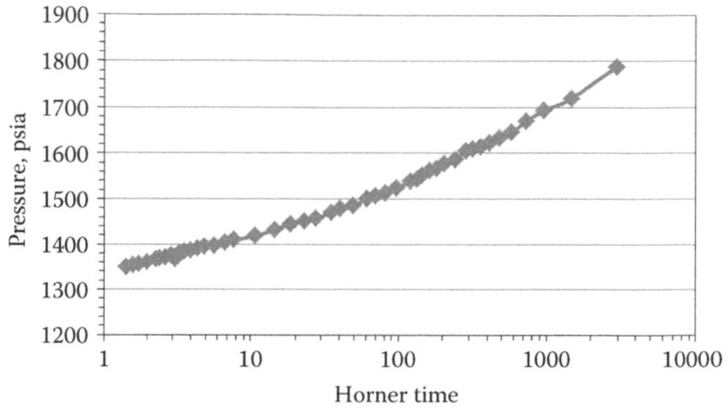

FIGURE 4.17 Semilog plot of bottom hole pressure as a function of Horner time. (Data from Seidle, J., Coal well pressure transient tests, in: *Fundamentals of Coalbed Methane Reservoir Engineering*. J. Seidle (ed.), PennWell, Tulsa, Oklahoma, pp. 185–215, 2011b.)

DST data are typically measured during four periods: a short initial flow period, followed by a short initial shut-in period, a longer flow period and a longer shut-in period (Gas Research Institute 1996). Coal permeability is acquired in the final flow and shut-in periods, with the previous two periods yielding other valuable characteristics. A representative example of the results of a DST is plotted in Figure 4.19.

DSTs exert less damage to wellbore, thereby reducing the wellbore effect. Despite that, due to the presence of gas and coal fines, a DST may be problematic in coal reservoirs (Seidle 2011b). In a coal seam that produces both gas and water, a DST produces water effective permeability instead of absolute permeability which can only be obtained if gas and water relative permeabilities and fluid saturations are known. Moreover, the coal fines, especially in open holes, can interfere with the DST tool, complicating the data interpretation.

4.4 DETERMINATION OF COALBED GAS-IN-PLACE

Methane gas was long considered to be a major threat to underground coal mining but now it is recognized as a valuable resource (Chen et al. 2012). Determination of gas total volume in place in a coal deposit is critical to the exploitation of the gas resources, because it not only affects the gas production design but also impacts the planning of the subsequent coal mining operations. Unlike conventional gas reservoirs, coal reservoir can hold large volume of gas by sorption due to the high organic matter component. The ratio of the sorbed gas phase volume to the total gas in place varies a lot from basin to basin, but it could be as high as over 90% (Gray 1987). The measurement of gas content in conventional reservoirs such as sandstones by logs is a benefit not yet available in coal seams without extensive calibration of the logs from previous core analyses. Therefore, coalbed gas contents in place are generally determined in the laboratories from coal samples taken from the field (Halliburton Company 2007). Direct method and indirect method are the two methods

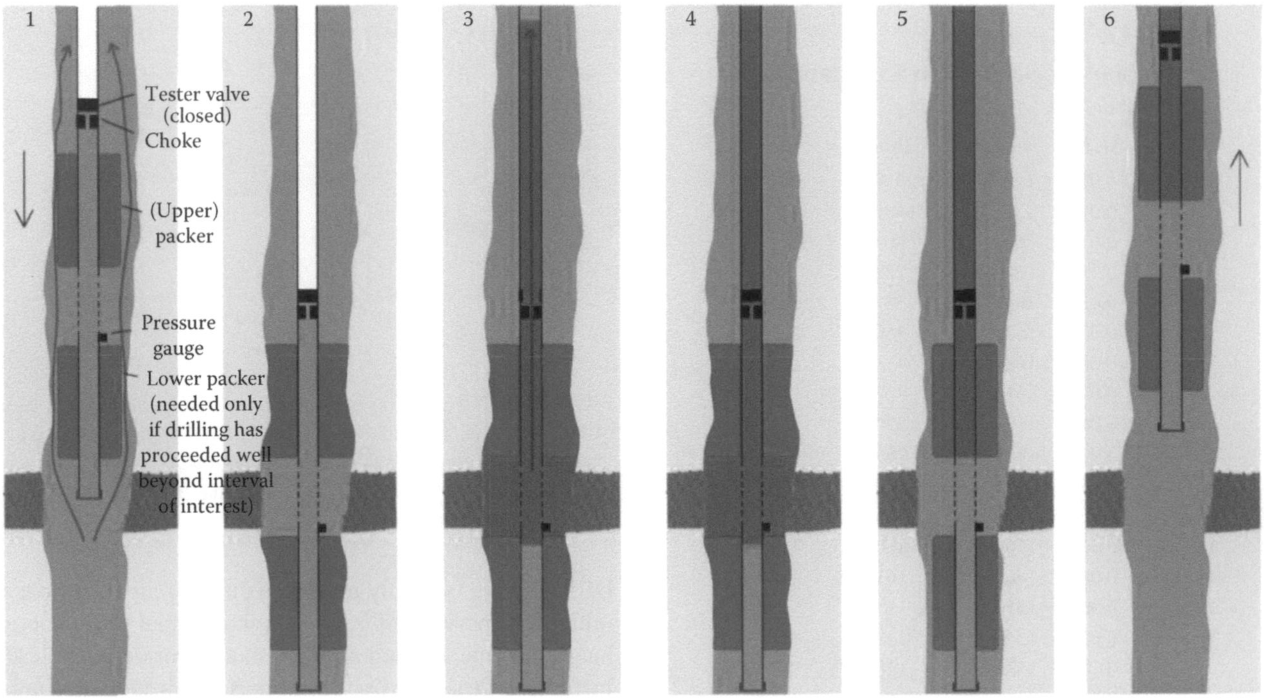

FIGURE 4.18 Procedure of DST measurement. (From Assaad, F. A., *Field Methods for Petroleum Geologists: A Guide to Computerized Lithostratigraphic Correlation Charts Case Study: Northern Africa*, F.A. Assaad (ed.), Springer, Berlin, Heidelberg, 2008.)

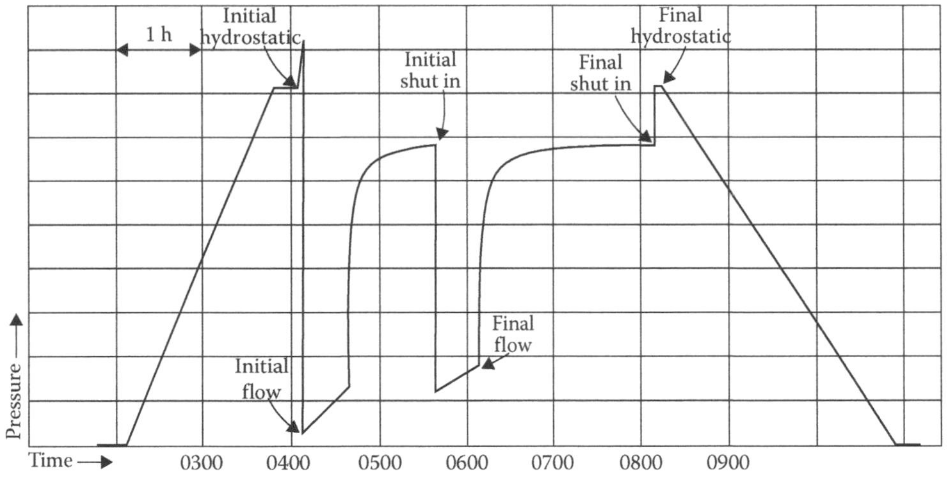

FIGURE 4.19 Typical DST chart. (Data from Oklahoma Geological Survey, *Well Testing*. http://www.ogs.ou.edu/pdf/Welltest.pdf, 2014.)

commonly used to determine gas contents, which will be discussed in detail below.

4.4.1 Direct Measurement of CBM Content in Place

Direct measurement of gas content in place is very difficult due to a number of uncertainties, such as the variation of reservoir conditions, the estimation of lost gas during the core retrieval process, and limitations of measuring methods. The current gas industry standard for determining gas content is desorption of whole core sample using the U.S. Bureau of Mines direct method (Seidle 2011c). Estimating the sorbed gas content of coal formation requires the measurements of three components: lost gas, desorbed gas in the canister, and residual gas. The relationship is given by the following equation (Gas Research Institute 1996):

$$G = G_R + G_C + G_L \tag{4.21}$$

where:

G is the gas content per unit weight of coal, m^3/t

G_R is the core residual gas, m^3/t. It is measured at the end of the desorption cycle by pulverizing the core or crushing the core. Note that all gas remaining adsorbed in coal below atmosphere pressure should not be considered because it would not be recovered in practice (Seidle 2011c)

G_C is the gas desorbed by the core in the canister, m^3/t. It is collected from the whole core in the canister at reservoir temperature and atmospheric pressure over a certain period of time in the canister. The use of reservoir temperature during core desorption is to give accurate assessment of gas to be desorbed from the coal formation. Note that metric standard temperature and pressure conditions are defined as 0°C and 101 kPa

G_L is the gas lost from the core in the coring process, m^3/t. It refers to gas desorbed from the core from the time the core is extracted from the formation to the time the core is placed in the canister and sealed

Please note that generally the standard volume gas per unit weight coal convention is used in coalbed methane (CBM) reservoir engineering, and when quantifying the gassiness of a particular coal reservoir, the unit of cubic meters per ton (m^3/t) is commonly used or standard cubic feet of gas per ton (scf/ton) if in the United States (Seidle 2011c).

The general procedure for evaluating gas content of a coal seam involves the following steps (Halliburton Company 2007):

- Remove cores from the coal seam with conventional coring equipment.
- Transport cores in core barrels rapidly to the surface. Record transit time.
- Place cores in canisters immediately upon reaching the surface.
- Measure desorbed gas over several days, weeks, or months until the rate drops below a certain threshold value. The desorbed gas volume at each step must be corrected to the condition of standard temperature and pressure.
- Estimate gas lost during cutting, retrieval, and surface handling prior to being hermetically sealed.
- Determine residual gas either from laboratory procedures or analytic methods.

To maintain the accuracy of direct method, it is critical to properly select the starting time of gas desorption at reservoir temperature. The time is assumed to be the moment the external pressure of the core equals to the desorption pressure. When a coal reservoir is gas saturated, the pressure of desorption is considered to be equal to the reservoir pressure; thus desorption begins when once fluid pressure around the core sample is less than the reservoir pressure. When coal seam is under-saturated, desorption does not occur until the removal of the drill collars; thus the estimate of lost gas volume can be very challenging. In this case, the determination of gas content needs to combine with the information of core collection (Gas Research Institute 1996).

4.4.1.1 Sample and Measurement Settings

The desorption test was conducted by ABC Desorption and Isotherm Company from Denver, Colorado, the United States at the depth of 383.5 m. Core diameter was about 7.62 cm (3 in.) using wireline coring, and bulk sample weight was 2177 grams. Ash and sulfur contents were about 4.13% and 2.66%, respectively.

4.4.1.2 Results and Discussion

The cumulative desorbed gas volume, corrected to standard temperature and pressure, is plotted in Figure 4.20. The result clearly shows that gas desorption happens very fast at the early stage of the desorption test, but it slows down gradually until the cumulative gas volume almost levels off. The trend line can be roughly described with an exponential function. In practice, gas desorption test can be terminated when gas emission rate falls below a certain threshold value, typically 0.05 $cm^3/g/day$ for a week (Seidle 2011c).

When gas desorption process from a coal sample is being analyzed, Fick's law of diffusion is commonly used, and for early short times, cumulative gas released from coal cores is generally considered to be proportional to the square root of time (Gas Research Institute 1996; Seidle 2011c). Therefore, a linear relationship between the cumulative gas volume at standard temperature and pressure condition and square root of time is expected. The lost gas volume can be estimated by extrapolating the linear trend line back to zero time, which is the start of desorption. For instance, the cumulative desorbed gas volume at the early stage of desorption is plotted in Figure 4.21 and fitted with a linear trend line. Extrapolating back to zero time of the trend line, the lost gas volume for this case can be yielded as around 1665 cm^3.

Similar to the approach of lost gas determination, the residual gas volume can be estimated by plotting the reciprocal time versus the cumulative desorbed gas, because according to Fick's law, at late time of gas desorption, the desorbed gas is linear to the reciprocal time (Mavor et al. 1990). Hence, straight line is

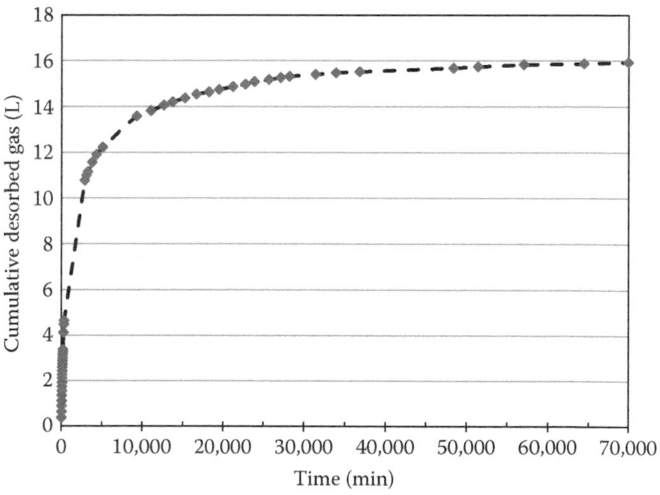

FIGURE 4.20 Cumulative desorbed gas. (Data from Seidle, J., Measurement of Coalbed Gas Content, In. *Fundamentals of Coalbed Methane Reservoir Engineering*, J. Seidle (ed.), PennWell, Tulsa, Oklahoma, 2011c.)

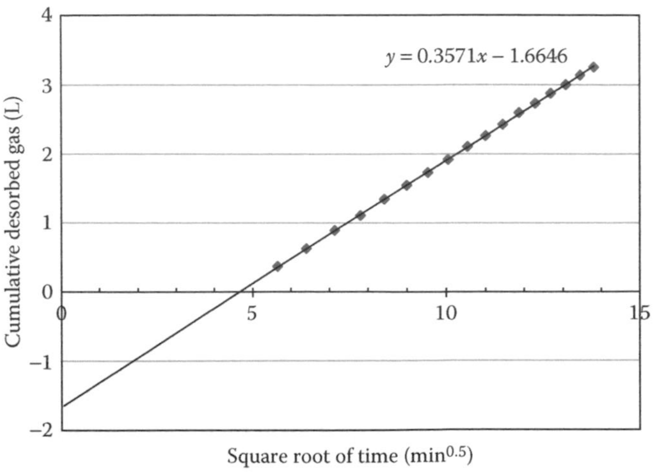

FIGURE 4.21 Estimate of lost gas. (Data from Seidle, J., Measurement of Coalbed Gas Content, in: *Fundamentals of Coalbed Methane Reservoir Engineering*, J. Seidle (ed.), Tulsa, Oklahoma, PennWell, 2011c.)

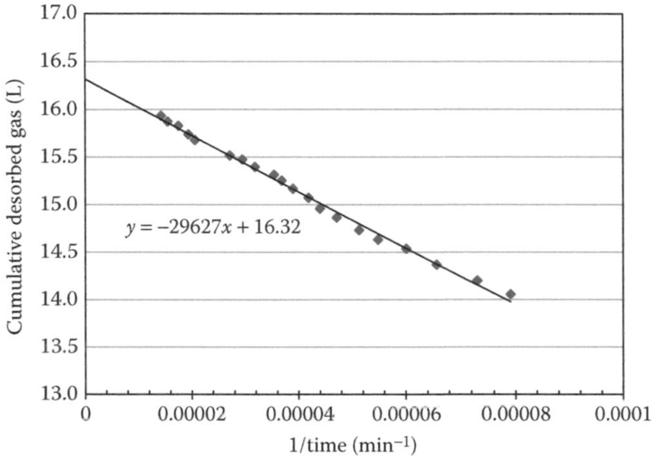

FIGURE 4.22 Estimate of residual gas. (Data from Seidle, J., Measurement of Coalbed Gas Content, in: *Fundamentals of Coalbed Methane Reservoir Engineering*, J. Seidle (ed.), PennWell, Tulsa, Oklahoma, 2011c.)

yielded when plotting cumulative desorbed gas against the reciprocal time, and the total desorbed gas can be calculated when intercepting the line to zero reciprocal time (Mavor et al. 1990). The residual gas is the difference between total desorbed gas and the cumulative desorbed gas. Take Figure 4.22, for instance, a trend line fit to the data give an intercept of 16,320 cm^3. Subtracting the cumulative gas volume of 15,931 cm^3, as shown in Figure 4.22, yields a residual gas volume of 389 cm^3.

4.4.2 Isothermal Adsorption of CH_4 on Coals

Different to direct measurement of gas content in place, indirect methods rely on equations, correlations, or laboratory isotherm measurements, assuming that the coal formation is gas saturated, to estimate gas content in place (Seidle 2011c). The isothermal adsorption is the common method used to estimate gas in place of a coal formation. This method involves the measurements of gas adsorption on the coal sample with varying pressure for a specific temperature, and is typically measured with pure component gases.

A number of models or theories have been developed to describe experimental coal gas adsorption data, such as linear adsorption isotherm, Langmuir equation and BET theory et al. (Pan 2012). Among them the Langmuir equation is probably the most commonly used one in CBM reservoir engineering largely due to the close fit of the adsorption data and simplicity, and as such, only this approach is introduced in this section.

The Langmuir theory was developed in 1918 (Langmuir 1918), and the major assumptions in the derivation of the equation include the following:

- One gas molecule is adsorbed at a single adsorption site.
- An adsorbed molecule does not affect the molecule on the neighboring site.
- Adsorption is on an open surface, and there is no resistance to gas access to adsorption sites.

The general Langmuir equation with the above assumptions is given as (Langmuir 1918) follows:

$$V = V_L \frac{p}{p + p_L} \tag{4.22}$$

where:

V is the gas content, cm^3/g

V_L is the Langmuir volume constant (cm^3/g), which represent the maximum amount of gas that can be adsorbed onto a given sample at infinite gas pressure

p is the gas pressure, MPa

p_L is the Langmuir pressure constant (MPa), which is the gas pressure at which the adsorption volume is equal to one-half the maximum gas volume

Using Equation 4.22, the isotherm of gas adsorption on coal with varying pressure values while keeping temperature constant can be constructed, which can be extrapolated to predict the maximum gas content at any pressure value. More importantly, the equation provides an important guide to the estimate of remaining gas volume for a specific coal formation during production process.

4.4.2.1 Sample

An example of this procedure is presented for an Australian coal sample from the southern Sydney Basin. The coal sample was bituminous coal from the Bulli seam and cored to 4.5 cm in diameter and 10.55 cm in length. The void volume of the sample is 81.97 ml and the weight of coal sample is 222.38 g (Pan et al. 2010). Methane gas was used in the tests.

4.4.2.2 Measurements Settings

Prior to gas adsorption, the void volume, in the cell is determined by injecting known quantities of helium from a calibrated gas injection pump, and measurements on a number of different pressures were conducted to check the consistency of void volume values.

The Gibbs excess adsorption is calculated directly from experimental quantities. For this case study, the quantity of methane gas, n_{inj}, injected from the gas injection pump into the cell is monitored. The injected gas will exist in two forms in the coal sample: adsorbed phase (n_{ads}^{Gibbs}), and free gas phase (n_{unads}^{Gibbs}). A mass balance is used to calculate the amount adsorbed phase as (Pan et al. 2010):

$$n_{ads}^{Gibbs} = n_{inj} - n_{unads}^{Gibbs} \tag{4.23}$$

The amount injected is calculated directly from pressure, temperature, and volume measurements of the pump, and the amount of unadsorbed gas can be determined based on the experimental conditions of the cell.

Equation 4.24 is used to calculated absolute adsorption from the measured Gibbs excess adsorption (Ruppel et al. 1974):

$$n_{ads}^{Abs} = n_{ads}^{Gibbs}\left(\frac{\rho_{ads}}{\rho_{ads} - \rho_{gas}}\right) \tag{4.24}$$

where:

ρ_{gas} is the gas phase density
ρ_{ads} is the adsorbed phase density

In CBM reservoir engineering, the sorbed gas density is usually assumed to be the density of the liquid at the atmospheric pressure boiling point. The density of adsorbed methane is about 0.421 g/cm^3.

4.4.2.3 Results and Discussion

The gas pressure and temperature in the pump were 10.25 MPa and 43.55°C, respectively. The compressibility value (Z) at this condition is 0.882212. The experimental pressure-volume-temperature (PVT) data are listed in Table 4.5.

The absolute adsorption volume at standard temperature and pressure conditions is plotted in Figure 4.23. Using Langmuir equation to fit the data gives p_L = 2.96 MPa and V_L = 25.37 m^3/t. Therefore, the adsorption character of this sample can be expressed using Langmuir equation as

$$V = 25.37\frac{p}{p + 2.96} \tag{4.25}$$

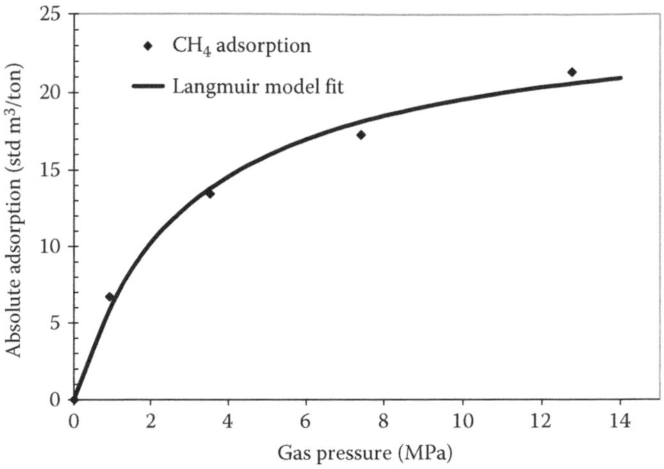

FIGURE 4.23 Absolute adsorption volume versus gas pressure at T = 44.3°C. (Adapted from Pan, Z. et al., *Int. J. Coal Geol.*, 82, 252–261, 2010.)

4.5 SUMMARY

This chapter describes the key properties needed to characterize a coal reservoir, including coal porosity, permeability, and gas-in-place measurement. The accurate quantitative determination of these parameters plays a critical role in evaluating and predicting coal reservoir production performance.

Seven different techniques for estimating coal formation porosity were introduced. The principle for each technique was explained with some specific examples, the strengths and limitations of each method were also provided in the context.

Coal permeability is probably the most important parameter in determining gas production performance. As such, this chapter spent a large proportion explaining two key approaches to estimate permeability values: laboratory measurements and field well testing. Regarding the laboratory measurements of permeability, its working principles were explained in detail and a real case study was provided. Two well-testing methods were also introduced: IFT and DST. These two commonly used methods were explained with the specific examples.

Accurate determination of gas content in coal seams not only helps to plan exploitation of gas resources but also helps to optimize mining system design when coming to the mining stage.

TABLE 4.5
Experimental Data and Calculation of Absolute Adsorption Mole Mass

Pump	Cell							
ΔV (ml)	P (MPa)	Z	T (°C)	N_{inj} (mol)	n_{unads}^{Gibbs} (mol)	n_{ads}^{Gibbs} (mol)	ρ_{gas} (g/cm^3)	n_{ads}^{Abs} (mol)
0	−0.086	0.99980	44.3	–	–	–	–	–
−20.3234	0.798	0.98811	44.3	0.40725	0.12498	0.28228	0.005518	0.28602
−52.6722	3.422	0.95395	44.5	1.05548	0.51339	0.54210	0.022376	0.57252
−89.1361	7.311	0.90887	44.3	1.78617	1.13694	0.64923	0.049441	0.73362
−137.91	12.67	0.86983	44.5	2.76353	2.04726	0.71627	0.088953	0.90816

This chapter listed both direct method and indirect method for gas content estimates. For the direct method, the difference between lost gas, desorbed gas, and residual gas was explained in detail and a set of data from field measurement was introduced. The principles of laboratory measurements to determine gas content were also provided and the calculation of adsorption volume for an Australian coal sample was also explained.

REFERENCES

Abell, A. B., Willis, K. L., Lange, D. A., 1999. Mercury intrusion porosimetry and image analysis of cement-based materials. *Journal of Colloid and Interface Science* 211, 39–44.

Aminian, K., 2006. *Evaluation of Coalbed Methane Reservoirs*. Petroleum & Natural Gas Engineering Department, West Virginia University.

Assaad, F. A., 2008. *Field Methods for Petroleum Geologists: A Guide to Computerized Lithostratigraphic Correlation Charts Case Study: Northern Africa*. Springer, Berlin, Heidelberg.

Auzerais, F. M., Dunsmuir, J., Ferréol, B. B., Martys, N., Olson, J., Ramakrishnan, T. S., Rothman, D. H., Schwartz, L. M., 1996. Transport in sandstone: A study based on three dimensional microtomography. *Geophysical Research Letters* 23, 705–708.

Bale Harold, D., Carlson Marvin, L., Kalliat, M., Kwak Chul, Y., Schmidt Paul, W., 1984. Small-angle X-ray scattering of the submicroscopic porosity of some low-rank coals, in: Harold H. Schobert (Ed.), *The Chemistry of Low-Rank Coals*. American Chemical Society, Washington, DC, pp. 79–94.

Bansal, R. C., Goyal, M., 2005. *Activated Carbon Adsorption*. Taylor & Francis Group, Boca Raton, FL,.

Bloembergen, N., Purcell, E. M., Pound, R. V., 1948. Relaxation effects in nuclear magnetic resonance absorption. *Physical Review* 73, 679–712.

Bo, L., Qiyan, F., Xiongdong, L., Lai, Z., Yue, S., 2006. Analysis to porosity of coal by mercury porosimetry, Sciencepaper Online.

Bonnie, J. H. M., Fens, T. W., 1992. *Porosity and Permeability from SEM Based Image Analysis of Core Material*. SPE Latin America Petroleum Engineering Conference, Caracas, Venezuela, pp. 8–11.

Brace, W. F., Walsh, J. B., Frangos, W. T., 1968. Permeability of granite under high pressure. *Journal of Geophysical Research* 73, 2225–2236.

Brunauer, S., Emmett, P. H., Teller, E., 1938. Adsorption of gases in multimolecular layers. *Journal of the American Chemical Society* 60, 309–319.

Calo, J. M., Hall, P. J., 2004. The application of small angle scattering techniques to porosity characterization in carbons. *Carbon* 42, 1299–1304.

Carr, H. Y., Purcell, E. M., 1954. Effects of diffusion on free precession in nuclear magnetic resonance experiments. *Physical Review* 94, 630–638.

Chen, Y., Li, D., 2006. Analysis of error for pore structure of porous materials measured by MIP. *Bulletin of the Chinese Ceramic Society* 25, 198–202.

Chen, Z., Lian, Y., 2003. Discussion on technical problems for injection/fall-off well test in the coalbed methane well. *Meitiandizhi Yu Kantan/Coal Geology and Exploration* 31, 23–26.

Chen, Z., Liu, J., Pan, Z., Connell, L. D., Elsworth, D., 2012. Influence of the effective stress coefficient and sorption-induced strain on the evolution of coal permeability: Model development and analysis. *International Journal of Greenhouse Gas Control* 8, 101–110.

Chen, Z., Pan, Z., Liu, J., Connell, L. D., Elsworth, D., 2011. Effect of the effective stress coefficient and sorption-induced strain on the evolution of coal permeability: Experimental observations. *International Journal of Greenhouse Gas Control* 5, 1284–1293.

Cheng, X., Ling, Y., 2008. Use well logging to obtain coal reservoir assessment parameters. *Energy Technology and Management* 4, 84–85.

Cherry, R., 1997. Magnetic resonance technology and its applications in the oil and gas industry. *Petroleum Engineer International* 70(3), 29–35. Other Information: PBD: Medium: X; Size.

Coates, G., Xiao, L., Prammer, M. G., 1999. *NMR: Logging Principles and Applications*. Halliburton Energy Services, Houston, TX.

Cohaut, N., Blanche, C., Dumas, D., Guet, J. M., Rouzaud, J. N., 2000. A small angle X-ray scattering study on the porosity of anthracites. *Carbon* 38, 1391–1400.

Cohen, M. H., Mendelson, K. S., 1982. Nuclear magnetic relaxation and the internal geometry of sedimentary rocks. *Journal of Applied Physics* 53, 1127–1135.

Compton, A. H., 1923. A quantum theory of the scattering of X-rays by light elements. *Physical Review* 21, 483–502.

Cox, D. O., Young, G. B. C., Bell, M. J., 1995. *Well Testing in Coalbed Methane (CBM) Wells: An Environmental Remediation Case History*. SPE Annual Technical Conference and Exhibition, Dallas, Texas, pp. 22–25, October 1995.

Davis, M. R., White, A., Deegan, M. D., 1986. Scanning electron microscopy of coal macerals. *Fuel* 65, 277–280.

Dragsdorf, R. D., 1956. Small-angle X-ray scattering. *Journal of Applied Physics* 27, 620–626.

Dunlap, M., Adaskaveg, J. E., 1997. *Introduction to the Scanning Electron Microscope: Theory, Practice & Procedures*. Edited by Facility for advanced instrumentation, U.C. Davis, Davis, CA.

Dunn, K. J., Bergman, D. J., Latorraca, G. A., 2002. NMR logging applications, in: K. J. Dunn, D. J. Bergman and G. A. Latorraca (Eds.), *Nuclear Magnetic Resonance—Petrophysical and Logging Applications*. Elsevier, Kidlington, Oxford.

Earlougher, R. C., 1977. *Advances in Well Test Analysis*. Henry L. Doherty Memorial Fund of AIME, Dallas, Texas.

Flannery, B. P., Deckman, H. W., Roberge, W. G., D'Amico, K. L., 1987. Three-dimensional X-ray microtomography. *Science* 237, 1439–1444.

Francis, W., 1961. *Coal: Its Formation and Composition*, 2nd ed. E. Arnold, London.

Fu, X., Qin, Y., Wang, G. G. X., Rudolph, V., 2009. Evaluation of gas content of coalbed methane reservoirs with the aid of geophysical logging technology. *Fuel* 88, 2269–2277.

Gan, H., Nandi, S. P., Walker Jr, P. L., 1972. Nature of the porosity in American coals. *Fuel* 51, 272–277.

Gane, P. A. C., Ridgway, C. J., Lehtinen, E., Valiullin, R., Furó, I., Schoelkopf, J., Paulapuro, H., Daicic, J., 2004. Comparison of NMR cryoporometry, mercury intrusion porosimetry, and DSC thermoporosimetry in characterizing pore size distributions of compressed finely ground calcium carbonate structures. *Industrial & Engineering Chemistry Research* 43, 7920–7927.

Gas Research Institute, 1996. *A Guide to Coalbed Methane Reservoir Engineering*. Gas Research Institute, Chicago, IL.

Giesche, H., 2006. Mercury porosimetry: A general (practical) overview. *Particle & Particle Systems Characterization* 23, 9–19.

Giffin, S., Littke, R., Klaver, J., Urai, J. L., 2013. Application of BIB–SEM technology to characterize macropore morphology in coal. *International Journal of Coal Geology* 114, 85–95.

Gray, I., 1987. *Reservoir Engineering in Coal Seams: Part 1-The Physical Process of Gas Storage and Movement in Coal Seams. SPE Reservoir Engineering* 2(1), 28–34.

Guo, P., Cheng, Y., 2013. Permeability prediction in deep coal seam: A case study on the No. 3 coal seam of the Southern Qinshui Basin in China. *The Scientific World Journal* 2013, 161457.

Gupta, R., 2007. Advanced coal characterization: a review. *Energy & Fuels* 21, 451–460.

Halliburton Company, 2007. *Coalbed Methane: Principles and Practices*. Halliburton, Duncan, OK.

Harpalani, S., 1987. *Influence of Microscopic Structure of Coal on Methane Migration*, Mine Ventilation Symposium (Third), The Pennsylvania State University, University Park, Pennsylvania, October 12–14, pp. 113–118.

Harpalani, S. S., 1999. *Compressibility of Coal and Its Impact on Gas Production from Coalbed Reservoirs*. American Rock Mechanics Association.

Harris, L. A., Yust, C. S., 1976. Transmission electron microscope observations of porosity in coal. *Fuel* 55, 233–236.

Hedenblad, G., 1997. The use of mercury intrusion porosimetry or helium porosity to predict the moisture transport properties of hardened cement paste. *Advanced Cement Based Materials* 6, 123–129.

Hinde, A., 2004. PRINSAS—A Windows-based computer program for the processing and interpretation of small-angle scattering data tailored to the analysis of sedimentary rocks. *Journal of Applied Crystallography* 37, 1020–1024.

Horner, D. R., 1951. *Pressure Build-up in Wells*. 3rd World Petroleum Congress, 28 May-6 June, The Hague, the Netherlands.

Ishaq, U.M., Bijaksana, S., Nurhandoko, B. E. B., 2009. Porosity and fracture pattern of coal as CBM reservoir, *Proceedings of the 3rd Asian Physics Symposium*, Bandung, Indonesia.

Jarzyna, J. A., Bala, M. J., Mortimer, Z. M., Puskarczyk, E., 2013. Reservoir parameter classification of a Miocene formation using a fractal approach to well logging, porosimetry and nuclear magnetic resonance. *Geophysical Prospecting* 61, 16.

Kalliat, M., Kwak, C. Y., Schmidt, P. W., 1981. Small-angle X-ray investigation of the porosity in coals. in: Bernar D. Blaustain and Bradley C. Bockrath (Eds.), New *Approaches in Coal Chemistry*. Sidney Friedman, American Chemical Society, Washington DC, pp. 3–22.

Karacan, C. Ö., Mitchell, G. D., 2003. Behavior and effect of different coal microlithotypes during gas transport for carbon dioxide sequestration into coal seams. *International Journal of Coal Geology* 53, 201–217.

Karacan, C. O., Okandan, E., 2001. Adsorption and gas transport in coal microstructure: Investigation and evaluation by quantitative X-ray CT imaging. *Fuel* 80, 509–520.

Kazimierz, T., Jacek, T., Stanislaw, R., 2004. Evaluation of rock porosity measurement accuracy with a helium porosimeter. *Acta Montanistica Slovaca* 9, 316–318.

Kenyon, W. E., 1992. Nuclear-magnetic-resonance as a petrophysical measurement. *Nuclear Geophysics* 6, 153–171.

Ketcham, R. A., Carlson, W. D., 2001. Acquisition, optimization and interpretation of X-ray computed tomographic imagery: Applications to the geosciences. *Computers & Geosciences* 27, 381–400.

Kleinberg, R. L., 1996. Utility of NMR T2 distributions, connection with capillary pressure, clay effect, and determination of the surface relaxivity parameter ρ2. *Magnetic Resonance Imaging* 14, 761–767.

Kleinberg, R. L., Farooqui, S. A., Horsfield, M. A., 1993a. T1/T2 ratio and frequency dependence of NMR relaxation in porous sedimentary rocks. *Journal of Colloid and Interface Science* 158, 195–198.

Kleinberg, R. L., Straley, C., Kenyon, W. E., Akkurt, R., Farooqui, S. A., 1993b. *Nuclear Magnetic Resonance of Rocks: T1 vs. T2*. SPE Annual Technical Conference and Exhibition, 3-6 October, Houston, Texas.

Kotlensky, W. V., Walker Jr, P. L., 1960. *Proceedings of the 4th Carbon Conference*. Pergamon Press, London, pp. 423–442.

Langmuir, I., 1918. The adsorption of gases on plane surfaces of glass, mica and platinum. *Journal of the American Chemical Society* 40, 1361–1403.

Levine, J. R., 1996. Model study of the influence of matrix shrinkage on absolute permeability of coal bed reservoirs, in: R. Gayer and I. Harris (Eds.), *Coalbed Methane and Coal Geology*, The Geological Society, London, pp. 197–212.

Li, S., Tang, D., Xu, H., Yang, Z., Guo, L., 2012. Porosity and permeability models for coals using low-field nuclear magnetic resonance. *Energy & Fuels* 26, 5005–5014.

Mahajan, O. P., Walker, P. L., 1978. Porosity of coals and coal products, in: Karr, C. J. (Ed.), *Analytical Methods of Coal and Coal Products*, Academic Press, New York.

Mancktelow, N. S., Grujic, D., Johnson, E. L., 1998. An SEM study of porosity and grain boundary microstructure in quartz mylonites, Simplon Fault Zone, Central Alps. *Contributions to Mineralogy and Petrology* 131, 71–85.

Mares, T. E., Radliński, A. P., Moore, T. A., Cookson, D., Thiyagarajan, P., Ilavsky, J., Klepp, J., 2009. Assessing the potential for CO_2 adsorption in a subbituminous coal, Huntly Coalfield, New Zealand, using small angle scattering techniques. *International Journal of Coal Geology* 77, 54–68.

Mavor, M. J., Owen, L. B., Pratt, T. J., 1990. *Measurement and Evaluation of Coal Sorption Isotherm Data*. SPE Annual Technical Conference and Exhibition, 23-26 September, New Orleans, LA.

Miedzinska, D., Niezgoda, T., Malek, E., Zasada, D., 2013. Study on coal microstructure for porosity levels assessment. *Bulletin of the Polish Academy of Sciences: Technical Sciences* 61, 499–505.

Moore, T. A., 2012. Coalbed methane: A review. *International Journal of Coal Geology* 101, 36–81.

Murphy, D. P., 1995. NMR logging and core analysis—Simplified. *World Oil* 216, 65.

Nakashima, Y., Yamaguchi, T., 2004. A Mathematica® program for three-dimensional mapping of tortuosity and porosity of porous media. *Bulletin of the Geological Survey of Japan* 55, 93–103.

Oklahoma Geological Survey, 2014. *Well Testing*. http://www.ogs.ou.edu/pdf/Welltest.pdf.

Pan, Z., 2012. Modeling of coal swelling induced by water vapor adsorption. *Frontiers Chemical Science and Engineering* 6, 94–103.

Pan, Z., Connell, L. D., Camilleri, M., 2010. Laboratory characterisation of coal reservoir permeability for primary and enhanced coalbed methane recovery. *International Journal of Coal Geology* 82, 252–261.

Prinz, D., Pyckhout-Hintzen, W., Littke, R., 2004. Development of the meso- and macroporous structure of coals with rank as analysed with small angle neutron scattering and adsorption experiments. *Fuel* 83, 547–556.

Rabi, I. I., Zacharias, J. R., Millman, S., Kusch, P., 1938. A new method of measuring nuclear magnetic moment. *Physical Review* 53, 318–327.

Radlinski, A. P., Boreham, C. J., Lindner, P., Randl, O., Wignall, G. D., Hinde, A., Hope, J. M., 2000. Small angle neutron scattering signature of oil generation in artificially and naturally matured hydrocarbon source rocks. *Organic Geochemistry* 31, 1–14.

Radlinski, A. P., Mastalerz, M., Hinde, A. L., Hainbuchner, M., Rauch, H., Baron, M., Lin, J. S., Fan, L., Thiyagarajan, P., 2004. Application of SAXS and SANS in evaluation of porosity, pore size distribution and surface area of coal. *International Journal of Coal Geology* 59, 245–271.

Radlinski, A. P., Radlinska, E. Z., 1999. The microstructure of pore space in coals of different rank, in: Mastalerz, M., Glikson, M., Golding, S. (Eds.), *Coalbed Methane: Scientific, Environmental and Economic Evaluation*. Springer, Dordrecht, the Netherlands, pp. 329–365.

Remeysen, K., Swennen, R., 2006. Beam hardening artifact reduction in microfocus computed tomography for improved quantitative coal characterization. *International Journal of Coal Geology* 67, 101–111.

Rodrigues, C. F., Lemos de Sousa, M. J., 1999. Further results on the influence of moisture in coal adsorption isotherms, *Proceedings of the 51st Meeting of the International Committee for Coal and Organic Petrology*. Romanian Journal of Mineralogy, Bucharest, Romania.

Rodrigues, C. F., Lemos de Sousa, M. J., 2002. The measurement of coal porosity with different gases. *International Journal of Coal Geology* 48, 245–251.

Rouquerol, J., Avnir, D., Fairbridge, C. W., Everett, D. H., Haynes, J. M., Pernicone, N., Ramsay, J. D. F., Sing, K. S. W., Unger, K. K., 1994. Recommendations for the characterization of porous solids (Technical Report). *Pure and Applied Chemistry* 66, 1739–1758.

Ruppel, T. C., Grein, C. T., Bienstock, D., 1974. Adsorption of methane on dry coal at elevated pressure. *Fuel* 53, 152–162.

Sakurovs, R., He, L., Melnichenko, Y. B., Radlinski, A. P., Blach, T., Lemmel, H., Mildner, D. F. R., 2012. Pore size distribution and accessible pore size distribution in bituminous coals. *International Journal of Coal Geology* 100, 51–64.

Schnablegger, H., Singh, Y., 2011. The SAXS Guide: Getting acquainted with the principles, 2nd ed. Anto Paar GmbH, Austria.

Seidle, J., 2011a. Coal permeability, in: J. Seidle (Ed.), *Fundamentals of Coalbed Methane Reservoir Engineering*. PennWell, Tulsa, Oklahoma, pp. 155–183.

Seidle, J., 2011b. Coal well pressure transient tests, in: J. Seidle (Ed.), *Fundamentals of Coalbed Methane Reservoir Engineering*. PennWell, Tulsa, Oklahoma, pp. 185–215.

Seidle, J., 2011c. Measurement of Coalbed Gas Content, in: J. Seidle (Ed.), *Fundamentals of Coalbed Methane Reservoir Engineering*. PennWell, Tulsa, Oklahoma.

Seidle, J. P., Kutas, G. M., Krase, L. D., 1991. *Pressure Falloff Tests of New Coal Wells*. Low Permeability Reservoirs Symposium, 15–17 April, Denver, CO.

Shen, J., Qin, Y., Wang, G. X., Fu, X., Wei, C., Lei, B., 2011. Relative permeabilities of gas and water for different rank coals. *International Journal of Coal Geology* 86, 266–275.

Speight, J. G., 2005. *Handbook of Coal Analysis*. J.D. Winefordner (Ed.), Vol. 166, John Wiley & Sons, Inc., Hoboken, NJ.

Straley, C., Rossini, D., Vinegar, H., Tutunjian, P., Morriss, C., 1997. Core analysis by low-field NMR. *Log Analyst* 38, 84–93.

Suuberg, E. M., Deevi, S. C., Yun, Y., 1995. Elastic behaviour of coals studied by mercury porosimetry. *Fuel* 74, 1522–1530.

Tang, L., Gupta, R., Sheng, C., Wall, T., 2005. The char structure characterization from the coal reflectogram. *Fuel* 84, 1268–1276.

Tang, J., Pan, Y., 2011. *Study of the Theory of Coalbed Methane NMRI, The Nuclear Magnetic Resonance Imaging for Coalbed Methane Reservoir Transportation*. Northeastern University Press, Shenyang, China.

Timur, A., 1969. *Pulsed nuclear magnetic resonance studies of porosity, movable fluid, and permeability of sandstones. Journal of Petroleum Technology* 21(06).

Torsæter, O., Abtahi, M., 2000. *Experimental Reservoir Engineering Labortory Work Book*. Norwegian University of Science and Technology, Trondheim, Norway.

Tricker, M. J., Grint, A., Audley, G. J., Church, S. M., Rainey, V. S., Wright, C. J., 1983. Application of small-angle neutron scattering (SANS) to the study of coal porosity. *Fuel* 62, 1092–1096.

Van Geet, M., Swennen, R., 2001. Quantitative 3D-fracture analysis by means of microfocus X-xay Computer Tomography (μCT): An example from coal. *Geophysical Research Letters* 28, 3333–3336.

Van Geet, M., Swennen, R., Wevers, M., 2000. Quantitative analysis of reservoir rocks by microfocus X-ray computerised tomography. *Sedimentary Geology* 132, 25–36.

Verhelst, F., David, P., Fermont, W., Jegers, L., Vervoort, A., 1996. Correlation of 3D-computerized tomographic scans and 2D-colour image analysis of Westphalian coal by means of multivariate statistics. *International Journal of Coal Geology* 29, 1–21.

Wang, G., Wang, K., Ren, T., 2014. Improved analytic methods for coal surface area and pore size distribution determination using 77K nitrogen adsorption experiment. *International Journal of Mining Science and Technology* 24, 329–334.

Wardlaw, N. C., McKellar, M., 1981. Mercury porosimetry and the interpretation of pore geometry in sedimentary rocks and artificial models. *Powder Technology* 29, 127–143.

Washburn, E. W., 1921. The dynamics of capillary flow. *Physical Review* 17, 273–283.

Wellington, S. L., Vinegar, H. J., 1987. X-ray computerized tomography. *Journal of Petroleum Technology* 39(08).

Wonik, T., Olea, R., 2007. *Borehole Logging, Environmental Geology*. Springer, Berlin, Germany, pp. 431–474.

Wood Jr, G. H., Kehn, T. M., Carter, M. D., Culbertson, W. C., 1983. *Coal Resource Classification System of the U. S. Geological Survey*. Dallas L. Peck (Ed.), U.S. Department of the interior, Denver, CO.

Xing, D., Yan, W., 2007. Analysis to pore structure of typical semicokes by mercury porosimetry. *Journal of North China Electric Power University* 34, 57–63.

Yao, S., Jiao, K., Zhang, K., Hu, W., Ding, H., Li, M., Pei, W., 2011. An atomic force microscopy study of coal nanopore structure. *Chinese Science Bulletin* 56, 2706–2712.

Yao, Y., Liu, D., 2012. Comparison of low-field NMR and mercury intrusion porosimetry in characterizing pore size distributions of coals. *Fuel* 95, 152–158.

Yao, Y., Liu, D., Cai, Y., Li, J., 2010a. Advanced characterization of pores and fractures in coals by nuclear magnetic resonance and X-ray computed tomography. *Science China Earth Sciences* 53, 854–862.

Yao, Y., Liu, D., Che, Y., Tang, D., Tang, S., Huang, W., 2009. Non-destructive characterization of coal samples from China using microfocus X-ray computed tomography. *International Journal of Coal Geology* 80, 113–123.

Yao, Y., Liu, D., Che, Y., Tang, D., Tang, S., Huang, W., 2010b. Petrophysical characterization of coals by low-field nuclear magnetic resonance (NMR). *Fuel* 89, 1371–1380.

Zhao, P., Lu, Q., Liu, Y., Xu, J., Jiang, H., Han, X., 2010. Significance of well testing of coalbed methane. *Well Testing* 19, 1–5.

Zhuang, H., 2013. Chapter 7—Coalbed methane well test analysis, in: Zhuang, H. (Ed.), *Dynamic Well Testing in Petroleum Exploration and Development*. Elsevier, Kidlington, Oxford, pp. 497–525.

Zhuang, H. N., 2012. *Dynamic Well Testing in Petroleum Exploration and Development*. Elsevier Science, Kidlington, Oxford.

Section II

Coal Mining and Production

5 Drilling and Blasting in Coal Mining

Celal Karpuz and Hakan Basarir

CONTENTS

Abstract: The basics features of both exploration (diamond) and blasthole drillings are introduced. The main operational parameters in drilling works are briefly described. The rock properties affecting the drillability and their relations to operational parameters are discussed. The basic features of both surface and underground works, and their design principles are described. The main types and properties of explosives are given. Initiation methods and devices such as electrical and nonelectrical systems are mentioned.

5.1 INTRODUCTION

Drilling is the act or process of making a circular hole in the rock with a drill. The main purposes of the drilling are higher core recovery in the case of core-drilling, higher penetration rate, and higher economy. Drilling serves many engineering works, each with its particular needs, such as mining, civil, petroleum, and natural gas engineering. In mining engineering, the fields of application can be prospecting, development, production, and grouting.

General drilling classification can be listed as follows:

1. Groundwater development/production
2. Mining engineering: exploration drilling including resource definition (for geological information including nature and strength of the materials and type of ore body such as vein, hydrothermal, disseminated, and large ore bodies), mine site investigation, rock bolt and foundation drilling, and blasthole drilling and drilling for access, ventilation, drainage, shaft, dewatering wells, disposal wells, and so on
3. Construction of building works
4. Oil, gas, and geothermal development/production holes
5. Subsurface storage and disposal access

Considering the energy source, drilling machines are divided into following groups: pneumatic drills (compressed air energy), hydraulic drills, electric motor-driven drills, internal combustion engine-driven drills, combined energy source, manpower, and steam power in early years.

The drilling system consists of three main parts such as drill (machine itself), rods or pipes, and bits. Penetration mechanism of the bit may be percussive, rotary crushing, and rotary cutting. Actually, the combination of percussion, rotating, and cutting action with geometric properties of the bit enables the rock penetration.

Depending on the size of the drilling, drills (drilling machines) can be divided into two groups as hand-held drills, that is, jackhammer, jackleg, and stoper and mounted drills, that is, crawler mounted and wheel mounted.

In this chapter, the basic features of the exploration drilling and blasthole drilling will be presented.

5.2 EXPLORATION (DIAMOND) DRILLING

5.2.1 Overview

Exploration drilling sometimes is also called *diamond drilling* or *diamond core drilling*. The drilling rig with required capacity turns the drill rods, by exerting a torque, which depends on the horse power and bit revolutions/min. The turning effect is transferred to the bit, which starts the cutting action under enough bit load. Essentially, a diamond drill has four requirements: (1) means for rotating the drill rods; (2) means for controlling the pressure on the bit and the downward motion of the drill rods and bit; (3) means for hoisting and lowering the drill rods; and (4) a pump for circulating water to the face of the bit (Cumming and Wicklund 1956).

In diamond core drilling, there are two basic applications. The first one is the conventional drilling, where all drill pipes together with the core barrel (in some cases, there may be double or triple core barrels) are removed to the surface, in each run. After removing the cores, those drill pipes and core barrel are lowered down to the hole to begin to the new run. The second one is the wire line drilling, which has many advantages over conventional drilling.

5.2.1.1 Wire Line Drilling

The main feature of wire line drilling (Cumming and Wicklund 1956) is the reduction in rod pulling. When the core barrel is full or a block occurs, an inner tube containing the core is detached from the core barrel assembly. The tube and core are pulled to the surface by a wire dropped down the line of drill rods. A latch or an *overshot assembly* snaps onto the top of the inner tube and is used for this purpose. The inner tube is rapidly hoisted to surface, within the drill rod string.

After the core is removed, the inner tube is dropped down into the outer core barrel and drilling resumes. To save time, a spare inner tube can be dropped into place and drilling can be resumed, while the core is removed from the original inner tube and it is cleaned and inspected.

The advantages of the wire line barrel over conventional system are particularly evident in deep work, and when drilling conditions allow for good footage per bit. Some of these advantages are (1) round-trip time for the retrievable inner barrel is only a fraction of the time for that of a string of drill rods, (2) lower round-trip time results in reduced *down time*, increased net drilling time, more round trips per shift, and higher footages per shift, (3) less caving in the hole—less core blocking, (4) longer core runs—higher core recovery, (5) longer bit life, (6) less fatigue for the operator, (7) less wear and tear on the drill motor and hoist, and (8) lower costs per foot of hole drilled.

The operation is as follows: drilling commences and proceeds in a customary manner. At the end of each core run, the string of drill rods and the outer barrel assembly remain in the hole, while the inner, core-retaining barrel is pulled up from the hole by use of the overshot assembly. The overshot assembly is lowered into the hole, inside the wireline rods, when retrieving the inner barrel. At surface, the overshot is released from the inner barrel and a spare inner barrel assembly is dropped into the hole and makes contact with the landing shoulder of the outer tube, locks in place, and drilling is resumed.

The string of drill rods is normally hoisted from the hole only when it is required to replace the diamond bit. It is generally necessary to keep the hole straight, when drilling vertically, and much time has been spent in determining proper bit contours and speed and rate of advance, designed to reduce the number of wedges required. Hydraulic feed machines are desirable because of better bit pressure control. The wire line hoist, for retrieving the inner barrel assembly, is a separate unit. As the depth of drill hole increases, then to save the energy consumption, the core sizes must also be decreased. In order to advance in the same hole, the drill pipes, bits, and casings (protect the hole from cavings) must be compatible (nesting with each other) with each other, and the core sizes generally follow a standard. A typical nesting is given in Figure 5.1.

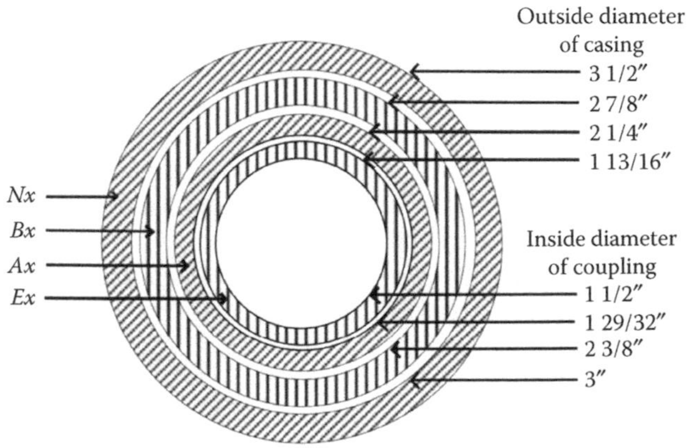

FIGURE 5.1 Nesting of different sizes of bits and casings. (From Cumming, J. and Wicklund, A. *Diamond Drill Handbook*, J.K. Smit & Sons Diamond Products, Toronto, Canada, 1956.)

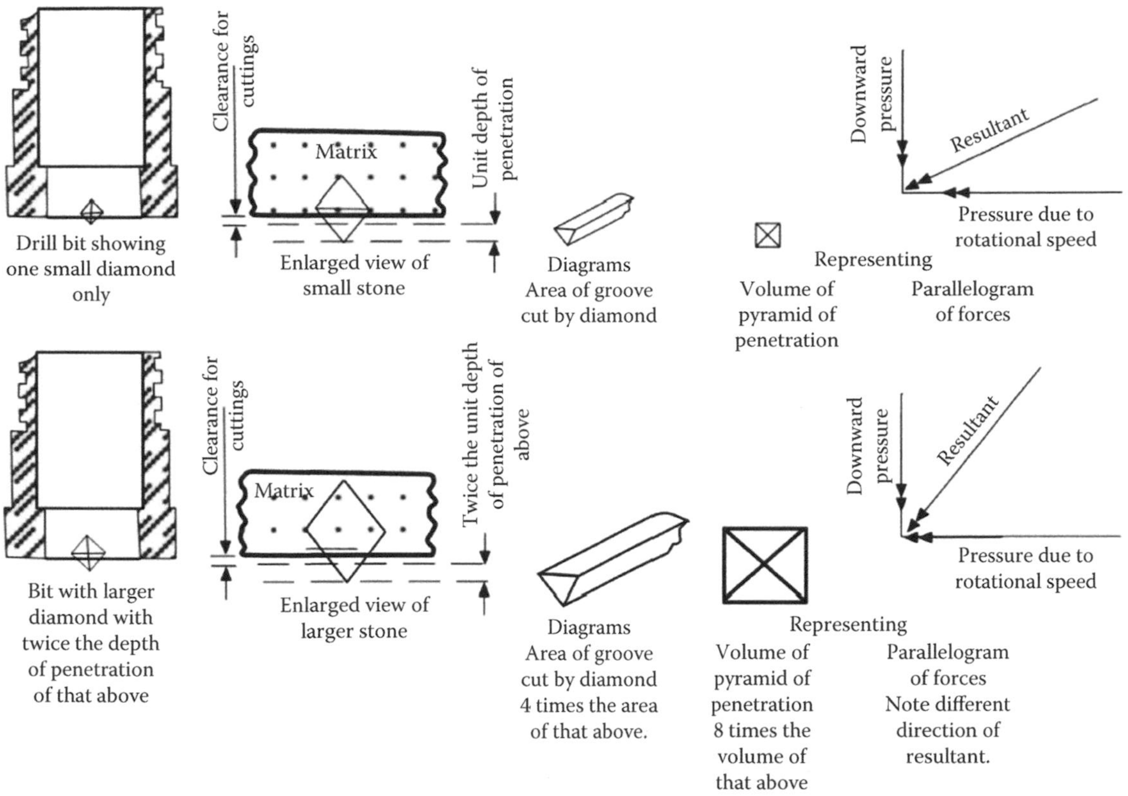

FIGURE 5.2 Cutting action of small- and large-size diamonds. (From Cumming, J. and Wicklund, A. *Diamond Drill Handbook*, J.K. Smit & Sons Diamond Products, Toronto, Canada, 1956.)

5.2.2 Cutting Mechanism

The basic principle in drilling, to be able to cause the propagation of cracks, is to subject the diamonds to pressures in excess of the strength of the rock. The active forces at the diamond-rock intact are twofold: force due to downward pressure and force due to torque exerted by the drilling rig. Forming the parallelogram of forces, one can find the active resultant force (Figure 5.2). The direction of the resultant force can be changed by the variation of speed and load. Thus, for a specific rock type, bit rotation and bit load are alternatively altered and a constant maximum rate of penetration is achieved. If the resultant pressure exerted on the rock surface is lower than the compressive strength of the rock, cutting will not commence and the diamonds are polished, without producing any cracks. Since the aim of core drilling is to achieve as much core as possible, the optimum penetration rate is not the practically achievable maximum penetration rate, but the maximum rate at which maximum core recovery with a minimum cost can be obtained.

5.2.2.1 Bits

In diamond drilling, bits are made of diamonds. Diamonds (pure carbon) are used for the cutting purposes, which are embedded into the matrix of the bit. The industrial diamonds are divided into four main categories (A, AA, AAA, and AAAA) according to their qualities. With wire line drilling equipment and especially in deep holes (<500 m), it is important to use high-quality stones. In diamond drilling, two kinds of diamond bits are utilized: surface-set and impregnated bits. The size of diamond stones set on a bit mainly depends on the type of formation to be drilled. Surface-set bits are mainly used for soft rock formations, while impregnated bits are preferred for harder rock formations. The weight of stones used per surface-set bit is expressed in terms of carats (1 carat = 0.202 g). For each carat, the number of stones used gives an idea about the size of the stones. This is expressed as stones per carat (spc). In surface-set bits, diamonds are embedded into the matrix and as the diamonds are eroded during drilling, they may be recuperated and sharpened for the next use. As the hardness and strength of rock increases, the size of the diamonds in bit decreases or spc values of the bit increases.

In impregnated core bits, the smaller diamond particles embedded to the matrix and as matrix is eroded during drilling, new diamonds are exposed and cut the rock. That is why it is most important that sufficient wear occurs to expose fresh diamonds to maintain penetration rates.

For surface-set diamond bits, the diamond particles set in the matrix generally cut with a ploughing action. In harder rocks, cracks are propagated via stress relaxation. These actions are illustrated, in Figure 5.2.

The specific energy concept is also very important in drilling and has also been studied by Miller and Ball (1990).

They related thrust, bit area, bit rotation, bit torque, and penetration rate with specific energy, introducing the following equation:

$$e = \frac{F}{A} + \frac{2\pi}{A} N \frac{T}{\text{PR}} \tag{5.1}$$

where:

e is the specific energy (kg/cm^2)
F is the thrust (bit load in kg)
A is the area of bit (cm^2)
N is the bit rotation (rpm)
T is the torque (kg-cm)
PR is the penetration rate (cm/min)

Miller and Ball (1990) have made studies relating to rock drilling with impregnated diamond bits and found a parabolic relation between thrust, specific energy, friction, and bit wear. The apex of the parabola shows the balanced bit wear with minimum specific energy; while the sides show either low-friction wear flats or high-friction stalling or seizure with higher specific energy values. This is shown in Figure 5.3.

By careful regulation of the operational parameters with respect to the properties of the rock cut, optimum penetration rates can be achieved. Reaching at optimum penetration rates needs very detailed information on operational parameters, such as bit load, bit rotation, mud circulation volumes, and pressures with bit wears and knowledge on the rock properties such as uniaxial compressive strengths, rock quality designations, discontinuity frequency, and quartz content. Some of these parameters are changeable, while some are not. Akun (1997) has tabulated these factors, as shown Table 5.1.

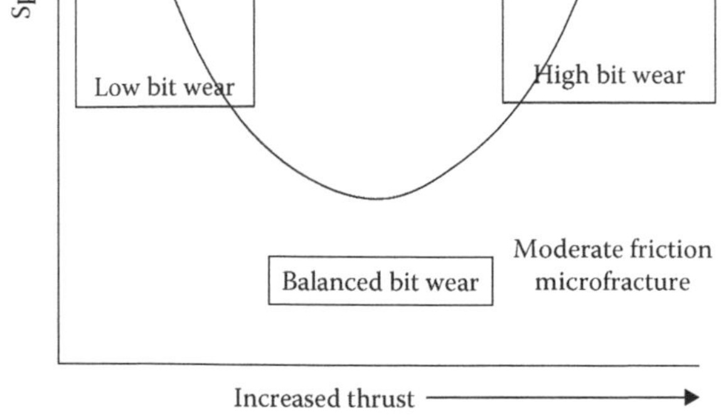

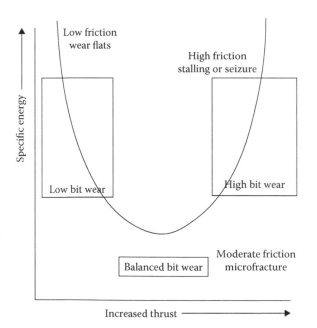

FIGURE 5.3 Thrust-specific energy and bit wear relations for impregnated bits. (From Miller, A. and Ball, A., Rock drilling with impregnated diamond microbits—An experimental study. *International Journal of Rock Mechanics and Mining Science Geomechanics Abstracts*, 27(5), 363–371, 1990.)

5.2.2.2 Operational Parameters in Diamond Drilling

Bit RPM: In diamond drilling, relatively high rates of rotation are applied. This rotative action cuts the rock continuously and results in penetration. Cutting speeds (C.S.) referred to the outside diameter of the bits are used as guidelines in drilling (peripheral speed). Cutting speeds V_s, of 180 to 540 ft/min (55–165 m/min) are considered for practical for surface-set drill bits. V_s-rpm relation in diamond drilling is important and these values surely depend on specific drilling conditions and in the case of surface-set bits, the diamond stones, which are embedded in the matrix, protrude a few thousands of an inch, so that very small cutting depths are obtained. The diamond stones on a surface-set bit are found by multiplying the number of carats by the average number of stones per carat. Those stones, which carry out actual cutting, are generally found by multiplying the total number of stones by 2/3, as advised by Christensen (1980).

The cutting speed commonly referred to either *outside diameter* (O.D.) or *mean bit diameter.*

$$\text{Cutting speed (peripheral, lineal)} = \text{RPM O.D. } \pi \tag{5.2}$$

For impregnated bits, cutting speed must be around 3–5 m/s. Higher cutting speed obviously causes higher penetration rate, but matrix wear is reduced, thus new diamonds are not exposed and the old diamonds become flat and polished. On the other side, in the case of low cutting speed, abrasion increases, especially with a high bit load, and diamonds are discarded before their useful life is completed, thereby reducing overall bit life.

For surface-set bits, cutting speed must be around 0.5–3.5 m/s. Since the cutting speed (peripheral) is related to the O.D. and I.D., cutting speed is different for outer diameter and inner diameter. Even in noncoring bits, centrally situated diamonds have zero cutting speed. Therefore, it must be kept in mind that for noncoring and wide kerf (cut) bits, C.S. differs too much. For example, for AQ bits, O.D. lineal speed is greater than I.D. speed by 76.6%. In this content, wireline bits have larger kerf; therefore, there is larger difference between lineal travel of the diamonds on I.D. and O.D. of the bit. Therefore, the presence of higher wear on the outside of a used wireline bit is an indication that the lineal speed on the ID of the bit is too low or at least not optimal. Therefore, possibly the diamonds on the intermediate diameter are the only stones cutting optimally.

5.2.2.3 RPM-Penetration Rate Relation

In the surface-set bits, the method of placing the stones into the matrix is so arranged that the stones continuously cut the same point. Thus, in one revolution, each cutting stone takes microscopic chips from the rock surface. This depth of cut per diamond per revolution is referred to as the *specific depth of cut.*

Christensen (1980) has derived the following equation, showing the relationship between penetration rate, number of cutting stones, specific depth of cut, and rotational speed for surface-set bits:

$$V_b = C \cdot a \cdot S \cdot n \tag{5.3}$$

TABLE 5.1
Some Important Parameters Affecting Drillability

Name of Parameter	Type of Parameter	Dependent On	Changeability	Seen At
Bit load	Operational	Bit diameter, carat load, SPC, diamond quality, matrix composition, and kerf area	Changeable	Bit
		Rig HP, hyd. cylinder areas, rock type, UCS, and RQD	Unchangeable	Rig
Bit rotation	Operational	Bit diameter and bit manufacture type	Changeable	Bit
		UCS, RQD, quartz content, discontinuity spacing, rock type, and rig HP	Unchangeable	Rock
		Length of drill string and length of drill rods	Changeable	Equipment
		Torque cap. of rods	Unchangeable	
Circulation volume	Operational	Pump cyl. diameter	Unchangeable	Pump
		Length of strokes		
		Annulus area	Changeable	Hole
		Crit. upward velocity	Unchangeable	
		Rock type	Unchangeable	Rock
		Bit waterways	Changeable	Bit
		Mud properties	Changeable	Mud
		Pump press. losses	Changeable	Hole
UCS	Rock	Geology	Unchangeable	Formation
RQD	Rock	Geology	Unchangeable	Formation
Discontinuity frequency	Rock	Geology	Unchangeable	Formation
Abrasivity	Rock	Geology	Unchangeable	Formation

Source: Akun, M.E., *Effect of Operational Parameters and Formation Properties on Drillability in Surface Set Diamond Core Drilling*. PhD thesis, Middle East Technical University, Ankara, Turkey, 1997.

where:

V_b is the penetration rate (m/h)
C is the conversion factor, 6×10^{-2}
a is the number of cutting stones on average diameter (usually 1/5 or 20% of total number of stones on a bit is taken)

For a certain rock formation and for a certain bit, a and S are theoretically constant. Therefore, the penetration rate becomes proportional to the rotational speed only.

Thus, C.a. S = constant and = C_1, and

$$V_b = C_1 n \text{ or, } \log V_b = \log n + \log C_1 \tag{5.4}$$

This relation shows that in a log–log system, the function of the penetration rate versus revolution is a straight line with 45° slope. It simply means that doubling of revolution will double the penetration rate under ideal conditions.

In order to find the rotational pressure, the active force and bit radius is necessary. The force in turn is calculated by using the following equations:

$$T = \frac{(60 \times 75 \times \text{HP})}{2\pi N} \tag{5.5}$$

where:

T is the created torque at the Chuck (kg-m)
HP is the horse power of the drill rig (HP)
N is the rotational speed of bit, rpm
$F = T/r$, where F is the rotational force (kg) and r is the bit radius (m)

Special emphasis should be given to factors that limit the use of rotation as a means of increasing penetration rates. However, there are some limitations: (a) torque capacity of rods, (b) hole straightness and stability, and (c) length of drill string.

Heinz (1985) investigated the effect of the length of drill string on the bit rotation and his suggestions have been tabulated in Table 5.2.

TABLE 5.2
Critical RPM for Drill Rods

Drill Rod	OD (mm)	ID (mm)	Rpm$_c$ for 6 m	Rpm$_c$ for 3 m
AQ	44.5	34.9	190	760
BQ	55.6	46.0	243	970
NQ	69.9	60.3	310	1241
HQ	88.9	77.8	397	1588
FQ	114.3	103.2	518	2070

Source: Heinz, W.R., *Diamond Drilling Handbook*, South African Drilling Association, Johannesburg, South Africa, 1983.

5.2.2.4 Bit Load

In surface-set bits, considering the fact that all the stones in the matrix may not be cutting considerably, small penetration rates are obtained. Hence, for the surface-set bits,

Bit load = number of active diamonds ×
load bearing capacity of a single dimond stone

Moppes (1986) formed the following empirical equation for surface-set bits.

$$BL = \frac{2}{3} \times CL \times SPC_{av} \times BL_{sp} \tag{5.6}$$

where:
BL is the bit load (kg)
CL is the carat weight
SPC_{av} is the average number of stones per carat
BL_{sp} is the supposed strength of diamond stone (kg/stone)

Here, the strength of the diamond stones is also important, and then the following must be satisfied.

$$\sigma_R \le \frac{P}{bf} \le \sigma_D \tag{5.7}$$

where:
σ_R is the strength of rock (kg/cm^2)
σ_D is the strength of diamond (kg/cm^2)
P is the bit load (kg)
b is the number of active diamonds (2/3 of total stones)
f is the contact surface (cm^2)

Bit load is also limited with the core barrel and the load transfer capacity of rods. Excessive load damages the bit stones. Low bit load with high RPM causes higher polishing. Impact loads causes more damage on stones.

Moppes (1986) suggested the following kg/stone values for different varieties of diamond as follows:

VM Bortz = 3.178 kg/stone, Congo = 2.270 kg/stone, and Carbonado = 6.810 kg/stone. In the case of the diamond, which is a perfect sphere, the contact surface remains constant throughout its life.

For impregnated bits,

$$BL = \text{kerf area, in cm}^2 \times 100 \text{ kg/cm}^2$$
$$\left(\text{kerf area is calculated using outer and inner diameters of bit}\right)$$

5.2.2.5 Circulation (Flushing) Medium

As drilling progresses, the formation is broken into small chips, which must be removed out of the drill hole for the continuation of drilling. The chip removal is made possible by the circulation of water or mud (together with chemical additives) through the drill rods, which carry the chips upward and outside the hole into the mud pits through the hole annulus.

The other function of drilling medium are (1) cooling of the diamonds and (2) to overcome the formation pressure in the hole, thereby preventing the small cavings from borehole wall (Cumming and Wicklund 1956).

Although being very strong, excessive heat can easily burn the diamonds, consisting of pure carbon. Thus, required flushing amount must be continuously applied. The minimum amounts necessary for cooling are much less than those required for the removal of cuttings from the hole. Therefore, studies are concentrated on the second item.

Annulus velocity, V (ft/min)

$$V = 170\sqrt{\frac{d_c(\gamma_r - \gamma_m)}{\gamma_m}} \quad \text{for spherical particles} \tag{5.8}$$

$$V = 133\sqrt{\frac{t_c}{d_c}}\sqrt{\frac{d_c(\gamma_r - \gamma_m)}{\gamma_m}} \tag{5.9}$$

where:
d_c is the average diameter of particles (in.)
t_c is the thickness of flat particles (in.)
γ_r is the specific gravity of rods
γ_m is the specific gravity of fluid

Experience has shown that annulus velocities should lie between 0.3 and 0.5 m/s, in order to prevent the disintegration of cores. The circulation rate necessary to create this velocity depends on the rods (conventional and wireline) and the annulus area (cased holes-uncased holes).Water is the common fluid employed in diamond drilling and bentonite is also added in some cases.

The factors affecting to cutting removal can be listed as follows: (1) size and shape of the cuttings, that is, spherical, flat, and so on; (2) specific gravity of rock; (3) specific gravity of the fluid; and (4) viscosity of the fluid (ft/min).

After determining the required annulus velocity, the required quantity, Q, can be calculated as follows:

$$Q_{mean} = \frac{V_{an}S}{1000} \cdot 60 \tag{5.10}$$

where:
Q is the average required water quantity, L/min
S is the annulus area, cm^2
V_{an} is the annulus velocity, m/s

Fluid pressure losses in diamond drilling at shallow depths are not very high. Therefore, few hundred psi pump pressures are usually more than enough. However, in some special cases, especially drilling in clayey formations, more pressure may be needed because of the sticky character of the formation.

There is a certain relation between pump pressure, delivery, and hydraulic horsepower (HP) of the pump.

$$HP = \frac{QP}{1714} \tag{5.11}$$

TABLE 5.3
Drilling Guidelines for Longyear Impregnated Bits

System	Fluid Volume Range, L/min	Rotation Speed, RPM	Penetration Rates for Series 1–9 ± 10%, cm/min	Penetration Rates for Series 10 ± 10%, cm/min	Indicative Bit Weight Range, kg
LTK46	10–13	2300	26	14	1360–4500
		1400	16	8	
		1000	11	6	
AQ	15–19	2000	22	12	910–2260
		1200	13	7	
		850	9	5	
LTK56	10–13	1700	19	10	910–1810
		1000	11	6	
		700	8	4	
BQ	23–30	1700	19	10	910–2260
		1000	11	6	
		700	8	4	
NQ CHD76	30–38	1350	15	8	1360–2720
		800	9	5	
		550	6	3	
HQ	38–45	1000	11	6	1810–3620
CHD101		600	7	4	
		400	4	2	
PQ CHD134	68–87	800	9	5	2260–4530
		500	6	3	
		350	4	2	

Source: Longyear, *Q-CQ-CHD Mini Manual*, Longyear, Mississauga, Canada, 1989.

where:

HP is the hydraulic horse power of the flush pump
Q is the delivery of the pump gallons/min
P is the discharge pressure of the pump in psi

Flush pumps are employed and capacity of pumps is generally 20–50 g/min at 500–1000 psi.

When impregnated bits are in use, Longyear (1987) suggested the following operational parameters (Table 5.3) for their different impregnated bit series bits, expressing rotational speeds in terms of rpm (revolutions per meter).

5.3 BLASTHOLE DRILLING

Blasthole drilling is utilized both for the production and development works of coal. The explosives are replaced in the blastholes and then detonation is started.

5.3.1 Percussive Drilling

In this system, the dominant force is the thrust force and there are two main drilling systems working with this principle: top hammer drilling equipment and down-the-hole (DTH) drilling (in the hole).

Percussion systems consist of four main drilling parameters, which utilize the energy to penetrate the rock (Sandvik-Tamrock 1999). These are as follows:

1. *Percussion power (percussion energy and frequency)*: It is mainly provided by piston, which is inside the drill and the main part converting the energy from its original form (fluid, electronic, pneumatic, or combustion engine drive) to mechanical energy to activate the system. The commonly used sources are pneumatic and hydraulic energies.

   Compared to pneumatic drills, hydraulic drills have higher percussion power and hence faster penetration rate is achieved. They are more efficient, more reliable, provide constant penetration rate, can easily be adjusted to different rock types, smoother drilling is obtained, more ergonomic, and more economic. Pneumatic drilling systems have the impact frequency of between 1600 and 3400 hits/min, while hydraulic systems have 2000–4500 hits/min.
2. *Feed force*: Feed force, sometimes called the *shank adapter*, is the force required to keep the shank in contact with the drill and bit and hence to keep the bit in contact with the rock. It ensures the maximum impact energy transfer from pistons to drill rods (pipes) and provides the rotation torque.

   Optimum feed force depends on the percussion pressure level, rock condition, hole depth, drilling angle, and the size and type of drill steels. Optimum feed pressure level is observed by monitoring the penetration rate, and for broken rock, the percussion power and feed force.

3. *Bit rotation*: The main purpose of bit rotation is to rotate the drill bit between consecutive blows. Because, after each blow, the drill bit must be turned to ensure both giving the chance to clean the chips by flushing medium and to allow the bit to meet the unbroken rock. As it is mentioned above, one of the main purposes of the drilling is to provide the maximum penetration rate; the factors affecting the optimum rotation to provide the maximum penetration are rock properties, drill frequency, bit type, and bit diameter. Tamrock (1984) suggested the following formula to calculate the rotation for top hammer drilling:

$$n = \frac{(S \times f \times 60)}{\pi d} \tag{5.12}$$

where:
n is the bit rotation (rpm)
S is the gauge button travel distance between consecutive blows (mm)
f is the impact frequency of the hammer (1/s)
d is the bit diameter (mm)

The rpm is closely related to the bit type (such as cross and button bits) and for the smaller hole sizes (35–51 mm), the cross bits need 5%–10% lower values compared to button bits, while for the larger hole diameters (76 mm and larger), cross bits need 5%–10% higher rpm values suggested by Tamrock (1984).

Insufficient (less) rpm values result in energy loss due to recutting the rocks and hence giving lower penetration rate, and excess rpm values result in excessive bit wear since rock cutting is dominantly carried out by rotation instead of the rpm-thrust combination.

4. *Flushing*: It is used both to remove the cuttings from the drill hole and to cool the drill bit. The flushing medium can be air, water, and foam. Insufficient flushing gives low penetration rate, and high bit wear since increase in recutting.

Higher flushing, in air flushing, causes bit and steel rods erosion by the "sand-blasting" effect. In water flushing, penetration rate decreases since before water acts as a cushion effect against the bit.

Bits are the end part of the drilling system and the bit applies the energy to rock to penetrate it. They are either made of strengthened metals or harder materials such as diamonds.

5.3.1.1 Types of Surface Blasthole Drills (Drilling Machines)

5.3.1.1.1 Top Hammer Drilling

In top hammer drilling, penetration is obtained by the combination of percussion (impact frequency and hence produced energy, which depends on air or hydraulic pressure, area of piston, stroke number of piston, and weight of piston) feed force, rotation, and flushing but the percussion is provided at the top of drill by means of a hammer.

The top hammer's piston hits the shank adapter and creates a shock wave, which is transmitted through the drill string to the bit. The energy is discharged against the bottom of the hole and the surface of the rock is crushed into drill cuttings. These are in turn transported to the surface by means of flushing air that is supplied through the flushing hole in the drill string. As the drill bit is rotated, the whole bottom area is worked over. The rock drill and drill string are arranged on a feeding device. The feed force keeps the drill bit constantly in contact with the rock surface in order to utilize the impact power to the maximum.

The top hammer equipment can work either by hydraulic rock drills or pneumatic rock drills. In the top hammer drills, percussion may reach 2000–4500 blows/min in hydraulic drifters. Rotation, which provides to turn the bit to a new position for a new blow, may be in the range of 80–250 rpm and flushing (dominantly air flow) velocity to remove the cuttings is around 12–20 m/s. It is suitable for medium-hard to hard rocks. The hole diameter is up to 9 in. Integral drill steel chisel bits (button or brazed insert bits) are the commonly used bit types. The drilling machines have different sizes such as starting from hand-held rock drills (6–7 atm. air pressures): hand-held rock drills with feed devices for manual operations, by the use of extension drill steels; light drills mounted on a feed device; light crawler-based drill rigs with boom, feed, and medium heavy drifter; and heavy crawler-based or wheel-based drill rigs with boom (or mast), feed, and heavy drifters.

5.3.1.1.2 Down-the-Hole Drilling

In this drilling system, percussive hammer works in the hole during drilling. It means that the power source drill piston directly strikes the bit. Penetration is obtained by the combination of percussion, feed, rotation, and flushing as similar to the top hammer drilling but the percussion is given at the bottom of the hole. Since drill piston directly strikes the bit, the energy lost is minimum compared to top hammer drilling. It also provides constant penetration rate irrespective to hole depth and better hole accuracy and less wear and damage to drill steels (rods) and bits. On the other side, DTH drills have poor mobility and low penetration rate (sometimes three times), since they need large separate compressor. The energy consumption is large and they need larger hole diameters compared to top hammer drills. Usual hole depths for blastholes are 12–20 meters, but down to 150 m drilling depth is possible.

DTH and top hammer may have both pneumatic hammer with pneumatic rotation motor and pneumatic hammer with hydraulic rotation motor.

In percussive system, up to 2100 rpm can be applied and minimum flushing air velocity should be 3000 fpm (14 m/s). But in the case of rock having higher specific gravity, or breaking into large chips or the ground is muddy or if there is excessive groundwater then 5000 fpm bailing velocity can be applied. Most common bit sizes are 4″–6 ½″, but sizes up to 12.5″ can also be used. Most common bit type is flat-faced button bit.

Comparison of the penetration rates of top hammer and down-the-hole drilling (Tamrock 1984) is given in Table 5.4.

TABLE 5.4
Comparison of Penetration Rates of Top Hammer and DTH Drills

	Top-Hammer Drill (Pneumatic)	DTH Drill
Type of rock	64 mm (2.5″)	152 mm (6″)
Barre granite	23 m/h (75 fph)	7.6 m/h (25 fph)
Granite (Idoha)	25 m/h (83 fph)	8.5 m/h (28 fph)
Limestone (Colorado)	18 m/h (59 fph)	6.1 m/h (20 fph)
Limestone (Pennsylvania)	20 m/h (67 fph)	6.7 m/h (22 fph)
Limestone (Iowa)	41 m/h (137 fph)	13.7 m/h (45 fph)
Andesite (Washington)	29 m/h (95 fph)	9.8 m/h (32 fph)
Quartzide (Minnesota)	13 m/h (42 fph)	4.3 m/h (14 fph)

Source: Tamrock, *Handbook on Surface Drilling and Blasting*, Tampere, Finland, 1985.

5.3.1.1.3 Rotary Drilling

Rotary crushing is a drilling method, which was originally used for drilling oil wells, but is nowadays also employed for blasthole drilling in large open pits and in harder species of rock.

All rotary crushing drilling requires high feed pressure and slow rotation. The relationship between these two parameters varies with the type of rock. For softer rock, lower pressure and higher rotation rate is applied. Suitable hole diameters may change from 5″ (127 mm) to 17″ 1/2 (441 mm) and up to 40–70 m deep holes can be penetrated. Small and large hole rotary cutting drills, especially applied for small-scale and metal mining, large hole rotary cutting is generally applied for coal mining. In rotary drilling, 1 cm/min in hard rock and 3 cm/min in soft rock penetration rate can be achieved.

Flushing: Bailing velocity (V) should be in the range of 30–45 m/s and can be calculated by using the following equation (Tamrock 1984):

$$V = \frac{183.3\text{CFM}}{D^2 - d^2} \tag{5.13}$$

where:

D is the bit diameter
d is the drill steed (rod) diameter
CFM is air requirement (ft^3/min).

Four key elements involved in rotary drilling:

1. Sufficient torque in the drill stem drive to turn the bit in any strata encountered.

$$T = \frac{\text{Rotary motor HP} \times 5250}{\text{RPM}} \text{ft} - \text{lb} \tag{5.14}$$

2. Sufficiently high load on bit (pull down force or thrust). For optimum penetration $P = (0.9 \times D) \times 1000$ lb.
3. Sufficient air-water volume to remove the chips and to cool the bit bearings.
4. Selection of a proper type of bit for the ground to be drilled.

Rotation is usually around 40–100 rpm for coal and soft rocks.

The rotary roller bits are usually categorized according to their use as *soft*, *medium*, and *hard* formation types. Generally, large tooth is for soft formation, whereas small tooth is for hard formation. Tricone bit is used in rotary crushing type of drilling.

Tricone bits for hard rotary crushing and drag bits for soft rock rotary cuttings, respectively, are applied for coal and its host rock. The rotary bit design parameters are given in Table 5.5.

The suggested operational parameters and bit types for rotary drilling system with respect to rock properties are presented in Table 5.6.

The relationship between penetration rate, operational parameters, and uniaxial compressive strength are given in Figures 5.4 and 5.5. These relations were derived from the research carried out on the rotary blasthole drilling operation for both drag and tricone bits for the coal measure rocks for the Turkish Coal Enterprises open cast mines (Karpuz et al. 1990).

TABLE 5.5
Rotary Bit Design Parameters

Type of Ground	Tooth or Insert Spacing	Tooth Depth or Insert Projection and Size	Cutting Action
Soft formations with low compressive strengths and high drillability: shales, unconsolidated sands, and calcites	High	Large: inserts extended chisel shaped	Mostly gouging and scraping by skew cone action, with little chipping and crushing
Medium formations: harder shales, limestones, sandstones, and dolomites	Medium close	Medium: inserts short or blunt chisel shaped	Partly by gouging and scraping but with significant chipping and crushing action, especially at harder end of type
Hard formations: siliceous limestones, hard sandstones, and porphyry copper ores	Close with low intermesh	Low: inserts domed, hemispherical, or conical	Mostly by chipping and crushing by cutter rolling action
Very hard rocks such as taconites and quartzites	Very close with low intermesh	Very low: insert hemispherical, conical, or ovoid	Nearly all excavation by true rolling action of cutters

Source: Tamrock, *Handbook of Surface Drilling and Blasting*. Painofaktorit, Tampere, Finland, 1984.

TABLE 5.6
Suggested Bit Type, RPM, and Thrust Values Based on Rock Properties

	Rock Type	Proper Bit Type	RPM	Thrust
1	Soft rock	Roller bits with large teeth drag bit	70–100	1000–3000 lb/in. of diameter (0.5–1.4 tons/in.)
2	Medium rock	Roller bits with medium size teeth	50–70	3000–5000 lb/in. of diameter (0.14–2.3 tons/in.)
3	Hard rock	Roller bits with small size teeth	40–60	4000–7000 lb/in. of diameter (1.8–3.5 tons/in.)

Source: Tamrock, *Handbook of Surface Drilling and Blasting*, Painofaktorit, Tampere, Finland, 1984.

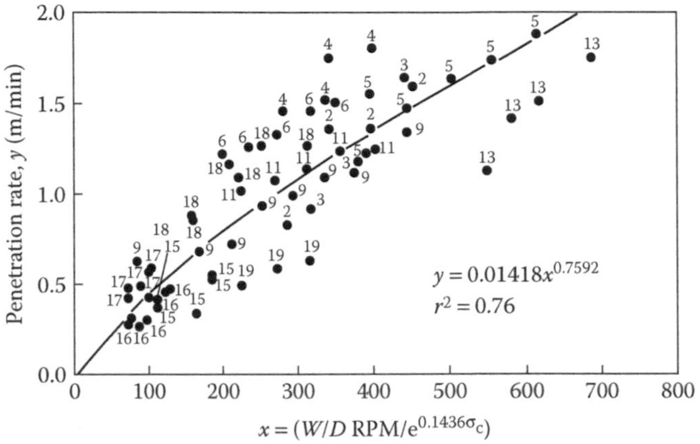

FIGURE 5.4 The interaction between rock properties and penetration rate. (Data from Karpuz, C. et al., Drillability studies on the rotary blast hole drilling of lignite overburden series. *International Journal of Surface Mining and Reclamation*, 4, 83–93, 1990.)

5.3.1.2 Types of Blasthole Drills (Drilling Machines)

Drilling machines and their applications can be shortly summarized as follows:

1. Handheld drills:
   a. *Jackhammer* (sinker): For general utility purpose such as anchor holes (short vertical holes for bolts and anchors), pin holes (short horizontal holes), and shaft sinking and winzing
   b. *Jackdrill* (jackleg): For mine development (in small tunnels and drifts) and production (room and pillar, cut, and fill stoping)
   c. *Stoper*: For mine development in raises, for production purpose of shrinkage, cut and fill stoping, and for supporting such as in rock bolting
2. Drifter mounted machines (drill jumbos): It is a heavy and powerful drill; so it is usually mounted on a wheel or crawler chassis; a *drill jumbo* is a drilling unit equipped with one or more rock drills and mounted on a conveyance. There are many types of jumbos. Jumbos range from single-drill and ring drills mounted on simple skeeds to sophisticated multiple drill units mounted on powered carriers and equipped with automatic controls. Drifters are mainly used for horizontal drilling (i.e., development of drifts, cross-cuts, and in driving tunnels).
3. DTH drills
4. Rotary blasthole drills

Types of drill steels and bits are summarized in Tables 5.7 and 5.8.

Drill holes are characterized by four factors such as diameter, length, deviation, and stability. Considering these factors, the suitable drilling equipment can be selected. Tamrock (1984) suggested the drill equipment selection guide for surface works, as given in Figure 5.6 and Table 5.9.

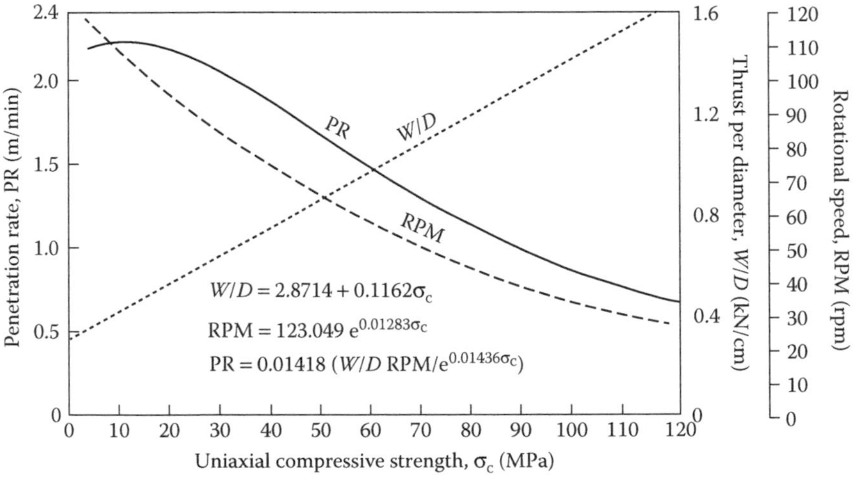

FIGURE 5.5 The relationship between penetration rate and operational parameters and uniaxial compressive strength. (Data from Karpuz, C. et al., Drillability studies on the rotary blast hole drilling of lignite overburden series. *International Journal of Surface Mining and Reclamation*, 4, 83–93, 1990).

TABLE 5.7
Types of Drill Steels and Bits

	Percussion Drilling
I.1. Single pass system	Hand held drills
	Light mounted drifters (light jumbos)
II.1. Multiple pass system	Heavy mounted drifters (heavy jumbos)
	Underground DTL drills
	Rotary Drilling
I.1. Single pass systems	Handheld drills
	Light rotary drills
II.2. Multiple pass system	Heavy jumbos
	Underground rotary blasthole drills
	Surface rotary blasthole drills

Source: Tamrock, *Handbook of Surface Drilling and Blasting*, Painofaktorit, Tampere, Finland, 1984.

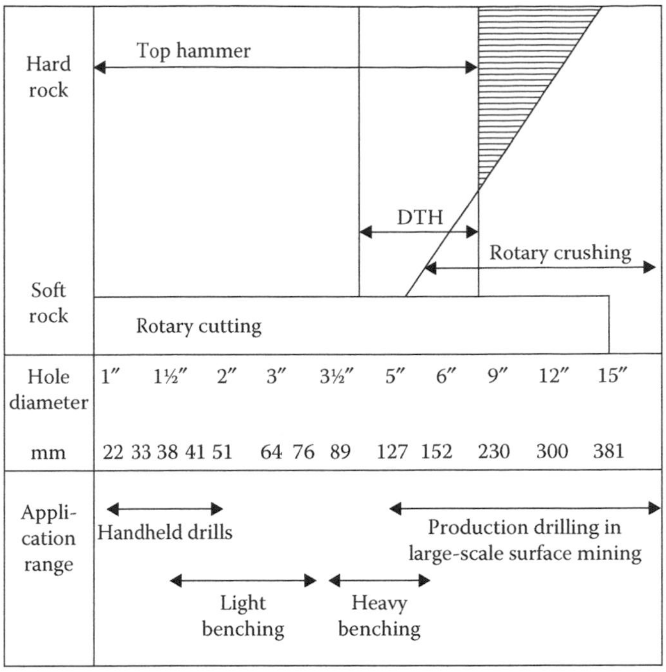

FIGURE 5.6 Drilling equipment selection guide. (From Tamrock, *Handbook of Surface Drilling and Blasting*. Painofaktorit, Tampere, Finland, 1984.)

5.4 BLASTING

5.4.1 Introduction

The use of explosives is the alternative to the mechanical way of rock breakage. Commercial rock breakage, which utilizes chemical explosives, is called *blasting*. The understanding of the blasting process (breakage mechanism) and its relation to the rock properties is the key concept to develop an efficient blasting design. And thus, the best results in terms of the desired particle size distributions, the placement of muck piles for ease of rock removal and handling, and control of ground vibrations, is achieved.

An explosive, or blasting agent, is a compound or a mixture of compounds, which, when initiated by heat, impact, friction, or shock, is capable of undergoing a rapid decomposition,

TABLE 5.8
Types of Drill Bits Used in Drilling Work

I. Percussion Drill Bits	**II. Rotary Drill Bits**
Medium hard to very hard rocks	Soft to very hard rocks
I.1. Chisel bit with integral steel, handheld drills, and light jumbos (5 = 23–42 mm, L = 0.8–3.7 m) can be resharpened, tungsten carbide (WC) inserts	II.1. Drag bits (WC inserts) (soft RX) small and large diameters <2 in., up to 17 in. handheld heavy rotary drill rigs light jumbos (also PDC) PDC up to 6″ in small<*P* drag bits
I.2. Brazed bits with extension steel tungsten carbide (WC) inserts 1. Cross-bit: insert located at 90° angles ϕ = 35–64 mm usually 2. X-bit: inserts located at 80°–100° ϕ = 51–76 mm usually—can be resharpened	II.2. Diamond bits (embedded in bronze matrix) Exploration drilling types: surface set and impregnated matrix wear Matrix: resist abrasion matrix wear diamond tones: 10–80 spc 80–1000 spc Rock: soft to medium and medium to very hard
I. 3. Button bits with extension steel top hammer + DTH drills φ = 51 to 204 mm Resharpening became possible in the last few years	II.3. Roller cone bits Three cones mounted on rolling bearings Soft rods: Hardened steel teeth Hard rods: Cemented WC inches Diameter: >5″ (127 mm)
I. 4. Polycrystalline diamond compacts Higher hardness than WC bits Lower bit wear and higher performance (Doubled bit life and 45% higher drilling rate) Susceptible to brittle fracture (limit the rock types to be drilled) More expensive (2 or more times)	

Source: Tamrock, *Handbook of Surface Drilling and Blasting*, Painofaktorit, Tampere, Finland, 1984.

TABLE 5.9
Blasthole Drilling Equipment Selection Chart

	Top-Hammer	Down the Hole	Rotary Crushing	Rotary Cutting
Rock Types	Medium to very hard	Medium to very hard	Soft to very hard	Soft to medium
Hole size range	Up to 9″	3″1/2 to 8″	5″ to 17″	1″ to 17″
Hole depth (max)	25 m	150 m	–	–
(normal)	15 m	20–60 m	70 m for blasthole drilling	70 m
Operational parameters				
Percussion	2000–3500 blows/min			
Rotation	80–250 rpm	15–30 rpm (minimum 0 and maximum 50 rpm)	Soft RX 70–100 rpm, medium RX 50–70, hard RX 40–60	70–1000 rpm
Flushing velocity (air/water)	12–20 m/s (2400–4000 fpm)	14–25 m/s (3000–5000 ft/m)	35–45 m/s (6000–9000 fpm)	30–45 m/s
Pulldown force (thrust)	–		Soft RX 1000–3000 lb/in of diameter Medium RX 3000–5000 lb/in Hard RX 4000–7000 lb/in	1000–3000 lb/in-diameter

Source: Tamrock, *Handbook of Surface Drilling and Blasting*, Painofaktorit, Tampere, Finland, 1984.

releasing tremendous amounts of heat and gas. The decomposition is a self-propagating exothermic reaction called as an *explosion*. The stable end products are gases that are compressed, under elevated temperature (as high as 4000°C) to very high pressures (up to 100,000 atm), and it is these pressures that cause the rock to be fragmented (Hartman 1992).

All commercial explosives are mixtures of carbon, hydrogen, oxygen, and nitrogen. The maximum energy release upon detonation occurs when the explosive mix is formulated for oxygen balance. An oxygen-balanced mixture is one in which there is no excess or deficiency in oxygen, such that the gaseous products formed are chiefly water vapor, carbon dioxide, and nitrogen. In actual blasting practice, small amounts of noxious gases such as nitric oxide, carbon monoxide, ammonia, methane, and solid carbon, are formed resulting in nonideal detonations. The work done by chemical explosives in the fragmentation and displacement of rock depends on the shock energy as well as the energy of the expanding gases.

Deflagration is the chemical burning of explosive ingredients at a rate well below the sonic velocity. It is associated with heat only and carries no shock. Deflagration occurs when less than ideal hole-loading conditions or explosive formulation are involved.

Blasting is an empirical approach, which means derived from experience; it needs always blast evaluation after blasting work and needs fine-tuning and modifications continuously, although there are some basic features. Then it is purely *site-specific work*, and it means that somebody cannot directly apply the same blasting work somewhere, even if the rock conditions are similar.

There are many empirical relationships proposed by different investigators used in the design of blasting works. This chapter is mainly based on the Ash's (1963 in Hartman 1992) approach for surface (bench) blasting works given in the SME handbook (Hartman 1992) and Nitro Nobel Company's (in Sweden) approach given their seminar notes and British approach for coal.

This chapter mainly includes short reviews of types and properties of explosives used for industrial blasting and the application principles of surface and underground blast designs, rather than breakage mechanism of the explosives.

5.4.2 Types of Industrial Explosives

Industrial explosives are classified as one of the following:

(1) Nitroglycerin based, (2) dry blasting agents, (3) water gels, (4) emulsions, (5) permissibles, (6) primers, and (7) boosters. Often, the difference among these products is formulation; however, product packaging and consistency can also change a classification. The U.S. Bureau of Mines (Hartman 1992) reported the relative consumption of these products as a percentage of total industrial explosive use and consumption by industry. Ammonium nitrate fuel oil (ANFO) and other powder explosives are called as *blasting agents*, and need other high explosives to initiate the blasting work. These consumptions are shown in Figure 5.7.

5.4.3 Properties of Explosives

The main characteristics of explosives are as follows:

Fume class: The fume class is a measure of the toxic gases in ft^3/0.44 lb (200 g) of unreacted explosive. The U.S. Bureau of Mines limits the volume of poisonous gases produced by permissible explosives (those used in underground coal and other gaseous mines) to 2.5.

Density: The density of an explosive is defined as the weight per unit volume or the specific gravity. Commercial explosives range in density from 0.5 to 1.7. Explosives with a density less than 1 will float in water. Therefore, in water-filled holes, an explosive with a density greater than 1 is required. However, for water-based explosives, this is not the case, and often the reverse is true. Density is most useful in determining the loading density or the weight of explosives one can load per unit length of borehole (in lb/ft or kg/m), and is calculated in English units as

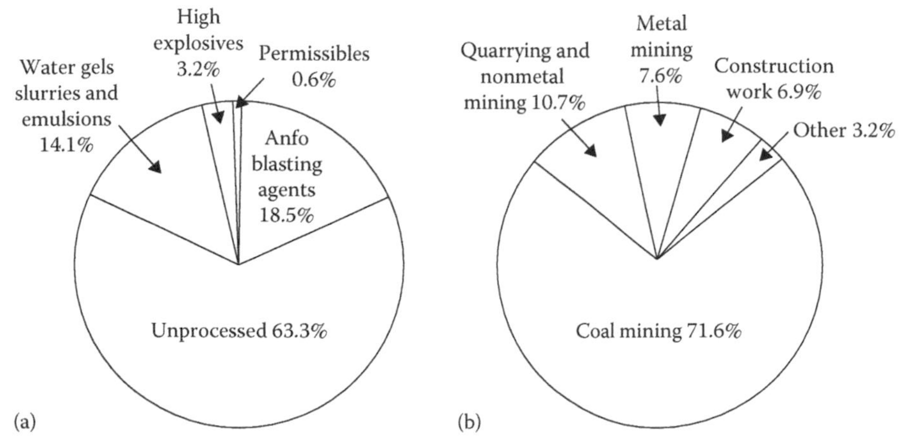

FIGURE 5.7 Industrial explosives and blasting agents sold in the United States for consumption in the United States (a) by classification for 1988 and (b) by use for 1987. (Data from Hartman, L., *SME Mining Engineering Handbook*. Englewood, CO: Society for Mining, Metallurgy and Exploration, 1992.)

$$\text{Loading density} = 0.3405\,\rho D^2 \qquad (5.15)$$

where:

ρ is density
D is explosive column diameter in inches, or can be directly calculated by the used volume of the borehole and the density of the explosive

Most ANFO has a poured density of about 0.78–0.85 g/cm^3. Pneumatic loading with some types of field air-loading equipment can increase the density of ANFO. At densities above 1.2 g/cm^3 the sensitivity of ANFO rapidly decreases. It is generally agreed that ANFO products with a density greater than 1.2 g/cm^3 will not detonate efficiently and may not initiate.

Water resistance: The ability of an explosive to withstand exposure to water for long periods of time without loss of strength or ability to detonate defines the water resistance. The presence of moisture in amounts greater than 5% dissolves chemical components in dry blasting agents and alters the composition of gases produced, contributing to the formation of noxious fumes and lower energy output. Explosive manufacturers individually rate products based on a relative basis as good, fair, or poor rating. Gelled granular products have good water resistance, and certain water-based mixtures have an excellent rating.

Ammonium nitrate prills have no water resistance and should not be used in the water-filled portions of a borehole. The emission of brown nitrogen oxide fumes from a blast often indicates inefficient detonation frequently caused by water deterioration, and signifies the need of a more water-resistant explosive or external protection from water in the form of a plastic sleeve or a water-proof cartridge or dewatering the hole by pumping out the water if possible.

Temperature effects (resistance to freezing): Extreme low temperatures affect the stability as well as the performance of explosives. The sensitivity and detonation velocity are hampered for certain water-based explosives at low temperatures, while dynamites can become dangerously unstable below freezing temperatures. Explosive manufacturers recommend the appropriate range of temperature for storage and use.

Detonation velocity: Detonation velocity is the speed at which the detonation front moves through a column of explosives. For high explosives such as dynamite, the strength of an explosive increases with detonation rate. For dry blasting agents and water-based explosives, field loading conditions greatly affect detonation velocity. Such conditions include borehole diameter, density, confinement within the borehole, presence of water, and other factors. The speed of detonation is important when blasting in hard, competent rock, where a brisance effect is desired for good fragmentation.

For most explosives, there is a minimum diameter D_{min} below which detonation velocity increases nonlinearly with increasing borehole diameter (Figure 5.8). Above D_{min}, the explosive has reached its steady-state velocity. At this point, all thermodynamic properties are at a maximum as the reaction front approaches a plane shock front. At diameters less than D_{min}, complete reactions do not take place, and less than

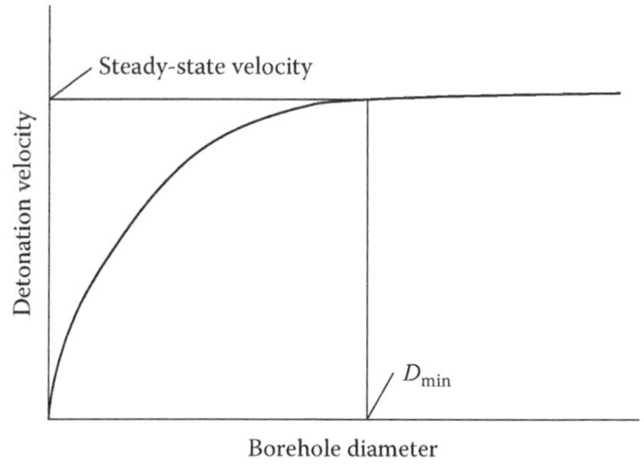

FIGURE 5.8 Generalized relationship between detonation velocity and borehole diameter. (From Hartman, L., *SME Mining Engineering Handbook*. Englewood, CO: Society for Mining, Metallurgy and Exploration, 1992.)

ideal energy and pressure evolve from the slower detonation rates. This represents a loss in terms of dollars spent on explosive energy.

Detonation pressure: The detonation pressure is the maximum theoretical pressure achieved within the reaction zone and measured at the C-J plane in a column of explosives. The actual pressure achieved is somewhat less than this maximum due to nonideal loading conditions always present in practice and due to certain explosive formulation. Most commercial explosives achieve pressures in the range of 0.29–3.48 × 10^4 psi (2–24 GPa).

Although detonation pressure is related to the temperature of the reaction, a number of simplifying formulas are available for estimating detonation pressure for granular explosives based on detonation velocity and density, such as (in English units):

$$P = 3.37\times10^{-3}\rho V^2 \quad (5.16)$$

where:

P is detonation pressure in psi
ρ is density
V is detonation velocity m fps

whereas in metric units:

$$P = 0.25\ \rho\ V^2 \quad (5.17)$$

In general, pressures after detonation within the borehole are estimated to be less than 30% of the theoretical detonation pressure. Borehole pressure is a function of confinement and the quantity and temperature of the gases of detonation. Borehole pressure is generally considered to play the dominant role in breaking most rocks and displacing all types of rocks encountered in blasting. This accounts for the success of ANFO and aluminized products, which yield low detonation pressures, but relatively high borehole pressures.

Borehole pressure: Borehole pressure is the maximum pressure exerted within the borehole upon completion of the explosive reaction measured behind the C-J plane. Such measurements cannot be made directly and are done during underwater tests performed for energy and strength determinations.

Sensitivity: The definition of explosive sensitivity is twofold. It includes sensitivity against accidental detonations in addition to the ease by which explosives can be intentionally detonated. From the standpoint of safety and accidental detonations, the sensitivity of an explosive to shock, impact, friction, and heat determines its storage and handling characteristics.

The term properly used to define the propagating ability of an explosive is *sensitiveness*. In this respect, tests such as the No. 8 strength blasting cap test, air-gap test, and the minimum critical diameter test are used. The cap sensitivity test measures the minimum energy required for initiation and is used to classify explosives (e.g., cap sensitive versus noncap sensitive products) or the ability to initiate an explosive directly with a standard cap.

If an explosive can be ignited by No. 8 cap: EXPLOSIVE
If it cannot be ignited by No. 8 cap: BLASTING AGENT

The No. 8 cap is an industry standard cap of specific dimensions and charge characteristics.

Strength: The strength of an explosive is a measure of its ability to break rock. The terms *weight strength* and *bulk strength* were useful many years ago when explosives were primarily composed of nitroglycerin cartridges, packaged in 50 lb (23 kg) boxes.

Absolute bulk strength in calories per cubic centimeter and absolute weight strength in calories per gram are computed from the heat liberated during the detonation and formation of gaseous end products. Absolute bulk strength and absolute weight strength can be computed from one another if density is known, and it is the volumetric basis of reaction heat, which correlates with energy. Other common strength terms are the relative weight strength and relative bulk strength, in which the relative measure of energy available per unit weight or volume of an explosive is compared to an equal weight or volume of the standard commercial explosive ANFO. The relative weight strength and relative bulk strength are computed as a percentage of that available from ANFO.

5.4.4 Rock Properties Affecting Blastability

Performance of particular explosive is controlled by several rock properties such as Young's modulus of the rock medium, uniaxial compressive strength, and unit weight of rock mass. Young's modulus of rock medium controls the capacity of rock mass to transmit energy. Strength properties of rock material control the ease of generating new fractures in the medium. The unit weight of the rock mass controls the energy required for displacing the fragmented rock and the energy transmissive properties of the intact rock medium.

Muftuoglu et al. (1991) studied the relationship between rock mass quality index, uniaxial compressive strength, tensile strength, P-wave velocity, and powder factor as shown in Figure 5.9.

In the Figure 5.10, an empirical correlation between rock mass material properties and explosive type suggested by Brady and Brown (2006) is shown. In the figure, the term *high brisance* is used to define explosives characterized by detonation velocities greater than about 5000 m/s. In the figure C_0 is the rock material strength. Although the graph can be used to assess the type of explosives, other explosive properties should also be considered.

For medium-soft rock ANFO-type explosives are more suitable. In many operations, it is advantageous to use dense high-velocity explosive in bottom of the borehole and ANFO as top load.

5.4.5 Initiation Methods and Devices

Initiators are devices containing high explosives that, upon receiving an appropriate mechanical or electrical impulse, produce a detonation or burning action. Initiators are used as components within a system of explosives and other devices to start

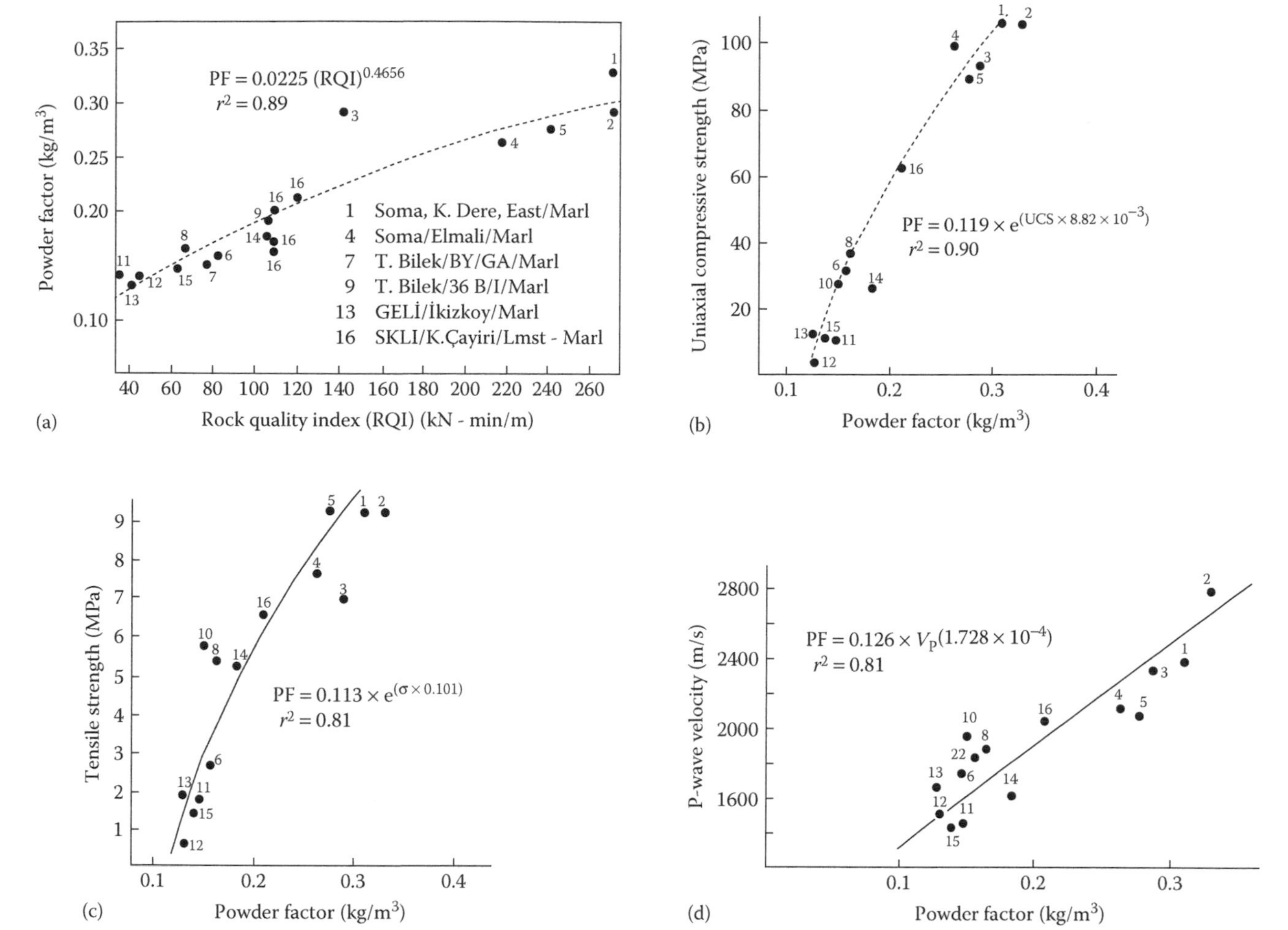

FIGURE 5.9 The relationships between rock properties and powder factor: (a) RQI and powder factor; (b) uniaxial compressive strength and powder factor; (c) tensile strength and powder factor; and (d) P-wave velocity and powder factor. (Data from Muftuoglu, Y. et al., Correlation of powder factor with physical rock properties and rotary drill performance in Turkish surface coal mines. W. Wittke (Ed.), *International Congress on Rock Mechanics* (pp. 1049–1051). ISRM, Aachen, Germany, 1991.)

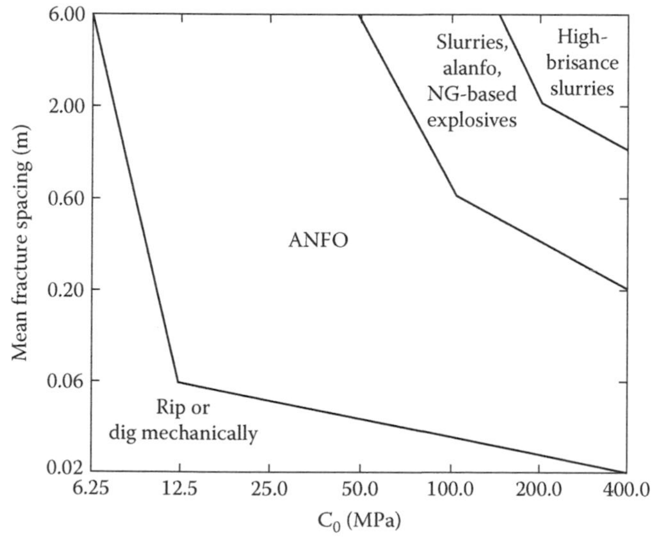

FIGURE 5.10 The correlation between rock mass material properties and explosive type. (Data from Brady, B. and Brown, E., *Rock Mechanics for Underground Mining*, Springer, Dordrecht, The Netherlands, 2006.)

the detonation of all other components. Initiation systems are either electric or nonelectric. Both in electric and in nonelectric firings, charges may be initiated instantaneously, implying that all the blastholes detonate at the same time. However, in delay blasting, time intervals are required between the detonation of various blastholes or even between decks within a blasthole.

The main initiation types, their energy properties, and in the hole components are presented in Table 5.10.

5.4.5.1 Electrical Initiation Systems

Electric blasting caps are one of the used method of initiation. Electrical energy (ac or dc) is sent through copper or iron lead wires to heat an internal-connecting bridge wire. This heat, in turn, starts a chain reaction of explosives burning within the metal cap shell, through a powder delay train. This process detonates a high-explosive base charge, igniting a cap-sensitive explosive. They are manufactured with an instantaneous (no delay train) time of initiation, or time delays (in milliseconds) used in delayed blasting practices. Time delays with intervals of 25, 50, and 100 ms are available

TABLE 5.10
Initiation Types and Energy Properties

Type of Initiation			Type of Energy Conveyed	Energy Distribution System	In the Hole Component
I. Electrical			Electric impulse	Copper or iron wires	Blasting cap (electric detonator)
II. Nonelectrical	Safety fuse and plain detonator		Heat (flame)	Safety fuse	Blasting cap (plain detonator)
	Detonating cord		High-energy explosive detonation	Detonating cord	Detonating cord
	Shock tubes	Nonel	Low-energy explosive detonation	Plastic tube	Special cap
		Herculet	Low-energy gas detonation	Tubing	Special cap

Source: Tamrock, *Handbook of Surface Drilling and Blasting*, Painofaktorit, Tampere, Finland, 1984.

for short- (ms) or long-period delays. Short delays are used in surface blasting operations, while longer delays are used underground where blasting conditions are more confined. The use of time delays in blasting enhances fragmentation and the control of ground vibrations. The new generation of high-precision detonators will contain an electronic circuit instead of pyrotechnical delay elements. The electrical circuits can be connected as series, series and parallel, and parallel.

5.4.5.2 Nonelectrical Initiation Systems

These systems use different types of chemical reactions ranging from deflagration to detonations as a means of conveying the impulse to nonelectric detonators, or as in the case of detonating cord, it is the initiator (Bhandari 1997). Most widely used nonelectrical initiation systems and the main advantages and disadvantages of electrical and nonelectrical initiation systems are presented in Table 5.11.

As it can be seen from the table, to select the most proper initiation system, many factors such as type of explosive, borehole temperature, geology, hydrostatic pressure, extraneous electricity and environment, and other constraints should be considered.

5.4.6 Surface Blasting (Bench Blasting) Design

In surface blast design, the empirical approach suggested by Ash (1963), is presented here. In surface blast designs, the required parameters are hole diameter, *d*; burden, *B*; hole spacing, *S*; charge weight, *W*; top-hole stemming length, *T*; and subgrade drilling depth, *J*. Design parameters are shown in Figure 5.11 and Table 5.12.

Borehole patterns are drilled square (Spacing/Burden [*S/B*] = 1) or rectangular ($S/B > 1$) on center or offset (staggered). The sequence of hole initiation timing, *S/B* ratio, actual timing between charge detonations, and number of blasthole rows determine the shape of the broken rock pile as well as the degree of rock fragmentation.

Borehole diameter and burden are perhaps the most important factors used in design. Burden values should be selected based on geology and explosive energy output. Usually hole diameter is set by the drill rig capacity, which is matched to the range of hole depths anticipated for the job. It is desirable to select a size that will provide an adequate powder factor, or specific charge for breakage while distributing the explosive evenly throughout the hole depth. Fragmentation and particle

TABLE 5.11
Comparison of Initiation Systems

		Nonelectrical			
Characteristics	**Electrical**	**Safety Fuse and Plain Detonator**	**Detonating Cord**	**Nonel**	**Herculet**
Timing ability	Precise	Limited	Precise	Precise	Precise
Blasting capacity	Unlimited	Limited	Unlimited	Unlimited	Limited
Blasting adaptability	Mainly underground blasting	Underground or surface blasting	ANFO and large diameter slurries	Most aspects	Most aspects
External electric hazard	Electricity and radio energy	None	None	None	Water and dirt contamination
Airblast	Nonexistent	Nonexistent	Limited	Nonexistent	Nonexistent
Fire risk	Nonexistent	Existent	Nonexistent	Nonexistent	Nonexistent
Means of checking	Instrumental	Visual	Visual	Visual	Instrumental
Craftsmanship	Skilled	Unskilled	Unskilled	Unskilled	Skilled

Source: Bhandari, S., *Engineering Rock Blasting Operations*, A.A. Balkema, Rotterdam, The Netherlands, 1997.

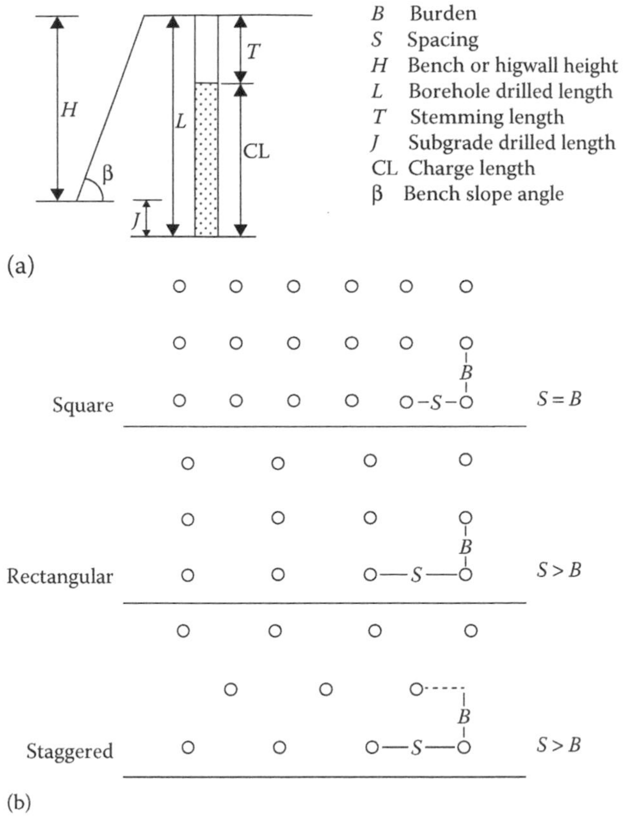

FIGURE 5.11 Surface blasting design parameters: (a) section view—terminology and (b) pattern array. (From Hartman, L., *SME Mining Engineering Handbook*, Society for Mining, Metallurgy and Exploration, Englewood, CO.)

TABLE 5.12
Empirical Relationships Proposed by Ash (1963)

Ash (1963)	$B = K\frac{D}{12}$	Using ANFO K = 22 for rock density <2.7 g/cm³ K = 30 for rock density >2.7 g/cm³
	$S = KsB$	Using slurry, dynamite, or other tech explosives *K* = 27 for rock density <2.7 g/cm³ *K* = 35 for rock density >2.7 g/cm³ Ks = 2.0 for simultaneous initiation, 1 for sequenced blasthole with long delay between 1.2–1.8 for sequenced blasthole with short delay
	$J = KjB$	Kj between 0.2 and 0.4
	$T = KtB$	Kt between 0.7 and 1

Source: Hartman, L., *SME Mining Engineering Handbook*, Society for Mining, Metallurgy and Exploration, Englewood, CO, 1992.

size distribution are a function of hole diameter and burden. The capacity of the excavation equipment dictates the required fragmentation. The charge length to charge diameter ratio for a cylindrical charge should be five or greater.

5.4.7 Powder Factor/Specific Charge

The ratio between the total weight of explosive detonated in a blast divided by the amount of blasted rock is described as

TABLE 5.13
Powder Factor for Different Surface Coal Mines

Major Removal Equipment	Geological Unit	Powder Factor (kg/m³)	Bench Height (m)
Large dragline	Shale	0.30	15
	Shale	0.35	23
	Shale	0.40	–
	Sandstone	0.35	18
Small dragline	Shale	0.20	6
	Shale	0.50	7.6
	Shale	0.35	18
	Sandstone	0.65	26
	Sandstone	0.35	9
Front-end loader	Shale	0.65	16
	Sandstone	0.40	20
	Sandstone	0.95	15

powder factor, or sometimes called *specific charge factor*. The unit of the factor can be kilogram per ton or kilogram per cubic meter. The following equation is used to calculate powder factor:

$$PF = \frac{(\text{Wlb})(27\text{ft}^3/\text{yd}^3)}{(\text{BSH})(\text{ft}^3)} \tag{5.18}$$

The term is generally used in empirical relations and describing performance of blasting. Powder factors (PF) range from 0.25 to 2.5 lb/yd³ (0.15 to 1.5 kg/m³) for surface blasting but average 0.5 to 1 lb/yd³ (0.3 to 0.6 kg/m³). Various powder factors for different surface coal mines are given in Table 5.13 (Bhandari 1997). Higher powder factors result in fine fragmentation and are required for small capacity removal equipment such as front-end loaders. Smaller powder factors result in coarser fragmentation and are typically used for rock removal using draglines and large shovels.

5.4.8 Underground Blasting

In underground coal mines, blasting operations can be conducted for different purposes such as in drifting (development), on longwall face for production, in cut coal, and in depillaring operations.

5.4.8.1 Underground Production Blasting Design

Blasting in underground coal mines requires special attention due to highly inflammable mixtures of methane, air, and explosible dust. In coal mines, to avoid gas or dust ignitions, only those explosives should be used in which the reaction is extremely fast and produces a short flame. Moreover, the reaction should produce a large quantity of inert gases, which are unable to enter into further reaction with the oxygen in the air. Such explosives are called *permitted* and/or *permissible explosives*. In most countries, permitted explosives are well described by their institution, that is, Mine Safety and Health Association (MSHA) in Canada, Ministry of Energy in the United Kingdom. The classification of permissible explosive suggested by the British safety in Mines Research Board is given in Table 5.14.

TABLE 5.14
Summary of the British Permitted Explosives Classifications and Tests

Group	Application	Tests
P1	Used for instantaneous blasting in undercut coal or relieved rock (rippings) near a coalface, but in British mines with minor exceptions, they are principally used for delay blasting in shafts and tunnels away from sources of gas.	1. Twenty-six shots of 142 g of explosive, inversely primed and un-stemmed, are fired into a methane/air mixture. Not more than 13 ignitions may occur. 2. Five shots of 795 g, directly primed and stemmed, are fired into a methane/air mixture. No ignitions may occur. 3. Five shots of 795 g, directly primed and stemmed, are fired into a coal dust/air mixture. No ignitions may occur.
P2	These are P1 explosives sheathed by sodium bicarbonate (a flame suppressant). Their use was discontinued many years ago.	
P3	Previously known as *equivalent to sheathed*, the flame suppressant is incorporated in the composition. Used mainly for blasting undercut coal and rock rippings by single shot-firing or instantaneous firing of up to six shots.	1. Twenty-six shots of 397 g of explosive, inversely primed and un-stemmed, are fired into a methane/air mixture. Not more than 13 ignitions may occur. 2. Five shots of 1020 g, directly primed and stemmed, are fired into a methane/air mixture. No ignitions may occur. 3. Five shots of 567 g, inversely primed and un-stemmed, are fired into a coal dust/air mixture. No ignitions may occur.
P4	Developed specifically for use in rock rippings with delay firing, where there is an inherent possibility of the charge firing into a gas-filled break or parting.	1. Twenty-six shots of 397 g of explosive, inversely primed and un-stemmed, are fired into a methane/air mixture. Not more than three ignitions may occur. 2. Five shots of the maximum permitted charge mass are fired into a methane/air mixture using the Break Test 1. No ignitions may occur. 3. Break Test 2 uses a gas mixture of 3.60% propane with air and nitrogen. This mixture is more easily ignited than methane/air. Some test shots are fired and the most hazardous charge not exceeding 227 g determined. Twenty-six shots are then fired at this mass and not more than 13 ignitions may occur. 4. Five shots of 30.5 cm length and 3.7 cm diameter are fired in methane/air in Break Test 2. No ignitions may occur.
P5	Designed for delay blasting in solid coal (i.e., not undercut).	1. Twenty-six shots of 567 g of explosive, inversely primed and charged to reach to 5 cm from the mouth of the cannon, are fired into a methane/air mixture. No ignitions may occur. 2. Five shots of 1020 g, directly primed and stemmed, are fired into a methane/air mixture. No ignitions may occur. 3. Five shots of 567 g, inversely primed and un-stemmed, are fired into a coal dust/air mixture. No ignitions may occur.
P6	Designed to meet both P4 and P5 test conditions.	These explosives must pass both the P4 and P5 tests.

5.4.8.2 Underground Development Blasting Design (Tunnel Blasting)

Tunnel blasting is much more complicated compared to bench blasting. Because when the tunnels are being blasted, the only free face available is the tunnel heading, that is, rock to be blasted is confined from all other sides. Therefore, the specific charge increases. The increase in the specific charge brings the possibility of greater damage to the remaining rock. Thus, the need for greater amount of scaling and heavy support also increases; this in turn results in higher costs. On the contrary, it is not required to damage the rock, and is expected that the rock (roof and walls) remains intact and stands up by itself, and uses less support. Also, the rock should be broken efficiently and economically by the blast and a well-fragmented muck (or ore) pile produced, which will be easy to remove, transport, store, and process. It is clear that *Less Damage!* and *Good Breakage*, conditions conflict each other. So, an optimum solution should be found.

The stopping can be compared well with bench blasting, but requires considerably larger charges. No theoretical basic information is available for the design of tunnel blasting rounds. The calculating principles are usually based on pure experience. Empirical relationships only serve as a first approximation to be taken in the design process. For each design situation, a trial-and-error approach is usually taken by experienced and qualified blasters.

Borehole patterns are selected based on rock type and size of the face (cross-sectional area of tunnel). Main parameters to be selected in design are cut type, powder factor, diameter and length of holes, total number of holes (or face area per hole), burden and spacing of holes, location and inclination of holes, charge per hole, stemming length, delay time, and initiation sequence.

The principle behind tunnel blasting is to create an opening at the central part of the heading, which is fired first to produce a free face (called *cut*) and then breaking the rock toward this opening (called *stoping*).

The common types are as follows (Figure 5.12): fan cut, plough or V-cut, instantaneous cut, and parallel hole (burn) cut.

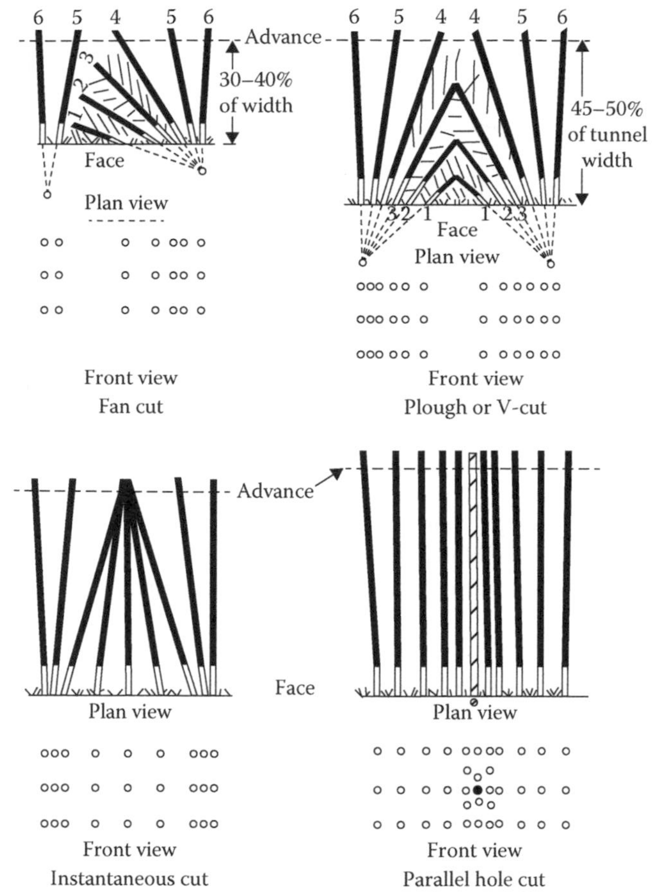

FIGURE 5.12 Various cuts used in underground drifts and tunnels. (From Nitro Nobel, Blasting Techniques, Nitro Nobel Blasting Techniques Department, Seminar notes, 1980.)

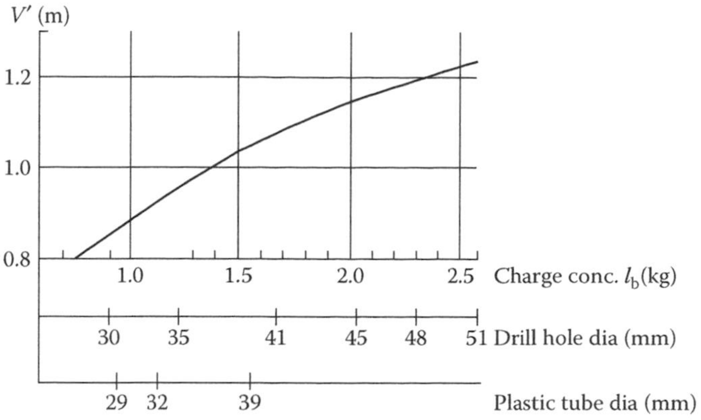

FIGURE 5.13 A diagram showing relationship between drilling and blasting parameters. (From Nitro Nobel, Blasting Techniques, Nitro Nobel Blasting Techniques Department, Seminar notes, 1980.)

Among those cuts, the design features of the commonly used V-cut and parallel hole cut will be given here.

5.4.8.2.1 Parallel Hole Cut

Parallel hole cuts consist of several parallel holes, some of which are left unchanged to provide a free face (relief hole) for the neighboring charged holes. Parallel hole cuts provide accurate drilling, especially in narrow openings, faster drilling, suitable for mechanization, better advance till 95% of drilled depth, longer advance per round, and good fragmentation.

In order to make blast design using parallel hole cut, a guideline for charging given in Table 5.15 can be used, where the concentration of the bottom charge is lb (kg/m) and hole depth is *H* (Nitro Nobel 1980).

TABLE 5.15
A Guideline for Charging the Holes

Square No	Height of Bottom Charge, h_b (xH)	Concentration of Column Charge, l_p (xl_b)	Uncharged Part, h_0
1	0.05	0.5	–
2	0.05	0.5	0.5
3	0.20	0.5	0.5
4	0.33	0.5	0.5

TABLE 5.16
Blasting Design Table

Hole Type	Burden (V), m, xV	Spacing (E), m, V	H_b, m, xH	L_p, kg/m, xlb	H_0, m, xV
Floor	1.0	1.1	1/3	1.0	0.2
Wall	0.9	1.2	1/6	0.4	0.5
Roof	0.9	1.2	1/6	0.3	0.5
Stoping	1.0	1.1	1/3	0.5	0.5
Stoping	1.0	1.2	1/3	0.5	0.5

The bottom charge concentration is obtained by using the hole diameter in Figure 5.13. The horizontal axis provides bottom charge concentration and vertical axis correspondence of the intersection of the vertical line from the charge concentration with the curve will give the *effective burden*, which determines the size of the *cut*. The number of rounds around empty is stopped when the last round burden reaches to *effective burden*. *Cut* is fixed at the center of the tunnel and then *stoping holes burden* and *spacings* are calculated using Table 5.16. Stoping holes are floor holes, wall holes, roof holes, stoping holes upward and horizontally and stoping holes downward.

Simple sketch showing parallel hole cut is shown in Figure 5.14. As shown in figure, the cuts consist of one or several empty large holes surrounded by the charged parallel holes with a small burden. Depending on the used bit diameter, the area of the opening can be in the range from 2 to 4 m^2. The advantages and disadvantages of parallel cut are given in Table 5.17.

5.4.8.2.2 V-Cut

In V-cut, the angled holes are drilled as shown in Figure 5.14. In order to have good blasting results the angle of inner plough should not be too acute. The advantages and disadvantages of V-cut are presented in Table 5.17.

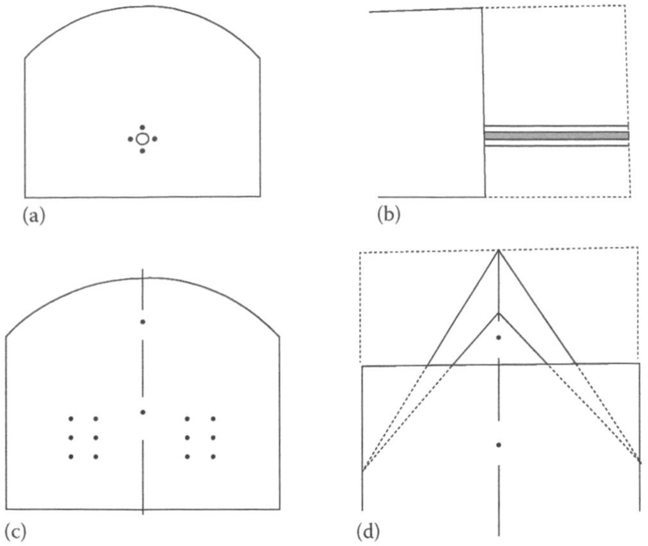

FIGURE 5.14 Sketches showing parallel and V-cuts: (a) parallel cut front view; (b) parallel cut side view; (c) V-cut front view; and (d) V-cut top view. (From Nitro Nobel, Blasting Techniques, Nitro Nobel Blasting Techniques Department, Seminar notes, 1980.)

TABLE 5.17
Advantages and Disadvantages of Parallel and V-Cut

Type of Cut	Advantages	Disadvantages
Parallel	Relatively simple design	To drill empty large hole special drill steel equipment required
	Same drill steel length is used for all holes	One rock drill is occupied during a relatively long time when drilling the empty large hole
	Can be used for both small and large cross sections	Great precision is required during drilling and charging the cut holes
V-cut	For all of the holes, the same type of drill steel equipment can be used	As the holes are angled the full length of drill steel set is not used
	Symmetry makes the drill easier, especially for mechanized equipment	A minimum width of the drift is required as the holes are drilled with certain angle

Since the holes are inclined, advance is dependent on tunnel width. Most typical advance for this cut is 45%–50% of tunnel width. It seems that if tunnel width is increased, advance will also increase. However, advance is also affected by drill hole deviation, for example, in 5 m long hole with ±5% deviation, the end of a hole may deviate ±25 cm, and this can increase burden at the bottom, and the blast will be less efficient. Also, the cooperation between the holes will be lost. Acute angle at the inner plough should be at least 60° and the deviations in the drilling should be minimum to have better application.

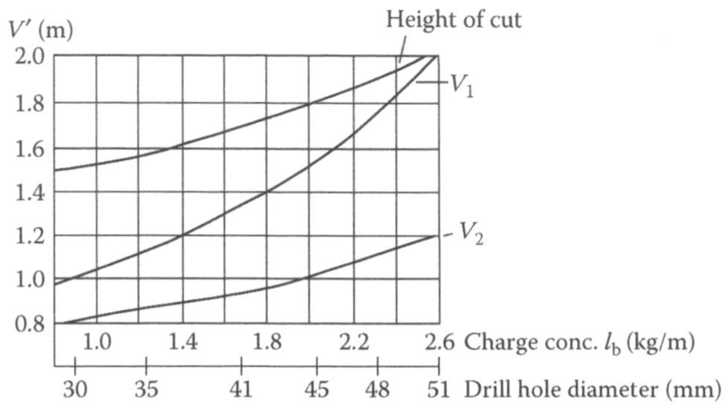

FIGURE 5.15 Design diagram for V-hole cut. (Data from Nitro Nobel, Blasting Techniques, Nitro Nobel Blasting Techniques Department, Seminar notes, 1980.)

V-cut design charts for a dynamite with weight strength $s = 1.0$ (78% of blasting gelatin).

In order to make blast design using a V-cut, first the height of the cut and the burdens V_1 and V_2 are found in the Figure 5.15.

The height of the cut provides three ploughs vertically as shown in the figure. The bottom charge concentration is again determined using Figure 5.13.

Uncharged part for the cut holes $h_0 = 0.3\ V_1$. To calculate burdens, spacings and charges of the remaining holes can be calculated again using Table 5.16. For all the remaining holes, the height of the bottom charge $h_b = H/3$, where H is the hole depth. The column charge concentrations $l_p = 40\%–50\%$ of lb. Uncharged part for the remaining holes $h_0 = 0.5\ V_2$.

REFERENCES

Akun, M. E. (1997). *Effect of Operational Parameters and Formation Properties on Drillability in Surface Set Diamond Core Drilling*. PhD thesis, Middle East Technical University, Ankara, Turkey.

Ash, R. (1963). The mechanics of rock breakage, standards for blasting design. *Pit and Quarry*, 1, 118–122.

Bhandari, S. (1997). *Engineering Rock Blasting Operations*. Rotterdam, the Netherlands: A.A. Balkema.

Brady, B., and Brown, E. (2006). *Rock Mechanics for Underground Mining*. Dordrecht, the Netherlands: Springer.

Christensen Diamond Products, C. (1980). *Diamond Bits and Their Use in Shallow Holes*. Wuppertal, Germany: CDPS.

Cumming, J., and Wicklund, A. (1956). *Diamond Drill Handbook*. Toronto, Canada: J.K. Smit & Sons Diamond Products.

Hartman, L. (1992). *SME Mining Engineering Handbook*. Englewood, CO: Society for Mining, Metallurgy and Exploration.

Heinz, W. R. (1985). *Diamond Drilling Handbook*. Johannesburg, South Africa: South African Drilling Association.

Karpuz, C., Pasamehmetoglu, A., Muftuoglu, Y., and Dincer, T. (1990). Drillability studies on the rotary blast hole drilling of lignite overburden series. *International Journal of Surface Mining and Reclamation*, 4, 83–93.

Longyear. (1989). *Q-CQ-CHD Mini Manual*. Mississauga, Canada: Longyear.

Miller, A., and Ball, A. (1990). Rock drilling with impregnated diamond microbits—An experimental study. *International Journal of Rock Mechanics and Mining Science Geomechanics Abstracts*, 27(5), 363–371.

Moppes, V. (1986). *Diamond Core Bits for Mining and Exploration*. Moppes, Italy.

Muftuoglu, Y., Pasamehmetoglu, A., and Karpuz, C. (1991). Correlation of powder factor with physical rock properties and rotary drill performance in Turkish surface coal mines. W. Wittke (Ed.), *International Congress on Rock Mechanics* (pp. 1049–1051). Aachen, Germany: ISRM.

Nitro Nobel, Blasting Techniques, Nitro Nobel Blasting Techniques Department, Seminar notes, 1980.

Sandvik-Tamrock. (1999). *Rock Excavation Handbook for Civil Engineering*. Tampere, Finland: Sandvik Tamrock.

Tamrock. (1984). *Handbook of Surface Drilling and Blasting*. Painofaktorit, Tampere, Finland.

Tamrock. (1985). *Handbook on Surface Drilling and Blasting*. Tampere, Finland.

6 Excavatability Assessment of Surface Coal Mine

Celal Karpuz and Hakan Basarir

CONTENTS

Abstract: In surface mining, the selection of appropriate excavation method becomes an important issue for most surface coal mines since it directly affects the productivity of the selected equipment. The basic definitions related to the excavatability are given. The mechanics of rock cutting and excavation are briefly introduced. Rock properties and equipment properties affecting the excavatability are presented. The main types of excavatability assessment methods such as single parameter, graphical, and grading methods are mentioned.

6.1 INTRODUCTION

The most common method of mine production is surface mining. In surface mines, the assessment of suitable method for loosening and breaking of ground rock and rock masses has crucial importance from both economic and technical viewpoints. Therefore, the reliable assessment of excavatability class for determining optimum excavating method and equipment is a very important process in mine planning. In case of wrong equipment selection, additional expenditures, production delay, and production loss may occur. Moreover, wrong estimation of this cost may lead to disagreement between clients and contractors.

The two main methods for breaking and loosening ground rocks are mechanical excavation and blasting methods. Large-scale mechanical excavation method can be subdivided into three groups as (a) direct digging, (b) ripping, and (c) digging with the help of blasting. The energy generated by the machines is transmitted to ground by means of a tyne or buckets in mechanical excavations, whereas in blasting, the energy is generated by blasting agents. In general, if the ground conditions allow easy/very easy excavation, then *direct digging* is considered as appropriate term to describe excavation conditions. In this case, bucket is directly crowded to ground to excavate it. *Ripping* is the term used to describe dominantly moderate to difficult excavation conditions. *Ground* is the term characterizing difficult to very difficult digging condition for which blasting is needed in general. Hadjigeorgiou and Poulin (1998) described diggability as the process of cutting and displacement by a blade or bucket. In some references, instead of *diggability*, the term *excavatability* is also used. Therefore, in this chapter, both these terms will be used for the same description.

Rippability is described as the process of breaking the harder ground by dragging tynes attached to a bulldozer. In other words, ripping is a method of loosening rock, during excavation using steel tynes attached to the rear of bulldozer. The tynes penetrate the ground as the bulldozer moves forward, and soil and rock are displaced by the tynes or rippers (MacGregor et al. 1994). When the physical and/or economical limits of ripping is reached or the excavation condition is very difficult, then *blasting* is the most appropriate term and method to loose and break the ground.

Mechanical excavation methods are more preferable to drilling and blasting method due to the following reasons:

- Under suitable conditions, if rock has diggable properties, ripping and direct digging are cheaper (30%–80%) than drilling and blasting as an overburden stripping method.
- In open casts, vibrations due to blasting can negatively affect the stability of slopes. In mechanical excavation, steps, levels, and slopes would not be disturbed, and hence slopes and levels can be formed regularly.

In blasting, the desired fragmentation size may not be obtained and sometimes secondary blasting is needed. However, in mechanical excavations, by adjusting the distance between the two runs, desired cut dimensions can be obtained easily.

For safety of operations and men, equipment and workers should be kept away before blasting; this leads to idle time and decreases the efficient of both equipment and workers.

Near urban regions to prevent the blasting damages, some strict regulations limit the use of blasting. Risks exist in supplying and stocking the blasting materials, which are classified as dangerous in general.

6.2 MECHANICS OF ROCK CUTTING AND EXCAVATION

Rock excavation methods, as it was mentioned before, can be subdivided into two groups: (1) cutting with mechanical tools and (2) excavation by explosives as shown in Figure 6.1. The main cutting tools are indenters and drag bits (Figure 6.2).

The tool strength of drag bits is weak compared to the indenters. Indenters are more widely used than drag bits, but they are less efficient than drag bits for the excavation of the same volume of rock due to higher energy consumption. Indenters break the rocks by applying forces normal to surface as shown in Figure 6.2. Examples of this type of

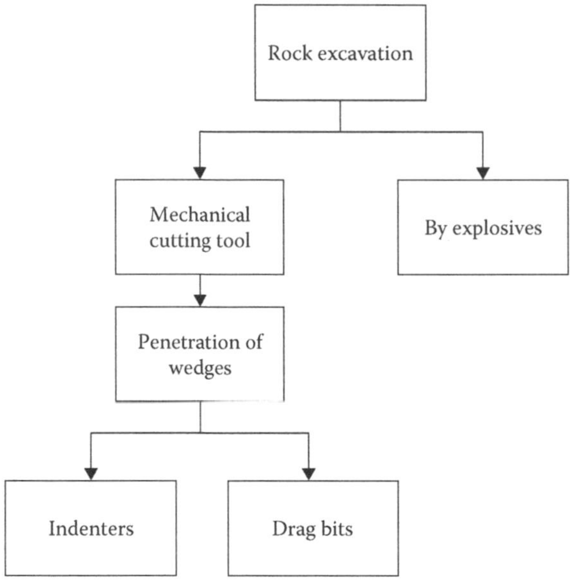

FIGURE 6.1 Rock excavation methods.

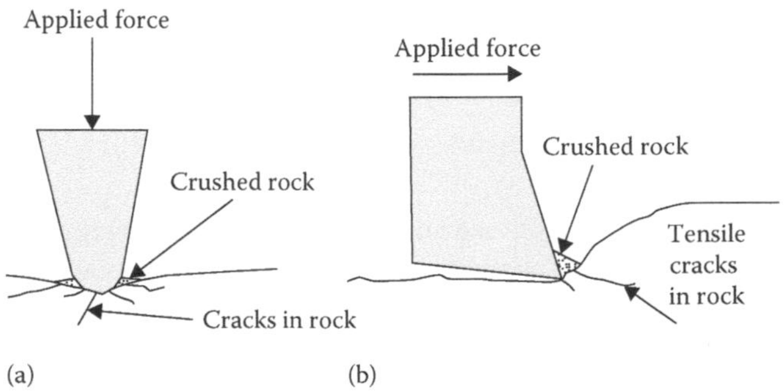

FIGURE 6.2 Mechanical cutting tools: (a) indenters and (b) drag bits. (From Hartman, L.H., *SME Mining Engineering Handbook*, Society for Mining, Metallurgy and Exploration, Englewood, CO, 1992.)

indenting action-based tools are roller cutters, disc cutters, rolling cone bits, and all percussive tools such as percussion drill bits, down the hole drilling drill bits, and high-energy impact bits.

With drag bits, the main cutting force is parallel to the rock surface as given in Figure 6.2. Rotary drill bits and picks can be considered as examples of drag bits. Mode of rock breaking in the form of tensile fractures is the same for both indenters and drag bits. Drag bits initiate a tensile fracture in a fairly direct manner, whereas an indenter generates tensile stress indirectly in the rock by crushing a portion of the rock mass right beneath the indenting tool.

The failure mechanism and stresses around indenters such as flat-bottomed punches, wedge indenters, and disc cutters are shown in Figure 6.3. In the indentation process of flat-bottomed punches (see Figure 6.3a), initially rock deforms elastically when σi is low; as stress increases (σii), Hertzian cracks initiate and intense comminution (crushing) takes place beneath the tool (σiii); and finally, major fractures are enforced from the zone of crushed rock beneath the cutting tool (σiv). Paul and Sikarskie (1965) developed a model assuming that failure along the fracture surface takes place when the Coulomb criterion is exceeded.

$$|\tau| = S_0 + \mu\sigma \tag{6.1}$$

where:

τ is the shear stress acting on the inherent fracture surface
μ is the coefficient of internal friction
S_0 is the inherent shear strength of the rock

For each subsequent loading with the wedge shaped indenters, indenter force increases and stress is transmitted to the major fracturing area through a zone of crushed rock produced from previous loading as shown in Figure 6.3b.

Evans (1974) used the same argument of Paul and Sikarskie (1965) for the disc cutters and the following equation was derived.

$$\frac{F_T}{d} = 2\times\frac{S_0\cos\phi\sin(\theta+\phi_f)}{\sin^2\{(\pi/4)-[(1/2)\cdot(\theta+\phi_f+\phi)]\}} \tag{6.2}$$

where:

F_T is thrust force
d is depth of cut
ϕ_f is friction angle between wedge and rock

As for the drag bits, although a shear force is applied to the rock by the bit, the rock is broken as the result of tensile cracking. To explain the drag bit force (wedge bits), Shuttleworth (in Hartman 1992) offered the following equations by adapting metal machining theory.

$$F_c = \frac{S_0 dw\cos(\phi_f-\alpha)}{\sin\phi\cos(\phi+\phi_f-\alpha)} \tag{6.3}$$

$$F_T = \frac{S_0 dw\sin(\phi_f-\alpha)}{\sin\phi\cos(\phi+\phi_f-\alpha)} \tag{6.4}$$

The formula implies a linear increase in drag bit forces with depth of cut. A monotonic decrease in forces is predicted with increasing rake angle, and it is observed that bit forces increase linearly with rock strength symbolized by S_0. Evans (1962) developed the following equation for a symmetric wedge-type cutting tool by proposing that failure is essentially tensile and by considering the limit equilibrium of the chip, a circular area, and the action of three principle forces.

$$F_c = \frac{2T_0 d\sin\theta}{1-\sin\theta} \tag{6.5}$$

For asymmetric geometry,

$$F_c = \frac{2T_0\sin(1/2)[(\pi/2)-\alpha]}{1-\sin(1/2)[(\pi/2)-\alpha]} \tag{6.6}$$

Specific cutting energy, required to excavate a unit volume of material, is another valuable parameter for mechanical rock cutting. The high specific energy values indicate low cutting efficiency; in other words, the specific energy is an inverse measure of cutting efficiency.

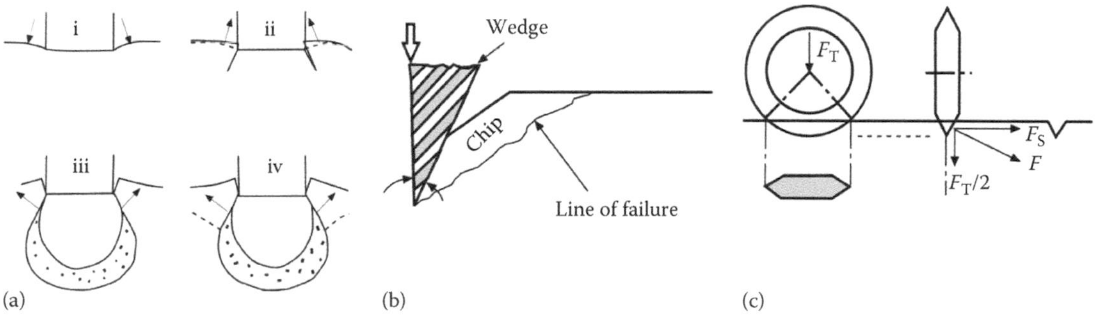

FIGURE 6.3 Failure mechanism and stresses of indenters: (a) flat-bottomed punches (From Wagner, H., and Schumann, E.H.R., *Rock Mech.*, 3, 185–207, 1971), (b) wedge indenters (From Paul, B., and Sikarskie, D.I., *SME-AIME Trans.*, 232, 372–383, 1965), and (c) disc cutter (From Roxborough, F.F., and Phillips, H.R., *Int. J. Rock Mech. Min. Sci.*, 12, 361–366, 1975).

6.3 FACTORS AFFECTING THE PHYSICAL LIMIT OF EXCAVATABILITY

In order to decide the most suitable method to be used, there are many factors that have to be taken into consideration. The main parameters are the type of projects, rock mass and rock material properties, extraction methods, production rate, cost, and environmental constraints. Pettifer and Fookes (1997) summarized them as follows: the characteristics of rock and its defects, dimension of pit, extraction methods, production rate, the equipment available and its condition and cost, environmental constraints, and the skill of operator.

As the factors affecting the economic limit of excavatability vary from site to site, it is more useful to concentrate on defining physical factors, which influence diggability, and to examine their effect on the productivity.

6.3.1 Characteristics of Rock Affecting Excavatability

Rock properties can be expressed as rock material and mass properties. The rock material and rock mass properties such as type, strength, degree of weathering, structure, fabric, abrasiveness, moisture content, cohesiveness, and seismic wave velocity used in excavatability determination by different investigators are reviewed and presented in Table 6.1.

Those rock mass and material properties used by different investigators are described in Section 6.3.1.1.

TABLE 6.1
Rock Material and Mass Properties Used by Different Researchers

Assessment	Rock Material Property					Rock Mass Property														
Method	UCS	PL	SHV	Ab	SE	Jcont	Jn	W	Ja	Jr	If	Jg	Jor	BS	SV	Gs	RQD	GSI	RMi	BV
Anon (1994)															+					
Franklin et al. (1971)		+									+									
Bailey (1975)															+					
Weaver (1975)	+							+			+	+	+		+					
Kirsten (1982)	+						+		+	+	+		+							
Abdullatif et al. (1983)		+									+									
Scoble and Muftuoglu (1984)	+	+						+			+			+						
Smith (1986)						+		+			+	+	+							
Singh et al. (1987)		+		+				+			+				+					
Anon (1987)															+					
Karpuz (1990)	+		+					+			+				+					
Hadjigeorgiou and Scoble (1998)		+					+	+												
MacGregor et al. (1994)	+						+	+		+	+			+	+	+				
Pettifer and Fookes (1997)		+									+									
Kramadibrata (1998)	+	+		+						+	+	+					+	+	+	+
Atkinson (1971)															+					
Bailey (1975)															+					
Bozdag (1988)		+									+									
Muftuoglu (1983)	+	+						+			+			+						
Basarir and Karpuz (2004)	+	+	+								+				+					
Basarir et al. (2008)					+															
Tsiambaos and Saroglou (2010)																		+		
Khamehchiyana et al. (2014)																			+	+

Notes: Ab, abrasiveness; BS, bed separation; BV, block volume; Gs, grain size; GSI, geological strength index; If, joint spacing; Ja, discontinuity alteration; Jcont, discontinuity continuity; Jg, joint gouge; Jn, joint set number; Jor, joint orientation; Jr, joint roughness; PL, point load strength; RMi, rock mass index; RQD, rock quality designation; SE, specific energy; SHV, Schore hardness; SV, Seismic velocity; UCS, uniaxial compressive strength; W, degree of weathering.

6.3.1.1 Rock Material Properties

The dominant rock material properties widely used in diggability assessment methods are strength of intact rock, abrasiveness, laboratory specific energy, and the rock fabric.

6.3.1.1.1 Intact Rock Strength

Point load strength index and uniaxial compressive strength (UCS) are widely used in the assessment of rock excavatability. There is a close connection between UCS and other rock properties such as porosity and hardness. Some researchers argued that tensile strength is more important and effective than compressive strength (Singh et al. 1986). Nevertheless, UCS is the most frequently used and suitable application in more massive and/or weaker materials. The lower UCS value, the higher digging production is expected.

Cohesion is the attraction force between the molecules of the rock and the friction angle; this is due to the resistance between the asperities of one on the other while the surface is shearing. Hardness as well as Schmidt hardness value is closely related to uniaxial compressive strength.

6.3.1.1.2 Abrasiveness

Abrasiveness is an important parameter regarding the excavating tool breakdown and operational cost. The wear of excavating tools due to abrasiveness of rock increases operational cost. The abrasiveness of rock increases expenditures. The abrasiveness index classification proposed by Singh (1983) is given in Table 6.2. The abrasiveness index can be assigned considering the properties of hard rock-forming minerals, angularity of hard minerals, strength of cementing material, Cerchar index, and rock toughness index. The toughness index used in the table is determined by the following equation:

$$T = \left(\frac{\text{UCS}^2}{2E}\right) \times 100 \tag{6.7}$$

where:

T is toughness index
UCS is uniaxial compressive strength
E is elasticity modulus

6.3.1.1.3 Specific Energy

Laboratory specific energy is the work done per unit volume of rock excavated. It is a commonly accepted measure of cutting efficiency, and when it is obtained under specified conditions, it provides a realistic and meaningful measure of ability to cut rock by road heading and tunnel boring machines (Fowell and Pyrcroft 1980). Specific energy is also used as a good indicator to estimate surface excavation machine performance such as bucket wheel excavator (BWE) as Koncagul (1997) emphasized.

6.3.1.1.4 Rock Fabric

The microstructural and textural features of rock material are described by means of rock fabric. In general, it is believed that course-grained rocks are more easily excavated than fine-grained rock, as they also have lower strength values.

6.3.1.2 Rock Mass Properties

Degree of weathering, rock structure-related features such as joints, bedding planes, lamination, cleavages and faults, and seismic wave velocity are the mostly used rock mass properties in assessing the excavatability.

6.3.1.2.1 Degree of Weathering

Hydrospheric and atmospheric conditions result in mechanical weathering of rock, leading to opening of discontinuities by rock fracture, opening of grain boundaries, and the fracture on cleavage of individual mineral grains. In addition to mechanical weathering, chemical weathering or chemical alteration may also be observed, resulting in chemical changes of the minerals. Both weathering types have reducing effect on the strength, density, and volumetric stability of the rock. The reduced strength due to the increase in weathering assists the excavation process (Hadjigeorgiou and Scoble 1988). The greater the degree of weathering, the easier it is to excavate the rock. ISRM weathering classification chart is given in Table 6.3.

TABLE 6.2
Abrasiveness Classification

Class	Cerchar Index	% Hard Rock Mineral	Angularity	Cementing Material	Toughness Index
Very low abrasiveness	<1.2	2–10	Well rounded	Non cemented or rock with 20% voids	<9
Low abrasiveness	1.2–2.5	10–20	Rounded	Ferruginous or clay or both	9–15
Moderate abrasiveness	2.5–4.0	20–30	Sub-rounded	Calcite or calcite and clay	15–25
High abrasiveness	4.0–4.5	30–60	Sub-angular	Silt clay or calcite with quartz overgrowths	25–45
Extreme abrasiveness	>4.5	60–90	Angular	Quartz cement or quartz mozale cements	>45

Source: Singh, R.N., *Testing of Rocks Samples from Underwater Trenching Operations of Folkstone for the Central Electricity Generating Board*, Nottingham, Unpublished report, 1–101, 1983.

TABLE 6.3
Weathering Classification Chart

Term	Description
Fresh	No visible sign of rock material weathering; perhaps slight discoloration on major discontinuity surfaces.
Slightly weathered	Discoloration indicates weathering of rock material and discontinuity surfaces.
	All the rock material may be discolored by weathering and may be somewhat weaker than its fresh condition.
Moderately weathered	Less than half of the rock material is decomposed and/or disintegrated to a soil.
	Fresh or discolored rock is present either as a discontinuous framework or as corestones.
Highly weathered	More than a half of the rock material is decomposed and/or disintegrated to a soil.
	Fresh or discolored rock is present either as a discontinuous framework or as corestones.
Completely weathered	All rock material is decomposed and/or disintegrated to soil. The original mass structure is still largely intact.

Source: ISRM, *Rock Testing Characterization Testing and Monitoring*, Pergamon Press, New York, 1981.

6.3.1.2.2 Rock Mass Structural Features

Structural features such as joints, bedding planes, lamination, cleavages, and faults affect the excavatability of rocks. The continuity and spacing of joints, strike and dip orientation, and the presence of gouge materials are of particular importance in digging. For example, regarding the rippabilities of rocks optimum discontinuity inclination is 45°.

6.3.1.2.3 Seismic Wave Velocity

Seismic velocity is one of the most widely used parameter for the assessment of rock excavatability. It depends on the most of rock properties effecting rock excavatability such as density, porosity, moisture content, degree of fracturing, and the weathering of the rock mass (Singh et al. 1986). Therefore, it is the most widely used parameter for the assessment of rock excavatability. Rock masses having lower wave velocity are more easily diggable than the others.

6.3.2 Equipment Properties Affecting Diggability

The main properties of excavating machine regarding excavatability can be outlined as follows: maximum power, power consumption, bucket capacity, and boom length.

Larger equipment will be able to apply a greater force but, as will be shown later on, when rocks become stronger, the ability to excavate them becomes more dependent on the presence of defects in the rock mass. If rock mass has poor quality, which is classified as directly diggable or easily rippable in this case, type and properties of equipment become less important. In other words, for the same quality of the rock mass, while high-capacity equipment can directly dig the rock, blasting may be needed for the low-capacity equipment.

TABLE 6.4
Operator Skill Coefficients

Operator Skill	Coefficient
Excellent	1.0
Average	0.75
Poor	0–0.60

6.3.3 Operator Skill

Operator skill is a very important parameter. For the same equipment, poor-experienced operator may decrease the equipment performance up to 35%–40%. The training of operators, especially for the larger capacity excavation machines, becomes more important, since their impact on the equipment productivity is great. If the operator skill is evaluated numerically in calculating the productivity, skill coefficients suggested by Anon (1994) are proposed to be taken into account. These are tabulated in Table 6.4.

6.4 EXCAVATABILITY ASSESSMENT METHODS

Many researchers have worked to improve a system used for determining excavatability of rock. It should be stressed here that all these methods are empirical methods. These systems can be grouped in two main parts: direct and indirect methods.

In cases where conclusive determination of excavatability is difficult by using available rock mass and material data or where equipment for trial test is readily available, a trial demonstration may more appropriately decide the issue: Such trial demonstration can be used to obtain good estimates of production for given equipment as in the case of rippability. If field trial or direct ripping runs cannot be conducted, then indirect methods become useful and in most cases, it is the unique way of determining rippability of rocks. As an example of direct method, Caterpillar Tractor Co., Peoria, Illinois, (Anon 1994) reported three methods to estimate ripper production according to availability and practicality of dozers. Volume by weight, volume by cross-sectioning, and volume by length are the subgroups of direct rippability classification methods.

In cases where there is absence of any equipment to conduct trials, the direct digging cannot be employed as a practical solution. For this reason, over the last 40 years, there have been numerous attempts to develop accurate methods for predicting excavatability.

Except the economical and operator parameters, rock properties are used basically for the prediction methods. These methods will be examined in the following sections.

Indirect methods can also be grouped into three parts: single parameter-based approximations, graphical methods, and grading methods. All types are mainly based on the physical and mechanical properties of rock mass and material.

6.4.1 Single Parameter Methods

These systems are purely based on a single parameter, such as uniaxial compressive strength, cutting resistance and specific energy, and seismic velocity.

6.4.1.1 Uniaxial Compressive Strength-Based Methods

Kolleth (1990) suggested an assessment system for excavators, scrapers, surface miners, and BWEs based on the compressive strength of rock samples. The proposed system is shown in Figure 6.4.

6.4.1.2 Cutting Resistance- and Specific Energy-Based Methods

The specific cutting resistance or specific separation force of intact rock (Fa) measured from laboratory tests with Orenstein and Koppel (O&K) wedge test ring has been widely used as an indicator for the assessment of performances and selection of BWEs (Hindistan 1997). Bolukbasi et al. (1991a) compiled the available BWE diggability criteria as presented in Table 6.5. Bolukbasi et al. (1991b) proposed BWE diggability criteria based primarily on cutting specific energy, independent from specimen size and rock anisotropy. Bolukbasi's method is given in Table 6.6.

Hindistan (1997) and Karpuz et al. (2001) proposed a digging difficulty classification system for blasted material at face and for the rehandling of piled material for the cable shovels, which have a dipper capacity of 20 yd^3 (Tables 6.7 and 6.8). Ceylanoglu (1991) and Karpuz et al. (1994) studied the effect of specific cutting energy on the performance parameters and proposed a classification system for power shovels based on the research carried out for stripping works of lignite mines as presented in Table 6.9. The rock mass and material properties used by Ceylanoglu (1991) are given in Table 6.10.

6.4.1.3 Seismic Velocity-Based Methods

These systems are purely based on measurement of seismic velocities of rock masses. For the estimation of rock excavatability (rippability), seismic velocity was started to be used by Caterpillar Co., Peoria, Illinois, in the 1958.

The basic of this method is to measure the ground movement by receivers that are away from the source. There are two types of seismic velocity measurement techniques: seismic refraction and seismic reflection. The refraction method is used and the first arrival movement pulses are taken into account.

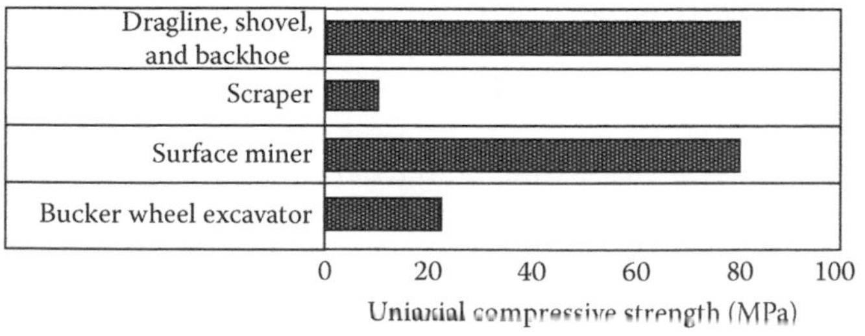

FIGURE 6.4 Excavatability assessment system. (Data from Kolleth, H., *Bulk Solids Handl.*, 10, 29–35, 1990.)

TABLE 6.5
Published BWE Diggability Criteria

Criteria	Class	Cutting Resistance, F_a, MPa
Highvale (Wade and Clark 1989)	Easy	<0.60
	Diggable	0.60–1.10
	Hard	1.10–1.40
	Marginal	1.40–1.80
	Undiggable	>1.80
Goonyella (O'Regan et al. 1987)	Easy	0.15–0.45
	Diggable	0.45–0.60
	Hard	0.60–0.75
	Marginal	0.75–1.00
	Undiggable	>1.00
Neyveli (Rodenberg 1987)	Easy	–
	Diggable	<1.10
	Hard	1.10–2.30
	Marginal	–
	Undiggable	>2.30
Canmet (Weise 1981)	Easy	–
	Diggable	<1.00
	Hard	1.00–1.50
	Marginal	1.50–2.40
	Undiggable	>2.40
Kozlowski (Kozlowski 1981 in Koncagul, 1997)	Easy	<0.17
	Diggable	0.17–0.36
	Hard	0.36–0.54
	Marginal	0.54–0.80
	Undiggable	>0.80
Krzanowski (Krzanowski et al., 1984 in Hindistan, 1997)	Easy	<0.27
	Diggable	0.27–0.90
	Hard	0.90–1.85
	Marginal	–
	Undiggable	>1.85

Source: Koncagul, O., Diggability assessment of bucket wheel excavators in Elbistan Lignite mines, PhD thesis, METU, Ankara, Turkey, 1997.

TABLE 6.6
Bolukbasi's Diggability Assessment for BWE

	Laboratory Specific Energy (MJ/cm^3)	
Rock Class	Minimum	Maximum
Easy	0.5	1.94
Diggable	1.12	3.72
Hard	1.73	4.81
Marginal	2.64	8.58
Undiggable	>2.64	>8.58

Source: Bolukbasi, N. et al., *Mining Sci. Technol.*, 13, 271–277, 1991b.

The success of the method depends on determination of suitable profiles and frequency. It is also dependent on the experienced person to analyze the results required. Although it is widely used, seismic refraction method may sometimes not represent real field conditions of the rock units and it

TABLE 6.7
Diggability Classification for Rehandling Material for 20 yd³ Cable Shovel

Digging Difficulty	Specific Digging Energy kWh/t	Specific Digging Energy kWh/m³	Hourly Digging Capacity t/h	Hourly Digging Capacity m³/h
Easy	≤0.041	≤0.100	>14,000	>6,000
Moderate	0.042–0.053	0.101–0.120	14,000–11,501	6,000–5,201
Mod. Difficult	0.054–0.065	0.121–0.140	11,500–9,000	5,200–4,400
Difficult	>0.065	>0.140	<9,000	<4,400

Source: Hindistan, M., Development of a computer based monitoring system and its usage for power shovels monitoring, PhD thesis, Ankara, Turkey, 1997.

TABLE 6.8
Diggability Classification for Blasted Material

Digging Difficulty	Specific Digging Energy kWh/t	Specific Digging Energy kWh/m³	Hourly Digging Capacity t/h	Hourly Digging Capacity m³/h
Easy	≤0.05	≤0.115	>10,500	>5,000
Moderate	0.051–0.08	0.116–0.16	10,500–8,001	5,000–4,001
Mod. Difficult	0.081–0.11	0.161–0.205	8,000–5,500	4,000–3,000
Difficult	>0.11	>0.205	<5,500	<3,000

Source: Hindistan, M., Development of a computer based monitoring system and its usage for power shovels monitoring, PhD thesis, Ankara, Turkey, 1997.

TABLE 6.9
Ceylanoglu's (1991) Classification System for Power Shovels

	Specific Digging Energy (kWh/m³)			
	Ease of Digging			
Dipper Capacity (yd³)	Easy	Moderate	Moderately Difficult	Difficult
10	≤0.235	0.236–0.300	0.301–0.390	≥0.391
15	≤0.210	0.211–0.275	0.276–0.345	≥0.346
20	≤0.185	0.186–0.250	0.251–0.315	≥0.316
25	≤0.155	0.156–0.230	0.221–0.290	≥0.291

Source: Ceylanoglu, A., Performance monitoring of electrical power shovel for diggability assessment in surface coal mines, PhD thesis, METU, Ankara, Turkey, 1991.

has certain drawbacks. These drawbacks can be outlined as follows.

Penetration stage is one of the most important stages during the ripping and it is independent from seismic velocity. According to Singh et al. (1987), presence of abrasive material does not affect the seismic velocity. However, during ripping, these abrasive materials sometimes become serious obstacles, by shortening the total ripper life.

Fresh boulders or rock columns in a matrix of completely weathered material are often difficult to excavate. But these fresh parts cannot be sensed by seismic velocity and they negatively affect rock rippabilities.

If the spacing between geophones are greater than the thickness of high-velocity layer or where the thickness of the high-velocity layer is less than one-third the thickness of the overlying layer, surface seismic methods may see through the layer (Bradybrooke 1988). The opposite problem to the thin high-velocity layer is the presence of low-velocity material below a high-velocity layer. In this case, the upper layer masks lower layers giving rise to interpretational problems. This can be overcome by up-hole shooting.

When two rock types with approximate mechanical properties are considered, their seismic velocities are near to each other but due to the differences in their water contents, their velocities are different from each other. The reason for this is the misleading of seismic refraction method. Since wet material transmits seismic waves faster, the porous rock with higher water content is seen as hard and compact, hard-to-rip material. For example, porous rock, dry and at 50%–70% saturation, may yield two different seismic velocities, 300 m/s for dry material and it may reach 1350 m/s if material is saturated.

Due to the orientation of main discontinuities with respect to shooting directions up to 1000 m/s reading differences can be observed for the same rock unit. The accuracy of the results of seismic velocity must be in the range of ±20%.

Seismic velocity-based methods are namely Caterpillar (Anon 1994), Komatsu (Anon 1987), Bailey (1975), Church (1981), and Atkinson (1971).

6.4.1.3.1 Caterpillar Method

Caterpillar Company first used the seismic refraction method in 1958 as a means of predicting rippability. The charts that Caterpillar produced show the rippable velocities of several different rock types for each model of bulldozer. In the charts, rocks are divided into categories of *rippable*, *marginal*, and *non-rippable* based on the seismic velocity. As an example, the chart recommended for D11N type dozer is given in Figure 6.5. But manufacturer does not clearly define these terms, and it is difficult to make practical use of this information.

It is clear that as the power rating and weight of dozer increases, rippability also increases.

6.4.1.3.2 Komatsu Method

Similar to Caterpillar, Komatsu Company (Anon 1987) produced charts for their own machines. An example graph prepared for D155 A dozer is shown at Figure 6.6.

TABLE 6.10
Rock Mass and Material Properties Used by Ceylanoglu

Case no	Formation	UCS (MPa)	Schmidt Hardness	Joint Spacing (m)	Number of Joints	Seismic P Wave Velocity (m/s)	Normalized Specific Digging Energy (kWh/m^3)
1	Fresh—slightly weathered marl	103.8	40	1.25	3	2800	0.446
2	Fresh—slightly weathered marl	91.4	41	1.30	3	2908	0.390
3	Fresh—slightly weathered marl	103.8	47	1.30	3	2800	–
4	Fresh—slightly weathered marl	97.1	52	1.50	3	2763	0.565
5	Fresh—slightly weathered marl	97.1	52	0.80	3	2763	–
6	Fresh—slightly weathered marl	88.0	53	1.25	3	2700	0.425
7	Fresh—slightly weathered marl	31.4	42	0.35	3	1855	0.262
8	Moderately weathered marl	15.0	–	0.30	3	1460	0.198
9	Fresh—slightly weathered marl	27.5	46	1.00	3	2440	–
10	Fresh—slightly weathered marl	36.5	49	1.50	3	1972	0.334
11	Fresh—slightly weathered marl	38.3	46	1.50	3	2430	0.377
12	Fresh—slightly weathered marl	11.2	28	0.50	3	1683	0.243
13	Fresh—slightly weathered marl	6.0	–	0.40	3	1683	0.236
14	Fresh—slightly weathered marl	3.9	30	1.20	3	1400	0.224
15	Conglomerate	28.0	–	–	–	3280	0.402
16	Moderately weathered marl	11.2	–	0.40	3	–	0.263
17	Fresh—slightly weathered marl	20.3	–	0.80	3	1860	0.309
18	Fresh—slightly weathered marl	2.1	27	1.00	3	1400	0.234
19	Limestone	86.3	52	3.00	3	2600	0.518

Source: Ceylanoglu, A., Performance monitoring of electrical power shovel for diggability assessment in surface coal mines, PhD thesis, METU, Ankara, Turkey, 1991.

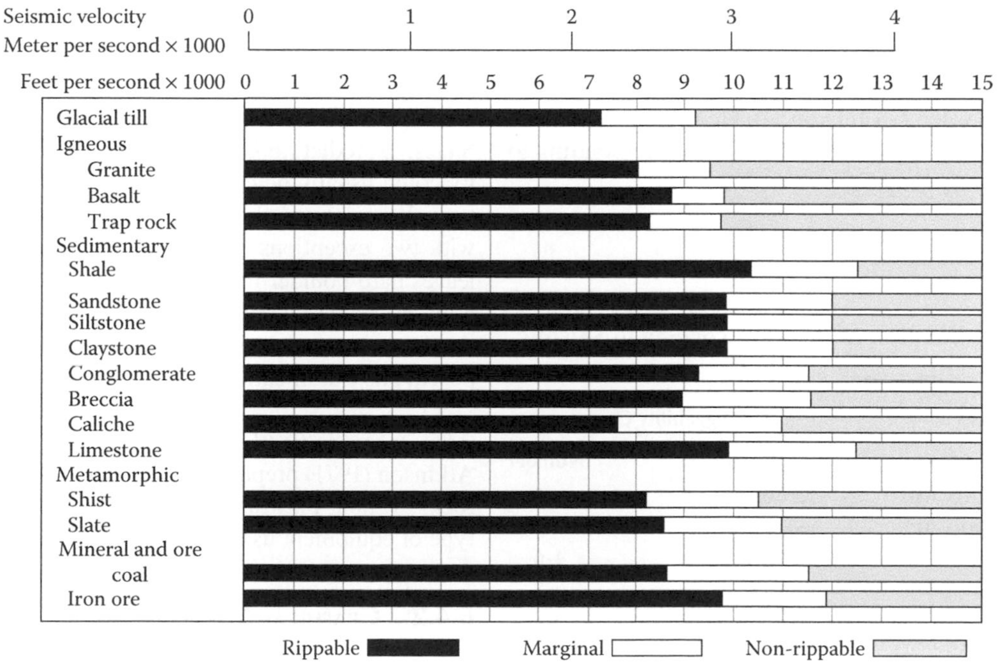

FIGURE 6.5 Rippability assessment chart recommended by Caterpillar Tractor Co. for Cat D11N type dozer. (Data from Anon, *Caterpillar Performance Handbook*, 25th edition, Caterpillar, Peoria, IL, 1994.)

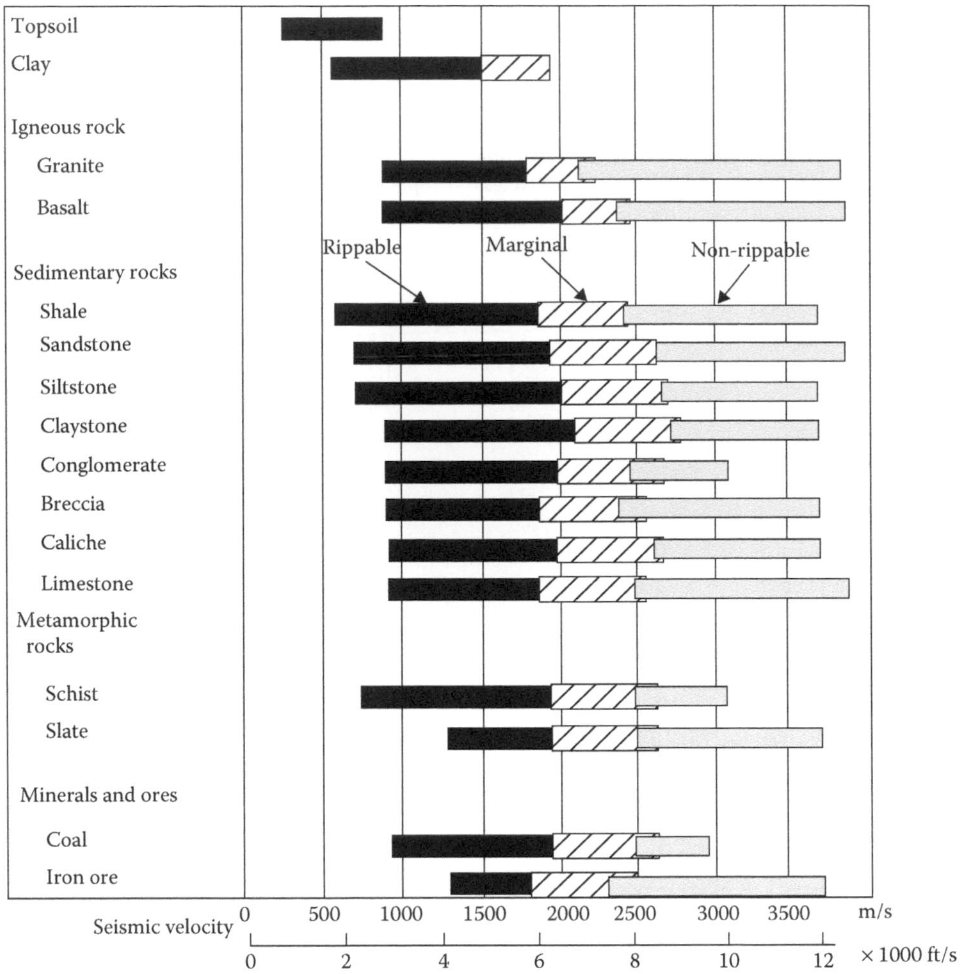

FIGURE 6.6 Rippability assessment chart recommended by Komatsu Ltd. for D155A ripper. (Data from Anon, *Specifications and Application Handbook*, 10th edition, Komatsu, Tokyo, Japan, 1987.)

6.4.1.3.3 Bailey's Method

Bailey (1975) proposed a rippability classification system based on P-wave velocity (Table 6.11). He also presented the description of each class and their numbers corresponding to each P-wave velocity ranges.

TABLE 6.11
Diggability Classification of Rocks according to Seismic Velocity

P-Wave Velocity		Diggability Class	
Ft/sec	**M/sec**	**Definition**	**Number**
1000–2000	305–610	Very easy	1–3
2000–3000	610–915	Easy	3–4
3000–5000	915–1525	Moderate	4–6
5000–7000	1525–2135	Difficult	6–8
7000–8000	2135–2440	Very difficult	6–8
8000–9000	2440–2743	Extremely difficult	8–10

Source: Bailey, A.D., *Highway Geol. Symp. Proc.*, 26, 135–142, 1975.

6.4.1.3.4 Church's Method

Church (1981) collected around 400 excavation sites' data, based on the seismic refraction method in the western United States, to predict rippability. It was reported that seismic velocity could usually determine the rippability regardless of rock type. It is claimed that there is a general weathering pattern with two exceptions. These are spherical weathering, which leaves hard boulder in weathered environment, and columnar jointing, which leaves hard columns in weathered media. For these exceptions, usage of reduced seismic velocity was proposed. His classification system is presented in Table 6.12.

6.4.1.3.5 Atkinson's Method

Atkinson (1971) prepared a chart showing the diggabilities of rocks without blasting, based on their P-wave velocities and type of equipment used as shown in Figure 6.7.

6.4.2 Graphical Methods

These are simple, practical, and quick methods. Because of their practicality, these methods are used especially when quick estimation of excavatability is required.

TABLE 6.12
Diggability Classification of Rock according to Seismic Velocity

Diggability Class	Seismic Velocity, m/s
Direct Digging	<458
Easy	458–1220
Moderately-Difficult	1220–1525
Difficult	1525–1830
Very Difficult	1830–2135
Blasting	>2135

Source: Church, H.K., *Excavation Handbook*, McGraw-Hill, New York, 1981.

The pioneer of these methods is proposed by Franklin et al. (1971) and remaining methods are the modified version of Franklin et al.'s (1971) method. These methods can be outlined as follows: Franklin et al. (1971), Church (1981), Pettifer and Fookes (1997), Bozdag (1988), Tsiambaos and Saroglou (2010), Khamehchiyana et al. (2014), and Basarir et al. (2008). Basarir et al. (2008) proposed a chart relating the specific energy of rock to hourly ripper production and rippability classes. Most recently, Khamehchiyana et al. (2014) and Tsiambaos and Saroglou (2010) proposed classification charts based on rock mass index and geological strength index, respectively.

6.4.2.1 Franklin et al.'s Method

In this method, two parameters are used: joint spacing (If) and point load index value ($I_{s(50)}$). Six construction projects are involved for preparing this chart. Rocks are divided into four groups with respect to their rippabilities. The rippability classes of rocks are given in Figure 6.8 without specifying equipment type.

6.4.2.2 Church's Method

Church (1981) proposed a relationship between seismic shock wave velocities and depth below ground surface for sedimentary, metamorphic, and igneous rocks for their minimum, average, and maximum degrees of weathering as shown in Figure 6.9.

6.4.2.3 Pettifer and Fookes's Method

Pettifer and Fookes (1997) modified the Franklin et al. (1971) graph based on 100 civil engineering and small-scale open-pit operations. They also utilized some data provided by different investigators. Classification system suggests the types of equipments such as Caterpillar dozers, cable and hydraulic shovels, and expected block sizes (Figure 6.10).

6.4.2.4 Bozdag's Method

Bozdag (1988) modified the Franklin's chart based on their research carried out at different sites of Turkish Coal Enterprises. Modification covered various dozer types, including the dozers manufactured by Caterpillar Company. Characteristics and assessment of different cases are shown in Figure 6.11.

6.4.2.5 Tsiambos and Saroglou's Method

Tsiambaos and Saroglou (2010) proposed another classification method for the assessment of ease of excavation of rock masses based on the Hoek et al.'s (1995) geological strength index and the point load strength of the intact rock. The data used for deriving the system originated from the excavation sites in Greece, mainly sedimentary and metamorphic

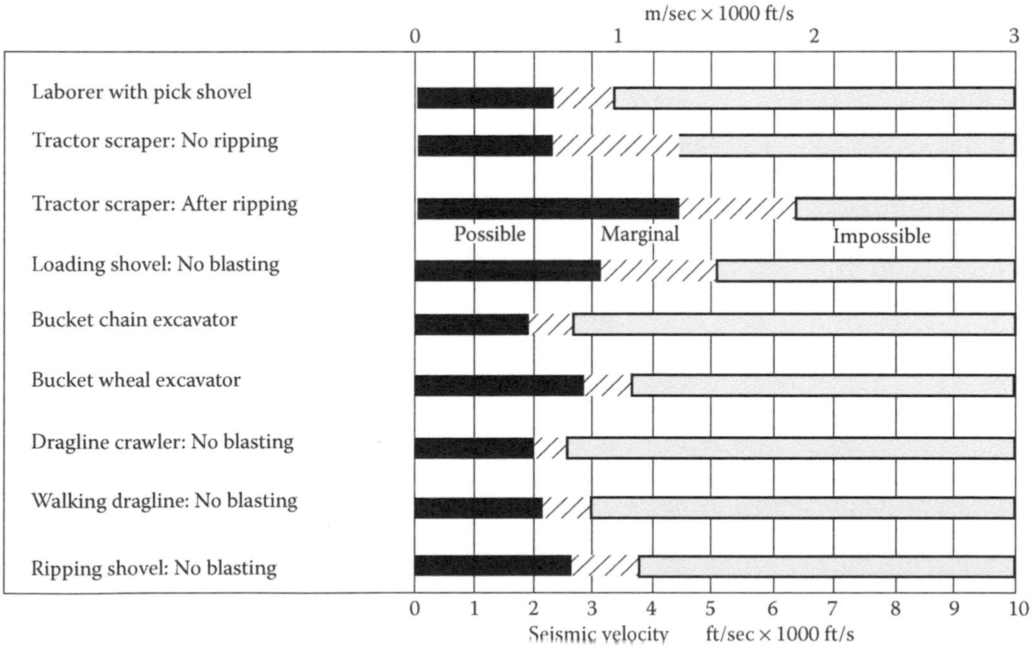

FIGURE 6.7 Seismic velocity method for determination of excavation possibilities. (Data from Atkinson, T., Selection of open pit excavating and loading equipment, *Transactions of the Institution of Mining and Metallurgy*, 80, A101–A129, 1971.)

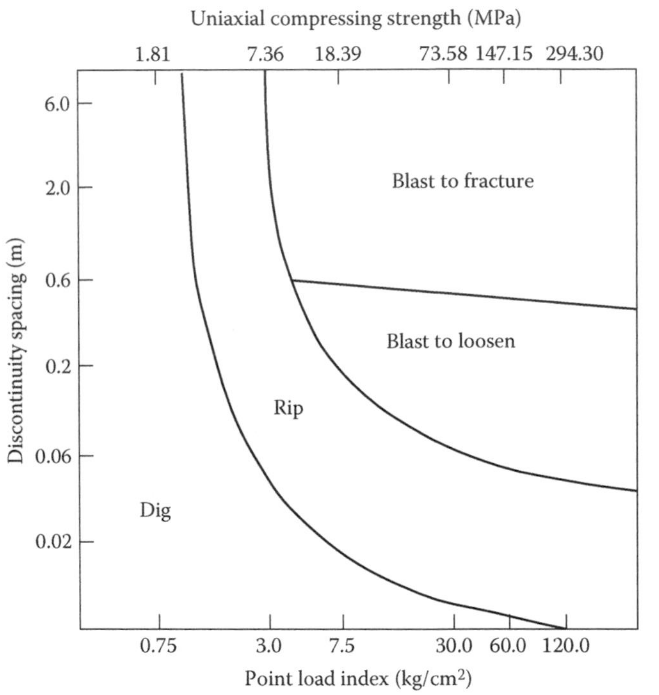

FIGURE 6.8 Rippability chart proposed. (Data from Franklin, J.A. et al., *Trans. Inst. Mining Metall. Sect. A*, 80, 1–9, 1971.)

rock masses. They noted that the method cannot be applied to heterogeneous rock masses and soft and/hard soils. Equipments covered in the field included CAT D6, D7, and D8 type dozers, hydraulic breaker, and hammer. They proposed two charts for the rocks based on a threshold point load strength ($I_{s(50)}$) value of 3 MPa; one included rocks with lower index value and the other was for the higher index value rocks as shown in Figures 6.12 and 6.13.

6.4.2.6 Khamehchiyana et al.'s Method

Khamehchiyana et al. (2014) reviewed previous systems and used data obtained from the surface excavations in dam and hydropower plant sites to propose a new classification system. The proposed system uses the Palmstrom's rock mass index and block volume to represent the rock properties (Figure 6.14).

6.4.2.7 Basarir et al.'s System

Basarir et al. (2008) proposed a rippability classification chart for coal measure rocks relating the specific energies to hourly production of different dozer types ranging from CAT D8 to CAT D11 (Figure 6.15).

6.4.3 Grading Methods

Grading-based classification systems relying on indirect diggability methods were proposed by many researchers. In these

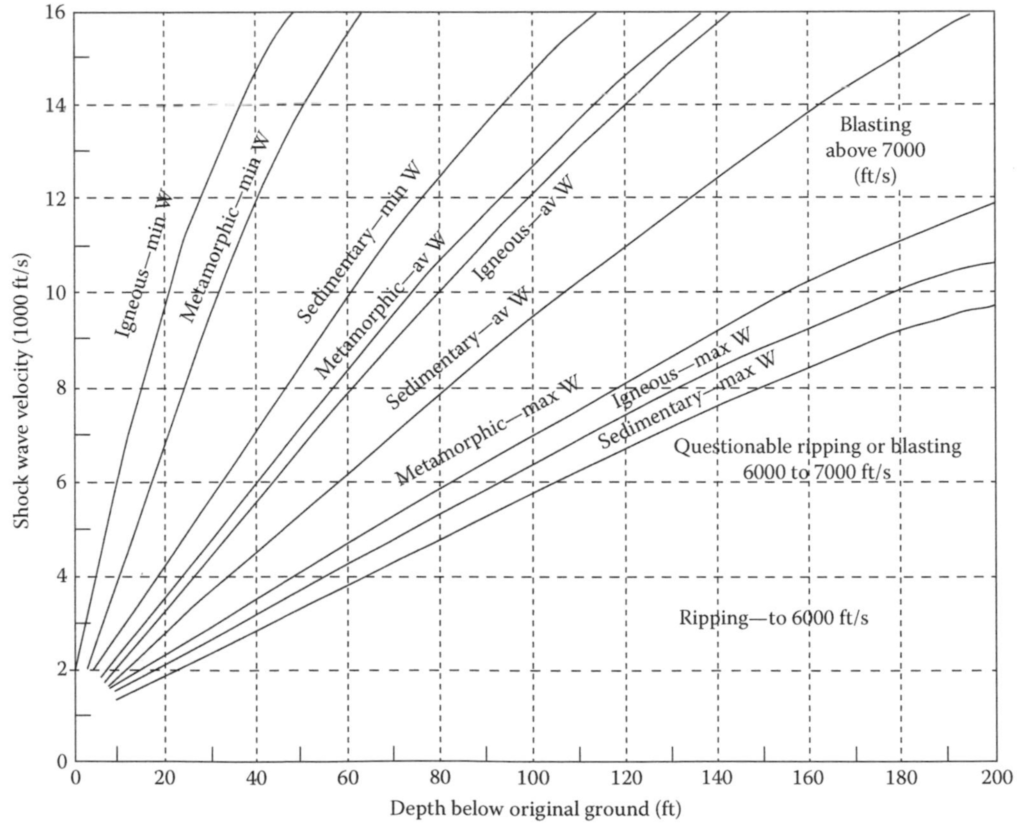

FIGURE 6.9 Church rippability classification system.

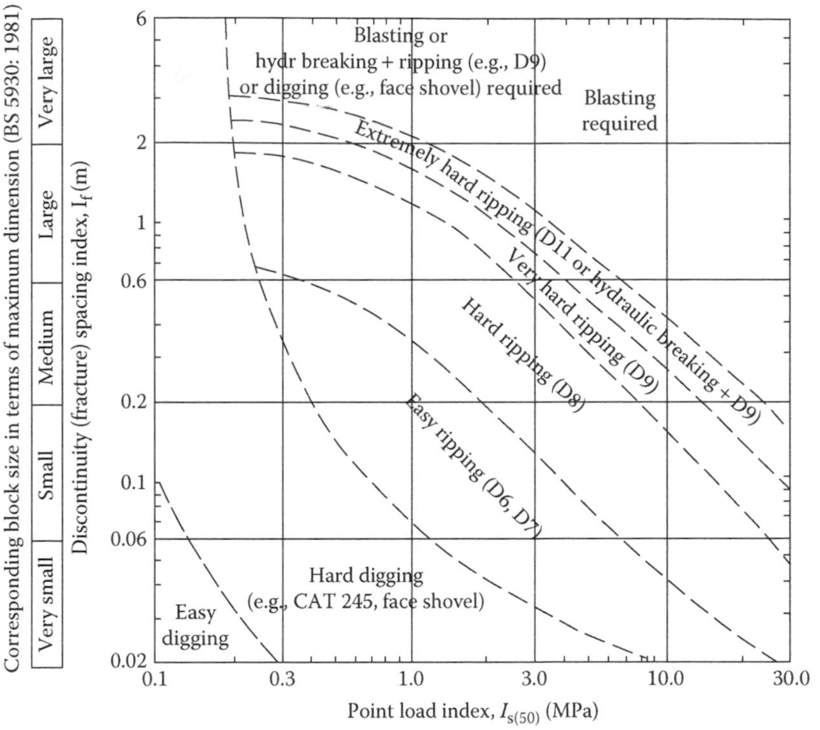

FIGURE 6.10 Rippability chart proposed. (Data from Pettifer, G.S. and Fookes, P.G., *Quart. J. Eng. Geol.*, 27, 145–164, 1997.)

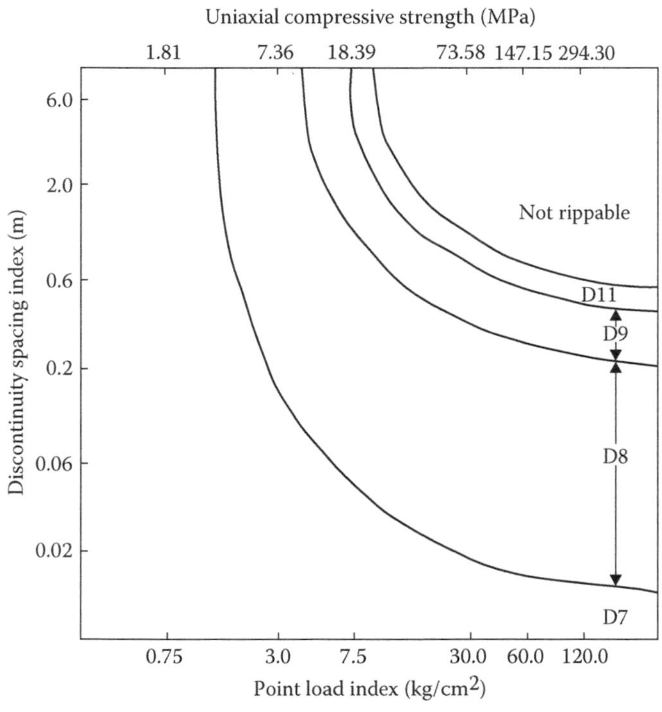

FIGURE 6.11 Rippability chart proposed. (Data from Bozdag, T., Indirect rippability assessment of coal measures rocks, MS thesis, METU, Ankara, Turkey, 86p., 1988.)

systems, rock mass and rock material properties are graded with respect to their importance in diggability. According to total grade that is obtained by summing up all parameters diggability of rock can be determined. The former of these methods is Weaver's (1975) method. Other methods were then developed by Kirsten (1982), Muftuoglu (1983), Smith (1986), Abdullatif and Cruden (1983), Singh et al. (1987), Karpuz (1990), Hadjigeorgiou and Scoble (1990), MacGregor et al. (1994), Kramadibrata (1998), Basarir and Karpuz (2004), and Dey and Ghose (2011). When those classification systems are closely reviewed, it is clearly noticed that the input parameters are different in each system and also the assigned grades to the same parameters show great variations from each other.

6.4.3.1 Weaver's Method

Weaver (1975) suggested a rippability prediction method based on Bieniawski's (1974) geomechanics classification system (RMR). Weaver (1975) considered seismic velocity, rock strength, hardness, weathering, and structure (discontinuities, weakness planes, their dip, and orientation) in determining rippability of rock. The parameters, their grades and rippability classes, and the corresponding selected equipment manufactured by Caterpillar Company, and their horse powers are given in Table 6.13.

6.4.3.2 Kirsten's Method

Kirsten's (1982) parameters were chosen and graded based on Barton et al.'s (1974) Q system. Then he defined an index *N*, called *excavatability index* (Equation 6.8). Effect of groundwater and stresses of ground were ignored in adopting the Q system. Excavatability index and corresponding rippability classes are given in Table 6.14.

$$N = \text{Ms}\frac{\text{RQD}}{\text{Jn}}\text{Js}\frac{\text{Jr}}{\text{Ja}} \tag{6.8}$$

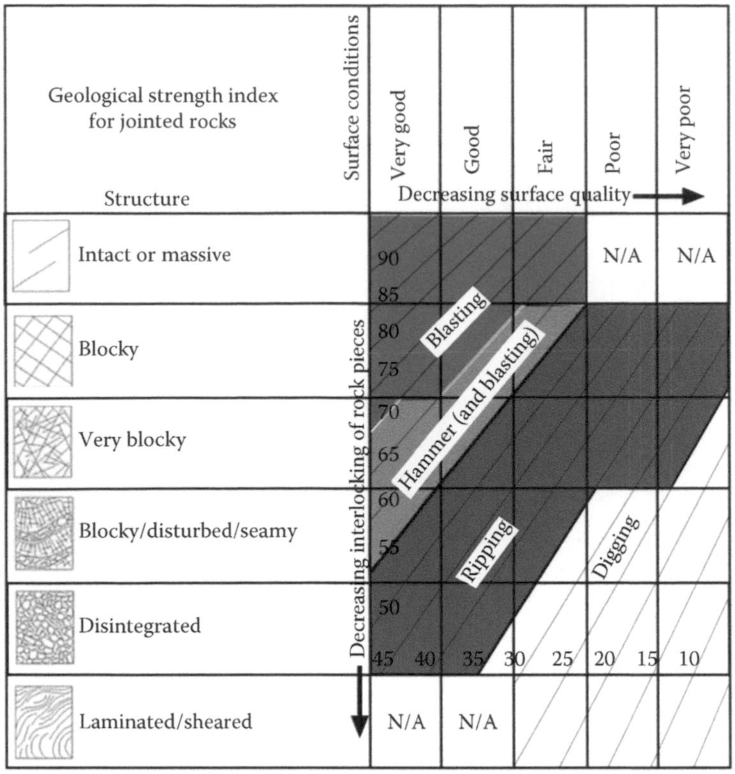

FIGURE 6.12 Geological Strength Index-based diggability classification for rocks with point load strength ≥3 MPa. (From Tsiambos, G. and Saroglou, H., *Bull. Eng. Geol. Environ.*, 13–27, 2010; Hoek, E. and Marinos P., *Tunnels and Tunnelling International*, Part 1, 32(11), 45–51, Part 2, 32 (12), 34–36, 2000.)

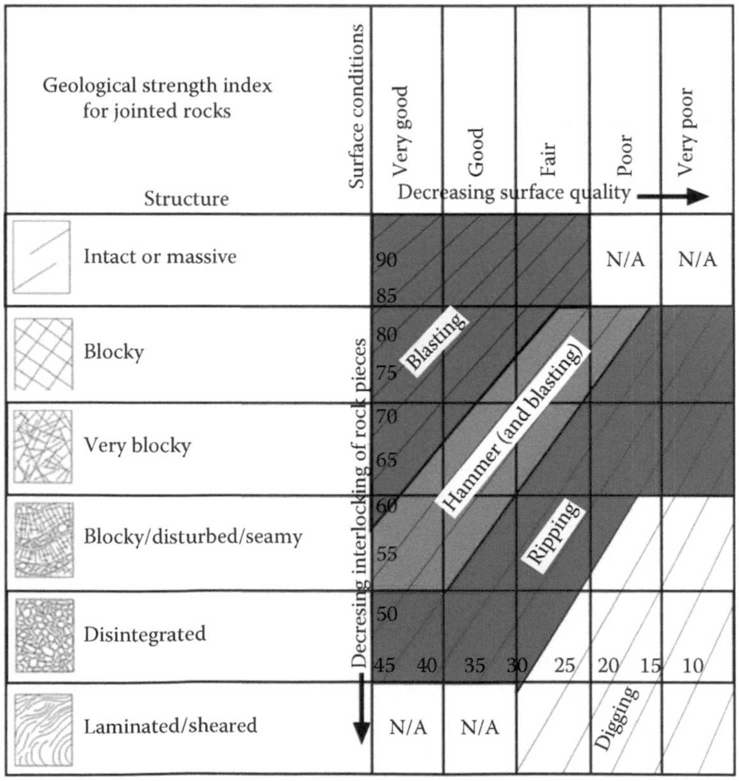

FIGURE 6.13 Geological Strength Index-based diggability classification for rocks point load strength <3 MPa. (From Tsiambos, G. and Saroglou, H., *Bull. Eng. Geol. Environ.*, 13–27, 2010; Hoek, E. and Marinos P., *Tunnels and Tunnelling International*, Part 1, 32(11), 45–51, Part 2, 32 (12), 34–36, 2000.)

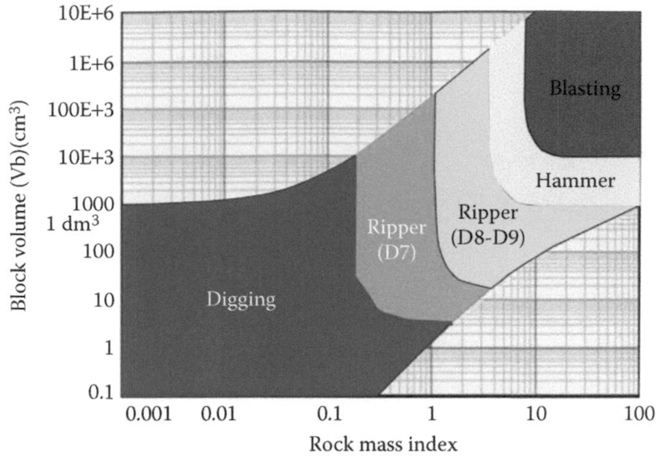

FIGURE 6.14 Excavatability assessment chart based on rock mass index and block size. (Data from Khamehchiyana, M. et al., *Geomechanics and Geoengineering*, 9, 63–71, 2014.)

where:

- Ms is the mass strength number
- RQD is rock quality designation
- Jn is the joint set number
- Js is the relative ground structure number
- Jr is the joint roughness number
- Ja is the joint alteration number

6.4.3.3 Muftuoglu's Method

Type of excavating equipment (ripper, dragline, cable, and hydraulic shovel) and ground conditions are both considered by Muftuoglu (1983), to predict diggability of rock such as sandstone and mudstone. Four main parameters form a basis for diggability index such as weathering, rock strength, joint spacing, and bedding spacing. Joint spacing is described as spacing between two orthogonal directions. Six site experiences are used as database for this classification system. The parameters, their grades, and the classification scheme are given in Tables 6.15 and 6.16.

6.4.3.4 Smith's Method

Smith (1986) prepared a systematic means of numerically weighing six rock parameters to produce a rippability rating chart, which is a modification of Weaver's system. Modified version of Weaver's (1975) rippability rating chart is given in Table 6.17. Rock hardness, rock weathering, joint spacing, joint continuity, joint gauge, and strike and dip orientation are the six parameters considered in Smith's (1986) classification system. There is not any field verification in his classification system. He recommended a method of correlating this rating with refraction seismograph method and tractor horsepower as shown in Figure 6.16.

6.4.3.5 Abdullatif and Cruden's Method

Abdullatif and Cruden (1983) used a database including 23 quarries involving different rock masses such as ball clay, China clay, dolerite, gravel, limestone, sandstone, and shale to develop the classification system. The excavatability of rock masses were evaluated by using point load strength index, Q system and rock mass rating (RMR) system. It was concluded that rock mass could be rated as directly diggable if RMR was less than 30. For the RMR ranging from 30 to 60, the rock mass was described as rippable. For the rock masses having RMR value greater than 60, blasting was suggested.

6.4.3.6 Singh et al.'s Method

Another rippability index for mining applications was suggested by Singh et al. (1987). Six road construction sites were used to develop the system. In this method, abrasiveness of rock, discontinuity spacing, seismic velocity, degree of weathering, and indirect tensile strength are taken into account as presented in Table 6.18. Singh et al. (1987) noted that rock mass can be classified as rippable if the shank penetrates to ground more than 0.6 m. with minimum forward speed of 2.5 km/h.

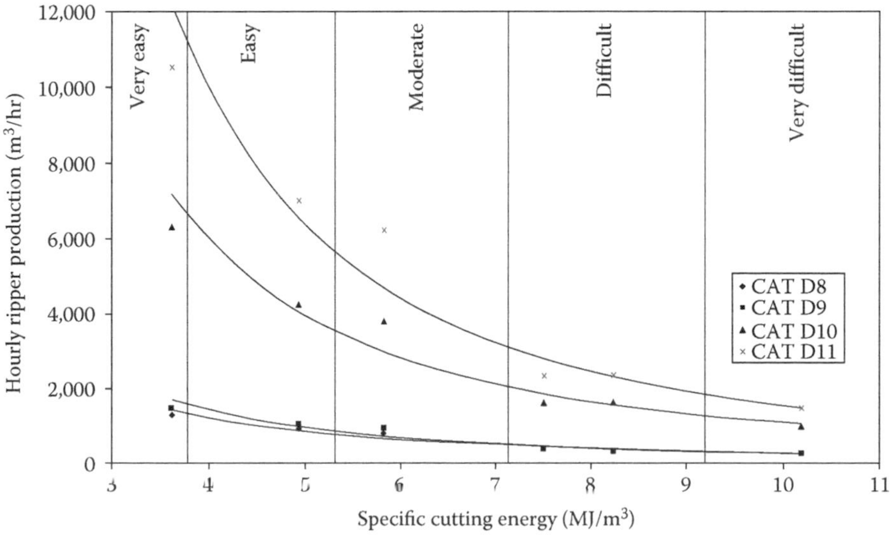

FIGURE 6.15 Rippability classification chart based on specific cutting energy. (Data from Basarir, H. et al., *J. Terramech.*, 45, 51–62, 2008.)

TABLE 6.13
Rippability Rating Chart

Rock Mass	I	II	III	IV	V
Description	**Very Good Rock**	**Good Rock**	**Fair Rock**	**Poor Rock**	**Very Poor Rock**
Seismic Velocity (m/sec)	>2150	2150–1850	1850–1500	1500–1200	1200–450
Rating	26	24	20	12	5
Rock hardness	Extremely hard rock	Very hard rock	Hard rock	Soft rock	Very soft rock
Rating	10	5	2	1	0
Rock weathering	Unweathered	Slightly weathered	Weathered	Highly weathered	Completely weathered
Rating	9	7	5	3	1
Joint spacing (mm)	>3000	3000–1000	1000–300	300–50	<50
Rating	30	25	20	10	5
Joint continuity	Noncontinuous	Slightly continuous	Continuous—no gouge	Continuous—some gouge	Continuous—with gouge
Rating	5	5	3	0	0
Joint gouge	No separation	Slight separation	Separation <1 mm	Gouge <5 mm	Gouge >5 mm
Rating	5	5	4	3	1
Strike dip and orientation *	Very unfavorable	Unfavorable	Slightly unfavorable	Favorable	Very favorable
Rating	15	13	10	5	3
Total rating	100–90	90–70 **	70–50	50–25	<25
Rippability assessment	Blasting	Extremely hard ripping and blasting	Very hard ripping	Hard ripping	Easy ripping
Tractor selection	–	DD9G/D9G	D9/D8	D8/D7	D7
Horse power	–	770/385	385/270	270/180	180
Kilowatts	–	570/290	290/200	200/135	135

Source: Weaver, J.M., *Civil Eng. South Afr.*, 17, 313–316, 1975.

TABLE 6.14
Excavation Index and Corresponding Rippability Classes

Excavation Index, N	Rippability
$N < 0.1$	Hand tools
$0.1 < N < 10$	Easy ripping, D6/D7
$10 < N < 1000$	Hard to very hard ripping, D8/D9
$N > 1000$	Extremely hard ripping to blasting, D10

Source: Weaver, J.M., *Civil Eng. South Afr.*, 17, 313–316, 1975.

6.4.3.7 Karpuz's System

Karpuz (1990) carried out a two-year research work on the diggability of coal measure rocks at different sites of Turkish Coal Enterprises. Karpuz (1990) used the following rock parameters in his assessment method: uniaxial compressive strength, joint spacing, P-wave velocity, weathering, and hardness. The suggested system considers both the ground conditions and equipment types such as ripper, cable, and rope shovel. The parameters used and their grades are given in Table 6.19; the classification scheme according to the grades is shown at Table 6.20.

TABLE 6.15
Diggability Index Rating Method

Class Parameter		I	II	III	IV	V
Weathering		**Completely**	**Highly**	**Moderately**	**Slightly**	**Unweathered**
Rating		0	5	15	20	25
Strength (MPa)	UCS	<20	20–40	40–60	60–100	>100
	$I_{s(50)}$	0.5	0.5–1.5	1.5–2.0	2.0–3.5	>3.5
Rating		0	10	15	20	25
Joint spacing (m)		<0.3	0.3–0.6	0.6–1.5	1.5–2.0	>2.0
Rating		5	15	30	45	50
Bedding spacing (m)		<0.1	0.1–0.3	0.3–0.6	0.6–1.5	>1.5
Rating		0	5	10	20	30

Source: Muftuoglu, Y.V., A Study of factors affecting diggability in British surface coal mines, PhD University of Nottingham, Nottingham, 1983.

TABLE 6.16
Diggability Classification according to Diggability Index Rating

Class	Ease of Digging	Index (W + S + J + B)	Excavation Method	Plant to Be Employed (Without Resort to Blasting) (With Examples)
I	Very easy	<40	1. Ripping	A. Ripper-Scraper Cat D8
			2. Dragline cast	B. Dragline >5 m^3 Lima 2400
			3. Shovel digging	C. Rope shovel >3 m^3 Ruston Bucyrus 71 RB
II	Easy	40–50	1. Ripping	A. Ripper-Scaraper Cat D9
			2. Dragline cast	B. Dragline >5 m^3 Marion 195
			3. Shovel digging	C. Rope shovel >3 m^3 Ruston Bucyrus 150 RB
III	Moderately difficult	50–60	1. Ripping	A. Ripper shovel/F.E. Ldr. Cat D9
			2. Shovel digging	B. Hydraulic shovel > 3 m^3 Cat 245
IV	Difficult	60–70	1. Ripping	A. Ripper shovel/F.E. Ldr. Cat D10
			2. Shovel digging	B. Hydraulic shovel > 3 m^3 Cat 245 or O&K RH240
V	Very difficult	70–95	Shovel digging	Hydraulic shovel > 3 m^3 Cat 245 or O&K RH40
VI	Extremely difficult	95–100	Shovel digging	Hydraulic shovel > 7 m^3 Demag H111, Poclain 1000 CK, P&H 1200, O&K RH75
VII	Marginal without blasting	>100	Shovel digging	Hydraulic shovel > 10 m^3 Demag H185/241, O&K RH300

Source: Muftuoglu, Y.V., A Study of factors affecting diggability in British surface coal mines, PhD thesis, University of Nottingham, Nottingham, 1983.

TABLE 6.17
Modified Version of Weaver's Rippability Rating Chart

Descriptive Classification	Very Good Rock	Good Rock	Fair Rock	Poor Rock	Very Poor Rock
Rock hardness*	Very hard Rock >= 70 MPa	Hard rock 70–25 MPa	Medium hard rock 25–10 MPa	Soft rock 10–3 MPa	Very soft rock >3 MPa
Rating	≥10	5	2	1	0
Rock Weathering	Unweathered	Slightly weathered	Highly weathered	Completely weathered	Completely weathered
Rating	10	7	5	3	1
Joint spacing (mm)	>3000	3000–1000	1000–300	300–50	<50
Rating	30	25	20	10	5
Joint continuity	Noncontinuous	Slightly continuous	Continuous—no gouge	Continuous—some gouges	Continuous—with gouges
Rating	5	5	3	0	0
J. Gouge (mm)	No Sep.	Slight Sep.	Separation < 1	Gouge < 5 mm	Gouge > 5 mm
Rating	5	5	4	3	1
Strike dip and Orient.	Very unfavorable	Unfavorable	Slightly unfavorable	Favorable	Very favorable
Rating	15	13	10	5	3

Source: Smith, H.J., Estimating rippability of rock mass classification, *Proceedings of the 27th US Symposium on Rock Mechanics*, University of Alabama, Tuscaloosa, AL, 443–448, 1986.

6.4.3.8 Hadjigeorgiou and Scoble's Method

Hadjigeorgiou and Scoble (1990) proposed an excavatability assessment system based on excavating index rating (Equation 6.9) of parameters such as block size, material strength, degree of weathering, and relative orientation of discontinuities. The parameters and the allocated grades are shown in Table 6.21; the definitions of excavating classes are given in Table 6.22.

$$\mathrm{EI} = (I_s + B_s) \cdot W \cdot J_s \quad (6.9)$$

6.4.3.9 MacGregor et al.'s Method

MacGregor et al. (1994) developed a method of predicting the productivity of bulldozers ripping in different rock types. They suggested that reasonable estimates of ripper productivity could be obtained from combinations of UCS, degree of weathering (WR), refracted seismic velocity (SV), grain size of the rock (GR), geological structure of the area (SR), number of defect sets (Ds), roughness of the defects (RR), and the defect spacing (DS). The following

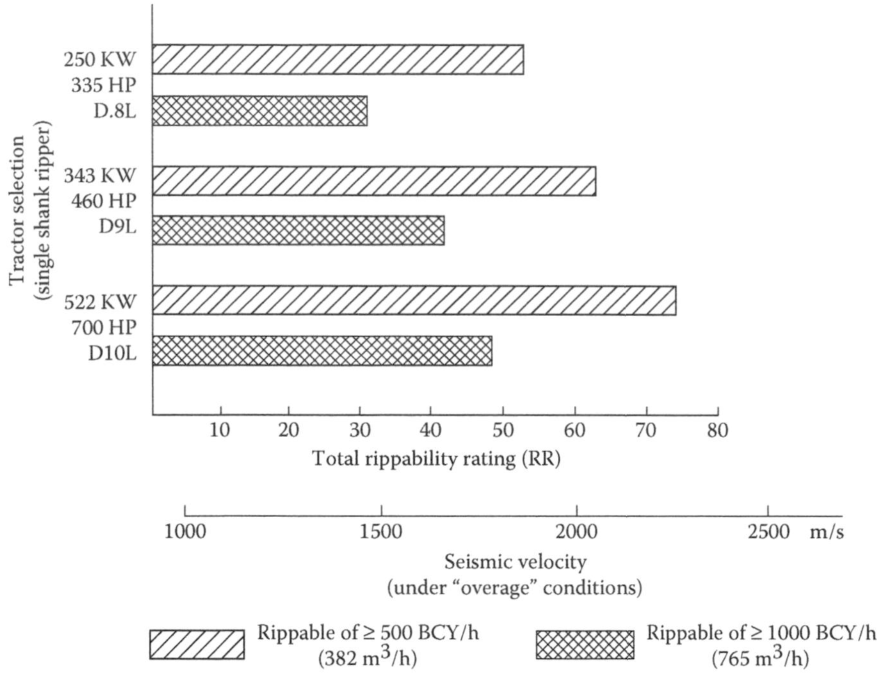

FIGURE 6.16 Correlation of rippability rating index with seismic velocity. (Data from Smith, H.J., Estimating rippability of rock mass classification, *Proceedings of the 27th US Symposium on Rock Mechanics*, University of Alabama, Tuscaloosa, AL, 443–448, 1986.)

TABLE 6.18
Rock Rippability Index

	Rock Class				
Parameters	**1**	**2**	**3**	**4**	**5**
UTS (MPa)	<2	2–6	6–10	10–15	>15
Rating	0–3	3–7	7–11	11–14	14–17
Weathering	Completely	Highly	Moderately	Slightly	Unweathered
Rating	0–2	2–6	6–10	10–14	14–18
Seismic velocity (m/sec)	400–1100	1100–1600	1600–1900	1900–2500	>2500
Abrasiveness	0–6	6–10	10–14	14–18	18–25
Rating	Very Low	Low	Moderately	Highly	Extremely
Disc. spacing (m)	0–5	5–9	9–13	13–18	18–22
Rating	<0.06	0.06–0.3	0.3–1.0	1.0–2.0	>2.0
Total rating	<30	30–50	50–70	70–90	>90
Rippability Asses.	Easy	Moderate	Difficult	Marginal	Blast
Recommended dozer	None—Class 1	Class 2	Class 3	Class 4	–
	Light duty	Medium duty	Heavy duty	Very heavy duty	–
Output (kW)	<150	150–250	250–350	>350	–
Weight (kg)	<25,000	25,000–35,000	35,000–55,000	>55,000	–

Source: Singh, R.N. et al., Development of new rippability index for coal measures excavation, *Proceedings of the 28th US Symposium on Rock Mechanics*, Balkema, Tuscon, AZ, 935–945, 1987.

equation is suggested for reliable estimation of ripping production.

$$\frac{\sqrt{Q}}{W} = 0.469 - 0.00321\text{UCS} + 0.023\text{WR} - 0.0205\text{GR} - 0.00011\text{SV} + 0.0535\text{RR} + 0.0524\text{DS} + 0.0114\text{SR} \quad (6.10)$$

where:

Q is productivity (m^3/h)

W is the weight of the dozer (ton).

Table 6.23 lists the coefficients of variables used in equation for different rock masses.

The ratings of different parameters are specified in MacGregor et al. (1994). If the calculated rippability is less

TABLE 6.19
Parameters and Their Ratings

	Excavation Class				
Parameters	**1**	**2**	**3**	**4**	**5**
UCS (MPa)	<5	5–20	20–40	40–110	>110
$I_s(50)$	0.2	0.2–0.8	0.8–1.6	1.6–4.4	>4.4
Rating	2	5	10	20	25
Joint space (cm)	<30	30–60	60–120	120–200	>200
Rating	5	10	15	20	25
P-velocity (m/sn)	<1600	1600–2000	2000–2500	2500–3000	>3000
Rating	5	10	15	20	25
Weathering	Completely	Highly	Moderately	Slightly	Fresh
Rating	0	3	6	10	10
Hardness (SHV)	<20	20–30	30–45	45–55	>55
Rating	3	5	8	12	15

Source: Karpuz, C.A., *Mining Sci., Technol.*, 11, 157–163, 1990.

TABLE 6.20
Diggability Classification Scheme

Excavation Class	Description	Rating	Excavation Method			When Blasting Necessary	
			Cable Shovel	**Hyd. Excv.**	**Ripping**	**Dril. Rate (m/min)**	**Spec. Charge (kg/m³)**
1	Easy	0–25	Direct dig.	Direct dig.	D7		
2	Medium	25–45	Blast req.	Direct dig.	D8/D9	1.48	130–220
3	Moderately	45–65	Blast req.	Blast req.	D9/D11	1.28	200–280
4	Hard	65–85	Blast req.	Blast req.	D11/Blast	0.57	280–350
5	Very Hard	85–100	Blast req.	Blast req.	Blast	<0.42	>350

Source: Karpuz, C.A., *Mining Sci., Technol.*, 11, 157–163, 1990.

TABLE 6.21
Excavation Index Scheme

Class	**I**	**II**	**III**	**IV**	**V**
$I_{s(50)}$ (MPa)	0.5	0.5–1.5	1.5–2.0	2.0–.3.5	>3.5
Rating	0	10	15	20	25
Block size (B_s)	Very small	Small	Medium	Large	Very large
J_v (joint/m³)	30	10–30	3–10	1–3	1
Rating	5	15	30	45	50
Weathering (W)	Completely	Highly	Moderately	Slightly	Unweathered
Rating	0.6	0.7	0.8	0.9	1.0
Relative ground structure (J_s)	V. favorable	Favorable	Slightly unfavorable	Unfavorable	V. unfavorable
Rating	0.5	0.7	1.0	1.3	1.5

Source: Hadjigeorgiou, J. and Scoble, M.J., Ground characterization for assessment of ease of excavation, *Proceedings of the International Seminar on Mine Planning and Equipment Selection*, Calgary, Alberta, Canada, 323–331, 1990.

TABLE 6.22
Definition of Excavating Classes

Class	Excavation Effort	Index Range
I	Very easy	<20
II	Easy	20–30
III	Difficult	30–45
IV	Very difficult	45–55
V	Blasting	>55

Source: Hadjigeorgiou, J. and Scoble, M.J., Ground characterization for assessment of ease of excavation, *Proceedings of the International Seminar on Mine Planning and Equipment Selection*, Calgary, Alberta, Canada, 323–331, 1990.

than 750 m^3/h it is likely that ripping will be difficult; if it is less than 250 m^3/h, ripping will be very difficult, and in many cases, blasting can be more appropriate and economical than ripping. Additionally, based on trials and the modeling at Nolan's quarry, MacGregor et al. (1994) estimated the limits of ripping in unjointed Hawkesbury Sandstone for Komatsu bulldozers based on UCS as shown in Table 6.24.

6.4.3.10 Kramadibrata's Method

Kramadibrata (1998) used RMR and Q systems to evaluate properties of rock masses in a limestone quarry, an open pit gold mine, and coal mines in Austria, Australia, and Indonesia. The excavatability index used was the same as Kirsten's (1982) index. Initially, the data compiled from the sites above were used to establish a link between excavation index, RMR, and Q values as shown in Figure 6.17 and 6.18.

6.4.3.11 Basarir and Karpuz's Method

Basarir and Karpuz (2004) proposed a rippability classification system based on the field trials at surface coal mines in Turkey. The system involved four quantitative rock properties such as UCS, seismic P-wave velocity, average discontinuity spacing, and Schmidt rebound hardness value. In their system, not only rock properties but also the dozer properties were considered. In the proposed method, rock properties are graded separately and rippability classes of rocks are determined according to the final grade. Appropriate dozer types and their productions are also estimated. Parameters and allocated grades are given in Table 6.25. The suggested rippability classes and hourly ripper productions are given in Table 6.26.

TABLE 6.24
Estimated Upper Limits of UCS

Class	Excavation Effort	Index Range
I	Very easy	<20
II	Easy	20–30
III	Difficult	30–45
IV	Very difficult	45–55
V	Blasting	>55

Source: MacGregor, F. et al., *Quart. J. Eng. Geol.*, 27, 123–144, 1994.

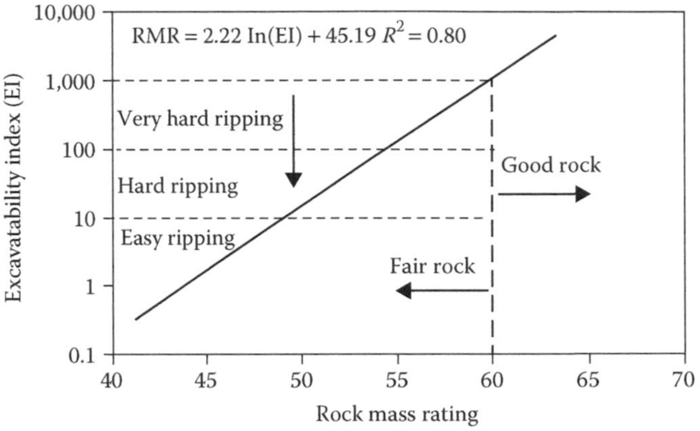

FIGURE 6.17 Relationship between excavation index and RMR. (Data from Kramadibrata, S., The influence of rockmass and intact rock properties on the design of surface mines with particular reference to the excavatability of rock, PhD thesis, Curtin University of Technology, Curtin, Australia, 1998.)

TABLE 6.23
Variables and Coefficients

Class	I	II	III	IV	V
$I_{s(50)}$ (MPa)	0.5	0.5–1.5	1.5–2.0	2.0–3.5	>3.5
Rating	0	10	15	20	25
Block size, B_s	Very small	Small	Medium	Large	Very large
J_v (joint/m^3)	30	10–30	3–10	1–3	1
Rating	5	15	30	45	50
Weathering, W	Completely	Highly	Moderately	Slightly	Unweathered
Rating	0.6	0.7	0.8	0.9	1.0
Relative ground structure, J_s	V. favorable	Favorable	Slightly unfavorable	Unfavorable	V. unfavorable
Rating	0.5	0.7	1.0	1.3	1.5

Source: MacGregor, F. et al., *Quart. J. Eng. Geol.*, 27, 123–144, 1994.

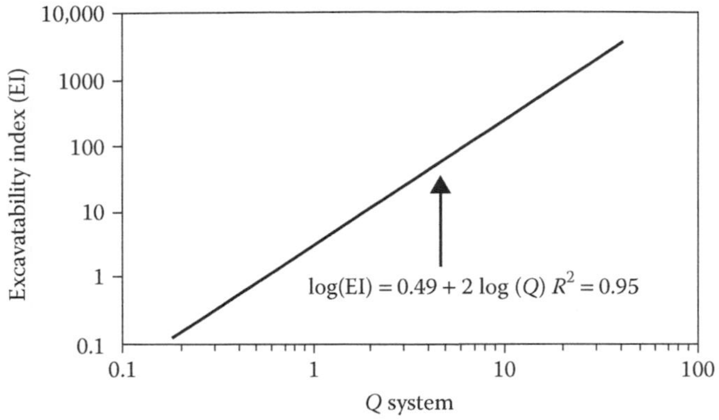

FIGURE 6.18 Relationship between excavation index and Q system. (Data from Kramadibrata, S., The influence of rockmass and intact rock properties on the design of surface mines with particular reference to the excavatability of rock, PhD thesis, Curtin University of Technology, Curtin, Australia, 1998.)

TABLE 6.25
Parameters and Grades

Variables	All Rock Types	Sedimentary	Igneous
Constant	0.469	0.866	−0.138
UCS (MPa)	−0.00321	−0.00736	
Weathering rating, WR	0.0230		0.112
Grain size rating, GS	−0.0205		−0.0599
Seismic velocity (m/sec), SV	−0.00011	−0.000119	−0.000084
Roughness rating, RR	0.0535	0.0496	0.106
No of defect sets, Ds	0.0524		
Structure rating, SR	0.0114		
Defect spacing, DS		−0.00004	−0.000225
R^2	58	52	85
Standard error	0.18	0.17	0.10

Source: Basarir, H. and Karpuz, C., *Eng. Geol.*, 74, 303–318, 2004.

6.4.3.12 Dey and Ghose

Dey and Ghose (2011) reviewed existing cuttability indices and proposed a cuttability index for surface miners. The proposed cuttability index is the sum of the ratings assigned to five parameters such as point load index, volumetric joint count, abrasiveness, direction of cutting with respect to major joint direction, and machine power. The parameters and allocated grades are given in Table 6.27. The assessment of excavatability by a surface miner is presented in Table 6.28.

Dey and Ghose (2011) proposed an equation to estimate the production rate of a surface miner.

$$L^* = \left(1 - \frac{\text{CI}}{100}\right) kM_c \tag{6.11}$$

TABLE 6.27
Rating of the Parameters of Rockmass Cuttability Classification

Class	I	II	III	IV	V
Point load index, $I_{s(50)}$	<0.5	0.5–1.5	1.5–2.0	2.0–3.5	>3.5
Rating, I_s	5	10	15	20	25
Volumetric joint count (no/m³)	>30	30–10	10–3	3–1	1
Rating (J_v)	5	10	15	20	25
Abrasivity (Cerchar)	<0.5	0.5–1.0	1.0–2.0	2.0–3.0	>3.0
Rating (A_w)	3	6	9	12	15
Direction of cutting respect to major joint direction	72°–90°	54°–72°	36°–54°	18°–36°	0°–18°
Rating (J_s)	3	6	9	12	15
Machine power (kW)	>1000	800–1000	600–800	400–600	<400
Rating (M)	4	8	12	16	20

Source: Dey, K. and Ghose, A.K., *Rock Mechanics and Rock Engineering*, 44, 601–611, 2011.

TABLE 6.26
Rippability Classes and Ripper Productions of Different Dozer Types

Grade	D8 Production, m³/h	Dozer Assedsed Class	D9 Production, m³/h	Dozer Assedsed Class	D10 Production, m³/h	Dozer Assedsed Class	D11 Production, m³/h	Dozer Assedsed Class
95–100	0	Blast	0–35	Blast	<150	Very difficult	<250	Difficult
85–95	<250	Very Difficult	<285	Very Difficult	<600	Difficult	<800	Easy
70–85	250–400	Difficult	285–450	Difficult	1,200–1,900	Easy	2,000–3,000	Very easy
55–70	400–900	Moderate	450–1,000	Moderate	1,900–4,300	Very easy	3,000–7,000	Very easy
20–55	900–1,300	Easy	1,000–1,500	Easy	4,300–6,000	Very easy	7,000–10,000	Very easy
0–20	>1,300	Very easy	>1,500	Very easy	>6,000	Very easy	>10,000	Very easy

TABLE 6.28
Assessment of Excavatability of Surface Miner Based on Cuttability Index

Cuttability Index	Possibility of Cutting
50 > CI	Very easy excavation
50 < CI < 60	Easy excavation
60 < CI < 70	Economic excavation
70 < CI < 80	Difficult excavation, may be not economic
CI > 80	Surface miner should not be deployed

Source: Dey, K. and Ghose, A.K., *Rock Mechanics and Rock Engineering*, 44, 601–611, 2011.

where:

L^* is the production or cutting performance (bcm/h)
M_c is the rated capacity of the machine (bcm/h)
CI is the cuttability index
k is a factor varying between 0.5 and 1 for taking into consideration the influence of specific cutting conditions; it is a function of pick array, pick shape, and atmospheric conditions

REFERENCES

Abdullatif, OM, and DM Cruden. The relationship between rock mass quality and ease of excavation. *Bulletin International Association of Engineering Geology* 28, 183–187, 1983.

Anon. *Caterpillar Performance Handbook*. 25th edition, Caterpillar, Peoria, IL, 1994.

Anon. *Specifications and Application Handbook*, 10th edition, Komatsu, Tokyo, Japan, 1987.

Atkinson, T. Selection of open pit excavating and loading equipment. *Transactions of the Institution of Mining and Metallurgy* 80, A101–A129, 1971.

Bailey, AD. Rock types and seismic velocity versus rippability. *Highway Geology Symposium Proceeding* 26, 135–142, 1975.

Barton, N, R Lien, and J Lunde. Engineering classification of rock masses for the design of tunnel support. *Rock Mechanics* 6(4), 189–239, 1974.

Basarir, H, and C Karpuz. A rippability classification system for marls in lignite mines. *Engineering Geology* 74, 303–318, 2004.

Basarir, H, C Karpuz, and L Tutluoglu. Specific energy based rippability classification system for coal measure rock. *Journal of Terramechanics* 45, 51–62, 2008.

Bieniawski, ZT. Geomechanics classification of rock masses and its application in tunneling. *Proceddings of the Third International Congress on Rock Mechanics*. International Society of Rock Mechanics, Denver, CO, 27–32, 1974.

Bolukbasi, N, O Koncagul, and AG Pasamehmetoglu. The effect of anisotrophy on the asessment of diggability for BWE. *International Journal of Surface Mining and Reclamation* 5, 107–111, 1991a.

Bolukbasi, N, O Koncagul, and AG Pasamehmetoglu. Material diggability studies for the assessment of BWE performance. *Mining Science and Technology* 13, 271–277, 1991b.

Bozdag, T. Indirect rippability assessment of coal measures rocks. MS thesis, METU, Ankara, Turkey, 86p, 1988.

Bradybrooke, JC. The state of art of rock cuttability and rippability prediction. *Proceedings of the 5th Australia-New Zealand Conference on Geomechanics*, Sydney, Australia, 13–42, 1988.

Ceylanoglu, A. Performance monitoring of electrical power shovel for diggability assessment in surface coal mines. PhD thesis, METU, Ankara, Turkey, 1991.

Church, HK. *Excavation Handbook*. McGraw-Hill, New York, 1981.

Dey, K, and AK Ghose. Review of cuttability indices and a new rock mass classification approach for selection of surface miners. *Rock Mechanics and Rock Engineering* 44, 601–611, 2011.

Evans, I. Relative efficiency of picks and discs for cutting rock. *Proceedings of the 3rd Congress on Advances in Rock Mechanics*, Washington, D.C., 1399–1405, 1974.

Evans, I. A theory of the basic mechanics of coal ploughing. *Proceedings of International Symposium On Mine Research*, Pergamon, London, Vol. 2, p. 761, 1962.

Fowell, RJ, and AS Pyrcroft. Rock mechaniability studies for the assessment of selecting tunneling machine performance. *Proceedings of the US Mechanics Symposium*, Rolla, MO, 149–158, 1980.

Franklin, JA, E Broch, and G Walton. Logging the mechanical character of rock. *Transactions of the Institute of Mining and Metallurgy, Section A* 80, 1–9, 1971.

Hajugeorgiou, JA, and R Poulin. Assessment of ease of excavation of surface mines. *Journal of Terramechanics* 35, 137–153, 1998.

Hadjigeorgiou, JA, and MJ Scoble. Ground characterization for assessment of ease of excavation. *Proceedings of the International Seminar on Mine Planning and Equipment Selection*, Calgary, Alberta, Canada, 323–331, 1990.

Hadjigeorgiou, JA, and MJ Scoble. Prediction of digging performance in mining. *International Journal of Surface Mining* 2, 237–244, 1988.

Hartman, LH. *SME Mining Engineering Handbook*. Society for Mining, Metallurgy and Exploration, Englewood, CO, 1992.

Hindistan, M. Development of a computer based monitoring system and its usage for power shovels monitoring. PhD thesis, Ankara, Turkey, 1997.

Hoek, E, PK Kaiser, and WF Bawden. *Support of Underground Excavations in Hard Rock*. AA Balkema, Rotterdam, the Netherlands, 1995.

Hoek, E, and P Marinos. Predicting tunnel squeezing. *Tunnels and Tunnelling International* Part 1, 32(11), 45–51. Part 2, 32 (12), 34–36, 2000.

ISRM. *Rock Testing Characterization Testing and Monitoring*. Pergamon Press, New York, 1981.

Karpuz, C. A classification system for excavation of surface coal measures. *Mining Science Technology* 11, 157–163, 1990.

Karpuz, C, MA Hindistan, and T Bozdag. A new method for determining the depth of cut using power shovel monitoring. *Journal of Mining Science* 37, 85–94, 2001.

Karpuz, C, AG Pasamehmetoglu, and A Ceylanoglu. Specific digging energy as a measure of diggability. *Proceedings of the 3rd International Symposium on Mine Planning and Equipment Selection*. Istanbul, Turkey, 1994.

Khamehchiyana, M, MR Dizadji, and M Esmaeili. Application of rock mass index (RMi) to the rock mass excavatability assessment in open face excavations. *Geomechanics and Geoengineering* 9, 63–71, 2014.

Kirsten, HAD. A classification system for excavation in natural materials. *The Civil Engineer in South Africa* 24, 293–308, 1982.

Kolleth, H. Overview of open pit mines for mining technologies with high outputs. *Bulk Solids Handling* 10, 29–35, 1990.

Koncagul, O. Diggability assessment of bucket wheel excavators in Elbistan Lignite mines. PhD thesis, METU, Ankara, Turkey, 1997.

Kramadibrata, S. The influence of rockmass and intact rock properties on the design of surface mines with particular reference to the excavatability of rock. PhD thesis, Curtin University of Technology, Curtin, Australia, 1998.

MacGregor, F, R Fell, and GR Mostyn. The estimation of rock ripping. *Quarterly Journal of Engineering Geology* 27, 123–144, 1994.

Muftuoglu, YV. A Study of factors affecting diggability in British surface coal mines. PhD thesis, University of Nottingham, Nottingham, 1983.

O'Regan, G, AL Davies, and BI Ellerly. Correlation of BWE performance with geotechnical properties of overburden at Gonyella mine, Australia. *Continuous Surface Mining* 381–396, 1987.

Paul, B, and DI Sikarskie. A preliminary theory of static penetration by a rigid wedge into a brittle material. *SME-AIME Transactions* 232, 372–383, 1965.

Pettifer, GS, and PG Fookes. A revision of the graphical method for assessing the excavatability of rock. *Quarterly Journal of Engineering Geology* 27, 145–164, 1997.

Rodenberg, JF. BWE working in extreme climatic and severe digging conditions. *Proceedings of the International Conference on Continuous Surface Mining.* Trans Tech Publications, Claustal, Germany, 1987.

Roxborough, FF, and HR Phillips. Rock excavation by disc cutter. *International Journal of Rock Mechanics and Mining Sciences and Geomechanics Abstract* 12, 361–366, 1975.

Scoble, MJ, and YV Muftuoglu. Derivation of a diggability index for surface mine equipment selection. *Mining Science and Technology* 1, 305–332, 1984.

Singh, RN. *Testing of Rocks Samples from Underwater Trenching Operations of Folkstone for the Central Electricity Generating Board*, Nottingham, Unpublished report, 1–101, 1983.

Singh, RN, B Denby, and I Egretli. Development of new rippability index for coal measures excavation. *Proceedings of the 28th US Symposium On Rock Mechanics.* Balkema, Tuscon, AZ, 935–945, 1987.

Singh, RN, AM Elmherig, and MZ Sunu. Application of rock mass characterization to the stability assessment and blast design in hard rock surface mining excavations. *Proceedings of the 27th US Symposium on Rock Mechanics*, Tuscaloosa, AL, 471–478, 1986.

Smith, HJ. Estimating rippability of rock mass classification. *Proceedings of the 27th US Symposium on Rock Mechanics.* University of Alabama, Tuscaloosa, AL, 443–448, 1986.

Tsiambaos, G, and H Saroglou. Excavatability assessment of rock masses using the Geological Strength Index. *Bulletin of the Engineering Geology and Environment* 69, 13–27, 2010.

Wade, NH, and PR Clark. Material diggability assessment for BWE stripping of plains coal. *International Journal of Surface Mining* 3, 35–41, 1989.

Wagner, H, and EHR Schumann. The stamp-load bearing strength of rock: An experimental and theoretical investigation. *Rock Mechanics* 3, 185–207, 1971.

Weaver, JM. Geological factors significant in the assessment of rippability. *Civil Engineering in South Africa* 17, 313–316, 1975.

Weise, H. Bucket wheel applicability study to plains coal mines. *CANMET*, 1981.

7 Surface Coal Production Methods and Equipments

Nuray Demirel and Celal Karpuz

CONTENTS

Abstract: This chapter presents an overview on six surface coal production methods and equipments used. Unit operations in both cyclic and continuous systems are introduced. Emphasis is given on shovel-truck combination and draglines as the main production machine for cyclic systems and bucket wheel excavators (BWE) as continuous mining systems in strip mining. Draglines and power shovels are also included as a cyclic system for inside dumping. Capacity calculations and corresponding equipment selections to satisfy the required production rate are also presented. Moreover, dozers and graders are covered as auxiliary equipments in surface coal mines.

7.1 INTRODUCTION

Surface coal mining operations require careful planning and effective scheduling to sustain the economic success of mining projects. Since large-scale production and stripping operations can only be achieved by uninterrupted flow of materials, appropriate selection, efficient operation, and well-planned maintenance of mining machines and equipments are critically important. These pieces of equipment are mainly used in excavation and loading, material haulage operations, and auxiliary operations, which support the main production operations.

Excavating-loading equipment used in surface mining is generally classified as cyclic or continuous depending on the continuity of unit operations. Cyclic machines operate in a cyclic manner in which several consecutive work phases are involved. Continuous machines combine rock breakage and materials handling functions, eliminating, in most cases, rock penetration and rock fragmentation; extraction and loading are performed in a single function, which is termed as *excavation*. Appropriate selection of the mining equipment and machinery is highly influenced by a number of key criteria that must be considered. These criteria fall into eight categories: technical, machine operation, geology and deposit characterization, digging and loading, productivity, maintenance, environmental impact, commercial considerations (Pfleider 1968). There are basically six different material excavation and haulage systems as given in Table 7.1.

For surface mining, shovels (electric and hydraulic) trucks, front-end loaders trucks, scrapers, graders, and draglines are the most commonly utilized pieces of equipment having a cyclic nature in operation. Among them, electric/hydraulic shovel-truck systems and front-end loaders truck systems are mostly used for coal excavation and overburden stripping and dumping to the outside of the pit. On the other hand, draglines and power shovels are cyclic systems used for inside dumping. Bucket-wheel excavators and bucket-chain excavators are continuous-type machines used in surface mining, their main use being on soft coals or overburden.

In this chapter, some of the main production machines and support equipments are covered to give an overview on commonly utilized mining machines. It includes BWEs as the main production machines in strip mining as continuous mining systems. For cyclic systems, shovels and front end loaders (FELs) are covered as the main production equipment, and dump trucks are covered as the material haulage units of cyclic systems. Draglines and power shovels and cut methods are included in the chapter as a cyclic system utilized in strip mining operations. Besides, main production and material haulage machines, dozers, and graders are covered as commonly used support auxiliary equipment in surface coal mining operations. For each piece of equipment, the main structures, available sizes and capacities, and productivities are presented. For detailed information, readers should consult the given references.

TABLE 7.1
Methods Used in Surface Coal Mining

	Material Excavation and Handling Methods
1	Draglines
2	Excavator or front-end loader and truck systems
3	Dozer and front-end loader
4	Shovel or grader
5	Shovel or loader, crusher, and conveyor
6	Bucket wheel excavator and conveyor

7.2 ELECTRIC AND HYDRAULIC SHOVELS

In most of the medium-to-large-scale open-pit mining operations, for dense, abrasive, badly-fragmented ground, heavy-duty electric and hydraulic shovels are used. Electric mining shovels are also called as *cable shovels* and *power shovels*. They are bucket-equipped machines, usually electrically powered, used for digging and loading earth or fragmented rock and for mineral extraction. Electric energy is furnished to the machine through a trailing cable and usually powers a large motor-generator set with multiple generators.

They are used principally for excavation of coal and removal of overburden in open cast mining operations. Figure 7.1 presents the use of a cable shovel-truck system in an open-strip mining operation. Once the thickness of overburden layer is decreased by stripping the upper parts of the overburden using electric mining shovel-truck systems to the point where the thickness of the overburden layer is not excessive, then either draglines or power shovels can be utilized for excavation and inside dumping of the waste material.

The work cycle of both hydraulic and electric mining shovels consists of five distinct phases: (1) the digging phase, (2) the swinging phase, (3) the dumping phase, (4) the returning phase, and (5) the positioning phase, sometimes required.

There are four different loading methods for shovels: single back-up method, double back-up method, drive-by method, and modified drive-by system. These four loading methods are distinguished from each other by three distinct features: the position of the shovel, the position of the truck loading, and travel routes of trucks. There are advantages and disadvantages associated with each shovel loading method. The decision made on the loading method depends on the mine planning and production requirements, the size of the equipment fleet, and the material characteristics.

FIGURE 7.1 Shovel-truck system in an open-pit mine.

In the selection of an excavator, the primary concern is the determination of the dipper capacity. For shovels, the dipper capacity can be estimated using Equation 7.1 (SME 1992).

$$C_d = \frac{Q \times t_s \times S}{3600 \times E \times F \times D \times A} \tag{7.1}$$

where:

C_d is the dipper capacity, in yd^3 or m^3
Q is the bank volume of the required production or stripping in an hour
t_s is the cycle time
E is the efficiency coefficient
F is the dipper fill factor (dipper efficiency)
D is the depth cutting coefficient
A is the swing angle correction factor
S is the swell factor

Dipper fill factor is highly correlated with the rock fragmentation efficiency. If the material is well-blasted, dipper fill factor ranges from 80% to 90% and if it is badly fragmented, dipper fill factor may drop to 75%.

Productivity of a shovel is determined by two important parameters: utilization and availability. Electric shovels can normally sustain 75%–80% availability and 80%–90% utilization of availability for a substantial portion of their nominal 20-year life. This amounts to 640 to 770 operating shifts annually, or 42,000 to 5,2000 operating hours for 21 shifts per week operation. Allowing for reasonable delays, the operating hours per shift range from 6.50 to 6.75 (Crawford and Hustrulid 1979).

Selection of required size and number of shovels depends on the required stripping and/or production output and operation factors such as working time, operating conditions, and rock fragmentation (Hartman 1987). Shovel cycle time changes depending on the operating conditions. Table 7.2 presents typical cycle times in seconds, measured in Turkish Coal Enterprises' lignite mines in Turkey, for various dipper capacities at different operating conditions (Karpuz 1990). The details of ease of excavation are given in Chapter 10.

Depth of cut correction is used when a shovel (or dragline) is working at a depth of cut other than optimum depth of cut, which is the depth of cut that produces the greatest output for a given shovel size and type of material. Cycle time correction factor is used to correct the cycle times for different depth of cuts. Table 7.3 presents the cycle time correction factors for different digging depths.

A hydraulic mining excavator, also referred to as a *hydraulic mining shovel*, is available in a face shovel configuration or as a backhoe. The hydraulic mining shovel uses diesel engines or electric motors to drive hydraulic pumps, motors,

TABLE 7.2
Shovel Cycle Times for Various Operating Conditions

C_d		Ease of Excavation[a]			
Yd³	M³	E	M	M-H	H
4	3	18	23	28	32
5	4	20	25	29	33
6	5	21	26	30	34
7	5.5	21	26	30	34
8	6	22	27	31	35
10	8	23	28	32	36
12	9	24	29	32	37
15	11.5	26	30	33	38
20	15	27	32	35	40
25	20	29	34	37	42

Source: Karpuz, C., *Mining Sci. Technol.*, 11, 157–163, 1990.

[a] E is easy, M is medium, M-H is medium to hard, and H is hard digging conditions.

TABLE 7.3
Cycle Time Correction Factors for Different Digging Depths

Optimum digging depth (%)	40	60	80	100
Cycle time correction factor	1.25	1.10	1.02	1.00

and cylinders that in turn actuate the motions required to dig and load material and propel the machine.

The key advantages of using hydraulic excavators over electric mining shovels are (P&H 2003) (1) smaller size yields a mobility advantage, (2) minimal assembly time, (3) has greater control of material discharge, (4) can dig layer by layer and remove large rocks from the digging face, and (5) generally less capital expenditure is required.

7.3 OFF-HIGHWAY MINING TRUCKS

Off-highway mining trucks are used as a haulage equipment integrated with loading and excavation units. They are also called as *dump trucks* (Figure 7.1). Dump trucks are classified into three main groups: rear dump trucks, tractor trailer trucks, and bottom dump trucks. Among them, rear dump trucks are predominantly used for the haulage of a variety of materials. Rear dump trucks, which can have a rigid frame or articulated body, can be electrically or mechanically driven with good maneuverability and flexible systems. The performance of trucks mainly depends on the payload capacity of the truck, haul road characteristics, and loading and dumping conditions. For long distances, they cannot be used very efficiently and economically.

Truck productivity is calculated as a function of truck theoretical capacity, cycle time of truck, and truck availability and utilization factors. Truck cycle time is the summation of individual time components involved in one full haulage cycle of truck and is given in Equation 7.2.

$$T = t_{te} + t_s + t_l + t_{tl} + t_d + t_w \quad (7.2)$$

where:
t_{te} is travel time empty
t_s is spot time at the shovel
t_l is loading time
t_{tl} is travel loaded time
t_d is dump time
t_w is waiting time, all in minutes

Time to spot, load, and dump can be estimated or calculated readily, but travel times are read from the rimpull-speed-gradability characteristics charts prepared for trucks. Load time can be calculated for the excavation equipment selected. Spot and dump times are usually estimated based on operating conditions. Table 7.4 presents spot and dump times for both trucks and trailers (Anon. 1981). Knowing cycle time T and theoretical capacity of truck (C_t) (in ton) and job efficiency factor (E), productivity of the truck (in ton/hour) (P_T) can be estimated using Equation 7.3 (SME 1992).

$$P_T = \frac{3600 \times C_t \times E}{T} \quad (7.3)$$

In order to maintain the full haulage fleet in operation even if breakdowns occur, spare units are usually purchased. For every five to six production units at the mine, one spare is provided (Hartman 1987).

In assessing the efficiency of any shovel-truck system, there are three main focus areas that must be investigated: (1) ore and waste production schedules, haulage routes, and operating conditions; (2) equipment productive capacities including availability and utilization in percent, productivity, and effects of interaction between shovels and haulage trucks on productivity; and (3) unit capital and operating costs. Generally, shovel performance is of primary concern due to the fact that truck performance is highly influenced by shovel choice and its performance (Crawford and Hustrulid 1979).

In shovel-truck systems, good match of the capacity and number of trucks per shovel plays an important role for the

TABLE 7.4
Spot and Dump Times for Trucks and Trailers

	Spot Times (min)		Dump Times (min)	
Conditions	Trucks	Trailers	Trucks	Trailers
Favorable	0.15	0.15	1.0	0.3
Average	0.30	0.50	1.3	0.6
Unfavorable	0.50	1.00	1.8	1.5

Source: Anon., *Production and Cost Estimation of Material Movement with Earthmoving Equipment*, Terex Corporation, Hudson, OH, 82 pp, 1981.

efficiency and productivity of the mine. For a proper match of shovel and truck, a truck size with ±5% of the calculated load should be selected. A good match of shovel truck ensures that the shovel does not wait for the trucks and the truck wait is not excessive (more than a few minutes). Therefore, for correct shovel-truck matching, both shovel time and truck idle time should be minimized. For a good match, the number of dipper fills range from three to six and the optimum is found to be four. That means the truck should be filled with four passes of shovel dipper fills. Inappropriate shovel-truck match affects not only production but also capital cost, operating costs, and maintenance costs. Particularly, shovel as a main production unit should not wait for the trucks. Synchronization of shovel truck system can be checked by Equation 7.4 (Hartman 1987).

$$t \le n\left(t_1 + t_s\right) \tag{7.4}$$

where:

n is number of trucks
t is the cycle time for one truck
t_1 is the loading time
t_s is the spot time

For a synchronized cycle, t must be less than the time required to spot and load the truck fleet (Hartman 1987). Table 7.5, as an example, presents the appropriate number of passes for different capacities of dippers and trucks for both electric and hydraulic shovels.

7.4 FRONT-END LOADERS

FELs are primarily used for loading trucks, hoppers, and/or conveyors. They generally need the loosening of the ground. They can also be utilized for cleaning around large excavators. FEL has been mainly restricted to weak overburden stripping and auxiliary duties. Compared with loading shovels, they are short-life machines not greatly affected by obsolescence (SME 1992). They can be crawler-mounted or trackless. Crawler-mounted FELs have stronger digging capability, low ground pressure, perform better on gradients, high maneuverability but slow travel speeds, and high maintenance in abrasive conditions. On the other hand, trackless FELs are very mobile, lower maintenance except for tires in sharp abrasive conditions, less maneuverability, well-suited to narrow benches, high travel speeds, and high ground pressures. Life spans of crawler-mounted FELs range from 7,000 to 12,000 h, and those of trackless FELs range from 1,500 to 5,000 h (CAT 2000).

When compared to shovels, FELs have extremely good mobility, are more versatile, can operate on moderate grades, have lower capital cost, require one operator only, and lumps are not trapped in bucket.

TABLE 7.5
Appropriate Number of Dipper Fills for Different Capacity of Dump Trucks

HMS Truck/Shovel Match-Up

Truck Rated Capacity	Shovel Capacity				
Metric Ton (US ton)	21 m³ (28 yd³)	25 m³ (33 yd³)	35 m³ (46 yd³)	44 m³ (57 yd³)	56 m³ (73 yd³)
154 (170)	5 pass	4 pass	3 pass		
172 (190)	5 pass	4 pass	3 pass		
186 (205)	5–6 pass	4–5 pass	3 pass		
218 (240)	6–7 pass	5 pass	4 pass		
231 (255)	6–7 pass	5–6 pass	4 pass		
290 (320)		7 pass	5 pass		
327 (360)					
363 (400)					

EMS Truck/Shovel Match-Up

Truck Rated Capacity	Shovel Capacity				
Metric Ton (US ton)	21 m³ (28 yd³)	25 m³ (33 yd³)	35 m³ (46 yd³)	44 m³ (57 yd³)	56 m³ (73 yd³)
154 (170)	5 pass	3–4 pass	3 pass	2 pass	
172 (190)	5 pass	4 pass	3 pass	2–3 pass	2 pass
186 (205)	5–6 pass	4–5 pass	3 pass	3 pass	2 pass
218 (240)	6–7 pass	5 pass	3–4 pass	3 pass	3 pass
231 (255)	7 pass	5–6 pass	4 pass	3 pass	2–3 pass
290 (320)		7 pass	5 pass	3–4 pass	3 pass
327 (360)		8 pass	5–6 pass	4 pass	3–4 pass
363 (400)		8 pass	6 pass	5 pass	4 pass

Source: P&H, *MinePro Services Peak Performance Practices, Excavator Selection*, Harnischfeger Corporation, Milwaukee, WI, 2003.

Note: HMS, Hydraulic Mining Shovel; EMS, Electric Mining Shovel.

7.5 DRAGLINES

Dragline is a large-scale mining equipment for removing overburden and interzone rock intervals in strip/open cast mining. Strip mining is only practical when the ore body to be excavated is relatively near the surface or the above thick overburden is removed by a shovel-truck system. The rock is removed along a long strip by dragging a bucket capable of holding up to 220 yd^3 until the top of the coal is exposed. Waste rock is deposited behind the active mining area on land under which the coal has already been removed. Walking draglines are the predominantly used type, and they have larger bucket capacities (Figure 7.2).

Figure 7.3 presents a typical walking dragline and its main components.

FIGURE 7.2 Dragline in an open castlignite mine.

Walking draglines cover an extensive range of bucket capacities, from approximately 7 to 168 m^3 (10 to 220 yd^3). They can employ 16,500-hp motors on the swing drive and have boom lengths varying from approximately 37–128 m (120–420 ft) (Humphrey 1990).

Some technical specifications of draglines are listed as follows (Humphrey 1990):

Bucket capacity	7–168 m^3 (10–220 yd^3)
Boom length	37–128 m (120–420 ft)
Boom angle	32
Pit length	300–3000 m
Pit width	25–60 m
Maximum allowable load	250,000 to 340,000 kg (550,000 to 750,000 lb)
Maximum working weight	6,580,000 kg (14,500,000 lb)
Power	48,500 hp (36,000 kw)

In a strip mine, the number of loading machines is limited and their reliability and flexibility are very important (Ebrahimi et al. 2003). The dragline is operated constantly throughout the year except for planned and unplanned maintenance actions. Therefore, if the dragline is idle for a few minutes, heavy losses will occur in terms of depreciation and production. Townson et al. (2003) claimed that the loss of revenue associated with dragline downtimes could be substantial, and in extreme cases, it could reach up to $1 million/day. Analysis of the production process requires the basic understanding of two job parameters, which include capacity and time, or meters and minutes, or loads and cycles. Thus, the mining selectivity, productivity, reliability, and flexibility are essential factors for loading machine selection (Ebrahimi et al. 2003). There are several different approaches to define dragline productivity. Reddy and Dhar (1988) expressed productivity as in Equation 7.5.

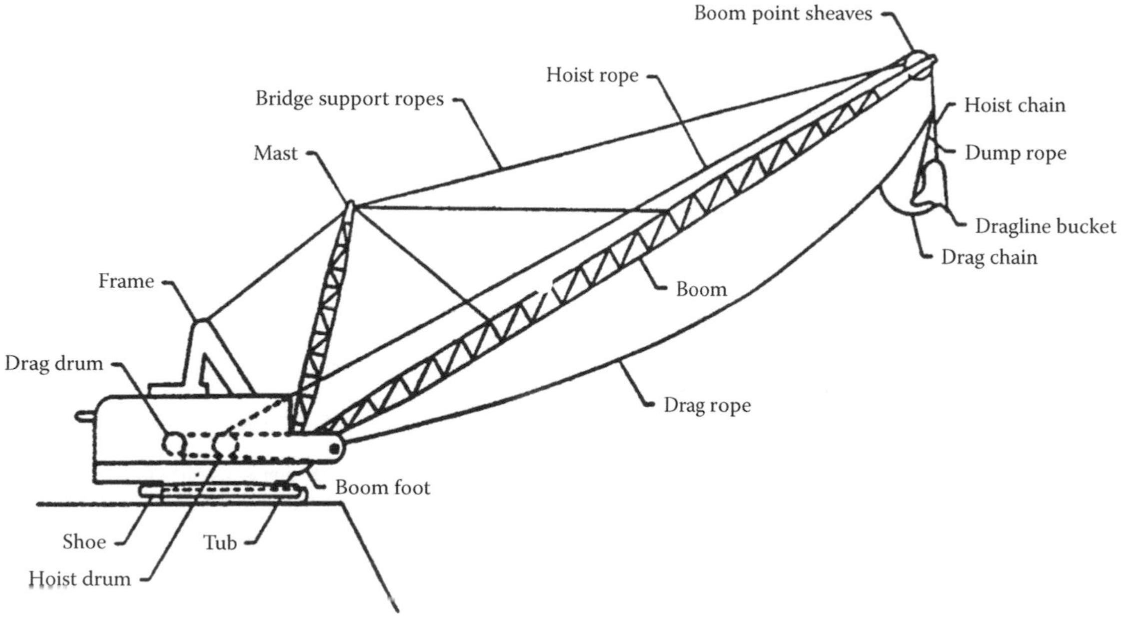

FIGURE 7.3 Schematic of a dragline showing major components. (From Bucyrus-Erie, *Bucyrus-Erie Surface Mining, Supervisory Training Programme*, Shovel/Truck, South Milwaukee, WI, 1979.)

$$\text{Productivity} = \text{material carried} \times \text{cycles/h} \quad (7.5)$$

From Equation 7.5, the average cycle time and bucket capacity are required to calculate average productivity of a dragline. Therefore, the necessary input data is the total time consumed to perform a complete cycle (one cycle comprises a number of time elements). A basic approach to estimating dragline production involves the use of a standard cyclic excavator equation such as Equation 7.6, which has been modified to yield monthly dragline output (Rai et al. 2000). It is obvious from Equation 7.6 that productivity increases by decreasing the amount of rehandle and cycle time.

$$O = \frac{B \times \text{BF} \times \text{HS} \times A \times J \times 3600}{(1+S) \times C \times (1+R)} \quad (7.6)$$

where:

- O is dragline output in m^3
- B is bucket capacity in m^3
- BF is bucket fill factor
- HS is hours scheduled per month
- A is availability
- J is the job efficiency factor
- S is swell factor
- C is cycle time in seconds
- R is rehandle amount

Table 7.6 presents a list of bulk densities, swell factors, and diggabilities for common materials.

Draglines, if properly used, have several advantages over shovel-truck systems. First of all, planning and execution of mining using dragline is simple unless multiple coal beddings and geology offers complexities. Also, handling of waste material is minimized due to the fact that draglines do not need extra haulage equipment. There are also several disadvantages due to high capital expenditures and high maintenance costs. This high maintenance cost makes production focus on a few dragline units. In addition to this, dragline mobility and flexibility is less efficient when compared to shovels.

One cycle of a typical dragline operation consists of five stages (Figure 7.4). At first, an empty bucket is placed in position for filling, then the bucket of the dragline is dragged toward dragline to be filled with overburden material. In the next step, the filled bucket is simultaneously hoisted and swung over spoil pile. Then the material is dumped on the spoil pile and finally the bucket is swung back and is simultaneously lowered to digging position.

Overburden stripping using draglines requires initially a key cut and then a production cut. For the key cut, a wedge of overburden adjacent to highwall is removed to form a free face for excavation. Then the dragline is positioned in line with the key cut, so that the material removed could form a spoil pile toe. In this case, a highwall slope and pit width is established and the third digging surface for production cut is provided. Designing the key cut is important to reduce swing angles for main cuts and to protect highway from disturbance during the digging of main cuts. Plan and cross-sectional views of initial key cut excavation is presented in Figure 7.5.

TABLE 7.6
Bulk Density, Swell Factor, and Diggability of Common Materials

Rock	Bank Density[a] t/m^3	Bank Density[a] lb/yd^3	Swell Factor	Fillability	Diggability[b]
Asbestos ore	1.9	3200	1.4	0.85	M
Basalt	2.95	5000	1.6	0.80	H
Bauxite	1.9	3200	1.35	0.90	M
Chalk	1.85	3100	1.3	0.90	M
Clay (dry)	1.4	2400	1.25	0.85	M
Clay (light)	1.65	2800	1.3	0.85	M
Clay (heavy)	2.1	3600	1.35	0.80	M-H
Clay and gravel (dry)	1.5	2500	1.3	0.85	M
Clay and gravel (wet)	1.8	3000	1.35	0.80	M-H
Coal (anthracite)	1.6	2700	1.35	0.90	M
Coal (bituminous)	1.25	2100	1.35	0.90	M
Coal (lignite)	1.0	1700	1.3	0.90	M
Copper ores (low-grade)	2.55	4300	1.5	0.85	M-H
Copper ores (high-grade)	3.2	5400	1.6	0.80	H
Earth (dry)	1.65	2800	1.3	0.95	E
Earth (wet)	2.0	3400	1.3	0.90	M
Granite	2.41	4000	1.55	0.80	H
Gravel (dry)	1.8	3000	1.25	1.0	E
Gravel (wet)	2.1	3600	1.25	1.0	E
Gypsum	2.8	4700	1.5	0.85	M-H
Ilmenite	3.2	5400	1.4	0.85	M
Iron ore 40% Fe	2.65	4500	1.4	0.80	M-H
Iron ore +40% Fe	2.95	5000	1.45	0.80	M-H
Iron ore +60% Fe	3.85	6500	1.55	0.75	H
Iron (taconite)	4.75	8000	1.65	0.75	H
Limestone (hard)	2.6	4400	1.6	0.80	M-H
Limestone (soft)	2.2	3700	1.5	0.85	M-H
Manganese ore	3.1	5200	1.45	0.85	M-H
Phosphate rock	2.0	3400	1.5	0.85	M-H
Sand (dry)	1.7	2900	1.15	1.0	E
Sand (wet)	2.0	3400	1.15	1.0	E
Sand and gravel (dry)	1.95	3300	1.15	1.0	E
Sand and gravel (wet)	2.25	3800	1.15	1.0	E
Sandstone (porous)	2.5	4200	1.6	0.80	M
Sandstone (cemented)	2.65	4500	1.6	0.80	M-H
Shales	2.35	4000	1.45	0.80	M-H

Source: Atkinson, T., Selection and sizing of excavating equipment. In *SME Mining Engineering Handbook*, 2nd ed., Vol. 2. Edited by H.L. Hartman, SME, Littleton, CO, 1992.

[a] These figures vary from location to location and tests should be made where possible. Allowance should be made for operation in wet conditions as density varies with moisture content.

[b] Diggability is based on shovel dippers.

FIGURE 7.4 Typical area mining dragline stripping operations.

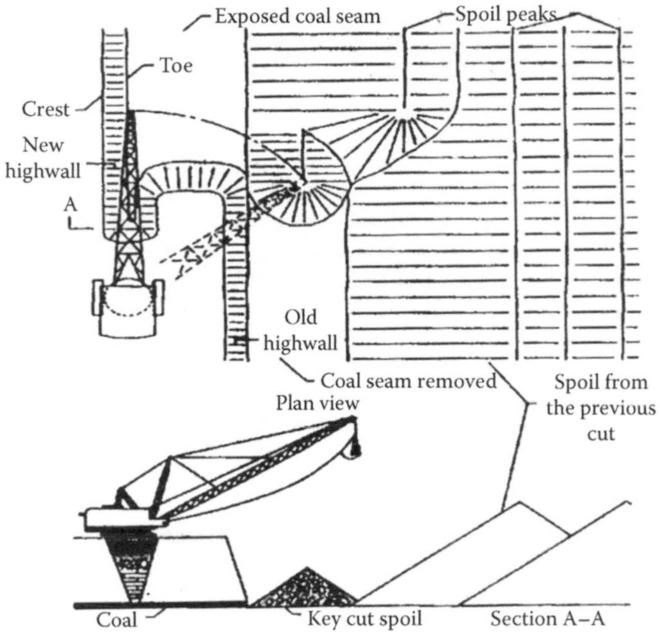

FIGURE 7.5 Plan and cross-sectional views of the removal of the initial key cut. (From Bucyrus-Erie, *Bucyrus-Erie Surface Mining, Supervisory Training Programme*, Shovel/Truck, South Milwaukee, WI, 1979.)

Initial key cut creation can be done using different methods:

1. *End cut (no rehandle)*: Dragline begins at one end of proposed pit. Overburden side casted to one side only. Swing angles limited to 60° (Figure 7.6).
2. *End cut (rehandle)*: Dragline spoils overburden on both sides of pit. Spoil in line of advance is rehandled. Fifty percent of material could be rehandled. Rehandle can be used to prepare dragline pads (Figure 7.7).
3. *Side cut (no rehandle)*: Dragline begins at one end of proposed pit. Overburden dumped far away from pit advance. Swing angles limited to 130° (Figure 7.8).
4. *Borrow pit (rehandle)*: Initial shallow pit parallel to waste side. Borrow pit waste set far away from box cut. Borrow pit large to cater for swell. Box pit material placed in borrow pit till adequate space for spoiling directly into pit (Figures 7.9 and 7.10).

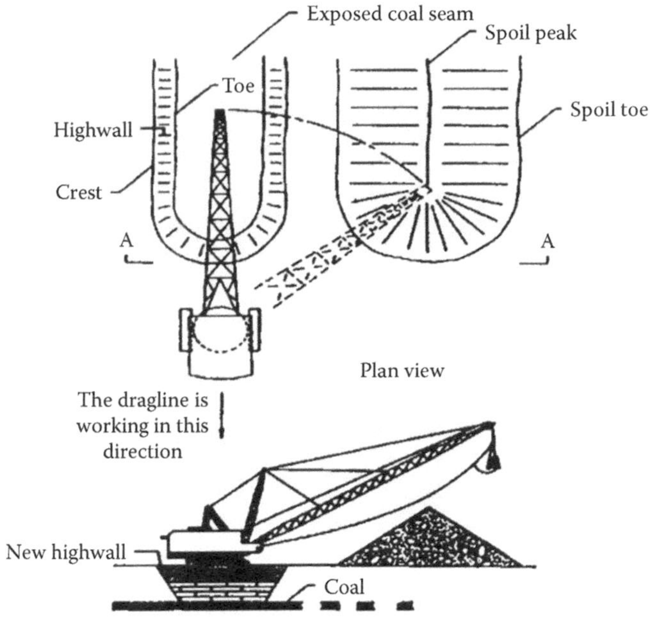

FIGURE 7.6 Plan and cross-sectional views of a box pit with end cut with no rehandle method. (From Bucyrus-Erie, *Bucyrus-Erie Surface Mining, Supervisory Training Programme*, Shovel/Truck, South Milwaukee, WI, 1979.)

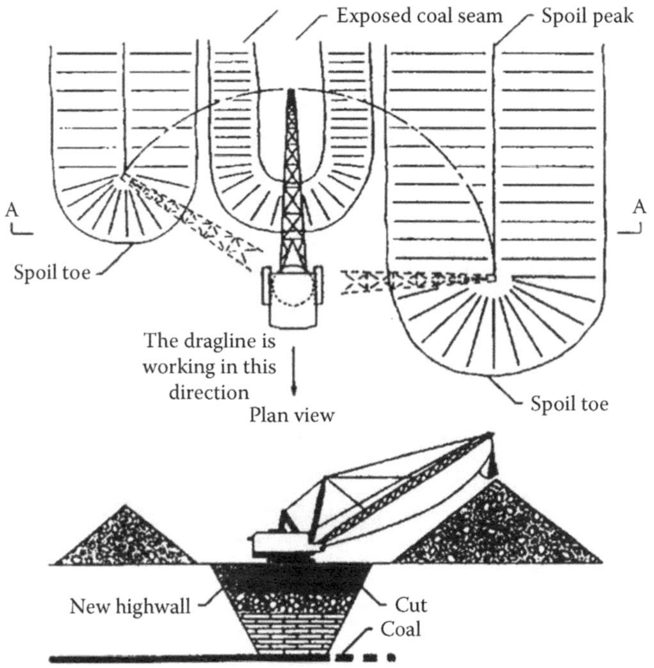

FIGURE 7.7 Plan and cross-sectional views of a box pit with end cut with rehandle method. (From Bucyrus-Erie, *Bucyrus-Erie Surface Mining, Supervisory Training Programme*, Shovel/Truck, South Milwaukee, WI, 1979.)

Dragline mining method selection depends on factors such as the bank height, nature of the overburden, highwall angle and stability, spoil pile angle of repose and stability, and the dragline dimensions.

Rehandling is applied when the reach factor of the dragline is not compatible with the depth of the overburden material. If

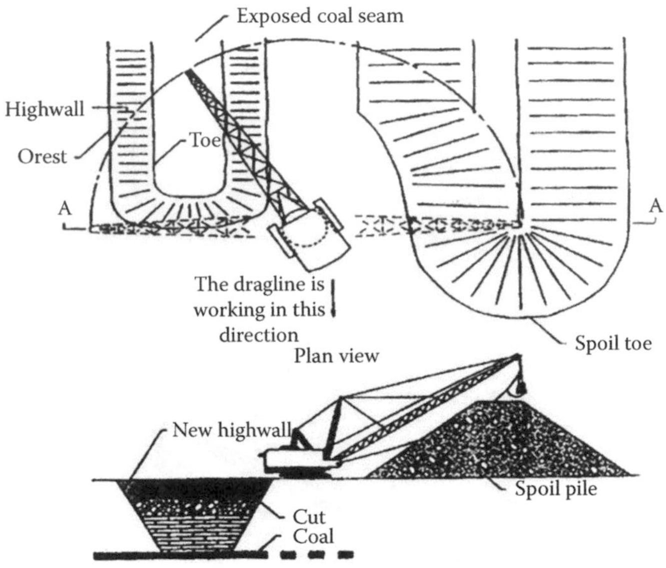

FIGURE 7.8 Plan and cross-sectional views of a box pit with side cut with no rehandle method. (From Bucyrus-Erie, *Bucyrus-Erie Surface Mining, Supervisory Training Programme*, Shovel/Truck, South Milwaukee, WI, 1979.)

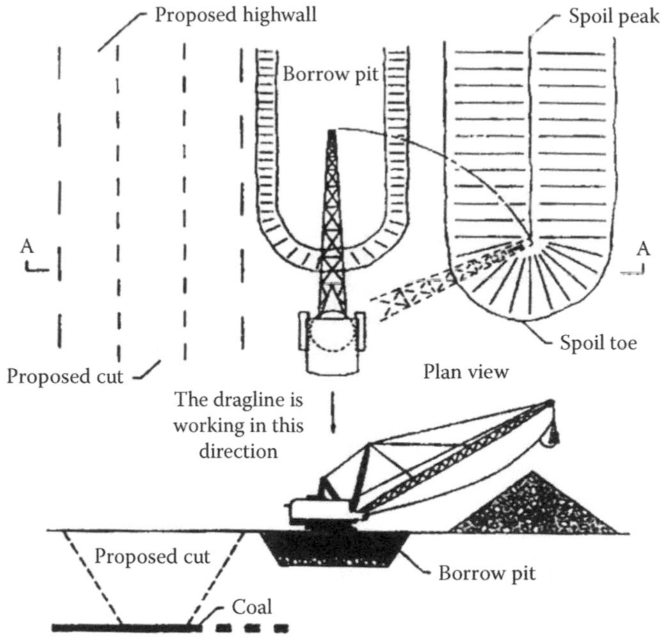

FIGURE 7.9 Plan and cross-sectional views of initial borrow pit excavation in preparation for the eventual box pit. (From Bucyrus-Erie, *Bucyrus-Erie Surface Mining, Supervisory Training Programme*, Shovel/Truck, South Milwaukee, WI, 1979.)

the reach factor is not sufficient, then rehandling, which is handling the stripped material twice to dump it further, is used. Since it increases the operating costs, it should be avoided or at least be kept at minimum when possible.

Production cut is advanced after forming the initial key cut by using remaining cut material. Plan and cross-sectional view of the final cut removal is illustrated in Figure 7.11.

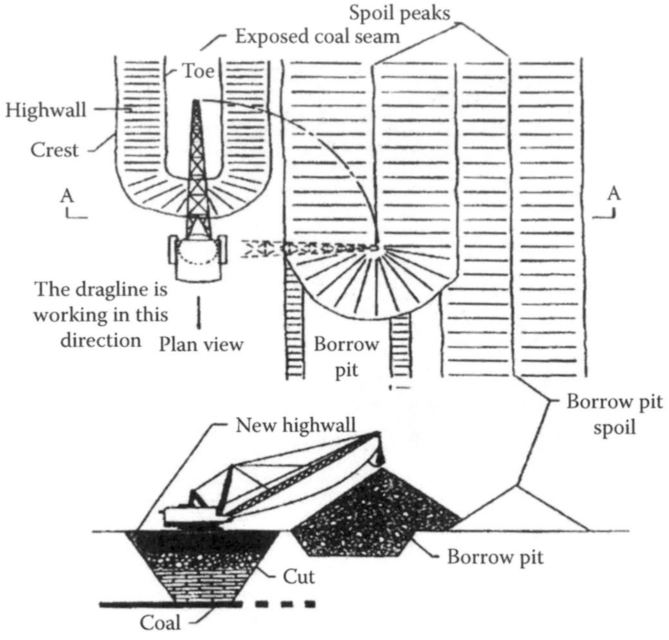

FIGURE 7.10 Plan and cross-sectional views of box pit being produced using the rehandle (borrow pit) method. (From Bucyrus-Erie, *Bucyrus-Erie Surface Mining, Supervisory Training Programme*, Shovel/Truck, South Milwaukee, WI, 1979.)

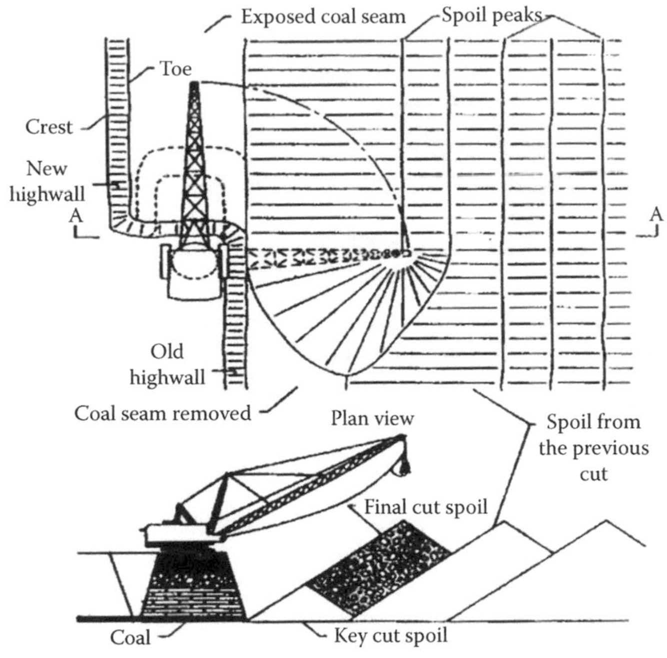

FIGURE 7.11 Plan and cross-sectional views of the final cut removal. (From Bucyrus-Erie, *Bucyrus-Erie Surface Mining, Supervisory Training Programme*, Shovel/Truck, South Milwaukee, WI, 1979.)

The classification of dragline mining methods is done according to the depth of overburden material as basic and advanced methods. The basic dragline mining method is a simple side casting. It is used when the reach factor and digging depth of the dragline is compatible with the depth overburden material or amount of excavated material. In the final cut removal, dragline utilizes its maximum load. In Figure 7.11, the type of production cut is called as a *simple side casting*, which is the most commonly utilized dragline mining method for a single coal bedding. The cycle of operations in the simple side casting consists of eight consecutive stages.

1. Top soil is removed and then placed in a stockpile to be used later for reclamation after the excavation of coal is finished
2. Subsoil is removed and then it is also placed in a separate stockpile to be used for reclamation after the excavation of coal is finished
3. Overburden is stripped using draglines to expose the coal to the surface and dumped on to the spoil pile
4. The exposed coal is then excavated using power shovel-truck systems (Figure 7.12)
5. The surface of spoil piles are leveled out using dozers and get it ready for placing vegetable soils, which were stripped at the beginning
6. Subsoil is placed on the surface of spoil piles for reclamation
7. Topsoil is placed on the surface of spoil piles for reclamation
8. Plantation and revegetation is done and the area is reclaimed

The simple side casting is advantageous because its planning and executing is simple, swing angles can be kept to a minimum, and there is no need for rehandling if digging depth is greater than overburden thickness. Also, reclamation can concurrently be done with mining. However, the capacity is finite and controlled by dragline geometry and overburden depth.

When the mineralization is deeper and/or reach of dragline is not sufficient, then more complex methods are used. These are extended bench, pull back, chop cutting, advance benching, terrace mining, and contour mining methods.

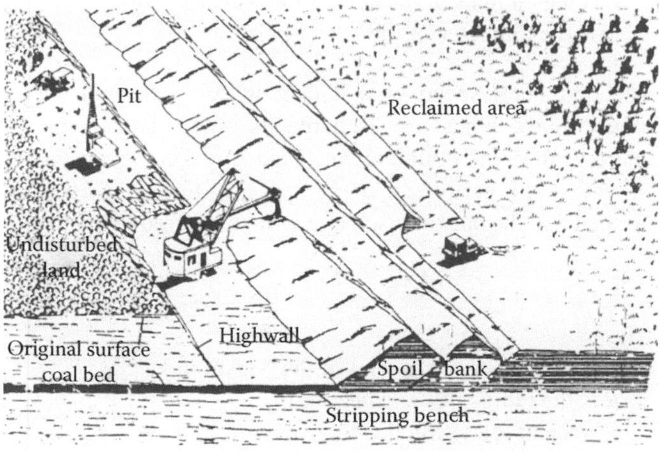

FIGURE 7.12 Typical area mining method with stripping shovel. (From U.S. Bureau of Mines, *Economic Engineering Analysis of US Surface Coal Mines and Effective Land Reclamation*, US Department of Commerce, NTIS PB-245-315/A8, 1975.)

7.6 BUCKET WHEEL EXCAVATORS

Surface excavation is done on a continuous basis with a variety of machines such as trenchers and ditchers, conveyor loaders, and bucket chain and BWEs. Since BWEs are most popular for surface coal mining, this section of the chapter is limited to discussing only BWEs.

Continuous mining systems consists of chain-like operations such as excavation of material, loading on to the conveyors, continuous material transportation, and dumping of material from the face to the dumping point. It can generally be utilized in excavation of large volumes of materials, excavation of soft minerals such as lignite, bituminous coal, oil sands, phosphate, clay, chalk, gypsum, and overburden. Continuous mining system is mainly utilized in open cast mines and is not suitable if hardness of materials is higher than the diggability of machines. Major components of any continuous mining systems include large or compact bucket chain excavators or BWEs, mobile transfer or bridge conveyors, large or compact stackers, cross-pit stackers, cross-pit conveyors, conveyor belt trippers, transfer stations, and shiftable and stationary belt conveyors. BWEs are the largest mobile machines used in mining (Figure 7.13).

BWE is a crawler or rail-mounted machine, with machine service weight exceeding 7000 tons, having a rotating wheel at the end of slewing boom. The crawler track assembly is made up of six crawler arranged in tandem with one group of nonsteering crawlers. The crawler track assembly supports the ring-shaped substructure in three points. The rotating wheel has evenly spaced buckets at the periphery of the wheel with various capacities. The wheel continuously rotates to excavate the material. The material excavated by the bucket wheel is transferred by the bucket wheel boom conveyor and then to the discharge conveyor in machine center. The material in buckets is discharged using gravity. The boom contains belt conveyor for material transport. Excavated material is fed via a transfer point inside the wheel, for example, a plough and/or a rotating disc to the belt conveyor system of the excavator

FIGURE 7.13 Bucket wheel excavator on site.

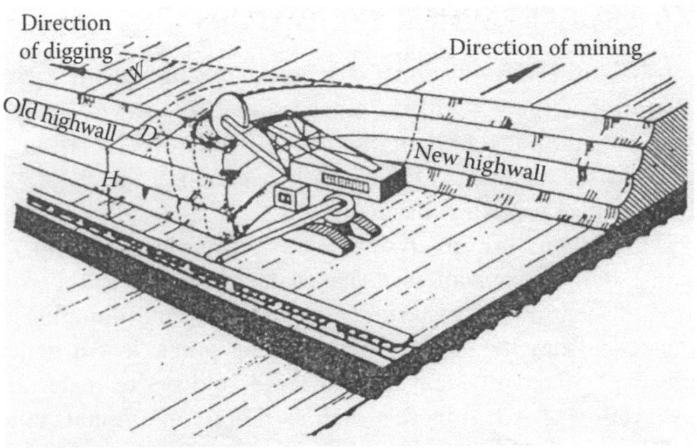

FIGURE 7.14 Block-digging operation with wheel excavator feeding onto a movable conveyor belt: H is the height of face, D is the depth of block, and W is the width of the block. (From Eskikaya, Ş. et al., *TMMOB Chamber of Mining Engineers of Turkey, Mining Engineering Surface Mining Handbook*, TMMOB Maden Mühendisleri Odası, Ankara, Turkey, 2005.)

TABLE 7.7
Bucket Capacities and Stripping Outputs of Boom-Type Excavators under Similar Operating Conditions

Shovels		Draglines		BWEs	
Bucket Size (m^3)	Output (m^3/h)	Bucket Size (m^3)	Output (m^3/h)	Bucket Size (m^3)	Output (m^3/h)
15	760	19	670	0.10	330
38	1500	23	800	0.15	840
50	1880	34	1190	0.25	1150
57	2180	46	1570	0.85	3180
61	2320	57	1990	1.35	4200
80	3060	69	2350	1.50	4280
96	3630	80	2750	4.00	8490
107	4070	92	3210		
138	5200	99	3440		
		168	5730		

Source: Pflieder, E.P., Planning and design for mining conservation section, *SME Mineral Processing Handbook*, Society of Mining Engineers, AIME, New York.

for discharge. The discharge conveyor transfers the material to a tripper car or to the bench conveyor. The discharge conveyor can be lifted, lowered, and swung with the respect to the superstructure. Theoretical output of BWEs can increase up to 14,000 bank m^3/h. BWE in operation and the rotating bucket are presented in Figure 7.14.

BWEs usually have high performance when they are utilized above the track level and have low performance when they are operated below the track level. BWEs are the most effective machine for mining large outputs in weak unconsolidated ground. BWEs operate in a continuous manner and swinging is not necessary. Since they have long discharge ranges, BWEs can be operated on a highwall bench or on a coal seam. BWEs can easily handle spoil with poor stacking characteristic and poor stability, and they can extend the range of shovel or dragline when operated in tandem. They can also be facilitated for land reclamation, as it dumps surface material back on top of the spoil pile. Another advantage is that BWEs can selectively mine faulted or intercalated ground. It is possible to accurately cut bands as thin as 10 cm at a cost of much reduced output. Due to lower impact loadings than comparable single-bucket machines, dynamic stresses and machine mass (service weight) can be reduced (Hartman 1987).

On the other hand, BWEs also have some disadvantages over other mining systems. BWEs usually have low availabilities and poor downward-digging capabilities and mobilities. They cannot dig hard materials. They need large maintenance crew. BWEs have high capital cost compared with output, and they can be susceptible to spoil slides and flooding. Also some surface preparation is required.

Table 7.7 lists the bucket capacities and outputs of shovels, draglines, and BWEs and allows us to compare these machines under similar operating conditions.

Productivity of BWEs can be estimated using Equation 7.7.

$$Q_t = I \times S_s \times 3600 \tag{7.7}$$

where:
Q_t is the theoretical capacity of the excavator (m^3/h)
I is the nominal bucket capacity (m^3)
S_s is the number of bucket discharges (s)

Number of bucket discharges is a function of cutting speed of the wheel (m/s), number of buckets around the wheel, and the size of the diameter and it can be found using Equation 7.8 (Pfleider 1973).

$$S_s = V_1 \times Z / \pi D \tag{7.8}$$

where:
V_1 is the cutting speed of the wheel in m/s
Z is the number of buckets in the wheel
D is the diameter of the wheel in m

The peripheral speed of a bucket wheel is limited by the ability of the wheel to discharge its bucket content on the chute against the counteracting centrifugal force. In theory, the maximum peripheral speed, given in Equation 7.9, must be such that the bucket discharge will be ensured (SME 1992).

$$M \times g = M \times \left(V_{max}\right)^2 / R \tag{7.9}$$

where:
M is the mass of material in the bucket in kg
R is the radius of the wheel in m
g is the acceleration due to gravity in m/s^2

Therefore, maximum peripheral speed can be obtained using Equation 7.10.

$$V_{max} = \sqrt{\left(g \times R\right)} \tag{7.10}$$

In practice, peripheral speed is kept between 40% and 60% of V_{max} and usually does not exceed 5 m/s to keep the wear on bucket's cutting knives or teeth at a minimum.

The peripheral speed is greatly dependent on the nature of material to be excavated. In principle, however, a higher peripheral speed will be decided upon if hard material is to be cut, in which case maximum output may not be attained. Another factor that affects the output of a BWE is the bucket-filling capacity. In hard ground, bucket fill factor ranges from 30% to 40% of the nominal capacity.

The relationship between digging resistance and hourly capacity of the BWE is given by Equation 7.11.

$$Q_1 / Q_2 = \left(k_2 / k_1\right)^2 \quad (7.11)$$

where:

Q_1 is the BWE hourly capacity in m^3 in soil with specific cutting resistance k_1

Q_2 is the BWE hourly capacity in m^3 in soil with specific cutting resistance k_2

Thus, the actual capacity of the BWE in any soil is given by Equation 7.12.

$$Q_a = I \times B_f \times S_s \times 3600 \quad (7.12)$$

where:

Q_a is actual capacity of the BWE in m^3/h

I is nominal bucket capacity in m^3

B_f is bucket filling capacity in the soil expressed as a fraction of the nominal bucket capacity

S_s is the number of bucket discharges per s

In soils with high cutting resistance, higher cutting speeds with lower bucket filling will result in a very small Q_a as compared to Q_t. In some cases, it may be as low as 20% of theoretical capacity of the excavator. In most BWE calculations, the ratio of bench height to wheel diameter is 2/3 at which machine performance is optimum. Lesser bench height means that full advantage of the wheel capacity is not used and higher bench height results in undercutting and lead to excessive spillage to be rehandled.

7.7 DOZERS

For support operations, dozers are utilized together with main production operation machines for various purposes. A few of the numerous application areas of dozers are as follows: land clearing and preparation, construction and maintenance of access roads and haul roads, cleanup tool for shovels or draglines, boost trucks from pit, tow-disabled vehicles, build dikes, relocate culvert pipe, relocate pumps, maintain spoil piles, move electric cables, bulldoze material to trap for backloading, ripping, pushing unconsolidated or rippable materials to other loading equipment, and push unit to assist scraper.

Dozers have two primary tools: the blade and the ripper. The blade is a heavy metal plate on the front of the tractor, used to push objects, and shoving sand, soil, and debris. Dozer blades usually come in three varieties: (1) a straight (S) blade, (2) a universal (U) blade, and (3) an S-U combination blade.

7.8 GRADERS

Graders have been integral pieces of support equipment to surface mining operations. They can be utilized in a wide range of areas, including reclaiming spoil piles, stripping top soil and subsoil in advance of overburden stripping, construction and maintaining haul roads, and constructing safety berms and drainage ditches. A variety of components convert the motor grader into a flexible machine. The attachment of an elevating conveyor enables the machine to take loose material from the trailing end of its blade, elevate it, and cast it into a hauling unit.

Motor graders are equipped with a movable blade known as *moldboard*, ranging from 3 to 4.9 m and up to 7.3 m wide. The world's current largest grader weighs 137,000 lb (62,142 kg) and boasts a 7.3 m moldboard. Moldboards are normally mounted to the grader in a circle-mounted configuration. The front wheels of motor graders can be tilted up to 20° so as to balance the machine horizontally when the moldboard is working at a vertical angle. Each motor grader has either a rigid frame or an articulated frame. An articulated frame is advantageous since it provides great maneuverability and versatility over rigid-framed graders. There are two types of articulated motor graders: graders with the articulated joint in front of the cab and graders with the articulated joint behind the cab, which enables the grader operator to have a better visibility.

REFERENCES

Anon., 1981., *Production and Cost Estimation of Material Movement with Earthmoving Equipment*, Terex Corporation, Hudson, OH, 82 pp.

Atkinson, T., 1992. Selection and sizing of excavating equipment. In *SME Mining Engineering Handbook*, 2nd ed., Vol. 2. Edited by H. L. Hartman. SME, Littleton, CO.

Bucyrus-Erie, 1979. *Bucyrus-Erie Surface Mining, Supervisory Training Programme*, Shovel/Truck, South Milwaukee, WI.

CAT, 2000. *Caterpillar Handbook*, Peoria, IL.

Crawford, J. T. and Hustrulid, W. A., 1979. *Open Pit Mine Planning and Design*, AIME Society of Mining Engineers, New York, 367 pp.

Ebrahimi, E., Hall, R. A., and Blackwell, G. H., 2003. Sizing equipment for open pit mining-a review of critical parameters, *Mining Technology* (Transactions of the Institution of Mining and Metallurgy A), Vol. 112, pp. A171–A179.

Eskikaya, Ş., Karpuz, C., Hindistan, M. A., and Tamzok, N., 2005. *TMMOB Chamber of Mining Engineers of Turkey, Mining Engineering Surface Mining Handbook*, TMMOB Maden Mühendisleri Odası, Ankara, Turkey.

Hartman, H.L., 1987. *Introduction to Mining Engineering*, John Wiley & Sons, Hoboken, NJ, 633 pp.

Humphrey, J. 1990. *The Fundamentals of the Dragline*, Marion Division, Dresser Ind. Inc., Marion, OH, pp 28.

Karpuz, C., 1990. A classification system for excavation of surface coal measures, *Mining Science and Technology*, Vol. 11, pp. 157–163.

Pfleider, E.P., 1968. *Surface Mining*, 1st ed., SME, New York.

Pflieder, E.P. 1973.Planning and design for mining conservation section, *SME Mineral Processing Handbook*, Society of Mining Engineers, AIME, New York.

P&H, 2003. *MinePro Services Peak Performance Practices, Excavator Selection*, Harnischfeger Corporation, Milwaukee, WI.

P&H, 2005. MinePro Services Peak Performance Practices, Excavator Selection, Harnischfeger Corporation, Milwaukee, WI.

Rai, P., Trivedi, R., and Nath, R., 2000. Cycle time and idle time analysis of draglines for increased productivity—A case study, *Indian Journal of Engineering and Materials Sciences*, Vol. 7, pp. 77–81.

Reddy, V. R. and Dhar, B. B., 1988. Dragline performance in open pit Indian coal mines. In *Mine Planning and Equipment Selection*. Edited by Singhal. Balkema, Rotterdam, the Netherlands, pp. 341–346.

SME, 1992. *SME Mining Engineering Handbook*2nd ed., Vol. 1. Edited by H. L. Hartman. Society for Mining, Metallurgy, and Exploration, Littleton, CO, 2161 pp.

Townson, P. G., Murthy, D. N. P., and Gurgenci, H., 2003. Optimization of dragline load. In *Case Studies in Reliability and Maintenance*. Edited by W. R. Blischke and D. N. Prabhakar Murthy, Wiley Series in Probability and Statistics, John Wiley & Sons, Hoboken, NJ, pp. 517–544.

U.S. Bureau of Mines, 1975. *Economic Engineering Analysis of US Surface Coal Mines and Effective Land Reclamation*. US Department of Commerce, NTIS PB-245-315/A8.

8 Strata Control for Underground Coal Mines

Celal Karpuz and Hakan Basarir

CONTENTS

Abstract: The design features of support systems for longwall panel roadways and of coal pillars are discussed in detail. The main support design methods such as empirical, numerical, and analytical methods are described. Ground-reaction curve approach for main gallery design is also presented. The widely used numerical modeling methods are presented. Monitoring techniques used to evaluate the performance of the designed support systems are given. Finally, the general guidelines for safe and economical support system and pillar design for underground coal mining are drawn.

8.1 INTRODUCTION

The purpose of this chapter is to draw general guidelines for safe and economical main haulage way and panel gateways support system and designing underground coal pillar. As it is known, the commonly used underground coal mining methods are longwall panels and room and pillars methods. The room and pillar method is mainly applied at shallow depths, while longwall panels are utilized for deeper seams. Lonwall mining methods have two different applications such as advance and retreat workings. While main haulage ways serve for both longwall panels and room and pillar methods, gateways are the vital parts of longwall panels and connected to main haulageways. Initially, general guidelines will be introduced regarding mine roadways such as maingate, tailgate, and main haulage gateways. In Section 8.4, general design principles of safe and economical pillar design will be introduced. In Section 8.5, the basic principles of numerical modeling methods used in both roadway and pillar design is introduced. In Section 8.6, instrumentation (monitoring), which is a must both to control and to improve the designed support systems and pillar, will be introduced.

The basic principles for support design of main haulage ways and panel roadways or gateways (tail- and maingates) are similar. The main difference between designing support system for haulage ways and panel roadways is the induced mining stresses due to applied mining method. Panel roadways are affected by the mining induced stresses sourced from the advancing mining face.

8.2 DESIGN OF SUPPORT SYSTEMS FOR MAIN HAULAGE WAYS

There are three basic design methods employed in designing haulage way support systems: empirical methods based on records of past experiences, analytical methods, and numerical methods. Numerical method is used not only for designing roadway support system but also in pillar design. As a common method, numerical method will be introduced in a separate section. Therefore, in this section only empirical and analytical methods will be introduced.

8.2.1 Empirical Methods

Rock mass classification evaluates the quality and expected behavior of rock masses based on the most important parameters that influence the rock mass quality. Rock mass classification systems are important as they provide a consistent means of describing quantitatively the rock mass quality. This in turn has led to the development of many empirical design systems involving rock masses. Many researchers have developed rock mass classification systems.

In empirical methods, the miner estimated, based on his experience, what kind of support element is needed, and if the applied support failed, it was rebuilt stronger. Terzaghi (1946) has written the rules for support selection. Terzaghi's classification system was followed by rock quality designation (RQD) classification (Deere et al. 1967), rock structure rating (RSR) (Wickham et al. 1972), geomechanics classification (Bieniawski 1989), rock quality index (Barton et al. 1974), modified rock mass rating (RMR) (M-RMR) (Unal 1996), and coal mine roof rating (CMRR) (CDC 2013).

8.2.1.1 Terzaghi's Rock Load Classification

Terzaghi's (1946) rock load classification is the earliest reference relating to rock mass classification for design of tunnel support. In this system, rock loads carried by steel sets are estimated on the basis of a descriptive rock classification. The vertical and side loads on the ribs are described in terms of the height of a loosened rock mass weighing on the steel rib. Figure 8.1 describes the rock load on the tunnel support system. Rock load for different rock masses are given in Table 8.1.

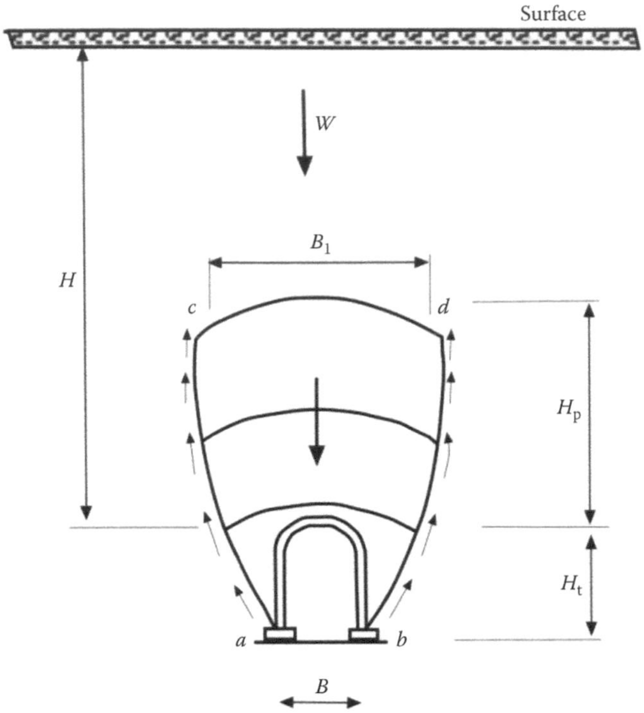

FIGURE 8.1 Rock load on support system.

8.2.1.2 Rock Quality Designation

To provide a quantitative estimate of rock mass quality from drill core logs Deere et al. (1967) developed RQD. RQD is the percentage of intact core pieces longer than 100 mm in the total length of core. It should be noted that for proper determination of RQD, the core size should be at least NX-size. After the development of RQD (Deere et al. 1967), as a means to describe the character or quality of rock mass, the correlations between Terzaghi's rock load and RQD were established. RQD has been used in many classification systems such as RMR (Bieniawski 1989), Q-system (Barton et al. 1974) and also in RSR (Wickham et al. 1972). Terzaghi's rock load height correlated with approximate RQD values by Deere et al. (1970) as shown in Figure 8.2.

8.2.1.3 Rock Structure Rating

RSR classification system was developed by Wickham et al. (1972) to describe the quality of rock mass and to select appropriate support. RSR is the first system making reference

TABLE 8.1
Rock Load for Different Rock Conditions

S. No.	Rock Condition	Rock Load, H_p (ft)	Remarks
1	Hard and intact	Zero	Light lining required only if spalling or popping occurs
2	Hard stratified or schistose	0–0.5 B	Light support, mainly for protection against spalls. Load may change erratically from point to point
3	Massive, moderately jointed	0–0.25 B	
4	Moderately blocky and seamy	0.25–0.35 $(B + H_t)$	No side pressure
5	Very blocky and seamy	0.35–1.1 $(B + H_t)$	Little or no side pressure
6	Completely crushed	1.1 $(B + H_t)$	Considerable side pressure. Softening effects of seepage toward bottom of tunnel require continuous support for either lower ends of ribs or circular ribs
7	Squeezing rock, moderate depth	1.1–2.1 $(B + H_t)$	Heavy side pressure; invert struts require circular ribs
8	Squeezing rock, great depth	2.1–4.5 $(B + H_t)$	
9	Swelling rock	Up to 250 ft	Circular ribs are required in extreme cases, use yielding support

to shotcrete support. RSR system yields in a numerical value by rating each of the components listed below:

1. Parameter A, geology: General appraisal of geological structure on the basis of (see Table 8.2):
   a. Rock type origin (igneous, metamorphic, or sedimentary)
   b. Rock hardness (hard, medium, soft, or decomposed)
   c. Geologic structure (massive, slightly faulted/folded, moderately faulted/folded, or intensely faulted/folded)
2. Parameter B, geometry: Effect of discontinuity pattern with respect to the direction of the tunnel drive on the basis of (see Table 8.3):
   a. Joint spacing
   b. Joint orientation (strike and dip)
   c. Direction of tunnel drive
3. Parameter C: Effect of groundwater inflow and joint condition on the basis of (see Table 8.4):
   a. Overall rock mass quality on the basis of parameters A and B combined
   b. Joint condition (good, fair, or poor)
   c. Amount of water inflow (in gallons/min/1000 ft of tunnel)

An example showing support estimates for 7.3 m diameter circular tunnel is shown in Figure 8.3.

8.2.1.4 Rock Mass Rating

The RMR rock mass classification system was initially developed at the South African Council of Scientific and Industrial

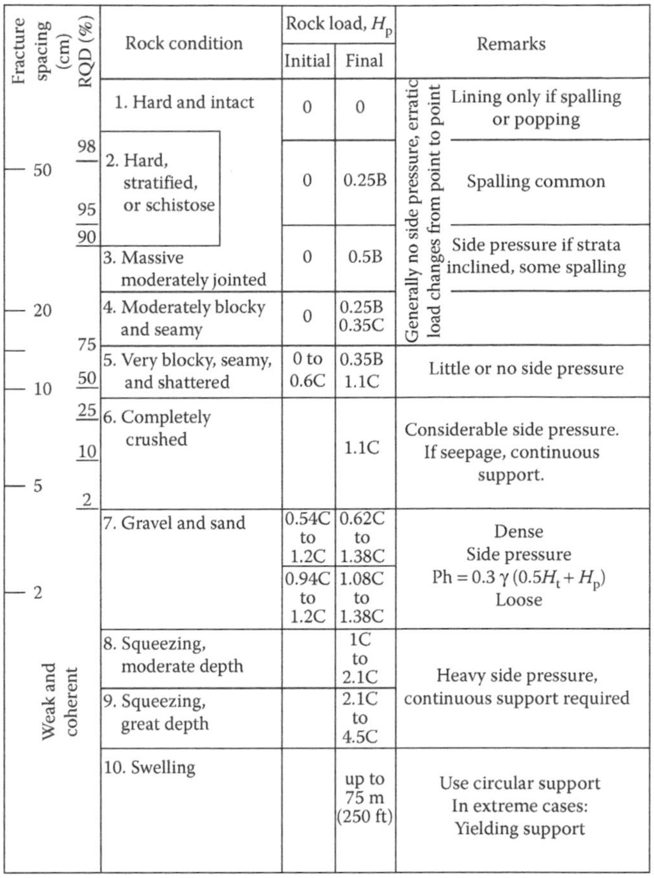

FIGURE 8.2 The correlation between RQD and rock load height. *Notes*: (1) For rock classes 4, 5, 6, and 7, when above groundwater level, loads reduce by 50%. (2) B is the tunnel width; $C = B + H_t$ = width + height of tunnel. (3) γ is the density of medium. (From Deere, D.U. et al., *Highw. Res. Rec.*, 39, 26–33, 1970.)

Research by Bieniawski (1974) on the basis of his experiences in shallow tunnels on sedimentary rocks. The system uses six parameters; RMR value is the summation of the assigned rating of these parameters. The parameters and allocated grades can be found in Bieniawski (1989).

By using RMR values, guidelines for excavation and support of 10 m span rock tunnels are given in Table 8.5.

Rock load carried by a support system is a function of rock mass condition and initial stress state. If the rock mass is not overstressed and squeezing is not occurring, the design load can be calculated using the following formula:

$$P_r = \frac{100 - \text{RMR}}{100} 10\sqrt{\frac{\text{Span}}{10}} \rho_r \gamma_r \tag{8.1}$$

where:

γ_r is a partial factor
ρ_r is the rock density

Stand-up time, unsupported roof span, and corresponding RMR values are shown in Figure 8.4.

The ranges of RMR and suggested bolt spacings are given in Table 8.6. Spot bolting is proposed for RMR values

TABLE 8.2
Parameter A: General Area Geology

	Basic Rock Type				Geological Structure			
	Hard	Medium	Soft	Decomposed	Massive	Slightly folded or faulted	Moderately folded or faulted	Intensively folded or faulted
Igneous	1	2	3	4				
Metamorphic	1	2	3	4				
Sedimentary	2	3	4	4				
Type 1					30	22	15	9
Type 2					27	20	13	8
Type 3					24	18	12	7
Type 4					19	15	10	6

Source: Wickham, G.E. et al., Support determination based on geological information, *Proceedings of the North American Rapid Excavation Tunneling Conference*, Society of Mining Engineers, American Institute of Mining Metallurgical and Petroleum Engineers, New York, 43–64, 1972.

TABLE 8.3
Parameter B: Joint Pattern, Direction of Drive

	Strike ⊥ to Axis Direction of Drive					Strike // to Axis Direction of Drive		
	Both	With Dip		Against Dip		Either Direction		
	Dip of Prominent Joints[a]					Dip of Prominent Joints		
	Flat	Dipping	Vertical	Dipping	Vertical	Flat	Dipping	Vertical
1. Very closely jointed, <2 in	9	11	13	10	12	9	9	7
2. Closely jointed, 2–6 in	13	16	19	15	17	14	14	11
3. Moderately jointed, 6–12 in	23	24	28	19	22	23	23	19
4. Moderate to blocky, 1–2 ft	30	32	36	25	28	30	28	24
5. Blocky to massive, 2–4 ft	36	38	40	33	35	36	24	28
6. Massive, >4 ft	40	43	45	37	40	40	38	34

Source: Wickham, G.E. et al., Support determination based on geological information, *Proceedings of the North American Rapid Excavation Tunneling Conference*, Society of Mining Engineers, American Institute of Mining Metallurgical and Petroleum Engineers, New York, 43–64, 1972.

[a] Dip, flat: 0°–20°; dipping: 20°–50°; and vertical: 50°–90°.

TABLE 8.4
Parameter C: Effect of Groundwater Inflow and Joint Condition

	Sum of Parameters A + B					
	13–44			45–75		
	Joint Condition[a]					
Anticipated Water Inflow gpm/1000 ft of Tunnel	Good	Fair	Poor	Good	Fair	Poor
None	22	18	12	25	22	18
Slight, <200 gpm	19	15	9	23	19	14
Moderate, 200–1000 gpm	15	22	7	21	16	12
Heavy, >1000 gp	10	8	6	18	14	10

Source: Wickham, G.E. et al., Support determination based on geological information, *Proceedings of the North American Rapid Excavation Tunneling Conference*, Society of Mining Engineers, American Institute of Mining Metallurgical and Petroleum Engineers, New York, 43–64, 1972.

[a] Joint condition: good = tight or cemented; fair = slightly weathered or altered; and poor = severely weathered, altered or open.

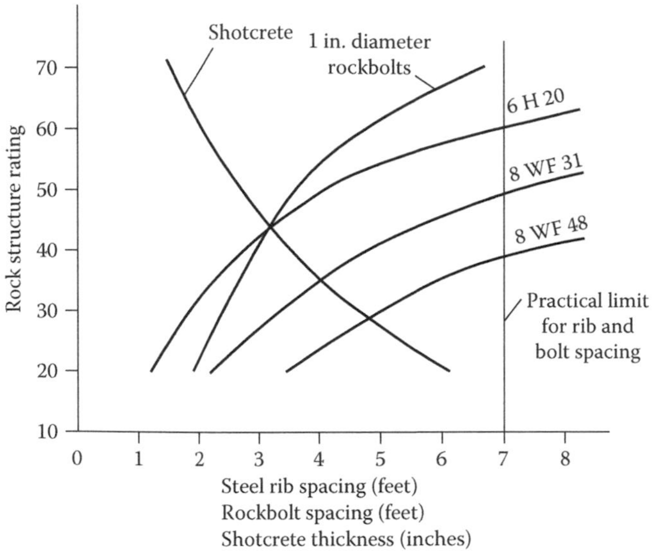

FIGURE 8.3 Rock structure rating support estimates for a 24 ft (7.3 m) diameter circular tunnel. (From Wickham, G.E. et al., Support determination based on geological information, *Proceedings of the North American Rapid Excavation Tunneling Conference*, Society of Mining Engineers, American Institute of Mining Metallurgical and Petroleum Engineers, New York, 43–64, 1972.)

above 85. The equation relating RMR to bolt length is also given in Table 8.6.

The capacity of each bolt divided by the area, it has to support, is assumed to be the support capacity of pattern bolt and can be calculated using Equation 8.2.

$$F_{bd} = \frac{F_b}{\gamma_b}\left(\frac{RMR}{85}\right)^{\frac{40}{RMR}} \tag{8.2}$$

where:

F_b is ultimate tensile capacity of bolt

γ_b is a partial factor

Based on the concept of shotcrete acting simply as an arch in compression, the design capacity of shotcrete support can be calculated using the formula (Equation 8.3) given by Lowson and Bieniawski (2013).

$$f_{cd} = \frac{f_{ck}}{\gamma_s}\left[0.2 + 0.8\left(\frac{RMR}{100}\right)^{3/2}\right] \tag{8.3}$$

where:

f_{ck} is shotcrete cylinder strength

γ_s is a partial factor

The capacities of the rock bolts and the shotcrete are considered to be additive (Lowson and Bieniawski 2013).

For construction purposes and the cost effectiveness of primary support system, ideal opening shapes and corresponding ground conditions have been proposed by Lowson and Bieniawski (2013) and are given in Table 8.7.

8.2.1.5 *Q*-System

The *Q*-system (Barton et al. 1974) is considered as an elaborate and the most detailed system for designing support system for underground opening. The system was developed in 1974 and has been modified twice since then. The first update in 1993 was based on 1050 examples mainly from Norway (Grimstad and Barton 1993). Based on more than 900 new examples from Norway, Switzerland, and India, the system was again updated in 2002. In a later update, analytical research with respect to the thickness, spacing, and reinforcement of reinforced ribs of sprayed shotcrete as a

TABLE 8.5
Guidelines for Excavation and Support System

Rock Mass Class	Excavation	Rock Bolts (20 mm Diameter, Fully Grouted)	Shotcrete	Steel Sets
I—Very good rock RMR: 81–100	Full face 3 m advance	Generally no support required except spot bolting		
II—Good rock RMR: 61–80	Top heading and bench 1–1.5 m advance; complete support 20 m from face	Locally, bolts in crown 3 m long, spaced 2.5 m with occasional wire mesh	50 mm in crown where required	None
III—Poor rock RMR: 21–40	Top heading and bench 1.0–1.5 m advance in top heading; install support concurrently with excavation, 10 m from face	Systematic bolts 4–5 m long, spaced 1–1.5 m in crown and walls with wire mesh	100–150 mm in crown and 100 mm in sides	Light to medium ribs spaced 1.5 m where required
IV—Very poor rock RMR<20	Multiple drifts 0.5–1.5 m advance in top heading; install support concurrently with excavation; shotcrete as soon as possible after blasting	Systematic bolts 5–6 m long, spaced 1–1.5 m in crown and walls with wire mesh Bolt invert	150–200 mm in crown, 150 mm in sides, and 50 mm on face	Medium to heavy ribs spaced 0.75 m with steel lagging and forepoling if required Close invert

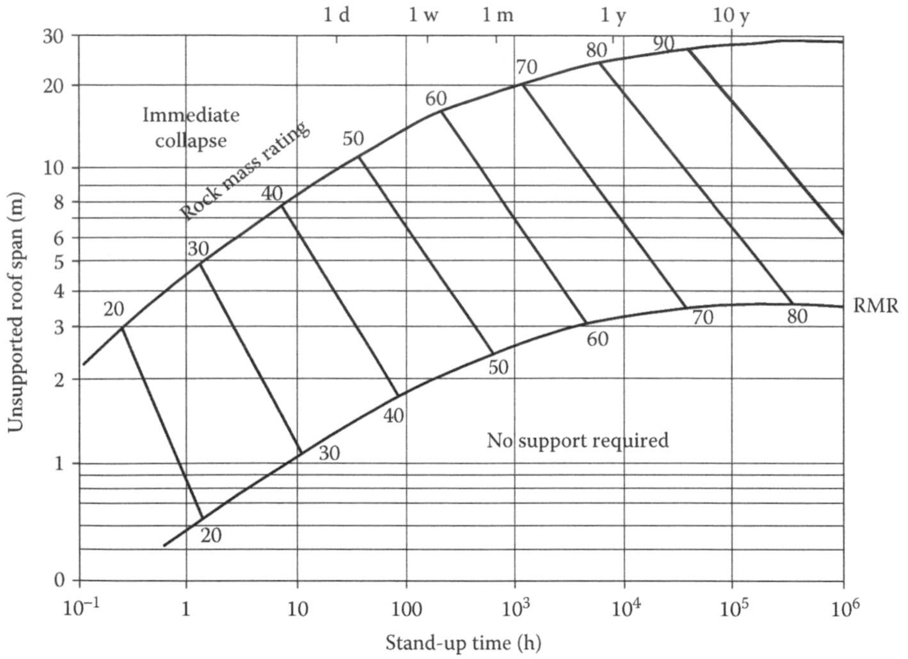

FIGURE 8.4 Stand-up time of an unsupported tunnel span, as a function of rock mass quality RMR. (Modified from Lowson and Bieniawski, Critical assessment of RMR based tunnel design practices: A practical engineer's approach, *Rapid Excavation and Tunneling Conference*, Design and Planning Session, Washington, DC, 2013.)

TABLE 8.6
RMR Ranges and Calculation of Bolt Length

RMR Range	Bolt Spacing	Bolt Length
20 < RMR ≤ 85	$S_b = 0.5 + 2.5\left(\frac{\text{RMR} - 20}{65}\right) \ldots m$	$\text{Span} = \frac{(L_b + 2.5)^{(\text{RMR}+25)/52}}{3.6} \ldots m$
10 < RMR ≤ 20	$S_b = 0.25 + \frac{(\text{RMR} - 20)^{1.5}}{140} \ldots m$	
RMR ≤ 10	$S_b = 0.25 \ldots m$	

function of load and the rock mass quality were included (Grimstad et al. 2002).

Q-value is calculated based on six parameters as follows:

$$Q = \frac{\text{RQD}}{J_n} \frac{J_r}{J_a} \frac{J_w}{\text{SRF}} \tag{8.4}$$

where:

RQD is rock quality designation
J_n is number of joint set
J_r is joint roughness number
J_a is joint alteration number
J_w is joint water reduction factor
SRF is stress reduction factor.

In fact, these six parameters are used to express the three main factors.

More detailed and up-to-date information about the classification of individual parameters used to obtain the tunneling quality index Q for a rock mass is supplied by NGI (NGI 2013).

8.2.1.5.1 Using the Q-System for Support Requirements

For the support design for underground openings, two decisive parameters are required in addition to the Q-value. These parameters are excavation support ratio (ESR) and span or height of the opening. ESR is used to express safety requirements depending on the purpose of excavation. High ESR values indicate that a lower level of safety will be acceptable, whereas low values indicate a high level of safety. ESR values are given in Table 8.8.

The *equivalent dimension* to be used in support chart is calculated by dividing span to ESR value. Having Q and equivalent dimension values, the support recommendation chart given in Figure 8.5 can be used.

The support chart is primarily designed for the crown and springlines of underground openings and caverns. As the

TABLE 8.7
Ideal Opening Shapes and Corresponding Ground Conditions

Ground	Shape	Comments
RMR > 50	D-shape with vertical or inclined sides and flat invert	Easiest to construct
30 < RMR < 50	Horseshoe with curved sidewalls	Reduces sidewall support costs
20 < RMR < 30	Horseshoe with curved sidewalls and curved invert	A curved shotcreted invert can be more economic than bolting the invert and/or an RC structural invert
10 < RMR < 20	Shape made up of three or more curves	Usually a three-curve comprising arch, haunch, and invert radii, or a 5-curve with arch, shoulder, sidewall, haunch, and invert radii
RMR < 10	Circular	

Sources: Bieniawski, Z.T., *Engineering Rock Mass Classification*. 1989. Copyright Wiley-VCH Verlag GmbH & Co. KGaA. Reproduced with permission. Lowson, A.R. and Bieniawski, Z.T., Critical assessment of RMR based tunnel design practices: A practical engineer's approach, *Rapid Excavation and Tunneling Conference*, Washington, DC, Design and Planning Session, 2013.

TABLE 8.8
ESR Values

Type of Excavation		ESR
A	Temporary mine openings	ca. 3–5
B	Vertical shafts[a]:	
	1. Circular sections	ca. 2.5
	2. Rectangular/square sections	ca. 2.0
C	Permanent mine openings, water tunnels for hydropower (excluding high pressure penstocks) water supply tunnels, pilot tunnels, drifts, and headings for large openings	1.6
D	Minor road and railway tunnels, surge chambers, access tunnels, and sewage tunnels	1.3
E	Power houses, storage rooms, water treatment plants, major road and railway tunnels, civil defense chambers, portals, and intersections	1.0
F	Underground nuclear power stations, railways stations, sports and public facilitates, and factories	0.8
G	Very important caverns and underground openings with a long lifetime, ≈100 years, or without access for maintenance	0.5

Source: NGI, Using the Q-system–Rock mass classification and support design. www.ngi.no/en/Contentboxes-and-structures/Reference-Projects/Reference-projects/Q-method/, 2013.

[a] Dependent of purpose, may be lower than given values.

walls of the opening require less support for high and intermediate Q-values ($Q > 0.1$), the height of wall must be used instead of the span. The actual Q-value is adjusted as shown in Table 8.9.

In cases where $Q < 1$, reinforced ribs of sprayed concrete is the preferred alternative to cast concrete. These ribs are constructed with a combination of steel bars (usually with a diameter of 16 or 20 mm). sprayed concrete, and rock bolts. The spacing between the ribs and their thicknesses, as well as the diameter of steel bars, and the number of ribs varies based on the dimensions of the opening and the rock mass quality.

Although forepoling is not indicated in support chart, generally it is suggested that forepoling should be used if the Q-value ranges from 0.1 to 0.6, depending on the span of the opening. The spacing between forepoles is normally around 0.3 m (0.2–0.6 m). In the support chart, energy absorption classes defined by Experts for Specialised Construction and Concrete Systems are specified for sprayed concrete. The guidelines for the defined classes are specified in the guidelines of Norwegian Concrete Association's publication no 7-2001 (NB 2011).

In some cases, there may be narrow weakness zones, that is, 0.5–3 m wide. In such cases, the quality of the surrounding rock will determine the necessary rock support as it is not usually convenient to design the support system based only on the Q-value of the zone. In order to decide the appropriate Q-value. both the Q-values of surrounding rock mass and weakness zone should be considered. Equation 8.5 can be used for such purposes.

$$\text{Log}\, Q_{\text{m}} = \frac{b \cdot \log Q_{\text{zone}} + \log Q_{\text{sr}}}{b+1} \tag{8.5}$$

where:

Q_{m} is the mean Q-value for weakness zone/surrounding rock
Q_{zone} is the Q-value for the weakness zone
Q_{sr} is the Q-value for the surrounding rock
b is the width of the weakness zone measured along the length of the excavation

The use of the equation should be well justified in cases of high Q-value of surrounding rock mass and/or thickness and acute angle of weakness zone along tunnel axis.

8.2.1.6 Coal Mine Roof Rating

The system was first introduced in 1994. It has been mainly developed to quantify geological description of coal measure rocks into an engineering value, which can be used for mining design purposes (CDC 2013). CMRR evaluates the properties of coal mine roof rock mass and determines its relative strength rating from 0 to 100.

CMRR can be determined from exploratory drill cores or from underground exposures. In any cases, four main parameters are required such as uniaxial compressive strength of intact rock, the intensity (spacing and persistence) of discontinuities such as bedding planes and slickensides, the shear strength (cohesion and roughness) of discontinuities, and the

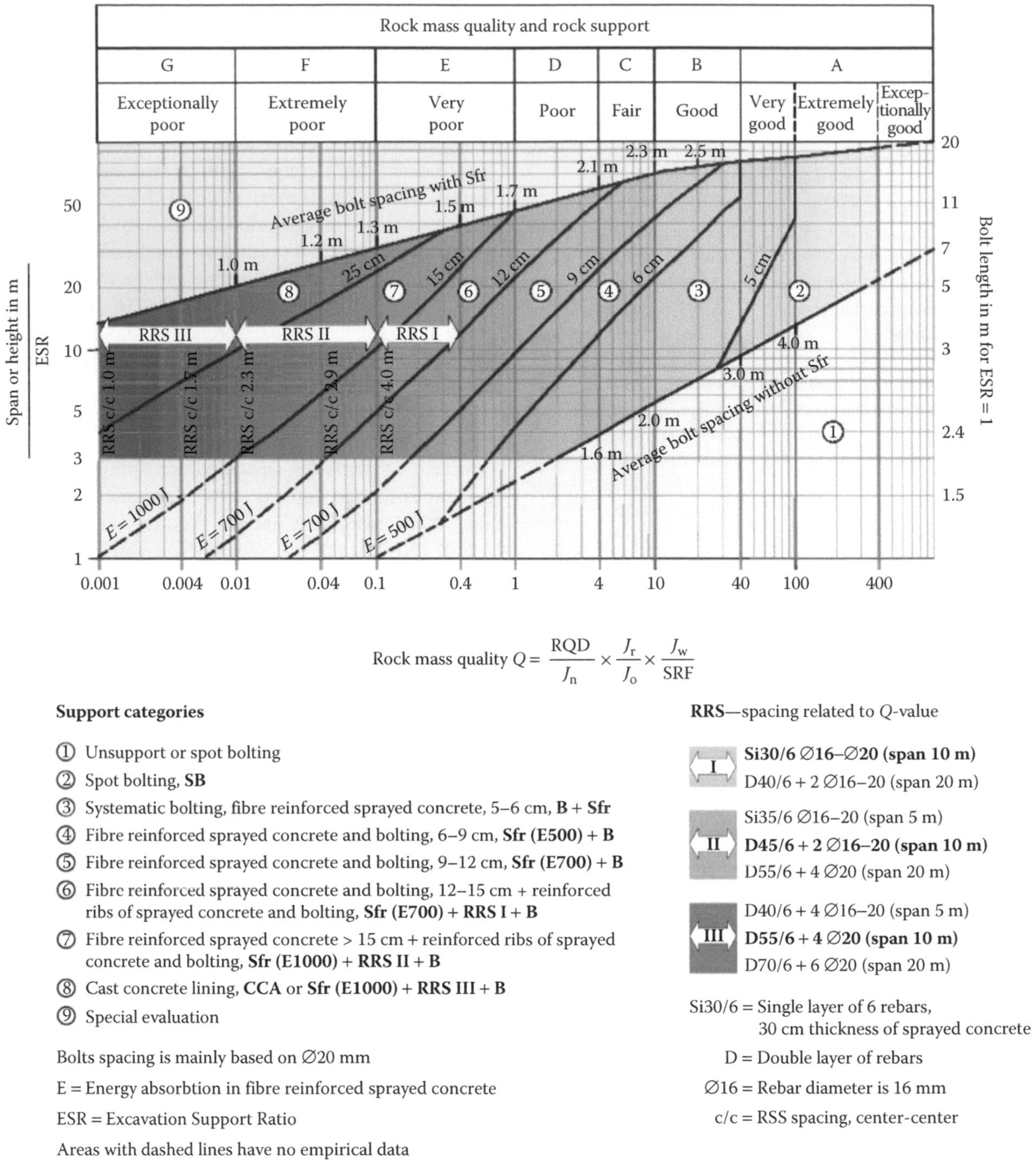

FIGURE 8.5 Support recommendation chart. (From NGI, Using the Q-system –Rock mass classification and support design. www.ngi.no/en/Contentboxes-and-structures/Reference-Projects/Reference-projects/Q-method/, 2013.)

TABLE 8.9
Conversion from Actual *Q*-Values to Adjusted *Q*-Values for Design of Wall Support

For rock masses of good quality	$Q > 10$	Multiply *Q*-values by a factor of 5
For rock masses of intermediate quality	$0.1 < Q < 10$	Multiply *Q*-values by a factor of 2.5. In cases of high rock stresses, use the actual *Q*-value
For rock masses of poor quality	$Q < 0.1$	Use the actual *Q*-value

Source: NGI, Using the Q-system –Rock mass classification and support design.www.ngi.no/en/Contentboxes-and-structures/Reference-Projects/Reference-projects/Q-method/, 2013.

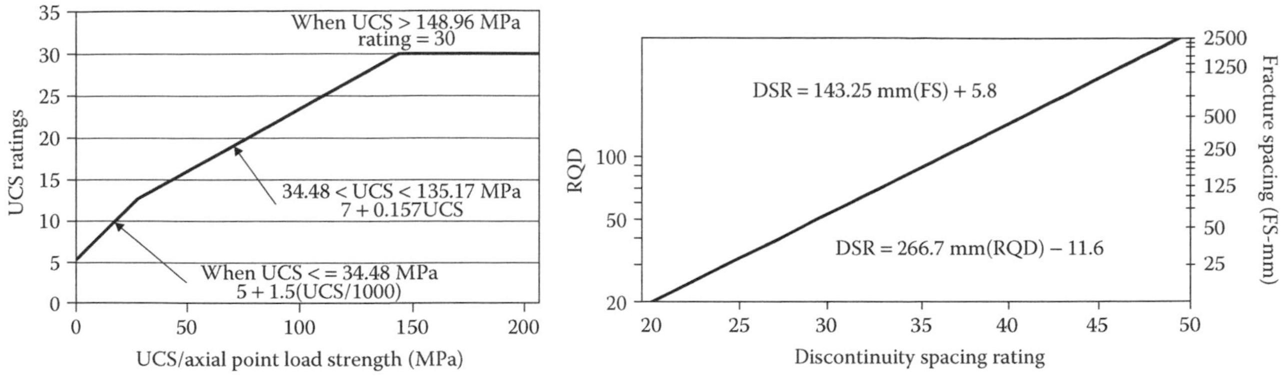

FIGURE 8.6 Ratings for strength and discontinuity spacing. (Data from Mark, C. and Molinda, G.M., The coal mine roof rating in mining engineering practice, *Coal 2003: Coal Operator's Conference*, University of Wollongong and the Australian Institute of Mining and Metallurgy, 50–61, 2003.)

moisture sensitivity of the rock. CMRR is calculated using a two-step process (Mark and Molinda 2003). As the first step, the mine roof is divided into lithologic and/or structural units and unit ratings are determined for each. In the second step, CMRR is determined by combining the unit ratings and applying appropriate adjustment factors. The second step is the same whether unit ratings are determined from core or underground. An example datasheet for underground data is supplied by CDC (2013).

Molinda and Mark (1996) presented the procedures for determining unit ratings from drill core. In this case, three types of information are required: uniaxial compressive strength, fracture spacing, and diametrical point load (an index of bedding plane shear strength) strength.

The graph showing the uniaxial compressive strength (UCS) rating and strength is given in Figure 8.6. Both fracture spacing and RQD can be used as input to consider fracture spacing. The graph showing the parameters and corresponding CMRR rating is given in Figure 8.6. The diametrical point load test rating values are shown in Figure 8.7.

The moisture sensitivity rating and immersion test results are given in Table 8.10. In cases where immerse test results are not available, two alternatives were mentioned by Mark and Molinda (2003). Moisture sensitivity can be estimated visually in underground exposures or slake durability test can be correlated with immersion test; in this case Figure 8.8 can be used as guideline.

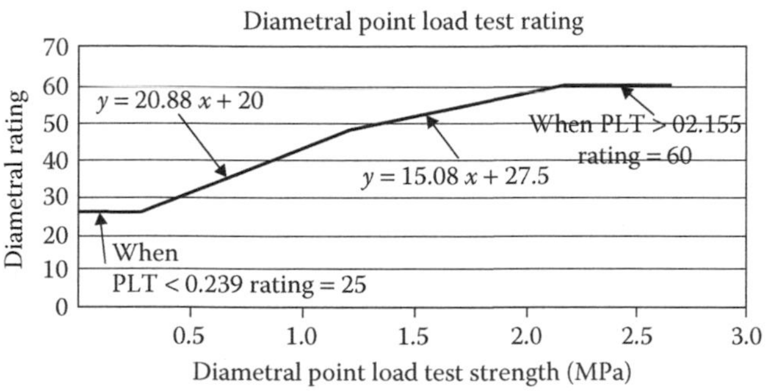

FIGURE 8.7 Diametrical point test and CMRR ratings. (Data from CDC, *Mining Product: CMRR—Coal Mine Roof Rating*, August 30, NIOSH (National Institute for Occupational Safety and Health), Pittsburgh, PA, 2013.)

TABLE 8.10
Moisture Sensitivity Ratings

Moisture Sensitivity	Immersion Index	Rating
Not sensitive	0–1	0
Slightly sensitive	2–4	−3
Moderately sensitive	5–9	−7
Severely sensitive	>9	−15

Source: Mark, C. and Molinda, G.M., The coal mine roof rating in mining engineering practice, *Coal 2003: Coal Operator's Conference*, University of Wollongong and the Australian Institute of Mining and Metallurgy, 50–61, 2003.

If there is only one unit in the roof, then the unit rating plus the groundwater adjustment will give final CMRR rating. If there are more than one unit in the roof, then final CMRR is calculated using thickness weighted average of the unit ratings. It should be noted that only the bolted interval is included in the average. Then the adjustments are applied to get final CMRR. The adjustments can be listed as strong bed adjustment, groundwater adjustment, ranging from 0 to 10 for large inflow and damp, surcharge adjustment (if the rock above the bolted interval has lower strength, then such deduction has to be made), and number of weak contacts: roof failure is often associated with major bedding contacts between rock units; for such conditions the adjustment is necessary.

To facilitate the entry storage and processing of field data, the CMRR program is designed and supplied to users. The software can be downloaded from the web page of *Centers for Disease Control and Prevention* (CDC 2013).

CMRR can also be used in the following application areas as described in the following sections: input for numerical models (Karabin and Evanto 1999), multiple seam mine design (Lou et al. 1997), and hazard analysis and mapping (Wuest et al. 1996) are some of the application areas.

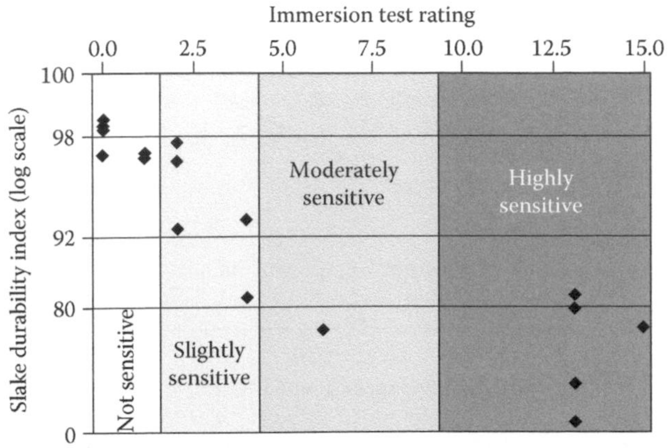

FIGURE 8.8 Slake durability and immersion test rating. (Data from Mark, C. and Molinda, G.M., The coal mine roof rating in mining engineering practice, *Coal 2003: Coal Operator's Conference*. University of Wollongong and the Australian Institute of Mining and Metallurgy, 50–61, 2003.)

8.2.1.7 Geological Strength Index

One of the widely used rock mass characterization system is the geological strength index (GSI) system (Hoek et al. 1995). Unlike rock mass classification systems such as RMR and Q, the GSI system does not suggest any support system directly. The GSI system is mainly utilized to provide the rock mass data required for numerical and analytical methods, such as deformability modulus, rock mass strength, cohesion, and friction angle of rock mass. Recently, more researchers have been using the system for deriving mass properties of coal and coal measure rocks (Pine et al. 2006; Simpson and Simpson 2006; Unver and Yasitli 2006; Hosseini et al. 2008; Lawrence 2009; Deisman et al. 2013).

Hoek et al. (2013) updated the original GSI chart proposed by Hoek and Marinos (2000). They proposed a GSI chart by adding scales to quantify each axis as shown in Figure 8.9. In the Figure 8.9 JCond_{89} represents the joint conditions defined by Bieniawski (1989) and RQD represents the rock quality designation. The value of GSI is given by the sum of predefined scales, that is,

$$\text{GSI} = 1.5\text{JCond}_{89} + \frac{\text{RQD}}{2} \tag{8.6}$$

Hoek et al. (2013) also suggested of alternative equations to obtain JCond_{89} and RQD values to be used in the equation as follows:

$$\text{RQD} = 110 - 2.5J_v \tag{8.7}$$

J_r and J_a included in the Q-system and described by Barton et al. (1974). As for the estimation of RQD, the following equations proposed by Priest and Hudson (1976), Palmstrom (1982), and Palmstrom (2005) can be used:

$$\text{RQD} = 100e^{0.1\lambda}(0.1\lambda + 1) \tag{8.8}$$

$$\text{RQD} = 110 - 2.5J_v \tag{8.9}$$

where:

λ is the average number of discontinuities per meter
J_v is volumetric joint count

By using the GSI system, it is possible to obtain the complete set of strength and deformability parameters of rock mass as long as UCS, GSI, and the material constant, m_i, are known.

8.2.2 Analytical Methods

The most widely used analytical method for designing supports in galleries is based on the ground-support interaction analysis, which is composed of ground reaction curve (GRC) and support reaction curves (SRC). By means of the method, the displacements and stresses around circular opening both elastic and inelastic material can be analyzed. The solution is in two dimension and longitudinal displacement profile is used for calibrating the relationship between support pressure, radial displacement, and the distance from the face. In other words, to relate wall displacement to the distance from the face, longitudinal displacement profile is used. In this approach, it is assumed that a circular tunnel with a radius of r_0 is subjected to hydrostatic stresses p_0 and a uniform support pressure p_i as shown in Figure 8.10. All of the terms are shown in Figure 8.11.

As it can be seen from Figure 8.11, deformation begins in front of the opening face and reaches its maximum value behind face. When the support is erected at L_0 distance behind face at the face, the displacement at the face is u_{s0}. When the applied support pressure or inner pressure is equal to in situ stresses ($p_i = p_0$), there will not be any displacement behind face. If the rock mass fails, then plastic zone around the opening (r_p) develops and inward plastic radial displacement (u_{ip}) occurs.

When the support pressure is larger than critical support pressure ($p_i > p_{cr}$), elastic displacement occurs. In the figure, the tolerable elastic deformation is indicated by u_{se}, and the maximum support pressure is p_{sm}. There will be some displacement before installation of support (u_{s0}), which corresponds to stiffness and the capacity of the support system. Ground reaction curve is intersected by SRC; at this point, radial displacement is u_{se} and support pressure is p_{se}. Therefore, the factor of safety for the used support system will be equal to the ratio between p_{smax} and p_{se}.

There are a number of methods to derive GRC and SRC. The common points of all the methods are assumption of circular tunnel, hydrostatic stress field. The main difference between the methods are the choice of the rock mass failure criterion and whether or not the rock mass dilates during failure. Although there are a number of methods, only two of the widely used methods will be introduced here (Duncan Fama 1993; Carranza-Torres and Fairhurst 1999). The equations for the derivation of GRC suggested by Duncan Fama (1993) and Carranza-Torres and Fairhurst (2000) are given in Tables 8.11 and 8.12, respectively.

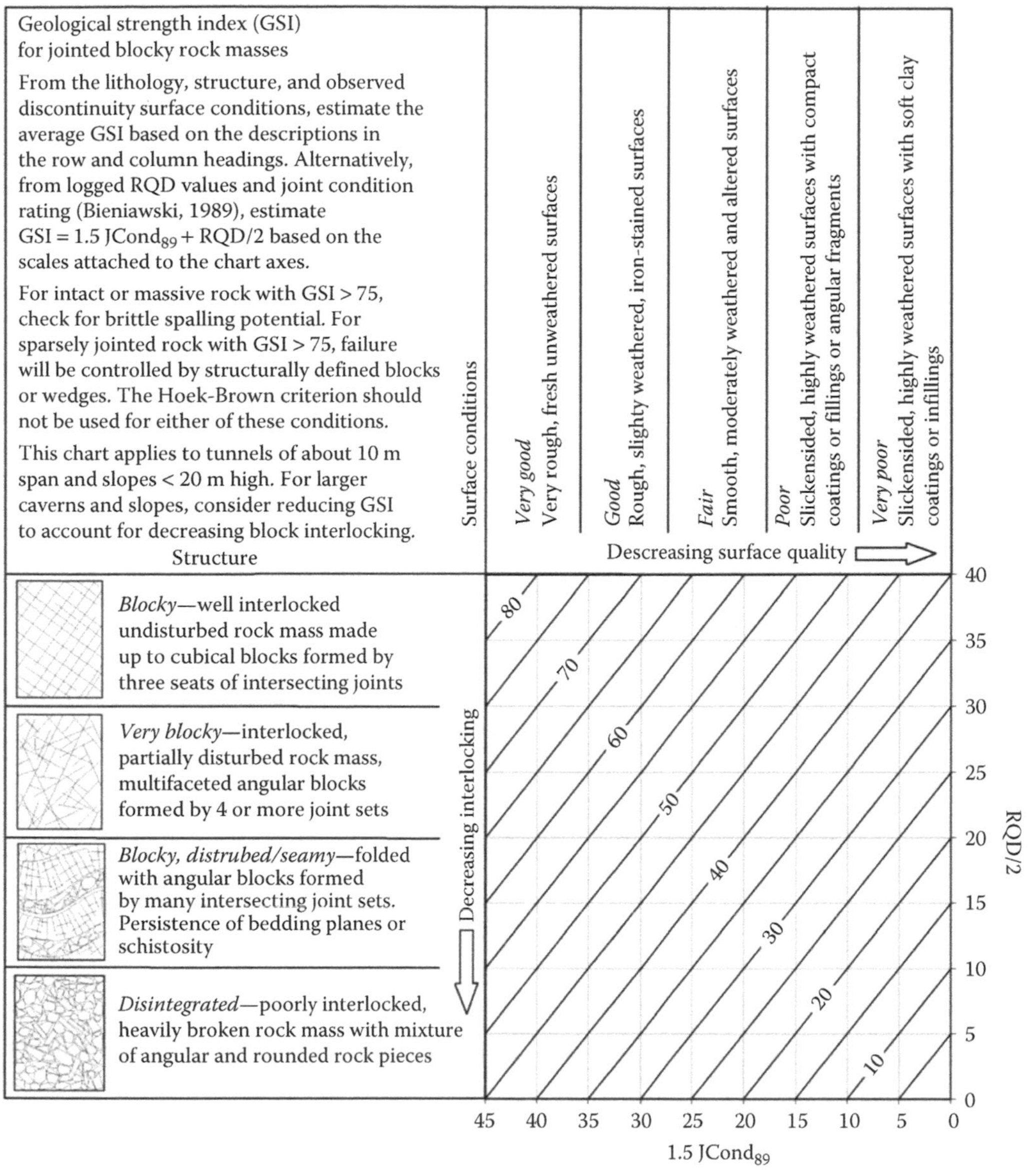

FIGURE 8.9 The basic structure of the GSI chart and quantification attempt. (Modified from Hoek, E., *Practical Rock Engineering*, http://www.rocscience.com/hoek/corner/Practical_Rock_Engineering.pdf, 2007.)

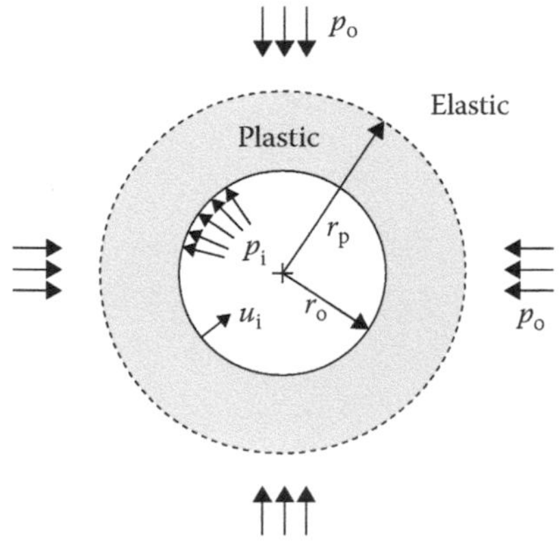

FIGURE 8.10 Basic terms used in ground reaction curve.

To construct SRC, the elastic stiffness K_s (MPa/m) and maximum support pressure are needed. The formulas for the construction of SRC of shotcrete, blocked steel set, and ungrouted bolts and cables were compiled and well presented by Carranza-Torres and Fairhurst (2000) (Table 8.13). For more practical use, the graph shown in Figure 8.12, supplied by Hoek and Brown (1980), can be used.

More recent formulas on the calculation of stiffness and maximum support pressure of steel set were presented by Oreste (2003). Another widely used support elements in underground coal mines is the yielding steel ribs or Taussenhousen (TH) ribs. Rodriguez and Diaz-Aguado (2013) suggested for the construction of support reaction curve of TH ribs.

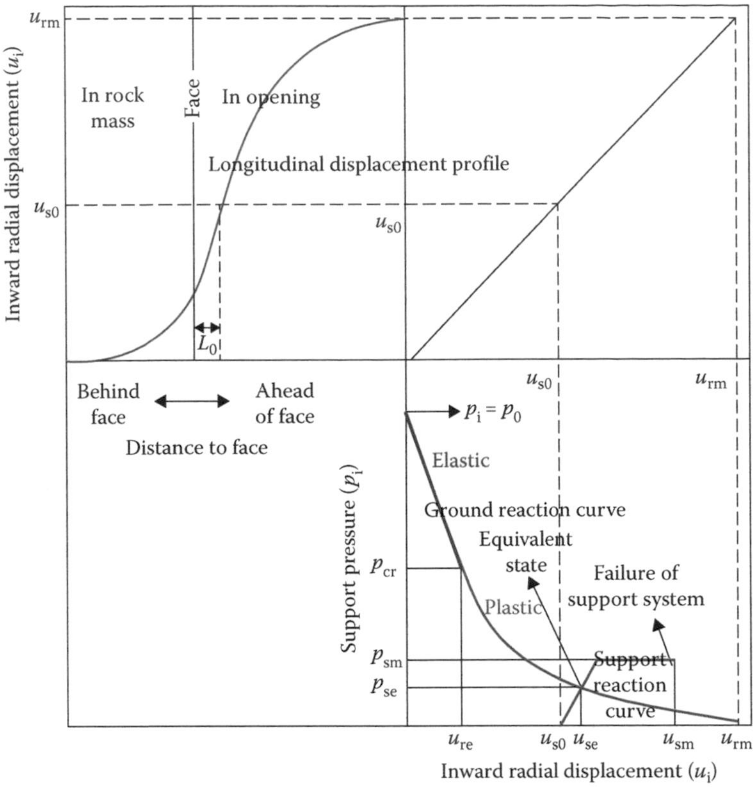

FIGURE 8.11 The terms used in analytical methods such as ground reaction curve, support reaction curve, and longitudinal displacement profile.

TABLE 8.11
Suggested Equations for Derivation of GRC

	Duncan Fama	Notes
Failure criterion	$\sigma_1' = \sigma_{cm} + k\sigma_3'$	σ_1' is the axial stress, where failure occurs, σ_3' is the confining stress, c' is the cohesive strength, and ϕ' is the angle of friction of rock mass
Rock mass strength	$\sigma_{cm} = \dfrac{2c' \cdot \cos\phi'}{1-\sin\phi'}$	
	$k = \dfrac{1+\sin\phi'}{1-\sin\phi'}$	E_m and ν are deformation modulus and Poisson's ratio of rock mass
Critical support pressure	$p_{cr} = \dfrac{2p_o - \sigma_{cm}}{1+k}$	
Elastic inward deformation	$u_{ie} = \dfrac{r_o(1+\nu)}{E_m} \cdot (p_o - p_i)$	
Plastic zone radius	$r_p = r_o \left\{ \dfrac{2[p_o(k-1)+\sigma_{cm}]}{(1+k)[(k-1)p_i + \sigma_{cm}]} \right\}^{1/(k-1)}$	
Plastic inward deformation	$u_{ip} = \dfrac{r_o(1+\nu)}{E}\left[2(1-\nu)(p_o - p_{cr})\left(\dfrac{r_p}{r_o}\right)^2 - (1-2\nu)(p_o - p_i) \right]$	

Source: Duncan Fama, M.E., Numerical modeling of yield zones in weak rock, in *Comprehensive Rock Engineering*, J.A. Hudson, 49–75, Pergamon, Oxford, 1993.

TABLE 8.12
Suggested Equations for Derivation of GRC

	Carranza Torres and Fairhurst	Notes
Failure criterion	$\sigma_1 = \sigma_3 + \sigma_{ci}\left(m_b \frac{\sigma_3}{\sigma_{ci}} + s\right)^a$	m_b, s, and a are Hoek-Brown constants and can be calculated using the suggestion by Hoek et al. (2002), where G_{rm} is the shear modulus of rock mass
	$p_i^{cr} = \left[P_i^{cr} - \frac{s}{m_b^2}\right] m_b \sigma_{ci}$	
Critical support pressure	$P_i^{cr} = \frac{1}{16}\left[1 - \sqrt{1 + 16S_0}\right]^2$	
	$S_0 = \frac{\sigma_0}{m_b \sigma_{ci}} + \frac{s}{m_b^2}$	
Elastic inward deformation	$u_r^{el} = \frac{\sigma_0 - p_i}{2G_{rm}} R$	
Plastic zone radius	$R_{pl} = R \exp\left[2\left(\sqrt{P_i^{cr}} - \sqrt{P_i}\right)\right]$	
Plastic inward deformation	$u_r^{pl} = \frac{(\sigma_0 - p_i^{cr})}{2RG_{rm}}\left[\frac{1-2\nu}{2} \cdot \frac{\sqrt{P_i^{cr}}}{S_0 - P_i^{cr}} + 1\right]\left(\frac{R_{pl}}{R}\right) + \frac{1-2\nu}{4(S_0 - P_i^{cr})}\left[\ln\left(\frac{R_{pl}}{R}\right)\right]^2$	
	$- \frac{1-2\nu}{2} \cdot \frac{\sqrt{P_i^{cr}}}{S_0 - P_i^{cr}}\left[2\ln\left(\frac{R_{pl}}{R}\right) + 1\right]$	

Source: Carranza-Torres, C. and Fairhurst, C., *Tunnel. Undegr. Space Technol.*, 15, 187–213, 2000.

TABLE 8.13
Maximum Support Pressure and Support Stiffness

	K_s (MPa/m)	p_s^{max}	Notes
Shotcrete	$K_{sc} = \frac{E_c}{(1-\nu_c)R} \frac{R^2 - (R - t_c)^2}{(1-2\nu)R^2 + (R - t_c)^2}$	$p_{sc}^{max} = \frac{\sigma_{cc}}{2}\left[1 - \frac{(R - t_c)^2}{R^2}\right]$	σ_{cc} is the uniaxial compressive strength of shotcrete or concrete (MPa), E_c is the Young's modulus for shotcrete or concrete (MPa), ν_c is the Poisson's ratio, t_c is the thickness of ring (m), R is the external radius of the support (m)
Blocked steel set	$\frac{1}{K_{ss}} = \frac{SR^2}{E_{ss}A_{ss}} + \frac{SR^4}{E_{ss}I_{ss}}\left[\frac{\theta(\theta + \sin\theta\cos\theta)}{2\sin^2\theta} - 1\right] + \frac{2S\theta t_B R}{E_B B^2}$	$p_{ss}^{max} = \frac{3}{2}\frac{\sigma_{ys}}{SR\theta} \frac{A_s I_s}{3I_s + DA_s\left[R - (t_B + 0.5D)\right](1 - \cos\theta)}$	B is the flange width of steel set and side length of the square block, D is the depth of steel section (m), A_s is the cross-sectional area of the section (m²), I_s is the moment of inertia of the section (m⁴), E_s is the Young's modulus of steel (MPa), σ_{ys} is the yield strength of the steel (MPa), S is the steel spacing along the tunnel axis (m), q is half the angle between blocking points (radians), t_s is the thickness of the block (m), E_s is Young's modulus for the block material (MPa), R is the tunnel radius (m)
Ungrouted bolts and cables	$\frac{1}{K_{sb}} = s_c s_l\left[\frac{4l}{\Pi d_b^2 E_s} + Q\right]$	$p_{sb}^{max} = \frac{T_{bf}}{s_c s_l}$	d_b is the bolt or cable diameter (m), l is the free length of the bolt or cable (m), T_{bf} is the ultimate load obtained from a pull-out test (MN), Q is a deformation load constant for the anchor and head (m/MN), E_s is the Young's modulus for the bolt or cable (MPa), s_c is the circumferential bolt spacing (m), and s_l is the longitudinal bolt spacing (m)

Maximum support pressures for various systems				
Support system/Tunnel radius	r_i 1 m 39 in.	r_i 2.5 m 98 in.	r_i 5 m 197 in.	r_i 10 m 394 in.
A. Shotcrete – 5 cm (0.05 m)/ 2 in. thick shotcrete. $\sigma_{c, conc}$ = 14 MPa/2000 psi after 1 day.	p_{smax} 0.65 MPa 95 psi	p_{smax} 0.27 MPa 39 psi	p_{smax} 0.14 MPa 20 psi	p_{smax} 0.07 MPa 10 psi
B. Shotcrete – 5 cm (0.05 m)/ 2 in. thick shotcrete. $\sigma_{c, conc}$ = 35 MPa/5000 psi after 28 day.	1.63 MPa 236 psi	0.68 MPa 99 psi	0.34 MPa 50 psi	0.17 MPa 25 psi
C. Concrete – 30 cm (0.3 m)/ 12 in. thick concrete. $\sigma_{c, conc}$ = 35 MPa/5000 psi after 28 day.	7.14 MPa 1036 psi	3.55 MPa 515 psi	1.93 MPa 279 psi	1.00 MPa 146 psi
D. Concrete – 50 cm (0.5 m)/ 19.5 in. thick concrete. $\sigma_{c, conc}$ = 35 MPa/5000 psi after 28 day.	9.72 MPa 1410 psi	5.35 MPa 775 psi	3.04 MPa 440 psi	1.63 MPa 236 psi
E. Steel sets – (6 I 12) space 2m/79 in. Blocket 2θ = 22 1/2°, σ_{ys} = 248 Mpa/36000 psi	0.61 MPa 88 psi	0.18 MPa 27 psi	0.07 MPa 10 psi	0.02 MPa 3 psi
F. Steel sets – (8 I 23) space 1.5 m/59 in. Blocket 2θ = 22 1/2°, σ_{ys} = 248 Mpa/36000 psi	1.59 MPa 230 psi	0.50 MPa 72 psi	0.18 MPa 27 psi	0.06 MPa 9 psi
G. Steel sets – (12 W 65) at 1 m/39 in. Blocket 2θ = 22 1/2°, σ_{ys} = 248 Mpa/36000 psi	7.28 MPa 1055 psi	2.53 MPa 3.66 psi	1.04 MPa 150 psi	0.38 MPa 55 psi
H. Very light rockbolts 16 mm/ 5/8 in. ϕ at 2.5 m/98 in. centers. Mechanical anchor. T_{bf} = 0.11 MN/25 000 lb	0.02 MPa 2.6 psi	0.02 MPa 2.6 psi	0.02 MPa 2.6 psi	0.02 MPa 2.6 psi
I. Light rockbolts 19 mm/ 3/4 in. ϕ at 2 m/79 in. Mechanical anchor. T_{bf} = 0.18 MN/40 000 lb	0.045 MPa 6.5 psi	0.045 MPa 6.5 psi	0.045 MPa 6.5 psi	0.045 MPa 6.5 psi
J. Medium rockbolts 25 mm/ 1 in. ϕ at 1.5 m/59 in. centers mechanical anchor. T_{bf} = 0.267 MN/60 000 lb	0.12 MPa 17 psi	0.12 MPa 17 psi	0.12 MPa 17 psi	0.12 MPa 17 psi
K. Heavy rockbolts 34 mm/ 1 3/8 in. ϕ at 1 m/39 in. centers. Resin anchor. T_{bf} = 0.345 MN/150 000 lb	0.34 MPa 49 psi	0.34 MPa 49 psi	0.34 MPa 49 psi	0.34 MPa 49 psi

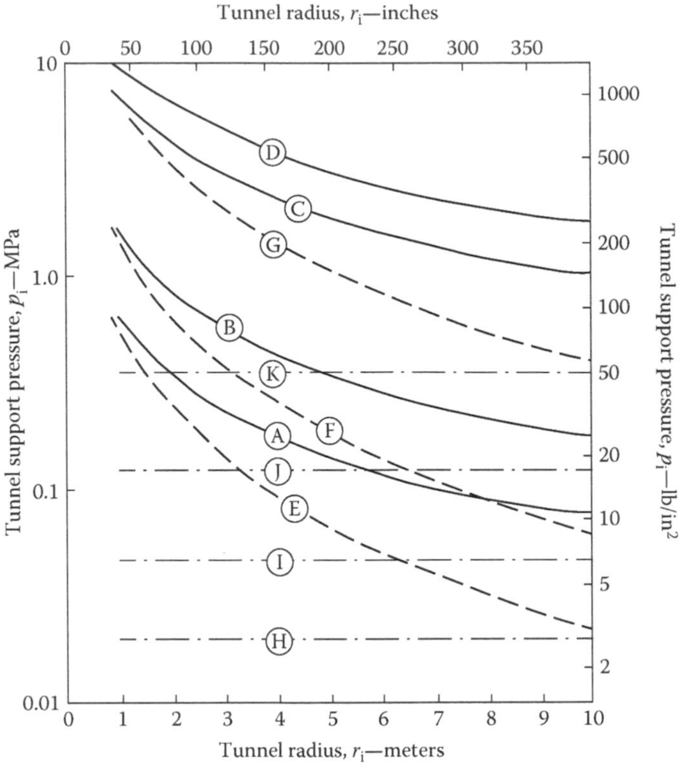

FIGURE 8.12 Support pressures of various systems. (Data from Hoek, E. and Brown, E.T., *Underground Excavations in Rock*, Institution of Mining Metallurgy, London, 1980.)

8.3 DESIGN OF SUPPORT SYSTEMS FOR LONGWALL PANEL ROADWAYS

The stability of the gate roadways is more complex than the stability of main haulage ways. The stability of gate roads is generally expressed in terms of convergence, which is defined as the sum of the roof lowering and floor heaving. Convergence can be expressed in terms of absolute displacement or percentage. The stability issues of the gateways are the result of the combined effect of number of parameters. Istanbulluoglu (1995) and Istanbulluoglu and Karpuz (1997) listed the parameters affecting the stability of panel roadways as presented in Figure 8.13. The parameters can be divided into two main groups such as controllable and uncontrollable parameters. Controllable parameters can be controlled during planning and designing phase, but others cannot be controlled such as geological parameters. Controllable parameters can be subdivided into two subgroups such as factors related to mining method and the factors related to support design.

For safe and economical design, the factors outlined in Sections 8.3.1 and 8.3.2 should carefully be considered.

8.3.1 Controllable Factors Related to Mining Methods

Peng et al. (1980) studied the relationship between panel width and side abutment pressure using finite element analysis. The results indicate that as the panel width increases the amount of side abutment pressure, consequently the amount of deformation in gateways increases. Peng and Chiang (1984) stated that the increase in panel width and panel number within the mining sequence result in increased gateway convergence.

Increasing seam height leads to increased vertical closure in gateways. The effect of seam height increases with increasing mine depth (Unver 1988).

In the report compiled in Nottingham University (Hazine et al. 1978), it was indicated that advance heading system showed the highest amount of gateway closure than conventional ripping formation method.

The gateway shape and size affect the amount of convergence. Hobbs (1968) reported that the order of preference roadway shape regarding the stability is hexagonal, circular, pentagonal, arched, and rectangular. Whittaker and Hodkinson

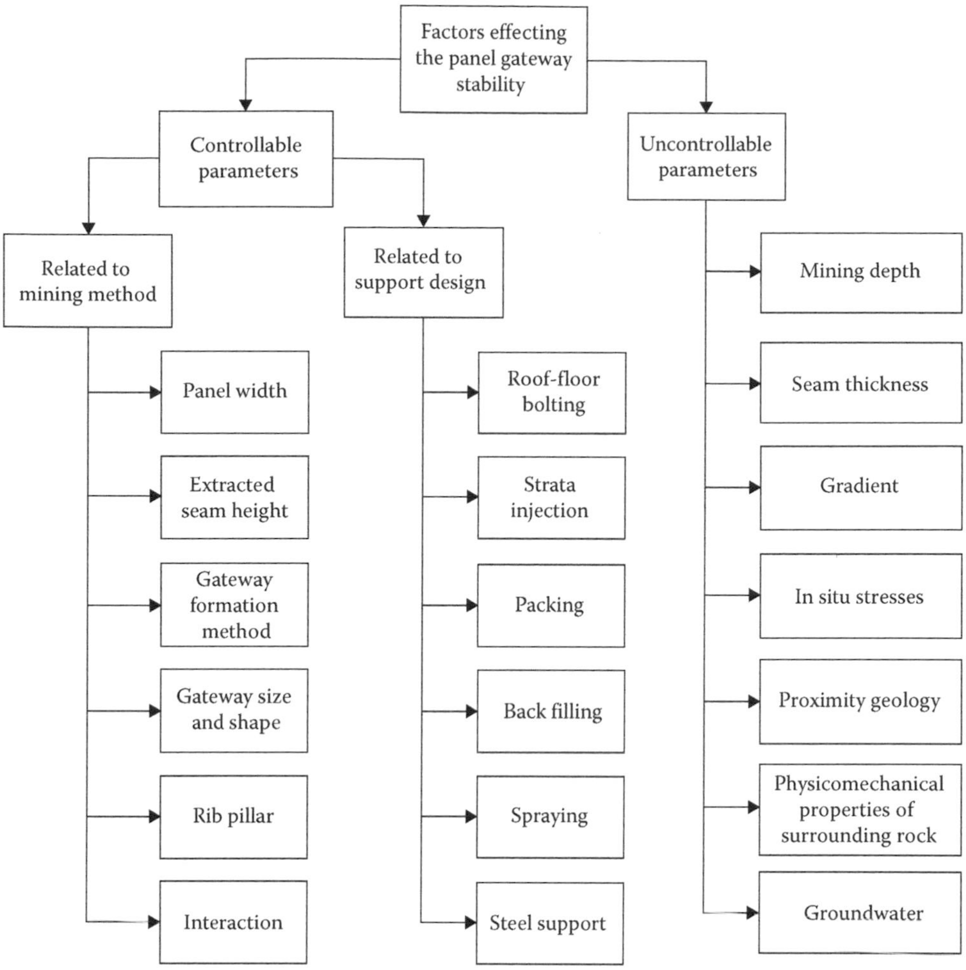

FIGURE 8.13 Factors affecting the stability of mine roadways.

(1971) used scale-model test to investigate the influence of size. They reported that the amount of deformation increases with the size of gateway. Width-to-height ratio of the gateway is another effective parameter. Studies indicated that for the same cross-sectional area, narrow roadway will stand much better. Afroz (1990) claimed that increase in the height of a roadway relative to its width results increase in the side closure. The formation of gateways by roadheader or similar machines reduces over breaking leading to less stability problems (Carr 1970). The design of rib pillar is one of the most important parameter of longwall layout planning, which has a great influence on the stability of gateways. To maintain the stability of roadways in longwall mining, a coal pillar of adequate dimension should be left between two longwall faces. On the other hand, increased pillar size reduces overall profit. Therefore, it is necessary to optimize coal recovery without jeopardizing gateway stability.

Most of the ground control difficulties are sourced from interaction. There can be different type of interaction. For example, the extraction of coal in the proximity of adjacent working or working in different seams as a consequence of multiseam mining can be considered as interaction. Interaction can also occur between the side abutment pressure of two adjacent roadways if they are separated by a narrow pillar. To explain the interaction phenomenon, the stress distribution around panel roadways should be analyzed.

Whittaker's (1974) study is the very first attempt for the investigation of vertical stress distribution around the longwall production panel. Concentrated high vertical stress around the rock surrounding the panel diminishes to virgin stress field by increasing distance from the excavation (Figure 8.14).

Whittaker and Singh (1979) considered an advancing longwall face at moderate depth in Britain and prepared a nomograph indicating the stress distribution around a longwall panel (Figure 8.15). Hudson (1993) commented on the validity of the nomograph for longwall retreat method.

The ratio of face length to depth, characteristics of immediate roof, and overburden and subsidence will surely alter the distribution or the distance where the stress levels equalized to virgin field stress. High stress concentration existing in front of the face may lead to use of special support systems, especially for thick coal seams. Stresses existing on pillar abutment have great influence on the stability of maingate and tailgate. Special analyses are required with respect to the pillar and roadway geometry for the analysis of the stability of maingate and tailgate roadways, since the nomograph of Whittaker and Singh (1979) suggests that the stresses developing around the panel

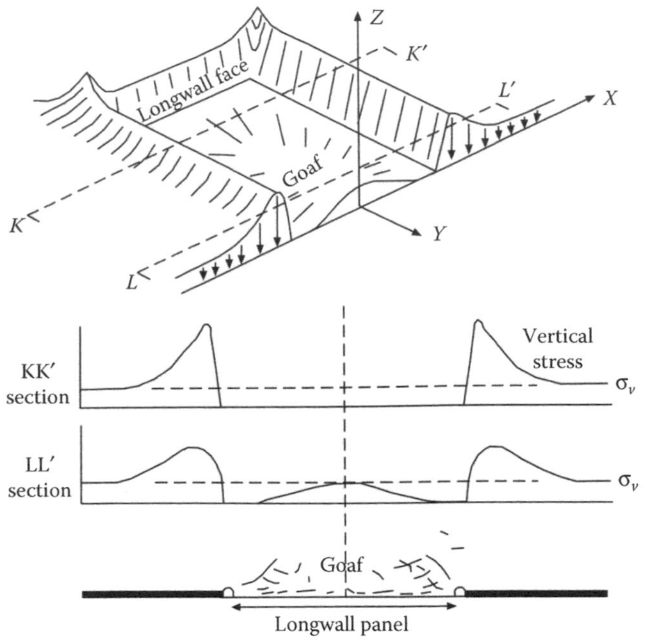

FIGURE 8.14 Vertical stress distribution in and around the longwall panel. (Modified from Whittaker, B.N., *Mining Eng.*, 134, 9–24, 1974.)

may increase up to four times of the in situ stress magnitude. Vertical stress distribution on pillar abutments was examined by several researchers (Wilson and Ashwin 1972; Carr and Wilson 1982; Wilson 1983; Suchowerska et al. 2013). Wilson studied on the estimation of the stress around the excavations by utilizing cavity expansion theory, which assumes that the openings are excavated in cylindrical form in a homogenous, isotropic, and elastoplastic medium. Wilson presumes that there exists a failed or yielded zone and an elastic zone in the pillar and a goaf region with a stress increasing linearly (Figure 8.16).

Around the goaf zone, vertical stress increases linearly until reaching the in situ stress value. Slope of the linear line is related to the material properties of the goaf. The figure is valid for a pillar, which can be allowed to be yielded but the roof and bottom are assumed to be stable and sustaining load.

8.3.2 Controllable Factors Related to Support Design

Most of the factors related to support design are explained in the previous section such as designing and selection of appropriate strata bolts, steel support, and shotcrete. In addition to these explained factors, backfilling and packing techniques will be explained in the following paragraph. Backfilling is

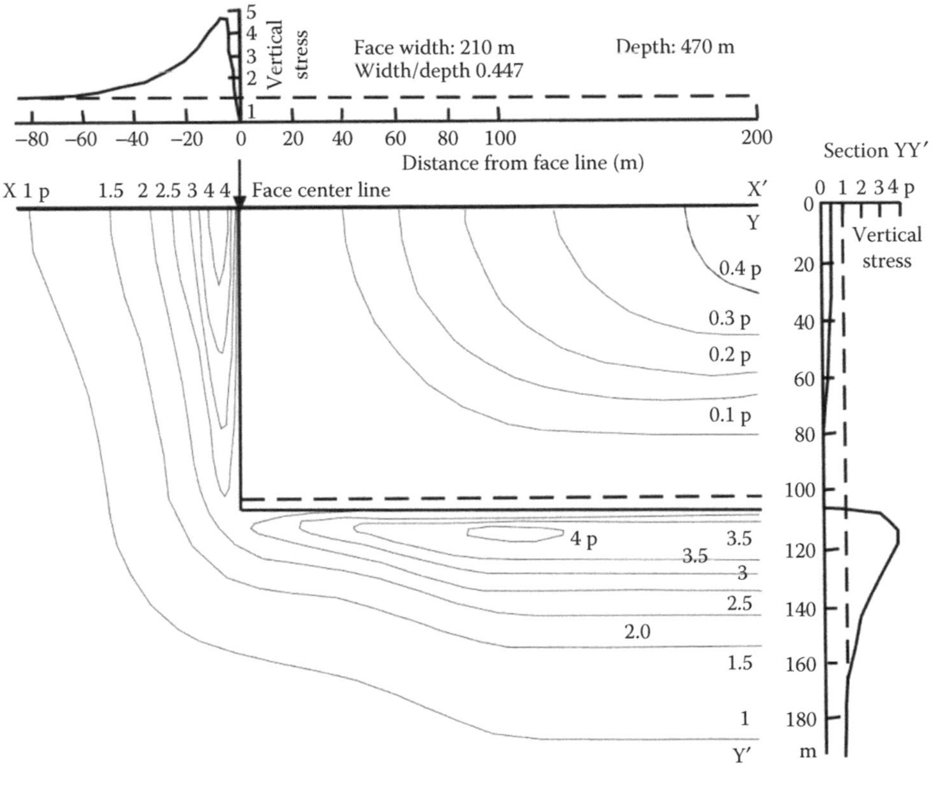

FIGURE 8.15 Nomograph indicating vertical stress distribution in and around the longwall panel. (Modified from Whittaker, B.N. and Singh, R.N., *Mining Eng.*, 139, 59–70, 1979.)

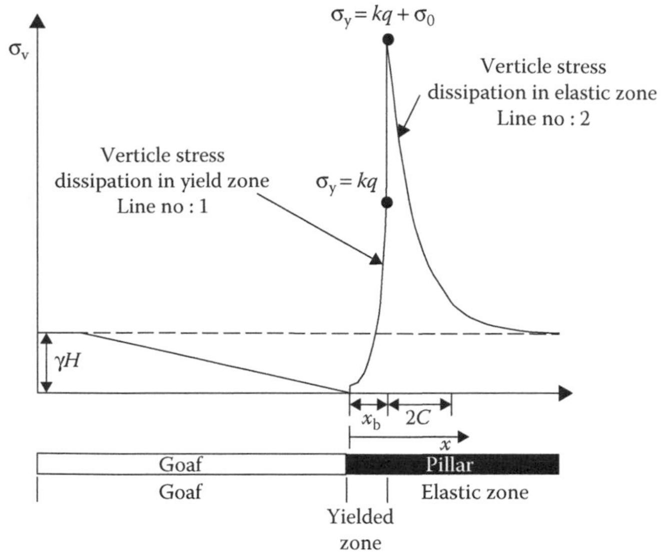

FIGURE 8.16 Variation of the vertical stress from goaf to pillar abutment. (Modified from Wilson, A.H., *Int. J. Mining Eng.*, 1, 91–187, 1983.)

used to fill the gaps between gate roadway and steel support. Several foamed grout are available to fill the gap. Mostley (1984) emphasized that total roof lowering was reduced by 50% in the backfilled areas of a gateway. Packing is used to prevent overbreakage around the face side and to provide early support to tilting roof toward the caved area. Therefore, packing is considered as one of the most important support element where panel gateways are drifted adjacent to goaf (Istanbulluoglu 1995). Clark and Newson (1985) summarized the purpose of face-side packs as follows: to provide support to the immediate strata around the roadway and to protect roadway support elements, to effect a breaking off line at the goaf side, to facilitate a convenient area for the disposal of unwanted debris produced during the face working, and to minimize air leakage across the waste area.

8.3.3 Uncontrollable Parameters

Localized faulting, for example, increases the roadway closure over a limited length of gate roadway and may require special support measures. Mudstones mostly observed at the roof of coal seam are weakened by water and may require special roadway forming methods and support measures. Afroz (1990) stated that gate roadways along the strike and parallel to the main cleat of the coal seam would have a more stable floor than those driven along the full dip and perpendicular to the main cleat. Mining depth is directly proportional to vertical stress and side abutment pressure which are two of the important sources of gate roadway deformation. Siddal and Gale (1992) reported that in general horizontal stress plays a major role in the stability of the roof and floor strata. Whereas vertical stress has a crucial role in rib side and pillar deformation.

8.4 PILLAR DESIGN

In any underground coal mine, two different pillar types are observed. The design of pillars has crucial importance on the stability of underground mine. The traditional pillar design methodology consists of three main steps such as estimation of pillar load, estimation of pillar strength, and calculation of safety factor.

8.4.1 Estimation of Pillar Load

The tributary area theory, also known as *extraction ratio approach*, is based on simple equilibrium analysis. The extraction ratio depends on the pillar layout; pillar load is calculated based on the geometry and dimensions of pillar and rooms as shown in Figure 8.17 (Hoek and Brown 1980). In Figure 8.17, W_p is the pillar width, R_s is the width of the room, and H is depth.

$$\sigma_p = \frac{A_t}{A_p}\sigma_z \tag{8.10}$$

$$\sigma_p = \left(\frac{1}{1-e}\right)\sigma_z \tag{8.11}$$

where:

σ_p average pillar stress
σ_z is the vertical stress
A_t is the total area
A_p is the pillar area
e is the extraction ratio

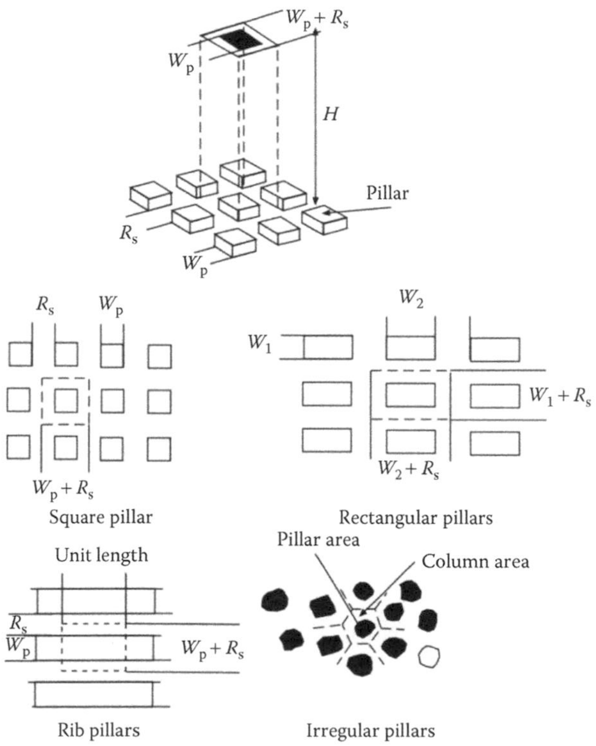

FIGURE 8.17 Different pillar types and dimensions. (Modified from Hoek, E. and Brown, E.T., *Underground Excavations in Rock*, Institution of Mining Metallurgy, London, 1980.)

TABLE 8.14
Pillar Stress on Different Shaped Pillars

For square pillars	$\sigma_p = \gamma H\left(1+\frac{R_s}{W_p}\right)^2$
For rectangular pillars	$\sigma_p = \gamma H\left(1+\frac{R_s}{W_1}\right)\left(1+\frac{R_s}{W_2}\right)$
For rib pillars	$\sigma_p = \gamma H\left(1+\frac{R_s}{W_p}\right)$
For irregular pillars	$\sigma_p = \gamma H\left(\frac{\text{Column area}}{\text{Pillar area}}\right)$

The stresses on pillars can be estimated using the formulas given in Table 8.14.

For describing pillar stress in terms of extraction ratio, the following expressions modified by Biron and Arioglu (1980) can be used:

$$\sigma_p = \frac{\sigma_v}{1-e} = \frac{\gamma H}{1-e} \cong 0.026\left(\frac{H}{1-e}\right) \quad (8.12)$$

For dipping seams, the above expression can be written as follows:

$$\sigma_p = \frac{\sigma_v \cos^2\alpha + \sigma_h \sin^2\alpha}{1-e} \quad (8.13)$$

where:
- σ_v is vertical stress
- σ_h is horizontal stress
- α is the dip of seam
- γ is unit weight of rock material
- e is extraction ratio, that is, the ratio of the extracted area to total area

For rectangular pillars, extraction ratio can be described as

$$e = 1 - \left[\left(\frac{W_1}{W_1+R_s}\right)\left(\frac{W_2}{W_2+R_s}\right)\right] \quad (8.14)$$

For square pillars, $W_p = W_1 = W_2$; therefore,

$$e = 1 - \frac{W_p^2}{(W_p+R_s)^2} \quad (8.15)$$

8.4.2 Pillar Strength Estimation

The strength of coal pillar is mainly affected by the size and shape of the pillar and geological factors. Bieniawski (1968) used the *critical size* concept to explain size effect. It is known that as the size increases, the strength decreases until critical specimen size. Shape effect is generally explained by using width (W) and height (h) of the pillar. Arioglu and Tokgoz (2011) outlined the following findings about the shape effect derived from South African and Australian database (Galvin et al. 1999). For a constant height pillar, strength increases with increasing width. Similarly, for a given pillar width, the strength increases with decreasing pillar height. The strength increases with increasing W/h ratio. Geology has significant effect on pillar strength such as strong floor and roof lead to increased pillar strength. Weak rock and bedding planes lead to limit the strength due to limited confinement.

In general, there are two main types of empirical equations for predicting coal pillar strength: linear form and power form. The respective equations are as follows:

$$S_p = \sigma_{cube}\left[A + B\left(\frac{W}{h}\right)\right] \quad (8.16)$$

$$S_p = K\left(\frac{W^a}{h^b}\right) \quad (8.17)$$

where:
- σ_p is the pillar strength
- σ_{cube} is the in situ coal strength incorporating *size effect*
- A, B, a, and b are the constants defining the *shape effect*

Very comprehensive review of the formulas regarding both size and shape effect is presented by Arioglu and Tokgoz (2011) and Jawed et al. (2013).

It is known that the strength decreases with increasing specimen size. The equations presented in Table 8.15 emphasize this phenomenon.

8.4.3 Factor of Safety

Factor of safety is simply defined as the ratio between the pillar strength and average pillar stress. Galvin et al. (1999) and Salamon and Munro (1967) suggested a factor of safety of 1.6 as an appropriate value for pillar designs.

8.5 NUMERICAL MODELING

Numerical modeling techniques are widely used for design purposes. The techniques follow the earliest analytical models dating back to closed-form solutions used to calculate stresses around a circular hole in a stresses plate (Kirsch 1898). With the development of microcomputer technology, numerical modeling techniques have become an important tool in underground coal mining. By means of the power of the computers and improved numerical modeling technique, now it is possible to perform a large number of complex analyses. Given the inherently inhomogeneous nature of rock mass, the possibility of conducting such large number of analysis can be considered as an asset.

The main advantages of numerical modeling compared to other methods such as empirical and analytical methods can be outlined as follows: complex engineering problems can be handled, accuracy of the analysis is satisfied, and fast

TABLE 8.15
Empirical Equations for Calculating Pillar Strength

Zern (1926)	$S_p = \sigma_{ci}\sqrt{\frac{W}{H}}$	σ_{ci} is the coal strength parameter ranging from 4.8 to 7.0 MPa
Greenwald et al. (1941)	$S_p = 0.67\sigma_{cube}\frac{\sqrt{W}}{H^{0.83}}$	σ_{cube} is the strength of unit cube coal sample
Obert-Duvall (1967)	$\frac{S_p}{\sigma_{ci}} = 0.778 + 0.222\left(\frac{W}{H}\right)$	σ_{ci} is the strength of coal sample having diameter to height ratio of 1
Salamon and Munro (1967)	$S_p = k_{SM}H^{\alpha}W^{\beta}$, kPa	$k_{SM} = 7.176$ kPa, $\alpha = 0.66$, $\beta = 0.46$. The value of k has to be evaluated by testing a specimen of size 30 cm.
Bieniawski (1968)	$S_p = \sigma_{cube}\left[0.64 + 0.34(W/h)\right]$, MPa	σ_{cube} is the strength of a 30 cm cube pillar specimen (MPa), $1 < W/h < 5$
Wilson (1972)	$S_p = 444\gamma(W/h)$, lb / in^2 $S_p = \frac{1333\gamma}{h}\left[W' - \left(\frac{W+W'}{2}\right) + \left(\frac{W}{3}\right)\right]$, lb / in^2	For the estimation of the square and rectangular pillar strengths
Borecki and Kidbinski (1972)	$S_p = 1.6\sigma_{lab5}W^{m}(W/h)^{0.5}$ $m = 7.57\times 0^{-4}\sigma_{lab5} - 0.4$	Where σ_{lab5} is the uniaxial compressive strength of cube sample with side length of 5 in. (kgf/cm^2), W and h are the width and height of pillar in centimeters.
Holland (1973)	$S_p = \sigma_{cube}\sqrt{W/h}$, psi $\sigma_{cube} = \frac{k}{6} = 0.166k = 0.166\sqrt{D}$ $\sigma_{cube} = 0.235\sigma_{ci}$	To use the equation D should be in the range from 2 to 4 in. diameter or the length of cube side dimension, σ_{cube} is the pillar coal strength incorporating critical specimen (generally ≥90 cm)
Logie and Matheson (1982)	$S_p = \sigma_{ci}\left[0.64 + 0.34\left(\frac{W}{h}\right)\right]^{1.4}$	Where σ_{ci} is the uniaxial compressive strength for square coal pillars with W/h ratio larger than or equal to 4.5
Wagner and Madden (1984)	$S_p = K\frac{2.5}{V^{0.07}}\left\{0.13\left[\left(\frac{R}{4.5}\right)^{4.5} - 1\right] + 1\right\}$, MPa	Where $V = (W_{eff})^2 h$ and $R = W/h$. In the equation, W_{eff} is the effective pillar width and K is a constant for South African coals, which is around 7.2 for $W/h \le 4.5$
Stacey and Page (1986)	$S_p = K\left[\frac{\sqrt{W_{eff}}}{h^{0.7}}\right]$	$W_{eff} = 2\left[\frac{W_1W_2}{W_1 + W_2}\right]$
Holland and Gaddy (1964)	$S_p = k\left[\frac{\sqrt{W}}{h}\right]$ $k = \sigma_{ci}\sqrt{D}$, psi	Where D is the size of laboratory specimen (inch) and σ_{ci} is the strength of laboratory test sample (psi)
Sheorey et al. (1987)	$S_p = 0.27\sigma_{clab}\left[\frac{\sqrt{W}}{h^{0.96}}\right]$, MPa	σ_{clab} is the strength of 2.5 cm cubes for any arbitrary in situ stresses and Indian coal beds
Sheorey (1992)	$S_p = 0.27\sigma_{ci}h^{-0.36} + \frac{\sigma_v}{150\gamma}\left[p + \frac{q}{H}\right]\left[\frac{W}{h} - 1\right]$, MPa $S_p = 0.27\sigma_{ci}h^{-0.36} + \frac{H}{150}\left[0.6 + \frac{150}{H}\right]\left[\frac{W}{h} - 1\right]$, MPa	
Mark and Bieniawski (1986)	$S_p = \sigma_{cube}\left[0.64 + 0.54\left(\frac{W}{h}\right) - 0.18\left(\frac{W^2}{Lh}\right)\right]$	σ_{cube} is the in situ coal strength and L is the pillar length for $W/h \le 6.0$
Madden (1991)	$S_p = 5.24\left(\frac{W^{0.63}}{h^{0.78}}\right)$	
Maleki (1992)	$S_p = 4700\{1 - \mathrm{EXP}[-0.339(W/h)]\}$, lb / in^2 $S_p = 3836\{1 - \mathrm{EXP}[-0.260(W/h)]\}$, lb / in^2	Equations for the situations when the pillar strength is governed by confinement control and structural control 3836 and 4700 are the design strength, which may range from 3000 to 5000 psi.

(Continued)

TABLE 8.15 (*Continued*)
Empirical Equations for Calculating Pillar Strength

Africa Van der Merwe (2003)	$S_p = k\frac{W}{h}$	For normal coal as in Withbank, South Rand, Utrecht, Spring-Withbank, and Free State Coal fields of South Africa suggested the following formula, with *k* in the range of 2.8 to 3.5 MPa, describing the strength of longwall chain pillar
Oraee et al. (2010)	$S_p = \sigma_{cube}\text{EXP}\left[-0.43 + 0.668\sqrt{W/h}\right]$	

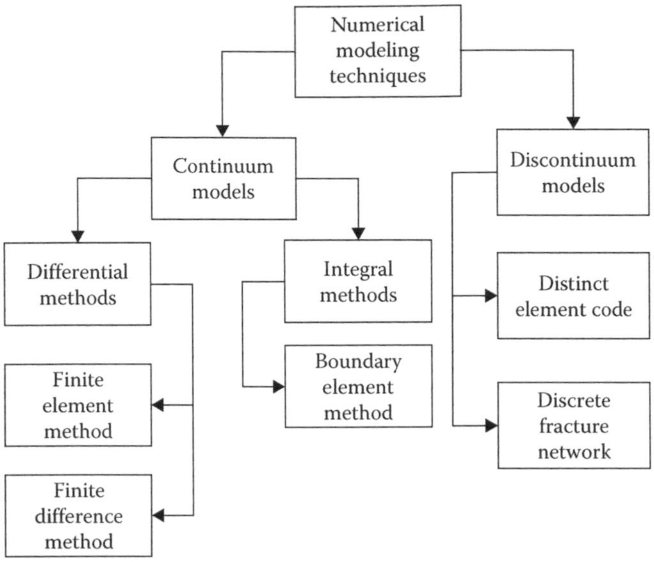

FIGURE 8.18 Numerical modeling techniques.

analysis is possible. However, care should be given to the following point: bad assumptions and wrong selection of input parameters may lead the user to make serious mistakes and misinterpretations.

The main groups of numerical modeling are the continuum and discontinuum methods. The main difference between these subgroups is that whether the displacement field is physically continuous or not. In continuum model, the medium is considered as continuity that is free of discontinuities and/or equivalent jointed rock mass. However, in discontinuum methods, the medium is assumed to be composed of a number of discrete blocks. The main groups of both continuum and discontinuum methods are given in Figure 8.18.

8.5.1 Continuum Methods

Continuum methods are also called as *differential methods* in which approximations are made throughout the problem domain. In differential methods, a problem is solved either by using equations of motion or by using equation of equilibrium. Depending on the used approximation, the method can be called either *finite difference* (explicit methods) or *finite element* (implicit methods) method. In practice, it is usually hard to distinguish these two methods (Hoek 2007). Another subgroup of the continuum method is the boundary element method or integral method. In this method, approximation is made on the boundary. Each of these methods has its own limitations and strengths. For example, finite difference method requires long time compared to others, whereas the method is very effective on nonlinear or large strain problems. Finite element method gives quick solution compared to finite difference method but requires longer processing time than boundary element method. Most importantly, the limitation of finite element method is that the method gives solution for small strain problems. On the other hand, finite element method can successfully handle complex material properties, boundary, and loading conditions. Boundary element method can solve problems in which medium is considered as homogeneous, isotropic/transversely isotropic and material is assumed to be linear elastic. These assumptions can be considered as unrealistic for some situations, whereas boundary element method has high computational efficiency compared to other subgroups of continuum methods. Further details on boundary element methods can be found in *Boundary Element Methods in Solid Mechanics* by Crouch and Starfield (1983).

The finite element method is well suited for solving heterogeneous and nonlinear material properties, but it can be considered as an inappropriate method for modeling infinite boundaries such as those that occur in underground excavation problems. In that case, different algorithms should be applied to handle infinite boundaries such as discretization of area beyond the area of influence and appropriate boundary conditions should be applied to the outer edges. Finite difference method, which was first applied to geomechanics by Cundall (1971), has the advantage where both geometric and material nonlinearities are accommodated by relatively increasing the computational time. Therefore, as the nonlinearity of the system increases, finite difference method becomes more attractive compared to the finite element method. Boundary element method can successfully model the tabular ore bodies, which are considered as homogeneous, isotropic/transversely isotropic, and material is assumed to be linear elastic. As long as the nature of the problem is suitable for the method, using boundary element method, the solution can be obtained with less computational time. In such situation, boundary element method becomes more efficient than other continuum methods.

8.5.2 Discontinuum Methods

Discontinuum methods can be subdivided into two groups: distinct element method and discrete fracture network method. In discontinuum methods, ground conditions can be described as discontinuous or blocky. In such case, the individual blocks may be free of rotate and translate and each block can be deformable and/or rigid. Although continuum models may include several individual joints, their use for such blocky conditions requiring explicitly modeling of many joints may result in relatively inefficient performance.

8.6 MONITORING IN UNDERGROUND COAL MINING

To evaluate the performance of the designed support elements and to improve the understanding of interaction between the designed rock support and rock mass, instrumentation studies are required.

Another utilization area of the instrumentation studies is the numerical modeling calibration study, which is a way of reliably estimating strength properties of rock mass and reduction of uncertainties (Sakurai 1997). The calibration studies include the comparison of modeling results with the field measurements. During calibration process, input parameters are modified in a systematic manner until a satisfactory agreement between the model results and field measurements is achieved. When the model is calibrated, then it can be applied to evaluate similar mining layout and geological conditions.

Instrumentation studies are widely used in underground coal mining. Chen et al. (2002) developed an instrumentation program for the determination of the stresses in front of advancing mining face. Seedman (2001) used the program to describe the failure mechanism in a drift. Shephard and Lewandowski (1993) tried to monitor the roof falls during coal pillar extraction in Australian underground coal mine. Zahl et al. (2002) monitored the changing stresses in the roadway around the advancing face. Bai et al. (2013) developed a measurement program to determine the extent of fractured zone due to advancing mining face.

The program can also be used to improve early warning system. The researchers claimed that increasing acceleration of roof displacement can be used as indicator of instability in gateways (Unal et al. 2001; Shen et al. 2008).

8.6.1 Widely Used Measurement Instruments in Underground Coal Mining

Although a wide variety of instrumentation tools is available, most of them have similar properties and measurement principles. The most widely used equipment have been described in the following paragraphs.

8.6.1.1 Stressmeters

Vibrating wire stressmeters are used for the long-term measurement of stress changes in rock. Vibrating wire transducers are used to measure the deformation of thick-walled steel ring preloaded into a borehole by a wedge and platen assembly. Changing rock stress leads to change in the load on the gage body, causing the body to deflect. This deflection is recorded as a change in tension and resonant frequency directly proportional to the change in diameter of the gage and therefore the change in stresses in the rock. The instrument is used to measure the change in stresses rather than the measurement of in situ stresses. Depending on the placement and the location of the device, the stress changes in different directions can be measured. By means of the sensor, the effect of the advancing face and mining induced stresses can be quantified.

8.6.1.2 Borehole Pressure Cell

Similar to stressmeters, these cells are used to quantify the stress changes. The cell is placed into a borehole and grouted. Then the amount of changing stresses are directly read from the cell dial gage. The cells can be considered as an alternative of stressmeters.

8.6.1.3 Support Load Cells

The load cells placed between the support element and rock mass directly measures the load on support element. Different designs are available to measure the load continuously on support elements such as steel set, rock bolts, and shield support.

8.6.1.4 Multipoint Borehole Extensometers

Multipoint borehole extensometers are used to measure lengths between one or more anchor points and a reference head point. From these measured lengths, movements due to fractures and deformations of rock around drift can be monitored. Different lengths are available.

8.6.1.5 Convergence Extensometers (Tape Extensometers)

Convergence in the drift can be measured by means of tape extensometer. Tape extensometers measure changes in the distance between selected two fixed points on the sidewalls of the drift. Tape extensometer provides high accuracy (0.25 mm) in deformation measurement. Although the accuracy is high, this extensometer cannot give detailed information about the absolute displacement. Therefore, it is difficult to identify the pattern of deformation.

8.6.1.6 Total Station

In recent years with the advent of geodetic instruments, it is possible to record absolute 3D displacements of a large number of fixed points accurately. With the introduction of electronic total stations, it is possible to control opening deformation during construction with in an accuracy of a few millimeters. This technique is widely used since it is simple, inexpensive, and functional.

REFERENCES

Afroz, A. Method to reduce floor heave and sides closure along the arched gate road. *Mining Science and Technology*, 10 (1990): 253–263.

Arioglu, E., and N. Tokgoz. *Hard Rock Mass and Coal Strength*. Istanbul, Turkey: Evrim Pubkisher, 2011.

Bai, Q.S., S.H. Tu, F. Wang, X. Zhang, H. Tu, and Y. Yuan. Observation and numerical analysis of the scope of fractured zones around gateroads under longwall influence. *Rock Mechanics and Rock Engineering*, 47 (2013): 1939–1950.

Barton, N., R. Lien, and J. Lunde. Engineering classification of rock masses for the design of tunnel support. *Rock Mechanics* 6(4) (1974): 189–239.

Bieniawski, Z.T. *Engineering Rock Mass Classification*. New York: Wiley, 1989.

Bieniawski, Z.T. Geomechanics classification of rock masses and its application in tunneling. *Proceedings of the 3rd International Congress on Rock Mechanics*. Denver, CO: International Society of Rock Mechanics, 1974, 27–32.

Bieniawski, Z.T. Notes on in-situ testing of the strength of coal pillar. *Journal of the South African Institute of Mining and Metallurgy*, 68 (1968): 189–236.

Biron, E., and E. Arioglu. *Design of Supports in Mines*. Istanbul, Turkey: Birsen Publisher, 1980.

Borecki, M., and A. Kidbinski. Coal strength and bearing capacity of coal pillars, 3–21. *Proceedings of 2nd Rock Mechanics Congress*. Belgrade, Serbia, 1972, 143–152.

Carr, T.L. Requirements controlling roadway design and construction. *Strata Control Symposium*, Paper No: 1. Nottingham, 1970, 1–3.

Carr, F., and A.H. Wilson. A new approach to the design of multi-entry developments for retreat longwall mining. *Proceedings of the 2nd International Conference on Ground Control in Mining*. Morgantown, WV: Society for Mining, Metallurgy & Exploration (SME), 1982, 1–21.

Carranza-Torres, C., and C. Fairhurst. Application of the convergence-confinement method of tunnel design to rock masses that satisfy the Hoek-Brown failure criterion. *Tunneling and Underground Space Technology*, 15 (2000): 187–213.

Carranza-Torres, C., and C. Fairhurst. The elasto-plastic response of underground excavations in rock masses that satisfy the Hoek-Brown failure criterion. *International Journal of Rock Mechanics and Mining Sciences*, 36 (1999): 777–809.

CDC. *Mining Product: CMRR—Coal Mine Roof Rating*. August 30, Pittsburg, CA: NIOSH, 2013. http://www.cdc.gov/niosh/mining/works/coversheet1812.html.

Chen, J., M. Mishra, E. Zahl, J. Dunford, and R. Thompson. Longwall mining-induced abutment loads and their impacts on pillar and entry stability. *Proceedings of the 21st International Conference on Ground Control in Mining*. Morgantown, WV, 2002, 11–17.

Clark, C.A., and S.R. Newson. A review of monolithic pumped packing systems. *The Mining Engineering*, 144 (1985): 491–495.

Crouch, S.L., and A.M. Starfield. *Boundary Element Methods in Solid Mechanics*. London: Allen and Unwin, 1983.

Cundall, P.A. Computer model for simulating progressive large scale movements in blocky rock systems. In *Rock Fracture, Proceedings of the Symposium on ISRM*. Nancy, France: International Society of Rock Mechanics, 1971, 2–8.

Deere, D.U., A.J. Hendron, F.D. Patton, and E.J. Cording. Design of surface and near surface construction in rock. *Failure and Breakage of Rock, Proceedings of the 8th U.S. Symposium on Rock Mechanics*. C. Fairhurst (Ed.), New York: Society of Mining Engineers, American Institute of Mining Metallurgical, and Petroleum Engineers, 1967.

Deere, D.U., R.B. Peck, H. Parker, J.E. Monsees, and B. Schmidt. Design of tunnel support systems. *Highway Research Record*, 39 (1970): 26–33.

Deisman, N., M. Khajeh, and R.J. Chalaturnyk. Using geological strength index (GSI) to model uncertainty in rock mass properties of coal for CBM/ECBM reservoir geomechanics. *International Journal of Coal Geology*, 112 (2013): 76–86.

Duncan Fama, M.E. Numerical modeling of yield zones in weak rock. In *Comprehensive Rock Engineering*, J. A. Hudson, 49–75. Oxford: Pergamon, 1993.

Galvin, J.M., B.K. Hebblewhite, and M.D.G. Salamon. UNSW coal pillar strength determinations for Australian and South African mining conditions. *Proceedings 2nd International Workshop on Coal Pillar Mechanics and Design*. Pittsburg, CA: NIOSH IC9448, 1999.

Greenwald, H.P., H.C. Howarth, and I. Hartman. Experiments on the strength of small pillars in the Pittsburgh bed. R.I. P 3575, MD, US Bureau of Mines, 1941.

Grimstad, E., and N. Barton. Updating of the Q system for NMT. *International Symposium on Sprayed Concrete*. Fagarnes, Oslo, Norway: Norwegian Concrete Association, 1993, 46–66.

Grimstad, E., K. Kankes, R. Bhasin, A. Magnussen, and A. Kaynia. Rock mass quality Q used in designing reinforced ribs of sprayed concrete and energy absorption. *International Symposium on Sprayed Concrete*. Davos, Switzerland, 2002, 134–142.

Hazine, H.I., B.N. Whittaker, and R.N. Singh. Evaluation and interpretation of performance and stability data of different face roadway design and operation with the aid of computer. NCB Report, London: National Coal Board (NCB), 1978.

Hobbs, D.W. Scale model studies of strata movement around mine roadway, Part 5: Roadway shape and size. MRE Report, No: 2325, 1968.

Hoek, E. *Practical Rock Engineering*, 2007. http://www.rocscience.com/hoek/corner/Practical_Rock_Engineering.pdf.

Hoek, E., and E.T. Brown. *Underground Excavations in Rock*. London: Institution of Mining Metallurgy, 1980.

Hoek, E., C. Carranza Torres, and B. Corcum. Hoek-Brown criterion—2002 edition. *NARMS-TAC Conference*. Toronto, Canada, 2002, 267–273.

Hoek, E., T.G. Carter, and M.S. Diederichs. Quantification of the geological strength index chart. *US Rock Mechanics/Geomechanics Symposium*. San Francisco, CA, 2013.

Hoek, E., P.K. Kaiser, and W.F. Bawden. *Support of Underground Excavations in Hard Rock*. Rotterdam, the Netherlands: AA Balkema, 1995.

Hoek, E., and P. Marinos. Predicting tunnel squeezing. *Tunnels and Tunnelling International*, 2000: Part I: November 2000, pp. 45–51; Part II: December 2000, pp. 34–36. http://www.rocscience.com/hoek/references/H1998d.pdf.

Holland, C.T. Mine pillar design. *SME Mining Engineering Handbook*. Society of Mining Engineers, AIME, New York, 1973: 13–69 and 113–118.

Holland, C.T., and F.L. Gaddy. The strength of coal in mine pillars. Proceedings of 6th US Symposium on Rock Mechanics. University of Missouri, Rolla, MO: 1964, 450–466.

Hosseini, S.M., F. Sereshki, and M. Ataei. Application of fuzzy GSI system for estimating strength of jointed rock masses—A case study from Eastern Alborz coalfield, Iran. *Proceedings of the 8th International Scientific Conference—SGEM*, 2008, 357–364.

Hudson, J.A. *Comprehensive Rock Engineering*. Vol. 4. Exeter: BPCC Wheatons, 1993.

Istanbulluoglu, Y.S. Strata control aspects at the gate roadways of OAL underground mine. PhD, Ankara, Turkey: METU, 1995.

Istanbulluoglu, S., and C. Karpuz. Convergence behaviour-packing relations of full extraction gate roadways of OAL lignite mine. *CIM Bulletin*, 90 (1997): 56–61.

Jawed, M., R.K. Sinha, and S. Sengupta. Chronological development in coal pillar design for bord and pillar workings: A critical appraisal. *Journal of Geology and Mining Research*, 5 (2013): 1–11.

Karabin, G.J., and M.A. Evanto. Experience with the boundary element method of numerical modeling to resolve complex ground control problems. *Proceedings 2nd International Workshop on Coal Pillar Mechanics and Design*. Pittsburg, PA: NIOSH IC-9448, 1999, 89–114.

Kirsch, G. Die theorie der elastizitat und die bedurfnisse der festigkeitslehre. *Veit. Diet. Ing.,* 1898: 797–807.

Lawrence, W. A method for the design of longwall gateroad roof support. *International Journal of Rock Mechanics & Mining Sciences*, 46 (2009): 789–795.

Logie, C.V., and G.M. Matheson. A critical review of the current state of the art design of mine pillars. *Proceedings of the 1st International Conference on Stability in Underground Mining*, Vancouver, Canada, n.d., 359–382.

Lou, J., C. Haycocks, and M. Karmis. Gate road design in overlying multiple seam mines. In *SME* Preprint. Denver, CO: SME annual meeting, 1997, 97–107.

Lowson, A.R., and Z.T. Bieniawski. Critical assessment of RMR based tunnel design practices: A practical engineer's approach. *Rapid Excavation and Tunneling Conference*. Washington, DC, 2013. Design and Planning Session.

Madden, B.J. A re-assessment of coal – pillar design. *The Journal of the South African Institute of Mining and Metallurgy*, 91(1) (1991): 27–37.

Maleki, H. In-situ pillar strength and failure mechanism for U.S. coal seams. *Workshop on coal pillar mechanics and design*. Santa Fe, U.S. Bureau of Mines IC 9315: 1992.

Mark, C., and Z.T.B. Bieniawski. An empirical method for the design of chain pillars for longwall mining. *Proceedings of 27th US Symposium on Rock Mechanics*, AIME, New York, 1986: 415–422.

Mark, C., and G.M. Molinda. The coal mine roof rating in mining engineering practice. *Coal 2003: Coal Operator's Conference*. University of Wollongong and the Australian Institute of Mining and Metallurgy, 2003, 50–61.

Molinda, G., and C. Mark. Rating the strength of coal mine roof rocks. USBM IC 9387: 36 pp, 1996.

Mostley, J.T.B. An investigation of roadway stability in a roadway of retreat working face using aqualight as a cavity filter above the supports. *MRDE Technical Memorandum* (84): 15, 1984.

Obert, l., and W.I. Duvall. Rock mechanics and the design of structures in Rock. New York: Wiley, 1967, p. 650.

NB. *Sprayed Concrete for Rock Support*. Publication no. 7. Oslo, Norway: Norwegian Concrete Association, 2011.

NGI. Norwegian Geotechnical Institute. 2013. Using the Q-system –Rock mass classification and support design. www.ngi.no/en/Contentboxes-and-structures/Reference-Projects/Reference-projects/Q-method/ (accessed June 2013).

Orea, K., B. Orea., and A.H. Bagian. Design optimization of longwall chain pillars. *29th International Conference on Ground Control in Mining*. Morgantown, WV: 2010, 1–4.

Oreste, P.P. Analysis of structural interaction in tunnels using the convergence confinement approach. *Tunneling and Underground Space Technology*, 18 (2003): 347–363.

Palmstrom, A. Measurements of and correlations between block size and rock quality designation (RQD). *Tunnels and Underground Space Technology*, 20 (2005): 326–377.

Palmstrom, A. The volumetric joint count—A useful and simple measure of the degree of jointing. *Proceedings of the 4th International Congress IAEG*, Vol V. New Delhi, India, 1982, 221–228.

Peng, S.S., and H.S. Chiang. *Longwall Mining*. New York: Wiley Interscience, 1984.

Peng, S.S., K. Matsuki, and W.H. Su. J-D Structural analysis of longwall panels. *Proceedings of the 1st US Symposium on Rock Mechanics*. Columbia, MO: University of Missouri, 1980, 44–56.

Pine, R.J., J.S. Coggan, Z. Flynn, and D. Elmo. The development of a comprehensive numerical modelling approach for prefractured rock masses. *Rock Mechanics and Rock Engineering*, 39 (2006): 395–419.

Priest, S.D., and J.A. Hudson. Discontinuity spacings in rock. *International Journal of Rock Mechanics and Mining Sciences & Geomechanics Abstracts*, 13 (1976): 135–148.

Rodriguez, R., and M.B. Diaz-Aguado. Deduction and use of an analytical expression for the characteristic curve of a support based on yielding steel ribs. *Tunneling and Underground Space Technology* 33 (2013): 159–170.

Sakurai, S. Lessons learned from field measurements in Tunneling. *Tunneling and Underground Space Technology*, 1997: 453–460.

Salamon, M.D.G., and A.H. Munro. A study of the strength of coal pillars. *Journal of South African Institute of Mining and Metallurgy*, 1967: 55–67.

Seedman, R. The stress and failure paths followed by coal mine roofs during longwall extraction and implications to tailgate support. *Proceedings of the 20th International Conference on Ground Control in Mining*, Morgantown, WV, 2001.

Shen, B., A. King, and H. Guo. Displacement, stress and seismicity in roadway roofs during mining-induced failure. *International Journal of Rock Mechanics and Mining Sciences*, 45 (2008): 672–688.

Sheorey, P.R., C. Bandopadhyay., M.N. Das., T.N. Singh., A.K. Biswas., R.K. Prasad., D. Barat., and R. Ramna. Optimisation of design of mine pillar parameters and feasibility of extraction of locked up coal below builtup structures, water logged areas and hard cover. *CMRS Report on partially funded coal S&T grant of Department of Coal*, Ministry of Energy Government of India, 1987.

Sheorey, P.R. Pillar strength considering in situ stress. *Proceedings of the Workshop on Coal Pillar Mechanics and Design*. USBM IC 9315, A.T. Iannacchione et al. (Santa Fe) (Eds.), 1992, 122–127.

Shepherd, J., and T. Lewandowski. Instrumentation of roof support for colliery pillar extraction. *Geotechnical Instrumentation and Monitoring in Open Pit and Underground Mining*. T. Szwedzicki (Ed.), Rotterdam, The Netherlands: Balkema, 1993, 409–416.

Siddall, R.G., and W. Gale. Strata control—A new science for old problem. *Mining Engineer*, 1992: 151, 341–347, 349–351, 353–356.

Simpson, J.V., and P.J. Simpson. Composite failure mechanisms in coal measures' rock masses—Myths and reality. *The Journal of the South African Institute of Mining and Metallurgy*, 106 (2006): 459–469.

Stacey, T.R. and C.H. Page. Practical Handbook for Underground Rock Mechanics. Clausthal-Zellerfeld, Germany: Trans Tech Publications, 1986, 144.

Suchowerska, A.M., R.S. Merifield, and J.P. Carter. Vertical stress changes in multi-seam mining under supercritical longwall panels. *International Journal of Rock Mechanics and Mining Sciences*, 61 (2013): 306–320.

Terzaghi, K. Rock defects and loads on tunnel supports. *Rock Tunneling with Steel Support*. R. V. Proctor and T. White (Eds.), Youngstown, OH: Commercial Shearing and Stamping Company, 1946, 17–99.

Unal, E. Modified rock mass classification: M-RMR system. *Milestones in Rock Engineering: A Jubilee Collection.* Z.T. Bieniawski (Ed.), Rotterdam, The Netherlands: Balkema, 1996, 203–223.

Unal, E., I. Ozkan, and G. Cakmakci. Modeling the behavior of longwall coal mine gate roadways subjected to dynamic loading. *International Journal of Rock Mechanics and Mining Sciences*, 38 (2001): 181–197.

Unver, B. Closure in longwall access roadways. PhD Thesis, Nottingham: University of Nottingham, 1988.

Unver, B., and N.E. Yasitli. Modelling of strata movement with a special reference to caving mechanism in thick seam coal mining. *International Journal of Coal Geology*, 2006: 227–252.

Van Der Merwe, J.N. New pillar strength formula for South African coal. *The journal of South African Institute of Mining and Metallurgy*, June, 2003: 281–292.http://www.saimm.co.za/Journal/v103n05p281.pdf.

Wagner, H., and B.J. Madden. Fifteen years` experience with the design of coal pillars in shallow South African collieries: an evaluation of the performance of the design procedures and recent improvements. *Design and Performance of Underground Excavations, ISRM Symposium*, Cambridge, 391–399, British Geotech. Soc.: London, 1984.

Whittaker, B.N. An appraisal of strata control practice. *Mining Engineering* 134 (1974): 9–24.

Whittaker, B.N., and D.R. Hodgkinson. The influence of size on gate roadway. *Mining Engineering*, 139 (1971): 62–73.

Whittaker, B.N., and R.N. Singh. Design and stability of pillars in longwall mining. *Mining Engineering* 139 (1979): 59–70.

Wickham, G.E., H.R. Tiedemann, and E.H. Skinner. Support determination based on geological information. *Proceedings of the North American Rapid Excavation Tunneling Conference*. New York: Society of Mining Engineers, American Institute of Mining Metallurgical and Petroleum Engineers, 1972, 43–64.

Wilson, A.H. An hypothesis concerning pillar stability. *The Mining Engineer*, 131 (1972): 409–417.

Wilson, A.H. The stability of underground workings in the soft rocks of the coal. *International Journal of Mining Engineering*, 1 (1983): 91–187.

Wilson, A.H., and D.P. Ashwin. Research into the determination of pillar size. *Mining Engineering*, 20 (1972): 409–417.

Wuest, W., M.J. DeMarco, and C. Mark. Review of applications of the Coal Mine Roof Rating (CMRR) for ground control planning and operations. *Mining Engineering*, 48 (1996): 49–55.

Zahl, E., J. Dunford, M. Larson, T. Brady, and J. Chen. Stress measurements for safety decisions in longwall coal mines. *Proceedings of the 21st International Conference on Ground Control in Mining*. Morgantown, WV, 2002, 45–52.

Zern, E.N. *Coal Miners' Pocketbook: Principles, Rules, Formulas and Tables*. New York: McGraw-Hill, 1926.

9 Longwall Production with Subsidence

C. Okay Aksoy and Halil Köse

CONTENTS

Abstract: Subsidence is a ground settlement that occurs on the surface after the underground mine excavation. The amount of the subsidence is particularly high for caving method applied in the thick coal seams. Subsidence amount depends on the thickness of bed, the slope of bed, and geological formations. This phenomenon also changed the width of the affected area. In particular, the control of the subsidence in the mines close to the residential area is very important. This section will be described in general and the effects of subsidence.

9.1 INTRODUCTION

The removal of materials from earth's crust by underground mining creates an obvious potential for ground movement and consequential deformation of the surface. The circumstances under which this may arise vary widely, the main parameters being

1. *The geometry of the mineral deposit*: This can range from thin stratified flat seams to steeply dipping irregular veins or lenses, and massive ore bodies large in all three dimensions.
2. *The method of mining*: There may be partial or total extraction, with or without artificial support, and caving of the roof or hanging wall may be undesired or deliberately induced.
3. *The nature of the mineral deposit and the overlying strata*: There is a wide variety in the physical characteristics, hydrology, geology, depth of cover, and other factors pertinent to ground behavior.

Despite significant advances in the science of rock mechanics in the last two decades, analysis and prediction of stress and strain in large rock masses remains formidably complex because of such factors as anisotropy, lack of homogeneity, and the presence of geological discontinuities. These problems have so far prevented the development of a unified phenomenological theory capable of predicting satisfactorily ground movement and surface subsidence in the wide range of mining situations described above.

The environmental importance of subsidence is related to three main factors:

1. The surface land area affected
2. The nature of the land uses within the affected areas
3. The type and magnitude of ground movement

A large majority of the investigations into surface subsidence so far undertaken have concentrated upon areally extensive mining in countries with high population density. Underground extraction of seam deposits, such as coal, normally requires that large areas are undermined if significant tonnages are to be produced. Particularly in Western Europe, this type of mining has for many years co-existed with important surface land uses. There has thus been strong pressure to devise techniques to predict and minimize surface subsidence damage, especially since collapse of the roof and overlying strata behind the working faces is normally an integral part of the longwall mining systems most often adopted. Other types of mining have seldom presented the same urgency to understand and control the subsidence mechanism. Many mining methods commonly used in steeply dipping and irregular deposits require that the structural integrity of the hanging wall is preserved, as far as possible, by the use of natural or artificial support. For those methods, which rely upon caving the hanging wall, it is often assumed that major surface disruption is inevitable and that the affected surface zone must be cleared of installations and effectively left derelict. There have been very few attempts to predict or measure subsidence for these types of mining methods.

9.2 PREDICTING SURFACE DEFORMATION

Mining one or more seams most commonly produces a relatively continuous surface deformation, which can be measured in terms of vertical and horizontal displacements. Various calculation techniques are used to predict these displacements, founded on extensive investigation in European coalfields. These are principally based on the complete extraction of stratified deposits in situations where the depth and area of excavation are large in relation to the seam thickness. Most techniques currently in widespread use rely upon an empirical approach, although there are continuing research efforts to base prediction upon theoretical or phenomenological considerations.

9.2.1 Empirical Methods

It is observed that surface deformation over a single extraction area in a flat seam has the following characteristics:

1. A subsidence trough is formed by the vertical displacements of surface points and this normally extends beyond the limits of the mined area.
2. Horizontal displacements occur with magnitudes and directions approximately proportional to the slope of the subsidence profiles.
3. If the extraction is geometrically regular in shape, the distribution of vertical and horizontal displacements is approximately symmetrical about the center.

Two concepts are fundamental to most empirical calculation techniques. The first is that vertical displacement has a maximum possible value (S_{max}) for a particular excavation and this full subsidence is determined from

$$S_{max} = a\,m \tag{9.1}$$

where:
m is the seam thickness
a is the *subsidence factor*, which varies with local conditions

The second concept is that full subsidence (S_{max}) only develops if a sufficient area of the seam, relative to the depth of mining, is extracted. The *critical area* is defined as the extraction area that produces full subsidence at one surface point only. Areas that produce no full subsidence are termed *sub-critical*, and those that produce full subsidence at more than one point are *super-critical*.

Figure 9.1 shows the typical profile of a subsidence trough with the vertical and horizontal displacements of surface points. The angle of draw (or limit angle) is the angle of inclination, measured from a horizontal axis, from the edge of the mine workings to the point of zero subsidence. It is a function of seam dip and local geology, and is commonly in the range of 25°–35° in Britain. In practice, an arbitrary small value of vertical displacement is often taken as the limit of subsidence.

Figure 9.2 shows typical distributions of strain, horizontal, and vertical displacements for critical, super-critical, and sub-critical extraction areas. Maximum tensile strain normally occurs approximately vertically over the edges of the extracted panel and horizontal displacement is zero where full subsidence has developed.

An empirical relationship in common use is that the maximum subsidence over a sub-critical area depends upon the size of that area in relation to its depth. Provided the length of the extracted area is at least 1.4 times the width, it is found in practice that the two-dimensional case can be considered adequate. In this case, the relationship states that for sub-critical widths, extraction areas having the same width-to-depth ratio produce the same maximum subsidence, other parameters being equal. Figure 9.3 shows a typical graph used for prediction by the National Coal Board (NCB 1975) based on the

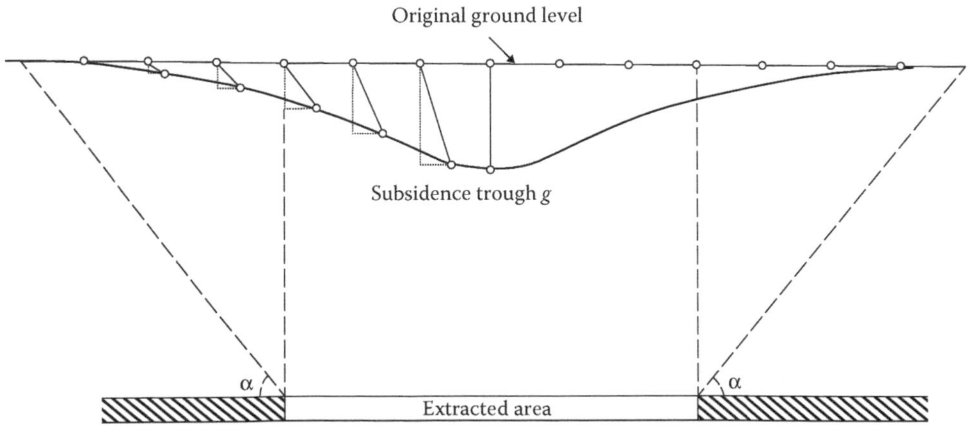

FIGURE 9.1 Profile of typical subsidence trough, α = limit angle.

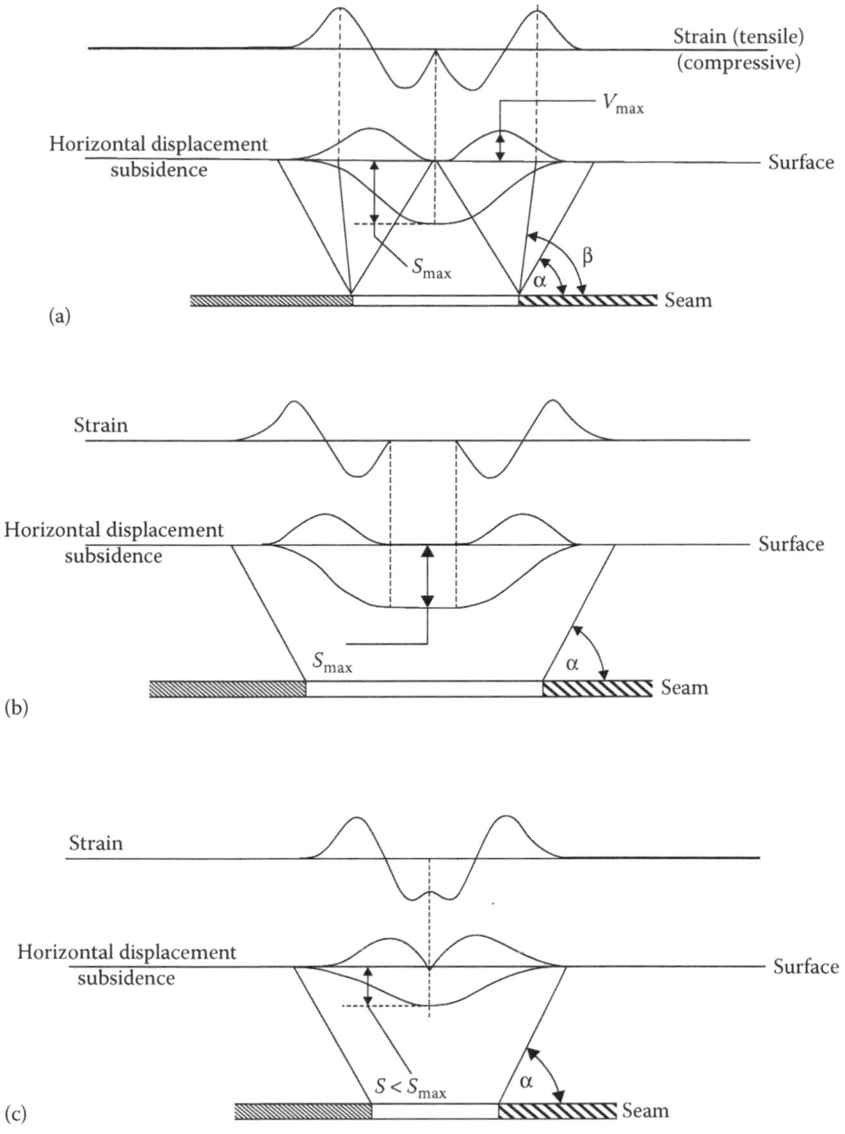

FIGURE 9.2 Critical (a), super-critical (b), and sub-critical (c) mining depth.

above principle. The graph assumes that there are no zones of special support and that the average panel width is used where panel sides are not parallel. An accuracy of ±10% is claimed for this method in the majority of cases in British coalfields.

The greatest possible subsidence S_{max} is found to occur at width-to-depth ratios exceeding 1–4, where the subsidence factor a is approximately 0.9 for full caving. Solid stowing reduces a to about 0.4–0.5.

Despite the fact that each subsidence process must obviously change the state of the affected rock mass, in practice, the principle of super-position is often found to be accurate within acceptable limits.

The subsidence factor, a, used in Equation 9.1, is determined by measuring the full subsidence over super-critical areas. Variations in a may be attributable to the type of packing (if any) used behind the working face, and additionally significant variations are found from one coalfield to another, indicating the influence of particular roof strata. Values of a determined in a number of major coal-mining countries and are summarized in Table 9.1 (Brauner 1973) and the value of break angle and limit angle for different ground conditions are given in Table 9.2 from the different countries, respectively (Aksoy et al. 2004).

Calculation methods, which generally follow the principles discussed above, predict the final subsidence of horizontal formations and may be extended to other factors such as horizontal displacements and dipping strata. Two general methods are in use and involve functions termed *profile* or *influence*.

Profile functions express mathematically the distribution of displacements over two-dimensional critical extraction areas. For super-critical widths, the central trough has a constant subsidence of S_{max}, and for sub-critical widths, the profile is determined from the critical case using empirical relationships. The method is normally restricted to rectangular extraction areas.

Influence functions apply the principle of super-position. The extraction area is considered as an infinite number of infinitesimal elements and likewise the subsidence trough is regarded as a composition of infinitesimal troughs produced by the extraction elements. The subsidence of any surface

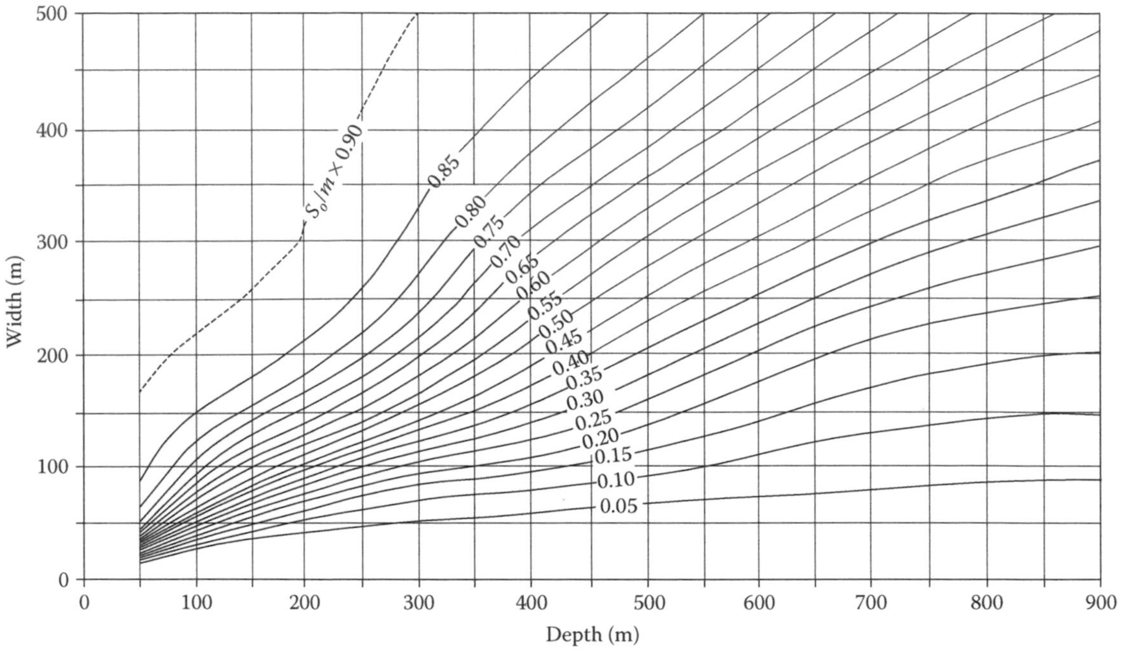

FIGURE 9.3 Relationship of subsidence to width and depth. (Data from NCB, *Subsidence Engineering Handbook*, London, 1975.)

point is the sum of the individual subsidence amounts due to each extraction element. This method imposes no restrictions on the geometric shape of the excavation area.

9.2.2 Physical Model Studies

Quantitative simulation of subsidence by the use of physical models has seldom been undertaken because of the difficulties of devising three-dimensional models, which accurately reflect the mechanical behavior of actual rock masses. Most experiments have therefore concentrated upon qualitative investigations. Problems such as the justification of the principle of super-position or the effects of previous workings can be studied with the use of models.

9.2.3 Theoretical Studies

Theoretical techniques consist essentially of replacing the rock mass with an idealized material, which deforms in

TABLE 9.1
Typical Value of Subsidence Factor, *a*

Location and Method of Packing	Subsidence Factor (*a*)
Britain: Roof caving or strip-packing	0.90
Solid stowing	0.45
Germany (Ruhr): Roof caving	0.90
Pneumatic stowing	0.45
Other solid stowing	0.50
France (Pas de Calais): Roof caving	0.85–0.90
Pneumatic stowing	0.45–0.55
Hydraulic stowing	0.25–0.35
Upper Silasia: Roof caving	0.70
Hydraulic stowing	0.12
U.S.S.R.: Roof caving	0.60–0.90
Pennsylvania	0.50–0.60

Source: Brauner, *Subsidence due to Underground Mining (in Two Parts)*, BuMines Circular, IC 8571 and IC 8572, U.S. Department of Interior, 1973.

TABLE 9.2
Values of Limit Angle Determined in Coal Mines for Various Ground Conditions

Draw Angle (Vertical Axis)	Ground Condition	Coal Site	Reference
45	Thick alluvial sediments	Limburg Coal Site, Holland	
41	Thick alluvial sediments	Trend River Valley, England	
35	Rock-covered shale and siltstone sequence	Midlands Coal Site, England	
23–39	Rock-covered shale and siltstone sequence		
25	Rock-covered sedimentary hard sandstone		
38–45	Claystone-limestone	Soma Eynez Coal District, Turkey	Aksoy et al. (2004)

Source: Onargan et al., *Determination of Break Angle From Subsidence Cracks in Soma-Eynez Coal Field of Turkey*, 7th National Mine Surveying Conference With International Participation, Proceeding Book, pp.189–201, 19th-25th June, Varna, Bulgaria, 2000.

accordance with the principles of continuum mechanics. Most of the traditional mathematical models of solids have been applied to the problems of subsidence. These include isotropic and anisotropy elasticities and viscoelastic and viscoplastic behaviors.

Up to the present time, this abstract model approach has had little practical significance. A great potential advantage of the method is that it can lead to a much deeper understanding of subsidence mechanisms than those achieved by empirical techniques. The current disadvantages are that application to actual problems requires simplifying assumptions, which seem to preclude realistic analytical solutions.

9.3 SUBSIDENCE DAMAGE

Knowledge of damage occasioned by subsidence is mainly based on stratified deposits, although other types of mining have been the subject of some study (Crane 1929).

In assessing subsidence damage, there may be confusion with *pseudo-mining* damage, which can be similar in effect but is not caused by mining. Foundation settlement due to the weight of the building or plaster cracks caused by bad construction techniques are common examples of *pseudo-mining* damage. Changes in the level of the groundwater table, which may be caused by mine pumping, can cause soil shrinkage and settlement of buildings. In localities where damage due to mining subsidence might be expected, it can be difficult to determine unequivocally the cause of such damage.

9.4 CONTROL OF SUBSIDENCE DAMAGE

The alleviation of subsidence damage may be undertaken either by precautionary measures on surface to protect installations or by modification of the mining method so as to minimize deformation of the surface (Voight and Pariseau 1970; Brauner 1973; NCB 1975).

9.4.1 Structural Precautions

The location of new installations is an important control measure. It is advisable to avoid areas of natural geological discontinuity, such as the outcrop of a fault, because of the higher probability of discontinuous deformation. Likewise, the hanging wall of a steeply dipping ore body, particularly if caved or unsupported, should be avoided.

9.4.2 Underground Precautions

Surface areas can be protected by leaving a safety pillar of mineral of sufficient dimensions or by controlled mining, which ensures that allowable deformations are not exceeded.

The size of safety pillars has often been determined by using a constant *angle of protection* often equated with the angle of draw (limit angle), as illustrated in Figure 9.4. This is now commonly regarded as unacceptable because most structures are only susceptible to differential movements and unnecessary sterilization of mineral may occur. This loss of mineral increases

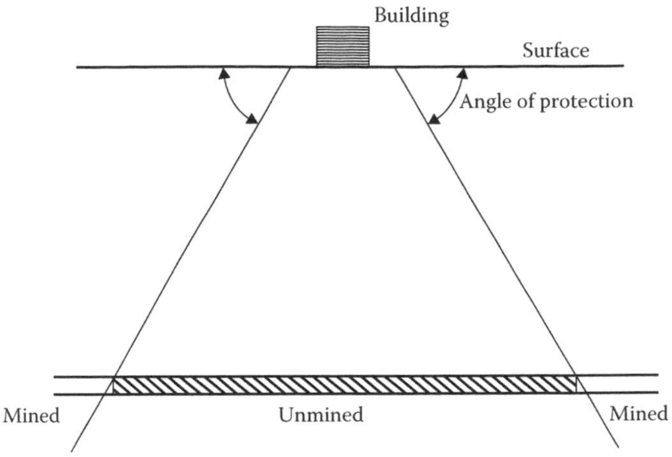

FIGURE 9.4 Safety pillar under constant angle.

with depth of working. Furthermore, adequate protection may not be ensured, particularly in multiseam mining, and other problems such as high stress concentrations underground can result. In workings where subsidence effects are predictable with reasonable accuracy, a better approach is to assess the allowable surface deformations and to undermine as far as possible without exceeding these deformations, subject to a reasonable safety factor. This method is particularly applicable to deep seams where a large loss of mineral may thus be avoided. In shallower workings, the loss of mineral is less severe and the potential surface deformations are more damaging and it may be economical to leave a relatively large safety pillar.

In seam mining, it may be possible to avoid adverse configurations, which produce maximum differential movements. These are illustrated in Figure 9.5.

An alternative to leaving a completely unmined safety pillar is to mine only part of the deposit, leaving the remainder in situ to reduce surface deformation. Mining systems of this type are usually termed *room and pillar* or *bord and pillar.* In European coalfields, there is a preference for relatively long narrow panels separated by permanent pillars, the width of both panels and pillars being of the order of 20%–30% of the depth of mining. With extraction ranging from 40%–70%, observed subsidences are 3%–20% of seam thickness (Brauner 1973). Filling or stowing of the excavated area reduces the potential for subsidence damage.

Control of the mining process can help to prevent the occurrence of subsidence damage. In longwall mining of coal under British trunk roads, for example, three-shift, seven days per week mining is employed during the critical period, thus avoiding the intermittent stresses arising from faces, which stand for one shift in three and at weekends.

9.5 SUBSIDENCE DAMAGE TO THE SHAFT

When working takes place in the safety zone, the outer cylinder of the shaft wall, enclosed as it is by the rock mass, is exposed to the highly varied movements of the strata through which it passes. If the inner safety zone is worked, the lower

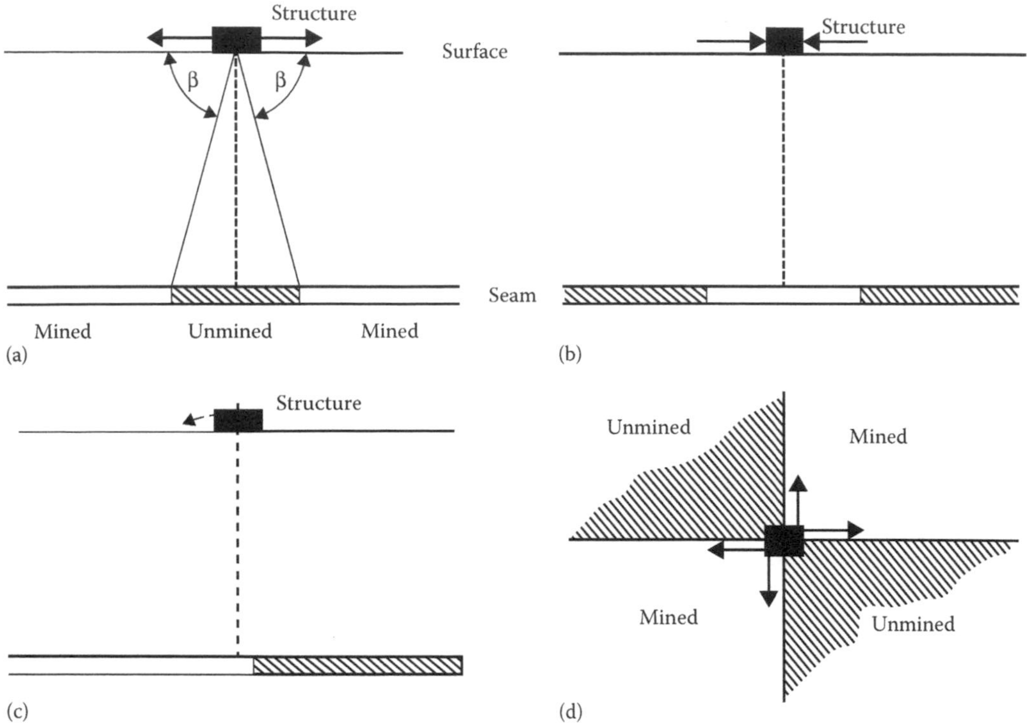

FIGURE 9.5 Unfavorable extraction layouts; β is the angle of break and arrows show direction of strain: (a) the surface structure is placed under maximum extension, (b) under maximum compression, (c) gives maximum slope, and (d) produce maximum distortion.

part of the shaft is forced somewhat upward into the mine excavation by the footwall beds, whereas the hanging wall or roof, to which the upper section of the shaft is bonded by mortar, concrete, or some general interlocking action, tends to drag it down with itself. In this process, considerable compaction damage to the wall and internal structures of the shaft often results at the working horizon. Higher up, as workings pass round the shaft, it is axially stretched and lengthened by the strata as they become perpendicularly relaxed. If mining takes place at the outer boundary of the shaft safety pillar, the part of the shaft nearest to the surface falls within the zone of vertical compression ahead of the face, a zone in which higher points in the rock mass subside more than lower ones. At the dividing line between the sandy surface layers, which slip down the shaft, and the solid rock beneath, which acts as a firm support for the shaft, a peak of axial compressive stress can build up in the shaft lining.

The horizontal components of strata movement have the effect of displacing the shaft outward, that is, away from the center of the workings both in the floor and in the main roof, and inward—toward the center—in the higher strata. The result is that the shaft structure, from the extraction level to the surface, leans toward the goaf and bends. This inclined position of the shaft starts to develop ahead of the approaching face and is at its maximum just after the passage of the face, gradually reducing toward the center line, where the shaft is once again vertical. In an inclined shaft, a somewhat higher rate of wear is noticeable in the hoisting installations, but a rupture of the shaft is unusual, given a regular pattern of strata displacement, since even a horizontal compressive stress of 2 mm/m implies that the walls of a 6 m diameter shaft will close in by only 12 mm. This degree of horizontal compression is largely taken up by the pressure arch in the rock mass encircling the shaft section and by the surrounding porous layer. If, however, the surrounding concrete has been tamped down and compacted, the rock pressure can produce a strain of 1–2 mm/m of diameter in the shaft section itself (Skuta 1979). Bending of the shaft axis is mostly slight, but possible bending stresses have to be taken into account in the construction of rigid linings. Horizontal shear movements in severely bent strata are always associated with fractures in shaft linings.

The method followed in Germany since the 1930s for calculating the amount of lengthening and shortening in mine shafts (Bals 1939) starts from the assumption that each section of the shaft, all down its length, sinks to exactly the same extent as the subsiding rock mass around it. Axial deformation of the shaft is therefore equated with vertical deformation of the rock mass. Table 9.3 shows the effects of mining on the shaft.

Although this simple method of calculation has served well for over three decades in the prediction of axial shaft stresses, the assumption that *shaft deformation equals strata deformation* is in fact valid only for the middle sections of the shaft. It does not hold good for the top and bottom sections, those nearest the surface and the workings. Measurements show that the shaft head frequently appears to *grow* out of the ground, which indicates that there is relative movement between the upper shaft structure and the surface zone of strata. When the inner shaft safety pillar is mined, a shortening of the shaft

TABLE 9.3
Effects of Mining on the Shaft

Effects and Damage	Cause	Manner of Transmission	Critical Factors	To Be Determined
Vertical compression and extension of shaft	Subsidence of rock mass of varying amplitude at depth	Frictional and adhesive shear stress	Vertical compression and extension of strata; shear force (wall area, effective length) deformation resistance (cross section, linear changes, and E modulus)	Transmission relationship of rock mass and shaft
Tilting and bending	Horizontal strata displacement, varying at depth	Close fit between shaft section and rock face	Sitting of shaft in relation to workings	Extend of strata displacement
Shearing fracture	Intense bending of strata	Shear force parallel to stratification	Degree of bending resistance of bending planes to shear	Shaft path
Horizontal pressure	Compaction of a weak bed	Transverse expansion	Vertical compressive pressure, degree of yield in strata, and compression	Degree of pressure on shaft

Source: With kind permission from Springer Science+Business Media: *Mining Subsidence Engineering*, 1983, 93-116, H. Kratzsch, trans. R. F. S. Fleming.

structure has been observed both above and below the mining horizon where, according to calculations, only a lengthening or, as in the footwall, no effects at all should occur. Ahead of the face the structure has been shortened even below (i.e., outside) the limit of influence of the workings, whereas by calculation, it should be unaffected. In one shaft, subsidence of 14 cm was measured as much as 300 m below extraction at the edge of the pillar (Skuta 1979). A shaft structure bound to the rock mass thus reacts to the influence of extraction differently from the rock mass itself.

The breaking strength of the shaft's close fit in the rock and its interlocking with it is of the order of τ_{Br} = 100 N/cm^2 = 1 MN/m^2. A break between the shaft wall and the rock face will accordingly occur almost as soon as the effects of extraction begin to be felt, that is, when axial deformation of the shaft lining amounts to 0.05 mm/m in the case of concrete bonding, or to 0.1 mm/m with mortar bonding. Once a rupture plane has been formed in this way, the local force transmissible to the shaft wall drops to an adhesive shear or frictional stress of around 10 N/cm^2 in solid rock, and 3 N/cm^2 in loose ground.

For the frictional or tangential shear force F_t acting on the shaft as the rock strata slide down it, the following equation can be formulated (Kratzsch 1983):

$$F_t = \tau A l \tag{9.2}$$

The transmitted shear force is met, by way of force of reaction, by the elastic deformation resistance D of the shaft lining:

$$D = \varepsilon E S 10^{-3} \tag{9.3}$$

The frictional force, which builds up and is transmitted to the shaft, thus increases linearly with shaft length. Correspondingly, the graph of shaft deformation will rise in a straight line until it reaches the magnitude of the strata deformation. Finally, after restatement, the required slope tan α of this line of shaft deformation in mm/m per running meter of shaft length:

$$\frac{\varepsilon}{l} = \tan\alpha = \frac{\tau A 1000}{ES} \tag{9.4}$$

In this way, a simple connection is found between the shear force and the degree of vertical extension or compression, E, in relation to shaft length. At the same time, however, though this is not completely true, the coefficient of friction and the strata pressure normal to the shaft section remains constant, while elastic, though not plastic, deformations after fracture of the lining cylinder are taken into account.

9.6 EXTENT OF VERTICAL EXPANSION AND CONTRACTION IN STRATA AND SHAFT

Vertical strata deformation can be ascertained from the subsidence of points in the rock mass calculated with an integration grid, by dividing the difference in subsidence (in mm) by the vertical distance between points (in m). Special integration grids for the direct calculation of expansion and contraction have also been designed, but they offer no great saving in labor or time as compared with deriving strata deformation from subsidence values (Kochmanski and Magdziorz 1972; Sauer 1975).

What appears very promising for shaft investigation is the finite elements method, if the shaft-lining cylinder is regarded as a vertical column of infinitesimal elements within the total mesh, to which is assigned a substitute elastic modulus for tubing, brickwork, and concrete, related to the shaft section as a whole (Köse 1978). The adhesive shear bond between the exterior wall and the rock face can be represented in model terms by a weak boundary layer. A gradual diminution of the elastic modulus in those elements of the boundary layer, which have reached a critical shear stress, enables the transition to kinetic friction to be modeled.

Calculate chiefly the situation at depth and maximum extend of vertical strata expansion or contradiction at the shaft and equate this value to axial shaft deformation (Knothe 1969).

During working, dynamic contraction reaches only 30% of its final value, whereas dynamic expansion is 60% greater than final expansion (Pielok and Sroka 1979).

9.7 ANALYSIS OF SHAFT STRESSES DUE TO MINING

At the beginning of the mine planning exercise, therefore, the section of the shaft to be protected and the permissible degree of deformation must be determined. The anticipated effect on the shaft will depend on the amplitude of subsidence in the workings and initially also by the added influence of deformation forces acting on the section of the shaft concerned. The type of deformation expected will depend on the horizontal location of the shaft in relation to the workings.

Interim stress on the shaft can differ greatly from this final stage of strata deformation, depending on the conduct of the mining operation and the treatment of the shaft lining at the mining horizon (whether it is slit and provided with a timber crib). A working, which proceeds at critical extraction width 2*R*, from the edge of the shaft pillar toward the shaft, and from one side only, has the effect at first of axially compressing the shaft head (Figure 9.6a). As the face advances, the zone of axial compression in the shaft lining extends to the mining horizon (curve 2). After the face has passed round the shaft, the middle and upper reaches of the shaft become extended (curve 3). The compression maximum, which develops at the mining horizon, up to the area of extension-free shaft at the center of a critical area (curve 4), damages the shaft lining if it has not, as is more usual, been opened up to seam height and had a yielding timber crib inserted (Figure 9.6b). On this basis, a face carried inward from one side stresses the shaft structure, first by compression and then by extension. Furthermore, the shaft becomes tilted in the meantime. With a face of less than 2*R* (critical width), a slight axial extension remains in the upper shaft lining after the shaft safety zone has been worked through, instead of curve 4.

The alternating stress on the shaft and its deviations from plumb can be avoided by working the shaft safety zone symmetrically and from the outer edge inward. Compression of the upper section of the shaft, down to just below the limit of mining influence, is greater than shown at Figure 9.6a (curve 1) because of the doubled mining influence. At position 2 of the

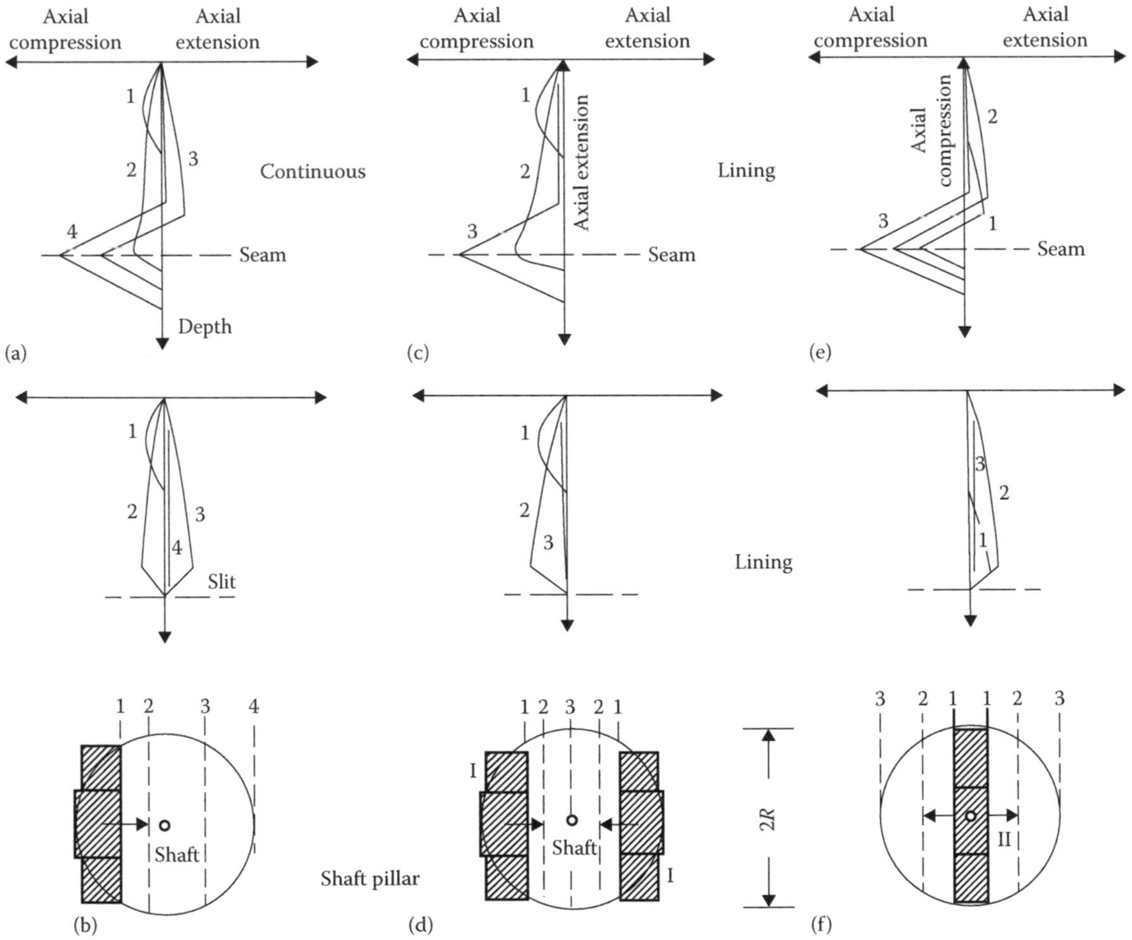

FIGURE 9.6 Stresses affecting a continuous shaft (top) and shaft which is *slit* at working horizon (center), with extraction proceeding from one side of shaft pillar, from both sides symmetrically, and from center of pillar symmetrically outward (bottom).

face, the lower section of the shaft is also axially compressed. After that the upper shaft structure becomes unstressed, but high compression remains above and below the workings (curve 3), unless the shaft lining has been slit (Figure 9.6d). Consequently, with symmetrical working from the edge of the pillar inward, at no time does the damaging stretching effect arise in the upper shaft structure, which consists of segments of tubbing to seal off the water-soaked sand beds of the overburden layer. With this mining procedure, stretching can be avoided throughout the length of a concrete shaft (which is vulnerable to tensile stress), provided it is slit at the mining horizon.

If the shaft pillar is mined symmetrically from the shaft outward—and there are many good mining reasons for doing this, not least the early uncovering of the shaft walling at the mining horizon, which is at risk—then either the upper section of a continuous lining (Figure 9.6e) or the whole shaft structure above a slit (Figure 9.6f) remains free of axial compression. Working on both sides has the effect of acting first on the lower part of the shaft (curve 1), which is axially compressed and extended (Figure 9.6e) or only extended (Figure 9.6f). In position 2 of the face, the extending action is affecting the shaft lining right up to the pit-bank level. With a critical extraction area, a slit shaft shows no linear change on completion of mining. A continuous lining, on the other hand, shows (as in Figure 9.6a and c) a dangerously high degree of axial compression at the working level. Since rupture cracks in the lining cannot be ruled out, this mining plan can only be followed in dry strata.

Shear fracture occurs at the shaft if the beds, which it intersects, are so severely bent by one-sided working (position 2 of the face in Figure 9.6b) that the bending shear stress rising parabolically at the center of group of beds overcomes the adhesive shear stress, or limiting friction, at a bedding plane.

Two points, standing on opposite side of the slip plane, are then displaced by a shear travel. No shaft lining, however strong, can withstand the shear movement of two beds. In inclined to steep-lying measures, however, shear movements can occur without causing bending (Sljacheckij 1974).

In the initial phases of mining, it is not possible to achieve a complete balancing-out of deformation at the shaft by causing axial extension and compression to coincide, both in time and in place. The compression resulting from working at the pillar edge develops at the shaft head (Figure 9.6d). Axial extension over a central working, on the other hand, starts in the bottom section of the shaft (Figure 9.6f). Simultaneous opening-up of symmetrical edge workings (I) and central working (II) therefore leads to compressing the shaft lining at the top and extending it at the bottom. It is only from position 2 onward that the extending and compressing effects on the shaft structure cancel themselves out. All the overlying strata then subside equally at the shaft side and take the lining cylinder down, without linear change, by the amount of the roof settlement.

For a face advance, which spares the shaft, it is usually sufficient to mine only the inner shaft protective zone from the point of view of balancing out expansions and contractions, in time and place, as much as possible. The protective zone in a seam, which at 700 m depth measures around 1000 m in diameter, is divided into three or four sections with a face length down the dip of 250 or 200 m, as the case may be. The outer segments of the circle, which will have only a slight compressional effect on the shaft head when mined later, are ignored. Where the protective area is very large, a limited approach such as this is all the more necessary to keep the operation within a manageable size and, by concentrating on the sectors having the strongest influence, to minimize the time during which the shaft is endangered (Kratzsch 1983).

The mining plans represented in Figure 9.7 envisage, in ground plan, *a* perimeter workings I starting first, followed slightly later by central working II in both directions; in ground plan *b*, central working I in both directions from eccentrically offset faces, followed by perimeter working II; and in ground plan *c*, extracting the shaft pillar *harmonically* on a checkerboard pattern (Lehman 1938; Bals 1939), which leaves the shaft practically stress-free. In a water-bearing surface rock, the initial compression of the shaft head expected under mining plans *a* and *b* is a desirable handicap to offset its elongation that will come later. The expansion in the lower roof strata caused by central working represents no threat to the operational security of the shaft in dry rock and does less damage to the loading stations than compression.

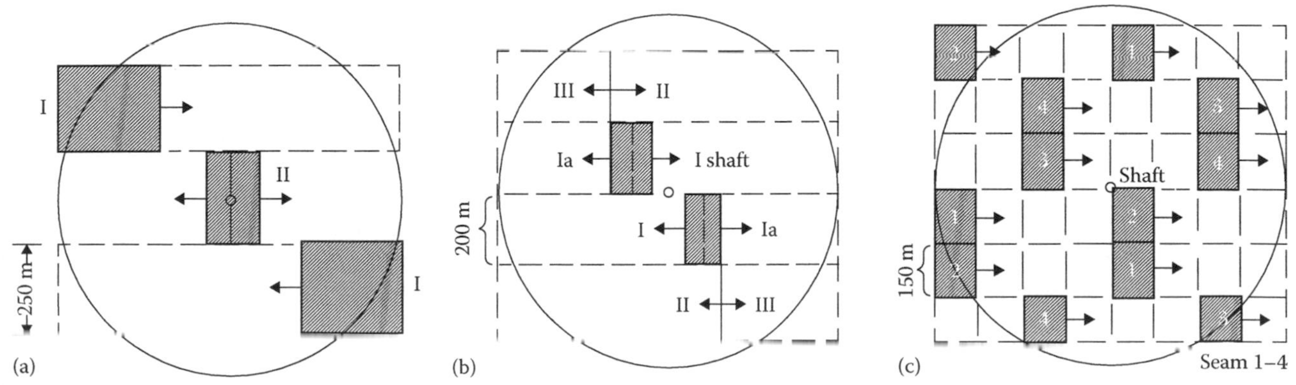

FIGURE 9.7 Examples of combined peripheral and central working of shaft's protective zone in one seam or more seams.

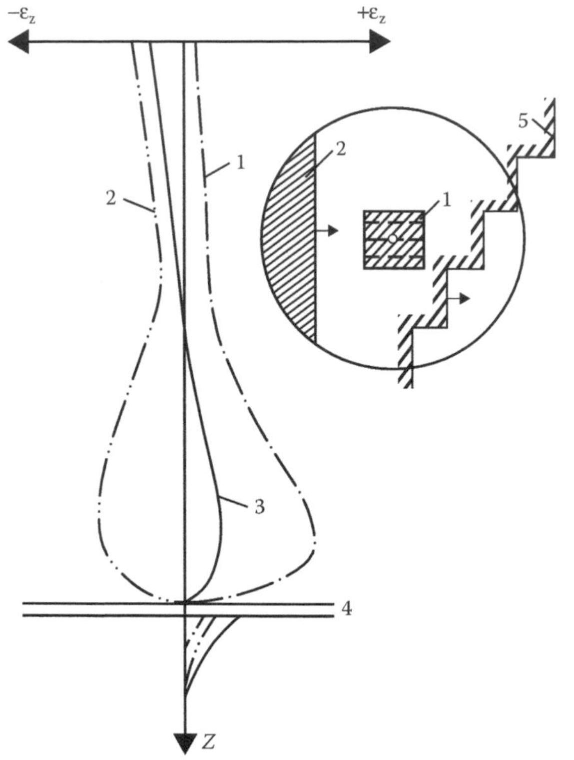

FIGURE 9.8 Extraction of shaft's protective zone in Polish coal-mining practice: 1—first of all, core zone is extracted and stowed; 2—extraction from pillar edge; 3—combined effects of 1 and 2; 4—seam; 5—staggered working face (stepped-face method). (After Kowalczyk, Z. et al., Abbau von Schactsicherheitspfeilern mit Anwendung Hydraulischen Grubenausbaus, Freiberg-Forsch-H. A., 578, pp. 89–101, 1976.)

As regards the rate of face advance, rapid peripheral working will reduce dynamic contraction, and rapid central working will reduce dynamic expansion. Rapid caving causes rapid subsidence and, consequently, since the stiffness of the bitumen depends on the rate of movement, an increase in the axial force transmitted to the lining cylinder by the slip layer.

A mining plan such as this, entailing symmetrically combined peripheral and central working to cancel out stresses, and capable of being enlarged to take in several seams, often runs into serious difficulties in practice.

A further mining technique that can be mentioned for protecting the shaft lining is stowing the goaf. With stowing—which should be considered for at least the inner shaft-protection zone—subsidence of the roof beds is only half as great as with caving. Vertical contraction and expansion of the rock mass, and axial deformation of the shaft cylinder, both also decrease to a similar extent. In other words, stowing protects the shaft lining very effectively from linear changes, and this applies equally to *low-subsidence* room-and-pillar working.

Over the goaf, the lower strata become significantly more relaxed than the upper ones. When subsidence is rapid and bending is severe, the bonding between roof beds can become loosened up to a height of 200 m, causing the lower shaft structure to be pulled apart in the early stages of central working. It has been suggested that this expansion in the main roof be kept within bounds by a *core zone*, formed either by leaving the shaft pillar unworked within a radius of 50 to 100 m of the shaft, or by stowing the same area with a very hard, unyielding material (Bals 1956).

When the core zone itself, 1 in Figure 9.8, is extracted, axial tensional stresses (curve 1) are set up in the shaft lining, which reduce the compressive effect (curve 2) of extraction starting at the edge of the protective zone (curve 3). In Polish coal mining, the core zone is stowed with flushed-in sand or with concrete, and this lessens strata subsidence down the length of the shaft cylinder. A staggered face is said to minimize the dynamic effects of extraction (Kowalczyk et al. 1976).

9.8 RECENT STUDIES

The subsidence quantity depends, among other things, on coal seam thickness and the depth of the coal seam. One of the mining methods used for thick coal seams is called *longwall top coal caving*. In this method, the coal seam is divided into slices. In each slice, some amount of coal is extracted from the face and coal above the powered supports is caved and collected onto a conveyor at the rear of the powered supports. Since coal seams mined by *longwall top coal caving* are usually very thick, substantial subsidence occurs at the surface. The magnitude of the subsidence effect depends on the coal seam thickness and the limit angles of the strata (Aksoy et al. 2004). Analytical calculations can be carried out to determine the area affected by the subsidence. The calculations for the limit angle, especially for deep mines, are notably complex (Whittaker and Pasamehmetoglu 1981; Yao et al. 1991; Singh and Singh 1998; Sheorey et al. 2000; Aksoy 2005). Recent studies mention that the real limit angles differ from thc values calculated with empirical methods (Marschalko et al. 2011, 2012a, b). These studies showed that the real limit angles develop over a long period of time. Even after underground coal production is complete the consolidation of the caved material takes place over an extended period and the subsidence area of effect increases. It is important to mention that the long-term subsidence effects are considerably less than the subsidence effects, which occur immediately after mining. In other words, shortly after the production has ended, large subsidence occurs and later, the long lasting consolidation of the caved material increases the subsidence area but the deformations are considerably lower than earlier. The presence of water makes the situation more serious and it can take years for the subsidence to end (Wang et al. 2009). Especially for countries like China where the thick coal seams are near residential areas, occurrence of long lasting subsidence may cause serious damage (Hur et al. 2004).

The analytical approaches used today make it hard to determine the long-term consolidation of the caved material (Whittaker and Pasamehmetoglu 1981). Analyses carried out using numerical modeling techniques give close results to the real values (Whittaker and Pasamehmetoglu 1981; Hur et al. 2004). Numerical modeling, which is commonly used in construction and mining industry, makes it possible to analyze long-term caving consolidation (Aksoy 2002). As mentioned above, the real area of effect of subsidence forms as a result of long-term caving consolidation. It is possible to determine the real subsidence area by using time dependent numerical consolidation analysis.

In recent studies, the area of effect of the subsidence after a long duration due to the coal production using the *longwall top coal caving* method in Manisa-Soma-Eynez region was determined using time-dependent numerical modeling. The subsidence area and amount are determined in the field. The results obtained from the field are consistent with the results obtained from the time-dependent numerical modeling.

In Soma region, especially in Eynez, there are many underground mines. These mines have been in continuous production for a long period resulting in wide subsidence areas. Even in some locations, there are overlapping subsided areas from different mining companies. After the long-term consolidation of the caved material, the subsidence ends. Determining the subsidence areas in advance is a complex problem including complicated operations. Numerical modeling can be used to predetermine the subsidence areas as it is used to predict various problems in construction and mining industry. The results obtained using the right methods in numerical modeling are consistent with the real results.

Subsidence starts to affect the surface after the initiation of underground caving process. The areal extent of the surface subsidence changes depending on the effect of geological formations above the production area. The areal extent just after the production is instant effect area. Consolidation of the cave-in material can continue for two or more years (Marschalko et al. 2011, 2012b).

9.9 CONCLUSION

Most of the available knowledge relates to trough subsidence over stratified deposits, particularly coal, worked by longwall methods. The prediction and control techniques in common use are based largely upon empirical investigation and measurement. The science of rock mechanics has enabled some progress on a phenomenological approach with an understanding of the behavior of the rock mass overlying an excavation. However, this method still yields mainly qualitative rather than quantitative assessment of subsidence phenomena. Continuing research, both empirical and theoretical, will undoubtedly aid the mining engineer in the dual objectives of maximizing the percentage extraction of mineral whilst minimizing surface disruption.

REFERENCES

Aksoy, C.O., (2005), Three-dimensional finite element analysis of an underground shaft at the hustas mine. Turkey. *CIM Bulletin*, September/October Issue, 1–5.

Aksoy, C.O., Kose, H., Onargan, T., Koca, Y., Heasley, K., (2004), Estimation of limit angle by laminated displacement discontinuity analyses in soma coal field, Western Turkey, *International Journal of Rock Mechanics and Mining Science*, 41, 547–556.

Aksoy, C.O., (2002), *Numerical Modelling for the Recovery of Protecting Pillars in Soma District*. PhD Thesis, Dokuz Eylul University, Graduate School of Natural and Applied Science, Izmir-Turkey, p. 107.

Bals, R., (1939), Abbau von Schachtsicherheitspfeilern, *Glückauf*, 75, 253–259.

Bals, R., (1956), Einwirkungen auf Schachte, Sammelwerk, *Glückauf*, 2, 709–757.

Brauner, G., (1973), *Subsidence due to Underground Mining (in Two Parts)*, BuMines Circular, IC 8571 and IC 8572, U.S. Department of Interior.

Crane, W.R., (1929), *Subsidence and Ground Movement in the Copper and Iron Mines Upper Peninsula*, BuMines, Bulletin 295, U.S. Department of Interior.

Hur, L., Yue, Z.Q., Wang, L.C., and Wang, S.J., (2004), Review on current status and challenging issues of land subsidence in China. *Engineering Geology* 76, 65–77.

Knothe, S., (1969), Bestimmung der Voraussichtlichen Abbaueinflüsse in Schachtsicherheitspfeilern, Przegl. Gorn. pp. 377–381.

Kochmanski, T. and Magdziorz, J., (1972), Neue Methcden der Berechnung von Bewegungen und Vermessungen der Tagesoberflache und des Gebirges als Falge untertagigen Abbaus, Freib, p. 527.

Köse, H., (1978), Programmierte Berechnung ger Abbaueinwirkung auf Schachte nabh dem Modell der Finiten Elemente, *Glückauf*, 39, 141–145.

Kowalczyk, Z., Jaunusz, and Pielok, (1976), Abbau von Schactsicherheitspfeilern mit Anwendung Hydraulischen Grubenausbaus, Freiberg-Forsch-H. A., 578, pp. 89–101.

Kratzsch, H., (1983), *Mining Subsidence Engineering* (Fleming, R. F. S., Trans.), Springer-Verlag, Berlin, Germany, pp. 93–116.

Lehman, K., (1938), PlanmaBige Abbauführung, *Glückauf*, 74, 321–332.

Marschalko, M., Yilmaz, I., Bednarik, M., and Kubecka, K, (2011), Variations in the building site categories in the underground mining region of Doubrava (Czech Republic) for land use planning. *Engineering Geology*, 122, 169–178.

Marschalko, M., Yilmaz, I., Bednarik, M., and Kubecka, K., (2012a), Influence of underground mining activities on the slope deformation genesis: Doubrava Vrchovec. Doubrava Ujala and Staric case studies from Czech Republic. *Engineering Geology*, 147–148, 37–51.

Marschalko, M., Yilmaz, I., Kristkova, V., Fuka, M., Bednarik, M., and Kubecka, K., (2012b), Determination of actual limit angles to the surface and their comparison with the empirical values in the Upper Silesian Basin (Czech Republic). *Engineering Geology*, 124, 130–138.

National Coal Board, (1975), *Subsidence Engineering Handbook*, London.

Onargan, T., Koca, M.Y., and KÖse, H., (2000), *Determination of Break Angle From Subsidence Cracks in Soma-Eynez Coal Field of Turkey*, 7th National Mine Surveying Conference With International Participation, Proceeding Book, pp.189–201, 19th-25th June, Varna, Bulgaria.

Pielok, J. and Sroka, A., (1979), Verringerung der Deformationen im Schachtausbau Durch Entsprechende Zeitlich-Raumliche Gestaltung des Abbaus, International Symposium on Mine Surveying (ISM), *Aachen*, 4, 97–114.

Sauer, A., (1975), Die Einflüsse von Durchbauungsgrad, Abbaukonzentration und Abbaugeschwindiykeit auf die Vorausberechnung von Bodenbewegungen, *Glückauf*, 36, 16–26.

Sheorey. P.R., Loui. J.P., Singh. K.B., and Singh, S.K., (2000)Ground subsidence observations and a modified influence function method for complete subsidence prediction, *International Journal of Rock Mechanics and Mining Sciences*, 37(5), 801–818.

Singh, K.B. and Singh, T.N, (1998), Ground movements over long-wall workings in the Kamptee coalfield. India. *Engineering Geology*, 50 (1–2), 125–139.

Skuta, E., (1979), Die Wirkung von Abbau und Grundwasserentzug auf dieTübbingschaacte des Steinkohlenbergwerkes, Emin Mayrisch, Markschedewesen, 86, pp. 130–140.

Sljacheckij, V. K., (1974), Verringerung der Abbauverluste an Kohle durc Schubsicherheitspfeiler, Bezopasnot Truda Promyslenosti, pp. 21–23.

Voight, B. and Pariseau, W., (1970), State of the predictive art in subsidence engineering, *Journal of the Soil Mechanics and Foundation Division, Proceedings of the American Society of Civil Engineers*, 96, 721–750.

Wang, G.Y., You, G., Shi, B., Yu, J., and Tuck, M., (2009), Long-term land subsidence and strata compression in Changzhou, China. Engineering Geology 104, 109–118.

Whittaker, B.N. and Pasamehmetoglu, A.G., (1981), Ground tilt in relation to subsidence in Longwall mining. *International Journal of Rock Mechanics and Mining Sciences and Geomechanics*, 18 (4), 321–329.

Yao, X.L., Whittaker, B.N., Reddish, D.J., (1991), Influence of overburden mass behavioural properties on subsidence limit characteristics. *Mining Science and Technology*, 13 (2), 167–173.

10 Coalbed Methane

Gouthami Senthamaraikkannan and Vinay Prasad

CONTENTS

Abstract: Coalbed methane (CBM) is an unconventional source of natural gas produced from coal seams. In recent years, it has gained a lot of attention and is being developed as a significant energy source in various parts of the world. However, high capital cost requirements and uncertainties arising from heterogeneity of coal beds are major hindrances in CBM production. In this chapter, we explore the origin, composition, and properties of CBM, its extraction methods, the availability and production of CBM worldwide, the economics involved in it, and the challenges faced by this industry.

10.1 INTRODUCTION

Most geological strata have gases associated with them. CBM refers to the large density of gases (varying from 0.0003 to 18.66 m^3/metric ton) stored in coal beds that have a multiporous structure. These gases are a mixture of methane (80%–99% of volume) and minor amounts of carbon dioxide, nitrogen, hydrogen sulfide, and sulfur dioxide. In the early nineteenth century, CBM was recognized as a major mine hazard because the presence of high-pressure gases in coal seams led to structural stress on mine beds and caused explosions and outbursts. To prevent this, these gases used to be drained or ventilated. However, the drainage of these gases into the atmosphere contributed to the greenhouse effect, since the global warming potential of methane is 25 times higher than that of CO_2, measured over a 100-year period (IFP Energies Report 2008). In the late twentieth century, CBM began to be produced from conventionally drilled wells for use as a fuel, and soon it became recognized as an important unconventional source of natural gas (Flores 1998).

Since CBM consists mostly of methane, it is similar to other sources of natural gas. Natural gas is a cleaner fuel compared to oil and coal, which are 30% and 50% more carbon-intensive, respectively. It produces no ash and emits negligible amounts of toxins such as mercury and oxides of nitrogen and sulfur during combustion; however, it needs compression or liquefaction to make it suitable for storage and transportation. The improved prospects of liquefied natural gas (LNG) projects have played a role in increasing its presence as a fuel in the international market (Stevens 2010).

Until 1990, only the Soviet Union consumed gas for its primary energy requirements. Today, the United States, Iran, and China are the world's largest consumers of gas along with Russia, while the United States, Canada, and Iran are the world's largest producers. The demand for gas has risen in the recent past, and this surge is expected to continue in the near future. Today, natural gas is employed as a fuel in the residential, commercial, transportation, and power sectors. The power sector is currently the largest user, with around 40% of the global gas demand (IEA 2014). The U.S. Energy Information Administration predicts that natural gas will surpass coal as the largest source of energy generation in the United States by 2035. In 2009, the global consumption of natural gas was estimated at 2940 BCM (a 21% share of the global primary energy mix) and this has been growing at a compound annual growth rate of ~3.5% (Ernst & Young 2010).

To meet this demand, gas is produced from conventional and unconventional reserves. Conventional gas production occurs in relatively highly porous and permeable geological formations such as sandstone, siltstone, and carbonates. Unconventional gas reserves are difficult to produce from, owing to their low permeability values (NRCAN 2014) (e.g., CBM, shale gas, tight gas, and gas hydrates). These unconventional resources were first exploited in the mid-1970s in the United States as gas demand increased and conventional gas reserves declined. This was assisted by legislative action that included the removal of wellhead price controls and restrictions on use of gas for electricity generation, and provision of incentives for development of new sources of natural gas, which led ultimately to the establishment of a large number of CBM wells by the 1990s. By the turn of the century, CBM production accounted for about 7% of the total dry gas production and 9% of the proven dry gas reserves in the United States.

The major factors that sustain interest in CBM are (1) increased safety and productivity in coal mining operations, (2) possible greenhouse gas mitigation, (3) the possibility of greenhouse gas capture and storage, and (4) earning emission credits for investing in emission reduction projects according to the Kyoto Protocol. Greenhouse gas capture and storage refers to carbon dioxide sequestration, which involves injection of carbon dioxide into deep coal seams. The sequestration of carbon dioxide in coal seams is doubly attractive as it tackles the emission of this important anthropogenic greenhouse gas along with simultaneous improvement in the recovery of CBM.

Overall, CBM today is a rapidly growing significant energy source. The International Energy Agency (IEA) reports that in 2008, CBM was the source of 10% natural gas production in the United States, 4% in Canada, and 8% in Australia, and there has been increased interest in CBM production in other countries with large coal reserves, especially India, China, Russia, and Indonesia, which have large coal reserves (Stevens 2010). China has made CBM as one of its 16 priority projects in its 11th five-year plan. The global CBM market volume by application is shown in Figure 10.1, with the power sector being the largest consumer. In 2013, 35.3% of the total CBM produced globally was used for power generation, with an estimate that the usage would grow at a compound annual growth rate

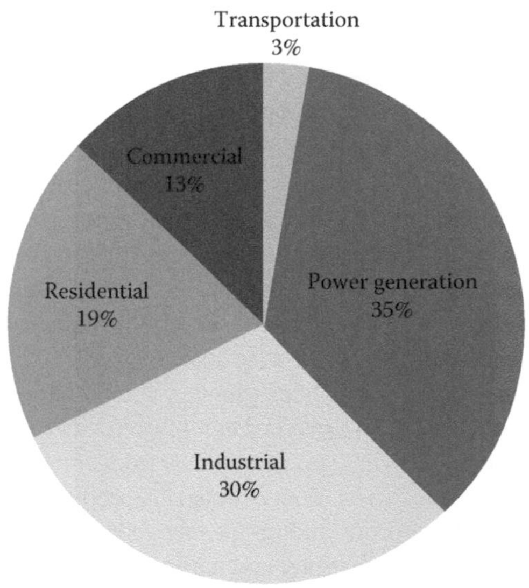

FIGURE 10.1 Coalbed methane utilization based on end-use application (worldwide).

of 8.5% from 2014 to 2020 (Grand View Research 2014b). However, large initial investment costs, long dewatering periods, and potential detrimental effects to the environment are major considerations in the planning and implementation of any CBM project.

10.2 ORIGIN, COMPOSITION, AND PROPERTIES OF CBM

Coal was formed initially through the biodegradation/oxidization of plant matter. Cellulose was rapidly degraded by aerobic bacteria and the resulting resistant biopolymers (lignin, cutan, and alganean) were transformed through the cleavage of bonds and reactions such as isomerization, modification of propyl structure, defunctionalization, and cross linking (Payne and Ortoleva 2001; Achyuthan et al. 2010). This transformation occurred under high temperature and pressure through cycles of subsidence and reemergence. As a result, coal layers are generally found interspersed between layers of fine-grained sediments (shales and limestones) and coarser sediments (siltstones and sandstones) due to the sedimentation of sand, silt and clay along over buried swamps. Sedimentation patterns determine the thickness and geometry of present-day coal beds. For example, the Black Warrior basin of Alabama has produced many thin coal beds (<1 in.–4 ft thick), while the San Juan basin has produced fewer but thicker coal beds (up to 70 ft thick) (U.S. EPA 2004a).

CBM associated with coal deposits are formed from the elimination of smaller molecules during the compaction and transformation of coal beds by thermogenic and biogenic processes. Based on the source or origin, CBM is classified as primary biogenic gas, thermogenic gas, and secondary biogenic gas. Gas eliminated during the aerobic microbial attack of cellulose is referred to as primary biogenic gas (Rice 1993). The elimination of small gas molecules from the deeper layers of coal seams due to thermal cracking leads to the production of thermogenic gas. Thermogenic gases are available in greater volumes compared to biogenic gases per unit volume of sediment (Broadhead 2000). As thermal processes dominate, microbial processes get suppressed under these extreme conditions. The microbes still survive nonetheless by feeding on the highly recalcitrant organic molecules for cellular maintenance. Such a geological habitat can become active again with changes in geochemical conditions such as brine incursion, basin uplift, basin cooling, the flow of associated groundwater or diffusion of thermogenic methane into coal deposits (Parkes et al. 2000; Ulrich et al. 2008). Biogenic gas generation resumes when conditions that are conducive for the anaerobic microbial attack of coal resume, such as the dilution of salinity levels and changes in the pH levels (Martini et al. 1998). In this instance, the biogenic gas produced is classified as secondary biogenic gas.

The source and origin of CBM gases can be determined through stable isotope composition, molecular composition, and the abundance of methane relative to higher hydrocarbons (Faiz and Hendry 2006). Isotopic signatures are used to deduce the composition, since different sources and sinks of methane have different affinities for ^{12}C and ^{13}C isotopes. For example, microbes preferentially attack the lighter isotopes; hence, biogenic gas is isotopically light.

$$\text{Isotopic signature}, \delta^{13}C = \left[\frac{\left(13_C/12_C\right)_{sample}}{\left(13_C/12_C\right)_{reference}} - 1\right] \times 1000‰$$

The abundance of methane relative to other higher hydrocarbons in CBM is given by the dryness index. For biogenic gas, the dryness index is near 1 due to near-complete breakdown during aerobic microbial attack; however, thermogenic gas usually has a reasonable amount of ethane and propane, leading to a dryness index of approximately 0.98 (Rice and Claypool 1981).

$$\text{Dryness index} = \left[\frac{\text{Amount of } CH_4}{\text{Amount of}\left(CH_4 + \text{higher hydrocarbons}\right)}\right]$$

Low-rank coals generally have lower gas content. Although higher rank coals usually produce more gas, their adsorptive capacity is lesser, thus causing gases to expel (Halliburton 2007). Apart from this, production is also difficult in higher rank coals due to slow desorption from narrow pore networks. Thus, commercial CBM production is usually pursued with coals of middle rank, which are typically sub-bituminous or bituminous coals with a range of volatilities.

10.3 CBM EXTRACTION

CBM is generally produced from relatively shallow coal beds (about 1000–1500 ft) as compared to conventional sources that are contained within sharply defined geological formations (at about 3000 ft). For a coal seam to be economically viable for CBM production, it must typically be at least 20 ft thick, produce 50–70 ft^3 of gas/ton of coal, contain 77%–87% of carbon (normally found in sub-bituminous coal), contain sufficient water pressure to hold the gas in place and have sodium bicarbonate as its dominant water chemistry as it helps to maintain the ideal pH required for anaerobic gas production (Ernst & Young 2010).

CBM exists in coal seams in three basic states: as free gas, as gas dissolved in the water in coal seams, and as gas adsorbed on the solid surface of coal (U.S. EPA 2013a). When a production well is drilled for CBM extraction, the hydrostatic pressure is reduced and the formation water starts flowing out. Following this, the free gas in the coal seams is produced. It is only later that the gas that is adsorbed in the porous spaces in the coal at almost liquid-like density (about 91%–95% of total gas content) is produced (EPA 2004a). In various studies, the transport of gas from these porous spaces to production wells is described as a multistep transport process consisting of gas diffusion in micropores, gas diffusion through partly blocked microfractures, gas flow through open un-mineralized microfractures, and finally, gas movement through the main fractures (Wei et al. 2007). This is usually simplified as a two-step transport process consisting of gas diffusion through micropores and gas flow through macropores. Referring to these two levels, coal beds are generally considered to have dual porosity. The micropores represent a coal matrix embedded in the macropores (or fractures) as shown in Figure 10.2.

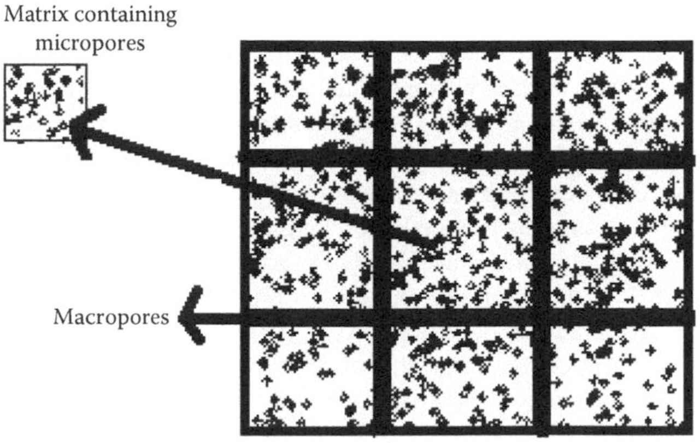

FIGURE 10.2 Dual porosity representation of a coalbed.

Conventional CBM recovery by reservoir pressure depletion typically recovers only 50% of the gas in place. The recovery of gases that are adsorbed in the coal matrix can be enhanced by the use of two techniques—(1) inert gas stripping using nitrogen injection and (2) displacement desorption employing CO_2 injection (White et al. 2005). This is called *enhanced CBM (ECBM) production.* Nitrogen injection ECBM (N_2-ECBM) works by decreasing the partial pressure of methane in the porous spaces of coal, thus promoting its desorption. Carbon dioxide injection (CO_2-ECBM) works on the principle that carbon dioxide is preferentially adsorbed in the coalbed matrix, subsequently causing methane desorption. The CO_2-ECBM technology is based on the existing knowledge of CO_2 sequestration in oceans, oil and gas fields, salt domes, and deep saline aquifers. Moderately permeable (about 1–5 mD) deep reservoirs with minimal faults, foldings, and discontinuities are ideal candidates for CO_2-ECBM. Based on the geological assessment, market potential, carbon dioxide supply potential, and site infrastructure costs, the major coal basins ranked highest to lowest for CO_2-ECBM potential (White et al. 2005) are (1) the Bowen basin (Australia), (2) Qinshui basin (China), (3) the Upper Silesian basin (Poland), and (4) the Cambay basin (India).

For high CBM production from wells, desorbed gases have to be able to flow out easily. This depends a great deal on the presence of preexisting natural fracture systems. In the absence of natural fractures (which is typically the case), artificial stimulation is performed. A common method of achieving this is hydraulic fracturing, where high-pressure fracturing fluids are injected with proppants (such as sand) into targeted coal zones. This procedure widens the naturally occurring planes of weakness (rarely creating new fractures), resulting in induced or enlarged fractures, which improves the connections for the flow of fluids to the production bore (U.S. EPA 2004a). Fracturing fluids are primarily water based, but may contain other substances, including acids and small quantities of hydrocarbons. Fracturing is also possible by the use of inert gases such as nitrogen and carbon dioxide or foams containing water and inert gases with a foaming agent. Proppants such as sand are generally added to the fracturing fluid, so that they penetrate into the seam and stay wedged, thus keeping the fractures propped open and allowing the fluid to flow easily (Gale and Freund 2001). Apart from hydraulic fracturing, the other common options to improve gas recovery for the commercial development of CBM are air or air-to-air mixture injection into well and electrothermal bed stimulation (Griffiths and Severson-Baker 2003).

A major technological advance that has enabled the commercial exploitation of CBM is directional drilling. CBM can be produced either from vertical or from horizontal production wells. Vertical wells are often targeted only in multiple coal seams. However, for drilling into methane-bearing coal seams interspersed by other geological formation layers, horizontal drilling is preferred. The directional drilling technique was developed to overcome practical difficulties in laying horizontal wells. Directionally drilled wells for CBM production usually drain more gas, thereby reducing the number of production wells required. In the document *Directional Drilling Technology* released by the U.S. Environmental Protection Agency, a directional well is defined as "a well bore that intersects a potentially productive formation and does not intentionally exit the formation for the remaining footage drilled" (U.S. EPA 2010a). This means that it is a well that is spudded like a conventional vertical well but deviates such that the well bore enters the formation roughly parallel to the bedding plane. Directional drilling is thus a technique that has evolved by taking the best elements of vertical boreholes and underground horizontal drilling techniques (U.S. EPA 2010a). In essence, a drill rig could be oriented in different directions on the same surface site to drill a succession of directional degasification holes. The use of directional drilling significantly reduces the cost of site preparation and production facilities as gas flow is centralized to one location.

Once completed, directionally drilled CBM wells typically cover an area sufficient just for a well head, pump, liquid separator and dehydrator, metering equipment, and maybe an optional water tank. The rest of the area can be re-vegetated and re-contoured. This mitigates or eliminates long-term surface disturbances (Ministry of Energy and Mines, BC). It also removes the need to fracture coal beds, which is a huge advantage as it saves groundwater from possible contamination by fracking fluids. This technique usually extracts 2 to 25 times more gas than vertical wells drilled in the same gas field, and this advantage can potentially offset the high expenses and general technical risks arising from the requirements for advanced geological equipment and constant monitoring of the placement of drill bits during directional drilling (Dogwood Initiative).

One of the primary requirements for directional drilling is that the site must have sufficient depth of cover over the target coal bed (Oyler and Diamond 1982). Two other critical coal characteristics that can make reservoirs unamenable to directional drilling are poor vertical permeability or impermeable streaks and variable formation depth and thickness. Apart from these reservoir limitations, a major operating constraint associated with directional drilling is dewatering from horizontal wells using conventional pumping systems designed for vertical wells (U.S. EPA 2010a). However, despite these concerns, it is indisputable that directional drilling has been a significant contributor to the commercialization of CBM production.

The life span of a CBM well typically lies between 5 and 15 years, with the maximum gas production achieved after about one to six months of water removal. The three stages of production (U.S. EPA 2010b) as shown in Figure 10.3 are as follows:

1. An early stage where formation water is pumped out, reducing underground pressure and thus allowing the release of natural gas.
2. A stable stage where natural gas production increases and water production decreases.
3. A late stage where gas production declines and water production remains low.

In the first few weeks or months, when only little methane or natural gas flows from the formation into the well bore, the gas is vented and later flared or incinerated for several weeks or months. When the gas production levels are high enough, the installation of a compressor station and pipelines for transportation to the markets becomes economic (Gale and Freund 2001). Since coalbed gas mostly consists of methane, it is usually suitable for direct introduction into a commercial pipeline with little or no treatment (Rice 1993).

The production dynamics of water over time are different for CBM production as compared to conventional gas extraction. In the case of CBM wells, a large quantity of water is produced initially as opposed to the case of conventional gas wells, where produced water increases over time. Since produced water management costs are a significant portion of operating costs, CBM projects begin with high

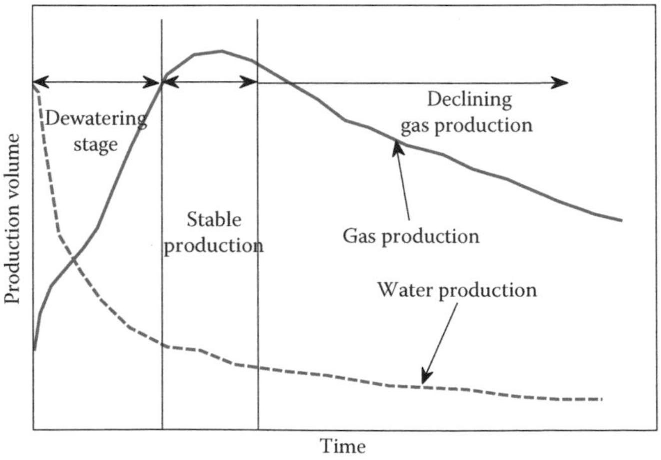

FIGURE 10.3 Three stages of production of coalbed methane.

operating costs, but these usually tend to diminish over time (U.S. EPA 2010b).

10.4 WORLDWIDE AVAILABILITY AND PRODUCTION OF CBM

Global natural gas reserves have steadily grown to 187.5 TCM towards the end of 2009 (Ernst & Young 2010). About 4% of this is unconventional, of which half is in North America. Its rapid growth for use in different energy sectors has projected an expectation that it will be on par with crude oil in the global energy landscape by 2040. The only significant barrier to this would be sustainability due to depleting conventional reserves and the difficulty and investment required in tapping unconventional reserves. In this scenario, CBM has the potential to become the significant resource in major coal-producing countries, especially the six countries that represent 80% of coal production in the world: China (39%), United States (19%), India (7%), Australia (7%), Russia (5%), and Indonesia (4%). In 1994, global CBM reserves were estimated by the U.S. Geological Survey at more than 7000 TCF (IFP Energies 2008). Figure 10.4 shows estimates of conventional and unconventional gas reserves (including CBM) around the world (CBM Asia Development Corp. report 2012b; U.S. EIA 2014a).

In the present day, the significant CBM producers in the world are the United States, Australia, Canada, and a few western European countries. The United States dominates CBM production, accounting for 61.8% of the total CBM produced globally in 2013, followed by Canada, which totaled 11.5% of the global production in the same year. The key driving force for the CBM market in North America comes from the need to become self-sustainable in its gas needs. CBM is also rapidly being developed in the Asia Pacific (especially China, India, and Indonesia) to meet the increased energy demand rising from these growing economies. Australia, with its vast resources and proximity to China, India, and Indonesia, is emerging as a leading exporter of CBM. Thus, the Asia Pacific region (Australia, China, Indonesia, and India) is expected to be one of the fastest growing markets, with an estimated compound annual growth rate of 14.9% from 2014 to 2020 (Grand View Research 2014b). The key market players investing in China, India, and Indonesia for the development of the vast proven and unproven resources are BG group, Arrow Energy, Origin Energy, PetroChina, Dart Energy, Great Eastern Energy, and Santos (Grand View Research 2014b).

A breakdown of the status of CBM reserves, production capacity, demand, and markets in these major regions are described in the subsequent sections.

10.4.1 United States and Canada

The United States dominates CBM production in the world, with nearly 5 BCFD production of CBM and 20 TCF produced to date. The production is expected to fall in the near future, however, due to resource maturity and depletion coupled with low gas prices (U.S. EIA 2014a).

There are 11 major coal basins in the United States, with production established in 10 of them. They are the San Juan basin, the Black Warrior basin, the Piceance Coal basin, the Uinta basin, the Central Appalachian basin, the Northern Appalachian

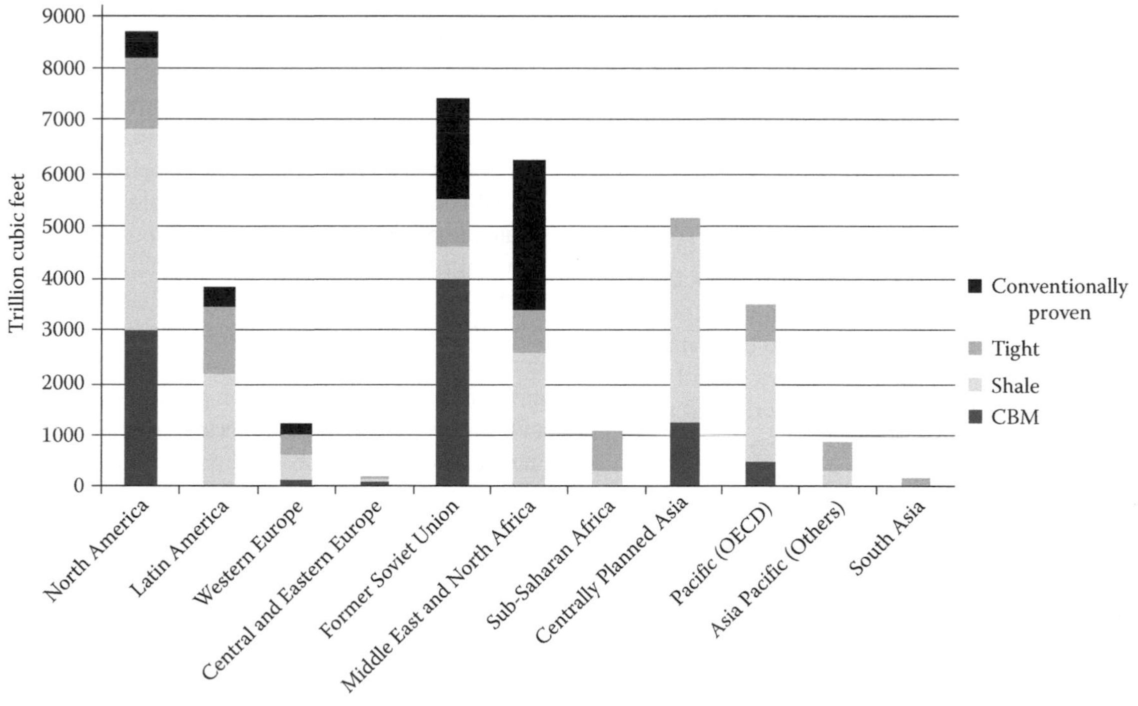

FIGURE 10.4 Global estimates of conventional and unconventional gas reserves, including coalbed methane. (Data from Stevens, P., The 'Shale gas revolution': Hype and reality, *Chatham House*, 2010; Ernst & Young, *Shale gas and coal bed methane: Potential sources of sustained energy in the future*, 2010.)

basin, the Western Interior coal region, the Raton basin, the Sand Wash basin, and the Washington coal regions (U.S. EPA 2004b). The San Juan basin is the most productive CBM basin in North America. The CBM production from these wells was over 800 BCF in 1996 and 925 BCF in 2000 (U.S. EPA 2004b). Among other high-productivity basins are the Powder River basin with 147 BCF, the Black Warrior basin with 112 BCF, the Uinta basin with 75.7 BCF, the Central Appalachian basin with 52.9 BCF, and the Raton basin with 30.8 BCF of gas production in 2000. The Sand Wash basin, which had 31 BCF of gas production in 1995, faced slow development in later years due to limited economic viability. The large depth of the CBM areas in the Piceance basin inhibits permeability and has therefore made it difficult to extract gas from this region. Another basin with very little production is the Northern Appalachian basin due to insufficient geological knowledge and CBM ownership issues. There was no gas production from the Washington coal regions as of 2000 due to the difficulty in interpreting the geological structure of these coal seams (U.S. EPA 2004b).

The CBM industry is relatively newer to Canada as compared to its neighbor, the United States. The regions of central and southern Alberta in Canada are known to contain one of the largest geographical extents of continuous coal in North America. In this region, individual coal seams are either laterally discontinuous or well correlated over large distances. All of them contain CBM to some extent, and each is potentially capable of producing a significant quantity of CBM. The CBM reserves recognized in Alberta by the Energy Resources Conservation Board are coals of the Horseshoe Canyon formation and the Belly River group with low gas content and low water volume, and the coals of the Mannville group with high gas content and high volume of saline water. The Alberta Geological Survey estimated that there is 500 TCF of gas in place within all of coal in Alberta (Alberta Energy). The estimate is accepted as an initial determination of Alberta's ultimate CBM gas in place with a large portion of the ultimate potential still remaining undetermined. The first production of CBM in Alberta was attempted in the 1970s; however, significant development with commercial production commenced only in 2002 in the Horseshoe Canyon region in Alberta. Alberta's CBM production was 7.2 BCM from 14,000 wells (not all active) in 2010, although the actual CBM production continues to be uncertain even today due to the commingling of CBM with conventional gas (Alberta Geological Survey). Producers in Alberta typically drill more horizontal gas wells and use multistage completion technology as it substantially improves well productivity. There is also a large coal resource in British Columbia, estimated at about 90 TCF. The cretaceous coals on Vancouver Island and at Telkwa are also structurally similar to CBM basins in the United States (Ryan 2003). Nova Scotia forms the third largest CBM reserve after Alberta and British Columbia. Pipeline systems in Alberta have been reconfigured to allow the low pressure intake of CBM, and gas ownership issues and regulatory controls have been addressed, even though there are no federal tax incentives to encourage investment in CBM technologies (Alberta Geological Survey).

10.4.2 Australia

Natural gas accounts for about 21% of the energy demand portfolio in Australia (U.S. EIA 2014b). In 2012, BREE (the Bureau of Resource and Energy Economics) projected that natural gas share in Australia's primary energy consumption would increase to 34% by 2050. Australia's abundant gas resources and its geographical proximity to major consumer markets (China, India, and Indonesia) have made it a leader of LNG supply for the Pacific basin. The large amounts of untapped CBM resources (also known as coal seam gas in Australia) have resulted in large investments in production in the recent past. Geoscience Australia has estimated that the total of the proven/probable commercial resources of CBM is 33 TCF as of 2012 (U.S. EIA 2014b). These resources are primarily located in the northeastern Queensland province in the Bowen and Surat basins. Commercial production from CBM fields in Australia began in 1996 in Queensland and New South Wales. CBM production of 246 BCF was totaled in 2012, which made up roughly 15% of the natural gas production in Australia, according to BREE (Bureau of Resources and Energy Economics, Australia 2013).

With key CBM developers aggressively exploring and drilling in several areas, it is anticipated by Geoscience Australia that resource distribution of natural gas will shift from offshore traditional gas production to CBM (or other sources) in the next two decades. Many of Australia's new gas field developments are tied to liquefaction projects. CBM-LNG projects are feasible due to the sizeable amount of gas reserves associated with coal production. The increased use of stream of CBM for LNG projects in Australia has attracted many Asian companies such as Sinopec, China National Offshore Oil Corporation, China National Petroleum Corporation in China and Tokyo Gas in Japan (U.S. EIA 2014b). The Queensland Curtis LNG project is the first CBM-LNG project in Australia. However, many of these projects that are under exploration have delayed schedules for production, with the main reason being the public resistance to CBM development due to the potential environmental impact related to produced water use and disposal, underground water contamination from methane, and fracking fluids; also, there are land rights issues in some cases. South Australia, which houses part of the Cooper basin, first published extensive guidelines encouraging investment for development of CBM projects with environmentally safe extraction practices. The low productivity of CBM wells is another important reason for delays in investment decisions on these CBM projects. Table 10.1 shows some of the proposed CBM-LNG projects in Australia and their estimated target dates.

10.4.3 Russia

The probable CBM resources in Russia are large, being estimated at approximately 84 TCM, which is about one-third of Russia's probable gas resources. The country's mining and geological conditions are similar to those in Australia, Canada, and the United States, and hence are favorable for

TABLE 10.1
Proposed Coalbed Methane and Liquefied Natural Gas (CBM-LNG) Projects in Australia

Project Name	Companies	Peak Output (Billion Cubic Feet/Year)	Target Date	Capital Cost
Queensland Curtis LNG	Train 1: BG 50% and CNOOC 50% Train 2: BG 97.5% and Tokyo Gas 2.5%	400	Fourth quarter 2014	$20.4 billion
Australia Pacific LNG	Origin Energy 37.5%, ConocoPhilips 37.5%, and Sinopec 25%	430	Third quarter 2015	$25.5 billion
Gladstone LNG	Santos 30%, Petronas 27.5%, Total 27.5%, and Kogas 15%	375	Fourth quarter 2015	$18.5 billion
Fisherman's Landing	LNG Ltd 67.5%, Inpex 17.5%, Kogas 10%, and CPC 5%	144	2016	$1.7 billion
Arrow LNG	Shell 50% and PetroChina 50%	384	2018	$24.2 billion

Source: Bureau of Resources and Energy Economics, Australia, *Gas Market Report*, 2013.

CBM development. Russia's first production facility was launched by Gazprom in the Taldinskoye coalfield, in southwestern Siberia in 2010. Gazprom is also pursuing expansions for CBM production from Kuzbass. The Kuzbass basin is estimated to have 13 TCM of gas, accessible at 1800–2000 m depth. It may reasonably be considered to be the world's largest explored CBM basin. It is projected that the annual CBM production in Kuzbass will be 4 BCM at the plateau period and 18 to 21 BCM in the long run (Gazprom 2010). The major barriers to CBM development in Russia are competition from large in-country proven gas resources with low-cost production capacity and state regulations that keep the sale price low for this large gas supply.

Similar to other countries, government aid has been proposed to increase the investment attractiveness of CBM projects. In November 2011, CBM was categorized as an independent mineral resource and included into the *Russian Classified Index of Natural Resources and Underground Waters*. Initiatives were also being taken to gradually increase the price of the natural gas supplied to industrial and residential users. CBM recovery is also listed in the policy passed for increase in energy supply from renewable power generation (Gazprom).

10.4.4 China

China's rapidly increasing demand for energy is making it one of the world's most influential energy markets. It is the largest producer and consumer of coal in the world, accounting for almost half of the world's coal consumption. Coal accounted for 69% of China's total energy consumption in 2011, and it had an estimated 126 billion tons of recoverable coal reserves in 2011, the third largest in the world behind the United States and Russia. China's CBM gas in place is estimated to be greater than 500 TCF. China is currently facing challenges in CBM exploitation due to poor geological conditions such as low gas saturation and low permeability in coal seams (U.S. EIA 2014c).

Most of China's CBM comes from basins in the north and northeastern portions of the country, the Sichuan basin in the southwest and the Junggar and Tarim basins in the west. FACTS Global Energy (FGE) reported that China produced about 441 BCF of CBM in 2012 from surface wells and coal mines and is now targeting about 700 BCF output by the end of 2015. China is also working on increasing its utilization rates from less than 40% to over 60% by the end of 2015. Apart from technical challenges and high development costs, CBM production in China also faces hurdles from the lack of pipeline infrastructure between coal mines and gas markets, and government regulations. However, the State Council in China has issued policy guidelines and financial incentives to encourage investment in CBM exploration and development. The first commercial CBM pipeline in China began operations in late 2009, linking the Qinshui basin with their west-east pipeline. Several more projects are under construction, with liquefaction plants and trucks also being installed to transport CBM to demand centers (U.S. EIA 2014c).

10.4.5 India

India's growing economy has driven its demand for energy, including gas (Global Methane Initiative 2010). Gas consumption rose from 36 BCM in 2005 to 51 BCM in 2009, with gas production rates of 29.9 BCM and 38.6 BCM in 2005 and 2009, respectively. The shortfall in production relative to consumption was compensated with imported natural gas and LNG. CBM production is therefore becoming economically attractive because of the added cost incurred in importing gas and LNG.

The Indian Ministry of Oil partnered with the U.S. Geological Survey (USGS) and Oil and Natural Gas Corporation to conduct a CBM resource assessment. The estimate was anywhere between 9 and 92 TCF of CBM resources (Ernst & Young 2010). The Directorate General of Hydrocarbons, Noida, India, estimated that there were 44 major coal and lignite fields in 12 states of India, covering an area of 35,330 km^2 and the Central Mine Planning and Design Institute Limited estimated that these basins contain 120 TCF of CBM depending on the rank of coal, depth of burial, and geotectonic settings. Beginning May 2001, the Indian government has offered four rounds of international bidding for the exploration and production of CBM from these basins, amounting to a resource potential of 65.1 TCF. As of 2010, more than 130 exploratory and 80 pilot test wells have been drilled in these basins (Global Methane Initiative 2010). The first commercial production of CBM was carried out by GEECL in June 2007, at Raniganj, with a production of 0.11 MMSCMD. Essar Oil and Reliance Industries Limited have CBM production active in West Bengal. The total CBM production in 2013 amounted to about 5.8 BCF (U.S. EIA 2014d). However, large-scale CBM production and utilization is hampered by lack of pipeline infrastructure and a nationally integrated system. The Gas Authority of India Limited and Reliance Gas Transportation Infrastructure Limited are the two most important companies operating gas pipelines in India for distribution from coal fields to end use markets.

10.4.6 Indonesia

Indonesia's gas market is characterized by a small but quickly growing domestic market and a large export market. The country's gas exports have fallen in the recent years due to an increase in domestic consumption, leading to the exploration of CBM to meet this shortage. The Indonesian government supported CBM development by providing incentivized production-sharing contract terms, which results in 45% earnings for contractors on an after tax basis, which has led to the establishment of many CBM exploration and development projects and a rise in gas prices (CBM Asia Development Corp. report 2012b).

Indonesia has an estimated total of 453 TCF gas-in-place, located mainly in the provinces of Sumatra and Kalimantan. The country's coal deposits are young in age (Miocene), extremely thick, low in ash content (<5%), are of relatively low thermal rank (sub-bituminous), reasonably highly permeable, contain low to moderate gas content, and are located at optimal depths (CBM Asia Development Corp. report 2012a). These characteristics and frequent strong gas kicks have been indicators of the rich CBM potential in these deposits. They are considered to be the world's best in undeveloped CBM potential today. Some of the basins in Indonesia are the Central Sumatra basin, the Ombilin basin, the South Sumatra basin, the Bengkulu basin, the Jatibarang basin, the Barito basin, the Kutai basin, the Berau basin, and the north Tarakan basin. The first commercial production in Indonesia was carried out by BP and ENI (VICO) in March 2011. The gas was then liquefied at the Bontang liquefaction facility for export to the high-priced north Asian markets. The project was the world's first export of CBM-to-LNG.

On the whole, increasing exploration and extraction on a global scale is expected to drive CBM production over the next six years. Figure 10.5 shows the projected CBM market volume trend in different regions up to 2020, with the demand in the Asia Pacific projected to match the demand in North America by 2020 (Grand View Research 2014b). Table 10.2 lists the major industrial stakeholders engaged in CBM activity in various parts of the world (CBM Asia Development Corp. report 2012b).

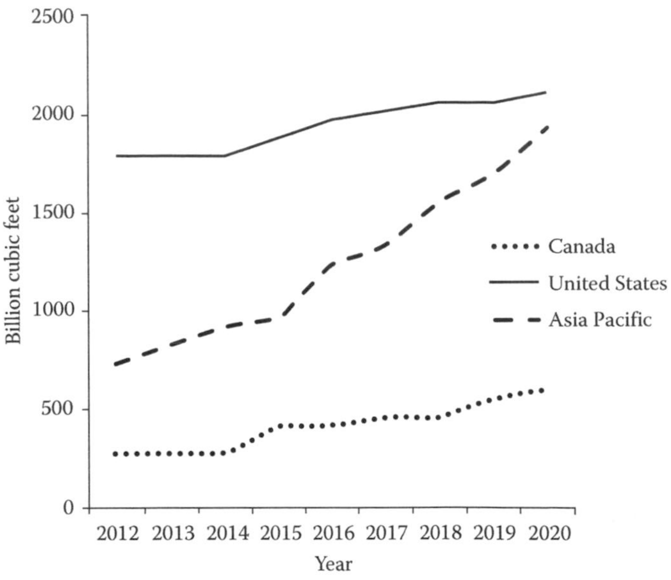

FIGURE 10.5 Projections of global trends in coalbed methane market demand up to 2020. (Data from Grand View Research, Coalbed methane market size, industry trends, regional outlook, company share and segment forecasts 2014 to 2020, 2014b. http://www.grandviewresearch.com/blog/coal-bed-methane-market/, Retrieved on July 21, 2015.)

TABLE 10.2
Companies Engaged in CBM Production Worldwide

Country	Company
United States	ConocoPhilips, BP, Chevron, Exxon Mobil, and Anadarka
Australia	ConocoPhilips, BG Group, Shell, Santos, Kogas, and Total
China	BHP, BP, ConocoPhilips, and Chevron: all tested CBM and left due to poor geology
India	Reliance Industries, Oil and Natural Gas Corporation, Great Eastern Energy Corporation, Essar Oil Ltd., BP, Arrow Energy, and Geopetrol
Indonesia	Santos, BP, ExxonMobil, Eni, and Total

Source: CBM Asia Development Corp. report, CBM around the world, 2012b. http://www.cbmasia.ca/CBM-Around-The-World, Retrieved on July 21, 2015.

10.5 ECONOMICS OF CBM PRODUCTION

The Crude Oil Windfall Profit Act (WPT) of 1980 enacted by Congress in the United States provided tax credits for the production of many unconventional sources of fuel, including CBM, shale gas, and biofuels (Soot 1991). Tax credits for unconventional gas wells and formations lasted until 1992, while credits for gas produced from these wells lasted through 2002. Even though exploration of CBM began much later than for tight or shale gas, the tax credits enabled CBM production to reach 1250 BCF in 1999. This development suggests that economic incentives from governments, such

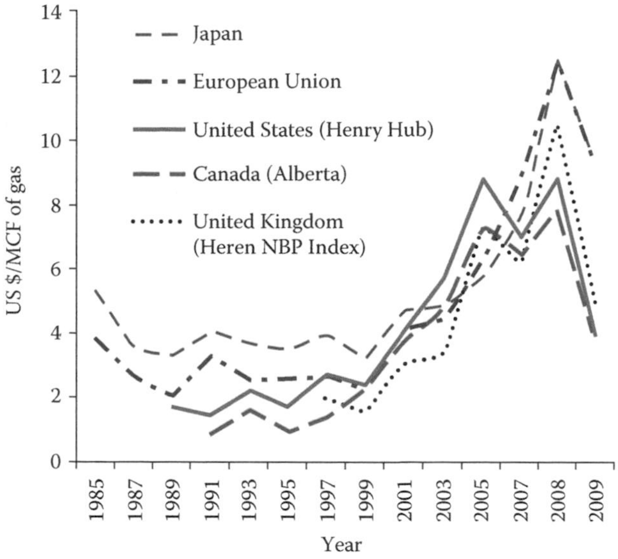

FIGURE 10.6 Global trends in gas prices from 1985 to 2009. (Data from Stevens, P., *The Shale gas revolution: Hype and reality*, Chatham House Report, page 6, Chatham House, London, 2010.)

as subsidies or tax credits, are necessary to make CBM production sustainable (Kuuskraa and Huge 2002).

Naturally, the profitability of CBM production also depends on the price of natural gas, and regional trends in prices greatly affect its economics. Figure 10.6 (Stevens 2010) shows the trends of gas prices around the world and indicates that prices have increased steadily from 1985 to 2009. The subsequent decrease in gas prices can be attributed to increased shale gas and LNG production, and this, coupled with relatively high production costs has brought into question the economic viability of CBM projects.

Based on a CBM survey conducted in 2008 and a review based on data from CBM-producing basins in the United States in 2010, it was concluded in a report by the U.S. Environmental Protection Agency that a large fraction of existing CBM projects are no longer economically viable, which is borne out by the trend of decreasing volume of CBM production shown in Figure 10.7 (U.S. EPA 2013b). Also, according to this study, wastewater management (using ion exchange or underground injection disposal) contributes significantly to the costs associated with CBM production.

Shale gas production has also provided significant competition to CBM production. Figure 10.8 shows the share of gas production from different sources in the United States from 1991, with projections up to 2039 (U.S. EIA 2014a). It is seen that shale gas production was 370 BCF in 1999, when a significant amount of CBM (about 1250 BCF) was already being produced from three basins. However, shale gas production has rapidly increased in the recent past (since 2005). In 2012, shale gas production was 9.7 TCF, accounting for over one-third of gas production in the United States (IEA 2014), and it is projected to be 19.8 TCF in 2040, accounting for approximately 53% of the projected gas production in the United States. The large amounts of gas produced from shale plays affect the development of the CBM market; it is projected that as shale gas production increases, CBM production would diminish and would resume growth again only after 2025. The increased production of shale gas was enabled by the development of horizontal drilling and hydraulic fracturing technologies, especially coil tube drilling. In addition, the estimates of shale gas resources, especially in the United States, have risen dramatically. In April 2009, the U.S. Department of Energy

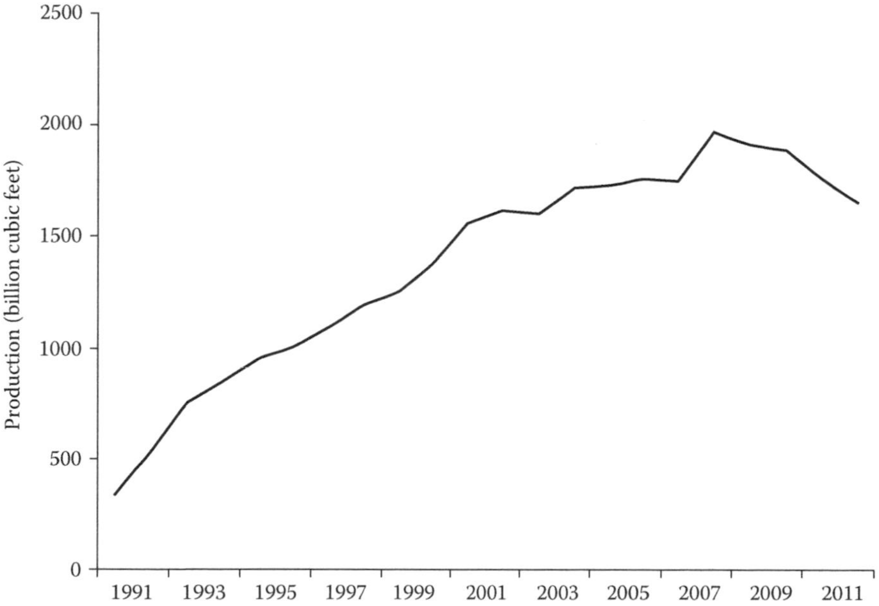

FIGURE 10.7 Production of coalbed methane in the United States. (Data from Stevens, P., *The Shale gas revolution: Hype and reality*, Chatham House Report, page 6, Chatham House, London, 2010; U.S. EPA, *U.S. Economic Analysis for Existing and New Projects in the Coalbed Methane Industry*, 2013b; U.S. EIA, U.S. Energy Information Administration, *U.S. Coalbed Methane Production In Natural Gas*, 2012, http://www.eia.gov/dnav/ng/hist/rngr52nus_1a.htm, Retrieved on July 21, 2015.)

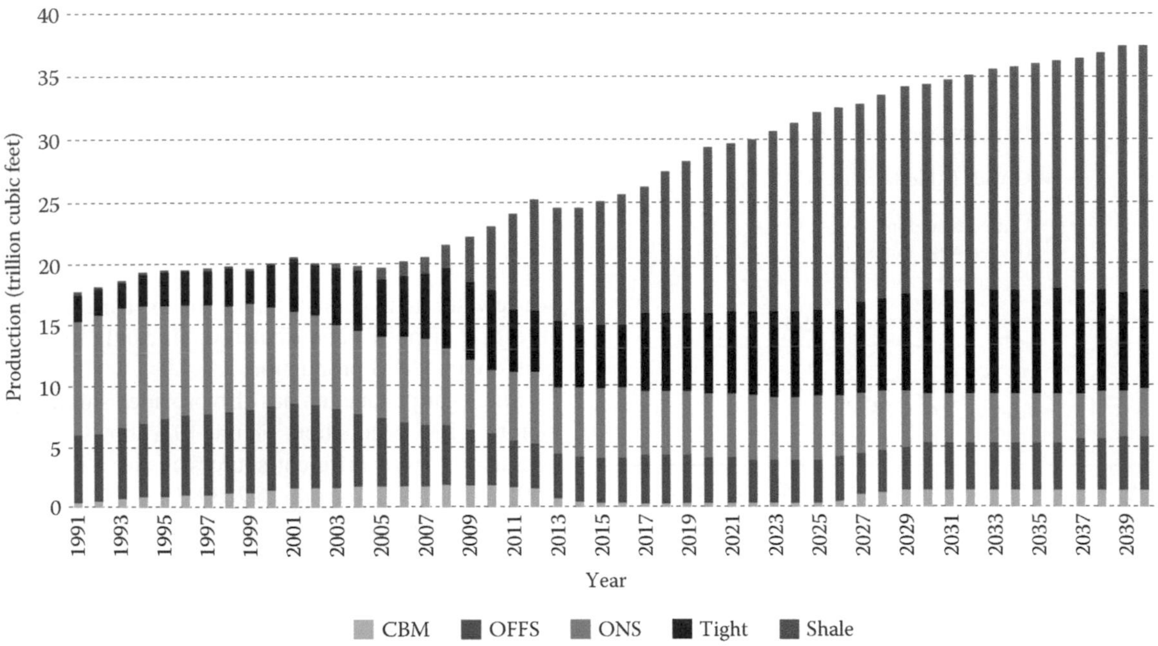

FIGURE 10.8 Historical and predicted shares of gas production in the United States. CBM—coalbed methane, OFFS—offshore, ONS—onshore, tight—tight gas, and shale—shale gas. (Data from U.S. EIA, *Annual Energy Outlook 2014 with projections to 2040*, 2014a. http://www.eia.gov/forecasts/aeo/pdf/0383(2014).pdf/, Retrieved on July 21, 2015.)

estimated that the Marcellus shale play has 262 TCF of recoverable resources, whereas the EIA suggested that the technologically recoverable resources were at 1744 TCF (Stevens 2010). The amount of water generated and its management is another key difference between CBM production and shale gas production. CBM deposits produce some of the highest water intensities (i.e., volume of water per unit of energy extracted) as compared to any other oil and gas source. On an average, CBM deposits produce 10 times the amount of water compared to shale gas, diverting a lot of its investment capital for water management rather than gas recovery.

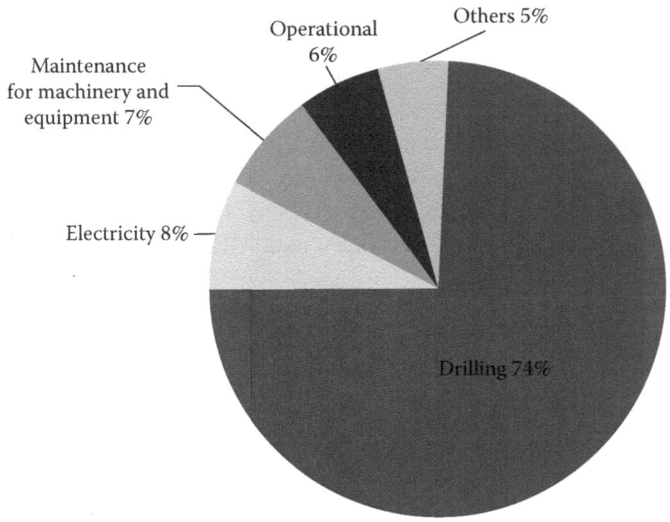

FIGURE 10.9 Production costs for coalbed methane, excluding water purification costs. (Data from Grand View Research, Coalbed methane (CBM) market analysis by application (Industrial, power generation, residential, commercial and transportation) and segment forecasts to 2020, http://www.grandviewresearch.com/industry-analysis/coal-bed-methane-industry, 2014a.)

10.6 CHALLENGES IN CBM PRODUCTION

The main issues affecting a CBM project are uncertainties in projected gas and water production due to the vast heterogeneity of coal beds, water and environmental management, the availability of gas and water pipelines, the availability of land and ownership issues, the difficulty of water and gas production, and the economics of gas demand and supply. The high investment requirement is another major hurdle. A large part of investment goes into the continued drilling of production wells, as they are usually of low productivity (producing about 100–500 MCF/day, much lesser than a conventional well). Figure 10.9 shows the share of different costs incurred in CBM production (excluding the water management costs). It is seen that drilling accounts for almost 74% of total cost incurred (U.S. EPA 2013b; Grand View Research 2014a).

The main concerns with CBM produced water are salinity, sodicity, and toxicity from various metals such as barium and iron. Apart from this, other trace pollutants such as potassium, sulfate, bicarbonate, fluoride, ammonia, arsenic, and radionuclides may also be present. Typically, water is either processed using chemical precipitation, ion exchange, reverse osmosis, and other desalination technologies or disposed through methods such as underground injection or evaporation/infiltration ponds.

Based on a survey of 112 CBM projects in the United States, it was estimated that adding water treatment using ion exchange would cause 24% of economically viable projects to become uneconomical, with an additional 26%–33% experiencing loss in gas production. Similarly, the addition of underground injection to meet environmental regulations would cause 27% of economically viable projects to become unviable, with additional 38%–44% experiencing production losses (U.S. EPA 2013a). Produced waters can also be considered for serving agricultural purposes such as irrigation or stock watering. This can however be done only after taking into account not just the quality of water, but also conditions such as soil mineralogy, texture, and sensitivity of plant species in the receiving areas (Veil et al. 2004).

Apart from water management issues, another serious environmental concern associated with water production in CBM fields is the drawdown of the underground water table, since coal seams are usually shallow and are located within/or close proximity to aquifers. Groundwater can also get contaminated from methane seepage through un-cemented annular spaces in production well casings, natural fractures, and stimulated fractures from hydrofracturing.

Finally, surface disturbances can also be caused by drilling of a large number of producing wells with very close spacing as compared to conventional wells. These can be controlled by good land use practices such as drilling multiple wells from a single location and aligning roads or pipelines along the natural field breaks using horizontal/directional drilling techniques (Alberta Energy).

10.7 CONCLUSIONS

The technology required for the production of CBM is relatively mature, and it is a viable source for natural gas, which is seen, at least in the short term, as a sustainable bridge in the transition from fossil-based to renewable energy. Directional drilling is one of the enabling technologies that has helped to enhance the recovery of gas from coal seams. CBM production can also be coupled with the injection of carbon dioxide in coal seams, which combines greenhouse gas sequestration with enhanced CBM recovery. However, CBM production is associated with high capital costs, mostly related to drilling and water management, and the heterogeneity of coal beds results in uncertainties in potential gas production. In addition, the competition from tight and shale gas production has led to significant uncertainty in the future of CBM production. However, CBM remains, and is projected to remain, a significant source of energy in the near future.

REFERENCES

Achyuthan, K.E., Achyuthan, A.M., Adams, P.D., Dirk, S.M., Harper, J.C., Simmons, B.A., Singh, A.K., Supramolecular self-assembled chaos: Polyphenolic lignin's barrier to cost-effective lignocellulosic biofuels, *Molecules*, 2010, 15 (12), 8641–8688.

Alberta Energy, *Coalbed Methane Development*, http://www.energy.alberta.ca/NaturalGas/753.asp, Retrieved on July 21, 2015.

Alberta Geological Survey, *Coal Bed Methane*, http://www.ags.gov.ab.ca/energy/cbm/, http://www.ags.gov.ab.ca/energy/cbm/, Retrieved on July 21, 2015.

Bureau of Resources and Energy Economics, Australia, *Gas Market Report*, 2013.

Broadhead, R.F., Petroleum geology of the McGregor range, Otero County, New Mexico, in Transactions Southwest Section AAPG Convention, Roswell Geological Society, pp. 40–55, 2002.

CBM Asia Development Corp. report, CBM in Indonesia, 2012a.

CBM Asia Development Corp. report, CBM around the world, 2012b, http://www.cbmasia.ca/CBM-Around-The-World, Retrieved on July 21, 2015.

Dogwood Initiative, *Coalbed Methane: Best Practices for British Columbia*, http://dogwoodinitiative.org/publications/reports/, Retrieved on July 21, 2015.

Ernst & Young, Shale gas and coal bed methane: Potential sources of sustained energy in the future, 2010.

Faiz, M., Hendry, P., Significance of microbial activity in Australian coal bed methane reservoirs—A review, *Bulletin of Canadian Petroleum Geology*, 2006, 54 (3), 261–272.

Flores, R.M., Coalbed methane: From hazard to resource, *International Journal of Coal Geology*, 1998, 35, 3–26.

Gale, J., Freund, P., Coal-bed methane enhancement with CO_2 sequestration worldwide potential. *Environmental Geosciences*, 2001, 8 (3), 210–217.

Global Methane Initiative, *Coal Mine Methane Country Profiles—India*, 2010. https://www.globalmethane.org/documents/toolsres_coal_overview_ch16.pdf, Retrieved on July 21, 2015.

Grand View Research, Coal bed methane (CBM) market analysis by application (Industrial, power generation, residential, commercial and transportation) and segment forecasts to 2020, http://www.grandviewresearch.com/industry-analysis/coal-bed-methane-industry, 2014a.

Grand View Research, Coal bed methane market size, industry trends, regional outlook, company share and segment forecasts 2014 to 2020, 2014b. http://www.grandviewresearch.com/blog/coal-bed-methane-market/, Retrieved on July 21, 2015.

Griffiths, M., Severson-Baker, C. *Unconventional Gas: The Environmental Challenges of Coalbed Methane Development in Alberta*, Pembina Institute for Appropriate Development, Drayton Valley, Alberta, Canada, 2003.

Halliburton, Chapter 3—Sorption, in *Coalbed Methane: Principles and Practices*, 2007. http://www.halliburton.com/public/pe/contents/Books_and_Catalogs/web/CBM/H06263_Chap_03.pdf, Retrieved on June, 2007.

Gazprom, Prospects for CBM production in Russia. http://www.gazprom.com/about/production/extraction/metan/, Retrieved on July 21, 2015.

Gazprom, *Gazprom launches Russia's first CBM facility*, February 12, 2010. http://www.gazprom.com/press/news/2010/february/article76350/, Retrieved on July 21, 2015.

IFP Energies report, *Coalbed Methane: Current Status and Outlook*, 2008. http://www.ifpenergiesnouvelles.com/Publications/Available-studies/Panorama-technical-reports/Panorama-2008/Coalbed-methane-current-status-and-outlook, Retrieved on July 21, 2015.

International Energy Agency (IEA), *FAQs on Natural Gas*, 2014. http://www.iea.org/aboutus/faqs/naturalgas/, Retrieved on July 21, 2015.

Kuuskraa, V.A., Hugh, D.G., Translating lessons learned from unconventional natural gas R&D to geologic sequestration technology, *Journal of Energy & Environmental Research*, 2002, 2 (1), 75–86.

Martini, A., Walter, L., Budai, J., Ku, T., Kaiser, C., Schoell, M., Genetic and temporal relations between formation waters and biogenic methane: Upper Devonian Antrim Shale, Michigan Basin, USA, *Geochimica et Cosmochimica Acta*, 1998, 62 (10), 1699–1720.

Ministry of Energy and Mines, *Coalbed Methane in British Columbia*, http://www.em.gov.bc.ca/mining/geoscience/coal/coalbc/cbm/pages/cbmbrochure.aspx, Retrieved on July 21, 2015.

Natural Resources Canada (NRCAN), *Natural Gas: A Primer*, 2014. http://www.nrcan.gc.ca/energy/natural-gas/5639.

Oyler, D.C., Diamond, W.P., *Drilling a horizontal coalbed methane drainage system from a directional surface borehole*, U.S. Department of The Interior, Bureau of Mines, RI 8640, Pittsburgh, PA, 1982.

Parkes, R.J., Cragg, B.A., Wellsbury, P., Recent studies on bacterial populations and processes in subseafloor sediments: A review, *Hydrogeology Journal*, 2000, 8 (1), 11–28.

Payne, D., Ortoleva, P., A model for lignin alteration—Part I: A kinetic reaction-network model, *Organic Geochemistry*, 2001, 32 (9), 1073–1085.

Rice, D.D., Composition and origins of coalbed gas, *Hydrocarbons from Coal: AAPG Studies in Geology*, B. E. Law and D. D. Rice (Eds.), 1993, vol. 38, 159–184.

Rice, D.D., Claypool, G.E., Generation, accumulation, and resource potential of biogenic gas, *AAPG Bulletin*, 1981, 65 (1), 5–25.

Ryan, B. A summary of coalbed methane potential in British Columbia, *CSEG Recorder*, November 2003, 32–40.

Soot, P.M., Tax incentives spur development of coalbed methane, *Oil & Gas Journal*, 1991, 89 (23), 40.

Stevens, P. *The Shale gas revolution: Hype and reality*, Chatham House Report, page 6, Chatham House, London, 2010.

U.S. Energy Information Administration (U.S. EIA), *U.S. Coalbed Methane Production In Natural Gas*, 2012. http://www.eia.gov/dnav/ng/hist/rngr52nus_1a.htm, Retrieved on July 21, 2015.

U.S. Energy Information Administration (U.S. EIA), *Annual Energy Outlook 2014 with projections to 2040*, 2014a. http://www.eia.gov/forecasts/aeo/pdf/0383(2014).pdf/, Retrieved on July 21, 2015.

U.S. Energy Information Administration (U.S. EIA), *Australia – International Energy and Data Analysis* 2014b. http://www.eia.gov/beta/international/analysis_includes/countries_long/Australia/australia.pdf, Retrieved on July 21, 2015.

U.S. Energy Information Administration (U.S. EIA), *China – International Energy and Data Analysis*, 2014c. http://www.eia.gov/beta/international/analysis_includes/countries_long/China/china.pdf, Retrieved on July 21, 2015.

U.S. Energy Information Administration (U.S. EIA), *India – International Energy and Data Analysis*, 2014d. http://www.eia.gov/beta/international/analysis_includes/countries_long/India/india.pdf, Retrieved on July 21, 2015.

U.S. Environmental Protection Agency (U.S. EPA), Characteristics of coalbed methane production and associated hydraulic fracturing practices, in *Evaluation of Impacts to Underground Sources of Drinking Water by Hydraulic Fracturing of Coalbed Methane Reservoirs*, 2004a. http://www.epa.gov/ogwdw/uic/pdfs/cbmstudy_attach_uic_ch03_cbm_practices.pdf, Retrieved on July 21, 2015.

U.S. Environmental Protection Agency (U.S. EPA), Summary of coalbed methane descriptions, in *Evaluation of Impacts to Underground Sources of Drinking Water by Hydraulic Fracturing of Coalbed Methane Reservoirs*, 2004b. http://www.epa.gov/ogwdw/uic/pdfs/cbmstudy_attach_uic_ch05_basins.pdf, Retrieved on July 21, 2015.

U.S. Environmental Protection Agency (U.S. EPA), *Directional Drilling Technology*, 2010a. http://www.epa.gov/cmop/docs/dir-drilling.pdf, Retrieved on July 21, 2015.

U.S. Environmental Protection Agency (U.S. EPA), *U.S. Coalbed Methane Extraction: Detailed Study Report*, 2010b. http://water.epa.gov/scitech/wastetech/guide/304m/upload/cbm_report_2011.pdf, Retrieved on July 21, 2015.

U.S. Environmental Protection Agency (U.S. EPA), *Technical Development Document for the Coalbed Methane (CBM) Extraction Industry*, 2013a. http://water.epa.gov/scitech/wastetech/guide/oilandgas/upload/cbmttd2013.pdf, Retrieved on July 21, 2015.

U.S. Environmental Protection Agency (U.S. EPA), *U.S. Economic Analysis for Existing and New Projects in the Coalbed Methane Industry*, 2013b. http://water.epa.gov/scitech/wastetech/guide/oilandgas/upload/cbmea2013.pdf, Retrieved on July 21, 2015.

Ulrich, G., Bower, S., Active methanogenesis and acetate utilization in Powder River Basin coals, United States, *International Journal of Coal Geology*, 2008, 76 (1), 25–33.

Veil, J.A., Puder, M.G., Elcock, D., Redweik Jr, R.J., A white paper describing produced water from production of crude oil, natural gas, and coal bed methane, prepared by Argonne National Laboratory for the US Department of Energy, National Energy Technology Laboratory, January, 2004. http://www.ead.anl.gov/pub/dsp_detail.

Wei, X., Wang, G., Massarotto, P., Golding, S., Rudolph, V., Numerical simulation of multicomponent gas diffusion and flow in coals for CO_2 enhanced coalbed methane recovery, *Chemical Engineering Science*, 2007, 62 (16), 4193–4203.

White, C.M., Smith, D.H., Jones, K.L., Goodman, A.L., Jikich, S.A., LaCount, R.B., DuBose, S.B., Ozdemir, E., Morsi, B.I., Schroeder, K.T., Sequestration of carbon dioxide in coal with enhanced coalbed methane recovery: A review, *Energy and Fuels*, 2005, 19 (3), 659–724.

11 Health and Safety Issues in Coal Mining

Nuray Demirel and Celal Karpuz

CONTENTS

Abstract: Health and safety issues are critically important in minimizing risks of accidents and occupational illnesses. This chapter serves as a comprehensive text to meet requirements of researchers, mining professionals, engineers, safety specialists, and students. The chapter is organized into five main sections: mine safety and health issues in underground coal mines, main hazards in coal mining and associated impacts on occupational health and safety, typical mine accidents, accident analysis and comparative costs, and effective mine safety and health management risk management including emergency preparedness and response.

11.1 INTRODUCTION

Historically, underground coal mining has been experiencing industrial accidents and occupational diseases and has been regarded as the most risky sector. It ranks first in accident frequency and severity rates among all other industries. Although there have been attempts to improve this situation through legal provisions and technological improvements, occupational health and safety in underground coal mining, especially in developing countries, is still an emerging issue.

According to the Turkish Statistical Institute (TSI 2007), accident rate of mining industry is 10.1%, while the average accident rates of different sectors, including agriculture, forestry, hunting, and fishing; manufacturing industry; electric, gas, and water; construction; and transportation is around 3.0%. Over the past decade, a number of accidents involving underground coal mining have resulted in fatalities and injuries to workers and others in mining workplaces. According to Mining Safety and Health Administration's statistics, in the United States alone there have been 289 fatalities since 2003 (MSHA Statistics 2014).

Although inherent risks exist because of potential for explosions, mine fires, often-confined nature of the workplace, and the proximity of dangerous equipment and personnel (Schofield et al. 2001), some of these risks can be managed by employing an effective mine safety and health management system. Minimizing the risk of accidents and diseases and sustaining work safety require commitment and cooperative effort involving governments, managements, and employees to improve three essential aspects of health and safety management such as the work environment, the person, and the systems. In this sense, these three essential components should integrate with each other and zero accident policy should be adopted and even extended beyond the workplace. Grayson and Watzman (2001) stated that government certainly plays an important role, through effective inspection, enforcement, legal provisions, health and safety knowledge, and technology development as the result of research. Inevitably, academia has a key function in conducting research and education to advance the current knowledge and expand the frontiers in mine health and safety.

This chapter attempts to cover comprehensively current issues to be addressed in coal mining industry, particularly in underground coal mining; it serves as a comprehensive text and endeavors to fulfill the essence of researchers, mining professionals, engineers, safety specialists, and students to adopt it according to their own requirements. The chapter has been organized as follows: Section 11.2 provides a brief overview about the mine safety and health issues in underground coal mines; Section 11.3 deals with the main stressors encountered in coal mining and associated impacts of these stressors on occupational health and accidents; Section 11.4 explains the typical underground mine accidents and safety issues encountered in underground coal mining; Section 11.5 presents accident analysis and costs, and Section 11.6 discusses two important components of effective mine safety and health management: risk management and emergency preparedness and response.

11.2 UNDERGROUND COAL MINE ATMOSPHERE

Marching toward continuous improvement in mine safety and health essentially requires a dedication to the anticipation, recognition, evaluation, communication, and control of all environmental stressors in, or arising from, the underground coal mine work place, which may result in fatality, injury, illness, impairment, or affect the well being of workers. These environmental stressors, to some extent, inherently exist in underground coal mines and are divided into five categories as (1) biological, (2) chemical, (3) physical, (4) ergonomic, and (5) psychosocial (AIOH 2014). These stressors and their associated potential hazards should be determined and evaluated to mitigate the available risks in underground coal mines.

Potential biological hazards usually stem from exposure to insects, molds, fungi, and bacteria and are not usually associated with mining activities. In mining industry, biological stressors have the least importance when compared to other stressors. Physical stressors, on the other hand, created by electromagnetic and ionizing radiation, noise, vibration, illumination, and extremes of temperature and pressure are critically important in the prevention of occupational diseases and accidents. Chemical stressors are airborne contaminants such as liquids, gases, dusts, fumes, mists, vapors, and skin irritants. Ergonomic stressors are caused by monotony (repetitive motion) and work pressure (fatigue). In order to achieve a high level of safety standards and minimize the risk of accidents and occupational illnesses, ergonomic principles should be applied in any workplace. Psychosocial stresses can be caused by upsetting events that happened recently. Such stressful instances may include fatal mine accidents in which colleagues are involved. Moreover, work pressure, fatigue, shift-work sleep disorder, and problems such as worry and inability to live up to a standard of performance may also be other stress factors involved in coal mining industry.

11.2.1 Mine Air

Mine air is a mixture of atmospheric air and mine gases. Atmospheric air sent to the mine through mine ventilation is called as *entry air* and the air used in underground and leave the mine is called as *return air*. Entry air is assumed to have the same content with atmospheric air. The distribution of gases in atmospheric air by volume is given in Table 11.1.

Besides these gases, water vapor is always present in various amounts in the mine air. The volumetric amount of water vapor in the mine air is around 1%.

The fresh air sent to the underground mine openings is contaminated by harmful gases and dust due to (1) harmful gas generated by the ore, coal, or rock strata, (2) oxidation of coal, (3) mining machineries, (4) drilling and blasting operations,

TABLE 11.1
Distribution of Gases in Atmospheric Air

Gas	Volume (%)
Nitrogen (N_2)	78.09
Oxygen (O_2)	20.95
Carbon dioxide (CO_2)	0.03
Rare gases[a]	0.93

Source: Hartman et al., *Mine Ventilation and Air Conditioning*, 3rd edition, John Wiley & Sons, New York, 1997.

[a] Argon, neon, krypton, xenon, helium, hydrogen, and ozone.

and (5) breathing of people. The contaminated air is classified into the following four groups (MLSS 2012):

- *Used air:* Air containing less than 20% O_2, also called as *suffocating air.*
- *Toxic air:* Air containing poisonous gas that affects human health and even poses threat to human life. These poisonous gases are carbon monoxide (CO), hydrogen sulfide (H_2S), nitrogen oxides (NO, NO_2, N_2O_3, etc.), sulfur dioxide (SO_2), and radon (Ra).
- *Explosive air:* Air containing explosive gases such as methane (CH_4), ethane (C_2H_6), propane (C_3H_8), butane (C_4H_{10}), hydrogen (H_2), and carbon monoxide (CO). The most dangerous among explosive gases is hydrocarbons (C_nH_{2n+2}), especially CH_4. The mixture of air and methane is explosive at right pressure and temperature conditions and can cause catastrophic damage.
- *Dusty air:* Air containing rock or coal dust. It can be explosive and detrimental to human health. It may cause an occupational disease called *pneumoconiosis.*

Besides atmospheric air, mine atmosphere also includes active mine gases, which are generated in various ways. Gases commonly observed in mine atmosphere are discussed in the following sections.

11.2.1.1 Nitrogen

Nitrogen (N_2) is a colorless, odorless, and tasteless gas. It is the largest constituent of mine air, comprising approximately 80% of it. Its main function is diluting O_2 in the atmospheric air, so that O_2 could be breathable. Main sources of N_2 are decomposition of organic substance, blasting agents, and emission from cracks in rock or coal.

11.2.1.2 Oxygen

Oxygen (O_2) is a colorless, tasteless, and an odorless gas. It can dissolve in water. It is an essential gas for breathing and also for combustion. Atmospheric air contains 20.95% of oxygen by volume. Its concentration decreases due to breathing; oxidation of organic and inorganic substances, such as timber, rock, and ore; mine fires; methane and dust explosions; emissions from engines; open flame lights; and emission of active mine gases from coal or rock strata.

If O_2 concentration decreases, a number of harmful effects are observed on human health. Harmful effects begin at an O_2 concentration of 17%. Table 11.2 presents the effects of O_2 depletion on human health. In general, oxygen deficiency leads to loss of mental alertness and distortion of judgment and performance within a relatively short time, without the person's knowledge and prior warning.

O_2 concentration may decrease up to 1%–3% after fires or explosions in locations where there is no or insufficient ventilation. In such places, breathing causes sudden death. Therefore, abandoned areas must be closely monitored and controlled.

At normal atmospheric conditions, a person at rest inhales 15–20 times and uses 8–9 L/min air. The amount of used air increases to 50 L/min when a person is active and increases to 100 L/min when a person is very active. Table 11.3 lists the amount of air inhaled, O_2 consumed, and respiratory quotient for different level of activities (Howard 1961).

TABLE 11.2
Effects of O_2 Depletion on Human Health

O_2 Concentration (%)	Health Impacts
17	Reduced night vision, increased breathing volume, and accelerated heartbeat
16	Dizziness; reaction time for new tasks is doubled
15	Poor judgment, poor coordination, abnormal fatigue upon exertion, and loss of muscle control.
10–12	Very faulty judgment, very poor muscular coordination, and loss of consciousness.
8–10	Nausea, vomiting, and coma
<8	Permanent brain damage
<6	Spasmodic breathing, convulsive movements, and death in 5–8 min

Source: *Oxygen Deficiency Hazard, Safety Booklet*, https://www.jlab.org/accel/safetylb/ODH-book.pdf.

TABLE 11.3
Respiratory Quotient for Different Level of Activities

Activity	Respiratory Rate (/min)	Air Inhaled (L/respir)	Air Inhaled (L/min)	Oxygen Consumed (L/min)	Respiratory Quotient
At rest	15	0.55	8.7	0.28	0.75
Moderate	30	1.75	52.5	1.96	0.90
Very active	45	2.50	100.0	2.80	1.00

Source: Howard, L.H., *Ventilation and Air Conditioning*, Ronald Press, New York, 1961.

A few practical rules dictating the minimum amount of O_2 that should be provided per person are as follows:

- 1–2 m^3/min/person in nongassy mines
- 3–6 m^3/min/person in medium gassy mines
- 20–25 m^3/min/person in mines with high concentration of CH_4 and other heavy gases.

Amount of O_2 in mines can be measured by safety lamp, chemical methods, physical methods, and electrochemical methods.

11.2.1.3 Carbon Dioxide

Carbon dioxide (CO_2) is colorless and odorless but has an acidic taste. Its specific gravity is 1.52 and is heavier than air. It causes suffocation. The maximum allowable concentration of CO_2 in the return air is 0.5%. Exposure to CO_2 concentration of 4% poses life threat. The main sources of CO_2 are breathing; burning of wood, coal, and petroleum; decay of wood; internal combustion engines; explosions; and mine fires. CO_2 concentration can be measured using chemical analysis in the laboratory, testing by flame safety lamp, and also using portable gas analyzers.

11.2.1.4 Carbon Monoxide

Carbon monoxide (CO) is a colorless, odorless, tasteless, toxic, and flammable gas produced by incomplete combustion of carbonaceous material. It is poisonous and explosive even at a very low concentration. Its explosibility range is 12.5%–74% in air. The specific gravity of CO is 0.967, which means it is slightly lighter than air. The causes of CO generation are underground mine fires, explosions, blasting, frictional heating prior to open burning, low-temperature oxidation, and internal combustion engines.

The effects of CO concentration in air on human health are given in Table 11.4 (Goldstein 2008).

TABLE 11.4
Effects of CO Concentration on Human Health

Concentration	Symptoms
35 ppm (0.0035%)	Headache and dizziness within 6–8 h of constant exposure
100 ppm (0.01%)	Slight headache in 2–3 h
200 ppm (0.02%)	Slight headache within 2–3 h; loss of judgment
400 ppm (0.04%)	Frontal headache within 1–2 h
800 ppm (0.08%)	Dizziness, nausea, and convulsions within 45 min; insensible within 2 h
1,600 ppm (0.16%)	Headache, tachycardia, dizziness, and nausea within 20 min; death in less than 2 h
3,200 ppm (0.32%)	Headache, dizziness, and nausea in 5–10 min. Death within 30 min
6,400 ppm (0.64%)	Headache and dizziness in 1–2 min. Convulsions, respiratory arrest, and death in less than 20 min
12,800 ppm (1.28%)	Unconsciousness after 2–3 breaths. Death in less than 3 min

Source: Goldstein, M., Carbon monoxide poisoning, *J. Emerg. Nurs.*, 34(6), 538–542, 2008.

CO acts as a type of asphyxiant by displacing the oxygen normally carried by the hemoglobin. The affinity of the blood for carbon monoxide is approximately 3000 times that for oxygen (Forbes and Grove 1954). Therefore, if the air breathed into the lungs contains only a small amount of carbon monoxide, the hemoglobin will absorb it in preference to the oxygen present.

For the detection of CO, chemical analysis in the lab, colorimetric detectors (change of color), thermal detectors, and mobile detectors are used.

11.2.1.5 Hydrogen

Hydrogen is colorless, odorless, tasteless, nontoxic, and the lightest of all gases found underground. The sources of underground hydrogen are the charging of batteries, the action of water or steam on hot materials, and the action of acid on metals. Hydrogen is also explosive within the range of 4%–74% in air. It can explode when the oxygen concentration in the air is as low as 5%.

TABLE 11.5
Effect of H_2S on Human Health

Concentration (ppm)	Effects
20	Accepted as safe for 8 h exposure
50–100	Marked eye irritation and respiratory irritation (1 h)
200	Dangerous after 1 h exposure
200–400	Marked eye and respiratory irritation (1 h)
400	Extremely dangerous after 0.5 h
400–900	Unconsciousness (0.5–1 h)
900–2000 or more	Acute poisoning, death in minutes

Source: National Coal Board, *"Noxious Gases Underground," A Handbook for Colliery Managers*, W.S. Cowell, Ipswich, 1970.

11.2.1.6 Hydrogen Sulfide

Hydrogen sulfide (H_2S) is an extremely poisonous gas; it has no color but a distinct smell of a rotten egg. However, H_2S destroys the smelling sense. It is explosive between 4.4% and 44.5%. Its specific gravity is 1.19. The main sources of H_2S are combustion of black blasting powder, blasting of sulfide ores, and dewatering of flooded area.

Table 11.5 tabulates the effect of H_2S on human health (National Coal Board 1970 in Guyaguler et al. 2005).

11.2.1.7 Sulfur Dioxide

Sulfur dioxide (SO_2) is nonflammable but a very poisonous gas and has a strong pungent sulfurous smell. It has a specific gravity of 2.26. The main sources of SO_2 are burning of iron pyrite and blasting of sulfide ores. It irritates eyes and throat. Breathing of air containing SO_2 severely damages lungs.

11.2.1.8 Nitrogen Oxides

Nitric oxide (NO), nitrogen dioxide (NO_2), nitrogen trioxide (NO_3), nitrogen tetraoxide (N_2O_4), and nitrous oxide (N_2O) are called as *nitrogen oxide* (NO_x) gases. Diesel engine exhausts are one of the sources for NOx generation Exposure to NO_x gases has a detrimental impact on human health. The threshold limit value of NO_2 is 5 ppm.

11.2.1.9 Methane

Methane (CH_4) is explosive, colorless, and odorless and has a specific gravity of 0.55. It may also cause suffocation when O_2 concentration of air decreases. CH_4 explodes within the range of 5%–15%. The strongest explosion takes place when CH_4 concentration is about 9.5%. Ignition temperature of CH_4 is between 650°C and 750°C. The main sources of CH_4 emissions are from the strata and sudden discharge mostly seen in narrow workings, short faces, geologically disturbed zones, and dry parts of the coal mine.

CH_4 in the above-mentioned zones should continuously be measured and closely monitored using automatic detectors and sensors, stationary and mobile methanometers, and digital methanometers.

11.2.1.10 Control of Gases Underground

Control of harmful gases underground is critically important due to the high risk of fatal accidents. The preferred order of the gas control measures is given as follows (Hartman et al. 1997):

1. Prevention
   a. Proper procedure in blasting
   b. Adjustment and maintenance of internal combustion engines
   c. Avoidance of open flames
2. Removal
   a. Drainage in advance of mining
   b. Drainage of bleeder entries
   c. Local-exhaust ventilation
   d. Water infusion in advance of mining
3. Adsorption
   a. Chemical reaction in internal combustion engine conditioner
   b. Solution by air-water spray in blasting
4. Isolation
   a. Sealing off abandoned workings of fire areas
   b. Restricted blasting or off-shift blasting
5. Dilution
   a. Local dilution by auxiliary ventilation
   b. Dilution by main ventilation airstream
   c. Local dilution by diffusers and water sprays.

Threshold limit values of harmful gases are tabulated in Table 11.6 (National Coal Board 1970).

The following simplified equation, Equation 11.1, is used to determine the amount of air required to dilute the harmful gases (Hartman et al. 1997).

$$Q = \frac{q}{\text{MAC} - b} - q \tag{11.1}$$

TABLE 11.6
Main Hazardous Gases and Their Threshold Limit Values

Hazardous Gas	Threshold Limit Value (ppm)	Explosion Range (%)
Carbon monoxide (CO)	50	12.5–74
Carbon dioxide (CO_2)	5000	Nonflammable
Nitric oxide (NO)	25	–
Nitrogen dioxide (NO_2)	5	–
Methane (CH_4)	–	5–15
Hydrogen sulfide (H_2S)	10	4–44
Sulfur dioxide (SO_2)	5	–
Hydrogen (H)	–	4–75

Source: National Coal Board, *"Noxious Gases Underground," A Handbook for Colliery Managers*, W.S. Cowell, Ipswich, 1970.

where:
Q is airflow (m^3/min)
q is harmful gas inflow (m^3/min)
b is gas concentration in intake air (%)
MAC is the maximum allowable concentration of the harmful gas

If time factor is involved, more exact solution can be obtained using Equation 11.2 (Hartman 1982).

$$\log\left[\frac{q - Q_x}{q - Q_{x0}}\right] = -\frac{T}{2.3} \cdot \frac{Q}{V} \tag{11.2}$$

where:
Q_x is gas concentration (%)
q is gas inflow (m^3/min)
Q is ventilating air (m^3/min)
T is time required (min)
V is volume of working area (m^3)
Q_{x0} is initial gas concentration

11.2.2 Dust

Dust is formed by the disintegration of materials by different processes, such as drilling and blasting operation, crushing, grinding, material handling, caving, use of explosives, cutting operation, loading and unloading, transportation or material handling, production operations, filling operations, cleaning after blasting, supporting operations, expanding galleries, and also operation of mine equipment and air movement.

Dust concentration is expressed in two ways: (1) gravimetric method (mg/m^3) and (2) counting method (number of particles/cm^3). According to particle size, coal dust is classified into the following four groups (Guyaguler et al. 2005):

- *Loose coal:* Any particle of coal that is too large to pass through a No.20 sieve
- *Coal dust:* Any particle of coal that can pass through a No.20 sieve

- *Float coal dust:* Any particle of coal that can pass through a No.200 sieve
- *Respirable coal dust:* Any particle of coal that is 5 μm or less in diameter

According to harmful effect, coal dust can be grouped as follows (Guyaguler et al. 2005):

- *Fibrogenic dusts:* Capable of producing fibrosis or scarring of the lung surfaces. Silica (quartz, cristobalite, tridymite, and chert), silicates (asbestos, talc, mica, and sillimanite), nearly all metal fumes, beryllium ore, tin ore, some iron ores, carborundum, and coal (bituminous and anthracite).
- *Carcinogenic dusts:* Asbestos, radon daughters (attached to any dust), arsenic, diesel particulate matter (a suspected carcinogen), and silica (a suspected carcinogen).
- *Toxic dusts:* Poisonous to body organs and tissue. Dust of ores of beryllium, arsenic, lead, uranium, radium, thorium, chromium, vanadium, mercury, cadmium, antimony, selenium, manganese, tungsten, nickel, and silver, principally the oxides and carbonates, mists and fumes of organic and other body-sensitizing chemicals.
- *Radioactive dusts:* Ores of uranium, radium, and thorium (injurious because of alpha and beta radiation) and dust with radon daughters attached (source of alpha radiation)
- *Explosive dusts:* Combustible when airborne. Metallic dusts (magnesium, aluminum, zinc, tin, and iron), coal (bituminous and lignite), sulfide ores, and organic dusts.
- *Nuisance dusts:* Little adverse effect on humans (gypsum, kaolin, and limestone).

Movement of dust particles depends on size, shape, specific gravity of dust, temperature, humidity of air, and airflow rate. Movement of dust can be classified into three main groups: Newton's motion, Stokes' motion, and Brownian motion. For the Newton's motion, dust particles drop down under the influence of gravity; for the Stokes' motion, dust particles settle down with constant velocity; for the Brownian motion, dust particles make zigzag motion, and never settle down (Lawrie 1967).

Dust particle size causing occupational diseases usually have a size range of 0.5–10 μm. Long-term exposure to mineral dust causes an occupational lung disease, called *pneumoconiosis*, in mineral-related industry, especially coal mining. Factors determining harmfulness of dust are composition, concentration, particle size, exposure time, and individual susceptibility. Depending on the type of mineral, pneumoconiosis can take different forms. Table 11.7 lists the common lung diseases caused by mineral dust (Lawrie 1967).

In underground coal mining, the preferred order of the dust control measures is given as follows:

1. *Prevention*: A cardinal rule of dust control is to prevent dust from occurring. It can be achieved by wetting the floor and by solidifying settled dust using chemicals. Water/steam infusion, foam infusion, and wet drilling techniques are commonly utilized to prevent dust from occurring at its source.
2. *Removal*: Wet and dry dust collectors can be used to clean and remove the settled dust. Wet dust collectors have high efficiency and moderate expenses.
3. *Suppression*: Water sprays, wet cutting, foam, deliquescent chemicals, rock dusting, and water-jet assisted cutting methodologies can be used for dust suppression.
4. *Isolation*: Enclosed cabs, enclosed dust generation, exhaust ventilation, off-shift blasting, control of airflow, separate split, spray fans, air curtains, control of personnel location, remote control, and unidirectional cutting are used for dust isolation.
5. *Dilution*: Main and local ventilation are the ways to dilute the dust from mine environment. A simplified equation (Equation 11.3) for finding the amount of fresh air to dilute mine air is given below. This steady-state equation for dust dilution is used when dust generation rate is constant or where an average value is known (Hartman et al. 1997).

$$Q = \frac{G}{\text{TLV} - B} \tag{11.3}$$

where:
Q is airflow (m^3/min)
G is dust generation (g/min)
TLV is the threshold limit value (mg/m^3)
B is the dust in the inflow air (mg/m^3)

Besides the above-mentioned dust control measures, personal protective devices such as standard respirators, powered positive-pressure respirators, and air helmets must always be used.

For successful dust control, the concentration of dust should closely be monitored using dust sampling techniques. Commonly utilized methods and instruments for dust sampling are filtration, sedimentation, centrifuging, scrubbing, precipitation by impact, electrostatic precipitation, thermal precipitation, optical properties of dusty air, and gravimetric

TABLE 11.7
Common Lung Diseases Caused by Mineral Dust

Mineral	Disease
Coal (especially hard coal)	Anthracosis
Silica	Silicosis
Asbestos	Asbestosis
Beryllium	Acute berylliosis
Iron oxide	Siderosis
Barium sulfate	Baritosis
Tin oxide	Stannosis

Source: Lawrie, W.B., *Some Aspects of Dust, Dust Sampling, the Interpretation of Results*, ILO, Occupational Safety and Health Service, No. 8, Geneva, Switzerland, 1967.

TABLE 11.8
Basic Dust Control Strategies and Methods

Strategy	Method	Expense	Efficiency
Prevention	Water/steam infusion	High	Moderate
	Foam infusion	High	Moderate
	Wet drilling	Low	High
Removal	Dust collectors—wet	Moderate	High
	Dust collectors—dry	Moderate	High
Suppression	Water sprays	Low	Moderate
	Wet cutting	Low	Moderate
	Foam	Moderate	Moderate
	Cutting variable optimization	Low	Moderate
	Deliquescent chemicals	Moderate	Moderate
	Rock dusting	Moderate	Moderate
	Water jet-assisted cutting	High	Moderate
Isolation	Enclosed cabs	Moderate	Moderate-high
	Enclosed dust generation	Moderate	Moderate
	Exhaust ventilation	Low	Moderate
	Blasting off-shift	Low	Moderate
	Control of airflow		
	Separate air split	Low	Moderate
	Spray fans	Low	Moderate
	Air curtains	Moderate	Moderate
	Control of personnel location		
	Remote control	Low	Moderate
	Unidirectional cutting	High	Moderate
Dilution	Main ventilation stream	Moderate	Moderate
	Local ventilation dilution	Low	Moderate

Source: Hartman, H., Ramani, R.V., Mutmansky, J.M., and Wang, Y.J., *Mine Ventilation and Air Conditioning*, 3rd edition. 1997. Copyright Wiley-VCH Verlag GmbH & Co. KGaA. Reproduced with permission.

dust sampler. Table 11.8 presents the current dust control measures and their effectiveness and drawbacks.

11.2.3 Heat

Heat is a type of energy that can be converted into other type, but it cannot be destroyed. In mines, especially electrical and mechanical energies are converted in to heat energy. The unit of heat is joule (J) and the unit of energy is watt (w = J/s). The effect of heat on human beings should be considered together with relative humidity, air velocity, and barometric pressure. Any discomfort about the work decreases efficiency, causes carelessness and inattentiveness, which lead to accidents, injuries, fatalities, and/or occupational diseases.

The main sources of heat in coal mining are heat from rock, heat due to autocompression of air, heat from power-operated machinery, oxidation, electric lights, blasting explosives, heat from rock movement, and heat from human beings.

11.2.4 Illumination

Illumination is the lighting of surfaces of objects in order to make the objects visible. Illumination plays an important role on efficiency in addition to contrast between the surroundings and tasks, which may be influenced by color and the presence or absence of glare (Table 11.9). It is still undecided exactly how much light is required for a particular job; however, it is known that at normal levels of illumination, the ability of the eye to see increases as the log of the illumination. A major issue in both surface and underground mining is the visibility from the operator's station and associated equipment design. The term *visibility* refers to how well the human eye can see something. Because human judgment is involved, the measurement of visibility is always subjective.

TABLE 11.9
Relationship between Illumination and Visual Performance

Increase in Illumination (cm/ft^2)		Increase in Visual Performance
From	To	(%)
5	10	10
10	20	10
20	50	12

Source: Poltev, M.K., *Occupational Health and Safety in Manufacturing Industries*, Mir Publishers, Moscow, Russia, 1985.

In mining, peripheral vision is very important for early recognition of conditions that might give a pre-warning of potential safety hazards in the peripheral field and performance of tasks that require knowledge of the relative spatial relationships among objects separated by significant distance. Lighting all surfaces in a miner's normal field of vision eliminates the tunnel vision effect of the narrow cap lamp beam and overcomes shadowing by machine or roof support structures. Each year, miners are killed and injured, equipments are damaged, cables are cut, and roof supports are knocked down because the operators of mobile equipment cannot see. Their line of sight is blocked due to the design of the equipment or canopies. It is also beneficial to a person's general psychology. Investigation into the relation between production and lighting has shown that adequate and uniform illumination may result in a maximum of production and a minimum of inefficiency.

Illumination standards require

- A uniform illumination of work surfaces.
- Absence of fluctuations and abrupt changes of illumination.
- Minimizing or eliminating of any usual discomfort.
- Elimination of any undesirable glare from illuminated surfaces in the direction of the eye.
- Illumination should be sufficient to provide safe working condition and provided in all surface structures, walkways, stairways, switch panel, loading and dumping sites, and working areas.

- Illumination that satisfies requirements of both health and economics is called *rational.*
- The luminous intensity (surface brightness) of surfaces that are in a miner's normal field of vision of areas in working place that are required to be lightened shall not be less than 0.06 fL or 0.21 cd/m^2.

11.2.5 Noise

Noise can be best described as unwanted sound. The intensity of sound depends on the amplitude of its constituent waves. The greater the amplitude, the greater will be the sound pressure transmitted. Table 11.10 tabulates typical sound pressure levels and their descriptions (Le Roux 1979).

It has been long recognized that individuals who have been exposed to high noise levels over a long period of time suffer from a hearing loss. However, the amount of damage that has occurred is not clear yet.

Noise pressure level in decibel is defined as the 10 times the log to the base 10 of the ratio of sound pressure to the reference pressure. Reference pressure is the lowest sound pressure that humans can hear, which is equal to 2×10^{-5} N/m^2.

The effect of noise depends on its type and is listed as follows (Murrell 1987):

- *Continues (broadband)*: May have an effect that is related to its intensity; it may cause deafness and may reduce working efficiency and interfere with communication. The extent of the damage depends on the susceptibility of the individual, the amount by which the noise exceeds the damage risk level, the length of exposure, and whether the noise is steady or intermittent (pneumatic hammer).
- *Continues (narrow band)*: If frequency is near the top of the spectrum, it may cause irritation and, indirectly, inefficiency.
- *Intermitted (regular)*: Have effects that differ little from those of continuous irritation.
- *Intermitted (irregular and unexpected)*: It *may cause* a *startle reaction*, which can be very disturbing.
- *Meaningful noise*: The influence will be related to a large extent to particular circumstances and is likely to depend on the nature of the noise and what it means to the hearer.

The impacts of noise on labor productivity and number of misdoings are shown in Figures 11.1 and 11.2 (Poltev 1985).

Hearing loss due to noise is called as *presbycusis.* Hearing may be conserved by two methods of approach: noise control

TABLE 11.10
Some Sound Pressure Levels and Their Description

Sound Pressure (Pa)	Sound Pressure (dB)	Source of Sound	Description
2×10^{-5}	0	Normal hearing level	
2×10^{-4}	20	Broadcast studio	Very little sound
6.3×10^{-3}	50	Low sound	
6.3×10^{-2}	70	Talking radio sound	Noisy
2×10^{-1}	80	Intensive traffic	Noisy
2.0	100	Drilling, lathe	Very noisy
6.3	110	Rock drilling	Very noisy
63.0	130	Jet engine	Unbearable

Source: Le Roux, W.L., *Mine Ventilation Notes for Beginners*, 3rd edition, The Mine Ventilation Society of South Africa.

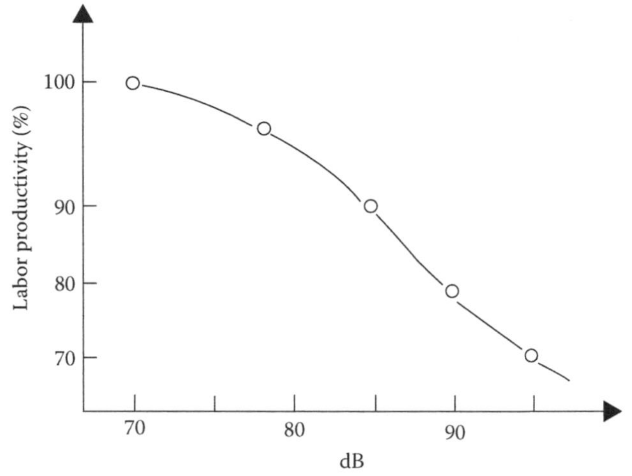

FIGURE 11.1 The effect of noise on productivity. (Data from Poltev, M.K., *Occupational Health and Safety in Manufacturing Industries*, Mir Publishers, Moscow, Russia, 1985.)

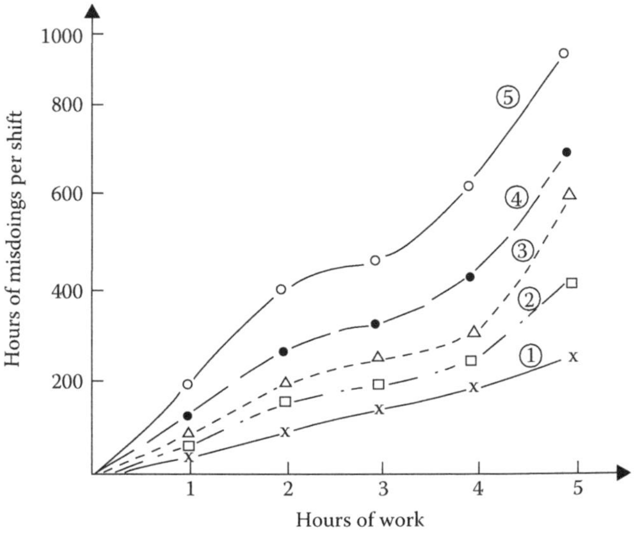

FIGURE 11.2 Number of misdoings relative to hour of work (for sound pressure levels of 1—76 dB, 2—78–80 dB, 3—85 dB, 4—90 dB, and 5—95 dB). (Data from Poltev, M.K., *Occupational Health and Safety in Manufacturing Industries*, Mir Publishers, Moscow, Russia, 1985.)

and personal protection. Effective noise control or reduction has three essential steps: (1) the reduction of the noise at the source, (2) the isolation of the equipment from the surrounding to prevent the noise from spreading, and (3) the absorption of noise to prevent either direct transmission or reflection from surrounding objects.

Apart from noise control strategies, personal protective measures should also be taken for hearing protection if individuals must work in irreducible noise levels, which exceed damage risk level.

Personal hearing protection devices are as follows:

- *Ear plugs*: Used in all frequencies up to sound pressure level of 110 dB.
- *Helmet (headpiece)*: Used up to 120 dB.
- *Noise stopper*: Same as ear plugs. It is a kind of ear defenders.
- *Ear muffs*: Used up to 125 dB above 600 c/s and used up to 115 dB below 600 c/s.

Critical noise levels and associated exposure times are given in Table 11.11.

If there are multiple noise sources in the work environment, then the cumulative noise level can be calculated using Equation 11.4 (Ministry of Environment and Forestry 2011).

$$SP_{\Sigma} = 20\log_{10}\left(\frac{SP_1}{10} + \frac{SP_2}{10} + \ldots + \frac{SP_n}{10}\right) \quad (11.4)$$

where:

SP_{Σ} refers to the cumulative sound pressure in decibels of different sources

SP_1, SP_2, and SP_n are the sound pressures from sources 1, 2, and n, respectively

TABLE 11.11
Permissible Noise Levels

Exposure Time (h)	Permissible Noise Level (dB)
8	90
6	92
4	95
3	97
2	100
1.5	102
1	105
0.75	107
0.50	110
0.25 or less	115

Source: Guyaguler, T. et al., *Occupational Health and Safety in Mining Industry*, Middle East Technical University, ODTU Basim İsligi, Ankara, 2005.

Alternative forms of decibel addition rely upon a few simple rules, which are as follows (Jensen et al. 1978):

1. When two decibel levels are equal or within 1 dB of each other, their sum is 3 dB higher than the higher individual level. For example, 89 dB + 89 dB = 92 dB; 72 dB + 73 dB = 76 dB.
2. When two decibel levels are equal or within 2 or 3 dB apart, their sum is 2 dB higher than the higher individual level. For example, 87 dB + 89 dB = 91 dB; 76 dB + 79 dB = 81 dB.
3. When two decibel levels are equal or within 4–9 dB apart, their sum is 1 dB higher than the higher individual level. For example, 82 dB + 86 dB = 87 dB, 32 dB + 40 dB = 41 dB.
4. When two decibel levels are equal or within 10 or more dB apart, their sum is the same as the higher individual level. For example, 82 dB + 92 dB = 92 dB.

11.2.6 Vibration

Like noise, vibration may be considered as another physical stressor in the working environment. Workplaces, where workers are exposed to excessive noise and vibration, experience increased compensation costs and lost work days through absenteeism, work inefficiencies, and early retirement (Poltev 1985).

Vibration is characterized by three parameters: (1) amplitude of displacement, (2) oscillatory velocity, and (3) acceleration. Vibration according to its type of oscillation can be classified into three classes as natural oscillation (no resistance to the motion, free oscillation), damp oscillation (there is resistance to the motion), and forced oscillation (there is external force to the motion, resonance). Vibration according to its frequency can be classified into three classes as low, medium, and high. Low frequency vibration has a frequency of 1–6 c/s; medium frequency vibration has a frequency of 6–60 c/s; and high frequency vibration has a frequency of 60 c/s.

Another classification based on frequency is as follows (Poltev 1985):

- Frequencies up to 6–8 Hz can be considered as natural oscillations.
- Frequencies up to 25 Hz are perceived as separate jolts and cause osteoarticular changes.
- Frequencies up to 250 Hz affect the nervous system and cause vascular reactions (spasms) and vibration sickness.

Movement with frequencies greater than 1 c/s can be seen in ships and vehicles and may have an amplitude about 10–90 cm. Its main effect is to produce motion sickness. If this movement is sudden and of large amplitude, there may be danger of injury through a person being thrown about. Frequencies greater than 30 c/s produce vibration.

Effects of vibration on human body vary depending on whether the whole body (general vibration) or part (local vibration) of it is involved. Under extreme conditions, vibration may cause displacement of internal organs of the body. General vibration from the jolting of the floor or operating platform or the operator's seat affects the whole body. Local vibration from the operation of hand tools and drills affect mostly hands of the operator, and it may interrupt blood supply to the blood vessels in hands and arms and may cause loss of sensitivity of the skin and lead to deformation and articular immobilization. Hand-arm vibration syndrome affect blood circulation, nerves, muscles, and bones in the hands and arms, leading to loss of sensation and grip and severe pain in the hands. It may even cause white finger syndrome. Sympathetic vibration of organs at low frequencies leads to nausea. Whole body vibration leads to lower back pain and spinal damage.

Prolonged effect on the body of a local or general vibration or both may cause an occupational disease known as *vibration sickness*. Vibration sickness develops gradually and for a long time does not affect the ability to work. The main indications are pain, weakness, increased sensitivity to chilling, cramps, whitening of the fingers, and decrease in skin sensitivity. Functional disorders of the nervous system such as rapid fatigue, headaches, and dizziness are seen. If vibration sickness progresses, disruption of the cardiovascular activity and of internal secretion and disturbance of metabolic processes are unavoidable. Besides physical effects, there are also psychological effects, including loss of concentration, which can cause accidents.

Prevention against vibration sickness consists of careful occupational screening for jobs, constant supervision by physicians, strict observance of protective measures for workers, reducing vibration at the source (e.g., engine vibration), isolation of vibration from the source to the operator seat, and modifying the vehicle suspension or seat suspension. Safe limits for vibration are tabulated in Table 11.12 (Poltev 1985).

Standard regulations for use of hand tools specify force of the hand pressure feed and mass of tools. The design of a hand tool or machine must ensure protection of the operator's hands against vibration. Suitable footwear with shock-absorbing soles affords protection to the worker from general vibration; safety boots made of leather and artificial or synthetic materials help in protecting persons against general vertical vibration and shock. Safety standards include also requirements for the protection of hands from harmful vibration, which include use of gloves with elastic, shock-absorbing, and special elastic devices to tackle with the vibration.

TABLE 11.12
Limit Values for an 8-Hour Exposure to Vibration

	Oscillatory Velocities' Safe Limits	
Geometric Mean of Octave Band Frequency (Hz)	**Effective Value (mm/s)**	**Pressure Level of Effective Value (dB)**
2	11.2	107
4	5.0	100
8	2.0	92
16	2.0	92
31.5	2.0	92
63	2.0	92

Source: Poltev, M.K., *Occupational Health and Safety in Manufacturing Industries*, Mir Publishers, Moscow, Russia, 1985.

11.3 TYPICAL UNDERGROUND COAL MINE ACCIDENTS

11.3.1 Mine Explosions

Explosion refers to a very rapid combustion process accompanied by a severe pressure gradient. Explosion advances at the speed of sound or faster. Explosions may or may not be associated with a mine fire, and they may occur as a result of an ongoing fire or it may lead to the development of a fire. Methane explosion is still the most important source of accident in coal mines. There have been major fatal accidents due to methane explosion. For example, on March 7, 1983, in Armutçuk, Zonguldak, Turkey, 103 miners lost their lives as a result of a methane explosion. Another tragedy occurred on March 3, 1992, and 263 miners were killed in Kozlu, Zonguldak, Turkey.

Methane explosion takes place with the combination of enough O_2 (higher than 12%), explosive gas methane (5%–15% CH_4), and an ignition source (minimum ignition temperature is 580°C). The explosion reaction is given below:

$$CH_4 + 2(O_2 + 4N_2) \rightarrow CO_2 + 2H_2O + 8N_2$$

In this reaction, one volume of methane reacts with a total of 11 volumes of gas mixture. When the methane concentration is 1/11 or 9.5% in volume, then the strongest explosion takes place. As a result of explosion, temperature reaches to 2150°C–2650°C (confined) and 1850°C (expand freely); pressure is increased 7–10 times, and two different waves, namely compressed air wave and flame wave, occur. Flame wave may cause further explosions.

After explosion, the area is filled with a mixture of hot gases (CO_2, N_2, and CO). Mine atmosphere becomes poisonous and unbreathable. Investigations show that two-third of the fatalities are due to CO poisoning or asphyxiation (O_2 starvation). In order to see the explosibility potential of gas mixtures, Coward triangle, which was developed by Coward and Jones (1952), is used (Figure 11.3).

In Coward triangle, air or O_2 percent is shown on y-axis and combustible gases are shown on x-axis from 0% to 100%. Point zero represents 100% blackdamp, which is known as *dead air* or *residual gas*, which contains CO_2 and N_2. All types of air, CH_4, and blackdamp mixture can be represented by a point on the triangle.

Some other explosive gases may take part in the explosion. Existence of coal dust makes methane explosion easier. When there is more than one explosive gas in the mine atmosphere, then the lower limit of explosion of the mixture is calculated by the Le Chatelier formula given in Equation 11.5 (Skochinsky and Komarov 1969):

$$X = \frac{100}{(P_1/N_1) + (P_2/N_2) + \ldots + (P_n/N_n)} \quad (11.5)$$

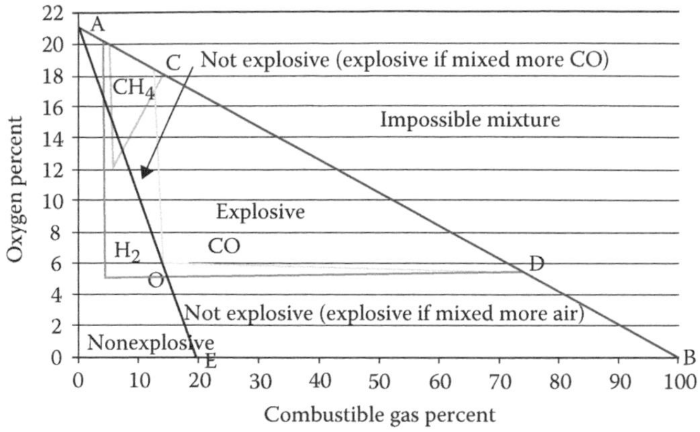

FIGURE 11.3 Coward triangle. (Data from Coward, H.F. and Jones, G.W., *Limits of Flammability of Gases and Vapors*, U.S. Bureau of Mines, Bulletin 503, U.S. Government Printing Office, 153pp., 1952.)

where:

P is explosive of gas concentration ($P_1 + P_2 + \ldots + P_n = 100\%$)

N is the lower explosion limit of corresponding gases

Actions to be taken for preventing methane explosion are listed below (Hartman et al. 1997):

1. *Prevention of accumulation of* CH_4
   a. Mechanical ventilation (fan) should be applied and natural ventilation should be avoided.
   b. Faces should be ventilated by the main fan.
   c. Ventilation design should be simple, and complex networks should be avoided.
   d. Ascending ventilation should be applied if dip angle $\alpha > 5°$.
   e. Proper and strong ventilation, doors, and regulators should be used.
   f. Leakage should be minimized.
   g. Proper and periodical gas and air monitoring velocity measurements should be conducted.
   h. If CH_4 reaches to 2%, working should be stopped.
2. *Prevention of ignition of* CH_4
   a. There should not be open flame.
   b. Battery lamps should be used for illumination.
   c. Blasting operations should be minimized, where:
      i. Permissible explosives should be utilized.
      ii. Proper stemming and stemming material should be designed.
      iii. Within 20 m distance before blasting, CH_4 concentration should be below 1%.
      iv. If needed, rock dusting should be applied.
      v. Delayed detonators should be used according to safety rules.
      vi. Flame-proof electrical equipment should be used.
3. *Minimizing effects of explosion*
   a. Ventilation sections should be designed separately.
   b. There should not be any leakage between upper and lower gateroads when explosion occurs.
   c. Dust explosion should be prevented after CH_4 explosion.
   d. Proper rescue team and equipment should be ready.
   e. Education about gases, explosions, rescue, and first aid should be given.

11.3.2 Outburst

Among other mine accidents, the most challenging one to tackle in underground workings are outbursts. Lama and Saghafi (2002) defined an outburst phenomena as an event in which coal or other rocks are ejected from an advancing face together with the emission of large amounts of gas. The major causes of outburst are stress disturbance, gas content, geological structures, and material properties of the coal. An unfavorable combination of these factors with mining methods may cause a disaster if not recognized at an early stage of mine development (Beamish and Crosdale 1998).

Gas outbursts can associate with coal and rock outbursts as well. Two gases are predominantly associated with coal seams: CH_4 and CO_2. CH_4 is generated as a result of the coalification process and/or microbial processes, while CO_2 is usually derived from an outside source such as magmatic activity (Juntgen and Karweil 1966). Gas compositions in the vicinity of these structures may vary from almost 100% CH_4 to almost 100% CO_2, if a coal field has been affected by igneous intrusions.

Instantaneous outbursts range in size from a few tons to thousands of tons of coal with corresponding gas volumes from tens of cubic meters to hundreds of thousands of cubic meters. CO_2 outbursts tend to be more violent, because of the greater sorption capacity for CO_2, although there is the added risk of a following explosion additional to a CH_4 outburst (Hargraves 1980).

Outbursts occur more in seams of high rank (bituminous, anthracites, and semi-anthracites), because high-rank coals have greater capacity to absorb gas at a given pressure, higher internal surface areas, lower porosities, and lower permeability (Lama and Saghafi 2002).

When coal is permeable and when stress levels are low and strength is high, high gas emissions occur without the ejection of coal or rock. Coal seams, which at lower stress levels show high gas emissions, invariably experience outbursts when the strength of coal is low or stress levels are high (Lama and Saghafi 2002).

There is a long history of outbursts of gas and coal in underground coal mining and they have occurred in most coal-producing countries. In the last 150 years, almost 30,000 outbursts have been recorded, and Bodziony and Lama (1996) had compiled the occurrence of outbursts in various countries as presented in Table 11.13. CO_2 bursts have been experienced only in Australia, Poland, Canada, Czech Republic, and France up to date (Lama and Saghafi 2002).

TABLE 11.13
Occurrence of Outbursts in Various Countries

Country	Mine Field(s)	Coal/Rock Burst	Gas Type	No. of Outbursts
Australia	Sydney basin	Coal	$CH_4 + CO_2$	>669
Belgium	Southern coalfield	Coal	CH_4	487
Bulgaria	Balkan	Coal	CH_4	250
Australia	Crows Nest Canmore Sydney	Coal and rocks	$CH_4 + CO_2$	411
China	Large number of coal fields	Coal and rocks	CH_4	>14,297
Czech Republic	Ostrava Slany Oslavany	Coal and rocks	$CH_4 + CO_2$	482
France	Various	Coal and other rocks	$CH_4 + CO_2$	>6814
Germany	Ruhr Ibbenbüren	Coal and other rocks	CH_4	359
Hungary	Mecsek	Coal and other rocks	CH_4	600
Japan	Hokkaido Kyushu	Coal	CH_4	920
Kazakhstan	Karaganda	Coal	CH_4	45
Poland	Upper Silesia	Coal	$CH_4 + CO_2$	1738
Rumania	Anima-Resica	Coal	CH_4	20
South Africa	Main Karoo	Coal	CH_4	5
Russia	Various	Coal and other rocks	CH_4	521
Taiwan	Taiwan	Coal	CH_4	60
Turkey	Zonguldak	Coal	CH_4	58
Ukraine	Donetsk	Coal and other rocks	CH_4	4689
United Kingdom	Various	Coal	CH_4	>219

Source: Bodziony, J. and Lama, R.D., *Sudden Outbursts of Gas and Coal in Underground Coal Mines*, 1996.

11.3.3 Mine Fires

Fire refers to a relatively slow combustion process without substantial pressure developments. Fire advances at a rate less than the speed of sound (343 m/s). Primary causes of mine fires are open flames; electricity and electrical equipment; mechanical friction; explosive dust, gas, vapor, and fluid; blasting; methane explosion; and spontaneous combustion. Fires can be open fires or concealed fires and/or oxygen-rich or fuel-rich fires.

Typical mine fires and their consequences are listed as follows:

- Carbon monoxide—incomplete combustion yields poisonous gas
- Carbon dioxide—complete combustion displaces oxygen
- Diesel particulate matter—unburned diesel fuel causes carcinogen matters
- Carcinogen products—products from the chemicals to treat flame-resistant belts and cables, however, will be released at high temperatures of a fire
- Smoke-unburned materials—can contain all of the above

In case of a fire, all people in the fire area should be notified and self-rescuer device should be used until fresh air is accessible. All firefighting equipments shall be maintained in a usable and operative condition. Chemical extinguishers shall be examined every six months and the date of the examination shall be written on a permanent tag attached to the extinguisher. Table 11.14 presents the color-coded symbols of fire extinguishers (Guyaguler et al. 2005).

Spontaneous combustion (sponcom) is also a type of fire, which is defined as the incomplete combustion. Sponcom is a

TABLE 11.14
Symbols of Fire Extinguishers

Symbol	Class of Fire	Type of Fire	Materials
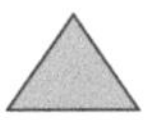	Class A	Ordinary combustible	Wood, coal, paper, cloth, and plastics
	Class B	Flammable liquids	Gasoline, diesel fuel, kerosene, and grease
	Class C	Electrical	Combustible materials in electrical equipment
	Class D	Metals	Magnesium and titanium

Source: Guyaguler, T. et al., *Occupational Health and Safety in Mining Industry*, Middle East Technical University, ODTU Basim İsligi, Ankara, 2005.

self-heating process of coal or any easily oxidizable substance such as sulfur due to auto-oxidation at atmospheric temperature. It leads to combustion of fuel at proper combinations of air movement, fuel, and material.

Factors affecting liability of coal to spontaneous combustion are surface area of coal, coal rank, volatile content, petrographical composition of coal, temperature and heat conductivity of coal, presence of pyrite, ash, moisture, concentration of oxygen in contact, and methane content of coal.

For the prevention of spontaneous combustion, providing careful and systematic mining operations is very important. In order to decrease the number of accidents, proper development work and coal mining techniques should be applied. Mine ventilation should be controlled closely and carbon monoxide and heat, which indicate the starting of fire, should be monitored.

The following should be satisfied as far as ventilation is concerned (Hartman et al. 1997):

- All active mine workings should be adequately ventilated.
- The mine ventilating pressure should not be very large.
- Short circuiting and uncontrolled circulation must be eliminated. Systematic air leakage tests by quantity surveys and smoke tube test should be conducted in workings of high-risk seams.
- All roadways in coal mines should be ventilated properly and abandoned roadways sealed off.
- Ventilation stoppings, doors, and regulators should be correctly placed.
- Air crossing should be airtight and should be constructed out of fire-proof material.
- Mine ventilation should be effectively supervised.

The most effective way to eliminate the risk of sponcom is methane drainage. Methane drainage is a process of collecting and removing gas mixtures from underground and bringing it to the surface. Various methane drainage techniques and their associated recoveries are tabulated in Table 11.15 (Hartman et al. 1997).

11.3.4 Blasting and Explosive Safety

Blasting and explosive safety is always of paramount concern to every mine involving blasting operations. Proper selection, handling, and use of explosives must be carefully planned and thoughtfully executed. As Flanagan and Santis (2001) stated that each individual involved in the process should be knowledgeable about the characteristics of the particular products being used and proper procedures. All members of the blasting crew should be experienced, and employing inexperienced and untrained blasters should always be avoided (Verakis 1988).

In surface mining, improper use of explosives and uncontrolled blasting may cause damage to structures and disturbance to people living and working in the blasting area. Three types of damage that can be caused by blasting include (Duncan and Mah 2005):

1. *Ground vibration*: Structural damage induced by shockwave spreading out from the blast area
2. *Flyrock*: Impact damage by rock ejected from the blast
3. *Airblast and noise*: Damage due to overpressure generated in the atmosphere

Figure 11.4 (a) and (b) represents hypothetical plot of measured particle velocity versus scaled distance from blast to determine attenuation constants k and β and typical blast vibration control diagram for residential structures, respectively (Duncan and Mah 2005).

The regulations permit only permissible explosives, approved sheathed explosive units, and permissible blasting units to be taken or used in underground coal mines. Black blasting powder, aluminum-cased detonators, and detonators with aluminum leg wires and safety uses are prohibited. Explosives, sheathed explosive units, and blasting units must meet certain design and performance requirements to be approved for use. There is a close relationship between peak particle velocity and the associated damage. Table 11.16 presents the peak particle velocity threshold damage levels.

TABLE 11.15
Methane Drainage Techniques

Method	Description	Methane Quality	Recovery
Vertical wells	Drilling from surface to coal seam several years in advance of mining	Recovers nearly pure methane	Up to 70%
Gob wells	Drilling from surface to a few feet above coal seam just prior to mining	Recovers methane that is sometimes contaminated with mine air	Up to 50%
Horizontal boreholes	Drilling from inside the mine to degasify the coal seam	Recovers nearly pure methane	Up to 20%
Cross-measure boreholes	Drilling from inside the mine to degasify surrounding rock strata	Recovers methane that is sometimes contaminated with mine air	Up to 0%

Source: Hartman, H., Ramani, R.V., Mutmansky, J.M., and Wang, Y.J., *Mine Ventilation and Air Conditioning*, 3rd edition. 1997. Copyright Wiley-VCH Verlag GmbH & Co. KGaA. Reproduced with permission.

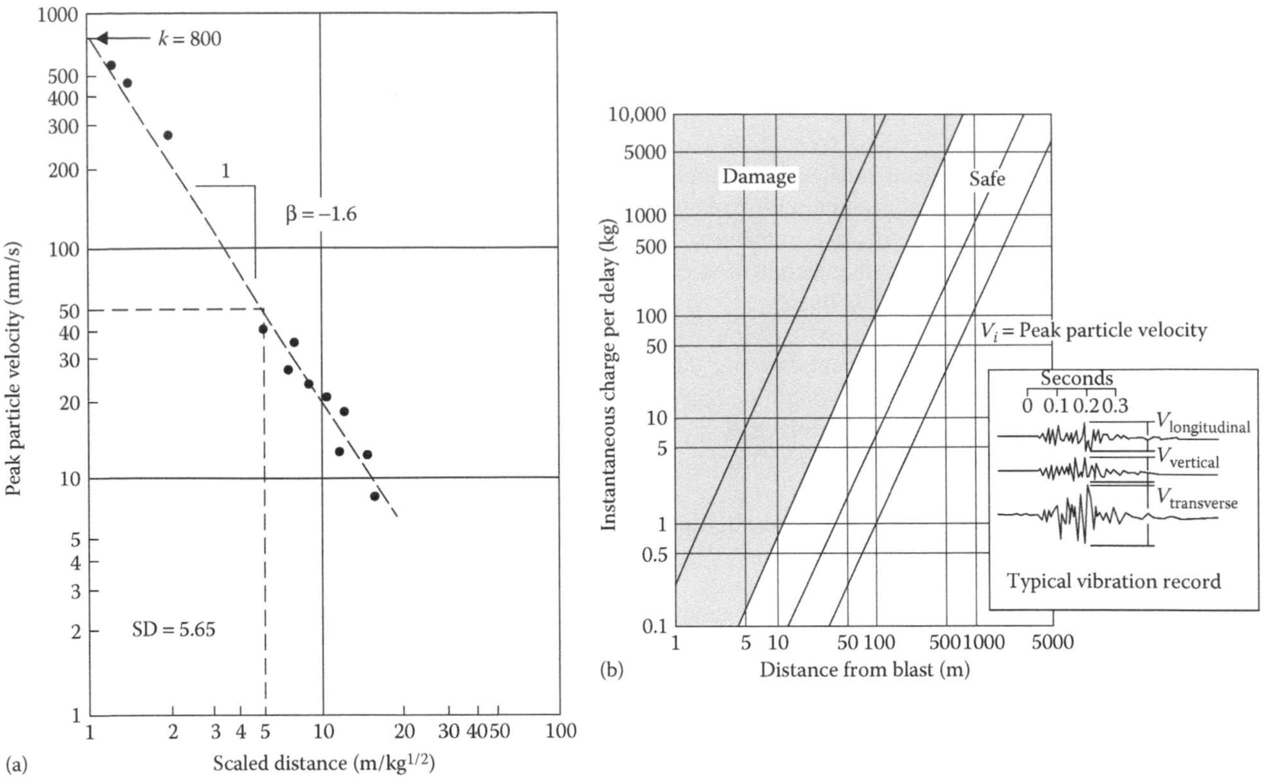

FIGURE 11.4 (a) Hypothetical plot of measured particle velocity versus scaled distance from blast to determine attenuation constants k and β and (b) typical blast vibration control diagram for residential structures. (Data from Duncan, C.W. and Mah, C.W., *Rock Slope Engineering*, 4th edition, Taylor & Francis Group, New York, 431 pp., 2005.)

TABLE 11.16
Peak Particle Velocity Threshold Damage Levels

Velocity (mm/s)	Effect/Damage
3–5	Vibrations perceptible to humans
10	Approximate limit for poorly constructed and historic buildings
33–50	Vibrations objectionable to humans
50	Limit below which risk of damage to structures is very slight (less than 5%)
125	Minor damage, cracking of plaster, and serious complaints
230	Cracks in concrete blocks

Source: Duncan, C.W. and Mah, C.W., *Rock Slope Engineering*, 4th edition, Taylor & Francis Group, New York, 431 pp., 2005.

Moreover, due to poor drill alignment and geologic conditions, flyrock problem can be encountered in surface mining. In order to control flyrock, stemming and burden dimensions should be designed carefully and blasting mats should be used (Duncan and Mah 2005). In addition to ground vibration and flyrock, air blast and noise may also be caused by improper blasting practices, especially close to the blasting area. Factors causing air blast and noise include blast holes, poor stemming, uncovered detonating cord, venting of explosive gases along cracks in the rock, and the use of insufficient burdens. Figure 11.5 illustrates response of structures and humans to sound pressure levels (Ladegaard-Pedersen and Dally 1975).

11.3.5 Haulage Safety

Material handling and powered haulage have been one of the leading causes of mine accidents. According to Mining Safety and Health Administration's accident statistics, more than 7000 accidents have been attributed between 1986 and 1995 related to conveyors in underground coal mines (MSHA 1996). Many of the accidents are caused by the head or tail pulleys, or return idler, walking around or passing over or under moving conveyors, slips and falls around or onto a moving conveyor. In some mines, belt conveyors are used for transportation of workers. It can be hazardous because of loose clothing and hair. Other mining equipments such as load-haul-dumps and continuous miners may also pose some threats in terms of safety due to confined workplaces, limited visibility, and lack of communication between operators and their surrounding (el-Bassiouni 1996 in Backer and Holt 2001).

In order to minimize the potential for material haulage-related accidents and injuries, safe working practices and procedures should be adopted by installing stop-start switches, guardrails, toe-boards, and stop cords around conveyors and conducting proper maintenance of machines and equipment.

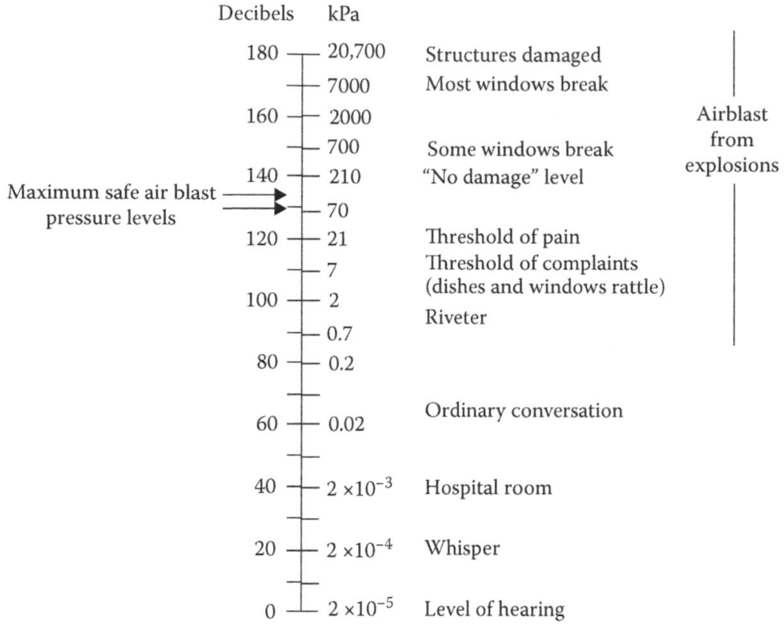

FIGURE 11.5 Response of structure and humans to sound pressure level. (Data from Ladegaard-Pedersen, A. and Dally, J. W., *A Review of Factors Affecting Damage in Blasting*, Report to the National Science Foundation, Mechanical Engineering Department, University of Maryland, 170 pp., 1975.)

11.4 ANALYSIS OF MINE ACCIDENTS

An accident is defined as an undesired and unplanned event that may result in personal injury and/or property damage. An incident is an occurrence or event that interrupts normal procedure. During an accident, a person's body comes into contact with or is exposed to some objects, other person, or substance, which is injurious; or the movement of a person causes injury or creates the probability of injury. An injury may be defined as a harmful condition sustained by the body as the result of an accident. It can take the form of an abrasion, a bruise, a laceration, a fracture, a foreign object in the body, a puncture wound, a burn, or an electric shock. Occupational disease is an unhealthy condition caused to person by exposure to unsafe working conditions.

11.4.1 Causes of Accidents

When the causes of accidents are analyzed, there are three levels of causes: basic causes, direct causes, and indirect causes. Basic causes are related mostly to management's safety policies and decisions, and lack of enforcements and legal provisions. The list of basic causes related to management's safety policies and decisions is as follows (Metzgar 2001 and Guyaguler et al. 2005):

- Health and safety policy is not written, not signed by top management, not distributed to each employee, and not reviewed periodically.
- Health and safety procedures do not provide for written manual, safety meeting, adequate housekeeping, preventive maintenance, safety inspection, accident investigations, job safety analyses, medical surveillance, and reports.
- Health and safety is not considered in the procurement of supplies, equipment, and services.
- Inadequate personal practices regarding employee selection, training, assignment, job observation, communication, assigned responsibility, and accountability.

The list of basic causes related to personal factors is given as follows:

- *Behavioral factors*: Accident repeater, risk taking, and lack of hazard awareness
- *Experience factors*: Insufficient knowledge, accident record, inadequate skills, and unsafe practices
- *Physical factors*: Size and strength
- *Mental factors*: Emotional instability, alcoholism, depression, and drug usage
- *Motivational factors*: Needs and capabilities
- *Attitude factors*: People, company, and job

The list of basic causes related to environmental factors is as follows:

- *Unsafe facility designs*: Mechanical layout, electrical systems, hydraulic systems, and air conditioning
- *Unsafe operating procedures*: Normal and emergency
- *Unsafe projections*: Physical plant, equipment, supplies, and procedures
- *Unsafe location factors*: Geographical area, terrain, surrounding, access roads, and weather conditions

Indirect causes are subclassified into two groups: unsafe acts and unsafe conditions. The following can be regarded as unsafe acts:

- Failing to use personal protective equipment
- Failing to warn coworkers
- Engaging in horseplay
- Lifting improperly or not following procedures
- Loading or placing equipment or supplies improperly
- Making safety devices inoperable
- Operating equipment without authority or at improper speed
- Serving equipment in motion or taking an improper working position
- Using drugs and alcoholic drinks
- Using defective equipment
- Using equipment improperly

On the other hand, unsafe conditions involve the following:

- Defective tools
- Equipment or supplies
- Fire and explosion hazards
- Excessive noise or radiation exposure
- Hazardous atmospheric conditions such as gases, dusts, fumes, and vapors
- Inadequate warning systems
- Inadequate supports or guards
- Poor illumination and ventilation
- Using defective equipment or using equipment improperly
- Ergonomic hazards
- Environmental hazards
- Inadequate housekeeping
- Blocked walkways
- Improper or damaged personal protective equipment
- Inadequate machine guarding

Direct causes of accidents are unplanned release of energy and/or hazardous material. Energy sources can be mechanical, such as machinery, tools, moving objects, compressed gases, and explosives; electrical sources can be uninsulated conductors and high-voltage sources; chemical sources are acids, fuels, bases, and reactive materials; thermal sources are flammable and nonflammable; radiation sources are noise, X-rays, lasers, microwaves, and radioactive materials. Hazardous materials as a direct cause of accidents in coal mining are compressed or liquefied gas, flames and hot surfaces, corrosive material, flammable solid, liquid, gas material, oxidizing material, poison, radioactive material, dust, and explosive. The causes of accidents are summarized in Figure 11.6.

Research studies revealed that 88% of the accidents are caused by human error, 10% of accidents are caused by machinery, and only 2% of accidents are caused inexplicably (Heinrich 1931).

11.4.2 Accident Analysis

Commonly used trends, which are usually evaluated in terms of accidents and injuries per man-shift, per million man-hours,

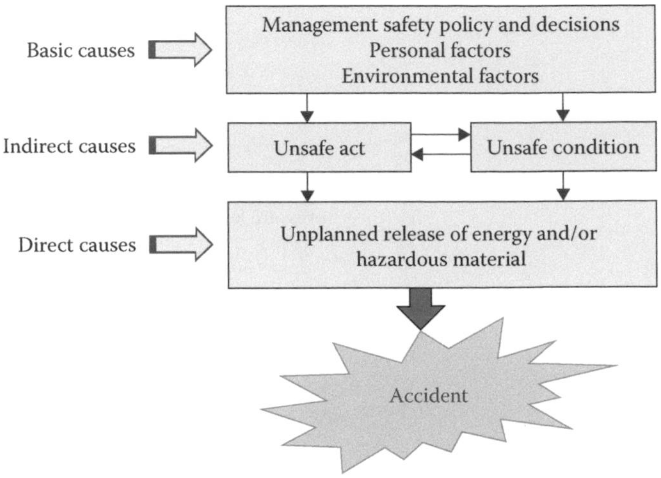

FIGURE 11.6 Causes of accidents. (Modified after Guyaguler et al., *Occupational Health and Safety in Mining Industry,* Middle East Technical University, ODTU Basim İsligi, Ankara, Turkey, 2005.)

and per million tons of material moved, to analyze accidents and compare the accident rates of various organizations are as follows (Equations 11.6 through 11.9):

- *Disabling injury frequency rating*: Stands for number of disabling injuries per hour for each million worker (Guyaguler et al. 2005):

$$\text{Disabling injury frequency rating} = \frac{\text{Number of disabling injuries} \times 1{,}000{,}000}{\text{Hours of exposure (or tonnage)}} \quad (11.6)$$

- *Fatality injury frequency rating*: Stands for number of fatalities per hour for each million worker (Guyaguler et al. 2005):

$$\text{Fatality injury frequency rating} = \frac{\text{Number of fatalities} \times 1{,}000{,}000}{\text{Hours of exposure (or tonnage)}} \quad (11.7)$$

- *Accident severity rate*: Stands for loss of working days due to occupational accidents per hour of each 1000 men (Guyaguler et al. 2005):

$$\text{Accident severity rate} = \frac{\text{Loss of working days due to accidents} \times 1000}{\text{Number of workers} \times \text{average working period}} \quad (11.8)$$

- *Accident frequency rate*: Stands for the number of accidents per hour for each million workers (Guyaguler et al. 2005):

$$\text{Accident frequency rate} = \frac{\text{Total number of accidents} \times 1{,}000{,}000}{\text{Number of workers} \times \text{average working period}} \quad (11.9)$$

11.4.3 Accident Reporting

Accident reporting is necessary for the statistical and economical analyses in which accident injury data are used. By having standard report form for the accidents for the period of time, the accidents can be analyzed considering the distribution of accidents according to accident types, job titles, part of the body injured, the procedure in which accident happened, age groups, days of week, type of shift, and working hours.

Incident report should contain the following information:

Part A—Introductory information: Names of the organization, section, and the mine.

Part B: Death, serious injury, fall, choked, gas or dust explosion, mine fire, explosives, falls of roof, transportation system (accident related to rope transportation, conveyor accidents, and rail system), person who recorded the accident, the recording date of the accident, and measures against its occurrence.

Part C: Surface/underground location, mining methods, date of accidents, hour of accident, starting hour of shift, all the reasons that caused the accident, equipment involved, names of the witnesses, number of reportable injury or sickness as a result of accident, name of the injured, date of birth, gender, insurance number, job title, injury details, other occupational diseases, type of work, present job experience, present mine experience, and so on.

11.4.4 Cost of Accidents

Cost of mine accidents are not limited by economical losses. There are consequences such as public opinions, industrial relations, and legal consequences, including improvement notices, prohibition notices, criminal prosecution, unlimited fines, and even imprisonment. These consequences cannot be easily externalized unlike economical consequences. However, mostly these hidden costs are overwhelmingly greater than unhidden costs.

Economic consequences can be itemized as costs to injured person, costs to company, and costs to nation. Costs to injured person include pain and suffering, mental strain, loss of earning, extra expenditure, loss of life, disablement, incapacitation, effect on dependants, and loss of leisure activities. Costs to company may be classified as insured costs, such as employer's liability and public liability, and uninsured costs, such as medical and first aid, loss of production, temporary staff, recruitment and training, overtime to make up, loss of sales, investigation and legal procedures, damage to equipment, replacement of parts/labor, effect on morale, rise of compensation amount, difficulty in supplying skilled people with right attitudes and motivation, strain in labor relationship, decrease in efficiency and productivity, loss of a key worker for a long period of time or even permanently, damage given to the plant, waste of materials, interruptions of work, disorganization, decreased efficiency, and production loss.

Moreover, the costs of accidents can be classified as direct and indirect costs. Usually indirect costs are not seen, but they account for the significant portion of the total costs. In some cases, the indirect costs of the accident may be four times of the direct costs.

Direct costs are as follows:

- Cash benefits payable under the related laws
- Medical benefits
- Disablement expense payable under the social insurance scheme
- Other benefits payable under the company's own scheme

Indirect costs are as follows:

- Time lost because of the injured employee
- Time lost because of other employees who stop work out of curiosity, out of sympathy, and to assist injured employee
- Cost of time lost by foremen, supervisor, or other executives when assisting the injured person(s); investigating the cause of accident; arranging for the injured employee's production to be continued by some other worker; selecting, training, or breaking-in a new worker to replace the injured man; and preparing accident reports required by law or attending hearings related to compensation claims
- Cost of time spent by the first-aid attendant and hospital department staff
- Cost due to damage to machine, tools, or other property or due to the spoilage of material
- Incident cost due to interference with production, failure to fulfill orders in time, payment forfeits, and other similar causes
- Cost of employer in continuing the wages of the injured worker who is not yet fully recovered
- Cost due to loss of profit or the injured worker's efficiency and production
- Cost of the subsequent injuries that occur in consequence of the excitement or weakened morale to the original accident
- Cost of light, rent, and other such items, which continue to be incurred while the injured employee continues to be nonproductive

11.5 MINE SAFETY AND HEALTH MANAGEMENT SYSTEM

11.5.1 Risk Management

Risk is defined as a measure of both the likelihood and the consequences of a hazard associated with a mining activity. Effective risk management essentially entails five stages: (1) identifying risks, (2) analyzing risks, (3) reducing or eliminating risks, (4) financing risks, and (5) administrating the process (Grayson and Watzman 2001).

There exists an inherent risk in coal mining industry because of potentials for explosions, mine fires, often-confined nature of the workplace, and the proximity to dangerous equipment. However, these risks can be managed by employing an effective mine safety and health management system.

Risk management is a proactive process that helps for continuous improvement in workplaces. It should be well planned and systematically cover all reasonably foreseeable hazards and associated risks.

The objective of risk management can be classified into two groups: preloss and postloss objectives (Brauer 1990). Preloss objectives include minimized expenditures for loss, reduced anxiety of loss, and meeting externally imposed obligations and social responsibility. On the other hand, postloss objectives include economic survival, continuity of operations, earnings stability, continued growth, and meeting social responsibility. Comprehensive explanation of the major stages of risk management with case studies was presented by Grayson and Watzman (2001) and Karmis et al. (2001), which readers can refer to for further information. It should be kept in mind that lack of good risk management practices results in ineffective use of resources.

11.5.2 Emergency Preparedness and Response

Emergency preparedness is a work of safety and health engineers and mine rescue teams, and they should be prepared for properly handling a violent incident in the workplace. It requires careful planning. The team should have only one mission—immediate response to accident—and should be chaired by a safety and health professional. Properly and adequately trained mine rescue teams are the heart and soul of emergency preparedness and response (Launhardt 2001). Emergency preparedness and response plan is designed as a guideline for all types of emergencies, including fires, explosions, entrapment of personnel, loss of electrical power, and natural disasters. Complete plan includes detailed information about the mine and operations.

Key issues in an emergency preparedness and response plan are as follows (Launhardt 2001):

1. *Identifying the potential for emergencies*: The type of emergency that is most likely to occur (roof collapse, mine fire, explosion, etc.) should be identified. Emergency preparedness should be focused on the identified potential emergencies according to the likelihood of hazards.
2. *Secondary escape ways and refuge chambers*: Secondary escape ways are required under the law. Reliability of secondary escape ways should be verified by mine management. Occasional testing should be done. Refugee chambers are required by law, while a secondary exit is being developed when mine personnel cannot reach the surface within an hour. Effective emergency preparedness requires regular inspections of refugee chambers and their contents.
3. *Ventilation reliability*: The ability to control mine ventilation is often the most important element in providing for the safety of personnel in a mine during evacuation. Emergency preparedness must include having up-to-date mine ventilation maps at headquarters for a mine emergency.
4. *Selecting equipment for emergency use*: Emergency preparedness must include equipment and supplies that will adequately protect personnel in any foreseeable emergency. Availability of emergency supplies should be determined. For instance, telephone service must include one phone system that has an emergency power supply to enable use if electrical power fails.
5. *Emergency response plan*: In an emergency, a mine needs its key personnel or their designees at a central location. Each key person must delegate responsibility under the emergency response plan when he/she is unavailable. Emergency response plan should incorporate a brief description of the role played by each person involved in the emergency response. Personnel not essential to the emergency must be excluded.
6. *Mine rescue*: Mine rescue teams are vital component of emergency preparedness and response. The team should be organized to handle the following tasks:
   a. Undergo trauma response training
   b. Handle media interaction
   c. Operate telephone and communication teams
   d. Develop and implement, as necessary, an emergency evacuation plan
   e. Establish a backup communication system
   f. Calm personnel after an incident
   g. Debrief witnesses after an incident
   h. Ensure that proper security procedures are established, kept up-to-date, and enforced
   i. Help employees deal with posttraumatic stress
   j. Keep employees informed about workplace violence as an issue, how to respond when it occurs, and how to help prevent it
7. *Ventilation planning*: It is essential that ventilation data are kept up-to-date and that current ventilation schematics are on hand at headquarters. The ventilation engineer must have the skills and training to quickly determine how changes in the mine that occur during an emergency will affect air flows throughout the active portions of the mine.
8. *Mine maps*: Up-to-date maps of all active portions of the mine must be on hand at headquarters. Maps of inactive mine openings must also be provided if there is a possibility of use by mine rescue teams. If there are adjoining mines, maps of such mines must also be at headquarters.
9. *Mine alarm system*: Alarms and signals to alert employees must be identified; this may include audio alarms, highly visible lights, and/or a public address system. Management and employees must know what actions to take when an emergency alarm is activated.

10. *Surprise drills*: It is always important to run drills without giving an advance notice. Stimulating fires in various parts of the mine and challenging personnel to respond is a key in firefighting emergency preparedness.

ACKNOWLEDGMENT

The authors express their special thanks to their families for their endless support and understanding throughout the preparation of this chapter.

REFERENCES

AIOH, Australian Institute of Occupational Health, http://www.aioh.org.au/index.aspx, Last accessed on July 2014.

Backer, R.R. and Holt, C.M.K. 2001. Haulage, In: Karmis, M. (Ed.), *Mine Safety and Health Management*, The USA, Society for Mining, Metallurgy, and Exploration, Inc. (SME), 453 pp.

Beamish, B.B. and Crosdale, P.J. 1998. Instantaneous outbursts in underground coal mines: An overview and association with coal type. *International Journal of Coal Geology* 35: 27–55.

Bodziony, J. and Lama, R.D., 1996. *Sudden Outbursts of Gas and Coal in Underground Coal Mines*, ACARP Project No. C4034, March, Australia, pp. 677.

Brauer, R.L. 1990. *Safety and Health for Engineers*. New York, Van Nostrand-Reinhold.

Coward, H.F. and Jones, G.W. 1952. *Limits of Flammability of Gases and Vapors*, U.S. Bureau of Mines, Bulletin 503, U.S. Government Printing Office, Pittsburgh, PA, 153 pp.

Duncan, C.W. and Mah, C.W. 2005. *Rock Slope Engineering*, 4th edition, Taylor & Francis Group, New York, 431 pp.

el-Bassiouni, S. 1996. *Continuous Haulage Systems*. Holmes Safety Association Bulletin. Mine Safety and Health Administration, pp. 10–11.

Flanagan, S.J.P. and Santis, L. 2001. Mining with explosives: Safety first. In: Karmis, M. (Ed.), *Mine Safety and Health Management*, The USA, Society for Mining, Metallurgy, and Exploration, Inc. (SME), 453 pp.

Forbes, J.J. and Grove, G.W. 1954. *Mine Gases and Methods for Detecting Them*. USBM M.C., No.33.

Goldstein, M. 2008. Carbon monoxide poisoning. *Journal of Emergency Nursing* 34(6): 538–542. doi:10.1016/j.jen.2007.11.014. PMID 19022078.

Grayson, R.L. and Watzman, B. 2001. History and overview of mine health and safety. Karmis, M. (Eds.), *Mine Safety and Health Management*, The USA, Society for Mining, Metallurgy, and Exploration, Inc. (SME), 453 pp.

Guyaguler, T., Karakas, A., and Gungor, A. 2005. *Occupational Health and Safety in Mining Industry*. Middle East Technical University, ODTU Basim İsligi, Ankara.

Hargraves, A.J. 1980. A review of instantaneous outburst data. Proc. The Occurrence, Prediction and Control of Outbursts in Coal Mines. The Australasian Institute of Mining and Metallurgy, Melbourne, pp. 1–18.

Hartman, H.L. 1982. *Mine Ventilation and Air Conditioning*, 2nd edition, John Wiley & Sons, New York.

Hartman, H., Ramani, R.V., Mutmansky, J.M., and Wang, Y.J. 1997. *Mine Ventilation and Air Conditioning*, 3rd edition, John Wiley & Sons, New York.

Heinrich, H.W. 1931. Industrial Accident Prevention. McGraw-Hill, New York.

Howard, L.H. 1961. *Ventilation and Air Conditioning*, The Ronald Press Company, New York.

Jensen, P., Jokel, C.R., and Miller, L.N. 1978. *Industrial Noise Control Manual*, revised Edition, U.S. Department of Health, Education, and Welfare, Public Health Service, Center for Disease Control, National Institute for Occupational Safety and Health, Division of Physical Sciences and Engineering, Cincinnati, Ohio, 357 pp.

Juntgen, H. and Karweil, J. 1966. Formation and storage of gases in bituminous coal seams. Part 1. Gas formation. *Erdoel Kohle Erdgas Petrochim* 19, 251–258.

Karmis, M. 2001. Mine Health and Safety Management, Society for Mining, Metallurgy, and Exploration, Inc., Littleton.

Ladegaard-Pedersen, A. and Dally, J.W. 1975. *A Review of Factors Affecting Damage in Blasting*. Report to the National Science Foundation. Mechanical Engineering Department, University of Maryland, 170 pp.

Lama, R. and Saghafi, A. 2002. *Overview of Gas Outbursts and Unusual Emissions*. Coal Operators Conference, pp. 1–15.

Launhardt, R.E. 2001. Emergency preparedness and response. In: Karmis, M. (Ed.), *Mine Safety and Health Management*, The USA, Society for Mining, Metallurgy, and Exploration, Inc. (SME), 453 pp.

Lawrie, W.B. 1967. *Some Aspects of Dust, Dust Sampling, the Interpretation of Results*, ILO, Occupational Safety and Health Service, No. 8, Geneva, Switzerland.

Le Roux, W.L. 1979. *Mine Ventilation Notes for Beginners*. 3rd edition, The Mine Ventilation Society of South Africa.

Metzgar, C.R. 2001. Causes and Effects of Loss. In: Karmis, M. (Ed.), *Mine Safety and Health Management*, The USA, Society for Mining, Metallurgy, and Exploration, Inc. (SME), 453 pages.

Ministry of Environment and Forestry. 2011. Environmental Noise Measurement and Assessment Guidelines, 106 pp.

MLSS, Ministry of Labor and Social Security of Turkey. 2012. Yeraltı ve Yerüstü Maden İşletmelerinde İş Sağlığı ve Güvenliği Rehberi, 140 pp.

Montgomery, H.E. 2008. *Oxygen Deficiency Hazard, Safety Booklet*. https://www.jlab.org/accel/safetylb/ODH-book.pdf, Last accessed on July 2014.

MSHA. Revised 1996. Metal-Nonmetal Monitor 1(8), 1 p.

MSHA Accident Statistics, http://www.msha.gov/stats/centurystats/coalstats.asp, Last accessed on July 2014.

Murrell, K. 1987. *Practical Ergonomics*, New York, John Wiley & Sons.

National Coal Board. 1970. *"Noxious Gases Underground" A Handbook for Colliery Managers*, Ipswich, W.S. Cowell.

Poltev, M.K. 1985. *Occupational Health and Safety in Manufacturing Industries*. Mir Publishers.

Skochinsky, A. and Komarov, V. 1969. *Mine Ventilation*, Moscow, Russia, Mir Publishers.

Schofield, D., Hollands, R., and Denby, B. 2001. Mine safety in the twenty-first century: The application of computer graphics and virtual reality. In: Karmis, M. (Ed.), *Mine Safety and Health Management*, The USA, Society for Mining, Metallurgy, and Exploration, Inc. (SME), 453 pp.

TSI 2007. Turkish Statistical Institute, http://www.sendika.org/2010/08/enerji-sektoru-ve-sendikal-mucadelenin-olanaklari-enerji-sen-yonetim-kurulu/, Last accessed on July 2014.

Verakis, H.C. 1988. *MSHA's New Regulations for Explosives Used in Coal Mines, Mine Safety and Health Administration*, Approval and Certification Center, 780–785.

Section III

Coal Conversion Processes

12 Coal Beneficiation

T. Gouri Charan

CONTENTS

Abstract: As it comes from the mine, coal is known as *run-of-mine* (ROM) coal and consists of a range of sizes from chunks to small particles mixed with some dirt and rocks. In most cases, this ROM coal needs some degree of preparation and beneficiation to meet certain market requirements as sizes, ash, sulfur, moisture, and heating values. *Coal beneficiation* is a generic term that is used to designate the various operations performed on the ROM coal to prepare it for specific end uses, such as feed to a coke oven or a coal-fired boiler or to a coal conversion process without destroying the physical identity of the coal. Coal beneficiation is now recognized as a combination of science, art, and engineering, recognized in its own right as a vital link between the production and marketing of coal.

The principal coal-cleaning processes used today are oriented toward product standardization and ash reduction, with increased attention being put on sulfur reduction in some of the countries. Coal preparation in commercial practice is currently limited to physical processes. In a modern coal-cleaning plant, the coal is typically subjected to (1) size reduction and screening, (2) separation from its impurities, and (3) dewatering and drying. A modern installation is a carefully designed assembly of component machines for handling, screening, cleaning, dewatering, and blending of coal, and for water clarification. The function of the modern plant is to produce the maximum yield of clean coal of suitable quality for the consumer at an economic cost. Up to the present, commercial practice has largely relied on physical coal-cleaning processes to beneficiate coals. Chemical, microbiological, and other novel coal beneficiation processes are of recent origin and still at various levels of process development.

12.1 BACKGROUND TO COAL PREPARATION AND CLEANING

The quality of mined coal is highly variable because coal is a heterogeneous material and because of the unavoidable incorporation of non-coal bands and possibly a certain amount of out-of-coal seam rock material. *Preparation* is the term used to describe the production of sized coals for different markets, with the use of separation processes to minimize the presence of mineral matter. *Cleaning* implies optimizing the separation processes to remove the maximum amount of unwanted impurities.

The processes used were simple and inexpensive but relatively inefficient. The principal purpose was to remove mineral matter, both lumps of stone that had been mined and smaller particles arising from coal breakage. This breakage was avoided, wherever possible, and fines cleaning were often not attempted. Washing water was commonly cleaned by settling in open circuits with no attempt to recover the fines. Consequently, waste piles and tailings ponds from past operations may contain enough organic coal to be worth rewashing and recovering. Because of the rising cost of coal mining, and the increased value of fuel generally, there has been strong interest in improving the recovery of organic coal during preparation. More importantly in terms of its effects on coal-cleaning practices, coal users have become increasingly aware of the cost to their operations of the presence of impurities in the coal feed. The Japanese steel industry is an example of a user that has set increasingly stringent specifications for the coals it buys. Power utilities around the world are becoming aware of the operating problems (and costs) that result from the presence of the mineral matter in the coal they use and from feedstock variability. New uses for coal are being developed and investigated and several of these are adversely affected by the presence of minerals (Konar et al. 1997).

All these factors have promoted extensive work on improved processes to clean coals. The main developments have been in the operation of the fines cleaning circuits of conventional coal preparation/cleaning plants, and improved processes to clean/separate material that has been milled, possibly to a very fine size, to promote liberation of the mineral matter.

Unit operations of coal preparation can broadly be divided into coal cleaning and three associated processes: (1) the pretreatment of feed coal, (2) subsequent treatment of products, and (3) the storage and loading of products. Coal cleaning is the key process in coal preparation to reduce mineral matter content. The introduction of mechanized, high-productivity extraction methods has resulted in run of mine coals that are finer, wetter, and dirtier than in the past, and have given rise to an increasing need to prepare or clean the material in some way before utilization.

Originally, coal preparation was confined to the hand picking of coarse discard from belts or tables and the crushing and screening of coal to get the appropriate size ranges required by the market at that time. Mechanical cleaning of coal to remove much of the impurity began in the second quarter of the nineteenth century, but it took more than 100 years for the practice to become widespread and develop into the multiple process preparation plants that are part of most current mining operations. Depending on the circumstances of production and consumption in each individual case, coal preparation may involve any combination of crushing, screening, and removal of a wide range of mineralogical contaminants from the various mined products.

12.2 NEED FOR COAL BENEFICIATION

Coal is defined as a sedimentary rock composed predominantly of solid organic materials with a greater or lesser proportion of mineral matter. It is derived from the accumulation of plant remains in sedimentary basins, and is altered to solid rock by heat and pressure applied during the basin's development. Its quality varies according to the content of ash, impurities, and volatile matter, which decreases as coal rank gets higher. It has a natural dark brown to black, graphite like appearance and is primarily used as a fuel. Types of coal according to increasing rank (in terms of hardness, purity, and heating value) are peat, lignite, sub-bituminous, bituminous, and anthracite. Globally, the reserves of sub-bituminous and bituminous coals are more, and the presence of inert material associated with the coal often referred to as *dirt* in various amounts is the prime concern (Narasimhan and Mukherjee 1999).

The need for coal beneficiation is more related to consumer demands than to the requirements of the producer, but it is also due to the development of greater environmental constraints than in the past. The introduction of mechanized, high productivity extraction methods has resulted in ROM coals that are finer, wetter, and dirtier than in the past, and given the rise to an increasing need to beneficiate or clean the coal in some way before use. The need to beneficiate coal is to address the following:

1. As a general rule, dirt in excess of specified limits affects the efficiency of utilization and the load factor, and it is more so when there are wide fluctuations in the quality and quantity of dirt.
2. The useless transport of dirt along with coal and its subsequent disposal as cinder or slag also cost money and results in some loss of sensible heat.
3. In thermal power stations, excess amount of ash in feed coal not only calls for additional capital investment but also causes frequent hazards in operation and maintenance.
4. In conventional iron and steel metallurgy, an increase in ash content of coke by 1% over a critical limit results in decrease of production by 3%–6% and involves an increase in coke consumption by 4%–5%.

12.3 HISTORY OF COAL PREPARATION

12.3.1 Global Scenario of Coal Preparation

The foundation for modern coal preparation technology was largely laid in Europe in 1915–1940. The use of coal throughout the world is increasing at a rapid rate due to worldwide energy shortages and increased demand for metallurgical, chemical, and synthetic fuel uses of coal. It is to be noted that not all of this coal will be washed in preparation plants. Only about one-third of the 3.3×10^9 metric tons of coal produced every year is at present cleaned.

The demand for coal preparation will grow along with increase in production for various reasons some of which are mentioned as follows:

1. Depletion of higher quality coal seams
2. Mechanized mining, which increases impurities in ROM coal
2. High cost of transportation, which makes it uneconomical to transport inert material
4. Market demands for higher quality coal
5. Higher mine costs, which make it imperative to improve coal washing techniques for optimized recovery
6. Environmental pollution control

The types of coal preparation plants being used today in various parts of the world are of almost endless varieties. Capacities may vary from as low as 70 to as high as 5,000 tph. They may use a simple screening and crushing operation or be equipped with very complex flow sheets. There is a wide variation in coal treatment that may be classified in accordance with level or degrees of preparation. Five levels of coal preparation may be summarized as follows:

Level I: Crushing and screening only
Level II: Coarse coal cleaning only

Level III: Coarse coal and simple fine coal cleaning (down to 0.5 mm)

Level IV: Coarse and fine coal cleaning plus cleaning of the minus 0.5 mm, with closed water recovery circuit.

Level V: Cleaning all sizes of coal, in multiple stages of sizing and crushing for maximum liberation, optimum yields, and producing two or more clean coal products, closed water recovery circuits.

The choice of process equipment involved in any of these levels will depend upon the type of coal being treated, the market requirements, and the economics or costs involved. In general, coal preparation, as practiced, is basically concerned with various densimetric or physical methods for the separation of impurities from coal.

There are many processes for cleaning coal in current use. (Osborne, 1988) Excluding the simple crushing and screening operations, the use of jig is still the most common. In the United States, based upon tonnage treated, 48% uses various types of jig followed by dense medium processes with 32%, Tables account for 11.5%, froth flotation 5% and other methods 3.5%. In the United Kingdom, some 60% of the coal is treated by jigs, 25% by heavy media methods, and 9% by other methods including froth flotation. In other parts of Europe, some 60% of coal is treated by jigs. In the erstwhile Soviet Union, jigging accounts for almost 50% of the washed coal produced, 25% is concentrated by heavy media, and approximately 15% is upgraded by froth flotation. The remaining 10% is cleaned by various methods, including trough washers and many other pneumatic cleaners. In Australia, about 45% of the coal is cleaned by jigs. The same percentage is cleaned by heavy media systems with the remaining 10% by froth flotation or other means, including tables.

In South Africa, where the coal is more difficult to treat, only 23% of the coal is treated by jigs, 64.5% by dense media methods, and 2.5% by froth flotation. In India, jigs account for 31% of total cleaning capacity, heavy media (including centrifugal separation) 66%, hydrocyclones and froth flotation 3% (Konar et al. 1997).

Thus, jigs are still the most common coal processing units closely followed by heavy-medium systems. In many cases, jigs are used as pre-washers and the final cleaning is done in heavy medium separators. Heavy media processes are rapidly gaining ground particularly for low relative density separations and for coal containing high percentage of middlings or near gravity materials (NGMs). Countries including Australia, Austria, Belgium, France, India, and South Africa have a decided preference for dense medium processes based either on gravity or on centrifugal separation. Among the countries where dense medium cyclone installations account for 8%–40% of total cleaning capacity, Australia tops the list followed by India, Japan, South Africa, Belgium, France, and the United Kingdom.

There has been an increase in the use of froth flotation not only to upgrade slurry forming fines below 0.5 mm but also to recover coal previously lost in tailings ponds. But all coals are not equally responsive to froth flotation. In the later situation, some of the newer plants are turning to water-only cyclones, spirals, and floatex separators.

12.4 COAL WASHABILITY

The extent of removal of free dirt or amenability of a coal to improvement in quality is known as *washability*. The normal procedure to assess the washability characteristics of a coal is to carry out float and sink tests in the laboratory on a representative sample of coal after crushing and screening it to proper size limits. Float and sink tests are universally used, which represent one of the basic coal-cleaning characteristics required in expressing the results of coal-cleaning tests. Most of the coal-producing countries have adapted the procedure of float and sink analysis as a laboratory method of producing basic data for designing coal-washing plant and subsequently for assessing its performance and efficiency (Konar et al. 1997).

The term *washability* implies the extent to which it is practicable to clean a coal, that is, to remove the dirt. Raw coal is primarily a mixture of pure coal substance (specific gravity: 1.25–1.29) and dirt (shale, sandstone, and other minerals of specific gravity: 2.70 and above). The pattern of distribution of this foreign matter in raw coal determines the washability characteristics of the coal.

Float and sink analysis is conducted by floating sized coal fractions in a series of liquid baths of increasing specific gravity generally ranging from 1.30 to 2.00. Various heavy media, including organic liquids, inorganic salts, and finely suspended solid particles for example, magnetite, are used to produce the desired specific gravities. The float and sink fractions in successive liquid baths of varying specific gravity are collected separately, dried and weighed, and subsequently analyzed for ash. This process has to be repeated until all the size fractions have been tested at desired relative densities. A float and sink analysis is performed on each individual size fraction then combined to obtain a complete float-sink profile for a wider size range. The percentage weights and the ash values of these density fractions separately are the basic data required for all information about the washability characteristics of the coal tested. The cumulative yield, ash sinks yield, and ash and characteristic yield of the fractions are then calculated from these data.

The specific objectives of the float and sink test are as follows:

- To know the cleaning potentialities of a coal
- To study the feasibility of a washery project
- To access the efficiency of different washing units
- To predict the practical yield

The standard washability curves namely, characteristics, floats, sinks, and yield gravity curves are usually drawn from the float and sink data to study the beneficiation characteristics of the coal under investigation. The procedure for construction of standard washability curves is well documented in various standards and text books (Osborne 1988). The brief inference, which may be drawn from the various curves, is explained as under.

12.4.1 Characteristics Curve

The basic data of laboratory float and sink test are directly used for construction of the characteristics curve. Characteristics curve is regarded as a parent curve that is constructed by plotting the determined ash percentage of individual gravity fractions against the yield of float up to mid-relative density of that fraction, expressed in percentages of the total coal. We must realize that the determined ash content of any particular specific gravity fraction represents the weighted average of ash contents coming from all the particles of those specific ranges.

12.4.2 Floats Curve

This curve is obtained by plotting the cumulative yield percent of floats at each specific gravity against the cumulative ash percent of floats at that specific gravity. As the float curve shows directly the recovery of clean coal at any desired ash content of the clean coal, it finds more frequent use in practice than the characteristics curve.

12.4.3 Sinks Curve

This curve is constructed by plotting cumulative yield percent of sinks at any specific gravity against the cumulative ash percent of sinks at that specific gravity. This is the complementary curve to total floats ash curve and gives directly the ash content of the total sinks corresponding to any yield level of the cleans or sinks. One may directly construct the sink curve by plotting the cumulative yield percent of sinks against the corresponding ash percentages from the basic float and sink data.

12.4.4 Yield Gravity Curve

The yield gravity curve shows the relationship between the specific gravity of separation and the total yield of clean coal, which floats at that specific gravity. The required specific gravity of separation, corresponding to any specified ash or yield level of clean coal may be provided by this curve.

12.4.5 Near Gravity Material

NGM in a coal is defined as the material lying between ±0.10 gravity of cut. The percentage of NGM at the operating density of cut, determines the easiness or difficulty in cleaning. The higher the percentage of NGM at the operating gravity the more difficult it becomes to clean the coal. Particles having specific gravity closer to the effective specific gravity of separation rise up (in case of float) or sink down (in case of rejects) at comparatively slower speed, and their course of movement is readily influenced by eddies or swirls as are encountered in commercial scale washing baths causing their misplacement of sink in floats or floats in sinks. An additional curve known as *NGM curve* may be drawn along with standard washability curves (Bird 1928).

12.4.6 Mayer's Curve or M-Curve

Till the year 1943, the use of standard washability curves was very popular. However, in 1943, Dr. F.W. Mayer of Germany developed a single curve, which could replace all the standard washability curves; it was named as *Mayer's curve* or simply *M-curve*. Almost all the requisite information provided by standard washability curves can be obtained from the single M-curve by precise geometrical construction.

The curve is constructed from float and sinks data by plotting the M-point values against cumulative weight percentage of floats on a vertical diagram in which the recovery of clean coal is directly read from the yield axis while the direction of vector represents the percent content of ash (Salama 1998).

This curve has the following principal advantages for use in practical coal washing.

- The most interesting feature of the curve is that with the help of one single curve, the entire requirement for the prediction of coal-cleaning results fulfilled by the standard washability curves can be met with.
- In a diagram of the same size as used for standard curves, the M-curve permits to read the figures of ash content with much greater accuracy.
- With the help of this curve the ash content of the middlings can also be determined with great accuracy, and so it is of utmost importance in cleaning practices.

12.4.7 Interpretation of Washability Data

The shape of characteristics curve of washability test characterizes the cleaning possibilities of a coal. If the curve is sharply angled, separation will be easy at relative density above the value at which the curve approaches the horizontal. If the curve is gently curved, then the coal is very difficult to washing or more truly the gradient at which the particles of coal are required to be separated is very much influenced by amount of NGMs present at the desired density of cut (Sanders and Brookes 1986).

Both shape and position of the characteristics curve are of significance insofar as the interpretation of the results is concerned. If the curve approaches a straight line and remains vertical, the coal can be absolutely unwashable in nature. If it is positioned in an inclined form and nearly takes the shape of a straight line right from its point of origin (corresponding to zero yield) up to the point of termination(corresponding to 100 yield) the coal supposed to be exceedingly difficult to wash without a clear cut point of separation. The greater concavity in the latter half of this curve implies better washability of a coal.

The floats ash curve directly shows the recovery of clean coal and it finds more frequent use in practice than the characteristics curve. The sinks curve is complementary to the floats ash curve corresponding to any yield level of the cleans or sinks. The yield gravity curve shows the relationship between the specific gravity of separation and the total yield of clean coal, which floats at that specific gravity. The required specific gravity of separation, corresponding to any specific ash or yield level of clean coal can be provided by this curve.

The advantage of drawing all the four curves on a common diagram are all the essential information required for studying the cleaning possibilities of a coal can be readily obtained by cross projection. For example, if one is interested to recover 15% ash clean coal, he/she has to read first from the floats curve the percent yield of cleans corresponding to 15% ash. Then from this yield point, a horizontal line is drawn to cut the sinks ash curve and the yield gravity curve. At the cut point of the sinks curve, the ash content of the sinks is read from the ash axis below and at the cut point of the yield-gravity cut, the required density of separation is read from the gravity axis above.

The standard washability curves help in estimating the recovery and quality of products. But for difficult-to-wash coals like that are mined presently in lower seams, conventional washability curves are too steep to provide sufficient information required for setting the cut points of different density separators in the plants. Mayer's curve is no doubt an improved representation of the washability data that explains in addition, the possibility of extracting middlings from an inferior grade coals or the blending of coals (Forrestor and Majumdar 1947).

When the combined calculations of the overall coal is made from the float and sink results of the constituent size fractions, the yields of different screen sizes should be taken into consideration. Then all the figures of fractional yields pertaining to the individual size groups are expressed based on total or composite coal by multiplying them with the corresponding yields of the screened fractions and dividing the products by 100. In the same way, the ash units of the individual fractions are all converted based on the total coal by multiplying them by the yield percentages of the screened fractions and dividing the products by 100. Then by simple addition of the corresponding figures of the different density fractions covering all the constituent size groups, the yield percentages as well as the weighs of ash (ash unit) of the overall coal are found out. Finally, the combined float and sink results of the overall coal are represented in the same form of table (Klima and Luckie 1986).

12.5 COAL BENEFICIATION

Coal beneficiation is a process by which impurities such as sulfur, ash, and rock are removed from coal to upgrade its value. Coal-cleaning processes are categorized as either physical cleaning or chemical cleaning. Physical coal-cleaning processes, the mechanical separation of coal from its contaminants using differences in density, are by far the major processes in use today. Chemical coal-cleaning processes are currently being developed, but their performance and cost are undetermined at this time. Therefore, chemical processes are not included in this chapter.

The scheme used in physical coal-cleaning processes varies among coal-cleaning plants but may generally be divided into four basic phases:

a. Size reduction or crushing
b. Screening or size classifications
c. Beneficiation
d. Dewatering

In the initial preparation phase of coal beneficiation, the raw coal is unloaded, stored, conveyed, crushed, and classified by screening into coarse and fine coal fractions. The size fractions are then conveyed to their respective cleaning processes.

12.6 SIZE REDUCTION

The particles of ROM coal may be up to 1 m in diameter when mined in open cast and up to about 30 cm in diameter when mined underground. The coal must be reduced in size before beneficiation. The optimum sizes are generally determined by float and sink tests to assess the size necessary for effective liberation of the coal from any shale particles.

Liberation is the process of releasing the individual components in the composite particles to form separate homogeneous fragments of coal and shale. The fragments in broken coal may include homogeneous particles made up entirely of coal or entirely of shale, and composite particles made up of coal and shale layers firmly bound together. Homogeneous particles are more readily separated from each other than composite particles. Composite particles normally exhibit intermediate characteristics and may be expected in a middling fraction (Subba Rao, 2003).

The degree of liberation may be defined as the proportion of liberated or free particles in relation to the total material. This value generally increases as the size of the particles present is reduced, partly because of preferential breakage along the planes of contact between components and partly because composites are less likely to occur in smaller sized materials.

Large coal is turned into more readily usable sizes by crushing and breaking. In general, the term *crushing* is applied to an indiscriminate reduction in size, while breaking implies size reduction in a machine designed to give the maximum possible yield of the desired sizes. The crux of the problem of size reduction of coal is to produce minimum fines.

12.6.1 Principles of Crushing

12.6.1.1 Impact

In crushing terminology, *impact* refers to the sharp, instantaneous collision of one moving object against another. Both objects may be moving, or one object may be motionless.

There are two variations of impact: gravity impact and dynamic impact. Coal dropped onto a hard surface such as a steel plate is an example of gravity impact. Gravity impact is most often used when it is necessary to separate two materials that have relatively different friability. The more friable material is broken, while the less friable material remains unbroken. Separation can then be done by screening.

Material dropping in front of a moving hammer (both objects in motion) illustrates dynamic impact. When crushed by gravity impact, the free-falling material is momentarily stopped by the stationary object. But when crushed by dynamic impact, the material is unsupported and the force of impact accelerates movement of the reduced particles toward breaker blocks and/or other hammers.

12.6.1.2 Attrition

Attrition is a term applied to the reduction of materials by scrubbing it between two hard surfaces. Hammer mills operate with close clearances between the hammers and the screen bars and they reduce by attrition combined with shear and impact reduction. Though attrition consumes more power and exacts heavier wear on hammers and screen bars, it is practical for crushing the less abrasive materials such as pure limestone and coal. Attrition crushing is most useful when material is friable or not too abrasive, when a closed-circuit system is not desirable to control top size.

12.6.1.3 Shear

Shear consists of a trimming or cleaving action rather than the rubbing action associated with attrition. Shear is usually combined with other methods. For example, single-roll crushers employ shear together with impact and compression. Shear crushing may be used when material is somewhat friable and has a relatively low silica content or for primary crushing with a reduction ratio of 6–1.

12.6.1.4 Compression

As the name implies, crushing by compression is done between two surfaces, with the work being done by one or both surfaces. Jaw crushers using this method of compression are suitable for reducing extremely hard and abrasive rock. However, some jaw crushers employ attrition as well as compression and are not as suitable for abrasive rock since the rubbing action accentuates the wear on crushing surfaces. As a mechanical reduction method, compression should be used if the material is hard, abrasive, and tough, if the material is not sticky or where the finished product is to be relatively coarse, or larger top size.

Based on the above principles, primary and secondary crushers of different sizes and capacities are manufactured by various firms and used as per the requirements by the customers (Lowrison 1974).

12.7 SCREENING AND CLASSIFICATION

Screening, also called *mechanical classification*, is a separation process that utilizes the differences in particle size. The particles that are smaller than screen opening pass through the screen, while larger particles either remain on the screen or fall off at a designated place.

Screening is affected by continuously presenting the material to be sized (the feed) to the screen surface, which provides a relative motion with respect to the feed. The screen surface can be fixed or movable. Agitation of the bed of material must be sufficient to expose all particles to the screen apertures several times during the travel of the material from feed end to the discharge end of the screen. At the same time the screen must act as a transporter for moving retained particles from the feed end to the discharge end. Particles of size more than the aperture size of the screen are retained and less size particles are passed through the apertures. Both the oversize and undersize particles are collected as overflow and underflow separately (Allen 1990).

In coal screening, it is desirable to keep breakage and production of fines during screening at a minimum. In addition, particularly for grading purposes, accuracy of sizing may be of somewhat greater importance than in ore screening, as the coal consumer may insist on getting a coal with the percentages of oversize and undersize held within rather narrow limits.

Screening is an important function in coal preparation plants, either dry or wet. The types of screens used for coal are fixed bar grizzlies and moving screens, that is, revolving Trommel screens, shaking or jigging screens, and vibrating screens. An important Dutch States Mines (DSM) development, the DSM *sieve bend* was first introduced to coal washing in 1954. The use of *sieve bend* is quite common today. Originally, it was used only for dewatering and classification of coal, but afterward, it was fully integrated with the HM cyclone process, making possible a considerable reduction in area of the expensive desliming and rinsing screens, or the application of a smaller number of these screens. They are now being used for sieving out ultrafines below 0.150 mm. The use of rapping devices, usually pneumatic, is found effective in preventing blinding (Konar et al. 1997).

Most screens are of the vibrating type and there are many makes and configurations. Many screen decks incorporate rubber or polyurethane screen surfaces. They are not only quieter but efficient screens for finer size screening and dewatering such as the Derrick-Linatex. This is a high frequency, low-amplitude screen dressed in the initial section with a sandwich screen and followed with a slotted natural rubber dewatering section. This section provides excellent dewatering performance on –20 mesh coal slurries without blinding. Additionally, a vacuum arrangement creating a pressure drop across the final section of the screen assists in lowering moisture content.

In erstwhile USSR, screen with a special belt-shaped screening surface and capacity up to 100 tph is in common use for primary sizing of coal. Some applications formerly delegated to sieve bends and vibrating screens in dewatering are now being performed by a stationary dewatering device, Vor-Siv, which is essentially a conical sieve bend developed in Polland. The high capacity and separation efficiency of the device have earned it a place in a number of new plants. In an attempt to overcome the screening problems at fine sizes with damp coals, Bretly in the United Kingdom has developed a rotating probability screen in 1973, which will produce a predominantly fine and predominantly coarse fraction within a range of size separation determined automatically by the process computer.

The state-of-the-art classification technologies generally can be divided into three groups: centrifugal classification, hydraulic classification, and screening classification.

Hydrocyclone is the most widely used centrifugal classification devices. Hydrocyclone utilizes centrifugal force, generated by converting the delivery head of slurry at the inlet volute into a spiraling passage through the cyclone, to separate coarser or greater mass particles from finer or lesser mass solids. Under the centrifugal force, coarser or greater mass particles move outward to the cylinder wall and then downward to the apex discharge. Majority of the liquid and very fine and light particles are drawn to the core due to less centrifugal force and are then forced upward to the overflow via the vortex finder (Dueck et al. 1998).

Hydrocyclone has been the principal unit of operation for fine coal classification for several decades due to its high mass and volumetric throughput capacity, small floor space requirement, and relatively effective classification. The advancements in cyclone structural design and also in the circuit design have greatly improved the size classification performance for this duty. However, the classifying cyclone has two fundamental limitations: (1) fine particles in the underflow due to hydraulic entrainment (fine light coal particles) and density effect (fine high-density tailings particles) and (2) coarse light fine particle in the overflow due to density effect. The misplaced fine particles (high-ash ultrafine fine coal/ clay) are often reported to the fine clean coal stream, ending up as contaminants. On the other hand, the misplaced coarse fine particles cannot be effectively recovered by flotation and therefore end up lost in the tailings (Plitt 1971).

Hydraulic classifiers are also used to achieve size classification for fine particles. The principle of operation of the hydraulic classifier is based on the concept of differential terminal settling velocity of solid particles of different size or mass. Due to the low classification efficiency and also the requirement of large floor space, it becomes less and less popular within the coal-processing industry (Mohanty et al. 2002).

Another method to achieve fine particle classification is to use screens. A static sieve bend screen is widely used in coal processing plants to separate heavy medium from coarse coal and also to dewater, partially to size, the fine coal spiral products. Slurry flows by gravity over the inclined screen surface, where the screen wires are mostly perpendicular to flow. The concave curved screen surface slices away layers of fine particles and slurry liquid. However, the sieve bend screen tends to be less efficient, low capacity, high maintenance, and high in operational cost. Recently developed screening technologies have been proven to be able to provide size classification at 150 μm or even finer while maintaining satisfactory performance. The Pansep screen technology showed that exceptional screen efficiencies could be achieved at a separation size of 45 μm. Meanwhile, Derrick Corporation, a well-known screen manufacturer, has commercialized a new fine coal-screening technology, known as *Stack Sizer*. Stack Sizer is fitted with Derrick Polyweburethane screen surfaces in the range of 1,000–45 μm. The linear motion provided to the screen decks by Derrick Corporation's Super G vibrating motors, together with an angle of inclination between 15 and 25 degrees, produces excellent screening efficiency with high oversize. The entrapped fines in the coarse overflow are released by repulping the initial overflow material with wash water in a trough between the upper and lower section of the screen deck. This repulping process allows the fines to find their way through the screen openings, thereby minimizing ultrafine bypass to the overflow product (Baojie et al. 2014).

12.8 COAL BENEFICIATION METHODS

The methods used in the mechanical preparation of coal are analogous in many ways to those used in the beneficiation of ores. The principles involved in coal beneficiation are similar to that of ore beneficiation. In some instances, the same machines could be used in both fields perhaps with slight modifications. An essential feature of coal beneficiation processes is that the treatment must be rapid and inexpensive, because coal is a cheap product. Accordingly, simple plants consisting of a few units, each capable of handling a large tonnage, are required.

Another factor important in coal beneficiation is that the coal, during all stages of its handling, from mining, through preparation, transportation, and finally to its delivery to the consumer, must undergo a minimum of degradation, because the value of coal depends on its particle size, and in general, the larger the particle size is, the better price the coal commands on the market.

Gravity separation principles form the basis of most of the coal beneficiation processes. In case of coal beneficiation, the valuable part, the coal, is light and the impurities to be removed by washing or cleaning are heavier, whereas in beneficiation of ores, the valuable mineral is heavy and impurities (gangue) is light.

The coals may be treated and upgraded by both dry and wet processes.

12.8.1 Dry Beneficiation

The dry beneficiation methods are based on the differences in physical properties between coal and mineral matters such as density, size, shape, lustrousness, magnetic susceptibilities, electrical conductivity, and frictional coefficient. Based on the difference in these particular properties, different types of equipments such as pneumatic jig, pneumatic table, sortex machine, tribo-electric separator, and air-dense medium fluidized bed separator, which are applicable to differently sized fractions, have been developed to beneficiate ROM coal.

12.8.1.1 Rotary Breaker

The relative friability of coal as compared to that of waste minerals has been exploited in coal preparation for many years. The simplest form of the separator is the rotary breaker devised by McNally and shown in Figure 12.1. It achieves size reduction by repeatedly raising the coal material and dropping it against strong perforated screen plates around the interior. Thus, it serves as the primary crusher for soft to medium-hard coals, deshaler, and screens the coal at a mine site as the lumps are broken down and pass through screen-sized openings. Moreover, it does not over crush the coal, thereby creation of excessive fines could be avoided. Various dimensions of rotary breakers treat varying capacities and a different level of product top sizes. The performance depends on several variables, such as drum diameter and drum length, effecting selective breakage between coal and rock, aperture size, percentage of open space, and prescreening of feed. In order to get a rational estimate of the height and number of drops required to refuse the dirt preferentially through a predetermined screen, it is required to study the differential breakage pattern of coals and stones through some simple laboratory methods, such as, drop breakage tests. Before feeding to rotary breaker, ROM coal is passed either through grizzly or scalping screen so that maximum part of naturally occurring undersize material is removed at an early stage (Bhattacharya 2006).

As described and discussed, rotary breaker may be termed as a *primary crusher, screen*, and *dry deshaler*. It is capable of discarding obvious dirt/stone and high ash content hard shales. But, the only use of rotary breaker is not the solution of coal preparation/upgradation. However, it may be used as preliminary crusher or preparation unit for the coals obtained from open cast mines. It not only increases the life of successive crushers but also improves the quality of feed coal to the beneficiation plant. Rotary breaker has proven to be a robust machine having very low operating costs, typically ranging from \$0.01/ton to \$0.04/ton, and a high capacity up to 2000 tph.

12.8.1.2 Ore Sortors

The sorting operation of coaly material can be carried out by different techniques such as optical, radioactivity, microwaves, and nuclear magnetic resonance. Coal beneficiation at a coarser size is possible by optical sorting when the coal does not contain a large amount of carbonaceous shale. This technique was applied by a British coal corporation that used a Gunson sortex device. In this process, the powerful lights from photoelectric cells illuminate the material. When the material passes under the cell, the amount of light reflected back is measured. If reflected light from a particle is sufficient, it means the material is not the coal and it is rejected from the conveyer by an air jet (Butel et al. 1993).

Coal sorting with an X-ray transmission usually consists of two components: detection and removal system as shown in Figure 12.2 (Ramana 2009). The sorting process may be divided into four interactive sub-processes: particle presentation, particle examination, data analysis, and particle separation. Powerful computers and increasingly sensitive X-ray scintillation counters enabled the development of high-performance sensor-based sorting machines. Feed preparation is more critical for sorters due the importance of surface characteristics and physical size of the particles; most sorters need a 3:1 or 2:1 ratio between the largest and smallest particle to be efficient. Once the particles have been properly prepared for sorting, they must be presented to the sensor. To operate

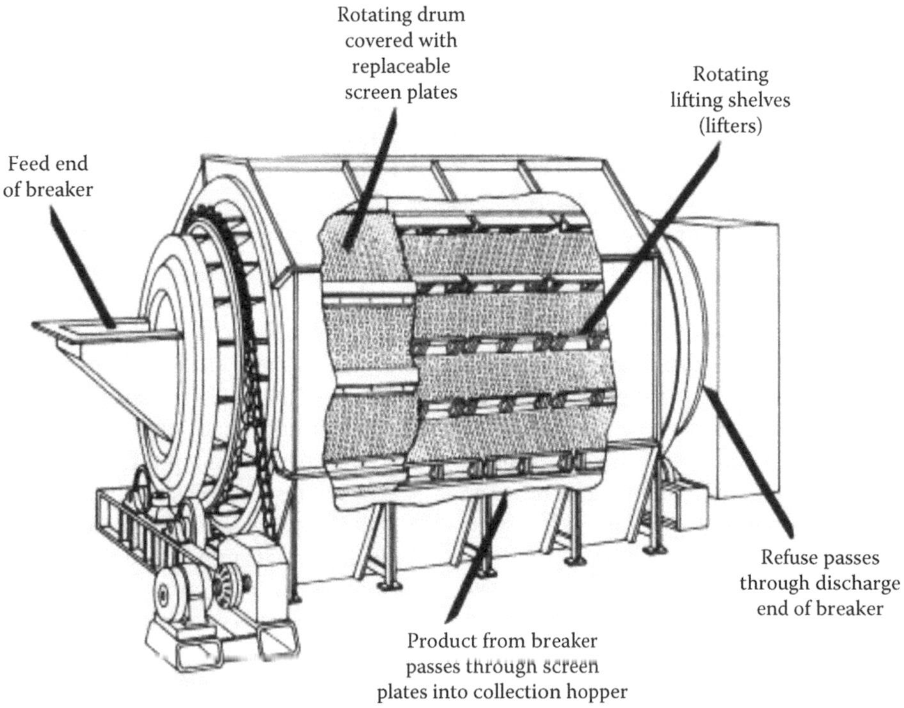

FIGURE 12.1 Rotary breaker. (From Osborne, D.G. *Coal Preparation Technology*, vol. 1, London, Graham & Trotman, 1988.)

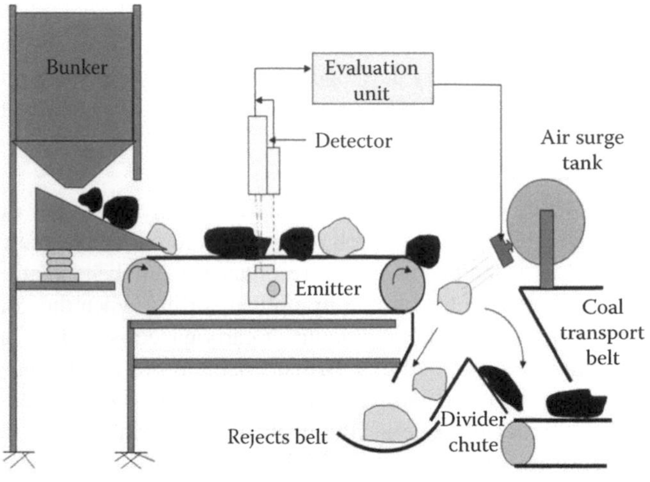

FIGURE 12.2 Ore sortor setup. (From Ramana, G.V. *CPSI J.*, 1, 35, 2009.)

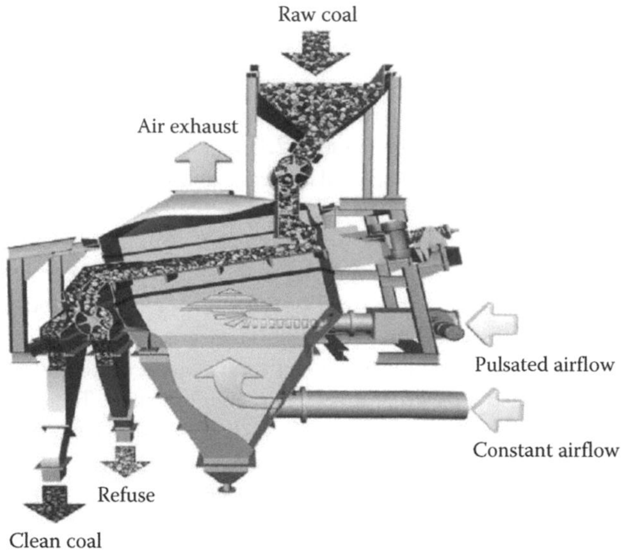

FIGURE 12.3 All air jig. (Courtesy of Allmineral Asia Pvt. Ltd., India.)

efficiently, the sensor must be able to analyze each single particle. As a result, feed rate and the materials handling methods are the critical components, with this most commonly being done by a conveyor belt or chute. The first industrial installation of a belt-type X-ray transmission sorting for coal has been in operation since 2004. Since 2010, chute-type sorters are in the South African market and, next to production, have been used for extensive test work (Robben et al. 2014).

The critical stage of examining the particle and determining whether material is valuable or barren, is done by a combination of sensor and processing unit. Once the decision has been made as to accept or reject a given particle, a mechanical device is required to physically sort. High-pressure jets of air are generally used to make this separation. Of all the components in a sorter, it is the choice of sensor that controls the design of a sorter.

12.8.1.3 Air Jigs

The allair®-jig uses the principles of jigging, which are also the basis for the design, and operation of conventional wet jigs. ROM coal consists of particles of comparable sizes but different densities. Eliminating particle friction and allowing the particles to be sorted according to the specific particle density can obtain stratification according to the particle density. In conventional wet jigs, this stratification is obtained by feeding the material across a screen and pulsating water upward and downward. allair-jigs carry out the separation process by pulsating air flowing through a layer of stratified material. The stratification begins by delivering constant airflow to illustrate or loosen the bed. Superimposed air is delivered to promote the stratification. Vibrating mechanisms assist in the transport of material across the jigging bed (Honaker 2007).

Air jigs as shown in Figure 12.3 use pulsed air to achieve particle stratification and separation through the hindered settling and consolidated trickling mechanisms. All mineral's air jig technology, known as an *allair-jig*, has been commercialized in the United States with the first 50 tph unit installed in an Ohio surface coal mine in 2001. The system provides high-density cut points as required in rock removal operation; however, the top particle size that it can treat is about 50 mm (Weinstein and Snoby 2007).

The allair-jig uses the several design features of the proven alljig wet jigs. The jig hutch design is optimized in order to achieve an even distribution of air. The feed star gate provides an even feed distribution over the jig width and an exclusion of air. The jigging frame is equipped with two vibration drives. The screen deck is made of stainless steel perforated plate. The pulse air valve controls the pulsed airflow, which is necessary for the separation. The material becomes stratified and fluidized by the pulsed airflow according to their specific densities. That means that low-density particles (coal) stratify on top of the material layer, while high-density particles (dilution rock) stratify in the bottom layer of the machine. A clean stratification depends mainly on the best possible stratification and fluidization. This is influenced by the particle size and density distribution of the feed material. As a consequence, different jigging stroke characteristics are required for different feed materials. At the discharge end of the jig a star gate continuously discharges the high-density material (dilution rock) from a reserve layer. The discharge control system maintains the thickness of the high-density material layer constant at the end of the jig. This discharge system speeds up or slows down according to the depth of the high density or refuse layer even when the amount of refuse in the feed changes over time. This maintains a consistent refuse layer, while discharging the low density or cleanly stratified coal. Thus, the amount of misplaced material is minimized and quality variations in the feed are automatically compensated (Sampaio and Aliaga 2008).

The allair-jig is designed to handle material up to a maximum particle size of 2″ (50 mm). The maximum feed capacity of the allair-jig depends on the particle size distribution of the feed material and reaches up to 100 t/h. The dedusting of the allair-jig is usually realized with a bag house type filter.

12.8.1.4 Air Dense Medium Fluidized Bed

Dry treatment of coal with an air dense medium fluidized bed is claimed to be an efficient separation method, which utilizes fluidization techniques into the field of coal preparation. It uses an air-solid fluidized bed as separating media, and thus, differs greatly from the conventional wet processes such as jigging, heavy medium separation, and coal flotation processes.

This technology was earlier studied in the former Soviet Union, the United States, and Canada, all in laboratory or pilot plant stages. China has recently developed up to 50 tph industrial scale proto-type for 50–6 mm coal. Air-dense medium fluidized bed (ADMFB) separator is effective for separation of 6–50 mm size fraction and the studies showed that the overall Ep value achieved was 0.03. Detailed results showed that the finer the feed coal particles, the lower the clean coal recovery as well as the separation efficiency. Vibrated ADMFB and magnetically stabilized ADMFB technologies were developed to provide dry separation of coal in 6–0 mm particle size range and dual-density ADMFB technology was developed and tested to achieve three-product separation (Luo and Chen 2001).

Air dense medium fluidization is a process in which tiny particles of media like magnetite, magnetic pearls, and mill scale are transformed from solid state into pseudofluid state by levitating with airflow, which is introduced from bottom of a vessel through air distributor located under the bed. The volume of air required for fluidization in 50 tph throughput unit is about 400 m^3/m^2 hr at an approximate pressure of 0.2 N/m^2 and fluidization velocity is varying from 1.8 to 1.9 cm/sec. The power consumption is reported as 0.5 kwh/ton of feed. It is claimed that the operating specific gravity ranges from 1.30 to 2.20. The particles of the feed with specific gravity lower than that of the medium floats to the top of the bed, where they are recovered at one end of the vessel and the heavier articles are rejected at the other end. It is claimed that the separation efficiency of the process is similar to that of the heavy media vessel or cyclones (Ep value, 0.05–0.07) with the following additional advantages (Luo et al. 2002):

1. It can be utilized in an arid region where water supply is scanty
2. It produces a dry product requiring no additional dewatering and coal slime treatment process
3. The airflow removes most of the surface moisture present in the feed
4. Magnetite medium can be recovered and loss of magnetite medium is less (0.5 kg/t of feed)

The limitations of the process, as visualized, are as follows:

1. The exhaust air carrying fines has to be passed through filters/electrostatic precipitator/cyclone to control the air pollution
2. The bed area increases significantly with the increase in feed rate, and the desired fluidization condition becomes difficult to achieve due to occurrence/initiation of the channeling phenomenon in the bed
3. There is a restriction on the top size of the feed coal (i.e., below 50 mm)
4. Limiting surface moisture (below 5%) required for the feed coal may restrict its applicability to some high moisture non-coking coals

12.8.1.5 FGX Dry Cleaning System

The FGX dry cleaning system employs the separation principles of an autogenous medium and a table concentrator. As may be seen in Figure 12.4, the feed is introduced into a surge bin from which the underflow is controlled using an electromagnetic feeder. Material from the surge bin is fed into a separation compartment at a predetermined mass flow rate. The separating compartment consists of a deck, vibrator, air chamber, and hanging mechanism. Upon introduction of feed coal into the separation chamber, a particle bed of a certain thickness is formed on top of the deck (Honaker et al. 2007).

A centrifugal fan provides air that passes through holes on the deck surface at a rate sufficient to fluidize and to transport the light particles. The presence of about 10%–20% material finer than 6 mm is needed to develop a fluidized autogenous medium particle bed. Low-density particles (such as coal) form the upper layer of particles that are collected along the front length of the table. The high-density particles (such as rock) maintain contact with the table surface where both vibration and the continuous influx of new feed material move the material along riffles toward the back side of the table and the narrow end of the table where the final refuse is collected. The separation process generates three product streams, that is, clean coal product, middlings, and tailing streams. Two dust collection systems are employed to clean the recycled air and to remove the dust from air before being emitted into the atmosphere (Honaker 2007).

The FGX separator provides a relatively efficient separation at high separation density values of around 1.8RD to 2.2RD. The typical probable error (Ep) value achieved was 0.25. However, if the middling stream is recycled to the feed

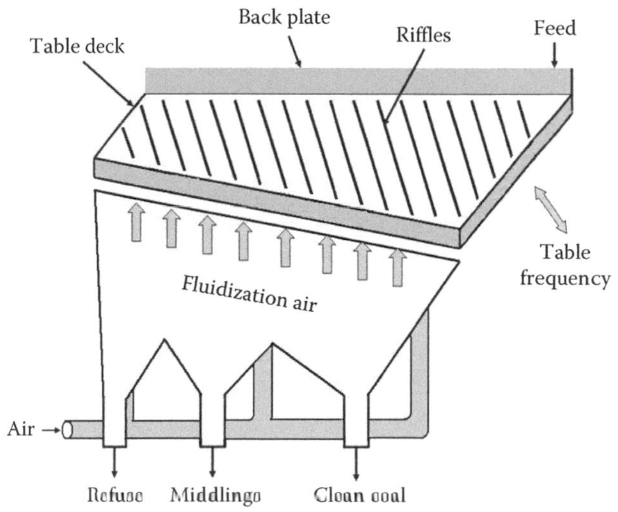

FIGURE 12.4 FGX separator. (Courtesy of FGX Sep Tech, LLC, USA.)

stream, process efficiency may improve due to the potential of capturing low-density particles that are mixed with high-density materials within the middle portion of the table discharge. On the other hand, recycling could overload the system with excessive near-density particles that could negatively impact separation performance. High separation efficiency, along with low cleaning costs, has resulted in the widespread application of the FGX dry separator in China. The first commercial installation of this technology in the United States took place in the year 2009 (Zhang et al. 2011).

12.8.2 Wet Beneficiation Processes

The wet processes have been well established and universally accepted for the beneficiation of coal. These are precise, efficient, economical, and viable. A wide range of particle size can also be treated by a single process. Hence, the wet processes have become attractive and all the coal-producing countries prefer to beneficiate the coals by wet processes.

Several technologies based on wet methods have been developed and quite a number of renowned manufacturers have been designing, fabricating, and marketing the washing equipment (washer) working on the same basic principle with different trade names.

Some of the wet beneficiation processes have been outlined in the following sections, which may be relevant to the present context.

12.8.2.1 Jig

Jigging was one of the oldest methods to be adopted for coal beneficiation, about 100 years ago. It is considered as an efficient washer for cleaning coals having NGM of about 20% or below (Ashish et al. 2011).

The separation of coal from shale is accomplished in a form of fluidized bed created by a pulsating column of water, which produces a stratifying effect on the raw coal. This is quite different in its effect from dense medium separation. This stratifying effect results in a definite order of deposition of all fragments contained in the bed. The main purpose of the rising and falling column is to create what is known as *dilation* or opening of the bed, and it is the extent to which this dilation may be controlled which governs the effectiveness of the separation.

During the pulsion, or rising part of the cycle, the bed is elevated en masse. But as the velocity decreases toward the end of the pulsion stroke, the bed begins to dilate, with the bottom ceasing motion first and the lowermost fragments commencing their descent. This produces an element of freedom of movement for all fragments signaling the commencement of the various principal effects leading to stratification. The most influential effects occurring during jigging are, in order of occurrence, as follows:

1. Dilation
2. Differential acceleration
3. Hindered settling
4. Consolidated trickling

Differential acceleration occurs, immediately following the end of the pulsion stroke. The theory has been advanced that, during this very brief period, the heavier fragments have a greater initial acceleration and velocity than lighter ones, even although their ultimate speeds may be the same, that is, if they were equisettling diameters. If the repetition of fall is frequent and the duration of fall is short enough, the distance traveled by dissimilar fragments should bear more resemblance to their initial acceleration than to their terminal velocities. Under such conditions, stratification would gradually occur as a result of differences in relative density alone.

The hindered-settling effect observed as occurring in jigging is a little different from that in sizing classifiers or trough washers. All things being equal, higher settling ratios are attainable during jigging than in other forms of classifier and this effect makes an important contribution in the separating mechanism of jigging.

Very small grains of coal and shale are likely to be the most influenced during the pulsion stroke; during dilation, the smallest particles may be pushed of the bed by the rising fluid, the coarser fragments being most influenced, first of all, by differential acceleration, and then by hindered settling, eventually bridging against one another and quickly becoming incapable of further movement. Fine fragments will remain free to move between the interstices of the coarse fragments. Aside from any velocity that may be imparted to these small fragments by the moving fluid, they are bound to settle under the influence of gravity in these interstitial passages. This phenomenon is described as *consolidation trickling*. There is no such thing as perfect separation by jigging even in concept because of the random nature of motion of the materials forming the bed as they pass along the jig unit.

The jig is divided into two compartments lengthwise: one completely sealed from the atmosphere—called the *air chamber*—and the other an open section, which receives the material to be separated and accommodates it during the stratification process. The water valve allows admission of *back water* at a level below that of the bed plate. The longitudinal section is further divided into several sections or compartments along the direction of flow. The purpose of this is to provide control over the separation as the material moves along the box, hence each of these sections has its own individual air and water controls. The two-elevator arrangement is the most common. In the first, moving along the direction of flow, the heavier shales are separated. In the second, lighter stones and any middlings are extracted. The plate that supports the coal and shale bed, usually referred to as the *bed plate* or *screen plate*, allows the water current to rise and fall and is usually perforated. Fine material inevitably percolates through the perforations to the hutch compartment and this is removed by screen conveyor, which delivers it to the bucket elevators (Fellensiek and Erdmann 1991).

Efficient collection of the product is of paramount importance. Clean coal overflows at the end of the box together with the majority of the flowing water.

12.8.2.1.1 Types of Jigs

There are basically two jigging principles: air pulsated or mechanical pulsated. For coal beneficiation, the air pulsated jigs are the most common. There are two different ways of air pulsating the jig: an air box on the side of the jig (Baum type) or where the air box is under the bed (Batac type).

The Baum jig, as shown in Figure 12.5, for coal was developed in Germany by Fritz Baum and was in operation in Germany since 1892. The movement of the water in the jig box was created with pressurized air. This is the principle distinction from the mechanically pulsated jigs, where movement of the water is induced through the up and down movement of a piston-like plunger connected to an eccentric driver. The air box of the Baum jig is on the side of the actual jig box. A U-tube like tub with two compartments is connected to each other, one is the air box, and the other is the actual jig box. The air box is sealed, and filled with air. Compressed air pressurizes the airbox in a short instance, and causes the water to move from the air box leg of the U-tube, through the bedplate, into the jig box. The material on the bedplate will move up and down with the movement of the water, and separation will take place (Brinkman and Helling 1964).

During the 1950s, Baum jigs were designed with widths up to 2.5 m, and it was felt that the maximum throughput and width had been reached. With wider jigs, the distribution of water and coal over the bed became less optimal. The development of the Batac jig overcame these limitations. The air box was placed under the bedplate in the jig box, and the jig box could be compartmentalized to handle high loads and still have a good separation (Figure 12.6).

At the end of the bedplate in each compartment, a gate routes the discards to the bottom and the product to either the next compartment or out to the product collection system. The discards are removed from the volume under the bedplate, or hutch, by means of a scroll or chain-bucket removal system.

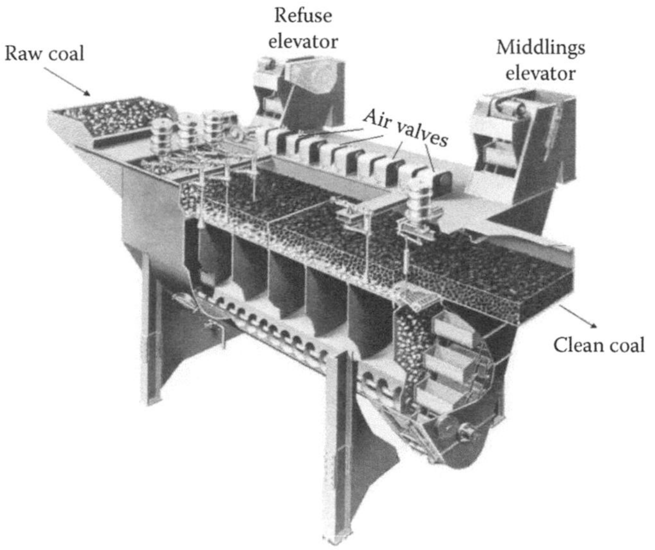

FIGURE 12.5 Baum jig. (From Subba Rao, D.V., *Coal—Its Beneficiation*, New Delhi, India, Emkay Publications, 2003.)

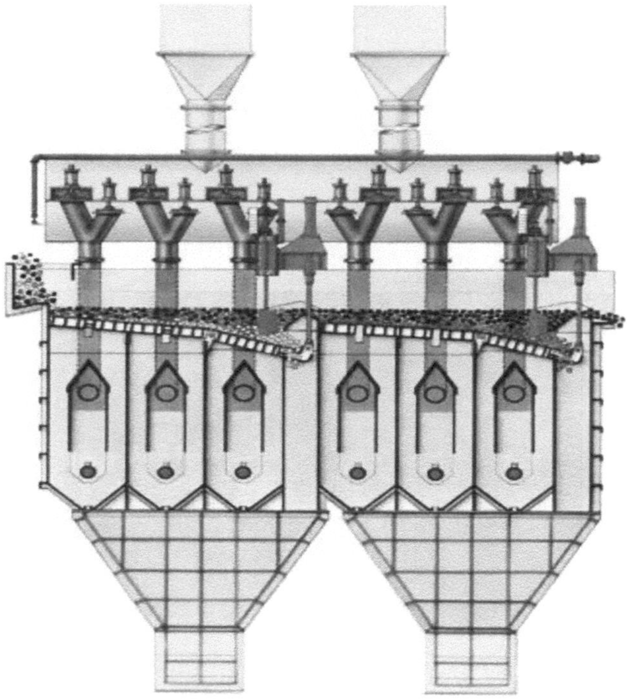

FIGURE 12.6 Batac Jig. (From Subba Rao, D.V., *Coal—Its Beneficiation*, New Delhi, India, Emkay Publications, 2003.)

The product flows of the final weir into a product collection system.

As discussed previously, the control of the pulse is of utmost importance (that, and the gate settings, are the two critical control parameters of a jig). Over the years, the various manufacturers of jig equipment have spent significant amounts of resources to develop their versions of the optimal control system and air supply system for their jig type.

Batac-type jigs are capable of handling high feed rates per single unit, up to 100 ton/h/m jig width is possible, and jigs have been manufactured to handle 700 ton/ h raw coal with good results (Sanders et al. 2002).

Jigs can be used for coarse coal and fine coal beneficiation, but the particle size range that can be handled in one jig is limited. This is due to the hindered settling and consolidated trickling actions during the downstroke, and the effect particle size has on these actions.

As mentioned before, the selection of the correct size range is important as this influences the separation efficiency (because of the hindered settling and consolidation effects). The size range treated also determines the frequency and amplitude of the pulsations required to achieve proper separation (Takakuwa and Matsumura 1954).

Batac-type jigs are capable of handling high feed rates per single unit, up to 100 ton/h/m jig width is possible, and jigs have been manufactured to handle 700 ton/h raw coal with good results.

KHD developed a jig as shown in Figure 12.7, especially suited to handle very large size particles, particularly suitable for ROM coal destoning, the ROM jig. The jig can handle large coal, 350 mm top size, with a bottom size of nominally

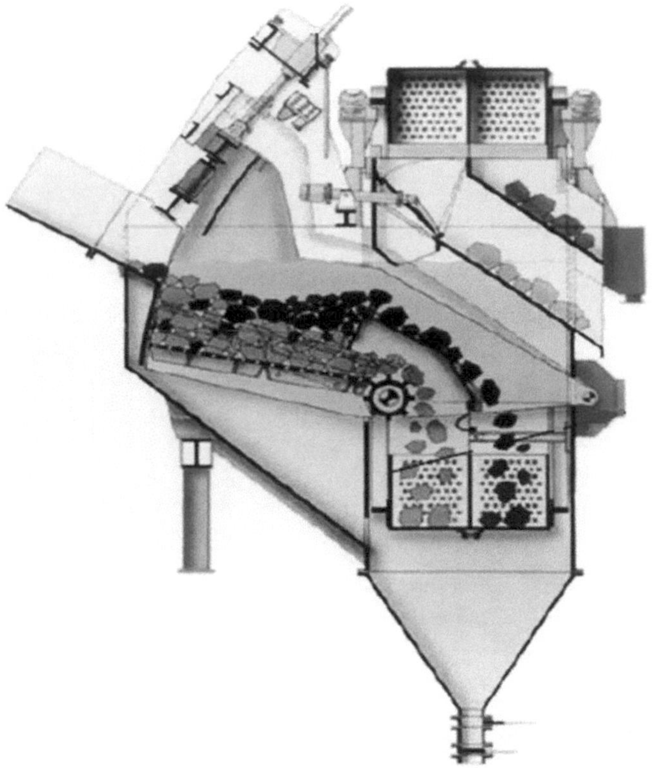

FIGURE 12.7 ROM jig. (From Osborne, D.G., *Coal Preparation Technology*, vol. 1, London, Graham & Trotman, 1988.)

40 mm. The water movement in this jig is not induced by air, but through the hydraulic lifting of the feed end of the screen. This is then allowed to lower again under gravity. The so induced jigging action causes the separation of discards from the product, the discards being on the bottom of the bed, the products on top (Florl and Heintges 1987).

The main benefit of the application of the ROM jig is the reduction of material to be treated in the next separation steps. Since the material removed is predominantly high mineral-content material, the abrasive tendencies of the product will be reduced and the downstream processes will suffer less mechanical wear, thus reducing maintenance costs.

The separation efficiency of the jig is greatly influenced by the amount of NGM (material of a relative density within 0.1 unit of the separation density). Material of density close to the separation density will reduce the sharpness of the separation as the three effects described above are no longer so distinct. The amount of NGM that can be tolerated in a jig is approximately 10%, rarely higher.

However, jigs separate well at higher densities where the amount of NGM is relatively low. In deshaling operations (the removal of stone and shale from a raw coal prior to final beneficiation), the separation density is high, generally above RD 1.8, and the amount of near density material is lower than 10%.

12.8.2.2 Dense Medium Baths

If a sample of raw coal is introduced into a solution of a predetermined density, a clean coal containing required ash content will be obtained in a float product, while the residue of the coal, the discard will sink. In commercial practice, four types of separating medium have been used. Organic liquids, dissolved salts in water, aerated solids, and suspensions consisting of fine solids suspended in water. Out of all the four, all the processes presently use suspension of finely divided solids in water, termed the *heavy medium*. A dense medium coal-washing unit consists usually of a separating vessel, feeding system for raw coal and removal of products, medium circulation, and finally medium cleaning and recovery system (Burt 1984).

There are two main categories of dense medium baths: deep and shallow. Deep baths (cones) have the advantage of a fairly quiescent pool area and longer residence time and consequently less prone to gravity changes by accidental addition of extra water. Barvoys, tromp deep bath, and chance cone come under this category. Shallow baths (troughs and drums) contain a relatively small volume of medium and when a stable or semi-stable medium is used, circulation rates can be comparatively low and take up considerably less area. Ridley-Scholes, Drewboy, Norwalt, Wemco, Teska, and so on come under this category (Leonard 1979).

12.8.2.2.1 Tromp Process

The *tromp process* was the first to introduce (about 1938) the use of magnetite suspension in dense-medium washing. The three-product tromp dense medium vessel, as shown in Figure 12.8, consists of shallow bath of mild steel plate. A scraper conveyor is provided at its upper section to remove clean coal. A dual-purpose scraper conveyor is arranged at its center and bottom sections for the separate removal of the middlings and refuse products. The raw coal is introduced at the top of the bath by means of a slow running balanced feeder screen through the chute. The clean coal floating on the top of the bath travels the length of the bath by laminar flow ant it is removed by the top scraper conveyor. The middlings are suspended in the bath below the clean coal conveyor

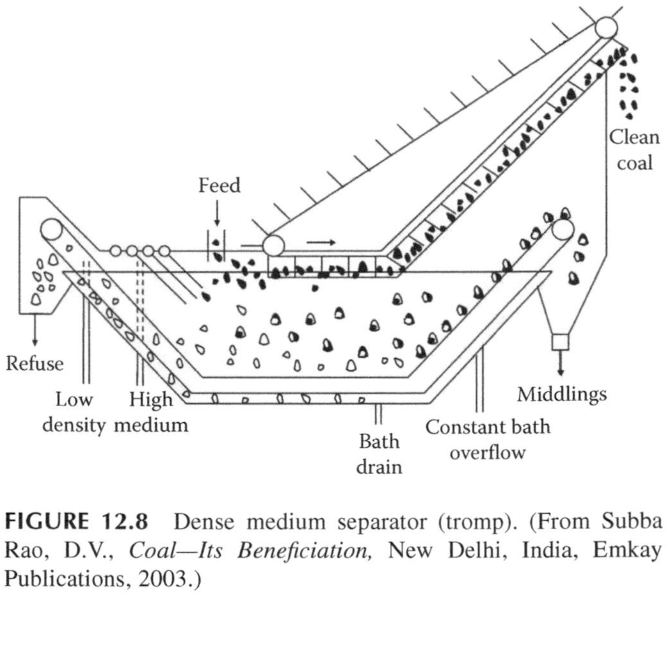

FIGURE 12.8 Dense medium separator (tromp). (From Subba Rao, D.V., *Coal—Its Beneficiation,* New Delhi, India, Emkay Publications, 2003.)

at the interface between the low-gravity medium and the high-gravity medium. They are removed by the combined action of the horizontal currents and the middlings strand of the conveyor. The refuse falls to the bottom of the bath where it is conveyed in the opposite direction of the middlings by the lower strand of the same conveyor.

The magnetite is ground to about −1/250 in. (0.1 mm) and added to water. The process makes use of an unstable suspension with horizontal currents of differing densities at intermediate levels. The process operates within the size range of 6–200 mm and in practice is used for raw coal down to 1/4 in. (6.4 mm). It gives a reasonably accurate three-product separation. This process may be used on any size of coal from 10 to 1/4 in. (254–6.4 mm) and for any specific gravity from 1.3 to 1.9. The grain size of the magnetite or pyrite is −0.1 mm. The quick settling of the magnetite particles gives a higher specific gravity in the lower layers of the wash box, which makes it possible to obtain three products: clean coal, middlings, and refuse (Aplan 1985).

12.8.2.2.2 Wemco Drum Separator

The Wemco drum separator (Figure 12.9) is the simplest example of a dense medium bath. Raw coal enters in the one end, into the magnetite/water mixture. Separation starts taking place immediately, with the heavier material sinking to the bottom, and the lighter material floating on the medium surface. The drum rotates, and internal scrolls move discards to a removal point in the middle of the drum, where they are removed with lifters into a removal chute. The product, the *floats*, moves with the medium out on the other end of the drum, over a drain screen to remove correct-density medium and over a rinse screen to remove the remainder of the medium. Medium is recovered, and returned to the bath. The discard material is also rinsed to recover more medium. If needed, the product and discards are further dewatered (separately) on dewatering screens before they are stacked for disposal (discards) or transported to the client (product). The maximum size of coal that can be handled in the Wemco drum depends on the size of the inlet, outlet, and lifters. Material with top size of 300–500 mm is not unusual. The Wemco drum can handle as much as 400 ton/h feed, approximately 20–25 ton/h/m^2 pool area, with very good separation efficiencies. High separation efficiency can be maintained through manipulation of the residence time in the vessel, either by supplying a large enough pool surface, or reducing the throughput of the raw coal. *Ecart probable moyen*, an indication of the sharpness of separation, values of 0.03 or better can be achieved.

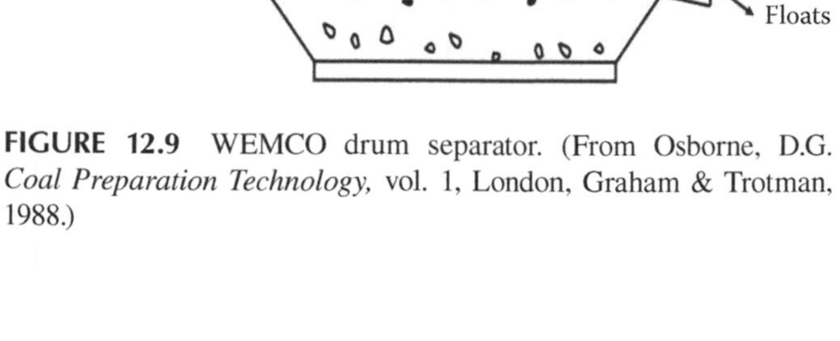

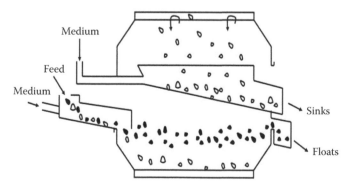

FIGURE 12.9 WEMCO drum separator. (From Osborne, D.G. *Coal Preparation Technology*, vol. 1, London, Graham & Trotman, 1988.)

No doubt, the dense medium baths are efficient and precise but the process is costly and requires highly skilled labor for its operation. One of the drawbacks of the process is that it cannot beneficiate coals below 10–15 mm; as such, additional small coal circuits such as HM cyclone, fine coal jigs, and so on are needed if this fraction of coal is required to be washed. The efficiency of dense medium baths is reflected in the range of Ep, which is between 0.025 and 0.06, depending on coal size and density of separation.

12.8.2.3 Processes Based on Centrifugal Force

The major limitation of dense medium vessels was found to be an alarming drop in cleaning efficiency with decrease in size of coal fed to the washer. The solution to the problem came with the development of the dense medium cyclone washer, first pioneered by M.G. Driessen and his colleagues at DSM and reported in 1945. In the 10 years that followed, the DSM cyclone technology became the single most important development in coal washing in the twentieth century, and will probably remain so. Hundreds of plants now utilize the DSM cyclone system and many hundreds more use cyclone types of washers, many of which are direct descendants of the original DSM design.

The DSM work pioneered the development of the familiar cylindroconical cyclone separator for sorting coal from waste, but other forms of cyclone separator have also emerged. All such forms can be categorized as follows:

1. Cylindroconical
2. Cylindrical

The first category includes the original DSM design and other similar types, including McNally cycloids, Krebs cyclone, and the Kilborn cyclone. It also includes other developments utilizing the conventional hydrocyclone profile, such as the Chinese electro-magnetic cyclone and the Japanese *swirl* cyclone (Zimmerman 1978). The original feed size range specified for the DSM design of 10 × 0.5 mm, with cyclone diameter ranging from 200 to 700 mm, has been expanded with practice to 50 × 0.5 mm and cyclone diameters up to 1,000 mm are now available.

The second category includes the Vorsyl separator designed and developed by British Coal in 1967, the Dyna whirlpool, an American development (1976) and the Tri-Flo separator, an Italian designed two-stage unit (1984). British Coal development has placed another type of separator in this category. This is the LARCODEM separator, which was specifically developed for treating coal sized 100 × 0.5 mm. It is similar in concept to the Vorsyl.

Heavy media cyclones (HMC) are in very common use today. There is trend toward increasing both the top size of the feed as well as feeding cyclones *zero* size particle. After much test work in Europe and in the United States, commercial plants have come into operation, treating feed sizes downs to zero, thus eliminating the usual screening at 0.5 mm. In South Africa, they have gone a step further in the use of HMC for cleaning sizes of 0.5 mm (Zimmerman 1978).

These trends show that the previous limits of sizes of coal that can be treated by heavy media systems will be extended, causing the preparation engineer to carefully study all options for the cleaning of coals. Various separators have been developed using the principles of centrifugal separation for the beneficiation of mineral matter. Following is a brief description of the commonly used ones.

12.8.2.3.1 Heavy Media Cyclone

The HMC, also called as *DSM cyclone*, was developed by M.G. Driessen and his colleagues at DSM and reported in 1945 (Driessen 1945). To create the dense medium, fine magnetite is added to water to form a suspension having a medium density between the solid densities of coal and rock. As a result, the coal floats, while the heavier mineral matter sinks through the medium.

Figure 12.10 shows a typical cyclone consists of a conically shaped vessel open at the bottom, called the *apex* or *spigot*, and connected to a cylindrical section, which has a tangential feed inlet. The top of this cylindrical section is closed with a plate, fitted through which is an axial pipe, extending into the body of the cyclone, know as the *overflow* or *vortex finder*.

In the HMC, a mixture of medium and raw coal enters tangentially near the top of the cylindrical section, thus forming a strong vertical flow. The high ash particles move along the wall of the cyclone due to the centrifugal force where the velocity is the least and is discharged through the underflow orifice or the spigot. The lighter washed coal moves toward the longitudinal axis of the cyclone due to the drag force where a high velocity zone exists and passes through the overflow orifice, or vortex finder, also called the *overflow chamber*. Centrifugal force also acts on the very fine magnetite medium and therefore the specific gravity of the medium will

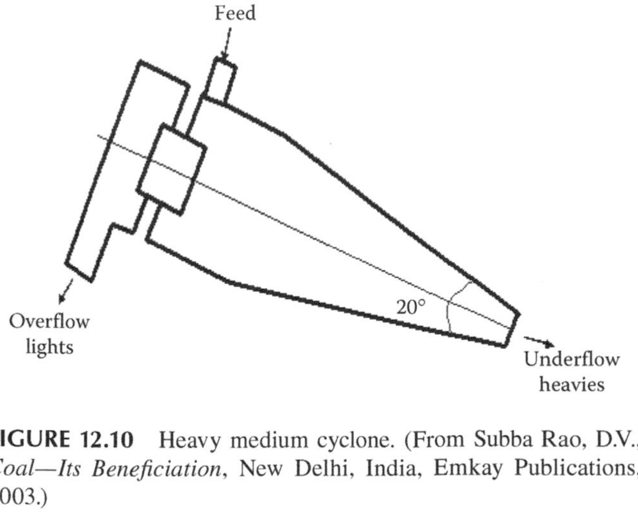

FIGURE 12.10 Heavy medium cyclone. (From Subba Rao, D.V., *Coal—Its Beneficiation*, New Delhi, India, Emkay Publications, 2003.)

increase toward the apex discharge opening. Therefore, the specific gravity of separation is higher than the specific gravity of feed. The capacity of HMC increases with its diameter. Cyclone diameters normally range from 200 mm up to 1 m, with the inlet pressure usually being higher for the smaller diameter units (Miller and DeMull 1985).

In recent times, cyclone diameters have increased to the point where 1,500 mm diameter cyclones are now being used in a number of coal-processing applications, particularly in Australia. The larger diameter also implies larger feed, overflow, and underflow openings and, as a result, allows larger feed particles to be fed to cyclones. Most cyclone manufacturers still adhere to the DSM recommendations, and Multotec *standard* cyclones are manufactured to these dimensions. There is, however, a demand for cyclones having a higher capacity, to be able to process larger feed particles, and furthermore to be able to handle higher amounts of sink material. The latter requirement is brought about by modern mining methods being less selective and including more roof and floor shale in the coal mined as well as more low-grade reserves being mined (Ruff 1984).

Cyclones were traditionally employed to process material between approximately 20 mm and 0.5 mm. The advent of large diameter cyclones now allows coarser material to be processed. The problems associated with pumping large particles, however, still limits the upper size range of material processed in dense medium cyclones to approximately 50 mm. Where sized products are required, it is still customary to install a dense medium bath and to process only the small material via dense medium cyclones. Dense medium cyclones, by virtue of their high separation efficiency, are the method of choice for processing difficult-to-process raw coals (Clarkson 1998).

12.8.2.3.2 Dyna Whirlpool Separator

Developed in the United States by the Minerals Separation Corporation of Arizona, this separator employs centrifugal forces in a cylindrical vessel. Other than in the DSM cyclone, the feed and the medium are not introduced together, but separately.

The vessel (Figure 12.11) is installed at an angle of approximately 25 degrees to the horizontal. The medium is introduced at the bottom, tangentially, and creates the vortex. This open vortex spirals upward toward the discards outlet. Raw coal is introduced at the top with a very small amount of medium (as to not disturb the existing open vortex). As the coal moves down, the centrifugal forces move the heavier material toward the wall in the outer spiral, and upward with the medium toward the discards outlet. The lighter material (coal) moves toward the inner spiral under centripetal forces, and with the rest of the medium move out toward the product outlet at the bottom of the separator.

Less horsepower will be required to supply the energy required to create the vortex, as this energy only needs to be applied to the medium that enters at the bottom. The feed enters at the top non-pressurized. This has an influence on the capital and operating costs of the process. There is also a medium-density gradient over the length of the separator. As the discards are moved to the outer spiral and up in the

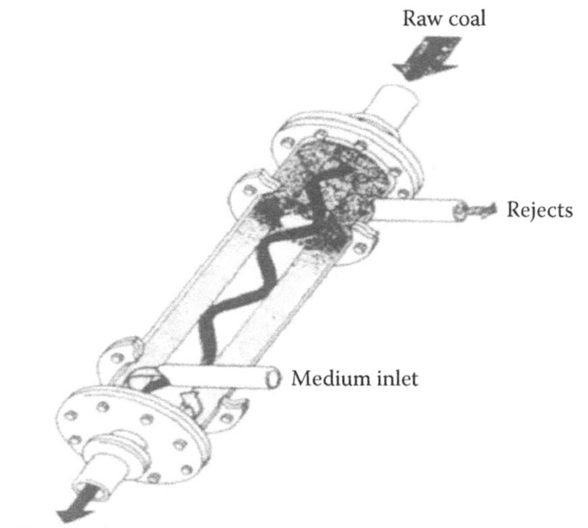

FIGURE 12.11 Dyna whirlpool separator. (From Osborne, D.G., *Coal Preparation Technology,* vol. 1, London, Graham & Trotman, 1988.)

separator, the density of the medium increases. Hence, the feed experiences a higher density at the entrance of the separator, which aids the separation.

As with all cyclonic separators, the material of construction for the separator is crucial since the more abrasive material (the discards) is moving with significant velocity against the wall of the vessel. Ceramic inserts or liners are used, as well as special nickel-based hard metal alloys to prevent excessive erosion. An advantage of the Dyna whirlpool's design is that the length of the vessel exposed to the harshest conditions is shorter than with the standard cyclones, as the highest concentration of discard material is closest to the outlet. Nevertheless, manufacturers prefer to fabricate the cyclones of wear-resistant material or incorporate liners, to extend vessel life. These separators are also capable of handling large-size coal particle (up to 100 mm).

12.8.2.3.3 Vorsyl Separator

This was developed by the Mining Research and Development Establishment of the British Coal in 1967 for the treatment of coal, sized from 50 to approximately 0.5 mm. This separator is vertically mounted, and has a separate discard removal section called the *vortex tractor*. The Vorsyl separator consists of vertically mounted completely cylindrical separating chamber with a cylindrical inlet at the top and an annular opening, the throat, at the bottom. The throat is encircled with a shallow shale chamber, which is provided with a tangential outlet and is connected by a short duct to a second shallow chamber known as *vortex tractor*. This is also a vertically mounted cylindrical chamber with a chamber with a tangential inlet for rejects and the medium and an axial outlet. A vortex finder is incorporated in a separating chamber eccentrically passing in a downward direction to the bottom of the chamber.

The feed consisting of deslimed raw coal, together with the separating medium of magnetite in water, is introduced tangentially, or more recently by an involute entry, at the top of the separating chamber, under pressure. Coal particles of specific gravity less than that of the medium passes into the clean coal outlet via the vortex finder, while the near-gravity particles and the heavier shale particles move to the wall of the vessel due to the centrifugal acceleration induced. The principles are the same as for the Dyna whirlpool, except that here the feed and medium are in co-current flow versus the counter-current flow in the DWP. A separate extractor is provided for the separation of the discards and the floats from the main vessel (Anon 1979).

12.8.2.3.4 Tri-Flow Separator

The Tri-Flo separator, as shown in Figure 12.12, combines two stages of dense medium cylindrical cyclones separation in a single unit operation that have been installed with a slope of 20° from horizontal. The cylindrical body of the Tri-Flo consists of two consecutive cylindrical chambers with an axial orifice. Each cylindrical chamber is equipped with an involutes media inlet and sinks discharges. The feed is sluiced with a small amount of dense medium and added to the first chamber of the vessel at atmosphere pressure, produced float and sink 1. The float from the first stage is the feed to the second chamber, at lower specific gravity, produced a sink 2 (middling) and the final float product (clean coal), Operation of Tri-Flo cyclones in general can be fed by gravity flow using head tank. Thus, there is no need pumping of feed coal. It can reduce energy and wear on the pumps and any degradation of the product. Additionally, the low-level feed entry makes a lower building and shorter feed conveyors, resulted in saving of space and cost.

Using dense medium cylindrical type cyclone separators in coal-cleaning plant, the raw coal can enter the separator at a much larger inlet compared with a dense medium conical cyclone. Typical top size for Tri-Flo separator is 45 mm for a 500 ID unit and 70 mm for a 700 mm ID unit. In Tri-Flo

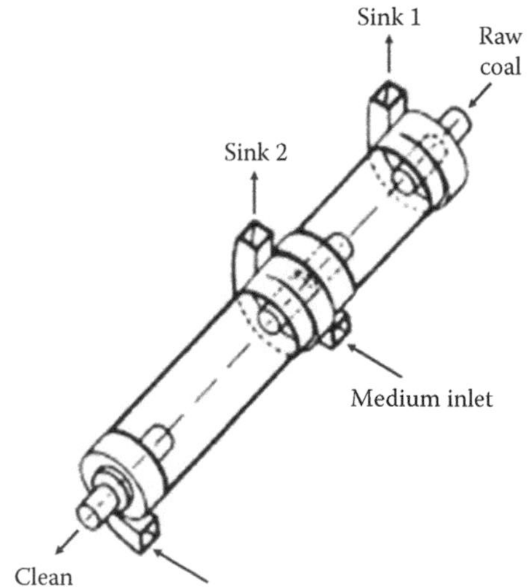

FIGURE 12.12 Tri-flo separator. (From Osborne, D.G., *Coal Preparation Technology,* vol. 1, London, Graham & Trotman, 1988.).

separator, there are two sinks outlets available. The second cleaner stage ensures that any remaining heavy materials or NGMs are rejected in the second stage, so it has high sinks capacity.

12.8.2.3.5 LARCODEM Separator

A further refinement of the Vorsyl separator was developed by Mining Research and Development Establishment, named the *large coal dense medium separator* or LARCODEMS. The unit, as shown in Figure 12.13, is also cylindrical, but mounted at an angle of approximately 30 degrees to the horizontal. Here, the medium flow and the feed material are in counter-current, like the DWP, and a separate discard extractor is provided analogue to the Vorsyl separator. The vortex tractor regulates the medium exit rate. This extractor is different from the one on the Vorsyl in that the outlet of the extractor of the LARCODEMS is off-center, which gives improved control. The medium split between discard and product outlet can be varied from 60/40 to 40/60, to suit the yields of product and discard, without detrimental effect on the separation. Furthermore, the separating efficiency is not materially affected when handling a feed with large percentages of discard material, which is an issue with dense medium cyclones. Efficiency characteristics can be maintained from discards yields as high as 100% and low as 20%.

The efficiency of the washer seems to be high as very attractive Ep values are claimed, which are given as follows:

Size Range (mm)	Ep (Separation Efficiency)
100.00–50.00	0.008
50.00–12.50	0.01
12.50–6.70	0.017
6.70–3.35	0.029
3.35–2.00	0.035
2.00–1.00	0.044

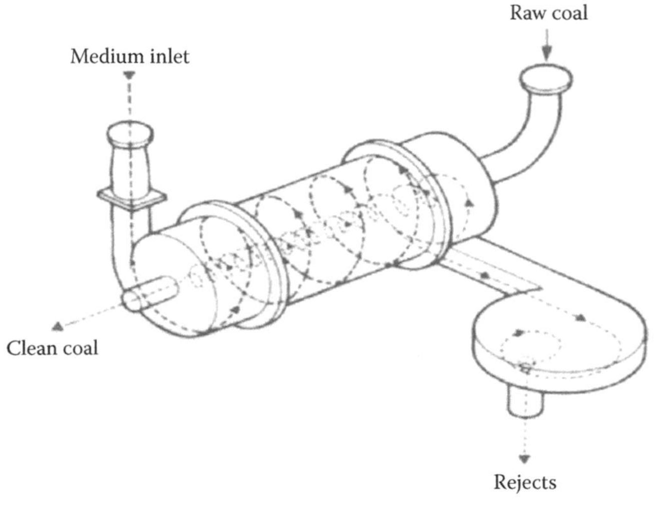

FIGURE 12.13 LARCODEM separator. (From Osborne, D.G., *Coal Preparation Technology*, vol. 1, London, Graham & Trotman, 1988.)

The previous paragraphs show that various technologies are available to beneficiate the coarse coal. The selection of the technology to use will have to be based on the achievable separation efficiency, the capital costs of the equipment, operating costs associated, and the effects misplaced material may have on the mass balances for whole washery.

12.8.3 Fine Coal Beneficiation

The need to crush coal to smaller sizes for better liberation of dirt for its subsequent removal by washing creates the problem of generation of more coal fines. Added to this, the adoption of mechanized mining has increased the quantum of fines generation in recent past and it will be increased day by day around the world. Up to 15% of ROM coal can be in the −0.5 fraction and economically this material showed an inferior product. Apart from the quantity, the qualities of these fines are enriched in coking properties with high percentage of ash. Beneficiation of coal fines, that is, coal of less than 0.5 mm in top size, was at one time not practiced in the world. Present trend in beneficiation shows that these fines must be beneficiated for getting a desirable quality of clean coal for coke-making purposes.

Normally, these fines are upgraded by conventional flotation cell in the washeries. Quality of feed coal to flotation plant having more fines, most of the flotation plants are either not running or running much below their capacity. Recovering additional clean from rich coal fines is possible by adopting different physical or physicochemical process. The upgradation of coal fines by some of the physical or physicochemical process has been discussed.

12.8.3.1 Physical Processes

12.8.3.1.1 Water-Only Cyclones

Cyclones were first developed at the DSM, Limburg in 1939; since that time, it has gained tremendous worldwide popularity, particularly in coal and mineral beneficiation industries for the separation of particles on the basis of size and specific gravity.

Unlike the HMC, the water-only cyclones differ in their design and appearance. The cone angle of the water-only cyclone is wide angled with an intruded angle of 120° or above. Besides this, the unit consists of long and large vortex finder. By virtue of this design, these units have gained the property of creating their own medium and hence they are also named as *autogenous cyclones*. Therefore, water-only cyclones are used for the separation of minerals based on specific gravity difference. The use of water-only cyclones for cleaning fine coal is gaining wide acceptance (Kim and Klima 1998).

In water-only cyclone, the feed in the form of slurry is introduced tangentially through the feed inlet, positioned at the cylindrical top of the cyclone, that is, just below vortex finder position. This tangential force imparts swirling motion to the pulp. This helps in the formation of vortex inside the cyclone with a low-pressure zone along the vertical axis. A stable bed is formed in the flat section of the cyclones with large cone angles, that is, 120°. Within the bed, the particles with high settling velocities are further separated according to specific gravity and size. The low-specific gravity particles migrate to the top of this bed and are stripped by the upward flow component near the air core interface. These particles then change of leaving through the vortex finder (depending on its location and the particle trajectory) or are returned to the bed (Rogers 1985).

12.8.3.1.2 Spiral

In a recent survey of worldwide coal preparation practices, it was reported that spiral concentrators treat 6% of the coal treated in processing plants. The application is typically associated with cleaning the 10.15 mm fraction of the plant feed (Holland-Batt et al. 1984).

The popularity of spiral concentrators is due to their simplicity, low cost, and ability to maintain minimal by-pass of low-density particles to the tailings stream. The principle of operation of spirals is simple. The equipment is more or less like a chute (Sivamohan and Forssberg 1985). The raw slurry, having 30%–40% solids content, is fed at the top of the spiral. As the slurry flows down along the spiral, it acquires certain velocity and centripetal forces, which start working on coal particles. The separation of cleans, middlings, and rejects takes place because of differences in their specific gravities. The reject particles get thrown toward the center of the spiral because of their high specific gravity. Similarly, the middlings stay in the center and the cleans collect toward the periphery of the spiral. At the bottom of the spiral, diverters/splitters are provided, which guide the three products (Hubbard et al. 1950).

Based on a typical coal-cleaning application, spirals effectively treat between 2 and 3 t/h per start at a recommended solid concentration of 30% by weight (Balderson 1982). The separation performance of a spiral is determined by both feed characteristics and operating parameters associated with the spiral. Feed characteristics include particle size-by-size weight and density distribution, volumetric flow rate and the solid concentration (Richards et al. 1985).

12.8.3.1.3 Reflux Classifier

The reflux classifier is a novel fluidized bed separator that has been in development since 2001. The reflux classifier consists of two sections: a lower vertical section and an upper inclined section divided into a number of parallel inclined channels. Within the inclined section, particles have only a relatively short vertical distance to fall before they settle against the lower surface of the channel, from where they can slide back down into the vertical section. This so-called Boycott effect produces an increase in the effective settling area compared to a conventional fluidized (or teetered) bed separator with the same footprint (Galvin et al. 2010).

In reality, particles that settle on the channel wall do not automatically slide back down into the vertical chamber. Some of them may be re-suspended. Although this reduces the capacity of the device as a clarifier, the re-suspension mechanism is density dependent, and so the ability of the device to separate particles based on their density is actually enhanced. Most early work on the reflux classifier used relatively wide channels under conditions of turbulent flow. However, by using the mechanism of a high shear rate coupled with laminar flow through relatively narrow channels, reflux classifiers with narrow channels were able to perform density separations on particles down to 38 μm in size, provided, of course, a commensurate reduction in solids throughput is applied. This has been confirmed at both laboratory and pilot scale (Galvin et al. 2005).

12.8.3.2 Physicochemical Processes

12.8.3.2.1 Froth Flotation

Froth flotation is a universally accepted process for the beneficiation of coal fines below 0.5 mm. This process utilizes the difference in surface properties of coal and its associated impurities for the separation. The coal surfaces being hydrophobic in nature are coated with some selective reagents, whereas the minerals of hydrophilic character remain water wet. The reagent coated coal particles are levitated to the surface of the flotation cell with the help of air bubbles and skimmed off (Figure 12.14). The impurities, that is, mineral matters are taken out as pulp from the bottom of the cell. The floated froth is collected and filtered (Fuerstenau 1976).

The modern mechanized mining methods utilized for winning the coal generate more coal fines and due to admixture of sand particles and other impurities, its quality deteriorates. The crushing of ROM coals to some suitable sizes for liberation and beneficiation also generate coal fines. Consequently, the beneficiation of coal fines, particularly when they are of coking variety, has become essential to recover the vitrinite-enriched coals, which are having the best coking properties (Fuerstenau et al. 1985). The coal slurry at a pulp density of 30%–35% is conditioned with diesel oil at the rate of (1.0–1.25 kg/t) of coal. The diesel oil termed as *collector* is dispersed, collided, and adsorbed on to the coal surfaces due to mechanical agitation in the conditioner. In some installations, the collector is emulsified

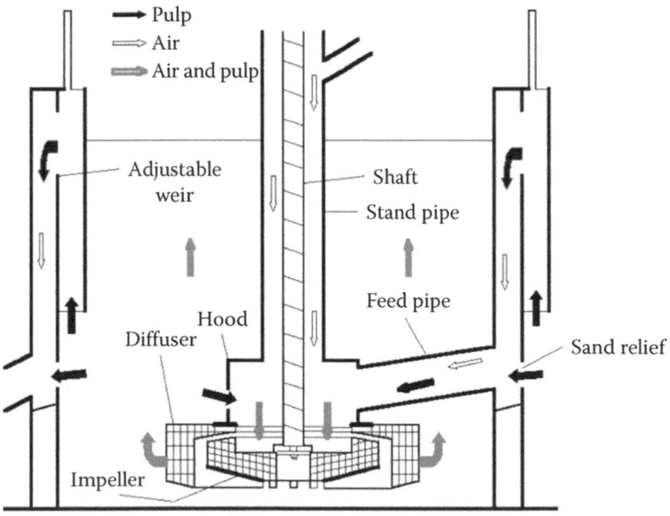

FIGURE 12.14 Typical flotation cell. (From Subba Rao, D.V. *Coal—Its Beneficiation*, New Delhi, India, Emkay Publications, 2003.)

before feeding to conditioner. The conditioned pulp is then diluted to the desired level (10%–15%) in the pulp density adjusting tank and then brought to the battery of flotation cells. Frother, normally pine oil or methyl isobutyl cyanate is added at the rate of 0.15–0.25 kg/t of coal. The conditioned coal particles attach with the air bubbles and reach to the top of the cells to be skimmed off. The remaining slurry is fed to the next cell and the same operation is repeated. Thus, sufficient residence time is provided to separate out the coal particles from its impurities (Wills and Napier-Munn 2006). The impurities in the form of tailings are taken out from the last cell. The beneficiated coal fines in the form of froth are collected and vacuum filtered.

One of the most significant discoveries in connection with flotation was that of Froment in 1902. It was particularly significant since it eventually led to flotation, as we know it today. It was not until 1918 that froth flotation was applied to coal beneficiation. Since then, the process has grown in popularity; until today, there is hardly a new preparation plant being designed or constructed that does not include froth flotation.

Much of the popularity of froth flotation in coal preparation was brought about by increasingly stringent water pollution control legislation. In addition to this legal or environmental problem, preparation plant operators were finding that because of the increased quantity of fines produced by modern mining and preparation methods, it was to their economic advantage to recover as much fine coal as possible.

Froth flotation continues to be the most widely accepted technology for the treatment of slurry forming fines (−0.5 mm), particularly when they are of coking variety. In recent years, a design trend of increasing volumetric capacity in individual machines has emerged with the result that most current installations employ machines of capacity between 8.50 and 14.2 m^3. There are, however, a number of larger machines (28.30 m^3) available, although many plant designers regard the 14.2 m^3 machines as being optimum when taking into account performance, capital, and operational costs (Eberts 1986).

Although most cells are mechanically agitated, often with some supplemental air, there are some new cells, which are strictly pneumatically actuated. Both the Column Flotation Company (Canada) and the Deister Concentrator Company (United States) have developed tall vertical cells up to 1.8 m^3 in cross section and over 12 m high. Air is introduced at the bottom of the column into a manifold that distributes the bubbles uniformly over the cross section of the unit with no mechanical agitation. A different type of machine called the *Heyl and Patterson cyclocell* has been developed in the United States in 1978 and has been on the market for many years as a strictly air agitated cell. Instead of mechanical impellers, this cell has submerged vortex chambers, which agitate the slurry and introduce air. The vortex chambers create cyclonic motion of the slurry, and low pressure is introduced at the center of the turbulent zone creating shearing of the incoming air into a multitude of fine bubbles.

One of the most sophisticated and beneficial devices now in use in Soviet coal preparation plants is flotation reagent metering device that measures the flow rate of the pulp to the flotation cells together with percent solids in the pulp and then delivers the required quantities of flotation frother and collector. This device gives a constant quality clean coal product with minimum loss of coal in the tailings product (Klassen and Mokrousov 1963).

12.8.3.2.2 Column Flotation

A significant development in flotation over the past few years has been the increasing industrial use of flotation columns. The column differs dramatically from conventional mechanical flotation machines, both in design and operating philosophy (Kawatra and Eisele 1987).

The basic principle, as depicted in Figure 12.15, involved in a flotation column is the counter current flow of air bubbles and solid particles and behavior similar to a plug flow reactor. The bubbles are generated by injecting air in the diffuser placed in the bottom of the column. Bubbles move upward in counter direction to downward flow of slurry. The attachment of hydrophobic mineral particles to the air bubbles take place in the lower enrichment section of the column between the feed point and air inlet known as *flotation zone*. The froth from flotation zone moves to cleaning zone (between interface and top of the column). The cleaning zone is a mobile-packed bubble bed that is contacted counter currently with wash water from the top of the column to remove the entrained gangue particles from the froth and send back to the flotation zone (Osborne and Foneca 1992). The design and operating philosophy of flotation column are totally different in comparison to conventional flotation cell (Rubinstein 1995).

In a typical flotation column, the feed slurry containing floatable particles is injected into the pulp zone for contact with the rising bubbles. Numerous efforts are aimed toward

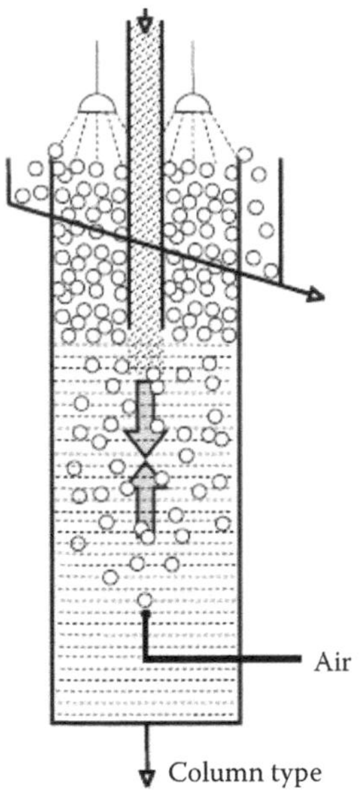

FIGURE 12.15 Column flotation.

increasing the probability of bubble-particle collision, reducing the degree to which hydrophobic particles are sheared off as bubbles transition from liquid to froth, maintaining an optimum ratio of bubble size to particle size. Problems in the pulp are high-liquid fraction, low number density of bubbles, and high turbulence. Even though the actual mineral collection depends on bubble-particle collision and their attachment, the overall flotation performance depends on the froth phase where separation occurs (Dobby et al. 1985).

The column is particularly attractive for applications involving multiple cleaning stages and can upgrade in a single stage compared with several stages of mechanical cells. This results in simpler, more controllable circuits. Importantly, the column itself is well suited to computer control.

Conventional columns provide better grade than the conventional flotation cells due to a deeper froth zone and employment of wash water to reduce the amount of hydrophilic material entrained in the rising froth that reports to the overhead product stream. Patil et al. (2010) showed that the columns with a longer froth zone will require only a relatively shallow pulp zone to obtain a low-ash clean coal at high yield. The relative balance between desired grade and recovery could be achieved by adjusting the location of the feed inlet to the froth section, combined with the extent of external reflux to the column.

12.8.3.2.3 Oil Agglomeration

Wet agglomeration processes are those in which the size enlargement occurs among particles suspended in a liquid phase. There are a number of ways in which this can occur, and perhaps the simplest is by the addition of electrolytes, which cause a reduction in zeta potential, with the resultant formation of coagulates. The next order of the process is the substitution of polymer flocculants for electrolytes, to create the collection of larger quantities of particles into floccules, which in some circumstances become quite massive in size (up to 5 mm). In the third order of process, that is, spherical agglomeration, finely divided solids are treated with a so-called *bridging liquid*, which preferentially wets the solid and is immiscible with the suspended liquid. On agitation, the bridging liquid becomes distributed over the exposed surfaces of the dispersed solid particles, and upon particle collision, the bridging liquid forms junctions (Farnand et al. 1961).

The *oil agglomeration* process is for the effective beneficiation of coal fines, finely ground high ash difficult-to-wash coals and washery middlings, which do not respond satisfactorily to the conventional processes, in laboratory and bench scales by conducting tests with coals from difficult sources, seams, and washeries. It was observed that the technique has three distinct merits:

- High yield of cleans with very low loss of carbonaceous matter through tailings
- Easy dewatering characteristics of cleans
- Improvements in the coking propensities of the cleans

When the natural coal fines or fine ground high ash coals/middlings in thickened slurry, under controlled pH, are agitated in a conditioning cell along with mineral oil (1%–2% by wt), the coal particles get preferentially coated with a thin layer of oil. These selectively coated coal particles along with non-combustibles and water are again agitated in an agglomeration cell in the presence of viscous/heavy petroleum oil (6%–7% by wt), whereby clean coal particles form dense compact spherical agglomerates and the mineral matters of the raw coal remain dispersed in water. These materials—agglomerates and tailings—are passed over bent sieve followed by a vibrating screen, whereby the water carrying the mineral matter passes through the aperture of the screens and agglomerates being bigger in size are separated and collected (Moza et al. 1976).

Thus, oil agglomeration is a process in which fines/ultrafines coal particles are bonded together to produce low ash and compact coarse sized agglomerates in liquid suspension by selective wetting (conditioning) and bridging with a second immiscible liquid. The process provides an attractive method of beneficiation and recovery of fine coal in the form of oil-bonded pellets.

12.8.3.2.4 Jameson Cell

In the 1960s flotation was introduced to treat fine coal using a bank of conventional mechanical flotation cells. These cells while partially successful in their application still sent a significant amount of fine coal to the tailings discharge and dumps. The early 1990s saw the introduction of Jameson cell technology to the coal industry. The fine bubble size, high intensity mixing, and froth-washing capabilities of the Jameson cell were perfectly suited to flotation of fine coal particles, and the cell was immediately able to achieve greater yields than the conventional flotation machines (Figure 12.16). In addition, the low fine ash product produced from the Jameson cell allows for greater flexibility in operating the coarser coal

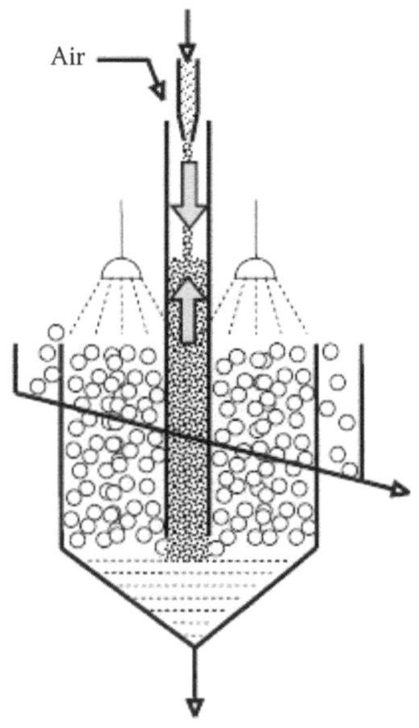

FIGURE 12.16 Jameson Cell.

circuits leading to significant improvements in the overall yields while still meeting product ash specification. Jameson cells have subsequently been installed in many coal preparation plants reducing the amount of fines being deposited in coal tailing dumps (Jameson and Manlapig 1991).

Jameson flotation cell is an innovative flotation cell developed as an alternative to column flotation. It is a high intensity and high efficiency flotation device, which uses induced air from the atmosphere. In this technology, air is absorbed from the atmosphere into the downcomer through the influence of the jet action caused by the highly pressurized pulp. This innovative flotation device can produce very fine bubbles (−0.3 mm) without a compressor or an impeller mechanism. When compared with the conventional cells, the main advantages of the Jameson cell include high-production capacity, excellent separation ability, no moving parts, simple construction, and low capital and operating costs (Hacifazlioglu 2011).

The introduction of Jameson cell technology into coal preparation plants has improved their overall environmental performance, improved mass recovery, and improved product quality. Additionally, Jameson cell technology has found exciting new applications such as recovering coal from tailings dams, which have both environmental and economic benefits. The ability to recover coal, while at the same time reducing the quantity of material impounded in tailings dams, has enormous benefits, economically and environmentally. Since the first installation, treating fine coal at Newlands Coal, Bowen Basin, in 1990, systematic reviews and assessments of the Jameson cell performance in the coal industry have resulted in a number of improvements and developments in the technology (Osborne and Foneca 1992). A decade after this first installation, research and development is still continuing to further advance the technology. Improvements include investigations into the jet velocity in the Jameson cell down comer, which has shown that by slowing the velocity the subsequent drop in cell turbulence has improved not only the recovery of fine coal particles but also the recovery of coarse coal (often the result of upstream plant upsets). Improvements such as these continue to be made to further advance the Jameson cell technology. Since the first coal industrial scale, Jameson cell installation at Newlands Coal Pty Ltd., in 1990, to treat coal fines, over 100 Jameson cells have been installed to service the global coal industry (Mohanty 2001).

12.8.4 Performance Evaluation of Washing Units

The primary objective of the majority of the coal-washing plants is selective separation of ash-forming materials from the combustible materials. Gravity concentration is one of the most important unit operations in many coal-washing circuits. Various types of gravity concentrators are available in the market and therefore, the coal washing plant designer's task is to select the most efficient type of separator to wash coals of specific characteristics. In order to carry this out, normally extensive laboratory data are first generated with a particular coal using various machines, and based on the performance of each machine, the desired washing circuit is designed. In an existing plant, the processing engineer's task is to optimize the overall plant performance (Gottfried 1978). Therefore, it is imperative that a coal preparation engineer evaluates the performance of coal-cleaning units accurately (Walters 1976). Some of the performance criteria to evaluate any gravity separator units are highlighted below.

12.8.4.1 Efficiency

Effectiveness of operation or accuracy of separation as measured usually by a comparison of actual and possible results.

12.8.4.2 Performance

A statement describing the scope and duty of a plant and often also the expected limits of results obtainable in plant operation in terms, for example, of the tonnage of coal treated per hour, the process used, the accuracy with which the separations are expected or the operations are controlled, and sizes produced or products obtained.

12.8.4.3 Theoretical Yield

The maximum yield of a product, as read from the washability curve with a specified percentage of ash or alternatively at a specified density.

12.8.4.4 Yield

Yield is the amount of the clean coal expressed as a percentage of the amount of that material in the quantity of feed treated.

12.8.4.5 Partition Curve or Distribution Curve

The curve indicating for each relative density or size fraction, the percentage of it, which is contained in one of the products of the separation, for example, the reject.

12.8.4.6 Partition Factor or Distribution Factor

The percentage of a relative density or size fraction recovered in one of the product of the separation, for example, the reject.

12.8.4.7 Ecart Probable (Mayen), Probable Error, EPM

One half of the difference between the densities corresponding to 29% and 75% ordinates (distribution factor) as shown in the partition curve.

Osborne (1988) recommends that the variation of EPM with particle size, equipment size, and separation density should be computed using a series of factor:

$$\text{EPM} = f1f2f3Es \tag{12.1}$$

where:

- $f1$ is a factor accounting for the variation of EPM with particle size
- $f2$ is a factor accounting for variation of EPM with equipment size
- $f3$ is a manufacturer's guarantee factor usually in the range 1.1–1.2
- Es is a standard function representing the variation of EPM with separation density for each type of equipment. Es for various types of coal washing equipment are as follows:

Dense-medium cyclone: $Es = 0.027\,{}^{p}50 - 0.01$
Dyna whirlpool: $Es = 0.15\,{}^{p}50 - 0.16$
Dense-medium bath: $Es = 0.047\,{}^{p}50 - 0.05$
Baum jig: $Es = 0.78({}^{p}50({}^{p}50 - 1) + 0.01)$
Water-only cyclone: $Es = 0.33^{p}\,50 - 0.31$
Shaking table and spiral concentrator: $Es = {}^{p}50 - 1$
For dense-medium cyclones $f1$ varies from 2 to 0.75 as particle size varies from 0.5 to 10 mm.
For dense-medium vessels, $f1$ varies from 0.5 for coarse coal to 1.4 for small coal.

Because both the Baum jig and the water-only cyclone are autogenous gravity separators, $f1$ is dependent on the size distribution and washability of the feed material. For jig values of $f1$ range from 3 to 0.5 as the average size of the feed material varies from 1 to 100 mm.

12.8.4.8 Imperfection

The ratio of the Ecart probable to partition density minus one.

12.8.4.9 Fraser and Yancy Efficiency (or Organic Efficiency)

$$\text{Fraser and Yancy efficiency (or organic efficiency)} = \frac{\text{Actual yield}}{\text{Theoretical yield}\begin{pmatrix}\text{at the same ash level}\\ \text{of the actual yield}\end{pmatrix}} \times 100 \quad (12.2)$$

12.8.4.10 Anderson Efficiency

$$\text{Anderson efficiency} = 100\begin{pmatrix}\text{Clean coal in sinks + Sinks in clean}\\ \text{coal at the effective density of cut}\end{pmatrix}$$

12.8.4.11 Ash Error

Difference between the ash percentage of clean coal from a separating process and that of the equivalent yield level indicated from washability curves.

12.8.4.12 Yield Reduction Factor

The yield reduction factor is defined as the percent reduction in yield for each 1% reduction in ash content at any selected ash level of the clean coal.

12.8.4.13 Recovery of Nonash Material

$$\text{Recovery of nonash material} = \frac{\text{Yield\%} \times (100 - \text{Concentrate ash\%})}{(100 - \text{Feed ash\%})} \quad (12.3)$$

12.8.4.14 Efficiency Index

$$\text{Efficiency index} = \frac{(\text{Recovery of nonash material} \times \text{Tailing ash\%})}{\text{Product ash\%}} \quad (12.4)$$

12.9 DEWATERING

The wet beneficiation of coal consumes vast quantities of water. After processing, the finished product must be dewatered. Conventional dewatering techniques utilize thickeners, dewatering screens, vacuum filters, centrifuges, and pressure (hyperbaric) filters to reduce product moisture to a lower value. The driving forces exerting dewatering range from 50 to >5,000 g (for centrifuges) or 0.3–200 bar (for filters). In general, the greater the driving force exerted by any machine, the lower the throughput and the higher the cost. Choice of the dewatering technology for a specific material is dependent on many factors such as particle size and initial moisture content. Dewatering is accomplished by many devices of which dewatering screens, centrifuges, and sieve bends provide a low-moisture product with +600 μm coal; however, their efficiency decreases rapidly with decreasing particle size. Dewatering of −600 μm fine coal is often accomplished using either vacuum or pressure filtration (Svarovsky 1990).

Where the particles are between about 8 and 0.5 mm, dewatering is mainly accomplished by using centrifuges. Centrifuges are the machines, which effectively create high gravity forces for purposes of dewatering coal. They find application in virtually every wet washing coal plant. The centrifuges developed for the coal industry are reliable, efficient machines. Their products are consistent, uniform, and easily handled. Properly centrifuged coal can be further dewatered only by evaporation of the moisture remaining on the coal.

Vacuum filtration is one of the dewatering processes widely used in the coal industry. It has substantial advantages, for example, low cost and easy to design and to maintain, over hyperbaric filtration and centrifugal filtration. The most significant advantage is that of continuous operation under relatively simple mechanical conditions. Continuous vacuum filter machines, particularly rotary drum filters and rotary disc filters, have long proved themselves to be capable of producing satisfactory performance in various commercial applications. Although the vacuum filtration process is accomplished over several steps, there are only two major stages: (1) the cake formation characterized by a consistent increase in cake thickness and single phase (water) flow through the cake and filter medium and (2) cake dewatering characterized by the flow of two phases (air and water). During the cake dewatering stage, water present in capillary pores is continuously removed until only discrete lenses of water are left, eventually leading to the lowest possible cake moisture, which is dependent on the cake configuration. In practice, however, this lower limit of cake moisture is not achievable even with extended cake dewatering time due to blockage of pores by micron or submicron particles (Osborne and Pierson 1990).

Standard disc or vacuum filters are still the most common devices for dewatering −0.5 mm fines. Steam-assisted vacuum filters have also been successfully employed in some installations, whereby the free moisture content of the slurry cake is reduced from the level of 25%–17%. But increasing attention is being given to belt filter press because of the problem of dealing with high ash tailings. This type of filter has recently been utilized for coal treatment, having been tested in sewage treatment and other industries over a period of almost 15 years (Purchas and Wakeman 1986).

The filter presses though expensive, is of frequent use in the United Kingdom and provides a positive means of filtering tailings of high clay or ash content. There are several pressure filters now in operation in U.S.A. The multiple roll band press filter of numerous configurations are now in the market for filtering difficult to dewater materials. They appear to hold considerable promise. The Arus-Audritz and the Magnum press filter are now in commercial use in the United States. The Polish and the Japanese have developed a rotary drum dehydrator. It should be mentioned that in practically all these types, the careful use of flocculating and conditioning agents is required (Schwalbach 1989).

In static clarifying devices, generally called *thickeners*, use of smaller diameters are being introduced with novel feed arrangements to improve their efficiencies. In 1962, the British have developed the *deep cone* thickeners of relatively small diameter but of deep conical shape. The standard thickener has undergone certain changes. Two newcomers for this job are the Lamella and Enviro-clear thickeners. The former uses a series of inclined plates in which the solids settle and slide to bottom collecting tank. The Enviro-clear is unique in that the feed is pumped into the bottom of the thickener and percolates up through the thickening zone. Both systems claim to be three or four times more effective than standard thickeners (Keane 1986).

Presently, typical fine coal-dewatering circuits produce filter cakes with moisture contents of the order of 20%–25%, which often are sufficiently high to prevent coal preparation plants from achieving overall product moisture that meets increasingly stringent customer specifications. Of the currently available techniques, hyperbaric filtration provides a low-moisture product for the fine clean coal slurries. It adds a significant cost to the price of clean coal, as it is energy intensive and costly capital expenditure process. Hyperbaric filters, which are heavily used in Europe and China, have shown the capability of producing filter cakes containing less than 20% moisture (Groppo et al. 1995).

Dewatering aids have been developed claiming increased adsorption on solids and measurements of the filtrate have shown very little decrease in surface tension but significant decrease in cake moisture. There are three principal mechanisms to aid the dewatering on using surfactants (Pearse and Allen 1983):

- By lowering the surface tension, thereby reducing the capillary forces between coal particles, thus making water removal easier.
- By rendering the coal surfaces hydrophobic. This is a dewetting effect resulting from surfactant adsorption at the water-coal interface causing the coal to be poorly wetted by the water.
- By flocculating the coal particles to give bigger aggregates having larger free spaces between them through which water may be more readily drained.

Gray (1958) used flocculants, oil, and surfactants and reported that each improved dewatering. The benefit of the oils may have been due to impurities of surface-active components in the oils, while evidence of surfactants only lowered the liquid/air interfacial tension. Dolina and Kominski (1979) used several surfactants during vacuum filtration and found that the residual moisture content of the filter cake decreased or increased depending on which surfactant was used. This is not surprising because solution pH and electrolyte content have a significant influence on surfactant adsorption. Nicol (1976) reported improvements in dewatering of coal using anionic surfactants, while Brooks and Bethell (1976) found that a cationic surfactant also improved dewatering. Keller et al. (1979) showed that surfactants (1) changing the pressure differential required for dewatering and (2) lowering the residual water content of the filter cake. The pressure differential required was correlated with a decrease in surface tension while the residual moisture content was related to surfactant adsorption at the solid/air interface. Cationic, anionic, and nonionic surfactants showed similar effectiveness at different dosages (Gala et al. 1983).

The use of surfactants as filtration aids shows potential benefits for lowering cake moisture, particularly in longer filtration cycles. There is contradictory evidence in the literature pertaining to the mechanism responsible for this improvement. Some evidence exists for surfactant adsorption increasing the hydrophobicity of the substrate. Additional research suggests that lowering the surface tension at the solid/liquid interface is the primary mechanism. Regardless of the mechanism responsible, removal of *free* or *surface* water is the primary objective and the addition of agglomerating, as well as surface tension modifying reagents can indeed reduce the moisture content of fine coal (Wen et al. 1993).

Surfactants used in the study by Firat et al. (2015) showed that they are all able to decrease surface tension of the coal slurry. Lowering viscosity is much more effective than other mechanisms on lowering cake moisture. It helps to remove water from coal surfaces reducing the capillary forces between particles. PEG-400 was adsorbed at the gas-liquid interfaces, causing a large reduction in surface tension and provides greater reductions in the residual cake moisture.

In recent years, the need for improved floc conditioning is being increasingly recognized. Operators are assessing options for improved and controlled mixing of flocculants into filter feeds, and also the use of mixed flocculant formulations (anionic and cationic) to produce smaller tighter flocs, which in turn can produce enhanced cake permeability without large increases in porosity. It has been demonstrated that enhanced fine coal dewatering can be achieved by a number of different approaches such as the use of flocculants and surfactants. Two distinct types of proprietary chemical additives are generally available for industrial use: flocculant filter aids and surfactant dewatering aids. Flocculants have been used in vacuum filtration to decrease the moisture of the product cake. This has been tried in centrifuges with minimal results, due to the weakness of the long polymer chain present in flocculants and the strong forces occurring in a centrifuge. These strong forces break apart flocculants before they can aid in dewatering. Even though flocculants cannot be effectively used in centrifuges, fuel oil has been well known to aid in dewatering. Fuel oil will adsorb onto the coal surface. This results in a much lower interfacial tension than at the coal-water interface. This lower interfacial tension results in lower pressure differences needed to dewater. Surfactants and a combination of surfactant coagulant can be used for dewatering of coal fines in centrifuge. Different combinations of reagents can also be used for reduction of moisture of fine as well as ultrafine coal in filtration (Hogg 2000).

12.10 RECENT DEVELOPMENTS IN COAL BENEFICIATION

12.10.1 Enhanced Gravity Separators

Enhanced gravity concentrators employ a centrifugal force to enhance the settling rate of particles. The use of centrifugal force to increase the settling rate of particles has been applied for many years for classification in hydrocyclones and for dynamic heavy medium separation (Luttrell et al. 1995). In a cyclone-type separator, centrifugal force is generated mainly by tangential feed entry mechanism, but in an enhanced gravity concentrator, the centrifugal force is generated by rotating the separating chamber itself. Therefore, in a cyclone, the separation of heavy and light particles occurs due to the formation of an air core as a result of a pressure drop, but in an enhanced gravity concentrator, the separation is primarily based on the relative settling velocity differential between the particles. To process ultrafine particles, cyclone-type separators are found to be inefficient, probably because of their inherent design constraints (Honaker et al. 1995).

12.10.1.1 Falcon Concentrator

The Falcon C-series unit is essentially a centrifugal sluice and consists of a smooth-surface truncated cone, which rotates at a very high speed. Feed slurry is introduced onto a spinning rotor at the bottom of the bowl and is accelerated up the cone wall by the centrifugal field (up to 300 Gs). A thin flowing film of slurry is formed along the cone wall and based on the hydraulic differences, the particles are stratified. Heavy particles sliding along the inner surface of the cone are discharged through the cone wall via small ports, while the light particles atop the stratified layer are discharged over the top of the cone lip (Honaker and Wang 1996).

12.10.1.2 Kelsey Jigs

This is the only known commercially available centrifugal jig and Chris Kelsey in Australia developed it. The Kelsey jig is essentially is a Harz jig placed vertically in a centrifugal field. It consists of a rotating bowl with a series of individual hutches wrapped around it. The hutches hold the pulse water and discharge concentrates through their spigots. Within this bowl, a wedge wire screen is placed to retain a ragging bed. A diaphragm whose stroke is controlled by a cam-motor assembly independent of the main drive provides the pulsation (Majumdar and Barnwal 2006).

Selective ragging materials, having intermediate density between the heavy and the light minerals likely to be separated, are placed over the wedge wire screen. The feed enters from the top and the centrifugal force, maximum 60 Gs imparted by the bowl rotation forces the slurry to make contact with the ragging material. The high-frequency sequential strokes of the pulse arms create an inward pulse of water through the ragging bed that causes the bed to dilate and contract. This, in turn, results in differential acceleration of the feed and the ragging particles according to their specific gravity. Low-density particles flow across the ragging material and overflow from the top of the unit, while high-density particles pass downward through the ragging screen and are discharged through the spigots in the concentrate launder. This is, in short, the way it is intended to operate (Yerriswamy et al. 2003).

Recovery of ragging material and the extra screening and handling involved in feed preparation are the drawbacks of this unit, especially when investment is concerned. However, the Kelsey jig allows the jigging of particles that are too fine to be recovered by conventional jigs.

12.10.1.3 Multigravity Separator

The Mozley multigravity separator (MGS) was invented and developed by Richard Mozley Limited with backing from the British Technology Group. The operating principle of the MGS *may be visualized as rolling the horizontal surface of a conventional shaking table into a drum, then rotating it.* The heart of this separator is a slightly tapered drum that rotates around its horizontal axis to generate a centrifugal field ranging from 8 to 22 G. Feed is introduced into the drum in the form of dilute slurry via static feed pipe located at the front of each drum. Due to the centrifugal field, heavy particles are stratified close to the drum surface and are carried upward by rotating scrapers toward the concentrate launder, while the light particles are carried by the flowing film toward the discharge end of the drum. Before the heavy particles enter into the concentrate launder, a small amount of wash water is added to wash out any entrained low-density particles (Cordingley 1997).

12.11 COKING COAL BENEFICIATION

When many kinds of bituminous coals are heated in an inert atmosphere, they soften and swell to form a plastic mass. The escaping volatiles pass through this plastic material, which then re-solidifies to a carbon-rich solid upon further heating. Coals that pass through a plastic stage on heating are called *caking coals*. Some caking coals resolidify on heating to form a hard, very strong, carbon-rich porous mass suitable for use as a reducing agent in the metallurgical industry. This material is called *coke*, and the special classes of caking coals that yield a satisfactory coke are known as *coking coals*. It is unfortunate that the words caking and coking are so similar, because they are not synonymous. All coking coals are necessarily caking coals, but not all caking coals are coking coals.

To be useful in the metallurgical industry, the coke needs to meet five criteria. To be a useful reducing agent, the coke must have very high carbon content. To keep the iron reasonably pure, the coke must have low contents of sulfur and ash. To provide ample heat, the coke must have a high content of fixed carbon and have a high calorific value. To let air pass through the fuel bed, but yet keep the fuel bed from being collapsed by the weight of iron ore, the coke must be quite porous and very strong. Finally, to help keep the cost of the iron low, the coke must be cheap.

Metallurgical coking coal is an important component in iron and steel production. It is converted into coke for use in blast furnaces as the heat source and reducing agent for converting iron ore into pig iron.

Only certain high-rank bituminous coals are classified as metallurgical coking coals. Of the one-third total world coal production currently cleaned, most is used to produce coke for the iron and steel industry, and the rest is almost entirely consumed in large utility boilers to generate electricity. Only 10%–13% of utility coal currently burned is cleaned to any degree. Thirty years ago, most coking coals were of such high quality as mined that only minimal cleaning was required; In the United Kingdom, West Germany, and the United States, large amounts of high-quality coking coal with ash content well below 5% were sent to the coke ovens without cleaning. Majority of the Gondwana coking coals are high in ash content and needs beneficiation prior to its use for metallurgical purposes (Narasimhan and Mukherjee 1999).

With depletion of higher quality coal seams and increased mechanization in mine, ROM coals have become progressively dirtier. Ash content in coking coal is required to be reduced to acceptable limit for the steel plant. Three product separations, that is, low ash clean coal with acceptable coking properties, medium ash steam coal, and discards, are becoming increasingly common where coking coals are to be obtained from dirtier ROM coals. Dense medium processing (baths, drums, and cyclones) is only considered to be applicable, as a *proven* technology, down to a bottom size of 0.5 mm, where the coal fines are mostly processed using spirals and physicochemical methods such as flotation/column flotation (Kalyan 1994).

12.12 THERMAL COAL BENEFICIATION

There is an urgent necessity for adopting some simple and economic schemes for preparing coal to meet the increasing demand of the power stations. From technoeconomic consideration also, it is becoming increasingly clear that coal for power station should be suitably processed and only clean coal/free of abrasive dirt is burnt in the boilers for thermal power generation. Apart from making economic utilization of the natural resource, it would also enable power sector to ensure reliable and efficient power generation, which is an imperative need of the day.

There are many significant direct and indirect gains in various forms, which the utilities and the consumers would be deriving by using adequately processed/cleaned coal in the power plants without significantly affecting the ultimate cost of energy.

A number of specific advantages may be obtained from beneficiating the thermal coal prior to its use and are summarized as follows (Goel 2011):

- The heating value of the cleaned coal is higher since the impurities are removed either partially or completely.
- Due to elimination of shale and abrasive material, there will be reduced wear and tear on the boiler plant and auxiliaries. This will reduce the incidence of forced outage.
- The beneficiated coal is more uniform in size, composition, and heating value, thereby resulting in more uniform and steady combustion than would otherwise be possible.
- By removing the sulfur impurities present in the coal beneficiation contributes to reduced slagging and fouling in the boiler combustion chamber. This increases the boiler's on-stream availability and reduces its maintenance costs.
- Removal of the associated mineral matter from the ROM coal can result in lower transportation costs, higher combustion efficiency, reduced ash disposal, and flue gas clean up requirements.

Recent studies conducted by the International Energy Agency have shown that 1% increase in ash (generally after passing the 10% ash level) results in a 1.2%–1.5% decrease in boiler availability and a decrease of 0.3% in boiler efficiency (Mark et al. 2000).

It is evident that the advantages thus gained by the utilities by way of improved and reliable performance ploughing in additional revenues will be far more than the extra cost that will be borne by them toward coal preparation.

To cut down the consumption of oil in existing oil-fired furnaces, coal-oil and coal-water mixture can be used for which low-ash coals recovered from non-coking coals by deep coal cleaning may be of great help. The purpose is to produce a fuel with similar incombustible residues to those of heavy oils.

Strong efforts are being made to develop new technologies suitable for converting coal into electric power with high efficiencies and low investment costs. Among these new technologies of coal utilization, fluidized bed combustion seems suitable, especially for smaller power plants. Out of the number of advantages of this system, the important one is its capability of burning: (1) the high ash as mined coal, (2) middlings/sinks, which are unacceptable in conventional firing systems, (3) rejects obtained from a 3-product washing system of a coking coal washery, and (4) rejects from non-coking coal washeries (Jimmy 2011).

12.13 COAL BENEFICIATION PRACTICE IN MAJOR COUNTRIES

12.13.1 Australia

Australia has abundant supplies of coal, mainly from Permian measures in the eastern part of the country. Annual production is around 420 Mt/y and coal is a major contributor to the Australian economy. The production from Australian coal mines is predominantly for the export market, although there is also a very large amount of coal used domestically for electricity generation (Swanson 2013). This means that coal quality must adhere to relatively tight specifications and therefore a very high proportion of coal production is washed. The coal beneficiation practice is as follows:

- ROM coal is sized to 50–60 mm prior to the wash plant.
- Plant feed is usually classified (deslimed) at between 0.7 and 2 mm on large multislope (banana) screens.
- For washing coarse coal, dense medium cyclones are almost universal—mostly 1000 mm in diameter or larger, with most new installations using 1150, 1300, or 1450 mm diameter units.
- There are still some plants with dense medium vessels (drums and baths) and a few with jigs.
- Coking coal plants traditionally process −0.5 mm w/w by froth flotation. Most new or upgraded plants producing a coking coal have employed a *mid* circuit with spirals or hindered bed separators (teeter bed separator or reflux classifier) treating the 2 + 0.3 mm fraction and the −0.3 mm processed by flotation (mostly Jameson or Microcel technology).
- For thermal coal plants, the fines are generally deslimed at about 0.1 mm, and then processed in spirals or a combination of spirals and teeter bed separator.
- Coarse and mid-size coal fractions are invariably dewatered in vibrating or scroll-type basket centrifuges, while the flotation product is dewatered on vacuum filters (mainly horizontal belt and disc) or screen bowl centrifuges.

12.13.2 China

Coal preparation capacity and the volumes of raw coal processing have rapidly increased. In 2005, China had 961 coal preparation plants and in 2012, the total number is over 2000. During the same period, total annual coal washing capacity has increased from 837 Mt to 2150 Mt. Thus, total coal processing rate has increased from 32% to 56.3% of country's total coal production. China has made rapid progress in the development of coal-processing technology and equipment manufacturing capacity (Zhang and Zhou 2013). A large number of advanced technology, excellent equipment, and super large-scale coal preparation plants have been built. Some of the examples are as follows:

- Max cyclone of 1500 mm
- Max column of 5000 mm
- Max filter plate 2500 × 2500 mm
- Max FGX 480 t/h (dry washing)

More than 61% of the plants are based on heavy medium coal-washing methods. Over these years, the following percentage changes in coal preparation methods have taken place:

Year	Heavy Medium	Jigging	Flotation	Others
1980	17	67	14	2
2010	61	25	11	3

Presently, thrust is also on upgrading of coal and high efficient and clean chemical conversion of coal. China's focus on development of coal-washing capacity is based on the benefits such as saving energy, environmental protection, saving in the transportation, and sustainable development of coal industry.

Accordingly, the coal-processing development targets of 2015 are set as follows:

- General capacity will be 2.65 Bt, raw coal selection more than 2.54 Bt, preparation rate more than 65%.
- By the end of 2015, preparation capacity will increase nearly 900 Mt/a than 2010, equally every year to increase nearly 180 Mt/a, total production of coal washing from 1.65 Bt in 2010 to 2.54 Bt, net to increase 890 Mt.

12.13.3 India

Indian coal industry is the world's third largest in terms of production and fourth largest in terms of proven coal resources. Coal deposits are of drift origin, high in ash content but low in sulfur. There are 20 major coalfields located in east and south-eastern quadrant of the country. Coal supply to major consumers located all over involves long rail hauling. Indian coal deposits are generally of high ash content varying from 24% to 50%. Due to predominance of open pit mining, *out-of-seam* dilution is high, which results in high ash and inconsistent quality of coal supplied to most consumers (Sachdev 2013).

Current washed coking coal production is ~7 Mt. The processes used for washing coking coal are jigs, heavy medium bath, and cyclones for the sizes up to 0.5 mm, while the coal fines are washed in spirals and conventional flotation cells. Production of washed thermal coal is ~36.3 Mt. The processes used for washing non-coking coals are barrel washers, jigs, heavy medium. baths, and heavy medium. cyclone. In most of the washeries, the fraction below 13 mm is not washed and it is mixed with the cleans of the coarser fraction.

12.13.4 South Africa

The South African economy depends heavily on coal, both as a source of foreign income and as a primary energy source. This dependence, coupled with South Africa's extensive coal reserves, indicates that the coal mining and processing industry is likely to continue to be prominent in the medium term despite global concerns of climate change resulting from coal-based power generation and might well survive into a *clean coal technologies* era. South Africa's coal is almost exclusively bituminous. The country has limited reserves of anthracite but no brown coal or lignite. Coal plays a vital role in South Africa's economy and coal requires processing. Coal processing is therefore an important part of the coal industry (Korte 2013).

- There are approximately 60 coal preparation plants in operation in South Africa and a number of new plants are under construction or in the planning phase.
- Largest plant is Grootegeluk complex—7000 tons/h.
- Most plants use dense medium (LARCODEMS, baths, and cyclones) and spirals.
- Many of the plants have two-stage washing to produce export coal (6000 kcal/kg) and thermal coal (21 MJ/kg).

12.13.5 The United States of America

The United States was the world's second largest producer and consumer of coal, behind China. It produces coal from three major coal regions located in the eastern, interior, and western regions of the country. The western deposits are largely composed of thick seams of sub-bituminous *compliance* coal that has inherently low sulfur content. These reserves have traditionally required little coal preparation other than simple crushing and screening. On the other hand, increased levels of contamination from out-of-seam dilution have begun to generate some interest in developing preparation facilities for these coals and some of the other higher rank coals in the western United States.

The eastern coalfields are largely dominated by high-rank bituminous coals in the Appalachian and Illinois coal basins. These coal seams have a high specific heat value that makes them very attractive for transportation and power generation. Also, nearly all of the metallurgical coking coal consumed in domestic steel production is mined from this region. Unfortunately, most eastern coal seams occur as thin bands of coal-bearing sediments mixed with sedimentary rock. Hence, these seams often require coal preparation facilities to separate marketable coal from unwanted waste rock (Laurila 2013).

Dry coal-cleaning technologies were prevalent in the industry throughout the early to mid-twentieth century. Major innovations and development of dry, density-based separators occurred during the 20-year period between 1910 and 1930. The US coal industry processed a significant amount of coal until about 1968 when production peaked at 25.4 Mt annually. Pennsylvania had the distinction of operating the largest dry cleaning plant with 14 units processing about 1400 tph of coal. By 1990, most of the dry cleaning plants were either closed down or their capacities were severely stunted due to federal dust exposure regulations that required the use of a significant amount of water prior to the processing plant to suppress dust.

The wet beneficiation processes usually practiced in the United States are gravity-based separators like dense medium cyclones for the coarser fractions, while both gravity and physicochemical processes are followed for beneficiation of the coal fines. Major emphasis is give for fine coal beneficiation and dewatering.

12.14 COAL DESULFURIZATION

12.14.1 Introduction

Coal sulfur mainly exists in three different forms: pyritic sulfur, organic sulfur, and sulfate sulfur. Pyritic sulfur, which comes from sulfide minerals, is the major component of coal sulfur. Organic sulfur, which can be found in the forms of mercaptan, sulfide or thio-ether, disulfide, or aromatic systems containing the thiophene ring, is chemically linked to the coal structure rendering physical coal-cleaning processes useless in its removal. Sulfate sulfur, which exists in coal due to weathering or oxidation, is present in very small amounts, usually less than 1%. Fresh coals usually contain almost no sulfate sulfur.

Pyritic sulfur in coal includes iron disulfides such as pyrite and marcasite (FeS_2), plus many other inorganic sulfides, that is, galena (PbS), chalcopyrite ($CuFeS_2$), arsenopyrite ($FeAsS$), and sphalerite (ZnS). The differences in the environment in which, pyrite is formed, results in large variations in its morphology and size distribution in coal structure. Its occurrences may be found as finely disseminated particles, framboids or groups of individual crystals, and amorphous or structure less pyrite.

12.14.2 Coal Desulfurization by Physical Methods

The physical coal beneficiation systems range from simple systems, for just removing coarse impurities, to sophisticated systems designed for more effectiveness in ash and sulfur removal. The methods of physical removal of ash-forming minerals and pyritic sulfur used today include gravity separation, magnetic separation, electrostatic separation, oil agglomeration, and flotation. In a commercial coal preparation plant, the cleaning process is usually confined to physical processes that are based on the difference in the specific gravity and on the difference in surface property of coal and its impurities. The raw coal is typically subjected to sizing and screening, specific gravity-based cleaning, dewatering, clarifying, and drying, respectively. However, a modern coal beneficiation plant does not employ a specific cleaning process. On the contrary, a number of different beneficiation processes are applied either sequentially or in various combinations.

Gravity separation, which takes the advantage of the differences in specific gravity between coal and mineral matters (ash-forming minerals and pyrite), are widely used in the coal preparation. Among gravity separation techniques, tables, jigs, and dense medium process are widely used (Beniuk et al. 1994; Honaker and Patil 2002; Honaker and Das 2004).

The HMCs or high-flow dense-medium vessels are one of the simplest and the most efficient techniques for physical coal beneficiation, particularly for the intermediate size coal. In this process, coal is premixed with the heavy medium of specific gravity intermediate between the coal and the mineral impurities and fed to the cyclone. The rate of floating or sinking for fine particles can be increased by using a cyclone or centrifuge. Yet, the effectiveness of this technique is limited by the particle size (Singh and Peterson 1977).

Despite its efficiency in ash removal for relatively coarse coal, they are not so useful in fine coal cleaning and coal desulfurization. The reasons are mainly that (1) gravity beneficiation for coal usually requires feed size larger than 0.5 mm and (2) sulfur minerals are usually finely disseminated in coal matrix and can be liberated only by grinding to a finer size. The sizes produced through grinding inhibit the separation of coal and pyrite using gravity techniques as the gravitational effects are lessened. The development of an enhanced gravity concentrator, which is successful to concentrate cassiterite, chromite, and so on, shows promise in fine coal treatment (Honaker and Patil 2002).

Since the magnetic properties of pyrite and several other minerals found in coal differ slightly from those of the organic matter, that is, coal is weakly diamagnetic while pyrite and the ash-forming minerals are weakly paramagnetic, there is a possibility of separating these components by magnetic methods. However, because the difference in magnetic susceptibility between the components is very small, the separation requires the combination of an intense magnetic field and a large field gradient. The wet high gradient magnetic separation has been applied successfully to fulfill this requirement (Lua and Boucher 1990). In exploratory laboratory measurements, high-field high gradient magnetic separation has removed as much as 74% of the mineral matter and 99% of the pyritic sulfur from micronized coals with mineral contents up to 16.39 wt% (Oder 1984).

Flotation separation, which is realized using the differences in surface properties of the minerals, is the most widely used physical separation technique in the mineral industry. In coal flotation, the hydrophobicity difference between coal and other minerals is the driving force to separate pyrite and ash-forming minerals from coal. A single-stage flotation process was developed in which *coal* was floated out and pyrite was depressed. Up to 90% of the pyritic sulfur content of bituminous coal could be removed at 75% coal recovery. The process was applied to three Canadian and two U.S. coals. Higher coal recoveries were obtained for low-sulfur coals; up to 94.4% coal recovery was possible with 18.2% pyritic sulfur removal. Sulfate sulfur, trace elements, and ash were also removed (Boateng and Phillips 1977).

In most operating coal-cleaning plants, a significant amount of pyrite is recovered in the froth during flotation of high-sulfur coal. Reducing the pyrite recovery first requires that the primary recovery mechanism should be identified, as different measures are required for reducing entrainment, locked-particle flotation, or true hydrophobic flotation. In the paper published by (Kawatra and Eisele 1992) evidence was presented, which suggests that hydrophobic flotation is not an important mechanism for recovery of liberated pyrite when the collector is a neutral oil, and that the bulk of the floated pyrite occurs either as a result of simple entrainment or by mechanical locking with floatable coal particles. Column flotation results are also presented which show that significant sulfur reductions can be achieved by reducing level of entrainment.

Chemical reagents always play a key role in improving the separation efficiency of flotation (Melo and Laskowski 2005). In sulfide flotation, the flotation behavior of pyrite is well studied and a series of effective pyrite depressants have been developed. However, these depressants do not show effectiveness in coal desulfurization. The reason is that although coal pyrite and mineral pyrite share the same overall chemical formula and crystal structure, their behavior is definitely different. One main cause of these differences appears to be the carbonaceous material in the coal pyrite. This material impregnates the pyrite structure and as a result lowers the apparent specific gravity, increases the porosity, and imparts a dark coloration to the coal pyrite (Kawatra and Eisele 1996).

Lots of efforts have been made to find effective coal pyrite depressants to improve the efficiency of coal desulfurization. It is known that to depress coal pyrite, a depressant molecule should adsorb specifically on the surface of pyrite and render the pyrite surface hydrophilic, and thus improve the efficacy of coal desulfurization The bacterium *Thiobacillus ferrooxidans* is capable of suppressing the natural floatability of pyritic sulfur in a conditioning time of 2 min. This is achieved by changing the surface properties of the pyrite from aerophilic to areophobic. Pyrite particles of various sizes can be suppressed with an efficiency of over 92% providing sufficient bacterial cells are present. The effect of bacteria on sulfur suppression of both synthetic coal mixtures and a high sulfur American coal were studied (Atkins et al. 1987).

REFERENCES

Anon. 1979. Vorsyl separator, mining research and development establishment information, *British Coal*, February, 1979.

Allen, T. 1990. *Particle Size Measurement*. 4th edition, London, Chapman & Hall.

Aplan, F.F. 1985 Heavy media separations, In: Weiss, N.L. (Ed.), *SME Mineral Processing Handbook*, pp. 4.1–4.16.

Ashish, Agarwal, Perminder, Singh, and Nikkam, Suresh (2011) Use of jigs in coal and mineral industries – an overview, *XII International Conference on Mineral Processing Technology*, CP 24 1271.

Atkins, A.S., Bridgewood E.W., and Davis, A.J. 1987. A study of the suppression of Pyritic sulphur in coal froth flotation by Thiobacillus ferrooxidans, *Coal Preparation*. 5: 1–13.

Balderson, G.F. 1982. Recent developments and applications of spiral concentrators, The Australian J.M.M, Mill Operators' Conference.

Baojie, Zhang, Paul, Brodzik, and Manoj, K. Mohanty. 2014. Improving fine coal cleaning performance by high-efficiency particle size classification, *International Journal of Coal Preparation and Utilization* 34: 145–156.

Beniuk, V.G., Vadeikis, C.A., and Enraght-Moony, J.N. 1994. Centrifugal jigging of gravity concentrate and tailing at Renison limited, *Minerals Engineering* 7: 577–589.

Bhattacharya, S. 2006. Rotary breakers: Prospects of application in India, *Proceedings of the 1st Asian Mining Congress*, January 16–18, 2006, Kolkata, India, 353–359.

Bird, B.M. 1928. *Interpretation offloat-and-sink, Anais, II International Conference on Bituminous Coal*, Vol 2, pp. 82–111.

Boateng, D.A.D. and Phillips, C.R. 1977. Desulfurization of coal by flotation of coal in a single-stage process separation science and technology, 12:71–86.

Brinkman, F. and Helling, H. 1964. The Baum Jig, *Gluckauf* 100: 1249–54.

Brooks, G.F. and Bethell, P.J. 1976. The development of a flotation/filtration reagent system for coal, *Proceedings of the 8th International Coal Preparation Congress*, Paris.

Burt, R.O. 1984. *Gravity Concentration Technology.* Elsevier, New York.

Butel, D., Howarth, W.J., Rogis, J., and Smith, K.G. 1993. Coal Sorting, *Coal Preparation.* 12: 203–214.

Clarkson, C. 1998. Efficiency of large diameter dense medium cyclones, *Mine and Quarry*, April 1998.

Cordingley, M.G. 1997. MGS developments and strategies in fine gravity separation techniques, In: *Innovation in Physical Separation Technologies*, Falmouth: IMM, pp. 75–82.

Dobby, G.S., Amelunxen, R., and Finch, J.A. 1985. Column flotation: Some plant experience and model development, *International Federation of Automatic Control* 259–263.

Dolina, L.F. and Kominski, U.S. 1971. Coke and chemistry, UUUR, No. 10. P. 16.

Driessen, M.G. 1945. The use of centrifugal force for cleaning fine coal in heavy liquids and suspensions with special reference to the cyclone washer, *Journal of the Institute of Fuel*, December 1945.

Dueck, J., Matvienko, O.V., and Neesse, T. 1998. Hydrodynamics and particle separation in the hydrocyclone. In: Two-phase Flow Modelling and Experimentation, Pisa.

Eberts, D.H. 1986. Flotation- choose the right equipment for your needs, *Canadian Mining Journal*, March: 25–33.

Farnand, J.R., Smith, H.M., and Puddington, I.E. 1961. Spherical agglomeration of solids in liquid suspension, *Canadian Journal of Chemical Engineering* 39(2): 94.

Fellensiek, E. and Erdmann, W. 1991. Jigging – historical and technical development, *Aufbereitungs-Technik* 32(11): 45–52.

Firat, Burat, Ayhan, A. Sirkeci, and GuVen, ONal. 2015. Improved fine coal dewatering by ultrasonic pretreatment and dewatering aids, *Mineral Processing & Extractive Metallurgical Review* 36: 129–135.

Florl, M. and Heintges, S. 1987. Applications of the ROM Jig at Emil Mayrisch Colliery, West Germany, *Mine and Quarry* December 1987.

Forrestor, C. and Majumdar, J. 1947. The washability of Indian coal, *Fuel Research Committee*, Report No. 1, 1947.

Fuerstenau, M.C. 1976. "Flotation" *The American Institute of Mining, Metallurgical and Petroleum Engineers*, Inc. A.M.Gaudin Memorial, pp. 1235–1259.

Fuerstenau, M.C., Miller, J.D., and Kuhn, M.C. 1985. *Chemistry of Flotation, Society of Mining Engineers.* New York, AIME, pp. 170.

Gala, H.B., Chiang, S.H., Klinzing, J.W. Tierney, J.W., and Wen, W.W. 1983. Effect of surfactant adsorption on the hydrophobicity of fine coal, *Proceedings of International Conference on Coal Science*, Pittsburgh, PA, pp. 260–263.

Galvin, K.P., Callen, A., Spear, S., Walton, K., and Zhou, J. 2010. Gravity separation of coal in the reflux classifier: New mechanisms for suppressing the effects of particle size. In: R.Q. Honaker (Ed.), *International Coal Preparation Congress.* Littleton, CO, Society of Mining, Metallurgy & Exploration (SME), pp. 345–351.

Galvin, K.P., Callen, A., Zhou, J., and Doroodchi, E. 2005. Performance of the reflux classifier for gravity separation at full scale, *Minerals Engineering* 18: 19–24.

Goel, 2011. Benefits of Using Washed Coal, *CPSI Journal*, April 2011: 12–18.

Gottfried, B.S. 1978. A generalization of distribution data for characterizing the performance of float-sink coal cleaning devices, *International Journal of Mineral Processing* 5: 1–20.

Gray, V.R., 1958. The dewatering of fine coal, *Journal of the Institute of Fuel* 31: 96–108.

Groppo, J.G., Sung, D.J., and Parekh, B.K. 1995. Evaluation of hyperbaric filtration for fine coal dewatering: Part 1, *Mineral Metallurgical Processing* 12: 28.

Hacifazlioglu, H. 2011. Recovery of coal from cyclone overflow waste coals by using a combination of Jameson and column flotation energy sources, Part A 33: 2044–2057.

Hogg, R. 2000. Flocculation and dewatering, *International Journal of Mineral Processing* 58: 223–236.

Holland·Batt, A.B, Hunter, J.L., and Turner, J.H. 1984. The separation of coal fines using flowing film gravity concentration, *Powder Technology* 40NI-3: 129–145.

Honaker, R.Q. 2007. *Workshop on Coal Beneficiation and Utilization of Rejects, Initiatives, Policies and Best Practices*, Ranchi, India, August 22–24,

Honaker, R.Q. and Das, A. 2004. Ultra-fine coal cleaning using a centrifugal fluidized-bed separator, *Coal Preparations* 24: 1–18.

Honaker, R.Q. and Patil, D.P. 2002. Parametric evaluation of a dense-medium process using an enhanced gravity separator, *Coal Preparations* 22: 1–17.

Honaker, R.Q., Saraconglu, M., Thompson, E., Bratton, R., Luttrell, G.H., and Richardson, V. 2007. Dry coal cleaning using the FGX separator. In: *Proceedings of Coal Preparations*, May 1–3, Lexington, KY, 61–76.

Honaker, R.Q., Paul, B.C., Wang, D., and Huang, M. 1995. Application of centrifugal washing for fine coal cleaning, *Minerals and Metallurgical Processing*, May: 80–84.

Honaker, R.Q., Wang, D., and Ho, K. 1996. Application of the Falcon concentrator for fine coal cleaning, *Minerals Engineering* 9(11): 1143–1156.

Hubbard, J.S., Brown E.W., and Welker, M. 1950. The Humphrey's spiral concentrator for cleaning minus 1/4 inch coal, *International Conference on Coal Preparation*, Paris, pp. E2.

Jameson, G.J. and Manlapig, E.V. 1991. Application of the Jameson cell. *International Conference on Column Flotation*, Sudbury, Canada, September 1–8: 672–687.

Jimmy, Yu. 2011. Simplified and high efficient coal process technology for high ash content Indian coals, *CPSI Journal* III, November 2011: 54–63.

Kalyan, Sen. 1994. Beneficiation of difficult to wash coals: Development and prospects, *12th ICPC*, pp 1485–1493.

Kawatra, S.K. and Eisele, T.C. 1987. Column flotation of coal, In: Mishra and Klimpel (Eds.), *Fine Coal Processing.* Park Ridge, NJ, Noyes Publications, Chapter 16, pp. 414–429.

Kawatra, S.K. and. Eisele, T.C. 1992. Removal of Pyrite in coal flotation, *Mineral Processing and Extractive Metallurgy Review* (1–4) 11: 205–218.

Kawatra, S.K. and Eisele, T.C. 1996. Pyrite recovery mechanisms in coal flotation, *International Journal of Minerals Processing* 50: 187–201.

Keane, J.M. 1986. Laboratory testing for design of thickener circuits. In: A.L. Mular and M.A. Anderson (Eds.), *Design and Installation of Concentration and Dewatering Circuits.* Littleton, CO, SME, pp. 498–505.

Keller, D.V., Stelma, G.J., and Chi, Y.M. 1979. Surface Phenomena in the Dewatering of Coal, *EPA* 600/7–79–008.

Kim, B.H. and Klima, M.S. 1998. Density separation of fine, high density particles in a water-only cyclone. *Minerals Metallurgical Processing* 15(4): 15–35.

Klassen, V.I. and Mokrousov, V.A. 1963. *An Introduction to the Theory of Flotation*, translated by J. Leja and G.W. Poling. London, Butterworth.

Klima, M.S. and Luckie, P.T. 1986. An interpolation methodology for washability data, *Coal Preparation* 1: 16S–177.

Konar, B.B., Banerjee, S.B., Chaudhuri, S.G., Choudhury, A., Das, N.S., and Sen, K. 1997. *Monograph-Coal Preparation*, Sen, Samir and Narasimhan, K.S. (Eds.), New Delhi, Allied Publishers.

Korte, G.J. de. 2013. *Country Report: Status of Coal Preparation in South Africa*. Istanbul, Turkey, ICPC.

Laurila, M.J. 2013. *Coal Preparation in the United States*. Istanbul, Turkey, ICPC.

Leonard, J.W. 1979. *Coal Preparation*. 4th Edition, New York, American Institute of Mining, Metallurgical, and Petroleum Engineers (AIME), p. 1165, Chapter 4.

Lowrison, G.C. 1974. *Crushing and Grinding*. Cleveland, OH: CRC Press.

Lua, A.C. and Boucher, R.F. 1990. Sulphur and ash reduction in coal by high gradient magnetic separation, *Coal Preparation* 8(1–2): 61–71.

Luo, Z., and Chen, Q. 2001. Dry beneficiation technology of coal with an air dense medium fluidized bed, *International Journal of Mineral Processing* 63(3): 167–175.

Luo, Z., Chen, Q., and Zhao, Y. 2002. Dry beneficiation of coarse coal using air dense medium fluidized bed (ADMFB), *Coal Preparation* 22: 57–64.

Luttrell, G.H., Honaker, R.Q., and Phillips, D.I. 1995. Enhanced gravity separators: New alternatives for fine coal cleaning, *12th International Coal Preparation Conference*, Lexington, KT, pp. 281–292.

Majumdar, A.K. and Barnwal, J.P. 2006. Modeling of enhanced gravity concentrators–present status, *Mineral Processing & Extractive Metallurgical Review* 27: 61–86.

Mark, A., Sharpe, D., Rao, N., and Dynys, A.. 2000. An effective coal process for deep cleaning Indian thermal coals, *Coal Preparation India, International Conference & Exhibition on Coal Beneficiation & Allied Subject*, New Delhi, March 9–11, 2000.

Melo, F. and Laskowski, J.S. 2005. Fundamental properties of flotation frothers and their effect on flotation, *Minerals Engineering* 46: 126–140.

Miller, F.G. Demull, T.J. Matoney, J.P. 1985. Centrifugal specific gravity separator. In: Weiss, N.L. (Ed.), *SME Mineral Processing Handbook*, pp. 16–27.

Mohanty, M.K. 2001. In-plant optimization of a full-scale Jameson flotation cell, *Minerals Engineering* 14: 1531–1536.

Mohanty, M.K., Palit, A., and Dube, B. 2002. A comparative evaluation of new fine particle size separation technologies, *Minerals Engineering* 15: 727–736.

Moza, A.K., Kini, K.A., and Sarkar, G.G. 1976. Basic studies on the applicability of the oil agglomeration technique to various coal beneficiation problem, *Proceedings of the 7th International Coal Preparation Congress*, Sydney, Australia, Paper No. H3.

Narasimhan, K.S. and Mukherjee, A.K. 1999. *Gondwana Coals of India*. Delhi: Allied Patnaik, NK, 2006.

Nicol, S.K. 1976. The effect of surfactants on the dewatering of fine coal, *Proceedings Australasian Institute Minerals Metallurgical* 260: 37–44.

Oder, R. R. 1984. The application of high field and high gradient methods to the magnetic separation of mineral matter from micronized coal separation, *Science and Technology* 19(11–12): 761–781.

Osborne, D.G. 1988. *Coal Preparation Technology*, vol.1, London, UK, Graham & Trotman.

Osborne, D.G. and Foneca, A.G 1992. Coal preparation: The past ten years, *Coal Preparation* 11: 115–143.

Osborne, D.G. and Pierson, H.G. 1990. Vacuum filtration Part I rotary vacuum filters. Part II horizontal vacuum bell filters. In: L. Svarovsky (Ed.), *Solid-Liquid Separation*. London, Butterworth-Heinemann, pp. 415–475.

Patil, D.P., Parekh, B.K., and Klunder, E.B. 2010. A novel approach for improving column flotation of fine and coarse coal, *International Journal of Coal Preparation and Utilization* 30: 173–188,

Pearse, M.J. and Allen, A.P. 1983. The use of flocculants and surfactants in the filtration of minerals slurries, *Filtration and Separation* 20: 22–27.

Plitt, L.R. 1971. The analysis of solid–solid separations in classifier, *CIM Bulletin* 64(708): 42–47

Purchas, D.B. and Wakeman, R.J. (Eds.). 1986. *Solid and Liquid Separation Equipment Scale-Up*. London, Uplands Press, p. 749.

Ramana, G.V. 2009. Dry beneficiation of coal–relevance for Indian coals, *CPSI Journal* 1(1): 35.

Richards, R.G., Hunter, J.L., and Holland-Batt, A.B. 1985. Spiral concentrators for fine coal treatment, *Coal Preparation* 1: 207–229.

Robben, Christopher, De Korte, Johan, Wotruba, Hermann, and Robben, Mathilde. 2014. Experiences in dry coarse coal separation using X-ray-transmission-based sorting, *International Journal of Coal Preparation and Utilization* 34: 210–219.

Rogers, J. 1985. The recovery of fine coal from secondary water-only cyclone underflow by spiral concentration at Gretley Colliery, Australia, *Proceedings of the Coal Preparation* 85 Conference; Lexington, KY.

Rubinstein, J.B. 1995. *Column Flotation: Processes, Designs, and Practices*. Basel, Switzerland, Gordon and Breach, pp. 300.

Ruff, H.J. 1984. New developments in dynamic dense-medium systems, *Mine and Quarry*, December 1984.

Sachdev, R.K. 2013. *Indian Coal Preparation – 2013 Report*, Istanbul Turkey, ICPC.

Salama, A.I.A. 1998. Coal washability characteristics index utilizing the M-curve and the CM-curve, *International Journal of Mineral Processing* 55: 139–152.

Sampaio, C.H., Aliaga, W., Pacheco, E. T., Petter, E., and Wotruba, H. 2008. Coal beneficiation of Candiota mine by dry jigging. *Fuel Processing Technology* 89: 198–202.

Sanders, G.J. and Brookes G.F. 1986. Preparation of Gondwana Coals, I. Washability Characteristics, *Coal Preparation*, Vol.3, Gordon & Breach Science Publishers S.A.

Sanders, G.J., Ziaja, D., and Kottmann, J. 2002. Cost-efficient beneficiation of coal by ROM jigs and BATAC jigs, *Coal Preparation* 22(4): 181–197.

Schwalbach, H. 1989. Horizontal belt vacuum filters-survey and state of the art, *Aufbereihmgs Technik* 8: 471–476.

Singh, S.P.N and Peterson, G.R. 1977. Survey and evaluation of current and potential coal beneficiation Processes. ORNL/TM-5953, Oak Ridge National Laboratories.

Sivamohan, R. and Forssberg, E. 1985. Principles of spiral concentration, *International Journal of Mineral Processing* 15: 173–181.

Subba Rao, D.V. 2003. *Coal – Its Beneficiation*. New Delhi, India, Emkay Publications.

Svarovsky, L. (Ed.) 1990. *Solid-Liquid Separation*. London, Butterworth, p. 750.

Swanson, R. 2013. *Australian Coal Preparation – 2013 Report*. Istanbul Turkey, ICPC.

Takakuwa, T. and Matsumura, M. 1954. Suggestions for the Improvement of the air-pulsated Jig, Paper A118, *International Coal Preparation Congress*, September 20–25, Essen.

Tromp, K.F. 1937. New methods of computing the washability of coals, *Gluckauf* 37: 125–131, 151–156; Excerpts in *Colliery Guardian* 154: 955–959, 1009.

Walters, A.D. 1976. A computer simulation model for coal preparation plant design and control, The Pennsylvania State University, Department of Mineral Engineering.

Weinstein, R. and Snoby, R. 2007. Advances in dry jigging improve coal quality, *Mining Engineering* 59: 29–34

Wen, W.W., Cho, H., and Killmeyer, R.P. 1993. The simultaneous use of a single additive for coal flotation, dewatering and cake hardening, In: B.K. Parekh and J.G. Groppo (Eds.), Proceedings of Processing and Utilization of High-sulfur Coals V, Elsevier, Amsterdam, pp. 237–249.

Wills, B.A. and Napier-Munn, T.J. 2006. Froth flotation, *Mineral Processing Technology*. 7th edition, Great Britain, Elsevier: Oxford.

Yerriswamy, P., Majumder, A.K., Barnwal, J.P., Govindarajan, B., and Rao, T.C. 2003. Study on Kelsey jig treating Indian coal fines, *Mineral Processing and Extractive Metallurgy* (Trans. Inst. Min. Metall. C), December, 112: C206–C210.

Zhang, B., Akbari, H., Yang, F., Mohanty, M. K., and Hirschi, J. 2011. Performance optimization of the FGX dry separator for cleaning high-sulfur coal, *International Journal of Coal Preparation and Utilization* 31: 161–186.

Zhang, S. and Zhou, S. 2013. *The Status and Prospects of Development of Coal Preparation in P.R.China.* Istanbul, Turkey, ICPC.

Zimmerman, R.E. 1978. The Japanese swirl cyclone, *Mining Engineering* 30: 189–193.

Zimmerman, R.E. 1978. Breakthrough in heavy media cyclone operation, *World Coal* 3: 19–21.

13 Coal Combustion for Power Production

Hisao Makino and Kenji Tanno

CONTENTS

Abstract: In this chapter, the coal combustion profile, the typical coal combustion reactor, and the power generation system using coal combustion are introduced. After that, the advanced combustion technologies for pulverized coal including the low NO_x combustion and the stable combustion at low burner load are explained. Finally, the combustion characteristics of many kinds of coal, such as low-rank coal with high moisture content, high ash content coal, and high-fuel-ratio coal, are introduced.

The best way to utilize coal as an energy source is converting it into thermal energy, which is generated by coal combustion. However, the coal combustion process is never simple due to the complex structure of coal, and is heavily influenced by the difference of coal property. Although there are several coal properties relevant to combustion, the elemental composition and the structure of combustible component are generally important. Moreover, there are several types of coal combustion system; hence, we have to choose an appropriate system according to coal property and capacity of combustor.

In this chapter, estimation methods of coal properties important for combustion, characteristics of combustion profile, and basic concepts of several types of coal combustor are introduced. The chapter also includes the outline of pulverized coal combustion, which is a typical example of the entrained flow combustor that is presented. The utilization process of the pulverized coal combustion in a thermal power plant is introduced. The latest coal combustion technologies utilized in a thermal power plant, such as low NO_x combustion, low-load combustion, and low grade coal combustion, are also introduced.

13.1 COAL PROPERTIES

Many kinds of analyzing methods for coal property were investigated for a very long time, and these methods were utilized to estimate coal characteristics for selecting a coal in a coal combustor. Almost all of these methods are applicable to other solid fuels, such as biomass. The most common coal properties for analyzing combustion characteristics are the proximate and ultimate analyses. Proximate analysis is used to quantify the amounts of incombustible matter (moisture and mineral matter) and the amounts of combustible matter (volatile matter and fixed carbon).

The amount of volatile matter is evaluated by measuring the weight loss in nitrogen atmosphere at 900°C±20°C and carbonization (coking) temperature in a coke oven, for 7 min. As it can be noticed from the analyzing method, the volatile matter is easily released into the gas phase and reacts with oxygen. Therefore, the amount of volatile matter is closely related to the ignition characteristics.

The amount of fixed carbon is evaluated by measuring the weight loss in air at 815°C±10°C for 1 h, and the residual is regarded as ash content. Therefore, fixed carbon is not easily combustible compared with volatile matter; hence, it burns slowly. The ratio of fixed carbon to volatile matter is called as *fuel ratio*, which indicates the degree of coalification. Fuel ratio is also used as a factor that represents level of combustibility. The low fuel ratio means the coalification has not progressed so much, and the low fuel ratio coal is easily combustible due to the high volatile matter content. In coal-fired power plants, although coals with fuel ratio from 0.8 to 3.0 are often used, there are combustors that are especially designed for higher fuel ratio coal. Recently, the utilization of various coals is becoming important in terms of reinforcement of fuel supply, resulting in a gradual increase in the utilization of high-fuel-ratio coals such as anthracite coal and very low fuel ratio coals such as sub-bituminous coal. Although fuel ratio is an important factor for analyzing combustibility of coal, devolatilization characteristics under the case of a real combustion system, in which a coal particle is rapidly heated up, is different from that in the proximate analysis. Therefore, comprehending such differences is important for investigating the coal combustion characteristics in detail. Ultimate analysis gives the amounts of carbon, hydrogen, nitrogen, oxygen, sulfur (combustible sulfur and total sulfur), and ash content. From the results of ultimate analysis, we can estimate the amount of theoretical combustion air, flue gas, carbon dioxide emission, SO_x emission, and conversion ratio from nitrogen in the fuel to NO_x in flue gas. Moreover, we can estimate the degree of coalification from the ratio of C to H or C to O. Table 13.1 shows the results of proximate and ultimate analysis for five kinds of bituminous coal.

13.2 COMBUSTION PROFILE OF COAL

Figure 13.1 shows the typical combustion profile of a single coal particle in pulverized coal combustion. The process consists of the combustions of volatile matter in gas phase (flamezone) and residual char, the latter of which starts on the surface of char particle (post-flamezone).

The pulverized coal and its transport air (primary air) are injected into the furnace through the burner to form a flame. Temperature of pulverized coal and primary air in the burner

TABLE 13.1
Example of Coal Property

Coal Name	Plateau	Wanbo	Warkworth	Newlands	Blair Athol
Proximate analysis [wt%]					
Moisture[1]	(5.9)	(3.5)	(4.5)	(2.5)	(7.9)
Volatile matter[2]	41.3	35.7	31.4	28.4	29.6
Fixed carbon[2]	48.8	54.6	57.5	56.4	62.5
Ash[2]	9.9	9.7	11.1	15.2	7.9
Fuel ratio [-]	1.18	1.53	1.83	1.99	2.11
Ultimate analysis [wt%][2]					
C	71.9	74.2	73.6	71.8	74.6
H	5.47	5.62	5.10	4.5	4.52
N	1.30	1.82	1.59	1.59	1.54
O	11.8	8.27	8.36	6.40	11.2
S (combustible)	0.41	0.42	0.35	0.48	0.23
Heating value (low) [MJ/kg][2]	28.7	29.6	28.3	28.1	28.0

[1] As received.
[2] Dry basis.

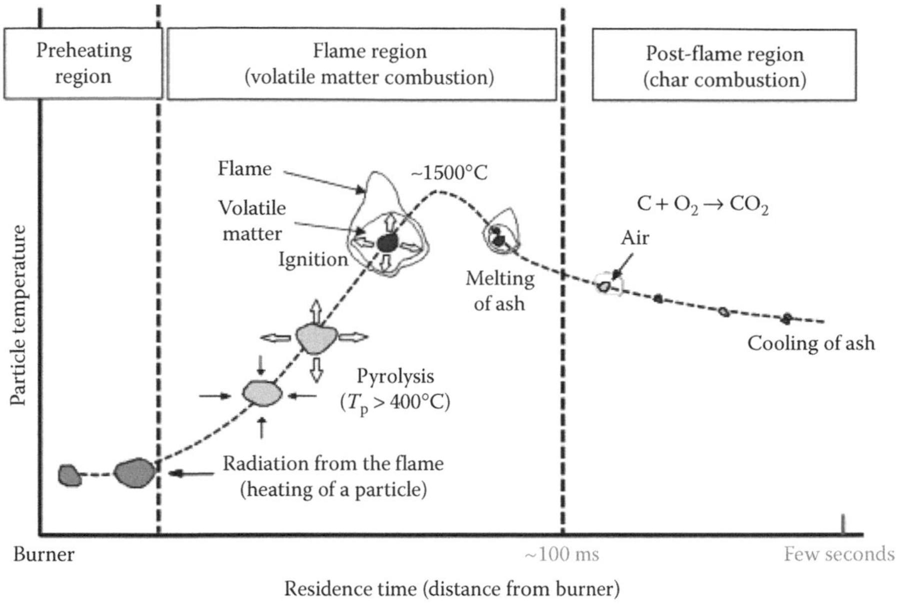

FIGURE 13.1 Combustion profile of a single coal particle in pulverized coal combustion.

is around 80°C. The coal particles are heated up in the furnace by radiation from the flame and the furnace wall. Moisture in coal particles is vaporized in this heating period. Volatile matter is decomposed; the decomposed gases ignite to combustion of volatile matter at around 400°C. Temperature of coal particles surrounded by the flame increases rapidly, reaching 1500°C and higher. Fixed carbon (char) reacts with the combustion air on its surface. A part of the char remains with the ash as unburned carbon.

13.2.1 Devolatilization and Ignition

The devolatilization is fast to finish within less than 100 ms. It is a very complicated phenomenon, and is affected by pulverized coal properties, temperature, and compositions of gas surrounding the coal particle. The higher heating rate from 10^2 K/s to 10^4 K/s increases the release of volatile matter by about 10% (Anthony et al. 1975; Kobayashi et al. 1977; Niksa et al. 1985). The actual heating rate in pulverized coal combustion is approximately 10^5 K/s (Williams et al. 2000). The devolatilization of a coal particle is often simulated by a first-order single reaction model, although other factors of the actual combustion influence the devolatization.

$$\frac{dV}{dt} = K_V\left(V^* - V\right)$$

$$K_V = A_V \exp\left(\frac{E_V}{RT_P}\right)$$

where:

V^* and V indicate the total volatile matter content in the particle and volatile mass released from the particle, respectively

R is the universal gas constant (=8.31 J/mol K)

Pre-exponential factor A_v and activation energy E_v are determined experimentally, for example, $A_v = 2021$ (1/s) and $E_v = 3.11 \times 10^4$ (J/mol) (Van Krevelen et al. 1951)

The sophisticated models such as FG-DVC, FLASH CHAIN, and CPD for coal devolatilization can predict masses of volatile matter and tar evolved at given heating rate and pressure, as well as product gas compositions (Niksa et al. 1985; Niksa and Kersten 1991a; Niksa 1991b).

The ignition temperatures of brown coals with high volatile matter and anthracite of low volatile content are around 250°C and 500°C, respectively. The ignition temperatures of most common bituminous coals are 300°C–400°C.

13.2.2 Combustion of Volatile Matter

The devolatilized gases react with combustion air in the gas phase. These reactions were simply modeled as follows. The two global reactions applied are

$$C_aH_bO_c + \alpha O_2 \rightarrow aCO + \frac{b}{2}H_2O$$

$$CO + 0.5O_2 \rightarrow CO_2$$

$$\alpha = \left(a + \frac{b}{2} - c\right)/2$$

where:

$C_aH_bO_c$ represents volatile stoichiometry

$$R_g = A_g \exp\left(-\frac{E_g}{RT_g}\right)\left[\text{Reactant}\right]^d\left[O_2\right]^e$$

where:

d and e are reaction orders in volatile and oxygen, respectively

These models must be combined with those of material flows. Since the burning rate of the devolatilized gases is faster than that of devolatilization, the soot particles are generated during this gaseous combustion only in case of poor mixing of the devolatilized gases and the combustion air.

13.2.3 Char Combustion

Two combustion processes of volatile matter and char are believed to begin simultaneously (Saito et al. 1987), although the duration of devolatilization and volatile matter combustion is 10–100 ms, while the char combustion continues for longer than 1 s.

The char burning rate is described by Field's model (Field 1969):

$$\frac{dC}{dt} = -\left(\frac{K_c K_d}{K_c + K_d}\right) P_g \pi D_p^2$$

$$K_d = \frac{5.06\times10^{-7}}{D_p}\left(\frac{T_p + T_g}{2}\right)^{0.75}$$

$$K_c = A_c \exp\left(-\frac{E_c}{RT_p}\right)$$

where:

C is the char mass
K_c and K_d are the chemical and diffusion rate coefficients
P_g is the partial pressure of oxygen in the bulk gas

This model assumes that the char burning rate is controlled by both the chemical reaction rate and the diffusion rate of oxygen to the surface of the char particle.

In the case of temperature lower than 1000°C, the char burning rate is mainly controlled not only by the chemical reaction rate, but also by the diffusion rate of oxygen in the high temperature region. If temperature of coal particle is higher than its melting point, the particle becomes spherical because of surface tension. A part of the remaining char is included in ash particles.

Field et al.'s model does not take account of the roles of ash on char burning rate. The ash covering the surface of char particles reduces the char burning rate at the later stage of the combustion. Carbon burnout kinetic model (Hurt et al. 1998; Sun and Hurt 2000) takes account of diffusion resistance of oxygen across molten ash layer.

13.3 COAL COMBUSTION SYSTEM

The combustion systems for coal are classified into the fixed bed, the fluidized bed, and the entrained bed (pulverized coal combustion) according to slip velocity between gas and particles as shown in Figure 13.2.

The fixed bed uses large coal particles (larger than 10 mm) and is called as *stoker combustion*. The lumpy coal is fed into a bed, where the grate is fixed in space, and combustion occurs with upward flow of air through the bed. Air velocity is so low that lump coal is fixed and burns in the bed. It is very difficult to control the combustion process, heat balance of the furnace, and air pollutant emissions. It is only used for small-scale furnaces.

13.3.1 Fluidized Bed Combustion

Air velocity is too low to fluidize the bed material (coal particles) in the fixed bed. As the air velocity is increased further, the coal particles separate from each other and the bed starts to fluidize where the particles have better contact with

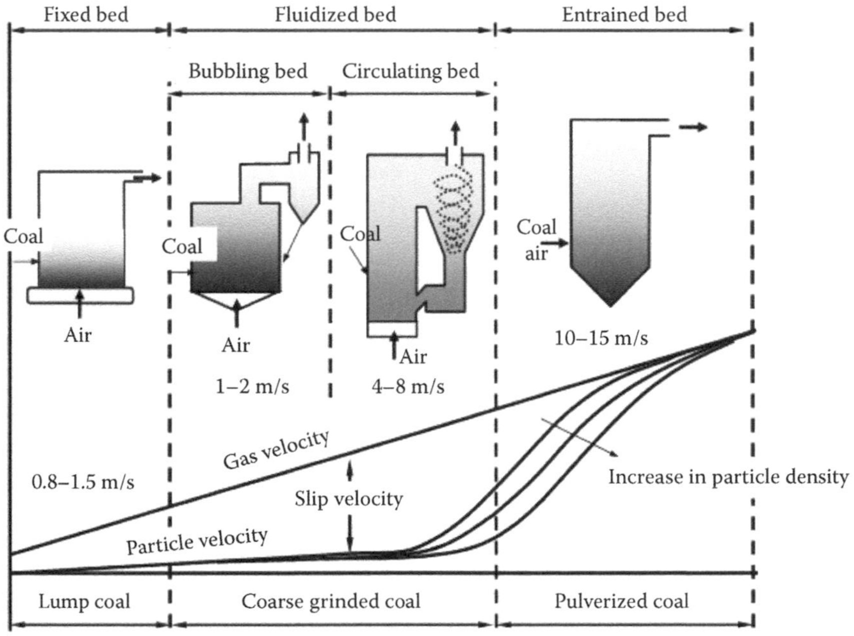

FIGURE 13.2 Coal combustion system.

air, and bed temperature is kept uniform due to good mixing of the particles. The bed material consists of coal ash, sand, and limestone.

The bubbling and circulating fluidized beds have been developed where gas velocities are 1–2 and 4–8 m/s, respectively. The fluidized bed combustion has the following characteristics:

1. As the combustion temperature is relatively lower (750°C–950°C) compared to other combustion systems, NO_x emission is low. The combustion efficiency, however, is relatively low.
2. The residence time of fuel particles in the bed is long, and heat capacity of the bed materials is large. These allow to burn low-grade fuel such as waste materials.
3. When limestone, dolomite, or other mineral is used as the desulfurization material, SO_x is removed in the furnace. This eliminates $DeSO_x$ equipment in flue gas treatment system.

Fluidized bed combustion is often applied for small- and middle-scale power generation.

13.3.2 Entrained-Bed Combustion

The pulverized coal particles of approximately 40 μm of median size are blown with primary air from the burner into the furnace to form a flame. The pulverized coal combustion is currently the most common method, because the combustion efficiency is higher than those of other systems, and its equipment is easy to scale up to 1000 MW class power plant boilers. The technologies developed on pulverized coal combustion are mainly focused in this chapter.

13.4 OUTLINE OF COAL-FIRED POWER PLANT

Heat generated from the combustion of coal or its derived fuels are transferred to steam, which rotate the steam turbine or directly utilized to rotate gas turbine. The rotation is transferred to the electro-magnetic rotors in the electric power generator to induce the electric power. Such a principle has been established in the early twentieth century, and the technological progress since then depends on how effective rotation of turbines is obtained through the combustion of coal and fuels.

The coal combustion to produce steam is the original form of power generation. The coal can be converted now into fuel gas (syngas) through the gasification, which is burnt in the gas turbine for power generation. The exhaust gas from the gas turbine is still very hot to produce the steam for steam turbine. Hence gas and steam turbines are combined in this system called as the integrated coal gasification combined cycle.

When the fuel gas is charged to the fuel cell, electricity is produced through the electrochemical combustion. The exhaust gas-flow from the fuel cell can be hot and rapid, sufficient for the gas turbine and the recovered steam to steam the turbine. Triple power generation in a series can be designed in the triple cycles (Integrated coal gasification fuel cell combined cycle). Such power generation schemes are illustrated in Figure 13.3.

Coal is currently combusted in the pulverized form by the entrained bed. The fluidized bed combustion of coal produces mainly the steam, but exhaust gas can be sent to the gas turbine if its pressure and temperature are high enough. This type of combined cycle produces small power at the gas turbine because of the limited kinetic energy of the exhaust gas without combustion.

1. Boilers and turbines
   The boiler is a combustor to recover the steam by heating pure water in the tube, which is installed within the boiler. It is very important to design its arrangement in the boiler as illustrated in Figure 13.4 (Makino 2007). The pressurized water is preheated to 300°C in the economizer installed at the flue gas line. The heated water becomes steam at the evaporator and the steam is next sent to the super heater in the boiler. Super-heated steam is sent to the high-pressure steam turbine. Steam coming out from the high-pressure turbine is sent to the reheater in the boiler and then sent to the intermediate pressure turbine.

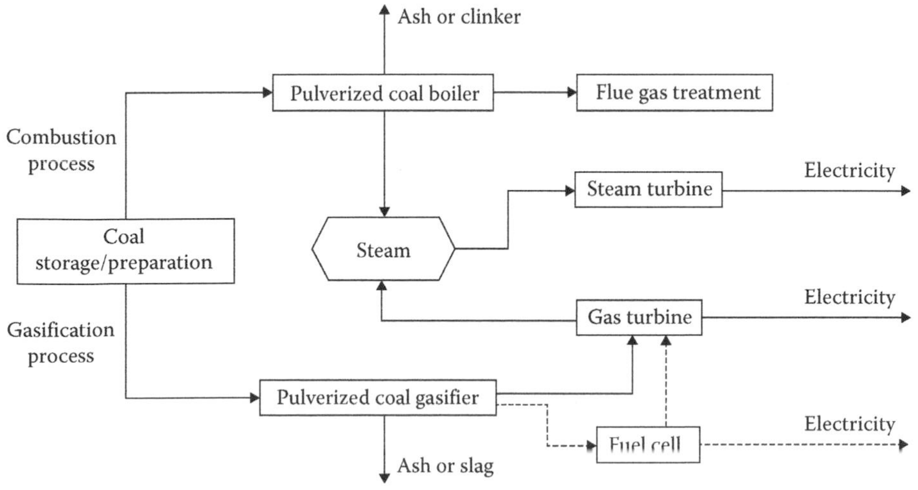

FIGURE 13.3 A total power generation system.

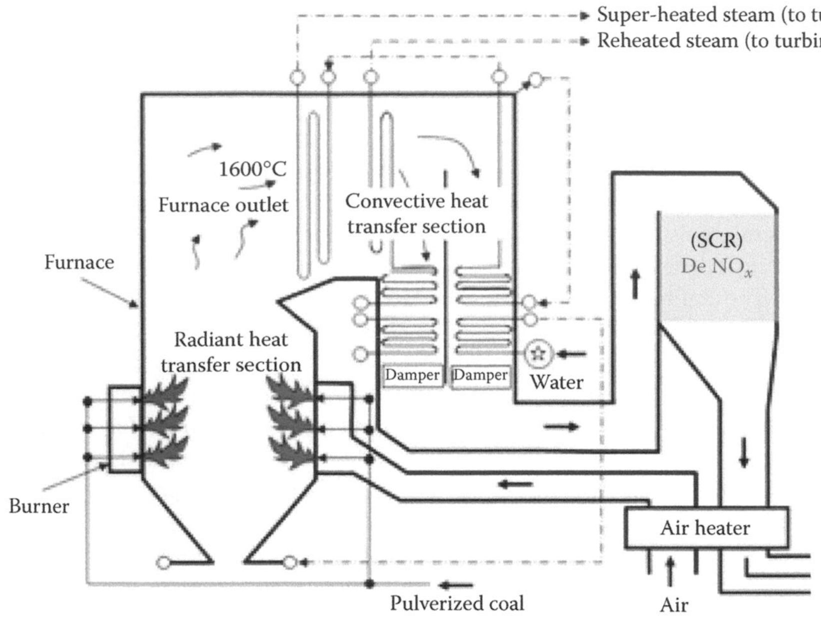

FIGURE 13.4 Schematic view of huge scale pulverized coal fired boiler.

Coal is combusted around 1600°C in the boiler and the flue gas leaves at 300°C. To recover the energy released by the 1300°C difference between the combustion and flue gases, the water is heated in the cascade manner as describe above. The steam can be heated by the flue gas of up to 1300°C. However, the temperature of the steam sent to the high-pressure turbine is now around 600°C (USC) and targeted to 700°C (A-USC) in future by developing the stability of material of steam tubes. Waste heat is utilized in the cascade manner. Flue gas is utilized to heat the water as described above and then the air for combustion. Thus, the energy of flue gas is transferred to the combustion system to utilize the waste heat to increase the thermal efficiency.

2. Power generation efficiency
   Power generation efficiency is calculated from combustion, heat-transfer, turbine, and generation efficiencies. The generation efficiency is usually very high, nearly unity; hence power generation efficiency is governed by the former three efficiencies. Combustion efficiency is governed by how much and how fast coal can be burned in the boiler. More than 99% of coal organic substance is burned within the boiler, discharging less than 5% unburned carbon in coal ash. Boiler efficiency is defined as how much energy of coal is transformed into that of the steam. Hence, combustion to produce the heat and its transfer define the efficiency of the boiler. Steam turbine efficiency is determined according to the Carnot cycle. Hence, higher temperature/higher pressure at the hot side and lower temperature/lower pressure at the cool side define the efficiency. The higher temperature and higher pressure of the steam are the key factors for the higher efficiency of the steam turbine. Figure 13.5 shows the trend and future prospect of steam temperature, pressure, and power generation efficiency. With the increase of steam temperature and pressure, power generation efficiency is increased. In addition, trend of increased steam temperature and pressure will continue into the future. The durability of transfer tube against the steam condition is restricted to the material properties. Hence, the higher efficiency in the past has been realized mainly due to better tube material, which can stand the steam conditions.

13.5 LOW NO_X COMBUSTION TECHNOLOGIES

Lower NO_x combustion has been achieved in spite of high temperature combustion. NO_x generated in pulverized coal combustion is classified into thermal NO_x and fuel NO_x. Thermal NO_x is formed from the nitrogen in combustion air, while fuel NO_x is formed from the nitrogen in the coal. The thermal NO_x involves two routes: Zeldovich NO_x and prompt NO_x.

The Zeldovich mechanism is described by following reactions:

$$N_2 + O \leftrightarrow NO + N$$

$$N + O_2 \leftrightarrow NO + O$$

$$N + OH \leftrightarrow NO + H$$

Zeldovich NO_x has a strong temperature dependence, being formed in high temperature region (>1500°C). Small amount of Zeldovich NO_x is generated in pulverized coal combustion.

Prompt NO_x is produced in flames surrounding coal particles. Prompt NO_x originates from HCN and CN, which are

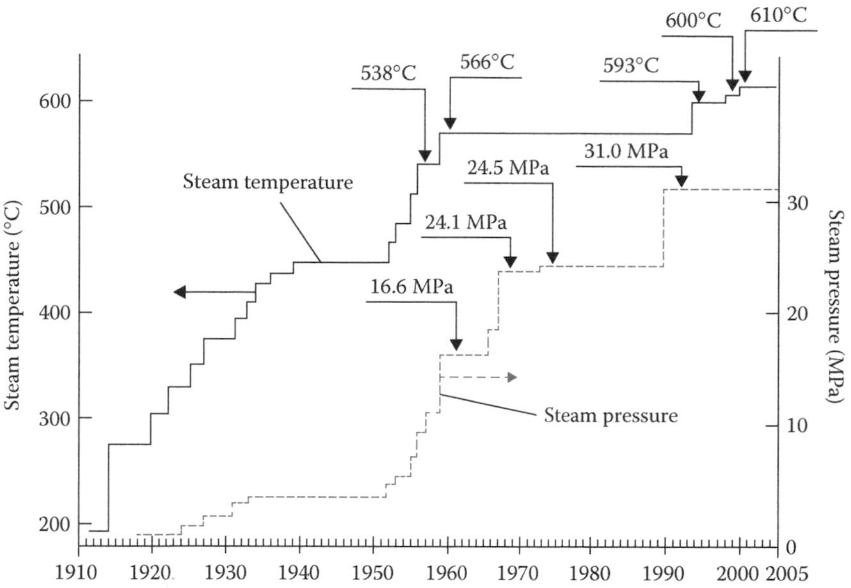

FIGURE 13.5 Historical progress of power generation steam condition and power generation efficiency of fossil power generation in Japan.

produced through the thermal reactions of hydrocarbon and nitrogen in air present in the flame. The quite small amount of prompt NO_x is formed in pulverized coal combustion.

NO_x in pulverized coal combustion is, therefore, mainly fuel NO_x. Twenty to sixty percent of nitrogen in coal is released into gas phase as volatile *N* species, which are mostly HCN and NH_3. Volatile N is converted to NO and N_2 during volatile matter combustion. The other nitrogen in coal remained is char (char N), which is transformed into N_2 or NO_x.

Several low NO_x combustion technologies have been developed. Figure 13.6 systematically shows the low NO_x combustion technologies (Makino and Kimoto 1994). The conventional low NO_x combustion technologies consist of conventional low NO_x burner, low air ratio combustion (low O_2 combustion), two-stage combustion, and flue gas recirculation. However, low air-ratio causes the increase of unburned carbon in fly ash. The advanced low NO_x combustion technology attempted to intensify the reduction of generated NO_x with unburned char.

13.5.1 Low NO_x Burner

Figure 13.7 shows the concept of the advanced low NO_x burner (Makino et al. 1997). A highly intensified internal recirculation zone is formed in the near-burner region, where the residence time of coal particles is extended. The initial stage of thermal decomposition of coal is therefore accelerated, thereby reducing the char. NO_x concentration in this internal recirculation zone is extremely high due to more production of volatile matter. However, NO_x is effectively reduced to N_2 in the downstream of the internal recirculation zone due to reducing environment—sub-stoichiometry ratio.

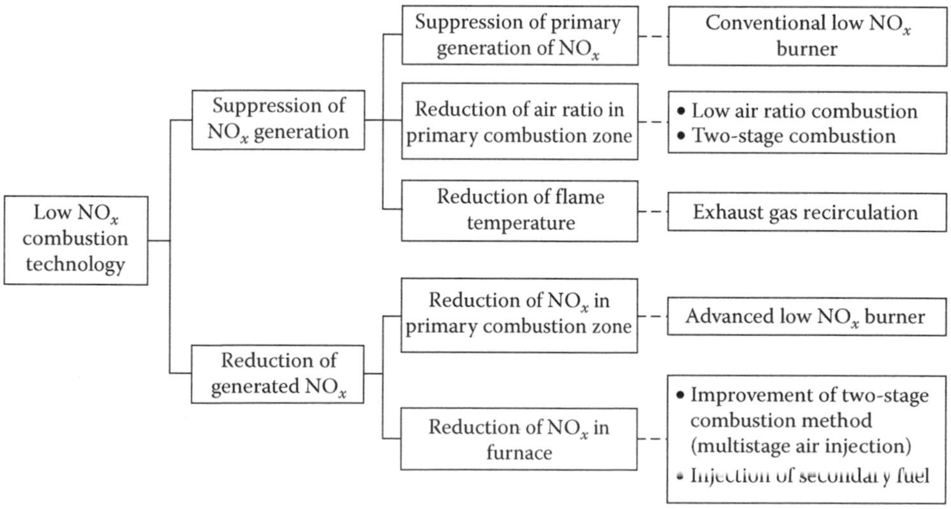

FIGURE 13.6 Low NO_x combustion technologies.

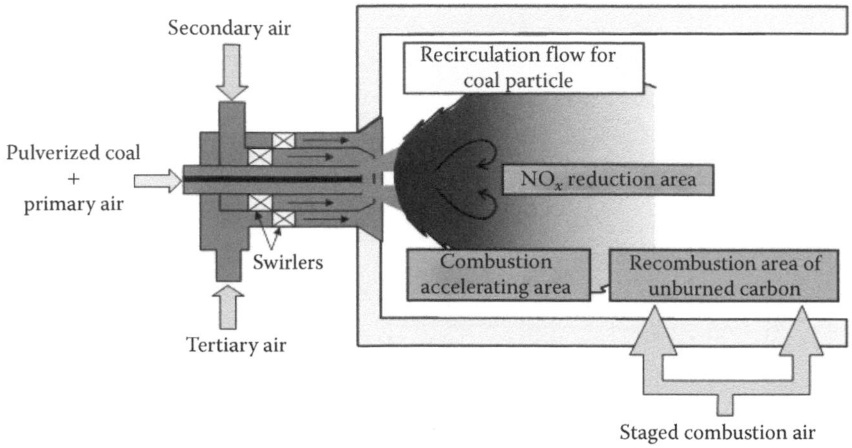

FIGURE 13.7 Concept of the advanced low NO_x Burner.

If two or more air injection ports are used for staged combustion, the NO_x decomposition in reduction flame is more efficient. This combustion method is called as a *multistage combustion method.* The conceptual design of the multistage combustion method is also presented in Figure 13.7. This method can suppress the regeneration of NO_x at the injection ports of staged combustion air.

The amounts of the unburned carbon in fly ash are compared in the multistages combustion with conventional and advanced burners, where the NO_x concentration is 100 ppm. Advanced low NO_x burner decreases the amount of unburned carbon content to about a half (14%) of that provided by the conventional low NO_x burner (30%). In the small-scale test furnace, the advanced low NO_x burner in the multistage combustion is found to decrease further unburned carbon to less than 10%, which corresponds to approximately 3% in the actual furnace of power generation.

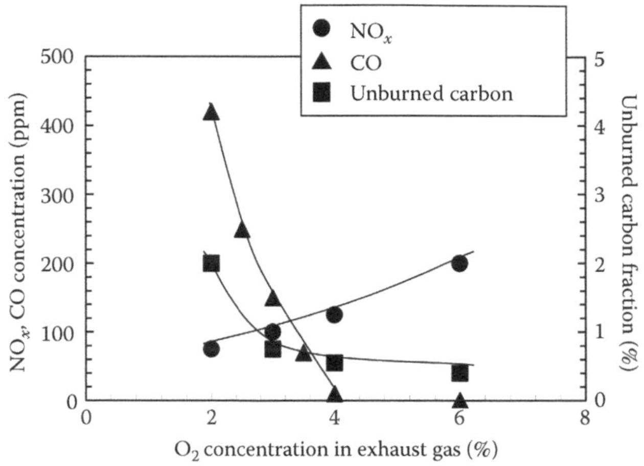

FIGURE 13.8 Relationship between O_2, NO_x, and CO concentration and unburned carbon content.

13.5.2 Low Air Ratio Combustion

In conventional coal combustion, air ratio (actual injected air/ideal injected air required for complete combustion) is set up to 1.0 in order to achieve high combustion efficiency. In pulverized coal combustion, air ratio is set between 1.15 and 1.25. However, increasing air ratio makes oxygen condition strong; consequently, NO_x emission increases. Therefore, NO_x is reduced by decreasing O_2 concentration as possible. Figure 13.8 shows the influence of O_2 concentration on NO_x, CO, and unburned carbon concentration (Makino and Kimoto 1994). NO_x concentration is decreased with the decrease of O_2 concentration. On the other hand, CO and unburned carbon concentration is increased and this tendency is typical below 4% O_2 concentration in flue gas.

13.5.3 Staged Combustion

To reduce the oxidization atmosphere at high temperature region of combustion flame near the burner, staged combustion method is applied. In staged combustion, the combustion air is divided and about 70% of combustion air is provided from burner part and about 30% is provided to the middle part of furnace. This method can keep the overall air ratio and reduce the primary air ratio near the burner. As NO_x is formed extremely near the burner, this method can reduce the NO_x concentration. For the combustion, efficiency in primary combustion zone near the burner is controlled by low primary air ratio, but unburned components from the primary combustion zone is combusted in the secondary combustion zone after staged air injection. So, overall combustion efficiency is kept at same level with usual combustion. The ratio of staged air, which is provided from the middle part of furnace, to the total combustion air is named *staged combustion air ratio.*

According to the increase in staged combustion air ratio, the oxidization condition of primary combustion area becomes weak and NO_x concentration is getting low as shown in Figure 13.9 (Makino et al. 1990). On the other hand, unburned carbon concentration in fly ash is increased with the increase of staged combustion air ratio. The injection point of staged combustion air is important for the NO_x reduction similar with the staged combustion air ratio. Figure 13.10 shows the influence

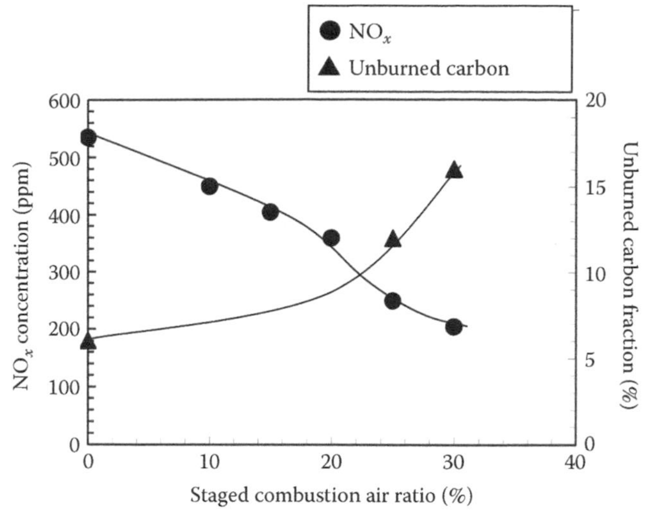

FIGURE 13.9 Relationship between NO_x concentration, unburned carbon fraction, and staged combustion air ratio.

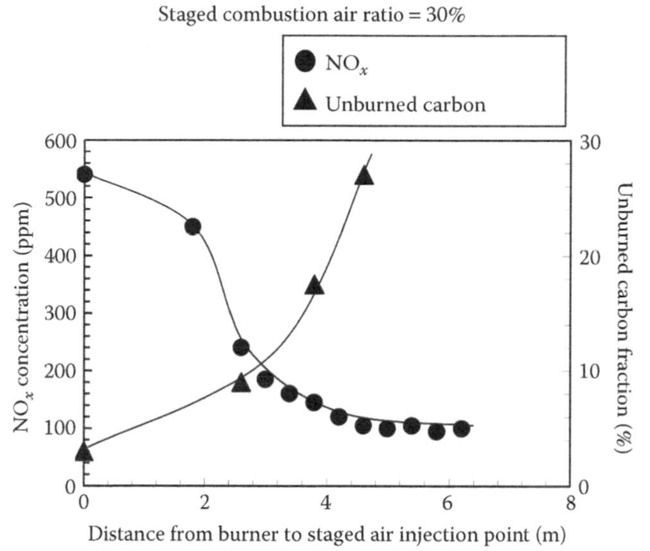

FIGURE 13.10 Relationship between NO_x concentration, unburned carbon fraction, and injection point of staged combustion air.

of the injection point of staged combustion air on NO_x concentration and unburned carbon concentration (Makino et al. 1990). NO_x concentration decreases with the increase of the distance of burner and injection point of staged combustion air. Because NO_x reduction area from burner to injection point of staged combustion air becomes wider. Of course, this wide reduction area causes the increase of unburned carbon concentration in fly ash. In Figure 13.10, the decrease of NO_x is remarkable between 2 m and 3 m of the distance from burner and ignition point of staged combustion air and the increase of unburned carbon concentration is remarkable over 3m of the distance from burner. So, for the simultaneous reduction of NO_x and unburned carbon, the optimum injection point of staged combustion air is about 3m from the burner in the combustion furnace used in the experiments.

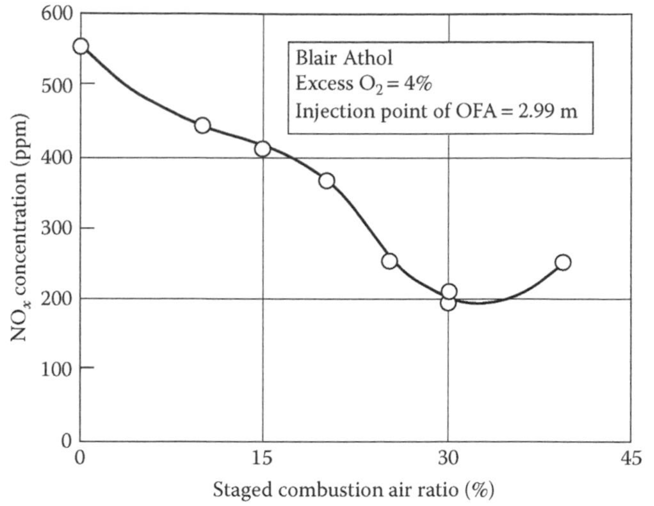

FIGURE 13.11 Relationship between staged combustion air ratio and NO_x concentration.

On the condition of optimum injection point, NO_x concentration increases over 30% of staged combustion air ratio shown in Figure 13.11 (Makino et al. 1990). This tendency is considered as follows. In the condition of 40% of staged combustion air ratio, NO_x concentration of primary combustion area is reduced remarkably due to low O_2 concentration atmosphere, but a lot of unburned matter remains before the injection point of staged combustion air. This unburned matter causes intense combustion by the injection of staged combustion air and regeneration of NO_x in this area is accelerated. For the control of this intense combustion and regeneration of NO_x, the division of staged combustion air is useful. This method is named *multistage air injection method.*

Figure 13.12 shows that the effect of the ratio of first stage air injection ratio to the total staged combustion air (Makino et al. 1990). The right side of this figure indicates all staged

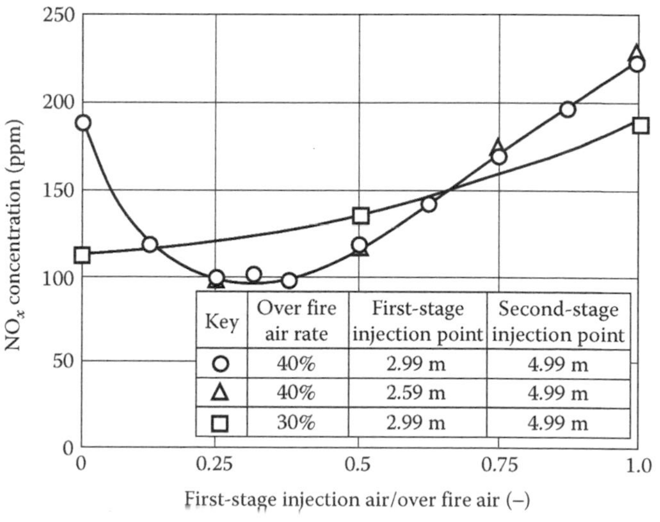

FIGURE 13.12 Relationship between air injection rate and NO_x concentration with multistage injection method.

combustion air is injected from the first stage of staged combustion air injection port (near burner), and the left side of this figure indicates all staged combustion air is injected from the second stage of staged combustion air injection port (far from burner). In the condition of 30% of staged combustion air ratio, NO_x concentration with the injection point of first stage is highest and the NO_x concentration decreases monotonously according to the decrease in the divided air ratio of first stage injection. On the other hand, the NO_x concentration has the minimum value on the condition of divided injection of staged combustion air at the 40% of staged combustion air ratio. This means that the control of the combustion on staged air injection point is very important for the NO_x reduction at the condition of high air ratio of staged combustion.

13.6 LOW-LOAD COMBUSTION

Pulverized coal-fired power stations are often required to reduce power generation at weekends and/or during nighttime. In order to reduce power generation of coal-fired power plant, it is necessary to decrease pulverized coal feed rate. The pulverized coal concentration in the primary air becomes lower by the decrease of coal feed rate, and it is very difficult to keep a stable flame under the condition of low pulverized coal concentration. For the improvement of flame stability at low load, concentrating pulverized coal to some part of burner is very effective. Several methods are proposed to concentrate the pulverized coal particles in the burner. Usually, swirling force of primary air is useful for the concentration of pulverized coal particles. Figure 13.13 indicates the effect of concentration of coal particles on combustion stability at low load. The concentrating coal particles can improve the minimum load of stable coal combustion to the half value of usual combustion method. On the condition without swirling force in primary air, the streamlined ring in the burner is useful. The streamlined ring shown in Figure 13.14 (Makino et al. 1999) can shift the pulverized coal particles to the outer region of the primary air nozzle and concentrate the pulverized coal particles. The ring is positioned near the burner outlet at low load conditions to increase the concentration of pulverized coal particles in the outer region of the primary air nozzle and positioned far from burner outlet at standard combustion load.

Figure 13.15 shows NO_x concentration and unburned carbon in fly ash of the burners with or without the streamlined ring (Makino et al. 1999). The burner with the streamlined ring is able to form a flame under 20% load, while 30% load is

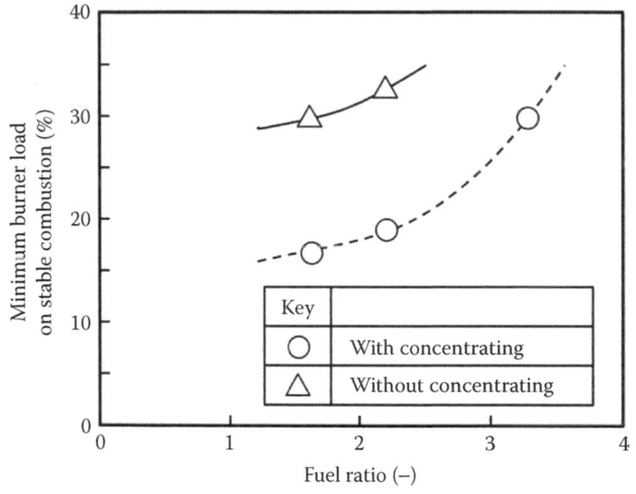

FIGURE 13.13 Effect of concentrating of coal particles on combustion stability at low load.

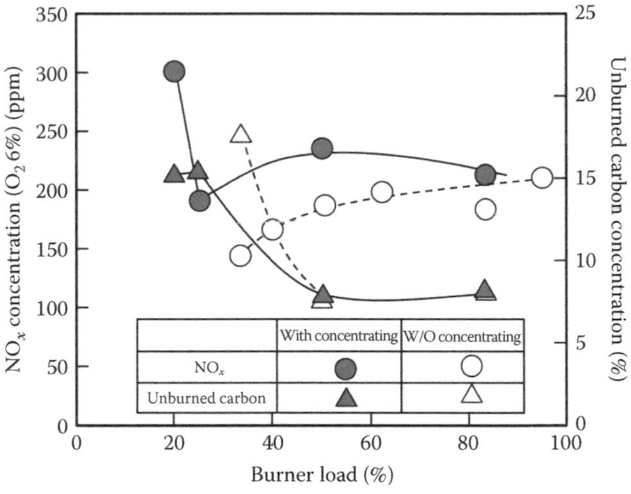

FIGURE 13.15 Relationship between NO_x and unburned carbon concentration and burner load with or without the streamlined ring.

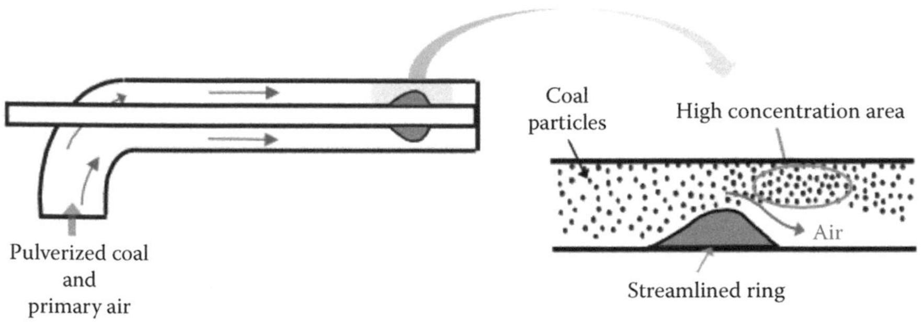

FIGURE 13.14 Schematics of coal concentrating mechanism of streamlined ring.

minimum load on the burner without the ring. The NO_x concentration on the burner with concentrating ring is higher than the NO_x concentration on the burner without ring at low load. On the other hand, the unburned carbon concentration in fly ash on the burner with concentrating ring is lower than the unburned carbon concentration in fly ash on the burner without ring at low load. This tendency of concentration of pulverized coal particles in the burner improves the combustion situation of pulverized coal burner, and combustion efficiency and NO_x concentration become higher.

13.7 COMBUSTION FOR LOW-GRADE COAL

For the stable coal supply for the pulverized coal-fired power station, the diversification of usable coal is very important. At present, the bituminous coal is used as a main coal in almost all power stations. For the diversification on usable coal, high moisture content coal, high-fuel-ratio coal, and high ash content coal are expected.

13.7.1 Combustion of High Moisture Content Coal

Low rank coals such as sub-bituminous coal and lignite are less coalified and contain high moisture. However, the minable reserve amount of such coals is next to that of bituminous coal. Therefore, for the diversification of coal, it is necessary to develop the utilization technology of high moisture coals.

13.7.1.1 Combustion Characteristics of Sub-Bituminous Coal

Pulverized coal combustion characteristics of sub-bituminous coal were evaluated with the combustion on air injection conditions to the burner optimized for bituminous coal. Figure 13.16 shows the distribution of the oxygen concentration in the furnace for Newlands (bituminous) and Wara (sub-bituminous) coal combustion (Ikeda et al. 2002). When Newlands coal was fired, the combustion flame was stable. The consumption of oxygen progressed gradually from the burner outlet. The reduction flame between the burner outlet and the injection port for two-stage combustion air was widely formed in the furnace. Distribution of the oxygen concentration for Wara coal combustion was then compared under the three levels of remaining moisture content. The total moisture content in the coal (Ct), is the sum of the vaporized moisture content (Cv) and remained moisture in coal (Cr). When Cr increased, the ignition worsened and the combustion flame became diffused, so that the reduction region was decreased.

Figure 13.17 shows the distribution of the NO_x concentration at the center axis in the furnace (Ikeda et al. 2002). When bituminous coal was fired, NO_x was formed rapidly at the burner exit. However, NO_x was immediately reduced at the reduction area between the burner outlet and the air injection port for two-stage combustion. In the case of sub-bituminous coal combustion, NO_x formation and decomposition were delayed. When Cr increased, NO_x formation was delayed. Therefore,

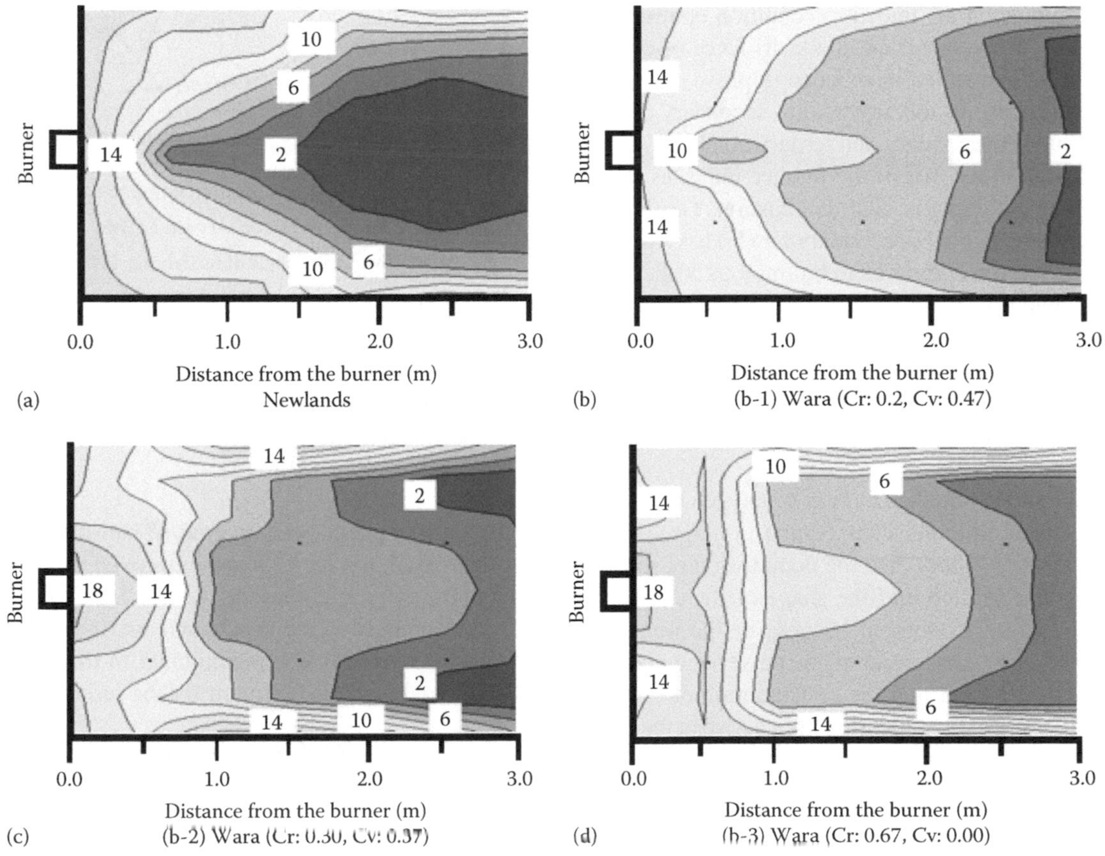

FIGURE 13.16 O_2 concentration in the furnace. (a) Bituminous coal case (b) sub-bituminous coal (Cr = 0.2) case (c) sub-bituminous coal (Cr = 0.30) case and (d) sub-bituminous coal (Cr = 0.67) case.

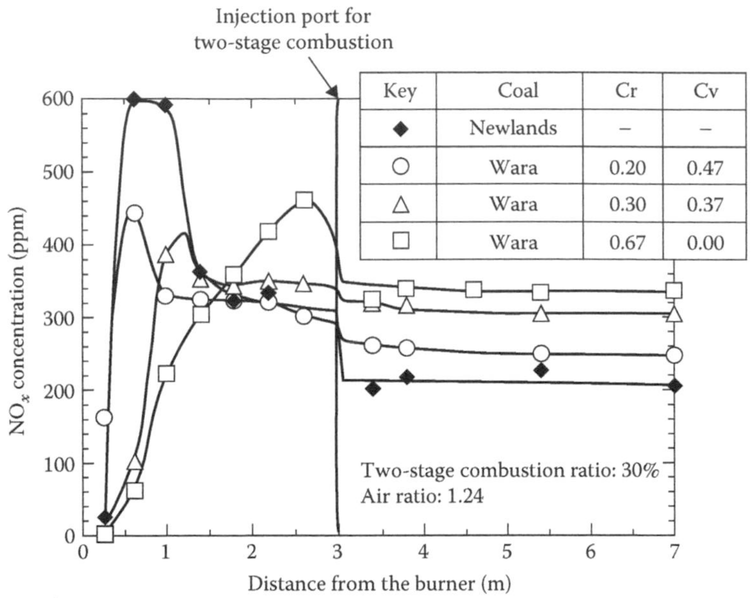

FIGURE 13.17 Influence of the remaining moisture content in coal on NO_x concentration in the center axis of the furnace.

NO_x concentration at the exit of the furnace increased, as Cr became higher, due to insufficient time for reduction.

13.7.1.2 Adjustment of Combustion Air Injection Conditions for Sub-Bituminous Coal

Since NO_x concentration at the exit of the furnace was high in the case of combustion air injection condition optimized for bituminous coal, the volume and the swirl force of combustion air should be controlled in order to improve combustion stability. The influence of the flow rate of secondary combustion air to the sum of secondary and tertiary combustion air on NO_x concentration at the exit of the furnace indicates tendencies similar with bituminous coal combustion. On the other hand, the swirl vane angle of secondary air (S_s) has a different effect on NO_x emission compared to bituminous coal combustion. The influence of S_s on NO_x concentration is shown in Figure 13.18 (Ikeda et al. 2002). When bituminous coal was fired, the optimum S_s was about 80°. The intensive swirl force was suitable for bituminous coal. When S_s is between 50° and 60°, the NO_x concentration at the exit of the furnace becomes the minimum value.

Figure 13.19 shows the distribution of oxygen at S_s of 80° and 54° for sub-bituminous coal combustion (Ikeda et al. 2002). When the S_s is about 50°, the combustion of sub-bituminous coal is accelerated and the shape of the combustion flame appeared to be moderated because of the weak swirl flow. As a result, the concentration of pulverized coal near the burner became higher and the consumption of oxygen is accelerated. Therefore, the reduction area between the burner outlet and the injection port for two-stage combustion air becomes wider and NO_x has sufficient time for reduction.

To improve ignition conditions at the burner outlet, the optimization of primary air volume is important. Figure 13.20 shows the relation between air/coal and NO_x concentration at the exit of the furnace (Ikeda et al. 2002). Here, air/coal is

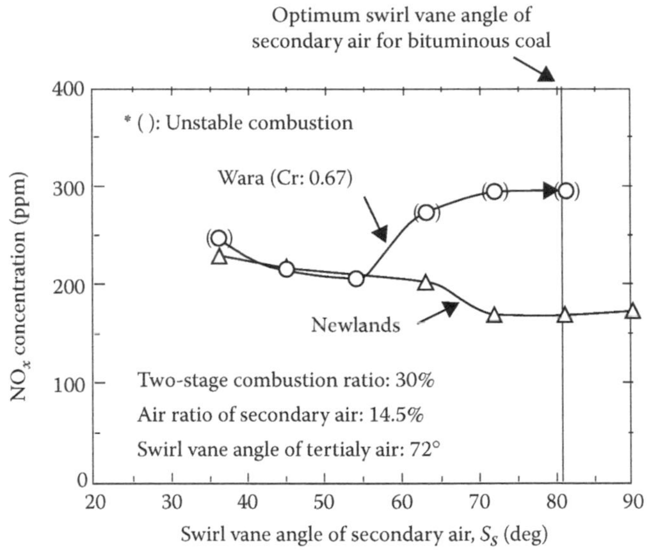

FIGURE 13.18 Relationship between the swirl vane angle of secondary air and NO_x concentration at the exit of the furnace.

mass ratio of primary air to coal. When the mass of primary air decreased, the ignition was improved and NO_x concentration at the exit of the furnace became lower. If air/coal is over 2.2, the combustion flame is blown out. If the air/coal is under 1.9, pulverized coal was precipitated in the primary air tube. The air/coal of 1.9 is optimum for the reduction of NO_x.

13.7.1.3 Emission Characteristics of NO_x and Unburned Carbon in Fly Ash

Figure 13.21 shows the influence of Cr on NO_x concentration at the exit of the furnace and the unburned carbon concentration in fly ash in the conditions with Cr of 0.20, 0.30, and 0.67, under the same Ct of 0.67 (Ikeda et al.

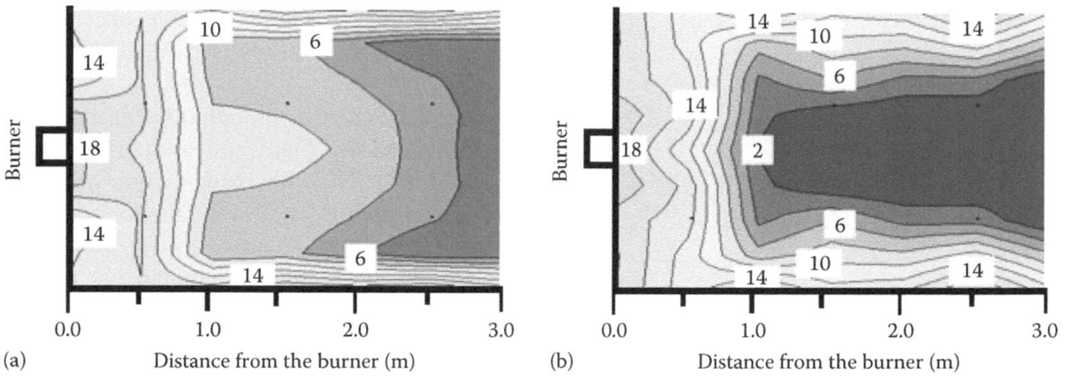

FIGURE 13.19 Relationship between the swirl vane angle of secondary air and O_2 concentration in the furnace (Cr: 0.67). (a) Condition optimized for bituminous coal and (b) condition optimized for sub-bituminous coal.

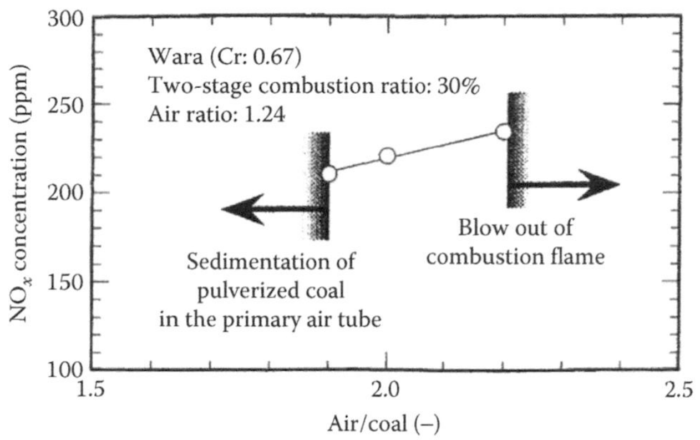

FIGURE 13.20 Relationship between air/coal and NO_x concentration at the exit of furnace.

Key	Remained moisture in coal (–)	Vaporized moisture (–)	Air/coal (–)	Ss (deg.)
○	0.20	0.47	2.2	81
●	0.20	0.47	2.2	72
△	0.30	0.37	2.2	81
▲	0.30	0.37	2.2	63
□	0.67	0.00	2.2	81
■	0.67	0.00	2.2	54
◆	0.67	0.00	1.9	54

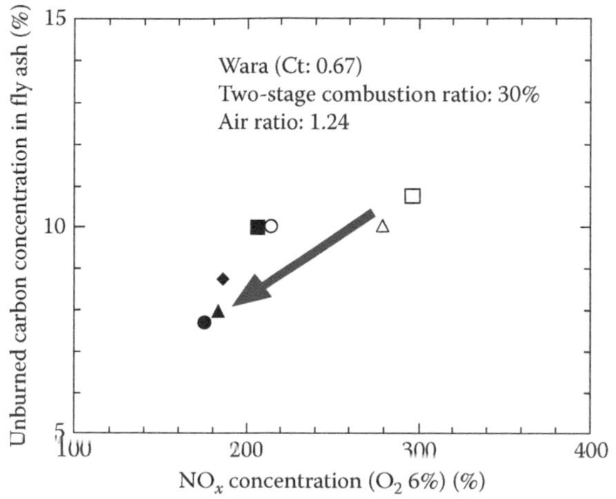

FIGURE 13.21 Influence of combustion air injection conditions on NO_x concentration and unburned carbon concentration in fly ash.

2002). When the S_s is optimized for sub-bituminous coal combustion, both NO_x concentration and the unburned carbon concentration decrease. When Cr becomes lower, NO_x concentration decreases. When Cr is high, the effect of NO_x reduction becomes higher because ignition is improved considerably by adjusting the combustion air injection conditions. When the air/coal ratio is 1.9, the emission of both NO_x and unburned carbon decrease further. By this optimization, the ratio of NO_x reduction is achieved about 40% with a Cr of 0.67.

13.7.1.4 NO_x Emission Characteristics in Blended Combustion

For the easy utilization of sub-bituminous coal, the combustion of its blend with bituminous coal is applied. Figure 13.22 shows the relationship between NO_x concentration at the exit of the furnace and the blend ratio of sub-bituminous coal (Ikeda et al. 2003). NO_x concentration at the exit of the furnace is different depending on the amount of remaining moisture in coal, Cr. In spite of Cr, NO_x concentration in blended combustion of sub-bituminous coal indicates the mean value estimated from those in non-blended combustion of bituminous coal and sub-bituminous coal, respectively. When the blend ratio of sub-bituminous coal increases, NO_x concentration in blended combustion approaches that in sub-bituminous coal combustion.

13.7.1.5 Unburned Carbon Concentration in Fly Ash in Blended Combustion

Figure 13.23 shows the relationship between unburned carbon concentration in fly ash and the blend ratio of sub-bituminous coal under the same conditions as in the case of Figure 13.22 (Ikeda et al. 2003). The unburned carbon concentration in fly ash in blended combustion is higher than that in non-blended combustion of each coal. However, unburned carbon concentration in fly ash is dependent on ash content. Even if combustion efficiency is high, unburned carbon concentration in fly ash is high in the case of low ash content coal. Therefore, it is very difficult to compare combustibility using unburned carbon concentration in fly ash. Then, the unburned fraction

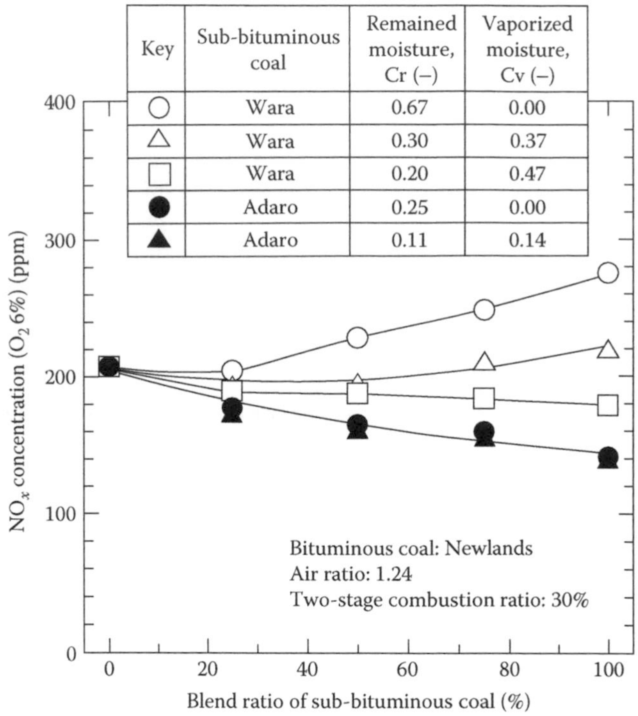

FIGURE 13.22 Relationship between NO_x concentration and blend ratio of sub-bituminous coal.

defined below is used for the investigation of combustion characteristics:

$$Uc^* = 100 - \eta = \frac{Uc}{100 - Uc} \times \frac{C_{Ash}}{100 - C_{Ash}} \times 100$$

where:

Uc^* (%) is the unburned fraction
η (%) is the combustion efficiency
Uc (%) is the unburned carbon concentration in fly ash
C_{Ash} (%) is the ash content in coal

Figure 13.24 shows the relationship between the unburned fraction and the bled ratio of sub-bituminous coal (Ikeda et al. 2003). The unburned fraction has a maximum value when the bled ratio of sub-bituminous coal is about 25%. This is presumably caused by the moisture in sub-bituminous coal, hindering the combustion process of the bituminous coal, because in blended combustion, moisture in sub-bituminous coal exists not only around sub-bituminous coal particles but also around bituminous coal particles. Then, as the partial oxygen pressure of circumstance of bituminous coal and the gas temperature decreases, combustion efficiency of bituminous coal becomes low. On the other hand, the partial oxygen pressure of circumstance of sub-bituminous coal in blended combustion becomes higher than that in sub-bituminous coal combustion. Then, the combustion efficiency of sub-bituminous coal seems to improve.

13.7.2 Combustion of High-Fuel-Ratio Coal

13.7.2.1 Combustion Characteristics of High-Fuel-Ratio Coal

In this section, the combustion characteristics of high-fuel-ratio coals are investigated under the non-staged combustion condition. To evaluate coal ignitability of high-fuel-ratio coal, the minimum burner load for stable combustion is examined by decreasing the coal feed rate. To keep the mass ratio of pulverized coal to primary air (=1:2.2), the primary air rate is decreased with decreasing coal feed rate. In order to consider the effect of the difference in burner type on the flame stability, the result for the CI-α burner is compared with that for the conventional low NO_x burner.

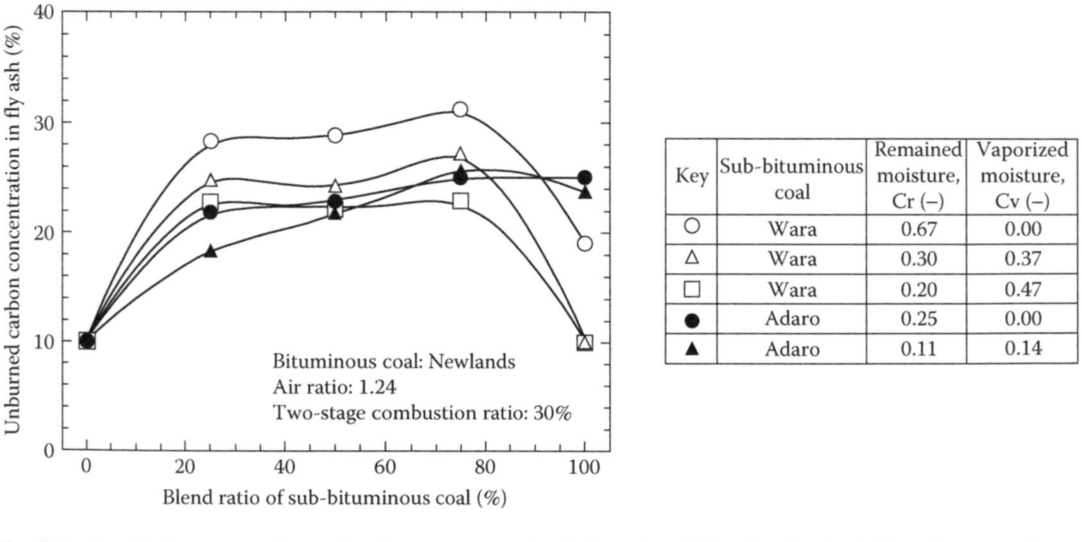

FIGURE 13.23 Relationship between unburned carbon concentration in fly ash and blend ratio of sub-bituminous coal.

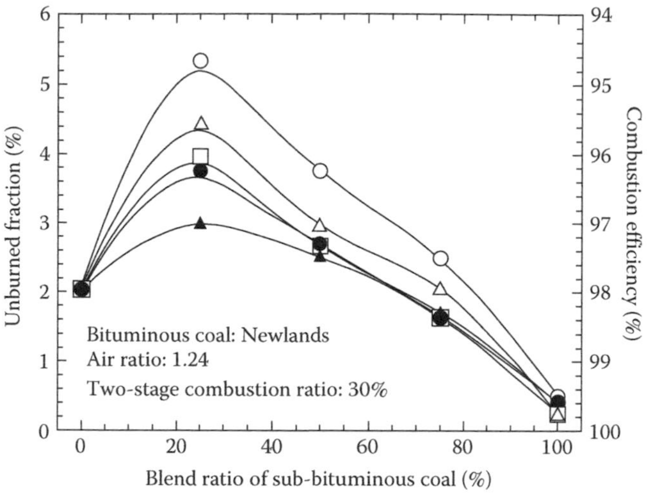

Key	Sub-bituminous coal	Remained moisture, Cr (–)	Vaporized moisture, Cv (–)
○	Wara	0.67	0.00
△	Wara	0.30	0.37
□	Wara	0.20	0.47
●	Adaro	0.25	0.00
▲	Adaro	0.11	0.14

FIGURE 13.24 Relationship between unburned fraction and blend ratio of sub-bituminous coal.

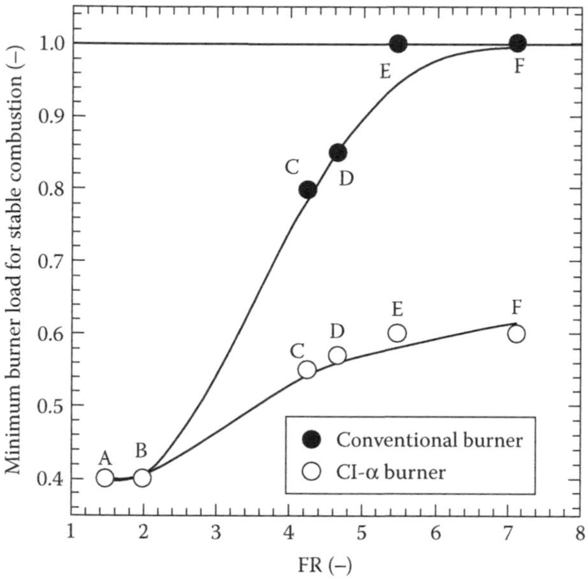

FIGURE 13.25 Effect of fuel ratio on the minimum burner load for stable combustion for non-staged combustion.

In general, the ignitability of high-fuel-ratio coals is poor because their volatile matter contents are low. Figure 13.25 shows the effect of fuel ratio FR on the minimum burner load for stable combustion under the non-staged combustion condition (Kurose et al. 2004). The minimum burner load was determined as follows. First, the combustion flame was stabilized at a burner load of 100%. Then, the coal feed rate was gradually decreased with decreasing air, and the minimum burner load was defined as a burner load when the ignition point apparently shifts to downstream. For comparison, the result with a conventional low NO_x burner is plotted in this figure. It is observed that although for both burners the minimum burner load rises with increasing FR, the value for the CI-α burner is much less than that for the conventional low NO_x burner at a certain FR. For a high-fuel-ratio range of FR > 5.5, the burner load for the conventional low NO_x burner could not be lowered below 100%, whereas the CI-α burner could be operated even for a lower burner load of 60%. This means that the CI-α burner is more suitable for the stable combustion of the pulverized coal than the conventional low NO_x burner for high-fuel-ratio coals. As verified by Kurose et al. (2004), the strong swirling flow generated by the CI-α burner produces a recirculation flow in the high-gas-temperature condition region close to the burner outlet. This increases the residence time of coal particles in this high-gas-temperature region and promotes the evolutions of volatile matter and char reaction. This effect decreases the value of the minimum burner load for high-fuel-ratio coals.

13.7.2.2 NO_x Emission and Combustion Efficiency

Figure 13.26 shows the relationship between the conversion of fuel-bound nitrogen to NO_x, CR [-], and three indexes consisting of fuel ratio FR [-], fuel-bound nitrogen FN [-], and fixed carbon FC [-] (Kurose et al. 2004). Six coals are combusted using the CI-α burner with a burner load of 100%. Here, CR is defined by

$$CR = \frac{C_{NOx}}{\left(\dfrac{2.24\times10^{-2}\times\dfrac{FN}{1.4\times10^{-2}}}{V_{dry}}\right)}\times10^{2}$$

where:

C_{NO_x} is the NO_x concentration at the furnace exit
V_{dry} is the rate of dry gas volume per feeding rate of coal [(Nm^3)/kg]
The denominator indicates the NO_x concentration assuming that all fuel N converts to NO_x

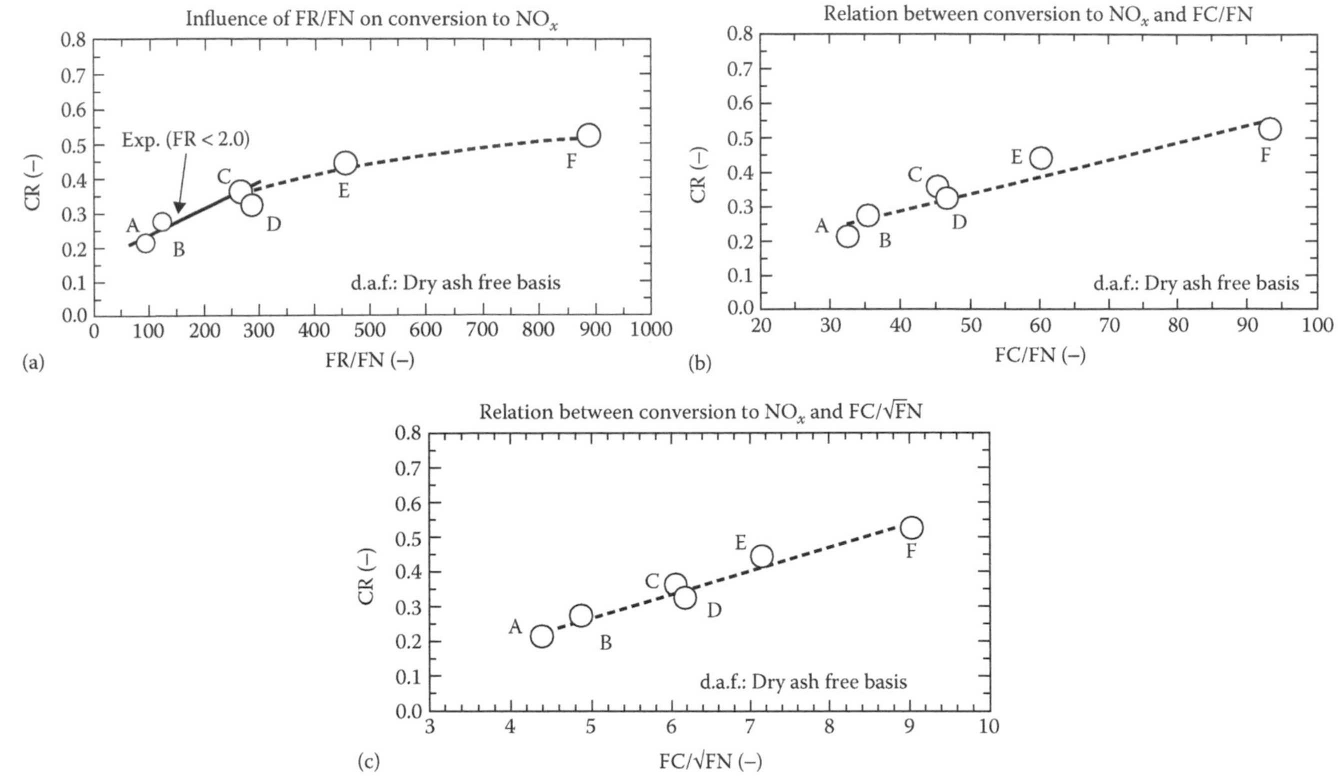

FIGURE 13.26 Relationship between CR and three indexes consisting of FR, FN, and FC. (a) Influence of FR/FN on conversion to NO*x* (b) relationship between conversion to NO*x* and FC/FN and (c) relationship between conversion to NO*x* and FC/FN.

In general, CR for bituminous coals with FR ≤ 2.5 was found to increase linearly with FR/FN. This suggest that CR is proportional to (1−VM)/(VM×FN), where VM is volatile matter content. It should be noted here that present VM and FC are given on a dry ash free basis to get rid of the influence of ash content on these values, because the ash content does not affect the combustion characteristics very much. Here, the reason why CR decreases with VM is that the NO_x reduction in the primary combustion region increases as VM increases. On the other hand, the reason why CR decreases with FN is that the increment of FN increases NO_x emission, which suppresses the conversion of N to NO_x to keep the state of balance between NO_x and N_2. For higher fuel-ratio coals with FR ≤ 2.5, CR similarly increases with FR/FN, but the increment becomes small and the deviation from the bituminous coals' linear correlation becomes large for higher FR/FN, as shown in Figure 13.26 (a). On the other hand, it is found in Figure 13.26 (b) and (c) that CR can be linearly correlated using fixed carbon FC instead of FR. CR increases almost linearly with FC/FN even for high-fuel-ratio coals with FR > 5.5 (FC/FN > 60), and the linearity is further improved by introducing FC/√FN. Here, the correlated parameter FC/FN is introduced on the basis of the inference that the denominator VM should be removed from FR because the variation of FR with VM becomes considerably large as VM decreases (FC increases) for

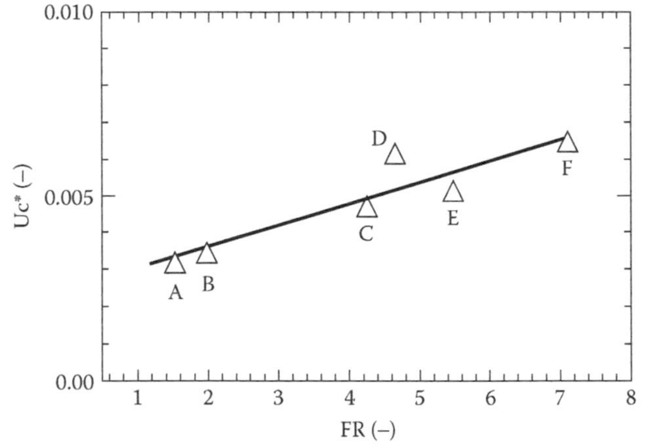

FIGURE 13.27 Relationship between Uc* and FR.

high-fuel-ratio coals. On the other hand, since the effect of FN on FC/FN becomes strong compared to that on FR for bituminous coal (for coals whose VM and FC are comparable), the other parameter FC√FN is employed to improve their relation.

The relationship between unburned carbon fraction Uc* and fuel ratio FR is shown in Figure 13.27 (Kurose et al. 2004). It is found that Uc* monotonically increases with FR not only for bituminous coals but also for high-fuel-ratio coals. The increment of the unburned carbon with FR is partially due to the char reaction rate being extremely low compared to the evolution rate of volatile matter and thereby the magnitude of the fixed carbon content mostly dominating the amount of residue carbon.

13.7.2.3 Effects of Staged Combustion on NO_x Emission and Combustion Efficiency

Effects of the staged combustion on NO_x emission and combustion efficiency are examined using the CI-α burner with a burner load of 100% and a staged combustion air ratio of 30%.

Figure 13.28 shows the effects of fuel ratio FR on NO_x reduction (R_{NO_x}) and unburned carbon fraction increment (R_{Uc^*}) at the furnace exit by the staged combustion (Kurose et al. 2004), R_{NO_x} [−] and R_{Uc^*} [−], which are defined by

$$R_{NO_x} = \frac{C_{NO_x\,nonstg} - C_{NO_x\,stg}}{C_{NO_x,\,nonstg}}$$

$$R_{Uc^*}\frac{U_{Ci,nonstg} - U_{Ci,stg}}{U_{Ci,nonstg}}$$

Here $C_{NO_{x,j}}$ and $U_{Ci,j}$ are the NO_x concentration and the unburned carbon fraction at the furnace exit [−], and the subscripts *nonstg* and *stg* indicate the non-staged and staged combustion, respectively. It is found that R_{NO_x} decreases and R_{Uc^*} increases with increasing FR. In other words, with increasing fuel ratio, the NO_x reduction effect due to the staged combustion weakens, whereas the unburned carbon fraction increment due to the staged combustion becomes significant.

To explain the reason for the above behaviors, the axial distributions of the gas temperature, and O_2 and NO_x concentrations, C_{O_2} and C_{NO_x}, in the furnace for coal A and coal E for the non-staged and staged combustions are shown in Figure 13.29. For both coals, the staged combustion depresses C_{O_2} in the primary combustion region, and also depresses C_{NO_x} in the whole region, because the deficiency of O_2 in the primary combustion region promotes NO_x reduction. It is found that the effects of the staged combustion on the C_{O_2} and C_{NO_x} distributions are weak for coal E. This is because the volatile matter content is low for coal E, and thereby O_2 concentration is relatively high in the primary combustion region. As a result, compared to low-fuel-ratio coals, the NO_x reduction effect due to the staged combustion for high-fuel-ratio coals deteriorates, as indicated in Figure 13.29 (Kurose et al. 2004). The unburned carbon fraction increment due to the staged combustion becomes significant for high-fuel-ratio coals, as indicated in Figure 13.28.

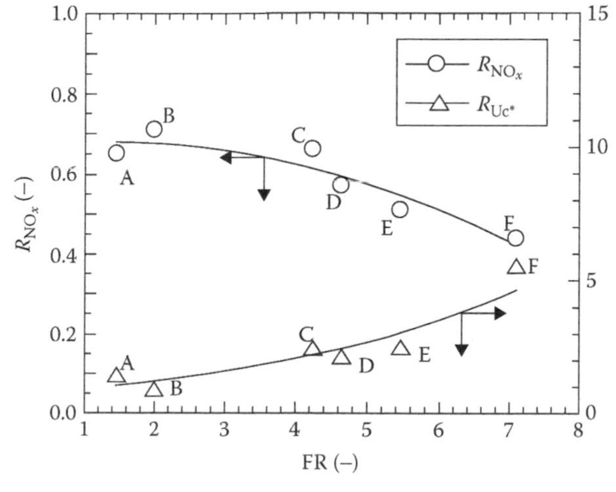

FIGURE 13.28 Effects of FR on NO_x reduction and unburned carbon fraction increment.

13.7.3 Combustion of High Ash Content Coal

13.7.3.1 Influence of Ash Content on NO_x and Unburned Carbon Concentration

For the utilization of high ash content coal, it is very important to analyze the influence of ash content on combustion characteristics. Figure 13.30 shows the distributions of gas temperature measured by Pt/Rt-Rh(13%) thermocouple, O_2 and NO_x concentration at the center of the furnace in the standard combustion of three high ash content coal (Kurose et al. 2001a). As the ash content increases, the gas temperature decreases while the O_2 consumption and the NO_x formation and reduction are delayed near the burner. Although the gas temperature and the O_2 concentration at the exit of the furnace tend to approach certain values independent of the ash content, NO_x concentration is higher for a high ash coal. This is because of the shortage of the NO_x reduction time because of the delay in NO_x formation.

Figure 13.31 shows the relationship between the conversion of fuel-bound nitrogen to NO_x, CR, and the index of the fuel ratio divided by the fuel-bound nitrogen, FR/FN, for high ash content coal, together with that for some typical bituminous coal with low ash content of 7.0%–18% (Kurose et al. 2001a).

CR for bituminous coal was found to increase linearly with FR/FN. On the other hand, although CR for high ash content coal with the ash content of 36 wt% is almost on the averaged line for bituminous coal, it tends to monotonously increase with the increasing ash content at a fixed FR/FN.

The relation between the unburned carbon fraction, Uc*, and the fuel ratio, FR, is shown in Figure 13.32 (Kurose et al. 2001a), together with those for some typical bituminous coals with low ash contents of 7.0%–18%.

It was shown that Uc* was proportional to FR for bituminous coals with low ash contents. Uc* of the high ash coal is found to be much higher than that of low ash coal. With increasing ash content, the unburned carbon fraction increases and the combustion efficiency decreases.

As mentioned above, the combustibility of the pulverized coal combustion is suppressed as the ash content in coal increases. One of the reasons of this is likely because the heat capacity of the ash increases with the increasing ash content.

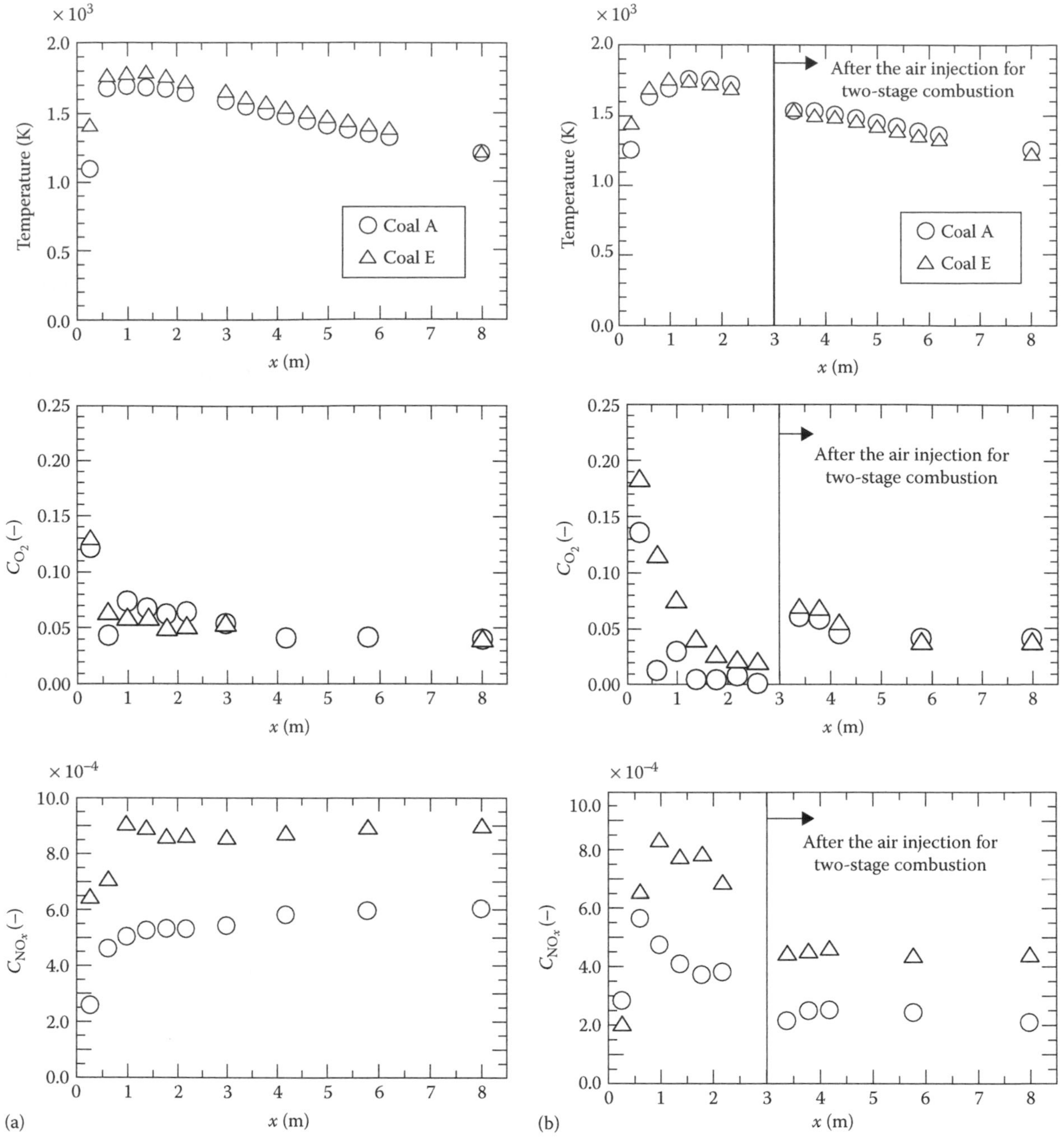

FIGURE 13.29 Axial distributions of gas temperature, C_{O_2} and C_{NO_x} in the furnace: (a) non-staged combustion and (b) staged combustion.

13.7.3.2 Influence of Ash Content on Combustion Profile

Figure 13.33 shows the gas temperature, O_2, and NO_x concentration in the staged combustion of high ash content coal (Kurose et al. 2001b). The O_2 concentration in the region before the staged air injection port is much less than that in the standard combustion condition, and the NO_x concentration at the furnace exit is also less because deficiency of O_2 promotes the NO_x reduction. The trends of the gas temperature, O_2, and NO_x concentration with the ash content are similar to those in the standard combustion. As the ash content increases, the gas temperature decreases, the O_2 consumption and the NO_x formation are delayed near the burner, and then NO_x concentration at the furnace exit increases.

Figure 13.34 shows the effect of the ash content on the NO_x concentration and the unburned carbon fraction at the furnace exit, together with those in the standard combustion (Kurose et al. 2001b). It is clearly observed that the

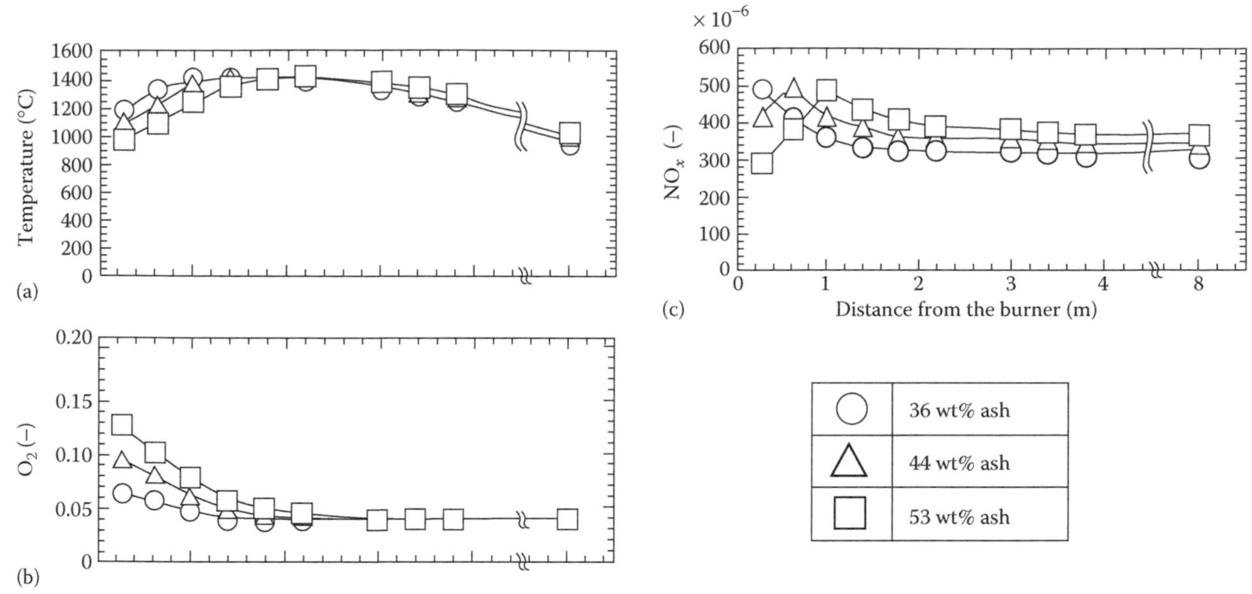

FIGURE 13.30 Axial distributions of (a) temperature, (b) O_2 concentration and (c) NO_x concentration.

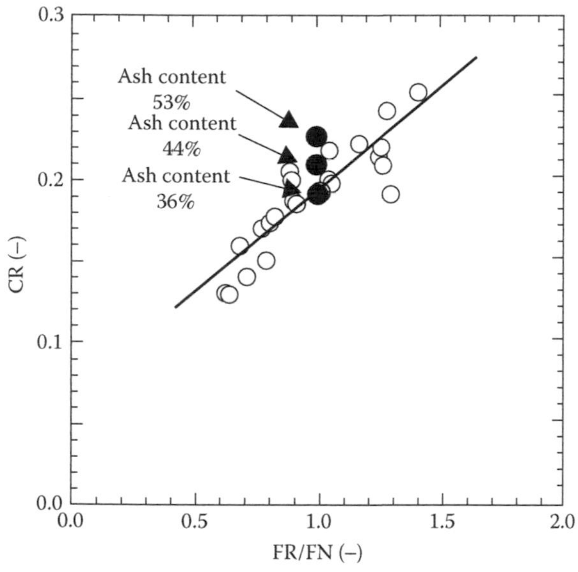

FIGURE 13.31 Relationship between CR and FR/FN under the standard combustion condition.

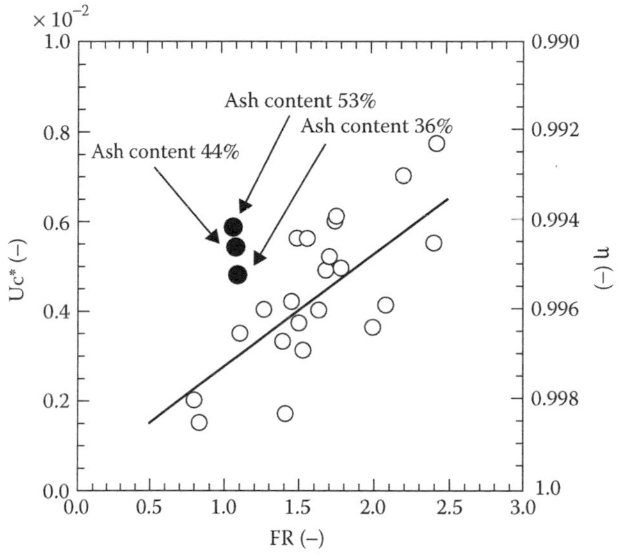

FIGURE 13.32 Relationship between Uc* and FR under the standard combustion condition.

staged combustion decreases the NO_x concentration and increases the unburned carbon fraction at the furnace exit. The decrease of the NO_x concentration is resulted from the promotion of the NO_x reduction, as mentioned above. On the other hand, the increase of the unburned carbon fraction is caused by the higher deficiency of the O_2 before the staged combustion air port. Similar to the results in the standard combustion, both the NO_x concentration and the unburned carbon fraction in the staged combustion increase with the ash content.

13.8 SUMMARY

In this chapter, the principle, science, and technology of coal combustion is introduced. At first, the relationship of coal property and combustion is explained, and combustion profile and basic concept of coal particle are described. After that, the concept of many kinds of coal combustor is explained and the significance of pulverized coal combustion is defined with the principle of thermal power engineering of pulverized coal utilization. At last, pulverized coal combustion technologies, including the low NO_x

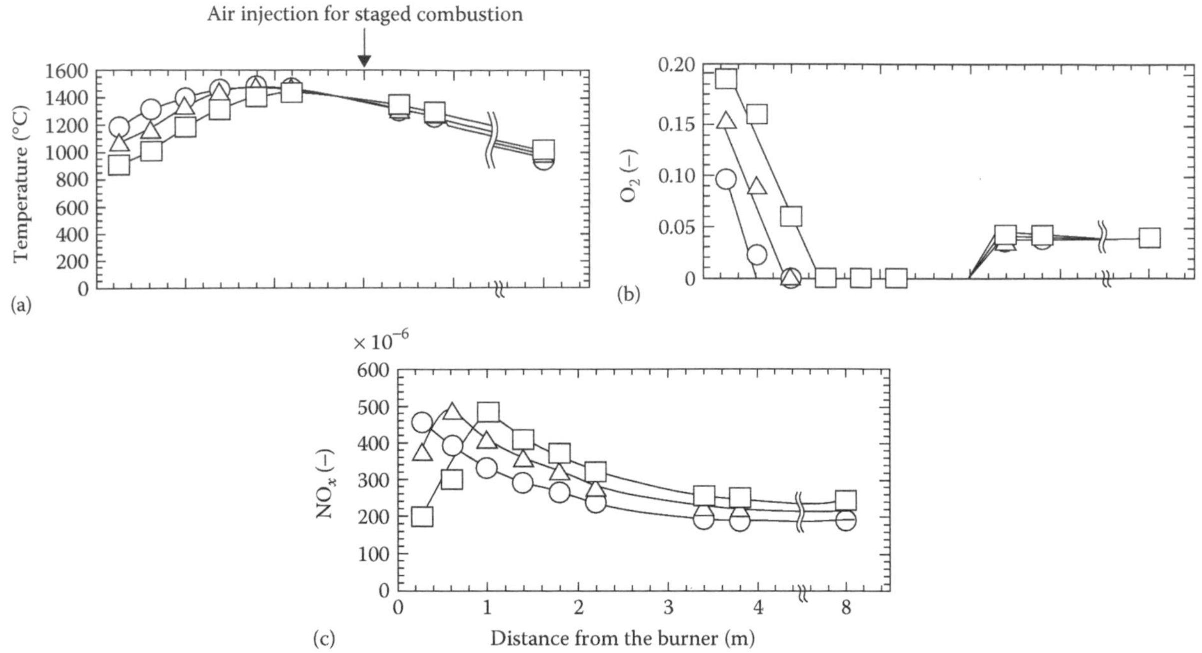

FIGURE 13.33 Axial distributions of (a) temperature, (b) O_2 concentration and (c) NO_x concentration at the center of furnace under the staged combustion.

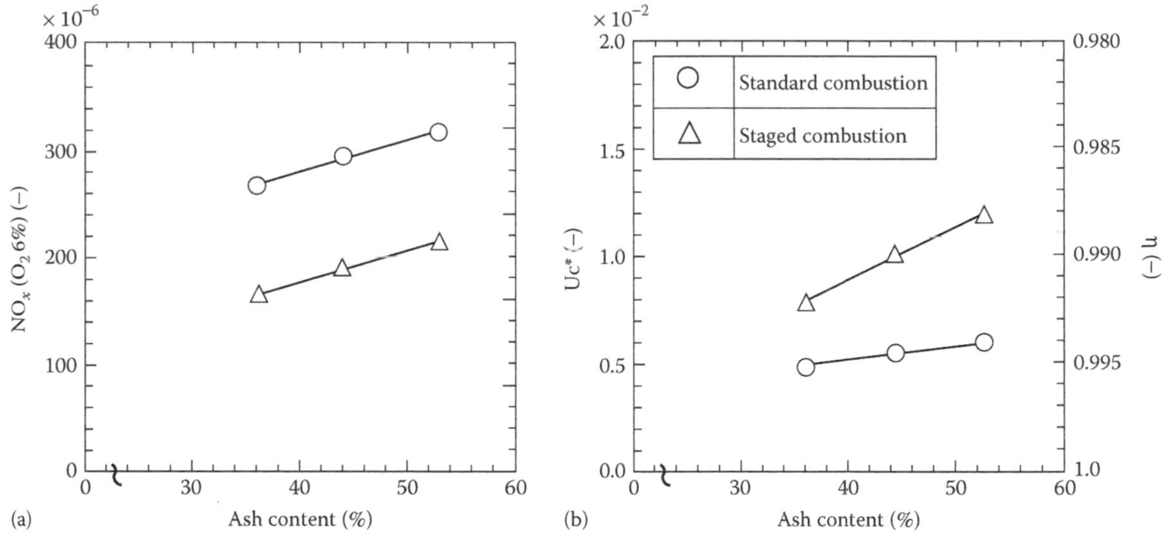

FIGURE 13.34 Effect of ash content on (a) NO_x concentration and (b) unburned carbon fraction at the furnace exit under the standard and staged combustion condition.

combustion and low-load combustion are described, and combustion technologies for low-grade coal including low-rank coal, high-fuel-ratio coal, and high ash content coal are explained.

REFERENCES

Anthony, D.B., Howard, J.B., Hottel, H.C. & Meissener, H.P., 1975. Rapid develatilization of pulverized coal. *Proc. of the Comb. Inst.* **15**, 1303–1317.

Field, M.A., 1969. Rate of combustion of size-graded fractions of char from a low-rank coal between 1200K and 2000K. *Comb. & Flame* **13**, 237–252.

Hurt, R., Sun, J.K. & Lunden, M., 1998. A kinetic model of carbon burnout in pulverized coal combustion. *Comb & Flame* **113**, 181–197.

Ikeda, M., Makino, H. & Kozai, Y., 2002. Emission characteristics of NO_x and unburned carbon in fly ash of sub-bituminous coal combustion. *JSME Int. J. Series B* **15**, 506–511.

Ikeda, M., Makino, H., Morinaga, H., Higashiyama, K. & Kozai, Y., 2003. Emission characteristics of NO_x and unburned carbon in fly ash during combustion of blends of bituminous/sub-bituminous coals. *Fuel* **82**, 1851–1857.

Kobayashi, H., Haward, J.B. & Sarofim, A.F., 1977. Coal Devolatilization at High Temperatures. *Proc. of the Comb. Inst.* **16**, 411–425.

Kurose, R., Tsuji, H. & Makino, H., 2001a. Effects of moisture in coal on pulverized coal combustion characteristics. *Fuel* **80**, 1457–1465.

Kurose, R., Ikeda, M. & Makino, H., 2001b. Combustion characteristics of high ash coal in a pulverized coal combustion. *Fuel* **80**, 1447–1455.

Kurose, R., Ikeda, M., Makino, H., Kimoto, M. & Miyazaki, T., 2004. Pulverized coal combustion characteristics of high-fuel-ratio coals. *Fuel* **83**, 1777–1785.

Makino, H., 2007. Series of Powder technology, Reaction of powder, The Society of Powder Technology Japan. *Huntai Kougaku Sensyo*, Huntaikougakkai. (in Japanese).

Makino, H. & Kimoto, M., 1994. Low NOx combustion technology in pulverized coal combustion. *Kagaku Kougaku Ronbunshu* **20**, 747–757. (in Japanese).

Makino, H., Kimoto, M., Kiga, T. & Endo, Y., 1997. Development of new type low NOx burner for pulverized coal combustion. *Karyoku-Genshiryoku Hatsuden* **48**, 702–710. (in Japanese).

Makino, H., Kimoto, M., Kiga, T. & Endo, Y., 1999. Development of advanced low NOx and wide range burner for pulverized coal combustion. *Karyoku-Genshiryoku Hatsuden* **50**, 790–798. (in Japanese).

Makino, H., Kimoto, M., Sato, M. & Ninomiya, T., 1990, *Nenryo-Kyokai-shi* **69**, 856–862. (in Japanese).

Niksa, S., Heyd, L.E., Russel, W.B. & Saville, D.A., 1985. On the role of heating rate in rapid coal devolatilization. *Proc. of the Comb. Inst.* **20**, 1445–1453.

Niksa, S. & Kersten, A.R., 1991a. FLASHCHAIN theory for rapid coal devolatilization kinetics. 1. Formulation. *Energ. & Fuels* **5**, 647–665.

Niksa, S., 1991b. FLASHCHAIN theory for rapid coal devolatilization kinetics. 2. Impact of operating conditions. *Energ. & Fuels* **5**, 665–673.

Saito, M., Sadakata, M. & Sakai, T., 1987. Measurements of Surface Combustion Rate of Single Coal Particles in Laminar Flow Furnace. *Combust. Sci. Tech.* **51**, 109–128.

Sun, J.K. & Hurt, R., 2000. Mechanisms of extinction and near-extinction in pulverized solid fuel combustion. *Proc. of the Comb. Inst.* **28**, 2205–2213.

Van Krevelen, D.W., Van Heerden, C. & Huntjens, F.J., 1951. Physiochemical aspects of the pyrolysis of coal and related organic compounds. *Fuel* **30**, 253–258.

Williams, A., Pourkashanian, M., Jones, J.M. & Skorupska, N., 2000, *Combustion and Gasification of Coal*, Taylor & Francis Group.

14 Coal Gasification

Mamoru Kaiho and Yoichi Kodera

CONTENTS

Abstract: A new stoichiometric approach was presented, and it was applied to the data by typical gasification processes. This approach is based on a stoichiometric analysis in the overall reaction formula of $CH_mO_n + \alpha O_2 + \beta H_2O \rightarrow \gamma H_2 + \delta CO + \varepsilon CO_2 + \eta CH_4$. Using the reported data of demonstration researches, basic stoichiometric relationship was described to show the principle of coal gasification governing composition of product gas, carbon conversion, and heat efficiency under the conditions such as coal composition and gasifying agents.

14.1 INTRODUCTION

Coal gasification is a versatile process to supply town gas in the past and feedstock for chemicals production as well as power generation. A stoichiometric approach is introduced to interpret the experimental results of coal gasification. This approach gives the theoretical basis of carbon conversion and cold gas efficiency based on a stoichiometric view. It also provides a theoretical value of adiabatic reaction temperature, which governs rate of gasification.

Coal gasification has been drawing attention for many years. Development of gasifiers and applications was shown in many publications (e.g., Bodle and Huebler 1981; Johnson 1981; Hebden and Stroud 1981; Higman and van der Burgt 2008). In the beginning of the history of modern industry, William Murdoch produced coal gas and used it for a gaslight in his house in 1792 (Meade 1921). This is the first successful demonstration of a practical use of coal gas. F. A. Winsor established London and Westminster Chartered Gas Light and Coke Company in 1812, and supplied coal gas as town gas for lighting and heating for public, domestic, and industrial uses (Hatheway and Doyle 2006). Town gas business spread over some countries in Europe and to other countries such as the United States and Japan within a few decades. Coal gas was produced by pyrolysis, and it was mixed with producer gas (Dowson and Larter 1907), water gas, and petroleum-derived gas to increase the heating value of coal gas. Coal gas had been used in developed countries until the 1960s. From about the end of World War II, many oil wells had been found and petroleum gas from refineries has been added to coal gas. During the 1960s, natural gas, which did not contain carbon monoxide, became a major source of town gas in many developed countries. Town gas suppliers looked for a new reliable source of fuel gas to meet strong demands of the increasing supply of town gas in the era of rapid economic growth. In those years, natural gas came into the market because many natural gas fields were discovered and transportation technology of liquefied gas using a large vessel was developed. Coal-derived fuel gas was losing its position as the main source of town gas.

In addition to coal gas and cokes, coal processing provides coal tar as a by-product. Application of coal tar components leads to the development of industrial organic chemistry in Germany in the nineteenth century. In the twentieth century, industrial production of ammonia was established by the Haber–Bosch process including the efforts of Badishe Anilin-und Soda Fabrik, BASF (Smil 2004). Industrial chemistry of coal developed along with conversion technologies such as liquefaction by Bergius and the Fischer–Tropsch method mainly in Germany, where petroleum production cannot be expected. New coal gasification processes were developed in this era. The Winkler process, one of the fluidized-bed gasification processes of coal, was developed in 1926. In 1936, the Lurgi process, a pressurized fixed-bed gasification process, was developed for converting coal into carbon monoxide and hydrogen under air, oxygen with or without steam. These gasification methods are called complete gasification. Production of ammonia and methanol was commercially conducted in many developed countries by using syngas that was produced by coal gasification. After World War II, coal chemical industry declined due to the significant development of the petroleum chemical industry in the fields of the production of hydrogen and other chemicals.

Coal gasification technologies, however, were developed in some cases. The Republic of South Africa faced trade embargo, including crude oil because of its notorious policy of apartheid. The SASOL process was developed for the production of gasoline and industrial chemicals by the technologies, including coal gasification using the Lurgi process. Since the first oil crisis in 1973, some developed countries have started many national projects on R&Ds of the effective use of coal, which has a long minable duration. Those efforts are now leading to the commercial power generation such as integrated coal gasification combined cycle (IGCC) in the United States, Europe, and Japan.

14.1.1 Purpose of Coal Gasification

Coal produces cokes, tar, and gaseous products upon pyrolysis. Powdery cokes were used for the production of hydrogen and carbon monoxide with a gas producer in the form of producer gas or water gas. The product gas has been used for various purposes such as town gas, industrial fuel, chemical synthesis, and power generation. Today, coal utilization is competing with many types of sources, such as petroleum, natural gas, atomic energy, renewable energy, and energy from wastes. Considering business environments, coal is suitable for a feedstock of mass consumption in power generation, hydrogen production, steal making, or C1 chemistry.

A variety of technical improvements had been done to improve the performance of a gasification process. In the Winkler process, reaction efficiency of coal with gasifying agents was increased by using smaller particles of coal. The higher pressure resulted in the higher reaction efficiency of coal with gasifying agents in fixed-bed gasification such as the Lurgi process. These processes still gave a small amount of tar despite that they are categorized as complete gasification. Considering cost reduction of gas production, a plurality of gasifier of over 2,000 ton/day capacity is possibly placed at a facility. In such case, we have to be careful for various emissions even if a gasifier discharges only small amounts of tar, char, or other by-products such as waste water and ash containing potential hazardous compounds. A new gasification process such as entrained-flow gasifier was developed to avoid the formation of tar and char and the elution of hazardous components from ash by agglomeration with the higher gasification efficiency.

14.1.2 Selection of a Gasification Process

When a gasification plant is supposed to be located near a coal field, a coal type would be fixed to use. In this case, suitable process and gasification conditions are determined by a coal type and the purpose of product gas utilization. When various types of imported coal are used in Japan and some European countries, a suitable process should be designed in order to achieve high efficiency.

It is not easy to produce a gaseous product of stable composition with optimizing feeding rate, oxygen-supplying rate, or steam-supplying rate. Based on the accumulation of analytical data, knowledge, and experiences, the mixing ratios of various types of coal were controlled to generate electricity at a constant output at coal-fired power stations in Japan. Different from combustion, operation conditions and gas compositions in gasification strongly depend on the type of coal. For an effective production of a gaseous product, each gasification process requires to fix a mixing ratio of coal in points of proximate analysis, ultimate analysis, calorific value, and other parameters.

14.1.3 Trends of a Technical Development of Coal Gasification

As mentioned earlier, coal gasification started with a production of flammable gas through pyrolysis in the nineteenth century. Later, powdery cokes were given as a waste in the production of cokes for metallurgy. Cokes were gasified in a fixed-bed gasifier, and the product gas of a mixture of carbon monoxide and hydrogen was used as fuel gas in factories and houses.

Industrial production of ammonia and methanol increased the demand of coal-derived gas. Later, complete gasification technology of coal using air, oxygen, and steam was developed. Different from coke gasification in the earlier history, complete gasification is a gasification of whole part of coal. Several types of gasification processes had been developed. A fluidized-bed gasifier such as the Winkler process was developed to increase a treatment capacity for the better economy by using the smaller size of coal. A fixed-bed reactor such as the Lurgi process under higher pressure was developed to increase the gasification efficiency of coal with a gasifying agent.

The rate of overall gasification varies by an exponential function of a gasification temperature. This means that the treatment capacity of a gasifier increases with temperature. However, it was difficult to perform gasification at a higher temperature than the melting range of ash from coal in the use of the gasifiers of the fixed bed and fluidized bed. Ash clinkering in those gasifiers disturbs a constant flow of gasifying agents through a coal layer in a gasifier. Thus, an entrained-flow gasifier such as the Koppers–Totzek process was developed. Coal was supplied in the form of powder and gasified with oxygen gas using a gasification burner. The gasification temperature was observed in the range of 1,330°C–1,927°C. Molten ash goes down to a water-quenching division in a gasifier, and the molten ash is recovered in the form of water-granulated slag. Gasification in an entrained-bed gasifier completes in the order of seconds. To accelerate gasification rates, the other processes such as Shell–Koppers and Texaco were developed. At the elevated temperature, the higher selectivity of carbon monoxide and hydrogen was achieved with suppressed formation of methane. Hydrogen consumption will increase in the near future because of the possible dissemination of hydrogen-fueled automobile and a fuel cell. The high-performance gasifier, especially entrained-bed gasifier, will be a major process in coal gasification.

The British Gas Corporation (BGC)/Lurgi process is an improved fixed-bed gasification process by removing a mechanical grate from the Lurgi gasifier. Ash is taken out from the bottom of a gasifier in the form of molten ash that is heated by combustion under the injection of pulverized coal with oxygen gas from a tuyer. This modification was an important solution of clinkering. Steam injection was reduced for controlling temperature, and the gasification at the higher temperature than the conventional fixed-bed gasifier was achieved to suppress water–gas shift reactions yielding carbon dioxide and water as well as methane formation reaction. This modified fixed-bed gasifier is also a promising technology for the future.

14.2 STOICHIOMETRIC APPROACH OF COAL GASIFICATION

Coal gasification is a chemical reaction in the presence of gasifying agents under a certain temperature and pressure. The yield of gaseous products, mainly hydrogen and carbon monoxide, is varied with the properties of coal and gasification conditions. Thus, theoretical interpretation of reactions and thermal balance of coal gasification is essential to maximize carbon conversion and cold gas efficiency with the optimization of gas composition. However, coal is a complex mixture of various types of compounds. Empirical approach has been a general way to optimize coal gasification in the practical sense of gasifier operation although reaction simulation was conducted by combining some reactions with assumed kinetic parameters.

The authors developed a new stoichiometric approach to understand coal gasification based on a simple description of gasification formula as follows:

$$CH_mO_n + \alpha O_2 + \beta H_2O \rightarrow \gamma H_2 + \delta CO + \varepsilon CO_2 + \zeta H_2O + \eta CH_4 + \theta CH_jO_k$$

Stoichiometric relationship among coal, gasifying agents, product gases, and by-products such as char and tar was clarified by constructing the stoichiometric approach of coal gasification. The theory helps to understand how gasifying agents and reaction conditions contribute to the product composition and the yields. It also gives a theoretical explanation on carbon conversion, cold gas efficiency, heat of gasification, adiabatic gasification temperature, and heat loss of a gasifier.

Some eighty demonstration results of various gasification processes were analyzed based on the theory after the reliability of the data from some 400 demonstrations were examined by the stoichiometric approach and the derived equations. Adiabatic gasification temperature, carbon conversion, and cold gas efficiency are the important values to assess the performance of a gasification process. Adiabatic gasification temperature is the essential driving factor to accelerate coal gasification. Carbon conversion and cold gas efficiency are the results of gasification achieved. Additionally, heat loss from a gasifier is another practical interest. Based on the theory, the features of typical gasification processes were evaluated in terms of adiabatic gasification temperature and heat loss.

Because of the clear stoichiometric relationship shown in the theory, governing factors such as contributions of oxygen and steam to the performance of gasification were clarified to suggest the optimized conditions. At the same time, the theory would help to improve the performance of coal gasification in the development of a gasification process.

14.2.1 Overall Equation of Coal Gasification

For a theoretical analysis of gasification, compositional formula was used for gasification formula in the stoichiometric approach. Equation 14.1 depicts gasification of coal in the presence of oxygen and steam. Coal is described as CH_mO_n, where m and n is the ratio of the number of atoms of H/C and O/C, respectively.

$$\underbrace{CH_mO_n}_{\text{Coal}} + \underbrace{\alpha O_2 + \beta H_2O}_{\text{Gasifying agents}} \rightarrow \underbrace{\gamma H_2 + \delta CO + \varepsilon CO_2 + \zeta H_2O + \eta CH_4}_{\text{Gaseous products}} + \underbrace{\theta CH_jO_k}_{\text{By-products}} \quad (14.1)$$

For the simpler mathematical handling, the by-products CH_jO_k were also described as CH_mO_n because of the assumed similarity in the chemical formula of coal and that of the by-products of char and tar in the following equation:

$$CH_mO_n + \alpha O_2 + \beta H_2O \rightarrow \gamma H_2 + \delta CO + \varepsilon CO_2 + \zeta H_2O + \eta CH_4 + \theta CH_mO_n \quad (14.2)$$

Letting $\beta = \gamma = \delta = \eta = \theta = 0$ in the above equation, the resulting formula, $CH_mO_n + \alpha O_2 \rightarrow \varepsilon CO_2 + \zeta H_2O$, becomes the reaction formula of complete combustion. Equation 14.2 is a general formula covering both gasification and combustion.

14.2.2 Yields of Hydrogen and Carbon Monoxide Based on Stoichiometry

Desired products of coal gasification are hydrogen and carbon monoxide. Methane is also produced in some cases. Theoretical yields of hydrogen and carbon monoxide can be determined based on the stoichiometry in the gasification formula (14.2).

Equation 14.2 gives the stoichiometric parameters of carbon, hydrogen, and oxygen as follows:

$$C: 1 = \delta + \varepsilon + \eta + \theta \quad (14.3)$$

$$H: m + 2\beta = 2\gamma + 2\zeta + 4\eta + m\theta \quad (14.4)$$

$$O: n + 2\alpha + \beta = \delta + 2\varepsilon + \zeta + n\theta \quad (14.5)$$

In coal combustion yielding only carbon dioxide and water, the theoretical requirement of oxygen, μ, is given as $\mu = 1 + 0.25m - 0.5n$ from the compositional formula of coal (CH_mO_n). Here, a parameter χ is defined as the oxygen ratio of α/μ. This ratio means that the ratio of oxygen amounts in gasification to the theoretical requirement of oxygen for complete combustion.

In the following mathematical transformation, the parameters χ and μ were introduced to the stoichiometric relationship in Equations 14.3 through 14.5 in order to understand the effect of oxygen amounts on the yields of hydrogen and carbon monoxide. Namely, 2 × (14.3) + 0.5 × (14.4) – (14.5) gives

$$(2 + 0.5m - n) - 2\alpha = (\gamma + \delta) + 4\eta + (2 + 0.5m - n) \quad (14.6)$$

Substituting μ and χ in the above equation gives

$$2\mu - 2\mu\chi = (\gamma + \delta) + 4\eta + \theta\mu \quad (14.7)$$

Then, Equation 14.8 is given as follows:

$$(\gamma + \delta) = 2\mu \times \{(1 - \theta) - \chi\} - 4\eta \quad (14.8)$$

The right hand of the above equation, the summed parameters $\gamma + \delta$, is the total yield of hydrogen and carbon monoxide. The term $1 - \theta$ means the total molar yield of gaseous product, which corresponds to carbon conversion. This stoichiometric relationship indicates that the total yield of hydrogen and carbon monoxide maximizes under the conditions of $\chi = 0$ with minimizing values of θ and η. It means that the gasification conditions can be optimized by supplying a rate of oxygen, and gasification can be monitored for methane formation under the conditions of the minimum formation of methane, tar, and char.

14.2.3 Yields of Carbon Dioxide and Water in Gasification

Stoichiometric approach was made to see the contribution of oxygen in coal and the other conditions to the formation of carbon dioxide and water, which are undesired products in coal gasification.

The total yield of carbon dioxide and water, $\varepsilon + \zeta$, is given as $(\varepsilon + \zeta) = (n - 1)(1 - \theta) + 2\alpha + \beta + \eta$ by subtracting Equation 14.5 from Equation 14.3. Substituting α with $\mu\chi$ gives the following equation:

$$(\varepsilon + \zeta) = (n - 1)(1 - \theta) + 2\mu\chi + \beta + \eta \quad (14.9)$$

The above equation indicates that the contribution to the formation of carbon dioxide and water by the O/C ratio of coal, carbon conversion, theoretical requirement of oxygen, oxygen ratio, oxygen supply from steam, and methane formation.

14.2.4 Theoretical Interpretation of Oxidation Process in Coal Gasification

The general formula of gasification of Equation 14.1 gives the stoichiometric relationship that was described as Equations 14.8 and 14.9. We considered the gasification of Equation 14.1 as a combination of two types of conversion: partial oxidation stage and secondary stage. In the partial oxidation stage, carbon and hydrogen in coal undergo oxidation to form water, carbon monoxide, and carbon dioxide with hydrogen formation by the reaction of carbon moiety with steam. The resulting hydrogen and carbon monoxide were assumed to react in water–gas shift reaction and methane formation in the secondary stage.

Assuming no by-product formation with $\theta = 0$, the partial oxidation stage can be depicted as follows:

$$CH_mO_n + \alpha O_2 + \beta H_2O \rightarrow \gamma H_2 + \delta CO + \varepsilon CO_2 + \zeta H_2O \quad (14.10)$$

The total yield of hydrogen and carbon monoxide, $\gamma + \delta$, is given as follows:

$$(\gamma + \delta) = 2\mu(1 - \chi) \quad (14.11)$$

The total yield of water and carbon dioxide, $\varepsilon + \zeta$, is given as follows:

$$(\varepsilon + \zeta) = (n - 1) + 2\mu\chi + \beta \quad (14.12)$$

Steam is often used as a gasifying agent. Reacted amount of steam, $\beta - \zeta$, is given as follows:

$$(\beta - \zeta) = -2\mu\chi + (1 - n) + \varepsilon \quad (14.13)$$

The yield of carbon dioxide, ε, is given as follows:

$$\varepsilon = (n - 1) + 2\mu\chi + (\beta - \zeta) \quad (14.14)$$

The value, ε, is always greater than 0. Equation 14.14 gives $(n-1) + 2\mu\chi + (\beta - \zeta) \geq 0$. Dividing the both sides of this equation by 2μ affords the following equation:

$$\left[\frac{0.5(1-n)}{\mu}\right] - \chi \leq \left[\frac{(\beta - \zeta)}{2\mu}\right] \quad (14.15)$$

Because the value of n is always less than 1 in the compositional formula of coal, the above equation indicates that $(\beta - \zeta)/2\mu > 0$ in the case of $\chi < 0.5(1-n)/\mu$. This means that water is consumed $(\beta > \zeta)$ during gasification in the case of $\chi < 0.5(1-n)/\mu$.

14.2.5 Graphical Representation of the Stoichiometry of Partial Oxidation Stage

As shown in Equations 14.11 through 14.13, the yields of flammable products (hydrogen and carbon monoxide), inflammable products (water and carbon dioxide), and steam consumption were expressed as liner functions of the oxygen ratio χ. These equations were drawn in Figure 14.1. The yields of flammable and inflammable products, and steam consumption were normalized by the required oxygen amount for complete combustion, μ, which has a liner relationship with slopes of ± 2 and 0.

The diagram in Figure 14.1 provides a convenient method to estimate the yield of flammable products in the function of oxygen ratio χ. It also gives the yield of inflammable products and steam consumption. It is noteworthy that the stoichiometric relationship at $\chi = 0$ gives the following results:

$$\gamma + \delta = 2\mu \quad (14.16)$$

$$\beta - \zeta = 1 - n \quad (14.17)$$

The condition of $\chi = 0$ means a theoretical model of an externally heated gasifier, in which there is no oxygen supply to the gasifier. These theoretical discussions would help to design a new gasifier using external heat source such as solar heat and nuclear power, or plan a reactor as a part of gasification processes.

When $\chi = 0.5(1 - n)/\mu$ in Figure 14.1, Equation 14.10 is simplified as follows:

$$CH_mO_n + 0.5(1 - n)O_2 \rightarrow 0.5mH_2 + CO \quad (14.18)$$

We use the reaction formula, Equation 14.18, as a basic equation to discuss coal gasification. The related terms are defined as follows:

Standard requirement of oxygen $O_{st} = 0.5\,(1 - n)$

Excess amount of oxygen $O_{ex} = \alpha - O_{st}$

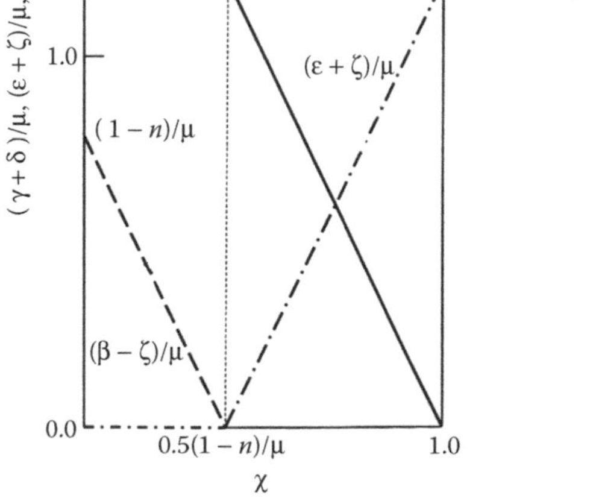

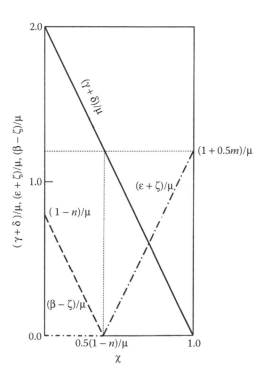

FIGURE 14.1 Linear relationship of the amounts of gaseous products and steam consumption with oxygen ratio χ.

Partial oxidation stage can be interpreted by using O_{st} and O_{ex}. When oxygen is supplied to a gasifier under the condition of $0 \leqq \alpha < 0.5\ (1-n)$, O_{ex} is less than zero and the amount of unreacted carbon corresponds to $-2O_{ex}$. In the case of a shortage of oxygen supply, steam will be used for oxidation of unreacted carbon of coal. When oxygen is supplied to a gasifier under the condition of $O_{st} < \alpha \leqq \mu$, O_{ex} is greater than zero, and CO and $0.5mH_2$ in Equation 14.18 react with an excess oxygen molecule to give combustion products, H_2O and CO_2, in the amount of $2O_{ex}$. The parameter O_{ex} can be an indicator in a practical operation of a gasifier to know the formation of unreacted carbon at an insufficient amount of oxygen supply and the formation of combustion products at an excess amount of oxygen supply.

14.2.6 Contribution of By-Products to the Stoichiometry of Coal Gasification

In the Sections 14.2.2 through 14.2.5, methane, tar, and char were neglected. Here the stoichiometric relationship was examined in gasification with the formation of by-products such as tar and char. When coal is gasified to gaseous products and a by-product of $\theta\ CH_mO_n$, steam supply and the yields of hydrogen, carbon monoxide, carbon dioxide, and steam are reduced at $(1-\theta)$. Letting the oxygen amount equal to A, which is smaller than α, gives the equation as follows:

$$\begin{aligned} CH_mO_n + AO_2 + (1-\theta)\beta H_2O \rightarrow (1-\theta)\gamma H_2 + (1-\theta) \\ \delta CO + (1-\theta)\varepsilon CO_2 + (1-\theta)\zeta H_2O + (1-\theta)\eta CH_4 \\ + \theta CH_mO_n \end{aligned} \quad (14.19)$$

Transposition of $\theta\ CH_mO_n$ on the left-hand side gives

$$\begin{aligned} (1-\theta)CH_mO_n + AO_2 + (1-\theta)\beta H_2O \rightarrow (1-\theta)\gamma H_2 \\ + (1-\theta)\delta CO + (1-\theta)\varepsilon CO_2 + (1-\theta)\zeta H_2O \\ + (1-\theta)\eta CH_4 \end{aligned} \quad (14.20)$$

Multiplying $(1-\theta)$ on both the sides of Equation 14.2 yields

$$\begin{aligned} (1-\theta)CH_mO_n + (1-\theta)\alpha O_2 + (1-\theta)\beta H_2O \rightarrow (1-\theta)\gamma H_2 \\ + (1-\theta)\delta CO + (1-\theta)\varepsilon CO_2 + (1-\theta)\zeta H_2O \\ + (1-\theta)\eta CH_4 \end{aligned} \quad (14.21)$$

Comparison of Equation 14.20 with the above equation gives $A = (1-\theta)\alpha$ and the following equation:

$$(1-\theta) = A/\alpha \quad (14.22)$$

Letting $\eta = 0$, Equations 14.8, 14.9, and 14.14 give the following equations, respectively:

$$(\gamma+\delta) = 2\mu \times [(1-\theta) - \chi] \quad (14.23)$$

$$(\varepsilon+\zeta) = (n-1)(1-\theta) + 2\mu\chi + \beta \quad (14.24)$$

$$\varepsilon = (n-1)(1-\theta) + 2\mu\chi + (\beta - \zeta) \quad (14.25)$$

The stoichiometric relationship of the above equations is now projected in Figure 14.2. Equations 14.11 through 14.14 of Figure 14.1 are also overlaid on it. The lines representing Equations 14.23 through 14.25 are the shifted lines of slopes of ±2 and 0 from Equations 14.11 through 14.13 by θ to the left, respectively. The intersection points of x and y axes with the lines of $(\gamma+\delta)/\mu$($\beta-\zeta$)$/\mu$ in Figure 14.2 are the values in Figure 14.1 that were multiplied by $(1-\theta)$. The required conditions of carbon dioxide formation, $0.5(1-n)/\mu - \chi \leqq (\beta-\zeta)/2\mu$, can be transformed as follows:

$$2\mu\chi + (\beta-\zeta) > (1-n)\ (1-\theta) \quad (14.26)$$

This means that carbon dioxide generates under the conditions of $\chi \geqq 0.5(1-n)\ (1-\theta)/\mu$, and it does not generate under the conditions of $\chi < 0.5(1-n)\ (1-\theta)/\mu$.

14.2.7 Secondary Reactions of Gasification

The initial products of hydrogen and carbon monoxide, which were obtained in partial oxidation stage, undergo secondary reactions such as water–gas shift reaction and other reactions as described in the following schemes:

$$CO + H_2O \rightarrow H_2 + CO_2 \quad (14.27)$$

$$CO + 3H_2 \rightarrow CH_4 + H_2O \quad (14.28)$$

$$4H_2 + 2CO \rightarrow C_2H_4 + 2H_2O \quad (14.29)$$

$$5H_2 + 2CO \rightarrow C_2H_6 + 2H_2O \quad (14.30)$$

Steam-induced shift reaction decreases the amount of carbon monoxide, whereas the same molar amount of hydrogen molecule generates at the same time. The total molar amounts of hydrogen and carbon monoxide, $(\gamma+\delta)$, are the same before and after a shift reaction. Partial oxidation as the initial stage of coal gasification is followed by shift reaction of carbon monoxide with water. During the shift reaction, the amount of water will decrease with the formation of carbon dioxide by the same molar amount. The total amount of water and carbon dioxide $(\varepsilon+\zeta)$ is kept at the same molar amounts. Similarly, the amount of carbon monoxide will decrease with the formation of hydrogen by the same molar amount. The total amount of hydrogen and carbon monoxide $(\gamma+\delta)$ is kept at the same molar amount. Thus, the reacted amount of water, $(\beta-\zeta)/\mu$, increases with the progress of shift reaction, but the values of $(\gamma+\delta)/\mu$ and $(\varepsilon+\zeta)/\mu$ are constant.

14.3 COLD GAS EFFICIENCY OF COAL GASIFICATION

Thermal efficiency is a heating value of a product gas divided by that of coal. Heating value is expressed by using a higher heating value (HHV) or a lower heating value (LHV). An HHV is the summed value of an LHV and a latent heat of condensation of steam that generates by combustion. A heating value of coal is

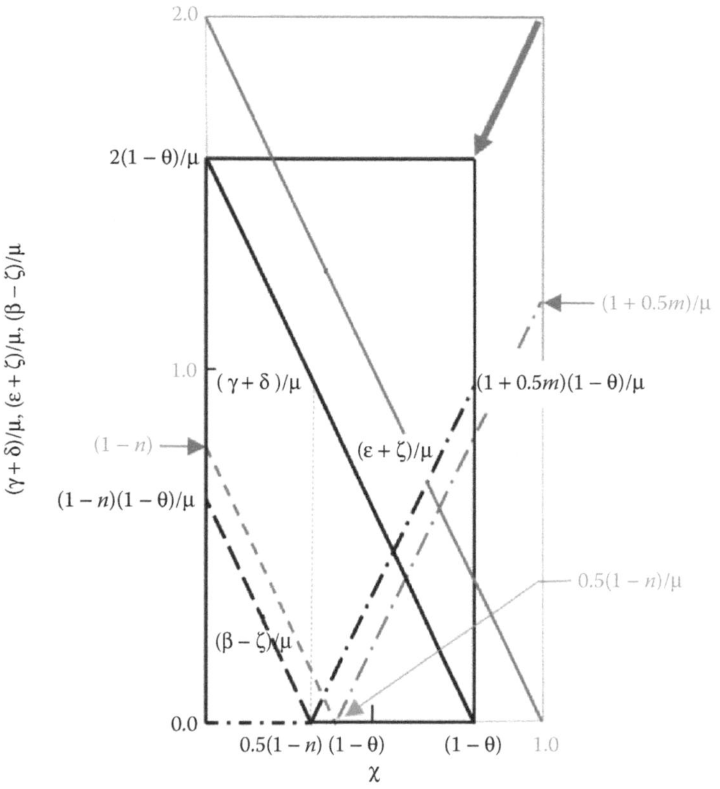

FIGURE 14.2 Relationship between the amounts of product gas and oxygen ratio.

determined with 1 g of powdered coal in a pressure vessel submerged in water at 25°C using a bomb calorimeter. The analytical data are given as HHV because the resulting steam in combustion condenses to release the heat of condensation. When latent heat of steam is not required to consider, LHV is worth discussing because the latent heat of steam cannot be converted into work. In coal gasification studies, both HHV and LHV are used and this makes us confused. Gasification is a chemical conversion process without the conversion step of heat into work. Thus, there is no necessity to use LHV. If thermal efficiency is calculated on an LHV basis, one should note it as LHV basis.

Same as carbon conversion, cold gas efficiency is the important indicator to measure an efficiency of coal conversion. It is obtained by mass balance data. There are no theoretical explanation to understand how cold gas efficiency relates to carbon conversion and the amounts of oxygen in gasifying agents and coal. In this section, we derive cold gas efficiency based on the stoichiometric approach and discuss suitable conditions to improve the performance of a gasifier.

14.3.1 Cold Gas Efficiency

As mentioned earlier, carbon conversion can be expressed as $(1 - \theta)$ in the following equation:

$$\underbrace{CH_mO_n}_{\text{Coal}} + \underbrace{\alpha O_2 + \beta H_2O}_{\text{Gasifying agents}} \rightarrow \underbrace{\gamma H_2 + \delta CO + \varepsilon CO_2 + \zeta H_2O + \eta CH_4}_{\text{Gaseous products}} + \underbrace{\theta CH_jO_k}_{\text{By-products}}$$

Cold gas efficiency (E) is expressed by HHV of coal divided by HHV of product gas as given in the following equation:

$$E = \frac{(\gamma h_{H2} + \delta h_{CO} + \eta h_{CH4})}{h_{coal}} \tag{14.31}$$

where h_{H2}, h_{CO}, h_{CH4}, and h_{coal} are the heat of combustion of H_2, CO, CH_4, and CH_mO_n (kJ/mol), respectively, and h_{H2} = −286.0 kJ/mol, h_{CO} = −283.1 kJ/mol, and h_{CH4} = −890.8 kJ/mol. The heating value of coal, h_{coal}, varies with the type of coal.

Because of the close values of h_{H_2} and h_{CO}, assuming $\gamma h_{H2} + \delta h_{CO} = (\gamma + \delta)h_{CO}$ and $(\gamma + \delta) = 2\mu \times [(1 - \theta) - \chi] - 4\eta$ in Equation 14.8 gives

$$\gamma h_{H2} + \delta h_{CO} = 2\mu \times h_{CO}\left[(1-\theta) - \chi\right] - 4\eta h_{CO} \tag{14.32}$$

Applying this equation to Equation 14.31 gives the following equation:

$$E = \frac{\{2\mu h_{CO}\left[(1-\theta) - \chi\right] + \eta(h_{CH4} - 4h_{CO})\}}{h_{coal}} \tag{14.33}$$

The above equation indicated that cold gas efficiency is determined by the difference between carbon conversion, oxygen ratio, methane formation, and HHV of coal. At the end on the right-hand side of the above equation, the term, $\eta(h_{CH4} - 4h_{CO})$, is an approximated form of a reaction heat $(h_{CH4} - h_{CO} - 3h_{H2})$ by approximating $h_{H2} \fallingdotseq h_{CO}$. The error range of E is estimated as less than 0.2%.

One can rationally estimate gasification conditions such as the oxygen ratio and methane contents to achieve targeted cold gas efficiency. And the demonstration data in the past can be evaluated in the points of cold gas efficiency in relation with carbon conversion and oxygen ratio.

14.3.2 Cold Gas Efficiency and the Property of Coal

Generally, the preferred goal of gasification is to achieve the higher processing rate of coal by elevating the operation temperature of a gasifier. Practically, it promotes ash melting to cause slag plugging, one of the serious problems in the operation of a gasifier. The higher operation temperature reduces the methane content in product gas as observed in gasification and is also known in chemical equilibrium studies. Equation 14.33 gives the following equation with decreasing methane content:

$$E = K\left[(1-\theta)-\chi\right] \tag{14.34}$$

where $K = 2\mu h_{CO}/h_{coal}$.

The above equation indicates that cold gas efficiency E is proportional to K, which is a specific constant depending on the coal rank. To say mathematically, cold gas efficiency increases under the case of coal CH_mO_n with the larger μ and smaller h_{coal}.

The term K was calculated by using ultimate analyses and heating values of 100 samples of coal, and the relationship of K with carbon contents (C%) was shown in Figure 14.3.

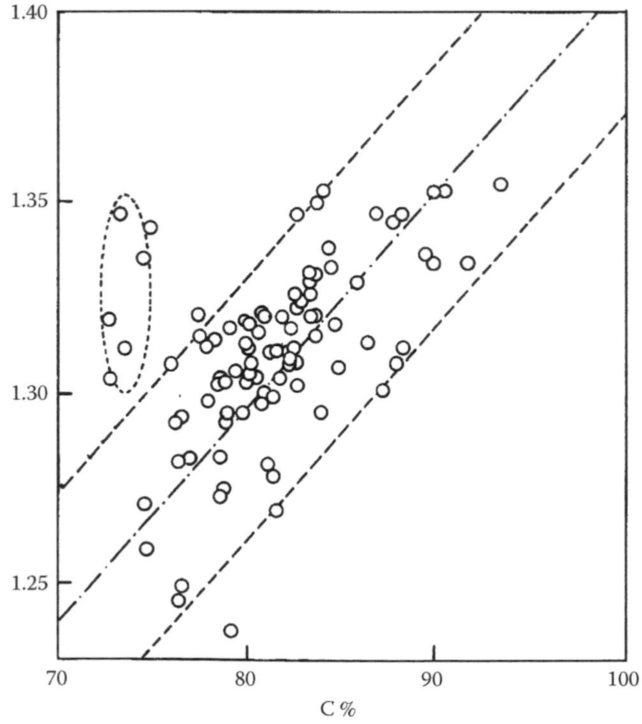

FIGURE 14.3 Relationship of K with C% of 100 specimens of coal.

Sub-bituminous and bituminous coals of the carbon content at 75%–85% are often used for gasification. The K values mostly distribute 1.28–1.33 and increase with an increase of the carbon content. Plots of lignite samples with C% of 70%–75% are surrounded by break line. The K values of lignite are bigger than those of anthracite and sub-bituminous coal by 0.02–0.05 due to the lower h_{coal}.

Table 14.1 shows the K values of typical organic compounds to examine the influence of molecular structure on the K values.

Except methanol and ethanol, the K values of various organic compounds are about 1.3, which are similar to the K values of coal. The K values of aliphatic hydrocarbons increase with the number of carbon atoms. For aromatic compounds, the K value increases by 0.02 in case of the difference between naphthalene (dicyclic compound) and benzene (monocyclic compound). The smaller increase of the K value is given in case of the difference between anthracene (tricyclic compound) and naphthalene. Methyl and hydroxyl groups decrease the K value and carboxyl groups increase it.

The observed number of aromatic nuclei of coal is mostly one in lignite and two in sub-bituminous coal, and the number is increasing with C% in the other types of coal. The contents of oxygen-containing functional groups decrease with an increase of C%. These observations explain the increase of K values with increasing C% in Figure 14.3. But the K values of lignite are quite different from the linear tendency of other types of coal. The higher K values can be explained

TABLE 14.1
***K* of Typical Organic Compounds**

Compound	Molecular Formula	Heating Value (kJ/mol)	μ (mol)	K (–)
Methane	CH_4	–890.8	2.0	1.27
Ethane	C_2H_6	–1561	3.5	1.27
Propane	C_3H_8	–2221	5.0	1.27
Butane	C_4H_{10}	879	6.5	1.28
Pentane	C_5H_{12}	–3538	8.0	1.28
Hexane	C_6H_{14}	–4165	9.5	1.29
Methanol	CH_3OH	–726.7	1.5	1.17
Ethanol	C_2H_5OH	–1368	3.0	1.24
Propanol	C_3H_7OH	2022	4.5	1.26
Butanol	C_4H_9OH	–2676	6.0	1.27
Pentanol	$C_5H_{11}OH$	–3318	7.5	1.28
Benzene	C_6H_6	–3269	7.5	1.30
Toluene	$C_6H_5CH_3$	–3958	9.0	1.29
Phenol	C_6H_5OH	–3057	7.0	1.30
Benzoic acid	C_6H_5COOH	–3232	7.5	1.31
Naphthalene	$C_{10}H_8$	–5159	12.0	1.32
α-Naphthol	$C_{10}H_7OH$	–4964	11.5	1.31
Anthracene	$C_{14}H_{10}$	–7062	16.5	1.32

Notes: $\mu = 1 + 0.25m - 0.5n$, $K = 2\mu h_{CO}/h_{coal}$.

by high contents of carboxylic group. There would be many other factors influencing chemical and physical properties of coal, especially with monocyclic and dicyclic aromatic moieties.

14.3.3 Estimation Procedure of Cold Gas Efficiency by Oxygen Ratio

Cold gas efficiency (1 – θ) can be given by Equation 14.34. However, it is difficult to monitor the amounts of by-products θ of tar and char. Some demonstrations lack the data such as the yields of by-products. This section describes an estimation procedure for cold gas efficiency (1 – θ) based on the amount of oxygen supply. Two pathways of coal combustion are described in Figure 14.4. A bold line shows combustion of coal with oxygen. A thin line shows partial oxidation of coal followed by combustion of gaseous products and by-products. Total amounts of oxygen for each pathway are the same as described in the following equation:

$$\mu = \alpha + \pi + \xi \tag{14.35}$$

where π and ξ are oxygen requirements for gaseous products and by-products, respectively.

Substituting ξ with μθ, dividing both sides of Equation 3.4 by μ, and applying α/μ = χ afford the following equation:

$$(1-\theta) - \chi = \frac{\pi}{\mu} \tag{14.36}$$

The above equation estimates carbon conversion $(1-\theta)$ as $\chi + \pi/\mu$. Applying the above equation to Equation 14.34 gives the following equation:

$$E = K(\pi/\mu) \tag{14.37}$$

The above equation can be used to estimate cold gas efficiency *E* when required data are not available.

By examining stoichiometry of the simple gasification formula, one can understand the relationship between cold gas efficiency and carbon conversion, and the effects of oxygen addition and elemental composition of coal on gasification.

14.4 INTERPRETATION OF THE GASIFICATION PROCESS BY THE STOICHIOMETRIC APPROACH

The developed stoichiometric approach explains the relationship of the oxygen ratio to the amounts of hydrogen, carbon monoxide, water, and carbon dioxide in product gas and steam consumption in gasification in Sections 14.2 and 14.3. Cold gas efficiency, theoretical requirement of oxygen, heat generation, and carbon conversion were described as the equations based on the theoretical relationships. Those equations are the effective tool to evaluate a gasifier and a gasification process in a demonstration project. The stoichiometric approach can

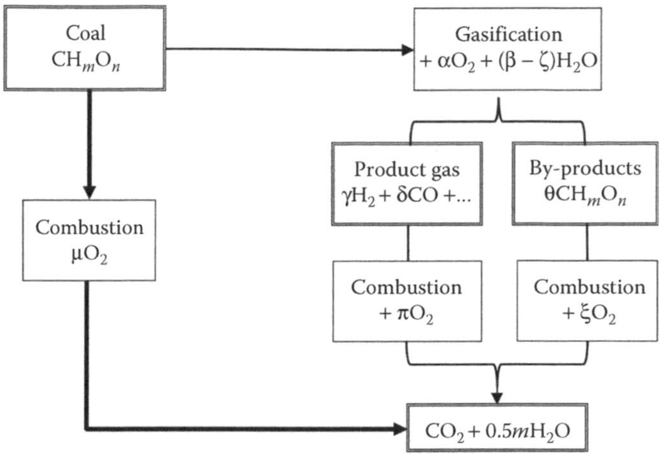

FIGURE 14.4 Schematic diagram of oxygen balance of gasification.

be used to improve a gasification process by applying gasification conditions and analytical data such as feeding rates of raw materials and gas composition.

In this section, coal gasification is described as a general formula in a simple and apparent manner as Equation 14.38. Stoichiometric relationship among the parameters was examined on a molar basis. Then, we will examine how the general gasification formula relates to the typical underlying reactions such as partial oxidation and the other secondary reactions under some assumptions or hypothetical idea to analyze demonstration data of various gasification processes in Section 14.5.

14.4.1 General Reaction Formula Representing Gasification

A general reaction formula, Equation 14.38, was defined to discuss a more general case of gasification of various feeds such as biomass and plastics:

$$CH_mO_n + \alpha O_2 + \beta H_2O \rightarrow \gamma H_2 + \delta CO + \varepsilon CO_2 + \eta CH_4 + \kappa C_2H_4 + \lambda C_2H_6 + \theta CH_jO_k \tag{14.38}$$

The balance equations of carbon, hydrogen, and oxygen are expressed by Equations 14.39 through 14.41, respectively:

$$(C): 1 = \delta + \varepsilon + \eta + 2\kappa + 2\lambda + \theta \tag{14.39}$$

$$(H): m + 2\beta = 2\gamma + 4\eta + 4\kappa + 6\lambda + j\theta \tag{14.40}$$

$$(O): n + 2\alpha + \beta = \delta + 2\varepsilon + \kappa\theta \tag{14.41}$$

Letting the total molar amounts of each gaseous component to be expressed as Σ as in the following equation:

$$\Sigma = \gamma + \delta + \varepsilon + \eta + \kappa + \lambda \tag{14.42}$$

The molar amounts of hydrogen, carbon monoxide, carbon dioxide, methane, ethylene, and ethane are expressed by the following equations:

$$H_2: \gamma = p\Sigma \tag{14.43}$$

$$CO: \delta = q\Sigma \tag{14.44}$$

$$CO_2: \varepsilon = r\Sigma \tag{14.45}$$

$$CH_4: \eta = s\Sigma \tag{14.46}$$

$$C_2H_4: \kappa = t\Sigma \tag{14.47}$$

$$C_2H_6: \lambda = u\Sigma \tag{14.48}$$

where the molar concentrations of hydrogen, carbon monoxide, carbon dioxide, methane, ethylene, and ethane are p, q, r, s, t, u, and $p + q + r + s + t + u = 1$.

The parameters, α through θ and Σ, are numerically obtained by Equations 14.39 through 14.48. Substituting the parameters, δ through θ, in Equations 14.43 through 14.48 yields the following equation:

$$1 = q\Sigma + r\Sigma + s\Sigma + 2t\Sigma + 2u\Sigma + \theta \tag{14.49}$$

The above equation yields

$$\Sigma = \frac{(1-\theta)}{(q+r+s+2t+2u)} \tag{14.50}$$

Applying the above equation to Equations 14.43 through 14.48 gives the yield of each gaseous product and by-product as the following equations:

$$H_2: \gamma = \frac{p(1-\theta)}{(q+r+s+2t+2u)} \tag{14.51}$$

$$CO: \delta = \frac{q(1-\theta)}{(q+r+s+2t+2u)} \tag{14.52}$$

$$CO_2: \varepsilon = \frac{r(1-\theta)}{(q+r+s+2t+2u)} \tag{14.53}$$

$$CH_4: \eta = \frac{s(1-\theta)}{(q+r+s+2t+2u)} \tag{14.54}$$

$$C_2H_4: \kappa = \frac{t(1-\theta)}{(q+r+s+2t+2u)} \tag{14.55}$$

$$C_2H_6: \lambda = \frac{u(1-\theta)}{(q+r+s+2t+2u)} \tag{14.56}$$

Substituting the parameters, η, κ, and λ in Equation 14.40 gives the amount of reacted steam, β, as the following equation:

$$\beta = \frac{(2\gamma + 4\eta + 4\kappa + 6\lambda + j\theta - m)}{2} \tag{14.57}$$

To obtain the parameter α, the balance equation of oxygen of Equation 14.41 is modified as follows:

$$\alpha = \frac{(\delta + 2\varepsilon - \beta + \kappa\theta - n)}{2} \tag{14.58}$$

The mathematical relationships mentioned above suggest that the parameters α through θ in Equation 14.38 can be expressed by a combination of the concentrations of gaseous products p through u in Equations 14.51 through 14.58.

14.4.2 Construction of the Detailed Stoichiometric Model Involving the Underlying Reactions

The general reaction formula of coal gasification was described in the previous Section 14.4.1. The parameters of the chemical species were determined in the stoichiometric relationship. In this section, stoichiometric models are constructed by combining the typical underlying reactions during gasification. One is the partial oxidation to give hydrogen and carbon monoxide. Another is a series of secondary reactions of the products by partial oxidation. Partial oxidation is the primary reaction of coal with oxygen and steam to yield hydrogen, carbon monoxide, carbon dioxide, and water. Secondary reactions include water–gas shift reactions of hydrogen and carbon monoxide and formation reactions of gaseous hydrocarbons such as methane. Based on these simplified schemes, stoichiometric models were constructed, and the contribution of gasifying agents to the overall gasification was examined. A stoichiometric model of partial oxidation was described in two cases of $O_{ex} > 0$ and $O_{ex} \leqq 0$. In the case of $O_{ex} > 0$, hydrogen and carbon monoxide react with excess oxygen to give water and carbon dioxide in the total amounts of $2O_{ex}$. In the case of $O_{ex} \leqq 0$, the reaction of steam in the amount of $-2O_{ex}$ with carbon of coal (CH_mO_n) gives $(0.5m - 2O_{ex})H_2 + CO$.

Formation of by-products such as gaseous hydrocarbons of C_2, tar, and char can be described as the secondary reaction. Those compounds were assumed to be generated by thermal decomposition and other reactions of hydrogen and carbon monoxide although tar would be generated in several types of reactions such as thermal recombination and hydrogenation. Here, the stoichiometric relationship on tar formation was examined in three cases: thermal recombination, hydrogenation, and synthetic reaction. By-products would be generated via various unknown reaction pathways including pyrolysis of coal. Mathematical or stoichiometric equations of the unknown pathways are quite difficult to construct. The stoichiometric relationship on the formation of by-product through pyrolysis is followed by thermal recombination: This is a hypothetical recombination of elements of iC, jH, and kO to form tar Ci HjOk as in Equation 14.59. The term, lH_2O, was added to the both sides of the equation for easy comparison of this equation with the equations of hydrogenation and synthetic reaction.

$$iC + jH + kO + lH_2O \rightarrow CiHjOk + lH_2O \tag{14.59}$$

Rewriting jH as $0.5jH_2$ gives the following equation:

$$0.5jH_2 + iC + kO + lH_2O \rightarrow CiHjOk + lH_2O \quad (14.60)$$

The above equation is the same as the equation expressing hydrogenation. The similarity means that we cannot distinguish reactions of tar formation by thermal recombination and hydrogenation. When steam reforming ($iC + iH_2O \rightarrow iH_2 + iCO$) is involved in the reaction of Equation 14.60, the following reaction equation is obtained:

$$(i + 0.5j)H_2 + iCO + kO + (l - i)H_2O \rightarrow CiHjOk + lH_2O$$

The further reaction of kH_2 with kO yielding kH_2O gives the following equation:

$$iCO + (i + 0.5j - k)H_2 + (l + k - i)H_2O \rightarrow CiHj\,Ok + l\,H_2O \quad (14.61)$$

The above equation is the by-product formation through synthetic reaction from gaseous compounds. Compatible transformation among Equations 14.59 through 14.61 suggests that tar formation pathway cannot be identified as one among the typical routes of thermal recombination, hydrogenation, and synthetic reaction. Similarly, precise pathways of the formation of ethane and ethylene are not known. A stoichiometric model through synthetic reaction was used for describing the formation of gaseous hydrocarbon and tar in Section 14.24.2.1:

14.4.2.1 Case 1: $O_{ex} > 0$

When O_{ex} is greater than 0, excess oxygen reacts with hydrogen and carbon monoxide on the right-hand side of the standard reaction formula ($CH_mO_n + 0.5(1 - n)O_2 \rightarrow 0.5mH_2 + CO$). The total amount of H_2O and CO_2 is $2O_{ex}$. Letting the standard amount of molecular oxygen as O_{st} and hydrogen consumption as x, partial combustion can be described as follows:

$$CH_mO_n + (O_{st} + O_{ex})O_2 \rightarrow (0.5m - x)H_2$$
$$+ (1 - 2O_{ex} + x)CO + xH_2O + (2O_{ex} - x)CO_2$$

In the secondary reactions including the shift reaction of yCO and yH$_2$O, the reaction formula can be expressed with the following equation:

$$CH_mO_n + (O_{st} + O_{ex})O_2 + (y - x)H_2O \rightarrow$$
$$(0.5m + y - x)H_2 + (1 - 2O_{ex} - y + x)CO + (2O_{ex} + y - x)CO_2$$

When z mol of methane generates through the reaction $3H_2 + CO \rightarrow CH_4 + H_2O$, the following equation is obtained:

$$CH_mO_n + (O_{st} + O_{ex})O_2 + (y - x - z)H_2O \rightarrow$$
$$(0.5m + y - x - 3z)H_2 + (1 - 2O_{ex} - y + x - z)CO \quad (14.62)$$
$$+ (2O_{ex} + y - x)CO_2 + zCH_4$$

When t mol of ethylene is generated by the reaction $4H_2 + 2CO \rightarrow C_2H_4 + 2H_2O$, the following formula is obtained:

$$CH_mO_n + (O_{st} + O_{ex})O_2 + (y - x - z - 2t)H_2O \rightarrow$$
$$(0.5m + y - x - 3z - 4t)H_2 + (1 - 2O_{ex} - y + x - z - 2t)CO$$
$$+ (2O_{ex} + y - x)CO_2 + zCH_4 + tC_2H_4$$

When u mol of ethane generates by the reaction $5H_2 + 2CO \rightarrow C_2H_6 + 2H_2O$, the following formula is obtained:

$$CH_mO_n + (O_{st} + O_{ex})O_2 + (y - x - z - 2w - 2u)H_2O \rightarrow$$
$$(0.5m + y - x - 3z - 4t - 5u)H_2 + (1 - 2O_{ex} - y + x - z - 2t - 2u)CO$$
$$+ (2O_{ex} + y - x)CO_2 + zCH_4 + tC_2H_4 + uC_2H_6$$

When v mol of tar generates, the following equation is obtained:

$$CH_mO_n + (O_{st} + O_{ex})O_2 + \left[y - x - z - 2t - 2u - (1 - k)v\right]$$
$$H_2O \rightarrow \left\{0.5m + y - x - 3z - 4t - 5u - \left[(1 - k) + 0.5j\right]v\right\}H_2 \quad (14.63)$$
$$+ (1 - 2O_{ex} - y + x - z - 2t - 2u - v)CO + (2O_{ex} + y - x)CO_2$$
$$+ zCH_4 + tC_2H_4 + uC_2H_6 + vCHj\,Ok$$

14.4.2.2 Case 2: $O_{ex} \leqq 0$

In case of $O_{ex} \leqq 0$, coal carbon at the amount of $-2O_{ex}C$ remains unreacted. Further reaction of carbon of $-2O_{ex}C$ with $-2O_{ex}H_2O$ to complete gasification yielding $-2O_{ex}(H_2 + CO)$ is described as follows:

$$CH_mO_n + (O_{st} + O_{ex})O_2 - 2O_{ex}H_2O \rightarrow (0.5m - 2O_{ex})H_2 + CO$$

When shift reaction $y(CO + H_2O \rightarrow H_2 + CO_2)$ is involved in the secondary reaction, the following formula is obtained:

$$CH_mO_n + (O_{st} + O_{ex})O_2 + (-2O_{ex} + y)H_2O \rightarrow$$
$$(0.5m - 2O_{ex} + y)H_2 + (1 - y)CO + yCO_2$$

When zCH_4 is generated by the reaction $z(3H_2 + CO \rightarrow CH_4 + H_2O)$, the following formula is obtained:

$$CH_mO_n + (O_{st} + O_{ex})O_2 + (-2O_{ex} + y - z)H_2O \rightarrow$$
$$(0.5m - 2O_{ex} + y - 3z)H_2 + (1 - y - z)CO + yCO_2 + zCH_4 \quad (14.64)$$

The formation of tC_2H_4 by the reaction $4H_2 + 2CO \rightarrow C_2H_4 + 2H_2O$ can be expressed by the following formula:

$$CH_mO_n + (O_{st} + O_{ex})O_2 + (-2O_{ex} + y - z - 2t)H_2O \rightarrow$$
$$(0.5m - 2O_{ex} + y - 3z - 4t)H_2 + (1 - y - z - 2t)CO$$
$$+ yCO_2 + zCH_4 + tC_2H_4$$

When uC_2H_6 is generated by the reaction $5H_2 + 2CO \rightarrow C_2H_6 + 2H_2O$, the following formula is obtained:

$$CH_mO_n + (O_{st} + O_{ex})O_2 + (-2O_{ex} + y - z - 2t - 2u)H_2O \rightarrow$$
$$(0.5m - 2O_{ex} + y - 3z - 4t - 5u)H_2 + (1 - y - z - 2t - 2u)CO$$
$$+ yCO_2 + zCH_4 + tC_2H_4 + uC_2H_6$$

When v mol of tar CH_jO_k is generated, the following equation is obtained according to Equation 14.61:

TABLE 14.2
Mathematical Results of the Relationship between the Observed Gas Concentration and the Stoichiometric Parameters of the Overall Formula for Coal Gasification

	$O_{ex} \leqq 0$	$O_{ex} > 0$
α	$O_{st} + O_{ex}$	$O_{st} + O_{ex}$
β	$-2O_{ex} + y - z - 2t - 2u - (1 - k)v$	$y - x - z - 2t - 2u - (1 - k)v$
γ	$0.5m - 2O_{ex} + y - 3z - 4t - 5u - [(1 - k) + 0.5j]v$	$0.5m + y - x - 3z - 4t - 5u - [(1 - k) + 0.5j]v$
δ	$1 - y - z - 2t - 2u - v$	$1 - 2O_{ex} - y + x - z - 2t - 2u - v$
ε	y	$2O_{ex} + y - x$
η	z	z
κ	t	t
λ	u	u
θ	v	v

Note: Overall formula for coal gasification: $CH_mO_n + \alpha O_2 + \beta H_2O \rightarrow \gamma H_2 + \delta CO + \varepsilon CO_2 + \eta CH_4 + \kappa C_2H_4 + \lambda C_2H_6 + \theta CH_jO_k$.

$$\begin{gathered} CH_mO_n + (O_{st} + O_{ex})O_2 + \\ \left[-2O_{ex} + y - z - 2t - 2u - (1-k)v\right] H_2O \rightarrow \\ \left\{0.5m - 2O_{ex} + y - 3z - 4t - 5u - \left[(1-k) + 0.5j\right]v\right\} H_2 \\ +(1 - y - z - 2t - 2u - v)CO + yCO_2 + zCH_4 + tC_2H_4 \\ +uC_2H_6 + vCH_jO_k \end{gathered} \tag{14.65}$$

Stoichiometric models were given as Equations 14.63 and 14.65. Table 14.2 shows the parameters α through ζ in Equation 14.38 and the stoichiometric relationship in comparison with the terms of Equations 14.63 and 14.65.

Using the mathematical results in Table 14.2, the experimentally observed concentrations, α through θ, of the compounds in gasification correlate to the stoichiometric parameters in the stoichiometric model. We can understand the contributions of various reactions to the increase or decrease of the products. The stoichiometric model derived here is based on the stoichiometric discussion of the standard reaction formula of the combination of partial oxidation and some secondary reactions ignoring some underlying reactions such as pyrolysis of coal and steam gasification of char.

14.5 PROCESS EVALUATION BY THE STOICHIOMETRIC APPROACH

There have been many reports on the experimental data of coal gasification. Some 400 data around the 1990s were examined with respect to material balance and heat balance between coal and product gas. Additionally, elemental analysis of coal, flow rate of gasifying agents, and gasification temperature should be mentioned in literatures. Some eighty data were selected as shown in Tables 14.3 through 14.11. Applying the theory in Sections 14.2 through 14.4 to these data, stoichiometry of gasification was determined in each process based on the general formula describing coal gasification, and cold gas efficiency and carbon conversion were calculated. Gasification results achieved in various processes were compared with each other to clarify the features of various gasification processes.

14.5.1 Estimation of the Values of the Parameters in the Coal Gasification Formula

Process data were analyzed by the theory explained in Section 14.4. As shown in Table 14.3, the formation of C_2 compounds such as ethylene is negligible, and C_2 compounds are ignored in the general gasification formula:

$$CH_mO_n + \alpha O_2 + \beta H_2O \rightarrow \gamma H_2 + \delta CO + \varepsilon CO_2 + \eta CH_4 \tag{14.66}$$

Letting observed concentrations of H_2, CO, CO_2, and CH_4 as p, q, r, and s, the following equations were obtained to give the parameters α and η by the mathematical procedure in Section 14.4:

$$O_2{:}\ \alpha = \frac{(-p+q+2r-2s)}{2(q+r+s)} + 0.25m - 0.5n \tag{14.67}$$

$$H_2O{:}\ \beta = \frac{(p+2s)}{(q+r+s)} - 0.5m \tag{14.68}$$

$$H_2{:}\ \gamma = \frac{p}{(q+r+s)} \tag{14.69}$$

$$CO{:}\ \delta = \frac{q}{(q+r+s)} \tag{14.70}$$

$$CO_2{:}\ \varepsilon = \frac{r}{(q+r+s)} \tag{14.71}$$

TABLE 14.3
Analysis of the Data Fixed-Bed Process

Entry		1	2	3	4	5[k]	6
Process	Name	Riley-Morgan	Riley-Morgan	Lurgi	BGC/Lurgi	Lurgi	GEGas
	Capacity	100 tons/day	67 tons/day	–	–	–	–
	Type	Fixed-bed	Fixed-bed	Fixed-bed	Slagging Fixed-bed	Fixed-bed	Stirred Fixed-bed
Coal	Coal rank	Bituminous	Bituminous	Bituminous	Sub-bituminous	Sub-bituminous	Sub-bituminous
	Feed rate (kg/h)[a]	2722	–	4718	12,005	–	599
	Calorific value (kcal/kg)[b]	7987	7883	8060	8060	5746	6835
Proximate analysis	Moisture (%)[a]	3.2	5.5	–	10.0	16.3	–
	Ash (%)[b]	4.9	7.5	8.0	7.3	23.1	15.0
	Volatile matter (%)[b]	37.8	32.6	39.4	40.8	–	–
	Fixed carbon (%)[b]	57.3	59.9	52.6	51.9	–	–
Ultimate analysis[b]	C (%)	80.8	80.0	78.3	78.3	58.7	67.4
	H (%)	5.4	5.1	5.2	5.2	4.4	5.0
	O (%)	7.8	5.9	6.2	6.2	12.0	7.4
	N (%)	0.6	0.6	1.3	1.3	1.1	1.3
	S (%)	0.6	0.8	1.7	1.7	0.8	3.8
	Ash (%)[b]	4.8	7.5	7.3	7.3	23.1	15.0
Elemental ratio	*m* (H/C)	0.802	0.765	0.797	0.797	0.899	0.890
	n (O/C)	0.072	0.055	0.059	0.059	0.153	0.082
	O_2 (kg/kg)[c]	–	–	0.644	0.604	–	–
Gasifying agent	Air (O_2 content) (kg/kg)[c]	2.83 (0.654)	3.09 (0.714)	–	–	1.53 (0.353)	2.62 (0.607)
	Steam (kg/kg)[c]	0.584	0.601	5.543	0.442	0.738	0.450
	Total O_2 (kg/kg)[c]	0.654	0.714	0.644	0.604	0.353	0.607
	O_2 (%)	15.7	16.0	6.1	43.5	11.8	16.4
Condition	Pressure (kgf/cm^2)	0.15	1.10	25.60	24.6ata	20.70	15.8–21.1Atg
	Temperature top (°C)	600	600	600	600	600	549
	Bottom (°C)	–	Max 1150	1371	1471	–	–
Product gas composition[d]	H_2 (%)	14.3	17.9	39.8	27.8	23.4	17.0
	CO (%)	22.0	21.1	24.6	60.6	17.4	23.9
	CO_2 (%)	7.5	8.9	24.6	2.6	14.8	6.7
	CH_4 (%)	2.3	2.0	8.7	7.6	5.1	3.2
	C_2H_4 (%)	–	–	–	–	–	–
	C_2H_6 (%)	0.9	0.5	1.1	0.4	0.7	–
	N_2+Ar (%)	53.0	49.6	1.2	1.0	38.6	49.2

(Continued)

TABLE 14.3 (*Continued*)
Analysis of the Data Fixed-Bed Process

Entry		1	2	3	4	5[k]	6
Gas yield in volume (Nm^3/kg)[c]		3.755	3.827	2.189	1.890	2.384	3.305
Heating value (kcal/Nm^3)		1472	1460	2971	3477	1843	1549
Cold gas efficiency[e] (–)		0.692	0.675	–	–	0.767	0.748
Carbon conversion[e] (–)		–	–	–	–	–	0.885
Coal	Heating value (kJ/mol)	–496.5	–494.9	–517.1	–517.1	–491.9	–509.5
Oxygen	μ (mol/mol)	1.165	1.164	1.170	1.170	1.148	1.182
	O_2 supplied A (mol/mol)	0.304	0.335	0.308	0.289	0.226	0.338
	χ (=α/μ) (–)	0.385	0.369	0.268	0.282	0.286	0.327
	O_{st} [=0.5(1–n)] (mol/mol)	0.464	0.473	0.471	0.471	0.424	0.459
	O_{ex} (=α–O_{st}) (mol/mol)	–0.015	–0.043	–0.157	–0.141	–0.096	–0.072
Water[e]	w (mol/mol)	0.482	0.501	4.719	0.376	0.838	0.445
	w + moisture (mol/mol)	0.509	0.549	4.719	0.471	1.059	0.445
Product gas composition[f]	p (H_2) (%)	31.0	35.9	40.7	28.2	38.1	33.5
	q (CO) (%)	47.7	42.3	25.2	61.4	28.3	47.0
	r (CO_2) (%)	16.3	17.9	25.2	2.6	24.1	13.2
	s (CH_4) (%)	5.0	4.0	8.9	7.7	8.3	6.3
Stoichiometric parameter[g]	α (O_2) (mol/mol)	0.449	0.430	0.314	0.330	0.328	0.387
	β (H_2O) (mol/mol)	0.193	0.301	0.588	0.210	0.451	0.248
	γ (H_2) (mol/mol)	0.449	0.559	0.686	0.393	0.628	0.504
	δ (CO) (mol/mol)	0.691	0.659	0.425	0.856	0.466	0.707
	ε (CO_2) (mol/mol)	0.236	0.279	0.425	0.036	0.397	0.198
	η (CH_4) (mol/mol)	0.072	0.062	0.150	0.107	0.137	0.095

(Continued)

TABLE 14.3 (*Continued*)
Analysis of the Data Fixed-Bed Process

Entry	1	2	3	4	5[k]	6
Heating value of gas (kJ/mol)	−388.3	−401.8	−450.2	−450.2	−433.8	−429.2
H_2O in product gas[h] (mol/mol)	0.378	0.315	4.142	0.311	0.785	0.228
N_2 (mol/mol)	1.577	1.503	0.023	0.014	0.997	1.456
Heat of reaction $(1-\theta)\ h_r$ (kJ/mol)	−73.85	−73.26	−66.49	−51.69	−35.83	−70.94
$H_r\ [=(1-\theta)\ h_r - w \times 43.99]$ (kJ/mol)	−95.01	−95.27	−273.81	−68.22	−72.65	−90.49
Sensible heat of products H_g (kJ/mol)	−57.27	−58.27	−121.35	−28.53	−57.59	−51.17
Heat loss $H_l = H_r - H_g$ (kJ/mol)	−37.74	−37.00	−152.46	−39.69	−15.06	−39.32
T_{ad} (°C)	938	927	1208	1274	735	910
H_2O generation from H_2 combustion[i] × (mol/mol)	0.000	0.000	0.000	0.000	0.000	0.000
H_2O-consumed in shift reaction y (mol/mol)	0.160	0.217	0.417	0.027	0.241	0.173
Cold gas efficiency[i] (−)	0.529	0.633	0.854	0.815	0.765	0.736
Carbon conversion[j] $(1-\theta)$ (−)	0.677	0.779	0.981	0.762	0.608	0.873
Reference	Bodle and Huebler 1981	Hebden and Stroud 1981	Bodle and Huebler 1981	Bodle and Huebler 1981	Hebden and Stroud 1981	Hebden and Stroud 1981

[a] Dry base
[b] Wet base
[c] kg per kg-dry coal
[d] H_2S and NH_3 are not involved in the gas composition
[e] Reported value
[f] The concentration of N_2 and Ar are not involved in the theoretical calculations
[g] Parameters in the equation $CH_mO_n + \alpha O_2 + \beta H_2O \rightarrow \gamma H_2 + \delta CO + \varepsilon CO_2 + \eta CH_4$
[h] Moisture + steam − β
[i] Produced by combustion with excess of O_2
[j] Calculated based on the stoichiometric theory
[k] Designed values based on the data in their development stage

TABLE 14.4
Analysis of the Data Fluidized Bed Process (1)

Entry			1	2	3	4	5	6	7
Process	Name		Winkler	Winkler	Winkler	Winkler	Winkler	Winkler	Winkler
	Capacity		–	–	–	–	–	–	–
	Type		Fluidized-bed	Fluidized-bed	Fluidized-bed	Fluidized-bed	Fluidized-bed	Fluidized-bed	Fluidized-bed
Coal	Coal rank		Bituminous	Lignite	Bituminous char	Bituminous char	Sub-bituminous	Lignite	Lignite
	Feed rate (kg/h)[a]		5160	6100	5000	2790	4500	–	–
	Calorific value (kcal/kg)[b]		6828	6525	6070	5890	6522	5731	4849
Proximate analysis	Moisture (%)[a]		7.0	0.0	4.0	5.2	2.5	8.0	8.0
	Ash (%)[b]		–	–	19.4	17.7	13.0	13.8	23.7
	Volatile matter (%)[b]		–	–	70.6	71.7	50.1	–	–
	Fixed carbon (%)[b]		–	–	10.0	10.6	36.9	–	–
Ultimate analysis[b]	C (%)		68.0	64.4	68.9	68.3	68.8	61.3	54.3
	H (%)		4.8	4.7	2.0	2.1	4.9	4.7	3.7
	O (%)		11.3	7.0	5.7	5.4	9.7	16.3	15.4
	N (%)		1.3	0.9	1.1	1.8	1.0	0.8	1.7
	S (%)		0.4	0.4	0.5	0.5	0.4	3.3	1.2
	Ash (%)		14.2	12.6	18.7	16.8	12.7	13.8	23.7
Elemental ratio	*m* (H/C)		0.847	0.876	0.348	0.369	0.855	0.920	0.818
	n (O/C)		0.125	0.082	0.062	0.059	0.106	0.199	0.213
	O_2	(kg/kg)[c]	0.549	0.506	0.465	0.461	0.434	–	0.424
Gasifying agent	Air (O_2 content)	(kg/kg)[c]	0.598 (0.138)	0.623 (0.144)	0.394 (0.091)	0.592 (0.137)	0.533 (0.143)	2.73 (0.630)	–
	Steam	(kg/kg)[c]	0.475	0.302	0.310	0.367	0.302	0.130	0.424
	Total O_2	(kg/kg)[c]	0.687	0.650	0.556	0.598	0.557	0.630	0.424
	O_2	(%)	33.5	37.6	38.3	33.8	35.7	19.4	36.0
Condition	Pressure	(kgf/cm^2)	3.90	3.90	3.90	3.90	3.90	~1.1	~1.1
	Temperature top (°C)		1100	1100	1100	930	1075	800–1000	800–1000
	Bottom (°C)		–	–	–	–	–	1000–1200	1000–1200
Product gas composition[d]	H_2 (%)		31.0	30.9	30.8	28.4	32.9	12.7	40.1
	CO (%)		32.0	32.5	30.2	34.7	31.4	22.7	36.1
	CO_2 (%)		16.0	13.6	19.6	16.4	13.9	7.8	19.6
	CH_4 (%)		2.0	1.5	0.2	0.4	1.3	0.7	2.5
	C_2H_4 (%)		–	–	–	–	–	–	–
	C_2H_6 (%)		–	–	–	–	–	–	–
	N_2 + Ar (%)		19.0	21.5	19.2	20.1	20.5	56.1	1.7

(Continued)

TABLE 14.4 (*Continued*)
Analysis of the Data Fluidized Bed Process (1)

Entry		1	2	3	4	5	6	7
Gas yield in volume (Nm^3/kg)[c]		1.957	1.900	1.339	1.951	1.631	2.993	1.402
Heating value ($kcal/Nm^3$)		2109	2073	1877	1959	2082	1144	2559
Cold gas efficiency[e] (–)		0.605	0.604	0.398	0.616	0.508	0.619	0.744
Carbon conversion[e] (–)		0.718	0.750	0.500	0.749	0.577	0.830	0.810
Coal	Heating value (kJ/mol)	–504.5	–509.1	–442.5	–433.3	–476.4	–469.7	–448.8
Oxygen	μ (mol/mol)	1.149	1.178	1.056	1.063	1.161	1.131	1.098
	O_2 supplied A (mol/mol)	0.379	0.378	0.303	0.328	0.304	0.080	0.293
	χ (=α/μ) (–)	0.394	0.381	0.417	0.408	0.358	0.458	0.326
	O_{st} [=0.5(1–n)] (mol/mol)	0.438	0.459	0.469	0.471	0.447	0.401	0.394
	O_{ex} (=α–O_{st}) (mol/mol)	0.002	–0.010	–0.029	–0.037	–0.031	0.118	–0.036
Water[e]	w (mol/mol)	0.466	0.313	0.300	0.358	0.293	0.141	0.521
	w + moisture (mol/mol)	0.539	0.313	0.340	0.412	0.310	0.236	0.627
Product gas composition[f]	p (H_2) (%)	38.3	39.4	38.1	35.5	41.4	28.9	40.8
	q (CO) (%)	39.5	41.4	37.4	43.4	39.5	51.7	36.7
	r (CO_2) (%)	19.8	17.3	24.3	20.5	17.5	17.8	19.9
	s (CH_4) (%)	2.5	1.9	0.2	0.5	1.6	1.6	2.5
Stoichiometric parameter[g]	α (O_2) (mol/mol)	0.439	0.449	0.440	0.434	0.416	0.518	0.358
	β (H_2O) (mol/mol)	0.277	0.275	0.448	0.382	0.334	–0.009	0.366
	γ (H_2) (mol/mol)	0.620	0.650	0.616	0.551	0.706	0.406	0.690
	δ (CO) (mol/mol)	0.639	0.683	0.604	0.674	0.674	0.727	0.621
	ε (CO_2) (mol/mol)	0.320	0.285	0.393	0.318	0.299	0.250	0.337
	η (CH_4) (mol/mol)	0.040	0.031	0.003	0.008	0.027	0.023	0.042

(*Continued*)

TABLE 14.4 (*Continued*)
Analysis of the Data Fluidized Bed Process (1)

Entry	1	2	3	4	5	6	7
Heating value of gas (kJ/mol)	–394.0	–407.0	–349.9	–355.6	–416.9	–342.5	–410.7
H_2O in product gas[b] (mol/mol)	0.300	0.081	0.031	0.123	0.066	0.243	0.328
N_2 (mol/mol)	0.380	0.452	0.384	0.390	0.440	1.957	0.029
Heat of reaction (1–θ) h_r (kJ/mol)	–96.03	–86.59	–64.32	–59.21	–43.99	–94.99	–31.71
H_r [=(1–θ)h_r – w × 43.99) (kJ/mol)	–116.49	–100.32	–77.50	–74.95	–56.84	–101.20	–54.58
Apparent heat of products H_g (kJ/mol)	–82.01	–74.28	–63.67	–56.08	–68.58	–84.87	–49.67
Heat loss $H_l = H_r - H_g$ (kJ/mol)	–34.48	–26.04	–13.83	–18.87	11.74	–16.33	–4.91
T_{ad} (°C)	1498	1437	1306	1200	912	935	869
H_2O generation from H_2 combustion[i] × (mol/mol)	0.001	0.000	0.000	0.000	0.000	0.041	0.000
H_2O-consumed in shift reaction y (mol/mol)	0.274	0.240	0.271	0.240	0.219	0.051	0.276
Cold gas efficiency[i] (–)	0.674	0.673	0.545	0.620	0.640	0.542	0.749
Carbon conversion[j] (1–θ) (–)	0.863	0.842	0.689	0.756	0.731	0.743	0.818
Reference	Furusawa et al. 1985	Furusawa et al. 1985	Furusawa et al. 1985	Furusawa et al. 1985	Furusawa et al. 1985	Hebden and Stroud 1981	Hebden and Stroud 1981

[a] Dry base
[b] Wet base
[c] kg per kg-dry coal
[d] H_2S and NH_3 are not involved in the gas composition
[e] Reported value
[f] The concentration of N_2 and Ar are not involved in the theoretical calculations
[g] Parameters in the equation $CH_mO_n + \alpha O_2 + \beta H_2O \rightarrow \gamma H_2 + \delta CO + \varepsilon CO_2 + \eta CH_4$
[h] Moisture + steam – β
[i] Produced by combustion with excess of O_2
[j] Calculated based on the stoichiometric theory

TABLE 14.5
Analysis of the Data Fluidized Bed Process (2)

Entry		1	2	3	4	5	6
Process	Name	CMRC	CMRC	CMRC	CMRC	CMRC	CMRC
	Capacity	40 tons/day	40 tons/day	40 tons/day	40 tons/day	40 tons/day	40 tons/day
	Type	Two-stage fluidized-bed	Two-stage fluidized-bed	Two-stage fluidized-bed	Two-stage fluidized-bed	Two-stage fluidized-bed	Two-stage fluidized-bed
Coal	Coal rank	Sub-bituminous	Sub-bituminous	Sub-bituminous	Bituminous	Sub-bituminous	Bituminous
	Feed rate (kg/h)[a]	1385	1300	1296	748	607	530
	Calorific value (kcal/kg)[b]	6533	6502	6445	7241	7054	7300
Proximate analysis	Moisture (%)[a]	3.8	5.1	5.2	6.5	3.6	5.2
	Ash (%)[b]	15.8	17.2	16.8	8.4	12.0	7.3
	Volatile matter (%)[b]	–	44.2	44.9	28.3	29.7	30.5
	Fixed carbon (%)[b]	–	38.7	38.3	63.2	58.3	62.2
Ultimate analysis[b]	C (%)	65.3	64.3	64.7	76.5	72.1	78.5
	H (%)	5.5	5.5	5.4	4.4	4.5	4.4
	O (%)	12.0	11.6	12.1	8.7	8.8	8.4
	N (%)	1.1	1.1	1.0	1.7	1.8	0.9
	S (%)	0.3	0.3	0.3	0.3	5.3	0.7
	Ash (%)	15.8	17.2	16.5	8.4	7.5	7.1
Elemental ratio	*m* (H/C)	1.011	1.026	1.002	0.690	0.749	0.673
	n (O/C)	0.138	0.135	0.140	0.085	0.092	0.080
Gasifying agent	O_2 (kg/kg)[c]	–	–	–	–	–	–
	Air (O_2 content) (kg/kg)[c]	3.37 (0.779)	3.29 (0.760)	3.15 (0.728)	4.68 (1.081)	4.57 (1.056)	4.807 (1.110)
	Steam (kg/kg)[c]	0.742	0.880	0.702	1.034	1.412	1.751
	Total O_2 (kg/kg)[c]	0.779	0.760	0.728	1.081	1.056	1.110
	O_2 (%)	15.5	14.7	15.4	15.5	14.0	13.2
Conditions	Pressure (kgf/cm^2)	20	20	19.7	20.0	20.0	20.2
	Temperature top (°C)	982	940	930	1037	987	965
	Bottom (°C)	978	960	954	1051	997	977
Product gas composition[d]	H_2 (%)	15.8	15.8	15.0	13.0	12.4	12.4
	CO (%)	15.1	15.1	14.7	13.9	9.9	9.3
	CO_2 (%)	12.1	12.1	12.6	12.5	14.5	14.9

(Continued)

TABLE 14.5 (*Continued*)
Analysis of the Data Fluidized Bed Process (2)

Entry		1	2	3	4	5	6
	CH_4 (%)	3.6	3.6	4.6	1.1	1.5	1.5
	C_2H_4 (%)	–	–	–	–	–	–
	C_2H_6 (%)	–	–	–	–	–	–
	N_2 + Ar (%)	53.4	53.4	53.1	59.5	61.7	61.9
Gas yield in volume (Nm^3/kg)[c]		3.866	3.740	3.655	5.080	4.782	5.020
Heating value (kcal/Nm^3)		1283	1283	1341	924	822	804
Cold gas efficiency[e] (–)		0.758	0.725	0.742	0.624	0.561	0.521
Carbon conversion[e] (–)		0.978	0.950	0.948	0.961	0.930	0.877
Coal	Heating value (kJ/mol)	–502.8	–507.8	–500.3	–475.6	–491.5	–467.2
Oxygen	μ (mol/mol)	1.184	1.189	1.181	1.130	1.141	1.128
	O_2 supplied A (mol/mol)	0.447	0.443	0.422	0.530	0.549	0.530
	χ (=α/μ) (–)	0.379	0.381	0.362	0.496	0.522	0.523
	O_{st} [=0.5(1−n)] (mol/mol)	0.431	0.433	0.430	0.476	0.454	0.460
	O_{ex} (=α–O_{st}) (mol/mol)	0.018	0.022	–0.003	0.104	0.142	0.130
Water[e]	w (mol/mol)	0.756	0.912	0.723	0.901	1.306	1.487
	w + moisture (mol/mol)	0.798	0.969	0.780	0.962	1.340	1.534
Product gas composition[f]	p (H_2) (%)	33.9	33.9	32.0	32.1	32.4	32.5
	q (CO) (%)	32.4	32.4	31.3	34.3	25.8	24.4
	r (CO_2) (%)	26.0	26.0	26.9	30.9	37.9	39.1
	s (CH_4) (%)	7.7	7.7	9.8	2.7	3.9	3.9
Stoichiometric parameter[g]	α (O_2) (mol/mol)	0.449	0.454	0.427	0.561	0.596	0.590
	β (H_2O) (mol/mol)	0.240	0.240	0.258	0.207	0.220	0.262
	γ (H_2) (mol/mol)	0.513	0.513	0.471	0.473	0.479	0.482
	δ (CO) (mol/mol)	0.490	0.490	0.460	0.505	0.382	0.362
	ε (CO_2) (mol/mol)	0.393	0.393	0.396	0.455	0.561	0.580
	η (CH_4) (mol/mol)	0.116	0.116	0.144	0.040	0.058	0.058

(Continued)

TABLE 14.5 (*Continued*)
Analysis of the Data Fluidized Bed Process (2)

Entry	1	2	3	4	5	6
Heating value of gas (kJ/mol)	−388.9	−388.9	−393.3	−314.0	−296.9	−292.1
H_2O in product gas[h] (mol/mol)	0.559	0.735	0.525	0.766	1.137	1.299
N_2 (mol/mol)	1.734	1.734	1.665	2.163	2.382	2.409
Heat of reaction $(1-\theta)\ h_r$ (kJ/mol)	−114.05	−117.02	−106.6	−153.27	−179.81	−157.78
$H_r\ [=(1-\theta)\ h_r - w \times 43.99]$ (kJ/mol)	−147.32	−157.10	−138.38	−192.86	−237.16	−223.10
Apparent heat of products H_g (kJ/mol)	−127.35	−126.61	−115.98	−153.03	−166.64	−168.53
Heat loss $H_l = H_r - H_g$ (kJ/mol)	−19.97	−30.49	−22.40	−39.83	−70.52	−54.57
T_{ad} (°C)	1117	1137	1086	1276	1351	1236
H_2O generation from H_2 combustion[i] × (mol/mol)	0.012	0.015	0.000	0.050	0.071	0.058
H_2O-consumed in shift reaction y (mol/mol)	0.368	0.355	0.391	0.284	0.326	0.346
Cold gas efficiency[i] (–)	0.770	0.747	0.776	0.624	0.556	0.562
Carbon conversion[j] $(1-\theta)$ (–)	0.996	0.976	0.988	0.945	0.921	0.898
Reference	Hozumi and Seki 1990	Hozumi and Seki 1990	Shimode et al. 1985	Hozumi and Seki 1990	Hozumi and Seki 1990	Hozumi and Seki 1990

[a] Dry base
[b] Wet base
[c] kg per kg-dry coal
[d] H_2S and NH_3 are not involved in the gas composition
[e] Reported value
[f] The concentration of N_2 and Ar are not involved in the theoretical calculations
[g] Parameters in the equation $CH_mO_n + \alpha O_2 + \beta H_2O \rightarrow \gamma H_2 + \delta CO + \varepsilon CO_2 + \eta CH_4$
[h] Moisture + steam – β
[i] Produced by combustion with excess of O_2
[j] Calculated based on the stoichiometric theory

TABLE 14.6
Analysis of the Data Fluidized Bed Process (3)

Entry		1	2	3	4	5	6
Process	Name	AIST	AIST	AIST	AIST	AIST	AIST
	Capacity	–	–	–	–	–	–
	Type	Fluidized-bed	Fluidized-bed	Fluidized-bed	Fluidized-bed	Fluidized-bed	Fluidized-bed
Coal	Coal rank	Sub-bituminous char	Sub-bituminous char	Sub-bituminous char	Sub-bituminous char	Sub-bituminous	Sub-bituminous
	Feed rate (kg/h)[a]	5.0	2.4	3.5	5.1	20	15
	Calorific value (kcal/kg)[b]	5994	5994	5994	5994	6985	7020
Proximate analysis	Moisture (%)[a]	3.9	3.9	3.9	3.9	6.0	6.0
	Ash (%)[b]	16.3	16.3	16.3	16.3	10.5	9.8
	Volatile matter (%)[b]	21.9	21.9	21.9	21.9	47.5	48.0
	Fixed carbon (%)[b]	61.8	61.8	61.8	61.8	42.2	42.3
Ultimate analysis[b]	C (%)	67.4	67.4	67.4	67.4	69.2	69.8
	H (%)	3.5	3.5	3.5	3.5	5.5	5.5
	O (%)	12.4	12.4	12.4	12.4	13.3	13.4
	N (%)	0.8	0.8	0.8	0.8	1.3	1.3
	S (%)	0.1	0.1	0.1	0.1	0.3	0.3
	Ash (%)	15.7	15.7	15.7	15.7	10.5	9.8
Elemental ratio	*m* (H/C)	0.623	0.623	0.623	0.623	0.954	0.946
	n (O/C)	0.138	0.138	0.138	0.138	0.144	0.144
Gasifying agent	O_2 (kg/kg)[c]	–	–	–	–	0.610	0.744
	Air (O_2 content) (kg/kg)[c]	2.77 (0.639)	3.73 (0.861)	3.00 (0.694)	2.41 (0.556)	–	–
	Steam (kg/kg)[c]	1.061	2.125	1.070	0.714	1.615	2.292
	Total O_2 (kg/kg)[c]	0.639	0.861	0.694	0.556	0.610	0.744
	O_2 (%)	13.0	10.9	13.3	14.2	17.5	15.5
Conditions	Pressure (kgf/cm^2)	7	5	5	5	5	5
	Temperature top (°C)	950	950	950	950	870	890
	Bottom (°C)	–	–	–	–	–	–
Product gas composition[d]	H_2 (%)	19.7	18.5	20.6	21.4	32.3	34.7
	CO (%)	11.8	7.0	9.2	11.6	16.4	20.5
	CO_2 (%)	15.4	20.4	19.7	16.3	32.5	31.3
	CH_4 (%)	1.4	0.8	1.0	1.2	11.9	7.0
	C_2H_4 (%)	–	–	–	–	–	–
	C_2H_6 (%)	–	–	–	–	–	0.8
	N_2 + Ar (%)	51.7	53.3	49.5	49.5	6.9	5.7

(Continued)

TABLE 14.6 (*Continued*)
Analysis of the Data Fluidized Bed Process (3)

Entry		1	2	3	4	5	6
Gas yield in volume (Nm^3/kg)[c]		3.600	4.379	3.597	2.999	1.647	1.577
Heating value ($kcal/Nm^3$)		1093	854	1004	1120	2615	2479
Cold gas efficiency[e] (–)		0.640	0.620	0.590	0.550	0.626	0.576
Carbon conversion[e] (–)		0.810	0.980	0.850	0.690	–	–
Coal	Heating value (kJ/mol)	–446.7	–446.7	–446.7	–446.7	–507.0	–505.3
Oxygen	μ (mol/mol)	1.087	1.087	1.087	1.087	1.167	1.165
	O_2 supplied *A* (mol/mol)	0.356	0.479	0.386	0.309	0.331	0.400
	χ ($=\alpha/\mu$) (–)	0.403	0.533	0.479	0.402	0.320	0.392
	O_{st} [$=0.5(1-n)$](mol/mol)	0.431	0.431	0.431	0.431	0.428	0.428
	O_{ex} ($=\alpha-O_{st}$) (mol/mol)	0.007	0.148	0.090	0.006	–0.054	0.029
Water[e]	w (mol/mol)	1.049	2.102	1.058	0.706	1.556	2.189
	w + moisture (mol/mol)	1.190	2.290	1.099	0.746	1.617	2.250
Product gas composition[f]	*p* (H_2) (%)	40.8	39.6	40.8	42.4	34.7	36.8
	q (CO) (%)	24.4	15.0	18.2	23.0	17.6	21.7
	r (CO_2) (%)	31.9	43.7	39.0	32.3	34.9	33.2
	s (CH_4) (%)	2.9	1.7	2.0	2.4	12.8	7.4
Stoichiometric parameter[g]	α (O_2) (mol/mol)	0.438	0.579	0.521	0.437	0.374	0.457
	β (H_2O) (mol/mol)	0.476	0.400	0.445	0.506	0.447	0.355
	γ (H_2) (mol/mol)	0.689	0.656	0.689	0.735	0.531	0.591
	δ (CO) (mol/mol)	0.412	0.248	0.307	0.399	0.270	0.348
	ε (CO_2) (mol/mol)	0.539	0.724	0.659	0.560	0.534	0.533
	η (CH_4) (mol/mol)	0.049	0.028	0.034	0.042	0.196	0.119

(Continued)

TABLE 14.6 (*Continued*)
Analysis of the Data Fluidized Bed Process (3)

Entry	1	2	3	4	5	6
Heating value of gas (kJ/mol)	–357.4	–282.8	–314.3	–360.7	–403.0	–373.6
H_2O in product gas[b] (mol/mol)	0.803	1.959	0.769	0.388	1.221	1.939
N_2 (mol/mol)	1.808	1.890	1.656	1.701	0.113	0.094
Heat of reaction (1–θ) h_r (kJ/mol)	–73.22	–136.06	–98.62	–61.39	–92.83	–115.81
H_r [=(1–θ)h_r – w × 43.99] (kJ/mol)	–119.32	–228.40	–145.11	–92.41	–161.18	–211.98
Apparent heat of products H_g (kJ/mol)	–127.81	–171.48	–123.97	–110.20	–89.99	–114.91
Heat loss $H_l = H_r - H_g$ (kJ/mol)	**8.49**	–56.92	–21.14	**17.79**	–71.19	–97.07
T_{ad} (°C)	894	1220	1090	814	1419	1491
H_2O generation from H_2 combustioni × (mol/mol)	0.002	0.058	0.032	0.003	0.000	0.017
H_2O-consumed in shift reaction y (mol/mol)	0.429	0.412	0.387	0.390	0.473	0.432
Cold gas efficiency[i] (–)	0.651	0.524	0.521	0.560	0.703	0.647
Carbon conversion[j] (1–θ) (–)	0.813	0.827	0.741	0.707	0.885	0.875
Reference	Honma et al. 1982	Honma et al. 1982	Honma et al. 1982	Honma et al. 1982	Kimura, Fujii 1976	Kimura, Fujii 1976

[a] Wet base
[b] Dry base
[c] kg per kg-dry coal
[d] H_2S and NH_3 are not involved in the gas composition
[e] Reported value
[f] The concentration of N_2 and Ar are not involved in the theoretical calculations
[g] Parameters in the equation $CH_mO_n + \alpha O_2 + \beta H_2O \rightarrow \gamma H_2 + \delta CO + \varepsilon CO_2 + \eta CH_4$
[h] Moisture + steam – β
[i] Produced by combustion with excess of O_2
[j] Calculated based on the stoichiometric theory

TABLE 14.7
Analysis of the Data Fluidized Bed Process (4)

Entry		**1**	**2**	**3**	**4**	**5**	**6***	**7***	**8***	**9***
Process	Name	Hygas	Hygas	Synthane	Synthane	Synthane	Tri-Gas	West Virginia University	PEATGAS	U-Gas
	Capacity	–	–	72,144 tons/day	–	–	–	–	50 tons/day	–
	Type	Multiple stage fluidized-bed	Multiple stage fluidized-bed	– Fluidized-bed	– Fluidized-bed	– Fluidized-bed	Three-stage fluidized-bed	Two-stage fluidized-bed	– Fluidized-bed	– Spouted-flow
Coal	Coal rank	Sub-bituminous	Sub-bituminous	Sub-bituminous	Sub-bituminous	Sub-bituminous	Bituminous	Bituminous	Lignite	Coke
	Feed rate (kg/h)[a]	–	–	–	2140	2520	–	–	1829	1890
	Calorific value (kcal/kg)[b]	6390	6395	6117	6193	6101	7939	7283	4726	7661
Proximate analysis	Moisture (%)[a]	11.0	11.0	7.5	6.9	5.2	1.4	3.4	50.0	5.0
	Ash (%)[b]	–	9.0	11.2	11.4	12.2	6.3	7.6	11.3	16.6
	Volatile matter (%)[b]	–	38.5	39.0	37.3	35.8	38.4	–	65.0	–
	Fixed carbon (%)[b]	–	52.5	49.7	51.3	52.0	55.3	–	23.7	–
Ultimate analysis[a]	C (%)	60.4	67.9	66.6	66.4	66.2	80.3	73.5	49.9	71.4
	H (%)	5.3	4.6	5.0	4.4	4.2	5.4	5.3	5.1	5.3
	O (%)	24.7	16.8	7.4	15.9	15.8	4.9	10.1	30.7	0.6
	N (%)	0.9	1.0	1.0	1.0	0.9	1.4	0.7	2.7	1.7
	S (%)	0.7	0.8	0.6	0.9	0.5	1.5	2.8	0.3	4.4
	Ash (%)	8.0	0.9	11.2	11.4	12.2	6.4	7.6	11.3	16.6
Elemental ratio	*m* (H/C)	1.053	0.813	0.901	0.795	0.761	0.807	0.865	1.226	0.891
	n (O/C)	0.307	0.186	0.083	0.180	0.179	0.046	0.103	0.461	0.006
Gasifying agent	O_2 (kg/kg)[c]	0.281	0.281	~0.346	0.358	0.206	–	–	0.175	–
	Air (O_2 content) (kg/kg)[c]	–	?	–	–	–	3.75 (0.867)	1.98 (0.457)	–	3.23 (0.747)
	Steam (kg/kg)[c]	1.236	1.236	1.081	1.557	1.181	0.112	0.300	0.747	0.326
	Total O_2 (kg/kg)[c]	0.281	0.281	0.35	0.358	0.206	0.867	0.457	0.175	0.747
	O_2 (%)	11.4	11.4	15.4	11.5	8.9	20.0	16.8	11.7	18.0

(*Continued*)

TABLE 14.7 (*Continued*)
Analysis of the Data Fluidized Bed Process (4)

Entry		1	2	3	4	5	6*	7*	8*	9*
Conditions	Pressure (kgf/cm^2)	83.00	86.00	42–70	42.30	42.30	17.60	5.9Atg	–	8.1–25.7
	Temperature top (°C)	927	299/649/927	425	816	781	650/1090/1150	760/1040	–	370–425
	Bottom (°C)	–	–	760–980	–	–	–	–	–	950–1050
Product gas composition[d]	H_2 (%)	25.7	30.2	32.8	33.7	39.5	16.0	19.8	22.2	14.8
	CO (%)	30.8	25.2	13.4	10.6	7.6	31.5	27.4	12.4	23.2
	CO_2 (%)	25.9	25.4	36.9	42.9	38.7	0.5	6.8	42.0	6.4
	CH_4 (%)	17.6	17.3	15.3	12.5	12.7	–	3.9	19.6	3.2
	C_2H_4 (%)	–	–	–	–	–	–	–	–	–
	C_2H_6 (%)	–	1.9	1.6	0.2	1.2	–	0.7	3.5	–
	N_2 + Ar(%)	–	–	–	0.1	0.3	52.0	41.4	0.3	52.4
Gas yield in volume (Nm^3/kg)[c]		0.738	0.827	–	1.311	0.901	4.447	2.723	0.886	3.819
Heating value (kcal/Nm^3)		3392	3645	3126	2572	2842	1445	1923	3494	1460
Cold gas efficiency[e] (–)		0.428	0.470	~0.57	–	–	0.808	0.730	–	–
Carbon conversion[e] (–)		0.465	–	0.750	0.635	0.573	0.959	–	–	–
Coal	Heating value (kJ/mol)	–531.7	–473.1	–461.4	–468.5	–463.0	–496.5	–497.8	–476.0	–539.2
Oxygen	μ (mol/mol)	1.110	1.110	1.184	1.109	1.101	1.179	1.165	1.076	1.220
	O_2 supplied *A* (mol/mol)	0.174	0.155	0.197	0.202	0.117	0.405	0.233	0.132	0.392
	χ (=α/μ) (–)	0.231	0.175	0.309	0.356	0.247	0.370	0.291	0.289	0.365
	O_{st} [=0.5(1–*n*)] (mol/mol)	0.347	0.407	0.459	0.410	0.411	0.477	0.449	0.270	0.497
	O_{ex} (=α–O_{st}) (mol/mol)	–0.091	–0.213	–0.093	–0.015	–0.139	–0.041	–0.110	0.042	–0.052
Water[e]	w (mol/mol)	1.364	1.214	1.082	1.563	1.189	0.093	0.272	0.998	0.304
	w + moisture (mol/mol)	1.446	1.287	1.132	1.638	1.245	0.105	0.304	2.334	0.354
Product gas composition[f]	*p* (H_2) (%)	25.7	30.8	33.3	33.8	40.0	33.3	34.2	23.1	31.1
	q (CO) (%)	30.8	25.7	13.6	10.6	7.7	65.6	47.4	12.8	48.7
	r (CO_2) (%)	25.9	25.9	37.5	43.0	39.3	1.0	11.7	43.6	13.4
	s (CH_4) (%)	17.6	17.6	15.5	12.5	12.9	0.0	6.8	20.4	6.7
Stoichiometric parameter[g]	α (O_2) (mol/mol)	0.256	0.194	0.366	0.395	0.272	0.436	0.339	0.311	0.445
	β (H_2O) (mol/mol)	0.293	0.547	0.515	0.492	0.718	0.097	0.293	0.219	0.201

(*Continued*)

TABLE 14.7 (*Continued*)
Analysis of the Data Fluidized Bed Process (4)

Entry	1	2	3	4	5	6*	7*	8*	9*
γ (H_2) (mol/mol)	0.346	0.445	0.500	0.511	0.668	0.500	0.519	0.301	0.452
δ (CO) (mol/mol)	0.415	0.371	0.204	0.160	0.129	0.985	0.719	0.167	0.708
ε (CO_2) (mol/mol)	0.349	0.374	0.563	0.651	0.656	0.015	0.178	0.568	0.195
η (CH_4) (mol/mol)	0.237	0.254	0.233	0.189	0.215	0.000	0.103	0.266	0.097
Heating value of gas (kJ/mol)	–427.7	–458.7	–408.4	–359.9	–419.2	–422.0	–443.9	–370.4	–416.2
H_2O in product gas[h] (mol/mol)	1.247	0.850	0.855	1.387	0.936	0.015	0.103	2.241	0.177
N_2 (mol/mol)	1.153	0.740	0.6	0.002	0.005	1.625	1.048	0.004	1.598
Heat of reaction (1–θ) h_r (kJ/mol)	–71.16	–12.26	–28.94	–55.89	–19.19	–70.17	–37.65	–44.99	–108.91
H_r [=(1–θ)h_r – w × 43.99] (kJ/mol)	–131.09	–65.58	–76.48	–124.57	–71.44	–74.26	–49.61	–88.83	–122.28
Apparent heat of products H_g (kJ/mol)	–116.05	–26.36	–34.49	–73.19	–55.82	–58.38	–56.54	–71.02	–33.34
Heat loss $H_l = H_r - H_g$ (kJ/mol)	–15.04	–39.22	–41.99	–51.38	–15.62	–15.88	**6.93**	–17.81	–88.94
T_{ad} (°C)	1029	661	843	1269	960	808	677	780	1178
H_2O generation from H_2 combustion[i] × (mol/mol)	0.000	0.000	0.000	0.000	0.000	0.000	0.000	0.014	0.000
H_2O-consumed in shift reaction y (mol/mol)	0.237	0.299	0.303	0.333	0.282	0.014	0.122	0.219	0.172
Cold gas efficiency[i] (–)	0.546	0.774	0.476	0.393	0.389	0.789	0.613	0.330	0.679
Carbon conversion[j] (1–θ) (–)	0.680	0.799	0.538	0.511	0.430	0.929	0.687	0.424	0.881
Reference	Furusawa et al. 1985	Hebden and Stroud 1981	Hebden and Stroud 1981	Haynes et al. 1977	Haynes et al. 1977	Hedben and Stroud 1981	Hedben and Stroud 1981	Bodle and Huebler 1981	Hedben and Stroud 1981

[a] Wet base
[b] Dry base
[c] kg per kg-dry coal
[d] H_2S and NH_3 are not involved in the gas composition
[e] Reported value
[f] The concentration of N_2 and Ar are not involved in the theoretical calculations
[g] Parameters in the equation $CH_mO_n + \alpha O_2 + \beta H_2O \rightarrow \gamma H_2 + \delta CO + \varepsilon CO_2 + \eta CH_4$
[h] Moisture + steam – β
[i] Produced by combustion with excess of O_2
[j] Calculated based on the stoichiometric theory
[k] Designed values based on the data in their development stage

TABLE 14.8
Analysis of the Data of Spouted Flow Process

Entry		1	2	3	4	5	6	7	8	9	10
Process	Name	NKK	NKK	NKK	NKK	NKK	NKK	NKK	NKK	NKK	NKK
	Capacity	3 tons/day	←	←	←	←	←	←	←	←	←
	Type	–	–	–	–	–	–	–	–	–	–
		Spouted-flow	Spouted-flow	Spouted-flow	Spouted-flow	Spouted-flow	Spouted-flow	Spouted-flow	Spouted-flow	Spouted-flow	Spouted-flow
Coal	Coal rank	Bituminous	Bituminous	Bituminous	Bituminous	Bituminous	Bituminous	Bituminous	Bituminous	Bituminous	Bituminous
	Feed rate (kg/h)[a]	123	112	112	30	30	50	50	70	80	100
	Calorific value (kcal/kg)[b]	7018	7018	7018	7018	7018	7018	7018	7018	7018	7018
Proximate analysis	Moisture (%)[a]	3.5	3.5	3.5	3.5	3.5	3.5	3.5	3.5	3.5	3.5
	Ash (%)[b]	9.3	9.3	9.3	9.3	9.3	9.3	9.3	9.3	9.3	9.3
	Volatile matter (%)[b]	33.5	33.5	33.5	33.5	33.5	33.5	33.5	33.5	33.5	33.5
	Fixed carbon (%)[b]	57.2	57.2	57.2	57.2	57.2	57.2	57.2	57.2	57.2	57.2
Ultimate analysis[b]	C (%)	75.4	75.4	75.4	74.6	74.6	74.6	74.6	74.6	74.6	74.6
	H (%)	4.7	4.7	4.7	4.6	4.6	4.6	4.6	4.6	4.6	4.6
	O (%)	9.4	9.4	9.4	9.3	9.3	9.3	9.3	9.3	9.3	9.3
	N (%)	1.7	1.7	1.7	1.7	1.7	1.7	1.7	1.7	1.7	1.7
	S (%)	0.5	0.5	0.5	0.5	0.5	0.5	0.5	0.5	0.5	0.5
	Ash (%)	9.3	9.3	9.3	9.3	9.3	9.3	9.3	9.3	9.3	9.3
Elemental ratio	*m* (H/C)	0.748	0.748	0.748	0.740	0.740	0.740	0.740	0.740	0.740	0.740
	n (O/C)	0.094	0.094	0.094	0.093	0.093	0.093	0.093	0.093	0.093	0.093
Gasifying agent	O_2 (kg/kg)[c]	0.782	0.753	0.740	0.938	0.740	0.474	0.622	0.486	0.463	0.459
	Air (O_2 content) (kg/kg)[c]	–	–	–	–	–	–	–	–	–	–
	Steam (kg/kg)[c]	0.337	0.259	0.426	1.055	1.249	0.266	0.525	0.274	0.260	0.258
	Total O_2 (kg/kg)[c]	0.782	0.753	0.740	0.938	0.740	0.474	0.622	0.486	0.463	0.459
	O_2 (%)	56.6	62.1	49.5	33.4	25.0	50.1	40.0	50.0	50.1	50.1
Conditions	Pressure (kgf/cm^2)	Atmospheric	Atmospheric	Atmospheric	–	Atmospheric	–	–	Atmospheric	–	–
	Temperature top (°C)	1180	1175	1175	1120	1129	1140	1170	1150	1140	1140
	Bottom (°C)	1600?	1600?	1600?	–	–	–	–	–	–	–

(Continued)

TABLE 14.8 (*Continued*)
Analysis of the Data of Spouted Flow Process

Entry		1	2	3	4	5	6	7	8	9	10
Product gas composition[d]	H_2 (%)	33.1	32.6	36.2	29.4	23.2	22.6	23.2	25.7	27.3	27.1
	CO (%)	48.3	44.8	40.1	22.4	21.3	22.9	23.3	28.5	26.9	27.7
	CO_2 (%)	14.8	17.9	19.3	18.4	18.1	15.6	16.8	13.5	14.8	15.9
	CH_4 (%)	1.3	2.4	2.2	0.7	0.7	3.6	4.1	2.6	2.8	2.8
	C_2H_4 (%)	–	–	–	–	–	–	–	–	–	–
	C_2H_6 (%)	–	–	–	–	–	–	–	–	–	–
	N_2 + Ar (%)	2.5	2.3	2.2	29.1	36.7	35.3	32.6	29.7	28.2	26.5
Gas yield in volume (Nm^3/kg)[c]		1.821	1.858	1.973	2.769	2.621	1.629	1.741	1.735	1.662	1.480
Heating value ($kcal/Nm^3$)		2601	2584	2532	1645	1422	1728	1806	1897	1917	1935
Cold gas efficiency[e] (–)		0.675	0.684	0.712	0.649	0.531	0.401	0.448	0.469	0.454	0.408
Carbon conversion[e] (–)		0.917	0.958	0.963	0.850	0.750	0.480	0.550	0.560	0.550	0.530
Coal	Heating value (kJ/mol)	–467.7	–467.7	–467.7	–472.6	–472.6	–472.6	–472.6	–472.6	–472.6	–472.6
Oxygen	μ (mol/mol)	1.140	1.140	1.140	1.139	1.139	1.139	1.139	1.139	1.139	1.139
	O_2 supplied A (mol/mol)	0.389	0.375	0.368	0.472	0.372	0.238	0.313	0.244	0.233	0.231
	χ (=α/μ) (–)	0.411	0.413	0.395	0.421	0.481	0.374	0.374	0.363	0.355	0.375
	O_{st} [=0.5(1−n)] (mol/mol)	0.453	0.453	0.453	0.454	0.454	0.454	0.454	0.454	0.454	0.454
	O_{ex} (=α–O_{st}) (mol/mol)	–0.016	0.018	0.003	0.027	0.095	–0.028	–0.028	–0.041	–0.050	–0.027
Water[e]	w (mol/mol)	0.298	0.229	0.377	0.943	1.116	0.238	0.469	0.245	0.232	0.231
	w + moisture (mol/mol)	0.330	0.261	0.409	0.975	1.149	0.270	0.502	0.277	0.265	0.263
Product gas composition[f]	p (H_2) (%)	33.9	33.4	37.0	41.5	36.7	34.9	34.4	36.6	38.0	36.9
	q (CO) (%)	49.5	45.9	41.0	31.6	33.6	35.4	34.6	40.5	37.5	37.7
	r (CO_2) (%)	15.2	18.3	19.7	26.0	28.6	24.1	24.9	19.2	20.6	21.6
	s (CH_4) (%)	1.3	2.5	2.2	1.0	1.1	5.6	6.1	3.7	3.9	3.8
Stoichiometric parameter[g]	α (O_2) (mol/mol)	0.469	0.471	0.450	0.480	0.548	0.426	0.426	0.413	0.404	0.427
	β (H_2O) (mol/mol)	0.179	0.202	0.284	0.372	0.245	0.338	0.340	0.324	0.369	0.335

(Continued)

TABLE 14.8 (*Continued*)
Analysis of the Data of Spouted Flow Process

Entry	1	2	3	4	5	6	7	8	9	10
γ (H_2) (mol/mol)	0.514	0.501	0.588	0.708	0.580	0.536	0.524	0.577	0.613	0.585
δ (CO) (mol/mol)	0.750	0.688	0.652	0.539	0.531	0.544	0.527	0.639	0.605	0.597
ε (CO_2) (mol/mol)	0.230	0.274	0.313	0.444	0.452	0.370	0.380	0.303	0.332	0.342
η (CH_4) (mol/mol)	0.020	0.037	0.035	0.017	0.017	0.086	0.093	0.058	0.063	0.060
Heating value of gas (kJ/mol)	−377.2	−371.1	−384.0	−370.3	−331.4	−384.0	−382.0	−397.7	−402.8	−389.9
H_2O in product gas[h] (mol/mol)	0.182	0.100	0.177	0.609	0.983	0.081	0.252	0.101	0.052	0.082
N_2 (mol/mol)	0.039	0.035	0.036	0.701	0.915	0.838	0.738	0.666	0.634	0.571
Heat of reaction $(1-\theta)\ h_r$ (kJ/mol)	−75.52	−77.38	−68.98	−101.29	−96.31	−49.95	−67.06	−41.03	−40.73	−41.58
$H_r\ [=(1-\theta)h_r - w \times 43.99]$ (kJ/mol)	−88.61	−87.44	−85.53	−142.71	−145.35	−60.39	−87.67	−51.79	−50.94	−55.31
Apparent heat of products H_g (kJ/mol)	−64.08	−59.56	−67.11	−117.87	−125.39	−78.70	−93.01	−72.99	−71.57	−68.62
Heat loss $H_l = H_r - H_g$ (kJ/mol)	−24.53	−27.88	−18.42	−24.84	−19.96	**18.31**	**5.34**	**21.20**	**20.63**	**13.31**
T_{ad} (°C)	1570	1641	1449	1325	1284	906	1111	854	849	947
H_2O generation from H_2 combustion[i] × (mol/mol)	0.000	0.008	0.002	0.015	0.035	0.000	0.000	0.000	0.000	0.000
H_2O-consumed in shift reaction y (mol/mol)	0.191	0.197	0.253	0.398	0.213	0.207	0.279	0.164	0.192	0.185
Cold gas efficiency[j] (–)	0.669	0.684	0.671	0.770	0.476	0.454	0.594	0.456	0.491	0.446
Carbon conversion[j] $(1-\theta)$ (–)	0.829	0.796	0.818	0.983	0.679	0.559	0.734	0.542	0.577	0.541
Reference	Endo et al. 1985	Endo et al. 1985	Endo et al. 1985	Bodle and Huebler 1981	Bodle and Huebler 1981	Bodle and Huebler 1981	Bodle and Huebler 1981	Bodle and Huebler 1981	Bodle and Huebler 1981	Bodle and Huebler 1981

[a] Wet base
[b] Dry base
[c] kg per kg-dry coal
[d] H_2S and NH_3 are not involved in the gas composition
[e] Reported value
[f] The concentration of N_2 and Ar are not involved in the theoretical calculations
[g] Parameters in the equation $CH_mO_n + \alpha O_2 + \beta H_2O \rightarrow \gamma H_2 + \delta CO + \varepsilon CO_2 + \eta CH_4$
[h] Moisture + steam − β
[i] Produced by combustion with excess of O_2
[j] Calculated based on the stoichiometric theory

TABLE 14.9
Analysis of the Data of Entrained Flow Process (1)

Entry		1	2	3	4	5	6	7	8
Process	Name	Koppers–Totzek	Koppers–Totzek	Koppers–Totzek	Koppers–Totzek	Koppers–Totzek	Koppers–Totzek	Shell–Koppers	Shell–Koppers
	Capacity				Nihon Suiso	Nihon Suiso	Nihon Suiso		
	Type	Entrained-flow	Entrained-flow	Entrained-flow	Entrained-flow	Entrained-flow	Entrained-flow	Entrained-flow	Entrained-flow
Coal	Coal rank	Bituminous	Bituminous	Bituminous	Lignite	Lignite	Lignite	Bituminous	Sub-bituminous
	Feed rate (kg/h)[a]	–	–	2900	3831	4069	2428	–	–
	Calorific value (kcal/kg)[b]	6455	7196	6026	4739	7963	5099	7292	6924
Proximate analysis	Moisture (%)[a]	2.0	2.0	1.0	8.5	8.2	6.1	10.0	35.0
	Ash (%)[b]	19.5	14.0	20.4	27.4	24.1	18.9	10.1	9.2
	Volatile matter (%)[b]	–	–	–	38.0	38.7	38.8	–	–
	Fixed carbon (%)[b]	–	–	–	3.5	37.2	42.3	–	–
Ultimate analysis[b]	C (%)	63.2	71.3	63.5	52.5	55.4	59.3	73.9	68.6
	H (%)	4.4	5.0	3.6	4.1	4.3	4.5	4.8	5.4
	O (%)	6.9	7.2	10.7	14.8	15.3	15.5	8.9	15.2
	N (%)	1.0	1.4	1.4	0.9	0.8	0.6	1.2	0.9
	S (%)	5.0	1.1	0.4	0.2	0.2	0.1	1.2	0.6
	Ash (%)	19.5	14.0	20.4	27.4	24.1	18.9	10.1	9.2
Elemental ratio	*m* (H/C)	0.835	0.842	0.680	0.937	0.931	0.911	0.779	0.945
	n (O/C)	0.082	0.076	0.126	0.211	0.207	0.196	0.090	0.166
Gasifying agent	O_2 (kg/kg)[c]	0.704	0.817	0.716	0.652	0.664	0.702	0.649	0.817
	Air (O_2 content) (kg/kg)[c]	–	–	–	–	–	–	–	–
	Steam (kg/kg)[c]	0.251	0.272	0.100	0.291	0.245	0.395	–	–
	Total O_2 (kg/kg)[c]	0.704	0.817	0.716	0.652	0.664	0.702	0.649	0.817
	O_2 (%)	61.2	62.9	80.1	55.8	60.4	50.0	100	100
Conditions	Pressure (kgf/cm^2)	Atmospheric	Atmospheric	–	–	Atmospheric	–	~30	–
	Temperature top (°C)	1816–1927	1816–1927	–	1329	1330	1411	1400–1590	1400–1590
	Bottom (°C)	–	–	–	–	–	–	1790–1980	1790–1980
Product gas composition[d]	H_2 (%)	35.3	35.5	27.1	31.0	32.8	28.0	31.5	30.2
	CO (%)	56.5	56.1	64.4	56.0	53.9	54.4	66.0	66.4
	CO_2 (%)	7.2	7.2	7.0	11.4	12.6	13.8	1.5	2.5

(*Continued*)

TABLE 14.9 (*Continued*)
Analysis of the Data of Entrained Flow Process (1)

Entry		1	2	3	4	5	6	7	8
	CH_4 (%)	–	–	0.1	0.1	0.1	0.1	0.4	0.4
	C_2H_4 (%)	–	–	–	–	–	–	–	–
	C_2H_6 (%)	–	–	–	–	–	–	–	–
	N_2 + Ar (%)	1.0	1.2	1.4	1.5	1.1	3.6	0.6	0.5
Gas yield in volume (Nm^3/kg)[c]		1.686	1.910	1.492	1.402	1.446	1.462	1.825	1.722
Heating value (kcal/Nm^3)		2794	2788	2793	2657	2648	2516	3004	2976
Cold gas efficiency[d] (–)		0.758	0.770	0.778	0.786	0.773	0.708	0.77(L)	0.77(L)
Carbon conversion[d] (–)		–	–	–	0.957	0.932	0.760	–	
Coal	Heating value (kJ/mol)	–513.3	–507.0	–476.9	–453.4	–429.1	–432.1	–495.7	–507.0
Oxygen	μ (mol/mol)	1.168	1.173	1.107	1.129	1.129	1.130	1.150	1.153
	O_2 supplied *A* (mol/mol)	0.418	0.430	0.423	0.466	0.449	0.444	0.329	0.447
	χ (=α/μ) (–)	0.383	0.383	0.419	0.426	0.422	0.463	0.365	0.385
	O_{st} [=0.5(1–*n*)] (mol/mol)	0.459	0.462	0.437	0.395	0.397	0.402	0.455	0.417
	O_{ex} (=α–O_{st}) (mol/mol)	–0.012	–0.013	0.027	0.087	0.080	0.121	–0.035	0.027
Water[e]	w (mol/mol)	0.265	0.254	0.105	0.370	0.295	0.444	–	–
	w + moisture (mol/mol)	0.286	0.273	0.116	0.487	0.402	0.517	0.100	0.523
Product gas composition[f]	*p* (H_2) (%)	35.7	35.9	27.5	31.5	33.0	29.1	31.7	30.4
	q (CO) (%)	57.1	56.8	65.3	56.9	54.2	56.5	66.4	66.7
	r (CO_2) (%)	7.3	7.3	7.1	11.6	12.7	14.3	1.5	2.5
	s (CH_4) (%)	0.0	0.0	0.1	0.1	0.1	0.1	0.4	0.4
Stoichiometric parameter[g]	α (O_2) (mol/mol)	0.447	0.449	0.464	0.481	0.476	0.523	0.420	0.444
	β (H_2O) (mol/mol)	0.137	0.139	0.042	–0.006	0.030	–0.042	0.086	–0.024
	γ (H_2) (mol/mol)	0.554	0.560	0.379	0.459	0.493	0.410	0.464	0.437
	δ (CO) (mol/mol)	0.887	0.886	0.901	0.829	0.809	0.797	0.972	0.958
	ε (CO_2) (mol/mol)	0.113	0.114	0.098	0.169	0.190	0.202	0.022	0.036
	η (CH_4) (mol/mol)	0.000	0.000	0.001	0.001	0.001	0.001	0.006	0.006

(Continued)

TABLE 14.9 (*Continued*)
Analysis of the Data of Entrained Flow Process (1)

Entry	1	2	3	4	5	6	7	8
Heating value of gas (kJ/mol)	−409.7	−411.1	−364.5	−367.0	−371.1	−343.9	−413.3	−401.7
H_2O in product gas[h] (mol/mol)	0.158	0.140	0.078	0.493	0.373	0.553	0.033	0.547
N_2 (mol/mol)	0.016	0.019	0.020	0.022	0.017	0.053	0.009	0.007
Heat of reaction $(1-\theta)\ h_r$ (kJ/mol)	−97.06	−92.68	−103.05	−84.54	−331.79	−75.37	−65.13	−106.20
$H_r\ [=(1-\theta)h_r - w \times 43.99]$ (kJ/mol)	−108.69	−103.85	−107.67	−100.77	−344.74	−94.88	−65.13	−106.20
Apparent heat of products H_g (kJ/mol)	−102.14	−102.66	−98.80	−90.49	−84.95	−95.77	−59.04	−94.89
Heat loss $E_1 = H_r - H_g$ (kJ/mol)	−6.55	−1.19	−8.87	−10.28	−259.79	0.89	−6.09	−11.31
T_{ad} (°C)	1936	1839	–	1464	–	1399	1536	1557
H_2O generation from H_2 combustion[i] × (mol/mol)	0.000	0.000	0.013	0.054	0.048	0.064	0.000	0.017
H_2O-consumed in shift reaction y (mol/mol)	0.105	0.109	0.053	0.049	0.076	0.031	0.016	−0.001
Cold gas efficiency[h](–)	0.743	0.777	0.696	0.784	0.815	0.675	0.653	0.792
Carbon conversion[j] $(1-\theta)$ (–)	0.931	0.958	0.912	0.969	0.943	0.849	0.783	nearly 1.00
Reference	Bodle and Huebler 1981	Bodle and Huebler 1981	Hebden and Stroud 1981	Kimura and Fujii 1976	Kimura and Fujii 1976	Kimura and Fujii 1976	Bodle and Huebler 1981	Bodle and Huebler 1981

[a] Wet base
[b] Dry base
[c] kg per kg-dry coal
[d] H_2S and NH_3 are not involved in the gas composition
[e] Reported value
[f] The concentration of N_2 and Ar are not involved in the theoretical calculations
[g] Parameters in the equation $CH_mO_n + \alpha O_2 + \beta H_2O \rightarrow \gamma H_2 + \delta CO + \varepsilon CO_2 + \eta CH_4$
[h] Moisture + steam – β
[i] Produced by combustion with excess of O_2
[j] Calculated based on the stoichiometric theory

TABLE 14.10
Analysis of the Data of Entrained-Flow Process (2)

Entry		1	2	3	4	5	6	7	8[k]	9[k]	10[k]	11[k]
Process	Name	UBE	TEXACO	TEXACO	TEXACO	TEXACO	HITACHI	HITACHI	Babcock & Wilcox	Mountain Fuel	Combustion Engineering	The Japan Institute of Energy
	Capacity	500 tons/day	–	–	–	–	0.5 tons/day	0.5 tons/day	1000 tons/day	0.5 tons/day	5 tons/day	–
	Type	Entrained-flow	Entrained-flow	Entrained-flow	Entrained-flow	Entrained-flow	Two-stage entrained-flow	Two-stage entrained-flow	Entrained-flow	Entrained-flow	Two-stage entrained-flow	Two-stage entrained-flow
Coal	Coal rank	Bituminous	Liquefaction residue	Liquefaction residue	Liquefaction residue	Liquefaction residue	Sub-bituminous	Sub-bituminous	Bituminous	Bituminous	Bituminous	Bituminous
	Feed rate (kg/h)[a]	20800	296	342	309	452	20.2	34.0	37.4t	20.5	4472	2083
	Calorific value (kcal/ kg)[b]	6910	7452	7387	8078	7960	6610	6776	7321	7071	7673	7018
Proximate analysis	Moisture (%)[a]	8.0	0.0	0.0	0.0	0.0	5.3	3.5	10.0	1.0	2.4	0.0
	Ash (%)[b]	15.1	–	–	–	–	15.2	14.5	8.3	10.6	8.6	9.1
	Volatile matter (%)[b]	47.2	–	–	–	–	43.1	27.9	–	–	39.0	33.5
	Fixed carbon (%)[b]	37.8	–	–	–	–	41.7	57.6	–	–	52.4	57.5
Ultimate analysis[b]	C (%)	66.8	71.2	71.0	78.4	78.5	65.6	69.7	76.4	70.1	77.5	74.7
	H (%)	5.0	5.4	5.4	5.8	5.5	5.2	4.0	4.8	5.6	5.3	4.6
	O (%)	7.3	2.0	2.6	3.7	4.6	12.6	9.7	6.7	11.7	4.6	9.4
	N (%)	1.7	0.8	0.8	0.9	0.8	1.2	1.6	1.0	1.4	1.4	1.7
	S (%)	4.2	1.7	1.6	0.0	0.0	0.3	0.5	2.7	0.6	2.7	0.4
	Ash (%)	15.0	18.6	18.5	11.1	10.5	15.2	14.5	8.3	10.6	8.6	9.1
Elemental ratio	*m* (H/C)	0.898	0.910	0.913	0.888	0.841	0.951	0.689	0.754	0.959	0.821	0.739
	n (O/C)	0.082	0.021	0.027	0.035	0.044	0.144	0.104	0.066	0.125	0.045	0.094
Gasifying agent	O_2 (kg/kg)[c]	0.965	0.812	0.801	0.899	0.856	0.842	0.872	0.978	0.831	–	0.821
	Air (O_2 content) (kg/ kg)[c]	–	–	–	–	–	–	–	–	–	5.04 (1.164)	–
	Steam (kg/kg)[c]	–	0.310	0.240	0.318	0.274	–	–	0.200	0.392	–	0.525
	Total O_2 (kg/kg)[c]	0.965	0.812	0.801	0.899	0.856	0.842	0.872	0.978	0.831	1.164	0.821
	O_2 (%)	51.4	59.6	65.3	61.4	63.8	100.0	100.0	73.4	54.4	20.9	46.8
Conditions	Pressure (kgf/cm²)	39—40.5	81.70	81.70	81.70	81.70	4	5	22.2ata	11.20	1.10	30
	Temperature top (°C)	1380–1450	1500	1500	1500	1500	1200–1600	1648/ 1993	870–980	1566	930	900
	Bottom (°C)	–	–	–	–	–	>1600	–	1650–1855	–	1760	1400

(Continued)

TABLE 14.10 (*Continued*)
Analysis of the Data of Entrained-Flow Process (2)

Entry		1	2	3	4	5	6	7	8[k]	9[k]	10[k]	11[k]
Product gas composition[d]	H_2 (%)	36.3	39.4	38.5	38.0	37.8	32.3	23.0	30.4	35.5	10.7	31.9
	CO (%)	42.5	54.9	57.9	57.0	57.6	55.1	51.0	59.6	46.7	24.5	52.4
	CO_2 (%)	20.9	5.5	3.4	4.8	4.1	12.4	9.2	7.5	14.9	4.1	14.1
	CH_4 (%)	–	0.1	0.1	0.1	0.2	0.2	0.2	–	1.2	–	0.2
	C_2H_4 (%)	–	–	–	–	–	–	–	–	–	–	–
	C_2H_6 (%)	–	–	–	–	–	–	–	–	–	–	–
	N_2+Ar(%)	0.3	0.1	0.1	0.1	0.3	–	16.6	2.5	1.7	60.7	1.4
Gas yield in volume (Nm^3/kg)[c]		1.985	2.154	2.114	2.337	2.336	1.737	1.893	2.075	2.076	5.086	2.012
Heating value (kcal/Nm^3)		2399	2880	2944	2901	2923	2679	2270	2738	2616	1071	2584
Cold gas efficiency[d] (–)		0.689	0.834	0.843	0.840	0.858	0.704	0.618	0.776	0.750	0.719	0.752
Carbon conversion[d] (–)		0.985	0.981	0.981	0.989	0.987	0.930	0.879	–	0.940	0.975	0.985
Coal	Heating value (kJ/ mol)	–519.6	–525.8	–522.9	–517.5	–509.5	–506.2	–488.6	–481.5	–506.6	–497.4	–471.8
Oxygen	μ (mol/mol)	1.184	1.217	1.215	1.205	1.188	1.166	1.120	1.156	1.177	1.183	1.138
Oxygen	O_2 supplied *A* (mol/ mol)	0.542	0.428	0.423	0.430	0.409	0.481	0.469	0.480	0.445	0.563	0.412
	χ (=α/μ) (–)	0.445	0.357	0.351	0.360	0.341	0.441	0.447	0.420	0.412	0.479	0.438
	O_{st} [=0.5(1−*n*)] (mol/ mol)	0.459	0.490	0.487	0.483	0.478	0.428	0.448	0.467	0.438	0.478	0.453
	O_{ex} (=α–O_{st}) (mol/ mol)	0.118	–0.055	–0.060	–0.217	–0.067	0.086	0.053	0.018	0.048	0.090	0.046
Water[e]	w (mol/mol)	0.666	0.290	0.225	0.270	0.233	–	–	0.174	0.373	–	0.469
	w + moisture (mol/ mol)	1.000 (slurry)	0.290	0.225	0.270	0.233	0.037	0.023	0.271	0.382	0.007	0.469
Product gas composition[f]	*p* (H_2) (%)	36.6	39.4	38.5	38.0	37.9	32.3	27.6	31.2	36.1	27.2	32.4
	q (CO) (%)	42.8	55.0	58.0	57.1	57.8	55.1	61.2	61.1	47.5	62.3	53.1
	r (CO_2) (%)	20.6	5.5	3.4	4.8	4.1	12.4	11.0	7.7	15.2	10.4	14.3
	s (CH_4) (%)	0.0	0.1	0.1	0.1	0.2	0.2	0.2	0.0	1.2	0.0	0.2

(*Continued*)

TABLE 14.10 (*Continued*)
Analysis of the Data of Entrained-Flow Process (2)

Entry		1	2	3	4	5	6	7	8[k]	9[k]	10[k]	11[k]
Stoichiometric parameter[g]	α (O_2) (mol/mol)	0.557	0.435	0.427	0.434	0.411	0.514	0.501	0.485	0.485	0.567	0.499
	β (H_2O) (mol/mol)	0.128	0.198	0.173	0.172	0.196	0.007	0.042	0.077	0.123	–0.036	0.116
	γ (H_2) (mol/mol)	0.577	0.650	0.626	0.613	0.610	0.477	0.381	0.453	0.565	0.374	0.479
	δ (CO) (mol/mol)	0.675	0.908	0.943	0.921	0.931	0.814	0.845	0.888	0.743	0.857	0.786
	ε (CO_2) (mol/mol)	0.325	0.091	0.055	0.077	0.066	0.183	0.152	0.112	0.238	0.143	0.212
	η (CH_4) (mol/mol)	0.000	0.002	0.002	0.002	0.003	0.003	0.003	0.000	0.019	0.000	0.003
Heating value of gas (kJ/mol)		–356.2	–444.9	–447.9	–438.0	–440.8	–369.7	–351.0	–381.1	–389.0	–349.7	–362.3
H_2O in product gas[h] (mol/mol)		0.875	0.095	0.054	0.100	0.038	0.030	–0.016	0.195	0.269	0.043	0.373
N_2 (mol/mol)		0.005	0.002	0.002	0.002	0.005	–	0.275	0.037	0.027	2.122	0.021
Heat of reaction (1–θ) h_r (kJ/mol)		–159.75	–80.43	–74.98	–79.81	–69.12	–128.51	–129.29	–100.08	–108.84	–147.36	–91.09
H_r [=(1–θ)h_r – w × 43.99) (kJ/mol)		–159.75	–93.18	–84.88	–91.69	–79.34	–128.51	–129.29	–107.75	–125.21	–147.36	–111.67
Apparent heat of products H_g (kJ/ mol)		–73.75	–85.35	–80.79	–83.74	–79.79	–59.00	–87.24	–46.50	–100.47	–102.54	–51.50
Heat loss $H_l = H_r - H_g$ (kJ/mol)		–86.00	–7.83	–4.09	–7.95	0.45	–69.51	–42.05	–61.25	–24.74	–44.82	–60.17
T_{ad} (°C)		2684	1630	1572	1635	1492	2691	2527	1915	1908	1302	1780
H_2O generation from H_2 combustion[i] × (mol/mol)		0.071	0.000	0.000	0.000	0.000	0.051	0.025	0.010	0.028	0.052	0.021
H_2O-consumed in shift reaction y (mol/mol)		0.158	0.089	0.055	0.076	0.066	0.062	0.069	0.085	0.159	0.015	0.120
Cold gas efficiency[i] (–)		0.667	0.833	0.849	0.838	0.861	0.683	0.672	0.783	0.704	0.698	0.634
Carbon conversion[j] (1–θ) (–)		0.973	0.983	0.991	0.991	0.995	0.936	0.936	0.990	0.918	0.993	0.825
Reference		Sueyama et al. 1985	Furusawa et al. 1985	Furusawa et al. 1985	Furusawa et al. 1985	Furusawa et al. 1985	Tanaka et al. 1988	Tanaka et al. 1988	Hebden and Stroud 1981	Bodle and Huebler	Hebden and Stroud 1981	Kimura, Fujii 1976

[a] Wet Base
[b] Dry Base
[c] kg per kg-dry coal
[d] H_2S and NH_3 are not involved in the gas composition
[e] Reported value
[f] The concentrations of N_2 and Ar are not involved in the theoretical calculations
[g] Parameters in the equation of $CH_mO_n + \alpha O_2 + \beta H_2O \rightarrow \gamma H_2 + \delta CO + \varepsilon CO_2 + \eta CH_4$
[h] Moisture + steam – β
[i] Produced by combustion with excess of O_2
[j] Calculated based on the stoichiometric theory
[k] Designed values based on the data in their development stage

TABLE 14.11
Analysis of the Data of Molten Bath Process

Entry		1	2	3	4	5	6[k]	7[k]
Process	Name	Sumitomo Metal	Sumitomo Metal	Sumitomo Metal	Sumitomo Metal	Sumitomo Metal	Kellogg	Kellogg
	Capacity	60 tons/day	60 tons/day	60 tons/day	60 tons/day	60 tons/day	–	–
	Type	Molten bath	Molten bath	Molten bath	Molten bath	Molten bath	Molten bath	Molten bath
Coal	Coal rank	Bituminous	Bituminous	Oil pitch	Bituminous	Bituminous	Sub-bituminous	Sub-bituminous
	Feed rate (kg/h)[a]	1.5~3.0t	1.5~3.0t	1.5~3.0t	1.5~3.0t	1.5~3.0t	–	–
	Calorific value (kcal/kg)[b]	7423	7071	8602	7423	7423	6975	6975
Proximate analysis	Moisture (%)[a]	2.2	2.2	0.1	2.2	2.2	7.25	7.25
	Ash (%)[b]	8.2	15.1	0.5	8.2	8.2	16.6	16.6
	Volatile matter (%)[b]	35.2	39.6	40.4	35.2	35.2	–	–
	Fixed carbon (%)[b]	56.6	45.3	59.1	56.6	56.6	–	–
Ultimate analysis[b]	C (%)	77.4	70.8	85.3	77.4	77.4	65.9	65.9
	H (%)	4.8	5.3	6.1	4.8	4.8	5.3	5.3
	O (%)	7.3	7.1	1.6	7.3	7.3	6.1	6.1
	N (%)	1.7	1.4	1.0	1.7	1.7	1.5	1.5
	S (%)	0.7	0.3	5.5	0.7	0.7	4.6	4.6
	Ash (%)	8.2	15.1	0.5	8.2	8.2	16.6	16.6
Elemental ratio	*m* (H/C)	0.744	0.898	0.858	0.744	0.744	0.965	0.965
	n (O/C)	0.071	0.075	0.014	0.071	0.071	0.069	0.069
Gasifying agent	O_2 (kg/kg)[c]	0.971	0.919	1.114	0.922	0.840	0.453	–
	Air (O_2 content) (kg/kg)[c]	–	–	–	–	–	–	3.22 (0.745)
	Steam (kg/kg)[c]	–	–	–	0.051	0.153	1.941	–
	Total O_2 (kg/kg)[c]	0.971	0.919	1.114	0.922	0.840	0.453	0.745
	O_2 (%)	100.0	100.0	100.0	91.1	75.6	11.6	20.9
Conditions	Pressure (kgf/cm²)	–	–		–	–	84.50	15.10
	Temperature top (°C)	1400–1600	1400–1600	1400–1600	1400–1600	1400–1600	927–982	927
	Bottom (°C)	–	–	–	–	–	–	–

(Continued)

TABLE 14.11 (*Continued*)
Analysis of the Data of Molten Bath Process

Entry		1	2	3	4	5	6[k]	7[k]
Product gas composition[d]	H_2 (%)	25.0	28.5	29.5	27.5	31.0	26.7	13.9
	CO (%)	66.0	62.5	63.5	63.5	60.0	30.0	29.1
	CO_2 (%)	4.5	4.5	3.0	4.5	4.5	23.6	2.6
	CH_4 (%)	–	–	–	–	–	18.9	1.2
	C_2H_4 (%)	–	–	–	–	–	–	–
	C_2H_6 (%)	–	–	–	–	–	–	–
	N_2+Ar (%)	4.5	4.5	4.0	4.5	4.5	0.8	53.2
Gas yield in volume (Nm^3/kg)[c]		2.071	1.968	2.427	2.122	2.234	1.660	3.744
Heating value (kcal/Nm^3)		2768	2768	2829	2768	2769	3522	1422
Cold gas efficiency[d] (–)		0.727	0.722	0.750	0.743	0.764	0.873	0.764
Carbon conversion[d] (–)		> 0.98	> 0.98	> 0.98	> 0.98	> 0.98	0.976	0.977
Coal	Heating value (kJ/mol)	–481.9	–501.6	–506.6	–481.9	–481.9	–531.7	–531.7
Oxygen	μ (mol/mol)	1.151	1.187	1.208	1.151	1.151	1.207	1.207
	O_2 supplied *A* (mol/mol)	0.470	0.487	0.490	0.447	0.407	0.258	0.424
	χ (=α/μ) (–)	0.439	0.428	0.421	0.418	0.389	0.243	0.398
	O_{st} [=0.5(1−*n*)] (mol/ mol)	0.465	0.463	0.493	0.465	0.465	0.466	0.466
	O_{ex} (=α–O_{st}) (mol/mol)	0.041	0.046	0.015	0.017	–0.020	–0.173	0.014
Water[e]	w (mol/mol)	–	–	–	0.044	0.132	1.964	–
	w + moisture (mol/mol)		0.021	0.001	0.063	0.151	2.043	0.938
Product gas composition[f]	*p* (H_2) (%)	26.2	29.8	30.7	28.8	32.5	26.9	29.7
	q (CO) (%)	69.1	65.4	66.1	66.5	62.8	30.2	62.2
	r (CO_2) (%)	4.7	4.7	3.1	4.7	4.7	23.8	5.6
	s (CH_4) (%)	0.0	0.0	0.0	0.0	0.0	19.1	2.6
Stoichiometric parameter[g]	α (O_2) (mol/mol)	0.505	0.508	0.508	0.481	0.445	0.293	0.480
	β (H_2O) (mol/mol)	–0.017	–0.024	0.015	0.032	0.109	0.408	0.013

(*Continued*)

TABLE 14.11 (*Continued*)
Analysis of the Data of Molten Bath Process

Entry	1	2	3	4	5	6[k]	7[k]
γ (H_2) (mol/mol)	0.355	0.425	0.444	0.404	0.481	0.368	0.422
δ (CO) (mol/mol)	0.936	0.933	0.955	0.934	0.930	0.413	0.884
ε (CO_2) (mol/mol)	0.064	0.067	0.045	0.066	0.070	0.326	0.080
η (CH_4) (mol/mol)	0.000	0.000	0.000	0.000	0.000	0.261	0.037
Heating value of gas (kJ/mol)	–366.6	–385.8	–397.5	–380.1	–401.0	–454.8	–404.0
H_2O in product gas[h] (mol/mol)	0.035	0.044	–0.013	0.033	0.051	1.684	0.927
N_2 (mol/mol)	0.064	0.067	0.060	0.066	0.070	0.011	1.617
Heat of reaction $(1-\theta)\ h_r$ (kJ/mol)	–107.91	–111.92	–106.09	–95.20	–74.69	–68.55	–113.46
H_r [$=(1-\theta)h_r - w \times 43.99$) (kJ/mol)	–107.91	–111.92	–106.09	–97.13	–80.48	–154.82	–113.46
Apparent heat of products H_g (kJ/mol)	–63.30	–67.83	–64.64	–65.27	–69.13	–104.48	–86.75
Heat loss $H_l = H_r - H_g$ (kJ/mol)	–44.61	–44.09	–41.45	–31.86	–11.35	–50.34	–26.71
T_{ad} (°C)	2455	2593	2446	2255	1621	1288	1183
H_2O generation from H_2 combustion[i] × (mol/mol)	0.020	0.028	0.009	0.008	0.000	0.000	0.008
H_2O-consumed in shift reaction y (mol/mol)	0.004	0.004	0.063	0.039	0.065	0.287	0.054
Cold gas efficiency[i] (–)	0.708	0.737	0.757	0.733	0.761	0.753	0.671
Carbon conversion[j] $(1-\theta)$ (–)	0.931	0.959	0.965	0.929	0.915	0.881	0.883
Reference	Institute of Applied Energy 1993	Institute of Applied Energy 1993	Institute of Applied Energy 1993	Institute of Applied Energy 1993	Bodle and Huebler 1981	Hebden and Stroud	Hebden and Stroud

[a] Wet base
[b] Dry base
[c] kg per kg-dry coal
[d] H_2S and NH_3 are not involved in the gas composition
[e] Reported value
[f] The concentrations of N_2 and Ar are not involved in the theoretical calculations
[g] Parameters in the equation of $CH_mO_n + \alpha O_2 + \beta H_2O \rightarrow \gamma H_2 + \delta CO + \varepsilon CO_2 + \eta CH_4$
[h] Moisture + steam – β
[i] Produced by combustion with excess of O_2
[j] Calculated based on the stoichiometric theory
[k] Designed values based on the data in their development stage

$$\mathrm{CH_4}:\ \eta=\frac{s}{(q+r+s)} \tag{14.72}$$

14.5.2 Carbon Conversion of Coal Gasification

The amount of oxygen supply is defined as A (mol/mol$_{\text{-coal}}$). When the oxygen supply is not enough to oxidize coal ($A < \alpha$), unreacted solid is obtained as a by-product. This by-product is described by the same formula as coal.

In this case, coal gasification formula is described as Equation 14.73 as a result of the following modifications:

$$\mathrm{CH}_m\mathrm{O}_n + A\mathrm{O_2} + (1-\theta)\beta\mathrm{H_2O} \rightarrow$$

$$(1-\theta)\gamma\mathrm{H_2} + (1-\theta)\delta\mathrm{CO} + (1-\theta)\varepsilon\mathrm{CO_2} + (1-\theta)\eta\mathrm{CH_4} + \theta\mathrm{CH}_m\mathrm{O}_n$$

Transposing the term $\theta\mathrm{CH}_m\mathrm{O}_n$ on the right-hand side to the left-hand side gives

$$(1-\theta)\,\mathrm{CH}_m\mathrm{O}_n + A\mathrm{O_2} + (1-\theta)\beta\mathrm{H_2O} \rightarrow$$

$$(1-\theta)\gamma\mathrm{H_2} + (1-\theta)\delta\mathrm{CO} + (1-\theta)\varepsilon\mathrm{CO_2} + (1-\theta)\eta\mathrm{CH_4} \tag{14.73}$$

The carbon conversion $(1 - \theta)$ equals to A/α when coal is gasified with oxygen amount A, which is less than α, with the formation of char $\mathrm{CH}_m\mathrm{O}_n$ in the amount of θ. The calculated carbon conversion in each gasification process is listed in Table 14.3. Some data with the by-product formation under the condition of $A < \alpha$ were omitted from Table 14.3 because they are theoretically unacceptable.

14.5.3 Cold Gas Efficiency of Coal Gasification

Cold gas efficiency was calculated by the following equation using a general formula (Equation 14.73) of coal gasification with the formation of by-product:

$$E = \frac{2\mu h_{\mathrm{CO}}\left[(1-\theta)-\chi\right] + \eta(h_{\mathrm{CH4}} - 4h_{\mathrm{CO}})}{h_{\mathrm{coal}}}$$

where $\chi = A/\alpha$.

The calculated value was compared with the reported value of cold gas efficiency in Table 14.3.

14.5.4 Oxygen Balance Considering Shift Reaction and Hydrogen Combustion

According to the discussion in Section 14.4.2, stoichiometric models as Equations 14.60 and 14.62 were derived to describe the stoichiometric relationship of gasification under the conditions of $\mathrm{O_{st}} = 0.5(1 - n)$ and $\mathrm{O_{ex}} = \alpha - 0.5(1 - n)$ in the cases of $\mathrm{O_{ex}} \leq 0$ and $\mathrm{O_{ex}} > 0$. Using Equations 14.60 and 14.62, the converted amount of hydrogen yielding steam (x) and the amount of steam consumption (y) by water–gas shift reaction were calculated.

$$\mathrm{O_{ex}} > 0\ :\ \mathrm{CH}_m\mathrm{O}_n + (\mathrm{O_{st}} + \mathrm{O_{ex}})\mathrm{O_2} + (y - x - z)\mathrm{H_2O} \rightarrow$$

$$(0.5m + y - x - 3z)\mathrm{H_2} + (1 - 2\mathrm{O_{ex}} - y + x - z)\mathrm{CO}$$

$$+ (2\mathrm{O_{ex}} + y - x)\mathrm{CO_2} + z\mathrm{CH_4}$$

$$\mathrm{O_{ex}} > 0 : \mathrm{CH}_m\mathrm{O}_n + (\mathrm{O_{st}} + \mathrm{O_{ex}})\,\mathrm{O_2} + (-2\mathrm{O_{ex}} + y - z)\mathrm{H_2O} \rightarrow$$

$$(0.5m - 2\mathrm{O_{ex}} + y - 3z)\mathrm{H_2} + (1 - y - z)\mathrm{CO} + y\mathrm{CO_2} + z\mathrm{CH_4}$$

In the case of $\mathrm{O_{ex}} \leqq 0$ in Equation 14.60, CO consumption due to water–gas shift reaction of steam is the same amount of $\mathrm{CO_2}$ formation on a molar basis. In the case of $\mathrm{O_{ex}} > 0$ in Equation 14.62, oxygen gas of the amount $\mathrm{O_{st}}$ reacts with coal to give $0.5m\mathrm{H_2}$ and CO. And further oxidation of oxygen gas ($\mathrm{O_{ex}}$) gives $\mathrm{H_2O}$ and $\mathrm{CO_2}$ at the total amount of $2\mathrm{O_{ex}}$. Assuming the formation of $\mathrm{H_2}$ and CO in the molar ratio of 0.5m:1 in the former conversion, $\mathrm{H_2}$ combustion amount, $x = 2\mathrm{O_{ex}}\ 0.5m/(1 + 0.5m)$, and steam consumption by water–gas shift reaction, $y = \varepsilon - 2\mathrm{O_{ex}} + m\mathrm{O_{ex}}/(1 + 0.5m)$, were used for the calculation in Tables 14.3 through 14.11.

14.5.5 Heat Balance and Estimation of Adiabatic Gasification Temperature

Heat of gasification, H_r (kJ/mol$_{\text{-coal}}$), of coal per unit structure ($\mathrm{CH}_m\mathrm{O}_n$) can be obtained based on the following equation:

$$H_r = h_{\mathrm{coal}} - (\gamma h_{\mathrm{H_2}} + \delta h_{\mathrm{CO}} + \eta h_{\mathrm{CH_4}} + \sigma h_{\mathrm{CH}_m\mathrm{O}_n})$$

where h_{coal}, $h_{\mathrm{H_2}}$, h_{CO}, $h_{\mathrm{CH_4}}$, and $h_{\mathrm{CH}_m\mathrm{O}_n}$ are the molar heats of combustion of coal, hydrogen, carbon monoxide, methane, and char, respectively. Heat of combustion of char was approximated to that of coal, and the calculation results of H_r were summarized in Table 14.3. The HHV of coal and latent heat of water volatilization are involved in the calculation, and the temperatures of air, oxygen, and water were assumed to be at the standard temperature upon supply (Tables 14.3 through 14.11).

As mentioned in Section 14.3, the heat balance of gasification is discussed on an HHV basis at the standard temperature, 298.1 K. If steam was supplied to a gasifier, heat input in heat balance is a summation of the apparent heat and latent heat of steam and the generated heat by gasification of coal. Heat output is the total amount of the apparent heat of a mixture of the gaseous products, nitrogen from air, and steam remained.

Gaseous components in a gasifier are a mixture of product gas, steam, and nitrogen. Solid residue including char is assumed to be similar to graphite. Table 14.12 summarizes the heat input and output in Figure 14.5. Generally, the higher adiabatic gasification temperature accelerates the rate of gasification, and the lower heat emission means low heat loss of a gasifier. Based on the balance of the heat input and output, adiabatic gasification temperature during gasification and heat emission from a gasifier can be calculated to evaluate the performance of the gasification process.

TABLE 14.12
Energy Balance of Gasification to Estimate Heat Loss of a Gasifier

Heat Input Including Heat Generation		Heat Output	
Coal	Calorific value Apparent heat[1] Heat of gasification H_r[2]	Gaseous products (H_2, CO, H_2O, CO_2, N_2, steam remained)	Apparent heat HHV
Air (nitrogen, oxygen, and others)	Apparent heat[1]	Solid residue	Apparent heat HHV
Steam	Apparent heat and latent heat[3]	Heat emission from a gasifier	

[1] Molar flow rate × specific heat capacity × (298.1 – 298.1) = 0.
[2] $H_r = (1 - \theta)H_r$.
[3] When w mol/mol$_{\text{-coal}}$ of water is supplied as steam, heat input is the latent heat given as w × 43.99 (kJ/mol).

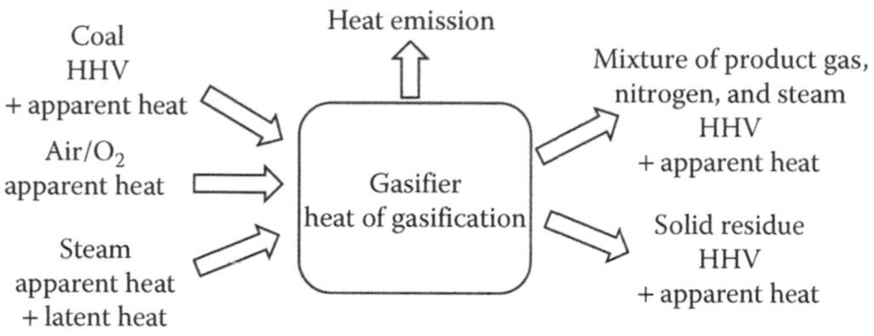

FIGURE 14.5 Heat balance calculation.

Then, an adiabatic gasification temperature is calculated as follows:

For heat input (heat of gasification and heat supply by steam addition),

1. Heat input by heat generation due to gasification of coal: $(1 - \theta)H_r$ kJ/mol$_{\text{-coal}}$, where the parameter H_r is a negative value.
2. Heat input by steam supply: $-w \times 43.99$ kJ/mol$_{\text{-coal}}$
3. Net heat input: $H_r = (1 - \theta)H_r - w \times 43.99$

For heat output,

4. Apparent heat of gaseous mixtures including gasification products, nitrogen from the air supplied and steam remained after gasification: (calculated heat capacity of the gaseous mixture) × (mass flow of the gaseous mixture)
5. Apparent heat of char

Then, the difference between the net heat input and the total apparent heat of gaseous mixture and char gives heat emission from a gasifier. The heat emission is considered to be a heat loss of the gasification process.

TABLE 14.13
Parameters of Heat Capacity (kJ/kg mol K)

Gas	a	$b \times 10^2$	$c \times 10^5$	$d \times 10^9$
H_2	27.144	0.9274	–1.381	7.645
CO	30.871	–1.285	2.789	–12.72
CO_2	19.796	7.344	–5.602	17.15
CH_4	19.252	5.213	1.197	–11.32
N_2	31.151	–0.1357	2.680	–11.86
H_2O	32.244	0.1924	1.056	–3.597

Heat capacity Cp_i of a gaseous component i at temperature T is given by the following equation:

$$Cp_i = a_i + b_iT + c_iT^2 + d_iT^3 \tag{14.74}$$

The parameters a–d are listed in Table 14.13. The heat capacity of char is approximated as that of graphite (Table 14.13).

Additionally, a = 2.673, $b \times 10^2$ = 0.2617, and $c \times 10^5$ = –1.169 were used as the parameters of heat capacity of char. Integral of Equation 14.75 from 298.1 K to T gives the heat requirement q_i to raise the temperature of a gaseous component i:

$$q_i = a_i\left[T - (298.1)\right] + \left(\frac{1}{2}\right)b_i\left[T^2 - (298.1)^2\right] + \left(\frac{1}{3}\right)c_i\left[T^3 - (298.1)^3\right] + \left(\frac{1}{4}\right)d_i\left[T^4 - (298.1)^4\right] \tag{14.75}$$

Letting the molar amount of a component i as n_i, the apparent heat Q of the component i is obtained by Σn_iq_i. An adiabatic gasification temperature T_{ad} was determined by seeking the temperature under the condition $H_r = Q$, and the obtained temperature is shown in Tables 14.3 through 14.11.

14.5.6 Examination of Experimental Data of Coal Gasification

Reported demonstration data of coal gasification were plotted by various parameters to clarify the features of the demonstration by various gasification processes. The calculated parameters α through η in the gasification formula (14.66) are listed in Table 14.3. The features of each gasification process are explained as follows.

In the most cases, demonstration researches were conducted by changing the supply rates of oxygen and steam to obtain their optimized rates, at which the best performance was given in terms of carbon conversion, cold gas efficiency, and product gas composition. The relationship between the supplied amount of O_2 (A mol/mol$_{\text{-coal}}$) and the molar amount of reacting with coal (α) is shown in Figure 14.6.

As explained in Section 14.5.2, the material balance of the demonstration data is invalid in the case of A > α, and the data plots are located above the line A = α. The bigger difference of the height of a plot and the height of the plot at the same A on the line means the lower carbon conversion because of the

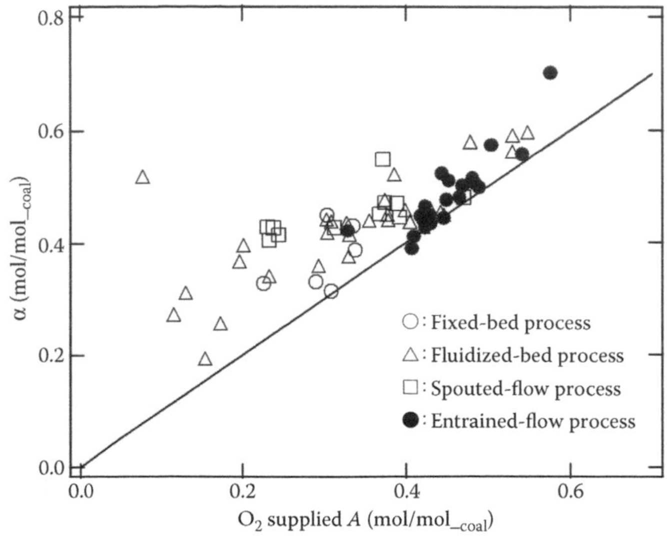

FIGURE 14.6 Relationship between the supplied amount of O_2 (A) and the reacted amount of O_2 (α).

relationship of carbon conversion $(1 - \theta) = A/\alpha$. In the region of $A <$ ca 0.4, many demonstration data by the fluidized-bed process locate apart from the line, which means low carbon conversion. In the region of $A >$ ca 0.4, many demonstration data of the entrained-flow process distribute close to the line, which means high carbon conversion.

Figure 14.7 shows the relationship between the supplied amount of steam [mol/mol$_{\text{-coal}}$] and the reacted amount of steam $(1 - \theta)\beta$. In the gasification by entrained flow process, $(1 - \theta)\beta$ shows a relatively small than that in the other processes. Typical entrained flow processes such as Shell–Koppers uses pulverized coal as a feed and oxygen as a gasifying agent without using steam. Steam source is moisture in coal. Some demonstrations were conducted under the conditions of $(1 - \theta)\beta < 0$ due to H_2 combustion by excess oxygen (O_{ex}).

In the gasification by fixed-bed and fluidized-bed processes, steam supply is larger than 0.3 and often larger than 5. The intention of increasing amount of steam would be the increase of cold gas efficiency by promoting steam gasification of $C + H_2O \rightarrow H_2 + CO$. Even if the large ratio of steam was supplied $(1 - \theta)\beta$ is about 0.2–0.4, which is not proportional to the supply amount of steam. As a result, large amount of steam would escape from a gasifier without reacting with coal.

Steam is consumed by both water–gas shift reaction with carbon monoxide and steam gasification with coal. Figure 14.8 shows the relationship between the observed amount of steam consumption $(1 - \theta)\beta$ and the steam consumption by steam gasification of coal. The distribution of the plots near the line of slope = 1 suggests that the dominant cause of steam consumption is water–gas shift reaction. In the case of steam consumption that is larger than the steam supply amount, the additional steam would be generated by hydrogen combustion and methane formation, and was consumed by water–gas shift reaction with carbon monoxide. The data under the line suggest that water–gas shift reaction was suppressed because steam gasification was preferred under the high temperature condition.

14.5.7 Confirmation of the Stoichiometry of Gasification

The total generation amount of hydrogen and carbon monoxide divided by theoretical oxygen requirement $(\gamma + \delta)/\mu$ was interpreted in the relation to oxygen ratio χ that was derived in Figure 14.1 by examining the data from the demonstrations. In this examination, methane was formed in the secondary

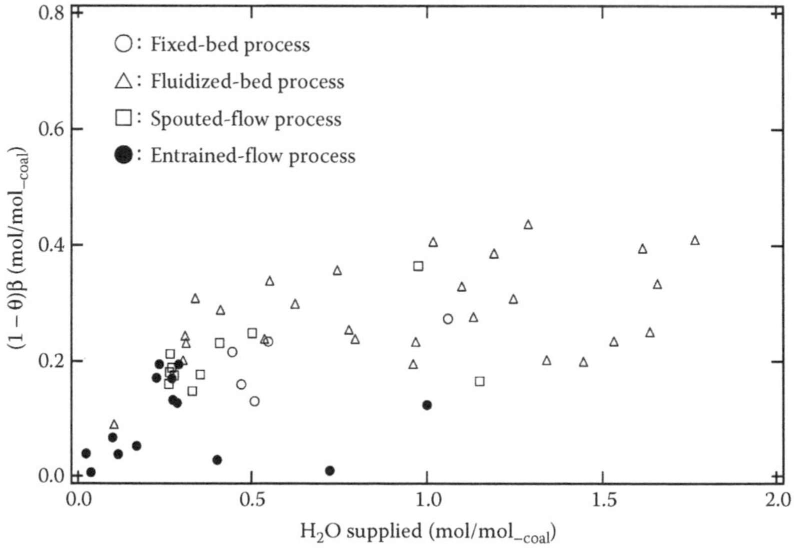

FIGURE 14.7 Relationship between the consumed amount of steam $(1 - \theta)\beta$ mol/mol$_{\text{-coal}}$ and the supplied amount of steam.

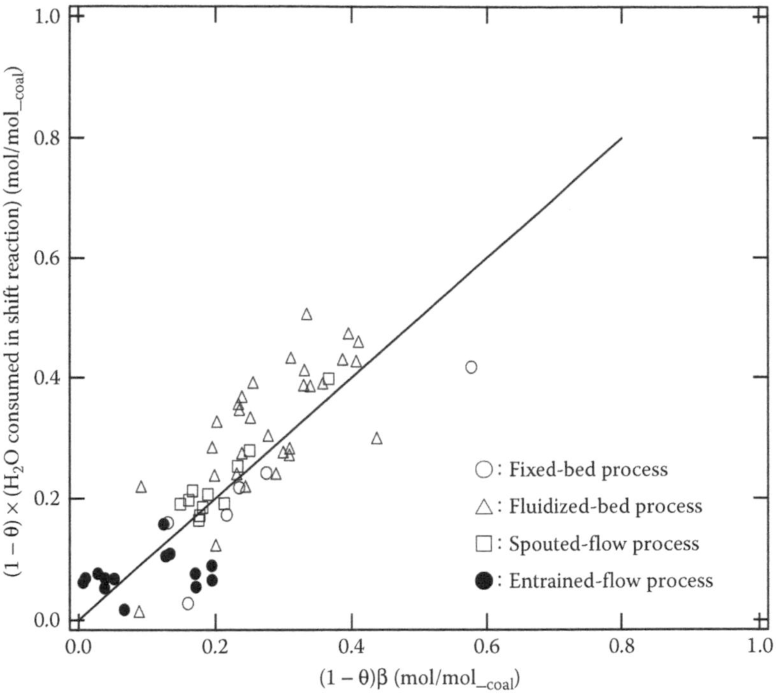

FIGURE 14.8 Steam consumption in shift reaction with consumed amount of steam $(1 - \theta)\beta$.

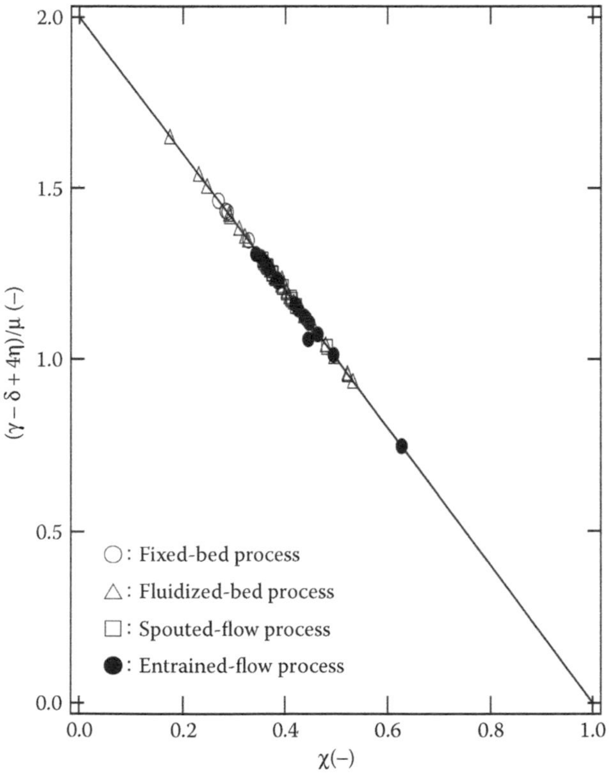

FIGURE 14.9 Relationship between $[(\gamma + \delta) + 4\eta]/\mu$ and χ.

reaction of $3H_2 + CO \rightarrow CH_4 + H_2O$. Figure 14.9 shows the total amounts of hydrogen and carbon monoxide divided by theoretical oxygen requirement $[(\gamma + \delta) + 4\eta]/\mu$ with the oxygen ratio χ. These data fit to the line with a slope of −2, which agrees to the stoichiometric approach.

14.5.8 Heat of Gasification, Adiabatic Gasification Temperature, and Heat Loss from a Gasifier

Heat of gasification has been unclear in the most demonstration projects. The stoichiometric approach provides heat of gasification H_r. Calculated values of heat of gasification H_r are listed in Table 14.3, and those values were plotted with χ $(=\alpha/\mu)$ in Figure 14.10.

Heat of gasification H_r has a tendency to increase with the increase of oxygen ratio. It also suggests that gasification becomes combustion with an increase of oxygen. Heat of gasification is also affected by a calorific value of coal.

Figure 14.11 shows $\chi \times h_{coal}$ with H_r. It shows the linear relationship between $\chi \times h_{coal}$ and $H_r = 103 + \chi \times h_{coal}$, which approximates heat of gasification in the function of calorific value of coal and oxygen ratio.

14.5.9 Adiabatic Gasification Temperature

Adiabatic gasification temperature T_{ad} is the theoretical maximum temperature during gasification. It will help to choose a suitable heat insulation and a cooling system of a gasifier. Generally, gasification simulation uses the observed temperature (T_g) of a gasifier. T_g is the observed temperature detected near a gasifier surface under heat emission to the surroundings. The real gasification temperature deep inside the gasifier would be the adiabatic gasification temperature T_{ad}, estimated by Equation 14.75.

The heat of gasification and the adiabatic gasification temperature are calculated based on the analytical values of demonstration operations. Figure 14.12 shows the adiabatic gasification temperature T_{ad} with heat of gasification

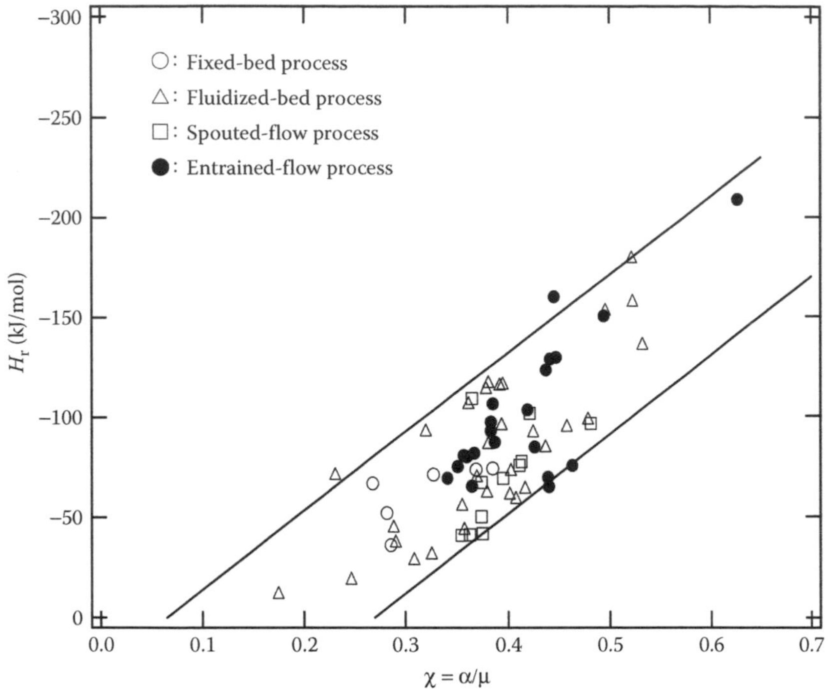

FIGURE 14.10 Heat of gasification H_r with oxygen ratio $\chi = \alpha/\mu$.

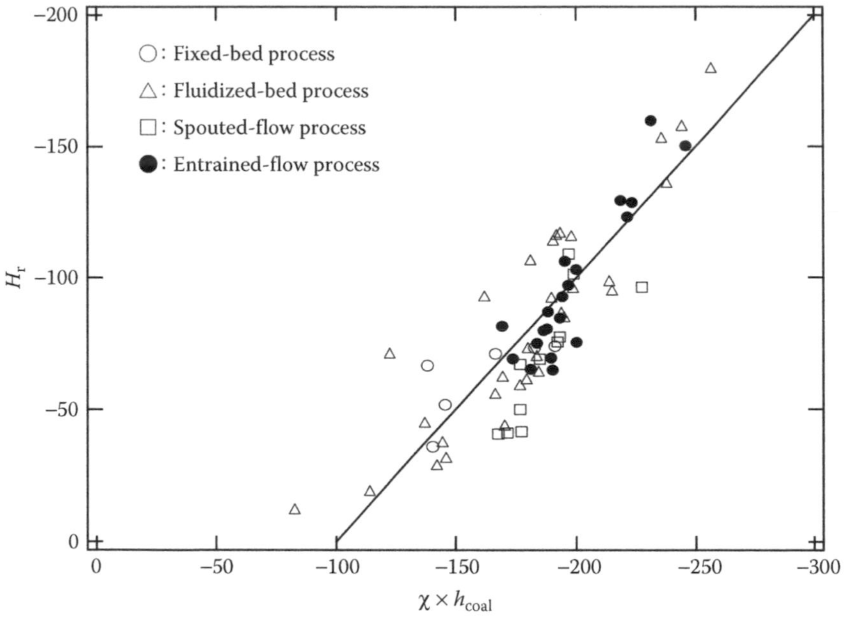

FIGURE 14.11 Heat of gasification H_r with $\chi \times h_{coal}$.

H_r. Most data of each gasification process are distributed along the lines indicating the tendency of proportional relationship between T_{ad} and H_r. The slopes of the lines are different from each other depending on the type and conditions of the gasification process. The calculated data of entrained-bed process (solid circle) are on the line of steep slope. Generally, the operation of this process is conducted at a high gasification temperature range in order to promote ash melting. The high operation temperature is achieved by increasing oxygen or air, and it increases the rate of gasification. However, the demonstration data by the fluidized-bed process (solid triangle) were along the line of gentle slope. A fluidized-bed process requires fluidizing gas, and large amounts of steam are used in the operations. Steam input increases H_r; however, the heat of gasification is spent to heat both the product gas from coal and the unreacted steam. As a result, the adiabatic gasification temperatures were relatively low for fluidized-bed processes.

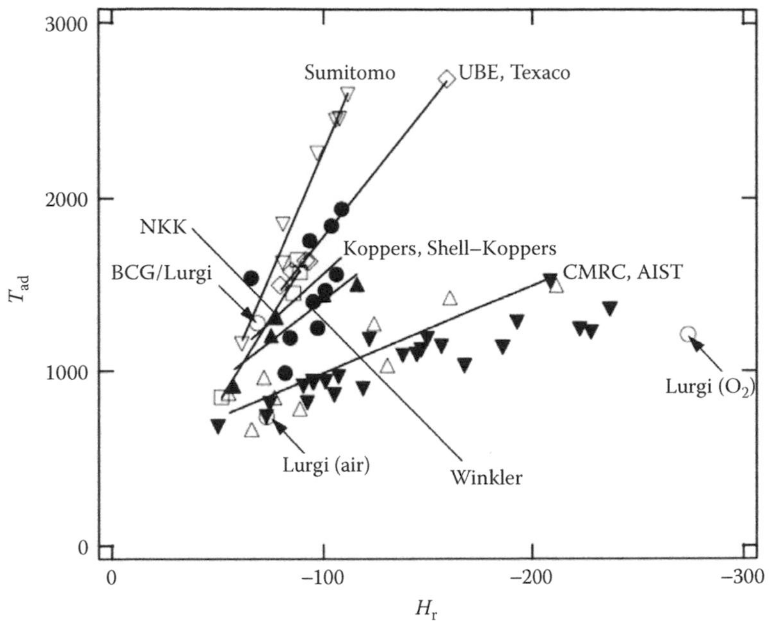

○: Demonstration data plots of the fixed-bed processes (Lurgi and BCG/Lugri processes with air-or oxygen-blow)

△: Fluidized-bed processes (CMRC and AIST processes) with a fitting line

▲: Fluidized-bed process (Winkler process) with a fitting line

▼: Demonstration data plots of the fluidized-bed process (Synthane and Hygas processes)

□: Spouted-flow process (NKK process) with a fitting line

●: Entrained- flow process (Koppers and shell–Kopper processes) with a fitting line

◇: Entrained- flow process (UBE and Texaco processes) with a fitting line

▽: Molten-bed process (Sumitomo process) with a fitting line

FIGURE 14.12 Relationship between H_r and T_{ad}. AIST, National Institute of Advanced Industrial Science and Technology; CMRC, Coal Mining Research Center.

The features of various gasification processes are discussed in Section 14.6.

14.5.10 Estimation of Heat Loss from a Gasifier

Heat loss is the important feature of the performance of a gasifier. For the heat loss per 1 mol of coal, H_l (kJ/mol$_{-coal}$), there is the relationship $H_l = H_r - H_g$, where H_g is an averaged apparent heat of a mixed gas of gasification products, by-products, nitrogen, and steam remained after gasification at a gasifier temperature T_g. Heat loss H_l per 1 mol of coal can be expressed as follows: (temperature inside a gasifier – atmospheric temperature) × (heat transfer coefficient) × (wall area of a gasifier). When a gasifier is a pressure-resistant vessel with a refractory material and heat insulation material in it, (temperature inside a gasifier – atmospheric temperature) = (T_{ad} − T_a), where T_a is the atmospheric temperature. When a gasifier is surrounded by refractory wall cooling with water, (temperature inside a gasifier – atmospheric temperature) = (T_{ad} − T_w), where T_w is the water temperature. Then, H_l can be expressed by the following equation:

$$H_l = (T_{ad} - 25) \times \begin{pmatrix} \text{heat transfer coefficient of} \\ \text{wall of a gasifier} \end{pmatrix} \times \left(\text{surface area of wall of a gasifier}\right) \times R_t \qquad (14.76)$$

where:

T is T_a or T_w

R_t is the averaged residence time of coal particles. The parameter R_t decreases with the smaller diameter of coal particles, the higher gasification temperature, and the higher operation pressure

According to rough estimation, the residence time of pulverized coal of 0.07 mm diameter (<200 mesh) in an entrained-bed gasifier would be several seconds, that of coal particles of up to 3 mm diameter in a fluidized-bed gasifier and that of lumped coal of up to 50 mm diameter in a fixed-bed gasifier would be around 20 min to over 1 h.

Figure 14.13 shows the calculated values of H_l in Table 14.3 with T_{ad}. Heat loss H_l is proportional to adiabatic gasification

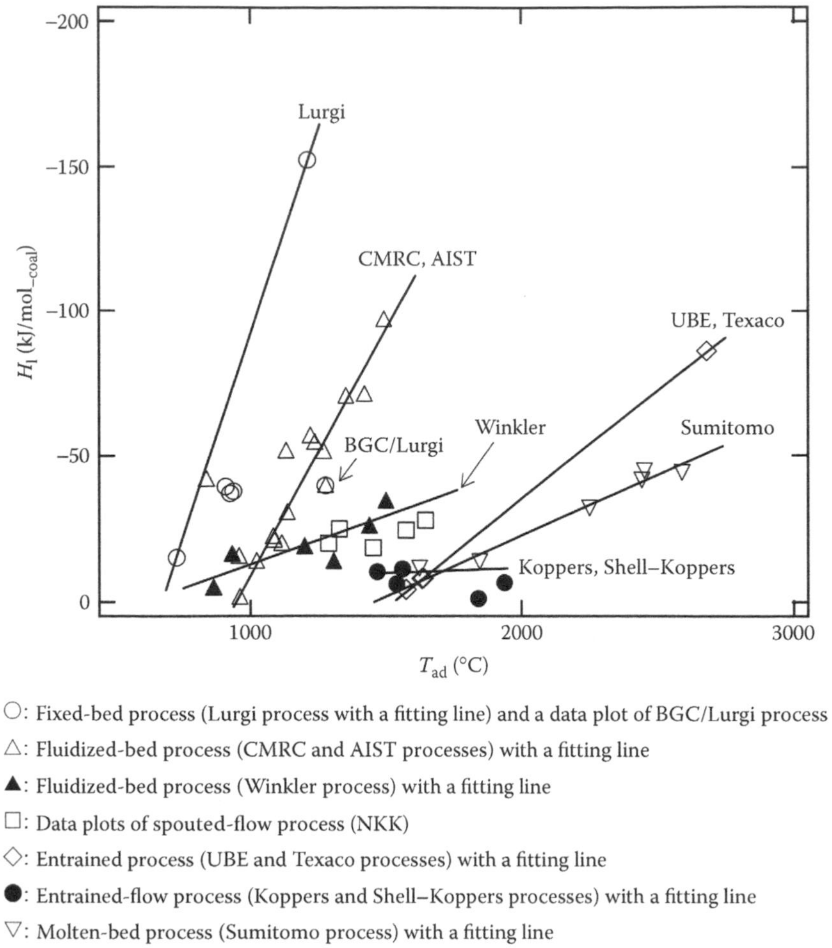

FIGURE 14.13 Heat emission H_l with T_{ad}. AIST, National Institute of Advanced Industrial Science and Technology; CMRC, Coal Mining Research Center.

temperature T_{ad}, and the proportional coefficients depend on the type of gasification process.

14.6 COMPARISON OF VARIOUS GASIFICATION PROCESSES

Effective utilization of coal had been drawing attention in many countries since the energy crisis in 1973, and many applications were proposed through technical innovations of coal conversion. Gasification was considered as the important core technology in coal conversion, which includes integrated combined cycle power generation, substitute natural gas, and C1 chemistry producing chemicals. Currently, there are some R&D projects involving large-scale gasification over 2000 ton/day.

The theoretical contribution of the supply amounts of gasifying agents and gas compositions to coal gasification can be analyzed in terms of calculated adiabatic gasification temperature, heat loss from a gasifier, and others. These calculated values give reliable values that indicate the performance of the gasification process such as carbon conversion and cold gas efficiency. The adiabatic gasification temperature can be used for estimation of a gasification rate. In this section, typical gasification technologies such as fixed-bed, fluidized-bed, spouted-flow, entrained-flow, and molten-bath processes are explained, and these processes are compared with each other in terms of carbon conversion and cold gas efficiency with explanations how high conversion or efficiency was achieved under the effects of oxygen and adiabatic gasification temperature.

14.6.1 Fixed-Bed Process

A fixed-bed gasifier is a shaft reactor with a rotating grate at the bottom. Lumped coal up to 70 mm in diameter is supplied from the top of a fixed-bed gasifier, and gasifying agents of oxygen and steam are supplied from the bottom. Oxygen with steam or air is supplied to a combustion zone of char, and char is oxidized to give gaseous products, steam, and carbon dioxide at high temperature. The mixed gas reacts with char in a reduction zone through the reactions of $C + H_2O \rightarrow H_2 + CO$ and $C + CO_2 \rightarrow 2CO$. The gaseous products are further moving upward to promote pyrolysis of coal at a pyrolysis zone and to make the incoming coal dry.

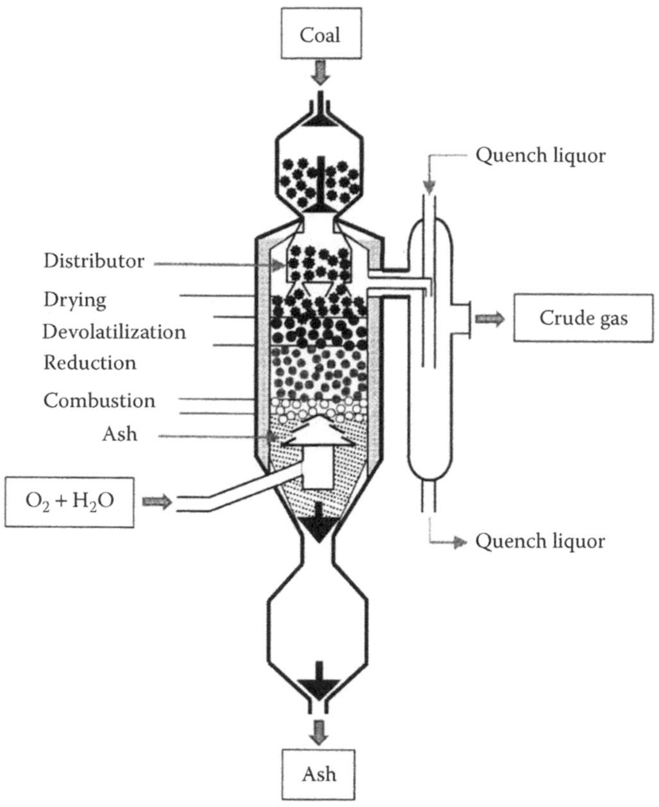

FIGURE 14.14 Schematic diagram of a Lurgi gasifier.

In this section, typical fixed-bed, Lurgi, and BGC/Lurgi processes are compared with each other. The Lurgi process has a fixed-bed gasifier under pressure (Figure 14.14). The BGC/Lurgi process is a modified Lurgi gasifier by removing a rotating grate and is equipped with tuyers that are injection nozzles of pulverized coal with oxygen. The operation temperature will be increased to promote ash melting for the easier removal.

In Table 14.12, the columns of Lurgi and BGC/Lurgi show the demonstration results by the use of the same coal using oxygen as a gasifying agent. There is another result by the use of air as a gasifying agent. Oxygen supply amounts were 0.308 in BGC/Lurgi, 0.298 in the Lurgi process with oxygen blow, and 0.226 mol/mol_{-coal} in the Lurgi process with air blow. Steam supply amounts were 0.471 in BGC/Lurgi, 4.719 in Lurgi with oxygen blow and 1.059 in Lurgi with air blow. There was a little difference between the Lurgi processes with oxygen blow and air blow. The BGC/Lurgi process gave less H_2, CO_2, and CH_4. But it gave more CO than the Lurgi process. The less steam supply would result in water–gas shift reaction being suppressed, and the less hydrogen would result in methane formation getting suppressed.

The heat of gasification H_r is the heat supply including the heat from both steam and enthalpy differences between coal and products. For the Lurgi process with oxygen blow, H_r was −273.81 kJ/mol_{-coal}, which is at about 4 times to −72.65 in the Lurgi process with air blow and −68.62 in the BGC/Lurgi process. The difference in H_r would be caused by the difference of the amounts of steam supply. The adiabatic gasification temperature of each process was 1208°C in the Lurgi process with oxygen blow, 1274°C in the BGC/Lurgi process with oxygen blow, and 735°C in the Lurgi process with air blow. The feeding rate of each process was 4718 kg/h in the Lurgi process with oxygen blow, 12005 kg/h in the BGC/Lurgi process, and 540–653 kg/h in the Lurgi process with air blow. Each feed rate multiplied by carbon conversion gives the actual gasification rate of coal in the operation. The actual rates of the Lurgi process with oxygen blow, the BGC/Lurgi process with oxygen blow, and the Lurgi process with air blow were 4628, 9148, and 328–397 kg/h, respectively. The gasification rate of the BGC/Lurgi process with oxygen blow (T_{ad} = 1274°C) was doubled to that of the Lurgi process with oxygen blow (T_{ad} = 1208°C) due to the higher adiabatic gasification temperature by 66 K. The adiabatic temperature in the Lurgi process with air blow was 735°C, which is 473 K lower than that in the Lurgi process with oxygen blow. The treat capacity of the Lurgi process with air blow was only one-thirteenth capacity of the Lurgi process with oxygen blow.

Ash amounts from the Lurgi process with air blow was 23.1 wt%, which was 15.1% bigger than ash from the Lurgi process with oxygen blow. In a fixed-bed gasifier, lumped coal is gasified without fluidizing. Coal surface is covered with a thick ash layer developed during gasification. Oxygen and steam would penetrate the ash layer and promote gasification at the coal surface. Supplied amounts of oxygen or steam influence on the rate of gasification rate, which will control carbon conversion. In the other gasification processes than fixed-bed process, the smaller coal particles are used as a feed, and ash layer would not develop so thick. As a result, ash would not give serious influence on the gasification rate in those gasification processes. Typically, the Lurgi process has been developed as a pressurized gasifier. The operation experiment indicated that the higher adiabatic gasification temperature was achieved to increase the gasification rate, or the treatment capacity of coal. It will be improved also in the future.

14.6.2 Fluidized-Bed Process

In a fluidized-bed gasifier, pulverized coal of around 3 mm in diameter is gasified under the suspended condition of coal by introducing a gasifying agent at a linear velocity 0.3–0.6 m/s. Pulverized coal and a gasifying agent are well mixed to result in uniform heating within a gasifier. There were various troubles such as tar formation, carryover of residual char, and clinker formation. This type of gasifier was not commercialized although a multistep fluidized-bed gasification process was proposed to solve those technical problems. Figure 14.15 shows a schematic diagram of the core gasifier of the Winkler process.

Operation data were collected in the processes of Winkler, CMRC (Coal Mining Research Center, Japan), AIST (National Institute of Advanced Industrial Science and Technology), Hygas, and Synthane. The CMRC process was a pilot project of IGCC (Integrated Gasification and Combined Cycle System), which is an air-blow gasification process. The AIST process is a bench plant to support the CMRC process. The Synthane and

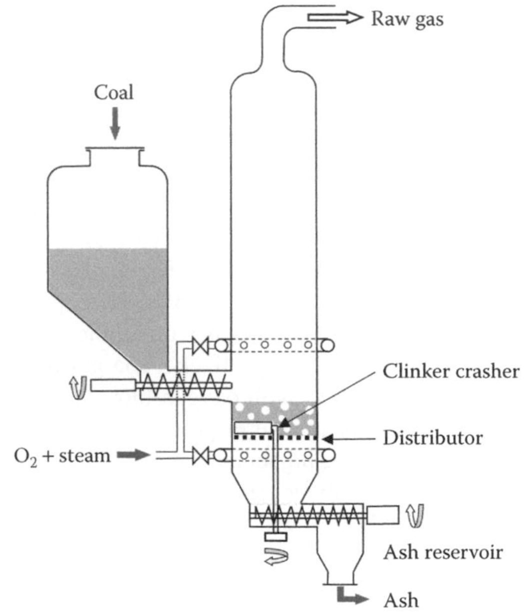

FIGURE 14.15 Schematic diagram of a Winker gasifier.

Hygas processes were the pilot projects for the production of a substitute of natural gas (SNG). Operation pressure was 1.1–3.9 kg/cm^2 in the Winkler process, 20 kg/cm^2 in the CMRC process, 3–7 kg/cm^2 in the AIST process, 83–86 kg/cm^2 in the Hygas process, and 42 kg/cm^2 in the Synthane process. To maintain fluidization, steam supply was increased under the high-pressure conditions. The Winkler process was conducted at a steam supply amount of 0.141–0.521 mol/mol$_{\text{-coal}}$ because of the relatively low operation pressure. In the other processes under the higher pressure, the CMRC and AIST processes were conducted at 0.706–2.189 mol/mol$_{\text{-coal}}$, the Hygas process was conducted at 1.214–1.364 mol/mol$_{\text{-coal}}$, and the Synthane process was conducted at 1.082–1.563 mol/mol$_{\text{-coal}}$.

The heat of gasification H_r is determined by (1 − θ) H_r − w 43.99, and H_r is increased with an increasing amount of steam supplied to a gasifier. The H_r was −54.58 to −116.49 kJ/mol$_{\text{-coal}}$ in the Winkler process, −62.41 to −237.16 kJ/mol$_{\text{-coal}}$ in the CMRC and AIST processes, and −65.80 to −131.09 kJ/mol$_{\text{-coal}}$ and −71.44 to −124.57 kJ/mol$_{\text{-coal}}$ in the Synthane process. The adiabatic gasification temperature T_{ad} was 916°C–1596°C in the Winkler process, 1086°C–1410°C in the CMRC process, 931°C–1548°C in the AIST process, 807°C–1515°C in the Hygas process, and 1073°C–1527°C in the Synthane process. The maximum T_{ad} reaches a melting range of ash, causing clinker trouble. The relationship between T_{ad} and H_r was shown in Figure 14.12. The slope of the line was about −10°C/kJ in the Winkler process, whereas the amount of unreacted steam was 0.031–0.328 mol/mol$_{\text{-coal}}$. For the CMRC and the AIST processes, in which the amount of unreacted steam was 0.215–1.959 mol/mol$_{\text{-coal}}$, the slope was −4°C/kJ. In the Hygas process, the amount of unreacted steam was 0.850–1.247 mol/mol$_{\text{-coal}}$. In the Synthane process, the amount of unreacted steam was 0.855–1.387 mol/mol$_{\text{-coal}}$, and the slope of the line was close to that of the CMRC and AIST processes.

Heat loss from a gasifier to the surroundings H_l (kJ/mol$_{\text{- coal}}$) was −4.91 to −34.48 in the Winkler process, −1.72 to −97.07 in the CMRC and AIST processes, −15.04 to −39.22 in the Hygas process, and 15.62–51.38 in the Synthane process. The slope of the line in Figure 14.13 was 0.05 in the Winkler and 0.2 kJ/°C in the other processes, namely, the CMRC, the AIST, Hygas, and Synthane. Carbon conversion (1 − θ) was 0.898–0.996 in the CMRC process, which is higher than that of the Winkler process (0.689–0.842). The high carbon conversion suggests that tar fraction decomposed effectively.

14.6.3 Spouted-Flow Process

As shown in Figure 14.16, spouted-flow gasifier is a gasification reactor with a conical bottom, where coal and gasifying agents are introduced. In this section, the demonstration results by NKK (currently JFE Holdings) are examined. Pulverized coal and gasifying agents of oxygen and steam are injected from the bottom of the gasifier. Coal is gasified in the fast stream traveling upward around the center of the gasifier. The resulting ash and residual char then go down near the inner surface of the gasifier, and those solid particles would circulate within the gasifier to undergo agglomeration until the formation of agglomerates is large enough to be discharged from the bottom.

The supply range of oxygen gas A is 0.368–0.047 mol/mol$_{\text{coal}}$, and the concentration of oxygen gas is 18.0%–56.6%. The calculated amount of reacted oxygen α was 0.445–0.548 mol/mol$_{\text{-coal}}$, and A was 0.365–0.481. The value of O_{ex} was −0.052 to 0.095 mol/mol$_{\text{-coal}}$, which was the similar range to that of fixed-bed gasifier. The amount of steam supply was 0.261–1.149 mol/mol$_{\text{-coal}}$, and the amount of unreacted steam was 0.100–0.983 mol/mol$_{\text{-coal}}$. Theoretical calculations gave H_r as −85.53 to −145.53 kJ/mol$_{\text{-coal}}$, H_g as −33.34 to −125.39 kJ/mol$_{\text{-coal}}$, and H_l as −18.42 to −88.94 kJ/mol$_{\text{-coal}}$. T_{ad} was calculated as 1178°C–1641°C. The obtained temperature range

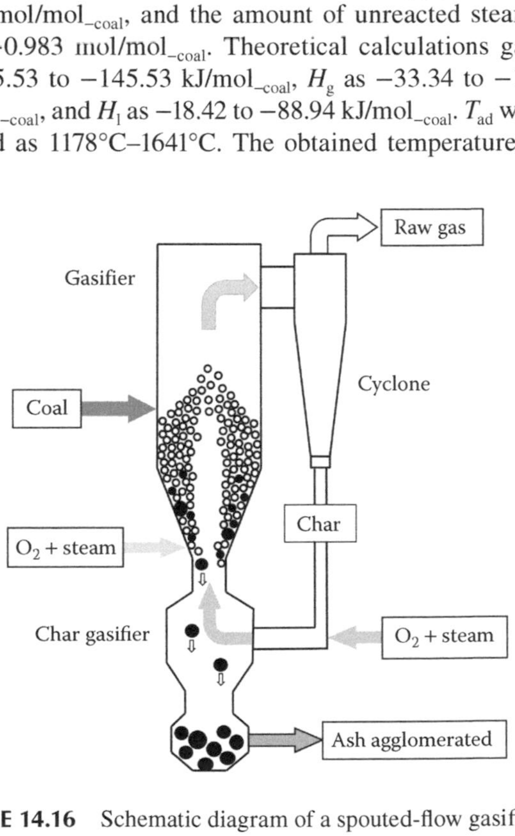

FIGURE 14.16 Schematic diagram of a spouted-flow gasifier.

was higher than the temperature range of the Winkler process (916°C–1596°C).

Comparing the temperature ranges in Figure 14.12, the T_{ad} of NKK's data is higher than that of the CMRC process and lower than that of the Winker process at the same H_r. The amount of steam supply is in the order of CMRC > NKK > Winkler. Figure 14.13 compares T_{ad} and H_l. The heat loss of the NKK process is near to that of the Winkler process. The NKK process is a laboratory-scaled gasifier. For ash agglomeration studies, propane gas (1.2–2.0 Nm^3/h) was combusted for heat supply. Steam supply was 0.262–1.149 mol/mol_{-coal}, and steam consumption in water–gas shift reaction was 0.191–0.398 mol/mol_{-coal}. The NKK process was conducted under relatively large A, T_{ad} (1356°C–1904°C) of this process was higher than that of the Winkler process (912°C–1647°C). The demonstration data of T_{ad} of the NKK process distribute near the line approximating the H_l–T_{ad} relationship of the Winkler process.

14.6.4 Entrained-Bed Process

Pulverized coal under 200 mesh is injected with oxygen through a burner. Data analyses were conducted for the operation data of the Texaco process, the Koppers–Totzek process (Figure 14.17), and the Shell–Koppers process. The Texaco process adopts a burner at the top of a gasifier, and coal is gasified in a downward flame. The Koppers–Totzek process uses plural burners with crossed flame. Shell–Koppers is a pressurized type of Koppers–Totzek process.

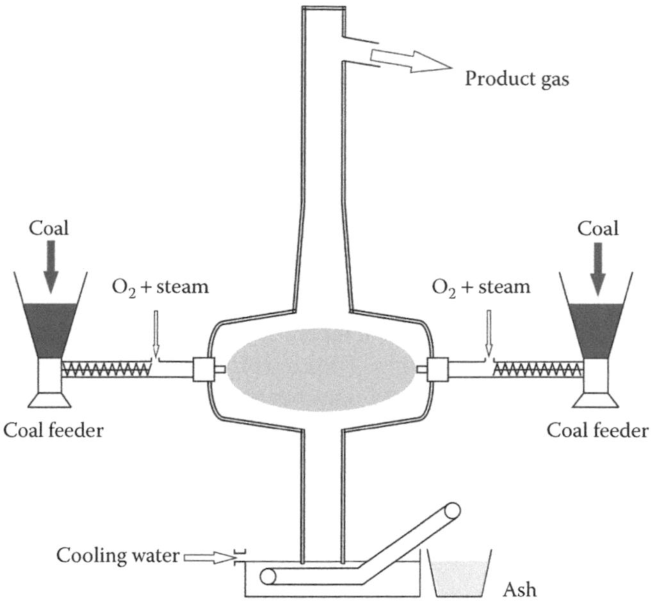

FIGURE 14.17 Schematic diagram of Koppers–Totzek gasifier.

The UBE process (Figure 14.18) is the similar process to Texaco process. Coal–water slurry is injected with a burner at the top of a gasifier with oxygen. The gasifier is a pressure-resistant vessel with a refractory material made from chromium compounds. Product gas and molten ash are sent to a water

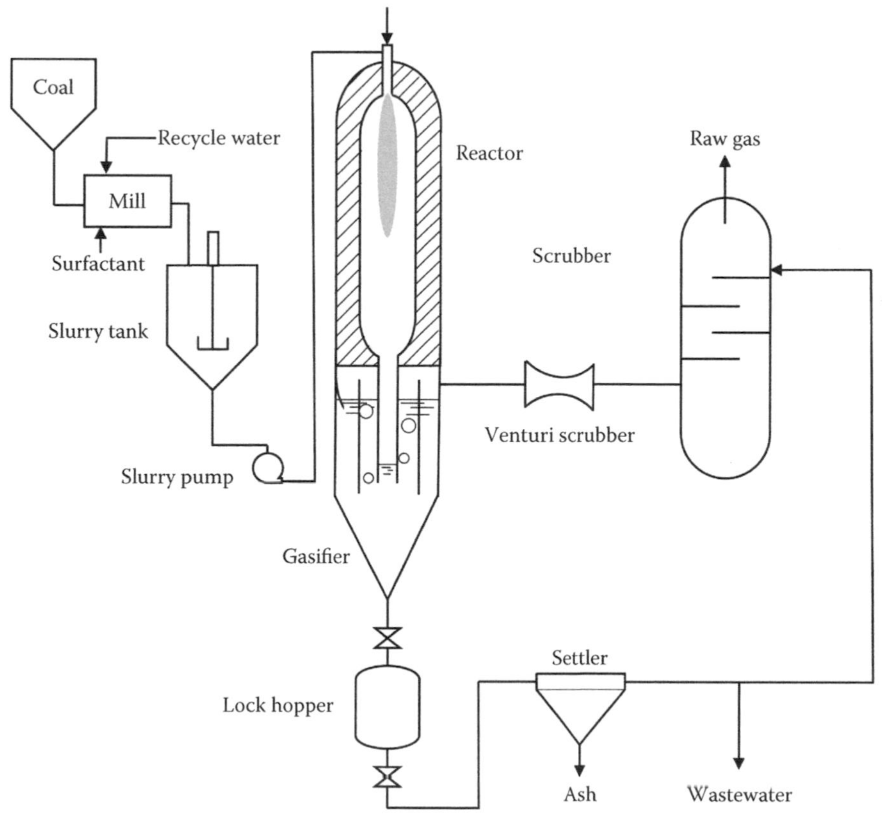

FIGURE 14.18 Schematic diagram of a UBE gasifier (modified TEXACO gasifier).

pool to remove dust with cooling. The observed and calculated values are as follows: $A = 0.542$ mol/mol$_{-coal}$, $\alpha = 0.557$ mol/mol$_{-coal}$, and $A/\alpha = 0.973$. Steam supply was 1.00 mol/mol$_{-coal}$, steam consumption $(1 - \theta)\beta = 0.125$ mol/mol$_{-coal}$, and unreacted steam 0.875 mol/mol$_{-coal}$. $H_r = -159.75$ kJ/mol$_{-coal}$, $H_g = -73.75$ kJ/mol$_{-coal}$, and $H_l = -86.00$ kJ/mol$_{-coal}$. The excess oxygen supply $O_{ex} = 0.118$ mol/mol$_{-coal}$ and the parts of the products, hydrogen and carbon monoxide, were burned. In the Texaco process, the use of steam at $O_{ex} = -0.055$ to -0.217 resulted in the formation of hydrogen and carbon monoxide. The adiabatic gasification temperature of the UBE process was higher (2684°C) than that of Texaco process (1492°C–1635°C) because the UBE process does not use steam that gives serious heat loss.

Four sets of data by the Texaco process were obtained under the conditions of supplying oxygen with steam: $A = 0.409$–0.430 mol/mol$_{-coal}$, $\alpha = 0.411$–0.435 mol/mol$_{-coal}$, and $A/\alpha = 0.983$–0.995. The steam supply is 0.225–0.290 mol/mol$_{-coal}$, $H_r = -79.34$ to -93.18 kJ/mol$_{-coal}$, $H_g = -79.79$ to 85.35 kJ/mol$_{-coal}$, and $H_l = -4.09$ to -7.95 kJ/mol$_{-coal}$. The operation of the Koppers–Totzek process was conducted under a normal pressure. For the acceleration of steam gasification, the burner was designed so that a flame of pulverized coal and oxygen was surrounded by steam flow. The product gas moves upward in the gasifier and escapes from an outlet. Ash sticks to a wall of a gasifier and molten ash downward to a water pool.

The Nippon Suiso process is a pilot model of the Koppers–Totzek process. For the operation conditions $A = 0.444$–0.466 mol/mol$_{-coal}$ and $\alpha = 0.476$–0.523 mol/mol$_{-coal}$ the steam supply was 0.295–0.444 mol/mol$_{-coal}$. Carbon conversion was 0.849–0.969, $H_r = -94.88$ to -344.74 kJ/mol$_{-coal}$, $H_g = -84.95$ to -95.77 kJ/mol$_{-coal}$, $H_l = -10.80$ to -259.79 kJ/mol$_{-coal}$, and $T_{ad} = 1399$°C–1464°C. During the initial period of the development, the data shows low T_{ad} possibly due to large steam supply.

In the modified gasifier, the Koppers–Totzek process, $A = 0.418$–0.430 mol/mol$_{-coal}$, $\alpha = 0.447$–0.464 mol/mol$_{-coal}$, and carbon conversion was 0.912–0.958 under the steam supply at 0.105–0.265 mol/mol$_{-coal}$. Calculated values are as follows: $H_r = -103.85$ to -108.69 kJ/mol$_{-coal}$, $H_g = -98.80$ to -102.66 kJ/mol$_{-coal}$, $H_l = -1.19$ to -8.87 kJ/mol$_{-coal}$, and $T_{ad} = 1839$°C–1936°C.

The Shell–Koppers process is the modified process of Koppers–Totzek under the higher pressure. The gasifier is a pressure-resistant vessel equipped with a heat recovery boiler in one reactor. It is also equipped with a counter burner. Product gas was injected to the outlet of a gasifier to circulate so that suspended molten ash was solidified for the easier collection. The gasifier has a membrane wall that consists of cooling water tubing. Gasification conditions and calculated values are as follows: $A = 0.329$ and 0.447 mol/mol$_{-coal}$, $\alpha = 0.420$ and 0.444 mol/mol$_{-coal}$, $A/\alpha = 0.783$ and 1.007, $H_r = -65.13$ and -106.20 kJ/mol$_{-coal}$, $H_g = -59.04$ and -94.89 kJ/mol$_{-coal}$, $H_l = -6.09$ and -11.31 kJ/mol$_{-coal}$, and T_{ad} was 1536°C and 1557°C.

In Figure 14.13, the operation data of the Koppers–Totzek and Shell–Koppers processes are in the region of high T_{ad} and low H_l, and the approximated line has the slope of nearly 0. Gasification performance was significantly increased in the Shell–Koppers process by increasing the operation pressure compared to the Koppers–Totzek process.

14.6.5 Molten-Bath Process

Molten-bath gasification is a gasification method of injecting coal and gasifying agents into a molten heating medium. Inorganic salts such as sodium carbonate, coal ash, and iron have been used as molten heating media. Two sets of results are analyzed in this section: (1) the data using molten iron by Sumitomo Metal and (2) the data using molten sodium carbonate by Kellog.

The Atgas method by Applied Technology Corporation is famous as a molten iron method. In Sumitomo Metal's demonstration, a laboratory-sized converter was used for coal gasification with injection of pulverized coal and oxygen. According to the report by the Atgas method, carbon moiety reacts with oxygen and steam to give hydrogen and carbon monoxide after it is dissolved in molten iron. The carbon content in molten iron is 3%–4% during the process. The gasification conditions and the results are as follows: $A = 0.407$–0.490 mol/mol$_{-coal}$, $\alpha = 0.429$–0.58 mol/mol$_{-coal}$, $A/\alpha = 0.915$–0.959, steam supply = 0.019–0.153 except gasification of oil pitch, unreacted steam = 0.033–0.051 mol/mol$_{-coal}$, $H_r = -80.48$ to -111.92 kJ/mol$_{-coal}$, $H_g = -63.30$ to -69.13 kJ/mol$_{-coal}$, $H_l = -11.35$ to -44.61 kJ/mol$_{-coal}$, and $T_{ad} = 1621$°C–2593°C. As shown in Figure 14.12, the adiabatic gasification temperatures of the data of molten-bath gasification linearly increased with H_r and the slope is close to the slope of the fitting line of the data by the entrained-flow process. Heat loss values increase with T_{ad}, and the slope of approximated line is close to that of the data-fitting line of the Winkler process and the NKK process in Figure 14.13.

14.7 CONCLUSIVE REMARKS

A new theory considering the stoichiometry of gasification was proposed to understand coal gasification. We described a general gasification formula with a simple compositional formula (CH_mO_n) representing coal. This enabled us to interpret the experimental results of coal gasification as a chemical conversion in rational ways on a molar basis rather than an empirical way. The compositional formula of coal was determined with elemental analysis, and it was used as an averaged molecular formula of coal. The observed values of the compositions of gaseous products were converted into the molar amounts of the products. The reliability of the analytical values in each demonstration project was confirmed by the mathematical relationship of the stoichiometric approach. Carbon conversion and cold gas efficiency were also defined by the underlying relationship among products in molar amounts under the stoichiometric approach.

Heat of gasification H_r was also calculated by the difference between the calorific value of feed and the total calorific value of products in molar basis. The calorific value of feed includes that of coal and latent heat of water vaporization. Heat of gasification H_r is a heat supply to raise the temperature of the gasification products. The maximum temperature was calculated as an adiabatic gasification temperature T_{ad}. This theoretically derived temperature could be a real gasification temperature in a gasifier, and it should be handled as the gasification temperature rather than the gasification temperature observed at a gasifier in demonstrations.

Adiabatic gasification temperature T_{ad} is an essential factor to increase the rate of gasification and otherwise treatment capacity of a gasifier because gasification rate generally follows the Arrhenius equation. Heat loss H_l is an indicator to evaluate the type of gasifier in relation to the gasification temperature, which directly affects an overall rate of gasification or the performance of the gasifier.

In these meanings mentioned above, the new stoichiometric approach on coal gasification would provide the insights of coal gasification. There are many types of gasification processes and attempts to improve them in terms of rate of gasification, carbon conversion, and cold gas efficiency. We are able to select a gasification system and to optimize the reaction conditions based on the process evaluation of the stoichiometric approach.

REFERENCES

Bodle, W. and Huebler, J. 1981. Coal gasification. In *Coal Handbook*, ed. by R. A. Meyers, Chapter 10, 493–713. New York: Marcel Dekker.

Dowson, J. E. and Larter, A. T. 1907. *Producer Gas*. New York: Longmans Green and Co. http://archive.org/details/producergas00dowsuoft (accessed on May 19, 2014).

Elliott, M. A. ed. 1981. *Chemistry of Coal Utilization, Second Supplementary Volume*. 1615. New York: Wiley Interscience Publication.

Endo, S., Mochizuki, T., Tanji, Y. and Suzuki, A. 1985. Paper presented at the 19th Autumn Meeting.

Furusawa, T., Kojima, T., Tokawa, S., Tanaka, S., Kawanish, T., Hada, H. 1985. Analysis of Coal Gastifier Performance, *Journal of the Fuel Society of Japan* 64: 716.

Hatheway, A. W. and Doyle, B. C. 2006. *Technical History of the Town Gas Plants of the BriTown Gas Plants of the British Isles*ngs of IAEG, http://iaeg2006.geolsoc.org.uk/cd/PAPERS/IAEG_564.PDF (accessed on May 19, 2014).

Haynes, W. P., Strakey, J. P., Santore, R. R. and Lewis, R. 1977. *Synthane Process Update* MID-'77.

Hebden, D. and Stroud, H. J. F. 1981. Coal gasification process. In *Chemistry of Coal Utilization, Second Supplementary Volume*, ed. by M. A. Elliot, Chapter 24, 1599–1752. New York: John Wiley & Sons.

Higman, C. and van der Burgt, M. 2008. *Gasification,* 2nd edition. Houston: Gulf Professional Publishing.

Honma, S., Tazaki, Y., Yumiyama, M., Takeda, S., Kitano, K., Yamaguchi, H., Kawabata, J., Mori. 1982. Coal Char Gastification by Single-Stage Fluidized Pressurized Bed. *Journal of the Fuel Society of Japan* 61: 998–1004.

Hozumi, S. and Seki, E. 1990. Low-caloric Gas Production from Coal for IGCC, Paper presented at the annual meeting of Coal Mining Research Center, Japan, 10th Sekitan Riyougijutu-Kenkyu Happyoukai, 245–281.

Institute of Applied Energy. ed. 1993. *Enerugi-Gijutu Data Handbook*. 441. Tokyo: The Institute of Applied Energy.

Johnson, J. L. 1981. Fundamentals of coal gasification. In *Chemistry of Coal Utilization, Second Supplementary Volume*, ed. by M. A. Elliot, Chapter 23, 1491–1598. New York: John Wiley & Sons.

Kikuchi, K., Suzuki, A., Mochizuki, T., Endo, S., Imai, T. and Tanji, Y. 1983. Ash-agglomerating Gasification of Coal in a Spouted Bed. *Journal of the Fuel Society of Japan* 62: 356.

Kimura, H. and Fujii, S. 1976. *Sekitan Kagaku to Kogyo*. 382. Tokyo: Sankyo Publishing.

Mantoux, P. 1961. *The Industrial Revolution in the Eighteenth Century*. London: Methuen & Co., Ltd.

Meade, A. 1921. *Modern Gasworks Practice*. London: Benn Brothers, https://archive.org/details/moderngasworks00meadrich (accessed on November 25, 2014).

Meyers, R. ed. 1981. *Coal Handbook*. 517. New York: Marcel Dekker.

Nogita, S., Koyama, S. and Morihara, A. 1986. Technology Developement of Hydrogen Production from Coal. *Sunshine Journal* 7: 70.

Scientific American. 1883. *Dowson's Gas Producer.* 230–231, April 14.

Shimode, M., Hozumi, S, Naka, M., Nakabayashi, Y., Nakahara, Y., Shirakawa, S., Iwahashi, Y., Seike, Y., Muraishi, K., Akiyama, H. 1985. Development of Fluidized Bed Coal Gasifier for Combined Cycle Power Plant. *MHI Technical Review* 22(3), 450–456.

Smil, V. 2004. Enriching the Earth: Fritz Haber, Carl Bosh, and the Transformation of Word Food Production. Cambridge: MIT Press.

Sueyama, T., Tsujino, T. and Chiba, Y. 1985. Ammonia Plant Using Coal Gasification Technology. *Kagaku-Souti* 27: 44.

Tanaka, S., Koyama, S., Morihara, A. and Nogita, S. 1988. Characteristics of Air-blown Gasification in an Entrained-bed Coal Gasifier. *Journal of Fuel Society of Japan* 67: 172.

NOTATION

A	an amount of oxygen supplied	mol
a_i	a parameter to estimate Cp_i	
b_i	a parameter to estimate Cp_i	
CH_jO_k	chemical formula representing by-products in stoichiometric approach	
CH_mO_n	chemical formula representing coal in stoichiometric approach	
Cp_i	heat capacity of a gaseous component i	J/K
c_i	a parameter to estimate Cp_i	
d_i	a parameter to estimate Cp_i	
E	cold gas efficiency	
h_{CH4}	heat of combustion of CH_4	kJ/mol
h_{CO}	heat of combustion of CO	kJ/mol
h_{Coal}	heat of combustion of CH_mO_n	kJ/mol
H_g	apparent heat of gas products	kJ/mol
h_{H2}	heat of combustion of H_2	kJ/mol
HHV	higher heating value	kJ/mol
H_l	heat loss per 1 mol of coal	kJ/mol
H_r	heat of gasification estimated from reaction formula	kJ/mol
i	component of gaseous product	
K	specific constant to evaluate a cold gas efficiency, $K = 2\mu\, h_{CO}/h_{Coal}$	
LHV	lower heating value	kJ/mol
m	number ratio of atoms of coal, hydrogen by carbon, H/C	

n	number ratio of atoms of coal, oxygen by carbon, O/C	
n_i	molar number of component i	mol
O_{st}	oxygen required for standard gasification, $O_{st} = 0.5\ (1 - n)$	mol
O_{ex}	excess amount of oxygen, $O_{ex} = \alpha - O_{st}$	mol
p	molar ratio of H_2 to whole gas produced	
Q	apparent heat of products calculated by $\Sigma n_i q_i$	kJ
q	molar ratio of CO to whole gas produced	
q_i	molar heat of a gaseous component i	J/mol
R_t	averaged residence time of coal particles in the reactor	s
r	molar ratio of CO_2 to the total gas produced	
s	molar ratio of CH_4 to the total gas produced	
t	molar ratio of C_2H_4 to the total gas produced	
T	temperature	K
T_a	atmospheric temperature of the air	K
T_{ad}	adiabatic gasification temperature	K
T_g	temperature of gas in a gasifer	K
T_w	temperature of water used to cool a gasifier wall	K
u	molar ratio of C_2H_6 to the total gas produced	
v	molar number of tar produced	mol
w	molar number of H_2O supplied as steam	mol
x	molar number of H_2 burned with excess O_2	mol
y	molar number of CO consumed by shift reaction	mol
z	molar number of CH_4 generated through the reaction $3H_2 + CO \rightarrow CH_4 + H_2O$	mol

GREEK SYMBOLS

α	molar number of O_2	mol
β	molar number of H_2O	mol
γ	molar number of H_2	mol
δ	molar number of CO	mol
ε	molar number of CO_2	mol
ζ	molar number of H_2O	mol
η	molar number of CH_4	mol
θ	molar number of by-products	mol
κ	molar number of C_2H_4	mol
λ	molar number of C_2H_6	mol
μ	molar number of oxygen for theoretical combustion $\mu = 1 + 0.25m - 0.5n$	mol
ξ	molar number of O_2 required for the combustion of by-products	mol
π	molar number of O_2 required for the combustion of gaseous products	mol
Σ	total molar amounts of gaseous components	mol
χ	oxygen ratio $\chi = \alpha/\mu$	

15 Coal Liquefaction and Processing

Arno de Klerk

CONTENTS

Abstract: Coal liquefaction processes turn coal into mineral-matter-free products. An overview of indirect and direct coal liquefaction is provided, with an emphasis on value addition by direct coal liquefaction, instead of just coal-to-liquids conversion in general. Coal chemistry is discussed from a conversion perspective. Oxidation, cracking, hydrogen transfer, and hydrogenation conversion pathways are explained, as well as how each influences coal conversion and product properties. Product properties are in turn related to refineability to chemicals and transportation fuels, namely, gasoline, jet fuel, and diesel fuel.

15.1 INTRODUCTION

The purpose of coal liquefaction is to turn coal, which is a solid, into a liquid product. There are primarily two reasons for doing so.

First, by turning the organic matter of coal into a liquid product, it is possible to separate the mineral matter from the organic matter. In this respect, it is an extreme form of coal cleaning. Even when the liquid coal product is so heavy that it turns solid at ambient conditions, the so-called ash-free coal is still an upgraded product. Ash-free coal can be used in applications such as integrated gasification combined cycle (e.g., Zhu and Frey 2012), because there are no ash-forming substances in the product. Using ash-free coal in conventional coal-fired power plants improves thermal efficiency, because energy is not expended to heat and convert the mineral matter to ash.

Second, by turning the organic matter of coal into an oil product, the coal-derived oil can be upgraded and refined to produce fuels and chemicals. Historically, coal tar was the main source of many aromatic chemicals (Ahland et al. 1982). There is also an historic precedent for the use of coal liquefaction to produce transportation fuels. A third of the German fuel production during World War II came from direct coal liquefaction (DCL) and indirect coal liquefaction (ICL) processes (Stranges 2007), and in the 1990s a quarter of the transportation fuel in South Africa was produced from coal and gas by ICL.

In this chapter, coal liquefaction will be discussed from the perspective of future production of fuels and chemicals from coal. The topic will be developed by first covering the basics by reviewing coal liquefaction processes in general. This will be followed by a discussion on the carbon efficiency of coal liquefaction. At the conclusion of these sections, the reader should have a basic understanding of coal-to-liquid processes. The remainder of the chapter will deal with a more in-depth discussion of coal conversion chemistry, the upgrading of coal liquids, and future challenges in DCL. Although the discussion will necessarily build on the vast literature that is already available, the objective is to provide a fresh view on the topic, rather than just a review of the literature. A more detailed discussion on synthesis gas-based coal-to-liquid processes is not provided, because only the synthesis gas generation is coal specific (Chapter 14) and not the downstream conversion.

15.2 OVERVIEW OF COAL LIQUEFACTION PROCESSES

There are two approaches of coal liquefaction: DCL and ICL. These approaches represent two fundamentally different viewpoints on how coal should be converted. For classification of coal liquefaction processes, the same classification system as used by Nowacki (1979) was adopted with slight modification (Figure 15.1).

During DCL, the molecules that constitute coal are progressively broken down to smaller molecules. Ideally speaking, the process is efficient, because the only work that must be performed is to break sufficient chemical bonds to turn the solid coal into liquid oil, some of which may be water soluble. The smaller molecules produced in this way retain aspects of the nature of the coal. The traditional DCL processes can be classified based on how they make use of temperature, solvents, H_2, and catalysts: coal pyrolysis, solvent extraction of coal, and catalytic coal liquefaction. A more recent development is the class of wet oxidative coal dissolution processes, which has not yet been fully developed as technologies. Different coals that are similarly converted will yield oils with somewhat different compositions. Irrespective of this, all coal-derived liquids have a high aromatic content and contain organic nitrogen, oxygen, and sulfur species. The further refining of the coal-derived oil is in principle analogous to the refining of conventional crude oil. However, in practice the coal-derived oil is more difficult to refine than conventional crude oil on account of its aromatic nature and associated low H:C ratio. Detailed descriptions of specific processes that were developed within each class of DCL technology can be found in literature (e.g., Richardson 1975; Nowacki 1979; Meyers 1984).

The first step in ICL is to gasify the coal (Chapter 14). During the gasification process, the coal is completely broken down to gaseous products. Considerable effort is required, because almost all chemical bonds in the coal are broken. Once the coal-derived synthesis gas is produced and cleaned, the clean synthesis gas that consists mainly of H_2 and CO is converted into a liquid product. The nature of the liquid product is determined by the technology employed to convert the synthesis gas. Two subclasses of ICL processes can be found in the industry: methanol synthesis (Bertau et al. 2014) and Fischer–Tropsch synthesis (Steynberg and Dry 2004; De Klerk 2011; Maitlis and De Klerk 2013). There are

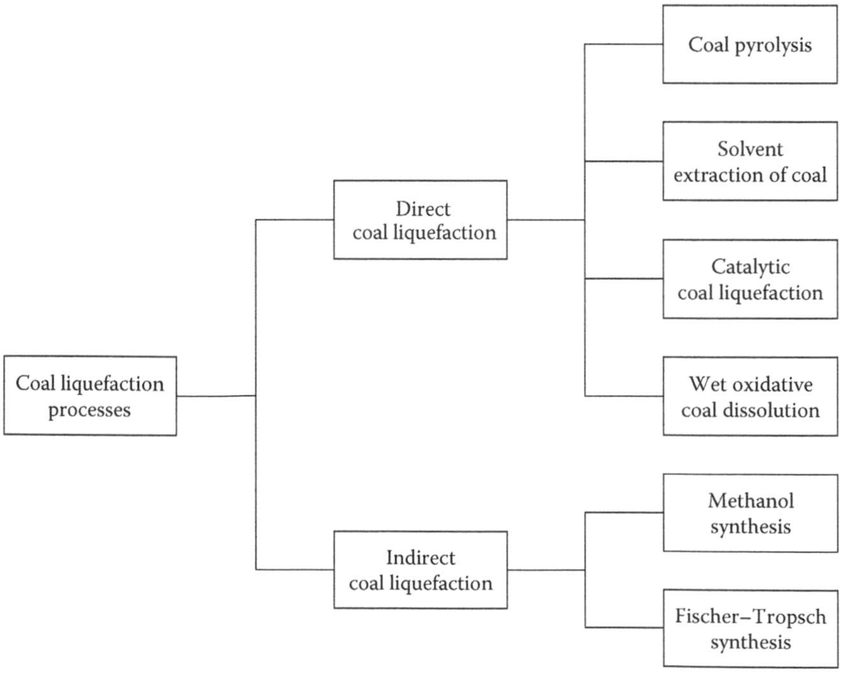

FIGURE 15.1 Classification of coal liquefaction processes.

also other variants on these two processes that are not shown in Figure 15.1, such as the Kölbel-Engelhardt synthesis and lower alcohol synthesis. Because the feed material is clean synthesis gas, the liquid products from indirect liquefaction contain only carbon, hydrogen, and oxygen. Further refining is required to produce fuels and chemicals, but the refining effort is less than that required for the conventional crude oil on account of the high H:C ratio of the liquid and the absence of sulfur and nitrogen species.

When the quality of the coal-derived oil does not matter, DCL is a more efficient strategy than ICL for coal-to-liquids conversion. However, the refining of the coal-derived oil from DCL is challenging. Unless heteroatom containing or aromatic chemicals are produced, oil quality is generally improved by increasing the H:C ratio and decreasing the heteroatom content. Improving the quality of oil from a DCL process to produce fuels requires hydrotreatment and the H_2 needed for hydrotreatment is obtained by coal gasification. Because coal gasification is the most demanding step in ICL, the efficiency difference between DCL and ICL is eroded as the hydrogen content of the oil from DCL is increased. Ultimately, if the H:C of the product is important, ICL is a more efficient strategy than DCL for coal-to-liquids conversion. Coal liquefaction should therefore not be evaluated independently and must be evaluated in the context of the intended downstream processing and intended application of the coal-derived liquids.

Each of the coal liquefaction process classes will be discussed to highlight the main elements of the different strategies for coal conversion. The reader is cautioned that for the sake of clarity much of the process complexity is hidden.

15.2.1 Coal Pyrolysis Processes

Coal pyrolysis processes obtain coal liquids from the coal by heating the coal in the absence of a solvent under an inert atmosphere. Simply put, the volatile matter in coal is recovered as a coal tar liquid, whereas the fixed carbon in coal is carbonized. The fixed carbon leaves the process with the pyrolized mineral matter as a solid coke-like product.

A generic coal pyrolysis process is shown in Figure 15.2. Coal of a particular size range is employed as coal feed (stream 1); the particle size range of coal depends on the specific coal pyrolysis technology. The first step is usually coal drying to remove excess moisture, but this can be integrated with the pyrolysis step. The coal feed must be heated for pyrolysis to take place and this is often accomplished by using a hot gaseous feed (stream 2). The typical pyrolysis temperature range is 425°C–550°C at near atmospheric pressure. The hot gas can be generated by combustion of the coal char or the coal gas produced by the process. In the pyrolysis reactor, the volatile matter obtained from the coal leaves the reactor as a vapor-phase product (stream 3), whereas the carbonized coal leaves as a solid product (stream 4). The vapor-phase product is cooled and phase separated, typically using a quench cooler. The coal gas (stream 5) contains the uncondensed product from the pyrolysis reactor. The condensed liquid (stream 6) must be further separated into an aqueous product phase, the

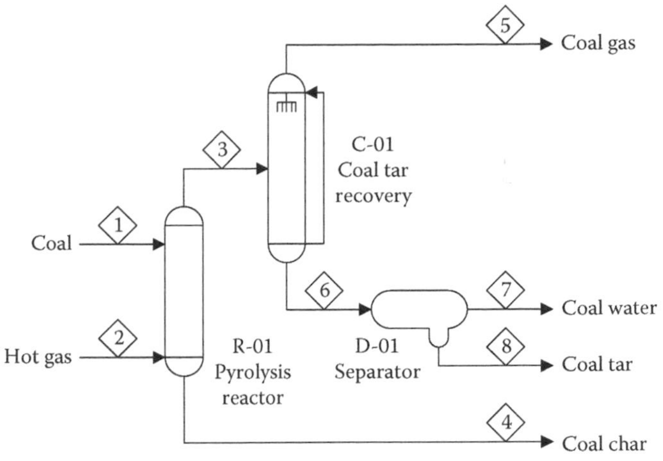

FIGURE 15.2 Generic coal pyrolysis process.

coal water (stream 7), and the organic product phase, the coal tar (stream 8). The coal tar is a very heavy oil product with a higher density than water and it may actually be a solid at ambient conditions.

15.2.2 Solvent Extraction of Coal Processes

Solvent extraction of coal processes are solvent-assisted coal pyrolysis processes. As the name suggests, a solvent is employed to facilitate the "extraction" of liquids from coal during pyrolysis. These processes are also sometimes referred to as thermal liquefaction or noncatalytic liquefaction processes. Solvent extraction of coal is not a physical extraction, although some physical extraction takes place at lower temperatures. Most of the liquid is produced by coal pyrolysis in a coal–solvent slurry at a high temperature and pressure. The solvent serves mainly as a source of transferable hydrogen and is therefore called a hydrogen-donor solvent. By transferring hydrogen from the solvent to the coal liquid, side reactions are suppressed and a higher yield of coal liquids can be obtained, because some fixed carbon is converted in addition to the volatile matter. In essence, the solvent hydrogenates the coal tar and in the process the coal tar dehydrogenates the solvent. Simply put, the solvent is a stoichiometric hydrogenation reagent that enables the coal pyrolysis process to produce more coal liquids than coal pyrolysis on its own. The solvent is regenerated to make this a continuous process. This can be accomplished by co-feeding H_2 to the process, or more efficiently, by hydrotreating the solvent with H_2 before it is recycled to the process, or both.

A generic solvent extraction of coal process is shown in Figure 15.3. It is useful to compare this process with coal pyrolysis to evaluate the increase in process complexity that is introduced by the solvent and the need for H_2.

Coal of a particular size range is employed. The coal must also be dried before use, because water present in the coal causes increased reactor pressure. The dried coal feed (stream 1) is mixed with the solvent (stream 2), which is actually a fraction of the coal liquids and not an externally supplied solvent.

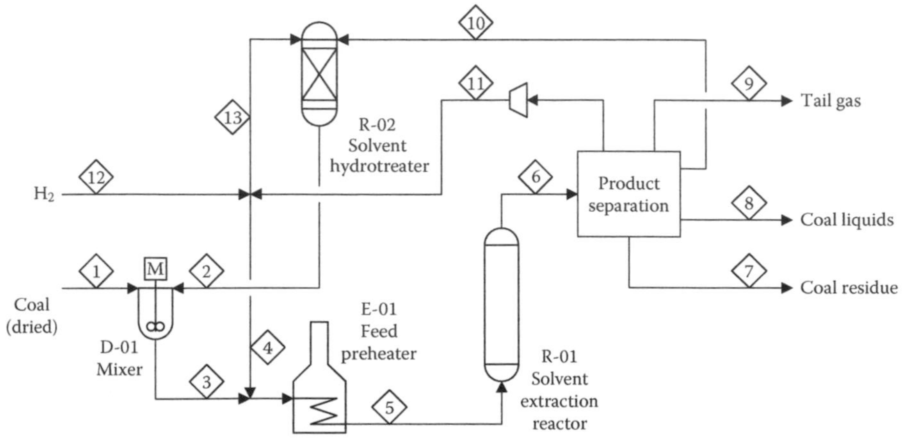

FIGURE 15.3 Generic solvent extraction of coal process.

Negligible physical coal dissolution takes place in the mixing tank. The coal–solvent slurry (stream 3) is then pumped to pressure and mixed with H_2 (stream 4). The mixture is preheated in a fired feed preheater to produce the reactor feed (stream 5). Typical operating conditions in the reactor are 400°C–500°C and 3–20 MPa. The reactor provides residence time and in many processes, it is operated as an up-flow plug flow reactor. The converted coal–solvent slurry (stream 6) must be separated into different products. Product separation, which comprises multiple steps, produces different product fractions. The coal residue (stream 7) contains the mineral matter and undissolved organic matter in the coal. The coal liquid (stream 8) is one or more distillation fractions of the dissolved coal. The tail gas (stream 9) contains the uncondensed gases and purge from the H_2 recycle. The solvent fraction (stream 10) of the coal liquids is sent to a high-pressure hydrotreater. In the hydrotreater, the solvent is hydrotreated to restore its hydro-donor properties and the hydrotreated solvent (stream 2) is returned to the mixer to prepare the coal–solvent slurry. Part of the H_2 is recycled (stream 11) to reduce the fresh H_2 feed (stream 12) requirements. The H_2 is supplied to the preheater feed (stream 4) and to the hydrotreater (stream 13).

Although the fresh H_2 feed (stream 12) is shown as a process input, hydrogen generation by coal gasification substantially contributes to the size and complexity of the solvent extraction of coal process. A block flow diagram of a generic coal gasification section to produce H_2 is shown in Figure 15.4. A more detailed description of the process and its utility requirements can be found in the chapter on coal gasification (Chapter 14).

In addition to the coal used for the solvent extraction of coal process, coal is also used as a feed for hydrogen production. In Figure 15.4, the coal feed (stream 1) is used as a feed for coal gasification (stream 2) and as an energy source for the steam boilers (stream 3). The steam boilers require water as a feed, but the freshwater feed (stream 4) must first be demineralized to produce boiler feed water (stream 5) that can then be turned into steam. Coal combustion provides energy for steam generation (Chapter 13). Steam is a widely used utility, but as shown, it is employed mainly as a feed for power generation (stream 6) and coal gasification (stream 7). Although power generation is shown as a separate unit, steam turbines

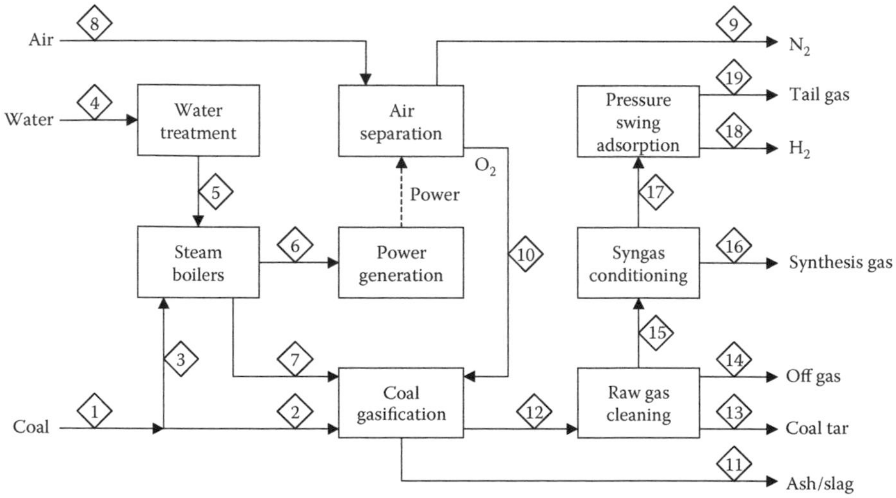

FIGURE 15.4 Generic coal gasification process section.

can be used to generate electric power, or to provide shaft work directly to the air separation unit (ASU). The ASU is needed to provide purified O_2 for gasification. Air (stream 8) is cryogenically separated into a purified N_2 stream (stream 9) and a purified O_2 stream (stream 10).

Coal gasification has three feed streams: coal (stream 2), steam (stream 7), and purified O_2 (stream 10). After gasification, the mineral matter in the coal feed leaves the coal gasifiers as either ash or slag (stream 11), depending on the gasification technology. The raw synthesis gas (stream 12) potentially contains coal tar, but mainly consists of water–gas shift products at equilibrium at the gasifier outlet temperature: H_2, CO, CO_2, and H_2O. It also contains contaminants such as H_2S and NH_3. The raw gas cleaning section is responsible for the separation of coal tar (stream 13) from the raw synthesis gas, as well as the removal of contaminant gases (stream 14). The extent of gas cleaning and the nature of the materials removed from the raw synthesis gas depend on both the gasification technology and the downstream process needs. The clean synthesis gas (stream 15) is sent for gas conditioning. Similar to gas cleaning, the extent and type of gas conditioning that must be performed depends on the downstream requirements. When the purpose is to produce H_2, gas conditioning will be a shift reactor to perform low-temperature water–gas shift conversion to maximize the H_2 content in the gas. Part of the gas can be produced as a synthesis gas (stream 16), but for H_2 production the conditioned gas (stream 17) will be sent for H_2 purification. Purified H_2 (stream 18) can be produced by pressure swing adsorption, the remainder becoming a low-pressure tail gas (stream 19).

From this simplified description, it is clear that the H_2 requirement of the solvent extraction of coal process is associated with much complexity.

15.2.3 Catalytic Coal Liquefaction Processes

Catalytic coal liquefaction is analogous to the solvent extraction of coal. It is also a solvent-assisted coal pyrolysis process. The main difference between catalytic coal liquefaction and solvent extraction of coal is that a hydrogenation catalyst is co-fed with the hydrogen-donor solvent and H_2 to the liquefaction reactor. By including a suitable catalyst in the feed to the coal liquefaction reactor, the hydrogenation efficiency is improved. Hydrogen transfer from the solvent to the pyrolyzed coal liquids is not the only type of hydrogenation that takes place in the reactor. The hydrogenation catalyst uses the co-fed H_2 to hydrogenate the coal liquids. The hydrogenation catalyst also hydrogenates the dehydrogenated hydrogen-donor solvent to restore the solvent's hydrogen-donor capability. The quality of the coal liquids produced is improved, because more hydrogen is added to the coal liquids during the liquefaction process. Furthermore, it was noted that improved performance due to added catalysts, even at short reactor residence times, exceeded the improvement in performance that could be obtained by better hydrogen-donor solvents (Derbyshire 1988). When co-feeding a homogeneous or finely dispersed catalyst with the process feed, catalyst recovery from the liquefaction products is challenging. For this reason, the use of a disposable catalyst is preferred in such applications. For the same reason, only iron-based catalysts are considered as hydrogenation catalysts in all of the more recent feasibility studies of catalytic coal liquefaction processes (Kaneko et al. 2012). An alternative approach is to employ larger catalyst particles in the liquefaction reactor so that catalyst–product separation is easier, albeit with a loss in catalyst efficiency due to increased mass transport resistance.

A generic catalytic coal liquefaction process is shown in Figure 15.5, which illustrates the use of a catalyst co-feed. The coal feed (stream 1), dispersed catalyst (stream 2), and solvent (stream 3) are mixed. The catalyst containing coal–solvent slurry (stream 4) is co-fed with H_2 (stream 5) to the fired feed preheater. The preheated feed (stream 6) is introduced to the liquefaction reactor. Typical operating conditions for catalytic coal liquefaction are 350°C–450°C and 10–30 MPa. Very high H_2 pressures are required due to the constraints imposed by the hydrogenation–dehydrogention equilibrium at these conditions. The liquefaction product (stream 7) is separated into various fractions by multiple separation units. This is a complex separation design. The solid coal residue and spent catalyst (stream 8) can be further processed to recover some of the catalyst, should that be required. The coal liquids (stream 9) are already partially

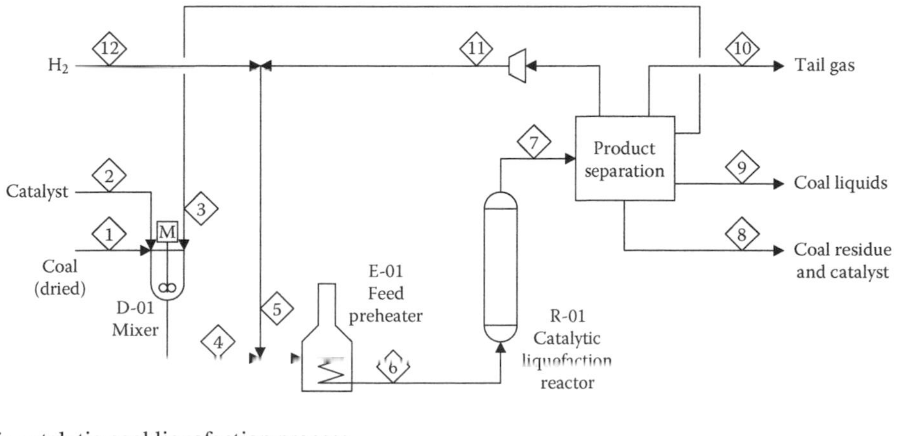

FIGURE 15.5 Generic catalytic coal liquefaction process.

upgraded and different product fractions may be produced by distillation, which is not shown. The solvent (stream 3) that is used to prepare the coal–liquid slurry is one of the coal liquid fractions. The uncondensed products are separated into a tail gas (stream 10) and a H_2 recycle (stream 11). The H_2 recycle reduces the fresh H_2 feed (stream 12) requirements.

Due to the improved hydrogenation efficiency during catalytic coal liquefaction, the H_2 requirements are higher for the solvent extraction of coal. Generating H_2 from coal is a complex process (Figure 15.4). The improvement in liquid quality and yield that can be obtained by catalytic coal liquefaction compared to the other DCL processes comes at the cost of additional H_2 capacity.

Another variation on catalytic coal liquefaction is substituting the hydrogenation catalyst for an acid catalyst. The process flow diagram for the liquefaction section is very similar to that shown in Figure 15.5. Zinc chloride ($ZnCl_2$) and stannous chloride ($SnCl_2$) were the two most successful catalysts in this category (Derbyshire 1988). Although these catalysts are sometimes referred to as "hydrocracking" catalysts, Zn and Sn are not very active for hydrogenation.

15.2.4 Wet Oxidative Coal Dissolution

Unlike the other DCL processes, wet oxidative coal dissolution has not been developed to pilot or commercial scale. Yet, there are encouraging reports in literature. Low-temperature wet oxidation of coal (Hayashi et al. 1997 and Mae et al. 1997) and oxidative hydrothermal dissolution of coal (Anderson et al. 2011) reportedly dissolve upward of 70% of the organic matter in coal. The products that are obtained from wet oxidative dissolution of coal are primarily water-soluble organics, extractable heavier organics, and gaseous products. The gaseous products are mainly CO and CO_2.

Oxygen, instead of hydrogen, is added to the coal liquids that are produced by an oxidative coal liquefaction process. The coal liquids are therefore rich in carboxylic acids and phenolics.

15.2.5 Methanol Synthesis Processes

Methanol synthesis from synthesis gas is an indirect liquefaction process. Bertau et al. (2014) provide an account of the state of the art in methanol synthesis.

Coal conversion takes place in the beginning of the process when the coal is gasified to produce synthesis gas (Figure 15.4). Sulfur-containing compounds must be removed from the synthesis gas, because they are methanol synthesis catalyst poisons. For methanol synthesis, a synthesis gas with an effective H_2:CO ratio of 2:1 is desired. Because CO_2 is also converted during methanol synthesis, the synthesis gas composition requirement is more often expressed as the stoichiometric number $(H_2 - CO_2)/(CO + CO_2) \approx 2$. These requirements are derived from the reaction stoichiometry for methanol synthesis from CO (Equation 15.1) and CO_2 (Equation 15.2).

$$CO + 2\,H_2 \rightleftharpoons CH_3OH\,,\ \Delta H_r = -91\ \text{kJ mol}^{-1} \quad (15.1)$$

$$CO_2 + 3\,H_2 \rightarrow CH_3OH + H_2O\,,\ \Delta H_r = -50\ \text{kJ mol}^{-1} \quad (15.2)$$

In Figure 15.4, the gas conditioning must therefore be performed in such a way that the main product is synthesis gas (stream 16) with the aforementioned composition requirements. During methanol synthesis, the synthesis gas is converted to methanol with around 99% selectivity (Lee 1990). Methanol synthesis from synthesis gas is an equilibrium-limited reaction and the single-pass conversion is limited by equilibrium. For this reason, methanol synthesis technologies benefit from producing dimethyl ether (H_3C—O—CH_3) instead, because it enables a higher single-pass conversion.

A generic synthesis gas-to-methanol conversion process is shown in Figure 15.6. The feed material is clean and conditioned synthesis gas (stream 1) from coal gasification. This feed is combined with the unconverted synthesis gas recycle (stream 2) to produce a combined feed (stream 3) with the required composition for methanol synthesis. The feed is preheated in a feed–product heat exchanger to the inlet temperature conditions for methanol synthesis. The preheated

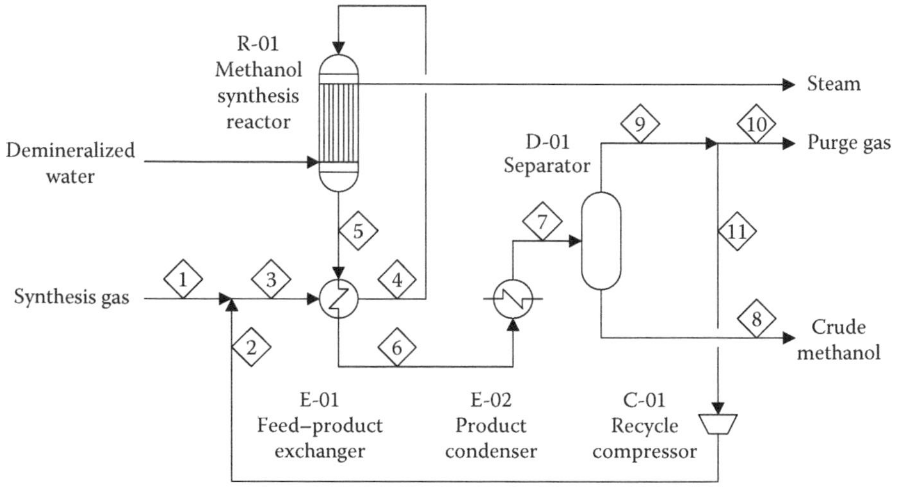

FIGURE 15.6 Generic methanol synthesis process.

synthesis gas feed (stream 4) is fed to the methanol synthesis reactor. Methanol synthesis is an exothermic reaction as shown in Equations 15.1 and 15.2. Heat management in the methanol synthesis reactor is important. There are different ways of managing the heat; the way shown in Figure 15.6 is heat removal by steam generation in a multitubular fixed-bed reactor. The reactor design and heat management strategy is technology dependent. Typical operating conditions for methanol synthesis are 230°C–260°C and 5 MPa and higher pressures. The hot reaction product (stream 5) is used to preheat the feed, thereby cooling down the reaction product (stream 6), which is further cooled to condense the methanol and produce a two-phase product in vapor–liquid equilibrium (stream 7). The vapor and liquid phases are separated to recover the crude methanol as liquid product (stream 8). The crude methanol contains some water and a small amount of side-products. The unconverted synthesis gas (stream 9) is split into two streams: the purge gas (stream 10) and the recycle gas (stream 11). The purge gas is a necessary loss from the process to control the buildup of inert gases in the recycle. The recycle gas is compressed to become the unconverted synthesis gas recycle (stream 2).

The crude methanol can easily be purified using two atmospheric pressure distillation columns to produce a final marketable product. In this respect, methanol synthesis is unique among the coal liquefaction technologies, because it is the only technology that produces a single-liquid product that can be marketed directly as an easily transportable final product without the need for extensive downstream refining.

15.2.6 Fischer–Tropsch Synthesis

Fischer–Tropsch synthesis, like methanol synthesis, is an indirect liquefaction process. Steynberg and Dry (2004), De Klerk (2011), and Maitlis and De Klerk (2013) provide accounts detailing Fischer–Tropsch synthesis, technology, and refining.

The first step in a coal-fed Fischer–Tropsch-based process is coal gasification (Figure 15.4) to produce the synthesis gas for the process. The synthesis gas must be cleaned to remove sulfur-containing species to very low levels, because sulfur-containing compounds are detrimental to Fischer–Tropsch synthesis. The synthesis gas composition requirements with respect to H_2, CO, and CO_2 depend on the type of Fischer–Tropsch catalyst that is employed and whether it is active for the water–gas shift reaction (Equation 15.3) or not.

$$CO + H_2O \rightleftharpoons CO_2 + H_2,\ \Delta H_r = -41\ \text{kJ mol}^{-1} \qquad (15.3)$$

When the Fischer–Tropsch catalyst is active for the water–gas shift reaction, there is little point in extensive conditioning of the synthesis gas, because the composition will change toward the water–gas shift equilibrium during the synthesis. Iron-based catalysts are water–gas shift active and can be operated with a wide range of H_2:CO ratios in the synthesis gas. When the Fischer–Tropsch catalyst is not active for the water–gas shift reaction, the synthesis gas should be conditioned to meet the stoichiometric requirements of the synthesis reactions. Cobalt-based catalysts are not active for the water–gas shift reaction.

Fischer–Tropsch synthesis involves CO polymerization and hydrogenation, which are very exothermic reactions. The heat of reaction is on the order of –140 to –160 kJ mol^{-1} CO converted. The catalyst and operating conditions determine the product distribution. The product is a syncrude that contains hydrocarbons and oxygenates with a wide boiling range; the main products are alkanes (Equation 15.4), alkenes (Equation 15.5), alcohols (Equation 15.6), carbonyls (Equation 15.7), and carboxylic acids (Equation 15.8).

$$nCO + 2n{+}1\ H_2 \rightarrow H(CH_2)_n H + nH_2O \qquad (15.4)$$

$$nCO + 2nH_2 \rightarrow (CH_2)_n + nH_2O \qquad (15.5)$$

$$nCO + 2nH_2 \rightarrow H(CH_2)_n OH + n-1\ H_2O \qquad (15.6)$$

$$nCO + 2n-1\ H_2 \rightarrow H(CH_2)_{n-1} CHO + n-1\ H_2O \qquad (15.7)$$

$$nCO + 2n-2\ H_2 \rightarrow H(CH_2)_{n-1} COOH + n-2\ H_2O \qquad (15.8)$$

A generic low-temperature Fischer–Tropsch synthesis process is shown in Figure 15.7. High-temperature Fischer–Tropsch synthesis is slightly less complicated, because no wax is produced. It is useful to compare Fischer–Tropsch synthesis with methanol synthesis to appreciate the relative increase in process complexity due to the production of a wide boiling range multiphase syncrude.

The synthesis gas feed from coal gasification (stream 1) is combined with the recycled synthesis gas (stream 2) to form the combined synthesis gas feed (stream 3) to the process. The combined synthesis gas feed is preheated by feed–product heat exchange and becomes the Fischer–Tropsch reactor feed (stream 4). Fischer–Tropsch synthesis is very exothermic and heat management is important in the reactor design. Reaction heat is removed by generating steam and the way in which this is accomplished depends on the Fischer–Tropsch technology employed. The most common reactor types are multitubular fixed-bed (shown in Figure 15.7), slurry bubble column, and fluidized-bed reactors. In the latter two reactor types, additional equipment for product–catalyst separation is required. The operating conditions depend on the Fischer–Tropsch technology and operating temperatures span the range 190°C–340°C. Operating pressure is usually in the range 2–2.5 MPa. During low-temperature Fischer–Tropsch synthesis, liquid and vapor phases are both present at reaction conditions. The liquid-phase product from Fischer–Tropsch synthesis is a paraffinic wax (stream 5). The wax leaves the reactor as a liquid-phase product, but it will congeal to form a solid wax product on cooling. The vapor-phase product from Fischer–Tropsch synthesis (stream 6) is stepwise cooled. Cooling is usually controlled in such a way that an oil product without any water is recovered from the partially

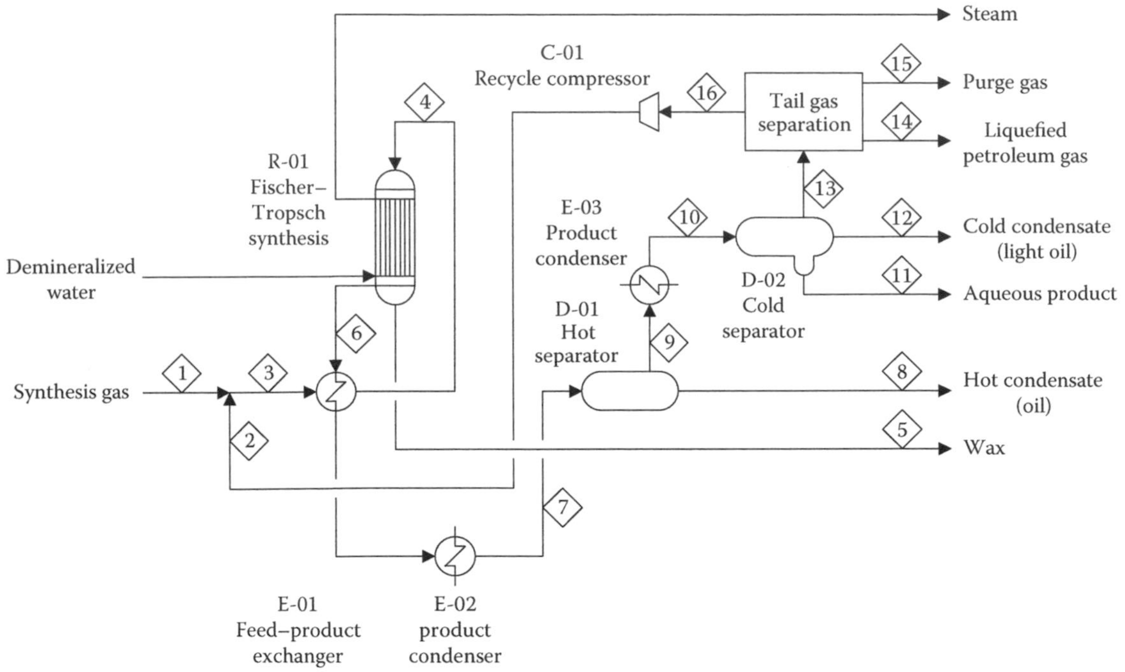

FIGURE 15.7 Generic low-temperature Fischer–Tropsch synthesis process.

condensed product (stream 7) at higher temperature, which also prevents congealing of the waxy fraction in the oil. The hot condensate (stream 8) is recovered by phase separation. The vapor-phase product (stream 9) is further cooled so that the water and lighter oil products that will not congeal on cooling can be condensed. The cooled three-phase product (stream 10) is phase separated. The aqueous product (stream 11) contains mainly water, with about 1% of dissolved oxygenates. The cold condensate (stream 12) is a light oil product consisting of hydrocarbons and some heavier oxygenates. The uncondensed vapor or tail gas (stream 13) contains the lightest fraction of hydrocarbons, the unconverted synthesis gas, and inert gases that were present in the synthesis gas feed from coal gasification. If the Fischer–Tropsch catalyst was active for the water–gas shift reaction, the uncondensed vapor will also contain much CO_2. In the tail gas separation section, additional light hydrocarbons can be recovered. Condensation under pressure can recover liquefied petroleum gas (stream 14) as a product. More complicated designs also involve CO_2 and lighter hydrocarbon recovery. The remaining gas is split between the purge stream (stream 15) to avoid inert gas buildup in the gas recycle (stream 16). The gas recycle is compressed to yield the recycled synthesis gas (stream 2).

The syncrude that is recovered from Fischer–Tropsch synthesis contains many different compound classes, primarily alkanes, alkenes, and alcohols, as well as some carboxylic acids, carbonyl compounds, and aromatics. Aromatics are produced only during high-temperature synthesis. These compounds are recovered in multiple product phases. For example, at ambient conditions, the syncrude from low-temperature Fischer–Tropsch synthesis is present in four phases: solid (wax), organic liquid (oil), aqueous liquid (oxygenates dissolved in water), and vapor (light hydrocarbons).

15.3 COAL-TO-LIQUIDS CARBON EFFICIENCY

The carbon efficiency, η_C, of coal liquefaction is the ratio of the carbon in liquefied coal product to the carbon in the organic matter of the coal (Equation 15.9):

$$\eta_C = \frac{(\text{mass C in coal liquids})}{(\text{mass C in organic matter of coal})} \tag{15.9}$$

For this measure to be meaningful, it must incorporate the energy cost associated with the process expressed on the mass of carbon converted to produce the coal liquids. This avoids the misrepresentation of process efficiency when technologies have different modes of energy input or different boundaries between units that are considered part of the process and part of the utility infrastructure.

Many other measures of process efficiency can also be employed, but they are all in some way or another related to the carbon efficiency. Carbon efficiency is important, because it captures economic and environmental performance in an easy to relate manner.

The carbon efficiency provides a measure of the economic viability of the process. There is a cost associated with the coal feed and there is an income related to the coal liquids. These values capture the monetary value attached to carbon quality. The gross carbon profit of the process, before operating and capital cost, can be calculated if the carbon efficiency is known (Equation 15.10).

$$\text{Gross carbon profit} = (\$/\text{kg C in coal liquids}) - (\$/\text{kg C in coal})\,(\eta_C)^{-1} \quad (15.10)$$

The cost of capital and the operating cost per unit mass of carbon production must be less than the gross carbon profit for the process to be economically viable. The higher the efficiency for the same level of upgrading of the carbon quality, the more likely it is that the process will be economically viable.

As an environmental measure, the carbon efficiency provides an indirect indication of the process impact. The carbon efficiency indicates how wasteful the transformation from one energy carrier (coal) to another energy carrier (coal liquids) is. Unless the societal benefit can justify the carbon cost, the process may not have the social license to operate. Currently the most prevalent expression of carbon loss is the CO_2 footprint of the process, which is also a measure of carbon efficiency. The main difference between the CO_2 footprint and the carbon efficiency is that not all carbon loss necessarily translates to loss as CO_2. For example, incomplete conversion of coal results in coal residue, which is a solid waste product.

Despite the value of carbon efficiency as a measure of process performance, it does not make a distinction between preventable carbon losses and inevitable carbon losses. Carbon that is lost due to the first and second laws of thermodynamics constitutes an inevitable loss of carbon. Similarly, carbon loss associated with reaction stoichiometry is also an inevitable loss of carbon. Other forms of carbon loss can in theory be prevented, but to be realistic, a distinction should be made between carbon losses associated with practical limitations of current technology and carbon losses due to design decisions. The amount of carbon that is lost and the form in which it is lost are both important for carbon management (Chapters 18 and 19). The main sources of carbon loss will be discussed so that the origin of carbon loss can be explained. In this way, carbon loss can be related to the different technologies that are employed for coal liquefaction and some design decisions that are associated with each.

15.3.1 Coal Conversion in Direct Liquefaction

The coal conversion in DCL technologies is incomplete and at least some of the organic matter remains associated with the mineral matter in the coal residue. The exact amount depends on the coal composition too. This makes a comparison of coal conversion between different DCL technologies meaningless unless the same coal was used as feed material.

Nevertheless, some indication of the carbon conversion in DCL can be obtained from the literature (e.g., Mzinyati 2007). A direct comparison of results published for the DCL of Illinois No. 6 coal is presented in Table 15.1.

Illinois No. 6 coal is a high volatile matter bituminous coal with the following contemporary analysis (Johnson et al. 1973): 42 wt% volatile matter, 48 wt% fixed carbon, and 10 wt% ash. The yields in Table 15.1 were expressed on moisture and ash-free (maf) basis, and the data were taken from different sources. The char-oil-energy-development (COED) process represents coal pyrolysis (Jones 1973). The Exxon donor-solvent (EDS) process represents solvent extraction of coal (Nowacki 1979 and Schlosberg 1985). The H-Coal process represents catalytic coal liquefaction (Johnson et al. 1973).

TABLE 15.1
Comparison of Product Yields from Illinois No. 6 Coal Reported for Different DCL Processes

	Yield (wt% maf)		
Product	Pyrolysis COED Process	Solvent Extraction EDS Process	Catalytic Liquefaction H-Coal Process
Coal residue	54.8	42.1	6.6
Oil products	21.7	47.9	73.7
Ammoniacal liquor	6.1	5.4	25.2
Gas	17.4	6.5	
H_2	0	−1.9	−5.5

The organic matter remaining in the coal residue is a carbon loss from the liquefaction reactor, but it is not necessarily a net loss to the overall coal liquefaction process. If the coal feed does not have high ash content, the coal residue can still be used for heating purposes, thereby reducing the additional amount of coal needed to produce energy. The coal residue can likewise be employed as a feed to a coal gasifier to produce H_2. In this respect, the low carbon loss associated with catalytic coal liquefaction (Table 15.1) might be suboptimal when the total process is considered, because the coal residue is no longer useful for heating or gasification purposes.

The coal conversion in the DCL reactor sets an upper bound on the carbon efficiency, but it is not indicative of the carbon efficiency of the process.

15.3.2 Coal Gasification Conversion

In a DCL facility, H_2 is prepared from coal that is a carbon-rich and hydrogen-poor feed material. In an ICL facility, coal gasification is the first step. Synthesis gas and H_2 are preferably prepared from natural gas, which is a hydrogen-rich feed material. However, if natural gas is abundantly available, it is less likely that coal liquefaction technology will be considered, because it is usually cheaper and less complex to construct gas-to-liquids facilities than coal-to-liquids facilities.

An aspect of coal gasification that is of interest here is coal consumption in relation to H_2 production. Adding H_2 to coal during DCL improves the yield and quality of the coal liquids, but H_2 production consumes coal. In ICL, the total coal feed is gasified to produce synthesis gas and the yield of synthesis gas with a H_2:CO equal to the consumption ratio during synthesis determines how much methanol or Fischer–Tropsch syncrude can be produced. The gasification technology selection and the composition of the coal affect the efficiency with which synthesis gas can be produced. This is explained in the chapter on coal gasification (Chapter 14).

In order to estimate the stoichiometry of coal consumption for synthesis gas production, some characteristics of the coal gasification process must be known (Higman and Van der Burgt 2008):

1. Cold gas efficiency based on the higher heating value (HHV) of the raw synthesis gas. This is the ratio of the HHV of the produced gas to the HHV of the coal. This enables a first estimate of the synthesis gas content of the raw gas.
2. The gasifier product selectivity is necessary to fully interpret the cold gas efficiency. Good assumptions can usually be made if the gasifier type is known. A high cold gas efficiency is not necessarily indicative of a high H_2 and CO yield, because the raw gas from fluidized-bed and fixed/moving-bed gasifiers may contain coal tar, as well as light hydrocarbon gases and methane in particular. If the raw gas composition is available, the synthesis gas yield is directly known and the cold gas efficiency is not needed.
3. Carbon conversion of the coal indicates how much carbon is lost due to incomplete conversion in the gasifier. This carbon loss is as a solid and from the point of view of carbon management, this is sequestered carbon. Good assumptions can usually be made if the gasifier type is known.

The gasifier also requires purified O_2 as a feed material. Any calculation of the carbon efficiency must take the energy consumption of this energy-intensive process into consideration. The energy required to produce 99.5% O_2 is 0.35 kWh per standard cubic meter of unpressurized oxygen (Häring 2008). In order to express this energy consumption on the same basis as the process feed, it must be converted on a coal consumption basis. Electric power generation from coal adds another efficiency factor to the calculation. The efficiency of power generation and transmission depends on the coal feed and the power generation technology. Nevertheless, for electric power generation from a power plant with 34%–35% thermal efficiency, an estimate of the overall efficiency, which includes transmission and other losses, is 25% (Pansini and Smalling 2002).

Due to the large number of variables involved in the calculation of the carbon cost of synthesis gas generation, it is better to illustrate coal gasification carbon efficiency by a specific example (Table 15.2). In this example, data from an entrained flow gasifier using an Illinois bituminous coal (Higman and Van der Burgt 2008) were employed to match the performance data in Table 15.1.

Due to the high outlet temperature of entrained-flow gasifiers, there is a negligible amount of hydrocarbon gases in the raw synthesis gas, carbon conversion is near complete and the water–gas shift (Equation 15.3) equilibrium favors CO and H_2O over CO_2 and H_2. The direct carbon loss due to entrained-flow gasification is therefore very little. The

TABLE 15.2
Synthesis Gas Generation Efficiency by Entrained-Flow Gasification of Illinois Bituminous Coal

Description	Entrained-Flow Gasifier Performance
Carbon conversion (%)	>99
Oxygen feed (kg O_2/kg coal, maf)	0.72
Steam feed (kg H_2O/kg coal, maf)	0.18
Dry, raw syngas (mol/kg coal, maf)	
H_2	33
CO	58
CO_2	0.9
Inert gases (N_2, Ar)	1.9
H_2S	1.4
CH_4	Trace
Calculated loss (kg C/kg C process feed)	
Incomplete conversion	0.001
CO_2 selectivity during gasification	0.016
Air separation for purified O_2	0.077
Steam generation for gasification	0.015
Water–gas shift for H_2:CO = 2:1	0.468

coal consumed to provide power for the ASU is 0.077 kg per kg coal in the process feed. The coal consumed to provide steam for the gasifier is 0.015 kg per kg coal in the process feed. Although these are inevitable and meaningful carbon losses, these losses are small compared to the inevitable stoichiometric carbon loss associated with water–gas shift conversion. Water–gas shift conversion to produce a synthesis gas composition with a H_2:CO molar ratio of 2:1 is 47% of the carbon in the process feed. The low H:C ratio of the coal feed cannot be adjusted to a high H:C ratio without sacrificing carbon.

Using the same example, it is possible to calculate the amount of coal that is required to produce H_2. Nearly complete CO conversion can be achieved when low-temperature water–gas shift conversion is performed to maximize the H_2 yield. Almost all carbon in the coal is rejected as CO_2. In the process, the carbon is rejected as CO_2 during water–gas shift conversion, and for utilities the carbon is converted to CO_2 during coal combustion to produce power for air separation and energy for steam generation. In total it requires 6 kg coal (maf) to produce 1 kg H_2, or differently put, 0.17 kg H_2 can be produced per kg coal converted. This is very close to the molar mass ratio of H_2 and C and reflects the dominant role of water–gas shift conversion (Equation 15.3) in the stoichiometry of the process.

When the coal gasification technology has a high yield of methane and light hydrocarbon gases, some process designs may incorporate a gas reformer to take advantage of these products. By including a gas reformer in the process, the cost and complexity are increased, but it can improve the carbon efficiency of the process. The production of light hydrocarbons in a coal liquefaction facility is therefore not necessarily a direct carbon loss. This aspect of coal gasification for synthesis gas generation is not apparent from the example provided.

TABLE 15.3
Thermal Efficiencies of Different DCL and ICL Processes Using a Bituminous Coal Feed

		Thermal Efficiency (%)	
Type	Process	Overall	Liquids Only
Solvent extraction of coal	SRC II	77	56
	CSF donor solvent	64	50
Catalytic coal liquefaction	Synthoil	68	65
	H-Coal	60	55
Fischer–Tropsch synthesis	Sasol 1	31	22
	Sasol 2	35	32

15.3.3 Product Hydrogen-to-Carbon Ratio

Although it is convenient to think of coal liquids as synthetic crude oils that can be refined like conventional crude oil, in reality the H:C ratio of the coal liquids will determine how onerous such refining will be. The DCL technologies produce coal liquids that retain some of the coal character and these coal liquids tend to have low H:C ratios. Additional hydrogen can be added during liquefaction by hydrogen disproportionation or H_2, but all forms of hydrogen come with a carbon cost. The ICL technologies can produce liquids with high H:C ratios, but the same carbon cost applies. Producing H_2 to adjust the H:C ratio from coal, which is a low H:C feed material, requires carbon to be sacrificed to meet stoichiometric requirements and to provide energy for the coal-to-liquids transformation.

The net carbon efficiency of a coal liquefaction process is therefore dependent on the nature of the feed and the products that are produced. The difference between the H:C ratio of the coal feed and the H:C of the desired product will affect the carbon efficiency of the process more than any other attribute.

This can also be seen from the disparity in the thermal efficiencies of DLC and ICL processes (Table 15.3) (O'Hara 1981). The overall thermal efficiencies for different DCL processes with bituminous coal are in the range 60%–77%, whereas the thermal efficiencies for Fischer–Tropsch-based ICL processes with bituminous coal are in the range 31%–35%. Based on liquid products only, the efficiencies are lower, but the thermal efficiency of DCL is still much higher than that of Fischer–Tropsch-based ICL. The key difference is the H:C ratio of the liquids. The liquid products from DCL typically has a H:C ratio on the order of 1.2:1, whereas the H:C ratio from Fischer–Tropsch-based ICL is on the order of 2:1.

15.3.4 Gas Loop Design

The gas loop design describes how synthesis gas flows through the process, whether that is to produce mainly H_2 for DCL processes, or whether it is to be used as synthesis gas in ICL processes. Complete single-pass conversion of H_2, or H_2 + CO, is impractical in any of the coal liquefaction processes. It is often desirable to recycle the unconverted H_2 and CO. Most gas loop designs can therefore be described as closed-loop designs, that is, H_2 and CO are not used on a once-through basis, but recycled.

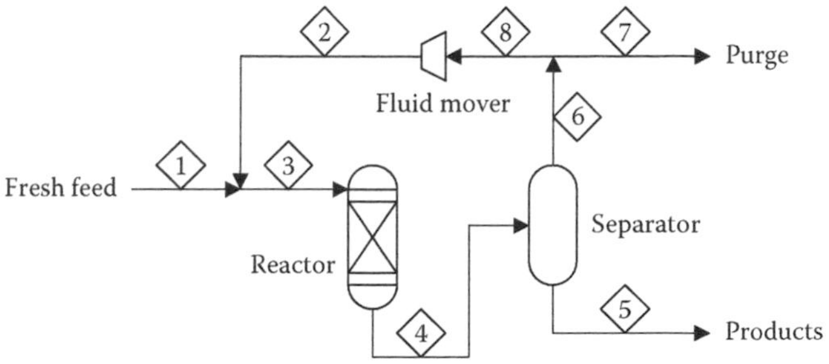

FIGURE 15.8 Elements of a basic gas loop design.

The design of a recycle loop for any reactive process stream involves six elements (Figure 15.8): a fresh feed stream, a reactor, a separator, a product stream, a fluid mover, and a purge. The purge can be omitted for an ideal process, where the separator is capable of performing a perfect separation between unreacted feed and all other materials. If this separation is not perfect, some of the unreacted material may be purged with the products even in the absence of a separate purge, or some of the products may never be able to leave the recycle stream of the process. In synthesis gas loops, the most common reason for purging is due to the presence of inert gases in the synthesis gas that are difficult to separate from the synthesis gas. The two most common inert gases found in the synthesis gas are N_2 and Ar. The N_2 is produced during coal gasification from the nitrogen-containing compounds in the coal. The Ar is introduced as a contaminant with the O_2 from the ASU.

The impact of inert gases in the synthesis gas feed can be illustrated in Figure 15.8. A simple component and material balance can be performed, where $x_{i,j}$ is the mass fraction of inert material in stream j and W_j is the mass flow rate of stream j:

$$x_{i,1}\ W_1 = x_{i,5}\ W_5 + x_{i,7}\ W_7 \tag{15.11}$$

$$W_1 = W_5 + W_7 \tag{15.12}$$

In the case of N_2 and Ar in unconverted synthesis gas, little of the inert material dissolves in the products, so that $x_{i,5} = 0$. Thus, $x_{i,1}{\cdot}W_1 = x_{i,7}{\cdot}W_7$. The concentration of inert material in the gas recycle is therefore higher than in the fresh feed, because $W_1 > W_7$, and hence, $x_{i,1} < x_{i,7} = x_{i,8} = x_{i,2}$. As the purge ($W_7$) becomes smaller, the concentration of the inert material in the purge ($x_{i,7}$) must increase, which is also the concentration of inert material in the recycle gas. The benefit of recycling unconverted synthesis gas is increasingly eroded as the concentration of the inert material in the recycle gas increases. Practically, some carbon loss due to the purging of inert material in the synthesis gas feed is inevitable.

15.3.5 Product Selectivity and Recovery

Highlighting the importance of product selectivity and recovery as a source of carbon loss is almost pointing out the obvious. Carbon that is present in the feed and that is converted to products that cannot be recovered, or that are not much more useful than the coal, represents a carbon loss. There are three common sources of carbon loss associated with product selectivity during liquefaction and product recovery after liquefaction:

1. Gaseous products. The design and location of the coal-to-liquids facility may not be conducive to the recovery and transport of light gaseous hydrocarbons to market. Light hydrocarbon gases have high H:C ratios. The loss of hydrogen as light hydrocarbon products is more detrimental to the overall carbon loss than the direct loss of carbon. For example, the indirect loss of carbon due to hydrogen in methane is equivalent to two times the direct carbon loss due to methane. In rare instances methane is the preferred product from coal gasification, like in the Great Plains Synfuels plant in North Dakota (Stelter 2001), which was designed specifically to produce synthetic natural gas from coal. In such instances the pipeline infrastructure to supply the market was developed as part of the facility.
2. Aqueous products. The water-soluble compounds, such as phenolics, light oxygenates, and ammonia, which are produced during liquefaction, are potentially valuable products. However, the efficient recovery of these compounds from the water can be challenging. The unrecoverable water-soluble compounds represent a carbon loss.
3. Oil-solid slurry products. Mixtures of coal liquids and coal residue from DCL processes, coal tar, and fine coal particulate fractions from some coal gasification technologies and syncrude–catalyst mixtures from DCL and ICL processes can present separation difficulties. In such cases, some of the liquid product might be lost due to incomplete oil–solid separation.

15.4 COAL CHEMISTRY FROM A CONVERSION PERSPECTIVE

The conversion chemistry of coal is central to DCL. Only DCL will be discussed. There is already much literature on this topic (e.g., Kabe et al. 2004; Li 2004; Van Krevelen 1993; Elliott 1981). This vast literature will not be reviewed and the aim is rather to discuss the fundamentals underlying coal conversion. The coal chemistry relevant to ICL takes place during coal gasification (Chapter 14) and ICL will not be discussed.

The nature and composition of coal was described in Chapter 1. In this section, the objective is to look at the chemistry again, but specifically from a conversion perspective. Coal conversion starts during coal storage and coal preparation. Weathering of coal by autoxidation is unavoidable and particularly troublesome for low-rank coal, where it may lead to spontaneous ignition of the coal (Given 1984). Coal preparation can involve chemical cleaning, which is a conversion process. Coal desulfurization is the main type of chemical cleaning practiced (Meyers 1977) and it is discussed in a separate chapter. The main focus will therefore be on the chemistry related to conversion during DCL and the upgrading of the coal liquids.

Coal chemistry is complex. It is not possible to classify coal molecules into neat categories, because coal molecules may contain more than one functional group. Rather than classifying the coal molecules, it is better to accept the diversity of functional groups within the coal molecules and identify the chemical and structural features that affect the conversion chemistry. The main structural features that influence coal conversion chemistry are as follows:

1. Acyclic aliphatic hydrocarbon structures
2. Cyclic aliphatic hydrocarbon structures
3. Aromatic hydrocarbon structures
4. Heteroatom modifications of structures (1–3)

The reason for this specific classification will become apparent as the impact of each structural feature on the conversion chemistry is discussed. Three main types of reaction will be discussed for each structural feature in coal. The reaction types are oxidation that is relevant to coal storage and to oxidative liquefaction, cracking that is relevant mainly to coal liquefaction, and hydrogenation and hydrogen transfer that are relevant to coal liquefaction and coal liquid processing.

15.4.1 Conversion Chemistry of Acyclic Aliphatic Hydrocarbon Structures

During the process of coal formation, long-chain hydrocarbons that were present in the original plant material are degraded to form oil, or light gases such as methane. Acyclic aliphatic hydrocarbon structures that are present in coal are therefore present mainly as alkyl groups attached to ring structures. The alkyl groups can be present as pendant groups, attached only on one end, or as bridging-groups, attached at both ends. The molecular environment of the alkyl group affects the chemistry of the acyclic aliphatic hydrocarbon structures. Waxlike compounds are found in appreciable quantities when the maceral groups, algenite or resinite, are present in the coal. The use of a Van Krevelen diagram with coal development lines is useful to determine the probability of finding acyclic aliphatic molecules in the coal (Van Krevelen 1993).

The near ambient temperature oxidative conversion of coal is referred to as weathering of the coal. Aliphatic C—H bonds, which are usually difficult to activate, are susceptible to low-temperature oxidation by O_2 in air (Emanuel 1965). This autoxidation reaction (Equation 15.13) is the chain initiation step that leads to oxygen addition to form a peroxyl radical (Equation 15.14) and propagation by forming a hydroperoxide (Equation 15.15):

$$R-H+O_2 \rightarrow R\bullet + HOO\bullet \quad (15.13)$$

$$R\bullet + O_2 \rightarrow R-OO\bullet \quad (15.14)$$

$$R\text{–}OO\bullet + R'\text{–}H \rightarrow R\text{–}OOH + R'\bullet \quad (15.15)$$

The hydroperoxide can in turn be decomposed to further propagate the free radical chain reaction and produce various products. The hydroperoxides have a rich chemistry (Swern 1970–1972).

The peroxide bond is weak and homolytic bond dissociation can readily take place at low temperature. The products from aliphatic hydrocarbon oxidation include alcohols, ketones, and carboxylic acids. The formation of carboxylic acids is of particular significance, because it can be accompanied by chain scission.

The oxidation of aliphatic structures and the scission of bridging groups by further oxidation make it possible to achieve near complete coal liquefaction. For example, it was reported that 70%–90% of the carbon mass in coal could be converted into water-soluble products by oxidation in the presence of water at temperatures from 220°C to 350°C (Anderson et al. 2011). This type of oxidative liquefaction was reportedly successful on all ranks of coal. Extensive oxidative degradation was also reported by many authors under basic, acidic, and near neutral conditions using a wide variety of oxidizing agents (Hayatsu et al. 1982 and Van Krevelen 1993). These oxidative conversions in the presence of water form the basis for the development of wet oxidation processes. The ability to oxidatively degrade coal in the presence of water indicates that at least some, if not most, of the acyclic aliphatic hydrocarbon structures in coal are present as bridging groups.

Coal liquefaction processes employ temperatures that are sufficiently high for thermal cracking to take place. Like autoxidation, thermal cracking involves free radicals as intermediate products. The C—C bond dissociation energy of an acyclic aliphatic structure depends on the molecular environment. At typical coal liquefaction temperatures, 375°C–450°C, there is sufficient thermal energy for appreciable homolysis of bonds with homolytic bond dissociation energy of 270 kJ mol^{-1} (65 kcal mol^{-1}) or less (Stein 1985). The rate of homolysis of stronger bonds will be much slower. Generally speaking, the higher the degree of substitution of the carbon atoms involved in the C—C bond, the lower the bond dissociation energy. This is due to the stability of the free radicals formed during bond scission (Equation 15.16).

$$\text{Stability: } \bullet CR_3 > \bullet CHR_2 > \bullet CH_2R > \bullet CH_3 \qquad (15.16)$$

The bond dissociation energy of a C—C bond that results in the formation of a benzylic free radical ($\bullet CH_2$—Ph) is even lower, because of the resonance stabilization of the lone electron of the carbon-centered free radical by the adjacent π-electrons of the aromatic. Values for the bond dissociation energies can be found in the literature (e.g., Blanksby and Ellison 2003; McMillen and Golden 1982). In an analogous way, a carbon-centered free radical next to a C=C can be stabilized by the adjacent π-electrons of the alkene. Resonance stabilization energies in coal chemistry can be estimated by the method proposed by Stein (1985). It should be noted that once a molecule becomes a free radical, many of the bond dissociation energies change. The adage that *a chain is only as strong as its weakest link* can be applied to the chemistry of free radical conversion.

The same principles apply to C—H bond scission, although C—H bonds in the same molecular environment as a C—C bond require more energy to be dissociated.

The susceptibility of acyclic aliphatic hydrocarbon structures in coal to thermal conversion is due to the prevalence of the structures as alkyl groups on aromatics. Such alkyl groups can be thermally cracked to produce benzylic free radicals, which explain the ease of thermal conversion. Alkanes (paraffins) are much more difficult to convert by thermal cracking. To put this into perspective, thermal cracking of pure *n*-hexadecane ($C_{16}H_{34}$) at 470°C for 1 h resulted in only 2% conversion to lighter products (Egloff 1937). Recent work in our laboratory also indicated that there is little synergistic thermal cracking of alkanes mixed with coal liquids. It is therefore fortunate that the acyclic aliphatic hydrocarbon structures in coal are present mainly as alkyl groups attached to ring structures.

Lewis acid catalysts have been employed for coal liquefaction (Derbyshire 1988). Acid catalysts usually require high temperatures to crack aliphatic molecules. Cracking and addition are competing reactions with addition reactions dominating at low temperatures and cracking reactions dominating at high temperatures. At high temperatures, catalysts such as $ZnCl_2$ and $SnCl_2$ assist cracking conversion. The use of acid catalysts with coal at low temperatures can also promote coal solubilization, but the improved coal dissolution is not due to cracking. The alkylation of the coal and disproportionation reactions taking place during alkylation can result in improved solubility of coal (e.g., Schlosberg et al 1978).

Whenever the coal conversion process involves a hydrogenation catalyst at temperatures where hydrogenation readily takes place, hydrogenation–dehydrogenation equilibrium is normally established. Aliphatic hydrocarbon structures with two or more adjacent carbon atoms are subject to the hydrogenation–dehydrogenation equilibrium. The introduction of alkenes (olefins) into the coal liquid opens the possibility for free radical addition reactions. This will be discussed in detail in Section 15.4.2.

15.4.2 Conversion Chemistry of Cyclic Aliphatic Hydrocarbon Structures

Coal contains many cyclic structures, but only some are cyclic aliphatic hydrocarbon structures. The process of coalification favors aromatization of cyclic aliphatic hydrocarbons. As before, a Van Krevelen diagram is useful to determine the likelihood of finding cyclic aliphatic molecules in the coal (Van Krevelen 1993). Most of the cyclic aliphatic hydrocarbon structures are present in association with aromatic structures.

The general chemistry of oxidation, cracking, hydrogenation, and hydrogen transfer that was discussed for acyclic aliphatics is equally applicable to cyclic aliphatics. However, there are two very important differences due to the cyclic nature of the aliphatics:

1. Bond scission reactions do not lead to the formation of two separate molecules. The two ends of the aliphatic chain that was broken remain in close proximity. This increases the probability of intramolecular

reactions. When intramolecular addition takes place after bond scission, the net effect may be that of ring-contraction. For example, a six-membered ring becomes a five-membered ring.

2. Hydrogen transfer reactions and catalytic dehydrogenation of six-membered cyclic aliphatic structures lead to the formation of aromatic structures. The additional resonance stabilization that can be gained by repeated loss of hydrogen provides a thermodynamic driving force for hydrogen loss that is not present in acyclic aliphatic structures.

The impact of the cyclic structure is particularly pronounced when it is attached to an aromatic ring. The additional resonance stabilization energy that is gained by carbon-centered free radicals in a benzylic position is found for all aliphatic hydrocarbon structures attached to an aromatic ring. Detailed discussions of the free radical chemistry of these reactions can also be found in the literature (e.g., Whitehurst et al. 1980; Stein 1985). However, when the aliphatic structure is cyclic, it can lead to intermediates and products that cannot be formed from acyclic aliphatic structures or that are more stable under reaction conditions than their acyclic analogs.

Tetralin is the prototypical hydrogen donor solvent used in coal liquefaction. Hydrogen transfer from the six-membered cyclic aliphatic structure of tetralin as shown in Figure 15.9a illustrates the process and why it readily takes place during coal liquefaction. The same hydrogen transfer reaction is possible from the five-membered structure of indane as shown in Figure 15.9b. The intermediate steps were not shown, but the intermediate nonradical product after two hydrogen atoms were transferred is indene. The carbon-centered free radical on the benzylic carbon of indene is resonance stabilized by the adjacent aromatic and adjacent olefinic π-electrons. Unlike the six-membered ring in tetralin, the five-membered ring in the indene free radical cannot transfer another hydrogen atom to become an aromatic. This has a profound effect on the chemistry that can take place.

Not only is the indene free radical readily formed and stable, but the olefinic group is also retained and not converted to an aromatic. Conditions giving rise to hydrogen transfer reactions will therefore also lead to the formation of alkenes that are stable under such conditions when five-membered cyclic aliphatic structures are present. The stability of the free

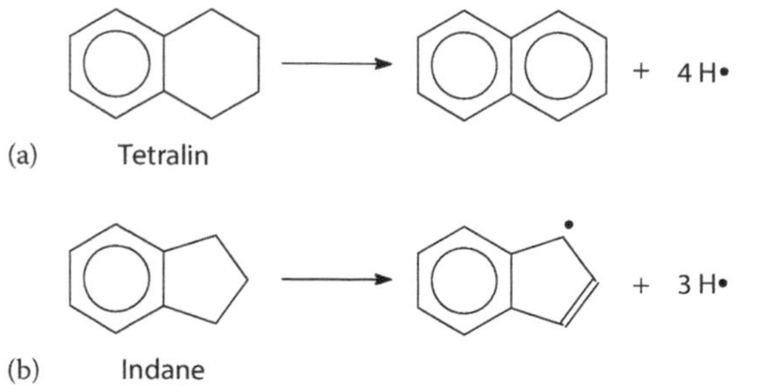

FIGURE 15.9 Hydrogen transfer from tetralin (a) and indane (b).

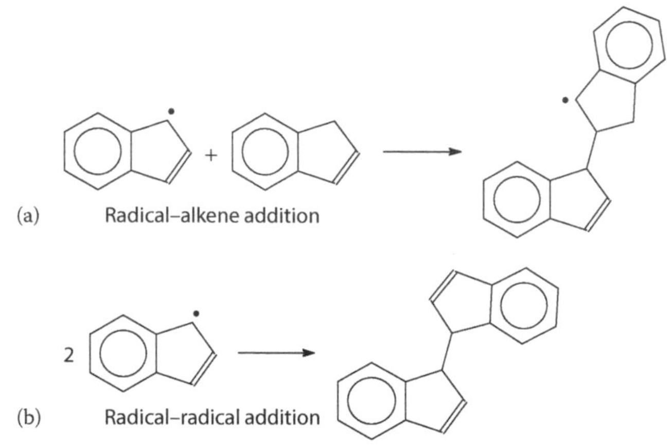

FIGURE 15.10 Free radical addition reactions illustrated by the indene radical addition to (a) an alkene and (b) another free radical.

radical and the alkene functionality increase the probability of free radical addition reactions. Free radical addition can take place by radical–alkene addition (Figure 15.10a) or by radical–radical addition (Figure 15.10b). Although the two types of free radical addition reactions were illustrated using just the indene radical, the five-membered cycloalkene radical structure can perform the same reactions with other alkene or radical species. Thus, the longevity of the radial intermediate and that of the alkene functional group are both detrimental to coal liquefaction.

The chemistry can be generalized to state that any process involving free radicals as intermediates will have an increased propensity for free radical addition if the material contains five-membered cyclic aliphatic structures. Free radical addition is favored by lower temperature. The propensity of free radical addition of indane and indene during autoxidation compared to other hydrocarbon classes was experimentally demonstrated (Siddiquee and De Klerk 2014).

In the presence of a hydrogenation catalyst, just like with acyclic aliphatic structures, the catalyst enables the hydrogenation–dehydrogenation equilibrium to be established at temperatures where hydrogenation readily takes place. The difference is that for six-membered cyclic aliphatic structures, the hydrogenation–dehydrogenation equilibrium that is pertinent at DCL operating conditions, is that between the cycloalkane and the aromatic and not between the cycloalkane and the cycloalkene. The equilibrium increasingly favors the aromatic as temperature is increased, H_2 pressure is decreased, or the number of ring structures or substituents are increased (Le Page 1987).

Much of the chemistry related to coal conversion provides pathways for the transformation of cyclic aliphatic hydrocarbon structures to aromatic structures. In this there is a warning and an opportunity for the improvement of coal liquefaction and processing of coal liquids. To be clear, coal naturally contains aromatic structures and the amount of aromatic carbon increases with coal rank, but it is not uncommon to find that the products from coal liquefaction have an even higher aromatic content than the coal feed. This sentiment was echoed by the remarks of Whitehurst et al. (1980) that the high aromaticity of coal products is not just due to the intrinsic properties of the coal, but also a consequence of coal conversion.

15.4.3 Conversion Chemistry of Aromatic Hydrocarbon Structures

The aromatic carbon content of lignite, sub-bituminous, and bituminous coals is typically in the range 60%–80%, although higher and lower values can be found. The aromatic hydrocarbon structures are unreactive at storage conditions. Under liquefaction conditions, the aromatics affect the chemistry, but for the most part the aromatic structures are thermally stable, but can participate in free radical reactions. It is worthwhile to bear in mind that when the concentration of aromatic C—H bonds in combination with reaction time increases, aromatic C—H bond scission may become noticeable, even though the probability of aromatic C—H bond dissociation is low. The amount of aromatic C—H bond scission is the low probability multiplied by the high concentration and reaction time. This is illustrated by the work of He et al. (1992), Khorasheh and Gray (1993), and Savage (1994).

The conversion of strong C—C bonds is another aspect of aromatic conversion that requires some explanation. When aromatics are linked by a single carbon bridge, as is found in diphenyl methane, this link is very refractory. Yet, some conversion of these compounds is found in coal, despite their extreme thermal stability in isolation. McMillen et al. (1987) forwarded an explanation for this observation in terms of *ipso*-hydrogenation of the aromatic carbon attached to the linking group (Figure 15.11). This type of chemistry is also found in the alkali metal conversion of aromatics (Pines and Stalick 1977), albeit with a carbanion intermediate. The chemistry for coal can be generalized by stating that any coal species that has strong reducing properties may reduce multinuclear aromatic systems to a radical species that have hydrogen transfer properties.

Oxidation of multinuclear aromatics to quinonoid compounds can take place at mild conditions, but further oxidation or hydrolysis that leads to ring opening does not take place readily. Hydrolysis is possible at higher temperature and under alkaline conditions (Hodge 1974). This is likely an important reaction in the dissolution of coal by wet oxidation. However, the extensive oxidative degradation of coal that is possible using various oxidizing agents and media (Hayatsu et al. 1982; Van Krevelen 1993) suggests that coal does not consist of large multinuclear aromatic clusters.

Under catalytic hydrogenation conditions, the aromatic structures can be hydrogenated to cyclic aliphatic hydrocarbon structures. During the hydrogenation of multinuclear aromatic structures, hydrogenation of each successive aromatic ring becomes progressively more difficult and thermodynamically less favorable (Le Page 1987). However, in coal the ring structures may be multinuclear, but the observation was made that if only the aromatic ring structures are considered, the aromatic rings are predominantly isolated as mono- and diaromatic rings (Whitehurst et al. 1980). The highly aromatic nature of the coal liquids from some DCL processes is therefore indicative of hydrogen disproportionation leading to the formation of aromatics from hydroaromatics, as was discussed in Section 15.4.2.

Hydrogen exchange between the gas-phase H_2 and a coal-solvent slurry takes place appreciably with or without catalysts, but hydrogenation that results in a net consumption of H_2 is increased considerably in the presence of catalysts (Kabe et al. 2004).

The catalysis of aromatic hydrogenation is vast. Hydrogenation of aromatics during catalytic coal liquefaction is performed by either dispersed catalysts or catalyst particles. In the case of dispersed catalysts that are not recoverable, the trade-off between hydrogenation activity and catalyst cost dictates what catalytically active materials can be profitably used. Most research effort was therefore directed at low concentration Mo-based catalysts and higher concentration Fe-based catalysts (Derbyshire 1988). Investigations dealing with packed and ebullated-bed hydrogenation catalysts developed along similar lines as hydrotreating catalysts for petroleum (Cusumano et al. 1978; Satriana 1982; Derbyshire 1988).

15.4.4 Conversion Chemistry Modified by Heteroatoms

Coal contains percentage levels of heteroatoms, mainly nitrogen, oxygen, and sulfur. The amount of these elements present in the organic matter of coal is reported as part of the ultimate analysis of the coal. Oxygen is usually the most abundant heteroatom and the oxygen content generally increases with a decrease in coal rank, that is, lignites have the highest oxygen contents.

Heteroatoms complicate the chemistry. Rather than dealing with all of the possible reactions, the approach taken here is to show how the heteroatoms modify the conversion chemistry of each of the main hydrocarbon structures that were discussed.

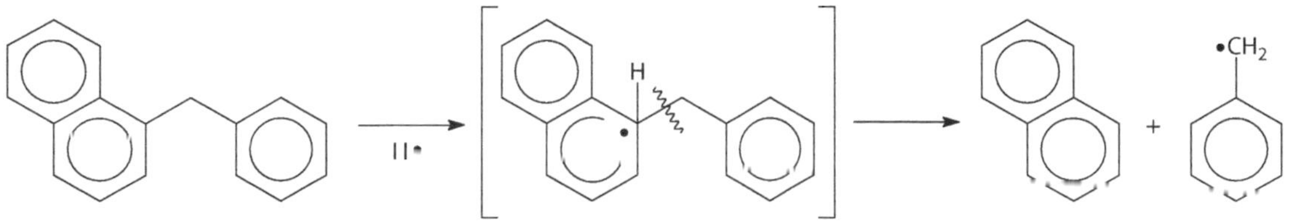

FIGURE 15.11 *Ipso*-hydrogenation and cleavage of strong aromatic bridging groups.

The insertion of oxygen or sulfur into an aliphatic hydrocarbon structure, irrespective of whether it is cyclic or not, changes the autoxidation behavior.

The aliphatic carbon adjacent to the ether becomes very susceptible to oxidation (Waters 1964). Once the free radical is formed (Equation 15.17), further oxidation to yield the peroxide may follow, or the radical may engage in different types of free radical reactions, including bond scission.

$$R{-}CH_2{-}O{-}R' + O_2 \rightarrow R{-}(\bullet CH){-}O{-}R' + HOO\bullet \quad (15.17)$$

Thioethers (sulfides), unlike ethers, are less susceptible to autoxidation, and in the absence of material that can form hydroperoxides, the thioethers do not react with oxygen in air at temperatures below 100°C (Barnard et al. 1961). However, hydroperoxides can readily oxidize thioethers to sulfoxides (Equation 15.18) and sulfoxides to sulfones (Equation 15.19).

$$R-S-R' + R''OOH \rightarrow R-(SO)-R' + R''OH \quad (15.18)$$

$$R-(SO)-R' + R''OOH \rightarrow R-(SO_2)-R' + R''OH \quad (15.19)$$

Secondary aliphatic amines (R_2NH) can be oxidized by hydroperoxides to produce hydroxylamines (Equation 15.20), which can in turn be oxidized further (Challis and Butler 1968).

$$R-(NH)-R' + R''OOH \rightarrow R-(NOH)-R' + R''OH \quad (15.20)$$

The presence of oxygen and sulfur in cyclic aliphatic structures changes the number of carbon atoms in the ring structure that is required to form an aromatic. The same principle of hydrogen transfer and dehydrogenation as discussed before applies to heteroatom-containing cyclics. The main difference is that the behavior of five-membered ether- and thioether-containing rings mimics the behavior of six-membered cyclic aliphatic hydrocarbons. Conversely, six-membered ether- and thioether-containing rings mimic the behavior of five-membered cyclic aliphatic hydrocarbons.

The homolytic bond dissociation energies of aliphatic C—C and C—O bonds are very similar, around 345 kJ mol^{-1} (82 kcal mol^{-1}), but thioether C—S bonds are weaker, around 305 kJ mol^{-1} (73 kcal mol^{-1}). Thioether bonds are therefore more susceptible to thermal cracking. Secondary aliphatic amines have a 10 kJ mol^{-1} lower C—N bond strength than C—C.

Hydrogen sulfide (H_2S) has a beneficial effect on coal dissolution, in both the presence and absence of catalysts (Derbyshire 1988). The exact role is not clear and the optimum partial pressure depends on the nature of the coal. There is an equilibrium between hydrogenation and sulfidation of some minerals. Sulfidation may also contribute to catalytic activity, as it does in the case of molybdenum sulfide acidity. The homolytic bond dissociation energy of H—SH is around 380 kJ mol^{-1} (91 kcal mol^{-1}), which makes H_2S a potential source of free radicals during coal liquefaction. It is likely that H_2S improves coal liquefaction through a combination of all of the above.

When the nitrogen, oxygen, or sulfur is incorporated into an aromatic, the heterocycle affects catalytic hydrogenation and autoxidation. The catalysis is analogous to that of aromatic hydrogenation. Literature reviews dealing specifically with coal can be found in the works by Cusumano et al. (1978), Satriana (1982), and Derbyshire (1988). Analogous heteroatom removal from coal liquids through the action of hydrogen donor solvents was reviewed by Shah and Cronauer (1979).

15.5 STRATEGIES TO IMPROVE DCL

15.5.1 Coal Preparation to Improve Liquefaction

The first step in most coal-based processes is a basic physical coal cleaning process, which takes place at or near the coal mine. The run-of-mine (ROM) coal is crushed, screened, and cleaned to remove most of the mineral matter associated with the coal. The coal is usually also dewatered to remove most of the surface water associated with it. These basic physical size reduction and cleaning steps of ROM coal are collectively called coal preparation or coal processing. Depending on the application, it may also include more extensive cleaning (e.g., Liu 1982).

The removal of pyrite containing minerals from coal is particularly important. There are some catalytic benefits of having pyrrhotite, $Fe_{1-j}S$, $0 \leq j \leq 1/8$, and possibly even iron pyrite, FeS_2 (Padrick and Granoff 1986). Iron is a weak hydrogenation catalyst and iron-based catalysts have been extensively investigated for catalytic coal liquefaction, as was pointed out before. However, iron pyrite is a source of sulfur, with around 50% of the pyritic sulfur being released on heating iron pyrite beyond 400°C (Bommannavar and Montano 1982). The sulfur released from the transformation of iron pyrite to iron pyrrhotite consumes hydrogen directly to produce H_2S, or it is transferred to the organic matter in coal to ultimately consume hydrogen during liquefaction and coal liquid upgradation. Using H_2 for such a purpose is costly. Even when coal is employed for combustion, the sulfur is released and the release of sulfur from iron pyrite in the coal increases gas cleaning cost. Sulfur removal from coal is a topic in its own right (e.g., Meyers 1977).

Coal preparation conducted close to the coal mine will usually optimize the size reduction to minimize the cost of producing a transportable coal. The production of coal fines is deliberately limited, unless the coal preparation is conducted close to a consumer of fine coal. Fine coal processing is challenging. However, many coal liquefaction processes require fine coal as feed, which is thought to decrease mass transport resistance and improve liquefaction yield. Practically, fine coal also facilitates coal–solvent slurry preparation and pumping. There is a cost involved in coal size reduction and there is a benefit in process strategies that can reduce the extent of coal size reduction before coal liquefaction (Haghighat and De Klerk 2014). Coal size reduction processing steps that are associated with DCL processes requiring fine coal as feed contribute to around 3% of the total capital cost of the facility (Nowacki 1979; Cavanaugh et al. 1984). It is therefore interesting to note that at least some literature indicate a less severe particle size requirement for liquefaction. For example, Berkowitz (1979) suggested a threshold of 3 mm as the size below which coal particle size will not influence liquefaction.

15.5.2 Coal Pretreatment to Improve Liquefaction

Coal is a very heterogeneous substance. Any coal liquefaction process that is proposed must deal with this heterogeneity. The aim of coal pretreatment is to modify the coal feed so that the overall coal liquefaction process will be more efficient. Various two-step liquefaction strategies have been proposed for DCL. The type of pretreatment step can be broadly classified into the categories outlined below (Okuma and Sakanishi 2004; Vasireddy et al. 2011):

1. *Coal drying*. Moisture in coal affects the operating pressure of the liquefaction reactor. It also reduces the H_2 partial pressure in processes that co-feed H_2. Because drying may also lead to pore collapse and other changes in the coal, drying does not necessarily increase liquid yield, even though it reduces the operating demands placed on the liquefaction reactor.
2. *Oxidation*. Oxidative weathering of coal takes place naturally, and with extended exposure to air the pyridine solubility of coal is reduced (Van Krevelen 1993). Oxidation decreases the effective H:C ratio of the coal, which ultimately degrades the liquefaction yield. Wet oxidation can be used for DCL, or it can be employed at lower severity as pretreatment. Oxidation can also be considered as a strategy for desulfurization of the organic matter in coal (e.g., Borah and Baruah 2001; Liu et al. 2008).
3. *Acid washing*. Acid washing is useful to remove metal cations from coal. Washing on its own would be able to reduce the dissolved salts present in the water trapped in the pore structure of the coal. By acidifying the water, the metal cations that are present as salts of weak acids, mainly carboxylates and phenolates, can be displaced and removed (Okuma and Sakanishi 2004). The nature of the cations affects pyrolysis behavior and in general higher yield is obtained during pyrolysis of acid-washed lignites (Morgan and Jenkins 1984). Removal of multivalent cations that can occupy bridging positions can directly improve coal liquefaction because intermolecular bridges are broken (Van Bodegom et al. 1984). This is illustrated by the acid washing of calcium benzoate (Figure 15.12).
4. *Hydrothermal treatment*. Steam treatment at high temperature increases the hydroxyl content of coal and improves liquefaction. The benefit of steam treatment is lost on exposure to oxygen and steam treatment must therefore be followed by liquefaction without exposure to air. Many explanations for the improved liquefaction have been offered in literature (Vasireddy et al. 2011) and it is likely a combination of different factors, at least some of which are related to coal drying and washing.
5. *Solvent preswelling*. The beneficial effect of solvent preswelling on coal liquefaction has been ascribed to the disruption of weak bonding, relaxing of the coal structure, and increase of macroporosity in coal. By extension the preswelling can also be used to perform physical extraction of coal liquids (e.g., Miura et al. 2001). The extent to which physical coal dissolution takes place is limited and physical dissolution from lower rank coals tends to be higher than from higher rank coals.
6. *Lower temperature thermal treatment*. Heating a coal–solvent slurry to temperatures below the liquefaction temperature and allowing the coal to soak was found to increase coal liquefaction yield. Many of the pretreatment procedures already discussed are effectively also conducted by such a low-temperature solvent soaking. In fact, lower temperature thermal treatment can also be regarded as a modification of the temperature-time relationship in DCL.

15.5.3 Temperature–Time History to Improve Liquefaction

The use of oversimplified reaction networks to show how coal is depolymerized to produce coal liquids, gas, and residue, or pre-asphaltenes, asphaltenes, and char (Shah et al. 1981) can create the impression that coal may be treated as a homogenous reagent. This is of course not true, because coal is complex and heterogeneous in nature. To capture the complexity of coal, a more elaborate and fundamentally grounded description of the reaction network is necessary (e.g., Alvarez et al. 2013).

Improvements in the coal liquid yield during DCL can be achieved by adhering to the principles that were established in the original Pott–Broche process for the solvent extraction of coal (Kiebler 1945). In essence, the Pott–Broche process controlled the temperature–time history of the coal, coal liquids,

FIGURE 15.12 Removal of intermolecular bridging by acid washing.

and coal residue. They appreciated the importance of the solvent, but from a practical point of view, the solvent is not really a process parameter that can be manipulated at will. Solvents and solvent combinations can be selected and controlled in laboratory studies, but process economics precludes the use of the same for industrial applications. The contribution of Pott and Broche was therefore to show how the variables that can be industrially manipulated, temperature and time, can be used in combination to obtain maximum coal liquid yield during the solvent extraction of coal. The same principles are also valid for other type of DCL processes.

To explain the contributions of Pott and Broche, let us start by looking at reaction temperature. The temperature determines the reaction classes that are capable of proceeding at a nonnegligible rate. The temperature determines whether there is sufficient energy available to overcome the activation energy, following the Arrhenius relationship. The Arrhenius relationship (Equation 15.21) describes the proportionality between the kinetic constant, k, to the activation energy needed for the reaction, E_a (J mol^{-1}), ideal gas constant, R (J mol^{-1} K^{-1}), and the reaction temperature, T (K).

$$k \propto \exp\left(\frac{-E_a}{RT}\right) \tag{15.21}$$

Viewed in terms of free radical decomposition only, the reaction temperature determines the strength of the chemical bonds that can be broken. In this instance, the bond dissociation energy is the activation energy. At lower temperature, some reactions do not occur. As the temperature is increased, progressively more reactions become possible. The maximum temperature of solvent extraction of coal that should not be exceeded according to the work of Pott and Broche is the temperature just below the onset of coal residue decomposition (Kiebler 1945).

Temperature also affects chemical and phase equilibria, solubility, diffusion rate, and physical properties such as density and viscosity. The coal dissolution process can be influenced through these variables. Coal has an inherently high free radical concentration, with values in the range of 5×10^{17} to 2×10^{19} spins per gram coal at 100°C as measured by electron spin resonance (ESR) spectroscopy (Petrakis and Grandy 1983). Mobility of these species is low, but at higher temperature mobility is increased. Although the increase in mobility will not affect the steric effects related to molecular structure, the increase in mobility will increase the probability of radical–radical interaction. Even without any thermally induced bond dissociation, coal is a reactive material. To put this into perspective, at 400°C–500°C the free radical concentration is increased only by tenfold over that at 100°C (Petrakis and Grandy 1983).

Once a reaction is possible, the actual rate of the reaction is influenced by temperature. If a reaction leads to the formation of stable products, there is less risk of undesirable side reactions taking place. If the reaction leads to reactive intermediates, the rate at which these intermediates are formed and how they are stabilized, affects selectivity. There is ample opportunity for undesirable side reactions to take place. The rate of formation affects the concentration of the reactive intermediates that must be stabilized per unit time, which creates the dependence between temperature and time.

Residence time must therefore be considered in relation to the reaction temperature. The amount of time spent at a given temperature, or the rate at which heating takes place, affects three aspects of the coal liquefaction process: conversion, mass transport, and product selectivity.

First, the conversion is determined by the residence time, because the conversion is the integral of reaction rate with respect to time. In the absence of reversible reactions, there is a monotonic increase in conversion with time. However, in practice, many coal conversion reactions are reversible, so that there is usually a maximum conversion, albeit incomplete conversion, even at infinite residence time.

Mass transport is the second process affected by residence time. Coal is not homogenous and as the coal dissolution proceeds, local concentration gradients develop. The rate at which these differences in concentration can be dissipated depends on the time and the relative ratio of the mass transport rate and the rate of reactions that causes the concentration differences. If there is insufficient time at a given temperature to dissipate these concentration differences, local depletion or enrichment of some species will occur. Of particular importance in coal liquefaction is the local concentration of reactive intermediates relative to the local concentration of compounds that will facilitate stabilization to form desirable stable products. Thus, selectivity is influenced by local concentration, which is in turn affected by the time available for mass transport in relation to conversion.

Selectivity is the third aspect to be influenced by residence time. The influence through mass transport has already been pointed out. In terms of the reaction, network, time also determines product selectivity through consecutive reactions. Even stable products at one temperature can become reactive products at a different temperature. Stable products may also not be immune to reaction with reactive intermediates. In any reaction of the type A → B → C, the residence time is very important when B is desirable and C is not.

These insights were implemented in the development of the Pott–Broche process for the solvent extraction of coal. The Pott–Broche process (Figure 15.13) employs stepwise extraction of coal. The coal (stream 1) and solvent (stream 2) are mixed and the coal–solvent slurry (stream 3) is heated and the hot slurry (stream 4) becomes the feed for solvent extraction. In each step, the reaction is allowed to proceed at a constant temperature and then the product (stream 5) is phase separated into coal extract (stream 6) and coal residue (stream 7). Only the coal residue (stream 7) and additional solvent (stream 8) are subjected to the next extraction step, which is conducted at a higher temperature. In this way, liquid products are recovered at each reaction temperature and only the coal residue is subjected to higher temperature. The highest temperature is a temperature just below the onset of coal residue decomposition, where the selectivity to gaseous products is meaningfully increased compared to the preceding reaction steps.

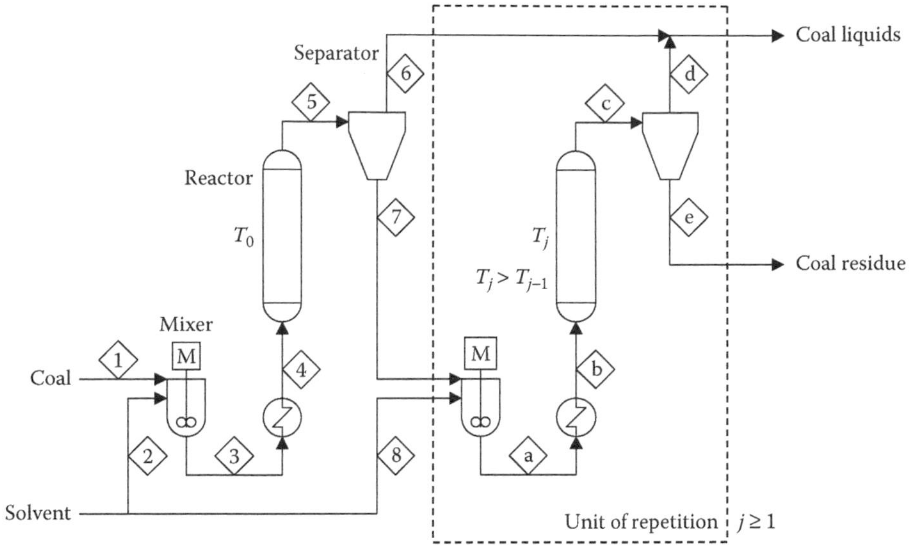

FIGURE 15.13 Pott–Broche process for the stepwise solvent extraction of coal.

Because the coal liquids are removed at the end of each extraction step, they are not subjected to further reaction, as is the case with many later DCL processes. In this way, the loss of coal liquids through side-reactions to undesirable products is avoided. However, there is a practical limitation to the number of reaction–separation steps that can be performed before the cost overtakes the benefit of improved liquid yield.

Other DCL process developments have employed the same insights, but in different ways. The use of short contact time liquefaction is of practical interest. Physical dissolution of coal is fast but limited, and liquid yield is mainly a function of temperature (Huang et al. 1996, 1998; Haghighat and De Klerk 2014). Initial reactive dissolution of coal is also fast and high liquid yields can be achieved within 2–3 min in the temperature range 375°C–450°C (Whitehurst 1980; Longanbach et al. 1980). On longer exposure at high temperature, the free radical species that are generated continue to react and liquid yield passes through a maximum. The amount of time needed to pass through maximum yield depends on the temperature. In an example given by Whitehurst (1980), liquid yield is lost almost immediately after reaching 450°C; at 425°C liquid yield is lost after 10 min, whereas liquid yield was not lost after more than 10 h at 400°C. By reducing the residence time, the impact of consecutive and other side-reactions is limited.

TABLE 15.4
Coal Liquid Products Produced by Different DCL Technologies before Further Upgrading

	Straight-Run Coal Liquid from DCL		
Property	**Pyrolysis COED Process**	**Solvent Extraction SRC-II**	**Catalytic Liquefaction H-Coal**
Coal	Illinois No. 6	Blacksville No. 2	Illinois No. 6
Density (kg/m^3)	1110	943	900
Density (°API)	–4	18.6	25.8
Viscosity at 38°C (cSt)	286[a]	2.2	1.6
Water content (wt%)	2	0.6	0.2
Ultimate analysis (wt%)			
C	80.5	84.6	87.0
H	7.0	10.5	11.4
N	1.2	0.9	0.5
S	2.0	0.3	0.3
O	9.2	3.8	1.8
Ash (µg/g)	700	40	90
H:C molar ratio	1.04	1.48	1.57

[a] Viscosity measured at 99°C; pour point is 46°C.

15.6 REFINING OF COAL LIQUIDS

Unless coal liquids will be employed as *ash-free coal* for heating applications, further refining of coal liquids from DCL is necessary to produce transportation fuels and chemicals. It was already pointed out that different types of DCL processes produce coal liquids of different quality. To illustrate the differences in coal liquid quality, the properties of coal liquids derived from different DCL process types are shown in Table 15.4 (Jones 1973; O'Rear et al. 1981).

The quality of the coal liquids that are produced by DCL increases with increasing degree of hydrogenation in the process. The increasing degree of hydrogenation is reflected by the increase in the H:C ratio of the coal liquids and the concomitant decrease in the heteroatom content. Yet, the quality of coal liquids is still poor compared to that of benchmark crude oils. It can therefore be anticipated that refining of coal liquids to transportation fuels and chemicals will be challenging.

A number of refining challenges can be highlighted based on the properties of the coal liquids from DCL in relation to the requirements for transportation fuels:

1. Low H:C ratio of the coal liquids. Transportation fuels typically have H:C ratios in the range 1.8–2.0. The coal liquids have H:C ratios in the range 1.0–1.6 and there is a clear need for further hydrogen addition to increase the H:C ratio of the coal liquids. It is therefore common to find one or more hydrotreating steps associated with the production of coal liquids. Hydrotreating can take place as part of the DCL process, as is the case with catalytic coal liquefaction. The low H:C ratio of coal liquids also reflects on the chemical nature of the coal liquids, which contain high concentrations of aromatics and cycloalkanes. The aromatic content of transportation fuels is regulated.
2. High heteroatom content of the coal liquids. Heteroatom (sulfur, nitrogen, and oxygen) removal is central to the upgrading of coal derived liquids to transportation fuels. The sulfur content is strictly regulated. The nitrogen content is usually not regulated, but the basic nitrogen containing compounds are detrimental to refining processes employing catalysts that rely on acid catalysis. Oxygenates are allowed in some transportation fuels, but there are limitations on the concentration and nature of the oxygenates.
3. Solids in the coal liquids. Solids may cause fouling or blocking of downstream units and are particularly detrimental to packed bed reactors, where the risk of causing a flow restriction is high. Although the presence of solids is not uncommon in oil products, for most oils the solids can be readily removed with water in the crude oil desalting unit, the remainder ending up in the vacuum residue. In the case of coal liquids, the solids in the coal liquids represent the solids that could not be removed even after multiple phase separation steps.
4. Water and oxygenates in coal liquids. There is water present in the coal liquids and the oxygenates in coal liquids will also produce water during refining. The catalysts employed in the refining of coal liquids must therefore be water tolerant, or water and oxygenates must be removed prior to processing in units with water-sensitive catalysts, such as catalytic naphtha reforming. In this respect, much can be learned from the catalysis related to biomass refining and refining of Fischer–Tropsch syncrude (e.g. De Klerk and Furimsky 2010). Because most conventional crude oils contain little oxygenates that are concentrated mainly in the heavy fractions, the impact of oxygenates on transportation fuels refining is an aspect that is easily overlooked.

General overviews on the refining of coal liquids from DCL processes can be found in the literature (e.g., Cusumano et al. 1978; Stiegel et al. 1981; Crynes 1981). It should be born in mind that transportation fuel specifications change over time. Some of the coal liquid-derived fuels that were acceptable transportation fuels in the past no longer meet fuel specifications and are no longer acceptable transportation fuels. The refining requirements have changed and the historic literature on the topic of coal liquid refining should therefore be reinterpreted in current context. Coal liquids can also be refined to chemicals.

The refining requirements for different coal liquid fractions to produce motor gasoline, diesel fuel, jet fuel, and chemicals will be discussed individually. To avoid overcomplicating the discussion, residue conversion processes will not be included in this discussion. The reader should nevertheless be aware that the atmospheric residue from the distillation of coal liquids that are produced by a DCL process must still be converted by one or more residue conversion processes in order to obtain the material that is in the boiling range for fuels or chemicals production.

15.6.1 Refining to Produce Motor Gasoline

Motor gasoline is refined from the naphtha fraction of oil. The distillation cut point depends on the individual refinery, but the distillation cut point for naphtha is mostly in the range 160°C–180°C. In older refining literature, the distillation cut point is often 204°C. When the distillation range is the expressed in terms of the carbon number range of the material in the naphtha fraction, naphtha typically contains C_4—C_{10} hydrocarbons.

In a conventional oil refinery (e.g., Gary et al. 2007), the naphtha is further separated into three fractions that are processed separately: C_4 hydrocarbons, the light naphtha that contains C_5—C_6 hydrocarbons, and the C_7 and heavier naphtha. The C_4 hydrocarbons are used for blending and alkylate production. The light naphtha is used for blending and hydroisomerization. The heavier naphtha is hydrotreated followed by catalytic naphtha reforming.

Despite hydrogen disproportionation reactions and hydrotreating, the heteroatom content of naphthas from DCL processes is high (Table 15.5) (Cavanaugh et al. 1984; Papso 1984).

TABLE 15.5
Properties of Straight-Run Naphtha from DCL of Illinois No. 6 Coal

	Straight-Run Naphtha	
Property	Solvent Extraction EDS Process	Catalytic Liquefaction H-Coal Process
Density (kg/m³)	820	770
Density (°API)	41	52.3
Elemental analysis (wt%)		
C	85.0	85.3
H	12.0	13.8
N	0.2	0.2
S	0.5	0.1
O	2.3	0.6
H:C molar ratio	1.69	1.94

Severe naphtha hydrotreating at hydrogen pressures of 7–10 MPa was required to reduce the sulfur content to below 10 µg/g (Caldwell and Eyerman 1979). Even then the oxygen content was on the order of 100–1000 µg/g, and in some instances the basic nitrogen content could not be reduced to less than 50 µg/g. With advances in hydrotreating catalysts made since then, near complete heteroatom removal is possible, but care must be taken with heat management in the naphtha hydrotreater. For example, the overall adiabatic temperature increase over multiple beds in an industrial coal tar naphtha hydrotreater used for fuels refining was reported to be 172°C (De Klerk 2005).

On hydrotreating, the octane number of coal tar naphtha is decreased to a research octane number (RON) of around 80–83 (De Klerk 2011), despite having an aromatic content of around 40%. This is due to the low octane numbers of the heavy alkanes in the naphtha. Severe hydrotreating of DCL naphtha fractions results in products with a RON of around 79–81 (O'Rear et al. 1981). Thus, the hydrotreated naphtha from DCL must serve as a feed to catalytic naphtha reforming to obtain a high octane number motor gasoline blend component.

The benefit of DCL naphtha is that it has a high cycloalkane content. A liquid yield of 89–90 vol.% can be expected from catalytic naphtha reforming of hydrotreated DCL naphtha to produce a RON 98 blend component (Fischer and Hildebrand 1981). However, there is a specification limit on benzene in motor gasoline. The feed material to the catalytic naphtha reformer must exclude cyclohexane, because cyclohexane is easily reformed to benzene. Even so, it might be necessary to reduce the benzene content of the final product, either by extraction or by aromatic alkylation (e.g., De Klerk and Nel 2008).

In conclusion, it can be said that the naphtha from DCL requires more severe hydrotreating than typical petroleum naphtha, but DCL naphtha can be refined in an analogous way to petroleum naphtha by employing standard fuel refining technologies. Coal-derived naphtha can be refined to an on-specification motor gasoline, or at least high octane number motor gasoline blend material.

15.6.2 Refining to Produce Diesel Fuel

Diesel fuel is refined from the distillate fraction of oil that is obtained by atmospheric distillation. The boiling range of distillate is from the naphtha cut point, say 180°C, to the bottom temperature of the atmospheric distillation column, which is usually around 340°C. Distillate contains mainly C_{11}—C_{22} hydrocarbons.

A conventional crude oil refinery straight-run distillate is normally just hydrotreated to produce diesel fuel, unless it is a very paraffinic crude oil that requires mild hydrocracking of the heavier distillate. Hydrotreating is necessary to reduce the sulfur content to 10–15 µg/g (10–15 ppm), which is the diesel fuel specification limit in most countries. Mild hydrocracking might be necessary to improve its cold flow properties.

Coal-derived distillates have a high heteroatom content (Table 15.6) (Cavanaugh et al. 1984; Papso 1984). Meeting the sulfur specification in a single hydroprocessing step is possible and requires a H_2 pressure of 12 MPa and below 400°C reaction temperature (Quignard et al. 2013).

TABLE 15.6
Properties of Straight-Run Distillate from DCL of Illinois No. 6 Coal

	Straight-Run Distillate	
Property	**Solvent Extraction EDS Process**	**Catalytic Liquefaction H-Coal Process**
Density (kg/m^3)	993	943
API gravity	11	18.5
Elemental analysis (wt%)		
C	89.0	88.4
H	9.0	10.1
N	0.3	0.5
S	0.2	0.1
O	1.3	1.0
H:C molar ratio	1.21	1.37

Diesel fuel quality is governed by the cetane number. Deep hydrotreating of the coal pyrolysis distillate at 18 MPa and over multiple catalyst beds can improve the cetane number to around 35 (De Klerk 2011). Unless there are blending options to increase the cetane number of the diesel fuel, the hydrotreated coal distillate is of insufficient quality for diesel fuel. When the hydrotreated distillate is of a poor quality, that is, low cetane number, hydrocracking may be considered. Hydrocracking is usually employed for vacuum gas oils, not straight-run distillates, because there is distillate yield loss as naphtha and gas are produced during straight-run distillate hydrocracking. Nevertheless, it is possible to reach a cetane number of 50 during high-pressure hydrocracking of coal distillates (Quignard et al. 2013). This is a notable achievement.

To better understand the challenge of producing diesel fuel of acceptable quality, it is useful to look at the molecular origin of cetane number. Cetane number is a measure of the ease of autoxidation leading to autoignition under conditions of high temperature and pressure. Although the relationship between cetane number and molecular structure is complex (e.g., Santana et al. 2006), some general statements about cetane number can be made. Aromatics have low cetane numbers and alkanes have high cetane numbers. The cetane numbers of cycloalkanes fall between that of aromatics and alkanes, but are closer to that of aromatics. Bearing this in mind, the main compounds that were identified in distillates from DCL are cyclic in nature (e.g., Aczel et al. 1978; Dooley et al. 1979). Examples of specific compounds in the distillates from DCL are shown in Figure 15.14 and Table 15.7 (Guin et al. 1977; Adesanwo et al. 2014).

Hydrodearomatization and some ring opening are needed to improve the cetane number of such aromatic distillates. The amount of H_2 required during refining of coal distillates is substantial.

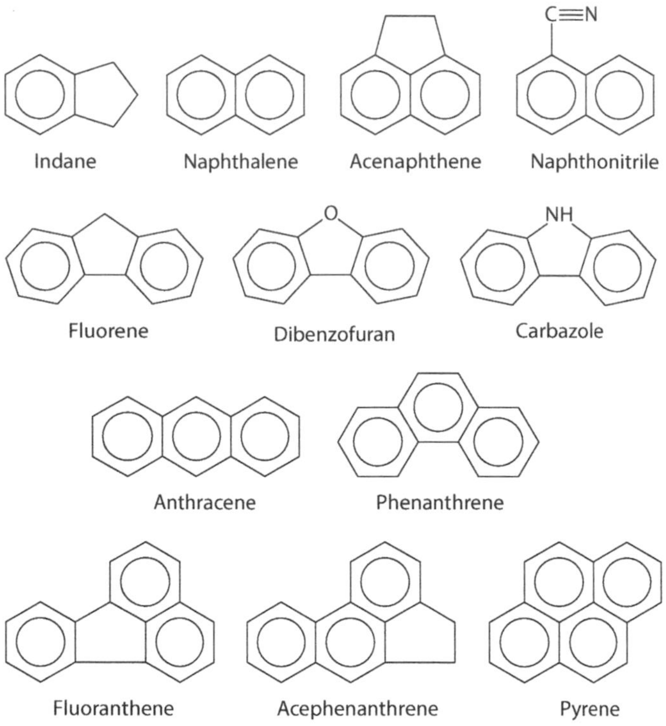

FIGURE 15.14 Structures of the most abundant compounds in coal distillate.

TABLE 15.7
Most Abundant Compounds (>1 wt%) in the Distillate Fraction from Coal Tar

	Distillate (wt%)	
Compound	**Creosote**	**Hydrotreated Distillate**
Indane	–	1.1
Naphthalene	5.1	2.5
Methylnaphthalene	1.7	1.3
Ethylnaphthalene	–	1.1
Acenaphthene	6.0	13.9
Naphthonitrile	2.4	–
Dibenzofuran	6.7	1.4
Methyldibenzofuran	1.7	–
Fluorene	10.3	2.3
Phenanthrene	18.6	20.0
Anthracene	4.3	2.3
Dihydroanthracene	2.4	–
Methylanthracene	1.4	2.0
Carbazole	2.2	1.6
Methylcarbazole	1.7	–
Acephenanthrene	2.5	–
Fluoranthene	1.0	15.3
Pyrene	2.6	13.0
Other compounds [a]	29.4	22.2

[a] Compounds present at concentrations <1 wt% in the distillate.

Another property of hydroprocessed coal distillates that may disqualify them from meeting diesel fuel specifications in some countries is density. The hydrotreated coal distillates have a density in the range 870–880 kg/m^3. Irrespective of this, there is a profitability impact related to the high density of the fuel. Transportation fuels are sold by volume and not by mass. The coal liquid refiner is therefore earning less money per unit mass of product due to the high density.

It was pointed out that there is an inherent synergy in co-refining or blending the products of DCL and ICL processes for the production of transportation fuels (Gray et al. 1990; De Klerk 2011). The distillates from Fischer–Tropsch synthesis have low density and high cetane number, whereas the distillates from DCL have high density and low cetane number.

In conclusion, it can be said that the distillate from DCL requires more severe hydroprocessing than petroleum distillates to be refined to diesel fuel. Coal distillate refining is very hydrogen intensive and hydrotreating alone is insufficient to meet cetane number requirements, so that hydrocracking is required. By doing so, a high-density distillate of acceptable cetane number can be produced. Thus, coal-derived distillates can be refined to on-specification diesel fuels, or at least to diesel fuel blend materials.

15.6.3 Refining to Produce Jet Fuel

Jet fuel is refined from the kerosene fraction, which overlaps in the boiling range with heavy naphtha and light distillate. A typical kerosene cut is from 160°C to 260°C. The boiling range is adjusted depending on refinery constraints, because jet fuel production affects motor gasoline and diesel fuel production. The carbon number range found in kerosene is mainly C_9—C_{14} hydrocarbons.

Jet fuel refining in a conventional oil refinery is rather straightforward. If the straight-run kerosene contains <3% naphthalene and the sulfur content is <0.3%, it may not even be necessary to hydrotreat the kerosene. The sulfur that is present as thiols (mercaptans) can be converted using a sweetening unit. If the naphthalene or sulfur content is too high, hydrotreating is required. The key jet fuel specification is freezing point temperature, which must be <–47°C for Jet A-1, the jet fuel that is used on international flights, or <–40°C for Jet A, that is used for shorter distance flights.

A special grade of jet fuel, JP-900, that is refined from coal liquids was under development in the United States (Balster et al. 2008). It required deep hydrogenation to remove heteroatoms and to remove naphthalene by hydrodearomatization. The two most challenging specifications to meet were hydrogen content and the density. Subsequently, Quignard et al. (2013) demonstrated that it was possible to produce a jet fuel that met all of the Jet A-1 specifications by high-pressure hydrocracking of coal liquids. The only property of the jet fuel that was arguably outside of the specification limit for a fully synthetic Jet A-1 was the aromatics content, which was too low. Unlike crude oil-derived Jet A-1, there is a minimum aromatic requirement of 8 vol.% for severely hydrotreated kerosenes and synthetic jet fuels. This requirement is set to avoid elasomer contraction that can lead to fuel leaks in fuel systems interchangeably employing crude oil and synthetic jet fuels. Coal liquid-derived kerosene has been used in commercial Jet A-1 blends for quite some time in South Africa (Roets et al. 2004).

It can be concluded that refining of kerosene from coal liquids to jet fuel is more demanding than refining of petroleum to jet fuel. High-pressure hydrocracking of coal liquids is needed to achieve Jet A-1 fuel specifications. Coal-derived kerosene can also be refined to jet fuel blend material.

15.6.4 Chemical Production

Falbe (1982) gives a general overview of chemical production from coal by various process routes. The review by Mochida et al. (2014) provides a more specific discussion on chemical production from DCL.

The coal tar industry was developed as a natural consequence of processing the by-products from the carbonization of coal to produce coke. Coke is employed as a reducing agent in the metallurgical industry. Coal carbonization is nothing but a coal pyrolysis process. The main difference between processes developed for DCL and that to produce coke for the metallurgical industry is the primary production objective. Coal pyrolysis to produce coal liquids is conducted at lower temperatures (<550°C) and it employs coals with a high volatile matter content to increase the yield of liquid products. In DCL processes, coke is the by-product. Coal carbonization to produce reactive coke (pyrolysis at 450°C–700°C) or hard coke (pyrolysis at >900°C) employs coking coals and the primary product is coke. In coal carbonization processes, coal liquids are the by-products. Typical product yields are shown in Table 15.8 (Ahland et al. 1982; Owen 1979).

TABLE 15.8
Product Yields from Different Coal Carbonization Processes

	Coal Carbonization (wt%, Dry)		
	Low Temperature		High
Product	Fast Heating	Slow Heating	Temperature
Gas	7.0	10.0	17.2
Ammoniacal liquor	7.5	8.1	2.5
Light oil	1.9	4.1	0.8
Coal tar	16.8	7.1	4.5
Coke	66.8	70.7	75.0

Moving-bed gasification technologies also perform coal pyrolysis in the coal preheating zone and low-temperature coal tar products can be recovered from the raw synthesis gas, just like from coal pyrolysis processes. Chemical production from coal tar is associated with ICL facilities, such as Sasol Synfuels (De Klerk 2011), and synthetic natural gas production facilities, such as the Great Plains Synfuels plant (Stelter 2001) that employ moving-bed coal gasifiers.

Not all DCL processes or all coals as raw materials are equally well suited for the production of chemicals. With increasing hydrogenation during DCL, the coal liquids become less aromatic and contain less heteroatom functional groups. The amount of hydrogen addition during DCL increases in the order: coal pyrolysis < solvent extraction of coal < catalytic coal liquefaction. The objective of increasing hydrogenation of the coal liquids is usually to produce fuels, not chemicals. Although hydrotreated coal liquids can still be used for chemical production, there is less diversity in the chemical products that can be recovered from the coal liquids.

The main product fractions and the classes of chemicals that can be recovered from coal pyrolysis product fractions are as follows:

1. Ammoniacal liquor. The water that is coproduced during coal pyrolysis forms an aqueous phase on condensation. Ammonia (NH_3) readily dissolves in the water and hence its name. The polar organic compounds partition between the organic and aqueous phases, and this partitioning is governed by the liquid–liquid equilibrium. Phenolic compounds, carboxylic acids, and some neutral oils are typically present. The concentration of these organic compounds is on the order of 1 wt% in the ammoniacal liquor. Detailed discussion on the recovery and treatment of the ammoniacal liquor can be found in the works of Wilson and Wells (1945) and Muder (1963).

2. Light oil. The light oil is essentially the naphtha fraction of the coal tar that is recovered from the gas after coal tar condensation. It is a very aromatic product that is particularly rich in benzene and toluene. This fraction is also referred to as benzole. The recovery of benzene–toluene–xylene (BTX) from the light oil is discussed in detail by Muder (1963).
3. Coal tar. This is the bulk of the liquid organic products obtained from coal carbonization. It is a wide boiling mixture. The ratio between the pitch, which is the atmospheric residue fraction of coal tar, and the coal tar distillates depends on the coal carbonization temperature. As the carbonization temperature is increased, the pitch fraction increases, mainly at the expense of tar acids and aliphatic compounds (Rhodes 1945). The coal tar distillates can be a source of phenolics and aromatics, as well as smaller amounts of pyridinic compounds. Descriptions of the composition and recovery of different coal tar chemicals can be found in the works of Rhodes (1945), Karr (1963), Aristoff et al. (1981), and McNiel (1981).

To conclude, an observation is made about the advantage of producing chemicals from coal liquids rather than producing transportation fuels. Many of the chemicals that can be recovered from coal liquids are aromatics, such as BTX, phenols, cresols, xylenols, naphthalene, and other aromatics. These compounds have inherently low H:C ratios. Producing products with a low H:C ratio is better matched to the properties of coal, which also has a low H:C ratio. Transportation fuels generally require a much higher H:C ratio. Although deep hydrogenation is possible to increase the H:C ratio of the coal liquids, producing H_2 erodes the liquid yield advantage that DCL processes have compared to ICL processes.

REFERENCES

Aczel, T., R. B. Williams, R. A. Brown and R. J. Pancirov. 1978. Chemical characterization of Synthoil feeds and products. In *Analytical methods for coal and coal products. Vol. 1*. ed. Karr, C. Jr., 499–540, Academic Press: New York.

Adesanwo, T., M. Rahman, R. Gupta and A. De Klerk. 2014. Characterization and refining pathways of straight-run heavy naphtha and distillate from the solvent extraction of lignite. *Energy Fuels* 28:4486–4495.

Ahland, E., G. Nashan, W. Peters and W. Weskamp. 1982. Low temperature carbonization and coking. In *Chemical feedstocks from coal*. ed. Falbe, J., 12–77, John Wiley & Sons: New York.

Alvarez, Y. E., B. M. Moreno, M. T. Klein, J. K. Watson, F. Castro-Marcano and J. P. Mathews. 2013. Novel simplification approach for large-scale structural models of coal: Three-dimensional molecules to two-dimensional lattices. Part 3: Reactive lattice simulations. *Energy Fuels* 27:2915–2922.

Anderson, K. B., J. C. Crelling, W. W. Huggett, D. Perry, T. Fullinghim, P. McGill and P. Kaelin. 2011. Oxidative hydrothermal dissolution (OHD) of coal and biomass. *Prepr. Pap.-Am. Chem. Soc., Div. Fuel Chem.* 56(2):310–311.

Aristoff, E., R. W. Rieve and H. Shalit. 1981. Low-temperature tar. In *Chemistry of coal utilization. Second supplementary volume*. ed. Elliott, M. A., 983–1002, John Wiley & Sons: New York.

Balster, L. M., E. Corporan, M. J. DeWittc, J. T. Edwards, J. S. Ervin, J. L. Graham, S-Y. Lee, S. Pale, D. K. Phelps, L. R. Rudnick, R. J. Santoro, H. H. Schobert, L. M. Shafer, R. C. Striebich, Z. J. West, G. R. Wilson, R. Woodward and S. Zabarnick. 2008. Development of an advanced, thermally stable, coal-based jet fuel. *Fuel Process. Technol.* 89:364–378.

Barnard, D., L. Bateman and J. I. Cunneen. 1961. Oxidation of organic sulfides. In *Organic sulfur compounds. Vol. 1*. ed. Kharasch, N., 229–247, Pergamon Press: Oxford.

Berkowitz, N. 1979. *An introduction to coal technology*. Academic Press: San Diego, CA.

Bertau, M., H. Offermanns, L. Plass, F. Schmidt and H.-J. Wernicke eds. 2014. *Methanol: The basic chemical and energy feedstock of the future. Asinger's vision today*. Springer: Heidelberg.

Blanksby, S. J. and G. B. Ellison. 2003. Bond dissociation energies of organic molecules. *Acc. Chem. Res.* 36:255–263.

Bommannavar, A. S. and P. A. Montano. 1982. Mössbauer study of the thermal decomposition of FeS_2 in coal. *Fuel* 61:523–528.

Borah, D. and M. K. Baruah. 2001. Kinetic and thermodynamic studies on oxidative desulphurisation of organic sulphur from Indian coal at 50–150°C. *Fuel Process. Technol.* 72:83–101.

Caldwell, R. D. and S. M. Eyerman. 1979. Chemicals from coal-derived synthetic crude oils. *Adv. Chem. Ser.* 179:145–158.

Cavanaugh, T. A., W. R. Epperly and D. T. Wade. 1984. EDS coal-liquefaction process. In *Handbook of synfuels technology*. ed. Meyers, R. A., 1.3–1.46, McGraw-Hill: New York.

Challis, B. C. and A. R. Butler. 1968. Substitution at an amino nitrogen. In *The chemistry of the amino group*. ed. Patai, S., 277–347, Interscience: London.

Crynes, B. L. 1981. Processing coal liquefaction products. In *Chemistry of coal utilization. Second supplementary volume*. ed. Elliott, M. A., 1991–2069, John Wiley & Sons: New York.

Cusumano, J. A., R. A. Dalla Betta and R. B. Levy. 1978. *Catalysis in coal conversion*. Academic Press: New York.

De Klerk, A. 2005. Adiabatic laboratory reactor design and verification. *Ind. Eng. Chem. Res.* 44:9440–9445.

De Klerk, A. 2011. *Fischer–Tropsch refining*. Wiley-VCH: Weinheim.

De Klerk, A. and E. Furimsky. 2010. *Catalysis in the refining of Fischer-Tropsch syncrude*. Royal Society of Chemistry: Cambridge.

De Klerk, A. and R. J. J. Nel. 2008. Benzene reduction in a fuels refinery: An unconventional approach. *Energy Fuels* 22:1449–1455.

Derbyshire, F. 1988. *Catalysis in coal liquefaction: New directions for research*. IEA Coal Research: London.

Dooley, J. E., W. C. Lanning and C. J. Thompson. 1979. Characterization data for syncrude and their implication for refining. *Adv. Chem. Ser.* 179:1–12.

Egloff, G. 1937. *The reactions of pure hydrocarbons*. Reinhold: New York.

Elliott, M. A. ed. 1981. *Chemistry of coal utilization. Second supplementary volume*. John Wiley & Sons: New York.

Emanuel, N. M. ed. 1965 *The oxidation of hydrocarbons in the liquid phase*. Macmillan: New York.

Falbe, J. ed. 1982. *Chemical feedstocks from coal*. John Wiley & Sons: New York.

Fischer, R. H. and R. E. Hildebrand. 1981. Refining if coal-derived syncrudes. *ACS Symp. Ser.* 156:251–267.

Gary, J. H., G. E. Handwerk and M. J. Kaiser. 2007. *Petroleum refining: Technology and economics*, 5th ed. CRC Press: Boca Raton, FL.

Given, P. H. 1984. An essay on the organic geochemistry of coal. In *Coal science*. Vol. 3. ed. Gorbaty, M. L., Larsen, J. W. and Wender, I., 63–252, Academic Press: Orlando, FL.

Gray, D., G. C. Tomlinson and A. ElSawy. 1990. The hybrid plant concept: Combining direct and indirect coal liquefaction processes. In *Indirect Liquefaction: Contractors' Review Meeting—Proceedings*. ed. Srivastava, R. D. and G. J. Stiegel, 299–316, US Dept. of Energy: Washington, DC.

Guin, J. A., A. R. Tarrer, W. S. Pitts and J. W. Prather. 1977. Kinetics and solubility of hydrogen in coal liquefaction reactions. In *Liquid fuels from coal*. ed. Ellington, R., 133–151, Academic Press: New York.

Haghighat, F. and A. de Klerk. 2014. Direct coal liquefaction: Low temperature dissolution process. *Energy Fuels* 28:1012–1019.

Häring, H-W. 2008. *Industrial gases processing*. Wiley-VCH: Weinheim, Germany.

Hayashi, J., Y. Matsuo, K. Kusakabe and S. Morooka. 1997. Depolymerization of lower rank coals by low-temperature O_2 oxidation. *Energy Fuels* 11:227–235.

Hayatsu, R., R. G. Scott and R. E. Winans. 1982. Oxidation of coal. In *Oxidation in organic chemistry*. Part D. ed. Trahanovsky, W. S., 279–354. Academic Press: New York.

He, S. J. X., M. A. Long, M. I. Attalla and M. A. Wilson. 1992. Methylation of naphthalene by methane over substituted aluminophosphate molecular sieves. *Energy Fuels* 6: 498–502.

Higman, C. and M. van der Burgt. 2008. *Gasification*, 2nd ed. Elsevier: Amsterdam, the Netherlands.

Hodge, P. 1974. Fragmentation reactions of quinones. In *The chemistry of the quinonoid compounds. Part 1*. ed. Patai, S., 579–616, John Wiley & Sons: New York.

Huang, H. and K. Wang, S. Wang, M. T. Klein and W. H. Calkins. 1996. Studies of coal liquefaction at very short reaction times. *Energy Fuels* 10:641–648.

Huang, H., K. Wang, S. Wang, M. T. Klein and W. H. Calkins. 1998. Studies of coal liquefaction at very short reaction times. 2. *Energy Fuels* 12:95–101.

Johnson, C. A., M. C. Chervenak, E. S. Johanson, H. H. Stotler, O. Winter and R. H. Wolk. 1973. Present status of the H-Coal® process. In *Symposium papers. Clean fuels from coal*. 549–575. Institute of Gas Technology: Chicago, IL.

Jones, J. F. 1973. Project COED (Char-Oil-Energy-Development). In *Symposium papers. Clean fuels from coal*. 383–402. Institute of Gas Technology: Chicago, IL.

Kabe, T., A. Ishihara, E. W. Qian, I. P. Sutrisna and Y. Kabe. 2004. *Coal and coal-related compounds* (Stud. Surf. Sci. Catal. 150). Elsevier: Amsterdam, the Netherlands.

Kaneko, T., F. Derbyshire, E. Makino, D. Gray and M. Tamura. 2012. Coal liquefaction. In *Ullmann's encyclopedia of industrial chemistry*. Wiley-VCH: Weinheim, Germany.

Karr, C. Jr. 1963. Low-temperature tar. In *Chemistry of coal utilization. Supplementary volume*. ed. Lowry, H. H., 539–579, John Wiley & Sons: New York.

Khorasheh F. and M. R. Gray. 1993. High-pressure thermal cracking of n-hexadecane in aromatic solvents. *Ind. Eng. Chem. Res.* 32:1864–1876.

Kiebler, M. W. 1945. The action of solvents on coal. In *Chemistry of coal utilization. Volume 1*. ed. Lowry, H. H., 677–760, John Wiley & Sons: New York.

Lee, S. 1990. *Methanol synthesis technology*. CRC Press: Boca Raton, FL.

Le Page, J. F. 1987. *Applied heterogeneous catalysis. Design–manufacture–use of solid catalysts*. Editions Technip: Paris, France.

Li, C.-Z. ed. 2004. *Advances in the science of Victorian brown coal*. Elsevier: Amsterdam, the Netherlands.

Liu, K., J. Yang, J. Jia and Y. Wang. 2008. Desulphurization of coal via low temperature atmospheric alkaline oxidation. *Chemosphere* 71:183–188.

Liu, Y. A. ed. 1982. *Physical cleaning of coal. Present and developing methods*. Marcel Dekker: New York.

Longanbach, J. R., J. W. Droege and S. P. Chauhan. 1980. Short-residence-time coal liquefaction. *ACS Symp. Ser*. 139:165–177.

Mae, K., T. Maki, J. Araki and K. Miura. 1997. Extraction of low-rank coals oxidized with hydrogen peroxide in conventionally used solvents at room temperature. *Energy Fuels* 11:825–831.

Maitlis, P. M. and A. De Klerk eds. 2013. *Greener Fischer-Tropsch processes for fuels and feedstocks*. Wiley-VCH: Weinheim, Germany.

McMillen, D. F. and D. M. Golden. 1982. Hydrocarbon bond dissociation energies. *Ann. Rev. Phys. Chem.* 33:493–532.

McMillen, D. F., R. Malhotra, G. P. Hum and S-J. Chang. 1987. Hydrogen-transfer-promoted bond scission initiated by coal fragments. *Energy Fuels* 1:193–198.

McNiel, D. 1981. High-temperature coal tar. In *Chemistry of coal utilization. Second supplementary volume*. ed. Elliott, M. A., 1003–1083, John Wiley & Sons: New York.

Meyers, R. A. 1977. *Coal desulfurization*. Marcel Dekker: New York.

Meyers, R. A. ed. 1984. *Handbook of synfuels technology*. McGraw-Hill: New York.

Miura, K., M. Shimada, K. Mae and H. Y. Sock. 2001. Extraction of coal below 350°C in flowing non-polar solvent. *Fuel* 80:1573–1582.

Mochida, I., O. Okuma and S-H. Yoon. 2014. Chemicals from direct coal liquefaction. *Chem. Rev.* 114:1637–1672.

Morgan, M. E. and R. G. Jenkins. 1984. Role of exchangeable cations in the rapid pyrolysis of lignites. *ACS Symp. Ser.* 264:213–226.

Muder, R. E. 1963. Light oil and other products of coal carbonization. In *Chemistry of coal utilization. Supplementary volume*. ed. Lowry, H. H., 629–674, John Wiley & Sons: New York.

Mzinyati, A. B. 2007. Fuel-blending stocks from the hydrotreatment of a distillate formed by direct coal liquefaction. *Energy Fuels* 21:2751–2762.

Nowacki, P. 1979. *Coal liquefaction processes*. Noyes Data Corp.: Park Ridge, NJ.

O'Hara, J. B. 1981. Liquid fuels from coal. In *Coal handbook*. ed. Meyers, R. A., 715–816, Marcel Dekker: New York.

O'Rear, D. J., R. F. Sullivan and B. E. Stangeland. 1981. Catalytic upgrading of H-Coal syncrudes. *ACS Symp. Ser*. 156:115–144.

Okuma, O. and K. Sakanishi. 2004. Liquefaction of Victorian brown coal. In *Advances in the science of Victorian brown coal*. ed. Li, C-Z., 401–457, Elsevier: Amsterdam, the Netherlands.

Owen, J. 1979. The coal tar industry and new products from coal. In *Coal and modern coal processing: An introduction*. ed. Pitt, G. J. and Millward, G. R., 183–204, Academic Press: London.

Padrick, T. D. and B. Granoff. 1986. Mineral matter catalysis of coal conversion. *ACS Symp. Ser.* 301:410–415.

Pansini, A. J. and K. D. Smalling. 2002. *Guide to electric power generation*, 2nd ed. Marcel Dekker: New York.

Papso, J. E. 1984. The H-Coal® process. In *Handbook of synfuels technology*. ed. Meyers, R. A., 1.47–1.63, McGraw-Hill: New York.

Petrakis, L. and D. W. Grandy. 1983. *Free radicals in coals and synthetic fuels* (Coal Sci. Technol. 5). Elsevier: Amsterdam, the Netherlands.

Pines, H. and W. M. Stalick. 1977. *Base-catalyzed reactions of hydrocarbons and related compounds*. Academic Press: New York.

Quignard, A., N. Caillol, N. Charon, M. Courtiade and D. Dendroulakis. 2013. DIRECT CTL: Innovative analyses for high quality distillates. *Fuel* 114:172–177.

Rhodes, E. O. 1945. The chemical nature of coal tar. In *Chemistry of coal utilization. Volume 2*. ed. Lowry, H. H., 1287–1370, John Wiley & Sons: New York.

Richardson, F. W. 1975. *Oil from coal*. Noyes Data Corp.: Park Ridge, NJ.

Roets, P. N. J., C. A. Moses and J. Van Heerden. 2004. Physical and chemical properties of fully synthetic aviation turbine fuel from coal. *Prepr. Pap.-Am. Chem. Soc., Div. Petrol. Chem.* 49(4):403–406.

Santana, R. C., P. T. Do, M. Santikunaporn, W. E. Alvarez, J. D. Taylor, E. L. Sughrue and D. E. Resasco. 2006. Evaluation of different reaction strategies for the improvement of cetane number in diesel fuels. *Fuel* 85:643–656.

Satriana, M. J. ed. 1982. *Hydroprocessing catalysts for heavy oil and coal*. Noyes Data Corp.: Park Ridge, NJ.

Savage, P. E. 1994. Are aromatic diluents used in pyrolysis experiments inert? *Ind. Eng. Chem. Res.* 33:1086–1089.

Schlosberg, R. H. ed. 1985. *Chemistry of coal conversion*. Plenum Press: New York.

Schlosberg, R. H., M. L. Gorbaty and T. Aczel. 1978. Friedel-Crafts isopropylation of a bituminous coal under remarkably mild conditions. *J. Am. Chem. Soc.* 100:4188–4190.

Shah, Y. T. and D. C. Cronauer. 1979. Oxygen, nitrogen, and sulfur removal reactions in donor solvent coal liquefaction. *Catal. Rev.-Sci. Eng*. 20:209–301.

Shah, Y. T., S. Krishnamurthy and R. G. Ruberto. 1981. Kinetic models for donor-solvent coal liquefaction. In *Reaction engineering in direct coal liquefaction*. ed. Shah, Y. T., 162–212, Addison-Wesley: Reading, MA.

Siddiquee, M. N. and A. de Klerk. 2014. Hydrocarbon addition reactions during low temperature autoxidation of oilsands bitumen. *Energy Fuels* 28:6848–6859.

Stein, S. E. 1985. Free radicals in coal conversion. In *Chemistry of coal conversion*. ed. Schlosberg, R. H., 13–44, Plenum Press: New York.

Stelter, S. 2001. *The new synfuels energy pioneers. A history of Dakota Gasification Company and the Great Plains synfuels plant*. Dakota Gasification Company: Bismarck, ND.

Steynberg, A. P. and M. E. Dry. eds. 2004. *Fischer–Tropsch technology* (Stud. Surf. Sci. Catal. 152). Elsevier: Amsterdam, the Netherlands.

Stiegel, G. J., Y. T. Shah, S. Krishnamurthy and S. V. Panvelker. 1981. Refining of coal liquids. In *Reaction engineering in direct coal liquefaction*. ed. Shah, Y. T., 285–381, Addison-Wesley: Reading, MA.

Stranges, A. N. 2007. A history of the Fischer-Tropsch synthesis in Germany 1926–45. *Stud. Surf. Sci. Catal*. 163:1–27.

Swern, D. ed. 1970. *Organic Peroxides. Vol. 1*. Wiley-Interscience: New York.

Swern, D. ed. 1971. *Organic Peroxides. Vol. 2*. Wiley-Interscience: New York.

Swern, D. ed. 1972. *Organic Peroxides. Vol. 3*. Wiley-Interscience: New York.

Van Bodegom, B., J. A. R. van Veen, G. M. M. van Kessel, M. W. A. Sinnige-Nijssen and H. C. M. Stuiver. 1984. Action of solvents on coal at low temperatures. 1. Low-rank coals. *Fuel* 63:346–354.

Van Krevelen, D. W. 1993. *Coal. Typology–Physics–Chemistry–Constitution*, 3rd ed. Elsevier: Amsterdam, the Netherlands.

Vasireddy, S., B. Morreale, A. Cugini, C. Song and J. J. Spivey. 2011. Clean liquid fuels from direct coal liquefaction: Chemistry, catalysis, technological status and challenges. *Energy Environ. Sci*. 4:311–345.

Waters, W. A. 1964. *Mechanism of oxidation of organic compounds*. Methuen: London.

Whitehurst, D. D. 1980. A new outlook on coal liquefaction through short-contact-time thermal reactions: Factors leading to high reactivity. *ACS Symp. Ser*. 139:133–164.

Whitehurst, D. D., T. O. Mitchell, and M. Farcasiu. 1980. *Coal liquefaction. The chemistry and technology of thermal processes*. Academic Press: New York.

Wilson, P. J., Jr. and J. H. Wells 1945. Ammoniacal liquor. In *Chemistry of coal utilization. Volume 2*. ed. Lowry, H. H., 1371–1481, John Wiley & Sons: New York.

Zhu, Y. and H. C. Frey. 2012. Integrated gasification combined cycle (IGCC) systems. In *Combined cycle systems for near-zero emission power generation*. ed. Rao, A. D., 129–161, Woodhead Publishing: Oxford.

16 Coal to Metallurgical Coke

Vinod K. Saxena and Hari P. Tiwari

CONTENTS

Abstract: The quality and cost of coke are the most important considerations for the smooth operation of blast furnace and technoeconomics of hot metal production. The quality of coke is primarily influenced by the intrinsic properties of coal/coal blend. The effort is being made worldwide to reduce the cost of coke by proper selection of coals and adopted coke-making technology to produce desired quality of coke. This chapter briefly describes the overview of coal carbonization, characterization of coal, coke-making technologies, effect of coal properties and operating parameters on coke quality, model for predicting coke strength after reaction (CSR), and yield of by-product during coking.

16.1 INTRODUCTION

The conservation of coking coal and reduction of coke cost are gaining much importance in the iron and steel industry. The effort is being made worldwide to increase the consumption of inferior grade coking coal/non-coking coal in coke-making processes. This is facilitated by proper selection of coals to produce desired quality of coke. The optimum proportion of blendable coal in any blend is ascertained primarily by CSR value of coke (Tiwari et al., 2014e).

The physicochemical characteristics of Indian and imported coals are different due to difference in organic microcomponents (macerals) mixed and dispersed with mineral matter. Considering the macerals distribution, it is difficult to utilize the potential of indigenous coal with imported coal in a blend properly. Also, the ash content of the coal charge influences the coke properties. Therefore, the right coal can be selected depending on the quality of other existing coals to produce good quality of coke. India has a high reserve of coal, out of which only 14% is of the coking variety and the remaining 86% is of poorly coking and non-coking type (Nandi et al., 1981 and Sreedhar et al., 1998). Therefore, it is beneficial to use poorly coking and non-coking coal in coke making by blending it with other suitable materials (Sofer and Zaborsky, 1981 and Jones, 1980). Indian coals are inferior in quality as compared to those of other steel plants abroad. The ash content of these coals varies in the range of 15%–17.5% (washed coal); the average reflectance in oil of vitrinite varies from 0.7% to 1.8% and are richer in inerts (Chaudhuri et al., 1990). High ash in coal associated with high ash coke translates to operating problems in blast furnace high slag volume and high coke rate (Tiwari et al., 2014e).

Coke is the major source of energy for producing hot metal through blast furnaces. The desired quality of coke to be acceptable for a blast furnace depends on the type and size of the furnace. In general, coke plays three major roles in a blast furnace namely, (a) as fuel providing the energy required for endothermic chemical reactions and for melting of iron and slag; (b) as reductant by providing reducing gases for reduction; and (c) as a strong permeable grid providing for passages of liquids and gases in the furnace, particularly in the lower part of the furnace and supporting the burden above it. As the coke moves toward the lower zone of a blast furnace, it degrades and generates coke fines, which affect the bed permeability and process efficiency of blast furnace. The rate at which coke degrades is mainly controlled by the solution loss reaction, thermal stress, mechanical stress, and alkali attack (Diez et al., 2002 and Sahajwalla et al., 2004). The blast furnace route will remain in next few decades in spite of the availability of any newer and more advanced iron and steel making technology. Auxiliary fuel can supply heat and reduce the iron oxides but cannot support the burden of iron and limestone in the furnace. Therefore, operation of large blast furnaces throughout the world has created the demand for better quality of coke, which is the major cost component of operation in the iron and steel industry.

Coke can be partially replaced by the injection of pulverized coal into the blast furnace, but it cannot completely eliminate the need of coke. Theoretically up to 40% of the coke can be replaced by coal injection. In general, better coke quality leads to more pulverized coal injection in blast furnace. Thus, even with injection of pulverized coal, blast furnaces will always require at least a minimum amount of high-quality coke (Price et al., 1997). The desired quality of coke for better utilization in a blast furnace depends on the type and size of the blast furnace. The control limits are constantly getting stringent as the quality of the coke has improved in recent years (Segers, 2004).

The important parameters influencing the productivity and the coke rate in blast furnaces are coke ash, cold strength (M_{40} and M_{10}), coke reactivity index (CRI), CSR, and size of the coke. For maintaining the above parameters of coke, the selection of coke-making technology is an important aspect because all coke-making technologies have inherent limitations.

It is a well-known fact that the selection of coal plays a key role in controlling the coke quality. In general, coke quality depends on the properties of the coal/coal blend and coking conditions. Therefore, the selection of coals for coal blend to produce desired coke quality is an important tool to reduce cost of coke and hot metal. Another option for controlling the hot metal cost is reduction in coke rate by injection of auxiliary fuels like coal as a pulverized coal injection through the tuyeres of a blast furnace.

The resources of prime coking coal is relatively less/small, therefore, to reduce cost of coal blend, semi-soft coking coal as well as poor coking coal needs to be used in coal blend. The quality of coal is a major issue that must be addressed by the coke producers in a holistic and inclusive manner. In iron making, coking coal is an important area to be addressed both for enhancement of its production and for its optimum use to maximize the yield of good quality coke that meets the requirements of steel plants. Low-volatile and medium-volatile coking coals are being mined in India, and they are used up to certain level in the coking process because the required heat is also an important parameter for carbonization. The incremental increase in coke production for self-sufficiency in coke for hot metal production is met mainly through imported low-ash coking coal. This is because of the limited capacity of coking coal and high ash content. This, however, leads to an increase in hot metal cost.

16.2 BRIEF HISTORY AND IMPORTANCE OF COAL

Coal is the world's most abundant and affordable fossil fuel. The other fossil fuels will play their part but coal's role will be a vital one. Coal makes a significant economic contribution to the global economy. It is mined commercially in more than 50 countries and used in more than 70 countries. Annual world coal consumption is about 5,800 Mt, of which about 75% is used as thermal coal for electricity production. This consumption is projected to nearly double by the year 2030 to meet the challenge of sustainable development and a growing demand for energy. Coal reserves are very large and will be widely available; more than 58% of the world's recoverable reserves are located in four countries such as the United States (27%), China (13%), India (10%), and Australia (8.7%). Approximately 53% of anthracite and bituminous coal, 30% sub-bituminous coal, and 17% lignite are estimated recoverable reserves. These four countries, taken together, accounted for about 64% of total world coal production in 2004, and consume about 65% of the total world production, with Japan (183 Mt) and South Korea (79 Mt) being the world's two largest coal importers for electricity generation and steel production. Coal produces 40% of the world's electricity, which is double the share of its nearest competitors (gas and hydro) and coal is an essential element in over 65% of the world's steel production. These proportions are expected to remain at similar levels over the next 30 years. The International Energy Agency (IEA) predicts that world energy demand will grow around 60% over the next 30 years, most of it in developing countries. China and India are very large countries in terms of substantial quantities of coal reserves; they account for 70% of the projected increase in world coal consumption. Strong economic growth is projected for both countries (averaging 6% per year in China and 5.4% per year in India from 2003 to 2030), and much of the increase in their demand for energy, particularly in the industrial and electricity sectors, is expected to be met by coal (http://www.asiapacificpartnership.org).

The first traces of coal use were found in northern China during the Neolithic era 10,000 years ago. Since then, coal has been used for multipurpose, but primarily as a fuel for heat and electricity. The Industrial Revolution in the eighteenth and nineteenth centuries in Britain led to the large-scale use of coal, creating the beginning of a dependency on coal that we are still stuck with today. In 1885, coal finally surpassed wood as the largest source of primary energy in the United States. For the next 65 years, coal was king until oil displaced it after World War II. In 1950, oil surpassed coal as the biggest source of the U.S. primary energy, mostly as the result of automobiles. For over six decades, oil's dominance as a fuel has been unmatched and nearly unchanged. In 2008, oil's share in the U.S. energy market was the same amount as it was in 1950: 38.4% (Macdonald, 2010). Beyond transportation, the United States and the rest of the world is heavily dependent on coal for electricity generation. In the United States, approximately 44% of electricity is produced with coal. Historically, empirical evidence has shown that as nations become wealthier, their energy demand rises. In most cases, this usually means a rise in demand for coal because coal is almost always the cheapest form of energy. For example, between 1990 and 2007, U.S. electricity generation jumped by 67.8% (Pedersen, 2011).

16.2.1 Definition of Coal

It is very difficult to suggest a definition of coal, which will be scientifically correct and of general acceptance. "Coal is a compact stratified mass of plants which has not undergone permineralization (which have in part suffered arrested decay to varying degrees of completeness) free from all save a very low percentage of other matter" (Stopes and Wheeler, 1918).

The above definition is only applicable for bituminous and anthracitic coals. It does not cover the wider range of all carbonaceous material. Due to deposition of impurities, the plant substance is being insufficiently free from inorganic matter, may grade into shales and other carbonaceous materials.

According to Lewes (1918),

> *The most satisfactory view to take of the composition of coal is that it is an agglomerate of the solid degradation products of vegetable decay, together with such of the original bodies as have resisted to a greater extent the action to which the material has been subjected (Lande and Mackay, 1924).*

16.2.2 Classifications of Coal

Coal is solid combustible sources of energy on earth. It consists of remains of various organic compounds, which were originally derived from ancient plant and have subsequently undergone changes in molecular and physical structures during the transition of coal (Speight, 2013). Geographical conditions, age, and origin of coal influences the quality of coal. This diversity of conditions and widespread occurrence of coal resulted in numerous classification system (Mushrush and Speight, 1995).

Based on the metamorphism nomenclature system, coals are divided in four major groups: anthracite coal, bituminous coal, sub-bituminous coal, and lignite coal. Peat is usually not classified as a coal and therefore is not included in this system. Anthracite coal is of the highest metamorphic rank. It is also known as *hard* coal and has a brilliant luster, whereas lignite is the lowest in the metamorphic scale. Lignite varies in appearance from brown to brown black. Other coals fall in between these two in descending order of metamorphism (Mushrush and Speight, 1995).

Coals are also classified according to the geological age in which they are believed to have originated. However, this method of classification is abandoned because of its least utility in scientific community.

Coals are further classified by their banded structure. According to this method of classification, the coal exists in two general forms: banded and non-banded. Since the banded structure persists in all types of coal from lignite to anthracite, there may be some merit in such a classification. However, its failure to take into account of elemental composition of coal is a serious bottleneck.

Coals are also classified by rank of coal. Coal contains significant proportion of carbon, hydrogen, and oxygen with lesser amount of nitrogen and sulfur. Thus, it is not surprising that several attempts have been made to classify coal based on elemental composition. Indeed, for coal below the anthracite rank and with oxygen content less than 15%, it is possible to derive relationships between the carbon content (C, %), hydrogen content (H, %), volatile matter, and calorific value of coals. In this method of ranking, coal rank is often equated directly with carbon content, thus rank progresses from high carbon coal to low carbon coal. Together with carbon content, other properties such as calorific value are also taken into consideration while assigning the rank.

Coal can be classified based on its heating value and elemental carbon. Commercially, in general, the price and quality of coal depends on heating values. It is generally measured in British Thermal Units (Btu)/calories (cal). The heating value of coal depends on the combustible matter, mainly carbon and hydrogen content. Thus, higher heat content coal is considered as more valuable coal. Beyond heating values, different types of coal are classified by the amount of impurities associated in the coal. For example, some types of coal have higher levels of sulfur and ash. Often, types of coal with higher levels of impurities are blended with low impurity coals (Pedersen, 2011).

16.3 TYPES OF COALS

In general, coals are classified into two categories: (1) non-coking coals and (2) coking/metallurgical coal. Coals that are not softened and resolidify during carbonization are called *non-coking coal*. Such type of coal is not suitable for coke making process, but it may be used in some other processes such as Corex, Roelt processes, and coal-based direct reduced iron plants. Non-coking coals generally have higher ash contents and are normally used in thermal power plants such as steam coals. Most of the non-coking coals are used for steam generation and other energy intensive industries. Coking coal is a type of coal that upon heating in the absence of air undergoes a transformation into plastic state, swells, and then solidifies to form a cake and is generally utilized in the metallurgical industries. The coking coal suitable for metallurgical purposes are further classified into three broad group based on the quality of coke produced and its physical properties: (1) the prime coking coals are those that, with or without beneficiation, by themselves produce coke of metallurgical specification by conventional carbonization process, (2) the medium coking coals are those that can, with or without beneficiation, produce slightly inferior coke, and when used in the blend with suitable matching coal can produce coke of metallurgical specification, and (3) the semi-coking/weakly coking coals are those that, with or without conventional beneficiation, are not capable of forming metallurgical grade coke on their own, except in blends with suitable matching coals (Mukherjee et al., 1982).

Globally coking coal is present in plenty in Australia, China, Canada, and the United States. In India, limited medium coking coal is available in the Gondwana region. Unfortunately, the bituminous coking coals are scarce worldwide accounting for only about 5% of the world's supply of coals (Bujnowska and Colling, 1992).

In the literature, poor coking coal is defined in various ways. One source classifies coals with less than 4 free swelling index (FSI) and 45 Roga index (RI) as poor quality coal (Ward, 1984). Schapiro and Gray (1964) classified high volatile coals with less than 6 FSI and reflectance less than 0.68%–0.85% as poor coking coal.

The limited reserve of the coking coal is a matter of major concern for coke makers in several countries, including India due to escalation in price and nonavailability of good coking coal in market. However, the major challenge of reducing the hot metal cost and subsequently preserve the available coking coal reserve is of great importance. The prime attention is being made to use less expensive coals, such as, widely available semi-soft coals, in blends to produce coke with desired quality. This has been undertaken at various steel plants of India by suitable blend of prime, medium, semi-coking, and non-coking coals in different proportion as suggested by expert bodies from time to time (Mukherjee et al., 1982).

16.4 CHARACTERIZATION OF COALS

Coal is a heterogeneous mixture of compounds with widely varying physical and chemical properties, and it is not expected that the result of any simple test will reveal all its important characteristics. Therefore, various tests can be performed on coals in order to obtain information on their quality and to determine their suitability as coal blend for coke making. The above information is useful, but by no means will it provide conclusive indication of the properties of the coal tested. The following typical analyses are widely used to characterize the coal samples to utilize the actual coking potential (CP) of coals to produce desired quality of coke:

1. *Physical properties*—Granulometry (screen analysis) and hardness (HGI test)
2. *Chemical properties*—Moisture content, ash content, volatile matter, fixed carbon, C, H, N, O, ash constituents, and alkalis
3. *Rheological properties*—Crucible swelling number/free swelling index (CSN/FSI), low temperature Gray-King assay (LTGK), Geisler's fluidity, Sapozhnikov plastometer, and Arnu or Ruhr dilatometer
4. *Petrographic properties*—Macerals analysis (vitrinite, semi-vitrinite, exinite, inertinite, and mineral matter), Reflectance of vitrinite (Ro avg.), and vitrinite distribution or v-step distribution.

All the above properties have been used with varying degrees of success to predict the coke quality, which can be expected from a coal and its potential in formulating suitable coal blends.

16.5 SELECTION OF COALS FOR COKE MAKING

The final selection of the coal charge used for coke making depends not only on the coke quality requirements and the coal processing conditions but also on any restrictions that may be imposed by safe coke oven operation. Generally, coals are typical heterogeneous materials and detailed analysis is difficult to get the representative data with reproducibility and repeatability. Based on actual coking potential of a particular coal, it is possible to group coals for coke making as *excellent/good/fair, and poor.*

The selection of coals plays a key role in controlling the coke quality. For prediction of coke quality, several mathematical models are available. These models are broadly divided into two groups. The first type of model focuses on the prediction of cold strength of coke and the second type of model is based on prediction of hot strength of coke. So far, no prediction model has reached universal application in the coke-making area. However, almost all coking plants have some form of a model based on coal rank, rheological properties, petrographic properties, and ash chemistry.

The ash content of the coal blend is one of the important parameters that will influence coke properties, especially the coke size, with other operating conditions remaining the same (Bernard et al., 1985). The ash content of the stamp-charged coal blend also influences the bulk density of the coal cake. Lower the ash content of the coal/coal blend, lower will be the bulk density of coal cake under same stamping energy (Karmarkar et al., 1997). Petrographic properties of coal/coal blend are another important parameter for assessing the carbonization potential of a coal/coal blend. In one of the attempts, the coke CSR is predicted based on vitrinite reflectance and the inert content (Miura Y, 1978). Study also reports that rank of the coal blend (mean Ro %) should be high for producing coke with high CSR (Chaudhuri et al., 1990). The pioneering works define the coal composition based on the optical properties and the *maceral concept* (Stopes, 1919 and 1935). This study provides a significant insight on the requirement of optimum proportion of reactive and inert components to produce desired quality of coke. The reactive maceral (comprising vitrinite, semi-vitrinite, and liptinite) first softens, goes through a plastic stage and then hardens, and during hardening assimilates the inerts (inerinite and mineral matter) to give a porous, fused, solid carbon material (Schapiro and Gray, 1960 and Diez et al., 2002).

The G-factor is also one of the tools for predicting coke quality, which is usually obtained from parameters derived from the Ruhr dilatometer test, which is a modification of the Audibert-Arnu dilatometer test (the softening and resolidification temperatures and the percentage of coal contraction and dilatation). Although the G-factor is considered additive property for coal blends, there is a limitation for blends composed of coals whose plastic ranges do not overlap sufficiently. It is generally used to predict the M40 value of coke (Gibson, 1972 and Chiu, 1982).

It is useful to understand the actual coking potential of individual property of coal on coke quality especially coke CSR. A comprehensive study of 24 coals has been conducted in 7 kg carbolite oven (laboratory scale oven). Figure 16.1 depicts the impact of coal properties such as ash, volatile matter (VM), CSN, LTGK, Gieseler fluidity, reflectance (Ro avg.), ratio of reactive:inert, V11–V12, ratio of SiO_2:Al_2O_3, and alkalinity index on coke CSR. The correlation coefficient of selected properties except Ro varies in the range of 0.04–0.32 and the correlation coefficient of Ro is 0.60. Results indicate that the rank of the coal is an important property of the coal/coal blend in deciding the coke quality ($R^2 = 0.60$). This relationship indicates that a coke with good CSR can be made from a coal with Ro of 1.1–1.3. Results show that the rheological properties of coals (CSN, LTGK, and Gieseler fluidity) were also not individually adequate for predicting the coke quality. However, these properties are not always proportional to the coking power of a coal/coal blend. Instead, they are essential factors for determining the quality of coke. Similarly, the other properties of coal like ash, VM, ratio of reactive:inert, V11–V12, ratio of SiO_2:Al_2O_3, and alkalinity index are also not individually adequate to predict the desired coke quality. Thus, it is clear from Figure 16.1 that a single property of coal is not adequate to make definite conclusion about the coke CSR. Based on the above study, authors have developed the unique approach namely composite CP (CCP) for assessing the suitability of coal/coal blend to simplify the selection criteria of coals for coke making (Tiwari et al., 2013 and Tiwari, 2014).

16.5.1 Composite Coking Potential

For predicting the desired quality of coke CSR from a particular coal/coal blend, it is useful to judge the CP of a particular coal/coal blend. The CP of coal depends on its individual properties, which is important for coke making to produce good quality of coke. In an effort to broaden the scope of coal selection, the authors have developed a novel CCP index. The CCP index takes into account the selected properties, namely, ash, VM, CSN, LTGK, Gieseler fluidity, (Ro avg.), ratio of reactive:inert, V11–V12, ratio of SiO_2:Al_2O_3 and alkalinity index of the coals and their proportions in a given coal blend. This methodology is necessary since each of these parameters represents different aspects of the coking phenomena with its varying importance. Some of these parameters also have inter-dependencies. Figure 16.2 shows the process flow for selection of coal through CCP model.

16.5.2 Development of the CCP Model

The prime objective of the CCP model is to optimize the proportion of prime hard coking coal in a coal blend that is influenced by various coking properties. Basically, CCP is a combination of statistical and mathematical methods that are useful for assessing the potential of coal/coal blends based on desired quality of coke. The calculation of CCP is based on the interpolation of the range of selected properties, which is used in this model. Experiments are performed with the variety of coal and subsequent charts are prepared, which depict various properties of the coal. For each property, a CP range is allocated to a permissible range. CP of individual property of each coal in a coal blend is found by using corresponding interpolation chart. Weightage is assigned to each property

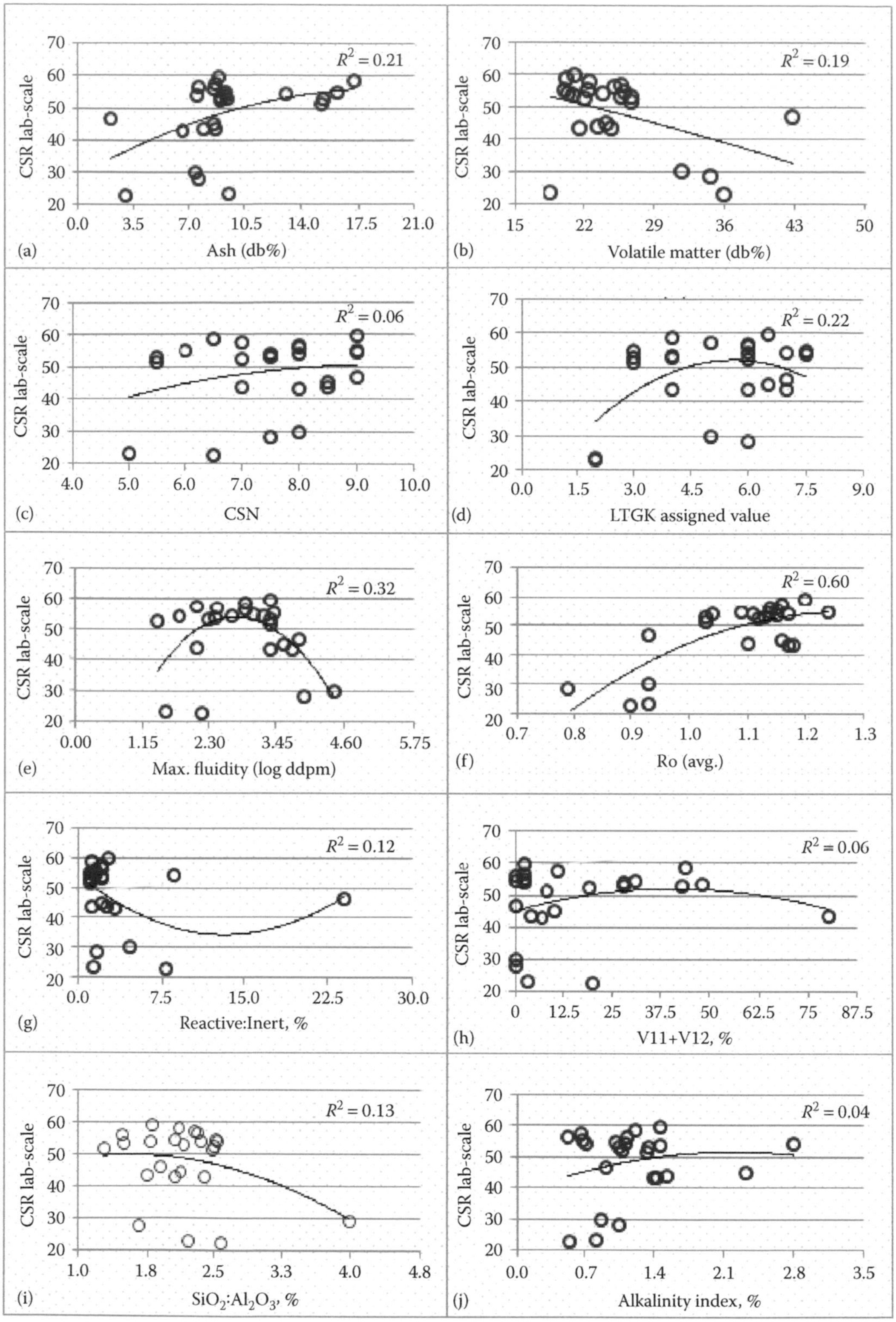

FIGURE 16.1 Effect of various coal properties on coke CSR. (a) Ash, (b) Volatile matter, (c) CSN, (d) LTGK assigned value, (e) Max. Fluidity (log ddpm), (f) Ro (avg.), (g) Reactive:Inert, (h) V11 + 12, (i) SiO_2:Al_2O_3 and (j) Alkalinity index. (Data from Tiwari HP, Banerjee PK and Saxena VK, A novel technique for assessing the coking potential of coals/coal blends for non-recovery coke making processes, *Fuel* 2013;107:615–622.)

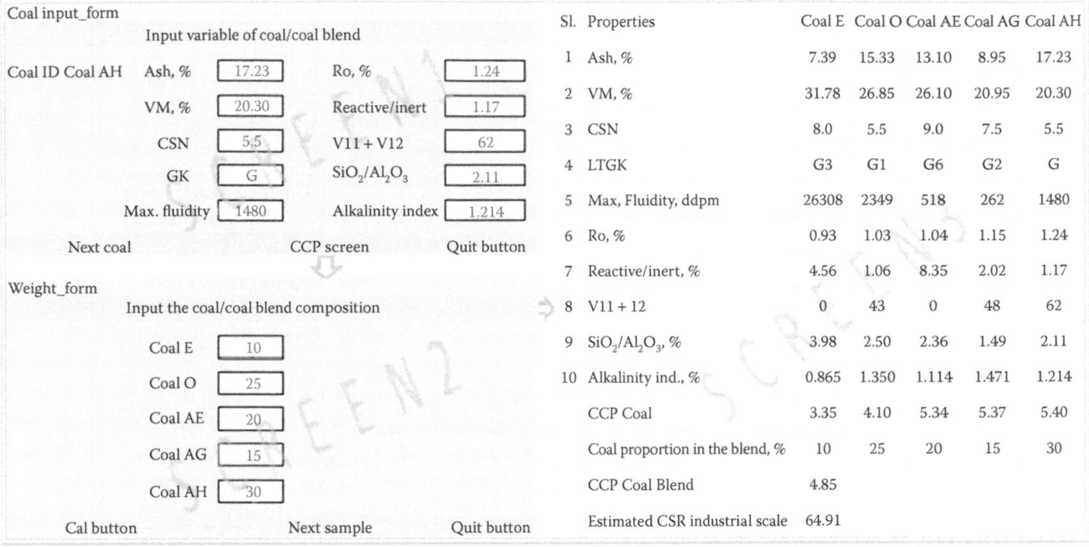

FIGURE 16.2 Process flow for selection of coals. (From Tiwari HP, Banerjee PK and Saxena VK, A novel technique for assessing the coking potential of coals/coal blends for non-recovery coke making processes, *Fuel* 2013;107:615–622.)

and thereafter CP is calculated for each coal. Mathematically, this can be expressed as follows:

$$R_{pm} = R_{pmj} + \frac{(R_{pmi} - R_{pmj}) \times (x_{pm} - x_{pmj})}{x_{pmi} - x_{pmj}} \tag{16.1}$$

where:

R_{pm} is the coking potential of *m*th property of *p*th sample of coal under experiment

x_{pm} represents the experimental value of the *m*th property of *p*th sample of coal under experiment

x_{pmi} is *i*th observed value of *m*th property of *p*th sample of coal under experiment

x_{pmj} is *j*th observed value of *m*th property of *p*th sample of coal under experiment

Based on above equation, the CCP of the individual coal as well as coal blend were calculated. Finally, CCP of coal can therefore be expressed as

$$\text{CCP of coal} = \frac{\sum_{m=1}^{10} (w_m \times R_{pm})}{100} \tag{16.2}$$

Therefore, CCP of coal blend can be expressed as

$$\text{CCP of coal blend} = \frac{\sum_{p=1}^{n} (wt_p \times \text{CCP of coal})}{100} \tag{16.3}$$

where:

wt_p is the proportion of individual coal used in coal blend

Thus, coefficient of CCP can be expressed as follows:

$$\text{CCP} = f(x, r) \tag{16.4}$$

where:

x is the coal property's value

r is the coal property's rank

The newly developed approach is very simple as compared to the old criteria, which is used for selection of coal for coke-making process to produce desired quality of coke. Figure 16.3a and b shows the process flow of existing and CCP model-based coal blend design criteria for non-recovery stamp-charged coke-making process. Figure 16.3a shows that in the existing selection process, input coals are characterized for chemical, rheological, and petrographic properties, and based on experience, a suitable coal blend is designed for carbonization tests in laboratory scale oven and evaluating the coke CSR. Based on laboratory, coke CSR suitable coal blend is used in plant, whereas coal blends, which are not able to produce targeted coke CSR, are further processed for making suitable coal blend for changing the proportion of individual coal. Figure 16.3b shows the coal blend selection based on newly developed CCP technique. In this unique approach, input coals are characterized as existing selection process and CCP value of input coals were calculated through the CCP model. According to prerequisite quality of coke, requisite CCP value of coal blend is designed by changing the proportion of input coals in the CCP model and estimates the coke CSR at industrial scale (Coke Plant). The CCP-based designed coal blend is directly implemented into the coke plant. It may be concluded that the CCP-based coal blend selection criteria is better than the old coal blend designing criteria (Tiwari et al., 2014e).

The CCP model has been used for selecting cheap coal blends with estimated plant coke CSR that would still comply with the minimum coke quality requirements of blast furnaces. This predicted coke CSR has been plotted against the actual coke CSR, as shown in Figure 16.4. Results show that the actual coke CSR is in close proximity with the predicted values of the CCP model with a correlation coefficient value of 0.86. CCP model may also be used for categorization of coals for better utilization. Based on the final outcomes, coals can be divided into two distinct groups: coals with CCP < 4, which give weak cokes after CO_2 gasification (CSR < 60) and coals with CCP > 4, which give

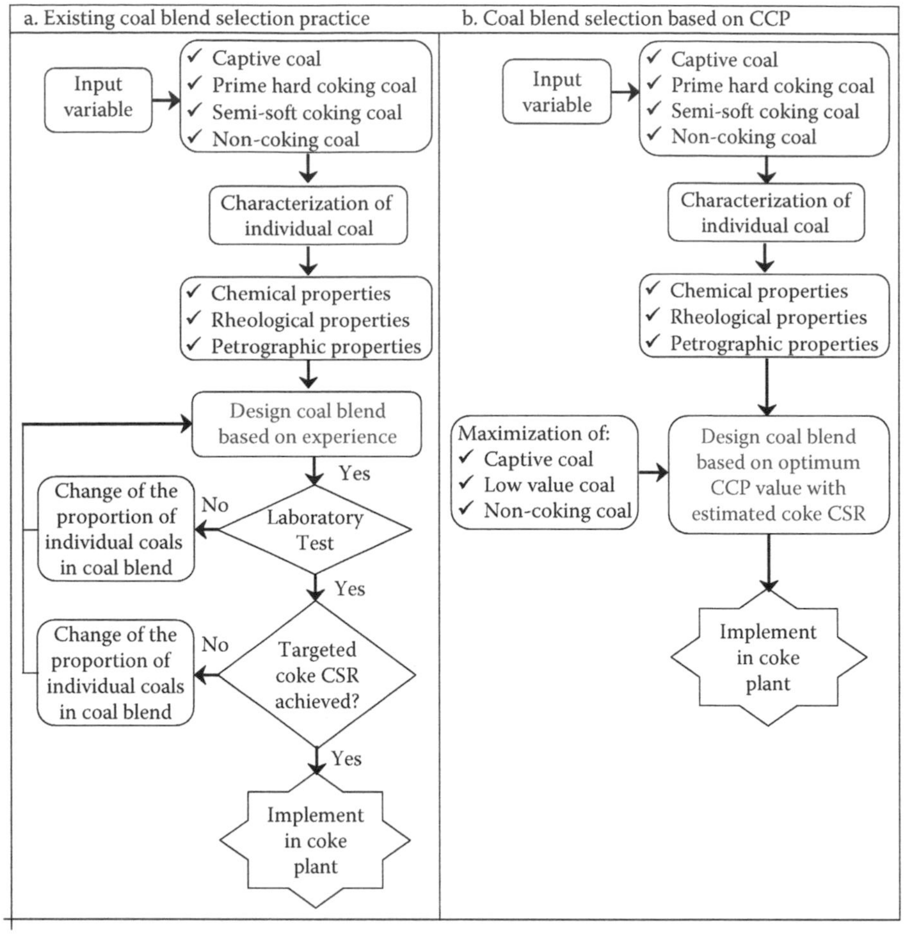

FIGURE 16.3 Process flow of existing and CCP-model based coal blend design criteria: (a) Existing coal blend selection practice and (b) coal blend selection based on CCP. (From Tiwari HP, Banerjee PK, Saxena VK et al., Efficient way to use of non-coking coals in non-recovery coke making process, *Metallurgical Research and Technology* 2014e;111:211 220.)

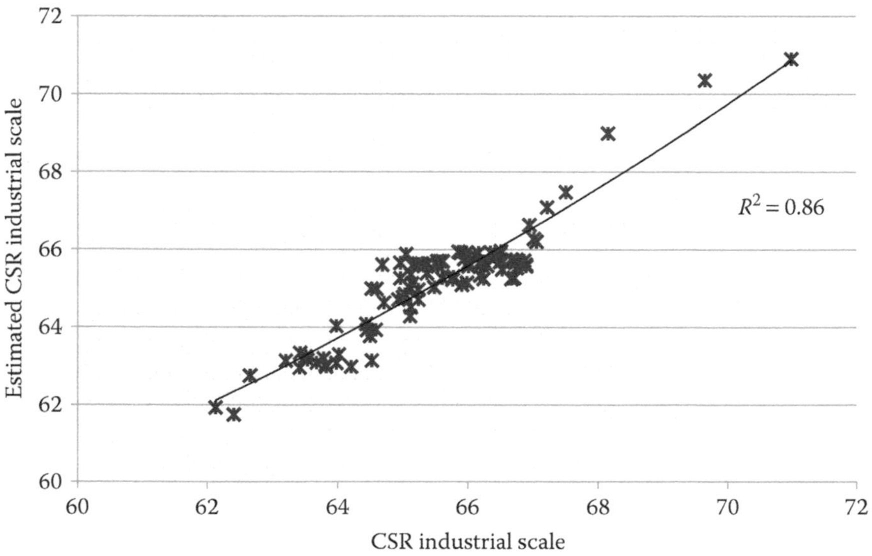

FIGURE 16.4 Relationship of CSR industrial scale vis-à-vis estimated CSR industrial scale. (From Tiwari HP, Banerjee PK and Saxena VK, A novel technique for assessing the coking potential of coals/coal blends for non-recovery coke making processes, *Fuel* 2013;107:615–622.)

strong cokes with CSR > 60. The accepted range of CCP of coal blend for producing ≥65.0 coke CSR in the non-recovery coke plant is ≥4.7. This model has been successfully implemented in the plant for selecting new coals and optimizing the blend composition (Tiwari et al., 2013 and Tiwari, 2014).

16.6 CARBONIZATION

It is defined as the thermal conversion of organic substances to carbon by the process of pyrolysis or destructive distillation of coal. The particles of coal on pyrolysis first releases water and absorbed gases and gradually become plastic, swell, and agglomerate. At a temperature of around 350°C, partial decomposition liberates some gaseous products within the plastic mass causing its swelling. Further rise in temperature of coal mass leads to its complete devolatilization and the left over residue, called *coke*, shrinks and hardens. The temperature range in which the coal becomes plastic, the maximum fluidity attained by the plastic mass, the amount of gaseous products evolved during coking, and the proportion of non-plastic component of coal are principal factors governing the quality of coke produced. In other words, aromatic growth takes place in which small aromatic and non-aromatic hydrocarbons present in coal are polymerized to form larger, complex aromatic polymer, which is then condensed by eliminating hydrogen and other side chains to form a solid residue called *coke*.

In coal, carbon is present in three forms: aromatic, hydroaromatic, and aliphatic. Of these, aromatic constitute the major portion, that is, 70%–85% of the total carbon, whereas, hydroaromatics and aliphatic constitute 10%–25% and 4%–5%, respectively. With increase in rank, aromaticity of coal increases and hydroaromaticity decreases. Coal pyrolysis involves two stages, that is, the primary stage up to 600°C and the secondary stage between 600°C and 1100°C. The primary stage of pyrolysis is the most important one, involving the major events, such as the softening of coal, development of plasticity and semi-coke formation, or accompanied by the release of primary volatiles. During this stage, the aromatic carbon fraction is effectively retained yielding semi-coke, while hydroaromatic carbon goes to the formation of primary tar, and the aliphatic structures contribute to the formation of gases. The secondary cracking leads to gas, low mass molecules, and polycondensation products.

16.6.1 Theory of Carbonization

A number of hypotheses have been set up to explain the plastic behavior of coal. Some of these have proved untenable in the light of knowledge derived from later experiments. According to Fitzgerald (1956) and Chermin and Van Krevelen (1957), decomposition may be distinguished into three successive reactions: formation of an unstable intermediate phase (metaplast), which is (partly) responsible for the plastification, and transformation of this intermediary into semi-coke and finally into coke:

$$\text{I. Coking coal} \xrightarrow{k_1} \text{metaplast } (M)$$

$$\text{II. Metaplast} \xrightarrow{k_2} \text{semi-coke}(R) + \text{primary volatile } (G_1)$$

$$\text{III. Semi-coke} \xrightarrow{k_3} \text{coke}(S) + \text{secondary gas } (G_2)$$

where:

k_1, k_2, and k_3 denote the respective reaction velocity constants

Reaction I is a de-polymerization reaction in which an unstable intermediate phase—metaplast—is formed, which is responsible for the plastic behavior of coal.

Reaction II denotes a cracking process in which tar is vaporized and nonaromatic groups are split off. This reaction is attended by re-condensation and formation of semi-coke. Toward the end of the process, the mass solidifies again.

Reaction III is a secondary degasification reaction in which the semi-coke units are welded together through evolution of methane and, especially at higher temperatures, of hydrogen, to yield the real coke (Fitzgerald, 1956).

16.6.2 Types of Carbonization

The process of carbonization can be classified into two types depending on the temperature.

Low temperature carbonization: The low temperature process is commercially practiced for the production of semi-coke, which is also called *soft coke* or *char*. For optimum yield, uniform heating is insured throughout the coal mass.

The various products of low temperature carbonization are semi-coke, low temperature tar, liquor, crude oil, and gas. The quality and yield of these products depend upon the process and the coal used. Low temperature carbonization (LTC) plants normally use low-rank coals and lignite. The properties of gas obtained by low temperature carbonization are in general specific gravity (referred to our) 0.6, calorific value 6500 K Cal/Nm^3 and Composition (in percent): CO_2 :4.0, C_mH_n :4.0, CO:7.0, H_2 :33, C_nH_{2n+2}: 45 and N_2 :7.0.

High temperature carbonization: In high temperature carbonization, coal is heated up to temperature of 900°C–1200°C (1600°F–2200°F). At these temperatures, practically all the volatile matter is driven off as gases, leaving behind a residue that consists principally of carbon with minor amounts of hydrogen, nitrogen, sulfur, and oxygen (which together constitute the fixed-carbon content of the coal). This residue is commonly known as *coke*. It has wide application in the iron and steel industry and other metallurgical industries.

16.6.3 Mechanism of Carbonization of Coal

The coal carbonization is the physicochemical processes, which depends on coal blend properties and transportation of thermal energy. The chemistry involved in the transformation of coal to coke is exceedingly complex. Much insight into the mechanism of coal carbonization has been reported by several researchers. In general, coal portion adjacent to the hot surface of the chamber wall becomes plastic and as the heat front advances to the next layer, the layer closest to the hot surface solidifies first. This sequence continues, layer by layer, until the

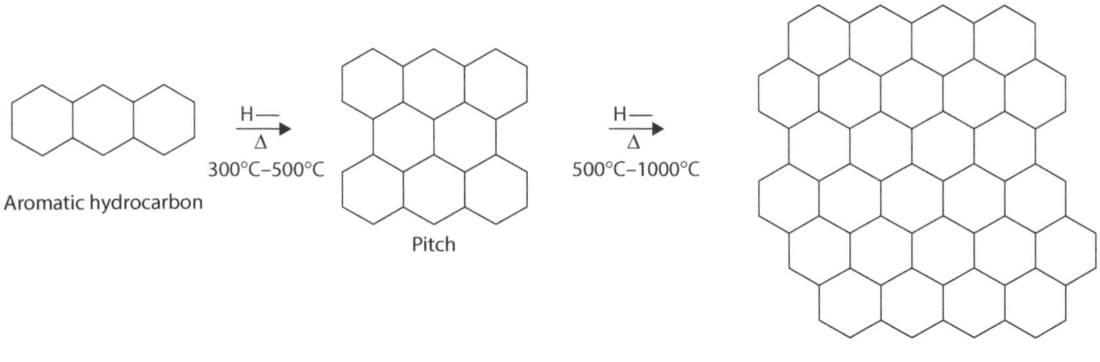

FIGURE 16.5 General reaction scheme for carbonization. (From Lewis IC, Chemistry of Carbonization, *Carbon* 1982;20:519–529.)

plastic layer reaches the center of the oven as layer near the hot surface gets transformed from semi-coke to coke during carbonization process. Gases and vapors from the inner layers escape through the hot coke, and fissures develop in the layer of semi-coke. Chemically, carbonization is a polymerization and aromatization process. It is considerable to explain a carbonization as parallel and consecutive of individual processes, which represent major reactions involving the pyrolysis of aromatic hydro carbons whereas overall process of carbonization is still exceedingly complicated (Lewis, 1982, Ruette et al., 1993 and Loison, 1989).

The major reactions involved in the pyrolysis of aromatic hydrocarbons include the following:

1. Rupture of C—H and C—C bond to form reactive free radicals
2. Molecular rearrangement
3. Thermal polymerization
4. Aromatic condensation
5. Elimination of side chains and hydrogen

In general, initially, a bond cleavage in the aromatic molecule results in the formation of a free radical intermediate. There are two types of radicals formed. An aromatic C—H bond ruptures to produce an σ radical and second, an aromatic C—C bond breaks to form a π radical. The formation of σ and π radicals allows us to analyze shape, size, and aromaticity effects in the energy formation of the radicals. The formation of σ radicals is a high-energy reaction. The bond dissociation energy is very high, and the σ radical intermediate form is very unstable and the free electrons are localized. The π radical intermediate form is comparatively more stable due to resonance.

$$(\circledcirc)_{n'} \rightarrow (\circledcirc)^{\bullet}_{n'} + H^{\bullet} \quad (\sigma - \text{radical}) \qquad (16.5)$$

$$(\circledcirc)_{n''} \rightarrow {}^{*}(\circledcirc)^{\bullet}_{n''} + H^{\bullet} \quad (\Pi - \text{radical}) \qquad (16.6)$$

After the formation of the intermediate radical, direct polymerization can take place. It initially involves the loss of hydrogen accomplished to internal hydrogen transfer. Thus, hydrogenated derivative of the original compound is formed with hydrogen added at the most reactive position in the molecules. There are two types of reactions that take place simultaneously in the pyrolysis scheme.

Cracking reaction: It involves the rupture of C—C bond producing two free radicals intermediate. The saturation of these radicals requires hydrogen. Cracking reaction produces components that are less polymerized and of which large proportion will be liquid at pyrolysis temperature.

Aromatization and condensation: It involves the formation of extensive aromatic groups by: (a) dehydrogenation of saturated rings and (b) recombination of aromatic groups with one another by formation of aromatic C—C bond. These reactions liberate hydrogen and lead to the production of solid carbon residue. It forms a large, thermally more stable aromatic molecule, which condenses along with elimination of water.

The condensation reaction is as follows:

$$R - OH + R'H \rightarrow R - R' + H_2O \qquad (16.7)$$

In this process, the next step is the thermal rearrangement, wherein the molecules rearrange among themselves to produce the most stable structure.

In an aromatic molecule, there are many sites for polymerization. However, the site at which the polymerization should predominantly occur can be predicted by reactivity parameters, such as, free valences, localized energies, and thermochemical kinetic analysis. Steric effects also play a role in the polymerization process.

The aromatic hydrocarbons condense to give either a condensed polymer or a non-condensed polymer. Both have almost the same molecular weights but differ in their structure and properties. Non-condensed polymers are non-planar in shape and their reactivity and ionization potentials change slowly with increasing polymerization, whereas, condensed polymers are planar and show marked changes with increasing size. One can envision carbonization by the scheme shown in Figure 16.5. Various stages occur during carbonization of coal/coal blend in laboratory and plant scale ovens as shown in Figure 16.6.

16.7 COKE-MAKING TECHNOLOGIES

Coal is converted into metallurgical coke broadly by two types of coke-making technologies: beehive coke making and the slot-type by-product coke-making technology. The operating

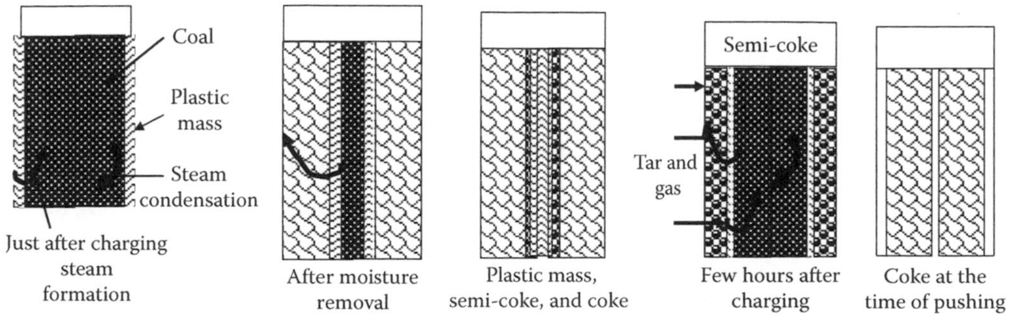

FIGURE 16.6 Various stages during carbonization.

conditions and heating rates of both recovery and non-recovery coke-making technologies are different due to asymmetry of heating. In the present context, emphasis has been given on non-recovery coke-making technology.

Producing metallurgical coke from other coke-making technology (except recovery and non-recovery coke making) has been proposed but is not yet fully commercialized till date. In one of the processes, pulverized coking or non-coking coal is dried and partially oxidized with steam or air in fluidized bed reactors to prevent agglomeration. The reactor product is carbonized in two stages at successively higher temperature to obtain char. The produced char is briquetted in a roll process with the help of binder (produced coal tar as a binder). After that the *green* briquettes are cured at low temperatures and carbonized at high temperature. After completion of carbonization process, hot coke is cooled in an inert atmosphere. This type of coke often called as *formed coke.*

16.7.1 Recovery Coke-Making Technology

The recovery ovens/slot ovens are normally tall and narrow chambers is called *recovery coke oven* or slot-type by-product coke oven in which coal mass is heated at a constant rate with the help of secondary fuel (blast furnace gas, coke oven gas, and Linz-Donawitz gas, or a combination of these gases) and operated with positive pressure. Normally, 10–100 ovens are grouped together in a battery. The ovens are run under high pressure and are associated with air pollution, such as gas leakage into atmosphere through the door (Walters, 1999 and Buss et al., 1999). The volatile matter liberated during the coking process and recovered as by-products are mainly coal tar, ammonia, sulfur, and coke oven gas.

16.7.2 Non-Recovery Coke-Making Technology

In the non-recovery/heat recovery coke plants originally referred to as beehive ovens, the coal mass is heated by burning the volatile matter inside the oven chamber by crown and sole flues and operated with negative pressure. The carbonization process takes place from the top by radiant heat transfer as well as conduction and from the bottom by conduction of heat through the floor by means of sole flue. This technique consumes the by-products, eliminating much of the air and water pollution associated with coke-making processes. A comparative cross-sectional view of recovery and non-recovery coke oven is shown in Figure 16.7.

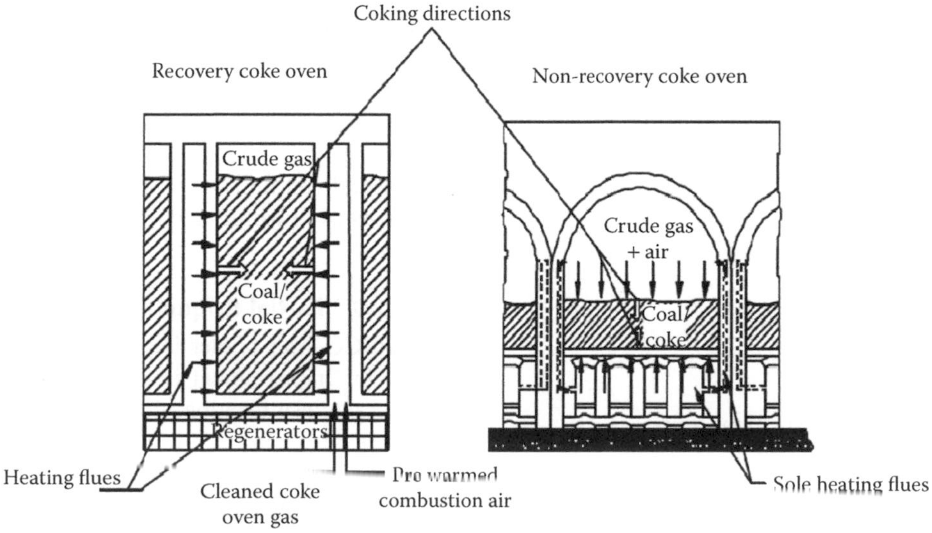

FIGURE 16.7 Cross-sectional view of recovery and non-recovery coke oven.

16.7.2.1 Process Description of Non-Recovery Oven

In non-recovery process, as soon as the coal is charged and comes into contact with the hot oven chamber and radiation heat from the oven crown, the coal mass ignites on the surface. The coal layer underneath the surface heats up and gas evolves. The raw gas thus produced is subsequently burnt in two stages and provides the heat required for carbonization. In the first stage, the raw gas burns in the presence of a controlled air volume at temperatures around ≥1000°C. The heat transfer into coal cake takes place mainly by radiation with the brick arch functioning as secondary heating surface. In this way, the coal is progressively carbonized from the surface of the coal mass down toward the oven sole. The required primary airflow enters through the oven doors and is controlled with the help of manually operated air dampers. As a result of the transfer of heat from the sole flue through the oven sole, carbonization progresses from the oven sole toward the surface of the coal mass. Identical carbonization rates from the top and bottom surfaces can be achieved by appropriately adjusting the primary and secondary airflows. The uniform coking rate from top and bottom will result in an even carbonization and hence improved coke quality and productivity (Tiwari et al., 2012, 2014c and Tiwari, 2014). Non-recovery ovens run under a negative pressure and all hydrocarbon is burnt within the oven, it eliminates leakage from doors and ports, a problem that generally prevails in the by-product ovens. The operating costs of non-recovery ovens are much lower and they can run with cheaper, comparatively high volatile (with >26%), low-ranked coal since there is no wall pressure occurs during coal carbonization.

16.8 TECHNOLOGIES FOR IMPROVING COKE QUALITY

The quality of coke depends upon factors such as properties of coal/coal blend, granulometry of charged coal/coal blend, methods of coal preparation—such as selective crushing, briquette blending, and partial briquetting of coal charge (PBCC), operating parameters—such as bulk density of charged coal, rate of heating and coke end temperature, and coke making technologies (top charged/recovery stamp charged/non-recovery stamp charged). The following technologies can be considered for improving coke quality (Tiwari, 2014):

- Selective crushing of coal
- PBCC
- Preheating of coal charge
- Stamp charging of coal
- Coke dry quenching process
- Solvent refining and formed coke

16.8.1 Precarbonization Technologies

Precarbonization technologies, such as selective crushing, preheating, briquette blending, and stamp charging, have been commercially implemented in different coke plant all over the world (Beck and Meckel, 1980). These technologies mainly aim to improve the bulk density of coal charge, and the resulting proximity of coal particles during softening phase may lead to stronger bond between coke cells and thereby improve coke strength.

16.8.1.1 Selective Crushing of Coals

Selective crushing aims at controlling the degree of crushing of the different constituents of coal. In general, reactive components of coal (primarily vitrinite) are softer constituents of coal, whereas inert content (inertinite and mineral matter) are hardest. In conventional coal crushing mills, reactive constituents get crushed to a relatively finer size as compared to inert. Hence, for producing good quality of coke, it is essential to crush the inert content into finer than the bulk coal mass, so during pyrolysis (softens stage) the former is assimilating in a better way, leading to improvement in the coke strength. This is achieved by selective crushing. In selective crushing, coal is crushed in two stages. The finer fractions of coal (rich in softener phase) are screened out after first stage of crushing and the oversized fraction is richer in inert and is crushed further in the second stage crusher. This approach not only saves the vitrinites from overcrushing but also increases the crushing fineness to almost close to 100% without excessive generation of zero fines (−3.2 mm fraction). Both these factors lead to the improvement in the coke quality. Thus, by progressive crushing and screening, the harder particles, which contain the high ash and more infusible components, are selectively crushed finer in order to effect uniform dispersal of the infusible mass in the coal charge. This technique provides help in preventing cracks and shrinkage, and hence improves the coke quality (Ghosh and Chatterjee, 2008).

16.8.1.2 Group-Wise Crushing of Coals

Group-wise crushing is a similar technique of selective crushing, which controls the degree of crushing of the different constituents of coal. It has been claimed to be the most suitable technique for multicomponent coal blend used in various plants of Steel Authority of India Limited and others in the iron and steel industry. Group-wise crushing envisages dividing the constituents of coals of the blend into *groups* based on their grindability, coking parameters, and petrographic analysis. Each group is then crushed separately. The aim is to decrease the heterogeneity in the different size fractions of coal blend and improving the coal properties of coarser size fractions by crushing them in groups to the required crushing levels.

16.8.1.3 Partial Briquetting of Coal Charge

The PBCC with binder is yet another commonly used technique for improvement in coke quality. The PBCC process enables partial utilization of weakly/non-coking coal in the form of briquettes along with coal charge. The technology involves charging about 30% coal blend in the form of briquettes. Briquettes are prepared using a binder (pitch and pitch + tar) up to 2.0%–3.0% of charge. Coke quality significantly improves as a result of increase in bulk density of charge.

16.8.1.4 Briquette Blending of Coal

In addition to enhance the bulk density of the coal charge, partial briquette blending has a distinct advantage in improving the quality of coke. When a mixture of coal fines and briquettes undergoes carbonization in an oven, the briquettes expand with marked evolution of gas during softening; this compresses the surrounding coal fines, promoting cohesion of the coal particles, and thus accelerates agglomeration among grains, both of the briquettes and of the fine coal. Hence, a more compact coke structure and improved coke quality is obtained.

Briquette blending is considered to be a very versatile tool where the relatively low grade coking coals are used for coke making. In this method, briquettes of coal were made by using coal tar, pitch, molasses etc. as binder, which utilize the beneficial effect of high density of charged coal to improve coke quality. The effectiveness of this technique depends primarily on the blending technique to distribute the briquettes uniformly in the charge of oven. A poor charging facility could cause segregation of briquettes and fine coal inside the oven and, obviously, too much of segregation would result in high wall pressure at a certain point.

16.8.1.5 Binderless Briquetting of Coal

In this technique, 100% of the coal is generally charged into any oven, briquetted just before taking it into the charging car. Since no binder is used in this technique, the briquettes are susceptible to disintegration and the fines generated by their disintegration during subsequent handling for charging into the oven make them behave like a partial briquette charge. The overall bulk density of the coal charge increases by 40–50 kg/m^3 in this process, which helps to improve the coke quality.

16.8.1.6 Preheating of Coal

The preheating technique is generally applicable for utilization of inferior grade coking coal in coke making. Since, dry coal have the higher flowability as compared to wet coal and, therefore, preheated coal is charged into the oven, it packed providing high charge bulk density which improves coke strength. This concept is utilized in preheating technology, in which the coal is preheated to 180°C–250°C, before charging it into the coke ovens. Improvement in bulk density of 100–150 kg/m^3 over the normal 700–750 kg/m^3 has been achieved by this method.

16.8.1.7 Stamp-Charged Coke-Making Technology

Stamp-charged coke making has been established as a versatile technology, which not only improves the coke properties that can be obtained from a given coal blend but also broadens the coal base for coke making, permitting the use of inferior coals without impairing the coke quality. It is a more energy-intensive process as compared to conventional top charged process because it needs extra quantity of water (3%–4%) to make a requisite strength of coal cake. Due to extra moisture content in coal blend, it needs additional thermal energy and this increases the coking cycle time. Out of all the techniques (Singh et al., 1983) available, namely, selective crushing, briquette blending, stamp charging, and preheating, it has been reiterated several times in the past that stamp-charged coke making would be the ideal technique to be adopted under Indian conditions, having about 50% of bituminous coal, even though the initial cost will be higher for the oven and other accessories (Haque, 1990).

The advantage of stamp-charged coke-making technology lies in its ability to use inferior coal blend for carbonization and still produce coke of superior property. The improvements of CSR in stamp-charged coke-making process are due to higher bulk density obtained through mechanical stamping of coal charge. Stamp-charged coke making produces denser coke of relatively low porosity, which makes the coke reactivity low (Loison et al., 1989).

16.8.2 Use of Additives/Binders

In stamp charging, stability of coal cake is of considerable importance as it ensures trouble free stamping and battery operations. The stamped coal cake must have certain stability in order to prevent breakage during charging by stamping machine into the coke oven chamber. The most important characteristics of stamped coal cake are its density, compressive strength, and shear strength. The strength of coal cake and hence its stability is due to the mechanical interlocking of grains, which is a function of the stamping energy, moisture content of the coal blend, and its surface area, which, in turn, is related to the granulometry of coal.

To preserve good quality of coking coal, preceding research work on the possibility of blending low-grade coals with abundantly available cheaper biomass and selective additive/binder to produce quality of coke (Graham and Wilkinson, 1980; Habermehl et al., 1981 and Samir and Zaborssky, 1981). Research efforts have also been focused to improve the coke quality through use of binders such as coal tar, coal tar pitch, phenolic resin, molasses, waste plastic, and nail. The addition of binder in coal blend not only improves the oven throughput and mechanical strength of coal cake but also improves the coke quality in terms of coke CSR, M40, and coke mean size (Tiwari et al., 2014a).

16.8.3 Coke Dry Quenching

Coke is generally cooled after completion of carbonization by two quenching methods: wet quenching and dry quenching. In the wet quenching process, water is used as a cooling media to cool the hot coke, which not only consumes large volume of water but also releases harmful pollutants, which are not permitted by environmental regulations.

In coke dry quenching (CDQ) process, red hot coke is cooled in the presence of inert gas, which subsequently saves the heat and is also eco-friendly. The CDQ process was originally developed on industrial scale in the former Soviet Union at the beginning of the 1960s (also known as the *Giprokoks process*). It was intended for application in coke oven plants located in regions that suffered from long periods of severe cold and where there was a strong need of energy for heating purposes (IPPC, 2001). In this process, sensible heat is recovered from red hot coke in the presence of inert atmosphere to produce steam for generation of electric power. Particulate matter is emitted during this process. These emissions can be avoided by spraying and/or transport in closed conveyors (IPPC, 2001). (http:/www.climatetechwiki.org/technology/coke-dry-quenching).

16.8.3.1 Features of Coke Dry Quenching

The generated waste heat is utilized for power generation and hence there is no *puff of white smoke* including dust because this process is operated under enclosed chamber. As a result, the working environment is improved. CDQ does not produces the greenhouse gas (CO_2), which an oil-burning boiler produce. Reduction of CO_2 by CDQ is nearly equivalent to 18T/h CO_2 that an oil-burning boiler produces when it generates 18MW electric power (equivalent to that by 100T/H CDQ).

By adopting CDQ process in coke making, coke quality in terms of hot strength and cold strength is improved by about 2%, respectively. As a result, the coke is free from surface pore due to aquatic gasification reaction and internal crack that may occur in wet quenching. In addition, brittle portions of coke are removed when it drops in the chamber. (http://www.jase-w.eccj.or.jp/technologies/pdf/iron_steel/S-10.pdf).

16.9 FACTORS CONTROLLING COKE CSR

The quality of coke depends on properties of coal/coal blend and operating parameters. Generally, a good quality of coke is made from good quality of coal/coal blend. The quality of coal/coal blend depends on its chemical, rheological, and petrological properties. Coal blend properties namely, volatile matter, CSN, maximum fluidity, dilation, vitrinite, inertinite, average reflectance, and operating parameters namely, fraction of −3.15 mm, fraction of +0.5 mm, stamping time, coke end temperature, and coking time are important parameters for controlling the coke quality. Since each of these properties and operating parameters represents different aspects of the coking phenomena with its varying importance, because of which it is important to maintain coke quality in terms of hot strength and cold strength. Literatures also reported that coke CSR depends about 70% on the coal/coal blend properties and 30% on the operating parameters (Nakamura et al., 1977).

16.9.1 Effect of Coal Properties on Coke Quality

In general, selection of coals plays a key role in controlling the coke quality. For prediction of coke quality for coke plant, several mathematical models are available. These models are broadly divided into two groups. The first type of model focuses on the prediction of cold strength of coke and the secondtype of model are for the on prediction of hot strength of coke. For predicting coke quality, almost all coke plants have some form of a model based on coal rank, rheological properties, petrology, and ash chemistry.

16.9.1.1 Chemical Properties of Coal/Coal Blend

Chemical analysis of coal/coal blend is useful for predicting coke chemistry, hot strength (coke reactivity index/coke strength after reaction), and coal strength (M40/M10), which directly affect the performance of blast furnace. Few important chemical properties are briefly explained below.

Ash content and VM are important properties for evaluating the properties of coals. Ash is the impurity component that decreases the rheological as well as agglomerating properties (Banerjee and Tiwari, 2008) of coals and increases viscosity of slag in blast furnaces. The ash content is the important parameter of the coal charge, which influences the coke size when other operating conditions remain the same (Prasad et al., 1996 and Nakamura et al., 1977).

VM is the light carbon component, which usually decreases with increase in coal rank. VM is closely related to the coke yield; with increase in VM, the yield decreases. On the other hand, VM is the most important parameter in deciding the blend for non-recovery ovens. Low VM coals generate less gas and hence the total thermal energy on its combustion may be insufficient for attaining the coking temperatures in the oven.

16.9.1.2 Mineralogy of Coal

Coals contain a wide range of minerals and the high temperature conditions involved in the coking process can cause various changes and lead to the formation of new phases. The major minerals present in coals are quartz, carbonates (calcite and dolomite), feldspars, sulfides (pyrite), and a number of phyllosilicates represented by clay minerals (e.g., kaolinite, montmorillonite, illite, and halloysite) and mica (muscovite). During the coking processes, some new mineral phases form and affect the properties of the coke and its behavior in blast furnace operation (Finkelmer, 1981).

The most common changes that occur to minerals in coke oven batteries are: (1) desulfurization, (2) decarbonation (loss of CO_2), (3) dehydration (loss of H_2O), (4) dehydroxylation (loss of OH^-), (5) polymorph transformations, (6) melting with the formation of an alkali and silica-rich liquid, and (7) transformation to another crystalline phase. The transformation (decomposition) of pyrite in an inert atmosphere proceeds through a multistep process in a sequence and is governed by the temperature at equilibrium conditions (Dam-Johansen et al., 2006).

$$\text{Pyrite}\left(FeS_2\right) \rightarrow \text{Pyrrhotite}\left(Fe_{1-x}S\right) \rightarrow \text{Troilite}\left(FeS\right) \rightarrow Fe$$

According to experimental data, dolomite, Ca $Mg(CO_3)_2$, will decompose directly to calcite, $CaCO_3$, in a CO_2 atmosphere, accompanied by the formation of MgO between 550°C and 765°C. The calcite will then decompose to CaO between 900°C and 960°C. In air, dolomite will decompose between 700°C and 750°C, with simultaneous formation of $CaCO_3$, CaO, and MgO. This may result in solid phases, periclase, MgO, lime, and CaO having higher density values (3.56 and 3.35 g/cm^3, respectively) than the parental calcite (2.71 g/cm^3) and dolomite (2.87 g/cm^3) (Engler, 1988).

Various minerals that contain H_2O and OH^- compounds dehydrate and dehydroxylate in a manner that depends not only on temperature but also on the heating rate. Halloysite, for example, dehydrates according to the following reaction:

$$Al_4(OH)_8(Si_4O_{10}) \times 4H_2O \rightarrow Al_4(OH)_8(Si_4O_{10}) + 4H_2O$$

But the temperature of dehydration varies between 99.7°C and 128.2°C depending on the heating rate (4°C–13°C/min). Thus, dehydration takes place at the temperatures corresponding to the first stage in the coking process, when the coal decomposes to form plastic layers.

Tata Steel (India) has proposed the following basicity index (BI) for optimizing the composition of rammed batch (Prasad et al., 1999).

$$BI = \frac{100 \times Ash(\%) \times (Na_2O + K_2O + CaO + MgO + Fe_2O_3)}{[(100 - VM) \times (SiO_2 + Al_2O_3)]}$$

Calculation of BI is based on the ash constituents of the ash, the yield of VM, and also the mass ratio of basic and acidic oxides present in the ash. This BI essentially characterizes the product of the ash coke's ash content and the BI of the batch (coke), expressed as a percentage (Sharma et al., 2014).

16.9.1.3 Rheological Properties of Coal/Coal Blend

Rheological properties such as CSN, LTGK assay, fluidity, dilation, and plastic layer thickness of coal/coal blend are important properties for producing the desired quality of coke. The rheological values of Indian coking coal cannot be compared with similar rank of imported coking coals; therefore, these properties will not be proportional to the coking power of blends containing coals from imported sources; instead, they are essential factors for determining the quality of the coke. Brief explanations of all important rheological properties are discussed separately (Banerjee and Tiwari, 2008 and Tiwari et al., 2014b).

16.9.1.3.1 Crucible Swelling Number

CSN is one of the key indicators for assessing the behavior of coal, which can be tested quickly, and is a reliable measurement to evaluate the coking behavior of coal for coke making. It does not need long time and precise preparations to analyze. For stamp-charged coke making, higher CSN values of blend, resulted high push force in the batteries. In general, coal blend having CSN value in the range of 5–6.5 is able to produce good quality of coke.

16.9.1.3.2 Gieseler Fluidity

The coking execution of coal is further assessed by the fluidity test. For high-strength coke, various coals used in the blend should have an almost identical temperature range of fluidity, otherwise, the coals will not be compatible as when one coal is in the plastic stage, the other has not yet become plastic or has resolidified into semi-coke. The fluidity of coal blend determines the bonding process during coke making. It significantly affects coke reactivity because it controls the size and shape of coke anisotropy and the nature of interfaces between reactive and inert macerals, which finally affects the CSR (Banerjee and Tiwari, 2008). It allows better wetting and subsequent bonding of coal particles during carbonization, as the latter undergoes polymerization and decantation. Considering the CSR requirement and better control on coal blend, desired operations range of maximum fluidity is 750–1000 ddpm. Lower fluidity results in weaker bonding during carbonization, thereby making the coke susceptible to breakage during transportation. Also, too high fluidity will make coke porous yet again imparting lower strength. Therefore, an optimum amount of fluidity is required in a coal blend to make a good coke.

16.9.1.3.3 Dilatometry

The dilatometric tests are based on the principle that when a column of coal is heated under a specified pressure at a fixed rate of heating, it contracts initially, when the coal begins to soften. This is followed by expansion because of the initial liberation of volatile matter, the extent of the plastic mass. The slope of the initial contraction and the extent of subsequent swelling, as indicated by the dilatometric curves obtained, are taken as characteristics of the coal. Prime coking coals are supposed to have more than 50% dilation. On the other hand, comparing dilatation values of Indian medium coking coal and imported coals of the same rank and inert content have different total Ruhr dilatation. The best predicted Ruhr dilatation of typical coal blend was 22.5%–25%.

16.9.1.3.4 Plastic Layer Thickness

PLT is an important property of coal, which directly affects the coalification processes. In coalification process, coal reaches an intermediate stage with reactive components that liquefy, flow, and bond during thermal decomposition in the absence of air with the presence of cracking and aromatization and condensation reactions.

During cracking, released aromatic liquid coalesce into anisotropic liquid entity, which is called *plastic layer* (and elastic). These plastic layers are deformed by gases, which releases through the molten materials. Vitrinite from high-rank coals release less volatile matter and decompose to more stable plastic material than low-rank coal. Consequently, higher rank coals have more viscous plastic phase with higher internal gas pressure in its pore than lower rank coals. Thus, the thickness of plastic layer can predict the rank as well as strength of the coke. Higher the value of the PLT, the better will be the assimilation of the coal particles during carbonization, and hence higher will be the coke strength. In general, coal should have a minimum of 22 mm PLT for producing higher strength coke.

16.9.1.4 Petrographic Properties of Coal/Coal Blend

During coalification, the original plant material is transformed into three main organic groups of the coal defined by the optical microscopy as *macerals*: vitrinite, liptinite, and inertinite.

There are two types of petrographic components: reactive component and inert component. Reactive components soften and melt upon heating, but inert components do not soften and melt upon heating. Inertinite that does not fuse yields the *inert maceral-derived component*. In the coking process, lump coke formed when particles are bonded with each other as inert components are wrapped in the reactive that have softened. In this case, inert component act to impart strength to lump coke.

The coal rank can be assessed according to the volatile matter, carbon content, or vitrinite reflectance. The coal rank increases with increasing carbon content, increasing vitrinite reflectance, and decreasing volatile matter yield. It is an important property of coal and has good impact on coke CSR. A positive impact of increased rank on CSR is considerable for coals below 1.10% mean max reflectance. With further increase in reflectance up to 1.6%, the effect of rank on CSR diminishes. Beyond a mean max reflectance of 1.6%, further increase in rank has negative impact on CSR (Hara et al., 1980). Another study reported that cokes reach the maximum CSR when their parent coals have the mean reflectance of 1.2%–1.3%, and the CSR of cokes decreases when parent coals have higher or lower rank, while the minimum value of coke reactivity index was in the region of prime coking coals as the coke reactivity index was found to be inversely related to coke CSR (Nakamura et al., 1978). For producing coke with good coke, CSR can be made from a coal blend with Ro of 1.125%–1.130%.

16.9.2 Effect of Operating Parameters on Coke Quality

To sustain the coke quality and production, it is essential to maintain the operating parameters. The operating parameters such as moisture content, bulk density, heating rate, stamping time, crushing fineness, and coking time play an important role in producing desired quality of coke.

16.9.2.1 Moisture Content

The moisture content throughout the coal charge influences the heating; on the one hand, a large amount of heat is required for the evaporation of water; on the other hand, thermal effects arise from the condensation of water. In addition, the distribution of moisture controls the bulk density distribution within the charge. Therefore, the final temperature of the coke is affected and hence it decreases the productivity (Tiwari et al., 2012 and 2014c). It was also noted that the moisture content of coal affects the coke quality in terms of cold and hot strength (Niekerk and Dippenaar, 1991).

Coal moisture in the range of 9.5%–10.5% for a given stamping time gives the optimum bulk density and helps to obtain desired coke CSR. However, an increase in moisture leads to increases in the bulk density of coal cake, which affects push force. Lower coal blend moisture (<9%) is detrimental as it leads to lower bulk density and hence, lower coke CSR and also leads to deposition of higher roof carbon (Agarwal et al., 2011).

16.9.2.2 Coal Granulometry

Coal granulometry is an important factor to ensure coal cake stability (in terms of compressive and shear strength) and to maintain bulk density. Higher level of coal fines are detrimental for the batteries as it leads to roof carbon formation and lower level of coal fines leads to higher proportion of +0.5 mm coal fraction. The course of crushing (+0.5 mm) ash material that is not embedded in the coal matrix is generally concentrated in the coarser fractions and is distributed in finer fractions to a large extent. They would remain coarser fractions and when such coal particles are carbonized, the coke shows dots of inerts (generally contain higher level of ash), which are distributed in the coke like *resins* in cake. These coarse inert particles from *nuclei* breakage fissures traverse the coke, thus leading to the deterioration in the physical properties of coke. When coal is crushed finer, the inertinite produced in finer grinding gets dispersed in the whole mass and improves the strength of coke.

16.9.2.3 Bulk Density

The bulk density of the charge is an important factor affecting the operation of an oven, its throughput, and coke quality. This parameter mostly depends on the granulometry of the charge and the addition of moisture/ binder in the charge. The bulk density in the chamber controls the local heat demand throughout the coke bed and also the operation of coke oven and coke quality. Increasing bulk density of coal charge during coke making decreased the porosity of coke and increased the apparent density of coke (Vogt et al., 1998; Graham and Wilkinson, 1978; Vander et al., 1996 and Tiwari et al., 2012); mechanical strength of formed coke also influenced by the bulk density of parent coal blend. It also depended on the moisture content and particle size of the coal/coal blend. Density of charge decreased with increasing moisture content until the moisture content of 10%; this effect was amplified by crushing coal to finer level (Tiwari et al., 2012).

16.9.2.4 Stamping Time

Stamping time is an important factor to ensure coal cake stability and maintain the bulk density of charged coal cake. Higher stamping time will help to improve bulk density for a given coal moisture and coal granulometry up to a maximum point and then it becomes constant.

16.9.2.5 Heating Rate

The heating rate of the coal cake influences the quality of coke, especially strength and the fissuring properties of coke. To maintain desired coke quality, the heating of the coal cake in a coke oven should therefore be uniform over the total length and height of the coal cake. Migration rate of plastic layer also influences the level of thermal stress in the resolidified mass and therefore, the level of fissuring. Heating rate is strongly associated with the pattern of heat supply during carbonization. An adequate heat requirement and pattern of heat supply during carbonization should be based on required coke quality with the optimum requirement of total heat consumption. The heating rate has positive impact on CSR and negative impact on M_{40} of coke with increase in the heating rate of the coal charge (Tiwari et al., 2012 and 2014c).

16.9.2.6 Coking Time and Coke End Temperature

The thermal condition prevailing during carbonization affects the coke strength characteristics. In general, it is crucial that a constant pushing schedule should be maintained to achieve rated productivity of coke oven. The flue temperature and the coking time adopted in a normal operation are not independent parameters but vary inversely. On the other hand, an increase in coking time increases the coke end temperature and affects the coke quality.

Coke end temperature is the temperature of coke as it reaches the quenching station and is measured by infrared pyrometer. Higher battery temperature leads to higher CET, apart from posing danger to silica bricks and chances of ash fusion, resulting in the generation of more fissures in coke due to faster rate of heating. As a result, the coke size drops and hence, lowers M_{40}.

16.10 EVALUATION OF COKE QUALITY

Metallurgical coke can be described as a porous solid material comprising an organic part, which is mainly carbon and small amounts of sulfur, nitrogen, hydrogen, and oxygen, and an inorganic part. The coke quality generally characterized by measuring its cold strength, hot strength and ash chemistry. The most often used and well-known tests are the coke reactivity index and coke CSR have done under standard conditions.

The coke sample is subjected to screen analysis on 100, 80, 50, 40, 30, 20, and 10 mm square size screens. After screen analysis, the coke sample is used for different analysis. The greater than 50 mm size coke obtained is tested for M_{40} and M_{10} following the method described in IS 1354: 1992. The coke reactivity index and CSR of the collected coke samples are assessed by the testing procedure using standard method adopted as per IS 4023: 1991.

16.11 ASSESSMENT OF THE BY-PRODUCTS YIELD OF COAL/COAL BLEND

Coking is a complex process, which produces coke and liquid and gaseous by-products. These by-products have several positive attributes when considered as a feedstock for aromatic chemicals, specialty chemicals, and carbon-based materials. Coke making by-products also created new opportunities for developing advanced polymer materials, incorporating aromatic and polyaromatic units in their main chains as value added organic chemicals. The behavior of coal during carbonization is quite complex and it is difficult to predict the outputs from the technologies in terms of by-products. The coke oven by-products, such as coke oven gas, coal tar, ammonia, and hydrogen sulfide, depend upon the physical and chemical properties of coal, operating technologies, and operational parameters like temperature, bulk density, pressure in the chamber, and the rate of carbonization (Schobert and Song, 2002; Tiwari, 2006 and Tiwari et al., 2014d).

16.11.1 Determination of By-Products by Jenkner Apparatus

By-products of coke oven plant can be estimated with fair degree of accuracy using Jenkner apparatus. The schematic diagram of the Jenkner setup is shown in Figure 16.8. In this test, coal samples are subjected to carbonization in a Jenkner apparatus and yield of by-products are determined. The coal sample is charged in the retort and the retort is kept inside the main electric furnace attached to a programmable heating control unit. A cracking column is fixed at the top of the retort and is provided with an independent heating system. The raw coal gas leaving from the top of the column passes through a water condenser and other separating units before exhausted through the outlet.

In top charging process, 1.2 kg crushed coal (80% ± 1% below 3.2 mm) having moisture content 7% ± 1% was slowly transferred into the retort, so that the bulk density of charging coal remains around the 0.75 t/m^3. In stamp charging technology, 1.5 kg crushed coal (90% ± 1% below 3.2 mm) having moisture content 10% ± 1% was carefully pressed in a suitable mould to get coal cake of bulk density 1.15 t/m^3. The coal cake was then charged into the retort carefully.

The carbonization furnace was first preheated to a temperature of 550°C. Once the temperature of the carbonization furnace reaches 550°C, the retort is slowly introduced into the furnace. The temperature of the carbonization furnace falls sharply to 350°C and then slowly rises to 550°C. The heating is subsequently controlled at the rate of 3°C/min till the temperature of the furnace reached 1050°C. The temperature of the cracking column is maintained at 850°C unless otherwise specified based on the test requirement. The carbonization temperature as well as the total carbonization time was kept constant for all the tests.

The typical properties of the coal samples and estimated yields of by-products are presented in Tables 16.1 and 16.2. Results show that yield of by-products varies from coal to coal and charging technologies. In general, yield of by-products is higher for top charging process as compared to stamp charging process (Tiwari, 2006 and Tiwari et al., 2014d).

It is observed that as the VM of coal increases, the coke oven gas and coal tar yield increases. It is also observed that the yields of coke oven gas and coal tar are higher in top charging as compared to stamp charging. This may be attributed to the difference in the density, porosity, and void density of the coal charges of the two charging processes. In case of top charging technology, the coal particles are loosely packed (bulk density: 0.75 t/m^3) in the retort, whereas for the stamp charging technology, the coal particles are compacted in the form of a cake (bulk density: 1.15 t/m^3) and charged in the retort. Escape of VM is fast through the loosely packed particles. However, the escape of VM is retarded in stamp charging mainly due to compactness of the coal particles. The VM may further decompose and produce carbon due to its longer residence time inside the hot coke.

It is observed from results that the NH_3 yield increases up to a certain level of VM (around 25%), beyond which the yield decreases with further increase in VM content. Similar

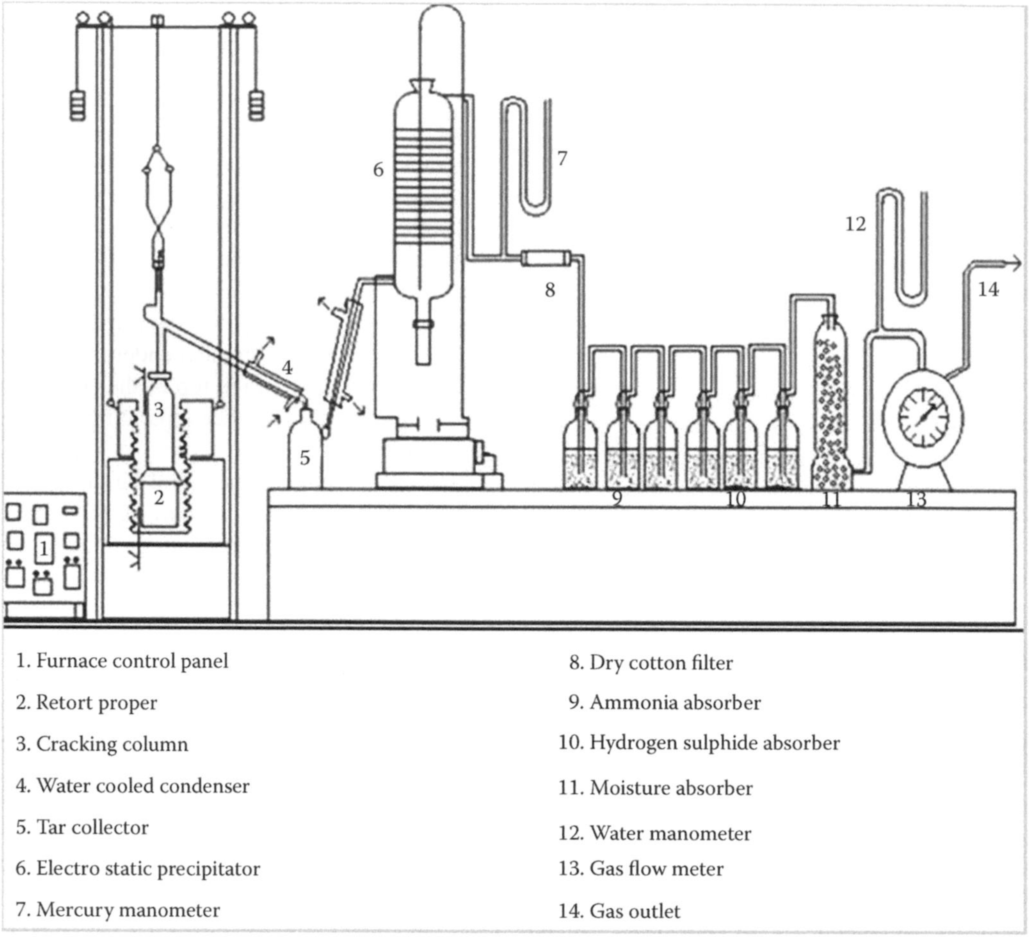

FIGURE 16.8 Schematic diagram of Jenkner apparatus. (From Loison R, Foch P and Boyer A., *Coke quality and production*, London: Butterworth, 1989.)

TABLE 16.1
Detailed Properties of Coals

Properties	Coal A	Coal B	Coal C	Coal D	Coal E	Coal F	Coal G	Coal H	Coal I	Coal J
Ash, % (adb)	9.57	9.10	10.85	16.10	13.42	12.70	9.30	14.70	8.41	7.10
VM, % (adb)	16.61	20.55	21.08	22.20	23.60	23.95	23.60	26.70	29.17	36.00
Carbon, % (daf)	86.90	86.55	87.43	86.98	87.77	86.48	87.49	86.11	81.06	87.51
Hydrogen, % (daf)	4.34	5.01	4.92	4.90	4.76	4.92	5.02	5.11	4.58	5.29
Nitrogen, % (daf)	1.82	2.00	1.60	2.00	2.02	2.02	1.99	1.99	1.84	2.18
Sulfur, % (daf)	0.90	0.71	0.44	0.66	0.52	0.81	0.05	0.76	0.26	0.67
Oxygen, % (daf)	6.04	5.75	5.61	5.46	5.04	5.76	5.00	6.03	12.26	10.35
Max. fluidity, ddpm	NIL	1821	34	2017	653	1957	482	2685	NIL	3913
(Ro avg.)	1.36	1.28	1.21	1.17	1.08	1.05	1.09	0.95	0.85	0.80

observations are made on the effect of VM content on H_2S yield. The H_2S yield is at a maximum when the VM content is of about 25%. It is interesting to note that the H_2S yield for stamp charging is higher as compared to top charging process. It may be due to the reason that the entrapped VM in coal cake react with elemental sulfur content of coal and produce an additional quantity of H_2S.

It has been found that the yields of coke oven gas and coal tar decrease with increase in the rank of the coals. The results are expected because of inverse relationship between the rank and VM content of coal. The results suggest that coal with lower rank, that is, high VM content and low reflectance of coal, would be more suitable for the production of higher yield of tar.

TABLE 16.2
By-Products Yield during Carbonization

Sample ID	Coal A	Coal B	Coal C	Coal D	Coal E	Coal F	Coal G	Coal H	Coal I	Coal J
					Coke Oven Gas Yield, Nm³/t					
Top charging	228.7	204.1	242.4	267.7	265.3	244.7	308.7	251.3	270.4	265.0
Stamp charging	202.1	204.0	236.1	268.2	255.0	228.6	304.1	217.2	236.1	230.5
					Coal Tar Yield, Kg/t					
Top charging	18.7	29.2	29.7	31.1	32.2	18.5	43.2	24.9	31.1	33.0
Stamp charging	14.3	19.5	23.0	23.3	23.8	14.1	36.6	19.5	25.9	28.0
					Ammonia Gas (NH_3) Yield, Nm³/t					
Top charging	2.27	8.38	7.50	10.92	5.56	0.66	2.65	5.53	8.17	6.53
Stamp charging	1.58	7.25	4.50	7.50	6.65	0.51	0.85	3.70	5.58	6.27
					Hydrogen Sulfide (H_2S) Yield, Nm³/t					
Top charging	2.46	1.88	2.03	3.51	3.01	3.20	1.70	2.20	3.51	3.01
Stamp charging	3.01	3.07	3.18	5.30	4.07	3.48	3.51	3.59	4.13	4.07

REFERENCES

Agarwal R, Tiwari HP, Roy A Choudhary PK, Sinha SK, Biswas B, Prasad D. A case study on impact of crushing fineness on cold strength of coke. *Proc. Int. Conf. Coal Coke, Jamshedpur, India* 2011:55–59.

Banerjee PK, Tiwari HP. Impact of beneficiation on coking properties of a typical Indian medium coking coal. *Proc. IMPC*, BICC, China 2008:1985–1993.

Beck KG, Mickel JF. Extension of coking-coal resources by new process technologies of coal preparation and carbonization. *Ironmaking and Steelmaking* 1980;7(3):111–115.

Bernard A, Duchene JM, Isler D. Etude de la gasification du coke. *Rev. Met. CIT* 1985;82:849–860.

Bujnowska B, Colling G. Coal tar pitch for improving coking properties of coal. *Proceedings 1st International Cokemaking Congress,* London 1992:142–146.

Buss WE, Merhof MA, Piduch HG Schumacher R, Kochanski U. The Thyssen still Otto/PACTI non-recovery coke making system. *Proceeding of 58th Ironmaking Conference*, Chicago, IL 1999:201–211.

Chaudhuri SG, Mukherjee AK, Roychoudhury KK. Relationship between physico-chemical and petrographic properties of Indian coking coals. *Proceeding National Seminar on Coal for Blast Furnace Coke and for Injection, Jamshedpur, India* 1990:65–76.

Chermin HAG, van Krevelen DW. The Theory of Metaplast. *Fuel* 1957;36:85.

Chiu YF. Study of coke petrography and factors attending coke reactivity. *Ironmaking and Steelmaking* 1982;9(5):193–199.

Dam-Johansen G, Hu K, Wedel S, Hansen JP. Decomposition and oxidation of pyrite. *Progress Energ. Combust. Sci.* 2006; 32(3):295–314.

Diez MA, Alvarez R, Barrioconals C. Coal for metallurgical coke production: Prediction of coke quality and further requirements for cokemaking. *Int. J. Coal. Geol.* 2002;50;389–412.

Engler P, Santana MW, Mittleman ML, Balazs D. Non-isothermal in situ XRD analysis of dolomite decomposition. *The Rigaku J.* 1988;5(2): 3–8.

Finkelmer RB. Modes of occurrence of trace elements in coal. US Geology survey. Open File Report (81–99), 1981:322.

Fitzgerald D. The kinetics of coal carbonizations in the plastic state. *Trans. Far. Soc.* 1956;52:362.

Ghosh A, Chatterjee A. *Ironmaking and Steelmaking: Theory and Practice*, PHI Learning Pvt. Ltd., New Delhi, 2008:155.

Gibson J. Dilatometry and the prediction of coke quality. *Coke Oven Managers Association* 1972:182–205.

Graham JP, Wilkinson HC. Coal properties, charge preparation and their influence on coke quality, *Ironmaking Conference Proceeding, Chicago,* 1978:421–436.

Graham JP, Wilkinson HC. Coke quality and its relation to coal properties. Charge preparation and carbonization practices. *Coke Oven Managers Association Year-Book*, Mexborough 1980:161.

Habermehl D, Orvwal F, Bever HD. In: *Chemistry of Coal Utilization* (Ed. MA Elliot), Second supplementary volume, John Wiley & Sons, New York 1981:317–368.

Haque R. Optimal utilization of Indian coals in iron making, *National Seminar on Coal for Blast Furnace Coke and for Injection, Jamshedpur, India* 1990:57–64.

Hara Y. Mikuni D, Yamanoto H, Yamanaki H. *The Assessment of Coke Quality with Particular Emphasis on Sampling Technique, Blast Furnace Coke: Quality, Cause and Effect*, McMaster University, Canada 1980:4–38.

Jones JL. Thermal Conversion of Solid Wastes and Biomass, *In: Symposium Series 130*. Washington, DC: American Chemical Society 1980:209–603.

Karmarkar RS, Tiwary M, Mishra KN Prasad HN, Banerjee KB, Krishnan SH. Study of variation of ash in coal blend for stamp charging on oven throughput. R&D Report no. 031/96, Tata Steel, 1997.

Lander CH, McKAY RF. *Low Temperature Carbonisation.* Ernest Benn Limited, London 1924:39.

Lewis IC. Chemistry of Carbonization. *Carbon* 1982;20(6):519–529.

Lewes V. The Carbonisation of Coal, Benn Brothers Ltd., London, 1918.

Loison R, Foch P, Boyer A. *Coke Quality and Production*. Butterworth, London, 1989.

Macdonald Gregor. Transition back to coal, Gregor.us, 2010, Annual. Web.<gregor.us>.

Miura Y. The science of cokemaking technology and its development in Japan. *The Coke Oven Managers' Association* 1978: 292–311.

Mukherjee AK, Chatterjee CN, Ghose S. Coal resources of India-Its Formation, distribution and utilization. *Fuel Sci. Tech.* 1982;1:19–34.

Mushrush GW, Speight JG. *Petroleum Products: Instability and Incompatibility*. CRC Press, Washington, DC, 1995:33.

Nakamura N, Togino Y, Tateoka T. *Behaviour of Coke in Large Blast Furnace. Coal, Coke and Blast Furnace*. The Metals Society, London 1977:1–18.

Nakamura N, Togino Y, Tateoka T, Behaviour of coke in large blast furnace. *Ironmaking Steelmaking* 1978;5:1–17.

Nandi BN, Ternen M, Belinko K. Conversion of non-coking coals to coking coals. *Fuel* 1981;60:347–53.

Niekerk WH Van, Dippenaar RJ, Blast-furnace coke: A coal-blending model. *J. S. Atr. Inst. Min. Metal.* 1991;91(2):53–61.

Pedersen Chris. Asia's spectacular coal demand: The new frontier for powder river basin coal. M. Sc. Thesis, New York, 2011.

Prasad HN, Karmakar RS, Tiwary M, Singh BK, Dhillon AS. Possibility of eliminating coke cutting in case of stamp charged coke. *Tata Search* 1996;2:52–57.

Prasad HN, Singh BK, Chatterjee A. Production of high CSR coke by stamp charging:Possibilities and limitations. *Coke Making Int.* 1999;2:50–59.

Price J, Granden J, Hampel K. *Microscopy, Chemistry and Rheology Tools to Determine Coal and Coke Characteristics. 1st McMaster's Cokemaking Course*. McMaster University, Hamilton, ON, 1997: 4.1–4.74.

Ruette F, Sierraalta A, Castells V, Lava M. Chemistry of carbonization-I. A theoretical study of free radical formation from starting materials. *Carbon* 1993;31(4):645–650.

Sahajwalla Veena, Hilding T, Oelreich Anne von, Gupta SK Bjorkman Bo, Wikstrom Jan-Olof, Fredriksson P, Seetharaman S. Structure and alkali content of coke in an experimental blast furnace and their gasification reaction. *Proc of AIS Tech*; vol. 1; 2004:491–500.

Samir SS, Zaborssky O. *Biomass Conversion Process for Energy and Fuel*. Plenum Press, New York 1981.

Schapiro N, Gray RJ. Petrographic classification applicable to coals of all ranks. *Proc. Illinois Min. Inst.* 1960:83–90.

Schapiro N, Gray RJ. The use of coal petrography in coke making. *J Inst. Fuels* 1964;11(30):9.

Schobert HH, Song C, Chemicals and materials from coal in the 21st century. *Fuel* 2002;81:15–32.

Segers M. Spatial variation of coke quality in the non-recovery beehive coke ovens. Master Thesis, University of Pretoria, Pretoria, 2004.

Sharma R, Tiwari HP, Banerjee PK. Producing high coke strength after reactivity in stamp charged coke making. *Coke and Chemistry* 2014;57(9):351–358.

Sofer SS, Zaborsky OR. *Biomass Conversion Processes for Energy and Fuel*. Plenum Press, New York 1981.

Singh K, Menon MVP, Raja K, Rao SK, JI. Min. Met. & Fuels, Special number on coal preparation 1983:538–4550.

Speight James G. *The Chemistry and Technology of Coal*. CRC Press, Boca Raton 2013.

Sreedhar I, Srikanth AV, Bhuyan BC, Choudhury R. Improvement of coking characteristics of non-coking Indian coals. *Ind. Chem. Engg.* 1998;40(1):27–29.

Stopes MC. On the four visible inrgadients in banded bituminous coals, studies in the composition of coals. *Proc. Roy. Soc.* London 1919;90(1):470–487.

Stopes MC. Petrography of bituminous coals. *Fuel Sci Pract.* 1935;14(1):4–13.

Stopes MC, Wheeler RV. Monograph on the constitution of coal, *Dept. Sci. Ind. Res.*, London, 1918.

Tiwari HP. Assessment of the yield of by-products during high temperature carbonization of coals, M. Tech Thesis, India, Ranchi, Mesra: BIT, 2006.

Tiwari HP, Saxena VK, Banerjee PK, Haldar SK Sharma R, Paul S. Study on heating behavior of coal during carbonization in Non-recovery. *Int. J. Metal Engg* 2012;1(6):135–142.

Tiwari HP. Studies on the effects of intrinsic parameters of coal and heat transfer rate on coke quality for non-recovery oven in coke making process, Ph D Thesis, India, Dhanbad: ISM, 2014.

Tiwari HP, Banerjee PK, Saxena VK. A novel technique for assessing the coking potential of coals/coal blends for non-recovery coke making processes. *Fuel* 2013;107:615–622.

Tiwari HP, Banerjee PK, Sharma R, Haldar SK, Kumar A, Roy A. Effect of binder on recovery stamp charged coke making process. *Coke Chem* 2014a;1(57):10–17.

Tiwari HP, Banerjee PK, Saxena VK, Haldar SK. Effect of Indian medium coking coal on coke quality in non-recovery stamp charged coke oven. *J. Iron Steel Res. Int.* 2014b;21(7): 673–678.

Tiwari HP, Banerjee PK, Saxena VK, Sharma R Haldar SK, Paul S. Effect of heating rate on coke quality and productivity in non-recovery coke making. *Int. J. Coal Prep Util* 2014c;34:306–320.

Tiwari HP, Banerjee PK, Sharma R, Haldar SK, Joshi PC. A comparative study of by-products yield from coke making processes in a Jenkner Apparatus. *Coke Chem* 2014d;57(5):192–198.

Tiwari HP, Banerjee PK, Saxena VK Sharma R, Haldar SK, Paul S. Efficient way to use of non-coking coals in non-recovery coke making process. *Metall. Res. Tech.* 2014e;111:211–220.

Vander T, Alvarez R, Ferraro M, Fohl J, Hofherr K, Huart JM Matilla E, Propson, Willmers R, Vd Velden B. Coke quality improvement. Possibilities and limitations, 3rd Int. *Ironmaking Congress Proceedings* 1996;Gent.:16–18.

Vogt D, Negro P, Isler D. Coke degradation in the blast furnace: Improvement of coke quality, *57th Ironmaking Conference* 1998:741–746.

Ward Colin. *Coal Geology and Coal Technology*. Blackwell Scientific Publications, Melbourne, 1984:345.

Walters EB. The Thyssen Still Otto/PACTI Non-recovery coke making system. *Ironmaking Conference Proceeding*, 1999:201–214.

www.climatetechwiki.org/technology/coke-dry-quenching

www.jase-w.eccj.or.jp/technologies/pdf/iron_steel/S-10.pdf

www.asiapacificpartnership.org/pdf/Projects/Coal%20Mining%20Task%20Force%20Action%20Plan%20030507.pdf

17 Coal-Based Products and Their Uses

Xian-Yong Wei, Zhi-Min Zong, Xing Fan, and Zhan-Ku Li

CONTENTS

Abstract: Coals are congenitally deficient as clean energy, but have overwhelming advantages as raw materials for fine organic chemicals and advanced carbon materials. In this chapter, coal-based products and their uses are reviewed. Many value-added organic products, especially condensed aromatics, can be obtained from coal extracts and soluble portions from coal thermal dissolution, catalytic hydroconversion, and oxidation. Such products have many important uses in the fields of medicine, food, synthetic plastic, rubber, protective coating, dye, superior electrical conductive polymeric conductor, and optoelectronic nanodevices. Directionally converting coals to the value-added organic products deserves investigation.

17.1 INTRODUCTION

Coals are typical heavy carbon resources with much higher contents of carbon, oxygen, sulfur, and nitrogen than other heavy carbon resources such as heavy petroleum, natural pitch, oil shale, and oil sand. At least because of their higher contents of carbon, oxygen, sulfur, and nitrogen, coals should not mainly be used as energy.

More attention was paid to the so-called clean coal technology. Although more or less nitrogen and sulfur can be removed before or after coal combustion, economical removal of CO_2 emitted from coal combustion is not feasible. In fact, economically removing most of the nitrogen- and sulfur-containing species is also difficult. Thus, the so-called clean coal technology does not facilitate low-carbon utilization of coals.

Less attention was paid to the limitation in coal reserve. Much larger amounts of coals were used as energy, especially in China, where more than three billion tons of coals were produced annually and most of them were used as energy sources directly or indirectly. As well known, most of coals were formed from plant remains at least 70 million years ago, and large-scale use of coals began in 1760, the beginning of the Industrial Revolution in Great Britain. According to the estimates of world coal reserves and consumption pattern of 2013, the world coal reserves are expected to be used up in about 135 years. The period of time is less 200 years even based on the most optimistic estimate. It is no doubt very tragic for human beings that such a precious fossil resource, the formation of which has taken more than 70 million years, disappears from our living earth within several centuries since 1760 (Figure 17.1).

Most of the people in China, including most of the researchers, insist the conventional viewpoints on coal utilization that coals are the most abundant fossil resources over the world, especially in China, so they should be used as cheap energy and that coals will also be the main energy in China during the coming decades. In fact, both coals and energy account for very small areas in natural resources (Figure 17.2). Coals resulting from solar energy of ancient age have mainly been used as energy during the past centuries. However, in light of sustainable development, the intersection between coals and energy should be as small as possible, especially for China, because huge expenses for coal mining and use as energy are being paid in China (Wei 2014):

1. More than 1000 miners die in various accidents annually to damage thousands of families; more than six billion tons of mine water are discharged annually to aggravate water shortage.
2. More than 10 billion m^3 of coalbed gas are discharged annually to aggravate greenhouse effect in addition to wasting of the methane resource.
3. Huge amounts of arable lands have been lost because of land subsidence and coal gangue accumulation.

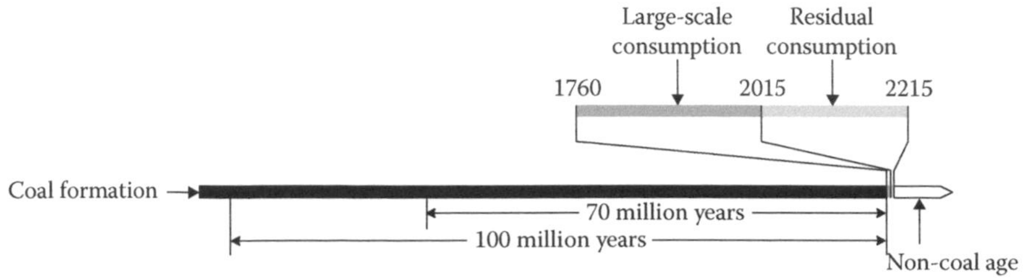

FIGURE 17.1 The periods of time for coal formation, large-scale consumption, residual consumption, and non-coal age.

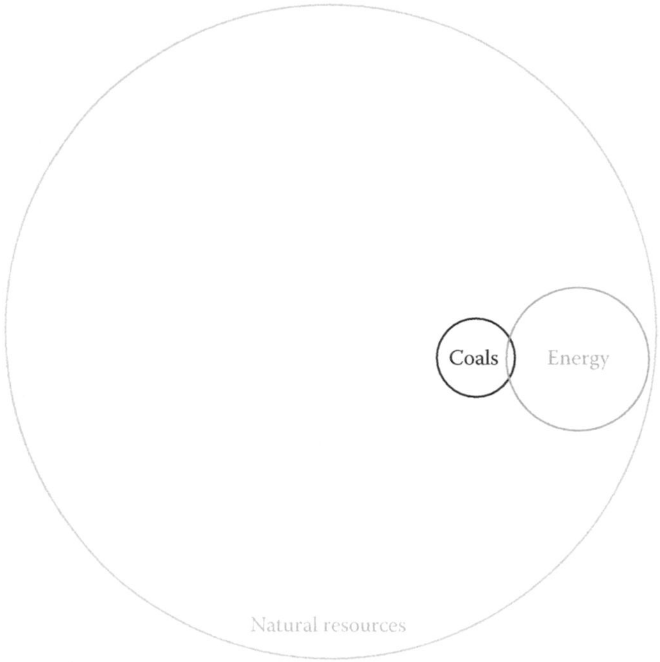

FIGURE 17.2 The relationship among resources, energy, and coals.

4. Direct coal combustion emits large amounts of harmful species, such as smoke and dust, CO_x, NO_x, and SO_x, leading to severe environmental pollution, especially haze (Figure 17.3).

Coals are congenitally deficient as clean energy but have overwhelming advantages as raw materials for fine organic chemicals and advanced carbon materials. The efficient utilization of coal resources should be clean, limited, and value added. To achieve efficient coal utilization, understanding molecular composition of organic matter in coals is indispensable. So far, molecular composition of organic matter in coals is far from clarification, leading to no breakthrough in both coal chemistry and coal chemical technology during the past decades. In conventional coal chemistry, non-separable and/or severely destructive methods are used for understanding coal structure. Using such methods, coal structure cannot be objectively revealed on molecular level. Molecular coal chemistry (MCC) aims at objectively revealing coal structure on molecular level using separable and nondestructive or less destructive methods, including sequential extraction, column chromatography, selective conversions under mild conditions, such as alkanolyses, oxidation, and catalytic hydroconversion (CHC), and separable analyses with multiple advanced instruments (Wei et al. 1999a, b, 2010b; Wang et al. 2001; Yuan et al. 2001). Information from MCC provides scientific basis for efficient utilization of coal resources. As shown in Figure 17.4, the main concern of MCC is molecular composition of organic matter in coals, that is, what kinds of organic compounds does organic matter in coals consist of? Understanding the molecular composition of organic matter in coals is crucial for reasonably utilizing organic matter in coals via technological development to achieve the industrialization of fine coal chemical technology (FCCT), bringing about huge economic benefits for human being with much smaller consumption of coals and much less environmental pollution, compared to conventional coal chemical technology and the so-called modern coal chemical technology (Figure 17.5). Therefore, the industrialization of FCCT will lead to a great revolution for coal utilization to incarnate real value of coals as black gold.

17.2 PRODUCTS FROM COAL EXTRACTS AND THEIR USES

Coal extracts usually contain numerous organic compounds in organic matter of coals. Most of the organic compounds are value added, but isolating these compounds from coal extracts is extremely difficult due to the compositional complexity of of coal extracts.

In conventional methods, coals and their derivates are sequentially extracted into oil, asphaltene, and preasphaltene in a Soxhlet extractor with an alkane (pentane, hexane, or heptane); benzene or toluene, and tetrahydrofuran (THF) or pyridine as solvents. Very long time is needed for exhaustive extraction with each solvent because of the difficulties in solvent permeation into the sample particles and solute dispersion from the sample particles in the Soxhlet extractor. In addition, coals and their derivates can only be roughly isolated with very limited kinds of solvents.

We extracted various coals with different solvents and developed a magnetically stirred extraction system with a polytetrafluoroethylene membrane filter (pore size of 0.45 μm). Relatively large amount of a coal sample can be extracted in the extraction system. Solvent permeation into the sample particles and solute dispersion from the sample particles become smooth by the magnetic agitation.

FIGURE 17.3 The examples of severe haze in China.

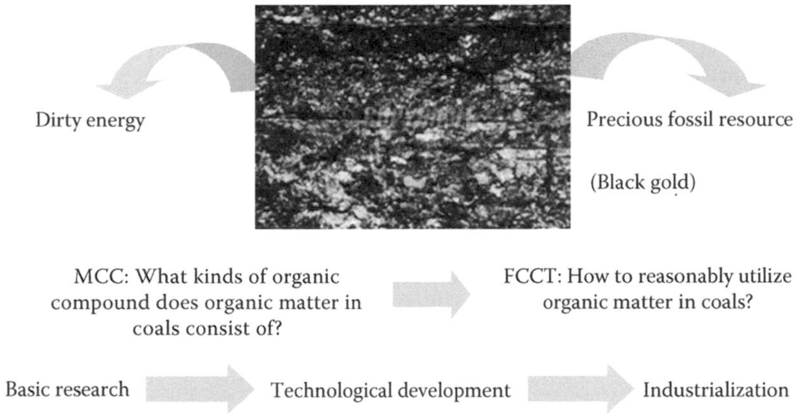

FIGURE 17.4 The relationship between MCC and FCCT.

As shown in Figure 17.6, Geting bituminous coal (GBC, collected from Geting Coal Mine, Shandong Province, China) can be sequentially extracted into extracts 1–5 (E_1–E_5) with petroleum ether (PE), carbon disulfide (CDS), methanol, acetone, and isometric CDS/acetone mixed solvent (IMCDSAMS) in the extraction system in a total extract yield of more than 10% (Figure 17.7) (Shi et al. 2013). According to the observation with a scanning electron microscope (SEM), particle sizes of the extraction residue are significantly smaller than those of GBC (Figure 17.8), suggesting that the sequential extraction resulted in significant destruction of GBC particles. According to the analysis with a gas chromatograph/mass spectrometer (GC/MS), the extracts consist of alkanes, arenes, and heteroatom-containing organic species (HACOSs). As displayed in Figure 17.9, E_1 and E_2 mainly contain arenes, E_3 and E_4 dominantly contain HACOSs, and E_5 consists of only arenes and HACOSs. Most of the arenes are condensed alkylarenes and condensed arenes. Condensed arenes were considered to be originated from small molecules in vitrinites and possibly formed during catagenesis (Davis et al. 1985). The relative content of alkanes in E_1 is obviously higher than that in any other extract. Most of the HACOSs contain oxygen, and some HACOSs contain both oxygen and nitrogen (Figure 17.10).

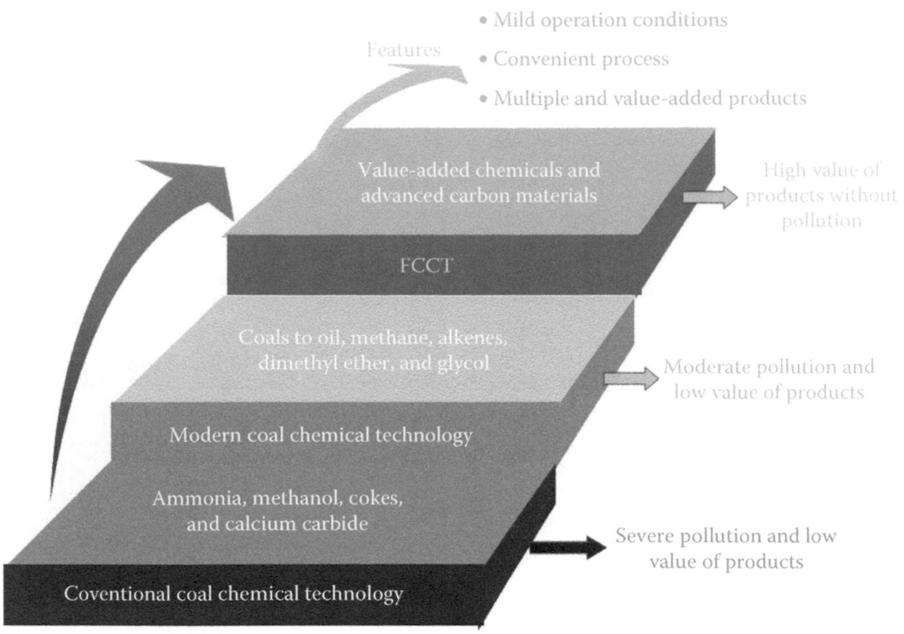

FIGURE 17.5 Scientific outlook on coal utilization.

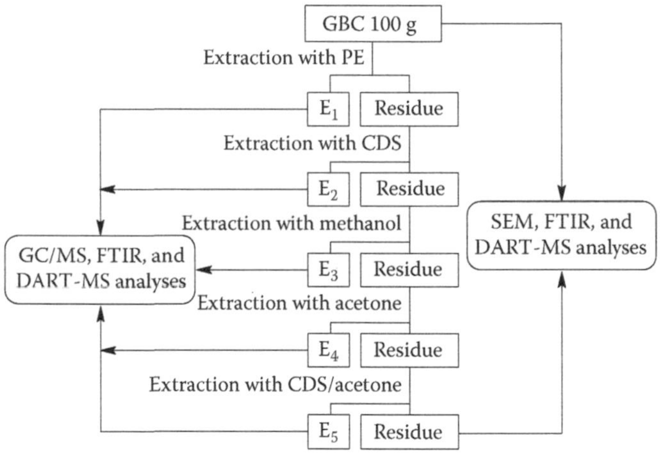

FIGURE 17.6 Procedure for fractional extraction of GBC and subsequent analyses.

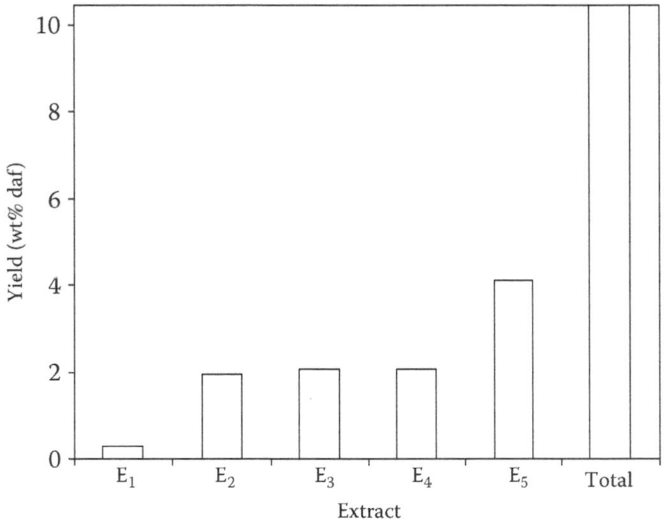

FIGURE 17.7 Yields of extracts from GBC.

The analysis with a Fourier transform infrared (FTIR) spectrometer shows that the extracts contain much more aliphatic moiety (AM)-rich species than GBC and its extraction residue (Figure 17.11), indicating that the AM-rich species in GBC are relatively easily extracted. AMs are present in alkanes, alkenes, alkylarenes, alkanols, alkanals, alkanones, alkanoic acids, esters, and many other HACOSs. There are remarkably different distributions of molecular mass among GBC and its extracts and among the extracts according to analysis with a direct analysis in real-time ionization source coupled to ion trap mass spectrometer (DARTIS/ITMS), as exhibited in Figure 17.12. As examples, some of the species detected were selected to understand their possible molecular structures by investigating their tandem mass spectra (TMS). As a result, a series of less volatile and/or strongly polar species were identified (Figures 17.13 through 17.16). Most of them cannot be identified with GC/MS.

E_2 from GBC is a complex mixture of numerous organic compounds (Figure 17.17) and thereby obtaining pure chemicals from the mixture is a huge challenge. Medium-pressure preparative liquid chromatography (MPLC) proved to be a potential and powerful approach for large-scale production in separation and purification of compounds in functional foods and pharmaceuticals (Sharma et al. 2003; Dev et al. 2006; Liang et al. 2012). Using MPLC, E_2 was eluted into four fractions (i.e., F_1–F_4 in Table 17.1), each of which contains a series of subfractions. In total, three condensed arenes, that is,

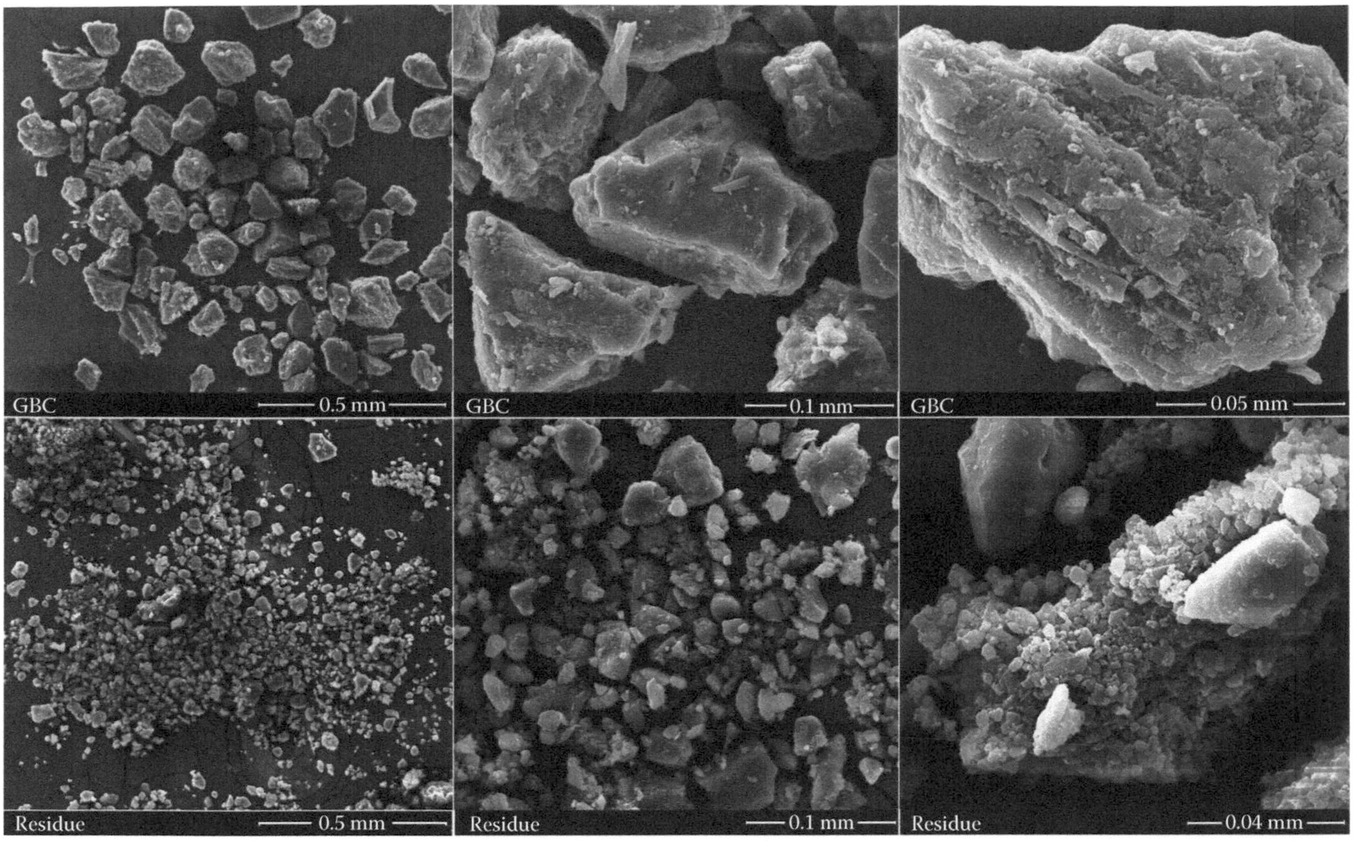

FIGURE 17.8 Scanning electron micrographs of GBC and its residue.

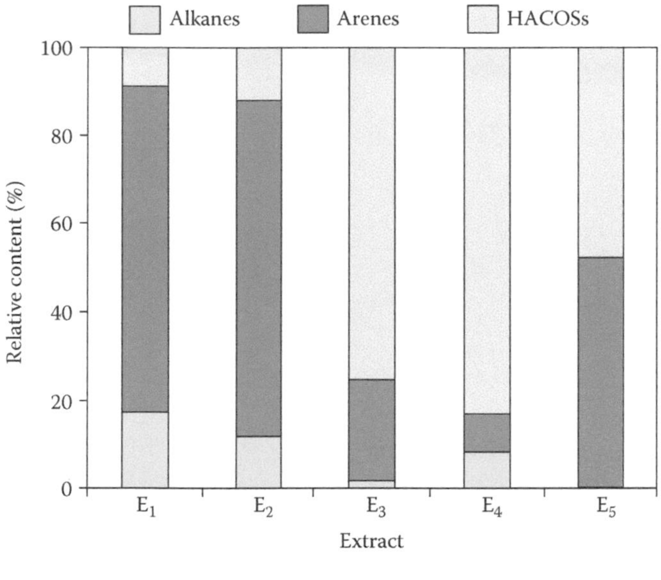

FIGURE 17.9 Distribution of group components in E_1–E_5 from GC/MS analysis.

benzo[*pqr*]tetraphene (BTP), benzo[*ghi*]perylene (BP), and coronene, and four dialkyl phthalates (DAPs), that is, bis(2 ethylhexyl) phthalate (BEHP), dioctyl phthalate (DOP), butyl 1-cyclopentylethyl phthalate (BCPEP), and dibutyl phthalate (DBP), were enriched into different subfractions in high purities, as demonstrated in Table 17.2 and Figures 17.18 through 17.25 (Shi et al. 2015).

Condensed arenes are used to synthesize superior electrical conductive polymeric conductor (Gama et al. 1992). Their derivates are applied in solar cell (Hiramoto et al. 1990; Wang et al. 2013; Yu et al. 2013), polymers (Tyutyulkov et al. 1992), and organic dye molecules (Olmsted III 1974; Lucenti et al. 2013; Yu et al. 2013). As a neurobehavioral toxicant (Obiri et al. 2013; Qiu et al. 2013), BTP is the first condensed arene identified as a cancer-causing compound (Harvey 1991). It was widely used to induce deoxyribonucleic acid (DNA) strand breaking in biochemistry research (Verhofstad et al. 2011; Vincent-Hubert et al. 2011). Dissolved BP in thiocyanate was used as fluorescent solute probes in different applications (Street et al. 1989). As a flat condensed arene, coronene was used as a fragment of graphene (Sato et al. 2011) and a ligand for DNA (Casagrande et al. 2009). In addition, *coronene nanodroplet* is a good solvent for dissolving C_{60} (Ghosh et al. 2010; Suzuki et al. 2010). By stacking coronene in single-wall carbon nanotube, a promising candidate for opto-electronic nanodevices can be formed (Dappe and Martinez 2013). Synthesizing such condensed arenes is an arduous and laborious work (Schmidt et al. 2011; Bedekar et al. 2013). Compared to synthesis, enriching condensed arenes directly from coals and their derivates could be more promising due to the high abundance of coals and their derivates.

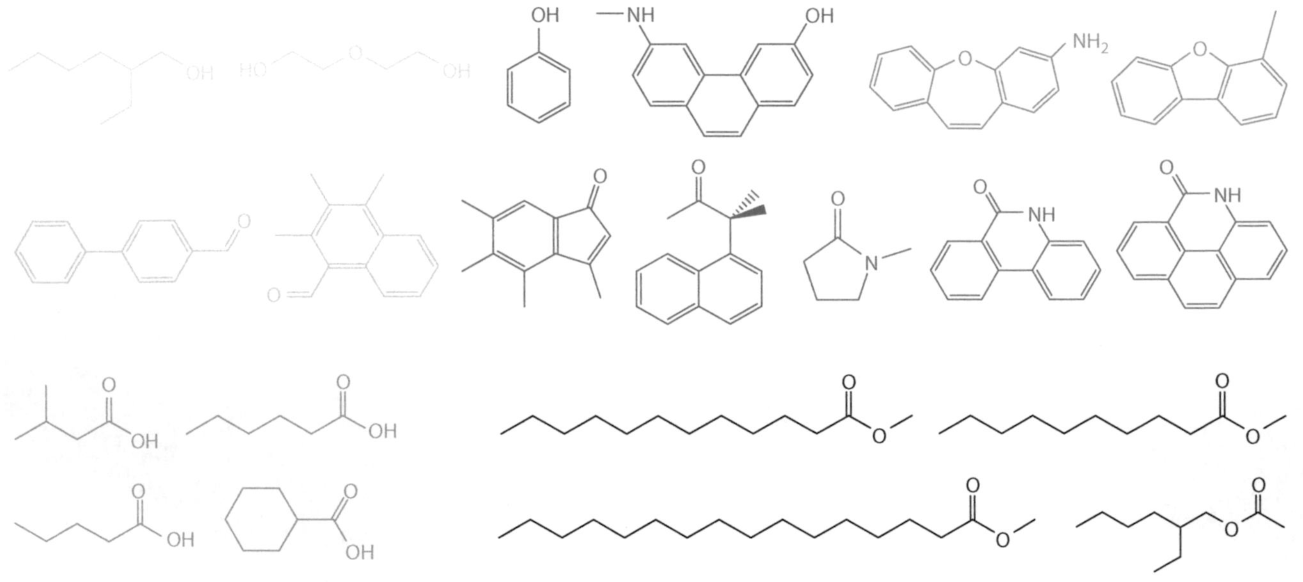

FIGURE 17.10 Typical HACOSs detected in E_3 from GC/MS analysis.

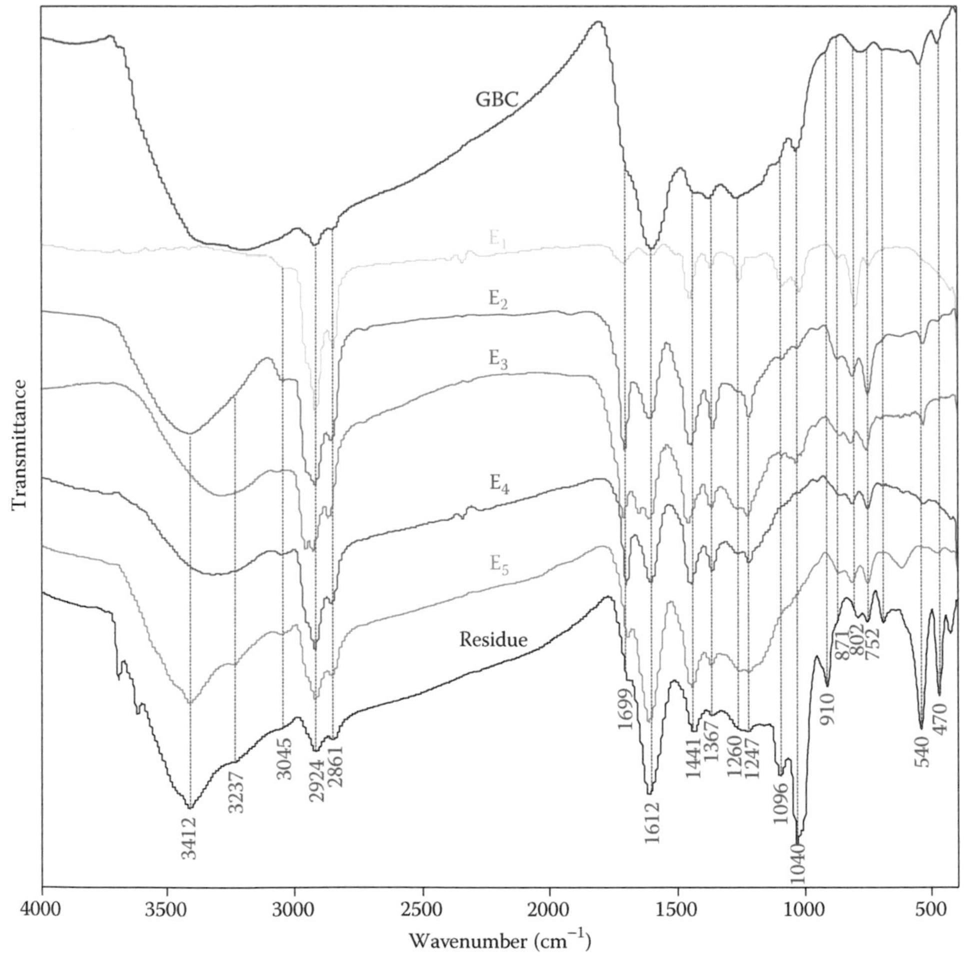

FIGURE 17.11 FTIR spectra of GBC along with its extracts and extraction residue.

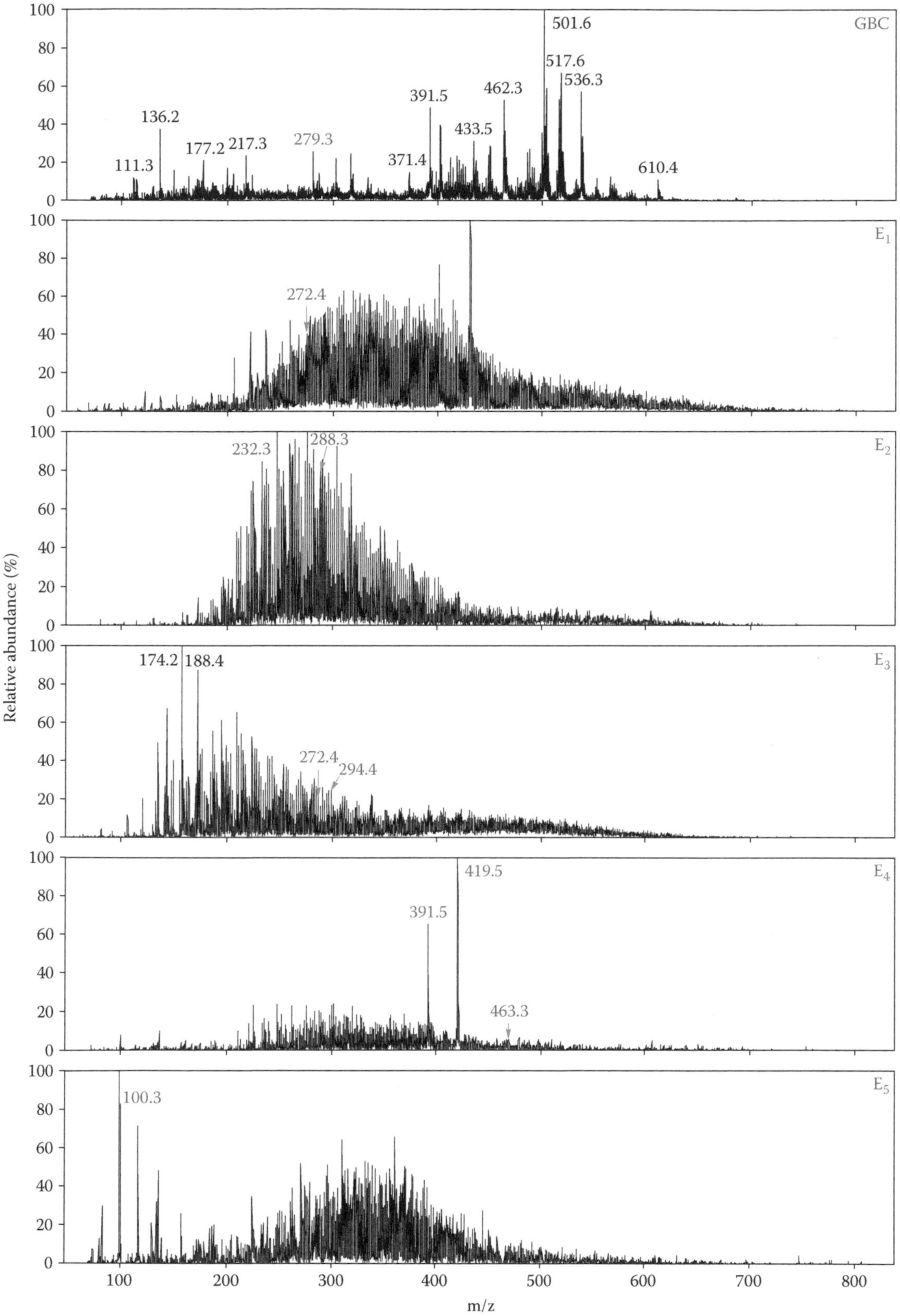

FIGURE 17.12 Mass spectra of GBC and its extracts from DARTIS/ITMS analysis.

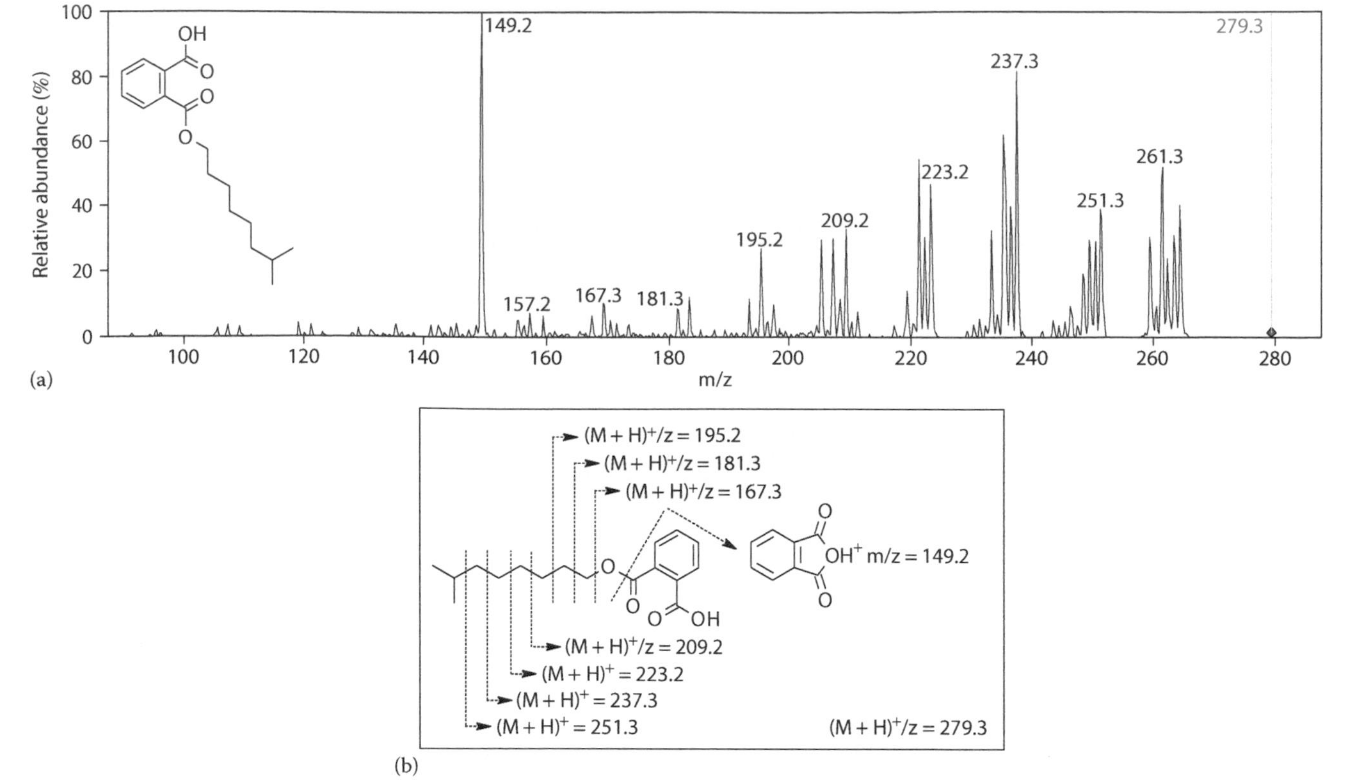

FIGURE 17.13 TMS of the ion at $[M + H]^+/z = 279.3$ from (a) DARTIS/ITMS analysis of GBC and (b) the corresponding fragmentation mechanism.

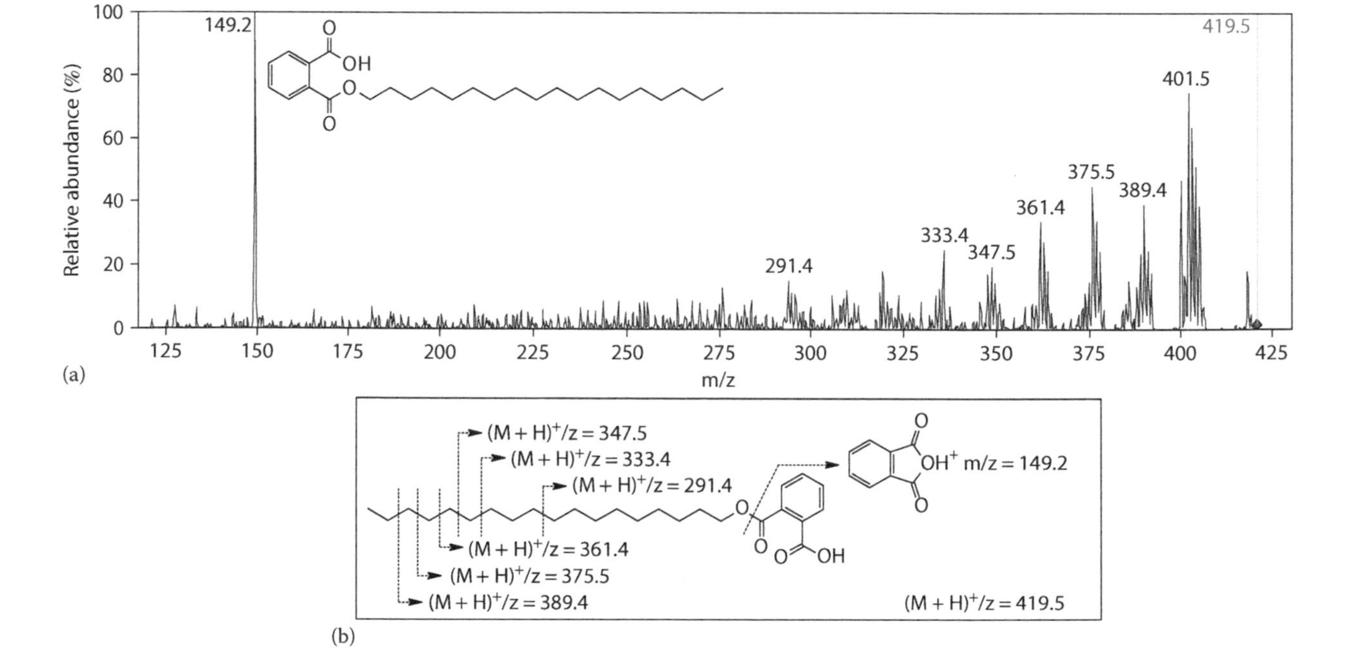

FIGURE 17.14 TMS of the ion at $[M + H]^+/z = 419.5$ from (a) DARTIS/ITMS analysis of E_4 and (b) the corresponding fragmentation mechanism.

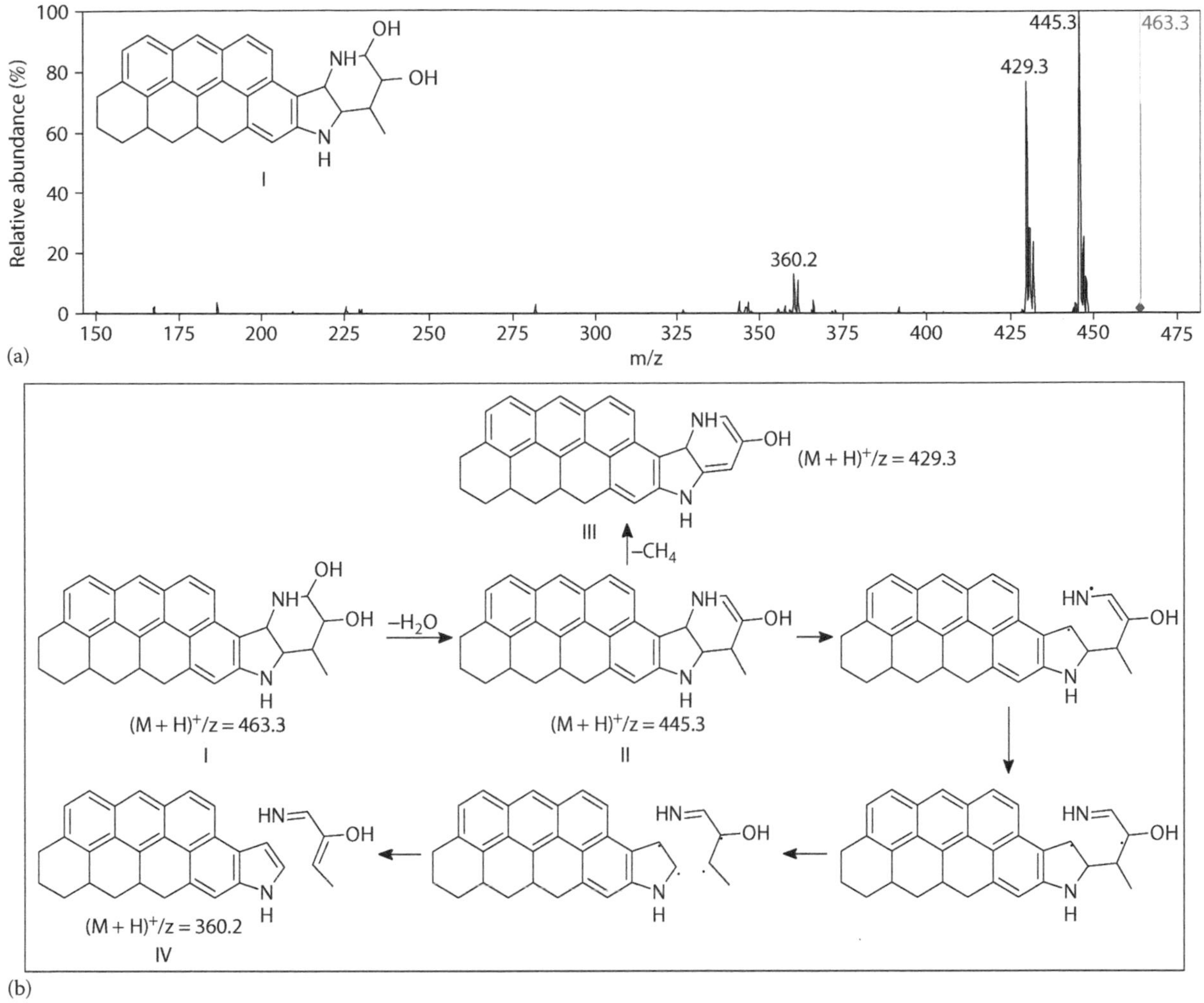

FIGURE 17.15 TMS of the ion at $[M + H]^+/z$ = 463.3 from (a) DARTIS/ITMS analysis of E_4 and (b) the corresponding fragmentation mechanisms.

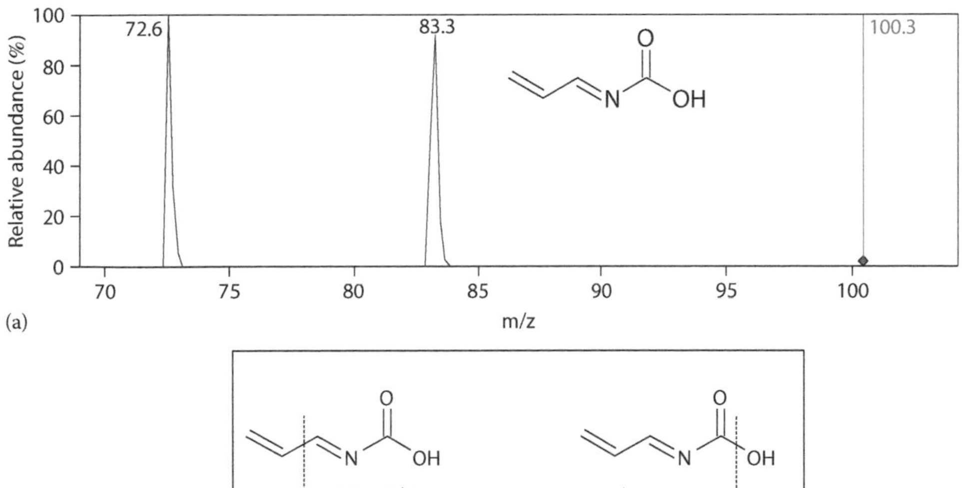

FIGURE 17.16 TMS of the ion at $[M + H]^+/z$ = 100.3 from (a) DARTIS/ITMS analysis of E_5 and (b) the corresponding fragmentation mechanisms.

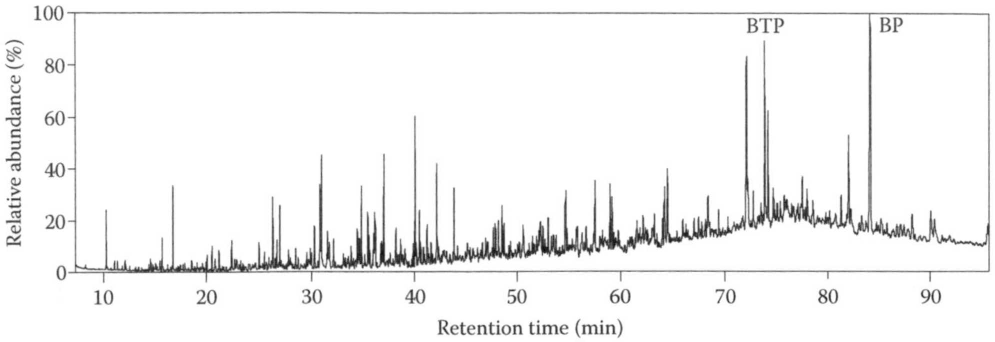

FIGURE 17.17 Total ion chromatogram (TIC) of E_2.

TABLE 17.1
Program of Variation of Solvent Concentrations during Gradient Elution

Fraction	Time (h)	Continual Variation of Solvent Concentration
F_1	0–4	From 100% PE to 100% CDS
F_2	4–8	From 100% CDS to 100% acetone
F_3	8–12	From 100% acetone to 100% methanol
F_4	12–	100% methanol

TABLE 17.2
Purities of Condensed Arenes and DAPs in the Corresponding Fractions

Fraction	Condensed Arene	Purity (Area%)	Fraction	DAP	Purity (Area%)
F_{2-22}	BTP	80.1	F_{4-17}	DOP	99.1
F_{2-38}	BP	72.2	F_{4-21}	BEHP	84.3
F_{2-39}	Coronene	96.7	F_{4-25}	BCPEP	99.8
			F_{4-29}	DBP	90.2

With total annual production of more than 25 Mt over the world (Bosnir et al. 2003), DAPs have many applications as plasticizers and stabilizers for cosmetic emulsifiers, adhesives, inks, hairsprays, lacquers, caulking, car parts, medical devices, tubes, films, insect repellents, floorings, paints, carpet backings, wood finishers, wallpaper, antifoaming agents, electric cables, and toys (Kavlock et al. 2002; Lehmann et al. 2004; Rahman and Brazel 2004). DAPs with alkyl carbon chains of C_4–C_6 are commonly used as plastic softer. As the most commonly used plasticizer, BEHP is

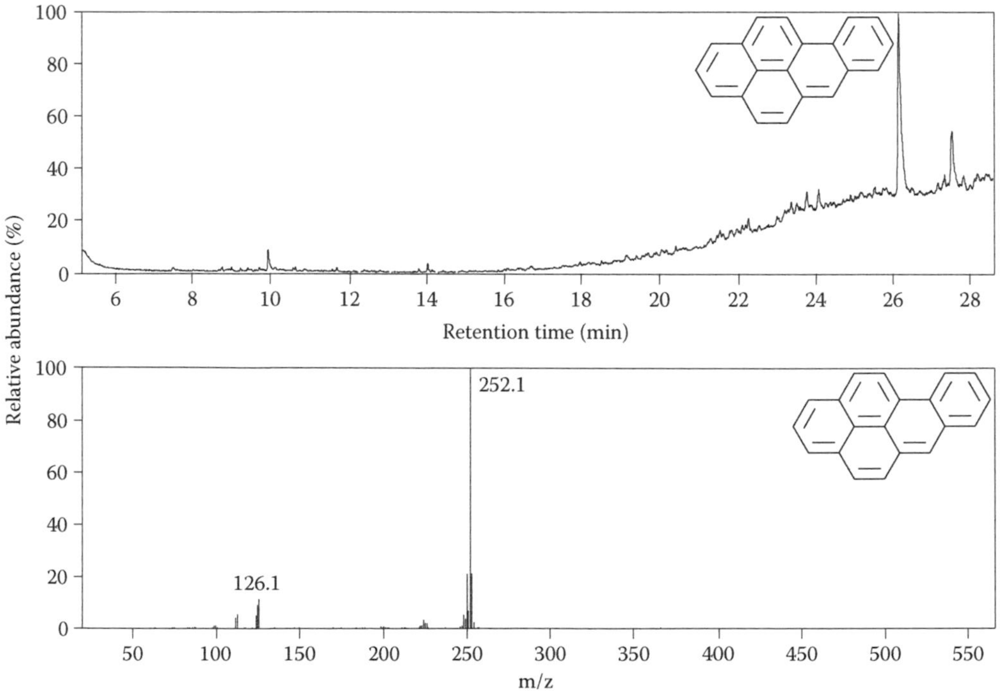

FIGURE 17.18 TIC of F_{2-22} and mass spectrum of BTP.

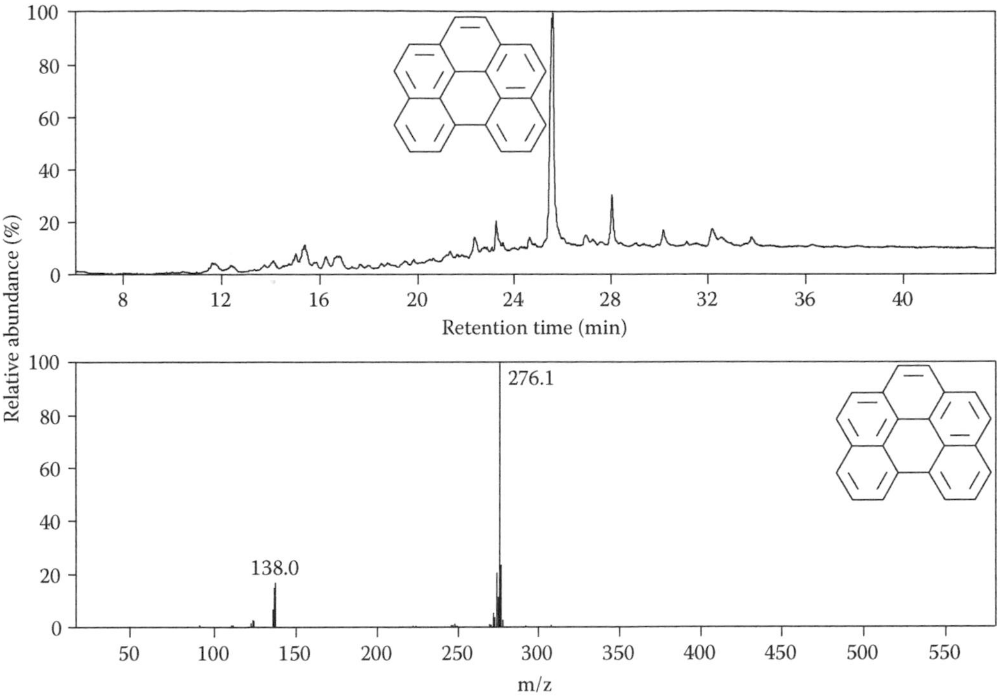

FIGURE 17.19 TIC of F_{2-38} and mass spectrum of BP.

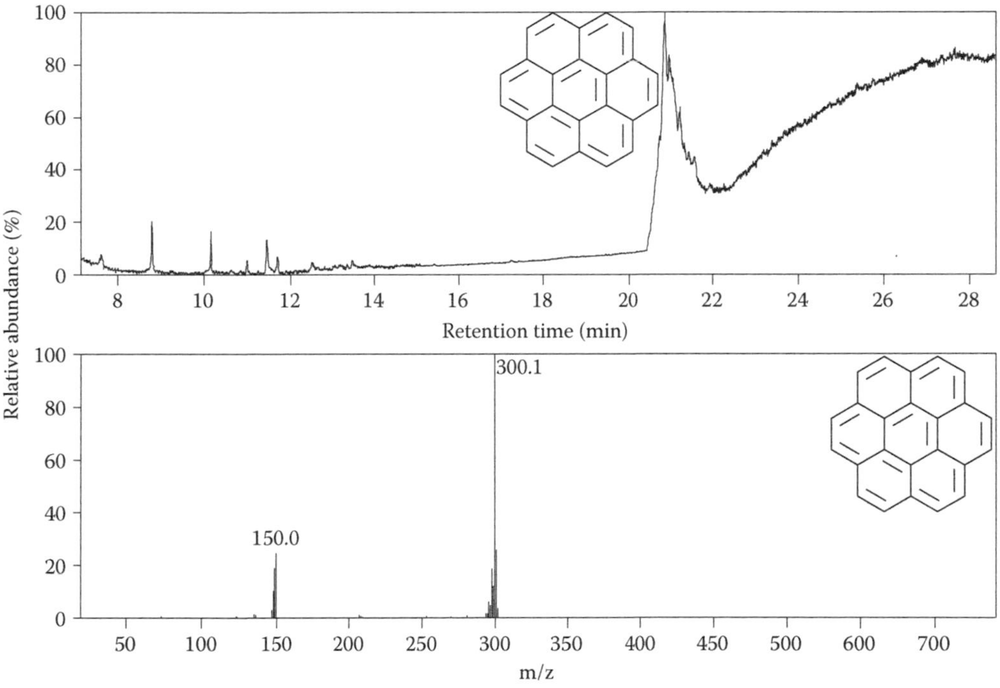

FIGURE 17.20 TIC of F_{2-39} and mass spectrum of coronene.

employed to impart flexibility of polyvinylchloride plastics worldwide (Rahman and Brazel 2004).

Using sequential extraction and subsequent column chromatography, pure BEHP, bis(2-ethylheptyl) terephthalate (BEHTP), methyl tetracosanoate (MTC), and methyl hexacosanoate (MHC) (Figure 17.26) were successfully isolated from Lingwu bituminous coal (LBC, collected from Lingwu Coal Mine, Ningxia Hui Autonomous Region, China) and identified by GC/MS, FTIR, ^{1}H, and ^{13}C nuclear magnetic resonance (NMR) analyses (Liu et al. 2009, 2010). These compounds could originally exist in LBC as biomarkers.

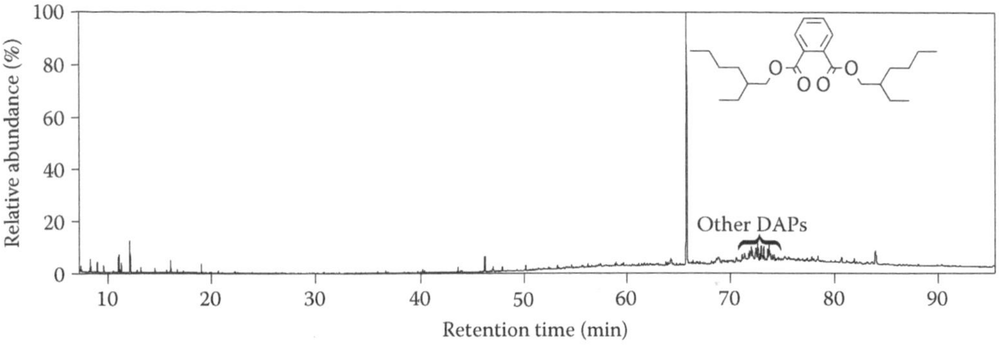

FIGURE 17.21 TIC of E_4.

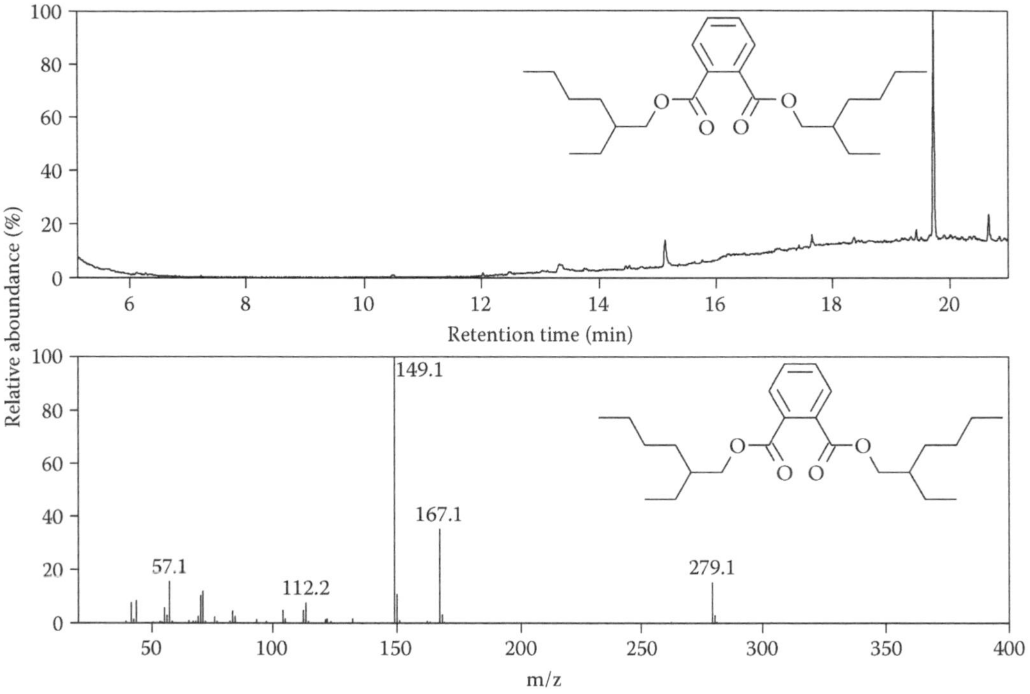

FIGURE 17.22 TIC of F_{4-21} and mass spectrum of BEHP.

Isometric CDS/THF mixed solvent was used to extract a subbituminous coal from Erdos, Inner Mongolia Autonomous Region, China. The extract was subsequently separated by silica-gel column chromatography. As a result, a series of long-chain normal alkanes, monoaromatic steranes, condensed arenes (especially a C_{10}-naphthalene and simonellite), and HACOSs in the extract were concentrated to large extents. Among them, abiettriene and retene were separated in high purities (Zong et al. 2009). Both simonellite and retene are known to be structurally specific higher-plant biomarkers (Grice et al. 2007), while abiettriene was seldom detected from fossil sources except for a previous investigation by our group (Ding et al. 2008a).

The combination of sequential extraction and subsequent column chromatography is also effective for enriching and isolating many other organic compounds from different coals. A series of long-chain normal alkanals with chain carbon number of C_{18}—C_{27} (Figure 17.27) was enriched (Cong et al. 2014d) and two novel condensed aromatic lactones (5*H*-phenanthro[1,10,9-*cde*]chromen-5-one and 4*H*-benzo[5,10]anthra[1,9,8-*cdef*]chromen-4-one, shown in Figure 17.28) were isolated as nearly pure compounds (Cong et al. 2014a) from Zhundong subbituminous coal (ZSBC, collected from Zhundong coalfield, Xinjiang Uygur Autonomous Region, China). Four arylhopanes (22-phenyl-30-norhopane, 22-[*o*-tolyl]-30-norhopane, 30-[5-methylthien-2-yl]hopane, and 30-[thien-2-ylmethyl]hopane, shown in Figure 17.29; Cong et al. 2014b) along with 14 cyclized hopanoids (Figure 17.30; Cong et al. 2015) were enriched, and 3-ethyl-8-methyl-2,3-dihydro-1*H*-cyclopenta[*a*]chrysene (Figure 17.31; Cong et al. 2014c) was isolated from Shengli lignite (SL, collected from Shengli coalfield, Xilinhot, Inner Mongolia Autonomous Region, China). A series of alkanamides (C_{15}—C_{28}) and three alkenamides (two C_{18} and one C22) were extracted with CDS from SL and enriched by

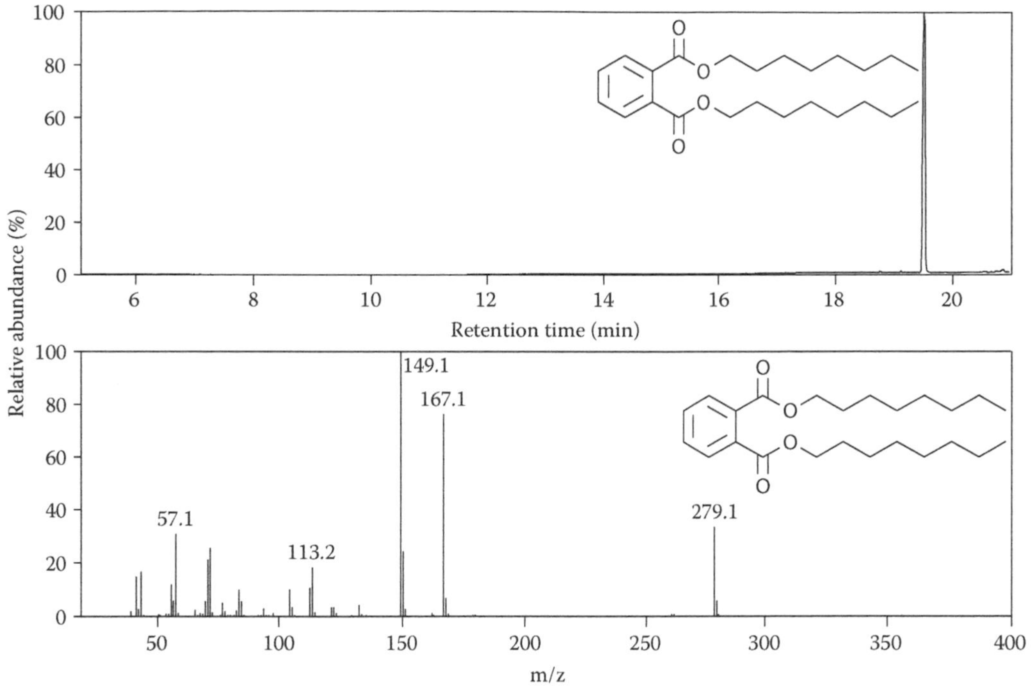

FIGURE 17.23 TIC of F_{4-17} and mass spectrum of DOP.

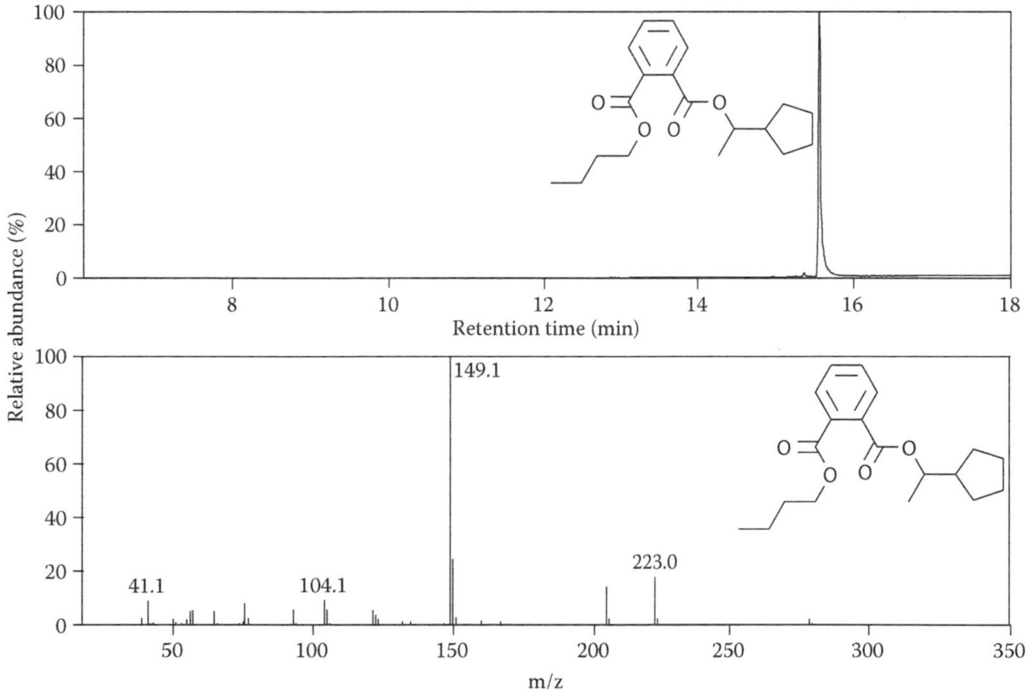

FIGURE 17.24 TIC of F_{4-25} and mass spectrum of BCPEP.

subsequent column chromatography (Figure 17.32) (Ding et al. 2008b). Strong interaction between C=S bond in CDS and C=O bond in the fatty acid amides (FAAs) could make the FAAs extractable with CDS from SL, although CDS is a weak polar solvent, whereas FAAs are strong polar compounds.

Chlorine and bromine in coals were suggested to occur as inorganic and organic species (Ren et al. 1999; Vassilev et al. 2000) and organic hydrochlorides (Shao et al. 1994; Huggins and Huffman 1995). Organochlorines and organobromines in coals deserve more attention because of their higher toxicities

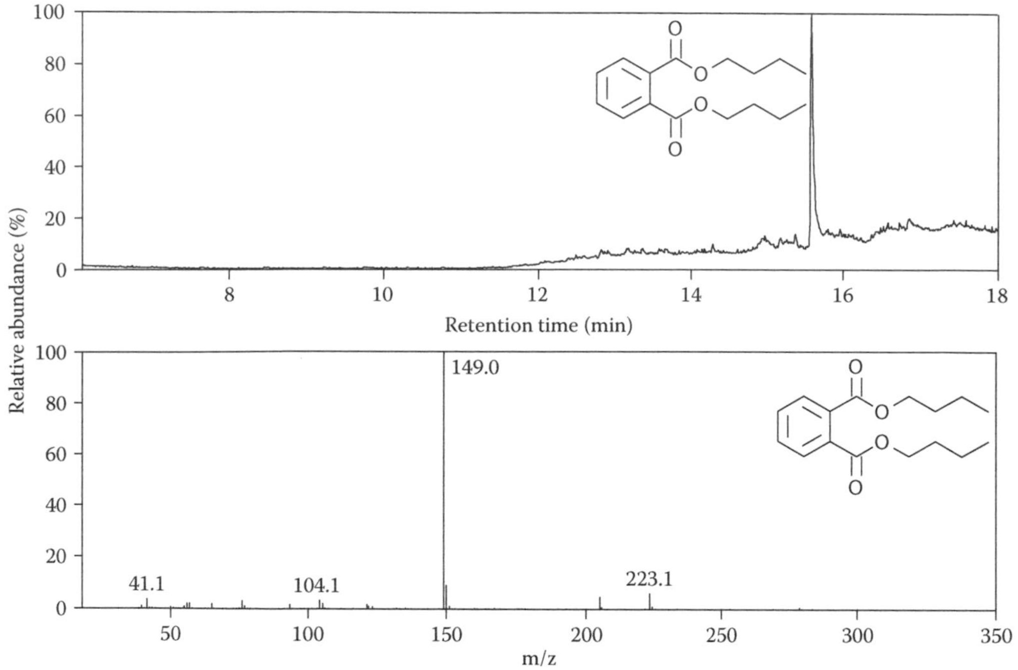

FIGURE 17.25 TIC of F_{4-29} and mass spectrum of DBP.

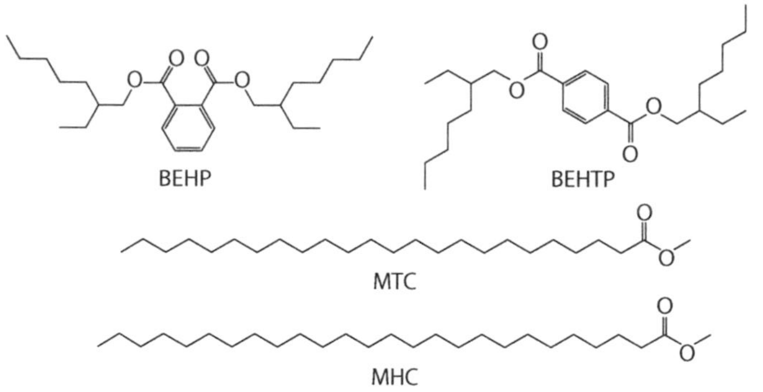

FIGURE 17.26 Four organic compounds isolated from LBC.

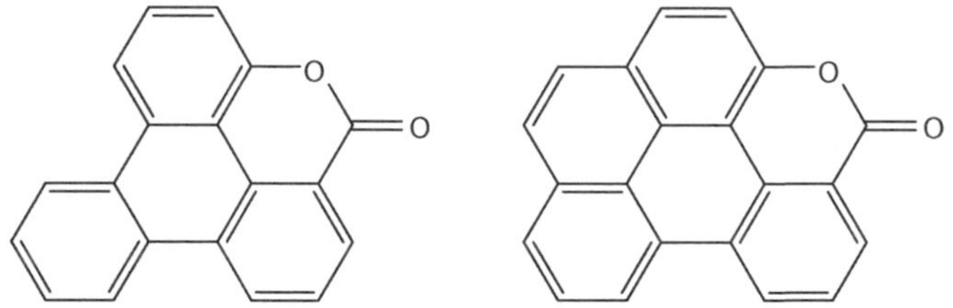

FIGURE 17.28 5*H*-Phenanthro[1,10,9-*cde*]chromen-5-one (left) and 4*H*-benzo[5,10]anthra[1,9,8-*cdef*]chromen-4-one (right) isolated from ZSBC.

and more significant relationship with geochemical implications than inorganic species. However, no reports were issued on the molecular structures of such species in any coals before the related findings of our group. In addition to GBC, 3 other Chinese coals, that is, Pingshuo bituminous coal (PBC, collected from Pingshuo Coal Mine, Shanxi Province, China), Shenmu-Fugu subbituminous coal (SFSBC, collected from Shenmu-Fugu coalfield, Shaanxi Province, China), and Datong bituminous coal (DBC, collected from Datong Coal Mine, Shanxi Province, China), were extracted sequentially with CDS, hexane, benzene, methanol, acetone, THF, and THF/methanol (1:3 vol/vol) mixed solvent, as depicted in Figure 17.33.

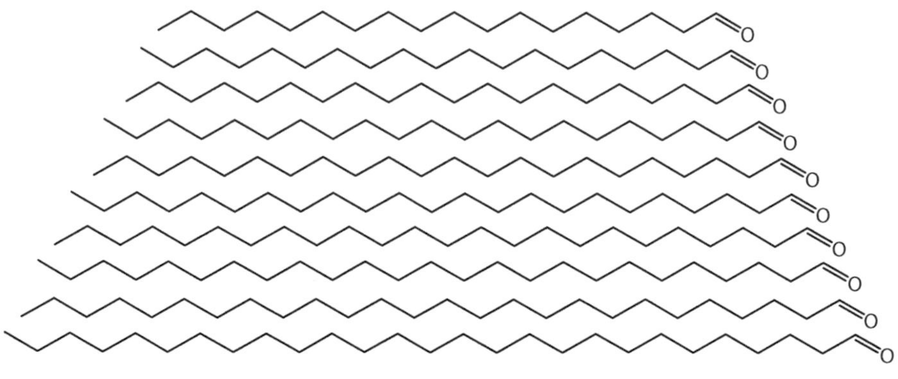

FIGURE 17.27 Long-chain normal alkanals enriched from ZSBC.

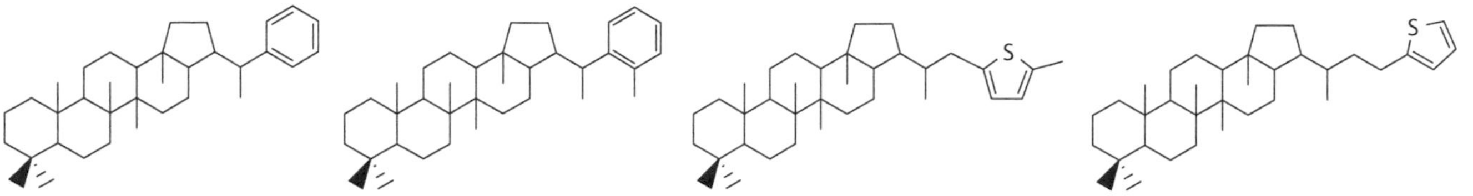

FIGURE 17.29 Four arylhopanes enriched from SL.

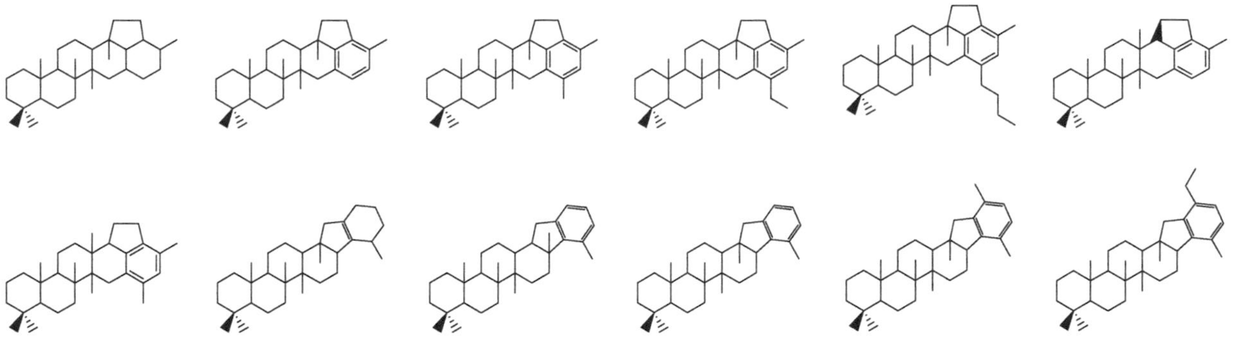

FIGURE 17.30 Cyclized hopanoids enriched from an extract of SL.

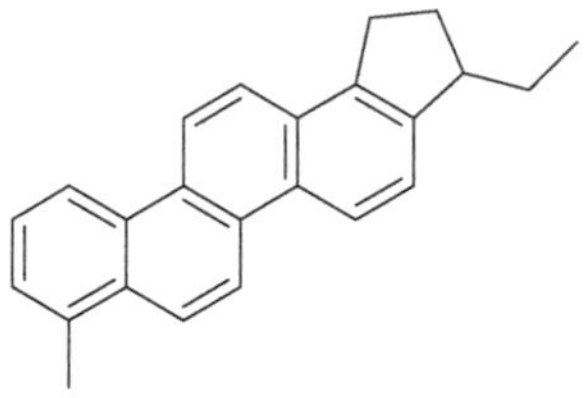

FIGURE 17.31 3-Ethyl-8-methyl-2,3-dihydro-1*H*-cyclopenta[*a*]chrysene isolated from SL.

In total, 28 extracts were obtained from the four Chinese coals. As listed in Table 17.3, six organochlorines and two organobromines were identified in eight extracts (Wei et al. 2004).

In addition to PBC, SFSBC, GBC, and DBC, Longkou lignite, collected from Longkou Coal Mine, Shandong Province, China, was also sequentially extracted according to the procedure depicted in Figure 17.33. A number of nitrogen-containing organic species (NCOSs) were identified with GC/MS in E_4 from the five Chinese coals (Wei et al. 2009). Hydrogen bonds O—H···N and O—H···π should be responsible for the effectiveness of methanol for extracting the NCOSs. Most of the NCOSs are toxicants for CHC of coals and their derivates (Furimsky and Massoth 1999, 2005). All the NCOSs are precursors of contaminant emitted during coal combustion (Kambara et al. 1995; Ohshima et al. 2000; Tomita 2001). On the other hand, they are important intermediates for synthesizing medicines, drugs, dyes, and many other fine chemicals. So, enriching and identifying NCOSs in coals are of great importance for clean and value-added use of coals.

Different macerals in the same coal may have different solubilities (Qin et al. 1997, 1998) and compositions. Vitrinite-rich sample (VRS) and inertinite-rich sample (IRS) were separated from SFSBC and PBC using sink-float method combined with hand-picking (Shu et al. 2002) and extracted with CDS. The results show that extract yields of the two maceral-rich samples from PBC are much higher than those from SFSBC and maceral-rich samples from the same coal, whereas the extract yield of VRS is higher than that of IRS (Figure 17.34); the VRSs contain much more AM along with more epoxide and ester moieties than the IRSs, whereas the contents of both free and associated hydroxyl groups and aromatic moiety in the IRSs are higher than the VRSs; total yield (TY) of oxygen-containing organic species (OCOSs) in IRS is higher than that in VRS for the same coal (Zhao et al. 2008).

Xianfeng lignite (XL) was sequentially extracted with PE, CDS, methanol, acetone, and IMCDSAMS under sonication to afford extracts 1–5 (E_1–E_5), as shown in Figure 17.35 (Liu et al. 2013). In total, 13.7% of organic matter in XL was extracted (Figure 17.36). The compounds detected in the extracts with GC/MS can be classified into alkanes, cyclanes, alkenes, cyclenes, alkylbenzenes, condensed arenes, alcohols, aldehydes, ketones, carboxylic acids (CAs), methyl alkanoates, ethyl alkanoates, methyl benzoates, ethyl benzoates, DAPs, and NCOSs. Different from the extracts from higher rank coals, such as GBC, the extracts from XL mainly consist of alkanes rather than condensed arenes (Figure 17.37). Taking E_1 as an example, the carbon number of alkanes continuously distributes from C_{13} to C_{33}, while the carbon number of alkan-2-ones ranges from C_{23} to C_{32} (Figure 17.38). Among other species, the relative contents of ketones, DAPs, alkenes and CAs, and NCOSs are the highest in E_2, E_3, E_4, and E_5, respectively. Phenols were not noteworthily detected in the extracts. As by-products from pyrolysis of low-rank coals, including lignites, low-temperature coal tars (LTCTs) are rich in phenols. Because of the absence of phenols in the extracts, phenols in LTCTs should result from pyrolysis of phenoxy group-containing macromolecules in low-rank coals. In other words, low-rank coals are rich in phenoxy group-containing moieties.

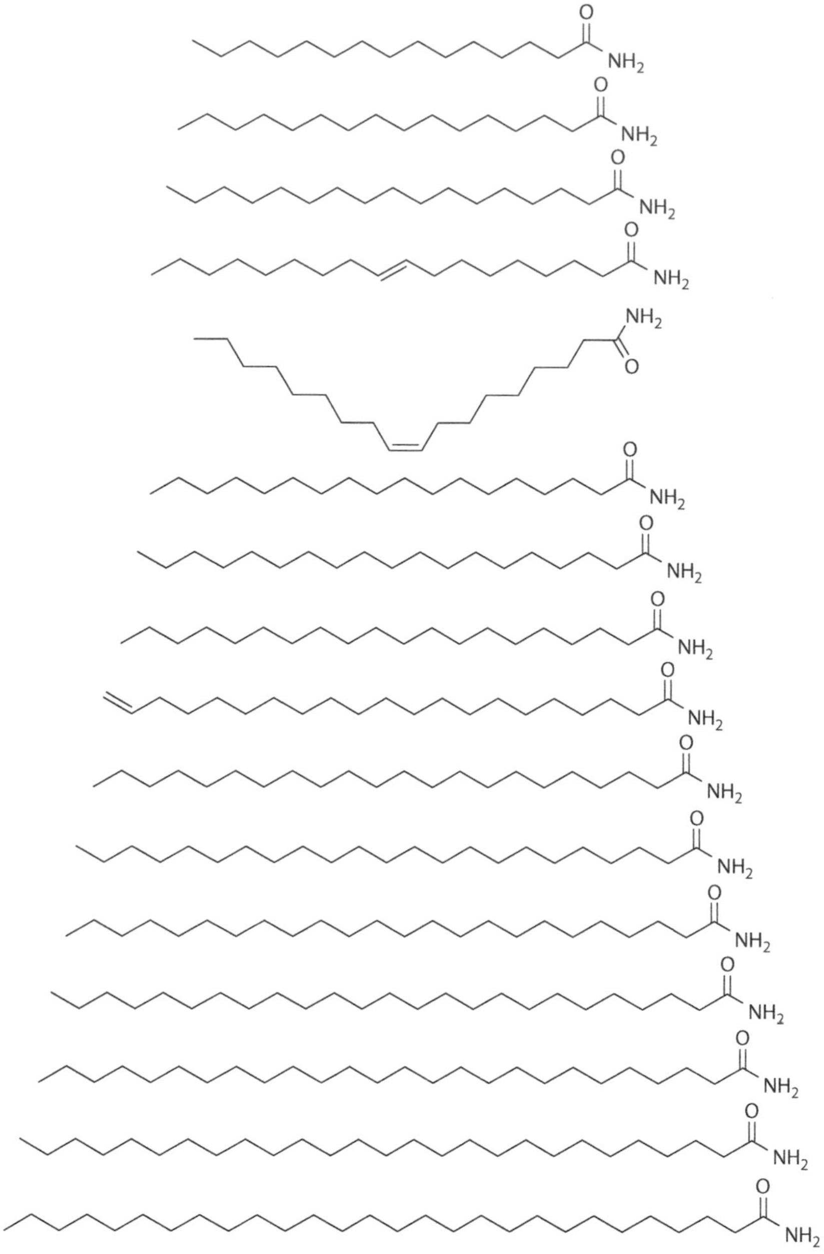

FIGURE 17.32 Alkanamides (C_{15}—C_{28}) and three alkenamides enriched from SL.

17.3 PRODUCTS FROM COAL THERMAL DISSOLUTION AND THEIR USES

In general, only small amount of organic matter can be extracted at room temperature except for the extraction of some bituminous coals with the famous isometric CDS/*N*-methyl-2-pyrrolidinone mixed solvent found by Iino et al. (1985, 1987, 1988). This finding has attracted continuous attention of coal researchers (Cai and Smart 1993; Mochida and Kinya 1994; Chervenick and Smart 1995; Gao et al. 1999; Opaprakasit and Painter 2004; Shui et al. 2006; Chen et al. 2011b; Qin et al. 2015). Investigation on thermal reaction of CDS with *N*-methyl-2-pyrrolidinone shows that there is a strong π-π interaction between C=S bond in CDS and C=O bond in *N*-methyl-2-pyrrolidinone, and association of the associated species with organic matter in coals was presumed to be an important reason for enhancing coal solubility in the mixed solvent (Zong et al. 2000, 2003). The results from calculation based on quantum chemistry also verified the π-π interaction between CDS and *N*-methyl-2-pyrrolidinone (Wang et al. 2004; Fu et al. 2006).

The mixed solvent was also used to investigate the effect of extractable substances on coal thermal dissolution (Wei et al. 1989) and prepare ashless coal extracts (Lee et al. 2007; Sönmez and Giray 2011). The high price and high

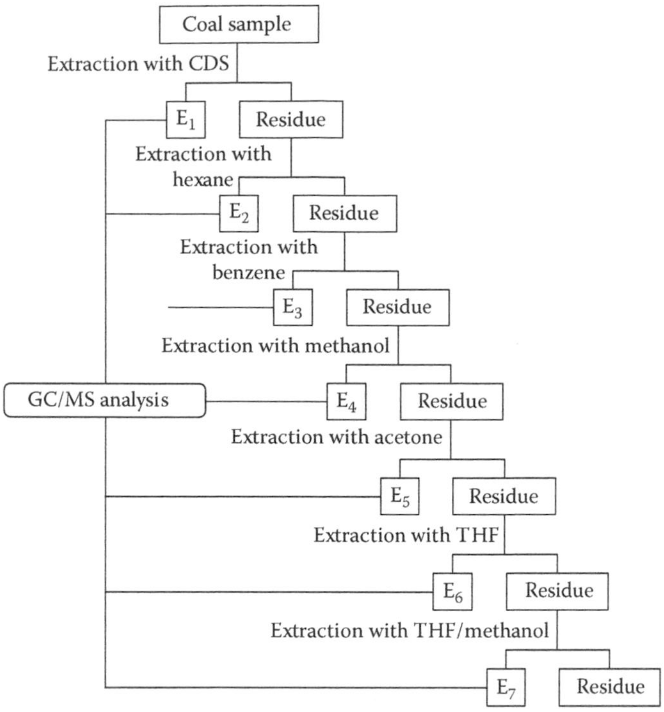

FIGURE 17.33 Sequential extraction of the four Chinese coals and subsequent GC/MS analysis of the extracts.

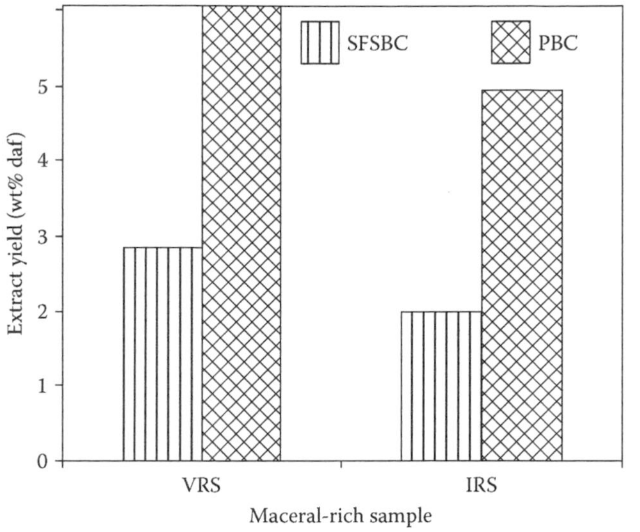

FIGURE 17.34 Extract yields of VRS and IRS from SFSBC and PBC.

boiling point of *N*-methyl-2-pyrrolidinone make the mixed solvent impractical. In fact, complete recovery of *N*-methyl-2-pyrrolidinone from the extract solution is impossible because of the formation of strongly associated molecular clusters between *N*-methyl-2-pyrrolidinone and organic species in coals (Liu et al. 2008).

Some other solvents, such as 1-methylnaphthalene (1-MN, Zhang et al. 2008) and crude methylnaphthalene oil (Masaki et al. 2004), with high boiling points and high viscosities were also used for coal thermal dissolution at high temperatures to prepare *HyperCoal*. The practical use of such a technology faces difficulties in hot filtration and solvent recovery. To overcome the difficulties, low-boiling point and low-viscosity solvents deserve consideration for coal thermal dissolution.

Huolinguole lignite (HL, collected from Huolinguole Coal Mine, Inner Mongolia Autonomous Region, China) was

TABLE 17.3
Organochlorines and Organobromines Identified in the Extracts from Four Chinese Coals

	Organochlorine and Organobromine Detected In				
Coal	**E_1**	**E_3**	**E_5**	**E_6**	**E_7**
PBC	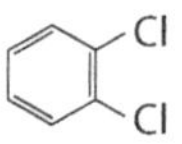				
DBC	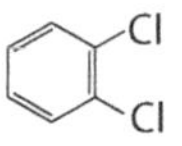	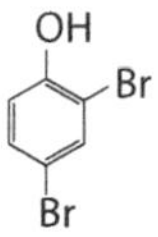 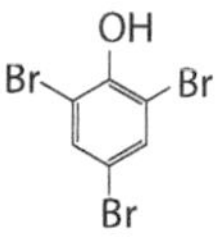		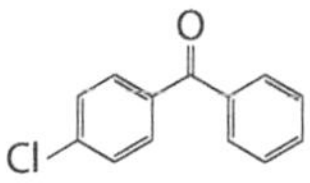	
GBC		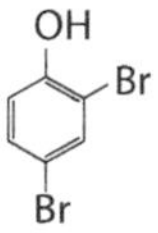 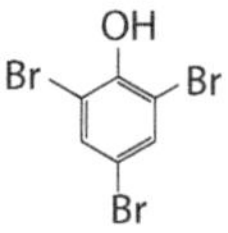			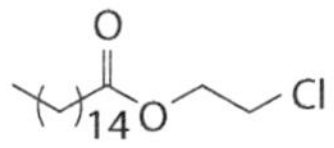
SFSBC		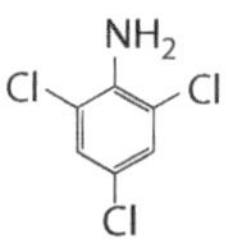 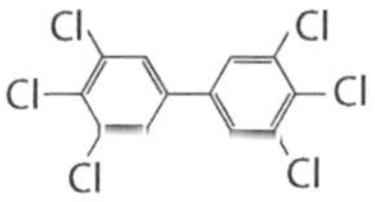	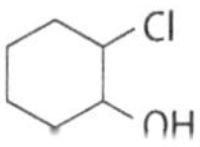		

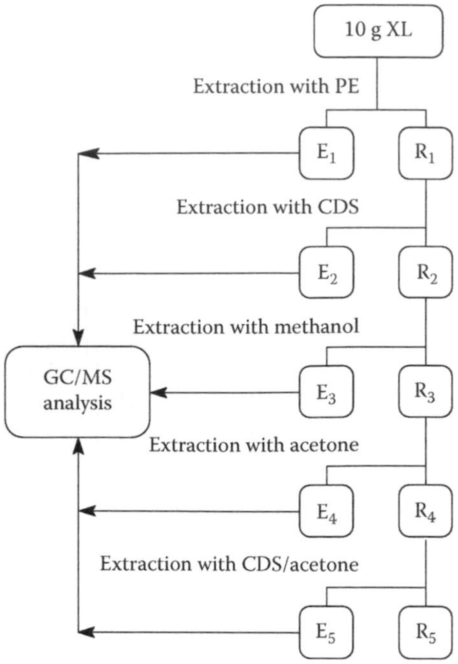

FIGURE 17.35 Sequential extraction of XL and subsequent GC/MS analysis of the extracts.

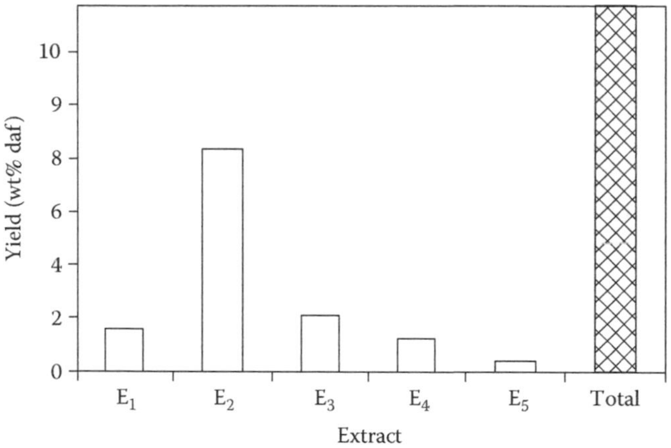

FIGURE 17.36 Yields of the extracts from XL.

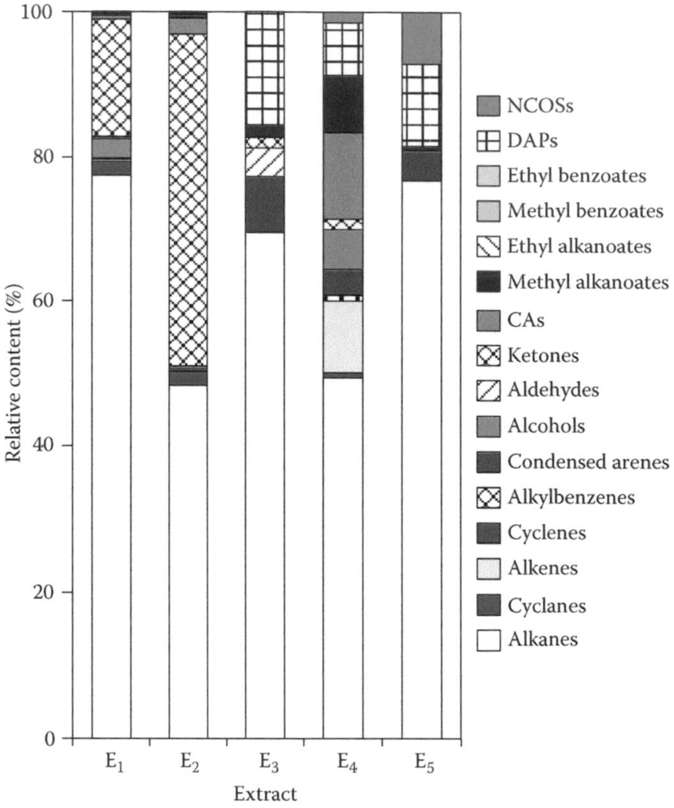

FIGURE 17.37 Distribution of group components in the extracts from XL.

sequentially dissolved at temperatures from 200°C to 330°C in methanol and ethanol, respectively (Lu et al. 2011). As Figure 17.39 shows, once-through yields (OTYs) of methanol-extractable portion (MEP) and ethanol-extractable portion (EEP) appreciably decreased with raising temperature from 200°C to 270°C and 240°C, respectively, but very rapidly increased for EEP and remarkably increased for MEP with further raising temperature. The significant increases in OTYs of MEP and EEP were also initiated from 270°C to 240°C and reached 23.0% and 55.3% at 330°C, respectively.

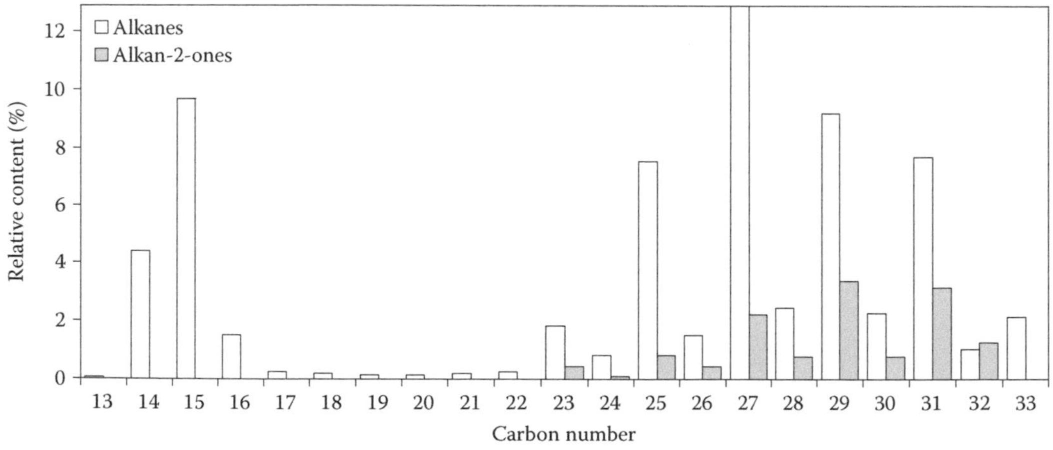

FIGURE 17.38 Distributions of alkanes and alkan-2-ones in E_1.

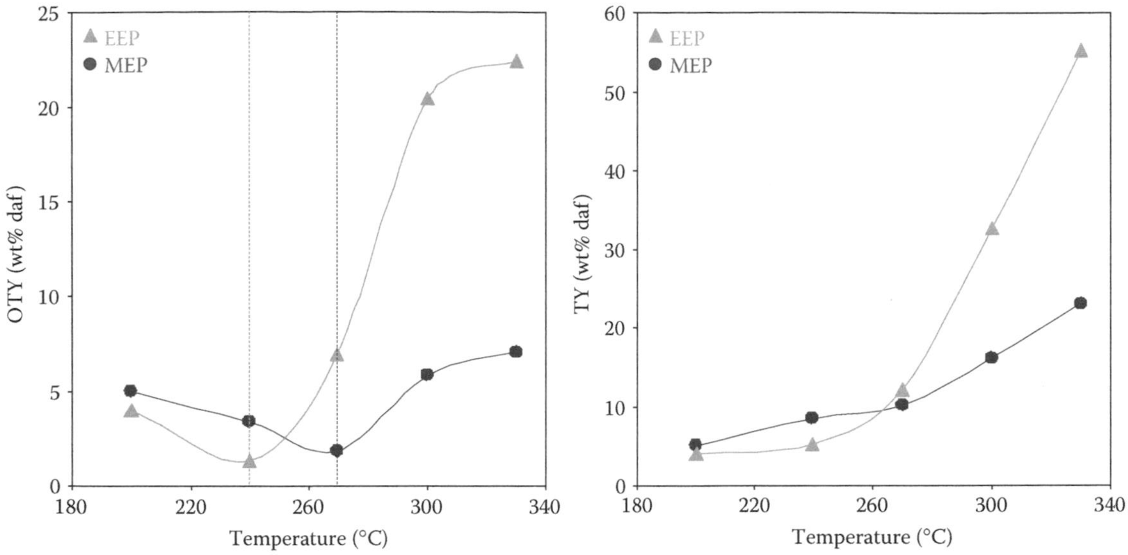

FIGURE 17.39 MEP and EEP yields at different temperatures.

At temperatures up to 240°C, OTYs of MEP are appreciably higher than those of EEP, whereas at temperatures higher than 270°C, OTYs of EEP are drastically higher than those of MEP. These results suggest that 270°C and 240°C could be initial temperatures (ITs) for methanol- and ethanol-induced decomposition of HL, respectively. In other words, thermal dissolution of HL predominantly proceeded via dissociation of intermolecular interactions in HL at temperatures lower than the corresponding IT, but significant cleavage of covalent bonds in HL proceeded at temperatures higher than the corresponding IT.

In total, 177 and 284 organic compounds were identified in MEP and EEP, respectively. They can be classified into alkanes, alkenes, alkyl arenes, non-substituted arenes, hydroarenes, methoxyarenes, ethoxyarenes, phenols, ketones, alkanols, CAs, methyl esters, ethyl esters, other esters, dialkyl ethers, amines, nitrocyclic aromatics, sulfur-containing organic species (SCOSs), and other species. Most of the compounds are OCOSs, and most of the dialkyl ethers contain an ethoxy group. MEP is rich in methoxyarenes and methyl esters, while EEP is rich in ethoxyarenes, ethyl esters, and dialkyl ethers, indicating that the reaction of methanol and ethanol with some macromolecular species in HL occurred. The high content (20.55%) of oxygen in HL and the detection of OCOSs as dominant compounds in MEP and EEP indicate that the main reactions should be methanolysis and ethanolysis. The nucleophilicity of ethanol is higher than that of methanol according to their pK_a values (ethanol of 15.9 and methanol of 15.5; Zhao and Zhang 1996). The nucleophilicity of ethanol than that of methanol and the larger molecular masses of the resulting $RCH_2OCH_2CH_3$ and $RCOOCH_2CH_3$ than those of the corresponding RCH_2OCH_3 and $RCOOCH_3$ resulted in the drastically higher OTYs of EEP than those of MEP. The contents of both oxygen and sulfur in both extractable portion and residue are significantly lower than those in HL, indicating that deoxygenation and desulfurization significantly proceeded during thermal dissolution of HL.

Compared to HL methanolysis at 310°C, SL methanolysis produced much more phenols, ketones, and alcohols but less arenes, alkanes, and esters; however, more than 50% of MEPs from methanolysis of both HL and SL are phenols and esters (Chen et al. 2011a). Phenols are important chemicals for synthesizing plastics, while esters can be used as both advanced lubricants and diesel oil. Therefore, methanolysis can be considered as a promising method for converting lignites to clean fuel and chemicals. Molecular dynamic simulations in terms of ReaxFF (reactive force field) provide a reasonable atomistic description of the initiation mechanism for the methanolysis (Chen et al. 2012).

The simulation of lignite-related model compounds, such as benzyloxybenzene, phenethoxybenzene, oxydibenzene, and anisole, using density functional theory further proved that ethanolysis proceeds much easier than methanolysis (Li et al. 2014b). For example, the rate constant of benzyloxybenzene ethanolysis is 10 times more than that of benzyloxybenzene methanolysis at the same temperature. Isopropanolysis proved to be more effective than ethanolysis for coal thermal dissolution because of the larger nucleophilicity of isopropanol than that of ethanol (Li et al. 2014c).

The extraction residue (R_5) mentioned in Figure 17.35 was subjected to sequential thermal dissolution in cyclohexane, benzene, 1-MN, methanol, and ethanol at 320°C (Figure 17.40; Liu et al. 2013). TY of the extracts from the sequential thermal dissolution reached 46.5% based on organic matter in XL (Figure 17.41), that is, in total 60.2% of organic matter in XL became soluble species by sequential extraction and subsequent sequential thermal dissolution.

The distributions of group components in the extracts from the sequential extraction are quite different from those in the extracts from the subsequent sequential thermal dissolution, according to comparison of Figure 17.37 with Figure 17.42.

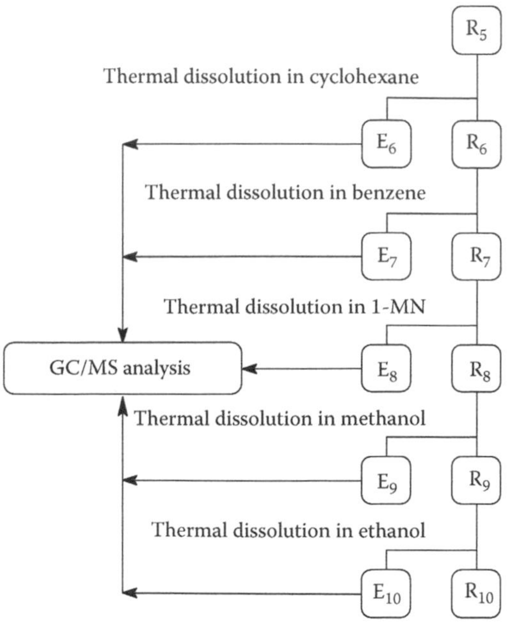

FIGURE 17.40 Sequential thermal dissolution of R_5 and subsequent GC/MS analysis of the extracts.

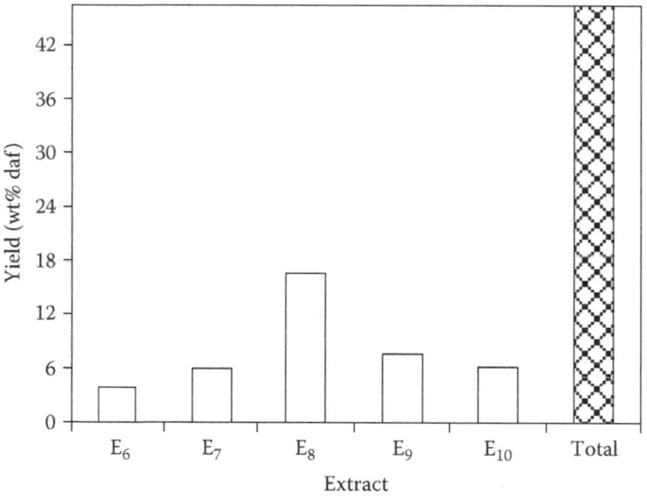

FIGURE 17.41 Yields of the extracts from R_5 thermal dissolution.

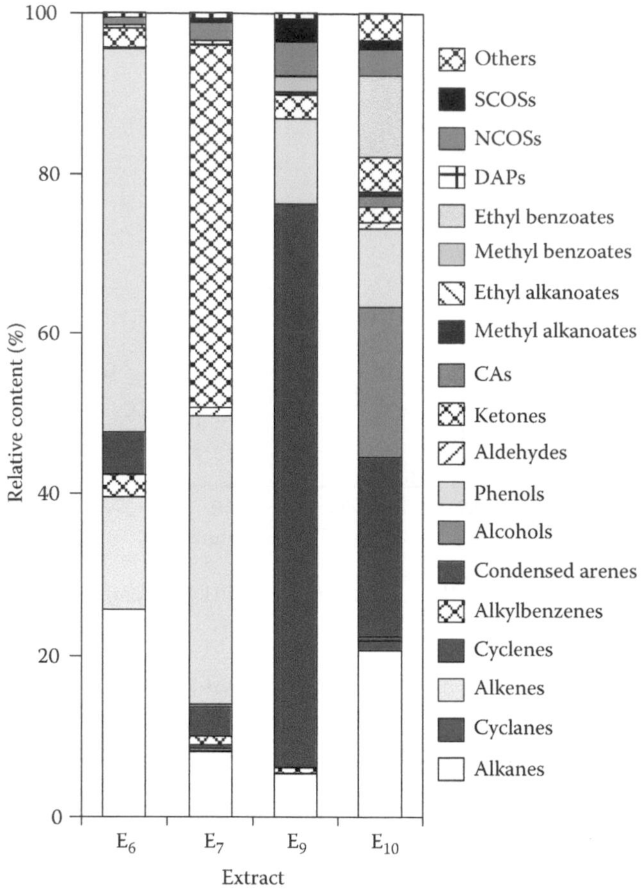

FIGURE 17.42 Distribution of group components in the extracts from R_5 thermal dissolution.

No species were detected in E_8 with GC/MS, although the yield of E_8 is much higher than that of other extracts, implying that E_8 consists of macromolecular species. In other words, covalent bonds in R_7 could not be significantly destroyed during thermal dissolution of R_7 in 1-MN. Phenols are the main group components in E_6 and secondary group components in E_7. They also appear in E_9 and E_{10}. These facts further suggest that the phenols resulted from the dissociation of weak covalent bonds connecting phenoxy groups in XL. Condensed arenes dominate the group components in E_9, implying that the methanolysis led to the cleavage of covalent bonds connecting condensed aryl or condensed arylalkyl groups.

Six series of homologues, that is, alk-1-enes, alk-2-enes, alk-3-enes, alkylbenzenes, *o*-alkyltoluenes, and alkyl-*p*-xylenes, were detected in E_6, as summarized in Figure 17.43. Some other series of homologues were mentioned in Figures 17.27, 17.31, and 17.37. A series of alkanes, methyl alkanoates (Zhou et al. 2013b), and methyl alkanones (Zhou et al. 2013a) were enriched from thermally soluble SL, and a series of alkyl phenyl carbonates (APCs) from heptyl phenyl carbonate to octacosyl phenyl carbonate were enriched from an extract of ZSBC (Zhou et al. 2014). 4*a*-Trimethyl-2,3,4,4*a*,10,10*a*-hexahydrophenanthren-9(1*H*)-one (HIPTMHHP) was isolated as a pure compound from the thermally soluble SL (Zhou et al. 2013b). APCs are protectants for selective synthesis of polyamines by reaction with carbamate (Pittelkow et al. 2002). Among them methyl phenyl carbonate improves the selectivity and yields of methoxycarbonylation of aliphatic diamines (Yoshida et al. 2005). HIPTMHHP was reported to have anti-inflammatory activity (Chao et al. 2005). It was isolated from the bark of *Celastrus orbiculatus* (Liu et al. 2010) and *Senecio cannabifolius* var. *integrilifolius* (Ma et al. 2009), but no reports were previously issued on the isolation of HIPTMHHP from any coal.

Much more other series of homologues were identified in R_5 (Liu et al. 2015), E_9 (Liu et al. 2014), and EEP from the

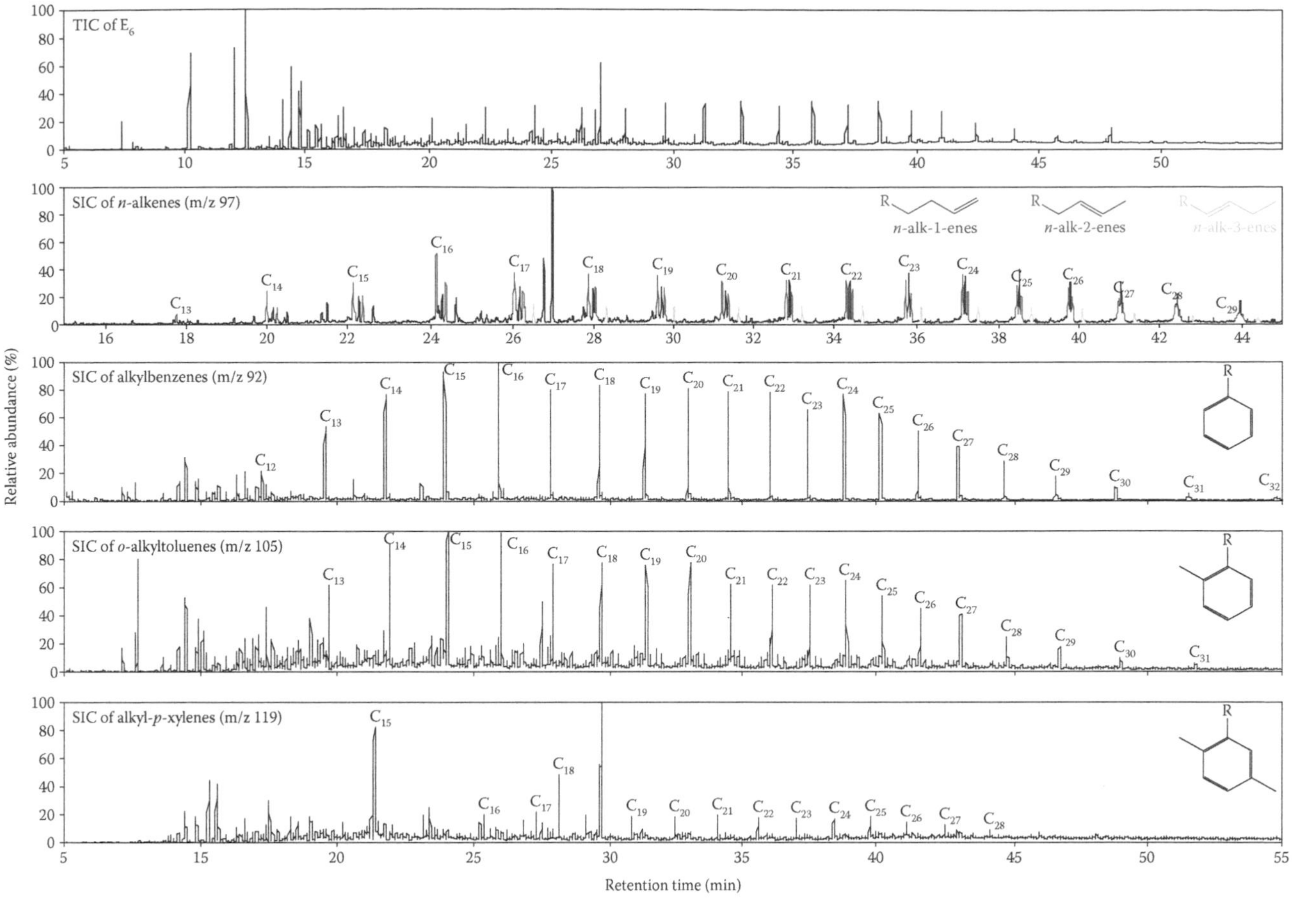

FIGURE 17.43 TIC of E_6 and selective ion chromatograms (SICs) of alkenes, alkylbenzenes, *o*-alkyltoluenes, and alkyl-*p*-xylenes in E_6.

ethanolysis of Zhaotong lignite, which was collected from Zhaotong Coal Mine, Yunnan Province, China (Li et al. 2014a). According to a number of experimental results, including aforementioned and undermentioned ones, homologues are generally present in coals and should be important portions of organic matter in coals. Hence, identifying different kinds of homologues in coals at the molecular level is of great importance both for coal chemistry and for value-added utilization of coals.

17.4 PRODUCTS FROM CHC OF COALS AND THEIR USES

CHC includes catalytic hydrogenation, hydrocracking, hydrodenitrogenation, hydrodeoxygenation, hydrodehydroxylation, and hydrodesulfurization. The development of highly active catalysts and the use of a proper solvent are crucially important for directional CHC of coals under mild conditions. Previous investigations of our group with coal-related model compounds (CRMCs) indicate that under mild conditions, metals catalyze biatomic hydrogen transfer, leading to the hydrogenation of unsaturated moieties (USMs), especially aromatic rings (Wei et al. 1991, 2003), while metal sulfides (or metal-sulfur system) (Wei et al. 1990, 1991, 1992a, b, c, d, 1993, 1995a, 2003; Ni et al. 2002) and activated carbon (AC) (Ni et al. 2003; Sun et al. 2005a, 2009b) promote monatomic hydrogen transfer (MAHT), and solid acids facilitate proton transfer (Yue et al. 2012a), resulting in the cleavage of some covalent bonds, especially bridged bonds. The reactivities of CRMCs toward catalytic hydrocracking not only depend on the hydrogen-accepting abilities of USMs but also are closely related to the stabilities of the leaving groups (Wei et al. 1990, 1992a, b; Ni et al. 2002). Quite different from conventional viewpoints, hydrogen-donating compounds, such as tetralin, 9,10-dihydrophenanthrene, and 9,10-dihydroanthracene (DHA), inhibit catalytic hydrocracking (Wei and Zong 1992e, 1993, 1995b) and thermolysis (Zong and Wei 1994) of CRMCs by strong adsorption on the catalyst surface and scavenge of active hydrogen species.

Coals themselves usually contain more or less extractable portions. Many organic compounds have been detected with GC/MS in the extractable portions, as mentioned above. It is difficult to discriminate GC/MS-detectable inherent organic compounds in coals with the resulting organic compounds from CHC of coals when using coals themselves as reactants. To avoid disturbance of the inherent organic compounds, using inextractable portions as the reactants for CHC is needed.

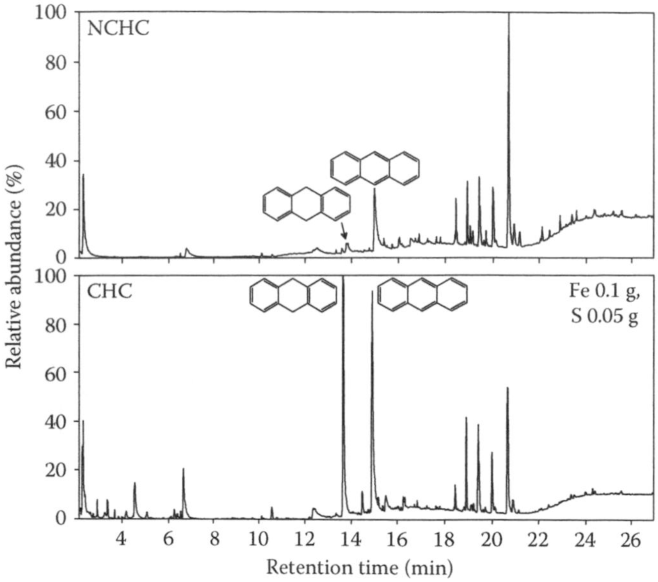

FIGURE 17.44 TICs of PEEPs from NCHC and CHC of THFMIEP from PBC—THFMIEP 1 g, cyclohexane 20 mL, initial H_2 pressure 5 MPa, 300°C, 4 h.

As Figure 17.44 displays, much more anthracene and DHA were detected in the PE-extractable portions (PEEPs) from CHC over Fe–S than that from non-catalytic hydroconversion (NCHC) of THF/methanol-inextractable portion (THFMIEP) from PBC (Wei et al. 2009). This result can be well interpreted by Fe-S-catalyzed MAHT to anthracene ring connected by PhX group (X denotes CH_2, NH, O, or S) on the benzene ring of the group where a macromolecular moiety is connected in the THFMIEP (Scheme 17.1). Isolating anthracene and DHA from the PEEPs should be much easier than from coal tar and coal liquefaction oil because of the very high contents of anthracene and DHA in PEEPs. The result

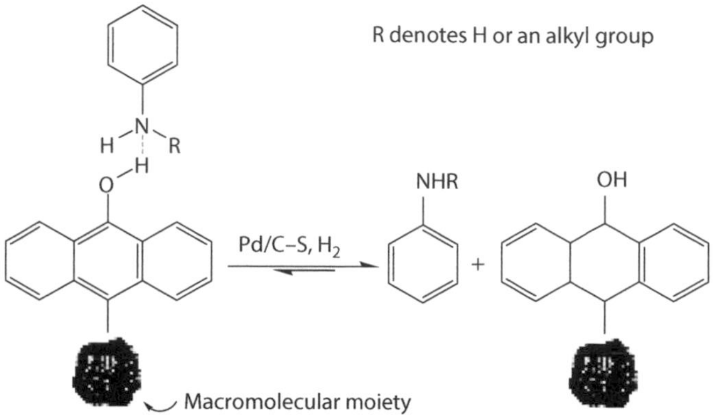

SCHEME 17.2 A possible reaction pathway for the release of aniline and *N*-alkylanilines from CHC of PBC- and DBC-derived THFMIEPs over Pd/C–S.

suggests that some species can be greatly enriched by CHC under mild conditions.

THFMIEPs from both PBC and DBC were also subjected to CHC over Pd/C at 300°C. A series of NCOSs were detected in the resulting PEEPs, and most of the NCOSs are aniline and *N*-alkylanilines (Wei et al. 2002). Aniline and *N*-alkylanilines may be associated strongly with condensed arenols by base–acid N—H bond in the THFMIEPs. Pd/C and the inherent sulfur could catalyze hydrogenation of the condensed aromatic ring (CAR) to convert the condensed arenols to diarylmethanols and thereby release aniline and *N*-alkylanilines (Scheme 17.2). Alternatively, anilinyl and *N*-alkylanilinyl groups could be connected with a CAR in the THFMIEPs. Pd/C and the inherent sulfur could catalyze MAHT to ipso-position of the CAR and release anilinyl and *N*-alkylanilinyl radicals, which accept H· from H· and/or H_2 to form aniline and *N*-alkylanilines (Scheme 17.3).

THF/CDS-inextractable portions from three Argonne Premium coals, that is, Pittsburgh No. 8 bituminous coal (P8BC), Upper

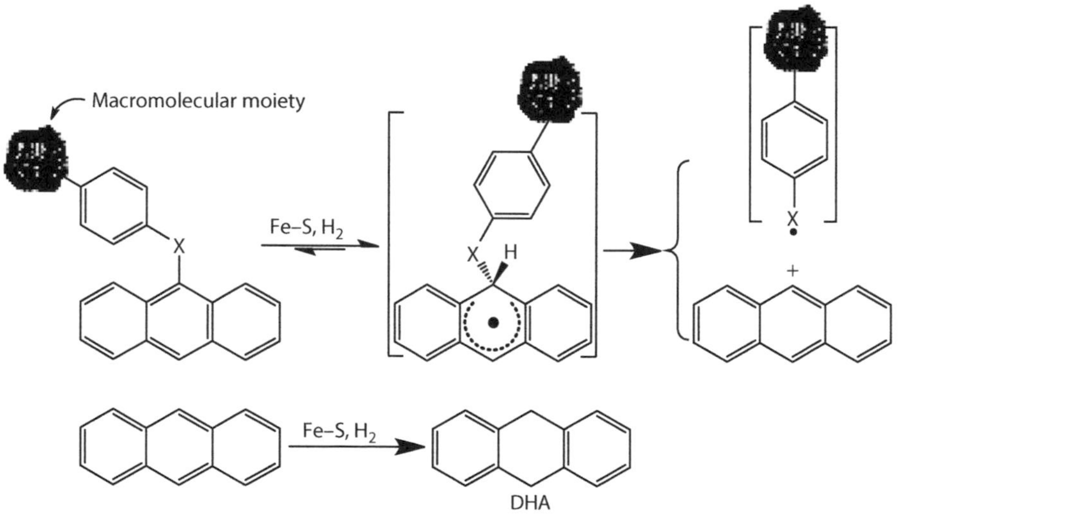

SCHEME 17.1 Reaction pathway for the formation of anthracene and DHA from CHC of the THFMIEP over Fe–S.

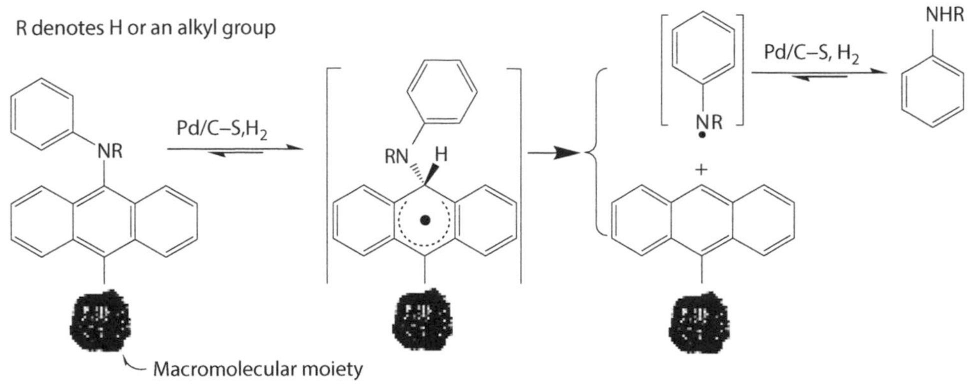

SCHEME 17.3 Another reaction pathway for the release of aniline and *N*-alkylanilines from CHC of PBC- and DBC-derived THFMIEPs over Pd/C–S.

Freeport bituminous coal (UFBC), and Pocahontas No. 3 bituminous coal (P3BC), were subject to NCHC and CHC over stabilized nickel at 300°C. According to GC/MS analysis, 22 SCOSs were detected in the resulting PEEP from the NCHC of UFBC, whereas only very small amounts of two SCOSs were detected in the resulting PEEP from the CHC of the inextractable portion from UFBC, suggesting that the thermally released SCOSs from the inextractable portion were effectively removed by the CHC over stabilized nickel (Sun et al. 2005b). A series of organochlorines and organoiodines were released from the NCHC and CHC of the inextractable portion from the three coals (Sun et al. 2007).

NCHC and CHC of THF/CDS-inextractable portion from P3BC were also investigated at 300°C using AC as the catalyst. A series of NCOSs and SCOSs were released from the CHC, but neither NCOSs nor SCOSs were released from the NCHC, suggesting that AC catalyzed the dissociation of nitrogen- and sulfur-containing moieties from some macromolecular species in the inextractable portion and subsequent release of the NCOSs and SCOSs (Sun et al. 2009a).

Three solid acids were prepared by impregnating the same volume of pentachloroantimony (PCA), trimethylsilyl trifluoromethanesulfonate (TMSTFMS), or isometric PCA and TMSTFMA into an AC. Di(1-naphthyl)methane (DNM) was used as a CRMC to evaluate their catalytic activities. The results show that bridged bond in DNM can be specifically cleaved over each catalyst to afford naphthalene and 1-MN under pressurized hydrogen at temperatures up to 300°C, while the new solid acid (NSA, i.e., PCA-TMSTFMS/AC) is significantly more active for DNM hydrocracking than the other two solid acids. The addition of mobile H^+ formed by heterolytically cleaving H_2 to ipso-position of DNM should be crucial step for DNM hydrocracking (Yue et al. 2012a). The NSA effectively catalyzed hydroconversion of SL and SFSBC (Lv et al. 2015) and proved to be active for CHC of IMCDSAMS-inextractable portion from LBC, especially for catalyzing the release of arenes and phenols from the inextractable portion, as shown in Figure 17.45 (Yue et al. 2012b). Unexpectedly, most of the release arenes are diphenylmethane (DPM). This is another example for releasing an

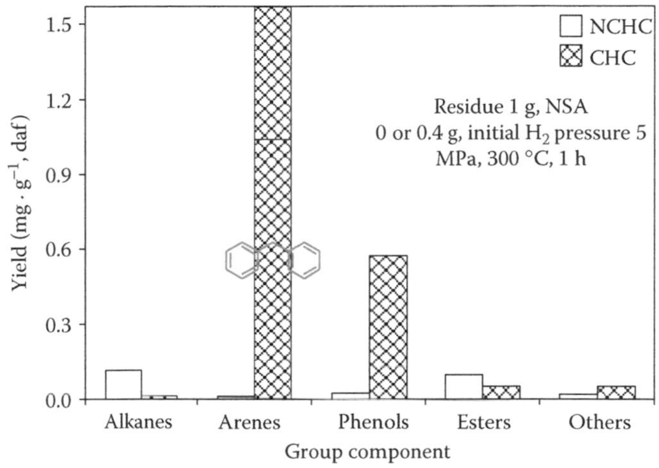

FIGURE 17.45 Distribution of group components in the PEEPs from NCHC and CHC of IMCDSAMS-inextractable portion from LBC.

arene in high relative content from CHC of a coal residue. The addition of mobile H^+ to ipso-position of CARs, which are connected with arylmethyl, diphenylmethyl, or phenoxy groups, in some macromolecular species of the inextractable portion could be responsible for releasing the arenes and phenols (Scheme 17.4). Fe–S/γ-Al_2O_3 is also active for CHC of LBC, especially for catalyzing the release of hydroarenes, arenes, and arenols and for completely removing 2,5-dimethylthiophene from LBC (Yu et al. 2014a), while Fe–S/ZSM effectively catalyzed hydroconversion of IMCDSAMS-inextractable portion from SL (Yu et al. 2014b).

More organic matter in GBC was catalytically converted into extractable portions over FeNi–S/γ-Al_2O_3 compared to NCHC, releasing more alkylmethylarenes and alkylarenols via the cleavage of —CH_2— and —O— linkages connecting CARs in GBC. Some series of homologues, such as alkylvinylphenols or alkyltetralinols, alkyltetralinols, and alkylalkenylketones, were detected in the extractable portions by analysis with an atmospheric solid analysis probe/time of flight mass spectrometer (Zhang et al. 2015).

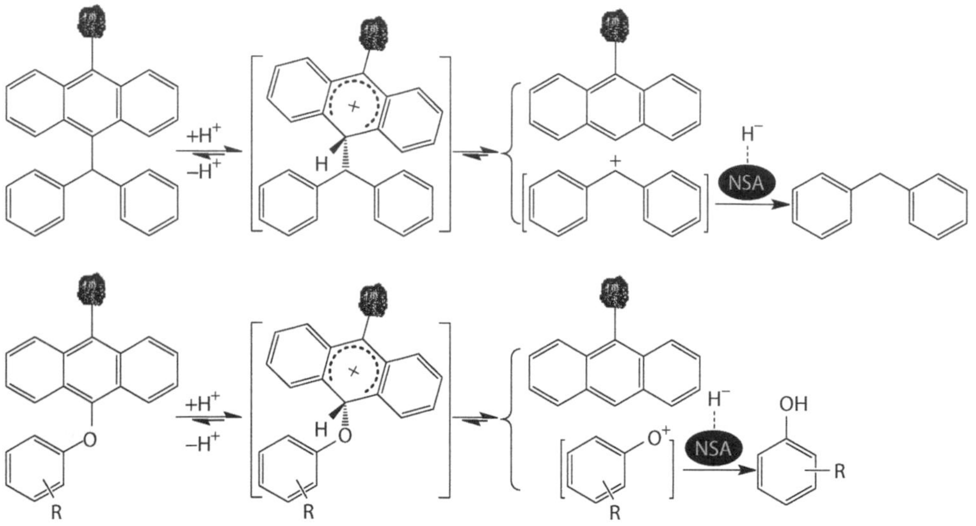

SCHEME 17.4 Reaction pathways for the release of DPM and phenols during CHC of IMCDSAMS-inextractable portion from LBC.

17.5 PRODUCTS FROM DIRECT COAL LIQUEFACTION (DCL) AND THEIR USES

DCL produces liquefaction oil (LO) as target products along with liquefaction residue (LR) and gaseous species as by-products. As the TY of gaseous species is very low, while LR yield is usually around 30% (Sugano et al. 2002), the efficient utilization of both LO and LR based on the understanding of molecular composition of LO and LR is important for improving economic benefits of DCL process.

A heavy LO (boiling point range of 360°C–380°C) from Heishan bituminous coal, which was collected from Heishan Coal Mine, Xinjiang Uygur Autonomous Region, China, was subjected to column chromatography and subsequent GC/MS analysis. The result shows that the heavy LO mainly consists of aromatics, including condensed arenes with two to four rings and heterocyclic aromatics with oxygen and nitrogen as main heteroatoms; the secondary group components are alkanes with C_{16}—C_{32} and alkenes with C_{16}—C_{26}; pyrene and methylpyrenes are dominant in the condensed arenes (Huang et al. 2009).

Like the parent coal, LR is also solid at room temperature. Based on this phenomenon, in a conventional viewpoint, LR is considered to consist of unreacted organic and inorganic matters in the parent coal along with the catalyst used. Theoretically, few organic compounds in coals remain unreacted under severe conditions (temperature up to 460°C, hydrogen pressure up to 20 MPa, and in the presence of a catalyst). Taking SFSBC as an example, its extract yields in both CDS and methanol are negligible compared to the extract yields of LR from SFSBC (Figure 17.46). This fact clearly indicates that at least most of the organic matter in LR from SFSBC is different from that in SFSBC itself (Wei et al. 2011). The excellent solubility makes LR useful as a potential precursor for preparing advanced carbon materials (Wei and Fan 2010a).

LR from SFSBC was exhaustively extracted with PE to afford PEEP, which was then eluted with PE through a silica gel-filled column. As a result, a number of 4- to 7-ring

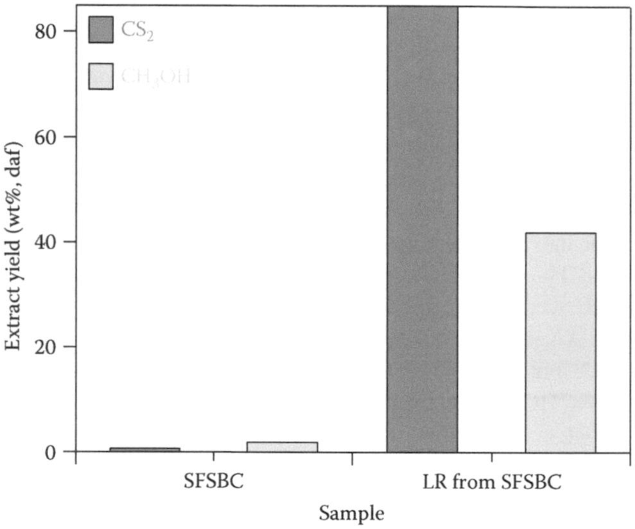

FIGURE 17.46 Extract yields of SFSBC and LR from SHSBC in CDS and methanol.

condensed arenes were enriched (Li et al. 2013). This result suggests that LR should be an important source of condensed arenes. Sequential extraction proved to be an effective approach for understanding detailed composition of LR (Li et al. 2015b). A number of nitrogen-containing homologues were detected from SFSBC-derived LR by positive-ion electrospray ionization Fourier transform ion cyclotron resonance mass spectrometry (Li et al. 2015a).

17.6 PRODUCTS FROM COAL OXIDATION AND THEIR USES

Coal oxidation is an effective method for understanding coal structures and also a very promising approach for obtaining CAs. The typical CAs from coal oxidation include alkanoic acids (AAs), alkanedioic acids (ADAs), and benzene carboxylic acids (BCAs), which are widely applied in the fine chemical field. For example, butyl butyrate from the esterification of butyric acid and butanol is usually used as a flavoring agent for candies, cookies, ice creams, and soft drinks and a solvent for shellacs, coumarone resins, and coatings (Wu et al. 2007). Succinic acid, as an intermediate for complex organic compounds, has been applied in the fields of medicine, food, synthetic plastic, rubber, protective coating, and dye (Li and Chen 1999). Adipic acid is an important organic chemical raw material and intermediate, which is mainly used to produce nylon-66, polyurethane synthetic resin, and plasticizer (Rao 2005). Compared with the aliphatic CAs, BCAs are more valuable but more difficult to be produced from the well-developed petrochemical industry. Phthalic acid and terephthalic acid are the important feedstocks for plasticizers (Zhang et al. 2007) and polyesters (Dong et al. 2008), respectively. Because of its good heat resistance, hydrolysis resistance, and chemical resistance, isophthalic acid could be used to produce polyesters, coatings, special fibers, hot melt adhesives, printing inks, and resin plasticizers (Cui 2000; Higashi et al. 2002). Trimesic acid is mainly used for producing the cross-linking agents of solid fuel as rocket boosters, alcohol-formaldehyde resins, water-soluble bakings, high-performance plastic plasticizers, and reverse osmosis membranes for water desalination (Liu et al. 2005). As a derivative of pyromellitic acid, pyromellitic dianhydride could be applied to synthesize high-performance polyimides, plasticizers, curing agents for epoxy resin, alkyd resin coatings, and matting agents (Cang and Yang 2007). As a significantly useful polydentate ligand, mellitic acid demonstrates quite flexible ligand properties during the synthetic reaction with the metal ions to form complexes (Kumagai et al. 2002; Karle et al. 2009). Moreover, the branch polyimide produced using mellitic acid dianhydride owns many particular characteristics (Wagner et al. 2006), which could be applied in the fields of aeronautics and other high technologies.

17.7 PROSPECTS FOR DIRECTIONAL COAL CONVERSION TO VALUE-ADDED PRODUCTS

Extensive investigations indicate that coals and their derivates are huge treasure houses, in which numerous value-added products are included. The most important issue is to find proper keys to the treasure houses. Previous investigations mentioned above proved the possibility for directional coal conversion to value-added products. For this purpose, developing highly active and renewable superacid and superbase is crucial for hydrocracking complex macromolecular species in coals to simple aromatics as value-added chemicals, achieving directional coal conversion under mild conditions and greatly facilitating subsequent separation (Figure 17.47). Elaborately and smartly using limited coal resources to create huge wealth should be an essential responsibility for coal chemists.

ACKNOWLEDGMENT

We are very grateful for Professor M. R. Riazi of Kuwait University for giving us such a chance to write this chapter. This chapter describes coal-based products and uses, mainly according to previous investigations of our group during the past few decades. The related investigations were financially supported in the past few decades by National Basic Research Program of China (Grants G1999022101, 2004CB217704, 2004CB217601,

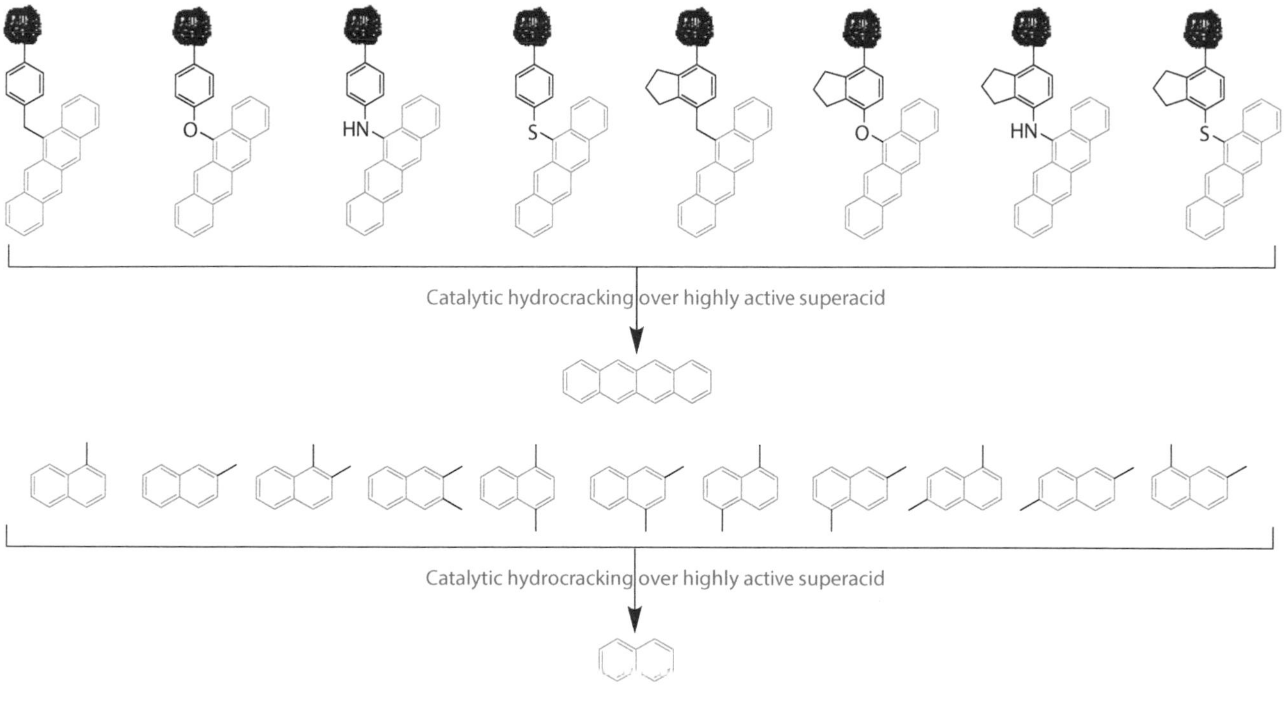

FIGURE 17.47 Examples for directionally cleaving bridged bonds connecting macromolecular moieties in coals to afford condensed arenes and for effectively eliminating side chain on aromatic rings in coal-derived small molecules to afford condensed arenes.

2005CB221204, and 2011CB201302), National High Technology Research and Development Program of China (Grant 2007AA06Z113), the fund from National Natural Science Foundation of China for Innovative Research Group (Grants 50921002 and 51221462), the Key Project of Coal Joint Fund from National Natural Science Foundation of China and Shenhua Group Corporation Limited (Grant 51134021), National Natural Science Foundation of China (Grants 29676045, 20076051, 90410018, 90510008, 20676142, 20776149, 50974121, 20936007, 51074153, 21276268, 21206187, 21206188, and 21306224), Strategic Chinese-Japanese Joint Research Program (Grant 2013DFG60060), the Fundamental Research Fund for the Doctoral Program of Higher Education (Grants 98029016, 20020290007, and 20120095110006), the Key Project of Chinese Ministry of Education (Grant 104031), Coal Science Foundation (Grant 93P410101), the fund from Chinese Ministry of Coal Industry for the Development of Science and Technology for Coal Industry (Grants 93 - 316, 96 - 306, and 97 - 312), Jiangsu Provincial Natural Science Foundation (Grants BK93136049, BK2011213, and BK20130171), the Program of the Universities in Jiangsu Province for Development of High-Tech Industries (Grant JHB05 - 33), the Fund from China Postdoctoral Science Foundation (Grants 2011M500975 and 2012T50501), the Fundamental Research Fund for the Central Universities (China University of Mining and Technology, Grants 2010ZDP02B03, 2010LKHX09, 2011QNA22, 2011QNA23, and 2014ZDPY34), and a Project Funded by the Priority Academic Program Development of Jiangsu Higher Education Institutions. We also acknowledge Professors Zhi-Hong Qin, Zhong-Hai Ni, and Ming-Jie Ding; Associate Professors Lin-Bing Sun, Zi-Wu Liu, Jing-Pei Cao, Yun-Peng Zhao, and Xiao-Yan Zhao; Lecturers Jun Zhou, Xiao-Ming Yue, and Bo Chen; Zi-Shuo Yao; Xing-Shun Cong; Rui-Lun Xie; Fang-Jing Liu; Yu-Gao Wang; Peng Li; Hai-Yun Lu; Shi-Chao Qi; Xiang Li; Zhu-Sheng Yang; Ying-Hua Wang; Li-Cheng Yu; Dong-Dong Zhang; Tie-Min Wang; Meng Qu; Wei-Tu Li; Mei-Xia Zhao; Tao-Xia Wang; Da-Ling Shi; Yin Zhou; Jing-Hui Lv; Jing Liu; Wei-Wei Ma; Lu-Lu Guo; Qing-Qing Teng; and many other members of our group for their contributions to this chapter.

REFERENCES

Bedekar, A. V., Chaudhary, A. R., Shyam Sundar, M., Rajappa, M. 2013. Expeditious synthesis of fluorinated styrylbenzenes and polyaromatic hydrocarbons. *Tetrahedron Letters* 54 (5): 392–6.

Bosnir, J., Puntaric, D., Skes, I., Klaric, M., Simic, S., Zoric, I. 2003. Migration of phthalates from plastic products to model solutions. *Collegium Antropologicum* 27 (Suppl 1): 23–30.

Cai, M. F., Smart, R. B. 1993. Quantitative analysis of *N*-methyl-2-pyrrodinone in coal extracts by TGA-FTIR. *Energy & Fuels* 7(1): 52–6.

Cang, L., Yang, X. D. 2007. Manufacture of pyromellitic dianhydride and its application in coating industry. *Paint and Coating Industry* 37 (12): 33–7.

Casagrande, V., Alvino, A., Bianco, A., Ortaggi, G., Franceschin, M. 2009. Study of binding affinity and selectivity of perylene and coronene derivatives towards duplex and quadruplex DNA by ESI-MS. *Journal of Mass Spectrometry* 44 (4): 530–40.

Chao, K. P., Hua, K. F., Hsu, H. Y. 2005. Anti-inflammatory activity of sugiol, a diterpene isolated from Calocedrus formosana bark. *Planta Medica* 71 (4): 300–5.

Chen, B., Wei, X. Y., Yang, Z. S., Liu, C., Fan, X., Qing, Y., Zong, Z. M. 2012. ReaxFF reactive force field for molecular dynamics simulations of lignite depolymerization in supercritical methanol with lignite-related model compounds. *Energy & Fuels* 26 (2): 984–9.

Chen, B., Wei, X. Y., Zong, Z. M., Yang, Z. S., Qing, Y., Liu, C. 2011a. Difference in chemical composition of the products from supercritical methanolysis of two lignites. *Applied Energy* 88 (12): 4570–6.

Chen, L., Yang, J., Liu, M. 2011b. Industrial and Engineering Chemistry Research. *Industrial and Engineering Chemistry Research* 50 (5): 2562–8.

Chervenick, S. W., Smart, R. B. 1995. Quantitative analysis of *N*-methyl-2-pyrrolidinone retained in coal extracts by thermal extraction G.C.-M.S. *Fuel* 74 (2): 241–5.

Cong, X. S., Zong, Z. M., Li, M., Gao, L., Wei, Z. H., Li, Y. et al. 2015. Enrichment and identification of cyclized hopanoids from Shengli lignite. *Fuel Processing Technology* 134: 399–403.

Cong, X. S., Zong, Z. M., Li, M., Zhou, Y., Gao, S. Q., Wei, X. Y. 2014a. Isolation and identification of two novel condensed aromatic lactones from Zhundong subbituminous coal. *Energy & Fuels* 28 (12): 7394–7.

Cong, X. S., Zong, Z. M., Wei, Z. H., Li, Y., Fan, X., Zhou, Y. et al. 2014b. Enrichment and identification of arylhopanes from Shengli lignite. *Energy & Fuels* 28 (11): 6745–8.

Cong, X. S., Zong, Z. M., Zhou, Y., Li, M., Wang, W. L., Li, F. G. 2014c. Isolation and identification of 3-ethyl-8-methyl-2,3-dihydro-1*H*-cyclopenta[*a*]chrysene from Shengli lignite. *Energy & Fuels* 28 (10): 6694–7.

Cong, X. S., Zong, Z. M., Zhou, Y., Li, M., Zhao, Y. P., Fan, X. et al. 2014d. Enrichment and analysis of long-chainnormal alkanals from Zhundong subbituminous coal. *Journal of Fuel Chemistry and Technology* 42 (3): 257–61.

Cui, X. M. 2000. Prospect on production, application and markets of isophthalic acid. *Advance in Fine Petrochemical* 1 (5): 44–6.

Dappe, Y. J., Martinez, J. I. 2013. Effect of van der Waals forces on the stacking of coronenes encapsulated in a single-wall carbon nanotube and many-body excitation spectrum. *Carbon* 54: 113–23.

Davis, M..R., Abbott, J. M., Gaines, A. F. 1985. Chemical structures of telocollinites and sporinites. 1. Differentiation between telocollinites and sporinites by the aromatic structures present in their pyridine extracts. *Fuel* 64 (10): 1362–9.

Dev, R. V., Babu, J. M., Vyas, K., Ram, P. S., Ramachandra, P., Sekhar, N. M. et al. 2006. Isolation and characterization of impurities in docetaxel. *Journal of Pharmaceutical and Biomedical Analysis* 40 (3): 614–22.

Ding, M. J., Zong, Z. M., Zong, Y., Ouyang, X. D., Huang, Y. G., Zhou, L. et al. 2008a. Group separation and analysis of a carbon disulfide-soluble fraction from Shenfu coal by column chromatography. *Journal of China University of Mining & Technology* 18 (1): 27–32.

Ding, M. J., Zong, Z. M., Zong, Y., Ou-Yang, X. D., Huang, Y. G., Zhou, L. et al. 2008b. Isolation and identification of fatty acid amides from Shengli coal. *Energy & Fuels* 22 (4): 2419–21.

Dong, Z. Z., Zhang, Z. Y., Chen, L., Wang, S. G. 2008. Research on non-isothermal crystallization kinetics of ply(ethylene terephalate). *Journal of Textile Research* 29 (3): 13–6.

Fu, X. B., Zhang, C., Zhang, D. J., Yuan, S. 2006. Theoretical evidence for the reaction of *N*-methyl-2-pyrrolidinone with carbon disulfide. *Chemical Physics Letters* 420 (1): 162–5.

Furimsky, E., Massoth, F. E. 1999. Deactivation of hydroprocessing catalysts. *Catalysis Today* 52 (4): 381–495.

Furimsky, E., Massoth, F. E. 2005. Hydrodenitrogenation of petroleum. *Catalysis Reviews—Science and Engineering* 47 (3): 297–489.

Gama, V., Henriques, R. T., Bonfait, G., Almeida, M., Meetsma, A., Van Smaalen, S. et al. 1992. (Perylene)Co(mnt)$_2$(CH$_2$Cl$_2$)$_{0.5}$: A mixed perylenecobalt complex as molecular and polymeric conductor. *Journal of the American Chemical Society* 114 (6): 1986–9.

Gao, H., Nomura, M., Murata, S., Artok, L. 1999. Statistical distribution characteristics of pyridine transport in coal particles and a series of new phenomenological models for overshoot and nonover shoot solvents welling of coal particles. *Energy & Fuels* 13 (2): 518–28.

Ghosh, A., Rao, K. V., George, S. J., Rao, C. N. 2010. Noncovalent functionalization, exfoliation, and solubilization of graphene in water by employing a fluorescent coronene carboxylate. *Chemistry* 16 (9): 2700–4.

Grice, K., Nabbefeld, B., Maslen, E. 2007. Source and significance of selected polycyclicaromatic hydrocarbons in sediments (Hovea-3 well, Perth Basin, Western Australia) spanning the Permian-Triassic boundary. *Organic Geochemistry* 38 (11): 1795–803.

Harvey, R. G. 1991. *Polycyclic Aromatic Hydrocarbons: Chemistry and Carcinogenicity*, Cambridge University Press, Cambridge.

Higashi, F., Hayashi, R., Yamazaki, T. 2002. Solution copolycondensation of isophthalic acid, terephthalic acid, 4,4′-dihydroxydiphenylsulfone, and bisphenols with a tosyl chloride/dimethylformamide/pyridine condensing agent. *Journal of Applied Polymer Science* 86 (10): 2607–10.

Hiramoto, M., Kishigami, Y., Yokoyama, M. 1990. Doping effect on the two-layer organic solar cell. *Chemistry Letters* 19 (1): 119–22.

Huang, Y., Wei, X. Y., Zhang, M. M., Lu, Y., Li, Y., Chen, F. J. et al. 2009. Composition of a heavy oil from direct liquefaction of Xinjiang Heishan coal. *Journal of Wuhan University of Science and Technology* 32 (5): 532–7.

Huggins, F. E., Huffman, G. P. 1995. Chlorine in coal: An XAFS spectroscopic investigation. *Fuel* 74 (4): 556–69.

Iino, M., Kumagai, J., Ito, O. 1985. Coal extraction with carbondisulfide mixed solvent at room temperature. *Journal of Fuel Society of Japan* 64 (3): 210–12.

Iino, M., Li, Q. T., Wei, X. Y. 1987. Extraction of coals with mixed solvents at room temperature. *Report of Asahi Glass Found Industrial Technology* 51: 121–8.

Iino, M, Takanohashi, T, Ohsuga, H, Toda, K. 1988. Extraction of coals with CS$_2$-*N*-methyl-2-pyrrolidinone mixed solvent at room temperature: Effect of coal bank and synergism of the mixed solvent. *Fuel* 67 (12): 1639–47.

Kambara, S., Takarada, T., Toyoshima, M., Kato, K. 1995. Relation between functional forms of coal nitrogen and NO$_x$ emissions from pulverized coal combustion. *Fuel* 74 (9): 1247–53.

Karle, I. L., Rajesh, Y. B., Ranganathan, S. 2009. Crystal engineering: A unique cyclic assembly of a 40 membered module composed from two alternating units each of benzenehexacarboxylic acid (mellitic acid, MA) and 2,5-bis-(4-pyridyl)-1,3,4-oxadiazole (4-BPO): Assembly of modules to macromolecules by intermolecular hydrogen bonding. *Journal of Chemical Crystallography* 39 (3): 201–8.

Kavlock, R., Boekelheide, K., Chapin, R., Cunningham, M., Faustman, E., Foster, P. et al. 2002. NTP center for the evaluation of risks to human reproduction: Phthalates expert panel report on the reproductive and developmental toxicity of di(2-ethylhexyl) phthalate. *Reproductive Toxicology* 16 (5): 529–53.

Kumagai, H., Oka, Y., Akita-Tanaka, M., Inoue, K. 2002. Hydrothermal synthesis and characterization of a two-dimensional nickel(II) complex containing benzenehexacarboxylic acid (mellitic acid). *Inorganica Chimica Acta* 332 (1): 176–80.

Lee, S., Kim, S., Woo, K., Jeong, S., Rhim, Y., Cho, H. 2007. Ashless coal preparation by using solvent extraction. In *Proceedings of the 24th Annual International Pittsburgh Coal Conference*, vol. 3, 1724–30, Johannesburg, South Africa, September 10–14, 2007.

Lehmann, K. P., Phillips, S., Sar, M., Foster, P. M., Gaido, K. W. 2004. Dose-dependent alterations in gene expression and testosterone synthesis in the fetal testes of male rats exposed to di(n-butyl)phthalate. *Toxicological Sciences* 81 (1): 60–8.

Li, C. L., Chen, Z. X. 1999. Preparation and uses of succinic acid. *Journal of Qinghai University* 17 (6): 193–6.

Li, P., Wei, X. Y., Sun, X. H., Lu, Y., Zong, Z. M., Mukasa, R. et al. 2013. The isolation of condensed arenes from Shenmu-Fugu coal liquefaction residue. *Energy Sources, Part A: Recovery, Utilization, and Environmental Effects* 35 (23): 2250–6.

Li, P., Zong, Z. M., Li, Z. K., Wang, Y. G., Liu, F. J., Wei, X. Y. 2015a. Characterization of basic heteroatom compounds in liquefaction residue from Shenmu-Fugu subbituminous coal by positive-ion electrospray ionization Fourier transform ion cyclotron resonance mass spectrometry. *Fuel Processing Technology* 132: 91–8.

Li, P., Zong, Z. M., Liu, F. J., Wang, Y. G., Wei, X. Y., Fan X. et al. 2015b. Sequential extraction and characterization of liquefaction residue from Shenmu-Fugu subbituminous coal. *Fuel Processing Technology* 136: 1–7.

Li, Z. K., Zong, Z. M., Yan, H. L., Wang, Y. G., Ni, H. X., Wei, X. Y. et al. 2014a. Characterization of acidic species in ethanol-soluble portion from Zhaotong lignite ethanolysis by negative-ion electrospray ionization Fourier transform ion cyclotron resonance mass spectrometry. *Fuel Processing Technology* 128: 297–302.

Li, Z. K., Zong, Z. M., Yan, H. L., Wang, Y. G., Wei, X. Y., Shi, D. L. et al. 2014b. Alkanolysis simulation of lignite-related model compounds using density functional theory. *Fuel* 120: 158–62.

Li, Z. K., Zong, Z. M., Yang, Z. S., Yan, H. L., Fan, X., Wei, X. Y. 2014c. Sequential thermal dissolution of Geting bituminous coal in low-boiling point solvents. *Energy Sources, Part A: Recovery, Utilization, and Environmental Effects* 36 (23): 2579–86.

Liang, S., Liang, Y., He, J. T., Ito, Y. 2012. Separation and purification of three flavonoids from Daphne GenkwaSieb. Et Zucc.: Comparison in performance between medium-pressure liquid chromatography and high-speed countercurrent chromatography. *Journal of Liquid Chromatography & Related Technologies* 35 (18): 2610–22.

Liu, F. J., Wei, X. Y., Gui, J., Wang, Y. G., Li, P., Zong, Z. M. 2013. Characterization of biomarkers and structural features of condensed aromatics in Xianfeng lignite. *Energy & Fuels* 27 (12): 7369–78.

Liu, F. J., Wei, X. Y., Wang, Y. G., Li, P., Li, Z. K., Zong, Z. M. 2015. Sulfur-containing species in the extraction residue from Xianfeng lignite characterized by X-ray photoelectron spectrometry and electrospray ionization Fourier transform ion cyclotron resonance mass spectrometry. *RSC Advances* 5 (10): 7125–30.

Liu, F. J., Wei, X. Y., Xie, R. L., Wang, Y. G., Li, W. T., Li, Z. K. et al. 2014. Characterization of oxygen-containing species in methanolysis products of extraction residue from Xianfeng lignite with negative-ion electrospray ionization Fourier transform ion cyclotron resonance mass spectrometry. *Energy & Fuels* 28 (9): 5596–605.

Liu, Z. W., Wei, X. Y., Zong, Z. M., Li, J. N., Xue, J. Q., Chen, X. F. et al. 2010. Isolation and identification of methyl alkanoates from Lingwu coal. *Energy & Fuels* 24 (4): 2784–6.

Liu, C. H., Zhang, C. Z., Zhang, X. B., Cai, W. F., Wang, Y., Xin, F. 2005. Synthesis of trimesic acid. *Chemical Industry and Engineering* 22 (2): 126–9.

Liu, C. M., Zong, Z. M., Jia, J. X., Liu, G. F., Wei, X. Y. 2008. An evidence for the strong association of *N*-methyl-2-pyrrolidinone with some organic species in three Chinese bituminous coals. *Chinese Science Bulletin* 53 (8): 1157–64.

Liu, Z. W., Zong, Z. M., Li, J. N., Chen, C. F., Jiang, H., Peng, Y. L. et al. 2009. Isolation and identification of two bis(2-ethylheptyl) benzenedicarboxylates from Lingwu coal. *Energy & Fuels* 23 (1): 588–90.

Lu, H. Y., Wei, X. Y., Yu, R., Peng, Y. L., Qi, X. Z., Qie, L. M. et al. 2011. Sequential thermal dissolution of Huolinguole lignite in methanol and in ethanol. *Energy & Fuels* 25 (6): 2741–5.

Lucenti, E., Botta, C., Cariati, E., Righetto, S., Scarpellini, M., Tordin, E. et al. 2013. New organic-inorganic hybrid materials based on perylene diimide-polyhedral oligomeric silsesquioxane dyes with reduced quenching of the emission in the solid state. Dyes and Pigments 96 (3): 748–55.

Lv, J. H., Wei, X. Y., Wang, Y. H., Yu, L. C., Zhang, D. D., Yue, X. M. et al. 2015. Light fraction from catalytic hydroconversion of two Chinese coals in cyclohexane over a solid acid. *Fuel Processing Technology* 129: 162–7.

Ma, H. Y., Wang, C. H., Yang, L., Zhang, M., Wang, Z. T. 2009. Chemical constituents of Senecio cannabifolius var. integrilifolius. *Chinese Journal of Natural Medicines* 7 (1): 28–30.

Masaki, K., Yoshida, T., Li, C., Takanohashi, T., Saito, I. 2004. The effects of pretreatment and the addition of polar compounds on the production of "HyperCoal" from subbituminous coals. *Energy & Fuels* 18 (4): 995–1000.

Mochida, I., Kinya, S. 1994. Advances in Catalysis, Academic Press, New York.

Ni, Z. H., Zong, Z. M., Zhang, L. F., Sun, L. B., Liu, Y., Yuan, X. H. et al. 2003. Synergic effect of sulfur on activated carbon-catalyzed hydrocracking of di(1-naphthyl)methane. *Energy & Fuels* 17 (1): 60–1.

Ni, Z. H., Zong, Z. M., Zhang, L. F., Zhou, S. L., Xiong, Y. C., Wang, X. H. et al. 2002. Reactivities of di(1-naphthyl)methane and hydrogenated di(1-naphthyl)methane toward hydrocracking over Ni-S. *Energy & Fuels* 16 (5): 1154–9.

Obiri, S., Cobbina, S. J., Armah, F. A., Luginaah, I. 2013. Assessment of cancer and noncancer health risks from exposure to PAHs in street dust in the Tamale Metropolis, Ghana. *Journal of Environmental Science and Health Part a-Toxic/Hazardous Substances & Environmental Engineering* 48 (4): 408–16.

Ohshima, Y., Wang, Y., Tsubouchi, N., Ohtsuka, Y. 2000. Approach to the iron-catalyzed formation process of N_2 from heterocyclic nitrogen in carbon by use of XRD and XPS methods. *Preprint Papers-American Chemical Society, Division of Fuel Chemistry* 45: 335.

Olmsted III, J. 1974. Oxygen quenching of fluorescence of organic dye molecules. *Chemical Physics Letters* 26 (1): 33–6.

Opaprakasit, P., Painter, P. C. 2004. Swelling of clays in *N*-methyl-2-pyrrolidinone/carbon disulfide mixed solvents. *Energy & Fuels* 18 (6): 1704–8.

Pittelkow, M., Lewinsky, R., Christensen, J. B. 2002. Selective synthesis of carbamate protected polyamines using alkyl phenyl carbonates. *Synthesis* (15): 2195–202.

Qin, Z. H., Chen, H., Yan, Y. J., Li, C. S., Rong, L. M., Yang, X. Q. 2015. FTIR quantitative analysis upon solubility of carbon disulfide/*N*-methyl-2-pyrrolidinone mixed solvent to coal petrographic constituents. *Fuel Processing Technology* 133: 14–19.

Qin, Z. H., Yuan, X. H., Yin, X. Q., Zong, Z. M., Wei, X. Y. 1998. Effects of macerals on the solubilities of some coals in CS_2-NMP mixed solvent. *Journal of China University of Mining & Technology* 27(4): 344–8.

Qin, Z. H., Zong, Z. M., Liu, J. Z., Ma, H. M., Yang, M. J., Wei, X. Y. 1997. Solubilities of litho types in carbon disulfide/*N*-methyl-2-pyrrolidinone mixed solvent. *Journal of Fuel Chemistry and Technology* 25 (6): 549–53.

Qiu, C. Y., Peng, B., Cheng, S. Q., Xia, Y. Y., Tu, B. J. 2013. The effect of occupational exposure to benzo a pyrene on neurobehavioral function in coke oven workers. *American Journal of Industrial Medicine* 56 (3): 347–55.

Rahman, M., Brazel, C. S. 2004. The plasticizer market: An assessment of traditional plasticizers and research trends to meet new challenges. *Progress in Polymer Science* 29 (12): 1223–48.

Rao, X. H. 2005. Production and demand of adipic acid and its production advance in the world. *China Petroleum and Chemical Industry* (7): 70–3.

Ren, D., Zhao, F., Wang, Y., Yang, S. 1999. Distributions of minor and trace elements in Chinese coals. *International Journal of Coal Geology* 40 (2–3): 109–18.

Sato, H., Kikumori, C., Sakaki, S. 2011. Solvation structure of coronene-transition metal complex: a RISM-SCF study. *Physical Chemistry Chemical Physics* 13 (1): 309–13.

Schmidt, C. D., Lang, N. N., Jux, N., Hirsch, A. 2011. A facile route to water-soluble coronenes and benzo[ghi]perylenes. *Chemistry-a European Journal* 17 (19): 5289–99.

Shao, D., Hutchinson, E. J., Cao, H., Pan, W. P., Chou, C. L. 1994. Behavior of chlorine during coal pyrolysis. *Energy & Fuels* 8 (2): 399–401.

Sharma, V., Walia, S., Kumar, J., Nair, M. G., Parmar, B. S. 2003. An efficient method for the purification and characterization of nematicidal azadirachtins A, B, and H, using MPLC and ESIMS. *Journal of Agricultural and Food Chemistry* 51 (14): 3966–72.

Shi, D. L., Wei, X. Y., Chen, B., Zong, Z. M. 2015. Enrichment of condensed arenes and dialkyl phthalates in the extracts from Geting bituminous coal. *International Journal of Oil, Gas and Coal Technology* in press.

Shi, D. L., Wei, X. Y., Fan, X., Zong, Z. M., Chen, B., Zhao, Y. P. et al. 2013. Characterizations of the extracts from Geting bituminous coal by spectrometries. *Energy & Fuels* 27 (7): 3709–17.

Shu, X. Q., Wang, Z. N., Xu, J. Q. 2002. Separation and preparation of macerals in Shenfu coals by flotation. *Fuel* 81 (3): 495–501.

Shui, H., Wang, Z., Gao, J. 2006. Examination of the role of CS_2 in the CS_2/NMP mixed solvents to coal extraction. *Fuel Processing Technology* 87 (3): 185–90.

Sönmez, O., Giray, E. S. 2011. Producing ashless coal extracts by microwave irradiation. *Fuel* 90 (6): 2125–31.

Street, K. W., Acree, W. E., Fetzer, J. C., Shetty, P. H., Poole, C. F. 1989. Polycyclic aromatic hydrocarbon solute probes. Part V: Fluorescence spectra of pyrene, ovalene, coronene, and benzo[*ghi*]perylene dissolved in liquid alkylammonium thiocyanate organic salts. *Applied Spectroscopy* 43 (7): 1149–53.

Sugano, M., Ikemizu, R., Mashimo, K. 2002. Effects of the oxidation pretreatment with hydrogen peroxide on the hydrogenolysis reactivity of coal liquefaction residue. *Fuel Processing Technology* 77–78 (Suppl 1):67–73.

Sun, L. B., Wei, X. Y., Liu, X. Q., Zong, Z. M., Li, W. 2009a. Release of organonitrogen and organosulfur compounds during hydrotreatment of Pocahontas (3 coal residue over an activated carbon. *Energy & Fuels* 23 (10): 5284–6.

Sun, L. B., Wei, X. Y., Liu, X. Q., Zong, Z. M., Li, W., Kou, J. H. 2009b. Selective hydrogen transfer to anthracene and its derivatives over an activated carbon. *Energy & Fuels* 23 (10): 4877–82.

Sun, L. B., Zong, Z. M., Kou, J. H., Cao, J. P., Yu, G. Y., Zhao, W. et al. 2007. Identification of organic chlorines and iodines in the extracts from hydrotreated Argonne Premium coal residues. *Energy & Fuels* 21 (4): 2238–9.

Sun, L. B., Zong, Z. M., Kou, J. H., Liu, G. F., Sun, X., Wei, X. Y. et al. 2005a. Activated carbon-catalyzed hydrogen transfer to α,ω-diarylalkanes. *Energy & Fuels* 19 (1): 1–6.

Sun, L. B., Zong, Z. M., Kou, J. H., Yu, G. Y., Chen, H., Liu, C. C. et al. 2005b. Thermal release and catalytic removal of organic sulfur compounds from Upper Freeport coal. *Energy & Fuels* 19 (2): 339–42.

Suzuki, K., Takao, K., Sato, S., Fujita, M. 2010. Coronene nanophase within coordination spheres: Increased solubility of C_{60}. *Journal of the American Chemical Society* 132 (8): 2544–5.

Tomita, A. 2001. Suppression of nitrogen oxides emission by carbonaceous reductants. *Fuel Processing Technology* 71 (1–3): 53–70.

Tyutyulkov, N., Karabunarliev, S., Müllen, K., Baumgarten, M. 1992. A class of narrow-band high-spin organic polymers II. Polymers with indirect exchange interaction. *Synthetic Metals* 52 (1): 71–85.

Vassilev, S. V., Eskenazy, G. M., Vassileva, C. G. 2000. Contents, modes of occurrence and origin of chlorine and bromine in coal. *Fuel* 79 (8): 903–21.

Verhofstad, N., van Oostrom, C. T. M., Zwart, E., Maas, L. M., van Benthem, J., van Schooten, F. J. et al. 2011. Evaluation of benzo(a)pyrene-induced gene mutations in male germ cells. *Toxicological Sciences* 119 (1): 218–23.

Vincent-Hubert, F., Arini, A., Gourlay-France, C. 2011. Early genotoxic effects in gill cells and haemocytes of *Dreissena polymorpha* exposed to cadmium, B[*a*]P and a combination of B[a]P and Cd. *Mutation Research-Genetic Toxicology and Environmental Mutagenesis* 723 (1): 26–35.

Wagner, S., Dai, H., Stapleton, R. A., Illingsworth, M. L., Siochi, E. J. 2006. Pendent polyimides using mellitic acid dianhydride. I. An atomic oxygen-resistant, pendent 4,4′-ODA/PMDA/MADA copolyimide containing zirconium. *High Performance Polymers* 18 (4): 399–419.

Wang, B., Wei, X., Xie, K. 2004. Study on reaction of *N*-methylpyrrolidine-2-thione with carbon disulfide using density functional theory. *Journal of Chemical Industry and Engineering (China)* 55 (4): 569–74.

Wang, H., Xu, X. Q., Shi, J. F., Xu, G. 2013. Application of ionic liquids with carboxyl and aromatic ring conjugated anions in dye-sensitized solar cells. *Acta Physico-Chimica Sinica* 29 (3): 525–32.

Wang, X. H., Zong, Z. M., Qin, Z. H., Wei, X. Y., He, L. T. 2001. The progress of organic spectroscopy applied to molecular coal chemistry. *Coal Conversion* 24 (1): 5–10.

Wei, X. Y. 2014. Molecular coal chemistry and fine coal chemical technology. In Keynote speech at *International Workshop on Clean Technologies of Coal and Biomass Utilization*. Anshan, China.

Wei, X. Y., Fan, M. H. 2010a. Advances in basic research and technology development of coal liquefaction. In *Clean Coal China Conference 2010*. Beijing, China, November 17, 2010.

Wei, X. Y., Gu, X. H., Zong, Z. M., Qin, Z. H., Wu, L., Wang, X. H. 1999a. Scientific basis of coal used as chemicals-separable and non-destructive analysis of the molecular structure of organic matter in coal. In *Proceedings of the 6th Symposium on Applied Chemistry of Chinese Chemical Society*, vol. 1, 77–81, Changzhou, China.

Wei, X. Y., Ni, Z. H., Xiong, Y. C., Zong, Z. M., Wang, X. H., Cai, C. W. et al. 2002. Pd/C-catalyzed release of organonitrogen compounds from bituminous coals. *Energy & Fuels* 16 (2): 527–8.

Wei, X. Y., Ni, Z. H., Zong, Z. M., Zhou, S. L., Xiong, Y. C., Wang, X. H. et al. 2003. Reaction of di(1-naphthyl)methane over metals and metal-sulfur systems. *Energy & Fuels* 17 (3): 652–7.

Wei, X. Y., Ogata, E., Futamura, S., Kamiya, Y. 1990. Thermal decomposition and hydrocracking of hydrogenated di(1-naphthyl)methane. *Fuel Processing Technology* 26 (2): 135–48.

Wei, X. Y., Ogata, E., Niki, E. 1991. Catalyses of Fe and FeS_2 on the reaction of di(1-naphthyl)methane. *Chemistry Letters* (12): 2199–202.

Wei, X. Y., Ogata, E., Niki, E. 1992a. FeS_2-catalyzed hydrocracking of α,ω-diarylalkanes. *Bulletin of the Chemical Society of Japan* 65 (4): 1114–19.

Wei, X. Y., Ogata, E., Niki, E. 1992b. FeS_2-catalyzed hydrocracking of diarylmethanes. *Sekiyu Gakkaishi* 35 (4): 358–61.

Wei, X. Y., Ogata, E., Niki, E. 1992c. FeS_2-catalyzed hydrocracking of di(1-naphthyl)methane. *Bulletin of the Chemical Society of Japan* 65 (4): 987–90.

Wei, X. Y., Ogata, E., Zong, Z. M., Niki, E. 1992d. Effects of hydrogen pressure, sulfur and FeS_2 on diphenylmethane hydrocracking. *Energy & Fuels* 6 (6): 868–9.

Wei, X. Y., Ogata, E., Zong, Z. M., Niki, E. 1993. Effects of iron catalyst precursors, sulfur, hydrogen pressure and solvent type on the hydrocracking of di(1-naphthyl)methane. *Fuel* 72 (11): 1547–52.

Wei, X. Y., Ogata, E., Zong, Z. M., Niki, E. 1995a. Inhibiting effects of hydrogen-donating compounds on 1,3-diphenylpropane thermolysis. *Coal Conversion* 18 (1): 67–70.

Wei, X. Y., Ogata, E., Zong, Z. M., Niki, E. 1995b. Promotional effects of molecular hydrogen and pyrite on 1,3-diphenylpropane thermolysis. *Journal of Fuel Chemistry and Technology* 23 (3): 231–5.

Wei, X. Y., Shen, J. L., Takanohashi, T., Iino, M. 1989. Effect of extractable substances on coal dissolution. Use of CS_2-*N*-methyl-2-pyrrolidinone mixed solvent for dissolution reaction products. *Energy & Fuels* 3 (5): 575–9.

Wei, X. Y., Wang, X. H., Zong, Z. M. 2009. Extraction of organonitrogen compounds from five Chinese coals with methanol. *Energy & Fuels* 23 (10): 4848–51.

Wei, X. Y., Wang, X. H., Zong, Z. M., Ni, Z. H., Zhang, L. F., Ji, Y. F. 2004. Identification of organochlorines and organobromines in coals. *Fuel* 83 (17–18): 2435–8.

Wei, X. Y., Yue, X. M., Sun, B., Liu, Z. W., Wang, Y. H., Zong, Z. M. 2011. Role of active species in direct coal liquefaction. In *The 6th Sino-US Joing Conference of Chemical Engineering*, November 7–10, 2011, Beijing, China (Keynote speech).

Wei, X. Y., Zong, Z. M. 1992e. Solvent effect on diphenylmethane hydrocracking. *Energy & Fuels* 6 (2): 236–7.

Wei, X. Y., Zong, Z. M., Qin, Z. H., Ji, Y. F., Liu, J. Z., Wu, L. et al. 1999b. Ideas and prospects of molecular coal chemistry. In *Proceedings of the 2nd Symposium on Chemical, Metallurgical and Materials Engineering Division of Chinese Academy of Engineering*, 623–8, Beijing, China.

Wei, X. Y., Zong, Z. M., Zhao, W., Li, B. M., Ni, Z. H., Sun, L. B. et al. 2010b. Molecular coal chemistry: scientific basis for efficient utilization of coal resources. In *BIT's 1st Annual World Congress of Well Simulation and EOR*, Chengdu, China.

Wei, X. Y., Zong, Z. M., Zhou, S. L., Ni, Z. H., Sun, L. B., Ma, Y. M. 2009. Monatomic and biatomic hydrogen transfer to coal-related model compounds. In *Proceedings of the 10th Japan-China Symposium on Coal and C1 Chemistry*, July 26–29, 2009, Tsukuba, Japan.

Wu, Y. H., Xiang, K. X., Hou, X. J., Wu, W. L. 2007. Progress of the development of catalytic synthesis of butyl butyrate in China. *China Surfactant Detergent & Cosmetics* 37 (3): 193–6.

Yoshida, T., Sasaki, M., Hirata, F., Kawamani, Y., Inazu, K., Ishikawa, A. et al. 2005. Highly selective methoxycarbonylation of aliphatic diamines withmethyl phenyl carbonate to the corresponding methyl *N*-alkyl dicarbamates. *Applied Catalysis A: General* 289 (2): 174–8.

Yu, C. C., Jiang, K. J., Huang, J. H., Zhang, F., Bao, X., Wang, F. W. et al. 2013. Novel pyrene-based donor–acceptor organic dyes for solar cell application. *Organic Electronics* 14 (2): 445–50.

Yu, L. C., Wei, X. Y., Wang, Y. H., Wen, Z., Zhang, D. D., Zong, Z. M. et al. 2014a. Catalytichydroconversion of Lingwu bituminous coal over Fe-S/γ-Al_2O_3. *International Journal of Oil, Gas and Coal Technology* 7 (4): 415–30.

Yu, L. C., Wei, X. Y., Wang, Y. H., Zhang, D. D., Wen, Z., Zong, Z. M. 2014b. Catalytic hydroconversion of extraction residue from Shengli lignite over Fe-S/ZSM-5. *Fuel Processing Technology* 126: 131–7.

Yuan, X. H., Xiong, Y. C., Zong, Z. M., Qin, Z. H., Wei, X. Y. 2001. Relationship between molecular coal chemistry and directional conversion of coal derivates. *Coal Conversion* 24 (1): 1–4.

Yue, X. M., Wei, X. Y., Sun, B., Wang, Y. H., Zong, Z. M., Fan, X. et al. 2012a. A new solid acid for specifically cleaving the C_{ar}—C_{alk} bond in di(1-naphthyl)methane. *Applied Catalysis A: General* 425–426: 79–84.

Yue, X. M., Wei, X. Y., Sun, B., Wang, Y. H., Zong, Z. M., Liu, Z. W. 2012b. Solid superacid-catalyzed hydroconversion of an extraction residue from Lingwu bituminous coal. *International Journal of Mining Science and Technology* 22 (2): 251–4.

Zhang, L., Kawashima, H., Takanohashi, T., Nakazato, T., Saito, I., Tao, H. 2008. Partitioning of boron during the generation of ultraclean fuel (HyperCoal) by solvent extraction of coal. *Energy & Fuels* 22 (2): 1183–90.

Zhang, W., Xu, Z., Pan, B., Lv, L., Zhang, Q., Zhang, Q. et al. 2007. Assessment on the removal of dimethyl phthalate from aqueous phase using a hydrophilic hyper-cross-linked polymer resin NDA-702. *Journal of Colloid and Interface Science* 311 (2): 382–90.

Zhang, D. D., Zong, Z. M., Liu, J., Wang, Y. H., Yu, L. C., Lv, J. H. et al. 2015. Catalytic hydroconversion of Geting bituminous coal over FeNi–S/γ-Al_2O_3. *Fuel Processing Technology* 133: 195–201.

Zhao, Y., Zhang, Z. Y. 1996. Reactivity of alcohols toward the phosphoenzyme intermediate in the protein-tyrosine phosphatase-catalyzed reaction: Probing the transition state of the dephosphorylation step. *Biochemistry* 35 (36): 11797–804.

Zhao, X. Y., Zong, Z. M., Cao, J. P., Ma, Y. M., Han, L., Liu, G. F. et al. 2008. Difference in chemical composition of carbon disulfide-extractable fraction between vitrinite and inertinite from Shenfu-Dongsheng and Pingshuo coals. *Fuel* 87 (4–5): 565–75.

Zhou, J., Zong, Z. M., Chen, B., Yang, Z. S., Li, P., Lu, Y. et al. 2013a. The enrichment and identification of methyl alkanones from thermally soluble Shengli lignite. *Energy Sources, Part A: Recovery, Utilization, and Environmental Effects* 35 (23): 2218–24.

Zhou, J., Zong, Z. M., Fan, X., Zhao, Y. P., Wei, X. Y. 2013b. Separation and identification of organic compounds from thermally dissolved Shengli lignite in a methanol/benzene mixed solvent. *International Journal of Oil, Gas and Coal Technology* 6 (5): 517–27.

Zhou, Y., Zong, Z. M., Zhao, Y. P., Cong, X. S., Fan, X., Dou, Y. Q., Sun, X. H., Wei, X. Y. 2014. Enrichment of oxygen-containing aromatics in an extract from Zhundong subbituminous coal. *International Journal of Oil, Gas and Coal Technology* 8 (3): 325–35.

Zong, Z. M., Peng, Y. L., Liu, Z. G., Zhou, S. L., Wu, L., Wang, X. H. et al. 2003. Convenient synthesis of *N*-methylpyrrolidine-2-thione and some thioamides. *Korean Journal of Chemical Engineering* 20 (2): 235–8.

Zong, Z. M., Peng, Y. L., Qin, Z. H., Liu, J. Z., Wu, L., Wang, X. H. et al. 2000. Reaction of *N*-methyl-2-pyrrodinonewith carbon disulfide. *Energy & Fuels* 14 (3): 734–5.

Zong, Z. M., Wei, X. Y. 1994. Effects of molecular hydrogen and hydrogen donor additives on 1,2-di(1-naphthyl)ethane thermolysis. *Fuel Processing Technology* 41 (1): 79–85.

Zong, Y., Zong, Z. M., Ding, M. J., Zhou, L., Huang, Y. G., Zheng, Y. X. et al. 2009. Separation and analysis of organic compounds in an Erdos coal. *Fuel* 88 (3): 469–74.

18 Disposal and Utilization of Coal Combustion, Gasification, and Coking Residues

L. Reijnders

CONTENTS

Abstract: Solid residues (ashes, gypsum, and chars) and semi-solid residues (tars) from coal combustion, coal gasification, and coking are discussed. Available data about their composition and volumes are presented. Coal-derived residues are in part disposed of and in part utilized. Main applications of these residues are in geotechnical works and building materials. Disposal and utilization of coal-derived residues may have workplace and environmental impacts. Workplace exposure to fly ashes and tars may be hazardous. Negative environmental impacts may be associated with a variety of substances leached from, or emitted by, coal-derived residues, including polycyclic aromatics, radon, and mercury.

18.1 INTRODUCTION

This chapter deals with solid residues and semi-solid residues (tars) from coal combustion, coal gasification, and coking. Section 18.2 discusses the generation of these residues, or by-products, and their definitions as used in this chapter. In Section 18.3, available data about composition and volumes of specified residues are outlined. Section 18.4 will deal with disposal and use of residues. The workplace and environmental impacts of coal combustion, gasification, and coking residues will be discussed in Section 18.5.

18.2 COAL COMBUSTION, GASIFICATION, AND COKING RESIDUES: GENERATION AND DEFINITIONS

Large volumes of coal are currently combusted, gasified, or coked (carbonized), and this gives rise to large amounts of residues.

Large-scale coal combustion serves mainly the generation of power, and it is expected that coal-based power generation will increase (Sarker et al. 2012). In power plants, co-combustion of pulverized coal with other fuels is also on the increase (Barnes 2010; Duan et al. 2012). Apart from large-scale coal combustion, there is also relatively small-scale *distributed* combustion of coal. The worldwide amount of ashes generated by distributed coal combustion is not documented. It is likely to be quite substantial, however, as, for example, in China distributed coal combustion in industry accounted for about 30% of overall coal consumption in 2009 (Hongjui et al. 2009). The residues of distributed coal combustion will not be considered here. The worldwide production of solid residues by large-scale plants that combust or co-combust coal has been estimated at 780 Mt (10^{12} g) in 2010 (Heidrich et al. 2013). These solid residues include fly ash including cenospheres, bottom ash, slag, flue-gas desulfurization residues such as gypsum, wastewater filtercake, fluidized bed combustion residue or ash, and mixtures thereof (e.g., Kosson et al. 2009; Hulburt et al. 2012). The definitions of these residues, as they are used in this chapter, are given in Table 18.1.

Coal gasification is on a smaller scale than coal combustion but is also expected to grow (Higman 2013). It may serve the production of fuels, chemicals, and power (Higman 2013; Higman and Tam 2014). Co-gasification of coal with other fuels (e.g., residues of the petrochemical industry, biomass, waste tires, sludge, and municipal waste) has been reported (Miccio et al. 2012; Higman and Tam 2014). Solid and semi-solid residues of coal gasification include fly ash, fluidized bed gasification ash, char, desulfurization residues, wastewater filtercake, and tar (Slaghuis et al. 1995; Pindoria et al. 1997; Seshadri and Shamsi 1998; Brage et al. 2000; Mahlaba 2006; Pinto et al. 2009; Meng et al. 2010; Galhetas et al. 2012; Li and Whitty 2012; Miccio et al. 2012; Emami-Taba et al. 2013; Kronbauer et al. 2013; Li et al. 2014; Higman and Tam 2014). For definitions of these residues as used in this chapter, see Table 18.1. The worldwide production volume of coal gasification residues is as yet not documented in the open literature.

TABLE 18.1
Types of Coal Combustion, Gasification, and Coking Residues and Their Definitions

Type of Residue	Definition Used in This Chapter
Bottom ash	Relatively coarse ash present at the base of coal combustion furnaces
Cenospheres	Hollow fly ash particles
Filtercake from wastewater treatment	Solids from wastewater treatment facilities of coal combustion and gasification plants
Fluidized bed combustion (boiler) or gasification ash	Ash from fluidized bed combustion or gasification, containing bed material
Fly ash	Relatively fine ash present in flue gas, as recovered by air pollution control devices (electrostatic precipitation, fabric filter)
Fly ash nanoparticles	Fly ash particles with a diameter <100 nm in at least one dimension
Flue-gas gypsum	Reaction product of limestone and sulfur oxides in flue gas
Slag (e.g., boiler slag)	Molten ash (e.g., from boilers)
Tar	Semi-solid or viscous liquid with high levels of cyclic organic compounds originating in coal gasification and coking

Coking of coal (heating at high temperatures while excluding oxygen) is largely linked to iron and steel production in blast furnaces and to the use of furnaces in foundries. Major residues of coking are coal tar and ashes (IARC 2012; Mu et al. 2012). In coking worldwide, coal tar is produced at a level of about 3%–4.5% of coke production, which would amount to 15–22.5 Mt (calculated on the basis of data in European Commission 2001; Li and Suzuki 2010; Jones 2011; Li et al. 2011).

18.3 COMPOSITION OF SPECIFIED COAL COMBUSTION, GASIFICATION, AND COKING RESIDUES AND THEIR VOLUMES

18.3.1 Ashes from Coal Combustion

The total worldwide production of coal combustion ashes from power plants may presently be estimated at over 700 Mt (extrapolated from Heidrich et al. 2013). About 80%–85% of the ashes from coal combustion–based power plants come in the category fly ash and the rest in the category bottom ash (Heidrich et al. 2013). Coal ashes from power plants have widely varying compositions. Most pulverized coal combustion ashes are alkaline, but some are acidic (Ram and Masto 2010). Pulverized coal combustion ashes predominantly consist of inorganic substances (see Table 18.2 for the elements involved).

The amounts of specific elements in fly ashes from pulverized coal-fired power plants vary widely. For instance, reported

TABLE 18.2
Characterization of Elements That Tend to Be Present in Ashes from Pulverized Coal-Fired Power Plants as Major (>0.1% by Weight) and Minor (<0.1% by Weight) Elements in Low Carbon Ashes

Major Element	Often Major Element	Major or Minor Element	Minor Element
Aluminum (Al)	Magnesium (Mg)	Chlorine (Cl)	Antimony (Sb)
Calcium (Ca)	Potassium (K)		Arsenic (As)
Iron (Fe)	Sodium (Na)		Barium (Ba)
Silicium (Si)	Sulfur (S)		Beryllium (Be)
Titanium (Ti)			Boron (B)
			Bromine (Br)
			Cadmium (Cd)
			Cesium (Cs)
			Chromium (Cr)
			Cobalt (Co)
			Copper (Cu)
			Fluorine (F)
			Iodine (I)
			Lead (Pb)
			Manganese (Mn)
			Mercury (Hg)
			Molybdenum (Mo)
			Nickel (Ni)
			Phosphorus (P)
			Polonium-210 (^{210}Po)
			Potassium-40 (^{40}K)
			Radium (Ra)
			Radon (Rn)
			Rare earths
			Rubidium (Rb)
			Selenium (Se)
			Thallium (Tl)
			Thorium (Th)
			Tin (Sn)
			Tungsten (W)
			Uranium (U)
			Vanadium (V)
			Zinc (Zn)
			Zirconium (Zr)

Sources: FitzGerald, T., Current issues in the regulation of coal ash, *World of Coal Ash Conference*, Lexington, KY, 2009; Izquierdo and Querol 2012; Peng, B. et al., *J. Coal Sci. Eng.*, 19, 387–391, 2013; Kalembkiewicz, J. and Chmielarz, U., *Fuel*, 122, 73–78, 2014; Sahu, S.K. et al., *J. Environ. Radioact.*, doi:10.1016/j.jenvrad.2014.04.010, 2014; Zacco, A. et al., *Environ. Chem. Lett.*, 12, 153–175, 2014.

concentrations of the major (>0.1% by weight) element Ti in pulverized coal combustion fly ashes vary between 0.13% and 1.26%, of the minor element Zr between 72.6 and 762 mg/kg (Kalembkiewicz and Chmielarz 2014), and of Ce between 151 and 1784 mg/kg (Mayfield and Lewis 2013). For minor elements such as As, Cd, Hg, Ni, Pb, and Y, the reported ranges are even wider, up to about 4 orders of magnitude (Baba et al. 2010; Pandey et al. 2011; Mayfield and Lewis 2013; Tian et al. 2013; Shaheen et al. 2014).

Fly ashes from pulverized coal-fired power plants may contain substantial amounts of carbon-based compounds (Reijnders 2005; Hower et al. 2010; Duan et al. 2012; McCarthy et al. 2013). In well-tuned power plants using pulverized coal combustion, percentages of carbon may range between a few tenths of a percent and a few percent, but in the absence of such tuning C percentages may be in the order of 10%–20% (Li and Xu 2009; Hower et al. 2010; McCarthy et al. 2013). When circulating fluidized bed technology is used, carbon contents in fly ash may range between about 5% and about 30% (Xiao et al. 2005).

To a large extent, C in fly ash from pulverized coal combustion may be present as deposit of char or soot (Hower et al. 2010; Pedersen et al. 2010). Carbonaceous compounds in coal combustion ashes may include polycyclic aromatic hydrocarbons, chlorinated dioxins and -benzofurans, organofluorocompounds, long-chain *n*-alkanes, steranes, pentacyclic terpanes, and methylsulfates (Reijnders 2005; FitzGerald 2009; Sahu et al. 2009; Tsubouchi et al. 2011; Masala et al. 2012; Ruwei et al. 2013; Pergal et al. 2014; Ribeiro et al. 2014). Concentrations of polycyclic aromatic hydrocarbons may be higher in bottom ash from pulverized coal combustion than in fly ash (Ruwei et al. 2013). Combustion ashes generated by co-combustion of pulverized coal with plastics, tire-derived materials, coal gasification wastes, and petroleum coke wastes have been found to contain chlorophenols, chlorobenzenes, polychlorinated biphenyls, chlorinated dioxins and benzofurans, polycyclic aromatics, benzene, and cyanides (FitzGerald 2009). Concentrations of polyaromatics in fly ashes may be much increased by co-combustion of coal and rubber tires, if compared with the combustion of coal only (Alvarez et al. 2004). The characterization of organic compounds in combustion ashes is as yet highly partial.

Concentrations of organic and inorganic substances in fly ashes from pulverized coal-fired power plants are in part dependent on the nature of the feed of these power plants. Concentrations and modes of occurrence of elements in feed and the presence of Al, S, and halogens matter, the latter to the extent that they generate gaseous halogens such as Cl_2 and HCl (Wang and Tomita 2003; Goodarzi 2006; Cao et al. 2008b; Dai et al. 2008). The arrangement for co-firing (direct, indirect, parallel) may impact ash composition (Dai et al. 2008), and it may also matter whether there is in-furnace or out-of-furnace blending of feed (Baek et al. 2014). Baek et al. (2014) found that for a blend of 40% sub-bituminous and 60% bituminous coal, in-furnace blending gave rise to lower C levels in ash than out-of-furnace blending. The character and operation of combustion plants are also important determinants of ash composition. These include temperature regime, residence time, gas velocity, interaction of flue gas with ashes, pollution controls, and, where applicable, fluidized bed materials (Rong et al. 2001; Wang and Tomita 2003; Xiao et al. 2005; Dai et al. 2008; Hower et al. 2010; Yao and Zhi 2010; Kalembkiewicz and Chmielarz 2012; Verbinnen et al. 2013; Wilczynska-Michalik et al. 2014). For instance, the relatively low temperatures in fluidized bed combustion may give rise to higher concentrations of leachable Cr^{6+} than conventional burning, which proceeds at higher temperatures (Verbinnen et al. 2013).

Changes in air pollution control may lead to changes in residue composition. Evidently, increased efforts to capture small fly ash particles (e.g., nanoparticles; see Table 18.1) from flue gas to reduce air pollution risk, increase the presence of such particles in the major coal combustion residue fly ash. Air pollution control efforts may also increase the amounts of volatile elements in fly ashes and wastewater streams from which wastewater filtercakes are derived (Clarke 1993). Efforts to reduce the emission of nitrogen oxides (NO_x) from coal-fired power plants may give rise to increases of unburned carbon in fly ash, especially when plants are retrofitted (Pedersen et al. 2008; Barnes 2010; Hower et al. 2010). NO_x removal by the use of selective and non-selective catalytic reduction and the use of ammonia to increase the efficiency of electrostatic precipitation of particulates may lead to substantial residues of ammonia and the ionic form thereof [mainly as $(NH_4)_2SO_4$] in fly ash (Wang et al. 2002; Barnes 2010). From such particulates, NH_3 may be released (Rubel 2002; Wang et al. 2002), and high ammonia levels in fly ashes may also increase the leaching of several heavy metals (Palumbo et al. 2007; Wang et al. 2007). SO_2 emission control with dry sprays of $Na_3HCO_3CO_3{\cdot}2H_2O$ (trona) in pulverized coal combustion plants may increase the leaching of As, S, Se, Mo, and V oxyanions and fluoride from fly ashes (Su et al. 2011; Dan et al. 2013).

The relevance of a variety of determinants of coal combustion residue composition can be further illustrated regarding the relatively well-studied element Hg. A main determinant of the amount of Hg in pulverized coal combustion fly ash is the amount of Hg in the feed of a power plant, whereas a relatively high concentration of Cl in the feed is correlated to relatively high Hg concentrations in fly ash (Hower et al. 2010). For the concentration of Hg in fly ash, the temperature at the ash collection hopper matters too. A lower temperature at the hopper is conducive to a higher concentration of Hg in fly ash (Hower et al. 2010). If electrostatic precipitation is replaced by fabric filters, the collection efficiency for Hg in fly ash is roughly doubled (Tian et al. 2014). The amount of Hg in pulverized coal combustion fly ash tends to be increased when the amount and surface area of carbon in fly ash increases (Hower et al. 2010). When active carbon injection is practiced, the amount of Hg in fly ash may increase substantially, as the efficiency of active carbon injection in capturing Hg may be up to about 90% (Tian et al. 2014). Mercury capture with wet flue-gas desulfurization scrubbers is estimated to be 30%–85% of gaseous mercury (Schuetze et al. 2012). Sulfite content in the slurry, pH, and oxygen concentrations are important determinants of mercury capture in wet desulfurization (Ochoa-Gonzalez et al. 2013). The presence of ammonia in fly ash, originating in NO_x emission control or use of ammonia in electrostatic precipitation, may lead to increased leaching of Hg from fly ashes (Wang et al. 2007).

Data about the composition of ashes from fluidized bed combustion are much fewer than data about ashes from pulverized coal combustion. Ceteris paribus, the concentrations of Ca and S in ashes from fluidized bed combustion tend to be higher than in pulverized coal combustion ashes. C contents in fly ashes from fluidized bed combustion tend to be higher than in the case of well-managed pulverized coal combustion, but the C content can be lowered by recirculation of fly ash and increased oxygen levels during combustion (Mei et al. 2014).

Whereas particle size matters much for the concentration of volatile minor elements in pulverized coal combustion fly ash, this is apparently less so in fluidized bed combustion fly ash (Weissman et al. 1983; Goodarzi 2006; Baba et al. 2010; Zhang et al. 2012; Zhao et al. 2014b). Also, differences in volatilization linked to differences in combustion temperature and CaO content lead to differences in composition of bottom ash and fly ash between pulverized coal combustion and fluidized bed coal combustion (e.g., Selcuk et al. 2006).

18.3.2 Ashes from Coal Gasification

As pointed out before, no data about worldwide ash production from coal gasification could be found in the open literature. Quantitative data have, however, been published about the major Sasol coal gasification plants in South Africa. Ginster and Matjie (2005) estimated the overall ash production from the Sasol coal gasification complex in Secunda at 7 Mt/year. A subdivision thereof is given by Mahlaba (2006), who reported that of the 1200 t ash per hour produced at the Sasol complex in Secunda, 200 t were fine ash (<250 μm), 700 t coarse ash (>250 μm), and 300 t fly ash.

The generation of gasification ashes is more complicated than in the case of coal combustion ashes, as their generation is partly linked with to the usage of gas, which is variable. This especially impacts the fate of relatively volatile components. Bunt and Waanders (2008, 2011) reported about Sasol-Lurgi fixed bed gasification that the volatile elements As, Cd, Hg, Se, and Pb partition preferentially to gas phase. Ilyushechkin et al. (2011) studying entrained flow gasification suggested that the elements As, B, Hg, F, Pb, Se, and V were highly volatile and Ba, Be, Li, Mn, Sc, Sr, Th, and Y were partially volatile.

Available data about the composition of ashes (including slags) from coal gasification are much less comprehensive than the corresponding data regarding pulverized coal combustion. Important determinants for the composition of ashes from coal gasification are feed composition and size, previous feeds, character and operation of the gasification plant including temperature regime, and, where applicable, fluidized bed materials (Wang and Tomita 2003; Bunt and Waanders 2008, 2009, 2011; Garcia et al. 2013; Ilyushechkin et al. 2014; Zeng et al. 2014).

Limited research into ashes from coal gasification noted alkaline, acid, and neutral ashes (Kim 2009). The distribution of elements among substances (speciation) in ashes formed by coal gasification tends to be different if compared with coal combustion ashes, which is relevant to element mobility, for example, to leaching behavior. For instance, Lopez-Anton et al. (2011) found that Hg in coal combustion fly ash was present as elemental Hg and $HgCl_2$, whereas in coal gasification fly ash it was present as HgS. Aineto et al. (2005) reported about the fly ash of an integrated coal gasification in combined cycle (IGCC) power plant that the contents of sulfides were higher than in pulverized coal combustion fly ashes. Also, it may be expected that fresh alkaline pulverized coal combustion fly ashes are relatively rich in mobile oxyanions of the elements Cr, Mo, Sb, V, and W (Izquierdo and Querol 2012), if compared with gasification ashes. Limited research comparing fresh coal gasification ashes (from integrated gasification combined cycle plants) with their counterparts from pulverized coal-fired power plants suggests that, except for Hg, cations present in coal gasification ashes tended to be more mobile than their counterparts in pulverized coal combustion ashes (Kim 2009).

An important component of gasification ashes tends to be carbon. Ginster and Matjie (2005) and Wagner et al. (2008) stated that in Sasol coarse ash the C content was 4%–7% and 3%–5%, respectively. Much of this C is present as particles with a high C content characterized as devolatilized coal dense anisotropic carbon, porous isotropic carbon, and remnant coal (Wagner et al. 2008). Bunt and Waanders (2011) reported for the Sasol-Lurgi fixed bed gasifiers that a reduction in coal feed size reduced the amount of carbon in discard ash.

Li and Whitty (2012) reported that in the practical operation of entrained-flow coal gasifiers, 30%–40% C in fine slags and fly ashes is common. This is in line with the findings of Wu et al. (2014). Pels et al. (2005) found similar concentrations of unburned carbon in fly ashes from circulating fluidized bed coal and biomass coal (co-)gasification and suggested the presence of polycyclic aromatics in the carbonaceous fraction of these fly ashes. Yoshiie et al. (2013) studied fly ashes from a drop tube coal gasification furnace. They found a bimodal size distribution of generated particles, in part peaking at >7.8 μm diameter consisting of ash with unburned carbon and at 0.5 μm diameter consisting of soot with mineral deposits. Tchapda and Pisupati (2014) noted high concentrations of unburned carbon in fly ashes from co-gasification of coal and biomass.

18.3.3 Ashes from Coking

During coking, ashes are transmitted to coke oven gas. These ashes contain mineral and carbonaceous constituents (Tsai et al. 2007; Kong et al. 2011). A relatively small part of coke oven ashes is emitted from coke ovens with fugitive gas into the wider environment, but mostly the fate of minerals in coke oven gas is dependent on what happens with coke oven gas not lost by fugitive emissions (Konieczynski et al. 2012; Mu et al. 2014). Razzaq et al. (2013) reported that in China most coke oven gas is directly emitted into the atmosphere and that 20% of Chinese coke oven gas is utilized as fuel. Usage of coke oven gas is far more common in OECD countries, where coke oven gas is often used as fuel but may also serve the production of methanol or syn-gas (Razzaq et al. 2013). When coke oven gas is used as fuel for power plants, coke oven gases are often treated, and in this case, ash-derived minerals may partially be found in tar water/condensate, Claus desulfurization units, aromatic solvents, ammonia liquor, and tar (Konieczynski et al. 2012; Razzaq et al. 2013). Ashes present in coke oven gas used as fuel may be collected in post-combustion particulate air pollution control facilities. No public worldwide data could be found about the amounts and fate of ashes originating in coking.

18.3.4 Desulfurization Residues

Desulfurization in coal gasification may be in situ (often with limestone or dolomite) or downstream, mainly with metal oxides, though the use of coal conversion residues has also been suggested (e.g., Meng et al. 2010; Garcia et al. 2011). Little data are available regarding the amounts and composition of desulfurization residues linked to coal gasification (Pinto et al. 2008; Meng et al. 2010; Razzaq et al. 2013). Much more is known about flue-gas desulfurization residues from coal-fired power plants. Wet, semi-dry, and dry technologies are available for desulfurization (e.g., Galos et al. 2003; Shi et al. 2011; Su et al. 2011; Dan et al. 2013).

Flue-gas desulfurization in pulverized coal-fired power plants is currently dominated by wet scrubbing with limestone, which generates suspensions of gypsum ($CaSO_4{\cdot}2H_2O$) (Leiva et al. 2010; Yan et al. 2014). Flue-gas desulfurization gypsum has been reported to be produced at a level of 35 Mt in China in 2008, >16 Mt in the United States in 2009, and 10.6 Mt in the European Union (EU) in 2010 (Pasini and Walker 2012; Galka 2013; Sun et al. 2014a). Apart from $CaSO_4{\cdot}2H_2O$, variable amounts of other compounds present in flue gas may end up in the suspension of gypsum, as the wet desulfurization system behaves as a scavenger (Alvarez-Ayuso et al. 2011; Schuetze et al. 2012; Sun et al. 2012, 2014a, b; Tian et al. 2014). As noted before, mercury capture with wet flue-gas desulfurization scrubbers is estimated to be 30%–85% of gaseous mercury (Schuetze et al. 2012). Tian et al. (2014) suggest that wet scrubbing for flue-gas desulfurization may capture substantial amounts of elements such as As, Cd, Cr, Ni, Pb, Sb, and Se.

A part of captured Hg may be re-emitted into air following the oxidation of $Hg^{2}+$ to zero-valent Hg (Wu et al. 2010). A part of the elements captured in wet scrubbers is transferred to desulfurization wastewater and a part ends up in desulfurization gypsum (Guan et al. 2009; Huang et al. 2013). Elements found in flue-gas desulfurization gypsum suspensions include, apart from Fe and Al, the following elements of potential environmental concern: As, B, Ba, Cd, Cr, Cu, F, Hg, Mo, Ni, ^{40}K, Pb, Sb, Se, Tl, and Zn (Leiva et al. 2010; Alvarez-Ayuso et al. 2011; Hansen et al. 2011; Paini and Walker 2012; see also Box 18.1). When good cleanup processes are used

BOX 18.1 ELEMENTS OF POTENTIAL ENVIRONMENTAL CONCERN WHEN EMITTED FROM RESIDUES

Antimony (Sb)
Arsenic (As)
Barium (Ba)
Beryllium (Be)
Boron (B)
Cadmium (Cd)
Chlorine (Cl)
Chromium (Cr)
Cobalt (Co)
Copper (Cu)
Fluorine (F)
Germanium (Ge)
Lead (Pb)
Lead-210 (^{210}Pb)
Lithium (Li)
Manganese (Mn)
Mercury (Hg)
Molybdenum (Mo)
Nickel (Ni)
Polonium-210 (^{210}Po)
Potassium-40 (^{40}K)
Radium (Ra)
Radon (Rn)
Rare earths
Selenium (Se)
Strontium (Sr)
Thallium (Tl)
Thorium (Th)
Tin (Sn)
Tungsten (W)
Uranium (U)
Vanadium (V)
Zinc (Zn)

Sources: Hopkins, W.A. et al., *Environ. Health Perspect.*, 114, 661–666, 2006; Barbosa, R. et al., *J. Hazar. Mater.*, 170, 902–909, 2009; FitzGerald, T., Current issues in the regulation of coal ash, *World of Coal Ash Conference*, Lexington, KY, 2009; Vejahati, F. et al., *Fuel*, 89, 904–911, 2010; Otter, R.R. et al., *Ecotoxicol. Environ. Safety*, 85, 30–36, 2012; Tsiridis, V. et al., *Ecotoxicol. Environ. Safety*, 84, 212–220, 2012; Yanusa et al., 2013; Izquierdo and Querol 2012; Szabo, Z. et al., *J. Environ. Radioact.*, 118, 64–74, 2013; Tripathi, R.C. et al., *Radiation Protect. Dosimetry*, 156, 198–206, 2013; Garrabrants, A.C. et al., *Chemosphere*, 103, 131–138, 2014; Grandjean, P. and Landrigan, P.J., *Lancet Neurol.*, 1, 330–338, 2014; Sahu, S.K. et al., *Microchem. J.*, 92, 92–96, 2014.

(e.g., Enoch et al. 1994; Leiva et al. 2010), levels of elements of potential environmental concern in the gypsum product output of power plants may well become low, whereas discharges to water are minimized.

Semi-dry and dry desulfurization processes used in pulverized coal-fired power plants generate dry desulfurization products. The use of $Na_3HCO_3CO_3{\cdot}2H_2O$ (trona) in dry desulfurization processes changes the composition of fly ash and, as noted before, may lead to increased leaching of a variety oxyanions and fluoride from fly ash (Su et al. 2011; Dan et al. 2013). When limestone is used in dry processes, gypsum is a component of desulfurization products, but other substances will also be present. These include $CaSO_3$, unreacted Ca-based adsorbents, and fly ash (Shi et al. 2011). Co-capture of Hg by spray dryer absorber and fabric filter is estimated to vary between <40% and up to 90% (Schuetze et al. 2012). In the case of fluidized bed combustion technology, mixtures of ashes, gypsum and other sulfates, and CaO may be produced (Galos et al. 2003).

18.3.5 Coal Tar

Coal tar originates in coking and coal (co-)gasification and is a viscous liquid or semi-solid (Li and Suzuki 2010). Coal tars and substances derived thereof may be collected as such (e.g., Razzaq et al. 2013) but may also be emitted into air (Lin et al. 2007; Mu et al. 2013, 2014; Razzaq et al. 2013) and discharged with wastewater (Zhang et al. 2006; Ghose 2007). Coal tar and derivatives thereof may end up in soils and sediments (e.g., Brown et al. 2006; Zhang et al. 2006; Kim et al. 2013).

Coking tars are complex mixtures of substances, with compositions dependent on coal composition, temperature regime, and time and pressure in coke ovens (Casal et al. 2008; Barriocanal et al. 2009; Li and Suzuki 2010; Dong et al. 2012; Morgan and Kandiyoti 2014). Substances present in coking tar number probably hundreds to over 10^3 and include aromatic hydrocarbons, phenols, amines, and S—, N—, and O— heterocyclics (Melbert et al. 2004; Brown et al. 2006; Li and Suzuki 2010; Maloletnev et al. 2014).

There is a variety of processes to fractionate and upgrade coking tars (Li and Suzuki 2010; Han et al. 2014).

Gasification tars are also complex mixtures, including linear (aliphatic and olefinic) and cyclic organic compounds, which may include phenols and heterocyclics such as N-aromatics, thiophenes, and benzofurans (Pindoria et al. 1997; Brage et al. 2000; Li et al. 2011; Schwarz et al. 2014). Dependent on the feed of, and the conditions in, the gasifier, the amount and composition of tar may vary strongly (Brage et al. 2000; Brown et al. 2006; Hernandez et al. 2013; Zeng et al. 2014). Co-gasification of coal with plastic wastes may increase tar generation, whereas co-gasification with pine woodchips may reduce the generation of tar (e.g., Pinto et al. 2009). Higher operating temperatures tend to reduce the amount of tar and increase the aromatization of tar, whereas low operating temperatures and high inputs of steam favor increased tar production and an increased concentration of phenols (Brage et al. 2000; Hernandez et al. 2013). Relatively low temperatures are also conducive to the formation of asphaltenes, which are rather cracked at elevated temperatures (Brage et al. 2000). A two-stage gasifier with a fluidized bed pyrolyzer (operating at 800°C–900°C) and a fixed bed char gasifier (operating at 1000°C–1100°C) were reported to produce tar with relatively high contents of low molecular weight components (Zeng et al. 2014). High temperature processing and a variety of catalysts (including dolomite, metal-impregnated char, zeolites, and Ni catalysts) have been reported to promote the destruction of (gasification) tar. It should be noted that de-activation of such catalysts is a matter of concern (Slaghuis et al. 1995; Seshadri and Shamsi 1998; Pinto et al. 2009; Miccio et al. 2012). There is also the option to hydrotreat and distill gasification tars. Products thereof include phenol, cresylic acid, naphtha, and distillate of diesel density (Higman and Tam 2014).

18.3.6 Chars

Chars may be formed in gasification at relatively low temperatures (Galhetas et al. 2012).

To the extent that they have been analyzed, chars are characterized by C and minerals contents in the order of tens of percents (Galhetas et al. 2012). High levels of polycyclic aromatic hydrocarbons may be present in chars (Zhou et al. 2009). Elements that may be present in chars also include N and S (Galhetas et al. 2012).

18.4 Fate of Coal Combustion, Coking, and Gasification Residues: Disposal and Utilization

Coal combustion, coking, and gasification residues are partly disposed of and partly applied. The worldwide utilization of all coal combustion residues generated by large-scale plants in 2011 has been estimated at about 53% of estimated worldwide production, with per country utilization rates varying between 10.6% and 94.6% (Heidrich et al. 2013), whereas for coal combustion ashes from power plants a worldwide utilization rate of marginally under 50% has been suggested (Izquerido and Querol 2012). The origin of coal combustion residues matters in this respect. Fluidized bed combustion ash is largely disposed of (e.g., Wang et al. 2006; Zhang et al. 2012), whereas bottom ash from pulverized coal combustion is often utilized (e.g., Goodarzi 2009). Regarding gypsum produced by wet flue-gas desulfurization, for China, the United States, and the EU, utilization rates of >70% have been estimated (Galka 2013; Lee et al. 2012; Yan et al. 2014). Utilization of gypsum containing residues from dry and semi-dry desulfurization processes in coal combustion is, however, very low (Shi et al. 2011).

The utilization and disposal rates of worldwide residues from coal gasification plants are not available in the open literature. Utilization of coal gasification ashes, however, has been documented (Ginster and Matjie 2005), but there seems to be no significant utilization of desulfurization residues linked to coal gasification (Meng et al. 2010).

Utilization of coal tar has been estimated at about 12 Mt, most of which originates in the coking of coal (Li et al. 2011). The worldwide production of coal tar in coking plants has been estimated at in the order of 15–22.5 Mt. The worldwide generation of coal tar in coal gasification could not be ascertained. The disposal of coal tar would thus presumably amount to at least 3–10.5 Mt.

No report could be found suggesting utilization of filter cake from wastewater treatment.

The distinction between application and disposal of coal combustion and gasification residues is not sharp. The distinction between *backfilling* of surface mines (utilization) and landfilling (disposal) is vague (e.g., Hulburt et al. 2012). Another example of the blurred boundary between disposal and application regards the adding of fly ash to agricultural soils. This might be viewed as disposal leading to *pollution* of farmland (e.g., Dellantonio et al. 2008; Tian et al. 2013), but fly ash from coal combustion has also been applied for intended beneficial agricultural purposes (e.g., Park and Chertow 2014; Ram and Masto 2014). *Geotechnical* applications of ashes include the inclusion in dams or dikes for ponds, lagoons, and surface impoundments of ashes. However, such dams or dikes may fail, leading to the release of ashes to surface water, which in turn may negatively affect water quality and cause aquatoxicity (e.g., Lemly and Skorupa 2012; Bevelhimer et al. 2014).

18.4.1 Disposal

Disposal of coal-derived residues may be uncontrolled (e.g., with wastewater into a river or a sea), wet or dry, with or without locational restrictions and in dumps without containment provisions, or subject to some sort of containment, e.g., dams, dikes, capping, liner systems (e.g., Wigmore and Kubrycht 1990; Reijnders 2005; Hulburt et al. 2012). Containments of coal-derived residues may be partial. For instance, hazardous substances may leach from lagoons, ponds, or surface impoundments (Lemly and Skorupa 2012; Shaheen et al. 2014). Soil layers capping coal ash landfills may be mixed with ashes due to soil tillage activities such as plowing (Dellantonio et al. 2008).

As coal-derived residues are there *forever*, the long-term fate of containment should be considered in dealing with ultimate impacts. For instance, surface impoundments may be subject to structural failure, and high-density polyethene (HDPE) linings may be subject to aging, damage, and failure near welded seams (e.g., Allen 2001; Rowe et al. 2009; Lemly and Skorupa 2012). Containment by solidification with cement may be vulnerable to freeze-thaw cycle, exposure to acid groundwater, and the activity of microorganisms (Coté and Bridle 1987; Rogers et al. 2003; Srivastava et al. 2008; Zhao et al. 2014a), whereas vitrification in metastable glass may be subject to changes in structure (e.g., Mazarenko and Yeo 1994). Quantitative estimates regarding the long-term fates of hazardous inorganic substances present in coal-derived residues are characterized by large uncertainties (Pettersen and Hertwich 2008).

18.4.2 Utilization of Coal Combustion and Gasification Residues

18.4.2.1 Major Applications of Combustion Ashes

Research has been done on actual and potential applications of coal combustion, coking, and gasification residues. Much thereof regards the application of ashes, flue-gas desulfurization gypsum and coal tar.

First, the applications of ashes from coal combustion will be considered, which represent the largest volume of coal residues.

The major applications of pulverized coal combustion ashes are associated with infrastructure and buildings, see Box 18.2.

The largest applications of pulverized coal combustion ashes in the United States and the EU are in geotechnical works (Park et al. 2014). But financially, most lucrative application of pulverized coal combustion fly ash tends to be in cement (Achternbosch et al. 2014). This application of fly ash, which lowers the clinker content of cement, is demanding in its requirements as to, for example, fineness, sulfate content, and carbon content (Li and Xu 2009; Stevens et al. 2009; Achternbosch et al. 2014). About 7% of coal combustion fly ash in the EU and a few percent of coal combustion fly ash in the United States and China meet the criteria for this application (Achternbosch et al. 2014).

Relatively high carbon levels in fly ash may negatively affect the performance of cement and concrete to which fly ash has been added (Gao et al. 1997; Külaots et al. 2002). This is linked to the adsorption by carbonaceous materials of air-entraining admixtures to concrete (Gao et al. 1997; Pedersen et al. 2008, 2010). The percentage of carbon, the surface area of carbonaceous material, and the surface chemistry of carbonaceous material have been identified as determinants of this impact (Pedersen et al. 2008, 2010). Percentages of C and loss of ignition have limited value in predicting the adsorption of air-entraining admixtures (Ahmed et al. 2014). Other tests do better in this respect (Stencel et al. 2009; Ahmed et al. 2014).

BOX 18.2 MAJOR APPLICATIONS OF PULVERIZED COAL COMBUSTION ASHES

- Geotechnical works such as road bases, road subbases, dikes, dams, and embankments (Asokan et al. 2005; Cao et al. 2008a; Ahmaruzzaman 2010; Blisset and Rowson 2012; Lemly and Skorupa 2012; Park and Chertow 2014).
- Building materials such as cement, concrete, aggregates, ceramics, asphalt, grout, and bricks (Asokan et al. 2005; Ahmaruzzaman 2010; Hulburt et al. 2012; Saraber et al. 2012; Park and Chertow 2014).
- Roofing granules and blasting grit (Park and Chertow 2014).

Standards for the application of fly ashes have suggested maximum levels of 5% or 6% C in fly ashes to safeguard satisfactory performance, but negative impacts have also been found at C percentages below 4% (Pedersen et al. 2008). Also, a maximum level of 2.5% of C in fly ash has been suggested for the incorporation in cement (Tao et al. 2009). When relatively large amounts of carbon are present in fly ashes, carbon levels may be reduced by increased oxidation during combustion, which has the negative side effect of increasing NO_x emissions (Pedersen et al. 2010), or by downstream oxidation of carbon (Cammarota et al. 2008; Hower et al. 2010; Duan et al. 2012; Wu et al. 2014). Also, flotation and rotary triboelectrostatic separator technology have been developed to separate out ash particles with high carbon concentrations (Li and Xu 2009; Tao et al. 2009).

The application of ashes from fluidized bed coal combustion would seem to be relatively difficult if compared with the application of ashes from pulverized coal combustion, and this is reflected in disposal being the dominant fate of fluidized bed coal combustion ashes (Montagnaro et al. 2009; Mejeoumov et al. 2010; Chindaprasirt and Rattanasak 2010; Zhang et al. 2012; Mei et al. 2014). Poor applicability for geotechnical purposes and in building materials is mainly linked to volume expansion and instability of the ash after application (Mejeoumov et al. 2010; Chindaprasirt and Rattanasak 2010; Zhang et al. 2012). Also, the limited presence of glassy materials in fluidized bed ashes and high C contents are disadvantages for the application in building materials (Montagnaro et al. 2009; Mei et al. 2014). Some research has been done on improving the applicability of ash from fluidized bed coal combustion. Weathered fluidized bed coal ash has been claimed to be sufficiently stable to be used as material in road base construction (Mejeoumov et al. 2010). Robl et al. (2010) have suggested removal of >200 mesh materials to improve the applicability of circulating fluidized bed coal combustion fly ash as pozzolanic additive to Portland cement concrete. Zhang et al. (2012) have suggested that autoclaving is conducive to the use of fluidized bed combustion fly ash in bricks. Furthermore, it has been suggested that fluidized bed coal combustion ash may be applied in geopolymers (*synthetic rock*) (Chindaprasirt and Rattanasak 2010) or may be recycled following steam hydration deactivation (Montagnaro et al. 2009).

18.4.2.2 Mine Site Rehabilitation

Application of coal combustion residues, mainly ashes, in mine site rehabilitation has been suggested by several authors (Asokan et al. 2005; Cao et al. 2008a; Ahmaruzzaman 2010; Shon et al. 2010; Blisset and Rowson 2012; Hulburt et al. 2012; Senapati and Mishra 2012; Yao and Sun 2012; Park and Chertow 2014; Park et al. 2014). Actual usage of coal combustion residues for this purpose is relatively important in the United States (>10% of the overall generation of coal combustion residues) but much less so in Australia, Canada, the EU, and Japan (Park et al. 2014). Ashes have been used for neutralizing acid mine drainage, soil management, backfilling, and engineering of covers.

18.4.2.3 Application of Coal Combustion Ash in Agriculture and Forestry

Use of coal combustion ashes in agriculture has been advocated as this may, for example, provide plant nutrients (e.g., Stout et al. 1997; Gupta et al. 2002; Asokan et al. 2005; Li et al. 2008; Basu et al. 2009; Ukwattage et al. 2013; Yanusa et al. 2013; Ram and Masto 2014; Masto et al. 2014). It should be noted, however, that availability of the major nutrient N in combustion ashes is zero or low, whereas the major nutrient P is not readily available (Shaheen et al. 2014). Also, the ash-induced increase in pH may lower P and Zn availability (Shaheen et al. 2014). Thus, when N and/or P limit crop yields, use of coal combustion ashes would not seem to make sense. Other suggested agricultural benefits of coal fly ash in agriculture are increasing the prevalence of beneficial microorganisms (e.g., Singh and Pandey 2013), modification of soil pH (by liming), and improvement of soil texture and nutrient-retaining abilities (Gupta et al. 2002; Sudha and Dinesh 2006; Gonzalez et al. 2009; Yanusa et al. 2013; Ram and Masto 2014). Improvement of water-holding capacity of agricultural soils by addition of coal combustion ashes has also been claimed but has been linked to very high ash inputs, which may also have negative impacts in other respects (Shaheen et al. 2014; Park et al. 2014). Reducing sodicity is another benefit claimed for ash addition, but high levels of sodium in coal combustion ashes may also have the opposite effect (Shaheen et al. 2014).

Negative impacts of the use of coal combustion ashes include potential phytotoxic effects (e.g., due to the presence of B) and increased levels of hazardous compounds in harvested materials caused by leached substances (e.g., Gonzalez et al. 2009; Tripathi et al. 2013; Yanusa et al. 2013; Shaheen et al. 2014). The estimated balances between benefits and negative impacts of the application of coal ashes may vary strongly, dependent on soil characteristics such as available nutrients and pre-existing environmental loads of hazardous compounds and on ash characteristics and amounts of ash added. Both advocacy of agricultural application and recommendations not to apply coal combustion ashes to agricultural soils can be found in the scientific literature (e.g., Gonzalez et al. 2009; Pandey et al. 2011; Shaheen et al. 2014). Actual application of coal combustion ashes for agricultural purposes is substantial in the United States and India (Tripathi et al. 2013; Park and Chertow 2014; Ram and Masto 2014). Beyond agricultural application, the use of coal combustion ashes has also been advocated as a soil management tool for forestry (Shaheen et al. 2014).

18.4.2.4 Application of Cenospheres

Cenospheres can be recovered from pulverized coal combustion fly ashes, in which they tend to be present in the 0.01–4.8% (w/w) range (Vassilev et al. 2004; Hirajima et al. 2010), being of varying size (Yu et al. 2012). A variety of wet and dry processes, and combinations thereof, which may be used as technologies for the separation of cenospheres from other fly ash particles, have been proposed and tested (e.g., Petrus et al. 2011; Li et al. 2014). A wide range of applications has been proposed for cenospheres, including insulating materials, concrete, fillers in polymers, paints, heat-resistant coatings, microwave absorption, refractory materials, sorbents, and support material for the photocatalytic degradation of pollutants (Vassilev et al. 2004; Asokan et al. 2005; Drozhzhin et al. 2005; Lu et al. 2013; Zhang et al. 2013; Zhiang et al. 2013). Applications of cenospheres in polymers and building materials are currently substantial.

18.4.2.5 Application of Magnetics Present in Ashes

Fine magnetite particles are used in coal cleaning (Groppo and Honaker 2009). It has been suggested that it might be possible to substitute those particles by fractions of fly ash or bottom ash fines with high concentrations of magnetics (Groppo and Honaker 2009). Separation technologies based on magnetic fields and concentrating spirals have been proposed to separate coal combustion bottom ashes with a high percentage of magnetics from other ashes (Groppo and Honaker 2009; Shoumkova 2011).

18.4.2.6 Applications of Coal Combustion Ashes in Environmental Technology

Many proposals have been made for the application of pulverized coal combustion fly ashes as adsorbent in environmental technology, whether or not after the activation or conversion to zeolites (Ahmaruzzaman 2010; Du Plessis et al. 2013). These applications include the cleaning of flue gases and the removal of a variety of substances from wastewater (Gonzalez et al. 2009; Ahmaruzzaman 2010; Garcia et al. 2011). Also, the use of coal combustion ashes in the immobilization of pollutants in solids has been suggested (e.g., Poon et al. 2004; Liu et al. 2009; Tomasevic et al. 2013; Nikolic et al. 2014; Osmanlioglu 2014). Actual utilization of coal combustion ashes in environmental technology seems to focus on the solidification and stabilization of wastes and is as such substantial in the United States (>3% of residue generation in 2012) but much smaller (<1% of residue generation) in Australia, Canada, the EU, and Japan (Park and Chertow 2014; Park et al. 2014). This application is contested, however, as freeze-thaw cycles; the activity of microorganisms and exposure to acid groundwater may negatively affect the containment of wastes (e.g., Coté and Bridle 1987; Rogers et al. 2003; Srivastava et al. 2008; Zhao et al. 2014a). Also, the use of coal combustion ashes in the neutralization of acid mine drainage has been noted (Park et al. 2014). In other environmental technologies, application seems to be much less successful, which may be due to such factors as limited removal efficiency, variable performance linked to variability in ashes, concern about the release or leaching of hazardous substances, and problems with disposal after utilization (Ahmaruzzaman 2010).

18.4.2.7 Mining Coal Combustion (and Gasification) Ashes

Coal combustion and gasification ashes may contain substantial amounts of minable elements. Al, Ca, Ga, Ge, V, Zn, Cr, Pb, Cd, Ti, lanthanides (rare earths), yttrium, and silica and/or substances derived thereof have been proposed for mining from coal-derived ashes (Font et al. 2005, 2007; Reijnders 2005; Gonzalez et al. 2009; Blisset and Rowson 2012; Seredin and Dai 2012; Chimenos et al. 2013; Mayfield and Lewis 2013; Achternbosch et al. 2014; Arroyo et al. 2014; Yao et al. 2014).

Technology has been proposed to generate the $CaCO_3$ polymorph vaterite from coal-derived fly ash with >25% CaO using $CaCl_2$ brine while adjusting pH at about 9.5 (Achternbosch et al. 2014). Other processes that have been proposed for the mining of coal-derived ashes are sinter processes, processes based on acid and alkaline leaching, on chlorination, on the use of chelators, and on reduction and oxidation (Font et al. 2005, 2007; Blisset and Rowson 2012; Arroyo et al. 2014; Yao et al. 2014).

Substantial work on Al extraction has been done in China, where an alumina content >30% in pulverized coal combustion ashes has been suggested to allow for the financial feasibility of Al mining (Yao et al. 2014). The Al content in fly ash from pulverized coal combustion in northern Shaanxi and Shanxi province and from southern Inner Mongolia has been reported to be 40%–50% (Yao et al. 2014; Yang et al. 2014). The building of factories extracting Al from coal combustion fly ash with relatively high contents of Al partly based on sintering and partly on acid leaching has been reported to be underway (Yang et al. 2014).

18.4.2.7.1 Utilization of Flue-Gas Desulfurization Gypsum in Building Materials

Most utilization of flue-gas desulfurization gypsum produced by wet technology (probably >50 Mt) regards the production of building materials. Both gypsum products and products based on mixtures of gypsum with other materials, including coal fly ash, are generated (Guo and Shi 2008; Telesca et al. 2013; Spadoni et al. 2014).

18.4.2.7.2 Utilization of Flue-Gas Desulfurization Gypsum in Agriculture

Substantial research has been done on the application of flue-gas desulfurization gypsum as soil amendment in agriculture. This has been stimulated by economic contraction of the application of flue-gas gypsum in U.S. building sector (Watts and Dick 2014). Application of flue-gas desulfurization gypsum as agricultural soil amendment in the United States has been estimated at about 5.5×10^{11} g in 2011 (Watts and Dick 2014). Flue-gas desulfurization gypsum may increase the availability of the plant nutrients Ca and S (Watts and Dick 2014). Also, it has been argued that amending agricultural soils with flue-gas desulfurization gypsum may improve soil aggregation, reduce P loss and Al toxicity, increase water infiltration, and be conducive to the reclamation of sodic soils (Park and Chertow 2014; Watts and Dick 2014).

18.4.2.8 Utilization of Coal Gasification Ashes

According to Ginster and Matjie (2005), coal ash from the Sasol gasification plants at Secunda was used as sub-base in roads, in brick manufacturing, as bedding material for paving bricks and floor screeds, in low-strength concrete, and as general fill material. This was stated to account for about 50% of all coal gasification ash produced at Secunda. Pels et al. (2005, 2006) have suggested that coal gasification fly ashes might be suitable for applications in asphalt. When the calorific value of the fly ashes is higher than 15 MJ/kg, utilization as fuel would be an option (Pels et al. 2005). Also, according to Pels et al. (2005), bottom ashes from fluidized bed gasification might be used for granulate production to be applied in concrete and infrastructural works as they do largely consist of sand and might replace virgin sand applications. They furthermore suggest that high levels (over 30%) of C in coal gasification ashes make such fly ashes unfit for direct applications in agriculture due to low nutrient value, hydrophobicity, and the presence of polycyclic carbons.

18.4.2.9 Utilization of Coal Tar

Coal tar as such, or after expelling substances with a boiling point <350°C, may be used as a fuel, for road materials and surface coatings. It is often further treated before utilization. Distillation is often applied, and other options for the treatment of coal tar include crystallization, cracking, reforming, catalytic upgrading, extraction, hydrogenation, and the use of sorbent (Li and Suzuki 2010). The use of catalysts for tar treatment tends to be vulnerable to poisoning by carbon deposits, sulfur, and chlorine (Li and Suzuki 2010). Major products of distillation are coal tar pitch and pesticidal impregnating agents for wood such as creosote. Creosote is a dense liquid containing over 289 compounds, including phenols, polycyclic hydrocarbons, and heterocyclics containing S, N, and O (Fraser et al. 2008). Applications of distilled coal tar pitch include use in road materials, coatings, electrodes for aluminum and electric arc steel production, and carbon fiber–reinforced composites (Sotoudehnia et al. 2010; Li and Suzuki 2010; Li et al. 2011). Other coal tar-derived products include BTX (mixture of benzene, toluene, and xylene), industrial oils, fuels, phthalic anhydride, acenaphthene, naphthalene, anthracene, pyrene, phenol, carbazole, thiophene, and quinolone (Li and Suzuki 2010).

18.5 WORKPLACE AND ENVIRONMENTAL IMPACTS

The disposal and utilization of coal-derived residues may have impacts on the use of natural resources and the exposure of humans and other living organisms to hazardous substances. The latter may be linked to the contamination of air, groundwater, surface water, and/or soil by coal-derived residues. Relatively much available research regards human exposure to coal-derived residues and substances originating in these residues, leaching, and soil contamination. Human exposure may be linked to dermal contact and airborne exposure to coal-derived residues or substances derived from such residues (e.g., Redmond 1983). Also, human exposure may be linked to leaching or volatilization and deposition of substances from coal-derived residues, followed by ingestion, for example, of contaminated food or soil (e.g., Ikarashi et al. 2005; Moret et al. 2007; Blokker et al. 2013; Zhao and Shi 2014). When elevated concentrations of radium (^{226}Ra) ^{232}Th, and/or ^{40}K are present in residues, there may be increased external irradiation with gamma rays, for instance, when such residues are included in building materials (e.g., Nemeth et al. 2000; Szabo et al. 2013). In this section, both workplace and environmental impacts of coal-derived residues will be discussed. Also, the impact of using coal-derived residues on the consumption of virgin natural resources and the potential contribution thereof to sustainability will be addressed.

18.5.1 Workplace Impacts

18.5.1.1 Workplace Impacts of Ashes

Airborne exposure to respirable coal-derived fly ash is possible in the working environment, linked to working in air-cleaning devices, storage facilities for fly ashes, the handling of fly ash, soils to which fly ash has been added, or secondary building materials with fly ash (Kleinjans et al. 1989; Borm 1997; Versluys et al. 2013). Wind erosion of soils to which fly ash has been added may also lead to workplace exposure (Park et al. 2014).

Available research suggests that workplace exposure to fly ash from coal combustion may cause cardiopulmonary dysfunction (Deering-Rice et al. 2012) and is associated with genotoxic risk (Kleinjans et al. 1989). It has been found that respirable coal-derived fly ash (from both coal combustion and coal gasification) does give rise to the generation of reactive oxygen species (Borm 1997; Van Maanen et al. 1999; Dwivedi et al. 2012), which might in turn be linked to inflammation and damage to DNA (Borm 1997; Van Maanen et al. 1999; Reijnders 2012). The number of respirable fly ash particles per cubic meter air, their surface area and surface characteristics, and the presence of hazardous compounds (e.g., polycyclic aromatics, Cr, Pb, Mo, and Ni, and radionuclides such as ^{40}K, ^{210}Po, ^{226}Ra, ^{232}Th, and ^{238}U) are likely to be determinants of hazards (Cohen et al. 2002; Bérubé et al. 2007; Reijnders 2012; Versluys et al. 2013; Bhangare et al. 2014; Sahu et al. 2014). When fly ash is handled, even when fly ash is contained in solid matrixes of, (for example, concrete), the emission of ammonia, volatile organics, and radon present in fly ash may occur and lead to human workplace exposure (e.g., Lindgren 2010; De Biase et al. 2014).

A hazardous substance that has drawn special attention in the context of occupational and environmental hazard of fly ash is vapor phase (zero-valent) mercury (Schuetze et al. 2012; Wang et al. 2014). When Hg is present in fly ash, the hazard may extend to cement kilns fed by such coal fly ash, which leads to loading of Hg in baghouse filter dust of cement kilns (Wang et al. 2014).

18.5.1.2 Workplace Impacts of Flue-Gas Desulfurization Gypsum

Some work has been published on the workplace impacts of flue-gas desulfurization gypsum. Studying the release of mercury vapor from flue-gas desulfurization gypsum, Schüetze et al. found that the emission of zero-valent Hg in flue-gas desulfurization scrubbers was strongly increased at pH values >7 and in the presence of Fe^{2+}. It has been found that mercury may be a workplace hazard in case of the utilization and disposal of flue-gas desulfurization gypsum (Gustin and Ladwig 2010; Sun et al. 2012, 2013, 2014a, b).

18.5.1.3 Workplace Exposure to Coal Tar and Derivatives Thereof

Worker exposure to coal tar and derivatives thereof may occur linked to coal conversion and to coal tar processing (e.g., Redmond 1983). Workplace exposure may also occur during the application of coal tar and its derivatives in products, handling such products, and recycling and disposal of products in which coal tar and its derivatives have been applied (Melbert et al. 2004; Ikarashi et al. 2005; Reinke and Glidden 2007; Fayerweather 2007). It should also be noted that knowledge about the toxicity of coal tar components is as yet partial (Schwarz et al. 2014).

Human exposure to coal tar is associated with an increased risk of damage to DNA, increased cancer risk, and damage to offspring (Schultz et al. 1983; Royer et al. 1983; Toraason et al. 2001; Ikarashi et al. 2005; Fayerweather 2007; Li et al. 2011; IARC 2012; Mahler et al. 2014). A number of studies have correlated workplace exposure to coal tar-derived substances to increased cancer risk, including lung cancer (Redmond 1983; Silverstein et al. 1985; Armstrong and Thériault 1996), bladder cancer, (Tremblay et al. 1995) and skin cancer (Redmond 1983). The workplaces studied included aluminum smelters with anodes partly consisting of coal tar derivatives, coke ovens, metal working, and tar distillation. Workplace exposure to coal tar-derived substances has also been correlated with an increased risk of ischemic heart disease (Ronneberg 1995; Burstyn et al. 2005; Friesen et al. 2010).

18.5.2 Environmental Impacts

18.5.2.1 Environmental Impacts of Airborne Exposure to Ashes, Substances Derived from Ashes, and Enhanced Gamma Irradiation

Storage of coal-derived fly ash in the open air and applications of fly ash in agriculture and soil management at mines may give rise to airborne environmental exposure (e.g., Park et al. 2014). And it may be that airborne exposure to respirable fly ash-derived particles may occur linked to wear, tear, and weathering of materials containing fly ash (Reijnders 2005; Shandilya et al. 2014). The hazards of such exposure have been outlined in section 18.5.1.1. Coal combustion and gasification residues may generate the airborne radionuclide radon (Rn). This may give rise to increased radiation doses following from the utilization of coal residues as construction material in buildings (e.g., Lokobauer et al. 1997; Reijnders 2005; Szabo et al. 2013). As noted before, application of coal-derived ashes with enhanced concentrations of radium (^{226}Ra), ^{232}Th, and/or ^{40}K in buildings may give rise to enhanced gamma irradiation of occupants thereof.

18.5.2.2 Environmental Impacts of Flue-Gas Desulfurization Gypsum

Limited research has been done regarding the environmental impacts of several elements of potential environmental concern as present in samples of flue-gas desulfurization gypsum (Sanchez et al. 2008; Alvarez-Ayuso et al. 2011; Gustin and Ladwig 2010; Pasini and Walker 2012; Sun et al. 2012, 2014a, b).

This research has added to the environmental concern about the presence of As, B, Be, Cd, Hg, Mo, Mn, Se, Tl, and Zn in desulfurization gypsum.

18.5.2.3 Risks of Environmental Exposure to Coal Tar and Derivatives Thereof

Human and ecosystem exposure linked to emissions into the environment of coal tar and its derivatives originating in the production, disposal, or application thereof have been noted (e.g., Melbert et al. 2004; Kramer et al. 2008; Moret et al. 2007; Mahler et al. 2012; Van Metre et al. 2014a, b; Blokker et al. 2013). Kramer et al. (2008) found significantly elevated rates of neuroblastoma in children correlating with increased ambient exposure to coal tar-related substances near a former gasification plant in Taylorville, Illinois. They suggested that environmental exposure to coal tar is a public health problem (Kramer et al. 2008). Environmental impacts of semi-volatiles in coal tar-derived sealant for pavements and creosote have been addressed by Van Metre et al. (2014a, b). These authors studied the volatilization of polycyclic aromatics following the application of coal tar-based pavement sealant. They estimated that in the United States new coal tar-based sealcoat applications could lead to an emission of polycyclic aromatics into air of about 1000 Mg/year, which would exceed the yearly emission of polycyclic aromatics by U.S. cars (Van Metre et al. 2014b). This might lead to the inhalation of polycyclic aromatics and to the deposition of polycyclic aromatics on agricultural produce and might also give rise to ecotoxicity (Mahler et al. 2014; Witter et al. 2014).

Moret et al. (2007) studied the release of semi-volatile polycyclic aromatics emitted into air from old creosoted railway ties and noted their deposition on olives grown in the vicinity of such ties. The potential for airborne exposure to soil particles and house dusts contaminated by coal tar and its derivatives has also been noted (White and Claxton 2004; Mahler et al. 2014). Such particles may give rise to mutagenesis (White and Claxton 2004).

18.5.2.4 Leaching

When residues of coal combustion and gasification, and products made thereof, are in contact with water, leaching of constituents thereof may occur. This may be relevant to the processing of residues (e.g., to generate cenospheres from fly ash) and to disposal and utilization of residues. Leaching may affect water quality aspects such as conductivity and hardness (e.g., Hwang and Latorre 2011). Also, a substantial number of specific constituents of residues is of potential environmental concern or hazardous on leaching. Hazards may regard impacts on humans and ecotoxicity (White and Claxton 2004; Reijnders 2005; Lemly and Skorupa 2012; Lemly 2014).

Most research about leaching regards ashes, but there is also research on leaching following disposal and utilization of coal tar and derivatives thereof.

18.5.2.4.1 Leaching from Ashes

The elements classified as being of potential environmental concern following emission thereof, including leaching from ashes, are in Box 18.1. Leaching of inorganic constituents depends on the composition of the residues, the speciation of elements, the absence or presence and nature of containment, pH, the flow and characteristics of water into which constituents may be leached, and the presence of (micro)organisms (Reijnders 2005; Thorneloe et al. 2010; Kosson et al. 2014). Changes in pH, weathering, and carbonation of coal combustion ashes and mineralogical changes such as the formation of amorphous phases and crystallization contribute to changes in leaching over time (Reijnders 2005; Akinyemi et al. 2013; Garrabrants et al. 2014; Kosson et al. 2014).

Inorganic constituents of potential environmental concern tend to be more abundant in fly ashes than in bottom ashes, and leaching of inorganic substances from combustion fly ashes is likely to be more problematical than leaching from combustion bottom ashes (e.g., Barbosa et al. 2009; Tsiridis et al. 2012). Co-combustibles may affect leachability. For instance, research by Stam et al. (2011) suggests that the percentage of Cr leaching from ashes may be much increased when there is co-combustion of coal and wood.

Leaching of organic substances from residues has been hardly studied (Park et al. 2014; Rowe 2014). Nevertheless, it should be borne in mind that several of the organic substances that may be present in fly and bottom ashes are hazardous and that these hazardous organic substances may be more abundant in bottom ash than in fly ash (e.g., Ruwei et al. 2013). Hazards of organic substances may regard humans but also other organisms. For instance, fish embryos may be negatively affected by relatively low concentrations of polycyclic aromatic compounds (Incardona et al. 2011).

Leaching may occur when residues are disposed of or applied. Nyale et al. (2014) studied leaching in a 20-year-old dump of hydraulically disposed fly ashes from coal combustion and coal gasification in South Africa. They found substantial losses due to leaching (estimated at 4.4%–27.4%) of the minor elements As, Cr, Cu, Mo, Ni, Pb, and Zn. Spadoni et al. (2014) investigated a case of groundwater pollution in Maharashtra (India), where concentrations of sulfate, Mo, and As exceeded safety limits. They linked high sulfate concentrations to infiltration water from factories that mix coal fly ash with gypsum for brick production. The high Mo and As concentrations were linked to the disposal of coal fly ash in ponds (Spadoni et al. 2014). Leaching of high levels of B, Cu, Mo, Se, and Zn following the agricultural application of coal combustion fly ashes has been linked to negative impacts on plant growth, which may in turn negatively affect agricultural yields (Yanusa et al. 2013). Low Cu : Mo ratios linked to mixtures of coal ash and agricultural soils may induce hypocuprosis in ruminants (Dellantonio et al. 2008). High levels of As in groundwater may negatively affect Scots pine (Lhotáková et al. 2013). Leaching of Se from coal ashes has been linked to damage to fish populations in Lake Sutton, North Carolina (Lemly 2014). Aquatic disposal of coal combustion residues has also been associated with negative impacts on aquatic biota including reduced hatching and lowered offspring viability of amphibians (Hopkins et al. 2006; Metts et al. 2013; Rowe 2014).

When coal ashes are integrated in products, substantial leaching during the product life cycle may also occur. For instance, substantial leaching of oxyanions may occur from bricks containing coal combustion fly ash (Reijnders 2005). Leaching may also occur linked to wear, tear, and weathering of products and after the end of life of products containing ashes, when they are, for instance, granulated (Reijnders 2005).

A variety of tests have been developed to assess leaching for regulatory purposes. These tests tend to focus on the leaching of inorganic constituents. Many of these tests are single-point extraction tests in which the leaching of a series of elements is measured (FitzGerald 2009; Thorneloe et al. 2010). The hazard as determined by such tests may well be at variance with the hazard as determined with aquatic and terrestrial ecotoxicity tests (Römbke et al. 2009). The concentrations of elements measured in those tests may also be very different from the concentrations as measured in the field (Twardowska and Szczepanska 2002; Reijnders 2005; FitzGerald 2009).

Also, sets of tests with extractions at different pHs and liquid–solid ratios have been developed (Thorneloe et al. 2010). The inclusion of different pHs is important because pH may strongly impact element mobility. For instance, at an alkaline pH, the leachability of elements of potential concern commonly present in coal combustion fly ashes as cations (e.g., Cd and Hg) may be reduced, whereas the leachability of elements commonly present as anions (e.g., As and Cr) may be increased (Izquierdo and Querol 2012).

Applying the set of tests of the Leaching Environmental Assessment Framework (LEAF) to a set of 76 residues of air pollution control residues from U.S. coal combustion power plants, Thorneloe et al. (2010) found that variability in metals leaching exceeded the variability of composition by several orders of magnitude. Differences in leaching from similar types of residues were large (Thorneloe et al. 2010).

It might moreover well be that in actual practice the concentrations of leached substances are beyond the range of results from the LEAF tests. In part, this possibility is linked to changes in the residues over time, such as in the case of fly ash, weathering, the formation of secondary minerals, and the dissolution of amorphous phases (Twardowska and Szczepanska 2002; Reijnders 2005). Also, differences between the outcomes of regulatory tests, leaching in the field may originate in microbial activity, the presence of organic compounds such as humic acids (which are not included in regulatory tests), and differences in ionic strength (Sandhu and Mills 1991; Idachaba and Mills 2003, 2004; Reijnders 2005).

Approaches have been suggested to reduce leaching of elements of potential environmental concern from coal consumption-derived residues in building materials and geotechnical applications. These include forced leaching before use (Reijnders 2005).

18.5.2.4.2 Leaching from Products Treated with Derivatives of Coal Tar

Leaching from products treated with derivatives of coal tar has been noted.

The disposal of creosote and its application to wood have been found to give rise to serious groundwater pollution (Ikarashi et al. 2005; Moret et al. 2007; Fraser et al. 2008).

The use of coal tar-based sealcoats for pavements has been found to increase the concentration of polycyclic aromatics and azaarenes in stormwater (Van Metre et al. 2009; Mahler et al. 2010, 2012, 2014). This in turn may give rise to ecotoxic effects in surface water (Witter et al. 2014).

18.5.3 Pollution of Soil and Sediments and Their Remediation

After emission into water or air, soils (including groundwater) and sediments may become sinks for substances present in coal tars and derivatives thereof (Brown et al. 2006; Moret et al. 2007; Fraser et al. 2008; Witter et al. 2014). Also, soils may be contaminated by direct addition of coal tar and its derivatives to soil (e.g., McGowan et al. 1996; Hatheway 2002; Antony and Wang 2006; Moret et al. 2007; Fraser et al. 2008). There have been substantial efforts to remediate soils, sediments, and groundwater contaminated with polycyclic aromatic hydrocarbons derived from coal tar, but high efficiencies of remediation, needed for minimizing present and future hazard (Simarro et al. 2013; Juhasz et al. 2014), remain a challenge especially in clayey soils and when contamination has aged (Birak and Miller 2009; Isosaari et al. 2007; Llado et al. 2013). High efficiencies of remediation for coal tar-derived substances tend to require technologies combining separation and destruction of polycyclic aromatics, while bioremediation technologies may well lead to only modest remediation efficiencies (McGowan et al. 1996; Hatheway 2002; Antony and Wang 2006; Gan et al. 2009; Simarro et al. 2013; Juhasz et al. 2014).

18.5.4 Impact on the Use of Virgin Natural Resources and Sustainability

It has been argued that the utilization of coal combustion and gasification residues contributes to sustainable production (e.g., Lee et al. 2012).

A problem with this claim is that *sustainability* as it is used in practice may have many meanings (e.g., Goodland 1995). The original meaning of (environmental) sustainability was closely linked to the concept of a steady-state economy, a type of economy that could continue indefinitely without negatively affecting resource stocks and the environment for future generations (Goodland 1995). In this sense, utilization of coal combustion and gasification residues cannot be considered sustainable, if only because coal stocks are practically finite and reduced by current coal combustion and gasification. Moreover, to be considered sustainable in the original sense, it should be shown that there are no future negative environmental impacts linked to the residues. Establishing the absence of future negative environmental impacts is severely hindered by the uncertainties regarding the future fate of hazardous substances (e.g., Pettersen and Hertwich 2008). Consideration of future environmental impacts may lead to limitations on the amounts of mobile substances of environmental concern in coal combustion and gasification residues to be applied. As noted before, a case for *cleaning up* coal-derived residues before application in building materials and geotechnical applications has been made.

Application of flue-gas gypsum in agricultural systems has been claimed to be sustainable (Watts and Dick 2014). A requirement for sustainable utilization of flue-gas desulfurization amendments would be that practices do not lead to increased concentrations of bioavailable hazardous substances in soils. In this context, a matter of concern regarding the agricultural application is the fate of minor hazardous elements present in flue-gas desulfurization gypsum. Of these elements, Hg has attracted most interest. Wang et al. (2013) have shown that the presence of Hg in flue-gas desulfurization gypsum may lead to increased soil mercury emissions to the atmosphere and increased mercury contents in (roots and leaves of) plants. A similar effect was noted by Briggs et al. (2014). Chen et al. (2014) found a (small) upward effect of flue-gas desulfurization amendments on Hg concentrations in earthworms. As noted before, the concentration of Hg and other elements of potential environmental concern in flue-gas desulfurization gypsum can be reduced by good cleanup practices.

Utilization of coal combustion residues can have environmental benefits. Such benefits can, for instance, originate in the actual replacement of virgin resources, when the environmental burden of these virgin resources is larger than the environmental burden of coal-derived residues. In this context, the availability of other non-virgin resources should also be considered. In the building sector, for example, large amounts of used gypsum are generated and linked to renovation and demolition, which in principle can be used to replace virgin gypsum resources. In practice, recycling of demolition-derived gypsum to new gypsum building materials appears to be very limited, and this may well be linked to the availability of flue-gas desulfurization gypsum. Similarly, aggregates made from ashes to be used in concrete may well compete with aggregates generated from end-of-life concrete (Lotfi et al. 2014). Regarding the use of coal fly ash in road construction, it has, for instance, been suggested that crushed concrete is less of a life cycle environmental burden (Mroueh et al. 1999).

It should be noted in this context that quantification of benefits of coal utilization residues by life cycle assessment is subject to uncertainty because the allocation of the environmental burden of generating coal combustion and gasification residues is controversial. Often this burden is valued as zero, because this valuation is often applied to wastes (e.g., Lee et al. 2012). However, in the case of utilization of residues, one might also be allocating the environmental burden in other ways, for example, on the basis of monetary value or physical units such as exergy (e.g., Marjeau-Bettez et al. 2014). Difference in allocation may give rise to divergence in quantitative results.

18.6 CONCLUDING REMARKS

There tend to be ranges of options for the disposal and utilization of residues from coal combustion, gasification, and coking. In view of workplace hazards and environmental burdens, decision making about what to do with such residues tends to lead to balancing acts, complicated by uncertainties about the long-term fate of hazardous substances present in the residues (Pettersen and Hertwich 2008). It would seem, however, that, for example, surface impoundments of ashes, discharges of residues into water, the use of coal tar-based sealcoats for pavements, and the agricultural application of fly ashes with high levels of mobile elements of potential environmental concern have to be evaluated negatively when workplace hazards and/or environmental burdens are considered.

REFERENCES

Achternbosch, M., U. Dewald, E. Nieke et al. 2014. Is coal fly ash a suitable alkaline resource for manufacturing new calcium carbonate-based cements. *Journal of Industrial Ecology*. doi:10.111/jiec.12147.

Ahmaruzzaman, M. 2010. A review on the utilization of fly ash. *Progress in Energy and Combustion Science* 36: 327–363.

Ahmed, Z.T., D.W. Hand, M.K. Watkins et al. 2014. Air-entraining admixture partitioning and adsorption by fly ash in concrete. *Industrial & Engineering Chemistry Research* 53: 4239–4246.

Aineto, M., A. Acosta, J.M. Rincón et al. April 11–15, 2005. Production of lightweight aggregates from coal gasification fly ash and slag. *World of Coal Ash*. Lexington, KY. http://www.flyash.info.

Akinyemi, S.A., W.M. Gitari, A. Akinlua et al. 2013. Chemical weathering and mobility of inorganic species in dry disposed ash. An insight from geochemical fractionation and physicochemical analysis. *Coal Combustion and Gasification Products* 5: 16–30.

Allen, A. 2001. Containment landfills: The myth of sustainability. *Engineering Geology* 60: 3–19.

Alvarez, R., M.S. Callén, and C. Clemente. 2004. Soil, water and air environmental impact from tire rubber/coal fluidized bed co-combustion. *Energy and Fuels* 18: 1633–1639.

Alvarez-Ayuso, E., A. Gimenez, and J.C. Ballesteros. 2011. Fluoride accumulation by plants grown in acid soils amended with flue gas desulphurisation gypsum. *Journal of Hazardous Materials* 192: 1659–1666.

Antony, E.J. and J. Wang. 2006. Pilot plant investigations of thermal remediation of tar-contaminated soil and oil-contaminated gravel. *Fuel* 85: 443–450.

Armstrong, B. and G. Thériault. 1996. Compensating lung cancer patients occupationally exposed to coal pitch tar volatiles. *Occupational and Environmental Medicine* 53: 160–167.

Arroyo, F., O. Font, J.M. Chimenos et al. 2014. IGCC fly ash valorisation. Optmisation of Ge and Ga recovery for an industrial operation. *Fuel Processing Technology* 124: 222–227.

Asokan, P., M. Saxena, and S.R. Asolekar. 2005. Coal combustion residues—Environmental implications and recycling potentials. *Resources, Conservation and Recycling* 43: 229–262.

Baba, A., G. Gurdal, and F. Sengunalp. 2010. Leaching characteristics of fly ash from fluidized bed combustion thermal power plant: Case study: Can (Cannakale-Turkey). *Fuel Processing Technology* 91: 1073–1080.

Baek, S.H., H.Y. Park, and S.H. Ko. 2014. The effect of the coal blending method in a coal fired boiler on carbon in ash and NO_x emission. *Fuel* 128: 62–70.

Barbosa, R., N. Lapa, D. Boavida et al. 2009. Co-combustion of coal and sewage sludge: Chemical and ecotoxicological properties of ashes. *Journal of Hazardous Materials* 170: 902–909.

Barnes, I. 2010. Ash utilisation-impact of recent changes in power generation practices. London: IEA Clean Coal Centre.

Barriocanal, C., M.A. Diez, and R. Alvarez. 2009. Relationship between coking pressure generated by coal blends and the composition of their primary tars. *Journal of Analytical and Applied Pyrolysis* 85: 514–520.

Basu, M., M. Pande, P.B.S. Bhadoria et al. 2009. Potential fly-ash utilization in agriculture: A global review. *Progress in Natural Science* 19: 1173–1186.

Bérubé, K., D. Balharry, K, Sexton et al. 2007. Combustion-derived nanoparticles: Mechanisms of pulmonary toxicity. *Clinical and Experimental Pharmacology and Physiology* 34: 1044–1050.

Bevelhimer, M.S., S.M. Adams, A.M. Fortner et al. 2014. Using coordination and clustering techniques to assess multimetric fish health response following a coal ash spill. *Environmental Toxicology and Chemistry*. doi:10.1002/etc.2622.

Bhangare, R.C., M. Tiwari, P.Y. Ajmal et al. 2014. Distribution of natural radioactivity in coal and combustion residues of thermal power plants. *Journal of Radioanalytical and Nuclear Chemistry*. doi:10.1007/s10967-014-2942-3.

Birak, P.S. and C.T. Miller. 2009. Dense non-aqueous phase liquids at former manufacture gas plants: Challenges to modeling and remediation. *Journal of Contaminant Hydrology* 105: 81–98.

Blisset, R.S. and N.A. Rowson. 2012. A review of the multi-component utilisation of coal fly ash. *Fuel* 97: 1–23.

Blokker, E.J.M., B.M. van de Ven, C.M. de Jongh et al. 2013. Health implications of PAH release from coated cast iron drinking water distribution systems in the Netherlands. *Environmental Health Perspectives* 12: 600–606.

Borm, P.J.A. 1997. Toxicity and occupational health hazards of coal fly ash (CFA). A review of data and comparison to coal mine dust. *Annals of Occupational Hygiene* 41: 659–676.

Brage, C, Q. Yu, G. Chen et al. 2000. Tar evolution profiles obtained from gasification of biomass and coal. *Biomass and Bioenergy* 18: 87–91.

Briggs, C.W., R. Fine, M. Markee et al. 2014. Investigation of the potential for mercury release from flue gas desulfurization solids applied as agricultural amendment. *Journal of Environmental Quality* 43: 253–262.

Brown, D.G., L. Gupta, T. Kim et al. 2006. Comparative assessment of coal tars obtained from 10 former manufactured gas plant sites in the Eastern United States. *Chemosphere* 65: 1562–1569.

Bunt, J.R. and F.B. Waanders. 2008. Trace element behavior in the Sasol-Lurgi MK IV FBDB gasifier. Part 1—The volatile elements: Hg, As, Se, Cd and Pb. *Fuel* 87: 2374–2387.

Bunt, J.R. and F.B. Waanders. 2009. Pipe reactor gasification studies of a South African bituminous coal blend. Part 1—Carbon and volatile matter behavior as a function of feed coal particle size reduction. *Fuel* 88: 585–594.

Bunt, J.R. and F.B. Waanders 2011. Volatile trace element behavior in the Sasol fixed bed dry-bottom (FBDB) gasifier treating coals of different rank. *Fuel Processing Technology* 92: 1646–1655.

Burstyn, I., H. Kromhout, T. Paratanen et al. 2005. Polycyclic aromatic hydrocarbons and fatal ischemic heart disease. *Epidemiology* 16: 744–750.

Cammarota, A., R. Chirone, R. Solimene et al. 2008. Beneficiation of pulverized coal combustion fly ash in fluidized bed reactors. *Experimental Thermal and Fluid Science* 32: 1324–1333.

Cao, D., E. Selic, and J. Herbell. 2008a. Utilization of fly ash from coal-fired power plants in China. *Journal of Zhejiang University Science A* 9: 681–687.

Cao, Y., H. Zhou, and J. Fan et al. 2008b. Mercury emissions during cofiring of sub-bituminous coal and biomass (chicken waste, wood, coffee residue, and tobacco stalk) in a laboratory-scale fluidized bed combustor. *Environmental Science & Technology* 42: 9378–9384.

Casal, M.D., M.A. Diez, R. Alvarez et al. 2008. Primary tar of different coking coal ranks. *International Journal of Coal Geology* 76: 237–242.

Chen, L., Y. Kost, X. Tian et al. 2014. Effects of gypsum on trace metals in soils and earthworms. *Journal of Environmental Quality* 43: 263–272

Chimenos, J.M., A.I. Fernadez, and R. del Valle-Zermeno. 2013. Arsenic and antimony removal by aqueous leaching of IGCC fly ash during germanium extraction. *Fuel* 112: 450–458.

Chindaprasirt, P. and U. Rattanasak. 2010. Utilization of blended fluidized bed combustion (FBC) ash and pulverized coal combustion as in geopolymer. *Waste Management* 30: 667–672.

Clarke, L.B. 1993. The Fate of trace elements during coal combustion and gasification: An overview. *Fuel* 72: 731–735.

Cohen, M.D., M. Sisco, K. Baker et al. 2002. Effect of inhaled chromium on pulmonary tract. *Inhalation Toxicology* 14: 765–771.

Coté, P.L. and T.R. Bridle. 1987. Long-term leaching scenario's for cement-based waste forms. *Waste Management and Research* 5: 55–66.

Dai, J., S. Sokhansanj, J.R. Grace et al. 2008. Overview of some issues related to co-firing biomass and coal. *Canadian Journal of Chemical Engineering* 86: 367–386.

Dan, Y., C. Zimmerman, K. Liu et al. 2013. Increased leaching of As, Se, Mo and V from high calcium coal ash containing trona reaction products. *Energy and Fuels* 27: 1531–1537.

De Biase, G., S. Loechel, T. Putzmann et al. 2014. Volatile organic compounds effective diffusion coefficients and fluxes estimation through two types of construction material. *Indoor Air* 24: 272–282.

Deering-Rice, C.E., M.E. Johansen, J.K. Roberts et al. 2012. Transient receptor potential vinilloid-1 (TRPV1) is a mediator of lung toxicity for coal fly ash particulate material. *Molecular Pharmacology* 81: 411–419.

Dellantonio, A., W.J. Fitz, H. Custovic et al. 2008. Environmental risks of farmed and barren coal ash landfills in Tuzla, Bosnia and Herzegovina. *Environmental Pollution* 153: 677–686.

Dong, J., F. Li, and K. Xie. 2012. Study on the source of polycyclic aromatic hydrocarbons (PAHs) during coal pyrolysis. *Journal of Hazardous Materials* 243: 80–85.

Drozhzhin, V.S., L. D. Danilin, I.V. Pikulin et al. April 11–15, 2005. Functional materials on the basis of cenospheres. *World of Coal Ash*. Lexington, KY, 9 pp.

Duan, L., D. Liu, X. Chen et al. 2012. Fly ash recirculation by bottom feeding on a circulating fluidized bed boiler co-burning coal sludge and coal. *Applied Energy* 95: 295–299.

Du Plessis, P.W., T.V. Ojumu, and L.F. Petrik. 2013. Waste minimization protocols for the process of synthesizing zeolites from South African coal fly ash. *Materials* 6: 1688–1793.

Dwivedi, S., Q. Saquib, A.A. El-Jedhairy et al. 2012. Characterization of coal fly ash nanoparticles and induced oxidative DNA damage in human peripheral blood mononuclear cells. *Science of the Total Environment* 437: 331–338.

Emami-Taba, L., M.F. Irfan, W.M.A.W. Daud. 2013. Fuel blending effects on the co-gasification of coal and biomass—A review. *Biomass and Bioenergy* 57: 249–263.

Enoch, G.D., W.F. van den Broeke, and W. Spiering. 1994. Removal of heavy metals and suspended solids from waste water from wet lime(stone)-gypsum flue gas desulphurization plants by means of hydrophobic and hydrophilic crossflow microfiltration membranes. *Journal of Membrane Science* 87: 191–198.

European Commission. December 2001. Best available techniques reference document on the production of iron and steel. Brussels, Belgium, p. 116.

Fayerweather, W.E. 2007. Meta-analysis of lung cancer in asphalt roofing and paving workers with external adjustment for confounding coal tar. *Journal of Occupational and Environmental Hygiene* 4 (S1): 175–200.

FitzGerald, T. May 4–7, 2009. Current issues in the regulation of coal ash. *World of Coal Ash Conference*. Lexington, KY.

Font, O., A. Querol, R. Juan et al. 2007. Recovery of gallium and vanadium from gasification fly ash. *Journal of Hazardous Materials A* 139: 413–423.

Font, O., X. Querol, A. Lopez-Soler et al. 2005. Ge extraction from gasification fly ash. *Fuel* 84: 1384–1392.

Fraser, M., J.F. Barker, B. Butler et al. 2008. Natural attenuation of a plume from an emplaced coal; tar creosote source over 14 years. *Journal of Contaminant Hydrology* 100: 101–115.

Friesen, M., P.A. Demers, J.J. Spinelli et al. 2010. Chronic and acute effects of coal tar pitch exposure and cardiopulmonary mortality among aluminum smelter workers. *American Journal of Epidemiology* 172: 790–799.

Galhetas, M., H. Lopes, and M. Freire. 2012. Characterization, leachability and valorization through combustion of residual chars from gasification of coals and pine. *Waste Management* 32: 769–779.

Galka, F. 2013. The gypsum market-usage of synthetic gypsum in Europe. *Presentation at ASHTRANS Europe*. Copenhagen, Denmark.

Galos, K.A., T.S. Smakowski, and J. Sziugai. 2003. Flue-gas desulphurisation products from Polish coal-fired power plants. *Applied Energy* 75: 257–265.

Gan, S., E.V. Lau and H.K. Ng. 2009. Remediation of soils contaminated with polycyclic aromatic hydrocarbons (PAHs). *Journal of Hazardous Materials* 172: 532–549.

Gao, Y., H. Shim, R.H. Hurt et al. 1997. Effects of carbon on air entrainment in fly ash concrete: The role of soot and carbon black. *Energy & Fuels* 11: 1497–1488.

Garcia, G., E. Cascarosa, J. Abrego et al. 2011. Use of different residues for high temperature desulphurisation of gasification gas. *Chemical Engineering Journal* 174: 644–651.

Garcia, G., J. Arauzo, A. Gonzalo et al. 2013. Influence of feedstock composition in fluidised bed co-gasification of mixtures of lignite, bituminous coal and sewage sludge. *Chemical Engineering Journal* 222: 345–352.

Garrabrants, A.C., D.S. Kosson, R. DeLapp et al. 2014. Effect of coal combustion fly ash use in concrete on the mass transport release of constituents of potential concern. *Chemosphere* 103: 131–138.

Ghose, M.K. 2007. Coke plant effluent treatment technology in India. *International Journal of Environmental Studies* 64: 253–268.

Ginster, M. and R.H. Matjie. April 11–15, 2005. Beneficial utilization of Sasol coal gasification ash. *World of Coal Ash*. Lexington, KY. http://www.flyash. Info.

Gonzalez, A., R. Navia, and N. Moreno. 2009. Fly ashes from coal and petroleum coke combustion: Current and innovative potential applications. *Waste Management & Research* 27: 976–987.

Goodarzi, F. 2006. Characteristics and composition of fly ash from Canadian coal-fired power plants. *Fuel* 85: 1418–1427.

Goodarzi, F. 2009. Environmental assessment of bottom ash from Canadian coal-fired power plants. *The Open Environmental & Biological Monitoring Journal* 2: 1–10.

Goodland, R. 1995. The concept of environmental sustainability. *Annual Review of Ecology and Systematics* 26: 1–24.

Grandjean, P. and P.J. Landrigan. 2014. Neurobehavioural effects of developmental toxicity. *Lancet Neurology* 1: 330–338.

Groppo, J. and R. Honaker.2009. Economical recovery of fly ash-derived magnetics and evaluation for coal cleaning. *Energeia* 20(8): 1–4.

Guan, B., W. Ni, Z. Wu et al. 2009. Removal of Mn(II) and Zn(II) ions from flue gas desulfurization wastewater with water-soluble chitosan. *Separation and Purification Technology* 65: 269–274.

Guo, X.L. and H.S. Shi. 2008. Thermal treatment and utilization of flue gas desulphurization gypsum as an admixture to cement and concrete. *Construction and Building Materials* 22: 1471–1476.

Gupta, D.K, U.N. Rai, R.D. Tripathi et al. 2002. Impacts of fly ash on soil and plant responses. *Journal of Plant Research* 115: 401–409.

Gustin, M. and K. Ladwig. 2010. Laboratory investigation of Hg release from flue gas desulfurization products. *Environmental Science and Technology* 44: 4012–4018.

Han, J., X. Wang, J. Yue et al. 2014. Catalytic upgrading of coal pyrolysis tar over char based catalysts. *Fuel Processing Technology* 122: 98–106.

Hansen, B.B., S. Kiil, and J.E. Johnsson. 2011. Investigation of the gypsum quality at three full-scale wet flue gas desulphurization plants. *Fuel* 90: 2965–2973.

Hatheway, A.W. 2002. Geoenvironmental protocol for site and waste characterization of former manufactured gas plants: Worldwide remediation challenge in semi-volatile organic wastes. *Engineering Geology* 64: 317–338.

Heidrich, G, H.-J. Feuerborn, and A. Weir. April 22–25, 2013. Coal combustion products: A global perspective. *Paper presented at the 2013 World of Coal Ash Conference*. Lexington, KY.

Hernandez, J.J., R. Ballesteros, and G. Aranda. 2013. Characterization of tars from biomass gasification; effect of the operating conditions. *Energy* 50: 339–342.

Higman, C. October 16, 2013. State of the gasification industry – the updated worldwide gasification database. *Paper presented at the Gasification Technologies Conference*. Colorado Springs, CO.

Higman, C. and S. Tam. 2014. Advances in coal gasification, hydrogenation and gas treating for the production of chemicals and fuels. *Chemical Reviews* 114: 1673–1708.

Hirajima, T., H.T.B.M. Petrus, Y. Oosako et al. 2010. Recovery of cenospheres from coal fly ash during a dry separation process: Separation estimation and potential application. *International Journal of Mineral Processing* 95: 18–24.

Hongjui, P., K. Pedersen, M. Jaccard et al. November 11–13, 2009. Sustainable use of coal and pollution control policy in China. *CCICED Policy Research Report presented at the CCICED 2009 Annual Meeting*, Beijing, pp 1–41.

Hopkins, W.A., S.E. DuRant, B.P. Staub et al. 2006. Reproduction, embryonic development, and maternal transfer of contaminants in the amphibian *Gastrophryne carolinensis*. *Environmental Health Perspectives* 114: 661–666.

Hower, J.C., C.L. Senior, E.M. Suuberg et al. 2010. Mercury capture by native fly ash carbons in coal-fired power plants. *Progress in Energy and Combustion Science* 36: 510–529.

Huang, Y.H., P.K. Peddi, C. Tang et al. 2013. Hybrid zero-valent iron process for removing heavy metals and nitrate from flue-gas-desulfurization wastewater. *Separation and Purification Technology* 118: 690–698.

Hulburt, C., T. Adamczyk, K. Prather et al. 2012. *Beneficial Use of Coal Combustion Residuals Survey Report*. Association of State and Territorial Waste Management Officials, Washington, DC, pp 1–25.

Hwang, S. and I. Latorre. 2011. Impact of manufactured coal ash aggregates on water quality during open pit restoration: 1. A statistical screening test. *Coal Combustion and Gasification Products* 3: 1–7.

IARC. 2012. Coke production. *IARC Monographs on the Evaluation of Carcinogenic Risk to Humans* 100F: 167–178.

Idachaba, M.A., K. Nyavor, and N.O. Egiebor. 2003. Microbial stability evaluation of cement-based waste forms at different waste to cement ratios. *Journal of Hazardous Materials* 96: 331–340.

Idachaba, M.A., K. Nyavor, and N.O. Egiebor. 2004. The leaching of chromium from cement-based waste form via a predominantly biological mechanism. *Advances in Environmental Research* 8: 483–491.

Ikarashi, Y., M. Kaniwa, and T. Tsuchiya. 2005. Monitoring of polycyclic aromatic hydrocarbons and water-extractable phenols and creosotes and creosote woods made and procurable in Japan. *Chemosphere* 60: 1279–287.

Ilyushechkin, A., D.G. Roberts and D.J. Harris et al. 2011. Trace element partitioning and leaching in solids derived from gasification of Australian coals. *Coal Combustion and Gasification Products* 3: 8–16.

Ilyushechkin, A., D.G. Roberts, and D.J. Harris. 2014. Characteristics of solid by products from entrained flow gasification of Australian coals. *Fuel Processing Technology* 118: 98–109.

Incardona, J.P, T.K. Collier, and N.L. Scholz. 2011. Oil spills and fish health: Exposing the heart of the matter. *Journal of Exposure Science and Epidemiology* 21: 3–4.

Isosaari, P., R. Piskonen, P. Ojala et al. 2007. Integration of electrokinetics and chemical oxidation for the remediation of creosote-contaminated clay. *Journal of Hazardous Materials* 144: 538–548.

Izquierdo, M. and X. Querol. 2012. Leaching behaviour of elements from coal combustion fly ash: An overview. *International Journal of Coal Geology* 94: 54–86.

Jones, A. January 2011. The world coke and coking coal markets. *Paper Presented at the Global Steel Conference*. New Delhi, India.

Juhasz, A.L., J. Weber, G. Stevenson et al. 2014. In vivo measurement, in vivo estimation and fugacity prediction of PAH bioavailability in post-remediated creosote contaminated soil. *Science of the Total Environment* 473–474: 147–154.

Kalembkiewicz, J. and U. Chmielarz. 2012. Ashes from co-combustion of coal and biomass. *Resources, Conservation and Recycling* 69: 109–121.

Kalembkiewicz, J. and U. Chmielarz. 2014. Functional speciation and leachability of titanium group from industrial fly ash. *Fuel* 122: 73–78.

Kim, A.G. 2009. Soluble metals in coal gasification residues. *Fuel* 88: 1444–1452.

Kim, Y.S., L.M. Nyberg, B. Jenkinson et al. 2013. PAH concentration gradients and fluzes through sand cap test cells installed *in situ* over river sediments containing coal tar. *Environmental Science Processes and Impacts* 15: 1601–1612.

Kleinjans, J.C.S, Y.M.W. Janssen, B. van Agen et al. 1989. Genotoxicty of coal fly ash, assesses in vitro in *Salmonella typhimurium* and human lymphocytes, and in vivo in an occupationally exposed population. *Mutation Research* 224: 127–134.

Kong, S.F., J.W. Shi, B. Lu et al. 2011. Characterization of PAHs within PM 10 fraction for ashes from coke production in Liaoning Province, China. *Atmospheric Environment* 45: 3777–3785.

Konieczynski, J., E. Zajusz-Zubek, and M. Jablonska. 2012. The release of trace elements in the process of coal coking. *Scientific World Journal*. doi:10.1100 / 2012 / 294927, 8 pp.

Kosson, D.S., A.C. Garrabrants, R. DeLapp et al. 2014. pH-dependent leaching of constituents of potential concern from concrete materials containing coal combustion fly ash. *Chemosphere* 103: 140–147.

Kosson, D., F. Sanchez, P. Kariher et al. 2009. *Characterization of Coal Combustion Residues from Electric Utilities—Leaching and Characterization Data.* Washington, DC: USEPA (EPA-600/R-09/151).

Kramer, S., M. Hawkins, and B.A. Ange. 2008. Elevated rates of neuroblastoma associated with exposure to coal tar from a former manufactured gas plant (MGP) site. *Annals of Epidemiology* 17: 726.

Kronbauer, M.A., Izquierdo, M., Dai S. et al. 2013. Geochemistry of ultrafine and nanocompounds in coal gasification ashes: A synoptic view. *Science of the Total Environment* 456–457: 95–101.

Külaots, I., R.H. Hurt, and E.M. Suuberg. 2002. Size distribution of unburned carbon in coal fly ash and its role in foam index. *ACS Fuel Chemistry Division Preprints* 47: 841–842.

Lee, J.C., S.L. Bradshaw, T.B. Edil et al. 2012. Quantifying the benefits of using flue gas desulfurization gypsum in sustainable wallboard production. *Coal Combustion and Gasification Products* 4. doi:10.4177/CCGP-D-11-00007.1.

Leiva, C., C.G. Arenas, L.F. Vilches et al. 2010. Use of FGD gypsum in fire resistant panels. *Waste Management* 30: 1123–1129.

Lemly, A.D. 2014. Teratogenic effects and monetary costs of selenium poisoning of fish in Lake Sutton, North Carolina. *Ecotoxicology and Environmental Safety* 104: 160–167.

Lemly, A.D. and J.P. Skorupa. 2012. Wildlife and coal waste policy debate: Proposed rules for coal waste disposal ignore lessons from 45 years of wildlife poisoning. *Environmental Science & Technology* 46: 8595–8600.

Lhotáková, Z., L. Brodský, L. Kupková et al. 2013. Detection of multiple stresses in Scots pine growing at post-mining sites using visible to near-infrared spectroscopy. *Environmental Science Processes & Impacts* 15: 2004–2015.

Li, Q, J. Chen and Y. Li. 2008. Heavy metal leaching from coal fly ash amended container substrates during *Syngonium* production. *Journal of Environmental Science and Health Part B* 43: 179–186.

Li, J., S.M. Iveson, A. Kiani et al. 2014. Recovery and concentration of buoyant cenospheres using an inverted reflux classifier. *Fuel Processing Technology* 123: 127–130.

Li, G. and K. Suzuki. 2010. Resources, properties and utilization of tar. *Resources, Conservation and Recycling* 54: 905–915.

Li, S. and K.J. Whitty. 2012. Physical phenomena of char-slag transition in pulverized coal gasification. *Fuel Processing Technology* 95: 127–136.

Li, Z., Y. Wu, Y. Zhao et al. 2011. Analysis of coal tar pitch and smoke extract components and their cytotoxicity on human bronchial epithelial cells. *Journal of Hazardous Materials* 186: 1277–1282.

Li, H. and D. Xu. 2009. The future resources for eco-building materials: II fly ash and coal waste. *Journal of Wuhan University of Technology-Materials* August 14: 667–672.

Lin, C., N. Liou, P. Chang et al. 2007. Fugitive coke oven gas emission profile by continuous line averaged open-path Fourier transform infrared monitoring. *Journal of the Air & Waste Management Association* 57: 472–479.

Lindgren, T. 2010. A case of indoor air pollution of ammonia emitted from concrete in a newly built office in Beijing. *Building and Environment* 45: 596–600.

Liu, W., H. Hou, C. Zhang et al. 2009. Feasibility study on solidification of municipal solid waste incinerator fly ash with circulating fluidized bed combustion coal fly ash. *Waste Management & Research* 27: 258–266.

Llado, S., S. Covino, A.M. Soalans et al. 2013. Comparative assessment of bioremediation approaches to highly recalcitrant PAH degradation in a real industrial polluted soil. *Journal of Hazardous Materials* 248–249: 407–414.

Lokobauer, N., Z. Franic, J. Sencar et al. 1997. Radon concentrations in houses around the Plomin coal-fired power plant. *Journal of Environmental Radioactivity* 34: 37–44.

Lopez-Anton, M.A., R. Perry, P. Abad-Valle et al. 2011. Speciation of mercury in fly ashes by temperature programmed decomposition. *Fuel Processing Technology* 92: 707–711.

Lotfi, S., J. Deja, P. Rem et al. 2014. Mechanical recycling of EOL concrete into high-grade aggregates. *Resources, Conservation and Recycling* 87: 117–125.

Lu, Z., W. Zhou, P. Huao et al. 2013. Performance of a novel TiO_2 photocatalyst based on the magnetic floating fly-ash cenospheres for the purpose of treating waste by waste. *Chemical Engineering Journal* 225: 39–42.

Mahlaba, S.J. December 2006. *Evaluation of Paste Technology to Co-Dispose of Ash and Brines at Sasol Synfuels Complex.* Thesis. University of Witwatersrand, Johannesburg, South Africa.

Mahler, B.J., P.C. van Metre, J.L. Crane et al. 2012. Coal-tar-based pavement sealcoat and PAHs: Implications for the environment, human health, and stormwater management. *Environmental Science and Technology* 46: 3039–3045.

Mahler, B.J., P.C. van Metre, and W.T. Foreman. 2014. Concentrations of polycyclic aromatic hydrocarbons (PAHs) and aza-arenes in runoff from coal-tar- and asphalt seal-coated pavement. *Environmental Pollution* 188: 81–87.

Mahler, B.J., P.C. van Metre, J.T. Wilson et al. 2010. Coal-tar-based parking lot sealcoat: An unrecognized source of PAH to settled house dust. *Environmental Science and Technology* 44: 894–900.

Maloletnev, A.S., A.M. Gyul'maliev and O.A. Mazneva. 2014. Chemical composition of the distillate fractions of coal tar from OAO Altai-Koks. *Solid Fuel Chemistry* 48: 11–21.

Marjeau-Bettez, G., R. Wood, and A.H. Stromman. 2014. Unified theory of allocations and constructs in life cycle assessment and input-output analysis. *Journal of Industrial Ecology.* doi:10.111/jie.12142.

Masala, S., C. Bergvall, and R. Westerholm. 2012. Determination of benzo[a]pyrene and dibenzopyrenes in Chinese coal fly ash certified reference material. *Science of the Total Environment* 432: 97–102.

Masto, R.E., T. Sengupta, J. George et al. 2014. The impact of fly ash amendment on soil carbon. *Energy Sources Part A* 36: 554–562.

Mayfield, D.B. and A.S. Lewis. 2013. Environmental review of coal ash as a resource for rare earth and strategic element. *2013 World of Coal Ash Conference.* Lexington, KY.

Mazarenko, G.F. and J. Yeo. 1994. Metastability, mode coupling and the glass transition. *Journal of Non-Crystalline Solids* 172–174: 1–6.

McCarthy, M.J., M.R. Jones, L. Zheng et al. 2013. Characteristics of long-term wet-stored fly ash following carbon and particle size separation. *Fuel* 111: 430–441.

McGowan, T.F., B.A. Greer, and M. Lawless. 1996. Thermal treatment and non-thermal technologies for remediation of manufactured gas plant sites. *Waste Management* 16: 691–698.

Mei, L., Q. Wang, X. Lu et al. 2014. Experimental study on combustion characteristics of residual carbon in fly ash at high concentration of oxygen in a circulating fluidized bed combustor. *Energy & Fuels* 28: 5534–5542.

Mejeoumov, G.G., C. Son, D. Saylak et al. 2010. Beneficiation of stockpiled fluidized bed coal ash in road base construction. *Construction and Building Materials* 24: 2072–2078.

Melbert, C., J. Kielhorn, and I. Mangelsdorf. 2004. *Coal Tar Creosote.* Geneva, Switzerland: World Health Organization.

Meng, X., W. de Jong, R. Pal et al. 2010. In bed and downstream hot gas desulphurization during solid fuel gasification. A review. *Fuel Processing Technology* 91: 964–981.

Metts, B.S., K.A. Bullmann, T.D. Turberville et al. 2013. Maternal transfer of contaminants and reduced reproductive success of southern toads (*Bufo [Anaxyris] terrestris*) exposed to coal combustion waste. *Environmental Science & Technology* 47: 2846–2853.

Miccio, F., G. Ruoppolo, S. Kalisz et al. 2012. Combines gasification of coal and biomass in internal circulating fluidized bed. *Fuel Processing Technology* 95: 45–54.

Montagnaro, F., M. Nobili, A. Telesca et al. 2009. Steam hydration-reactivation of FBC ashes for enhanced *in situ* desulphurization. *Fuel* 88: 1092–1098.

Moret, S., G. Purcaro, and L.S. Conte. 2007. Polycyclic aromatic hydrocarbon (PAH) content of soil and olives collected in areas contaminated with creosote released from old railway ties. *Science of the Total Environment* 386: 1–8.

Morgan, T.J. and R. Kandiyoti. 2014. Pyrolysis of coal and biomass: Analysis of thermal breakdown and its products. *Chemical Reviews* 114: 1547–1607.

Mroueh, U, P. Eskola, J. Laine-Ylijoki et al. 1999. *Life Cycle Assessment of Road Construction.* Helsinki, Finland: Finnish National Road Administration.

Mu, L., L. Peng, J. Cao et al. 2013. Emissions of polycyclic aromatic hydrocarbons from coking industries in China. *Particuology* 11: 86–93.

Mu, L., L. Peng and X. Liu. 2012. Emission characteristics of heavy metals and their behavior during coking processes. *Environmental Science and Technology* 46: 6425–6430.

Mu, L, L. Ping, X. Lu et al. 2014. Characteristics of polycyclic aromatic hydrocarbons and their gas/particle partitioning from fugitive emissions in coke plants. *Atmospheric Environment* 83: 202–210.

Nemeth, C., J. Somlai, and B. Kanyar. 2000. Estimation of external irradiation of children due to the use of coal slag as building material in Tatabanya, Hungary. *Journal of Environmental Radioactivity* 51: 371–378.

Nikolic, V., M. Komljenovic, N. Marjanovic et al. 2014. Lead immobilization by geopolymers based on mechanically activated fly ash. *Ceramics International* 40: 8479–8488.

Nyale, S.M., C.P. Eze, R.O. Akinyeye et al. 2014. The leaching behaviour and geochemical fractionation of trace elements in hydraulically disposed weathered coal fly ash. *Journal of Environmental Science and Health A* 49: 233–242.

Ochoa-Gonzalez, R., M. Diaz-Somoano, and M.R. Martinez-Tarazona. 2013. The capture of oxidized mercury from simulated desulphurization aqueous solutions. *Journal of Environmental Management* 120: 55–60.

Osmanlioglu, A.E. 2014. Utilization of coal fly ash in solidification of liquid radioactive waste from research reactor. *Waste Management & Research* 32: 366–370.

Otter, R.R., F.C. Bailey, A.M. Fortner et al. 2012. Trophic status and metal bioaccumulation differences in multiple fish species exposed to coal ash-associated metals. *Ecotoxicology and Environmental Safety* 85: 30–36.

Palumbo, A.V., J.R. Tarver, L.A. Fagan et al. 2007. Comparing metal leaching and toxicity from high pH, low pH and high ammonia fly ash. *Fuel* 86: 1623–1630.

Pandey, V.C., J.S. Singh, R.P. Singh et al. 2011. Arsenic hazards in coal fly as hand its fate in Indian Scenario. *Resources, Conservation and Recycling* 55: 819–835.

Park, J. and Chertow, M.R. 2014. Establishing and testing the 'reuse potential' indicator for managing wastes as resources. *Journal of Environmental Management* 137: 45–53.

Park, J.H., M. Edraki, D. Mulligan et al. 2014. The application of coal combustion by-products in mine site rehabilitation. *Journal of Cleaner Production.* doi:10.1016/j.jclepro.2014.01.049.

Pasini, R. and H.W. Walker. 2012. Estimating constituent release from FGD gypsum under different management scenarios. *Fuel* 95: 190–196.

Pedersen, K.H., A.D. Jensen, and K. Dam-Johansen. 2010. The effect of low-NO_x combustion on residual carbon in fly ash and its adsorption capacity for air entrainment admixtures in concrete. *Combustion and Flame* 157: 208–216.

Pedersen, K.H., A.D. Jensen, M.S. Skjoth-Rasmussen et al. 2008. A review of the interference of carbon containing fly ash with air entrainment in concrete. *Progress in Energy and Combustion Science* 34: 135–154.

Pels, J.R., D.S. de Nie, and J.H.A. Kiel. October 2005. Utilization of ashes from biomass combustion and gasification. *Proceedings of the 14th European Biomass Conference & Exhibition*. Paris, France, pp 17–21; also Report ECN-RX-05-182, available from ECN, Petten, the Netherlands.

Pels, J.R., D.S. de Nie, and J.H.A. Kiel. 2006. GASASH. ECN, Petten, the Netherlands. Report ECN-06–038.

Peng, B., L. Li, D. Wu. 2013. Distribution of bromine and iodine in thermal power plant. *Journal of Coal Science & Engineering* 19: 387–391.

Pergal, M.M., D. Relic, Z.L. Tesic et al. 2014. Leaching of polycyclic aromatic hydrocarbons from power plant lignite ash—Influence of parameters important for environmental pollution. *Environmental Science and Pollution Research* 21: 3435–3442.

Petrus, H.T.B.M., T. Hirajima, Y. Oosako et al. 2011. Performance of dry-separation processes in the recovery of cenospheres from fly ash and their implementation in a recovery unit. *International Journal of Mineral Processing* 98: 15–23.

Pettersen, J. and E.G. Hertwich. 2008. Critical review: Life cycle inventory procedures for long term release of metals. *Environmental Science & Technology* 42: 4639–4647.

Pindoria R.V., A. Megaritis, L.N. Chatzakis et al. 1997. Structural characterization of tar from a coal gasification plant. *Fuel* 76: 101–113.

Pinto, F., R.N. André, C. Franco et al. 2009. Co-gasification of coal and wastes in a pilot- scale installation I: Effect of catalysts in syngas treatment to achieve tar abatement. *Fuel* 88: 2392–2402.

Pinto, F., H. Lopes, R.N. André et al. 2008. Effects of catalysts in the quality of syngas and byproducts obtained by co-gasification of coal and wastes. 2: Heavy metals, sulphur and halogen compounds abatement. *Fuel* 87: 1050–1062.

Poon, C.S., X.C. Qiao, and Z.S. Lin. 2004. Effects of flue gas desulphurization sludge on the pozzolanic reaction of reject-fly-ash-blended cement pastes. *Cement and Concrete Research* 34: 1907–1918.

Ram, L.C. and R.E. Masto. 2010. An appraisal of the potential use of fly ash for reclaiming coal mine spoil. *Journal of Environmental Management* 91: 603–617.

Ram, L.C. and R.E. Masto. 2014. Fly ash for soil amelioration: a review on the influence of ash blending with inorganic and organic amendments. *Earth-Science Reviews* 128: 52–74.

Razzaq, R., C. Li, and S. Sang. 2013. Coke oven gas: Availability, properties, purification, and utilization. *Fuel* 113: 287–289.

Redmond, C.K. 1983. Cancer mortality among coke over workers. *Environmental Health Perspectives* 53: 67–73.

Reijnders, L. 2005. Disposal, uses and treatments of combustion ashes. *Resources, Conservation and Recycling* 43: 313–336.

Reijnders, L. 2012. Human health hazards of persistent inorganic and carbon nanoparticles. *Journal of Materials Science* 47: 5061–5073.

Reinke, G. and S. Glidden. 2007. Case study of worker exposure to coal tar containing paving materials on a routine paving project in Iowa. *Journal of Occupational and Environmental Hygiene* 4 (S): 228–232.

Ribeiro, J., T.F. Silva, J.C.M. Filho et al. 2014. Fly ash from coal combustion—An environmental source of organic compounds. *Applied Geochemistry*. doi:10.1016/j.apgeochem.2013.06.014.

Robl, T., K. Mahboub, W. Stevens et al. 2010. Fluidized bed combustion ash utilization: CFBC fly ash as a pozzolanic additive to Portland cement concrete. *Special Technical Proceedings of the Second International Conference in Sustainable Construction Materials and Technologies*. http://www.claisse.info/Proceedings.htm.

Rogers, R.D., J.J. Knight, C.R. Cheeseman et al. 2003. Development of test methods for assessing microbial influenced degradation of cement-solidified radioactive and industrial waste. *Cement and Concrete Research* 33: 2069–2076.

Römbke, J., T. Moser, and H. Moser. 2009. Ecotoxicological characterization of 12 incineration ashes using 6 laboratory tests. *Waste Management* 29: 2475–2482.

Rong, Y., D. Gauthier, and G. Flamant. 2001. Fate of selenium in coal combustion: Volatilization and speciation in flue gas. *Environmental Science & Technology* 35: 1406–1410.

Ronneberg, A. 1995. Mortality and cancer morbidity in workers from aluminum smelter with prebaked carbon anodes—Part III: Mortality from circulatory and respiratory diseases. *Occupational and Environmental Medicine* 52: 255–261.

Rowe, C.L. 2014. Bioaccumulation and effects of metals and trace elements from aquatic disposal of coal combustion residues: Recent advances and recommendations for further study. *Science of the Total Environment* 485–486: 490–496.

Rowe, R.K., S. Rimal, and H. Sangam. 2009. Aging of HDPE geomembrane exposed to air, water and leachate at different temperatures. *Geotextiles and Geomembranes* 27: 137–151.

Royer, R.E., C.E. Mitchell, R.L. Hanson et al. 1983. Fractionation, chemical analysis and mutagenicity testing of low-BTU coal gasifier tar. *Environmental Research* 31: 460–471.

Rubel, A.M. 2002. Forms of ammonia on SCR, SNCR, and FGG combustion ashes. *ACS Fuel Chemistry Division Preprints* 47: 834–835.

Ruwei, W., Z. Jiamei, L. Jingjing et al. 2013. Levels and patterns of polycyclic aromatic hydrocarbons in coal-fired power plant bottom as hand fly ash from Huainan, China. *Archives of Environmental Contamination and Toxicology* 65: 193–202.

Sahu, S.K., R.C. Bhangare, P.Y. Ajmal et al. 2009. Characterization and quantification of persistent organic pollutants in fly ash from coal fueled thermal power stations in India. *Microchemical Journal* 92: 92–96.

Sahu, S.K., M. Tiwari, R.C. Bhangare et al. 2014. Enrichment and particle size dependence of polonium and other naturally occurring radionuclides in coal ash. *Journal of Environmental Radioactivity*. doi:10.1016/j.jenvrad.2014.04.010.

Sanchez, F., D. Kosson, R. Keeny et al. 2008. Characterization of coal combustion residues from electric utilities using wet scrubbers for multi-pollutant control. Washington, DC: US Environmental Protection Agency, EPA/600/R-08/077.

Sandhu, S.S. and G.L. Mills. 1991. Mechanisms of mobilization and attenuation of inorganic contaminants in coal ash basins. *ACS Symposium Series* 468: 342–364.

Saraber, A., R. Overhof, T. Green et al. 2012. Artificial lightweight aggregates as utilization for future ashes. *Waste Management* 32: 144–152.

Sarker, M., B. Chowdhury, and I. Hossain. 2012. Power generation from coal—A review. *Journal of Chemical Engineering* 27(2): 50–54.

Schuetze, J., D. Kunth, S. Weissbach et al. 2012. Mercury vapor pressure of flue gas desulfurization scrubber suspensions. Effects of pH level, gypsum and iron. *Environmental Science & Technology* 46: 3008–3013.

Schultz, T.W., J.N. Dumont, and M.V. Buchanan. 1983. Toxic and teratogenic effects of chemical class fractions of a coal gasification electrostatic precipitator tar. *Toxicology* 29: 87–99.

Schwarz, M.A., A. Behnke, M. Brandt et al. 2014. Semipolar polycyclic aromatic compounds: Identification of 15 priority substances and the need of regulatory steps under REACH regulation. *Integrated Environmental Assessment and Management.* doi:10.1002/ieam. 1526.

Selcuk, N, Y. Gogebakan, and Z. Gogebakan. 2006. Partitioning of trace elements during pilot-scale fluidized bed combustion of high ash content lignite. *Journal of Hazardous Materials B* 137: 1698–1703.

Senapati, P.K. and B.K. Mishra. 2012. Design considerations for hydraulic backfilling with coal combustion products (CCPs) at high solids concentrations. *Powder Technology* 229: 119–125.

Seredin, V.V. and S. Dai. 2012. Coal deposits as potential alternative sources of lanthanides and yttrium. *Journal of Coal Geology* 94: 67–93.

Shaheen, S.M., P.S. Hooda, and C.D. Tsadilas. 2014. Opportunities and challenges in the use of coal fly ash for soil improvements—A review. *Journal of Environmental Management* 145: 249–267.

Shandilya, N., O. le Bihan, M. Morgeneyer. 2014. A review on the study of the generation of (nano)particles aerosols during the mechanical solicitation of materials. *Journal of Nanomaterials* 289108: 16 pp.

Sheshadri, K.S. and A. Shamsi. 1998. Effects of temperature, pressure, and carrier gas on the cracking of coal tar over a char-dolomite mixture and calcined dolomite in a fixed bed reactor. *Industrial Engineering and Chemistry Research* 37: 3830–3837.

Shi, L., P. Xu, K. Xie et al. 2011. Preparation of a modified flue gas desulphurization residue and its effect on pot sorghum growth and acidic soil amelioration. *Journal of Hazardous Materials* 192: 978–985.

Shon, C., A.K. Mukhopadhyay, D. Saylak et al. 2010. Potential use of stockpiled circulating fluidized bed combustion ashes in controlled low strength material (CLSM) mixture. *Construction and Building Materials* 24: 839–847.

Shoumkova, A.S. 2011. Magnetic separation of coal fly ash from Bulgarian power plants. *Waste Management & Research* 29: 1078–1089.

Silverstein, M., N. Maizlish, R. Park et al. 1985. Mortality among workers exposed to coal tar pitch volatiles and welding emissions. *American Journal of Public Health* 75: 1283–1287.

Simarro, R, N. Gonzalez, L.F. Bautista et al. 2013. Assessment of the efficiency of *in situ* bioremediation techniques in a creosote polluted soul: Change in bacterial community. *Journal of Hazardous Materials* 262: 158–167.

Singh, J.S. and V.C. Pandey. 2013. Fly ash application in nutrient poor agricultural soils: Impact on methanotrophs population dynamics and paddy fields. *Ecotoxicology and Environmental Safety* 89: 43–51.

Slaghuis, J.H., A.M. Ooms, and H.B. Erasmus. 1995. Thermal processing of unused waste products: The sasol perspective. *ACS Division of Fuel Chemistry Preprints* 40: 87–91.

Sotoudehnia, M.M., A.K. Soltani, A. Maghsouipour et al. 2010. The effect of modification of matrix on densification efficiency of pitch based carbon composites. *Journal of Coal Science & Engineering* 16: 498–414.

Spadoni, M., M. Voltaggio, E. Sacchi et al. 2014. Impact of the disposal and re-use of fly ash on water quality: The case of the Koradi and Khaperkheda thermal power plants (Maharashtra, India), *Science of the Total Environment* 479–480: 159–170.

Srivastava, S., R. Chaudary, and D. Khale. 2008. Influence of pH, curing time and environmental stress on the immobilization of hazardous waste using activated fly ash. *Journal of Hazardous Materials* 153: 1103–1109.

Stam, F., R. Meij, H. te Winkel et al. 2011. Chromium speciation in coal and biomass co-combustion products. *Environmental Science & Technology* 45: 2450–2456.

Stencel, J.M., H. Song, and F. Cangialosi. 2009. Automated foam index test: Quantifying air entrainment agent addition and interaction with fly ash cement admixtures. *Cement and Concrete Research* 39: 362–370.

Stevens, W., T. Robi, and K. Mahboub. May 4–7, 2009. The cementitious and pozzolanic properties of fluidized bed combustion fly ash. *World of Coal Ash Conference.* Lexington, KY.

Stout, W.L., M.R. Daily, T.L. Nickeson et al. 1997. Agricultural uses of alkaline fluidized bed combustion ash. *Fuel* 76: 767–769.

Su, T., H. Shi, and J. Wang. 2011. Impact of trona-based SO_2 control on the elemental leaching behavior of fly ash. *Energy & Fuels* 25: 3514–3521.

Sudha, J. and G. Dinesh. 2006. Fly ash as a soil ameliorant for improving crop production—A review. *Bioresource Technology* 97: 1136–1174

Sun, M., G. Cheng, R. Lu et al. 2014a. Characterization of Hg^0 re-emission and Hg^2+ leaching potential from flue gas desulfurization (FGD) gypsum. *Fuel Processing Technology* 118: 28–33.

Sun, L., L. Feng, D. Yuan et al. 2013. The extent of the influence and flux estimation of volatile mercury from the aeration poll in a typical coal fired power plant equipped with seawater flue gas desulfurization system. *Science of the Total Environment* 444: 559–564.

Sun, M., J. Hou, G. Cheng et al. 2014b. The relationship between speciation and release ability of mercury in flue gas desulfurization (FGD) gypsum. *Fuel.* doi:10.1016/j.fuel.2014.02.012.

Sun, M., J. Hou, T. Tang et al. 2012. Stabilization of mercury in flue gas desulphurization gypsum from coal-fired electric power plants with additives. *Fuel Processing Technology* 104: 160–166.

Szabo, Z., P. Völgesi, H.E. Nagy et al. 2013. Radioactivity of natural and artificial building materials—A comparative study. *Journal of Environmental Radioactivity* 118: 64–74.

Tao, D., M. Fan, and X. Jiang. 2009. Dry coal fly ash cleaning using rotary triboelectrostatic separator. *Mining Science and Technology* 19: 0642–0647.

Tchapda, A.H. and S.V. Pisupati. 2014. A review of thermal co-conversion of coal and biomass/waste. *Energies* 7: 1098–1148.

Telesca, A., M. Marroccoli, D. Calabrese et al. 2013. Flue gas desulphurization gypsum and coal fly ash as basic components of prefabricated building materials. *Waste Management* 33: 628–633.

Thorneloe, S., D.S. Kosson, F. Sanchez et al. 2010. Evaluating the fate of metals in air pollution control residues from coal-fired power plants. *Environmental Science and Technology* 44: 7351–7356.

Tian, H.Z., L. Lu, J.M. Hao et al. 2013. A review of key hazardous trace elements in Chinese coals: Abundance, occurrence, behavior during coal combustion and their environmental impacts. *Energy and Fuels* 27: 601–614.

Tian, H., K. Liu, J. Zhou et al. 2014. Atmospheric emission inventory of hazardous trace elements from China's coal fired power plants- temporal trends and spatial variation characteristics. *Environmental Science and Technology.* doi:10.1021/es404730j.

Tomasevic, D.D., M.B. Dalmacija, M.D. Prica et al. 2013. Use of fly ash for the remediation of metals polluted sediment—green remediation. *Chemosphere* 92: 1490–1497.

Toraason, M., C. Hayden, D. Marlow et al. 2001. DNA strand breaks, oxidative damage, and 1-OH pyrene in roofers with coal pitch dust and/or asphalt fume exposure. *International Archives of Occupational and Environmental Health* 74: 396–404.

Tremblay, C.B. Armstrong, G. Thériault et al. 1995. Estimation of developing bladder cancer among workers exposed to coal tar pitch volatiles in the primary aluminum industry. *American Journal of Industrial Medicine* 27: 335–348.

Tripathi, R.C., S.K. Jha, L.C. Ram et al. 2013. Fate of radionuclides present in Indian fly ashes on its application as soil ameliorant. *Radiation Protection Dosimetry* 156: 198–206.

Tsiridis, V., M. Petala, P. Samaras et al. 2012. Environmental hazard assessment of coal fly ashes using leaching and ecotoxicity tests. *Ecotoxicology and Environmental Safety* 84: 212–220.

Tsai, J.H., K.H. Lin, C.Y. Cheng et al. 2007. Chemical constituents in particulate emissions from an integrated iron and steel facility. *Journal of Hazardous Materials* 147: 111–119.

Tsubouchi, N., H. Hayashi, A. Kawahima et al. 2011. Chemical forms of fluorine and carbon in fly ashes recovered from electrostatic precipitators of pulverized coal-fired plants. *Fuel* 90: 376–383.

Twardowska, I. and Szczepanska, J. 2002. Solid waste: Terminological and long term environmental risk assessment problems exemplified in a power plant fly ash study. *Science of the Total Environment* 285: 29–51.

Ukwattage, N.L., P.G. Ranjith, and M. Bouazza. 2013. The use of coal combustion fly ash as soil amendment in agricultural lands (with comments on its potential to improve food security and sequester carbon). *Fuel* 109: 400–408.

Van Maanen, J.M.S, P.J.A. Borm, A. Knaapen et al. 1999. In vitro effect of coal fly ashes. *Inhalation Toxicology* 11: 1123–1141.

Van Metre, P.C., B.J. Mahler, and J.T. Wilson. 2009. PAHs underfoot: Contaminated dust from coal-tar sealcoated pavement is widespread in the United States. *Environmental Science and Technology* 43: 20–25.

Van Metre, P.C., M.S. Majewski, B.J. Mahler et al. 2014a. Volatilization of polycyclic aromatic hydrocarbons for coal-tar-sealed pavement. *Chemosphere* 88: 1–7.

Van Metre, P.C., M.S. Majewski, B.J. Mahler et al. 2014b. PAH volatilization following application of coal-tar-based pavement sealant. *Atmospheric Environment* 51: 108–115.

Vassilev, S.V., R. Menendez, M. Diaz-Somoano et al. 2004. Phase-mineral and chemical composition of coal fly ashes as a basis for their multicomponent utilization. 2. Characterization of ceramic cenospheres and salt concentrates. *Fuel* 83: 585–603.

Vejahati, F., Z. Xu, and R. Gupta. 2010. Trace elements in coal: associations with coal and minerals and their behavior during coal utilization—A review. *Fuel* 89: 904–911.

Verbinnen, B., P. Billen, M. van Coninckxloo et al. 2013. Heating temperature dependence of Cr (III) oxidation in the presence of alkali and alkaline earth salts and subsequent Cr (VI) leaching behavior. *Environmental Science & Technology* 47: 5858–5863.

Versluys, K., J. Bakker, P. Janssen et al. 2013. *Quick Scan on Human Health Risks of Working with Soil That Contains Recycled Building Materials*. Bilthoven, the Netherlands: RIVM.

Wagner, N.J., R.H. Matjie, J.H. Slaghuis et al. 2008. Characterization of unburnt carbon present in coarse gasification ash. *Fuel* 87: 683–691.

Wang, H., H. Ban, D. Golden et al. 2002. Ammonia release characteristics from coal combustion fly ash. *ACS Fuel Chemistry Division Preprints* 47: 836–838.

Wang, H., N. Bolan, M. Hedley et al. 2006. Potential uses of fluidized bed boiler ash (FBA) as a liming material, soil conditioner and sulfur fertilizer. In *Coal Combustion Byproducts and Environmental Issues*. Sajwan, K.S. et al. (eds.), New York: Springer, pp 202–215.

Wang, J., H.J. Hayes, T.G. Townsend et al. 2014. Characterization of vapor phase mercury released from concrete processing with baghouse filter dust added cement. *Environmental Science & Technology*. doi:10.1002/es4044962.

Wang, K., W. Orndorff, Y. Cao et al. 2013. Mercury transportation in soil using gypsum from flue gas desulphurization unit in coal fired power plant. *Journal of Environmental Science* 25: 1858–1864.

Wang, J. and A. Tomita. 2003. A chemistry on the volatility of some trace elements during coal combustion and pyrolysis. *Energy and Fuels* 17: 954–960.

Wang, J., T. Wang, H. Mallhi et al. 2007. The role of ammonia on mercury leaching from coal fly ash. *Chemosphere* 69: 1586–1592.

Watts D.B. and W.A. Dick. 2014. Sustainable uses of FGD gypsum in agricultural systems: introduction. *Journal of Environmental Quality* 43: 246–252.

Weissman, S.H., R.L. Carpenter, and G.J. Newton. 1983. Respirable aerosols from fluidized bed coal combustion. 3. Elemental composition f fly ash. *Environmental Science & Technology* 17: 65–71.

White, P.A. and L.D. Claxton. 2004. Mutagens in contaminated soil: A review. *Mutation Research* 567: 327–345.

Wigmore J.W. and Kubrycht. 1990. Sea lagoons in Hong Kong for the disposal of pulverized fuel ash. *Waste Management & Research* 8: 405–417.

Wilczynska-Michalik, W., R. Moryl, J. Sobczyk et al. 2014. Composition of coal combustion by-products: the importance of combustion technology. *Fuel Processing Technology* 124: 35–43.

Witter, A.E., M.H. Nguyen, S. Bidar et al. 2014. Coal-tar-based sealcoated pavement: A major PAH source to urban stream sediments. *Environmental Pollution* 185: 59–68.

Wu, C., U. Cao, Z. Doing et al. 2010. Mercury speciation and removal across full-scale wet FGD systems at coal-fired power plants. *Journal of Coal Science & Engineering* 16: 82–87.

Wu, S., S. Huang, Y. Wu et al. 2014. Characteristics and catalytic actions of inorganic constituents from entrained-flow coal gasification slag. *Journal of the Energy Institute*. doi:10.1016/j.joei.2014.04.001.

Xiao, X., H. Yang, H. Zhang et al. 2005. Research on carbon content in fly ash from circulating fluidized bed boilers. *Energy and Fuels* 19: 1520–1525.

Yan, L., X. Lu, Q. Wang et al. 2014. Research on sulfur recovery from the byproducts of magnesia wet flue gas desulfurization. *Applied Thermal Engineering* 65: 487–494.

Yanusa, I.A.M., V. Monoharan, B. Harris et al. 2013. Differential growth and yield by canola *(Brassica napus L.)* and wheat *(Triticum aestivum L.)* arising from alterations in chemical properties of sandy soils due to additions of fly ash. *Journal of the Science of Food and Agriculture* 93: 995–1002.

Yao, Y. and H. Sun. 2012. Characterization of new silica alumina-based backfill material utilizing large quantities of coal combustion byproducts. *Fuel* 97: 329–336.

Yao, D. and X. Zhi. 2010. The transformation and concentration of environmental hazardous trace elements during coal combustion. *Journal of Coal Science & Engineering* 16: 74–77.

Yao, Z.T., M.S. Xia and P.K. Sarker. 2014. A review of the alumina recovery from coal fly ash with a focus in China. *Fuel* 120: 74–85.

Yoshiie, R., Y. Taya. T. Ichiyanagi et al. 2013. Emissions of particles and trace elements from coal gasification. *Fuel* 108: 67–72.

Yu, J., X. Li, D. Fleming et al. 2012. Analysis on characteristics of fly ash from coal-fired power stations. *Energy Procedia* 17: 3–9.

Zacco, A., L. Borgese, A. Gianoncelli et al. 2014. Review of fly ash inertisation treatments and recycling. *Environmental Chemistry Letters* 12: 153–175.

Zeng, X., P. Wang and H. Li. 2014. Pilot verification of a low-tar two-stage coal gasification process with a fluidized bed pyrolyzer and fixed bed gasifier. *Applied Energy* 115: 9–16.

Zhang, J., H. Cui, B. Wang et al. 2013. Fly ash cenospheres supported visible-light-driven $BiVO_4$ photocatalyst: Synthesis, characterization and photocatalytic application. *Chemical Engineering Journal* 223: 737–746.

Zhang, W., J. Ma, S. Yang et al. 2006. Pretreatment of coal gasification wastewater by acidification. *Chinese Journal of Chemical Engineering* 14: 398–401.

Zhang, Z. J. Qian, C. You et al. 2012. Used of circulating fluidized bed combustion fly ash and slag in autoclaved brick. *Construction and Building Materials* 13: 109–116.

Zhao, J., G. Cai, D. Gao et al. 2014a. Influences of freeze-thaw cycle and curing time on chloride ion penetration resistance of sulphoaluminate cement concrete. *Construction and Building Materials* 53: 305–311.

Zhao, Z., Q. Du, G. Zhao et al. 2014b. Fine particle emission from and industrial coal-fired circulating fluidized-bed boiler equipped with a fabric filter in China. *Energy & Fuels* 28: 4769–4780.

Zhao, B. and S, Shi. 2014. Modeled exposure assessment via inhalation and dermal pathways to airborne semivolatile organic compounds (SVOCs) in residences. *Environmental Science & Technology* 48: 5691–5699.

Zhiang, J., J. Yu, P. Wang et al. 2013. A review on R&D status and application of fly ash cenospheres in microwave absorption. *Advanced Materials Research* 634–638: 1886–1889.

Zhou, H., B. Jin, R. Xiao et al. 2009. Distribution of polycyclic aromatic hydrocarbons in fly ash during coal and residual char combustion in a pressurized fluidized bed. *Energy and Fuels* 23: 2031–2034.

19 Environmental Issues in Coal Utilization and Mitigation Measures

Deepak Pudasainee and Rajender Gupta

CONTENTS

Abstract: Coal is one of the most polluting sources of energy. Environmental impacts are noticed during extraction, transportation, storage, preparation, and utilization of coal. Combustion of coal releases NO_x, SO_x, particulates, toxic metals, greenhouse gases, waste water, and residues into the environment. The use of appropriate technology and adoption of proper management practices can reduce the amount of pollutants released into the environment and their adverse impacts to the human beings and the ecosystem. There has been a distinct success in controlling NO_x, SO_x, and particulates emission from coal combustion flue gas; however, controlling CO_2 and volatile metals are still challenging.

19.1 BACKGROUND

Coal is widely available and cheaper source of energy compared to nuclear, natural gas, and oil. It has been utilized for residential, commercial, and industrial heating and power generation since long. Environmental impacts are noticed during coal extraction, transportation, storage, preparation, and utilization. Various aspects of coal such as composition, worldwide reserves, mining, processing, utilization, and future consumption trends are presented in earlier chapters. In this chapter, our major focus is to highlight the environmental impacts of coal utilization, mainly due to coal combustion in large power plants, and the mitigation measures to reduce their impact on the environment and to human beings. The following are some examples of environmental impacts during different stages of coal use:

- Extraction of coal from mine can cause landscape pollution, particulate emission, soil erosion, noise and vibration, water pollution, and impact local biodiversity.
- After coal is extracted from the mine, often the place is abandoned. Subsidence above mining area may occur.
- Acid mine drainage, due to reaction of water and sulfur bearing rocks, coming from coal mining area pollutes water bodies.
- Coal mining activities can also have secondary harmful effects due to leaching of acid and trace elements from residual materials.
- When the exposed coal gets in contact with water, trace metals get dissolved and contaminate drinking water resources. Some of these metals (e.g., Hg, Cd, Pb, and As) are very toxic to wildlife and human beings.
- The combustion of coal releases several harmful pollutants: particulates, NO_x, and SO_x. Release of NO_x and SO_x may cause acid rain.
- The combustion of coal releases toxic heavy metals, including mercury (Hg), which is neurotoxic.
- Fly ash contaminated with harmful pollutants generated from coal combustion often causes solid waste disposal problem.
- Combustion of coal releases greenhouse gases (GHGs) such as N_2O and CO_2 into the atmosphere.

19.2 EMISSION FROM COAL COMBUSTION

Currently, huge amount of coal is combusted in coal-fired power plants to generate electricity. Combustion of coal releases several air pollutants such as NO_x, SO_x, and particulates and GHGs such as N_2O and CO_2 into the atmosphere. In addition, volatile organic compounds (VOCs), toxic compounds, halides, and dioxins are emitted in lesser quantity. Once emitted into the atmosphere, these pollutants undergo transformation, transportation, and are finally deposited onto the earth surface and water bodies via dry and/or wet deposition. The harmful effect of them in the environment has been reported worldwide.

The emission of air pollutants from coal-fired power plants has been one of the major environmental problems. The concern over particulate, NO_x, and SO_x emission control has increased, influenced by the stringent regulation worldwide. Reducing NO_x and SO_x has been one of the major technical and regulatory challenges over the decade, and to some extent, technological control has been achieved. Some developed countries such as Germany, Japan, and South Korea have massively reduced the emission of such pollutants by the application of advanced air pollution control devices (APCDs) and other measures. Emission limits are stringent in the developed countries, and they have the technology in place to cope with this, whereas developing countries lack technologies (due to either technical know-how or the economy) and emission limits are less stringent. Emission standards for coal-fired power plants in China, the European Union, and the United States are compared in Table 19.1. This gives some insight into the regulatory standards and the technological advancements in developing and developed world.

To control emission of these pollutants into the atmosphere, it is essential to understand the pollutants'. formation and emission behavior. For this reason, schematic diagram of a typical modern coal-fired power plant with APCDs configuration is presented in Figure 19.1. Most of the modern coal-fired power plants are installed with selective catalytic reactor (SCR) for NO_x removal, electrostatic precipitator (ESP) for particulate removal, and flue-gas desulfurization (FGD) for SO_x removal. There are older plants that are with only ESP, or ESP plus FGD. Most of the power plants in developing countries lack advanced APCDs.

ESP removes suspended particles in gas stream using an electrostatic force. ESPs work on a principal of migration of charged particle in an electric field. The particles in gas stream are charged and exposed to an electric field from where they are removed. SCR is a post-combustion NO_x emission control technology. In SCR, NO_x is reduced to N_2 and H_2O through a series of reactions with ammonia or urea injected into the flue gas. An FGD system controls SO_x where lime or limestone is introduced into the flue gas to react with SO_2 and water; these form salt by-products that are separated and removed. The descriptions of each of these APCDs are presented in Section 19.3.

19.2.1 PARTICLE EMISSION

Coal combustion is a major sources of particulate matter (PM) emission into the atmosphere. PM consists of microscopic solid particles or liquid droplets. PM emission from the combustion of coal is resulted from the mineral portion. Poor combustion conditions lead to the formation of soot. Fine particulates with aerodynamic diameter less than 10 μ (PM_{10}) and less than 2.5 μ ($PM_{2.5}$) may cause severe human health effects. The health effects of ambient fine particulates have been correlated with human mortality rate. In the United States, National Ambient Air Quality Standard (NAAQS) for PM_{10} and $PM_{2.5}$ is established and regulated. Understanding characteristics of particle size distributions (PSDs) is critical in understanding mechanisms and devising control technology.

PM emitted from incinerators and boilers can be classified into two categories: (1) ultrafine PM under 0.1 μm with

TABLE 19.1
Emission Standards for Coal-Fired Power Plants in China, the European Union, and the United States

Pollutants	Plant Types	China	European Union[a]	United States[b]
Particulate matter (mg/m^3)	New and existing	30	50, with an exception of 100 for low-quality coal, such as lignite	22.5
NO_x (mg/m^3)	New	100	500 until December 31, 2015; then 200	117
	Existing	100 (built 2004–2011) 200 (built before 2004)	500 until December 31, 2015; then 200	117 (built after 2005) 160 (built 1997–2005) 640 (built 1978–1996)
SO_2 (mg/m^3)	New	100	200	160 (built after 2005)
	Existing	200 (28 provinces) 400 (4 provinces with high S coals)	400	160 (built 1997–2005) 640 (built 1978–1996)
Hg (mg/m^3)	New	0.03	–	0.001 (bituminous, gangue) 0.005 (lignite)
	Existing	0.03	–	0.002 (bituminous, gangue) 0.006 (lignite)

Source: World Resources Institute, China adopts world-class pollutant emissions standards for coal power plants, http://www.chinafaqs.org/files/chinainfo/China%20FAQs%20Emission%20Standards%20v1.4_0.pdf, June 2012.

[a] For power plants >500 MW in size.

[b] Units in the standards have been converted to concentrations.

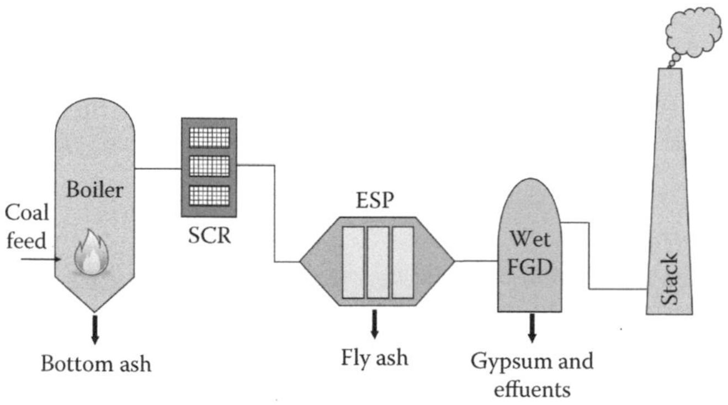

FIGURE 19.1 Schematic diagram of a typical modern coal-fired power plant with APCDs: SCR, ESP, and wet FGD.

nuclei formation (homogenous) and growth by condensation (heterogenous) and coagulation at high temperature (Linak et al. 2000) and showing a mode in submicron size range and (2) soot type of coarse PM known to be formed mechanically such as the generation of unburned products in poor combustion conditions. Ultrafine or submicron PMs are composed of metals that vaporize in the hot zones of flame and form in to coarser sizes but not bigger than 1 µm due to their coagulation mechanism in the downstream with decreasing temperatures.

The mechanism of particle formation in a combustion system is shown in Figure 19.2. The physical form and the partitioning of trace elements in combustion flue gas depend on several factors as follows (Ratafia-Brown 1994; Xu et al. 2003):

1. Heterogeneous condensation on the existing fly ash particles and heat exchange on surfaces
2. Physicochemical adsorption on fly ash particles
3. Species with high vapor pressure at typical boiler exit temperature continues to remain in vapor phase
4. Homogeneous (flue gas constituents) and heterogeneous (fly ash) chemical reactions between trace elements
5. Homogeneous condensation (nucleation) and coalescence as submicron aerosols

19.2.2 NO_x Emission

NO_x refers to oxides of nitrogen. NO_x emissions from coal combustion are mainly composed of nitric oxide (NO), with lesser amount of nitrogen dioxide (NO_2). In general, NO_x is composed of about 95% NO, about 5% NO_2, and <1% N_2O. NO and NO_2 are precursors of acid rain and form photochemical smog and ground level ozone, while N_2O is a GHG that causes global warming. NO_x during coal combustion is formed from two basic routes, that is, thermal conversion of N_2 (air injected) and oxidation of nitrogen (present in coal). In pulverized coal boilers, fuel N_2 contributes about 80% of total NO_x emission and the rest 20% is thermal. Thermal NO_x formation has shown that the NO_x concentration is exponentially dependent on temperature and is proportional to nitrogen concentration in the flame, the square root of oxygen concentration in the flame, and the gas residence time (Lim et al. 1979).

Formation of NO from N_2 (i.e., thermal NO) takes place via the following reaction pathways (Zeldovich 1946; Miller and Bowman 1989). Equation 19.1 has high activation energy,

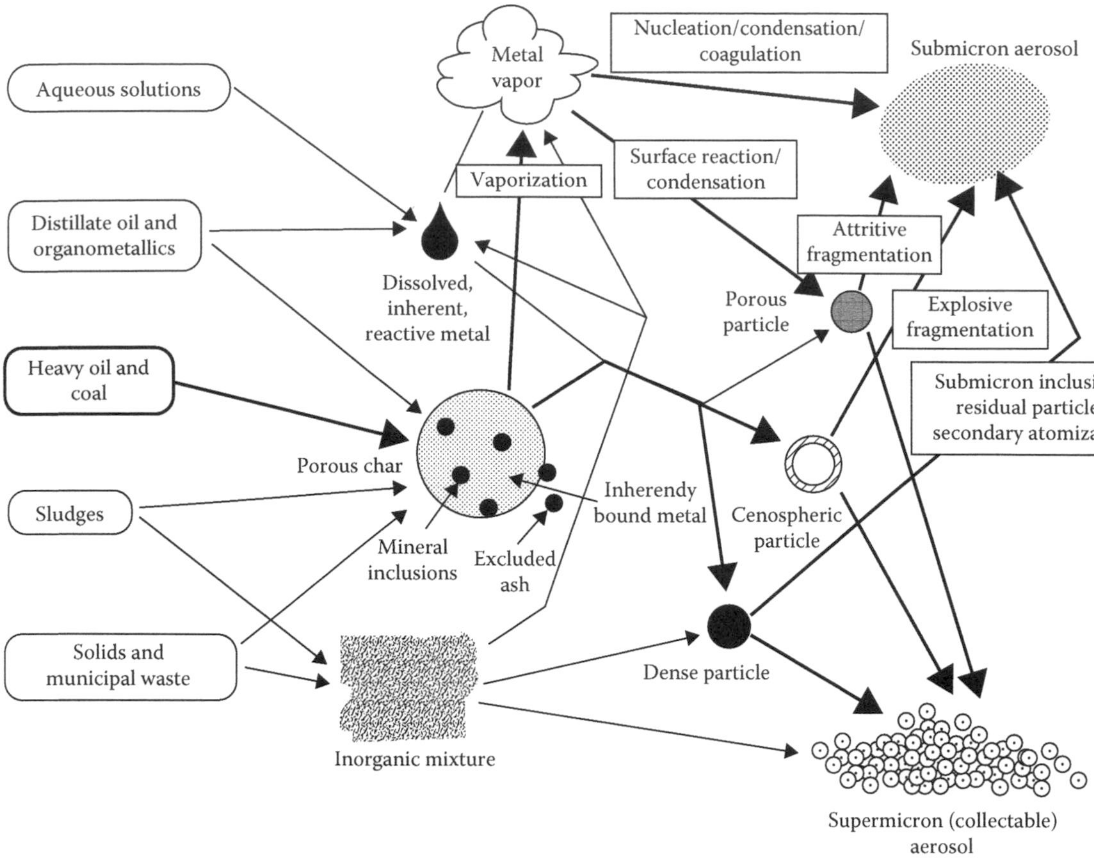

FIGURE 19.2 Particle formation mechanisms in combustion system. Bold arrows and lines show the specific paths for coal. (Reprinted from *Fuel Process. Technol.*, 39, Linak, W.P. and Wendt, J.O.L., Trace metal transformation mechanisms during coal combustion, 173–198, Copyright 1994, with permission from Elsevier.)

and formation of thermal NO is most important at temperatures above 1300°C. NO may be formed from N_2 through prompt NO (Fenimore 1971), initiated by attack of the CH radical on the N_2 triple bond (Glarborg et al. 1986; Miller and Bowman 1989).

$$O + N_2 \rightleftharpoons NO + N \quad (19.1)$$

$$N + O_2 \rightleftharpoons NO + O \quad (19.2)$$

$$N + OH \rightleftharpoons NO + H \quad (19.3)$$

$$CH + N_2 \rightleftharpoons NCN + H \quad (19.4)$$

Formation of N_2O is initiated as follows (Malte and Pratt 1975):

$$O + N_2 + M \rightleftharpoons N_2O + M \quad (19.5)$$

The nitrogen compounds thus formed (NCN, N_2O) are reactive, and depending on the reaction conditions, they may subsequently be oxidized to NO or yield back N_2.

Formation of fuel NO during combustion is well known. As combustion begins, the N in fuel is distributed between the volatiles and the solid char matrix. For bituminous coals, the volatile N consists mostly of tar compounds, which decay to hydrogen cyanide (HCN) or soot nitrogen at high temperatures. In case of low-rank coals, light nitrogen species may be released from the solid matrix. HCN is oxidized to NO by intermediate amines, depending on stoichiometry and fuel N concentration; it may be converted to N_2 by the recombination of NO with another nitrogen-containing species. The remaining char N and soot N undergo heterogeneous oxidation to NO and N_2, or they may evolve as light components such as HCN in high temperature. NO thus formed may be recycled by HC radicals to cyanide or reduced to N_2 by surface reactions on char or soot (referred from Glarborg et al. 2003).

Figure 19.3 shows results of the relative yields of fuel NO and thermal NO in pulverized coal combustion. In selected combustion experiments, the air was replaced by nitrogen-free synthetic air mixture ($O_2/Ar/CO_2$), so that fuel NO was formed. The NO formation from volatiles contributes to most of the fuel NO, whereas char N oxidation was a minor source. The high levels of NO formed from fuel N inhibit the thermal NO formation; the initiating step (Equation 19.6) is the reverse of Equation 19.1. Equations 19.1 and 19.6 have high activation energy, and due to the interaction between the two NO formation mechanisms, temperatures in excess of 1920°C may be required for a significant contribution from thermal NO in coal-dust flames (Pershing and Wendt 1976).

$$NO + N \rightleftharpoons O + N_2 \quad (19.6)$$

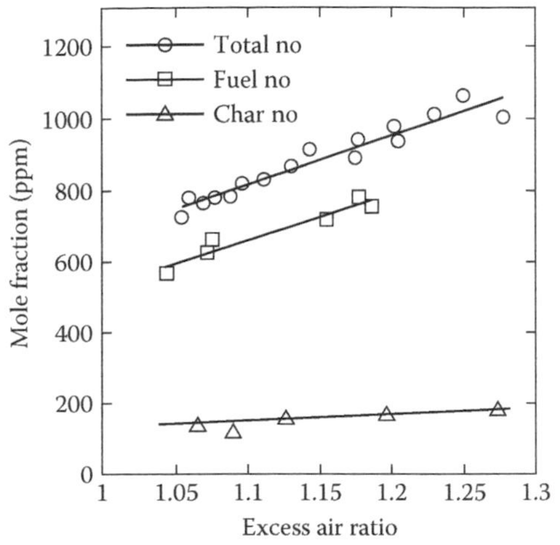

FIGURE 19.3 Source of NO_x emissions in turbulent diffusion pulverized coal flames. (Reprinted from *Prog. Energy Combus. Sci.*, 29, Glarborg, P. et al., Fuel nitrogen conversion in solid fuel fired systems, 89–113, Copyright 2003, with permission from Elsevier.)

19.2.3 SO_x Emission

SO_x refers to sulfur oxides. Combustion of sulfur-containing coal results in SO_x formation. Flue gas from coal combustion mainly consists of sulfur dioxide (SO_2) and lesser amount of sulfur trioxide (SO_3) (U.S. EPA 1993). SO_x emission during coal combustion is resulted from the oxidation of organic and pyritic sulfur present in the coal. When coal is combusted, on an average, about 95% of the sulfur present in coal is emitted as SO_2.

$$S + O_2 \rightarrow SO_2 \qquad (19.7)$$

In the presence of excess amount of oxygen and favorable conditions such as temperature (800°C), SO_2 is further oxidized to SO_3.

19.2.4 Emission of Greenhouse Gases

Coal mining and utilization can result GHG emission in two ways: firstly, due to release of methane (CH_4) into the atmosphere during mining, and secondly, emission of GHGs during coal combustion. Combustion of fossil fuels results in the emission of GHGs, mainly CO_2, CH_4, and N_2O (Wojtowicz et al. 1993; U.S. EPA 2008). As the carbon content in coal is the highest among fossil fuels, the emission of CO_2 from coal combustion is significant. Globally, fossil fuel–fired power plants are the largest anthropogenic emission sources contributing about one-third of CO_2 emissions. In 2010, 43% of CO_2 emissions from fuel combustion were produced from coal, 36% from oil, and 20% from gas (IEA 2012). Annual emissions of CO_2 have grown between 1970 and 2004 by about 80%, from 21 to 38 Gt, and represented 77% of total anthropogenic GHG emissions in 2004 (IPCC 2007b).

CO_2 capture and storage has become an important global topic due to the significant and continuous increase of CO_2 concentrations in the atmosphere, basically due to fossil fuel combustion, mainly coal. Between 2009 and 2010, CO_2 emissions from coal combustion increased by 4.9% and represented 13.1 Gt of CO_2 (IEA 2013). The CO_2 emission level in 2050 is estimated to be twice the 2007 CO_2 emission levels, given that no measures are taken to reduce the emissions (IEA 2010). The increasing trends of CO_2 in the atmosphere can lead to severe consequences including rise in sea levels, which causes displacement of human establishments, extreme weather events, and changes to rainfall patterns, resulting in droughts and floods affecting food production, affecting biodiversity, increasing human diseases, and increasing mortality (IPCC 2007a). Influence by these carbon capture and sequestration (CCS) has become a key solution to GHGs emission reduction.

19.2.5 Emission of Trace Metals (Except Mercury)

Coal contains wide varieties of elements—almost all the elements in the Periodic Table (Raask 1985; Swaine and Goodarzi 1995). Elements in coal can be divided into three groups: (1) major elements (C, H, N, S, O) that are in concentrations above 1000 ppm; (2) minor elements (Si, Al, Fe, Ca, Mg, and sometimes also the alkali elements [Na and K] and halogens [F, Cl, Br, I]) that are in concentrations between 100 and 1000 ppm; and (3) trace elements that are in concentrations below 100 ppm. The most important trace elements in coal are Be, B, K, Cr, Mn, Fe, Co, Ni, Cu, Zn, As, Se, Br, Mo, Cd, Sn, Sb, Ba, Hg, Tl, and Pb.

Table 19.2 shows the mean values of trace elements content in coal from the United States, Britain, Australia, and China, resulting from more than 1500 coal samples (Raask 1985; Swaine 1990; Swaine and Goodarzi 1995; referred from Xu et al. 2003). As seen from the table, most of the trace elements are concentrated less than 50 ppm, which varies in coal depending on the coalification processes, origin, geological formation, and many other factors; some of these effects are clearly noticeable here.

Trace elements such as Hg, Cd, Pb, and As are great threat to the environment and human beings. Minor elements (e.g., Na, K, V, Zn) may cause corrosion problems within the combustion and incineration facility, which lead to fouling of turbine blades, poison catalysts (mainly As), or sorbents downstream of the boiler. Many of the toxic metals emitted from power plants belong to persistent, bio-accumulative, toxic (PBT) chemicals and leads to serious human health effects (NET 2004). U.S. EPA has identified 67 hazardous compounds and chemicals in the flue gas emitted from power plant smokestacks. Those are known or suspected neurotoxins, developmental toxins, and probable human carcinogens (U.S. EPA 1998).

19.2.5.1 Distribution of Trace Elements in Combustion Flue Gas

Trace elements emission during coal combustion to a great extent is affected by their occurrence modes in coal. The elements that are associated with coal organic and sulfide fractions tend to vaporize first and are easily adsorbed by fine particles during flue-gas cooling. Trace elements associated with isolated mineral matters have higher possibility of

TABLE 19.2
Arithmetic Mean Values (ppm) of Concentrations of Trace Elements in Coal from the United States, Britain, Australia, and China

Element and Chemical Symbol	Coals													
	United States		Britain			Australia	China							For Most Coals
	(1)	(2)	(3)	(4)	(5)	(6)	(7)	(8)	(9)	(10)	(11)	(12)	(13)	
						>50 ppm								
Barium (Ba)		150	70–300		142	70–300								20–1000
Boron (B)	102	50	30–60			30–60								5–400
Fluorine (F)	61	74	150		114	150								20–500
Manganese (Mn)	49	100	130		84	130								5–300
Phosphorus (P)	71		–			–								10–3000
Strontium (Sr)	37	100	100			100								15–500
Titanium (Ti)	700	800	900	63		900								10–2000
Zinc (Zn)	272	39	25			25								5–300
						10–50 ppm								
Arsenic (As)	14	15	1.5		18	1.5	14.5	9.9	12.1	21.0	11.0	9.6	13.9	0.5–80
Cerium (Ce)	11		–			–								–
Bromine (Br)	15		–			–								–
Chlorine (Cl)			150			150								50–2000
Chromium (Cr)	14	15	6		34	6	36.8	25.4	21.6	30.4	26.0	12.0	74.0	0.5–60
Copper (Cu)	15	19	15	48		15	27.5	33.4	31.4	21.6	23.3	19.5	32.1	0.5–50
Lead (Pb)	35	16	10	48	38	10	20.9	18.1	12.2	29.4	22.8	22.7	24.4	2–80
Lithium (Li)		20	20			20								1–80
Nickel (Ni)	21	15	15		28	15	13.9	18.6	17.1	17.0	12.4	9.3	24.9	0.5–50
Rubidium (Rb)	14													2–50
Vanadium (V)	33	20	20		76	20	76.5	100.0	70.8	54.2	48.5	38.3	109.0	2–100
Zirconium (Zr)	72	30	100			100								5–200
						1–10 ppm								
Antimony (Sb)	1.3	1.1	0.5		3.1	0.5								0.05–10
Beryllium (Be)	1.6	2	1.5		1.8	1.5	3.1	2.6	1.8	1.3	1.9	1.3	2.5	0.1–15
Cadmium (Cd)	2.5	1.3	0.08	0.24	0.4	0.08	0.19	0.15	0.29	0.30	0.10	0.08	0.25	0.1–3
Cesium (Cs)			1.3			1.3								0.3–5
Cobalt (Co)	9.6	7	4			4	8.5	9.5	11.6	7.4	6.7	5.6	10.8	0.5–30
Gallium (Ga)	3.1		4			4								1–20
Germanium (Ge)	6.6		6	6.8	5.1	6	1.95	1.48	0.47	0.40	0.94	0.63	0.93	0.5–50
Iodine (I)	2.0		–			–								–
Lanthanum (La)	6.9		16			16								–
Molybdenum (Mo)	7.5	3	1.5		<2	1.5								0.1–10
Niobium (Nb)		3	–			–								1–20
Scandium (Sc)	2.4	3	4			4								1–10
Selenium (Se)	2.1	4.1	0.8		2.8	0.8								0.2–4
Thallium (Tl)			–			–								<02–1
Thorium (Th)	2.0		2.7	3.9		2.7								0.5–10
Uranium (U)	1.6	1.8	2		1.3	2								0.5–10
						<1 ppm								
Mercury (Hg)	0 2	0.18	0.1	0.2		0.1								0.02–1
Silver (Ag)	0.2		<0.1			<0.1								0.02–2
Tantalum (Ta)	0.15													

Source: Xu, M. et al., *Fuel Process. Technol.*, 85(2–3), 215–237, Copyright 2003, with permission from Elsevier.

Note: (1), 101 samples of mainly Illinois bituminous coals; (2), 799 samples of bituminous and sub-bituminous coals; (3), 23 samples of bituminous coals; (4), 231 samples of bituminous coals; (5), number of samples of not stated, bituminous coals; (6), 452 samples of New South Wales bituminous coals; (7), Qingshan bituminous coal; (8), Heshan bituminous coal; (9), Laiyang anthracite; (10), Jiafu anthracite; (11), Henan lean coal; (12), Huangshi lean coal; (13), Shaoguan lean coal.

remaining in the ash. Other factors affecting release of trace elements are combustion environment (oxidizing, reducing), temperature and pressure, halogen concentration, flue-gas chemistry, and so on. Trace elements emission from coal-fired power plants also depends on the configuration of APCDs.

Trace metals species in off-gas may be removed during their progression along the fuel-gas cleaning path or may remain in the vapor state and thus pass through the APCDs and finally released into the atmosphere. Figure 19.4 shows the behavior of trace elements in coal combustion process with particulate and SO_2 control devices. Metals with higher boiling point such as As, Be, Co, Cd, Cr, Mn, Ni, Pb, and Sb are attached with particles and are removed in particulate control device. So installation of highly efficient PM control devices promotes the co-beneficial control of metals species. Metals having relatively lower boiling point such as Hg and Se are mostly distributed into gaseous phase that pass through the conventional APCDs and released into the atmosphere.

It is to be noted that with the trend of stringent regulation to particulate, NO_x, and SO_x emission control, coal-fired power plants have been implementing the various measures to comply with the regulatory limit such as blending fuels, changing APCDs configuration, and using low NO_x burner; these affect the trace elements formation, distribution, and emission behavior. Some aspects of the effect of these on Hg distribution and emission behavior are discussed in Section 19.2.6.

Distribution of metals during coal combustion varies depending on their boiling and melting temperatures. To explain the distribution behavior during combustion, the trace elements have been classified into three classes (Meij 1989) (Figure 19.5). Class I elements do not volatize during combustion and are distributed mainly in bottom ash and fly ashes. Class II elements vaporize but are found in fly ashes after condensation on particulates and nucleation mechanism due to decreasing temperature. Significant portion of these fine particles are in the submicron size class where duct control system is less effective. Class III elements vaporize and condense partly within the system (Kema 1997).

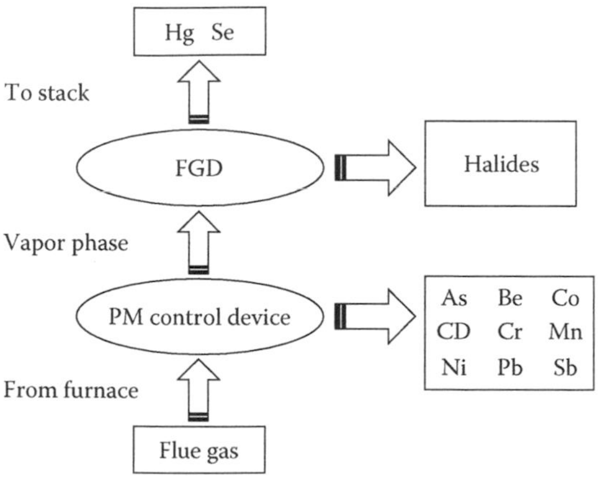

FIGURE 19.4 Behavior of trace elements in coal combustion flue gases. (Redrawn from CCT, *Clean Coal Today*, U.S. Office of Fossil Energy, Department of Energy, US DOE No. 25/Spring, 1997.)

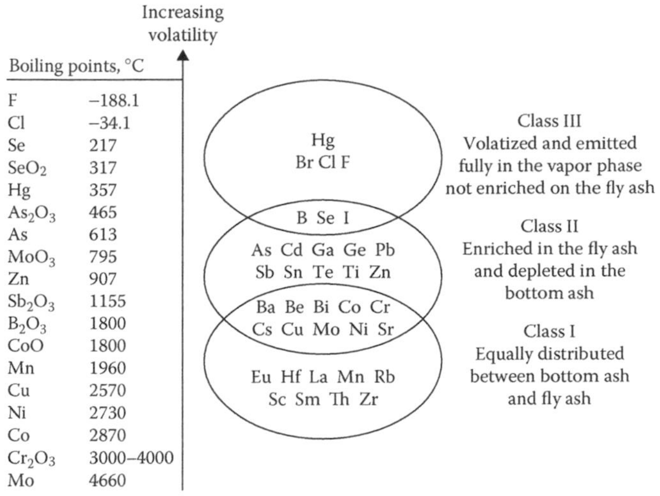

FIGURE 19.5 Categorization of trace elements during combustion based on volatility behavior. (Reprinted from *Fuel Process. Technol.*, 39(2), Ratafia-Brown, J.A., Overview of trace elements partitioning in flames and furnaces of utility coal-fired boilers, 139–157, Copyright 1994, with permission from Elsevier.)

The classification of trace elements into three classes (Figure 19.5) is based on the concept of relative enrichment. The relative enrichment factor of an element in ash is defined by relative enrichment index (REI), which reflects the change in concentration of elements (trace metals here) in combustion products. REI is also used to distinguish the different classes of elements. REI is calculated using the following formula (Huggins and Goodarzi 2009):

$$\text{REI} = \frac{\begin{array}{c}\text{Concentration of element in ash}\\ \times \text{ Ash percentage in the feed fuel}\end{array}}{\text{Concentration of element in feed fuel} \times 100}$$

For class I elements, REI ~ 1—these elements are equally distributed into bottom ash and fly ash. For class II elements, REI < 0.7 for bottom ashes and ~1.3–4 for fly ashes—these elements are usually enriched in fly ash and depleted in bottom ash. For class III elements, RE << 1 for bottom ashes and REI >> 10 for fly ashes—these elements are usually volatized and emitted fully in the vapor phase, not enriched in the fly ash. These numbers are based on pulverized coal combustion (dry bottom), with ESP (Kema 1997).

19.2.6 Emissions of Mercury

Trace quantity of Hg naturally occurs in coal (Table 19.2). Hg content in coal and raw materials differs depending on place of origin, formation, types, sulfide content, and so on. Hg in coal is primarily associated with inorganic mineral components such as pyrite (FeS_2), sphalerite (ZnCdS), cinnabar (HgS), and selenides such as ferroselite ($FeSe_2$) and also with organic components (Goodarzi and Swaine 1993; Finkelman 1994). The average Hg content in coals varies from 0.02 to 1 mg/kg (Swaine 1990; Sloss 1995). Concentration of Hg in coal combustion flue gas ranges 1–30 μg/m³. Even though Hg concentration in flue gas is lower, due to the large amount of coal consumption, combustion facilities emit significant amount of Hg into the atmosphere. Burning coal accounts for 46% of total global Hg emissions (IEA 2014b).

Hg enters into coal combustion facilities mainly through feed coal and to less extent from materials (e.g., $CaSO_3$). Once entered into the combustion system, Hg experiences transformation and removal within APCDs, and the remaining is emitted into the atmosphere. Among toxic heavy metals emitted from combustion facilities, Hg possesses greater concern in terms of controlling because it is volatile and passes through most of the conventional APCDs.

19.2.6.1 Mercury Speciation in Flue Gas

Hg speciation in flue gas refers to its distribution into various chemical forms. Hg compounds in combustion flue gas are speciated into three main forms: elemental (Hg^0), oxidized (Hg^{2+}), and particle bond (Hg_p) (Pacyna and Munch 1991). These all together comprise total Hg.

1. Particulate Hg (Hg_p): associated with particles (ash) and can be collected in particulate control devices
2. Gaseous elemental Hg (Hg^0): relatively non-reactive and insoluble in water and thus usually escapes emission control equipment and emitted into the atmosphere
3. Reactive gaseous or oxidized Hg (Hg^{2+}): soluble in water and can be removed in wet APCDs, such as in wet FGD

The property of various forms of Hg differs. Understanding Hg speciation is important to evaluate formation, distribution, transportation, deposition, and its impact in the environment. Further, it guides the emission control techniques and the policies to be implemented. Hg speciation in flue gas is a complex phenomenon primarily depending on the input composition (fuel, waste, and raw materials), availability of oxygen and chloride, affinity of Hg with fly ash, combustion temperature, flue gas residence time, process configuration, and so on.

Figure 19.6 shows the physicochemical transformation of Hg during coal combustion and in the resulting flue gas. During combustion, Hg in coal decomposes and converts mainly into Hg^0 (g), which is thermodynamically stable. As flue gas passes through the cleaning systems, temperature decreases and Hg^0 undergoes several reactions with flue-gas components transforming into Hg^{2+} and Hg_p. At lower concentration, such as in combustion flue gas, $Hg_2{}^{2+}$ compounds are unstable; thus, major oxidized form of Hg are assumed to be Hg^{2+} compounds (Aylett 1975). Hg compounds in combustion flue gas depend on the coal's chlorine concentration. Higher portion of Hg is speciated into Hg^{2+} in flue gas when the combusted coal has high chlorine concentration. During combustion, chlorine in coal is mainly released as atomic chlorine, which further forms HCl or Cl_2, as shown in Equations 19.8 through 19.11 (Kellie et al. 2005). The oxidation of Hg^0 by Cl atom is more favorable than by Cl_2 because the reactivity of Hg^0 with Cl atom (1.0×10^{-11} cm³/molecules/sec) is higher than that of Cl_2 (2.6×10^{-18} cm³/molecules/sec) (Parisa et al. 2002).

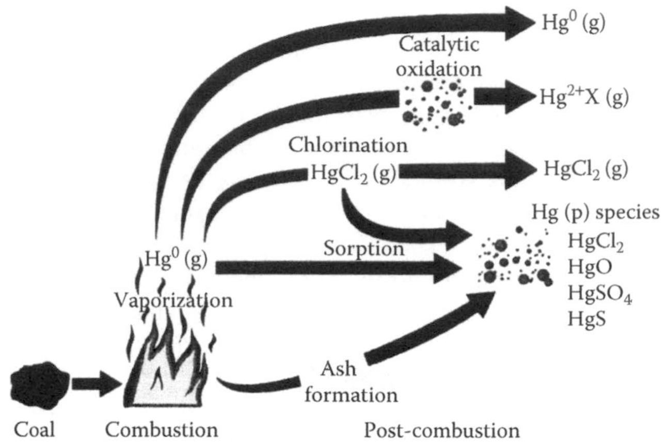

FIGURE 19.6 Hg transformations during coal combustion and in the resulting flue gas. (Reprinted from *Fuel Process. Technol.*, 65–66, Galbreath, K.C. and Zygarlicke, C.J., Mercury transformations in coal combustion flue gas, 289–310, Copyright 2000, with permission from Elsevier.)

$$Cl^{\bullet} + H^{\bullet} \leftrightarrow HCl \tag{19.8}$$

$$2Cl^{\bullet} \leftrightarrow Cl_2 \tag{19.9}$$

$$4Cl^{\bullet} + 2H_2O \leftrightarrow 4HCl + O_2 \tag{19.10}$$

$$4HCl + O_2 \leftrightarrow 2Cl_2 + 2H_2O \tag{19.11}$$

More Hg^{2+} is formed at lower temperatures (<475°C) and at higher chlorine content in coal, as shown in Equations 19.12 through 19.15 (Kellie et al. 2005). Further, chlorine in flue gas has been reported to be responsible for Hg oxidation and chemisorption on fly ash.

$$Hg + Cl^{\bullet} \leftrightarrow HgCl \tag{19.12}$$

$$HgCl + Cl_2 \leftrightarrow HgCl_2 + Cl^{\bullet} \tag{19.13}$$

$$Hg + Cl_2 \leftrightarrow HgCl_2 \tag{19.14}$$

$$2Hg + 4HCl + O_2 \leftrightarrow 2HgCl_2 + 2H_2O \tag{19.15}$$

Other potential mechanisms of Hg transformation is interaction with fly ash particles transforming into Hg^2 + X (g) (X = Cl_2 or O) and Hg^0 (g) and $HgCl_2$ (g) to Hg_p. There are several other parameters that affect the transformation of Hg compounds in flue gas. Hg speciation and oxidation in flue gas varies with flue-gas composition and APCDs type. Literatures show that coal sulfur, chlorine, and Hg concentration influence Hg emission during combustion (U.S. EPA 2002; Kellie et al. 2005).

19.2.6.2 Environmental Concerns of Mercury

Hg is a toxic air pollutant. Hg pollution is a global environmental problem. Once emitted into the atmosphere, Hg enters the global circulation pattern and travels long distance before deposited on earth surface. Residence time of Hg in atmosphere varies from few days to a year depending on its chemical form. Oxidized Hg (Hg^{2+}) and particulate Hg (Hg_p) reside in the atmosphere only for few days and deposit near the emission sources, whereas elemental Hg (Hg^0) has a lifetime of up to a year (Schroeder and Munthe 1998) and thus can travel to a long distance. In aquatic ecosystem, Hg is transformed into highly toxic methyl-Hg and bio-accumulates in living organisms posing serious health hazard.

Environmental concern over Hg grew after epidemics of methyl-Hg poisoning through fish consumption that occurred in Minamata City, Japan, in 1956. Wastewater contaminated with methyl-Hg was released from the chemical factory into the bay. Methyl-Hg bio-accumulated in shellfishes and fishes. Humans and wildlife were severely affected by consuming contaminated fishes and shellfishes. Cat, dog, pig, and human deaths continued over more than 30 years.

Hg^0 enters into human body through breathing as well. It crosses the blood–brain barrier and the placenta (Baldwin and Marshall 1999). In aquatic ecosystems, Hg is transformed into methyl-Hg compounds through microbial activities and bio-accumulated in food chain. Methyl-Hg causes nervous disorders, cancer, brain damage, unconsciousness, and even death (Zahir et al. 2005). Inorganic Hg mainly causes poisoning by ingestion. It can be absorbed in the gastrointestinal tract and skin. Acute exposure to inorganic Hg by the oral route may result in nausea, vomiting, and severe abdominal pain. The major effect from chronic exposure to inorganic Hg is kidney damage.

Because of these environmental and health concerns, controlling Hg from anthropogenic sources has received increasing scientific and regulatory interests worldwide. Due to toxic, persistent, and globally transporting behavior of Hg, *Minamata Convention*, a global legally binding treaty to prevent Hg emissions and release, was signed in October 2013. The scope of the new treaty is to ban production, export, and import of Hg-containing products; phase down the use of dental filling using Hg amalgam, provide some provisions for artisanal and small-scale gold mining operation, and so on. The treaty aims to control Hg emissions from major emission sources such as coal-fired power plants; however, emission limit value has not been set at the present.

19.2.7 Release of Wastewater

Wastewater may be released during extraction and utilization of coal through several pathways: washing, runoff from mining, discharge from combustion plants, and so on. Acid mine drainage, due to reaction of water with sulfur-bearing rocks, coming from mining area pollutes water bodies. Coal washing generates wastewater contaminated with particles and toxic elements. Sulfuric acid is formed when coal gets wet and dissolves toxic metals, which is very toxic to aquatic life and contaminates drinking water sources.

In coal fired power plants, water is used in steam production, cooling, and pollution control devices (e.g., wet APCDs). With once-through cooling systems, once cold water circulates through, it is released back into the lake, river, or ocean. This water is hotter (by up to 20°F–25°F) (UCS 2014) than the incoming water, creating *thermal pollution*. Coal power plants may add chlorine to their cooling water; once released into the natural water bodies, it decreases the algae growth. Coal-fired power plants also release wastewater (cooling and waste from processing and pollution control devices). Effluents generated from leaching of coal combustion residues (CCRs) typically have high concentrations of toxic elements.

Coal-fired power plants that discharge their coal ash and FGD wastewaters have a significant effect on water quality of receiving waters. Ruhl et al. (2012) have reported that even low concentrations of some contaminants, such as As with concentrations below health benchmarks at the National Pollution Discharge Elimination System (NPDES), can become problematic because As is retained in suspended sediments and remobilized with environmental changes in reduced bottom and pore waters.

19.2.8 Solid Waste

With the increase in coal-based power generation, the amount of fly ash generated from power plant is increasing worldwide. It has been estimated that worldwide, about 750 million tons of coal fly ash (CFA) is produced per year (Pandey and Singh 2010). As coal is a complex mixture of organic and inorganic matter, CFA, that is, generated at 1200°C–1700°C, is one of the complex matrix to characterize. Approximately 316 individual minerals and 188 mineral groups have been identified in different fly ash (Vassilev and Vassileva 2005). Fly ash from pulverized coal combustion can be considered as pozzolan (siliceous or siliceous and aluminous materials). The major components in fly ash are silica, alumina, magnetite, ferrous oxide, and calcium oxide. Fly ash contains macronutrients P, K, Ca, and Mg and micronutrients Zn, Fe, Cu, Mn, B, and Mo for plant growth. Fly ash from coal-fired power plants is contaminated with toxic elements (as discussed next), which restricts its disposal in landfill.

Figure 19.7 shows the potential environmental impacts of coal combustion residues. These residues may be bottom ash (from boiler), fly ash (from particulate control devices), waste sludge coming from FGD, and so on. Coal residues can be disposed or used for beneficial purposes. Depending on the disposal methods, they may create problems in the biosphere, hydrosphere, lithosphere, and atmosphere. On the other hand, after meeting all the requirements if these residues are utilized properly, they can have several advantages in agriculture.

19.2.8.1 Leaching of Toxic Elements from CFA

During high temperature coal combustion, mineral matters undergo phase transformation causing trace elements in the original coal matrix susceptible to leaching (Jones 1995). Some elements contained in fly ash may release when ash comes in contact with water. The leaching behavior of the major and trace elements from CFA has been recently reviewed by Izquierdo and Querol (2012). They have reported that a large number of elements are tightly bounded in fly ash and may not be easily released into the environment, regardless of

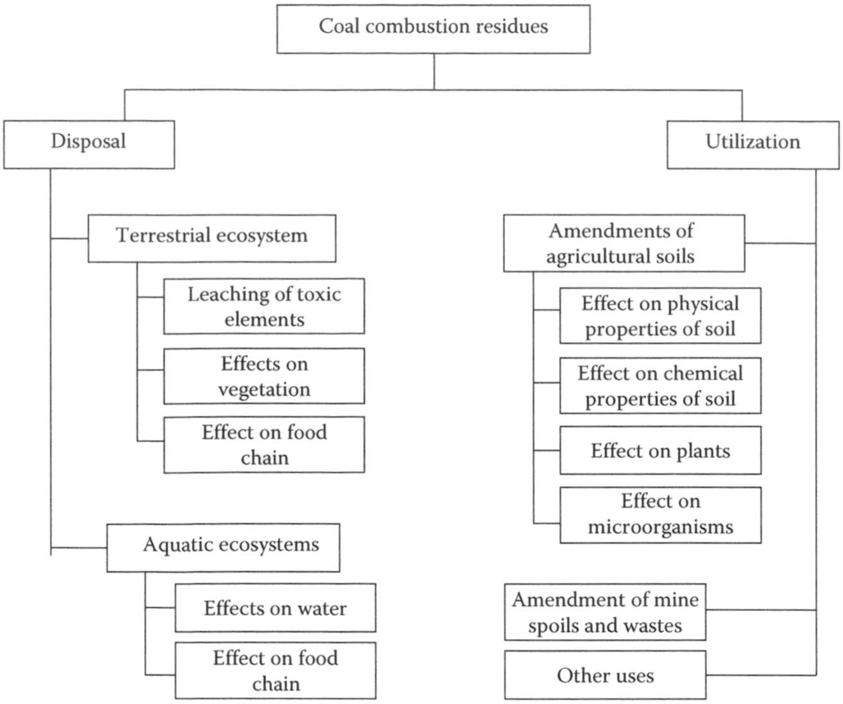

FIGURE 19.7 Schematic diagram of the potential environmental impacts of coal combustion residues. (Redrawn from Carlson, C.L. and Adrino, D.C., *J. Environ. Qual.*, 22, 227–247, 1993.)

the nature of the ash. The ratio between Ca and S determines the pH of the water–ash system and plays a major role in the leachability of most elements in fly ash. Whereas the alkalinity contributes to lessening the leachability of a large number of heavy metals, it also enhances the mobility of a few oxyanionic species: As, Cr, Mo, Sb, Se, V, and W. Elcments such as Be, Cd, Co, Cu, Fe, Mg, Mn, Ni, Pb, rare earth elements, Si, Sn, Th, Tl, U, and Zn have minimum solubility in pH range 7–10. On the other hand, oxyanionic-forming species such as As, B, Cr, Mo, Sb, Se, and V have maximum leachability in pH range 7–10 (Izquierdo and Querol 2012).

Table 19.3 shows the chemical composition of ash generated from bituminous, lignite, and sub-bituminous coal. CFAs have a bulk chemical composition containing a variety of metal oxides in the order SiO_2 > Al_2O_3 > Fe_2O_3 > CaO > M gO > K_2O > Na_2O > K_2O. This table shows that fly ash from lignite and sub-bituminous coals has a higher CaO content and lower loss of ignition (LOI) than ash from bituminous coal. In general, fly ash from sub-bituminous and lignite coals is found to be having CaO, MgO, and SO_3 and lower SiO_2 and Al_2O_3 compared to bituminous and anthracite coals (Blissett and Rowson 2012). Fly ash from bituminous and lignite coal contained less than 10% CaO in total, often consisted of mainly aluminosilicate glass, and usually do not contain any crystalline compounds of calcium.

Due to the pozzolanic properties of ash and the binding capacity, fly ash has been mostly used in cement manufacturing and as a substitute of materials in the construction industry. The geotechnical properties of fly ash (e.g., specific gravity, permeability, internal angular friction, and consolidation

TABLE 19.3
Normal Range of Chemical Composition for Fly Ash Produced from Different Coal Types

Component (wt%)	Bituminous	Sub-Bituminous	Lignite
SiO_2	20–60	40–60	15–45
Al_2O_3	5–35	20–30	10–25
Fe_2O_3	10–40	4–10	4–15
CaO	1–12	5–30	15–40
MgO	0–5	1–6	3–10
SO_3	0–4	0–2	0–10
Na_2O	0–4	0–2	0–6
K_2O	0–3	0–4	0–4
LOI	0–15	0–3	0–5

Source: Reprinted from *Prog. Energy Combus. Sci.*, 36, Ahmaruzzaman, M., A review on the utilization of fly ash, 327–363, Copyright 2010, with permission from Elsevier.

characteristics) make it suitable for use in construction activities (Ahmaruzzaman 2010).

In the ASTM classification, class F fly ash has a combined SiO_2, Al_2O_3, and Fe_2O_3 content of greater than 70% compared to greater than 50% for class C fly ash. Class F ash is regarded as a pozzolanic (siliceous, or a siliceous and aluminous) material that has no intrinsic cementitious property. In a very finely divided form, it will chemically react with $Ca(OH)_2$ at ordinary temperatures and in the presence of moisture to form compounds exhibiting cementitious

TABLE 19.4
U.S. and European Standards for Fly Ash to Be Used in Concrete

Class	$SiO_2 + Al_2O_3 + Fe_2O_3$ (%)	SO_3 (%)	Moisture (%)	LOI (%)
ASTM C618l				
C	>50	<5	<3	<6
F	>70			<12
Class	**$SiO_2 + Al_2O_3 + Fe_2O_3$ (%)**	**SO_3 (%)**	**Reactive Silica (%)**	**LOI (%)**
EN 450-1				
A	>70	<3	>25	<5
B				2–7
C				4–9

Source: Reprinted from *Fuel*, 97, Blissett, R.S. and Rowson, N.A., A review of the multi-component utilisation of coal fly ash, 1–23, Copyright 2012, with permission from Elsevier.

TABLE 19.5
Trace Element Content in 23 European Coal Fly Ashes

Element	Trace Element Composition (ppm) 25th Percentile	Median	75th Percentile
As	40	55	97
B	135	259	323
Ba	639	1302	1999
Be	6	8	12
Cd	1	2	2
Co	30	35	48
Cr	137	148	172
Cu	73	86	118
Ge	3	7	15
Hg	0.2	0.2	0.3
Li	150	185	252
Mo	7	11	13
Ni	87	96	144
Pb	59	80	109
Rb	50	108	147
Sb	4	4	8
Se	6	7	13
Sn	7	8	10
Sr	384	757	1647
Th	25	30	37
U	9	12	18
V	202	228	278
Zn	123	154	175

Source: Reprinted from *Fuel*, 84(11), Moreno, N., Querol, X., Andres, J.M., Stanton, K., Towler, M., Nugteren, H., Janssen-Jurkovicovád, M., and Jonese, R., Physico-chemical characteristics of European pulverized coal combustion fly ashes, 1351–1363, Copyright 2005, with permission from Elsevier; Reprinted from *Fuel*, 97, Blissett, R.S. and Rowson, N.A., A review of the multi-component utilisation of coal fly ash, 1–23, Copyright 2012, with permission from Elsevier.

properties (Blissett and Rowson 2012). The ASTM and the European Standards classifications to distinguish CFA types are suitable for use as a cement replacement (see Table 19.4). However, some (e.g., Manz 1999; Vassilev and Vassileva 2007) argue that this classification is not based on performance and can be controversial because in practice many class C ashes can meet the performance requirements of class F ashes.

Table 19.5 shows the concentrations of trace elements in 23 European CFAs. Most of the nonvolatile metal elements are collected with fly ash or bottom ash while volatile elements such as Hg are mostly speciated in flue gases. Pb, Cd, Zn, Ni, Be, and Cr are removed significantly by the pollution control devices. Intermediate volatile metals such as As and Se are emitted to the atmosphere to a large extent (Germani and Zoller 1988). The fly ash differs with coal type and also its ability to adsorb Hg. Fly ash containing higher unburned carbon contents may have significant catalytic and sorptive properties.

19.3 MITIGATION MEASURES AND SUSTAINABLE UTILIZATION OF COAL

19.3.1 Control of Particulate Emissions

The most commonly used devices for controlling particulate emissions are cyclone, ESP, bag filters, and wet scrubbers. In coal-fired power plants, ESP and bag filters are mostly used, so the descriptions on former control technologies are less prioritized here. The selection of the particulate control devices depends on a number of factors such as plant condition, regulatory requirements, efficiency required, and economics. In some cases, these pollution control devices are used in series to increase the removal efficiency.

19.3.1.1 Cyclone

Cyclones are relatively simple and most widely used APCDs for removing PM from gas streams. They are good for coarse particles, mainly for the removal of 50 μ or larger particles. Particles are deposited by inertial deposition. Schematic diagram of cyclone is shown in Figure 19.8. Particulate-rich air passes through the tangential inlet. When gas stream in cyclone changes direction (spins in a vortex) as it flows in cyclone, suspended particles tend to keep moving in their original direction due to their inertia, which leads particles to centrifuge toward the walls. Particles hit the wall and move down toward the narrowed base of the cyclone and are collected for removal. The cleaned gas in the middle of the cyclone moves up and exits from the top. Proper measures are to be taken to suppress the re-entrain of the collected dust in the cyclone.

Each cyclone design has a standard efficiency curve known as a *grade efficiency curve*, which shows what percentage

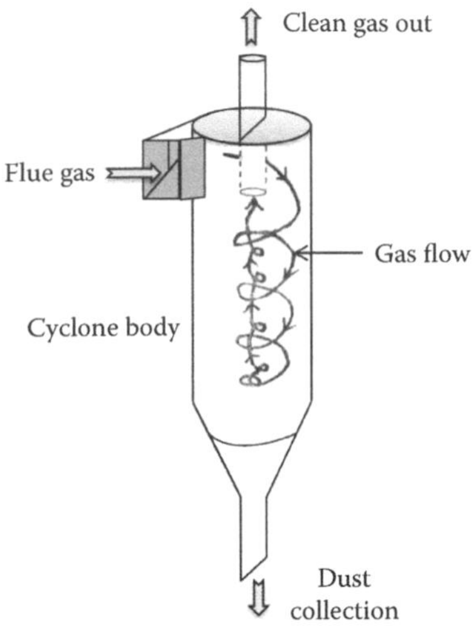

FIGURE 19.8 Schematic diagram of cyclone.

of each size range will be removed from the incoming flue-gas stream. Cyclone efficiency is determined mainly by the flue-gas inlet velocity (v), the cyclone diameter (D), and the cleaned gas outlet diameter relative to that of the cyclone diameter (d/D). Cyclone collection efficiency increases with increasing (1) particle size, (2) particle density, (3) inlet gas velocity, (4) cyclone body length, (5) number of gas revolutions, and (6) smoothness of the cyclone wall.

In cyclone, cut diameter is defined as the diameter of a particle for which the efficiency curve has the value of 0.5, that is, 50%. Cut diameter is one of the important parameters evaluating collection efficiency of cyclone. The smaller the cut diameter in a cyclone, the better is its dust performance. Lapple (1951) developed the semi-empirical relationship to calculate a 50% cut diameter, *dpc*, which is the diameter of particles collected with 50% efficiency.

19.3.1.2 Electrostatic Precipitators

ESP removes particles in gas stream using an electrostatic force. ESPs can be operated with high collection efficiency and a low pressure drop. They work on a principal of migration of charged particle in an electric field. The particles in gas stream are charged and exposed to an electric field. The resulting electrostatic force on the particles causes them to migrate toward collecting electrodes, where they are collected. Schematic diagram of cyclone is shown in Figure 19.9. The overall process in ESP can be classified into three distinct steps: (1) ionization of contaminated air flowing between the electrodes, (2) charging, migration, and collection of particulates, and (3) removal of particles from plate. ESP can achieve collection efficiency >99%. Submicron particles (diameter <1 μm) can also be collected effectively. Fly ash resistivity plays an important role in dust layer breakdown and the ESP efficiency. Other factors affecting the ESP performances are fuel type, coal grindability, coal and ash composition, and

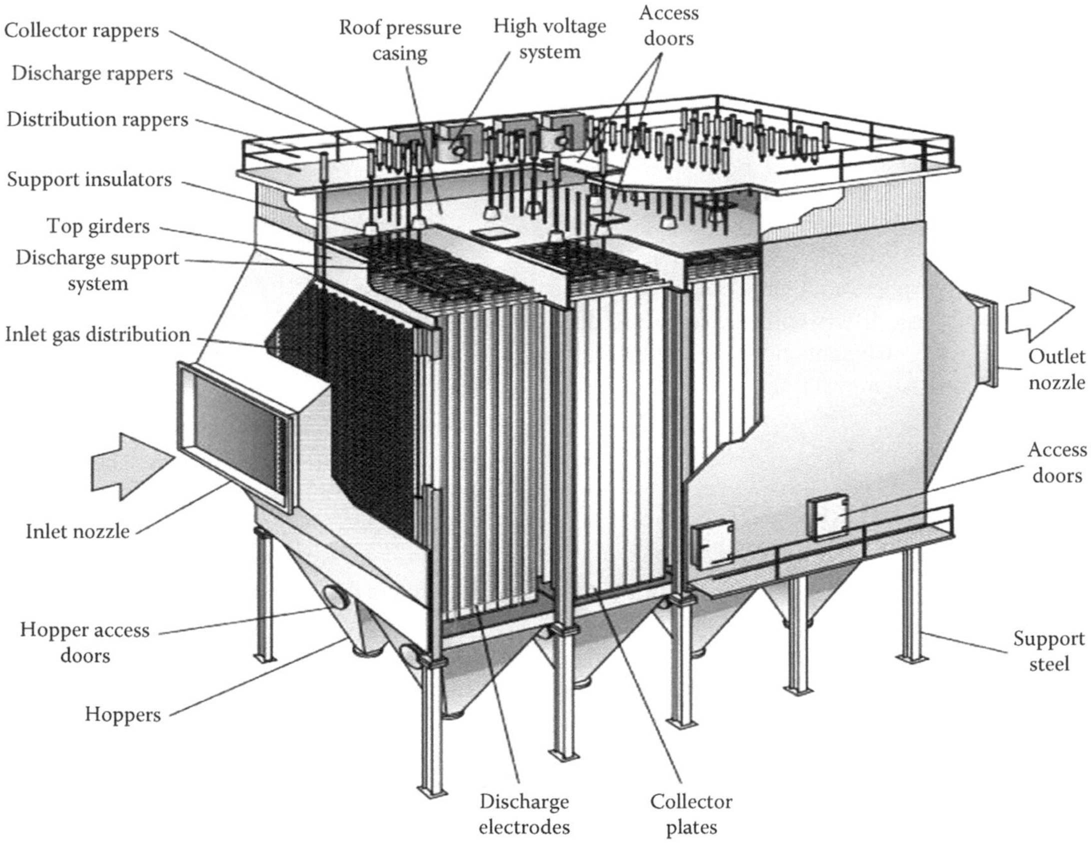

FIGURE 19.9 Schematic diagram of dry electrostatic precipitator. (From B&W, http://www.babcock.com/products/Pages/Dry-Electrostatic-Precipitator.aspx, 2014.)

other factors that affect the resistivity such as coal sulfur content and moisture.

The collection efficiency (η) of ESP is given by (Deutsch 1922)

$$\begin{aligned}\eta &= 1 - \exp(-\omega_e f) \\ &= 1 - \exp\left(\frac{-\omega_e A}{Q}\right)\end{aligned} \quad (19.16)$$

where:

ω_e is the migration velocity (m/s)
$f = A/Q$ is the specific collection area (s/m)
A is the area of the collecting electrode (m^2)
Q is the gas flow rate (m^3/s)

The collection efficiency is affected by many factors including the properties of dust particles and the geometry of the electrodes. Many studies suggest the amendment of the theoretical collection efficiency (from Mizuno et al. 2000).

$$\eta = 1 - \exp\left(\frac{-\omega_e LK}{v_0 b}\right) \quad (19.17)$$

where:

L is the length of the collecting electrode along the gas stream (m)
v_0 is the gas velocity (m/s)
b is the separation between the discharge and the collecting electrode (m)
K is a correction factor determined from actual measurements

$$t_0 = \frac{L}{v_0} \quad (19.18)$$

The detention time t_0 and the gas velocity v_0 are important factors for determining the ESP performance. Usually, t_0 and v_0 are designed to be ~10 s and 0.5–2 m/s, respectively.

ESP can be classified into cylindrical type and plate type based on the collecting electrodes; vertical gas flow and horizontal gas flow based on the direction of gas flow; one stage and two stage based on electrodes geometry; and dry type and wet type based on the use of scrubbing solution (water). Wet ESPs are also used to remove fumes, sulfuric acid mist. In the recent decades, conventional ESPs have been modified, and many new designs have been proposed for increasing collection efficiency, particularly for particles in the submicron size range that has been reviewed by Jaworek et al. (2007).

19.3.1.3 Fabric Filters or Bag Filters

Particulates are captured as flue gas rich in particles and are passed through the bag filter. The important filtering mechanisms are three aerodynamic capture mechanisms: direct interception, inertial impaction, and diffusion (Miller 2011). Electrostatic attraction may also play a role with certain types of dusts or dust fiber combinations (Wark et al. 1998). Dusts collected on the filter surface are periodically removed by blowing air in opposite direction through the filter, particles fall to a collection hopper. The dust layer on the filter itself can act as filter cake enhancing the removal efficiency. Some of the advantages of fabric filters (FF) are high collection efficiency in a broad range of particles sizes, flexibility in design installation of various filter media, and ability to handle a variety of solid materials (Wark et al. 1998). During filtration, particles may be attracted to or repulsed by filters due to Coulombic and polarization forces. Particles larger than 1 mm are removed by impaction and direct interception, whereas particles from 0.001 to 1 mm are removed mainly by diffusion and electrostatic separation (Wark et al. 1998).

Particulate removal efficiency of bag filters is higher (99.9%), which is good even for the fine particles. A FF normally achieves emission levels below 5 mg/Nm^3. The filtration process can be divided into three distinct time stages (Bustard et al. 1988):

1. Filtration by a clean FF, which occurs only once, that is, at the first use
2. Establishment of a residual dust cake, which occurs after many filtering and cleaning cycles
3. Steady-state operation, where the quantity of particulates removed during the cleaning cycles is equal to that collected during each filter cycle

Air-to-cloth (A/C) ratio and pressure drop across the filters are the two key parameters in sizing and operating baghouses. The A/C ratio determines the size of the baghouse and consequently the cost. The pressure drop is an important parameter because it determines the cost and energy requirements. In addition, flue-gas temperature, dew point, moisture content, PSD, and fly ash composition affect the performance of baghouses (Bustard et al. 1988).

The three basic types of baghouses are reverse-gas, shake-deflate, and pulse-jet, which are distinguished by the cleaning mechanisms and by their A/C ratio. For illustration, reverse-gas-type baghouse is shown in Figure 19.10. Reverse-gas type is the most traditional FF. It operates at low A/C ratio ranging from 1.5 to 3.5 ft/min (Soud 1995). Dusts collected on the filter surface are periodically removed by blowing air in opposite direction through the filter. The dislodged particles fall to a collection hopper from where it is removed away.

19.3.1.4 Wet Scrubbers

Particulates are controlled by passing a gas stream through solution (mostly water) or spraying liquid into gas stream. Commonly used scrubbers are packed towers, spray tower, and spray chambers. Wet collection devices used for particulate control are venture scrubbers, bubbling scrubbers, spray towers, and wet ESPs. Wet scrubbers are highly efficient in removing coarse particles, efficiency reaching up to 99%, but their removal efficiency for fine particles is poor. Besides particle collection, wet scrubbers have benefit control of acid gases, bases, and other elements that are soluble in water and/or scrubbing solution.

Cyclones and mechanical collectors are less efficient in particle removal; however, their operating cost is less. As they are less efficient, they alone are not best-available technologies; however, they can be used in a pre-cleaning stage in the

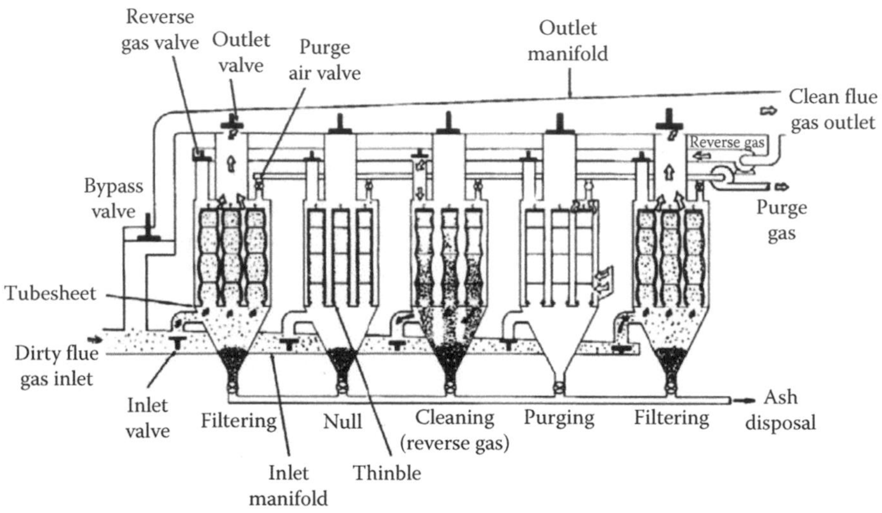

FIGURE 19.10 The reverse-gas-type baghouse showing the flue gas and cleaning air flows during the various cycles of operation. (Reprinted from *Fabric Filters for the Electric Utility Industry, Volume 1 General Concepts*, Bustard, C.J. et al., Copyright 1988; *J. Environ. Qual.*, 22(2), Carlson, C.L. and Adriano, D.C., Environmental impacts of coal combustion residues, 227–247, 1993, referred from Miller 2011, with permission from Elsevier.)

flue-gas cleaning steps. For removing PM from off-gases from new and existing combustion plants, the use of an ESP or a FF is considered to be the best-available technology.

19.3.2 Control of NO_x

A significant progress has been made in reducing NO_x emissions from pulverized coal combustion facilities. NO_x emission can be reduced by selecting low nitrogen fuel, using low NO_x combustion technology, controlling fuel–air mixing rates, and optimizing combustion temperatures. However, these measures may not be enough to meet the regulatory emission limit for which post-combustion NO_x control technologies such as selective catalytic reactor (SCR) or selective non-catalytic reactor (SNCR) are used. Both these technologies reduce NO_x to N_2 and H_2O through a series of reactions with ammonia or urea injected into the flue gas.

19.3.2.1 Selective Catalytic Reactor

SCR is a post-combustion NO_x emission control technology. SCR system has been used commercially on coal-fired power plants burning mainly low-sulfur coal and some medium-sulfur coal in Japan since 1980 and in Germany since 1986. During the 1990s, SCR demonstration and full-scale systems was installed in U.S. coal-fired power plants burning high-sulfur coal (IEA 2014a).

Among the available technologies, SCR is the most effective in reducing NO_x emissions (70%–90%). It is widely used in coal-fired power plants due to its higher efficiency, selectivity, and economic practicability. In the SCR process, ammonia (NH_3) is most commonly used for reducing NO_x to N_2 and H_2O. The optimum temperature for SCR operation is between 300°C and 400°C. Most of the catalysts used in commercial coal-fired power plants are metal oxides such as titanium dioxide–supported vanadium pentoxide (TiO_2/V_2O_5). SCR in general can achieve NO_x emission reductions over 80%–90%. The schematic representation of SCR is shown in Figure 19.11.

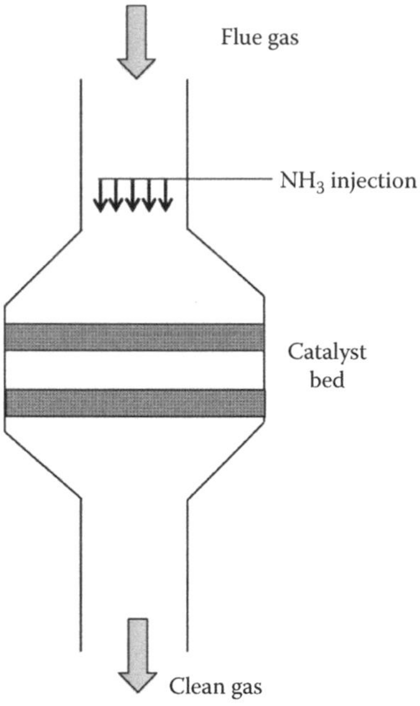

FIGURE 19.11 Schematic diagram of the selective catalytic reactor.

The stoichiometric chemical reaction for SCR system is as follows (Forzatti et al. 2001). At reaction temperature between 250°C and 450°C, in excess oxygen condition (to enhance NO–NH_3 reaction), Equation 19.19 proceeds rapidly on catalyst. The reaction proceeds in one-to-one NH_3-to-NO molar ratio. Here *selective* refers to the ability of ammonia reacting selectively with NO_x instead of being oxidized by oxygen to form N_2, N_2O, and NO.

$$4NH_3 + 4NO + O_2 \rightarrow 4N_2 + 6H_2O \quad (19.19)$$

$$2NH_3 + NO + NO_2 \rightarrow 2N_2 + 3H_2O \quad (19.20)$$

$$8NH_3 + 6NO_2 \rightarrow 7N_2 + 12H_2O \quad (19.21)$$

While burning sulfur-containing fuel, SO_2 is produced along with less SO_3. SO_2 can be further oxidized to SO_3 over the catalyst.

$$2SO_2 + O_2 \rightarrow SO_3 \quad (19.22)$$

SO_3 produced reacts with ammonia and water in flue gas and produces ammonium sulfate and sulfuric acid.

$$2NH_3 + SO_3 + H_2O \rightarrow (NH_4)_2 SO_4 \quad (19.23)$$

$$NH_3 + SO_3 + H_2O \rightarrow NH_4HSO_4 \quad (19.24)$$

Ammonium sulfate thus formed can accumulate on the catalyst as well as in the air pre-heater downstream of SCR and cause corrosion and pressure drop problems. At temperature less than 200°C, ammonium nitrate may also be formed from NO_2 and NH_3 and deposited as solid or liquid (Forzatti et al. 2001).

There are three typical configuration of SCR systems installation at coal-fired power stations:

1. High dust position: In this configuration, SCR is installed after boiler (as in Figure 19.1). This does not require particulate emissions control prior to the denitrification process. This is the most widely used configuration.
2. Low dust position: In this configuration, SCR is installed after particulate removal. Mostly hot-side ESP is installed prior to NO_x control. It has the advantage of less catalyst degradation by fly ash.
3. Tail end position: In this configuration, SCR is installed at the downstream end of the APCDs configuration. This type of SCR has been used primarily with wet bottom boilers with ash recirculation to avoid catalyst degradation caused by As poisoning (IEA 2014a). Especially, this configuration is more favored retrofit installations (due to SCR space requirements) between the economizer outlet and ESP.

19.3.2.2 Selective Non-Catalytic Reactor

In SNCR systems, ammonia or urea is injected into the flue gas in an appropriate temperature range (900°C and 1100°C). At temperature above 1000°C, the thermal decomposition of ammonia occurs and the NO_x removal rate decreases. The NO_x reduction decreases below 1000°C, and ammonia slip may increase. SNCR can reduce emissions of NO_x by 30%–50%. Because of higher stoichiometric ratios, either ammonia or urea, SNCR processes require 3–4 times reagent as of that of SCR systems to achieve similar NO_x reductions (IEA 2014a).

The unreacted ammonia in flue-gas stream can react with SO_3 to form ammonium bisulfate, which precipitates at air heater operating temperatures and can cause air heater fouling and plugging.

SNCR technology was commercially used in oil- or gas-fired power plants in Japan in the middle of the 1970s. In 1980s, SNCR systems have been used commercially in coal-fired power plants in west Europe and since the early 1990s in the United States. Besides these, combined SO_2/NO_x removal technologies are emerging, most of which are in developmental stage.

19.3.3 Control of SO_x

SO_x emission can be reduced by two approaches: first by reducing the sulfur content in fuel or using low-sulfur content fuel and second by using control technologies. The former one is more in use. Commonly used methods for removing SO_x from flue gas are wet FGD, dry FGD, spray dry scrubber, dry sorbent injection, and wet sulfuric acid process with sulfur recovery such as H_2SO_4. FGD is widely used in coal-fired power plants to control SO_2 emission. The FGD system can be classified into wet or dry, depending on the amount of water used to spray the reagent into the flue gas. Wet FGD systems are the most widely used ones. They can achieve SO_2 control of 90%–98% (DePriest and Gaikwad 2003).

Wet FGD systems use an alkaline reagent such as calcium-, sodium-, and ammonium-based sorbents that are injected in the scrubber to react with SO_2. Conventional wet FGD systems mostly use lime or limestone process in situ forced oxidation to remove SO_2 and form a gypsum by-product. Use of limestone is favored by its wide availability and low cost (see Figure 19.12). The reactions occurring in the wet FGD are much more complex, which include a combination of gas–liquid, solid–liquid, and liquid–liquid ionic

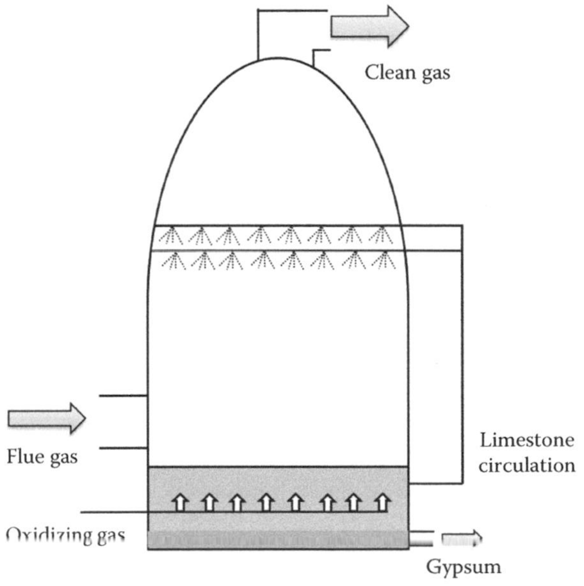

FIGURE 19.12 Schematic diagram of limestone-based wet FGD.

reactions. The overall simplified chemical reactions in a wet FGD system with a limestone reagent are as follows (Miller et al. 2006):

SO_2 reaction:

$$CaCO_3(s) + SO_2(g) + ½H_2O(l) \rightarrow CaSO_3 \cdot ½H_2O(s) + CO_2(g) \quad (19.25)$$

Sulfite oxidation:

$$CaSO_3 \cdot ½H_2O(s) + ½O_2(g) + \frac{3}{2}H_2O(l) \rightarrow CaSO_4 \cdot 2H_2O \quad (19.26)$$

In a lime/limestone wet scrubber, $CaCO_3$ reacts with SO_2 and H_2O and form calcium sulfite ($CaSO_3 \cdot ½H_2O$). An in situ forced oxidation system converts calcium sulfite ($CaSO_3 \cdot ½H_2O$) to calcium sulfate ($CaSO_4 \cdot 2H_2O$) or gypsum, which has low solubility in water and can be removed by filtering. Wet scrubbers operate at lower temperatures; thus, volatile trace elements in flue gas are condensed and removed in scrubber solutions.

Another widely used technology is limestone scrubbing with forced oxidation (LSFO). Limestone slurry is used in an open spray tower with in situ oxidation to remove SO_2 and form calcium sulfate (gypsum). The major advantages of this process compared to the aforementioned one is that in this process gypsum is formed, which is easier for dewatering, more economical disposal of the scrubber product solids, and decreasing scaling on the tower walls (Miller 2011). The other technologies such as the limestone/wallboard (LS/WB) gypsum FGD process, magnesium-enhanced lime process, limestone with dibasic acid, and sodium-based scrubbers are widely used (not discussed here). Regenerative FGD processes regenerate the alkaline reagent and convert SO_2 to a usable chemical by-product.

1. The Wellman–Lord process: This process uses sodium sulfite to absorb SO_2, which is then regenerated to release a concentrated stream of SO_2. Most of the sodium sulfite is converted to sodium bisulfite by reaction with SO_2, as in the dual alkali process. Some of the sodium sulfite is oxidized to sodium sulfate.
2. Regenerative magnesia scrubbing: In this process, MgO in the slurry is used in the same way limestone or lime is used in the lime scrubbing process. The magnesium oxide process is regenerative; on the other hand, lime scrubbing normally is a throwaway process.

Dry FGD technology includes lime or limestone spray drying; dry sorbent injection in furnace, economizer, duct, and hybrid methods; and circulating fluidized bed scrubbers. The wastes generated from dry FGD are dry and easy to handle and dispose than the wet one.

Spray dry scrubbers are also known as semi-dry FGDs, which are mainly suitable for power plants burning low- to medium-sulfur coal. This is the second most widely used method to control SO_2 emissions from coal-fired power plants. The sorbent used in the spray drying process is lime (CaO); however, hydrated lime [$Ca(OH)_2$] can also be used. In this process, slurry consisting of lime is atomized/sprayed into flue gas, presented in the reactions as follows (Miller 2011):

$$CaO_3(s) + H_2O(l) \rightarrow Ca(OH)_2(s) + \text{heat} \quad (19.27)$$

SO_2 in the flue gas reacts with $Ca(OH)_2$ (s) to yield $CaSO_3 \cdot ½H_2O$ (s), shown as follows:

$$Ca(OH)_2(s) + SO_2(g) \rightarrow CaSO_3 \cdot ½H_2O(s) + ½H_2O(v) \quad (19.28)$$

$$Ca(OH)_2(s) + SO_3(g) + H_2O(v) \rightarrow CaSO_4 \cdot 2H_2O(s) \quad (19.29)$$

SO_2 removal efficiency is affected by its concentration in the flue gas, flue-gas temperature, the size of the atomized slurry droplets, and so on. The product from the spray dryer is a dry solid that can be collected in dry particle collection devices.

Several dry injection processes have been developed demonstrating satisfactory SO_2 removal. These technologies can be easily retrofitted to existing facilities and are associated with low capital cost. Broadly these technologies can be classified into (1) furnace sorbent injection; (2) economizer sorbent injection; and (3) duct sorbent injection. Several combinations and hybrid of these processes are also present. Calcium-based sorbents are widely used followed by sodium compounds.

19.3.4 Control of GHGs

The public concern over the climate change has increased due to the increasing amount of CO_2 in the atmosphere. The development of new technologies for CO_2 emission control has been prioritized. CO_2 from coal combustion can be captured by three methods: pre-combustion capture in integrated gasification combined cycle, oxy-fuel combustion, and post-combustion captures of CO_2 from flue gas (Herzog 2001; IPCC 2005). The schematic diagrams of these technologies are presented in Figure 19.13a, b, and c. In pre-combustion capture, carbon is removed from the fuel prior to combustion. In oxy-fuel combustion, coal is combusted in an oxygen and CO_2-enriched environment. In post-combustion, coal is combusted normally in a boiler and CO_2 is removed from the flue gas. The flue-gas composition released from air combustion consists of N_2, NO_x, CO_2, and SO_2, whereas in oxy-fuel combustion flue gas mainly consists of CO_2 with lesser amount of N_2, SO_2, and O_2. The concentration of CO_2 in conventional coal combustion is about 13%–15%, which is difficult to capture. On the other hand, in oxy-fuel combustion, CO_2 concentration may be up to 95% (Hu et al. 2000) and can easily be captured.

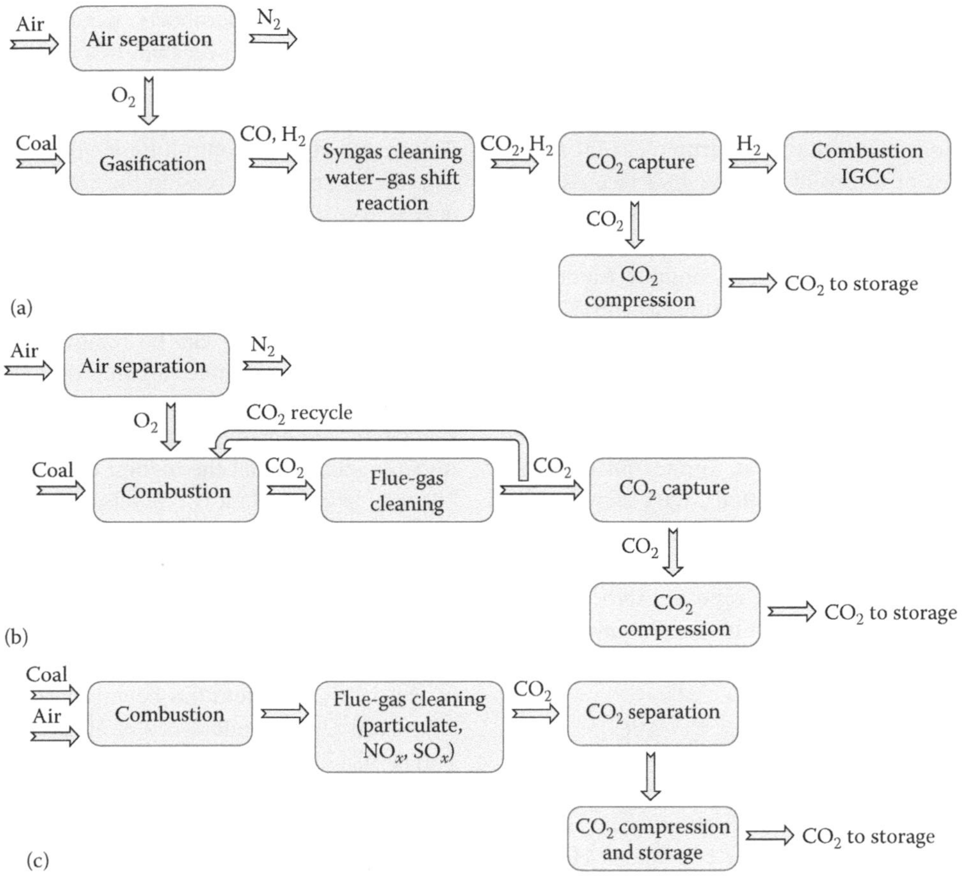

FIGURE 19.13 Flow diagram of (a) pre-combustion, (b) oxy-fuel combustion, and (c) post-combustion of CO_2 capture.

However, at present, these technologies are still in developmental stage and have not been applied in coal-fired power plants because of the following reasons (Samanta et al. 2012): (1) They have not been demonstrated in large-scale plants; (2) the required parasitic loads to supply both power and steam to the CO_2 capture plant would reduce power generation capacity approximately by one-third; and (3) if successfully scaled up, these technologies would not be cost-effective at their present state of the art.

Numerous numbers of CO_2 capture technologies using sorbent, solvent, and membrane are available. A wide range of materials has been studied as CO_2 sorbents through chemical reactions and physical absorptions. Amine-based chemical absorption processes such as monoethanolamine (MEA), methylmonoethanolamine (MMEA), diethanolamine (DEA), dimethylmonoethanolamine (DMMEA), diglycol-amine (DGA), N-methyldiethanolamine (MDEA), and 2-amino-2-methyl-1-propanol (AMP) have been widely used for several years for CO_2 capture from gas streams in natural gas, off-gas from refinery and synthesis gas processing (Astarita et al. 1983; Kohl and Nielsen 1997). The gas streams in these processes are at a high pressure. The major challenges to use the similar system for CO_2 capture from coal-fired power plants are limited due to the large volumetric flow rates of flue gas at atmospheric pressure with CO_2 at low partial pressures and in the temperature range of about 100°C–150°C (Samanta et al. 2012).

The solvent mostly used as discussed earlier is amine based. Amines react with CO_2 to form carbamate and bicarbonate (Equations 19.30 and 19.31):

$$2RNH_2 + CO_2 \leftrightarrow RNH^+ + RNHCOO^- \quad (19.30)$$

$$RNH_2 + H_2O + CO_2 \leftrightarrow RNH_3{}^+ + HCO_3{}^- \quad (19.31)$$

The modern process uses 20–30 wt% aqueous MEA for post-combustion CO_2 capture. The use of this technology in coal-fired power plants is limited because this process is highly energy intensive and solvent used in the process gets degraded with SO_x and NO_x in flue gas. Further, significant oxygen partial pressure in the flue gas is another challenge for implementing amine absorption process for CO_2 capture from flue-gas stream.

Aqueous ammonia process is similar in operation to amine systems. The main difference with amine system is the precipitation of crystalline solid salts at temperatures below 80°C and the relatively high vapor pressure of ammonia, which requires regeneration of the solvent at very high pressures to minimize ammonia slip (Miller 2011). In the absorber column, CO_2 is absorbed by reacting with $(NH_4)_2CO_3$ to form ammonium bicarbonate (Equation 19.32). Ammonium bicarbonate partly precipitates in the absorber solution.

$$(NH_4)_2 CO_3 + CO_2 + H_2O \leftrightarrow 2NH_4HCO_3 \quad (19.32)$$

CO_2 from coal combustion flue gas may be captured using a variety of solid physisorbents. The selective adsorption of CO_2 is caused by van der Walls attraction between the CO_2 molecules and adsorbent surface, as well as by pole–ion and pole–pole interactions between the quadruple of CO_2 and the ionic and polar sites of the solid adsorbent surface (Ruthven 1984). CO_2 can be captured using solid physisorbents such as (1) activated carbons, (2) carbon nanotubes, (3) carbon molecular sieves, (4) zeolites, and (5) metal organic frameworks (MOFs). Chemical sorbents for CO_2 capture include (1) regenerable alkali-metal carbonate-based sorbents and (2) amine functionalized solid sorbents.

Chemisorbents have an advantage over physisorbent to overcome the energy penalty in the base case: the MEA process. Several chemisorbents, such as amine functionalized sorbent (both impregnated and grafted), have shown promise to meet the desired working capacity target in simulated flue-gas conditions. Mostly, chemisorbents are found to have higher CO_2 selectivity. However, the reported sorbents are in the early development stage and require further investigation before they can be applied in the commercial plant.

19.3.5 Control of Toxic Metals

Best-available technology to reduce the emissions of heavy metals is generally the application of high efficiency particulate removal devices such as ESPs or bag filters. However, Hg and Se mainly present in the vapor phase in the operating temperature of most of the air pollution control devices in coal-fired power plants. For ESPs or bag filters operated in combination with FGD techniques, such as wet limestone scrubbers, spray dryer scrubbers, or dry sorbent injection, an average removal rate of Hg is 75% (50% in ESP and 50% in FGD) and removal of about 90% can be achieved in additional presence of a high dust SCR (U.S. EPA 1997).

19.3.5.1 Control of Mercury

Controlling Hg from coal combustion flue gas poses problem because Hg mainly exists in gaseous form at the operating conditions of the most of the conventional APCDs and is not sufficiently controlled. Powdered activated carbon injection has been applied for Hg control in coal combustion facilities. The novel Hg control approaches such as chemical additives, photochemical process, and non-carbon sorbents are being tested for Hg removal from flue gas in combustion facilities. In addition, sorbent injection technologies can be involved in multipollutant control technology. Multipollutant control technologies usually employ post-combustion flue-gas treatment to remove NO_x, SO_x, and acid gases. Acid gases are removed in the downstream wet ESPs or scrubbers, which also increase the removal of Hg from the flue gas.

In general, Hg^{2+} increases across dry APCDs. More of Hg tends to convert into Hg^{2+} in dry APCDs. In wet APCDs, Hg tends to be less oxidized due to the absorption of Hg^{2+} in scrubber solution. Major portions of Hg attached in particulates are removed across particulate control devices such as in cyclone, ESPs, and bag filters. Hg^{2+} decreased across the wet APCDs such as wet scrubbers, wet towers, and wet FGD. Hg removal in ESPs is lower than FFs due to lesser contact time between PM and Hg compounds in flue gas and high temperature condition particularly at hot-side ESPs. FF system is more effective in controlling Hg due to the formation of fly ash filter cake and increasing the contact time. The filter cake acts as fixed-bed reactor enhancing heterogeneous oxidation and adsorption of Hg.

Wet scrubbers operate relatively in lower temperature; thus, the volatile trace elements are condensed from the vapor phase in flue gas and can be removed in scrubber solution. As Hg^{2+} is water soluble, it can simultaneously be removed in scrubber solution. Hg removal in wet FGD depends on Hg speciation, the upstream of APCDs configuration. Field testing data indicate that the highest co-benefit of Hg control for bituminous-fired plant is on facilities equipped with FGD for SO_2 control and FF for PM control (U.S. EPA 2002).

As discussed in Section 19.2, typically the present-day coal-fired power plants are installed with SCR, ESP, and FGD to control NO_x, particles, and SO_x, respectively. Some fraction of Hg is controlled in APCDs as a co-benefit control. Oxidation of Hg^0 by SCR system has been reported in coal-fired boiler (Chu et al. 2003; Pudasainee et al. 2009, 2010) and in pilot-scale tests (EPRI 2000). Flue-gas Hg oxidation across an SCR system in plants burning bituminous coal is reported to be higher (30%–98%) (Laudal et al. 2002; Chu et al. 2003) than in plants burning sub-bituminous coal (0%–26%) (Machalek et al. 2003).

Mass distribution of Hg within coal-fired power plants with SCR + cold-side (CS) ESP + wet FGD and CS ESP + wet FGD configuration was compared by Pudasainee et al. (2012). The result from commercial power plant showed that 43.0% of Hg was removed in ESP fly ash, 49.4% was removed in FGD by-products and effluents, 3.9% was removed in boiler bottom ash, and 3.7% entering into the power plant was released into the atmosphere. In the SCR + CS ESP + wet FGD configuration, a major portion of Hg was removed in APCDs and less Hg was emitted into the atmosphere, whereas in CS ESP + wet FGD configuration, more Hg was emitted into the atmosphere due to less oxidation of Hg and more remission in FGD.

19.3.6 Wastewater Management

The main sources of effluent streams in coal-fired power plants in general are (1) cooling water, (2) wastewater from FGD, (3) wastewater from slag flushing and ash transport, (4) wastewater from washing of boilers, air preheater, and precipitator, and (5) run-off water from coal storage areas. The conventional wastewater treatment process can be summarized in the following five major steps:

1. pH adjustment: The pH of FGD wastewater/scrubber must be increased to precipitate heavy metals.
2. Coagulation/flocculation/precipitation: The addition of coagulation additive allows the agglomeration and aid precipitation.

Additional treatment: In some cases, for example, when metals concentration is higher, it is usual to add some organic sulfide (e.g., TMT 15) for precipitating heavy metals as sulfides, which is more effective than using hydroxide.
3. Sedimentation: The treated wastewater is allowed for sedimentation.
4. Filtration: Most of the slurry is dewatered by filtration.
5. Neutralization: pH is neutralized with the addition of acids.

The amount of wastewater generation from coal combustion facilities can be reduced to some extent by

- Maximizing the re-circulation of polluted wastewater in wet flue-gas treatment systems
- Use of semi-dry or dry sorption systems, which reduce the amount of wastewater generated
- Using the boiler drain water in the scrubber
- Proper waste storage and handling (i.e., roofed enclosures)

To reduce the impact of wastewater on the environment and human beings:

- Any surface run-off from the mining storage should be collected and treated before discharging into the natural water bodies.
- The wastewater before entering the river stream should be treated to remove chemicals and metals and to decrease the amount of solid matter. The wastewater must meet the effluent standards.
- The treatment plant in general need to adjust the pH and remove metals and solid particles.

19.3.7 Solid Waste Management

The solid wastes from coal combustion are highly contaminated with toxic elements that need to be properly handled, disposed, or reused for beneficial use. The higher fraction of fly ash is disposed of in landfill, which is costly and/or anticipated to be banned for dumping in the future. The fly ash from coal combustion can be utilized for beneficial purposes such as

1. Used in construction of roads and embankments, structural fill, and so on.
2. Use in cement manufacturing: Fly ash can be used as raw materials for the production of concrete in the manufacture of cement.
3. Use in agricultural field: Fly ash contains essential elements for plant growth, macronutrients P, K, Ca, and Mg, and micronutrients Zn, Fe, Cu, Mn, B, and Mo. After careful checking for contaminants, this can be applied in agricultural field as fertilizer.
4. Used as a low-cost adsorbent for the removal of NO_x, SO_x, organic compounds, metals from flue gas, and wastewater. The properties of fly ash, such as particle size, porosity, surface area, and bulk density, make it suitable for use as an adsorbent.
5. Used as light-weight aggregate.
6. Use for synthesis of zeolite: The high percentage of silica, alumina, and magnetite in fly ash enables its use for the synthesis of zeolite, alum, and precipitated silica.

19.4 SUMMARY

Coal has been utilized for residential and industrial heating since long. Environmental impacts are noticed in every steps of coal use: extraction, transportation, storage, preparation, and utilization. Combustion of coal releases several pollutants into air, water, and soil. These impacts can be mitigated with the adoption of proper technological and environmental management practices.

Air pollutants emission from coal-fired power plants has been one of the major environmental problems worldwide. There have been worldwide efforts (technological and regulatory) to control emission of particulates, NO_x, and SO_x. Some developed countries have significantly reduced the emission of such pollutants by the application of advanced APCDs and regulatory measures. The most commonly used devices for controlling particulate emissions are cyclone, ESP, bag filters, and wet scrubbers. The selection of the particulate control devices depends on number of factors such as plant condition, regulatory requirements, efficiency required, and economics.

NO_x emission can be reduced by selecting low nitrogen fuel, using low NO_x combustion technology, controlling fuel–air mixing rates, and optimizing combustion temperatures. However, these measures may not be enough to meet the regulatory emission limit, for which post-combustion NO_x control technologies such as SCR or SNCR are used. Both these technologies reduce NO_x to N_2 and H_2O through a series of reactions with ammonia or urea injected into the flue gas. SO_x emission can be reduced by either reducing sulfur content in fuel or using low-sulfur content fuel and by using control technologies. Commonly used methods for removing SO_x from flue gas are wet FGD, dry FGD, spray dry scrubber, and dry sorbent injection.

The uses of advanced APCDs not only control the targeted pollutants but also have co-beneficial control over other pollutants, such as trace metals are controlled in APCDs that are currently in use in many coal-fired power plants. Control technologies for trace metals are still in developmental stage. Best-available technology to reduce the emissions of heavy metals in general is the application of high efficiency particulate removal devices such as ESPs or bag filters. However, Hg and Se mainly present in the vapor phase in the operating temperature of most of the APCDs in coal-fired power plants. Activated carbon injection has been applied for Hg control in coal combustion facilities. The novel Hg control approaches such as chemical additives, photochemical process, and non-carbon sorbents are being tested for Hg removal from flue gas in combustion facilities. Multipollutant control technologies usually employ post-combustion flue-gas treatment to remove

NO_x, SO_x, and acid gases. CO_2 from coal combustion can be captured by three methods: pre-combustion capture in integrated gasification combined cycle, oxy-fuel combustion, and post-combustion capture of CO_2 from flue gas. However, at present, these technologies are still in development stage and have not been applied in commercial coal-fired power plants.

The amount of wastewater generation from coal combustion facilities to some extent can be reduced by maximizing the re-circulation of polluted wastewater in the treatment systems, switching to semi-dry or dry sorption systems, and so on. To reduce the impact of wastewater on the environment and human beings, any surface run-off from the mining storage should be collected and treated before discharging into the natural water bodies. With the increase in coal-based power generation, the amount of fly ash generated from power plant is increasing worldwide. Coal residues can be disposed or used for beneficial purposes. Depending on the disposal methods, they may create problems in the biosphere, hydrosphere, lithosphere, and atmosphere. On the other hand, after meeting all these requirements if these residues are utilized properly such as in construction, in cement kiln to prepare concrete, and as low-cost sorbents, they can have several advantages.

19.5 FUTURE RESEARCH NEEDS

Even though studies on environmental impact of coal utilization have been the subject of extensive research over the last decades, there are several issues that need further research; some of which are the following:

- Influenced by the stringent regulations facilities are changing the fuel, blending, and/or adding APCDs, the effects of which have to be assessed well in advance; for example, installation of SCR upstream of APCDs has a co-beneficial effect on Hg removal.
- Most of the regulatory control of NO_x and SO_x has been targeted to large industries; however, it is equally important to control it from small boilers and residential heating appliances.
- Most of the sorbents for CO_2 capture have been evaluated in simulated flue gas. The performances have to be evaluated in the real flue gas to see the effect of NO_x, SO_x, and particulates.
- Various ways to recycle fly ash for beneficial purposes have to be explored and applied.
- The pollutants formation and release behavior from emerging technologies such as coal gasification have to be studied.

REFERENCES

Ahmaruzzaman, M., 2010. A review on the utilization of fly ash. *Progress in Energy and Combustion Science* 36, 327–363.

Astarita, G., Savage, D.W., Bisio, A., 1983. *Gas Treating with Chemical Solvents*, John Wiley & Sons, New York.

Aylett, B.J., 1975. *The Chemistry of Zinc, Cadmium, and Mercury. Pergamon Texts in Inorganic Chemistry*, Vol. 18, Oxford: Pergamon Press.

Baldwin, D., Marshall, W., 1999. Heavy metal poisoning and its laboratory investigation. *Annals of Clinical Biochemistry* 36, 267–300.

Blissett, R.S., Rowson, N.A., 2012. A review of the multi-component utilisation of coal fly ash. *Fuel* 97, 1–23.

Bustard, C.J., Cushing, K.M., Pontius, D.H., Smith, W.B., Carr, R.C., 1988. *Fabric Filters for the Electric Utility Industry, Volume 1 General Concepts*, Electric Power Research Institute, California.

Carlson, C.L., Adriano, D.C., 1993. Environmental impacts of coal combustion residues. *Journal of Environmental Quality* 22 (2), 227–247.

CCT, 1997. *Clean Coal Today*, U.S. Office of Fossil Energy, Department of Energy, US DOE No. 25/Spring.

Chu, P., Laudal, D.L., Brickett, L., Lee, C.W., May 19–22, 2003. Power plant evaluation of the effect of SCR technology on mercury. In *Proceedings of the Combined Power Plant Air Pollutant Control Symposium—The Mega Symposium*, Washington, DC.

DePriest, W., Gaikwad, R., 2003. Economics of lime and limestone for control of sulfur dioxide. In *Proceedings of Combined Power Plant Air Pollutant Control Mega Symposium*, Washington, DC.

Deutsch, W., 1922. Bewegung und Ladnng der Elektrizitatstrager im Zylinderkondensator, *Annalen der Physik* 168, 335.

EPRI, Electric Power Research Institute, 2000. *Pilot-Scale Screening Evaluation of the Impact of Selective Catalytic Reduction for NO_x on Mercury Speciation.* EPRI Report No. 1000755, December.

Fenimore, C.P., 1971. Formation of Nitric Oxide in Premixed Hydrocarbon Flames. *Proceedings of the Combustion Institute* 13, 373–379.

Finkelman, R.B., 1994. Modes of occurrence of potentially hazardous elements in coal: Levels of confidence. *Fuel Processing Technology* 39, 21–34.

Forzatti, P., 2001. Present status and perspectives in de-NOx SCR catalysis. *Applied Catalysis A*: General 222, 221–236.

Galbreath, K.C., Zygarlicke, C.J., 2000. *Fuel Processing Technology* 65–66, 289–310.

Germani, M.S., Zoller, W.H., 1988. Vapor-phase concentrations of arsenic, selenium, bromine, iodine and mercury in the stack of a coal-fired power plant. *Environmental Science & Technology* 24, 1079–1085.

Glarborg, P., Jensen, A.D., Johnsson, J.E., 2003. Fuel nitrogen conversion in solid fuel fired systems. *Progress in Energy and Combustion Science* 29, 89–113.

Glarborg, P., Miller, J.A., Kee, R.J., 1986. Kinetic modeling and sensitivity analysis of nitrogen oxide formation in well stirred reactors. *Combustion and Flame* 65,177–202.

Goodarzi, F., Swaine, D., 1993. Chalcophile elements in western Canadian coals. *International Journal of Coal Geology* 24, 281–292.

Herzog, H.J. 2001. What future for carbon capture and sequestration? *Environmental Science & Technology* 35, 148A–153A.

Hu, Y., Naito, S., Kobayashi, N., Hasatani, M., 2000. CO_2, NO_x and SO_2 emissions from the combustion of coal with high oxygen concentration gases. *Fuel* 79, 1925–1932.

Huggins, F., Goodarzi, F., 2009. Environmental assessment of elements and polyaromatic hydrocarbons emitted from a Canadian coal-fired power plant. *International Journal of Coal Geology* 77, 282–288.

IEA, 2010. *Energy Technology Perspectives 2010– Scenarios & Strategies to 2050*, International Energy Agency, France p 650.

IEA, International Energy Agency, 2012. *CO_2 Emissions from Fuel Combustion Highlights* (2012 Edition), International Energy Agency, France.

IEA, 2013. *Tracking Clean Energy Progress 2013: IEA Input to the Clean Energy Ministerial*. OECD/IEA, Paris, France.

IPCC, 2005. *Working Group III of the Intergovernmental Panel on the Climate Change (IPCC)*. Carbon Dioxide Capture and Storage, Intergovernmental Panel on the Climate Change (IPCC), Cambridge University Press, Cambridge.

IPCC, 2007a. Summary for policymakers. *In: Climate Change 2007: The Physical Science Basis*, Cambridge University Press, Cambridge.

IPCC, 2007b. *Working Group III, I. Fourth Assessment Report of the IPCC*. Cambridge University Press, Cambridge.

Izquierdo, M., Querol, X., 2012. Leaching behaviour of elements from coal combustion fly ash: An overview. *International Journal of Coal Geology* 94, 54–66.

Jaworek, A., Krupa, A., Czech, T. 2007. Modern electrostatic devices and methods for exhaust gas cleaning: A brief review. *Journal of Electrostatics* 65, 133–155.

Jones, D.R., 1995. The leaching of major and trace elements from coal ash. In: Swaine, D.J., Goodarzi, F. (Eds.), *Environmental Aspects of Trace Elements in Coal*. Springer, the Netherlands.

Kellie, S., Cao, Y., Duan, Y., Li, L., Chu, P., Mehta, A., Carty, R., Riley, J.T., Pan, W.P., 2005. Factors affecting mercury speciation in a 100-MW coal-fired boiler with low-NO_x burners. *Energy & Fuels* 19, 800–806.

Kema, 1997. *Behaviour, Control and Emissions of Trace Species by Coal-Fired Power Plants in Europe*, Report 83428.SP.08 97P01.07B KEMA, Arnhem, the Netherlands.

Kohl, A.L., Nielsen, R.B., 1997. *Gas Purification*, 5th ed, Gulf Publishing Company, Houston, TX, 1997.

Laudal, D.L., Brown, T.D., Nott, B.R., 2000. Effects of flue gas constituents on mercury speciation. *Fuel Processing Technology* 65-66, 157–165.

Lapple, C. 1951. Processes use many collector types. *Chemical Engineering* 58, 144–151.

Lim, K.J., Milligan, R.J., Lips, H.I., Castaldini, C., Merrill, R.S., Mason, H.B., December 1979. *Technology Assessment Report for Industrial Boiler Applications: NOx Combustion Modification*, EPA-600/7-79-178f, U.S. Environmental Protection Agency, Research Triangle Park, NC.

Linak, W.P., Miller, C.A., Wendt, J.O.L., 2000. Comparison of particle size distributions and elemental partitioning from the combustion of pulverized coal and residual fuel oil. *Journal of Air and Waste Management Association* 50, 1532–1544.

Linak, W.P., Wendt, J.O.L., 1994. Trace metal transformation mechanisms during coal combustion. *Fuel Processing Technology* 39 (2), 173–198.

Machalek, T., Ramavajjala, M., Richardson, M., Richardson, C., Dene, C., Goeckner B., Anderson, H., Morris, E., May 19–22, 2003. Pilot evaluation of flue gas mercury reactions across an SCR unit. *Proceedings of the Combined Power Plant Air Pollutant Control Symposium—The Mega Symposium*, Washington, DC.

Malte, P.C, Pratt, D.T., 1975. Measurement of atomic oxygen and nitrogen oxides in jet-stirred combustion. *Proceedings of Combustion Institute* 15, 1061–1070.

Manz, O.E., 1999. Coal fly ash: A retrospective and future look. *Fuel* 78 (2), 133–136.

Meij, R., 1989. Tracking trace elements at a coal fired plant equipped with a wet flue gas desulphurization facility. *Kema Science and Technology report* 7 (5), 269–355.

Miller, B.G., 2011. *Clean Coal Engineering Technology*. Elsevier, Amsterdam.

Miller, J.A., Bowman, C.T., 1989. Mechanism and modeling of nitrogen chemistry in combustion. *Progress in Energy and Combustion Science* 15, 287–338.

Mizuno, A., October 2000. Electrostatic precipitation. *IEEE Transactions on Dielectrics and Electrical Insulation* 7 (5), 615–624.

Moreno, N., Querol, X., Andres, J.M., Stanton, K., Towler, M., Nugteren, H. et al., 2005. Physico-chemical characteristics of European pulverized coal combustion fly ashes. *Fuel* 84 (11), 1351–1363.

National Environmental Trust, NET, USA, 2004. *Beyond mercury*, Report.

Pacyna, J.M., Munch, J., 1991. Anthropogenic mercury emissions in Europe. *Water, Air, and Soil Pollution* 56, 51–61.

Pandey, V.C., Singh, N., 2010. Impact of fly ash incorporation in soil systems. *Agriculture, Ecosystems & Environment* 136, 16–27.

Parisa, A., Khalizov, A.A., Gidas, A., 2002. Reactions of gaseous mercury with atomic and molecular halogens: Kinetics, product studies, and atmospheric implications. *Journal of Physical Chemistry A* 106, 7310–7320.

Pershing, D.W., Wendt, J.O.L, 1976. Pulverized coal combustion: the influence of flame temperature and coal composition on thermal and fuel NO_x. *Proceedings of the Combustion Institute* 16, 389–399.

Pudasainee, D., Kim, J.H., Seo, Y.C., 2009. Mercury emission trend influenced by stringent air pollutants regulation for coal-fired power plants in Korea. *Atmospheric Environment* 43, 6254–6259.

Pudasainee, D., Kim, J.H., Yoon, Y.S., Seo, Y.C., 2012. Oxidation, reemission and mass distribution of mercury in bituminous coal fired power plants with SCR, CS-ESP and wet FGD, *Fuel* 93, 312–318.

Pudasainee, D., Lee, S.J., Lee, S.H., Kim, J.H., Jang, H.N., Cho, S.J., Seo, Y.C., 2010. Effect of selective catalytic reactor on oxidation and enhanced removal of mercury in coal-fired power plants. *Fuel* 89, 804–809.

Raask, E., 1985. The mode of occurrence and concentration of trace elements in coal. *Progress in Energy and Combustion Science* 11 (1), 97–118.

Ratafia-Brown, J.A., 1994. Overview of trace elements partitioning in flames and furnaces of utility coal-fired boilers. *Fuel Processing Technology* 39 (2), 139–157.

Ruhl, L., Vengosh, A., Dwyer, G.S., Kim, H.H., Schwartz, G., Romanski, A., Smith, D., 2012. The impact of coal combustion residue effluent on water resources: A North Carolina example. *Environmental Science & Technology* 46, 12226–12233.

Ruthven, D.M., 1984. *Principles of Adsorption and Adsorption Processes*, Wiley, New York.

Samanta, A., Zhao, A., Shimizu, G.K.H., Sarkar, P., Gupta, R., 2012. Post-combustion CO_2 capture using solid sorbents: A review. *Industrial Engineering & Chemistry Research* 51, 1438–1463.

Schroeder, W.H., Munthe, J., 1998. Atmospheric mercury—An overview. *Atmospheric Environment* 32 (5), 809–822.

Sloss, L.L., 1995. Mercury-emissions and effects—the role of coal. IEA Coal Research IEAPER/19.

Soud, H.N., 1995. *Developments in Particulate Control for Coal Combustion*, IEA Coal Research, London.

Swaine, D.J., 1990. *Trace Elements in Coal, (M), Butterworth*, London, 1990.

Swaine, D.J., Goodarzi, F., 1995. *Environmental Aspects of Trace Elements in Coal, (M),* Kluwer, Dordrecht, the Netherlands.

U.S. EPA., April 1993. Emission Factor Documentation for AP-42 Section 1.1—Bituminous and Subbituminous Coal Combustion, U.S. Environmental Protection Agency, Research Triangle Park, NC.

U.S. EPA, 1997. Mercury study report to congress; EPA-452/R-97-003; U.S. EPA Office of Air Quality Planning and Standards, U.S. Government Printing Office, Washington, DC.

U.S. EPA, 1998. A study of hazardous air pollutant emissions from electric utility steam generating units: final report to congress; EPA-453/R-98-004a; U.S. EPA Office of Air Quality Planning and Standards, U.S. Government Printing Office, Washington, DC.

U.S. EPA, 2002. Control of mercury emissions from coal-fired electric utility boilers. US Environmental Protection Agency, EPA-600/R-01-109.

US EPA, 2008. Climate Leaders Greenhouse gas inventory protocol core module guidance, Direct Emission from Stationary Combustion Sources. EPA430-K-08-003.

Vassilev, S.V., Vassileva, C.G., 2005. Methods for characterization of composition of fly ashes from coal-fired power stations: A critical overview. *Energy & Fuels* 19 (3), 1084–1098.

Vassilev, S.V., Vassileva, C.G., 2007. A new approach for the classification of coal fly ashes based on their origin, composition, properties, and behaviour. *Fuel*, 86 (1011):1490–1512.

Wark, K., Warner, C.F., Davis, W.T., 1998. *Air Pollution Its Origin and Control*, 3rd ed., Addison Welsey Longman, Menlo Park, CA.

Wojtowicz, M.A., Pels, J.R., Moulijn, J.A., 1993. Combustion of coal as a source of N_2O emission. *Fuel Processing Technology* 34, 1–71.

Xu, M., Yan, R., Zheng, C., Qiao, Y., Han, J., Sheng, C., 2003. Status of trace element emission in a coal combustion process: A review. *Fuel Processing Technology* 85(2–3), 215–237.

Zahir, F., Rizwi, S.J., Haq, S.K., Khan, R.H., 2005. Low dose mercury toxicity and human health. *Environmental Toxicology and Pharmacology* 20, 351–360.

Zeldovich, Y.B., 1974. The oxidation of nitrogen in combustion and explosions. *Acta Physicochem USSR* 21, 577–628.

INTERNET REFERENCES

B&W, 2014. http://www.babcock.com/products/Pages/Dry-Electrostatic-Precipitator.aspx.

http://www.chinafaqs.org/files/chinainfo/China%20FAQs%20Emission%20Standards%20v1.4_0.pdf (June 2012).

http://www.netl.doe.gov/technologies/coalpower/ewr/coal_utilization_byproducts/pdf/mercury_%20FGD%20white%20paper%20Final.pdf.

IEA, 2014a. http://www.iea-coal.org.uk/site/ieacoal/databases/ccts/selective-catalytic-reduction-scr-for-nox-control (accessed September, 2014).

IEA, International Energy Agency, 2014b. http://www.iea.org/techinitiatives/fossilfuels/cleancoal/ (assessed August 13 2014).

Miller, C.E, Feeley, T.J., Aljoe, W.W., Lani, B.W, Schroeder, K.T., Kairies, C. et al. 2006. Mercury capture and fate using wet FGD at coal-fired power plants. DOE/NETL mercury and wet FGD R & D.

UCS, 2014. http://www.ucsusa.org/clean_energy/coalvswind/c02d.html (assessed August 11, 2014).

World Resources Institute, 2012. China adopts world-class pollutant emissions standards for coal power plants. http://www.chinafaqs.org/files/chinainfo/China%20FAQs%20Emission%20Standards%20v1.4_0.pdf (accessed June, 2012).

Section IV

Global Issues and Trends

20 Carbon Management in the Coal Industry

Ali Elkamel, Zarook Shareefdeen, and Raymond Yeung

CONTENTS

Abstract: In recent years, coal has become the largest contributor to carbon dioxide (CO_2) emissions in the energy sector. The increasing concern regarding global warming has led to the research and development of more effective carbon management technologies for different coal-fueled power plants. The Kyoto Implementation Act gives countries initiatives and a plan to reduce carbon emissions to avoid further global warming and disruption of ecosystems around the world. This chapter explores different carbon management technologies that organizations implement to increase efficiency and decrease emissions. The topics include ultra-supercritical pulverized coal combustion, integrated gasification combined cycle, oxy-fueled combustion, and fluidized bed combustion. However, mitigation of CO_2 still largely relies on carbon capture methods such as amine scrubbing, Selexol, and Rectisol processes. This chapter also discusses recent advances in carbon capture such as the use of novel sorbent materials as well as gas separation using membrane technology. Sequestration is another major aspect of carbon management. The site for sequestration often includes geological storage in aquifers, ocean storage, oil reservoirs for enhanced oil recovery, or coal beds for methane recovery. While carbon management comes at a high cost, the captured carbon can be recycled for industrial applications.

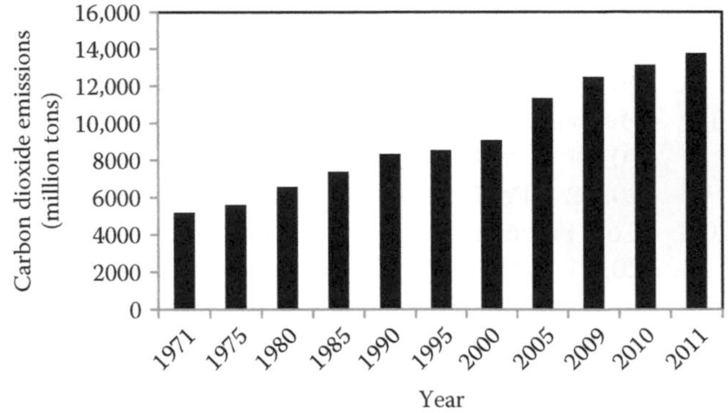

FIGURE 20.1 Coal-based emissions from different sources. (Data from International Energy Agency, CO_2 emissions—Sectoral approach, *CO_2 Emissions from Fuel Combustion*, International Energy Agency, France, 53, 2013.)

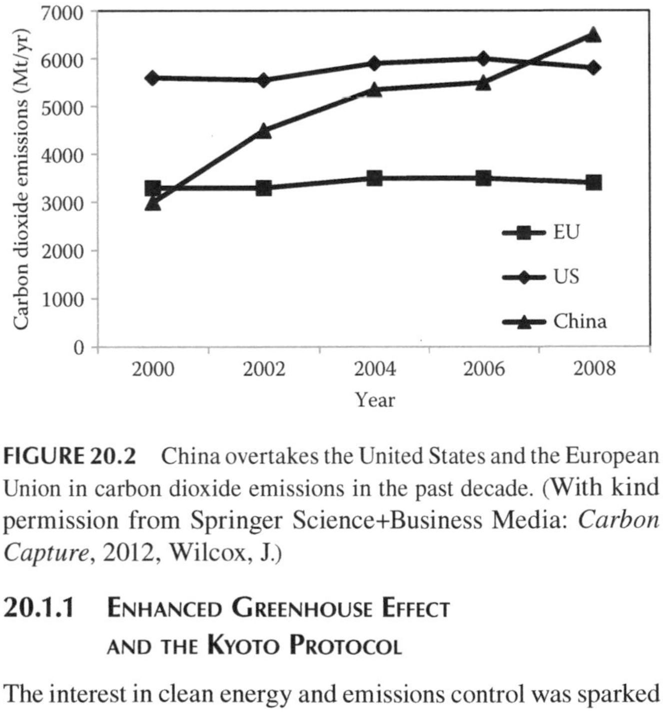

FIGURE 20.2 China overtakes the United States and the European Union in carbon dioxide emissions in the past decade. (With kind permission from Springer Science+Business Media: *Carbon Capture*, 2012, Wilcox, J.)

20.1 INTRODUCTION

Carbon management is a critical aspect of the coal industry because coal releases the most carbon dioxide (CO_2) per unit of energy generated. In 2011, the combustion of coal resulted in the release of 13.7 Gt of CO_2, while the combustion of oil and gas released 11.1 and 6.3 Gt of CO_2, respectively. Because of its high carbon content, coal contributes around 44% of global CO_2 emissions (International Energy Agency 2013).

Figure 20.1 shows that coal-based carbon emissions have increased between 2009 and 2011. Developing countries are joining the energy race and have contributed greatly to CO_2 emissions. After 2006, China's CO_2 level (see Figure 20.2) has surpassed the United States as the largest contributor to CO_2 emissions in the world.

20.1.1 Enhanced Greenhouse Effect and the Kyoto Protocol

The interest in clean energy and emissions control was sparked by the concern over global warming due to increase in CO_2 concentration levels in the atmosphere (see Figure 20.3). Between 1990 and 2010, CO_2 levels have increased by 11 billion tons, and because of this, the temperature has increased by roughly 0.8°C (Doyle 2014).

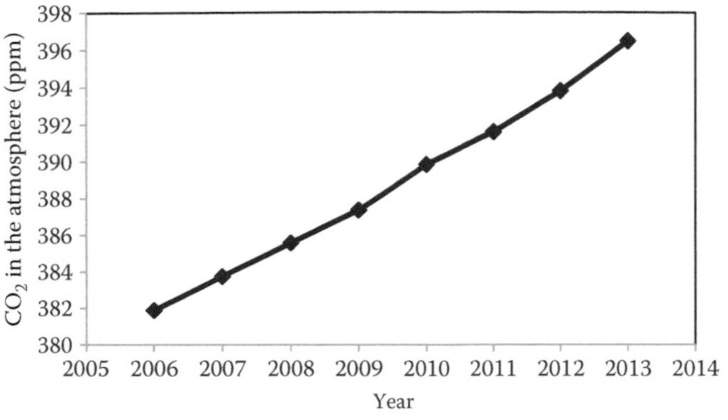

FIGURE 20.3 Carbon dioxide concentration (ppm). (Data from International Energy Agency, CO_2 emissions—Sectoral approach, *CO_2 Emissions from Fuel Combustion*, International Energy Agency, France, 53, 2013.)

The increase in temperature over the past decades has raised many issues within the ecosystem. For example, global warming created a lack of water and thus an increase in droughts in the Canadian Boreal Forests. Moreover, the balance of insects has been thrown off. Pests that were supposed to die in the winter months can now survive the milder temperatures and hence destroying forests at an increasing rate (Wells et al. 2008).

Global warming also causes the polar ice to melt, which disrupts agriculture in the prairies and petroleum recovery at the oil sands. With the polar ice caps shrinking, the albedo effect (coefficient of light reflected) will be reduced, allowing the earth to capture more heat (Borenstein 2014). To understand the proportionality of the situation, it has been estimated that the change in the average temperature of earth will range between 1.0°C and 2.1°C per trillion tons of CO_2 emitted (Matthews et al. 2009).

20.1.2 Effects of Carbon Dioxide on Humans

Concentrations of CO_2 in the atmosphere are roughly 390 ppm or 0.0038%, which does not affect human health. However, there are situations where CO_2 levels are higher, and workers are protected under regulations. For example, there is a limit of 5000 ppm/8 h exposure by the American Conference of Governmental Industrial Hygienists (ACGIH). Concentrations upward of 40,000 ppm will cause symptoms of intoxication, and extended exposure will be life threatening. A disorder called acidosis is the result of hazardous exposure to CO_2 (Charles 2005).

20.1.3 Kyoto Protocol

The concerns over the enhanced greenhouse effect resulted in the Kyoto Protocol Implementation Act (KPIA). With the KPIA, developing countries agreed to reduce their carbon footprint to 5% below 1990 levels within 2008 and 2012 (Environment Canada 2012). Countries have agreed to place a ceiling on global warming to less than 2°C increase when compared to pre-industrial era. To prevent reaching this 2°C limit, it was estimated that CO_2 emissions have to be limited to between 30 and 50 billion tons by 2030 (Doyle 2014). Prior to the KPIA, there was the United Nations Framework Convention on Climate Change (UNFCCC), but the plan fell through because it was a nonbinding agreement with voluntary goals. Despite having good intentions, many of the fossil fuel–rich countries are refusing to ratify the protocol, and even the countries that have ratified are far behind the target set for emission reduction (Al-Fattah et al. 2012).

20.1.4 Concept of Carbon Capture and High-Efficiency Power Plants

The increasing concern regarding the greenhouse effect and global warming sparks a race for researchers and engineers to develop a viable method to mitigate the effects of CO_2.

Developing processes such as ultra-supercritical pulverized coal combustion (PCC), integrated gasification combined cycle (IGCC), or fluidized bed combustion (FBC) seek to increase plant efficiency. These new technologies offer cleaner and more efficient energy but face challenges such as corrosion, complexity, air separation, and high capital costs. From Table 20.1, it can be observed that carbon capture creates a drastic drop in efficiency of power plants. With relatively low efficiencies already, the energy needed for capture will make plant operations not feasible.

At extreme operating temperatures and pressures, ultra-supercritical PCC can achieve increased efficiencies, but the drawbacks are high capital and maintenance cost. It is also difficult to find materials that can withstand the extreme environment inside an ultra-supercritical power plant unit. An alternate option is the IGCC. The plant converts the coal into syngas, which would drive both a gas engine and a steam engine. This technology produces clean energy and CO_2 at high pressure, making compression for transport a less-intensive

TABLE 20.1
Comparing Efficiency of Different Coal-Fueled Power Plants

Technology	Efficiency (%)	Efficiency with Carbon Capture (%)	Cost of CO_2 Mitigated USD/ton
Subcritical pulverized coal combustion	42–43	30–32	35–45
Ultra-supercritical pulverized coal combustion	43–45	32–35	40–51
Integrated gasification combined cycle	>43	32–34	18–28
Oxy-fuel combustion	37–44	36–38	26–45

Source: Mills, R., *Capturing Carbon: The New Weapon in the War against Climate Change*, C. Hurst & Co, London, p. 71, 197, 2011.

TABLE 20.2
Comparison of Chemical Absorption Technologies That Are Commercially Used

	Chemical Absorption		
Technology	**MEA**	**Secondary, Tertiary Amines**	**Chilled Ammonia**
Advantages	• High loading capacity for CO_2 • Low capital cost • Less solvent needed	• Low energy required for solvent regeneration • Less prone to degradation	• High pressure CO_2 released so compression for transport is reduced • High loading capacity
Disadvantages	• High energy required for solvent regeneration • Solvent degradation • High rate of solvent loss	• Low loading capacity for CO_2 • High capital cost • High amount of solvent needed	• Cooling of flue gas

TABLE 20.3
Comparison of Commercially Used Physical Absorption Technologies

	Physical Absorption	
Technology	**Selexol**	**Rectisol**
Advantages	• High selectivity toward CO_2	• Less steam required for regeneration • CO_2 is more soluble in cold temperatures • Low temperatures and low pressures reduce solvent loss • Low energy consumption
Disadvantages	• Gas streams are constantly heated and cooled • Low pressure CO_2 stream emitted thus high compression is required for transport	• Energy required to sustain low temperatures • Methanol have high vapor pressure

process. Due to its complexity in operation, it is not practical to retrofit existing plants with this technology. Commercially used carbon capture method in PCC and IGCC plants is chemical/physical absorption. Amine scrubbing is a well-developed and widely used method that is easily retrofitted into existing plants (see Table 20.2). The problem with amine scrubbing is the parasitic load that it has on the power plant. For IGCC plants, solvents such as Selexol and Rectisol are used because they are more suited for pre-combustion capture (see Table 20.3).

The carbon capture process requires extensive treatment of the flue gas before it can be captured by a solvent that translates into decreased plant efficiency. Oxy-fuel combustion can simplify the carbon capture process. An air separation unit (ASU) supplies oxygen to the combustion chamber where complete combustion will occur. The resulting flue gas is composed of water vapor and CO_2. After condensing the water vapor out, the CO_2 stream can be sent for transport with minimal treatment. The details of these technologies with respect to carbon management are discussed in the subsequent sections.

20.2 OVERVIEW OF THE COAL INDUSTRY

20.2.1 Coal as an Energy Source

Coal has been one of the most important sources of fuel for energy production as it supplies roughly 40% of the world's electricity (World Coal Institute 2005). Its extensive use is due to its abundance compared to other sources of fuel such as fossil fuel or natural gas. The stability of coal production also makes it a relatively inexpensive way to produce energy. From Figure 20.4, it can be observed that coal is among one of the

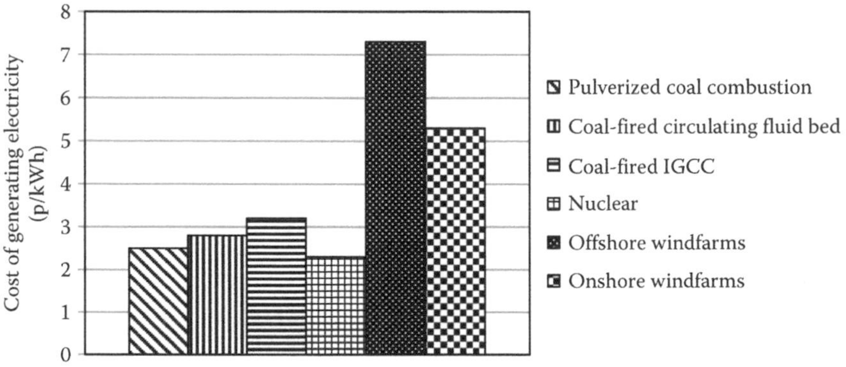

FIGURE 20.4 Price comparisons of different sources of energy including the cost of carbon management. (Data from Royal Academy of Engineering, *The Cost of Generating Electricity*, Royal Academy of Engineering, London, 8, 2004.)

TABLE 20.4
Projection of Energy Distribution

	Share in Electricity Generation (%)			
	Organization for Economic Cooperation and Development		Developing Countries	
Electricity Source	**2002**	**2030**	**2002**	**2030**
Coal	38	33	45	47
Oil	6	2	12	5
Gas	18	29	17	26
Nuclear	23	15	2	3
Hydro	13	11	23	16
Other renewables	3	10	1	3

cheapest sources of energy, especially compared with renewable sources. The figure compares different coal combustion technologies such as pulverized fuel (PF), circulating fluidized bed (CFB), and IGCC. Although some of those methods have high costs, they can be more efficient and have a smaller carbon footprint, which will be discussed later in the chapter.

Despite the continuous efforts of developing renewable energy sources, coal still dominates the energy market; in fact, it is predicted that coal-dependent electricity is to increase from 16,074 to 31,657 TWh from 2002 to 2030 (Stefaniak and Twardowska 2006). Table 20.4 shows a projection of energy distribution between fuel sources in the first three decades of the twenty-first century.

20.2.2 Production of Energy

The types of coal that are significant to energy production are sub-bituminous and bituminous coal. Bituminous coal used for power generation is classified as thermal or steam coal that has low moisture content and high energy content. The coal is ground into a fine powder to maximize the surface area for combustion at which point it is burned in a combustion chamber. The coal powder is mixed with air and pressurized before being forced into the chamber. The heat generated is then used to convert water into superheated steam to move numerous sets of turbines. The turbines then generate electricity, which is then delivered to the end user (Kaplan 2008). Steam that exits the turbine is condensed to be fed back into the boiler. This technology is well developed, but research in coal production is still being done to improve efficiency and to reduce greenhouse gas emissions. Temperatures outside of the flame zone remain at extremely high temperatures exceeding 3000°F; thus, an important aspect of power plants is to harness as much of this heat energy as possible to be used within the system. Along the way, heat will be extracted from the gases to improve efficiency. First, the heat is used to convert water to steam and then the steam gets superheated. After the superheated steam passes through early stages of the turbines, it goes through a re-heater. Finally, excess heat from the furnace is also used to pre-heat water that is entering the boiler as well as pre-heating the air entering the furnace (Hayes 2005). All these processes are meant to maximize the heat energy generated from the combustion of coal (Figure 20.5).

20.3 COAL PROCESSING METHODS AND CARBON DIOXIDE EMISSIONS

There are different ways of processing coal prior to its combustion to produce energy. Each of these methods has their own benefits with regard to cost effectiveness, carbon footprint, and efficiency. The following methods are discussed: (1) PCC (traditional, supercritical, ultra-supercritical PCC), (2) IGCC, (3) FBC, and (4) oxy-fuel combustion. The processing method and technology behind a power plant plays an essential part in carbon management. The goal is to make the process more efficient and also make the gas stream easier to clean so that the carbon capture process will be less energy intensive.

20.3.1 Pulverized Coal Combustion

PCC is a well-developed technology for coal-powered generators. There are many of its kind implemented around the world in developed and developing countries. Pulverization is needed to maximize the surface area for combustion. The coal powder

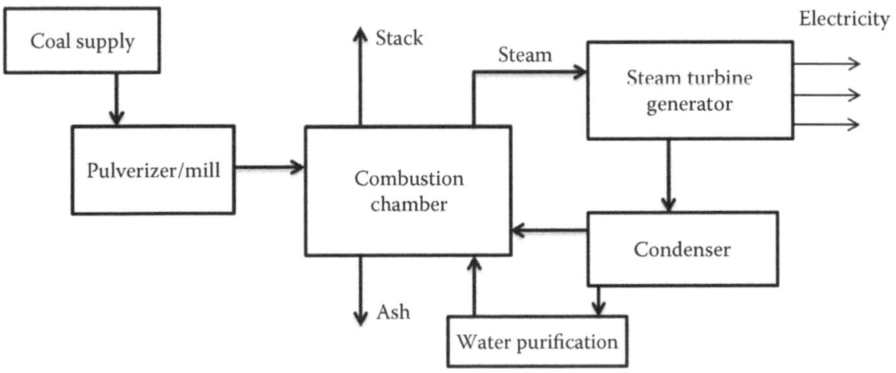

FIGURE 20.5 Simplified diagram of coal-fueled generators. (From World Coal Institute, *Coal Resource, a Comprehensive Overview of Coal*, World Coal Institute, London, 2005.)

TABLE 20.5
Comparison of Different PCC Technologies

Steam Cycle	Ultra-Supercritical	Supercritical	Subcritical
Pressure (MPa)	25–34	22–24	16–17
Steam temperature (°C)	701–759	596–656	537–546
CO_2 emissions (tons/MWh)	1.05	1.09	1.12

Source: Zhang, D., Introduction to advanced and ultra-supercritical fossil fuel power plants. *Ultra-Supercritical Coal Power Plants Materials, Technologies and Optimisation*, Editors: Zhang D, Woodhead Publishing, Cambridge, 4, 2013.

can be sprayed into the furnace as a mist that helps drive the reaction to complete combustion. The coal powder is sprayed into the furnace at roughly a 2:1 air to coal ratio at temperatures of around 65°C–95°C depending on the type of coal used as fuel. The velocity of the mixture has to be monitored as well because low velocity will result in a buildup of powder in the tubes while a high velocity will apply too much stress on the bends and inner surfaces of the tubes. There are generally three types of pulverized coal technology currently in use.

At operating pressures and temperatures of an ultra-supercritical plant, the efficiency is typically in the high 40% to low 50% range. Also, it can be observed from Table 20.5 that the emissions are less than the other two technologies. An issue with current ultra-supercritical plants is the availability and financial feasibility of acquiring advance materials for boiler walls and piping that are able to withstand the high pressure and temperature. More commonly, industries use supercritical pulverized coal (SCPC) technology, which typically have a first law efficiency (percent of thermal energy converted to electrical energy) of 38.5%. Difficulties surround supercritical and ultra-supercritical technology in finding a material that can combat the effects of corrosion under such harsh conditions (Zhang 2013).

20.3.1.1 Current Ultra-Supercritical Pulverized Coal Combustion

Most countries are leaving behind the technology of subcritical combustion due to its low efficiency and are now investing in supercritical or ultra-supercritical combustion. The International Energy Agency predicted that global usage of supercritical and ultra-supercritical technology will increase from 20% (2008) to 28% (2014).

In November 2012, an ultra-supercritical steam power plant started operation in Lunen, Germany. As a collaborative effort between Trianel Kohlkraftwerk Lunen GmbH und Co, Siemens, and IHI Corporation, an 800 MW coal-fueled power plant has been created with efficiency greater than 45%. Operating conditions are 28 MPa and 610°C with CO_2 emissions of less than 800 g CO_2/kWh (Bewerunge et al. 2009).

Another example of an ultra-supercritical unit is in Arkansas called the John W. Turk Power Plant. It is America's first power plant equipped with such technology. Despite strong legal opposition during construction, the 600 MW power plant commenced commercial production in December 2012 costing $2.1 billion. The power plant is one of the most efficient in the United States due to its extreme operating conditions of 25.3 MPa and 607°C allowing it to have an efficiency of about 40%. American Electric Power/Southwestern Electric Power Co. claims that emissions will be 320,000 tons of CO_2 less than other conventional PCC technology as well as using 180,000 tons less in coal per year.

20.3.2 Integrated Gasification Combined Cycle

Coal-based IGCC is a less-developed technology compared to PCC, but the benefits to the environment by IGCC are far greater. The following will give a description of the gasification of coal and the idea behind a combined cycle.

20.3.2.1 Gasifier Technology

IGCC is a very versatile technology as it uses liquid or solid fuel and converts it into a more manageable gaseous form. It first converts fossil fuel, coal in this case, into synthesis gas (syngas), which is composed mainly of hydrogen (H_2) and carbon monoxide (CO). Limited amounts of oxygen create a combustion reaction with some of the coal, generating heat and CO_2, which will trigger the gasification process where the rest of the coal is converted into H_2 and CO. One of the challenges of gasifier technology is the ASU. Partial oxidation of the coal requires pure oxygen to be fed into the gasifier that is supplied by the ASU. A parameter to be considered is the degree of integration, which is the term defining how much exhaust air from the gas turbine is supplied to the ASU. One hundred percent integration means all the air that the ASU uses is from the gas turbine, leading to maximized efficiency. However, partial integration will maximize power output; thus, a balance must be kept (Maurstad 2005). The following reactions characterize the conversion from coal (carbon) and water (or steam) to syngas (H_2 and CO) (Higman and Burgt 2008):

$$C+\frac{1}{2}O_2 \leftrightarrow CO\ -111\frac{MJ}{kmol} \tag{20.1}$$

$$C+H_2O \leftrightarrow CO+H_2+131\frac{MJ}{kmol} \tag{20.2}$$

Using coal as fuel, a chemical equation can be written as:

$$C_nH_m+\frac{n}{2}O_2=nCO+\frac{m}{2}H_2 \tag{20.3}$$

The fuel still contains trace amounts of sulfur and nitrogen. The quantity is not enough to affect the fuel composition, but it will still have to be isolated later on due to its adverse effects on the environment.

20.3.2.2 Combined Cycle

This technology is considered a combined cycle because it uses two turbines (a gas turbine and a steam turbine) to generate electricity. The syngas that is produced by the gasification

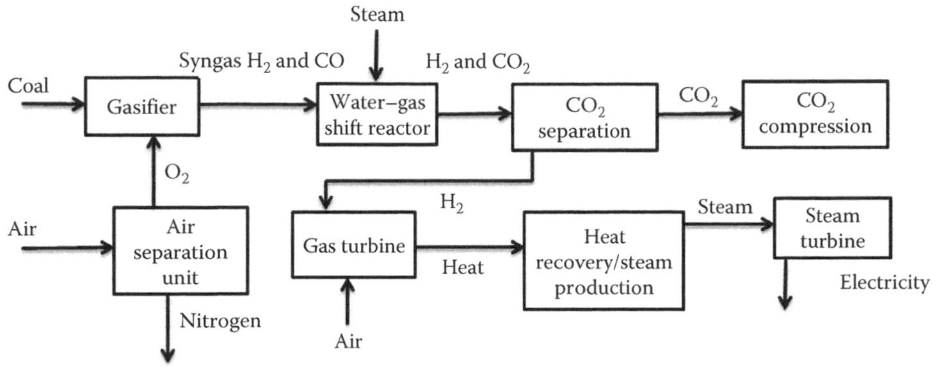

FIGURE 20.6 Block diagram of integrated gasification combined cycle. (From Abanades, J. et al., *Intergovernmental Panel on Climate Change Carbon Dioxide Capture and Storage*, Editors: Coninck H, Davidson O, Loos M, Metz B, Meyer L, Cambridge University Press, New York, 105–179, 2005.)

process is used to propel a gas turbine for electricity. The heat exiting the gas turbine is in great excess so it is captured for converting water to superheated steam. The steam converts its heat energy to mechanical energy via a steam turbine. A generator then uses the mechanical energy generated to produce electricity. The use of both a gas turbine and a steam turbine is a defining characteristic of the combined cycle. Processes such as the IGCC are still in its developing stages, but the technology proves to be more efficient and environmentally friendly than PCC (Figure 20.6).

20.3.2.3 Current IGCC Plants

Due to high costs and technological difficulties, IGCC plants have not dominated the energy market; however, they still remain as one of the cleanest sources of coal energy as proven in a few power plants around the world. One of the most recent developments in IGCC plants is spearheaded by Duke Energy in Indiana, named the Edwardsport Power Station. As of 2011, this power station has been taken offline to be replaced with an IGCC process and has resumed production in June 7, 2013, to become the largest IGCC power plant in the United States. With a final price tag of $3.5 billion, the Edwardsport power plant has missed its budget of under $2 billion as set in 2007. There has been huge opposition by citizens toward the IGCC project due to its high cost and large deviation from intended budget. However, the government has shown its support by providing up to $460 million in tax credits and incentives. Within months of starting up, the power plant has encountered numerous problems, causing it to operate below maximum power; a few difficulties even resulted in the temporary closure of the plant. The 618 MW power plant uses 1.7–1.9 million tons of coal per year. With six times the output compared to the former plant, the company claims that there will be 70% less emission of sulfur dioxide, nitrogen oxide, and other contaminants as well as half the amount of CO_2 emitted.

Another recent development in IGCC is the Kemper Project located in Mississippi. It is a 582 MW coal-fueled IGCC power plant that will be supplied with 160 million tons of coal over the course of 40 years. As of October 2013, a 60 mile pipeline was completed to deliver the captured CO_2 to oil fields for enhanced oil recovery (EOR). The proposed budget is $2.4 billion and is predicted to be completed late in 2014. The main deterrent in IGCC technology is the high capital costs, and this has been proven in the two existing plants in the United States.

20.3.3 Oxy-Fuel Combustion

A third method for producing coal-fueled energy is through oxy-fuel combustion. While PCC and IGCC technologies require the separation of CO_2 from the flue gas or syngas stream, oxy-fuel combustion is unique because it produces a highly pure CO_2 that will require little or no separation. Rather than combusting coal using air, oxy-fuel combustion drives toward complete combustion by feeding pure oxygen into the combustion chamber. Thus, the reaction products will be water vapor and CO_2. Ideally, once the water vapor is condensed, the flue-gas stream will contain only CO_2, which can be compressed and transported without further treatment, but often the gas still needs to be cleaned. In which case, a single- or double-flash distillation will easily make the flue-gas purity upward of 95 mol% CO_2 for transport. Besides the benefit of not needing an absorbing/scrubbing tower, the oxygen supplied into the combustion chamber can enter at low pressures, hence eliminating the need for compression. The challenge with this technology is the ASU. Although the current technology of ASU is well developed with regard to achieving high purity oxygen, the problem is still the volume of oxygen produced; oxy-fuel combustion plant requires around 20,000 tons of oxygen per day (Dillon and Wheeldon 2011).

20.3.3.1 Current Oxy-Fuel Power Plants

Oxy-fuel combustion technology is very difficult to achieve, and it comes at a high cost; thus, it is used only on smaller power plants. One of such project is the Callide Oxy-Fuel Project at Queensland, Australia. Budgeted at AUD 208 million, the 30 MW oxy-fuel combustion power plant was announced to be successful in April 2012. The Callide Oxy-Fuel project is a collaborative effort among CS Energy, Xstrata Coal, Australian Coal Association, Schlumberger, IHI, Mitsui, and J-Power with funding from the Commonwealth government's Low Emissions Technology Demonstration Fund, the Queensland government, and the Japanese government. This project is a good example of the possibility of retrofitting old power plants with a new technology. A budget has been allocated to evaluate the emissions and efficiencies of the plant over the span of 2 years to November 2014. The plant is to capture 90% of the CO_2 amounting up to 60,000 tons of CO_2/year for geological storage. Captured carbon will be sequestered near the Denison Trough via an estimated two trucks per day. The plant is currently in a demonstration phase until November 2015 and is predicted to commence commercial production by 2020.

20.3.4 Fluidized Bed Combustion

FBC is typically implemented with PCC, IGCC, or oxy-fuel combustion to improve combustion or oxidation reactions. Air enters a chamber through the bottom and runs through a bed of pulverized coal to create turbulence. This gas flow is said to have *fluidized* the coal because the turbulent bed behaves as if it were a fluid rather than a solid (Abanades et al. 2005). FBC is a cleaner method of combustion, but it sacrifices efficiency because of its low operating temperatures and the energy needed to fluidize the bed. Emissions are controlled by adding limestone into the bed to remove sulfur compounds. Furthermore, due to low combustion temperatures, there is no formation of nitrous oxide. FBC technology is also beneficial because it can use a variety of fuels such as low-grade coal, biomass, or waste-derived biomass fuels.

20.3.4.1 FBC Case Study

One of the most promising technologies using FBC is the Sargas process. It uses high pressure FBC to yield a high efficiency and effective carbon capture. The combustion process operates at a pressure of 1.2 MPa and temperatures of 850°C–880°C. The benefit of combusting at low temperature is that nitrogen oxides will not form in excess. The Sargas process can capture >90% of the emitted CO_2 at an efficiency of 40%. The capture technology that uses hot potassium carbonate solution for absorption is called the Benfield process (Hetland 2008).

20.4 TREATMENT AND SEPARATION OF CARBON DIOXIDE IN SYNGAS/FLUE GAS

As discussed in the previous section, different combustion or coal processing techniques were used to generate power, and each of those technologies has its own method for carbon capture. IGCC requires pre-combustion capture and PCC requires post-combustion capture, while oxy-fuel combustion aims to eliminate the need for capture altogether. In the next section, different capture technologies are discussed.

Pre-combustion stage: Pre-combustion capture is only used for IGCC processes and is typically less energy intensive compared to post-combustion capture. After the gasification of coal into syngas, it undergoes a water–gas shift (WGS) reaction and is converted into CO_2, which can be separated using various methods. Due to complexity in operation and technology, pre-combustion is difficult and costly to build or retrofit.

Post-combustion stage: Post-combustion capture is used to capture CO_2 in flue gas before it reaches the stack. It is usually implemented on traditional PCC or retrofitted to existing plants. Post-combustion capture is more common than pre-combustion technology and is also not as integrated into the plant as pre-combustion and oxy-fuel combustion. Currently, the challenge is the parasitic energy requirement of the capture process. In some cases, the addition of the capture process lowers the plant efficiency drastically; thus, researchers are trying to increase the plant efficiency or decrease the energy required for capture (Herzog et al. 2009).

20.4.1 Gasification and the Water–Gas Shift Reaction

Gasification is part of the IGCC process as mentioned earlier where it converts coal into syngas. By doing so, the CO_2 can be isolated prior to combustion. It was noted that gasification will result in H_2 and CO gases so the first step in pre-combustion capture is to convert CO to CO_2. The CO_2 is then separated for transportation. Although this technology is not used commercially yet, it shows promising prospects toward clean coal-fueled energy. The syngas produced from coal is more concentrated (almost 40 mol%) and also at higher pressure than flue gas from other plants (12–15 mol%). These two factors make extraction of the CO_2 much easier. The conversion of CO to CO_2 takes place via the WGS reaction (Fisher et al. 2011):

$$CO\ (g) + H_2O\ (g) \rightarrow CO_2\ (g) + H_2\ (g) \quad (20.4)$$

Steam is taken from the steam cycle of the plant and mixed with CO at a 2:1 molar ratio (steam: CO). The ratio will ensure the complete conversion of all CO into CO_2. Catalysts can be added into the reaction to improve the WGS. Like other processes, sulfur concentration will have adverse effects on catalysts such as iron-chromium or nickel-chromium; however, Haldor Topsoe has developed a sulfur-tolerant catalyst called the SSK catalyst, which is a compound containing cobalt and molybdenum (Hendriks 1994, p. 115).

Conversion of CO to CO_2 is maximized at lower temperatures, as shown in Figure 20.7.

Two heat exchangers are used to cool the hot syngas down to an optimal temperature for CO conversion. After the conversion to CO_2, a separation process such as absorption or adsorption is used to separate the syngas into a CO_2 stream and a H_2 stream.

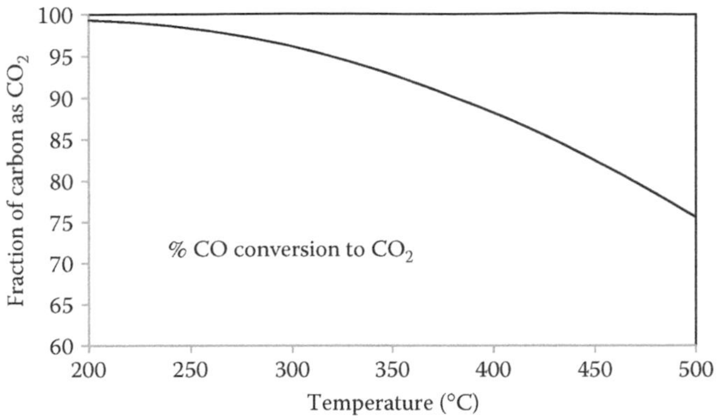

FIGURE 20.7 Carbon dioxide conversion with respect to temperature. (From Hendriks, C., *Carbon Dioxide Removal from Coal-Fired Power Plants*, PhD thesis, Utrecht University, Utrecht, The Netherlands, 118, 1994.)

20.4.2 Absorption Using Amine Scrubbers

The absorption of CO_2 into a solvent is typically preferred for post-combustion capture. The technology is well established compared to other methods and has easy layouts for retrofitting existing power plants. Chemical absorption forms chemical bonds with the targeted gas while physical absorption forms weak Van der Waals forces. In the case of amines, it is considered a chemical absorption process. Flue gas entering the scrubbing towers must be cooled because this exothermic reaction operates best under low temperature and pressure. Lean solvent (solvent that does not contain CO_2) is mixed with the CO_2-rich stream in a counter-current fashion where the CO_2 is absorbed into the amine. After the solvent has absorbed the CO_2, it is fed to a stripper where the solvent and CO_2 are separated. Heat provided by steam will vaporize the CO_2-rich solvent, hence regenerating the solvent for absorption again. As the solvent rises in the stripping tower, it condenses, liberating CO_2. The regenerated solvent is cooled and condensed before being sent to the absorber again, and the liquid water is recycled back into the stripping process. This process produces a concentrated CO_2 that can be sequestered or otherwise recycled (Hendriks 1994, pp. 20–50):

$$R_1—NH_2 + CO_2 \rightarrow R_1—\overset{H_2}{N^+}—CO_2^- \quad (20.5)$$

The most widely used chemical for absorption is monoethanolamine (MEA). Industries typically use a 30% MEA solution, which requires about 3.6 GJ/ton of CO_2 to be recycled back to the absorption tower (Bhown et al. 2011). The process consumes a large amount of energy, which will lower the plant efficiency drastically. A common method of less-energy-demanding processes for the regeneration of MEA is to increase the concentration of solvent. However, this will require the addition of corrosion inhibitors to minimize the effects of increased concentration. Another viable method is to alter the solvent using secondary or tertiary amines that requires less energy for regeneration but also has a lower loading capacity for CO_2. Besides energy costs associated with the regeneration of MEA, the degradation and loss of the solvent is a problem. MEA is an extremely volatile solvent so it has to be constantly replenished. Additionally, MEA reacts with impurities such as NO_2, SO_2, and O_2. These side reactions will create a buildup of by-products that contribute to corrosion (Hertwich et al. 2009).

An absorption column is easily implemented for existing plants, but it reduces the efficiency because it requires heat by steam extraction from the power plant. Simulations have shown that the usage of absorption technology will reduce the efficiency of the plant by 9.5% for 90% removal of CO_2. Other solvents are secondary amines such as diethanolamine or tertiary amines, have a higher capital cost, and have less of an effect on the plant's efficiency because it requires less energy for regeneration when compared to MEA (Hendriks 1994, pp. 25–50).

A different approach in absorption technology is the use of chilled ammonia. It utilizes a slurry of ammonium carbonate and ammonium bicarbonate suspended in an ammonia solution at temperatures of 0°C–10°C. This operation allows the slurry to have an extremely high loading capacity of CO_2 and also reduces the loss of ammonia via exiting flue gas. However, this process requires intensive cooling of flue gas. The solvent is regenerated at temperatures of >120°C and pressures of 2 MPa. The captured CO_2 stream is at high pressures, which will make it easier to compress for transportation (Rackley 2010, pp. 112–113). The Benfield process (see section 20.3.4), a type of pressurized FBC, utilizes hot potassium carbonate as an absorbent. This chemical is more stable and requires less regeneration energy when compared to amines. The energy consumption by the Benfield process is around 70–90 MJ/kg of CO_2 (Hetland 2008).

20.4.2.1 Current Amine-Based Carbon Capture Power Plants

The use of amines for post-combustion carbon capture is very common and has been implemented in many plants around the world. For example, there is a 660 MW supercritical PCC technology power plant in Brindisi, Italy, where MEA scrubbing process is used. The absorption chemical is 20% w/w MEA, which is to capture 85% of the carbon emitted that amounts to 1.0–1.5 Mt CO_2/year. The plant commenced production in 2010 (Barbucci 2009).

The 20 MW power plant in West Virginia, USA uses chilled ammonia, absorbing 110 kt CO_2 per year with carbon sequestration. The plant claims that it can achieve a capture rate of 75%–90% to produce a CO_2 stream of >99.9%. However, it stopped producing power in 2011 after a plan for further development was put off for economic reasons (Mills 2012).

More recent progression in child ammonia is located in Canada with Project Pioneer. A 450 MW supercritical power plant went online in 2011 and is now being considered for retrofitting with a chilled ammonia carbon capture process. The plant is to capture 90% of the CO_2 from the flue gas stream, which is estimated to be equivalent to 1 Mt of CO_2. According to the plant's proposal, construction began in 2013 and full commercial operation should commence in 2015. Part of the captured CO_2 will be injected into a saline formation while the rest of it will go toward EOR in Alberta (Mills 2012).

An IGCC plant in Puertollano, Spain, decided to use the solvent methyldiethanolamine for chemical absorption. The installed unit only removes 2% of the CO_2 emitted by the 335 MW IGCC power plant so it does not effectively mitigate the release of carbon into the atmosphere. Costing €3 million, the plant releases streams of CO_2 and H_2 with purity upward of 99% (Mills 2012).

20.4.3 Physical Absorption

While chemical solvents such as MEA is optimized under low pressures, physical solvent has a higher capacity for CO_2 at high pressures. Therefore, physical absorption is rarely used for post-combustion capture because of the large amount of flue gas that has to be compressed. Instead, this method is preferred for pre-combustion capture of CO_2 from a syngas stream. Also pre-combustion capture generally has a high concentration of CO_2, which makes physical absorption a better candidate than chemical. Solvents that are used commercially include Selexol (mixture of dimethyl ethers and polyethylene glycol), Rectisol (chilled methanol), Purisol (*N*-methyl-2-pyrrolidone), and Fluor (propylene carbonate).

20.4.3.1 Selexol Process

This process was developed by Allied-Signal (currently Honeywell) and is used by over 50 units in the world as of 2010. Syngas exiting the gasifier is first cooled to a temperature below 50°C to be fed into an absorber tower where it reacts with Selexol in a counter-current fashion. The process is very similar to that of amine absorption (see Section 20.4.2) where the solvent is regenerated by lowering pressure and increasing temperature, hence liberating a stream of CO_2. Selexol has a high selectivity toward CO_2 when compared to H_2, which helps minimize the energy required to separate the two gases later on. The CO_2 stream exits the capture process at atmospheric pressure, which means it requires a very high compression ratio to prepare it for transportation. This process of compressing, cooling, and heating is reflected in lowered plant efficiency (Fisher et al. 2012). A simulation done by National Energy Technology Laboratory has shown that the use of Selexol for carbon capture decreases the plant's efficiency from 39% to 32.6% (Fisher et al. 2012).

20.4.3.2 Rectisol Process

Another physical absorption technology is called the Rectisol process, which is also used to remove acid gas (CO_2). It functions as same as Selexol except it uses cold methanol between the temperatures of –40°C and –62°C. After gasification, the syngas enters the Rectisol unit to remove CO_2 and H_2S as well as other trace amounts of contaminants. The cleaned syngas (CO and H_2) undergoes the WGS reaction, which creates a gas containing around 33% CO_2. The gas then goes through another Rectisol unit to reduce the CO_2 concentration down to 3%. Maintaining low temperatures requires a lot of cooling energy, but the Rectisol process requires less steam energy for solvent regeneration. Methanol has a high vapor pressure, which means it has to operate at extremely low temperatures to reduce solvent loss (Burr and Lyddon 2008).

20.4.3.3 Purisol Process and Fluor Process

Purisol is a popular choice for IGCC plants with the highest selectivity for H_2S over CO_2. For power plants with low emissions of H_2S, the Fluor process is used. It is a mixture of propylene carbonate that does not dissolve syngas very well but has a very good loading capacity for CO_2. It has low vapor pressure so solvent loss is not as big of an issue as the other solvents. Operating temperatures of the Fluor solvent is between –18°C and 65°C. If the temperature is above that range, then the solvent reacts irreversibly with water and CO_2 (Burr and Lyddon 2008).

A simulation was done by Burr and Lyddon (2008) to compare the performance of the solvents. The Rectisol process requires the highest capital cost but has the least net power requirement. However, it was shown that Selexol is less costly to operate when compared to Rectisol. The Rectisol process is more versatile with gas absorption and cleaning, while Selexol and Fluor processes are only effective for removing CO_2. A comparison of data is made in Table 20.6.

20.4.3.4 Case Study

The case study about Kemper Project mentioned in section 20.3.2 includes a Selexol process unit for acid gas removal. The capture process removes 65% of the carbon emitted, which is around 3 Mt CO_2/year. The captured gas is compressed and used toward EOR. In Czech Republic, a 400 MW power plant is fitted with a two-stage Rectisol unit. The first is used to remove sulfur products, and the second is used to remove CO_2. The plant has been operating for over 16 years and is said to reduce the stream of syngas to ~5% CO_2 (Mills 2012). There are many coal-based IGCC plants with carbon capture in the process of development, and a few of them are summarized in Table 20.7.

TABLE 20.6
Comparison of Net Power and the Solubility for Different Solvents

	Selexol	Rectisol	Fluor	Purisol
Net power required (kW)	5826.9	1164	6232.6	5826.9
CO_2 solubility @ 25°C (L CO_2/L solvent)	3.628	3.179	3.398	3.304

Source: Burr, B. and Lyddon, L., *A Comparison of Physical Solvents for Acid Gas Removal*, Bryan Research and Engineering, Bryan, TX, 2008.

TABLE 20.7
Summary of Planned Coal-Fired IGCC Power Plants with Carbon Capture Technology

Project	Start-Up Year	Power Output	CO_2 Capture	CO_2 Captured (Mt/yr)
Texas Clean Energy Project	2015	411	Rectisol	2.7
Hydrogen Energy California Project	2016	390	Rectisol	32
Southern California Edison, Utah	2017	500	Selexol	3.5
Mississippi Power, Kemper County	2014	582	Selexol	3.5
C.GEN, Killingholme, UK	2016	520–570	Selexol	3.65
2CO Don Valley Project, UK	2016	900	Selexol	4.5–5

Source: Mills, S., *Coal-Fired CCS Demonstration Plants*, IEA Clean Coal Centre, 2012.

20.4.4 Adsorption

Adsorption is another method for carbon capture where the gas molecules are attached to the surface of the sorbent. This means that there will be less liquid waste that has to be processed by the plant. It is versatile when it comes to coal-fired power plants because chemical sorbents are good for post-combustion capture while physical sorbents are ideal for pre-combustion capture. Just like absorption, adsorption can happen through strong chemical bonds, which are covalent interaction between the sorbent and CO_2, or weaker physical bonds, which are Van der Waals or electrostatic forces. The difference between chemical and physical adsorption is typically their operating temperature with physical sorbents running between 25°C and 100°C and chemical sorbents operating in the range of 400°C–600°C. Depending on the material used, adsorption can work under a wide range of conditions, pressures, and temperatures. Ideally, the sorbent material will have a high selectivity, high loading capacity, and low energy requirement for regeneration.

There are two general types of system that adsorption can operate under: fixed-bed and moving-bed. In a fixed-bed system, the feed gas enters a column of sorbent where the capture takes place in a very limited space, as shown in Figure 20.8. This means that there is a large amount of sorbents either waiting to adsorb CO_2 or waiting to be regenerated. This result in low efficiency, and also a large amount of sorbent must be held to keep the process continuous. Also, the heat must be managed accordingly because there is a relatively small mass transfer zone (MTZ). The MTZ is the section of the column where CO_2 is adsorbed into the sorbent material.

Because of efficiency and practical issues, the moving-bed process is typically preferred over the fixed-bed system. In fluidized bed technology, feed gas is introduced in the middle of

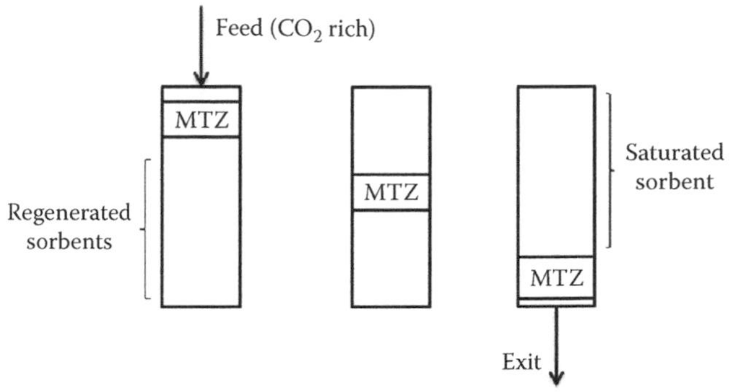

FIGURE 20.8 Fixed-bed processes for adsorption. (From Rackley, S., *Carbon Capture and Storage*, Elsevier, Oxford, 141, 2010.)

the column where intake air will force the feed gas toward the top. Sorbent adsorbs CO_2 in a counter-current fashion as the feed gas rises. The cleaned gas will exit the top of the column while the CO_2-rich sorbent will fall down into the desorption section. A heating element will liberate the adsorbed CO_2, and the regenerated sorbent will be recycled back to the top of the column.

This process minimizes the amount of sorbent that is required because all gas-rich sorbents are regenerated immediately. Sorbent regeneration can be done through numerous methods:

1. Pressure swing adsorption (PSA)—decrease in pressure
2. Temperature swing adsorption (TSA)—increase in temperature
3. Vacuum swing adsorption (VSA)—a vacuum is created
4. Electric swing adsorption (ESA)—heating by electricity

For post-combustion capture, flue gases are at a lower pressure, and for pre-combustion capture, syngas is at a higher pressure. PSA is preferred for pre-combustion because it can desorb the material at atmospheric pressure. VSA is preferred for post-combustion capture because it will eliminate the need to compress the flue gas.

Development of adsorption technology is mainly revolved around material science where researchers are trying to find a material that operates at the ideal conditions and can also have a high loading capacity for CO_2. Some materials include activated carbon, zeolite, and hydroxide compounds (Rackley 2010). Currently, adsorption technology is only performed in a laboratory setting with no commercial usage due to the fact that adsorption technology cannot handle large amounts of CO_2. Typical adsorption technology can clean gas streams with ~1% of the target gas. However, CO_2 in flue-gas streams in coal industry is many times higher (Morgan-Sagastume et al. 2005).

20.5 CARBON DIOXIDE TRANSPORTATION

Upon separation and capture of CO_2 at a power plant, it must be transported to the point of sequestration or processing for other applications. The use of pipelines is the most prominent method for transport, but marine transportation has been developed for overseas destinations. CO_2 is typically transported as a gas or liquid because the solid phase has limited industrial applications and also the energy required to keep as a solid is too costly.

20.5.1 Pipeline Transport

Transporting CO_2 at atmospheric pressure will be too costly due to large equipment needed to handle massive volume of gas. Pumping CO_2 as a subcooled liquid allows it to be transported through smaller pipes because the liquid is denser. However, subcooled fluids are more susceptible to fluctuation in the environment so they are preferred to transport through areas with cooler climates. Currently, CO_2 is transported in its dense phase, which is to say subcritical temperature and supercritical pressures with a density of around 700 kg/m^3 (Al-Fattah et al. 2012, pp. 139–157).

An issue around pipeline transport is the purity of CO_2. The composition of the gas is regulated and tested constantly because water, sulfur, and nitrogen compounds will cause corrosion problems. According to the regulation of the U.S. Department of Transportation, CO_2 must be at a purity of 90 mol% to be transported. However, many pipelines would have even higher standards upward of 95 mol% for EOR applications. Leak is another major issue with pipelines as it imposes environmental risks and also elevated CO_2 levels to surrounding residences. Crack arrestors made of glass fiber are often wrapped around a pipe at intervals to prevent corrosion. Also, valves are placed at intervals so that in the case of a leak, the flow of gas can be stopped. Many of the case studies mentioned in previous sections had a CO_2 pipeline constructed for sequestration (Doctor et al. 2005).

20.5.2 Marine Transport

Liquefaction is the process to treat CO_2 before it is loaded onto a ship. The purpose of liquefaction is to turn gaseous CO_2 into a high pressure, low temperature liquid, typically at –52°C and 6–7 bars. Liquefaction requires large storage facilities because the process is continuous, but the ship has a finite amount of space. There has to be two storage facilities, one for loading and one for unloading. In both cases, storage tanks are filled with dry, pressurized CO_2 to prevent the introduction of humid air or dry ice into the system. With these extra storage facilities, marine transport is quite costly in the long run and not to mention the CO_2 emission during transport (Al-Fattah et al. 2012, pp. 139–157).

20.5.3 Cost of Carbon Transport

The cost of carbon transport in pipelines varies with different projects as it depends on volume, distance, labor costs, and maintenance cost of a pipeline system. It was determined that 2%–5% of a carbon capture and sequestration project is allocated to carbon transport. It is a small part of the overall cost, but it still proved to be costly at \$2–7/ton CO_2 over 200 km. A study led by Rotterdam Climate Initiative has successfully concluded that sharing transport resources between carbon-emitting sources will significantly decrease the cost of carbon transport. As with any collaborative projects, a multiuser transport system has the problem of fairly allocating the shared cost of maintenance, decommissioning, and development (Doctor et al. 2005).

20.6 SEQUESTRATION

Upon carbon capture and transport, the gas must be sequestered such that it does not impact the environment anymore or recycled to compliment another industrial process. Possible solutions to long-term storage of CO_2 would be geological storage and ocean storage.

20.6.1 Geological Storage

One prominent option is to store the CO_2 underground as a supercritical fluid. This makes sure that a large volume of gas can be placed into a small space. The main concept in geological storage is to inject CO_2 into a underground cavity where rocks seal to ensure proper storage. Materials such as sandstone, shale, limestone, and salt are also suitable for containing CO_2. The amount of CO_2 that can be injected into a cavity is determined by the following factors: (1) size of reservoir, (2) porosity, (3) depth of storage (which affects pressure, temperature, and density), and (4) fraction of reservoir that can be filled (Hendriks 1994, pp. 177–200).

20.6.2 Aquifers

An aquifer is a permeable rock layer underground that contains water or occasionally gas and oil. Sometimes there is an impermeable surface over the aquifer making it an ideal reservoir for CO_2. As seen in Figure 20.9, the density of CO_2 increases exponentially with depth and then plateaus at a

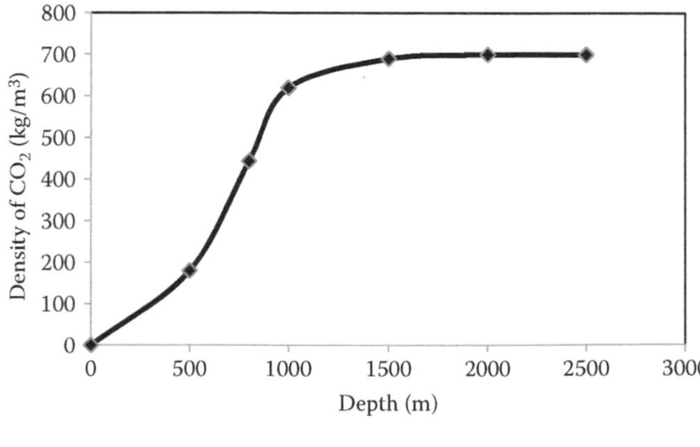

FIGURE 20.9 CO_2 density versus reservoir depth. (From Benson, S. and Cook, P., Underground geological storage. *Intergovernmental Panel on Climate Change Carbon Dioxide Capture and Storage*, Editors: Coninck H, Davidson O, Loos M, Metz B, Meyer L, Cambridge University Press, New York, 195–319, 2005.)

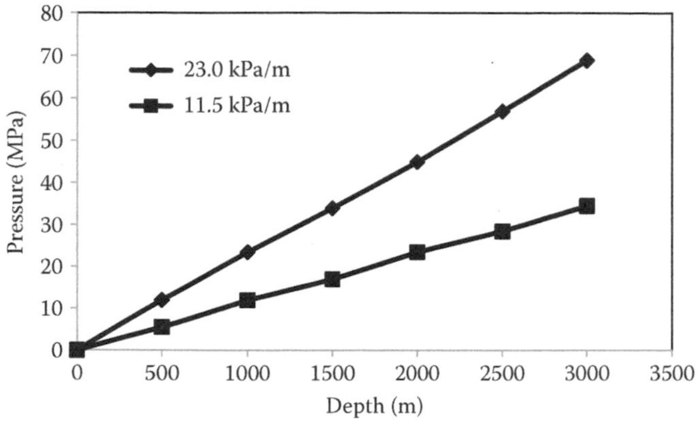

FIGURE 20.10 Correlation between reservoir pressure and depth. (Data from Rackley, S., *Carbon Capture and Storage*, Elsevier, Oxford, 262, 2010.)

certain level. It is important to inject the gas to a reasonable depth with a great enough density such that the maximum amount of gas can be held. At supercritical conditions, 420 m^3 of CO_2 will only take up 1 m^3 of space, allowing a lot more gas to be sequestered within a reservoir. However, reservoir pressure limits the depth of a well because too-deep reservoir will require extra compression at the surface. Figure 20.10 shows the correlation between reservoir pressure and depth of reservoir.

The Denison Trough is an example of an ideal geological storage location because of rare seismic activity, large storage potential, and cap rock formation for secure carbon storage. Risk assessment is a key component to geological storage because companies need to figure out how well CO_2 is going to stay within the ground and the long-term effects along with it. Excess supercritical CO_2 will be left in the reservoir, and diffusion will continue for long periods of time. Entrapment of the gas is distributed between the oil phase and the rest within water or other mineral compounds. A concern about ground storage is that CO_2 can seep into bodies of water and be released suddenly. For example in 1986, a plume of CO_2 erupted from Lake Nyos, Cameroon, and swept across a nearby village claiming 1700 lives (Evans et al. 1993).

20.6.3 Enhanced Oil Recovery

As oil wells struggle to keep up with the world's growing energy demand, EOR has proven to be one of the most important applications for CO_2 storage. A typical oil well can only extract 30%–40% of its oil, leaving much of it unattainable, but with EOR, extraction can increase 20% of the original amount of oil. When a reservoir's pressure is depleted, a method called liquid CO_2 flooding will be used to extract the remaining oil. CO_2 supplied by a pipeline is pumped into the reservoir (see Figure 20.11). Below 800 meters, CO_2 becomes a supercritical fluid and will dissolve with the oil that will cause expansion. The viscosity of the mixture will decrease as a result of expansion, making extraction of the remaining oil possible. Any CO_2 that comes up through the production well is recycled back into the injection process. The return of oil from CO_2 injected is around 10 barrels of oil per 3–6 tons of CO_2 (Mills 2011).

One of the largest EOR operations in Canada is located at the Weyburn oil fields in Saskatchewan. After the implementation of EOR in 2000, the production levels at Weyburn has risen 60% to 28,000 barrels/day. If the predicted operation holds, CO_2 injection will turn in $12 billion in the next 30 years. From 2000 to 2009, this plant has injected 10 million tons of CO_2 into an estimated 55 million ton capacity reservoir. The EOR project at Weyburn proves to be a successful method of sequestering CO_2. The second and third largest EOR projects in Canada are Midale Field and Ponoka/Chigwell, respectively (Brawn and Lewis 2009, pp. 9–20).

20.6.4 Enhanced Coal-Bed Methane Recovery

CO_2 can also be sequestered into coal seams for methane recovery to help offset the cost of carbon management. The recovery of methane is less explored when compared to oil because 60%–70% of the original methane reservoir can be extracted without the use of CO_2. The concept behind enhanced cold-bed methane recovery is very similar to EOR; CO_2 is injected into un-mineable coal beds, which force the methane out of a production well. This is possible because of coal's greater affinity toward CO_2 compared to methane. This method is effective in storing CO_2 because it takes 100 tons of CO_2 to release 3–5 tons of methane (Mills 2011).

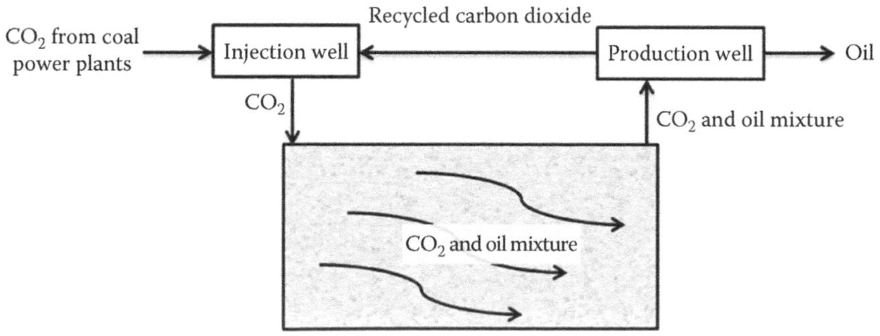

FIGURE 20.11 Diagram for enhanced oil recovery using carbon dioxide flooding. (From Mills, R., *Capturing Carbon: The New Weapon in the War against Climate Change*, C. Hurst & Co, London, 102, 2011.)

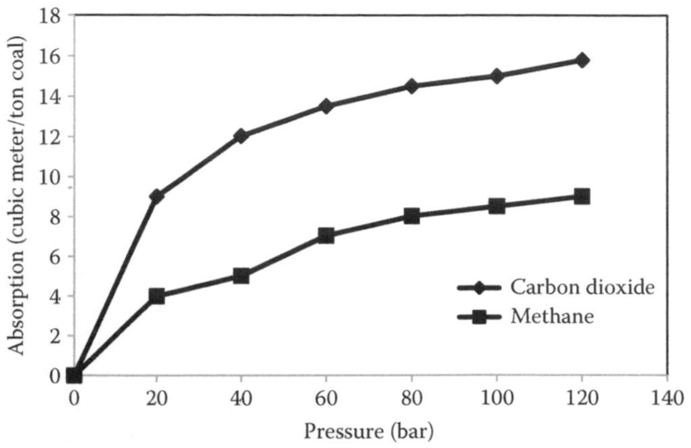

FIGURE 20.12 Absorption of different gases into Coal for enhanced cold-bed methane recovery. (Data from Bergen, F., *Enhanced Coal Bed Methane Recovery with CCS: Limitations and Possibilities*, Netherlands Organizations for Applied Scientific Research, https://www.iea.org/media/workshops/2012/ukraine/Van_Bergen.pdf. Accessed April 9, 2014.)

Figure 20.12 shows that coal as a higher affinity for absorbing CO_2 when compared to methane.

20.6.5 Ocean Storage

An alternative to geological storage would be ocean storage, which has the largest carbon storage potential. The conditions of which depend on depth, pressure, and temperature. At temperatures of 0°C–10°C and pressures of 4–5 MPa or depths of 400–500 m, CO_2 will liquefy; however, at these depths, any CO_2 that vaporizes due to a pressure drop will float back to the surface. To avoid this issue, CO_2 has to be pumped to a depth of 3 km. CO_2 from the atmosphere is absorbed into the ocean through wind turbulence and wave action. When deep water is driven toward the surface, CO_2 is released into the atmosphere (Rackley 2010, pp. 267–274).

At depths of about 3000 m, stored CO_2 in oceans will form lakes. It was simulated that 1% of sequestered carbon will leak in a century and 6% after two centuries. At this rate, it will not mitigate the effects of CO_2 in the long term. Another fear around ocean storage is what happens when something does go wrong. If a major leak breaks out or the environmental impact of sequestering CO_2 is too extreme, it will be almost impossible to reverse the process (Mills 2011). Because of limited knowledge of the interaction between mass storage of CO_2 in the ocean, the development of any marine storage has been restricted to small-scale testing.

20.7 RECYCLING AND CARBON DIOXIDE APPLICATIONS

The topics mentioned earlier are all about the capture and sequestration of CO_2, which is a cost for the power plant. However, CO_2 does have its applications in different industries, which can generate revenue to help offset the cost of carbon management.

20.7.1 Algae, Biomass, and Biofuel

Algae can capture CO_2 through the process of photosynthesis, which will in turn release oxygen back into the environment. Also, algae can be used to produce biomass and biofuel such as diesel or jet fuel. Table 20.8 compares the oil yield of different sources of biofuel. A benefit of growing algae over other crops is that it does not require potable water. The energy density of algae-based biofuel is comparable with processed fuel as well. Table 20.9 compares algae biofuel with other prominent fuel choices in different industries. However, this method of CO_2 capture will help reduce CO_2 levels but will not eliminate the effects of CO_2 on the environment because the burning of biofuel also results in the release of greenhouse gases (Mills 2011).

One of the issues with algae growth is the large amount of land that is required to capture CO_2 released from power plants. The chemical equation that describes the conversion of water and CO_2 by the algae into simple sugars used is given by (Wilcox 2012):

$$CO_2 + H_2O + 8 \text{ photons} \rightarrow \text{simple sugars} + O_2 \quad (20.6)$$

A simulation study showed that a massive 880 hectares of algae will be required to capture 70% of the emitted CO_2 from a 50 MW power plant. It is important to note that many commercial power plants have a much higher power output than 50 MW (Brune et al. 2009).

Algae can be produced in open pond or photo bioreactor. Algae in an open pond grows in a circular track, and a paddle

TABLE 20.8
Oil Yield of Different Biofuel Crops

Crop	Oil Yield (L/ha yr)
Soybean	446
Canola	1190
Palm	5950
Microalgae	58,700

Source: With kind permission from Springer Science+Business Media: *Carbon Capture*, 2012, Wilcox, J.

TABLE 20.9
Energy Density versus Types of Fuel

Fuel	Energy Density (MJ/L)
Diesel	35.8
Algae-based biofuel	34.0
Gasoline	29.8
Propane	23.2
Ethanol	21.1
Hydrogen	8.4

Source: With kind permission from Springer Science+Business Media: *Carbon Capture*, 2012, Wilcox, J.

wheel is used to ensure mixing for proper gas transport, algae growth, and nutrient distribution. When space is an issue, there is an option for more-expensive photo bioreactor. Thus, a combination of both pond and reactor technology can be used to optimize the process (Wilcox 2012). While biomass can be co-fired with coal or burned on its own, the potential of converting into fuel is also very captivating amid the energy crisis. The reaction to form biofuel is called *transesterification* where the extracted lipids of algae (typically triglycerides) react with methanol or ethanol to produce glycerol and methyl esters with a strong base as catalysis. The fuel is then separated by settling due to difference in density (Rackley 2010).

20.7.1.1 Case Study

One of the major strides toward carbon capture on algae to produce biomass or biofuel is done by an Australian company named Algae Tech Ltd. On January 23, 2014, the company announced that it received millions of dollars of financial backing from Reliance Industrial Investments and Holdings Limited (RIIHL). They will be opening biofuel plants in India as well as making the world's first algal pond for carbon capture at Australia's largest coal-fired power plant (Flitzpatrick-Napier 2014).

20.7.2 Carbon Dioxide in Methane Production

Methane is largely used as a fuel for many different applications that can help offset the cost of carbon capture at both the pre-combustion and post-combustion stages. Methane can be produced using CO_2 through the Sabatier process or by using CO via the Fischer–Tropsch process. CO_2 reacts with H_2 in the presence of a catalyst to produce water and methane. The process is operated usually under high pressure and temperature. It is suggested that methane can be used to generate electricity, which will create a closed loop where CO_2 will be reused and produced. In Germany, Audi announced that it will build a plant where surplus energy from renewable sources will be used to convert CO_2 into methane. It predicted that the plant will convert 2800 metric tons of CO_2 to produce 1000 metric tons of syngas (H_2, CH_4) (Rousseau 2013).

20.7.3 Hydrogen Storage

In the emerging field of fuel cells, H_2 is needed for many processes, and one of the challenges that researchers face is the proper storage of this flammable gas; thus, CO_2 is being investigated for properties to store H_2 safely and reversibly. The process involves the hydrogenation of CO_2 into formic acid and then the decomposition of formic acid to release H_2 and CO_2. Both hydrogenation and decomposition relies on a catalyst:

$$\text{Hydrogenation of carbon dioxide:} H_2 + CO_2 \rightarrow HCO_2H \quad (20.7)$$

$$\text{Decomposition of formic acid: } HCO_2H \rightarrow H_2 + CO_2 \quad (20.8)$$

An experiment done by Yuichiro Himeda used CO_2 as a storage material for H_2 with a bi-pyridine rhodium complex as the catalyst. The greatest benefit from the use of this process is that it produces CO-free H_2. The experiment showed that during decomposition of formic acid, the H_2 gas was released continuously at a pressure greater than 4 MPa. The rate of H_2 gas produced could be controlled by reaction temperature (Himeda 2010).

20.8 CURRENT RESEARCH AND DEVELOPMENT

The research and development of carbon capture methods is mostly revolving around optimization and material science. Absorption is currently the preferred technology in this field, but it requires a large amount of energy. In an ideal situation, the capture process conditions should be very similar to that of the entering gas stream so that minimum amount of energy is required to change the gas condition. In all cases, researchers are trying to find a material or chemical that has high selectivity for CO_2, low regeneration costs, and not prone to degradation or corrosion.

20.8.1 Novel Sorbents

20.8.1.1 Zeolite

Zeolite is an aluminosilicate compound that can be used as a physical adsorbent. Challenges that face this technology are the high energy requirement for sorbent regeneration, the energy needed to lower flue-gas temperature and sorbent degradation. Despite these drawbacks, zeolite is still a material of interest for adsorption because of its high capacity for CO_2. Research done by Fisher et al. (2011) has shown that zeolite 13X shows an adsorption capacity of 2.5–3 mol CO_2/kg at ambient temperature. However, zeolite is a low-temperature sorbent. At temperatures of around 120°C, the adsorption capacity drops to about 0.65 mol CO_2/kg. The presence of water degraded the sorbent but regained to its full capacity when regenerated at temperatures of 350°C. A simulation comparing the Selexol chemical and zeolite compounds resulted in efficiencies of 36.5% and 37%, respectively. The authors compared adsorption process using zeolite with the other processes and stated that zeolite can potentially be more effective than the commercially available Selexol process (Fisher et al. 2011).

20.8.1.2 Metal Hydroxides

Metal hydroxide is another material that can potentially compete with current carbon capture chemicals. In a study done for four hydroxide compounds—$Fe(OH)_2$, $Ni(OH)_2$, $Co(OH)_2$, and $Zn(OH)_2$—all four compounds are within 1% thermal efficiency to that of Selexol (Couling et al. 2012). The downside is that metal hydroxides are very costly and will degrade in the presence of water vapor (Fan et al. 2013). The most efficient compound is iron hydroxide, and its data is compared with Selexol in Table 20.10.

20.8.2 Membrane Separation

The use of membranes for gas separation is a very promising field of research because it compensates for the problems associated with current MEA or Selexol process. The energy load from

TABLE 20.10
Comparison of Efficiency: Selexol versus $Fe(OH)_2$

	Selexol (Cold Cleanup)	$Fe(OH)_2$
Thermal power input (MW)	1688	1733
Net power output (MW)	550	555.3
Thermal efficiency	32.6%	32.0%

Source: Couling, D. et al., *Industrial & Engineering Chemistry Research*, 51(41), 13473–13481, 2012.

carbon capture is significantly less than the previously discussed methods because membranes do not require intensive heating and cooling. Challenges facing the use of membranes are degradation at high temperatures and reduced performance in the presence of impurities (Algieri et al. 2011, p. 232).

Characteristics of a membrane can be broken down generally into two aspects: (1) selectivity—ratio of permeability values in different gas species (CO_2/N_2 or CO_2/H_2) and (2) permeability—flux of CO_2 passing through a membrane at a specified temperature and pressure ($Nm^3/m^2/bar/h$). High selectivity for CO_2 whether in CO_2/N (flue gas) or CO_2/H_2 (syngas) is important because this would determine how many times the gas have to pass through the membrane unit. A typical 600 MW coal-fired power plant releases 3 million tons of CO_2 a year at a flow rate of 500 m^3/s at normal conditions. This is why permeability is also an important factor; the membrane must be able to allow a large amount of gas or else the area needed will be too great (Zhang et al. 2013).

20.8.2.1 Pre-Combustion Membrane Capture

Syngas exiting a WGS reactor is composed of mainly H_2 and CO_2 at high temperatures and high pressures. For this reason, metal or inorganic materials are used because they are thermally stable while operating in temperatures upward of 300°C. The use of membranes can enable the separation of CO from the syngas produced by gasification. During this process, syngas is divided into a H_2 stream and a CO stream, at which point the H_2 stream proceeds to the gas turbine while the CO is converted to CO_2.

There are two types of membranes used for IGCC, and they are polymer membranes and metallic membranes (Hendriks 1994). Of the metallic membranes, palladium and silver membranes have potential for syngas cleanup. The metallic membrane is highly selective for H_2 and is usually implemented with the WGS reaction. This means that H_2 will be continuously removed as it is being produced. This will result in a higher conversion rate of CO to CO_2 because the constant removal of H_2 will prevent the system to reach equilibrium. Not only is a higher conversion rate beneficial, but also the pressure of the exiting CO_2 is no less than 10 MPa, making compression for transport a less-energy-demanding process (Rackley 2010, p. 183).

20.8.2.2 Post-Combustion Membrane Capture

Post-combustion CO_2 separation using membranes is a better alternative to the currently used amine solvents. It does not require replenishing or sorbent regeneration, and also the parasitic load on the overall plant efficiency is much lower than that of amine absorption process. An issue is that flue gas from power plants has a relatively low concentration of CO_2 (10%–20%), resulting in a large amount of gas to be processed. The driving force of gas through a membrane is the difference in pressure between the feed and permeate streams. Thus, with low concentration of CO_2, energy will have to be used to either compress the feed or vacuum the permeate side to make membranes viable (Algieri 2011, p. 233). A simulation done by Favre suggested that CO_2 selectivity has to be 150 to reduce an incoming flue-gas stream of 15% to get a capture ratio of 0.8. However, membranes with a high selectivity generally have low permeability. Even with a highly selective membrane, it was unable to achieve the goal of 2 GJ/ton CO_2 (Favre 2011).

A solution is to implement a multistage process; however, the problem with multistage membrane separation is that capital and operating cost will be significantly higher due to recompression pumps that have to be installed between the membrane units. A simulation done by Zhang et al. (2013) analyzed both a single-stage and a double-stage membrane separation for post-combustion carbon capture. Their simulation reflected the same idea that higher concentration of CO_2 will result in much lower energy consumption. For the two-stage analysis, it was discovered that with a membrane permeability of 2 $Nm^3/m^2/bar/h$, selectivity of 70, and CO_2 purity of 90%, the cost will be \$50/ton CO_2 captured. If the permeability was increased to 6 $Nm^3/m^2/bar/h$, then the cost drops to \$43/ton CO_2 captured. The cost analysis showed that it was less expensive than MEA absorption, but the energy consumed by membranes remain to be higher (Zhang et al. 2013).

In a different research, simulation was a two-stage membrane capture system, which also included a recycled CO_2 stream and an air sweep stream (see Figure 20.13). The flue gas enters the first membrane unit and the outlet CO_2 stream is sent to the second membrane unit, while the permeate stream is sent to an air sweep unit. In the air sweep unit, outlet air and CO_2 are sent back to the combustion chamber. Rest of the gas is released to the stack. The CO_2-rich stream is sent to compression for transportation. The proposed system operated on a membrane with a selectivity of 50 and achieved promising

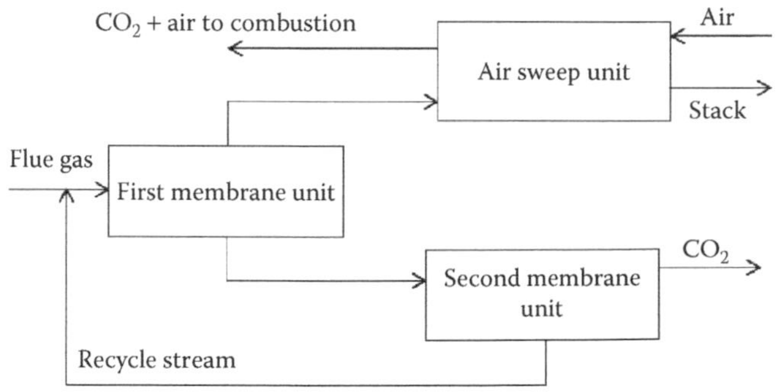

FIGURE 20.13 Two-staged membrane carbon capture systems with air sweep. (From Rubin, E. and Zhai, H., *Techno-Economic Assessment of Polymer Membrane System for Post combustion Carbon Capture at Coal-Fired Power Plants*, Environmental Science and Technology, 47, 3006–3014, 2013.)

results. It increased the incoming flue-gas stream from 12% to 18%, hence increasing the driving force across the membrane. By doing so, it decreased the energy requirement and cost by about a third (Rubin and Zhai 2013).

20.8.2.3 Membrane Contactors

Another use for membranes is a nonselective and nonreactive membrane contactor, which is used to facilitate amine absorption. It enhances gas–liquid absorption so that the size of the capture facility will be smaller and, hence, less costly. There is no stand-alone membrane system operating currently because the technology cannot handle the volume of gas that has to be processed. However, hybrid systems like the use of membrane contactors with absorption can be a viable alternative to conventional carbon capture processes (Puri 2011).

Typically, a hydrophobic hollow fiber membrane is used so that the CO_2 molecules can diffuse through the membrane and react with the absorbent. The application of hydrophobic hollow fiber membrane can reduce the reactor size by 63%–65% with mass transfer rate 2–4 times greater than the conventional chemical absorption (Puri 2011).

In this type of technology, the research is mostly focused on the absorbent used rather than the membrane material. In a study done by Fang et al. (2013), they simulated the hollow-fiber membrane contactor with MEA, DEA (diethanoloamine), MDEA (N-methyldiethanolamine), and also different blends of these chemicals. The paper focused on the effects of concentration and mixture proportion on different removal ratio and mass transfer coefficients (Fang et al. 2013). Another research involves the simulation of membrane contactors with the absorbent being an ionic liquid, 1-butyl-3-methylimidazolium dicyanamide. At temperatures of 100°C and pressures of 250 psig, the membrane was able to produce a stream containing ~88.2% CO_2 (Chau et al. 2014).

20.9 ASH MANAGEMENT

The technology behind capturing ash from flue gas is a well-developed process and is widely used in coal power plants around the world. Methods include electrostatic precipitator (ESP), fabric filters, mechanical collectors, and Venturi scrubbers; however, ESP and fabric filters are the most promising and developed technologies for this application. Collection efficiency often exceeds 99.5% removal. Due to sulfur dioxide being a criteria pollutant controlled by NAAQS (National Ambient Air Quality Standard), many power plants are switching to low-sulfur coal for combustion. However, low-sulfur coal tends to release more ash and particulate matter in flue gases. To solve this problem, flue-gas conditioning is used to improve efficiency in ESP processes (Phadke et al. 2009).

The effectiveness of ash removal is dependent on the resistivity of the suspended particles, which is to say how well it can accept and release electric charge. Sulfur trioxide is the most common gas conditioning agent used worldwide due to its low cost and low energy consumption. Typically, a gas conditioning process will have a catalytic reactor where sulfur or sulfur dioxide is converted to sulfur trioxide. The optimal concentration of sulfur trioxide is 5–25 ppm in the flue-gas stream (Phadke and Trivedi 2009).

For fabric filters, it has an operational temperature of around 120°C–180°C, and it can remove >99.5% of ash and solid particulates in the flue gas. Fabric filter parameters depend on the type of coal burned, plant size, and technology used for combustion. Other parameters include the pressure drop and the ratio between air and cloth. Pressure drop can be reduced, and efficiency can be increased by flue-gas conditioning.

A common method used for ash collection is a wet scrubbing technology called the *Venturi scrubber*. A Venturi scrubber has three sections: converging section, throat, and diverging section. The flue gas enters from the converging section and is accelerated due to the narrowing of the device. At the throat, the flue gas is at extremely high velocities, at which point, liquid is injected into the system to remove particles. The high velocity particles atomize (to make into an aerosol) the liquid and produce tiny droplets. The slurry containing the ash exits through the diverging section (Bettega et al. 2012).

20.10 CONCLUSION

Coal will continue to play a major role in supplying the world's ever-growing demand for energy, and with it, carbon emissions would increase sharply if it is not properly controlled. Highly efficient plants equipped with IGCC, ultra-supercritical PCC, and oxy-fuel combustion technology are being tested and implemented around the globe to further improve the environment. These technologies suffer from high capital cost and technological limitations but have a much lower carbon footprint than traditional PCC plants. New methods and solvents for carbon capture are being investigated to improve the traditional amine scrubbing towers. New processes such as adsorption by zeolite and membrane separation are being explored to have a high loading capacity for CO_2 while being able to regenerate the sorbent with minimal amount of energy. As international regulations continue to make strides toward limiting carbon emissions, there should be better carbon capture systems implemented to provide cleaner and more cost-effective energy.

REFERENCES

Abanades J, Soltanieh M, Thambimuthu K, Capture of CO_2. *Intergovernmental Panel on Climate Change Carbon Dioxide Capture and Storage*, Editors: Coninck H, Davidson O, Loos M, Metz B, Meyer L, (New York, Cambridge University Press, 2005) 105–179.

Al-Fattah S, Barghouty M, Bureau G et al., *Carbon Capture and Storage: Technologies, Policies, Economics, and Implementation Strategies.* (Rotterdam, the Netherlands, King Abdullah Petroleum Studies and Research Center, 2012) 139–157.

Algieri C, Barbieri G, Drioli E, Zeolite membranes for gas separation. *Membrane Engineering for the Treatment of Gases*, Editors: Drioli E, Barbieri G, (Royal Society of Chemistry, 2011) 231–233.

American Electric Power/Southwestern Electric Power Co, Turk JW, Jr. *Power Plant*, (AEP) Accessed April 24, 2014, http://www.slcatlanta.org/AL2013/presentations/AL2013_EE_Franklin.pdf.

Barbucci P, *The Enel's CSS Strategy and Projects*, (Edinburgh, The Enel Group, 2009).

Benson S, Cook P, Underground geological storage. *Intergovernmental Panel on Climate Change Carbon Dioxide Capture and Storage*, Editors: Coninck H, Davidson O, Loos M, Metz B, Meyer L (New York, Cambridge University Press, 2005) 195–319.

Bergen F, *Enhanced Coal Bed Methane Recovery with CCS: Limitations and Possibilities*, Netherlands Organizations for Applied Scientific Research, Accessed April 9, 2014, https://www.iea.org/media/workshops/2012/ukraine/Van_Bergen.pdf.

Bettega R, Coury J, Goncalves J, Guerra V, Pressure drop and liquid distribution in a venturi scrubber: Experimental data and CFD simulation. Industrial & Engineering Chemistry Research, (American Chemical Society, 2012) 51(23), 8049–8060.

Bewerunge J, Cziesla F, Senzel A, *Lunen—State of the Art Ultra Supercritical Steam Power Plant Under Construction*, (Cologne, Siemens, 2009) Accessed April 7, 2014, http://www.energy.siemens.com/ru/pool/hq/power-generation/power-plants/steam-power-plant-solutions/coal-fired-power-plants/Luenen.pdf.

Bhown A, Freeman B, Analysis and status of post combustion carbon dioxide capture technologies. (American Chemical Society, 2011) *Environmental Science Technology* 45(20), 8624–8632

Borenstein S, Melting Summer ice is turning Artic 8 per cent darker, which makes Earth Warmer, study says (Washington, DC, The Canadian press, 2014). Accessed February 18, 2014, http://ca.news.yahoo.com/melting-summer-ice-turning-arctic-8-per-cent-202814761.html.

Brawn B, Lewis I, Background. *Enhanced Oil Recovery Through Carbon Capture and Storage*, (Alberta Economic Development Authority, 2009) Accessed July 20, 2015, http://www.assembly.ab.ca/lao/library/egovdocs/2009/aleda/173910.pdf.

Brune DE, Lundquist TJ, Benemann JR, Microalgal biomass for greenhouse gas reductions: Potential for replacement of fossil-fuels and animal feeds. (2009) *Journal of Environmental Engineering* 135(11), 1136–1144.

Burr B, Lyddon L, *A Comparison of Physical Solvents for Acid Gas Removal*, (Bryan, TX, Bryan Research & Engineering, 2008).

Charles J, Heilman R, *Metabolic Acidosis*, (Wayne, PA, Turner White Communications, 2005).

Chau J, Jie X, Obuskovic G, Sirkar K, Enhanced pressure swing membrane absorption process for CO_2 removal from shifted syngas with dendrimer-ionic liquid mixture as absorbent, Industrial & Engineering Chemistry Research, (American Chemical Society, 2014) 53(8), 3305–3320.

Couling D, Das U, Green W, Analysis of hydroxide sorbents for CO_2 capture from warm syngas, Industrial & Engineering Chemistry Research, (American Chemical Society, 2012) 51(41), 13473–13481.

Dillon D, Wheeldon T, Economic comparison of oxy-coal, pre- and post-combustion CCS. Oxy-Fuel Combustion for Power Generation and Carbon Dioxide (CO_2) Capture, Editor: Ligang Zheng, (Cambridge, Woodhead Publishing, 2011) 18–19

Doctor R, Palmer A, Transport of CO_2. *Intergovernmental Panel on Climate Change Carbon Dioxide Capture and Storage* (New York, Cambridge University Press, 2005) 179–195.

Doyle A, Time running out to meet global warming target: U.N. report, (Reuters, 2014) Accessed July 20, 2014 5, http://www.reuters.com/article/2014/04/06/climate-un-idUSL5N0MW3MA20140406.

Environment Canada, *A Climate Change Plan for the Purposes of the Kyoto Protocol Implementation Act*, (Gatineau, Canada, Environment Canada, 2012), https://ca.news.yahoo.com/time-running-meet-global-warming-target-u-n-110118064--finance.html.

Evans WC, Kling GW, Tuttle ML, Tanyileke G, White LD, *Gas Buildup in Lake Nyos, Cameroon: The Recharge Process and its Consequences*. (Great Britain, Pergamon Press, 1993) Vol 8, 207–221.

Fan M, Hu X, Kenarsari S, Lin Y, Jiang G, Use of a robust and inexpensive nanoporous tio_2 for pre-combustion CO_2 separation, (American Chemical Society, 2013) *Energy & Fuels*, 27(11), 6938–6947.

Fang M, Luo Z, Wang Z, Wei C, Yu H, Experimental and modeling study of trace co_2 removal in a hollow-fiber membrane contactor, using CO_2 loaded monoethanolamine, (American Chemical Society, 2013) 52(50), 18059–18070.

Favre E, Simulation of polymeric membrane systems for CO_2 capture. *Membrane Engineering for the Treatment of Gas*, Editors: Drioli E, Barbieri G, (Royal Society of Chemistry, 2011) Vol 1, 29–54.

Fisher E, Losch J, Shen M, Siriwardane R, *CO_2 Capture Utilizing Solid Sorbents*, (Morgantown, WV, National Energy Technology Laboratory) Accessed July 20, 2015, https://www.netl.doe.gov/publications/proceedings/04/carbon-seq/039.pdf.

Fisher J, Siriwardane R, Stevens R, Zeolite-based process for CO_2 capture from high pressure, moderate-temperature gas streams, (American Chemical Society, 2011) *Industrial & Engineering Chemistry Research* 50(24), 139672–13968.

Fisher J, Siriwardane R, Stevens R, Process for CO2 capture from high-pressure and moderate-temperature gas streams. (American Chemical Society, 2012) *Industrial & Engineering Chemistry Research* 51, 5273–5281.

Flitzpatrick-Napier S, Algae. Tec signs carbon capture biofuel deal with Australia's largest coal-fired power company, (Algae. Tec, 2014) Accessed April 2, 2014, http://algaetec.com.au/news-room/press-releases/.

Hayes B, Power plants. *Infrastructure: A Guide to the Industrial Landscape* (New York, W.W. Norton & Company, 2005) Vol 5, 186–227.

Hendriks C, *Carbon Dioxide Removal from Coal-Fired Power Plants*, PhD thesis, Utrecht University, Utrecht, the Netherlands, 1994.

Hertwich E, Singh B, Veltman K, Human and environmental impact assessment of post combustion CO_2 capture focusing on emissions from amine-based scrubbing solvents to air, (American Chemical Society, 2009) *Environmental Science Technology* 44, 1496–1502.

Herzog H, Hatton A, Meldon J, *Advanced Post-Combustion CO_2 Capture* (Clean Air Task Force, 2009) 5–8.

Hetland J, *Assessment of a Fully Integrated SARGAS Process Operating on Coal with Near Zero Emissions*, (SINTEF Energy Research, 2008) http://www.cpi.umist.ac.uk/eminent2/Publications/148%20Hetland.pdf.

Higman C, Burgt M, Thermodynamics of Gasification. *Gasification*, (Oxford, Elsevier, 2008) 10–14.

Himeda Y, Utilization of carbon dioxide as a hydrogen storage material: Hydrogenation of carbon dioxide and decomposition of formic acid using iridium complex catalysts. *Advances in CO_2 Conversion and Utilization*, Editor: Yun Hang Hu, (Washington, DC, American Chemical Society, 2010) 141–153.

International Energy Agency, CO_2 emissions—Sectoral approach, *CO_2 Emissions from Fuel Combustion*, (France, International Energy Agency, 2013), Accessed July 20, 2015, http://www.iea.org/publications/.

Kaplan S, *Power Plants: Characteristics and Cost*, (Congressional Research Service, 2008), Accessed July 20, 2015, https://www.fas.org/sgp/crs/misc/RL34746.pdf.

Matthews D, Gillett P, Stott A, Zickfeld K, *The Proportionality of Global Warming to Cumulative Carbon Emissions*, (Macmillan Publishing, Nature Publishing, 2009), Accessed June 11, 2009, http://www.cccma.ec.gc.ca/papers/ngillett/PDFS/nature08047. pdf.

Maurstad O, *An Overview of Coal based Integrated Gasification Combined Cycle (IGCC) Technology*, (Cambridge, MA, Massachusetts Institute of Technology, 2005).

Mills R, *Capturing Carbon: The New Weapon in the War against Climate Change*, (London, C. Hurst & Co, 2011).

Mills S, *Coal-Fired CCS Demonstration Plants*, (IEA Clean Coal Centre, 2012), http://www.iea-coal.org/documents/83086/8635/Coal-fired-CCS-demonstration-plants,-2012,-CCC/207.

Morgan-Sagastume J, Revah S, Methods of odor and VOC control. *Biotechnology for Odor and Air Pollution Control*, Editor: Shareefdeen Z, Sing A, (New York, Springer, 2005), 35.

Phadke R, Trivedi S, Flue gas conditioning, *Electrostatic Precipitation, 11th International Conference on Electrostatic Precipitation*, Editor: Yan K, (Hangzhou, China, Zhejing University Press, 2009).

Puri P, Commercial application of membrane in gas separation, *Membrane Engineering for the Treatment of Gases*, (Royal Society of Chemistry, 2011) Vol 1, 215–244.

Rackley S, *Carbon Capture and Storage*, (Oxford, Elsevier, 2010).

Rousseau S, *Audi's New E-Gas Plant Will Make Carbon-Neutral Fuel*, (Popular Mechanics, 2013). Accessed April 19, 2014, http://www.popularmechanics.com/cars/news/auto-blog/audis-new-e-gas-plant-will-make-carbon-neutral-fuel-15627667.

Royal Academy of Engineering, *The Cost of Generating Electricity*, (London, Royal Academy of Engineering, 2004) 8.

Rubin Es, Zhai H, *Techno-Economic Assessment of Polymer Membrane System for Post combustion Carbon Capture at Coal-Fired Power Plants*, Environmental Science & Technology, (American Chemical Society, 2013) 47(6), 3006–3014.

Stefaniak S, Twardowska I, Coal and coal combustion products: Prospects for future and environmental issues, *Coal Combustion By-Products and Environmental Issues*, Editors: Alva A, Punshon T, Sajwan K, Twardowska I (New York, Springer, 2006) 13–20.

Wells J, Lefkowitz S, Chavarria G, Dyer S. Danger in the Nursery. *Impact on Birds of Tar Sands Oil Development in Canada's Boreal Forest* (New York, National Resource Defence Council, 2008) Vol 3, 21–22.

Wilcox J, *Carbon Capture*, Springer, New York, 2012.

World Coal Institute, *Coal Resource, A Comprehensive Overview of Coal*, (London, World Coal Institute, 2005).

Zhang D, Introduction to advanced and ultra-supercritical fossil fuel power plants. *Ultra-Supercritical Coal Power Plants Materials, Technologies and Optimisation*, Editors: Zhang D (Cambridge, Woodhead Publishing, 2013) 4.

Zhang X, He X, Gundersen T, *Post-combustion Carbon Capture with a Gas Separation Membrane: Parametric Study, Capture Cost, and Exergy Analysis*, (Energy Fuels, 2013) 27(8), 4137–4149.

21 Carbon Sequestration in Coal-Fired Combustion/Gasification Processes

Rajender Gupta, Deepak Pudasainee, and Bill Gunter

CONTENTS

Abstract: Combustion and gasification of coal emit CO_2 into the atmosphere. The carbon capture technologies, post-combustion, pre-combustion, and oxy-fuel combustion are in demonstration and pilot plant phases, whereas chemical looping combustion is currently in very early stage of development. Certain technologies have been commercially implemented, for example, enhanced oil recovery (storage) and production of urea (use). Currently, there are nine large-scale carbon capture and storage (CCS) projects in development/execute stage and 13 projects in the operation stage. Economic capture of CO_2 from flue gas, monitoring of the CCS site, maintaining a geological standard, and commercialization of the new CCS technologies are the major challenges.

21.1 INTRODUCTION

Fossil fuels are the major source of power generation. The total shares of the world's primary energy supply in 2012 were 32% coal, 29% oil, and 21% natural gas. The remaining 18% was composed of nuclear, hydro, geothermal, solar, tide, wind,

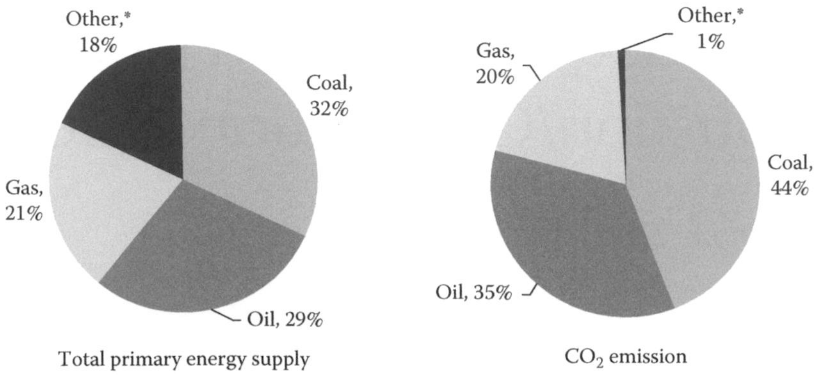

FIGURE 21.1 World's total primary energy supply and CO_2 emissions from different fuel types. *Note*: Other* includes nuclear, hydro, geothermal, solar, tide, wind, biofuels, and waste. (Data from IEA, *International Energy Agency, CO_2 Emissions from Fuel Combustion Highlights*, 2014.)

biofuels, and waste (IEA 2014) (Figure 21.1). Fossil fuels combustion and gasification release carbon dioxide (CO_2) into the atmosphere, which is the most important anthropogenic greenhouse gas (GHG). Worldwide CO_2 emissions from different fuels are shown in Figure 21.1. Coal combustion emits the largest amount of CO_2 into the atmosphere. In 2012, coal shared 29% of the total world primary energy supply; however, it contributed 44% of the global CO_2 emissions.

CO_2 concentrations in the atmosphere have been increasing significantly over the past century, compared to the pre-industrial concentration (about 280 ppmv). The concentration in 2013 (396 ppmv) was nearly 40% higher than in the mid-1800s (IEA 2014). The contribution of CO_2 in global radiative forcing is estimated to be about 1.66 W/m^2. It is the largest contributor (IPCC 2007). If no measures are taken to control CO_2 emissions, the emission level in 2050 is estimated to be twice the 2007 level (IEA 2010). The 4th IPCC report (IPCC 2007) warns that serious effects of warming have become evident and that the cost of reducing emissions would be far less than the damage they will cause. Due to increasing GHGs concentration in the atmosphere, several impacts are observed at the local and global scales. The global average surface temperature (the average of near-surface air temperature over land and sea surface temperature) has increased since 1861. It has increased over the twentieth century by about 0.6°C. Scientific projections indicate that climate change could affect the health and wellbeing of human beings in a number of ways such as loss of several endangered species, smog and heat waves, extreme weather events (more tornadoes/floods, increasing ice melt, etc.), diseases, disbalance in water distribution, and lack of food sources.

Geological storage of CO_2 is only one of a number of options to reduce GHG buildup in the atmosphere that have been summarized by Kaya and subsequently modified by Gunter et al. (1998) in the following generalized global equation for CO_2 emissions to the atmosphere:

$$GHG_a = POP \times \left(\frac{GGP}{POP}\right) \times \left(\frac{BTU}{GGP}\right) \times \left(\frac{GHG_t}{BTU}\right) - CM$$

where:

- GHG_a is GHG global emissions into the atmosphere
- GHG_t is total GHG emissions
- POP is global population
- GGP is gross global product
- GGP/POP represents the global standard of living
- BTU is British thermal units representing global energy consumption
- BTU/GGP represents the global energy intensity
- GHG/BTU represents the carbon intensity of energy production
- CM is contribution from the carbon management options

As population control and standard of living are the most difficult to reach global agreement on, the targets for GHG emission reductions are focusing on energy intensity (efficiency), carbon intensity (less carbon-intensive fuels such as natural gas, nuclear, wind, solar, biomass, hydro, geothermal, and hydrogen), and carbon management (chemical, biological, oceanic, and geological sinks). Efficiency improvements are the most desirable to reduce CO_2 emissions as they improve the economics of a process and consume less energy. Less carbon-intensive fuels are also desirable, but the cost (excluding other fossil fuels) is generally higher, and currently they have difficulty competing with carbon management of fossil fuels to reduce CO_2 emissions.

There is an agreement among scientists and policy makers that stabilizing the atmospheric concentration of GHGs at below 450 ppm of carbon dioxide equivalent (CO_2-eq) is consistent with a near 50% chance of achieving the 2°C target, and this would help to avoid the worst impacts of climate change (OECD/IEA 2013). Figure 21.2 shows CO_2 emissions in the absence of any action to cope with climate change and the delayed 450 scenario, that is, actions taken after 2019. The *450 scenario* (not discussed here) assumes a 50% chance of limiting global temperature increase to 2°C by the implementation of growing number of co-ordinated action against climate change from 2014, whereas the *delayed 450 scenario* is the prediction of what will happen if the early action to mitigate climate change is not taken. In the *new policies scenario*,

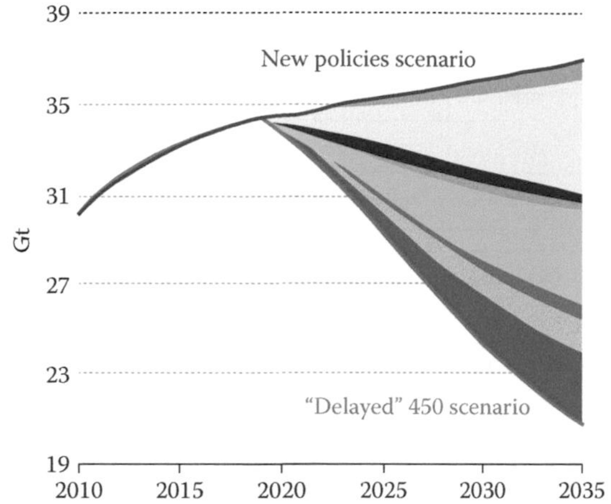

CO_2 abatement	2025	2035
Demand	5%	5%
End-use efficiency	27%	31%
Power plant efficiency	11%	3%
Fuel and technology switch	2%	2%
Renewables	25%	26%
Biofuels	5%	5%
Nuclear	9%	9%
CCS	15%	20%
Total (Gt of CO_2)	6.2	16.4

FIGURE 21.2 World's energy-related CO_2 emissions abatement in new policies scenario for a *delayed* 450 scenario. (Data from IEA, 2013b).

CO_2 emission keeps increasing to its annual emissions of about 37 Gt/yr, and in the *delayed 450 scenario*, it reduces to about 21 Gt/yr by 2035.

CCS is considered as one of the major reduction measures, which is more suitable to stationary sources. It includes four primary steps: CO_2 capture, compression, transport, and storage (Figure 21.3). Carbon enters from the fossil fuels. Once the carbon-containing fuel is combusted or gasified, CO_2 is released in the flue-gas stream. The CO_2 thus released can be captured by several technologies (pre-combustion capture, post-combustion capture, oxy-fuel combustion, and chemical looping combustion). After capturing, CO_2 is compressed for transportation and finally sent for storage.

There is a high potential for CO_2 reduction by increasing plant efficiency. There are a number of advanced technologies such as supercritical boilers, taking electricity generation efficiency from 35% to about 45% or higher. Other technologies included integrated gasification combined cycle (IGCC), and integrated gasification fuel cells take the efficiency above 50%. A single approach, however, cannot meet the desired CO_2 emission reductions. One need to take a more holistic approach to achieve the required target in reduction of CO_2 emissions including CCS in geological sites on a time scale needed to get renewable energy resources and other alternatives in place.

In this chapter, the overview of carbon sequestration in coal-fired combustion and gasification processes is presented. The recent state-of-the-art technologies in CO_2 capture, separation, transport, utilization, and storage are discussed in the next three sections followed by challenges and future works in this field.

21.2 CARBON CAPTURE

The flue gases from a boiler or a gasification process have less than 10%–15% of CO_2. However, CO_2 concentration may be higher in the resulting off-gas after the water–gas shift reaction in the gasification process. It is required to have a concentrated stream of CO_2 before it can be compressed and transported.

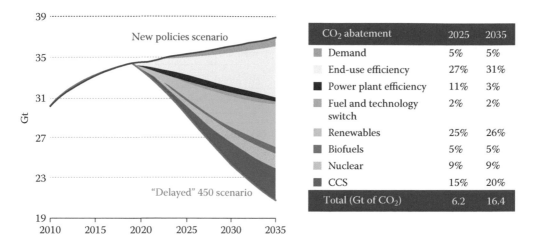

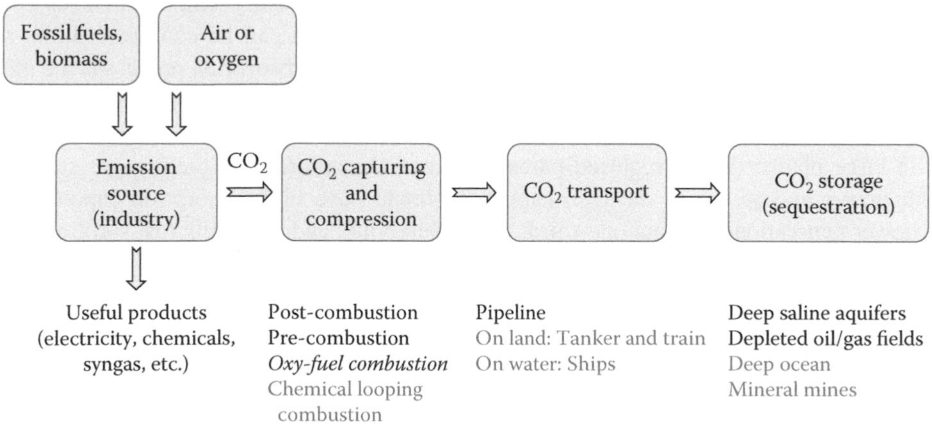

FIGURE 21.3 Schematic presentation of CCS system with available technical options. Technologies related are listed below each step. Solid texts represent technologies that are in commercial improvement phase; italics represent pilot plant and demonstration phase; faint texts represent laboratory-scale study and/or evaluation phase. (Redrawn from Rubin, E.S. et al., *Progr. Energy Combus. Sci.*, 38, 630–671, 2012.)

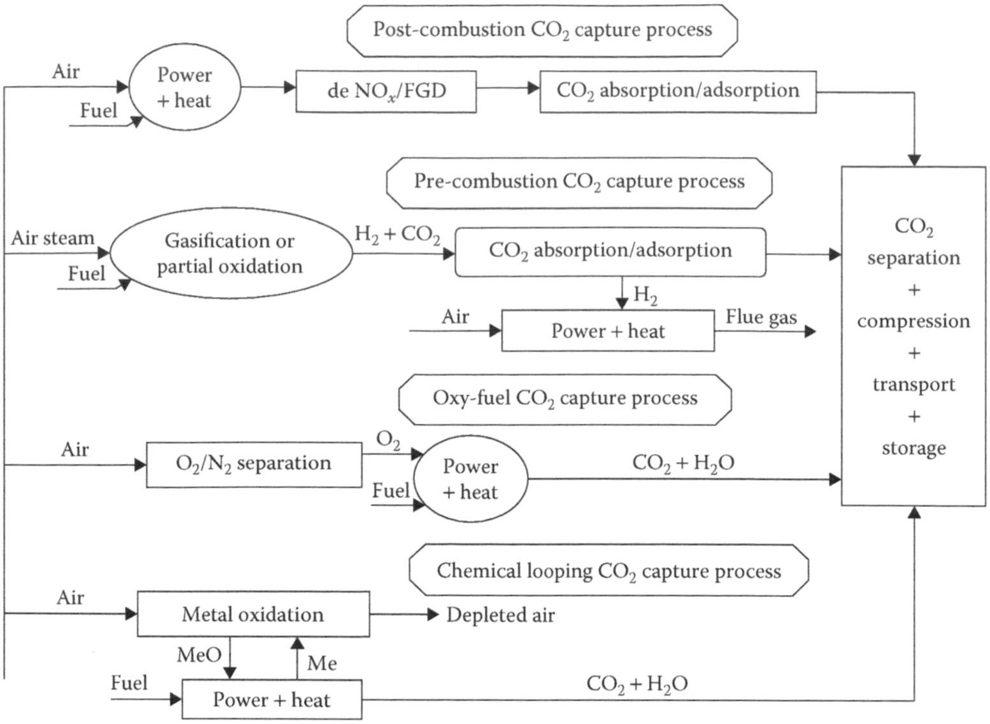

FIGURE 21.4 CO_2 capture options. (From Leung, D.Y.C. et al., *Renew. Sust. Energy Rev.*, 39, 426–443, 2014.)

If there are more than a few percent of noncondensable gases (e.g., N_2), the compression of CO_2 becomes excessively expensive. In the total process of CCS, obtaining the desired concentrated stream of CO_2, termed as *carbon capture*, constitutes more than 60%–70% of the total costs of CCS. It is, therefore, important to have a cost-effective capture to form a relatively concentrated CO_2 stream, and then compressing it to high pressures. There are several technological options for capturing CO_2 from combustion and gasification facilities such as pre-combustion capture, post-combustion capture, oxy-fuel combustion, and chemical looping combustion.

The overview of these capture processes and systems are presented in Figure 21.4. Among these technologies, post-combustion and oxy-fuel combustion can be applied to both coal- and gas-fired plants, whereas pre-combustion is mainly applied to coal gasification plants. Considering the present status, these approaches are still not ready for implementation on coal-based power plants because (1) they have not been demonstrated in large plants; (2) the required parasitic loads to supply both power and steam to the CO_2 capture plant would reduce power generation by about one-third; and (3) if successfully scaled up, they would not be cost effective (DOE/NETL 2010; Samanta et al. 2012).

21.2.1 Post-Combustion Capture

Post-combustion capture is of specific importance as it can be just an add-on to an existing power plant. In this approach, CO_2 is captured from combustion flue gas. The flue gas from combustion is treated for dust, NO_x, and SO_x removal followed by CO_2 capture. As stationary power-producing sources are the dominant sources (47%) of total anthropogenic CO_2 emissions into the atmosphere (IPCC 2005), most of the current research has focused on controlling CO_2 emission from such sources, utilizing post-combustion capture. The most important parameters for post-combustion CO_2 capture are flue-gas volume, CO_2 concentration, degree of CO_2 separation, solvent volume stream, purity of the separated CO_2 (Markewitz et al. 2012), and pressure.

The processes most widely used currently are chemical absorption using liquid solvents (Figure 21.5). CO_2 absorption in solvents, mainly amine, has been applied in industries and is a widely used separation technique. Solvents can be regenerated by increasing temperature and/or depressurization. CO_2 thus separated can be compressed and sent for storage. The amine-based absorption process is the most mature among the technologies available; however, it is not yet cost effective to be applied in industry. Cost-effective technology is essential for CO_2 capture from the flue-gas streams. The ideal sorbents should have high adsorption capacity, high CO_2 selectivity, durability, and relatively fast sorption and desorption kinetics (Samanta et al. 2012).

The following are some of the major issues in applying solvent absorption for post-combustion capture:

Solvent degradation by SO_x and NO_x: After reaction with SO_x and NO_x present in flue gas, amine forms salts that have to be separated. Salt formation can be prevented by lowering SO_2 concentration to about 10 ppm (Rao and Rubin 2002), which is a

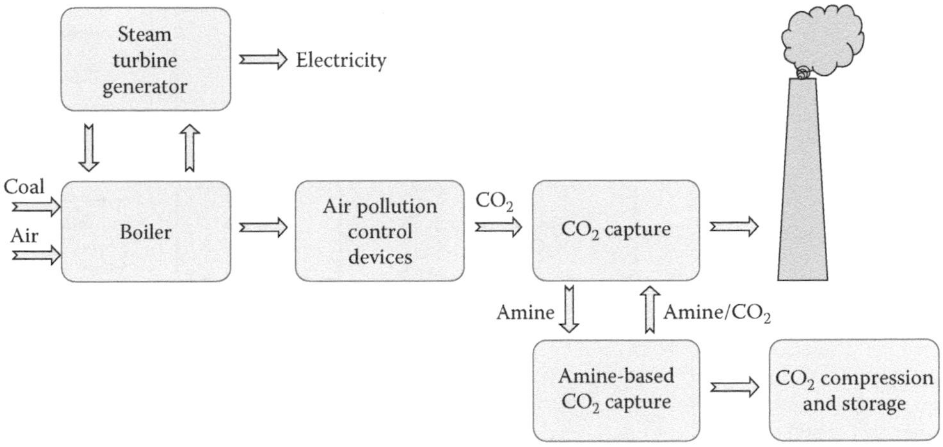

FIGURE 21.5 Schematic representation of a coal-fired power plant with post-combustion CO_2 capture using an amine-based scrubbing system.

very difficult task because current U.S. regulations for a coal-fired power plant specify 160 mg/m^3 for new facilities built after 2005 (IEA 2013a). Thus, new solvents tolerant to SO_x, NO_x, and flue-gas components are essential to overcome this problem.

Decomposition of amine in the presence of oxygen: The relatively higher concentration of oxygen in flue gas causes the decomposition of amine. For the cost-effective and the efficient capture of CO_2 from the flue gas, this issue needs to be addressed.

Higher regeneration cost: The energy consumption (low-pressure steam) amounts to about 4 GJ per ton CO_2 captured, about 40%–50% of the entire low-pressure steam is needed for the regeneration step (Irons et al. 2007; Markewitz et al. 2012). Reducing additional power consumption for solvent regeneration cost is a challenge to overcome.

Loss of ammonia: Ammonia's higher volatility compared to monoethanolamine is a concern during CO_2 capture. To minimize ammonia vapor emissions during the absorption and to improve CO_2 absorption, flue gas must be cooled. Process optimization to increase CO_2 loading and use of various engineering techniques to eliminate ammonia vapor losses from the system during operation has been proposed (e.g., Yeh et al. 2005).

There have been significant research efforts in developing new solvents of higher solubility with more tolerance to inherent pollutants such as SO_x and NO_x and lower regeneration costs. CO_2 has very high solubility in ionic liquids, but ionic liquids are more effective as physical adsorption at lower temperatures and highly dependent on CO_2 concentration. Many of these developments are proprietary in nature. Higher regeneration costs are directly related to the heat of evaporation. To overcome this, the development of new solid sorbents for CO_2 capture from flue gas with higher performance and desired economics has been carried out.

As CO_2 concentration in combustion flue gas is low (e.g., 10%–15% from coal-fired power plants), the energy penalty and the related cost of capture, sequestration, and transport are increased (Visser et al. 2008; Olajire 2010). It has been estimated that the inclusion of post-combustion CO_2 capture in coal- and oil-fired power plant would increase the cost of electricity by 65% and 32%, respectively (Kanniche et al. 2010).

21.2.2 Pre-Combustion Capture

Pre-combustion capture involves capturing CO_2 from syngas (mainly H_2 and CO_2) with lesser amounts of CO, H_2, and O_2. While, for example, Kemper IGCC plant uses the physical solvent process (Selexol) for CO_2 capture, many industries are using the Benfield process. Pre-combustion capture can be applied in IGCC power plant (Figure 21.6). In pre-combustion capture, the fuel (mainly coal and natural gas) is pretreated before combustion. Coal is gasified to produce syngas consisting mainly of CO and H_2 (Equation 21.1). After gasification, syngas undergoes the water–gas shift reaction, which converts CO to CO_2 and more H_2 is formed in the process (Equation 21.2). On the other hand, natural gas can be reformed to syngas according to Equation 21.3. Again, the H_2 and CO_2 concentration can be further increased by water–gas shift reaction.

$$\text{Gasification: } \text{Coal} \rightarrow CO + H_2 \quad (21.1)$$

$$\text{Water–gas shift: } CO + H_2O \rightarrow H_2 + CO_2 \quad (21.2)$$

$$\text{Reforming: } CH_4 + H_2O \rightarrow CO + H_2 \quad (21.3)$$

CO_2 removal in K_2CO_3 solution using the Benfield process is a standard method of removing CO_2 from syngas. This process occurs at high temperature (120°C) and pressure (3000 kPag). The advantages of CO_2 removal in K_2CO_3 solution over amine-based solvents are that absorption can occur at high temperatures that reduces parasitic load on the power station. In addition, cost of K_2CO_3 is low, less toxic,

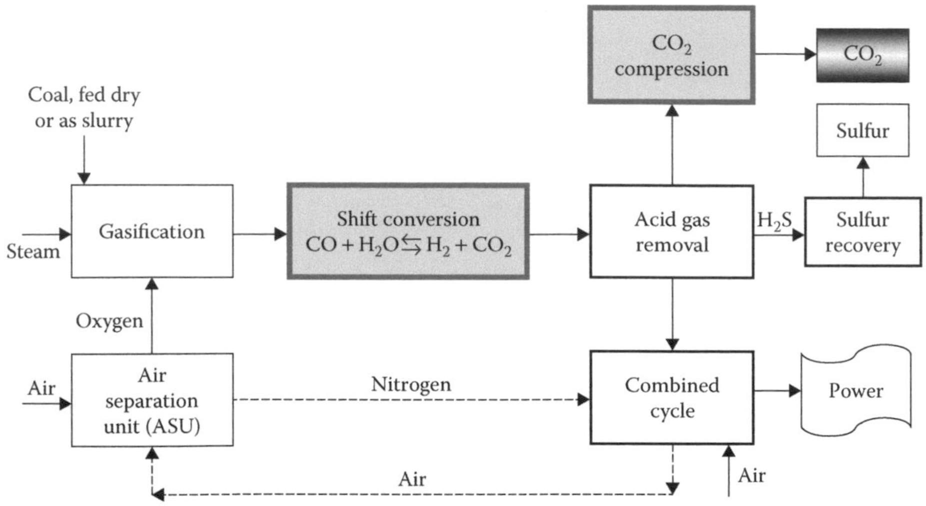

FIGURE 21.6 Schematic representation of IGCC (pre-combustion capture) process. The additional unit operations for carbon capture are shown in bold. (Reprinted from *Proc. Combus. Inst.*, 31, Wall, T.F., Combustion processes for carbon capture, 31–47, Copyright 2007, with permission from Elsevier.)

less corrosive, and less prone to degradation effects that are commonly seen with amine-based sorbents at high temperatures and in the presence of oxygen and other gas components (Smith et al. 2012).

21.2.3 Oxy-Fuel Combustion

Oxy-fuel combustion was first introduced in 1982 for obtaining CO_2-rich flue gas for enhancing oil recovery (Wall et al. 2009). It is the process of burning fuel with pure oxygen with recycled CO_2-rich flue gas (except particles), instead of O_2/N_2 as in air combustion systems. The major driving forces behind increasing interest in oxy-fuel combustion technology are the production of higher amount of CO_2 that can be separated and stored, and reduced cost of emission control, mainly for NO_x. The concentration of CO_2 in air coal combustion is about 13%–15%, which is difficult to capture, whereas in oxy-fuel combustion, CO_2 concentration may be up to 95% (Hu et al. 2001); this can easily be captured and stored.

Oxy-fuel combustion has several advantages over air combustion (Figure 21.7):

1. Concentration of CO_2 in flue-gas emission from oxy-fuel combustion is higher than air combustion, which can be more easily captured.
2. In oxy-fuel combustion, the volume of flue-gas emission, NO_x emission per unit of fuel consumption, is reduced (Liu and Okazaki 2003).
3. Reduction of flue-gas treatment cost due to decreasing flue-gas volume in oxy-fuel combustion systems.

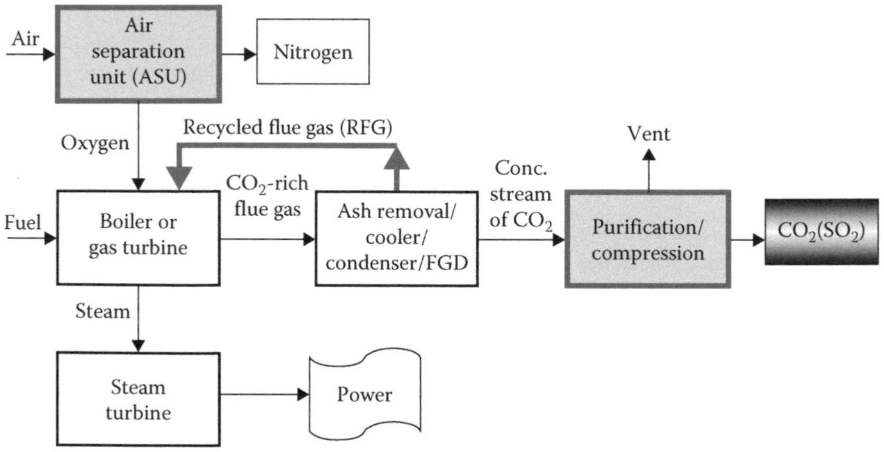

FIGURE 21.7 Oxy-fuel technology for power generation with CO_2 capture and storage. The additional unit operations are shown in bold. (Reprinted from *Chem. Eng. Res. Des.*, 87, Wall, T. et al., An overview on oxyfuel coal combustion-state of the art research and technology development, 1003–1016, Copyright 2009, with permission from Elsevier.)

4. Potential increase of boiler thermal efficiency due to reduction of flue-gas volume (Tan et al. 2002). Higher combustion efficiency of the boiler due to higher concentration of O_2 (Kimura et al. 1995; Hu et al. 2001).

21.2.4 Chemical Looping Combustion

Chemical looping combustion system employs two interconnected fluidized bed reactors: air and fuel reactors (Figure 21.8). In this process, oxygen is transported by an oxygen carrier from air reactor to the fuel in the fuel reactor.

The difference of this method with conventional combustion is utilization of these oxygen carriers in the system so there is no direct contact between air and fuel. The fuel and metal oxide (Me_xO_y) are introduced in the fuel reactor. The fuel and the metal oxide react (Equation 21.4). The reduced metal oxide, Me_xO_{y-1}, is transferred to the air reactor where it is re-oxidized (Equation 21.5). The reduced metal oxide will receive oxygen again from the air and then the cycle is repeated. Nitrogen and unused oxygen exit from the air reactor. So as a continuous looping system, this procedure will be repeated to combust fuel and achieve CO_2 with high purity. The products of the fuel reactor comprise CO_2 and H_2O. After condensation of water, a high concentration CO_2 stream will be obtained. Depending on the metal oxide and fuel used, Equation 21.4 can be endothermic or exothermic, whereas Equation 21.5 is exothermic.

$$(2n+m)Me_xO_y + CnH_2m \rightarrow (2n+m)Me_xO_{y\text{-}1} + mH_2O + nCO_2 \quad (21.4)$$

$$Me_xO_{y-1} + \frac{1}{2O_2} \rightarrow Me_xO_y \quad (21.5)$$

Chemical looping combustion process allows nitrogen-free combustion of fuel without requiring an expensive air separation unit (ASU). The product gas mainly consists of CO_2 and H_2O. CO_2 released is sequestration ready with little or no energy penalty. However, it is to be noted that all these capture technologies (pre-combustion, post-combustion, oxy-fuel combustion, and chemical looping combustion) need additional energy input for gas separation, capture, conditioning, and compression of CO_2.

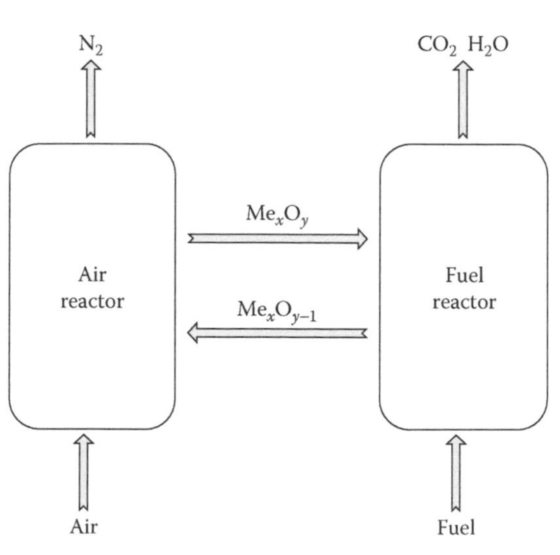

FIGURE 21.8 Schematic illustration of chemical looping combustion. (Me_xO_y/Me_xO_{y-1} means recirculating oxygen carrier material.)

21.3 CO_2 TRANSPORTATION

Once the carbon is captured from flue gas, it is compressed and transported for storage. Safe and reliable transportation of CO_2 is an important step in CCS process. CO_2 can be transported from the capture site to storage site by various means: (1) pipelines, (2) on land: transport via truck and rail, and (3) on water: ship transport. Figure 21.9 shows comparative costs for these three types of transportation.

1. *Pipeline*— CO_2 pipelines on land and under the sea have been operational worldwide. The existing pipelines in the United States are shown in Figure 21.10. The cost of pipeline transportation depends on the difficulty of terrain and whether it is onshore or offshore. The costs do decrease with increases in tonnages and distance transported. The cost of transportation is mainly due to high costs of infrastructure, and therefore, increases in distance and tonnages decrease the cost per ton slightly. The capital cost of pipelines on ground is usually cheaper than that under sea; however, the onshore capital costs can increase dramatically with difficulty in terrain. The comparative costs are presented in Figure 21.9. Transporting CO_2 through pipelines has been the most widely used means of transporting CO_2. Several thousands of kilometers of pipelines have been installed for transporting CO_2 around the world.

   At present, there are about 50 CO_2 pipelines operating in the United States, which transport approximately 68 million tons (Mt)/yr of CO_2 (GCCSI 2014).

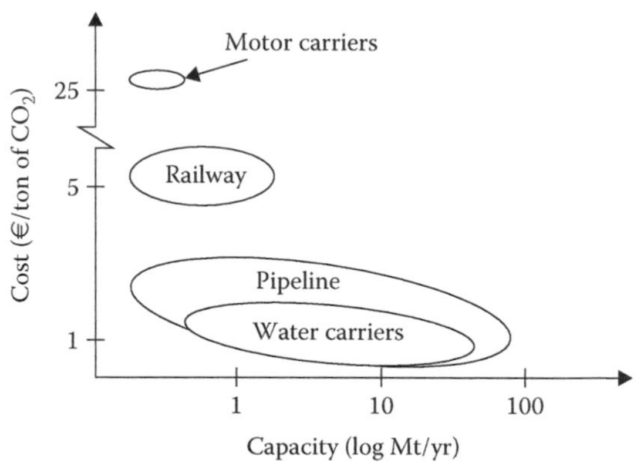

FIGURE 21.9 Comparative costs of transportation by different modes. (Reprinted from *Energy Convers. Manag.*, 45, Svensson, R. et al., Transportation systems for CO_2-application to carbon capture and storage, 2343–2353, Copyright 2004, with permission from Elsevier.)

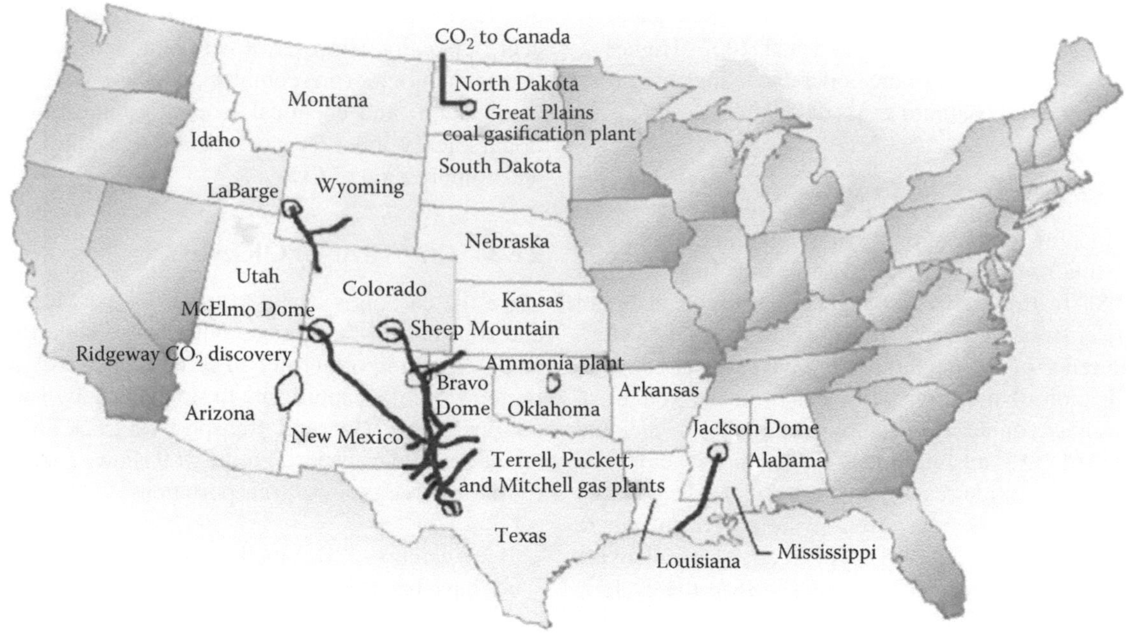

FIGURE 21.10 Existing pipelines for CO_2 transport in the United States. (From *Oil and Gas Journal*, http://www.ogj.com/, 2014.)

There is a concern that CO_2 is heavier than air, and in a pipeline accident, it may not disperse vertically.

2. *On land—transport via truck and rail*: Transportation by trucks and rail can be realized in small scale and when the capture and storage site is not very far and the quantity is small, probably used for piloting of storage sites or enhanced oil recovery (EOR). For example, in Callide oxy-firing plant, CO_2 was transported to storage site by trucks.
3. *On water—ship transport*: Transport of CO_2 via ship can be carried out in many sites where capture and storage site are connected to water bodies. Shipment of CO_2 has been carried out on a small scale in Europe, transporting food-quality CO_2 (about 1000 tons) from the emission site to the coastal distribution terminals (GCCSI 2014). There have been discussions in two large shipping companies to build large tankers similar to those for NG.

21.3.1 Challenges in CO_2 Transport

- Cost associated with CO_2 transport in general is \$10–15 per ton of CO_2 per 1000 km, which is not so expensive compared to capture costs.
- Large upfront costs of transportation by pipeline or tankers: 150,000 km pipeline is needed in North America alone.
- Risks of pipeline rupture, even though not so severe, need to be reduced.
- Trans-boundary transport–legal problems for hazardous waste.

21.4 CARBON STORAGE

Carbon management falls into five categories: geological storage, biomass storage, ocean storage, mineral mines, and producing bulk chemicals (e.g., fertilizers). The latter, although attractive, in most cases, is at too small-scale or storage period to appreciably lower CO_2 emissions. Mineral mines storage at the earth's surface has been proposed, with ultramafic and serpentinitic rocks being the primary targets. Storage in oceans, although practical, suffers from affecting the pH of the oceans, which may be harmful to life in the oceans, as well as the risk of leakage as the ocean is a liquid that has poor sealing capacity due to changing circulation patterns as well as potential upsets (abrupt overturns), as seen in one lake in Africa. Biomass storage is temporary, the length of time depending on the type of biomass (e.g., forests versus agriculture). Geological storage of CO_2 can be permanent ranging from millions to hundreds of millions of years, has huge capacities (measured in gigatons of CO_2), and therefore is the best long-term storage solution. However, pure geological storage is expensive once the *low hanging fruit* has been utilized; the costs being in the range of \$100/ton of CO_2 when the total cost is considered (capture, transport, and storage) and is unlikely to be enacted on a large scale without appropriate policies in place. This is why interest has turned to CCUS (carbon capture, utilization, and storage), primarily in CO_2 EOR, one of the *low hanging fruit* even though saline aquifers have the largest potential for geological storage. In between these two extremes are depleted oil and gas reservoirs.

These geological storage options are illustrated in Figure 21.11, which is a block diagram showing the sources of CO_2 from gas and oil processing, and electricity production

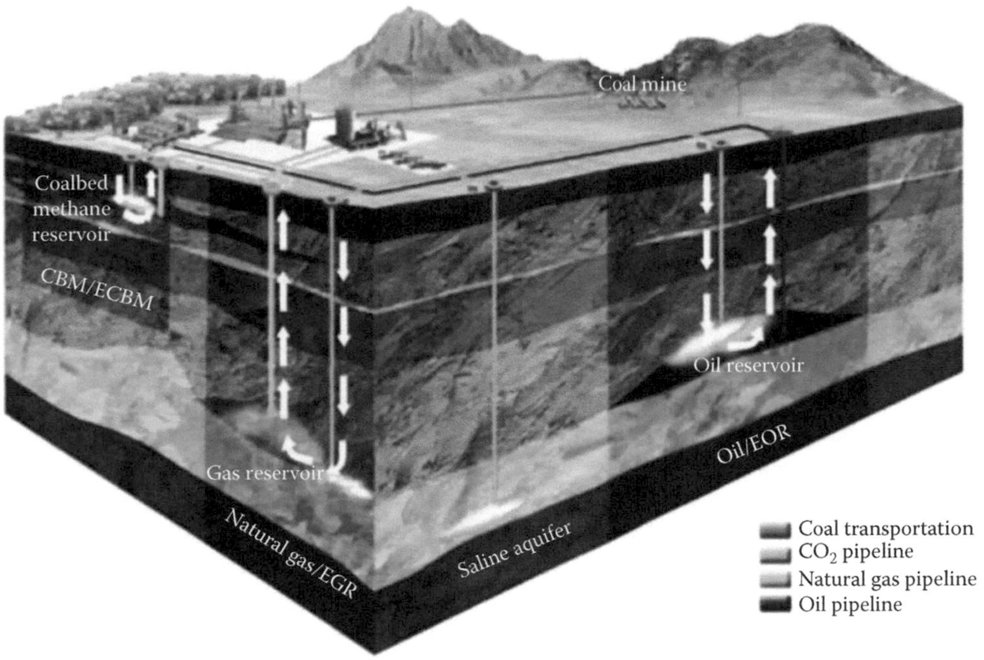

FIGURE 21.11 CCS and CCUS in resource production and CO_2 storage. CBM = coalbed methane, ECBM = enhanced coal bed methane, EGR = enhanced gas recovery, and EOR = enhanced oil recovery. (Reprinted from Asian Development Bank, *Prospects for Carbon Capture and Storage in Southeast Asia, a Regional Analysis*, Regional Technical Assistance program (RETA) #7575—for Determining the Potential for Carbon Capture and Storage in Southeast Asia, Asian Development Bank, Philippines,167 pages, Copyright 2013. With permission.)

from coal. CO_2 is shown being injected into shallow coals where it is strongly sorbed, releasing the methane sorbed to the coal that is produced. Another source of natural gas is from the natural gas reservoirs. Waste CO_2 is injected into the bottom of the natural gas reservoir to provide a pressure drive to recover more of the natural gas. This process has yet to be proved commercially and suffers from premature breakthrough of the CO_2 as it is miscible with the natural gas and is highly mobile. On the other hand, CO_2 EOR has been a commercial technology since 1972 and currently has over 100 commercial projects worldwide with most of them located in the United States. Figure 21.11 depicts the CO_2 being injected into the top of a reservoir where oil being the denser phase is produced by gravity displacement. However, CO_2 EOR has been successfully used in thinner reservoirs using a horizontal drive to become miscible with the oil and to drive the oil to the production wells. One of the challenges is the high mobility of the CO_2 with respect to the oil and water in the reservoir, resulting in early breakthrough of the CO_2. For storage in saline aquifers, the CO_2 is injected at the base of the aquifer so that it has a better chance to dissolve in the water. Dissolution of CO_2 in water at high pressures has a negative partial molal volume so that the CO_2-enriched water has a higher density than the formation water and will sink to the bottom of the aquifer over time. However, this process takes thousands of years to complete, and the CO_2 will mainly remain for a long time in a supercritical plume, which will tend to form at the top of the aquifer.

Although not evident in Figure 21.11, at the pore scale, microscopic trapping increases the permanence of storage and in some cases is the only trapping mechanism (e.g., mineral mines). These microscopic traps are geochemical in nature (Gunter et al. 2000). Geochemical traps consist of solubility traps, ionic traps, mineral traps, sorption traps, and phase traps. CO_2 will dissolve in water or oil, the amount proportional to the partial pressure of CO_2, and then will travel with the hydrous and oil phases rather than being confined to movement of the supercritical CO_2 plume. Any drop in pressure would cause the CO_2 to boil out of the hydrous and oil phases and revert back to its gaseous/supercritical form. However, once the CO_2 is dissolved in water, it forms a weak acid and will react with the rock at different rates. If Ca, Mg, or Fe are available in the minerals and the pH conditions are correct, carbonate minerals can precipitate (mineral trapping), which is one of the most secure forms of storage as evidenced by the abundance of carbonate rocks in the geological record, many of which are formed at the surface in shallow water. The formation of CO_2 hydrates may also become an important form of CO_2 storage in low temperature–moderate pressure environments such as that exist in the ocean floor sediments, particularly where methane hydrates are found throughout the world. Also under appropriate conditions, if K or Na is available in the minerals that make up the sediments, the CO_2-charged water will increase in K and Na with a corresponding increase in the dissolved bicarbonate ion (ionic trapping).

This form of dissolved CO_2 is more stable than dissolved CO_2 as the CO_2 will not be released by a pressure drop. Sorption of CO_2 onto minerals or solids depends on the minerals or solids' surface properties and is most important for organic phases such as coal. Chemically, it behaves similarly to solution in water as the amount adsorbed is dependent on the partial pressure of CO_2. Finally, phase trapping takes place in individual pores as multiphase fluid systems exhibit residual saturations due to the interaction of the various phases in the pore and the hydrodynamics of the system. When a new fluid phase such as CO_2 is introduced into a pore, part of the CO_2 is stripped out and becomes immobile to the flow system. Typically for CO_2, this residually trapped CO_2 can occupy 5% of the pore space or higher and is probably the most important contributor to the microscopic trapping that takes place in a slowly moving saline aquifer in sandstone or carbonate rocks. For coal aquifers, the most important trapping process is by adsorption of the CO_2. For basalt aquifers, it is probably mineral trapping.

Storage of CO_2 in the subsurface at high pressures is a process that has already been developed by mother nature. Natural CO_2 reservoirs exist in the geological column at a number of sites in the world that have been stable for millions of years and have been used in the United States as a source of CO_2 for CO_2 EOR. In addition, often natural gas reservoirs contain substantial amounts of acid gas (CO_2 and/or H_2S) that has to be separated from the produced natural gas to meet sale specifications. North America has pioneered acid gas disposal in depleted oil and gas reservoirs and saline aquifers in over 50 projects (Bachu and Gunter 2004) since 1990 to meet regulations limiting their emissions to the atmosphere. Although the technology was pioneered to meet H_2S emission regulations, as CO_2 was part of the H_2S fraction, it was more economical to dispose of them together than to separate CO_2 from H_2S. As leakage of H_2S is more serious than CO_2 from these underground reservoirs as regards human health, the regulations developed were quite rigorous. Consequently, these acid gas disposal sites represent an industrial analog for CCS. The only difference is the scale as the commercial geological storage projects would inject over a megaton of CO_2 per year (~>3000 tons/day) while most of the acid gas disposal projects would be regarded as pilots (<100 tons/day) in the geological storage sense. Their importance to CCS is that they have demonstrated that safe storage of CO_2 in the subsurface is possible over a number of years.

21.4.1 Enhanced Oil Recovery and Depleted Oil Reservoirs Storage

There is a continuum between CO_2 EOR and depleted oil reservoir storage of CO_2. In an EOR project, approximately one-third of the purchase CO_2 is captured and stored in the reservoir even though a goal of the project is to minimize CO_2 purchase. The injected CO_2 that is produced during oil recovery is recycled and re-injected to help produce more oil, and it represents approximately two-thirds of the CO_2 injected (Faltinson and Gunter 2010). The total CO_2 injected is the sum of the purchased and recycled CO_2. A life cycle analysis of a CO_2 EOR project shows that if the CO_2 captured in the reservoir during oil production is subtracted from the CO_2 life cycle emissions of the oil produced, the EOR produced oil has a significantly reduced footprint compared to oil produced by other means, approaching that of natural gas. If there is an economic value attached to storing CO_2, and if the government permits CO_2 credits for storing CO_2 during EOR, then integrated EOR CCS projects would be common and more oil would be recovered from EOR projects as the money received for storing CO_2 would offset the cost of CO_2 injection and allow more CO_2 to be injected during oil recovery. For example, currently in Alberta, the government charges \$15/ton to companies that are responsible for excess CO_2 emissions. If this CO_2 was captured (instead of being emitted) and used for EOR, there would be a cost saving. At the end of the EOR phase of the project, the production would cease, the producing wells would be converted to injectors, and the pure CO_2 storage phase (in depleted oil reservoirs) would commence until the original reservoir pressure is reached. Obviously, this would be more attractive than saline aquifer storage in the short term if the cost of the additional monitoring requirements and the care (e.g., remediation and abandonment) of the additional wells in the oil reservoir were not excessive.

Other EOR opportunities may appear in the future due to the properties of CO_2 (Gunter and Longworth 2013). CO_2 is highly mobile compared to the oil. EOR processes are based on increasing the capillary number (which reduces the interfacial tension between two phases) and/or lowering the mobility ratio between two phases by increasing the viscosity of the CO_2. Once the interfacial tension is lowered between oil and CO_2 to zero, as it is at higher pressures, the two phases become completely miscible. This latter property changes the properties of the oil and allows production of oil reservoir where even the primary residual oil can be produced. Recently, interest has been expressed in highly saturated residual oil zones (ROZs), which cannot be produced on primary or water floods but can be produced by CO_2 EOR, and in tight oils where low permeabilities (microdarcy range) restrict production to that allowed by massive induced hydrofracking. Currently, more than 90% of the oil is left behind after primary production that might be able to be produced by CO_2 EOR with substantial CO_2 storage occurring in the process.

21.4.2 Depleted Gas Reservoirs Storage Including Hydrates

Gas reservoirs are more abundant and have greater capacity than oil reservoirs. Gas reservoirs are produced to much lower pressures before they are depleted, and thus they generally can accept CO_2 at higher supply rates than oil reservoirs. The CO_2 is generally miscible with the natural gas (e.g., methane). Enhanced production of natural gas by injection of CO_2 has not been demonstrated commercially. Gas reservoirs generally have fewer well penetrations than oil reservoirs and are more attractive in that respect. If CO_2 EOR is not contemplated for an oil reservoir prior to pure storage, then gas reservoirs are more attractive for storage.

Gas hydrates are a future opportunity, particularly beneath the ocean floor where temperatures are low and pressures are high. The formation of CO_2 hydrates by combining with water in the pore space can cause plugging, reducing the permeability, and would be a self-sealing process that would minimize the amount of CO_2 stored. An EOR process similar to that for oil reservoirs can be envisioned where CO_2 is injected into a methane hydrate reservoir and forms a CO_2 hydrate releasing the methane that can be produced (Uddin et al. 2008). The technology has several hurdles to overcome because the methane hydrates are found to form in the pore space of the sediments, blocking the pores and making access for the CO_2 difficult. First, primary production of methane from methane hydrates has to be produced in a commercial setting.

21.4.3 Enhanced Coalbed Methane, Methanogenesis, and Depleted Coalbed Methane Storage

The production of coalbed methane (CBM) is an established commercial industry throughout the world. Because the methane is adsorbed onto the coal matrix but the flow of the fluids (including methane) takes place in fractures, the production of CBM is different to conventional gas reservoirs. The methane has to desorb from the coal in the matrix and diffuse to the coal cleats where Darcy flow takes place. Thus, the spacing of the fractures or cleats and the size of their opening is paramount to determining the success of commercial operation. When CO_2 is injected into the coal seams, it is more strongly adsorbing compared to methane and will displace the methane off of the coal and allow it to flow into the cleats where it can be produced. The stripping process is very efficient utilizing the same principles that allows gases to be separated in a gas chromatograph column and producing very sharp fronts so that most of the methane would be swept out of the reservoir before CO_2 breaks through to the production well. The problem is that sorption of the CO_2 onto the coal causes it to swell, which lowers the permeability of the cleats and in a worst-case scenario plugs any flow into and out of the CBM reservoir. Sweet spots can be found where the interplay of pressure and the concentration of CO_2 will maintain the permeability of the cleats (e.g., by injecting a CO_2/N_2 mixture). However, currently a commercial demonstration (Wong et al. 2010) does not exist that has been successful, although piloting has been carried out (Gunter et al. 2008).

A follow-up strategy has been proposed based on converting the stored CO_2 to methane in a depleted CBM reservoir by methanogenesis. Natural bacteria consortia that are present in the coal and have previously used the coal as an energy source producing a methane product as part of their waste (an analog to the methane production seen in landfills today) may be able to react with the CO_2 and regenerate the methane (Budwill et al. 2005), provided the right nutrients are available. This could be visualized as a cyclic process with a 4-step cycle: primary production of CBM, injection of CO_2 for enhanced recovery of methane, injection of nutrients, and regeneration of the methane. This has only been tested at the lab scale.

21.4.4 Saline Aquifer Storage

Sedimentary basins are composed of a mixture of consolidated sediments and their associated pore space (which is occupied predominantly by formation water) and are found distributed evenly throughout all continents of the world (Hitchon et al. 1999). Approximately 10% of the volume of a sedimentary basin is formation water, which forms the large saline aquifers that define the flow dynamics in the basin. Figure 21.12 shows the sedimentary basins in Canada. Approximately half of Canada is formed of sedimentary basins. The most mature basins are the Alberta and Williston basins, which together form the continental Canadian Western Sedimentary basin and contain most of the oil and gas production in Canada. Basins off the coast of Nova Scotia and Newfoundland are in the early stages of development. Other sedimentary basins in Canada, particularly those in the north, have yet to be developed. Along with oil and gas development comes detailed knowledge of the geology of the sediments and hydrodynamics of the saline aquifers of the basin, exactly the information needed to identify opportunities for storage of CO_2 in the subsurface in carbonate, sandstone, and basalt aquifers. Although the focus in the past has been on the exploration of the world's sedimentary basins for oil and gas, by far the predominant fluid in all these basins is formation water. Consequently, the capacity of saline aquifers to store CO_2 is much greater than oil and gas reservoirs.

Figure 21.13 is a schematic showing the distribution of regional aquifers in the Alberta Sedimentary basin. In fact, saline aquifers are the conduits through which oil and gas have been transported in the past to finally reside in stratigraphic and structural traps (Gunter et al. 2004). This is shown in Figure 21.13 by the abundant reefs in the Middle and Upper Devonian carbonates (shown as the local stratigraphic highs), which have formed stratigraphic traps for the oil and gas. The flow of these regional aquifers is slow measured in centimeters per year. The regional seals of these aquifers are often the low permeability shales that abound in the basin. Consequently, CO_2 injected in the deep part of the aquifers would have to travel several hundred kilometers to reach the surface over a time frame of millions of years. This is known as hydrodynamic trapping even though continuous movement of the CO_2 plume would take place. In this longer time frame, other forms of trapping that are more secure and at both a smaller microscopic (geochemical) and a larger geological (stratigraphic and structural) scale could take place and immobilize the CO_2 (Hitchon 1996).

However, saline aquifers represent the most expensive means of storage of CO_2 because there are no economic offsets such as the production of additional oil and gas, and as most of the saline aquifers are close to their initial pressures in the subsurface, storage occurs at higher pressures when compared to depleted oil and gas fields. Counterarguments against the use of depleted oil and gas fields for CO_2 storage are their higher risk of leakage of CO_2 because of the greater number of well penetrations that represent potential weak points in the seals.

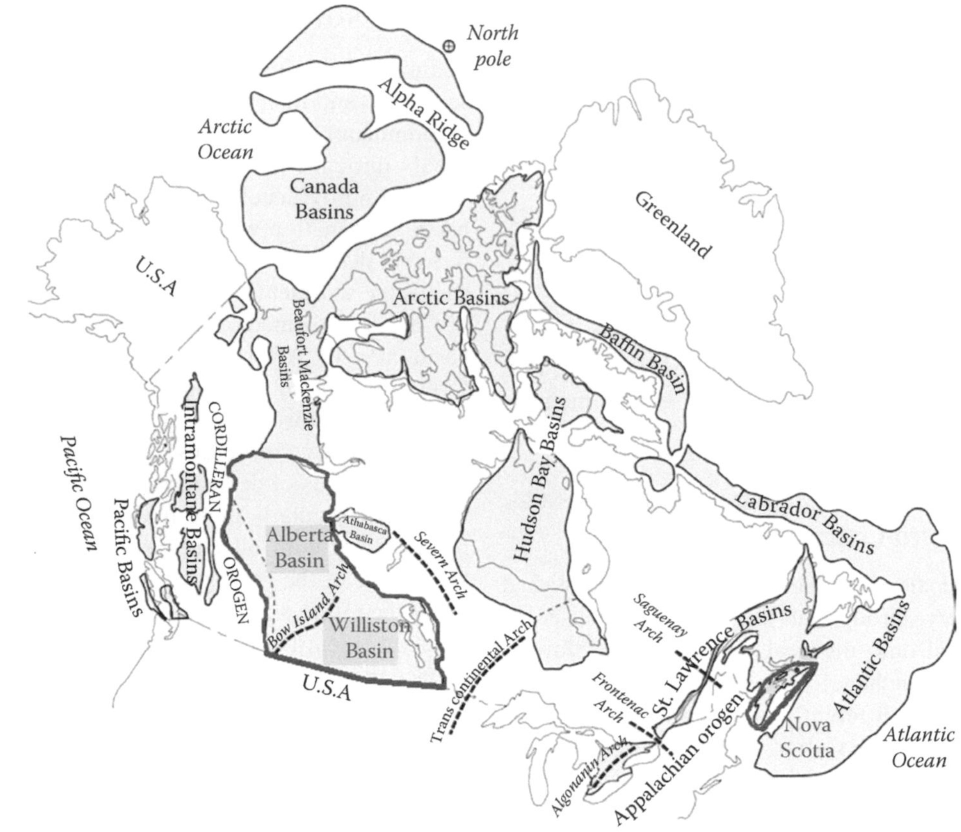

FIGURE 21.12 Map of Canada showing its sedimentary basins suitable for CO_2 storage. (With kind permission from Springer Science+Business Media: *Environ. Geol.*, Screening and ranking of sedimentary basins for sequestration of CO_2 in geological media in response to climate change, 44, 2003, 277–289, Bachu, S.)

The three most technical components to select a subsurface geological site for CO_2 storage are sufficient capacity to store the intended CO_2 volume, sufficient injectivity to receive the CO_2 at the supply rate, and containment to avoid CO_2 leakage due to incompetent seals or leaky wells. Injectivity depends on permeability, pressure distribution in the reservoir, thickness of the reservoir, and not exceeding fracture pressure. Hydrodynamic trapping and lateral containment in saline aquifers have already been discussed. Vertical containment is dependent on the competency of the

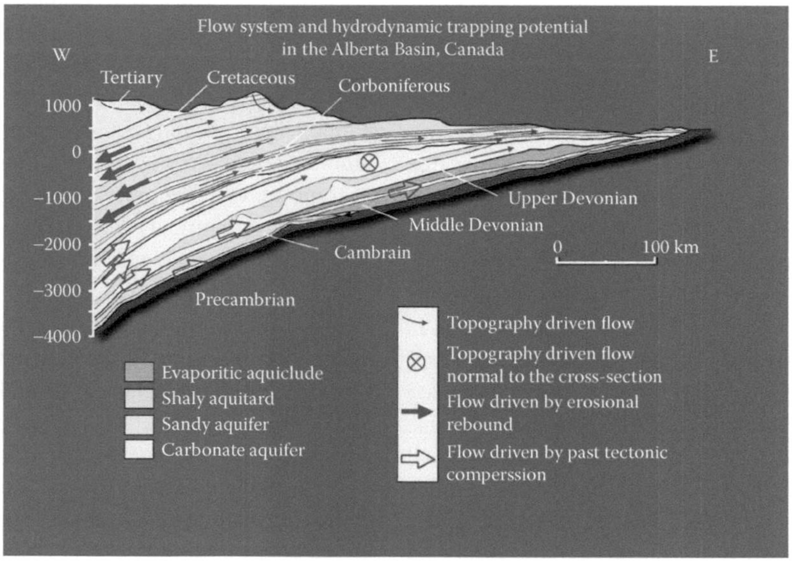

FIGURE 21.13 A schematic cross section of the central-southern part (including Alberta Basin) showing hydrodynamics of the regional saline aquifers. (Reprinted from Bachu, S., *Bull. Can. Petrol. Geol.*, 47(4), 455–474, 1999. With permission.)

seal and penetrations by faults, fracturing, and wellbores. Capacity depends on being able to access the pore space in a reservoir. A theoretical capacity for a geological storage unit can be calculated based on knowledge of the pore space volume and is a starting point. A more realistic capacity volume estimate can be made using technical considerations such as the physical attributes dictated by the geological model of the reservoir (e.g., geological heterogeneity, fluid saturations, injectivity, and containment), which is known as the effective capacity. Finally, a practical capacity is determined when legal, regulatory, and economics are considered (Bachu et al. 2007).

21.4.5 Storage in Mineral Mines

Mineralization of CO_2 for the storage in mineral mines is getting more attention as a CCS method to store CO_2 at the surface of the lithosphere. Mineral carbonation is the fixation of CO_2 as stable carbon in minerals, for example, calcite ($CaCO_3$), dolomite ($Ca_{0.5}Mg_{0.5}CO_3$), magnesite ($MgCO_3$), and siderite ($FeCO_3$) (Metz et al. 2005). As previously mentioned, in the mineral carbonation process, CO_2 is chemically reacted with calcium- and/or magnesium-containing minerals to form stable carbonate materials. These carbonates are thermodynamically more stable, and therefore, it is considered as one of the most attractive technologies for permanent and safe storage of CO_2 at the surface in mineral mines. It can be applied in situ or ex situ. CO_2 storage in mineral mines is favored with magnesium-based silicates $xMgO{\cdot}ySiO_2{\cdot}zH_2O$ (e.g., serpentinites) because they are available in huge amounts worldwide and are capable of binding all fossil fuel–bound carbon (Lackner and Ziock 2000).

A schematic conceptual diagram of CO_2 sequestration in minerals is presented in Figure 21.14. In the case of silicate rocks, CO_2 storage in mineral mines can be carried out in either ex situ in a chemical processing plant after mining and pre-treating the silicates or in situ by injecting CO_2 in geological formations (mainly aquifers). CO_2 from the power plant is transported to a carbonation reactor and reacted with minerals at appropriate conditions until the desired degree of carbonation is reached. Then carbonated minerals and residues in the aqueous CO_2 solution are separated. The CO_2 is recycled, useful materials are collected, and carbonated materials and residue are transported to the mine site (Olajire 2013). The economics and energy consumption of this process are tentative as it involves a huge mining operation where the crushed rock is reacted with CO_2 at elevated temperatures to form carbonates. Possibly some valuable by-products may be formed from this process as these rock types often contain small amounts of valuable metals to offset the storage costs.

It should be noted that the storage in minerals is more permanent and has a very large capacity as it also takes place in deep geological formations (carbonates, sandstones, and basalts) in contact with aquifers as well as on the surface in mineral mines.

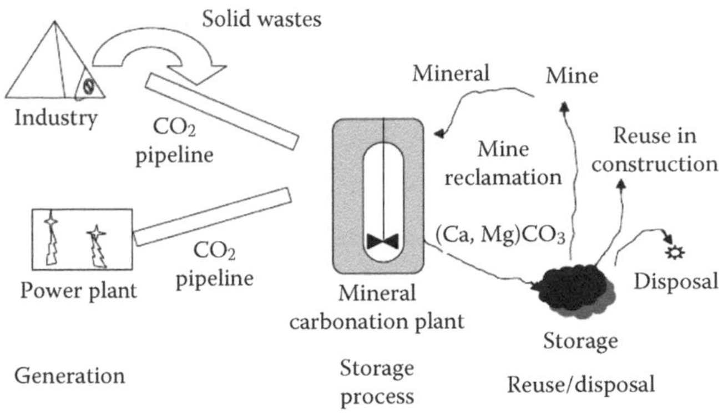

FIGURE 21.14 Conceptual diagram of CO_2 sequestration in mineral. (Reprinted from *J. Petrol. Sci. Eng.*, 109, Olajire, A.A., A review of mineral carbonation technology in sequestration of CO_2, 364–392, Copyright 2013, with permission from Elsevier.)

21.5 CURRENT STATUS OF CCS—COAL RELATED

21.5.1 Large-Scale Integrated Projects

The Global CCS Institute (GCCSI) provides an annual summary of large-scale integrated projects (LSIPs) worldwide. Overall in 2014, there are 9 projects in the development/execute stage and 13 projects in the operation stage. Of these, the Boundary Dam coal-fired power station in Saskatchewan is the first LSIP that applies CCS in a power station at large scale, although others are expected to come on line over the next few years (Figure 21.15). Other operating LSIPs are found in sectors where CO_2 is routinely separated from other gases such as synthetic natural gas production, natural gas processing, hydrogen production, and fertilizer production. Geographically, most projects are concentrated in North America as are the volumes of CO_2 stored and are tacked on to EOR projects as there is a thriving sales market for CO_2 for EOR. It can be seen in Figure 21.15 that the three operating pure CO_2 storage projects prior to 2014 are in either saline aquifers (Sleipner and Snohvit) or a depleted gas reservoir (In Salah). Those in saline aquifers were driven by a Norwegian tax on offshore CO_2 emissions of \$50/ton of CO_2 vented. However, the GCCSI expects that over the next 5 years, the ratio of new operating pure storage to EOR LSIPs will be close to 1 based on Figure 21.15.

21.5.1.1 Sask Power Boundary Dam—Post-Combustion CCS

The Boundary Dam coal-fired power station in Saskatchewan, Canada, is the first commercial scale post-combustion coal-fired CCS project, which was started in September 2014 (Stéphenne 2014). The coal-fired power plant capacity is 139 MW gross, 110 MW net, retrofit. At full capacity, the plant captures over 1 million metric tons of CO_2/yr, 90% CO_2 captured from the 139 MW coal-fired units. The captured CO_2 emissions are compressed and transported by pipelines to Cenovus Energy where CO_2 is used for EOR activities in the Weyburn oil field (presented in section 21.5.1.3). Transportation will be via

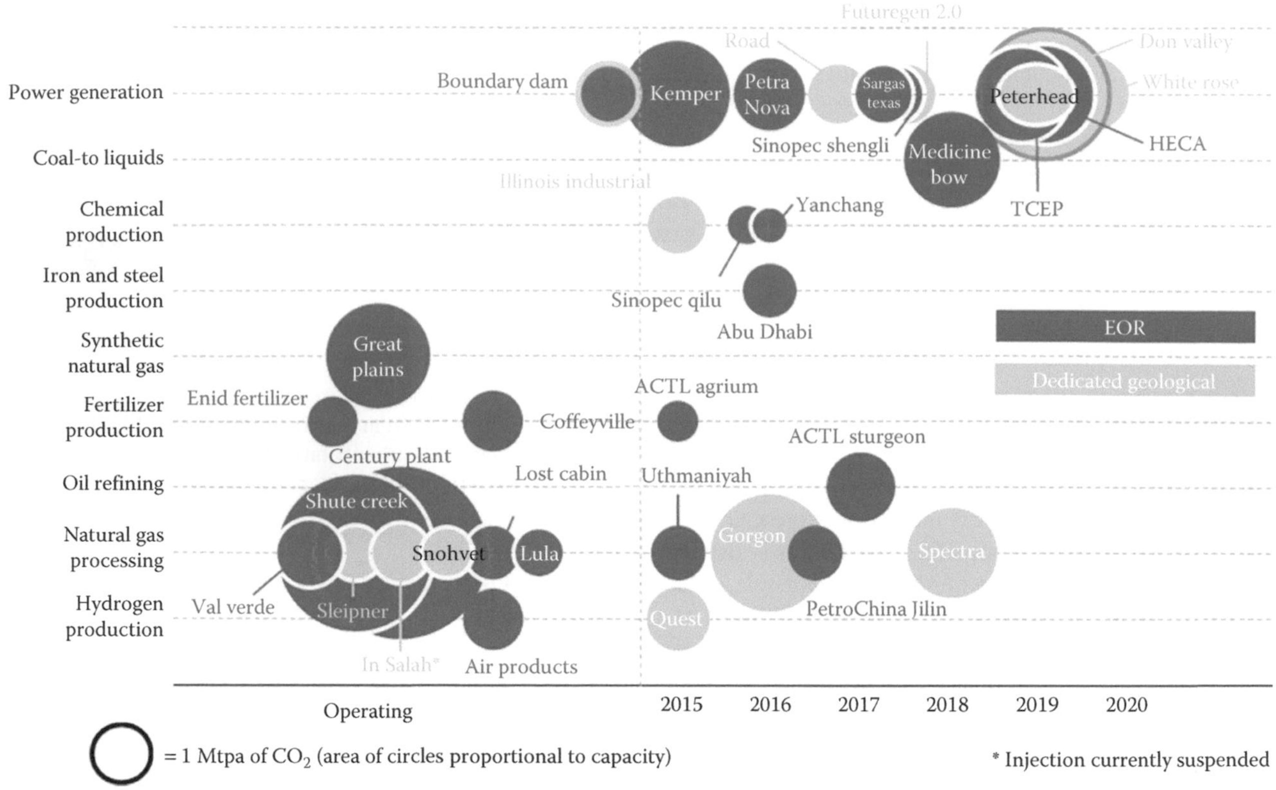

FIGURE 21.15 Expected operation dates for large-scale CCS projects. (Reprinted from GCCSI, *The Global Status of CCS: 2104 [Summary Report]*, Global CCS Institute, Melbourne, Australia, 14 pages, Copyright 2014. With permission.)

66 km pipeline. Excess CO_2 is to be used at the Aquistore, saline aquifer storage project 2 km away.

The schematic diagram of Cansolv process in the SaskPower ICCS Project is presented in Figure 21.16. The process includes the selective heat integration with Shell Cansolv's innovative combined SO_2 and CO_2 capture system, which reduces energy requirements associated with carbon capture. The technology uses regenerable amines to capture SO_2 and CO_2, and no direct waste by-products are generated.

Successful implementation of this project may reduce emissions of CO_2 by 1 Mt/yr. The successful operation of this project will be a milestone for the fossil fuel power industry worldwide, demonstrating and realizing large-scale CO_2 capture and storage.

21.5.1.2 Kemper County Energy Facility (Formerly Kemper County IGCC Project) (http://www.mississippipower.com/; www.globalccsinstitute.com; www.zeroco2.no; www.sequestration.mit.edu)

The construction of Kemper County Energy Facility (Figure 21.17) started in 2011 and is scheduled to come in commercial operation in 2016. The capital cost of the project as of January 2015 is \$6.1 billion. The lignite-fired plant located on Southwestern Kemper County, Mississippi, has the capacity of 582 MW. The IGCC plant consists of two major components: lignite gasification (including CO_2 capture) and combined-cycle power generation. Pre-combustion requires Transport Integrated Gasification (TRIG) technology to be used for capturing 3.5 Mt of CO_2/yr. The carbon capture and sequestration in the plant aims to reduce at least 65% of the facility's CO_2 emissions. Captured CO_2 will then be transported by pipeline for use on EOR projects to be used for EOR.

The facility is utilizing coal gasification technology called TRIG, which is a coal gasification process that can utilize low rank coals, including lignite, which was developed jointly by Southern Company and KBR in partnership with the U.S. Department of Energy. The gasification systems consist primarily of lignite handling, gasification, and syngas cleaning. At its full capacity, the gasifier would convert 12,500 tons/day of lignite to produce syngas. The CO_2 will be captured utilizing the physical solvent process (Selexol), which can reduce CO_2 emissions by up to 67% by removing carbon from the syngas during the gasification process. Syngas will be cleaned to remove particulate, S, Hg, and NO_x before using as fuel in the combined-cycle power-generating units. Thet facility will also produce about 135,000 tons/yr of H_2SO_4 and about 20,000 tons/yr of NH_3.

21.5.1.3 Development of a CCUS Commercial EOR Project: Weyburn Project (Cenovus)

Great Plains Synfuels Plant, North Dakota, is the only commercial-scale coal gasification plant in the United States that manufactures natural gas. The plant started CO_2 sequestration

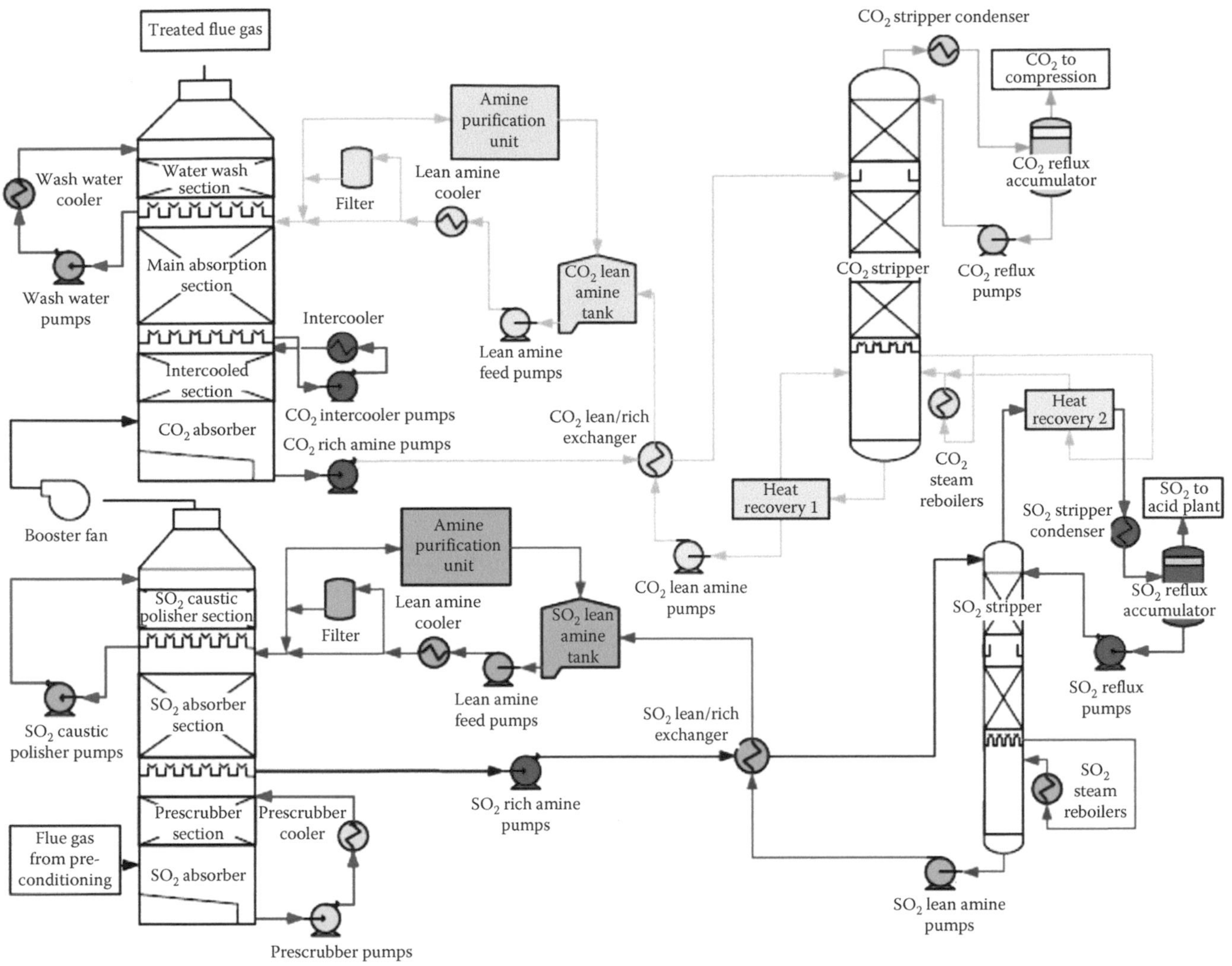

FIGURE 21.16 Process configuration of the SaskPower BD3 CCS Project. (Reprinted from *Start-Up of World's First Commercial Post-Combustion Coal Fired CCS Project: Contribution of Shell Cansolv to SaskPower Boundary Dam ICCS Project,* Stéphenne, Energy Procedia 00 (2013) 000–000, Copy right 2014, with permission from Elsevier.)

in 2000, and the project continues to inject around three Mt/yr of CO_2. Dakota Gas (subsidiary of Basin Electric Power Cooperative) captures and sells CO_2 produced at the plant to two places—Cenovus, operator of the Weyburn oil field, and Apache Energy, operator of the Midale field—and transports it through a 205 mile pipeline to Saskatchewan, Canada, to be used for EOR. Dakota Gas currently captures about three Mt of CO_2/yr—around 8000 tons a day (www.zeroco2.no).

The Weyburn field occupying an area of approximately 25 by 25 km contains approximately 1.4 billion barrels of oil and is similar in size to the Redwater oil pool and has evolved through primary and water flood production since 1955 with both vertical and horizontal infill wells drilled starting in 1987 and has been continuously CO_2 flooded since 2000. The geology of the pool is contained in the tilted Mississippian Midale Beds of the Charles formation consisting of a lower permeable limestone *Vuggy* zone overlain by the tighter dolostone *Marly* zone in turn capped by the impermeable Midale Evaporite and underlain by the Frobisher Evaporite. The Vuggy unit was formed mainly in a marine lagoonal environment affected by carbonate shoal development and is 10–22 m thick with permeabilities averaging 20 md with approximately 15% porosity. The Marly unit, 0–10 m thick, has higher porosity (averages 26%) but lower permeability and is the target of the infill horizontal wells and the CO_2 flood because it has been largely unswept by the water flood. The reservoir is fractured and is truncated in the north by the Mesozoic unconformity, which forms the trap for the oil pool.

The reservoir is approximately at 1450 m depth at 63°C with an initial pressure of 14 MPa, varying between 8 and 19 MPa during water flooding, and during CO_2 flooding exceeding the MMP of 15–17 MPa. Current injection rates are approximately 5 Mt of CO_2 per year with half of that being recycled CO_2. Oil production rates are 28,000 barrels per day with 18,000 barrels being incremental (Hitchon 2012). To date, about 20 Mt of CO_2 have been stored at Weyburn, which is approximately 5 Mt more than that based on earlier predictions (Figure 21.18), with the final CO_2 stored during EOR operations is also larger, expected to be 30 Mt. If the supercritical CO_2 plume rapidly equilibrates with the water and the oil in the reservoir, only

FIGURE 21.17 Kemper County Energy Facility under construction at a glance. (From www.mississippipower.com.)

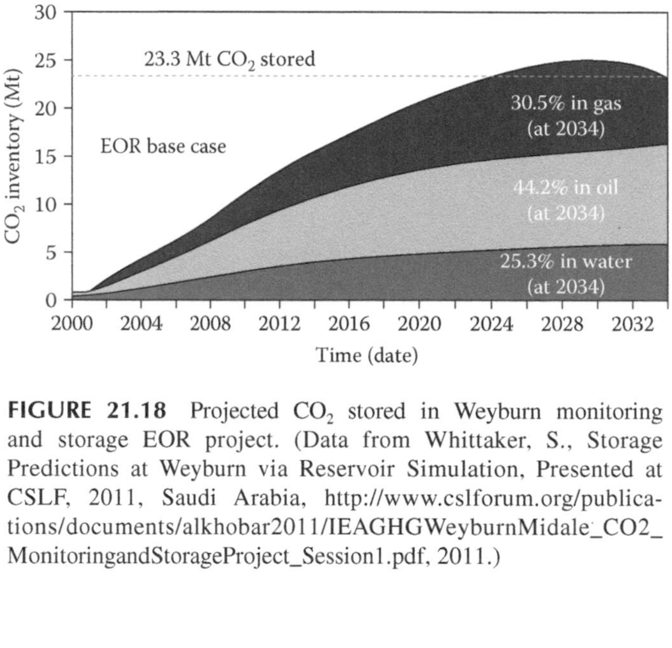

FIGURE 21.18 Projected CO_2 stored in Weyburn monitoring and storage EOR project. (Data from Whittaker, S., Storage Predictions at Weyburn via Reservoir Simulation, Presented at CSLF, 2011, Saudi Arabia, http://www.cslforum.org/publications/documents/alkhobar2011/IEAGHGWeyburnMidale_CO2_MonitoringandStorageProject_Session1.pdf, 2011.)

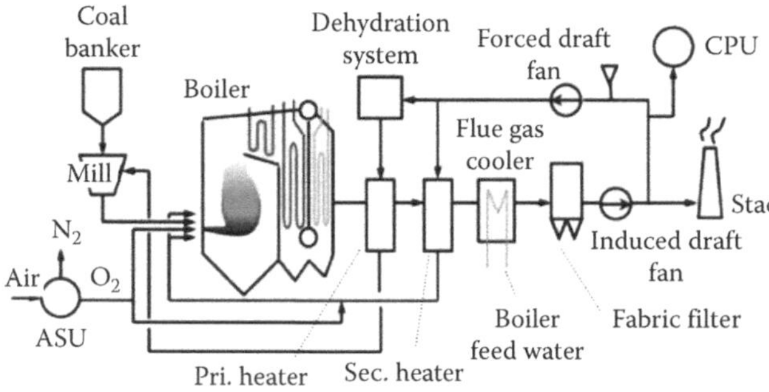

FIGURE 21.19 Callide oxy-fuel process (Callide A boiler) (ASU—Air Separation Unit). (Reprinted from *Energy Proc.*, 63, Komaki, A. et al., Operation experiences of oxyfuel power plant in callide oxyfuel project, 490–496, Copyright 2014, with permission from Elsevier.)

30% of the stored CO_2 will remain in the gas phase with the bulk of the CO_2 dissolved in the oil (44%) and the remainder dissolved in the water (25%). An additional 25 Mt of CO_2 could be stored by maintaining injection after cessation of EOR operations for storage purposes only and even more CO_2 could be stored if the CO_2 operations were extended into the whole of the oil field (Hitchon 2012).

21.5.1.4 Callide—Oxy-Fuel Combustion and Carbon Storage Demonstration Site (www.callideoxyfuel.com; www.globalccsinstitute.com)

The Callide oxy-fuel project, a first of its kind, at central Queensland, Australia, is a carbon capture storage demonstration site of an existing coal-fired power station. The project budget is nominally \$240 million. This is the 30 MW retrofit of CS Energy by IHI. The facility has 2 × 330 t per day ASUs and a 75 t per day CO_2 capture plant. Figure 21.19 shows the schematic of the Callide oxy-fuel process including ASU, CO_2 purification unit, and gas recirculation. The Callide oxy-fuel boiler has three mills to pulverize coal, two of which are required for normal full-load operation. The coal is dried and transported by recycled flue gas. O_2 from the ASUs is mixed with recycled flue gas after the secondary air heater, and a part of O_2 can be supplied to the burner flame area directly.

The plant was started in 2012 and as of March 2013 had achieved nominally 5500 h of operation in oxy-combustion mode and 2500 h of the CO_2 capture plant. The project has demonstrated more than 85% CO_2 capture rates from the flue-gas stream, producing a high-quality CO_2 suitable for geological storage. Removal of toxic gaseous emissions (such as SO_x, NO_x, particulates, and trace elements) from the flue-gas stream has been demonstrated.

21.5.1.5 Development of a Commercial Saline Aquifer Project—Quest Project (Shell)

The Quest Project aims to capture CO_2 produced at the Scotford Upgrader and to transport, compress, and inject the CO_2 for permanent storage at a saline formation in Alberta, Canada. It is proposed to remove CO_2 from the process gas streams of the three hydrogen manufacturing units. A commercially proven activated amine process is to be used to produce at least 95% CO_2 purity. The CO_2 will then be compressed and transported through a 12-inch diameter pipeline to a location approximately 80 km north of the Scotford Upgrader (Shell 2012).

Shell is developing a saline aquifer storage project in central Alberta where approximately 1 Mt of CO_2 will be stored annually in the basal Cambrian aquifer (Shell 2011). The stratigraphic section is shown in Figure 21.20. The Basal Aquifer site was chosen because of the few well penetrations (currently less than 10 well penetrations in the area of interest for the storage complex because there are no oil or gas reservoirs) compared to the Redwater Reef shown at D on Figure 21.20, where there are over 1000 wells completed in the reef (as it is the third biggest oil reservoir in Canada). Both reservoirs are targets for CO_2 storage although the Redwater Reef is the site of an EOR pilot owned by ARC Resources. Theoretical capacity of the Redwater Reef is 1 Gt of CO_2 (Gunter et al. 2009) immobilized by a stratigraphic trap, although it is connected to the regional Cooking Lake Aquifer at its bottom. As the Basal Aquifer is also regional in extent, its capacity is much larger than the Redwater Reef, but it is open to flow and relies on hydrodynamic trapping for the CO_2. The defined area of interest for the storage complex is 40 townships (60 × 70 km). It can hold 25 Mt of CO_2 within 3 km of the injection well. However, the elevated pressure zone formed by the CO_2 injection is much larger but is confined within the area of interest. Sufficient injectivity (20–500 md permeability) has been demonstrated by water injection. Sufficient capacity has been calculated based on logs (fine to coarse-grained sandstone sheet deposit, porosity ranges from 10% to 19%, 35–46 m thick, net to gross of 0.76–0.97) and static and dynamic transport models. The pressure (20.5 MPa) and temperature (60°C) define a density of the CO_2 plume of 0.73 g/cc. The Basal Aquifer has a primary seal in the middle Cambrian shale, and above it, secondary seals are formed by the Lotsberg salts, and it lies unconformably on the pre-Cambrian basement. One of the comprehensive monitoring plans to date for any commercial geological storage demonstration has been developed using a

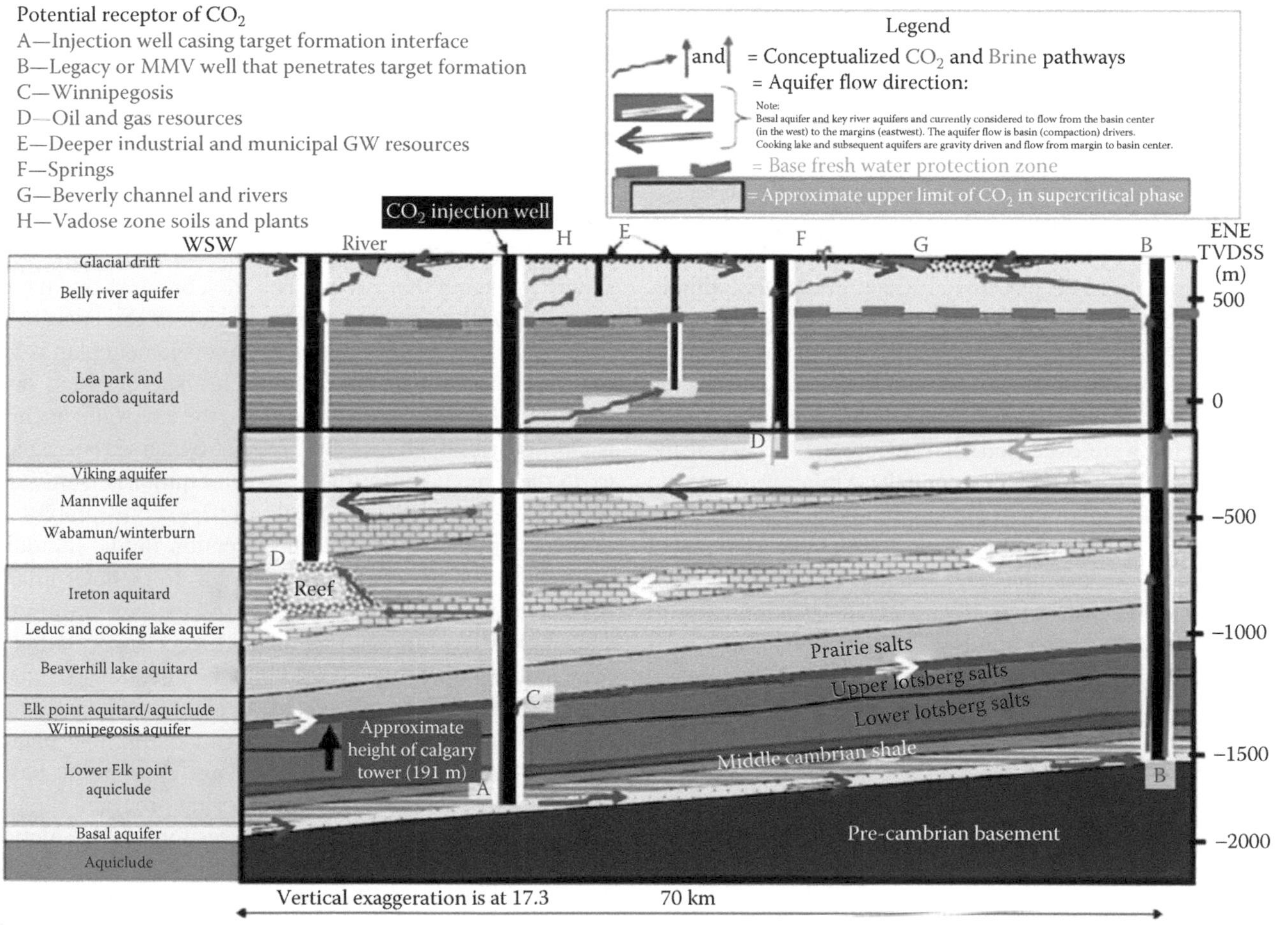

FIGURE 21.20 Geological storage: Quest CO_2 paths. (From Shell, Shell public documents submitted to Alberta Government, http://www.shell.ca/en/aboutshell/our-business-tpkg/upstream/oil-sands/quest/about-quest.html, 2011.)

combination of InSAR, time-lapse seismic, observation wells for temperature, pressure, water composition, and microsesmic, and at the surface line of sight CO_2 flux monitoring and soil chemistry. This has been integrated into the project's risk management plan.

21.5.2 Small-Scale Projects

Small-scale projects are geological storage pilots that inject less than 100,000 tons of CO_2. In 2013, 44 projects were identified globally with 18 in deep saline aquifers, 11 in ECBM, 9 in EOR, 4 in depleted oil and gas fields, and 2 in basalts with most of them located in North America and Australia (Cook et al. 2013). Most of the projects inject less than 15,000 tons as they are not intended to proceed to commerciality in this early period of CCS development and the CO_2 sources could be small and temporary where proven capture technologies are used. Generally, the pilots are designed as a research project to test the viability of an improved or new technology for geological storage and focus on monitoring and transport modeling. However, the need still exists for pilots preceding commercial demonstrations to lower the risks. A number of monitoring techniques have been tested in these pilots including 2D and 3D surface seismic, airborne EM and spectral imaging, boomer/sparker profiling, bubble stream chemistry, bubble stream detection, cross-hole EM, ERT and seismic, downhole fluid chemistry and pressure and temperature, ecosystem studies, Eddy covariance, ESP, fluid geochemistry, geophysical logs, ground-penetrating radar, high-resolution acoustic imaging, IR diode lasers, surface EM and ERT, long-term downhole pH, microsesmic, multibeam echo sounding, multicomponent surface seismic, non-dispersive IR gas analysis, satellite interferometry, seabottom EM and gas sampling, seawater chemistry, sidescan sonor, single-well EM, soil gas concentrations, surface gas flux, surface gravimetry, tiltmeters, tracers, VSP, and well gravimetry. Monitoring and transport modeling were used in an iterative fashion over time to build a risk assessment framework and identify risks, develop risk responses, apply risk response when trigger monitoring measurements were exceeded, and review the success of the risk response. Also, programs around these pilots were used to feed into legal/regulatory, socio/environmental, and government issues. Capacity building was also a strong component of most pilots. An example of a roadmap constructed for a pilot in Indonesia is shown in Figure 21.21, including the soft issues addressed and the go/no-go decision points (i.e., the stage gates).

21.6 CHALLENGES IN CCS

21.6.1 A Geological Storage Standard

Focus initially has been to define a standard that applies to natural subsurface reservoirs without production of hydrocarbons. It defines the necessary steps in identifying and utilizing the storage reservoir. Storage is targeted in reservoirs at depths generally more than 1000 m due to the density behavior of CO_2 with respect to the geothermal and geobarometric gradients in sedimentary basins. To better understand the steps that have to be taken to select and utilize a geological storage reservoir in a sedimentary basin, a number of studies have been done. They all have much in common, starting with site screening and selection (identify stage) followed by detailed site characterization (select stage) followed by design (design stage) and then development (execute stage) leading into the operational period (operate stage) and ending with the post-injection or abandon period (consisting of closure and post-closure stages), after injection of CO_2 has ceased. A typical example is provided in the Canadian Standards Association report Z741-12 *Geological Storage of Carbon Dioxide* (2012).

In site screening and selection or identify stage in a sedimentary basin, the basin is analyzed, and geographical, geological, and hydrogeological criteria are used to set the boundaries for potential sites in consultation with stakeholders and in accordance with applicable laws and regulations. Estimates of capacity and injectivity are made based on available data. Containment and contamination risks are assessed from knowledge of active faults, existing wells, and potential for intersection of other valuable resources. The availability of the reservoir, willingness of the operator, and economics will play a major role in the selection. A good example of this approach in developing countries is reported by the Asian Development Bank (2013) in their report on *Prospects for CCS in Southeast Asia*.

Once the site screening and selection has been completed for a sedimentary basin, detailed site characterization (select stage) takes place. Performance assessment criteria are set by which the development of the storage project can be evaluated. Data is collected to ensure that the site has sufficient capacity to accept the anticipated final volume of CO_2 at the desired supply rates and that the site has the containment characteristics that will ensure effective retention of the injected CO_2. Risk assessment is done and a risk management plan is developed to be confident that the selected site does not impose unacceptable risks to other resources, the environment, human health, project operators, and project owners. Data collected to do these analyses for the reservoir/aquifer is from seismic surveys, well tests, geophysical wireline, core samples, fluid samples, production fluids, and injection fluids. In addition, the integrity of the confining strata has to be determined for both primary and secondary seals and the penetrating wells. This data is used to build geological, geochemical, and geomechanical models, which feed into the flow transport models for the reservoir, any associated regional aquifer present, the seals, and the associated wells (both existing and planned). All this information is used to collect appropriate baseline information to be used in the monitoring program.

In design (design stage), procedures for health, safety, and environment protection programs are established. Protocols for integrated functioning of the project operator and subcontractors are made. Appropriate materials and methods are selected for development of the site including wellsite design, drilling operations procedures, well infrastructure,

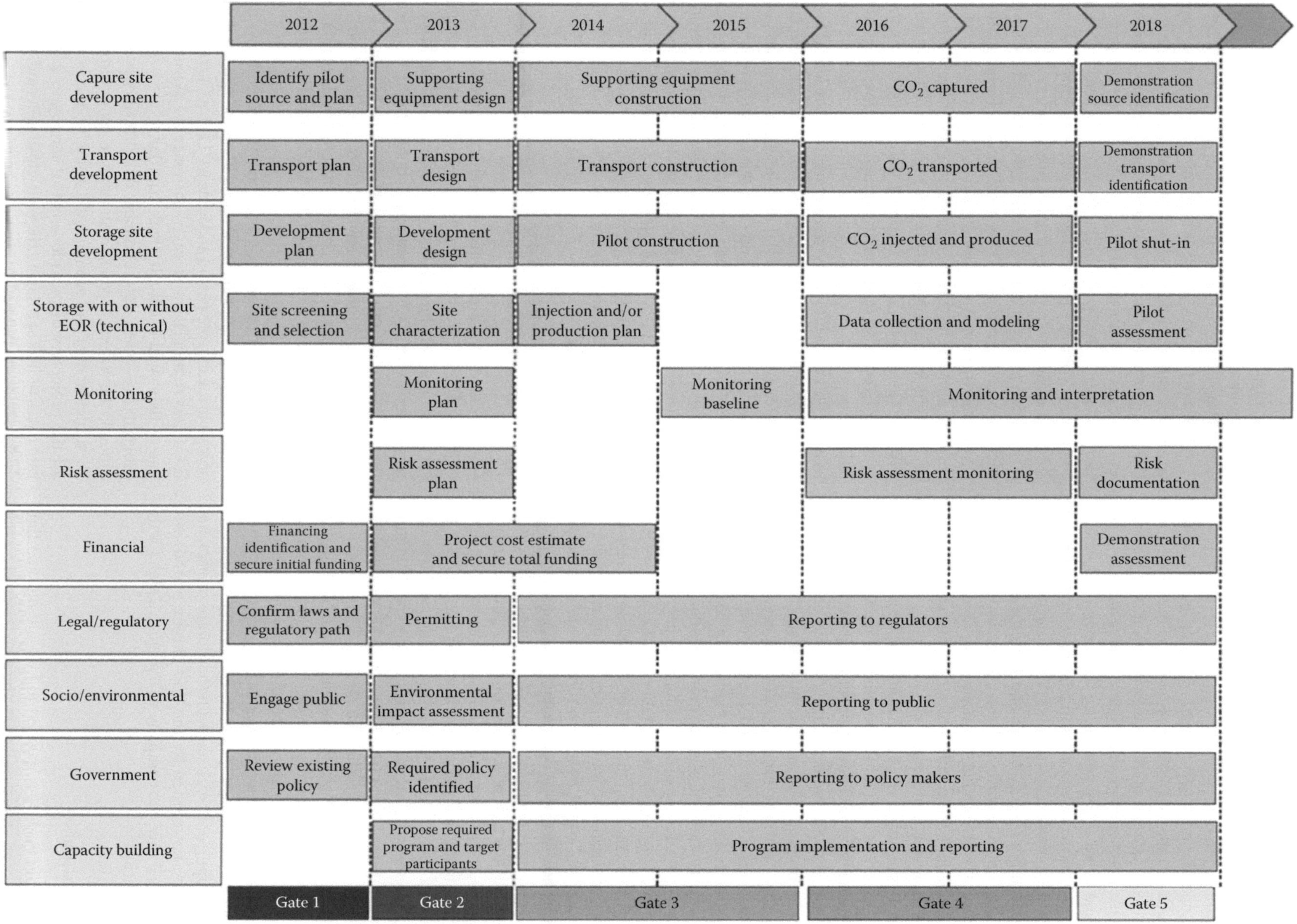

FIGURE 21.21 Roadmap constructed for a hypothetical Indonesian storage pilot. (Reprinted from Asian Development Bank, *Prospects for Carbon Capture and Storage in Southeast Asia, a Regional Analysis*, Regional Technical Assistance program (RETA) #7575—for Determining the Potential for Carbon Capture and Storage in Southeast Asia, Asian Development Bank, Philippines, 167 pages, Copyright 2013. With permission.)

facility construction, and monitoring hardware installation. Final operational procedures and maintenance programs are developed for monitoring and continuous improvement of the complete integrated storage system over the project life cycle. In development (development stage), wells are drilled, and surface facilities are built.

During operations (operate stage), injection of CO_2 at the supply rate is maintained, and movement of the CO_2 in the storage reservoir is monitored to be forewarned of any deviations from the planned injection scheme. These are evaluated, assessed, and acted on by comparing the monitoring data to the predictions of the reservoir fluid transport model—in the envelope of the risk management plan. This is an iterative process between the transport model and the monitoring data. As monitoring data becomes available, the transport model is improved to correctly predict the monitoring observations by changing the geological, geochemical, or geomechanical data that the transport model depends on.

After the storage objectives of the site have been met, injection ceases, but the CO_2 plume continues to move due to the pressure gradients formed during injection of the CO_2. These may take hundreds of years to disperse, and consequently, a monitoring program during the post-injection period is important. The post-injection period (abandon stage) is split into two sub periods: closure and post-closure. The reason for this is that as CO_2 storage is planned to be permanent, the risk always remains of leakage even a million years from now. In our society, tenure of companies is generally less than 100 years, so even though the liability of CO_2 leakage lasts much longer, the company may not be around to address the leakage. Governments are looking at assuming responsibility for the storage site in the post-closure period after appropriate due diligence has been completed. The requirements to obtain a *closure* certificate from the government will depend on showing that the risk of leakage of CO_2 from the storage site is very low over the long term. This will be accomplished by a closure monitoring period from which the data can be interpreted in a risk sense. In the closure period, archives and attendant systems will be established to ensure the future public availability of the project data and knowledge. The company will demonstrate that the storage complex has the appropriate monitoring systems in place and will specify criteria for well abandonment, well inspection, continued monitoring that meets regulatory requirements, and continuing the progressive reduction of uncertainties regarding plume fate. Decommissioning (or schedule for decommissioning), all surface equipment associated with the storage project, and wells (which are not needed for the post-closure period) will be completed. Finally, the company will prepare a plan for the post-closure period.

21.6.2 Monitoring

Monitoring technologies are needed to measure injected fluid volumes, flow rates, injection temperature and pressure, composition of injectant, spatial distribution of the CO_2 plume and of the elevated pressure zone associated with it, pressure within the storage complex, well integrity, leakage outside of the storage complex, integrity of the confining zone, extent of formation water displacement, pressure change in the aquifers above and below the storage units, potential induced seismicity including microseismic, geochemical changes, and contamination of competitive resources (Chalaturnyk and Gunter 2005).

There are two extreme types of monitoring based on sample size for the subsurface: point versus volume. Point monitoring is accomplished through the placement of observation wells and is a direct measurement (e.g., fluid composition, pressure, temperature) at that point in the formation where the well is completed. Volume measurements are less sensitive, are indirect, and measure a property that has to be interpreted in terms of a related property and is usually accomplished by surface surveys over time (e.g., 4D seismic, InSar). To compare the two methods, point measurements can be instantaneous and are very sensitive to any changes. However, if the well is not in the right location, the CO_2 plume may have bypassed the well and the event was not recorded. In contrast, comparison of seismic response over time will track the movement of the bulk of the plume over time, but because of its limited sensitivity, it will not yield early warning signals for leakage. The two methods are complimentary and in an ideal sense could be used together. For example, the placement of some observation wells could be delayed until the seismic predicted the movement of the plume and could be placed ahead of the moving front of the plume to verify the seismic interpretation. However, this is expensive and has to be integrated in the project budgets. In addition, drilling of an observation well into the plume introduces another potential weak point into the confining seals, and whether the reward exceeds the risk has to be assessed.

21.6.3 Policy/Regulations

Considerations for a legal and regulatory framework around CCS are classification of CO_2 in terms of being a pollutant, surface and subsurface rights for CO_2 transport and storage, legal liability of CCS operations and for stored CO_2, environmental protection, CO_2 transport, health and safety, EOR, and foreign direct investment for CCS credits. Policy and regulations will not be the same in different jurisdictions. An example of this can be found in the Asian Development Bank report (2013).

One issue that has already been mentioned is the requirements for the government to accept the company's handoff to complete the closure period and issue a *Closure Certificate*. As it is likely that once the storage sites enter the closure period, one of them will eventually leak. Similar to the *orphan well* program in Alberta, each company will probably have to post a bond, the value of which will have to be determined but be sufficient that the accumulated value of the bonds from all the projects will provide a pot of money to address the rare cases of leakage during the post-closure period. This also addresses *ownership of the pore space* where the CO_2 plume will reside. It makes sense that if the government is going to accept the ultimate liability for storage, it should own the pore

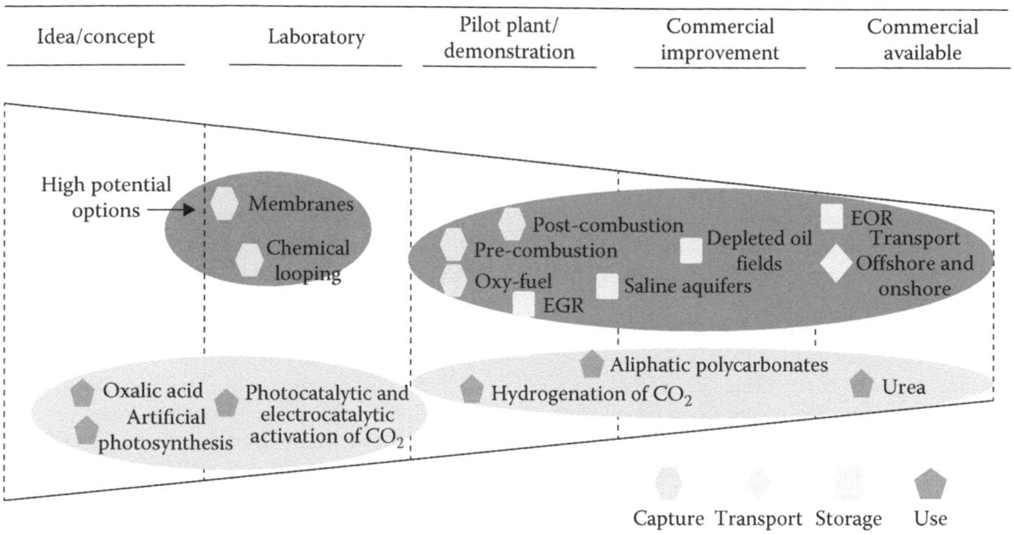

FIGURE 21.22 Innovation phases of technologies for capture, transport, storage, and use of CO_2. (Redrawn from Markewitz, P. et al., *Energy and Environ. Sci.*, 5, 7281–7305, 2012.)

space where the CO_2 resides. If it does this, it can also then be in a position to lease the pore space for CO_2 storage ensuring a stream of revenue to the government for CO_2 storage.

21.6.4 Public Perception

Negative public perception can *kill* any project. Particularly, because geological storage is easily related to oil and gas production and usage as one of the major sources of CO_2 emissions, any solution that proposes to solve the GHG venting by using oil and gas technology and storing the CO_2 in the same sedimentary basins that have produced oil and gas will be frowned upon by the public (who are not experts in the field of oil and gas). Public engagement is needed well in advance of any CCS project as public awareness of CCS is low. A robust communication and engagement strategy must be put in place locally with the communities where the project will take place, nationally with policy makers and industry, and broadly within scientific, social, and civic communities.

21.6.5 Commercializing New CCS Technologies

As previously discussed, the most expensive portion of CCS is the capture of CO_2 and is currently the most limiting factor in the commercialization of CCS. Various CO_2 capture technologies have been developed, and these are in different stages of innovation. The overview of current developmental stage of the various carbon captures, transport, use, and storage technologies is presented in Figure 21.22. The carbon capture technologies—post-combustion, pre-combustion, and oxy-fuel combustion—are in both demonstration and pilot plant phases. Certain technologies have been commercially implemented such as EOR (storage) and production of urea (use). Some technologies such as CO_2 membranes (capture) and artificial photosynthesis (capture) are in a very early stage of technical development or in the conceptualization, whereas chemical looping combustion—one of the high potential options—is currently in very early stage of technical development. It is to be noted here that CO_2 storage is for longer period; however, use of chemicals such as urea fertilizer may not reduce carbon emissions that decompose within several years releasing CO_2 back to the atmosphere. Even though many technologies are being developed and some are evolving for CCS, the real challenge is to commercialize the application of these technologies in industries in a cost-effective and environmental friendly manner.

REFERENCES

Asian Development Bank. 2013. *Prospects for Carbon Capture and Storage in Southeast Asia, a Regional Analysis*, Regional Technical Assistance program (RETA) #7575—for Determining the Potential for Carbon Capture and Storage in Southeast Asia, Asian Development Bank, Philippines, 167 pages.

Bachu, S. and Gunter, W.D. 2004. Acid gas injection in the Alberta basin: A CO_2 storage experience, In: Baines, S.J. and Worden, R.H. (Eds.), *Geological Storage of Carbon Dioxide*. Geological Society, London, Special Publications 233, 225–234.

Bachu, S., Bonijoli, D., Bradshaw, D., Burruss, R., Holloway, S., Christensen, N.P., and Maathiassen, O.P. 2007. CO_2 storage capacity estimation: Methodology and gaps. *International Journal of Greenhouse Gas Control*, 1, 430–443.

Bachu, S. 2003. Screening and ranking of sedimentary basins for sequestration of CO_2 in geological media in response to climate change. *Environmental Geology* 44, 277–289.

Bachu, S. December 1999. Flow systems in the Alberta Basin: Patterns, types and driving mechanisms. *Bulletin of Canadian Petroleum Geology* 47(4), 455–474.

Budwill, K., MarcBustin, M., Muehlenbachs, K., and Gunter, W.D. 2005. Characterization of a subsurface coalbed methanogenic culture and its role in coalbed methane recovery and CO_2 utilization. *Proceedings of the 7th International Conference on Greenhouse Gas Control Technologies: Volume II-Part II*, M. Wilson, T. Morris, J. Gale, and K. Thambimuthu (Eds.), Elsevier, the Netherlands, 2213–2216.

Canadian Standards Association. 2012. *Geological Storage of Carbon Dioxide; Standard Z741-12*, CSA Group, Ontario, Canada, 80 pages.

Chalaturnyk, R.J. and Gunter, W.D. 2005. Geological storage of CO_2: Time frames, monitoring and validation of technology, *Proceedings of the 7th International Conference on Greenhouse Gas Control Technologies, Volume I*, E.S. Rubin, D.W. Keith, and C.F. Gilboy (Eds.), Elsevier, the Netherlands, 623–631.

Cook, P., Causebrook, K., Michael, K., and Watson, M. 2013. *Test Injection Experience to Date and Best Practice*. IEAGHG report, United Kingdom, 231 pages.

DOE/NETL. December 2010. *Carbon Dioxide Capture and Storage RD&D Roadmap*, US Department of Energy.

Faltinson, J. and Gunter, W.D. 2010. *Canadian Unconventional Resources and International Petroleum Conference*, Calgary, Canada. CSUG/SPE paper #137730, 8 pages.

Global CCS Institute. 2014. *The Global Status of CCS: 2104 (Summary Report)*, Global CCS Institute, Melbourne, Australia, 14 pages.

Gunter, W.D. and Longworth, H. 2013. *Overcoming the Barriers to Commercial CO_2-EOR in Alberta, Canada*. Report prepared for Alberta Innovates—Energy and Environment Solutions, Edmonton, Canada, 90 pages.

Gunter, W.D., Bachu, S., and Benson, S. 2004. The role of hydrogeological and geochemical trapping in sedimentary basins for secure geological storage for carbon dioxide, In: Baines, S.J. and Worden, R.H. (Eds.), *Geological Storage of Carbon Dioxide*, Geological Society, London, Special Publications, 233, 129–145.

Gunter, W.D., Bachu, S., Buschkuehle, M., Michael, K., Ordorica-Garcia G., and Hauck, T. 2009. Reduction of GHG emissions by geological storage of CO_2: Anatomy of the Heartland Aquifer Redwater Carbon Capture and Geological Storage Project (HARP), Alberta, Canada, *International Journal of Climate Change Strategies and Management* 1(2), 160–178, Emerald Publishing Group Limited.

Gunter, W.D., Perkins, E.H., and Hutcheon, I., 2000. Aquifer disposal of acid gases: Modelling of water-rock reactions for trapping acid wastes, *Applied Geochemistry* 15, 1085–1095.

Gunter, W.D., Wong, S., Deng, X., Cech, R., Andrei S., and Macdonald, D. 2008. *Recommended Practices for CO_2-Enhanced Coal Bed Methane Pilot Tests in China*, Geology Publisher House, Beijing, China 275 pages.

Gunter, W.D., Wong, S., Cheel, D.B., and Sjostrom G. 1998. Large CO_2 sinks: Their role in the mitigation of greenhouse gases from an international, national (Canadian) and provincial (Alberta) perspective, *Applied Energy* 61, 209–227.

Hitchon, B. (Ed.) 1996. *Aquifer Disposal of Carbon Dioxide: Hydrodynamic and Mineral Trapping—Proof of Concept*, Geoscience Publishing, Sherwood Park, Alberta, 165 pages.

Hitchon, B., Gunter, W.D., Gentzis, T., and Bailey, R. 1999. Sedimentary basins and greenhouse gases: A serendipitous association, *Energy Conversion and Management* 40, 825–843.

Hitchon, B. 2012. *Best Practices for Validating CO_2 Geological Storage: Observations and Guidance from the IEA GHG Weyburn – Midale CO_2 Monitoring and Storage Project*, Geoscience Publishing, Alberta, Canada, 353 pages.

Hu, Y.Q., Kobayashi, N., and Hasatani, M. 2001. The reduction of recycled-NO_x in coal combustion with O_2/recycled flue gas under low recycling ratio. *Fuel* 80, 1851–1855.

IEA. 2013. *IEA Clean Coal Centre Advances in Multi-Pollutant Control*. Anne M Carpenter. United Kingdom, CCC/227.

IEA. 2013. *International Energy Agency. Redrawing the Energy-Climate Map*. World energy outlook special report, United Kingdom.

IEA. 2014. *International Energy Agency. CO_2 Emissions from Fuel Combustion Highlights, United Kingdom.*

IEA. 2010. *Energy Technology Perspectives 2010—Scenarios and Strategies to 2050*, United Kingdom, p. 650.

IPCC. 2005. IPCC special report on carbon dioxide capture and storage. In: B. Metz, O. Davidson, H. C. de Coninck, M. Loos, and L. A. Meyer (Eds.), *Prepared by Working Group III of the Intergovernmental Panel on Climate Change*. Cambridge University Press, Cambridge, 442 pages.

IPCC. 2007. Summary for policymakers. In: Solomon (Ed) *Climate Change 2007: The Physical Science Basis*, Cambridge University Press, Cambridge.

Irons, R., Davison, J., Sekkapaan, G., Gibbins, J. 2007. *Proceedings of the 3rd International Conference on Clean Coal Technologies for Our Future*, Cagliari, Sardinia.

Kanniche, M., Gros-Bonnivard, R., Jaud, P., Valle-Marcos, J., Amann, J.M., and Bouallou, C. 2010. Pre-combustion, post-combustion and oxy-combustion in thermal power plant for CO_2 catpure. *Applied Thermal Engineering* 30, 53–62.

Kimura, N., Omata, K., Kiga, T., Takano, S., and Shikisima, S. 1995. The characteristics of pulverized coal combustion in O_2/CO_2 mixtures for CO_2 recovery. *Energy Conversion and Management* 36, 805–808.

Komaki, A., Gotiu, T., Uchida, T., Yamada, T., Kiga, T., and Spero T. 2014. Operation experiences of oxyfuel power plant in callide oxyfuel project. *Energy Procedia* 63, 490–496.

Lackner, K. and Ziock, H., 2000. From low to no emissions. *Modular Power Systems* 20(3), 31–32.

Leung, D.Y.C., Caramanna, G., and Maroto-Valer, M.M. 2014. An overview of current status of carbon dioxide capture and storage technologies. *Renewable and Sustainable Energy Reviews* 39, 426–443.

Liu, H. and Okazaki, K. 2003. Simultaneous easy CO_2 recovery and drastic reduction of SO_x and NO*x* in O_2/CO_2 coal combustion with heat recirculation. *Fuel* 82, 1427–1436.

Markewitz, P., Kuckshinrichs, W., Leitner, W., Linssen, J., Zapp, P., Bongartz, R., Schreiber, A., and Muller, T. 2012. Worldwide innovations in the development of carbon capture technologies and the utilization of CO_2. *Energy & Environmental Science* 5, 7281–7305.

Metz, B., Davidson, O., de Coninck, H., Loos, M., and Meyer, L. (Eds.) 2005. Cambridge University Press, New York, 431 pages.

Olajire, A.A. 2010. CO_2 capture and separation technologies for end-of-pipe application—A review. *Energy*, 35, 2610–2628.

Olajire, A.A. 2013. A review of mineral carbonation technology in sequestration of CO_2. *Journal of Petroleum Science and Engineering* 109, 364–392.

Rao, A.B. and Rubin, E.S.A. 2002. Technical, economic, and environmental assessment of amine-based CO_2 capture technology for power plant greenhouse gas control. *Environmental Science & Technology* 36, 4467–4475.

Rubin, E.S., Mantripragada, H., Marks, A., Versteeg, P., and Kitchin, J. 2012. The outlook for improved carbon capture technology. *Progress in Energy and Combustion Science* 38, 630–671.

Samanta, A., Zhao, A., Shimizu, G.K., Sarkar, P., and Gupta, R. 2012. Post-combustion CO_2 capture using solid sorbents: A review. *Industrial & Engineering Chemistry Research* 51, 1438–1463.

Shell. 2011. Shell public documents submitted to Alberta Government. Available from http://www.shell.ca/en/aboutshell/our-business-tpkg/upstream/oil-sands/quest/about-quest.html.

Shell. March 2012. *Quest Carbon Capture and Storage Project*, Annual Summary Report—Alberta Department of Energy, Shell Canada, 82 pages.

Smith, K.H., Anderson, C.J., Tao, W., Endo, K., Mumford, K.A., Kentish, S.E., Qader, A., Hooper, B., and Stevens G.W. 2012. Pre-combustion capture of CO_2 -results from solvent absorption pilot plant trials using 30 wt% potassium carbonate and boric acid promoted potassium carbonate solvent. *International Journal of Greenhouse Gas Control* 10, 64–73.

Stéphenne, K. 2014. *Start-Up of World's First Commercial Post-Combustion Coal Fired CCS Project: Contribution of Shell Cansolv to SaskPower Boundary Dam ICCS Project.* Energy Procedia 00 (2013), Elsevier, the Netherlands, 000–000.

Svensson, R., Odenberger, M., Johnsson, F., and Stromberg, L. 2004. Transportation systems for CO_2-application to carbon capture and storage. *Energy Conversion and Management* 45, 2343–2353.

Tan, Y., Douglas, M.A., and Thambimuthu, K.V. 2002. CO_2 capture using oxygen enhanced combustion strategies for natural gas power plants. *Fuel* 81, 1007–1016.

Uddin, M., Coombe, D., Law, D., and Gunter, W.D. 2008. Journal of Energy Resources Technology, Numerical Studies of Gas Hydrate Formation and Decomposition in a Geological Reservoir, Journal of Energy Resources Technology, Transactions of the ASME, New York, Vol. 130, 032501-1–032501-14.

Visser, E., Hendricks, C., Barrio, M., Molnvik, M.J., deKoeijer, G., and Liljemark, S. 2008. Dynamics CO_2 quality recommendations. *International Journal of Greenhouse Gas Control* 2, 478–484.

Wall, T., Liu, Y., Spero, C., Elliott, L., Khare, S., Rathnam, R., Zeenathal, F. et al. 2009. An overview on oxyfuel coal combustion-state of the art research and technology development. *Chemical Engineering Research and Design* 87, 1003–1016.

Wall, T.F. 2007. Combustion processes for carbon capture. *Proceedings of the Combustion Institute* 31, 31–47.

Wong, S., Macdonald, D., Andrei, S., Gunter, W.D., Deng, X., Law, D., Ye, J., Feng, S., Fan, Z., and Ho, P. 2010. Conceptual economics of full scale enhanced coalbed methane production and CO_2 storage in anthracitic coals at South Qinshui Basin, Shanxi, China, *International Journal of Coal Geology* 82, 280–286.

Yeh, J.T., Resnik, K.P., Rygle, K., and Pennline, H.W. 2005. Semi batch absorption and regeneration studies for CO_2 capture by aqueous ammonia. *Fuel Processing Technology* 86(14–15), 1533–1546.

INTERNET REFERENCES

Callide Oxyfuel Project—Lessons Learned. Callide Oxyfuel Project and Global CCS institute. 2014. http://www.globalccsinstitute.com/publications/callide-oxyfuel-project-lessons-learned.

GCCSI, 2014. http://www.globalccsinstitute.com/content/how-ccs-works-transport.

http://www.callideoxyfuel.com/what/callideoxyfuelproject.aspx.

http://www.flickr.com/photos/mississippipower/10727408034/in/photostream/.

http://www.globalccsinstitute.com/project/kemper-county-energy-facility.

http://www.globalccsinstitute.com/publications/callide-oxyfuel-project-lessons-learned.

http://www.zeroco2.no/projects/kemper-county-igcc.

https://sequestration.mit.edu/tools/projects/kemper.html.

http://www.zeroco2.no/projects/the-great-plains-synfuels-plant.

Oil and Gas journal, 2014. http://www.ogj.com/.

Whittaker, S., 2011. CO_2 Storage Predictions at Weyburn via Reservoir Simulation. Presented at CSLF, 2011, Saudi Arabia. http://www.cslforum.org/publications/documents/alkhobar2011/IEAGHGWeyburnMidale_CO2_MonitoringandStorageProject_Session1.pdf.

22 Coal Company Valuation, Production, and Reserves

Mark J. Kaiser

CONTENTS

Abstract: The primary determinants of the value of a coal company are its cash flows and earnings, which is dependent on the quantity and quality of the coal it produces and coal selling prices; its future production potential, which is described by its reserves; and its inventory of capital assets, mining equipment, infrastructure, and acreage. The purpose of this chapter is to establish the relationship between the factors that determine company value for a cross-section of publicly traded North American coal companies at the end of 2010. We review the consolidation trends of North American companies up to this time and the factors that impact company value. Companies vary in the size of their reserves and production, earnings, number and type of mines, use of technology and automation, areas of operation, sales markets, degree of diversification, and financial structure. We construct regression models for U.S. and Canadian coal companies according to production, reserves, financial structure, and coal selling price based on 2010 data. We show that reserves and assets are strong indicators of market capitalization and enterprise value. Large producers exhibit a robust correlation between production and reserves, and small companies trade at a premium relative to large producers. We infer effective enterprise value and market cap for a sample of private companies. The methodology is general but the results will change with the period of evaluation.

22.1 INTRODUCTION

The value of a coal company and/or property is intended to reflect the worth of the company and/or property on the open market. According to the 2005 International Valuation Standards, worth is defined as "the value of property to a particular investor, or class of investors, for identified investment objectives." The market value of a property is defined as the "estimated amount for which a property should exchange on the date of valuation between a willing buyer and a willing seller in an arms-length transaction after proper marketing wherein the parties had each acted knowledgeably, prudently, and without compulsion … reflecting the collective perceptions and actions of a market" (IVS 2005).

The primary value of any company is its cash flows and earnings, which is dependent on the quantity and quality of the product that it provides, and the product sales price. Production is derived from reserves and the inventory of capital assets, mining equipment, infrastructure, and acreage. Reserves lie below the surface and have not yet been produced but are economically and technically viable to extract. In North America, the U.S. Securities and Exchange Commission (SEC), the U.S. Geological Survey (USGS), the Society for Mining, Metallurgy, and Exploration (SME), the Ontario Security Commission (OSC), Toronto Stock Exchange (TSE), and Canadian Security Administration (CSA) provide guidelines on resource classifications and company requirements to list on their stock exchanges.

Any member of society with enough money can buy shares of a public company, but a private company has only a few owners whose shares are not offered to the public. To estimate the value of a public company, there are four basic valuation techniques commonly employed—book value of assets, discounted cash flow, price earnings multiple, and market value—which can vary considerably depending on the assumptions used (Damodaran 2002; Abrams 2010). For a private company, the first three methods are not an option because detailed financial information is not publicly released.

In 2011, coal accounted for 25% of all mining and metal deals globally, worth $42 billion in value and surpassing mergers and acquisitions in the gold, copper, iron ore, and oil and gas industries according to Ernst & Young LLP (Miller and Maher 2012). Coal is one of the world's cheapest fuels for generating electricity, and the industry is the most highly fragmented of all commodities,* leaving room for consolidation. Mergers provide larger coal-reserves bases and can lower costs for infrastructure, equipment, and other inputs like diesel fuel. Prices for high-grade metallurgical (coking) coals were at historic levels in 2010, but since 2011 prices for all coal grades has fallen by nearly a third, significantly impacting the financial health of many companies. Every significant coal company has cut jobs and circa 2015 a half dozen have declared bankruptcy.

The purpose of this chapter is to review the factors that impact the market value of a coal company and establish the relationships that exist between reserves, production, assets, and financial metrics for a cross-section of public companies. We fix the time of assessment on December 31, 2010, to coincide with the release of production and reserves data. By fixing the time of assessment, we eliminate the impact of market price swings on company valuation, but the model results are linked to the evaluation date. The relationship between market capitalization, enterprise value, and company attributes is useful in understanding industry structure (Chapman 1983), performing due diligence (Antill and Arnott 2000; Howard and Harp 2009), and revealing the relationship between investment requirements and production output (Rutledge 2011).

The outline of the chapter is as follows. We begin by reviewing the top coal producers in the United States in 2010 and summarize the U.S. coal market and consolidation trends that have impacted the industry through 2010. The factors that influence company value are described followed by the model formulation. Summary statistics are presented followed by the definition of the valuation functional and model results. One-factor and multifactor regression models are presented for enterprise value and market capitalization, and we infer the effective values for a sample of private companies. We conclude with a discussion of the estimation challenges and limitations of analysis.

22.2 MAJOR U.S. COMPANIES

The top 25 coal producers in the United States in 2010 are shown in Figure 22.1. The largest companies are public, and only one company, BHP Billiton, is headquartered outside the

* Coal is produced in 24,000 mines around the world, and the top 50 coal miners accounted for 47% of 2010 production. By comparison, the top 50 copper miners controlled 83% of production (Miller and Maher 2012).

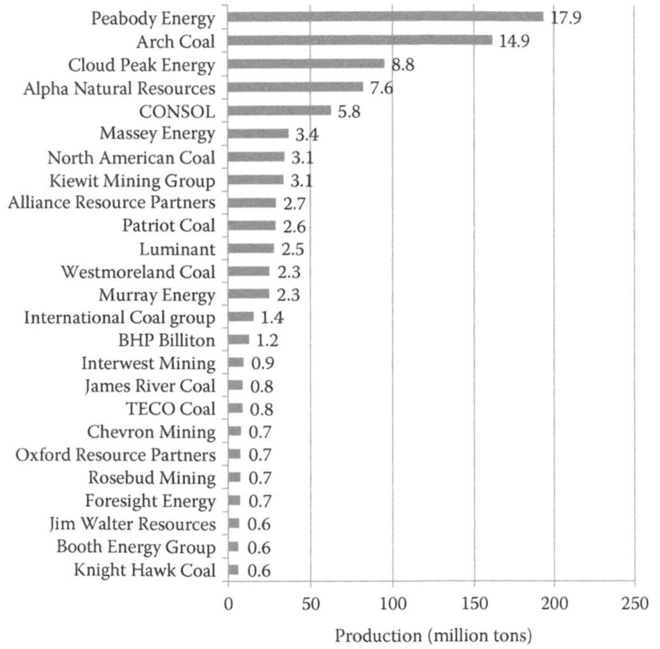

FIGURE 22.1 Top 25 U.S. coal producers (2010). Percentage of total U.S. coal production is shown at the end of bar. (Data from National Mining Association, *Coal Producer Survey*, May 2011, www.nma.org/pdf/members/coal_producer_survey2010.pdf, 2010.)

United States. Private coal companies tend to be much smaller than publicly traded companies and are more numerous.

Peabody Energy is the world's largest public coal company with surface and underground operations throughout the United States and Australia. In 2010, Peabody produced 194 million tons of coal and reported proven and probable reserves of 9 billion tons. It is the only domestic producer that has significant operations outside the United States.

Arch Coal is the second largest U.S. public coal company with production of 161 million tons and proven and probable reserves of 4.4 billion tons in 2010. Surface mines represent 87% of 1.9 billion tons assigned reserves and 68% of 2.5 billion tons unassigned reserves. Arch Coal classified 82% of its 2010 reserves as compliance coal.

Cloud Peak Energy, Alpha Natural Resources, and CONSOL Energy round out the top five U.S. producers in 2010. Cloud Peak Energy operates solely in the Powder River Basin, and all of its 970 million tons reserves are assigned compliance coal. Alpha Natural Resources produced 82 million tons in 2010 and acquired Massey Energy in a $7.1 billion cash and stock transaction in 2011.

CONSOL Energy is a multifuel energy producer and energy services provider with coal and gas business units. In 2010, CONSOL produced 62 million tons of coal from underground mines using longwall mining and held an estimated 4.4 billion tons of proven and probable reserves. It also produced 127 billion cubic feet of gas and holds 4.7 trillion cubic feet of proved and probable natural gas reserves and is the only public coal company with significant gas operations. In 2009, CONSOL acquired the E&P business of Dominion Resources for $3.5 billion.

22.3 COAL MARKETS AND CONSOLIDATION TRENDS

The U.S. coal market is divided into two broad categories—steaming or thermal coal used for the generation of heat and electrical power, and metallurgical or coking coal used in the production of iron and steel. Coking coal is the most expensive and is generally sold on evergreen contracts with volumes and prices set for 1 year. Hard coking coals are the highest priced coals in the United States. The selection of a coal for power generation is primarily one of cost subject to the coal's particular qualities, such as ash and sulfur content. Spot markets for thermal coal are more prevalent than for coking coals but only comprise a small portion of the market. Generally, the relative competitiveness of coal to other fuels or other coals is evaluated on a delivered cost per heating value unit basis. Coal with low moisture and ash content has high heat content and generally commands higher prices because less coal is needed to generate a given quantity of electric power.

One of the primary characteristics of all fossil fuel industries is that the asset base is continuously being depleted and must be replaced through development or acquisition. The high degree of fragmentation of the industry has led to consolidation at different points in time and for different reasons. During the 1970s, oil, steel, and utility companies were actively diversifying into coal and uranium, and by 1980, most leading coal companies were subsidiaries[*] or affiliated with larger companies; for example, Peabody Coal was affiliated with Newmont Mining, Williams, Boeing, Fluor, Bechtel, and Equitable; Arch Mineral was affiliated with Ashland Oil; Consolidation Coal was owned by Conoco/Dupont; Island Creek Coal was owned by Occidental; and Old Ben was owned by Sohio/BP (Chapman 1983). Arco, Exxon, Mobil, Phillips, Shell, Sun Oil, and other petroleum companies eventually left the coal sector, and today, only Chevron continues to produce coal. Internationally, over the past two decades there has also been a change of state ownership to privatization in the Former Soviet Union, Russia, India, and China.

The U.S. coal industry has consolidated over the past three decades, but despite many years of consolidation, the coal industry remains fragmented. The share of production from the top 15 producers grew from 43% in 1984 to 70% in 1998 to 77% in 2010 (Table 22.1). The top 5 producers in 1984 controlled 24% of the market, compared to 47% in 1998 and 53% in 2010. The largest market shares have also increased. In 1998, Peabody Energy and Arch Coal produced 15.1% and 9.4% of the total U.S. production, respectively; in 2010, Peabody Energy and Arch Coal produced 17.6% and 13.8%, respectively.

[*] A company that is controlled through the ownership of voting stock, or a corporate joint venture in which a corporation is owned by a small group of businesses for the benefit of the members of the group, is referred to as an operating subsidiary.

TABLE 22.1
Major U.S. Coal Producers—1984, 1998, and 2010

	1984			1998			2010		
Rank	Company	Production[a] (Million Tons)	Percent (%)	Company	Production[a] (Million Tons)	Percent (%)	Company	Production[a] (Million Tons)	Percent (%)
1	Peabody Energy	64	7.4	Peabody Energy	169	15.1	Peabody Energy	194	17.9
2	CONSOL	47	5.3	Arch Coal	105	9.4	Arch Coal	156	14.9
3	Amax Group	41	4.7	Kennecott	103	9.2	Cloud Peak	95	8.8
4	Texas Utilities	29	3.3	CONSOL	74	6.6	Alpha Natural	82	7.6
5	ARCO	24	2.8	American Coal	70	6.3	CONSOL	62	5.8
6	A.T. Massey	24	2.7	AEI Resources	51	4.6	Massey Energy	37	3.4
7	Exxon	23	2.7	A.T. Massey	38	3.4	N. American Coal	34	3.1
8	N. American Coal	19	2.1	N. American Coal	32	2.8	Kiewit Mining	34	3.1
9	Kerr-McGee	17	1.9	Texas Utilities	28	2.5	Alliance Resource	29	2.7
10	Nerco	16	1.9	PacifiCorp	22	2.0	Patriot Coal	29	2.6
11	Utah International	15	1.7	Pitt. & Midway	22	2.0	Luminant	28	2.5
12	Pittston	15	1.7	Triton Coal	17	1.5	Westmoreland	25	2.3
13	Old Ben	15	1.7	MAPCO	16	1.4	Murray Energy	25	2.3
14	AEP	14	1.5	BHP Minerals	16	1.4	Int. Coal Group	16	1.4
15	Pitt. & Midway	13	1.5	Kiewit Mining	15	1.3	BHP Billiton	13	1.2
Subtotal		376	43.0		778	69.6		857	79.6
Other producers		499	57.0		340	30.4		227	20.4
Total		875	100.0		1,118	100.0		1,084	100.0

Source: National Mining Association, *Coal Producer Survey*, May 2011, www.nma.org/pdf/members/coal_producer_survey2010.pdf, 2010.

[a] Overseas production is excluded.

22.4 FACTORS THAT IMPACT COMPANY VALUE

22.4.1 Reserves

22.4.1.1 Recoverable Reserves

Recoverable reserves are defined by SEC Industry Guide 7 (SEC 1992) as that part of a mineral deposit that could be economically and legally extracted or produced at the time of the reserve determination.* Recoverable reserves are calculated based on the area in which mineable coal exists, with coal seam thickness and average density determined by lab testing of drill core samples. Uncertainty is characterized according to distance to drilling holes, and U.S. Geological Survey Circular 891 (Wood et al. 1983) is widely used for public disclosure (Olea et al. 2011). Recoverable reserves are classified into proven (measured) and probable (indicated) reserves categories and are reported only for coal seams that are controlled by ownership or leases. Resources are not considered reserves because no economic constraints have been applied. The resource, or some part of it, may or may not become a reserve at a future date.

22.4.1.2 Proven (Measured) Reserves

Proven (measured) reserves are defined by SEC Industry Guide 7 as reserves for which (1) quantity is computed from dimensions revealed in outcrops, trenches, workings, or drill holes; (2) grade and/or quality are computed from the results of detailed sampling; and (3) the sites for inspection, sampling, and measurement are spaced so closely and the geologic character is so well defined that size, shape, depth, and mineral content of reserves are well established. Spacing of points of observation for confidence levels is based on guidelines in the U.S. Geological Survey Circular 891. Typically, estimates for proven reserves are based on observation points that are no greater than 0.5 mile apart. Calculations are adjusted to account for coal that will not be recovered during mining† and for losses that occur if the coal is processed after mining. The degree of variance from reserves estimate to tons produced varies with the continuity of the coal seam and the nature of the topography.

22.4.1.3 Probable (Indicated) Reserves

Probable (indicated) reserves are defined by SEC Industry Guide 7 as reserves for which quantity and grade and/or quality are computed from information similar to that used for proven reserves, but the sites for inspection, sampling, and measurement are farther apart, between 0.5 and 1.5 miles, or are otherwise less adequately spaced. Estimates of probable reserves have a moderate degree of geologic assurance and require further exploration to place into the proven reserves category. The degree of assurance, although lower than that for proven reserves, is high enough to assume continuity between points of observation.

22.4.1.4 Assigned and Unassigned Reserves

Assigned reserves are coal that is planned to be mined at an operation that is currently operating, is currently idled, or for which permits have been submitted and plans are eventually to develop the operation. Unassigned reserves are likely to be mined in the future but have not yet been designated for mining by a specific operation. They represent coal reserves that require significant investment (e.g., new mineshafts, mining equipment, plant facilities) before operations can begin. Assigned and unassigned coal reserves are classified within proven and probable reserves categories.

22.4.1.5 Leased and Owned Reserves

Coal companies generally do not own the land on which mining operations are conducted but have leases from third-party landowners. Federal coal leases are administered‡ by the U.S. Department of Interior Bureau of Land Management (BLM) under the Federal Coal Leasing Amendment Act of 1976. State and private coal leases have varying renewal terms and conditions but generally convey mineral rights in exchange for a per ton fee or royalty payment of a percentage of the gross sales price, subject to minimum payments, for the economic life of production.

22.4.2 Production

Production is the causal result of reserves and is an important measure of performance because production and sales price determines gross revenue and, when combined with costs, the cash flow and profitability of a mine. Companies report mine production in their annual reports, and the Energy Information Administration (EIA) has summarized annual production data for major§ mines since 1984 (EIA 2009).

Operators control production and generally produce at maximum safe rates, but differences arise in how coal is produced depending on the mine operations and market conditions. Companies may temporarily idle mines due to contract expiration, demand for and price of coal, depletion of economically

* Oil and gas reserves are defined using similar language, but because oil and gas are fugacious (movable) and the physical attributes of the asset class are much less certain, there are significant differences in reserves assessment and uncertainty. The relative uncertainty of oil and gas reserves is characterized by reference to deterministic categories—proved, P1 ("much more likely than not"); probable, P2 ("as likely than not"); and possible, P3 ("possible, but not likely")—or in probabilistic terms. If probabilistic methods are used, there should be at least a 90% probability that the quantities of proved reserves actually recovered will equal or exceed the estimate. For probable and possible reserves, the exceedance probabilities are 50% for probable and 10% for possible reserves (SPE 2007; SEC 2008).

† Recovery rates vary by company experience depending on the type of mine

‡ All BLM coal leasing is done competitively except in cases where a party holds a "prospecting permit" issued prior to the Federal Coal Leasing Amendment Act of 1976 or where coal lands are added to existing leases (Vann 2011). After the lease sale is given, BLM fields bids and announces the highest bid, which is awarded if all applicable lessee requirements are made. BLM does not accept any bid for less than the fair market values. Coal leases are for initial terms of 20 years with automatic extension as long as commercial quantities of coal are produced annually. Rental payments are not less than $3 per acre, and lessees are also required to make a royalty payment of at least 12.5% in amount or value of coal recovered from the leased land. Parties who have held a federal coal lease for 10 or more years that has not produced commercial quantities may not acquire any other federal leases.

§ The EIA currently defines major mines as mines that produce more than 4 million tons annually, but the threshold used to identify major mines has changed over time.

recoverable reserves, loss of transportation routes (e.g., due to flooding), availability of experienced labor, etc.

At the mine, production is not strongly correlated with reserves because of life cycle and other factors, but at a corporate level, production and reserves are expected to be more closely aligned because mines of many different sizes, types, and ages are aggregated, and the process of aggregation smooths out the production and life cycle variations of development. Companies with large reserves holdings are expected to produce more than companies with small reserves holdings, for all else equal.

22.4.3 Cash Flow and Capital Expenditures

The cash flow statement of public companies provides information on sources and uses of cash, with sections for operations, investing activities, and financing activities. Cash flow from operations consists of net income after taxes plus depreciation and other noncash expenses. Investing activities include the net effect of buying and selling property, plant, and equipment. Financing activities include the net effect of issuing and purchasing company stock, issuing and paying off debt, and paying dividends.

Major sources of cash include cash flow from operations, sales of assets, and proceeds from issuing debt or equity. Primary uses of cash include making capital expenditures, paying dividends, purchasing company stock, and paying off debt. Capital expenditures represent the value of assets acquired in the time period net of depreciation and also include investments and advancements to unconsolidated affiliate companies. Companies with large cash flows and earnings per share are worth more and are expected to be valued higher by the market than companies with smaller cash flow and earnings.

22.4.4 Cost

22.4.4.1 Capital and Operating Cost

Developing a mine requires a large initial investment and a continuing operating expenditure to produce cash flow. The capital cost to develop a mine depends on the mine type (underground or surface), mine method (longwall, shortwall, continuous),* production rate, equipment requirements, existing infrastructure (power, water, roads, rail, etc.) and processing plants, topography (terrain), climate, geologic complexity, time of construction, and related factors. Land acquisition is part of capital expenditures.

Production costs are the cost to operate and maintain the mine and related equipment after the coal has been found and developed for production. Each deposit has its own extraction, processing, and transportation cost. Extraction costs depend on the size of deposit, seam thickness, amount of overburden (if surface mined) or depth (if underground mined), continuity and angle of seam, inflow of water and methane generation, labor availability, and cost.

Operating costs often exhibit scale economies, meaning that the greater the production rate, the lower the cost on a unit basis (Runge 1998). Coal deposits deplete, and the cost of mining rises as a deposit approaches exhaustion. Costs are also impacted by productivity and state and federal regulatory scrutiny. Operating costs include wages and employee benefits, supplies, maintenance and repair, purchased coal, outside services,† royalties, and production and severance taxes. Production cost is usually a significant percentage of the commodity selling price, which is one of the distinguishing characteristics between coal and oil and gas operations.

22.4.4.2 Preparation and Blending

Surface mined coals typically require no additional preparation and after crushing can be shipped directly to customers. Coal extracted from underground mines often contains impurities such as rock, shale, and clay and occurs in a wide range of particle sizes that require treatment to ensure a consistent quality (Thomas 2002). Coal washing is the process where ash and waste are removed from run-of-mine coal to increase its value and meet specifications. Preparation plants remove impurities from run-of-mine coal and blend various coals and coal qualities to meet customer requirements (Arnold et al. 2007). Processing cost depends on the type of coal and the requirements of the user.

22.4.5 Coal Price

22.4.5.1 Contract Type

The price of coal is negotiated with utilities and steel companies, and long-term contracts are common. A typical coal contract has specific ranges for Btu content, moisture content, ash content, sulfur content, and volatile organic matter. Contracts come in many forms from fixed price to cost plus to prices renegotiated at predetermined intervals (Table 22.2). Joskow (1985) provides a detailed overview of contracts in the coal industry (see also Joskow 1987; Kerkvilet and Shogren 2001; Kozhevnikova and Lange 2009). Price risk varies with the contract type. A small percent of coal transactions occurs in spot markets where prices fluctuate on a daily/weekly basis.

* Access to coal is made by open cut or underground mining methods, depending on the thickness and depth of the seam. Differences in geology lead to variations in the techniques and layouts of mining and in the use of capital and other inputs (Hartman and Mutmansky 2002). Infrastructure is needed to support the project, including transport, power, water facilities, and processing plant. Underground mining methods include longwall and shortwall mining and continuous ("room-and-pillar") mining. Longwall mining systems typically have a lower variable cost compared with continuous mining and can achieve higher productivity levels but is only effective for large blocks of medium-to-thick coal seams (Hartman and Mutmansky 2002). Continuous mining is often used to mine smaller coal reserves or thin seams. Surface methods include truck-and-shovel/loader mining, dragline mining, and highwall mining. Contour and highwall mining is used where removal of all the overburden is either uneconomical or impossible. Underground mining has higher labor and capital costs than surface mining because of labor benefits and health care, specialized mining equipment, ventilation requirements, and related factors.

† Contract mining is sometimes employed in situations where the owner-operator has a limited capital budget.

TABLE 22.2
Coal Contract Types

Contract Type	Description
Fixed price	Price is fixed over the life of the contract.
Base price plus escalation	Different components of the price escalate (or de-escalate) as a function of changing economic conditions based on indices.
Price tied to market	Price tied to the price of coal being sold in a particular market.
Cost-plus contract with a fixed fee provision	Purchaser agrees to pay all producer's costs plus a management fee. Some contracts provide for payment of both a management fee and a profit. This contract has a fixed fee provision.
Cost-plus contract with an incentive fee	Purchaser agrees to pay all producer's costs plus a management fee. Some contracts provide for payment of both a management fee and a profit. This contract has an incentive fee provision; that is, a variable fee that is tied to various productivity and cost reduction incentives.
Price renegotiation	The price is renegotiated at predetermined intervals, usually 1 year. This type of contract, frequently known as an evergreen contract, may also contain provisions for price adjustments between renegotiations.

22.4.5.2 Coal Quality

Coal is classified on the basis of rank according to volatile matter, fixed carbon, and heating value (Schobert 1987; Thomas 2002). Heating value indicates the amount of heat derived from burning coal. High Btu coal is generally defined as coal delivered with an average heat value of 12,500 Btu per pound or greater. Coals are sometimes characterized as compliance or noncompliance coal in reference to sulfur dioxide emissions standards in Title IV of the Clean Air Act. A compliance coal will produce emissions that meet the current regulation limit without further cleanup or application of scrubbing technology*. Compliance coal is coal that, when burned, emits 1.2 lb or less of sulfur dioxide per million Btu. Most compliance coal and low-sulfur coal are suitable for steam coal markets in the United States, and a portion may also be used as metallurgical coal. Coking coal must meet stringent quality specifications.

22.4.5.3 Location

The location of the mine relative to the customer and alternatives available to the customer is an important factor in coal prices because transportation cost is usually borne by the customer. Eastern underground coal is costlier at the mine mouth, but its transportation costs are lower in regional markets, involving shorter hauls to consumers by rail and low-cost barge. Low-cost surface mined western coal is shipped primarily by rail over greater distances and involves a larger transportation cost. Powder River Basin coal with its lower energy content, lower production cost, and often greater distance to the consumer sells at a lower price than North and Central Appalachian coal that has a higher energy content and is often located closer to the end user.

22.4.6 Assets

Production is derived from reserves and capital assets, mining equipment, infrastructure, and acreage. The total assets of a company refer to the book value of the assets and include mining equipment and infrastructure, loading facilities, processing plants, support facilities, property, and buildings. Historical costs are applied in a company's financial statement rather than current market value. Reserves are not reported on the balance sheet but are reported separately according to conventions prescribed by the regulatory agency of the country where the company is listed.

22.4.7 Capital Structure

The capital structure of a company refers to the combination of securities employed (debt and equity) to raise and maintain the capital assets required to conduct business. Debt carries an obligation to repay the principal by a specified date plus a given rate of interest. Equity is in common and preferred stock.† The degree to which a firm is funded by loans is known as leverage and is usually expressed as the debt equity ratio (D/E). Considerable variation exists in the capital structure of coal companies and their D/E ratio. A company's D/E ratio is influenced by their access to capital markets, business strategy, and operational performance. A high debt ratio signifies a high risk for payment problems that can impede business growth and lead to liquidation. High indebtedness may prevent a company from obtaining additional financing to fund future working capital, capital expenditures, acquisitions, or other general corporate requirements.

22.4.8 Diversification

Coal development occurs throughout the United States in a variety of geologic environments, and companies pursue different strategies to balance and optimize their asset portfolio over time. Regional distinctions in geology, transportation routes, customers, contract pricing, and coal quality have caused market and contract pricing to develop by region. The business strategy of a few companies includes diversification into non-coal resources to stabilize and to grow out cash flows, but to date, most U.S. producers focus exclusively on coal operations. In North

* Electricity generators are able to use noncompliance coal by using emissions reduction technology, using emission allowance credits, or blending with low-sulfur coal.

† Preferred stock is a hybrid of financing between debt and common stock. Like debt, preferred stock carries a fixed commitment to make periodic payments, but failure to make the dividend payment does not result in bankruptcy.

TABLE 22.3
Model Variables and Definitions

Variable	Notation	Definition	Source
Market capitalization	CAP	Number of outstanding shares multiplied by stock price on December 31, 2010	Bloomberg
Enterprise value	EV	Theoretical takeover price of company on December 31, 2010	Bloomberg
Proven and probable reserves	R	Proven and probable coal reserves reported on December 31, 2010	Form 10-K and U.S. SEC
Production	P	Annual coal production in 2010	Form 10-K and U.S. SEC
Reserves to production ratio	R/P	Ratio of proven and probable reserves to 2010 production	
Debt to equity ratio	D/E	Ratio of debt to equity on December 31, 2010	Bloomberg
Price	PRICE	Weighted average selling price of coal in 2010	Form 10-K and U.S. SEC
Asset	A	Book value of total assets in 2010	Bloomberg

America, the coalbed methane business is run mostly by oil and gas companies, but a few new entrants include coal companies.

22.5 SUMMARY STATISTICS

22.5.1 Data Source

A summary of the model variables and data sources are provided in Table 22.3. All data was obtained using Bloomberg and 2010 company annual reports. Market capitalization, enterprise value, and D/E ratio were determined from Bloomberg for December 31, 2010. Reserves, production, asset value, and the weighted average price of coal sold in 2010 were collected from annual reports. R/P was computed based on reported data.

22.5.2 Sample Set

Twenty-two coal companies with headquarters in North America were identified. The list includes 18 U.S. and 4 Canadian* companies as shown in Table 22.4 and covers a range of large and small producers. In Table 22.5, a summary of the operating areas/basins of each company is depicted along with business segments other than coal mining.

Most companies in the sample only produce domestically, but Peabody Energy has operations in Australia, and three companies (SouthGobi Resources, L&L Energy, and Sinocoking) operate exclusively outside the United States. Companies that do not produce coal as their primary business operation were not considered. A few companies are diversified outside coal operations, but these activities are usually small.† Westmoreland has the largest degree of diversification in the sample with about 20% revenue from activities outside mining.‡

22.5.3 Descriptive Statistics

In 2010, the 22 largest public companies held proven and probable coal reserves of 22.4 billion tons, produced 736 million tons (about 70% U.S. production), and had a R/P ratio of 30 (Table 22.4). The combined market capitalization of the sample was \$59.4 billion, and the enterprise value totaled \$67.5 billion with reported book value of assets of \$46.8 billion. The D/E ratio of the sample averaged 71% and varied from 1% (L&L Energy) to 159% (Alliance Resource Partners).

Five companies lead by Peabody Energy each held reserves more than 1 billion tons and produced more than 50 million tons during the year. Large reserves companies produce more than small reserves holders, and a strong correlation exists between production and reserves (Figure 22.2).

Companies are classified as *large* producers if they have more than 350 million tons proven and probable reserves and produced more than 10 million tons in 2010; *small* producers are defined to have less than 350 million tons reserves and annual production less than 10 million tons. Our sample set splits into 8 large and 14 small companies. A statistical summary of the reserves, production, and financial indicators of the two classes is shown in Table 22.6.

The average market cap and enterprise value of large companies is about 6 times greater than small companies, large companies have assets about 13 times greater than small companies, and average reserves and production positions of large companies are about 30 times larger than small companies. The average R/P ratios for both company classes are around 35, but small companies exhibit a significantly larger variation. The D/E ratios also differ by class, with large companies having a higher debt to equity ratio (76%) relative to small companies (43%). The average sales price of coal differ significantly between the two classes and reflect variation in coal quality, location, and contract type.

* A number of Canadian companies such as Compliance Energy, Pacific Coal, Galway Resources, Elgin Mining, and NovaDx Ventures were excluded because they do not provide complete information about their operations and are considered high-risk micro- and small-cap companies. These companies are listed in the TSX Venture Exchange, which do not satisfy the Toronto Stock Exchange's listing criteria.

† For example, Hallador Energy owns a 45% equity interest in Savoy Energy (a private oil and gas company with operations in Michigan) and a 50% interest in Sunrise Energy, which plans to develop coalbed methane reserves; however, in 2010 less than 1% of Hallador's revenue was from activity outside mining.

‡ Westmoreland focuses on mine-mouth operations. In 2010, Westmoreland produced approximately 25 million tons of coal and generated 1.6 million MW of electricity from two power plants in North Carolina.

TABLE 22.4
North American Public Coal Company Operational and Financial Metrics (2010)

Company	Reserves (Million Tons)	Production (Million Tons)	R/P (yr)	D/E (%)	Market Cap[a] ($ Million)	EV ($ Million)	Total Assets[a] ($ Million)	Sale Price[b] ($/Ton)
Peabody Energy	9,013	218	41	59	17,287	18,771	11,363	27
Arch Coal	4,445	156	29	72	5,701	7,228	4,881	20
Alpha Natural Resources	2,254	82	27	28	7,234	7,433	5,179	46
Patriot Coal	1,865	29	65	54	1,762	2,023	3,810	66
CONSOL Energy	4,401	62	71	119	11,023	14,477	12,071	61
Cloud Peak Energy	970	95	10	134	1,414	1,788	1,915	14
Alliance Resource Partners	697	29	24	159	2,415	2,797	1,501	52
Westmoreland Coal	390	25	16	N/A[c]	133	365	750	20
Rhino Resource Partners	320	4	77	15	289	326	359	74
James River Coal	271	9	31	115	704	807	785	80
Walter Energy	194	8	24	28	6,793	6,668	1,658	166
Grande Cache Coal	151	2	97	28	1,010	1,071	434	173
SouthGobi Resources	117	3	38	36	2,248	1,984	962	35
Oxford Resource	94	8	12	98	502	607	261	44
Hallador Energy	73	3	24	22	293	310	187	43
L&L Energy	38	0.2	156	1	224	250	227	112
Xinergy	19	2	11	150	196	260	153	85
America West Resources	16	0.3	52	N/A	65	88	23	32
Americas Energy	10	0.1	97	15	19	21	25	64
Kentucky Energy	10	0.03	296	N/A	0.5	4	6	73
Royal Coal	9	0.4	21	N/A	26	43	16	76
Sinocoking	2	0.2	10	55	96	139	190	104
All	22,361	736	30	71	59,434	67,460	46,757	34

[a] As of December 31, 2010.
[b] Reported on 2010 annual reports or computed by dividing coal segment revenue by tons sold.
[c] D/E ratio is not applicable due to negative book value of equity.

22.6 MARKET CAPITALIZATION AND ENTERPRISE VALUE

22.6.1 Definition

The market capitalization of a publicly traded corporation represents the number of shares issued and outstanding multiplied by the per share price at a point in time. Enterprise value is the theoretical takeover price needed to acquire the company and is computed as the market value of equity, debt, and minority interest minus cash and investment.

22.6.2 Functional Specification

Market capitalization and enterprise value was regressed against reserves, reserves to production ratio, debt equity ratio, and average coal selling price using the following relation:

$$V = a + b \cdot R + c \cdot \frac{R}{P} + d \cdot \frac{D}{E} + e \cdot PRICE \qquad (22.1)$$

where:

V represents the market capitalization CAP and enterprise value EV in U.S. dollars
R represents the volumes of proven and probable coal reserves in million tons
R/P is the proven and probable reserves to annual production in years
D/E is the debt equity ratio defined as the book value of short-term borrowing and long-term debt to the book value of equity
PRICE is the weighted average sales price in $/ton sold in the fiscal year or computed by dividing coal segment revenue by tons sold.

22.6.3 Expected Signs

Reserves, assets, and production are expected to be primary indicators of company value and to be positively correlated with market capitalization and enterprise value. Reserves represent the inventory of the company, and larger reserves are expected to be associated with higher valuation and longer

TABLE 22.5
Operating Areas of North American Public Coal Companies (2010)

Company	Headquarter	Operating Areas	Other Business Segments
Peabody Energy	United States	United States, Australia	
CONSOL Energy	United States	Powder River Basin, Western Bituminous region, Central Appalachia	
Arch Coal	United States	Northern and Central Appalachia, Powder River Basin	
Alpha Natural Resources	United States	Appalachia, Illinois Basin	
Cloud Peak Energy	United States	Pennsylvania, West Virginia, Virginia, Ohio	Electricity
Alliance Resource Partners	United States	Powder River Basin	
Patriot Coal	United States	Illinois, Indiana, Kentucky, Maryland, Pennsylvania, West Virginia.	
Westmoreland Coal	United States	Pennsylvania, Appalachian Basin	Electricity
Rhino Resource Partners	United States	Appalachia, Illinois Basin, Western Bituminous	
James River Coal	United States	Eastern Kentucky, Southern Indiana	
Walter Energy	United States	Alabama	
Grande Cache Coal	Canada	Smoky River Coalfield	
SouthGobi Resources	Canada	Mongolia	
Oxford Resource	United States	Northern Appalachia, Illinois Basin	
Hallador Energy	United States	Indiana	Oil and gas exploration
L&L Energy	United States	China	
Xinergy	Canada	Kentucky, West Virginia	
America West Resources	United States	Utah	
Americas Energy	United States	Southeastern Kentucky	
Kentucky Energy	United States	Kentucky	
Royal Coal	Canada	Kentucky	
Sinocoking	United States	China	Electricity

Source: Companies' 2010 annual reports.

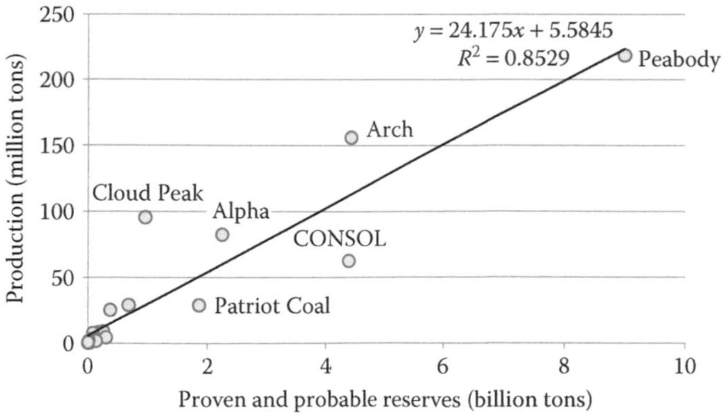

FIGURE 22.2 Proven and probable reserves and annual production (2010).

mine life. For all other things equal, high R/P ratios indicate that production and cash flows are weighted to the future, and thus, high R/P ratios should be negatively correlated with value, but conversely it can be argued that high R/P values are a signal of production longevity which is a positive factor of value, and so the expected sign is indeterminate.

Highly leveraged firms have higher fixed charges in the form of interest payments relative to discretionary outlays such as dividend payments. The higher the D/E ratio, the greater the financial risk, and the lower the expected market cap and enterprise value. The average selling price of coal indicates product value and, because cash flows are determined by production and price, are expected to be positively correlated with company value.

Companies that are geologically and geographically diversified, or have significant oil and gas resources, also point to less risk and greater potential upside that may contribute to a larger valuation. The ability to detect the impact of diversification on company value is expected to be limited, however, because of the primary domestic operations of the companies in the sample.

22.7 MODEL RESULTS

22.7.1 Enterprise Value

Enterprise value is strongly correlated with reserves (Figure 22.3) and assets (Figure 22.4), and less so with production (Figure 22.5). All the relations exhibit statistically significant positive correlations. The highest correlation is

TABLE 22.6
Statistical Summary of North American Company Sample (2010)

Class	Reserves (Million Tons)	Production (Million Tons)	R/P (yr)	D/E (%)	Market Cap ($ Million)	EV ($ Million)	Total Assets ($ Million)	Sale Price ($/Ton)
Large	3004 (2699)	87 (65)	35 (21)	76 (45)	5871 (5482)	6860 (6206)	5184 (4054)	31 (19)
Small	95 (101)	2.8 (3.1)	34 (75)	43 (46)	890 (1735)	898 (1682)	378 (449)	88 (42)
All	1153 (2148)	33 (56)	34 (63)	71 (49)	2702 (4311)	3066 (4902)	2125 (3384)	34 (41)

Note: Large (small) producers have more (less) than 350 million tons reserves and produced more (less) than 10 million tons in 2010.

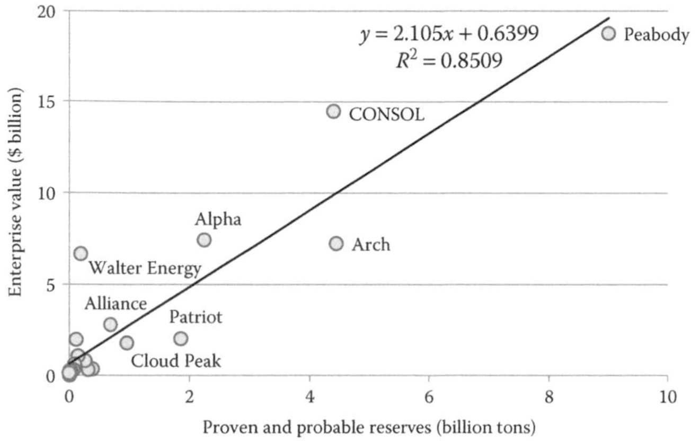

FIGURE 22.3 Enterprise value and proven and probable reserves (2010).

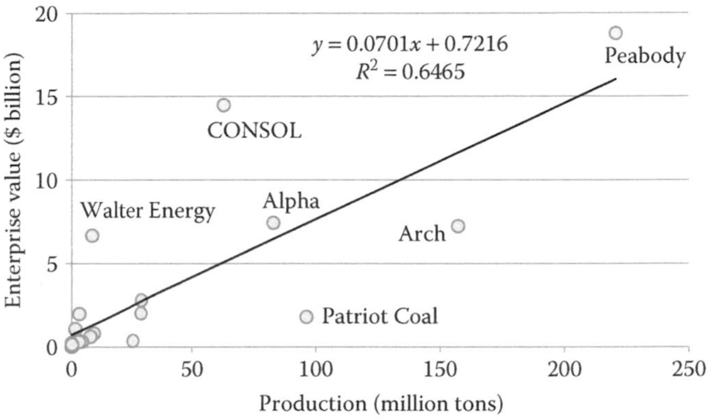

FIGURE 22.5 Enterprise value and annual production (2010).

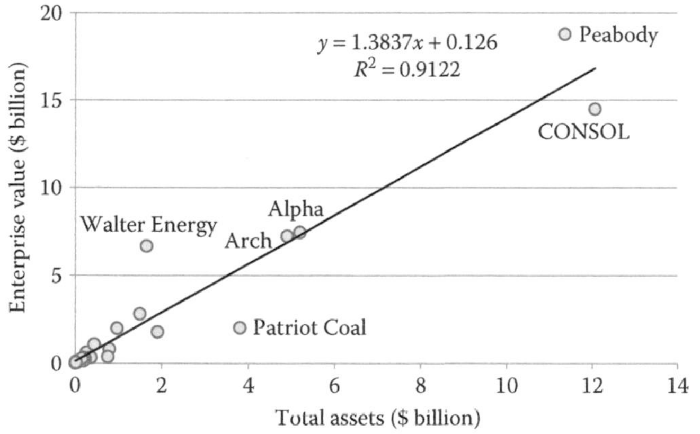

FIGURE 22.4 Enterprise value and total assets (2010).

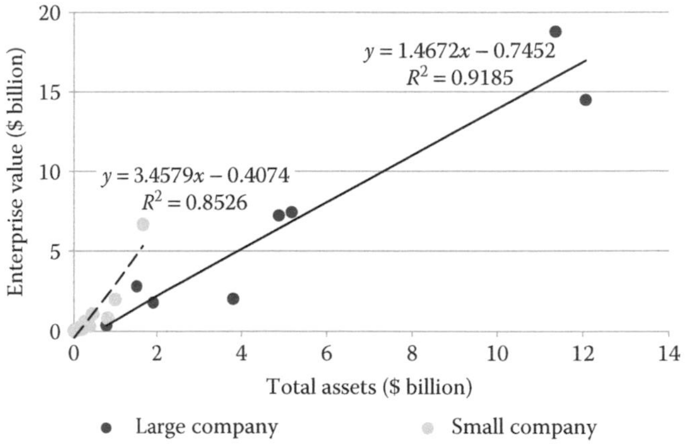

FIGURE 22.6 Enterprise value and total assets (2010).

between enterprise value and the book value of assets reflecting the direct and primary role of equipment and infrastructure in generating cash flow. Production is subject to greater variability across company because it is influenced by idled mines, contract expiration, reserves depletion, and availability of labor and transport routes.

22.7.2 Large versus Small Producers

Enterprise value and assets are strongly correlated for large and small producers (Figure 22.6). Large producers exhibit a markedly stronger correlation between enterprise value and reserves (Figure 22.7) relative to small companies (Figure 22.8), reflecting the heterogeneous nature of small producers operations and financing structure.

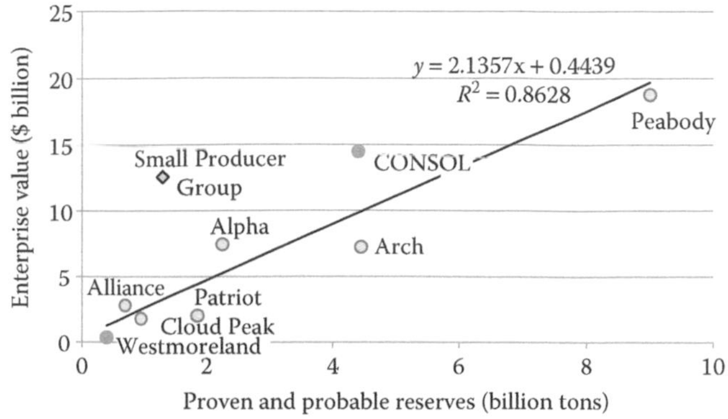

FIGURE 22.7 Enterprise value and proven and probable reserves—large companies (2010). Companies with multiple business segments denoted by dark circles.

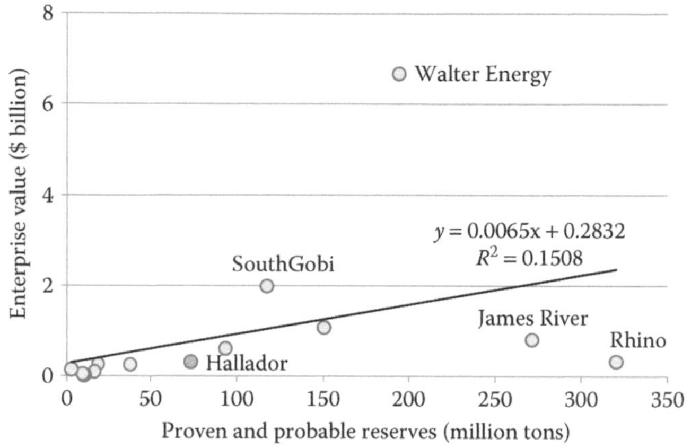

FIGURE 22.8 Enterprise value and proven and probable reserves—small companies (2010). Companies with multiple business segments denoted by dark circles.

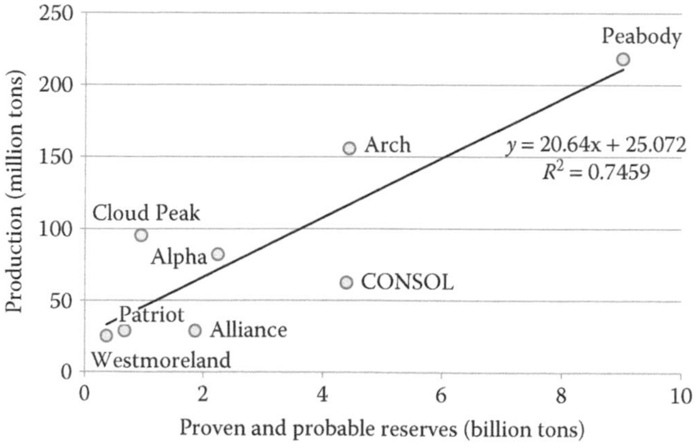

FIGURE 22.9 Proven and probable reserves and annual production—large companies (2010).

Large producers exhibit robust correlations between production and reserves (Figure 22.9) relative to small producers where the correlation is less robust (Figure 22.10). This trend is similar to the enterprise value relations observed for large and small companies.

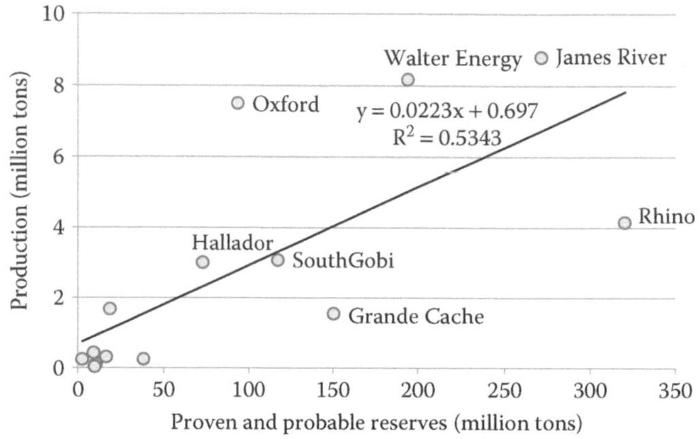

FIGURE 22.10 Proven and probable reserves and annual production—small companies (2010).

22.7.3 Consolidated Small Producers

Small companies are expected to be valued on a proportional basis similar to large companies but as a group are priced at a premium to the average enterprise value. In Figure 22.7, the small companies of the sample are consolidated in terms of their reserves and enterprise value. The composite small producer group falls above the large company average curve and indicates a higher value. CONSOL, for example, has reserves of 4.4 billion tons and an enterprise value of about $15 billion. The small producer group has reserves of 1.3 billion tons and an enterprise value of about $12.5 billion. Market premium is a reflection of the market's expectation that the companies might make new discoveries, increase their reserves base and production, and so on relative to the average group performance, but are also an indicator of overpricing.

22.7.4 Market Capitalization

Market capitalization is a function of the same variables as enterprise value, and because market capitalization is strongly correlated with enterprise value (Figure 22.11) for

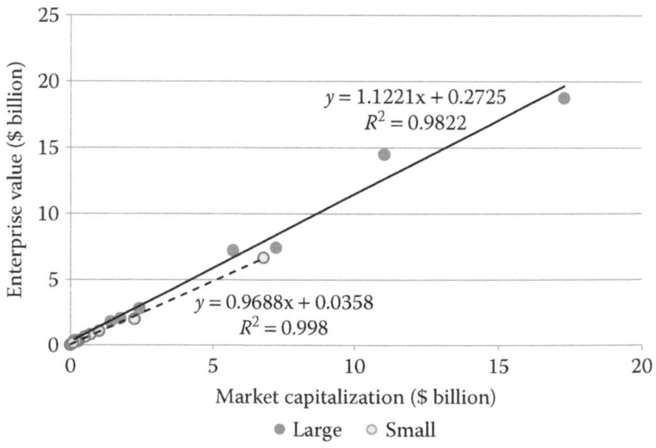

FIGURE 22.11 Enterprise value and market capitalization for large and small producers (2010).

both large and small producers, it is not necessary to present separate graphs for the market cap models.

Production and reserves are strongly correlated to market cap, and both provide reasonable estimates of valuation. Companies are often considered to value reserves over production and the empirical evidence generally supports this notion, but it is clear that both reserves and production are important variables in company valuation.

22.7.5 Regression Models

In Table 22.7, regression results for market capitalization and enterprise value are depicted. Reserves are the most significant explanatory variable, and while the PRICE coefficient is of the expected sign, it is only statistically significant across the entire sample. The R/P coefficients are indeterminate.

Large producers yield stronger correlations than small producers, which are due in part to the sample size relative to the number of factors. For small sample sizes, high R-squared values are suspect, and coefficients of interest may not be individually statistically significant. As long as the coefficients are jointly significant, however, the estimating equation can be utilized with benefit because the purpose of the regression model is to estimate value and is not necessarily concerned about the significance of each control variable.

22.7.6 Private Companies

If a private coal company *went public*, what market capitalization and enterprise value would we expect? Using data from the 2010 National Mining Association Coal Producer Survey, the regression relations were used to estimate enterprise value and market cap for five private companies as shown in Table 22.8.

Murray Energy, the largest private coal company in the United States, is estimated to have an enterprise value of \$2.4 billion and a market cap of \$2.1 billion in 2010. Enterprise value and market caps for Usibelli Coal, Rosebud Mining, TECO Coal, and Trapper Mining are also estimated (Figure 22.12). There is no good way to *validate* these numbers, however, because only a small number of transactions occur each year that can be used for comparison, and sales prices often deviate from corporate valuation. Companies that are bought and sold provide an indicator of market value, but discounts and premiums are common during negotiation, which will further distort the formula-derived value.

TABLE 22.7
Regression Results of North American Public Coal Companies (2010)

CAP or EV (\$ Million) = a + b·R (Million Tons) + c·R/P (yr) + d·PRICE (\$/Ton)

	Large		Small		All	
Parameter	CAP	EV	CAP	EV	CAP	EV
a	−3217 (−0.96)	−2220 (−0.80)	−884 (−0.72)	−750 (−0.75)	−747 (−0.96)	−716 (−0.79)
b	2.4 (4.4)	2.3 (4.3)	4.4 (0.87)	4.0 (0.92)	2.0 (10)	2.2 (11.4)
c	−100 (−1.0)	−44 (−0.40)	−5.6 (−0.83)	−5.0 (−0.88)	−5.7 (−0.90)	−5.3 (−0.83)
d	138 (1.3)	95 (0.86)	20 (1.7)	19 (1.9)	23 (2.1)	22 (2.2)
R^2	0.83	0.82	0.10	0.21	0.84	0.86

TABLE 22.8
Estimated Valuations of Five Private Coal Companies (2010)

Company	Reserves (Million Tons)	Production (Million Tons)	R/P (yr)	Enterprise Value (\$ Million)	Market Cap (\$ Million)
Murray Energy	832	25	34	2391	2114
Usibelli Coal Mine	700	2	350	2113	1872
Rosebud Mining	544	7	75	1785	1587
TECO Coal	270	9	31	1208	1085
Trapper Mining	23	2	10	688	633

Source: National Mining Association, *Coal Producer Survey*, May 2011, www.nma.org/pdf/members/coal_producer_survey2010.pdf, 2010.

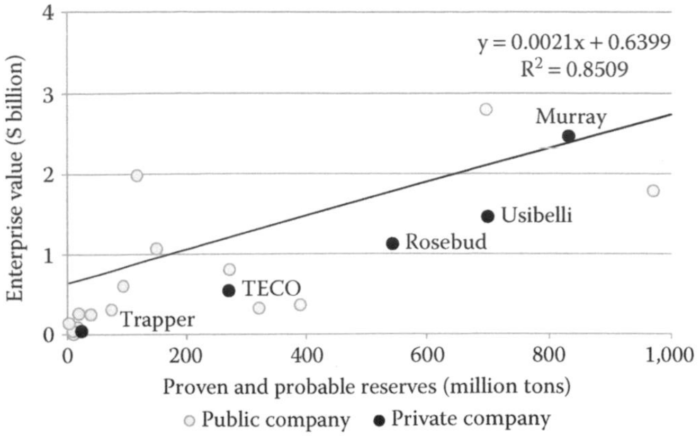

FIGURE 22.12 Estimated enterprise value of North American private coal companies (2010).

22.8 CONCLUSIONS

Constructing robust market capitalization and enterprise value models are subject to a variety of user preferences and estimation issues. In this analysis, the time of assessment was fixed to coincide with the annual release of reserves and production data, and it is clear that the model results will change over time and are only valid for the time period of the assessment. The methodology, however, is completely general. However, it is likely that the general conditions and relative positions of companies will not change significantly barring major merger and acquisition activity, and only the slopes of the regression relations will change. Over a longer period of time, more variation is expected to occur.

A large number of factors may impact market capitalization and enterprise value that were not included in the analysis, including but not limited to the cost of production, earnings, management quality, and exploration potential. Some of these factors can be quantified, but many cannot. Factors that are not reported or disclosed limit the use of regression models, but this limitation is relevant only if the excluded factors are significant. In most cases, these factors are believed to be secondary. We believe that we identified the primary factors of company value, but specialized metrics may also prove useful, and this is an area of potential future research.

All the major public coal companies in North America were enumerated, which significantly reduced sample selection bias, and although some of the measured factors are uncertain (i.e., reserves), they are reported according to U.S. GAPP and are believed to be reasonably consistent for evaluation purposes. The robust results of the model fits provide evidence of the validity of the model assumptions and expectations. There is no good way to validate the valuation of private coal companies, but the empirical relations provide a formal and objective means to benchmark these companies.

22.9 POSTSCRIPT

What a difference five years can make. In 2015, America's coal industry has struggled with oversupply and tepid global demand that has pushed the price of coal used in steelmaking to an 11-year low, off nearly a third since 2011. A plunge in natural gas prices has prompted power plant operators to switch fuels, increasing competition, and tougher environmental rules and labor issues have increased cash burn. Every significant coal company has cut jobs and a half dozen have declared bankruptcy 2014–2015, prompting a crisis that has impacted local economies throughout coal country.

In 2015, the nation's three largest publicly listed coal companies were still Peabody Energy Corp, Alpha Natural Resources Inc, and Arch Coal Inc, but their total market capitalization collapsed to below $1 billion (Miller and Frosch 2015).

Alpha is in talks to obtain financing for a potential bankruptcy filing in August 2015. Patriot Coal Corp filed for Chapter 11 bankruptcy protection in May 2015 (Brickley and Jarzemsky 2015), and Walter Energy Inc was negotiating with creditors in July 2015 after it stopped a bond payment.

REFERENCES

Abrams, J.B. 2010. *Quantitative Business Valuation*. New York: Wiley.

Antill, N., and R. Arnott. 2000. *Valuing Oil and Gas Companies*. London: Woodlands Publishing.

Arnold, B.J., M.S. Klima, and P.J. Bethell. 2007. *Designing the Coal Preparation Plant of the Future*. Littleton, CO: EME.

Brickley, P., and M. Jarzemsky. 2015. Patnot to sell mines to Blackhawk in debt-fueled deal, *Wall Street Journal* June 4, B3.

Chapman, D. 1983. *Energy Resources and Energy Corporations*. Ithaca, NY: Cornell University Press.

Damodaran, A. 2002. *Investment Valuation*. New York: Wiley.

EIA. 2009. *Coal Industry Annual*. Washington, DC: EIA.

Hartman, H.L., and J.M. Mutmansky. 2002. *Introduction to Mining Engineering*, 2nd Ed., Hoboken, NJ: John Wiley & Sons.

Howard, A.W., and A.B. Harp. 2009. Oil and gas company valuations, *Business Valuation Review* 28(1): 30–35.

International Valuation Standard 2005, 7th Ed., Washington, DC: International Valuation Standard Committee.

Joskow, P. 1985. Vertical integration and long-term contracts: The case of coal-burning electric generating plants, *Journal of Law, Economics, and Organization* 1(1): 33–80.

Joskow, P. 1987. Contract duration and relationship specific investments: Empirical evidence from coal markets, *American Economic Review* 77(1): 168–185.

Kerkvilet, J., and J. Shogren. 2001. The determinants of coal contract duration for the Powder River Basin, *Journal of Institutional and Theoretical Economics* 157: 608–622.

Kozhevnikova, M. and I. Lange. 2009. Determinants of contract duration: Further evidence from coal-fired power plants, *Review of Industrial Organization* 34: 217–229.

Miller, J. and D. Frosh. 2015. Coal industry slump puts states in a bind, *Wall Street Journal* July 18–19, A3.

Miller, J.W., and K. Maher. 2012. Coal, the hot commodity for deals, *Wall Street Journal* March 6, C12.

National Mining Association. 2010. *Coal Producer Survey*. May 2011. Available at: www.nma.org/pdf/members/coal_producer_survey2010.pdf.

Olea, R.A., J.A. Luppens, and S.J. Twealt. 2011. Methodology for quantifying uncertainty in coal measurements with an application to a Texas lignite deposit, *International Journal of Coal Geology* 85(1): 78–90.

Runge, I.C. 1998. *Mining Economics and Strategy*. Littleton, CO: Society for Mining, Metallurgy, and Exploration.

Rutledge, D. 2011. Estimating long-term world coal production with logit and probit transforms, *International Journal of Coal Geology* 85(1): 23–33.

Schobert, H.H. 1987. *Coal—The Energy Source of the Past and Future*. Washington, DC: American Chemical Society.

SPE/WPC/AAPG/SPEE. 2007. *Petroleum Resources Management System*. Available at: http://www.spe.org/industry/docs/Petroleum_Resources_Management_System_2007.pdf#redirected_from=/industry/reserves/prms.php.

Thomas, L. 2002. *Coal Geology*. Chichester: John Wiley & Sons.

US Securities and Exchange Commission. 1992. *Industry Guide 7: Description of Property by Issuers Engaged or to be Engaged in Significant Mining Operation*. Washington, DC.

US Securities and Exchange Commission. 2008. *Modernization of the Oil and Gas Reporting Requirements*. Confirming Version (proposed rule), 17 CFR Parts 210, 211, 229, and 249, Release Nos. 33-8995; 34-59192; FR-78; File No. S7-15-08, RIN 3235-AK00. Available at: http://www.sec.gov/rules/final/2008/33-8995.pdf.

Vann, A. 2011. *Energy Projects on Federal Lands: Leasing and Authorization*. Washington, DC: Congressional Research Service.

Wood, G.J. Jr, T.A. Kehn, M.D. Carter, and W.C. Culbertson. 1983. *Coal Reserve Classification System of the U.S. Geological Survey*. USGS Circular No. 891, Washington, DC.

23 Future Trends in Coal Technology

Arno de Klerk

CONTENTS

Abstract: The development of coal technology, as well as changes that influenced technology development, was used as the basis for predictions about the future. Change-drivers were classified and their impacts were analyzed under the headings of technical, economic, environmental, and political change drivers. Future trends were discussed in the fields of coal preparation, coal combustion and after treatment, coal gasification, coal carbonization, and coal liquefaction.

23.1 INTRODUCTION

Questions about the future of coal technology were posed by many, and it is interesting to read the predictions with hindsight. Making predictions about the future is for the most part an educated guess. It is ironic that it is often the past that serves as guide for the future. Looking back over history, one can identify cause-and-effect combinations that shaped coal technology. It is a reasonable expectation that future trends in coal technology would also be shaped in an analogous way. One type of prediction that cannot be made in this way is that of disruptive change. Disruptive change is a change where either the need to change or the means of change is not yet known or anticipated.

The approach that was taken in this chapter was to look at the history of coal development. In the literature, shortcomings in coal technology were noted, which were needs that anticipated change. In some cases, predictions were made about the future of coal technology, outlining anticipated trends and expected changes. Change drivers, which are the needs and trends of the present, form the factual basis for this chapter.

The scope of the chapter is further limited to the coal technology that is employed subsequent to coal mining, but it excludes coal mining technology. Coal mining is important, but it is a subject of which the author has limited firsthand knowledge or experience.

Coal technology in each field downstream of mining will be discussed under a separate heading. The disparity in coverage and detail in each of the fields is mostly reflective of the maturity and nature of the technologies. Mature coal technologies have converged on a set of accepted practices that will gradually improve over time. Changes in mature technologies are usually disruptive and unanticipated. Less-mature coal technologies still have some technical hurdles that need to be overcome, and trends are more easily discerned. Irrespective of the maturity of the coal technology, there are also change drivers that cut across the discipline and that force the development of coal technology in new directions. It is useful to start off by looking at these change drivers.

23.2 CHANGE DRIVERS IN COAL TECHNOLOGY

23.2.1 Technical Change Drivers

Future trends due to technical change drivers are the easiest to predict. These are changes in response to specific technical shortcomings or problems that have already been identified. Specialists in any field of coal technology will be quick to point out the main challenges in their respective fields. It is reasonable to anticipate that over time these challenges will be addressed. For this reason, future trends that are driven by technical needs are the easiest to predict.

23.2.2 Economic Change Drivers

Coal is both a source of energy and a raw material for production. In both instances, coal must compete with other raw materials for these markets. The cost, desirability, and availability of competing raw materials change over time. The economics affecting the use of coal and the development of coal technology respond to these changes.

The growth in the use of coal as energy source is responding to the increasing need for energy. The twentieth century saw a rapid increase in crude oil consumption, which was only moderate in the 1970s. The twenty-first century, like the nineteenth century, is dominated by the growth in coal consumption as primary energy resource (Figure 23.1) (Enerdata 2014).

Most of the recent growth in coal consumption is due to the increase in coal use in China. It is likely that economic growth in India will also largely be based on coal. In both countries, coal is a locally abundant energy source, and it highlights the importance of local demand and supply. China and India are also the two most populous countries in the world, accounting for over one-third of the global population. The global economic outlook of coal is and will continue to be influenced by the decisions made in these two countries.

There are also more general trends that are unlikely to change in the short to medium term that will affect the economics of coal: (1) The human population of the planet is growing; (2) each human consumes energy; (3) energy efficiency cannot exceed ideal thermodynamic efficiency; (4) standard of living and economic activity are partly related to energy use; and (5) humans strive to improve their standard of living.

Taken together, these trends indicate that there will be a growing need for energy. It is likely that coal will remain part of the mixture of energy sources that will be relied on to meet this need for energy.

23.2.3 Environmental Change Drivers

All human activities have an environmental impact. Mining and use of coal is no different. The nature of the environmental impact is determined mostly by the end use. Coal mining leads to concentrated and localized impact, but the impact of coal use is spread out. All the elements in coal came from the environment but are not returned to the environment in the same form. Through coal use, these elements are returned to the environment in a different form. The change in the chemical state of the elements and the impact that this has on the ecosystem, are central environmental change drivers influencing coal technology.

Coal and coal technology suffer the burden of hundreds of years of knowledge and experience with coal (Chapter 19). It is a burden, because the risks associated with coal mining, the use of coal, and the conversion of coal are better defined than for most substances in common use. These risks are translated into legislation and controls. We know that these controls are necessary. In this respect, coal serves as a model resource that has long since passed the stage where *ignorance is bliss*. Legislation, controls, and an increasing sense of environmental responsibility will continue to be change drivers that shape coal technology.

23.2.4 Political Change Drivers

Political decisions can have a significant impact on how and in what direction coal technology develops. The direct embodiment of the political influence on the development of coal technology is through legislation and tax. Legislation and taxation can be employed either to support or inhibit coal use or to promote or dissuade specific applications of coal. The motivation for political decisions affecting coal is usually due to strategic considerations (e.g., energy security), economic expediency (e.g., jobs and foreign exchange), or public opinion (e.g., pollution that is real or perceived and pressure groups promoting a competing material).

More often than not there is an uneasy relationship between current economic prosperity, current environmental impact, and future accountability. Economic prosperity can buy political power and influence. It is not uncommon to find that such political power and influence is wielded to ensure continued economic prosperity, despite environmental impact (Ross and Amter 2010).

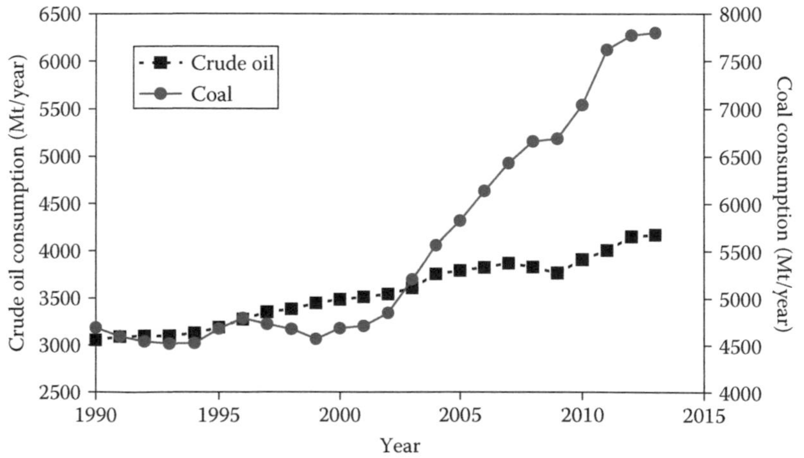

FIGURE 23.1 Global annual consumption in millions of tons (Mt) of crude oil and coal.

23.3 FUTURE TRENDS IN COAL PREPARATION

There are a couple of common themes in coal preparation: size reduction and organic–mineral matter separation. The latter includes desulfurization and coal drying. All these steps are necessary to improve coal use. It can therefore be anticipated that work will continue in all these areas. Improvement in the fundamental understanding of size reduction or coal drying will lead to incremental improvement in these technologies, but radical change in these technologies is not foreseen. The same cannot be said for the field of organic–mineral matter separation. Organic–mineral matter separation is governed by interfacial forces and coal chemistry. This is an area of coal preparation where it is anticipated that additional fundamental insight might lead to a breakthrough and drastic change in coal preparation technology.

It is interesting that organic–mineral matter separation, and specifically the chemical comminution of coal, was highlighted before as a future trend for development and change (Bevan 1981). By the same token, it also highlights the difficulty in making advances.

Chemical comminution of coal involves the use of a chemical reagent/solvent that weakens the interaction of the organic matter with the mineral matter. It is effective on coarser coal, because the coal particles are also fragmented through the action of the reagent. The chemicals that were reported to be the most active for comminution of coal are ammonia and methanol. More generally, they are the class of compounds that have a nonbonding pair of electrons that are able to dissolve and swell the coal at near ambient temperatures (Weiss et al. 1982). The benefit of chemical comminution of larger coal particles was clear, but little benefit was found for finer particles (Mahadevan 1991). In practice, ammonia and methanol are only effective in swelling the coal, so that liberation of mineral matter occurs by fragmentation, not very different from mechanical separation. Investigations in this field petered out.

Despite the apparent failure of chemical comminution of coal as it was originally envisioned, it is appealing as a concept, and it has not been fully explored. Viewed in a broad context, it is also a coal desulfurization strategy (see also Chapter 12). Coal desulfurization is driven mostly by environmental change drivers.

Effective separation of mineral matter is key to the removal of small crystals of iron pyrite (Figure 23.2) that appears to be difficult to be removed by mechanical means. Successful chemical comminution of coal, whether directed at mineral matter in general or desulfurization in particular, should result in meaningful additional desulfurization of the coal. In fact, Van Krevelen (1993) predicted that coal desulfurization may in future become as important as field of study as coal combustion, gasification, carbonization, and liquefaction.

Philosophically, coal preparation should be a core area of research and development, because many of the problems created by coal use is the consequence of not removing potential environmental contaminants from the coal before use. Environmental and political change drivers may in future make it more attractive to prevent pollution from coal use than to clean up pollution from coal use. It is speculated that the future trend in coal preparation will be that of chemical cleaning of the coal to prevent some contaminants entering the coal conversion or combustion process and thereby preventing the emission of the contaminants as consequence of downstream coal use.

23.4 FUTURE TRENDS IN COAL COMBUSTION AND AFTER TREATMENT

Coal combustion is a source of power and in many cases electric power. The second law of thermodynamics governs the amount of work that can be produced and the energy that is inevitably lost as heat. The usefulness of the combustion energy to generate work is determined by the temperature at which the energy is available. From this perspective, the main efficiency driver in coal combustion is combustion temperature, although there are many other technology areas that are also important to make coal combustion successful (Chapter 13).

The adiabatic temperature increase during coal combustion determines temperature driving force and the amount of energy that can be transferred at the maximum steam temperature of boiler operation. The maximum steam temperature of boiler operation is limited mainly by the material of construction, which should have sufficient mechanical strength at the

FIGURE 23.2 Small exposed iron pyrite crystals in Boundary Dam lignite.

operating temperature to contain the ultrahigh pressure steam. Further advances in the use of coal combustion are in this respect subservient to the development of materials that can be employed at more severe conditions. However, this does not preclude improvements in coal combustion, but it only limits the maximum steam temperature for power generation.

The adiabatic temperature increase during coal combustion is determined by the composition of the coal and the oxidant feed. The maximum adiabatic temperature increase during coal combustion is affected mostly by removal of material from the coal and oxidant feeds that do not participate in combustion and not so much by coal combustion technology. Preparation of ash-free coal and oxy-firing are two examples of improving the quality of the feed materials to improve coal combustion. In both cases, inert materials are removed from the feed. Interest in oxy-firing is also stimulated by gas companies that are potential providers of air separation technology to provide purified oxygen as oxidant feed.

Just like coal combustion can be improved by supplying cleaner feed materials, coal combustion after treatment is improved by using cleaner feed materials. If a potential contaminant is not present in the feed, it does not present an after-treatment problem.

The challenges in after treatment can roughly be divided into four categories: particulate matter emissions, major contaminants (sulfur in particular), trace element emissions, and carbon dioxide. The first three can all be mitigated by deeper coal cleaning. The latter, CO_2, is an inherent product of coal combustion.

It is anticipated that after-treatment technology will continue to evolve and improve. It is a necessary consequence of both environmental and political change drivers. Seen in isolation, the development of improved technology to increase the efficiency of after treatment will also improve the environmental performance of coal combustion. This seems to be a general future trend and an essential trend considering the increasing growth in coal use (Figure 23.1).

The emission of CO_2 from coal combustion has become the subject of public debate. The need, or not, to address CO_2 emissions will be determined by political change drivers in future. It is not clear whether the current trend to look at CO_2 capture and sequestration (Chapter 21) will persist or whether the predicted disastrous impact of CO_2 emissions on the environment is indeed a valid prediction.

Philosophically, the loss of efficiency from coal combustion when CO_2 capture and sequestration is necessary is difficult to justify when alternative high-density energy sources are available that are not carbon-based, such as nuclear power. Just like spent nuclear fuel, spent carbon (i.e., CO_2) is persistent over a long period of time. The impact of poor CO_2 containment following on CO_2 capture and sequestration is potentially disastrous, both at the point of release and globally, for the same reasons cited to justify sequestering the CO_2 in the first place. The need or not to capture and sequester CO_2 also affects the economics of power generation. The economic impact of higher-cost electric power makes public and political response uncertain. Even science is not value neutral, and political forces can and do manipulate science (e.g., Ross and Amter 2010). The environment and climate are complex systems. In complex systems, one might be able to describe behavior, but predictability is challenging and becomes more challenging when science and politics are intertwined.

A case study is presented to illustrate the challenge of predicting environmental outcomes of emission control strategies. In complex systems like the environment, potential pollutants do not act independently. This was indeed the case with sulfur and mercury emissions. Control over the release of sulfur and mercury from coal-fired power plants became increasingly stringent over time. The intent was to reduce acid rain that was caused by sulfur oxides and to reduce mercury levels in food due to the uptake of mercury liberated by coal combustion. Yet, when the sulfur and mercury emissions were both decreased by better flue-gas after-treatment technologies, it was observed that the mercury content in Norwegian fish increased (Hongve et al. 2012). This was not the intended or the anticipated outcome. It turned out that the decrease in sulfur emissions, which reduced acid rain, led to a decrease in water acidity. Decreased water acidity facilitated the uptake of mercury by fish. Thus, despite the reduced mercury emissions and mercury in the environment, the mercury content in fish still increased.

23.5 FUTURE TRENDS IN COAL GASIFICATION

Coal gasification is an enabling technology for many downstream processes. The synthesis gas that is produced by coal gasification can be applied in many different ways. Potential applications of synthesis gas include synthesis gas separation to produce pure hydrogen (H_2), synthetic or substitute natural gas (SNG) production, methanol synthesis, Fischer–Tropsch synthesis to produce liquid fuels and chemicals, and the use of synthesis gas as part of integrated gasification combined cycle (IGCC) processes. The synthesis gas from coal gasification is usually cleaned to remove sulfur-containing compounds, nitrogen-containing compounds, and trace elements before use. By cleaning the synthesis gas before use, it avoids the after-treatment challenges associated with typical coal combustion processes. Coal gasification and synthesis gas cleaning are expensive technologies, but they are the gateways to value addition to coal.

Despite the maturity of gasification as a process, many technical challenges remain (Chapter 14). One just has to look at the practical issues that need to be considered in gasification technology to appreciate some of the difficulties. Some of the important practical considerations pointed out by Higman and Van der Burgt (2008) are pressurization of coal and depressurization of ash, reactor containment, corrosion, and trace compound reactions. Due to the extreme conditions and solid handling involved in coal gasification, studying coal gasification is difficult and costly.

The future trend in coal gasification technology development is likely to be very similar to the recent past, with the field broadly advancing. Gradual improvements in the understanding of gasification and gasification technology will be made, but there is not a single aspect of the developments in coal gasification that can be highlighted as a clear future trend.

23.6 FUTURE TRENDS IN COAL CARBONIZATION AND COAL LIQUEFACTION

Coal carbonization and coal liquefaction processes are aimed at adding value to the coal beyond its combustion energy value. Coal coking produces coke and coal liquids as main products, and it is practiced industrially on large scale to support the metal industry. Coal liquefaction produces coal liquids, and it is an alternative route to crude oil as raw material for the production of transportation fuels and chemicals.

Coke production exploits the inherently high aromatic content of higher rank coals to produce an even more aromatic-rich product, namely coke. As a technology class, it is very well suited for the upgradation of coal. The quality of the coke is sensitive to the coal quality, so that advances made in coal preparation technology will also improve the operation of coal carbonization processes. The processes to produce coke for metallurgical applications are mature technologies, but it is likely that carbonization technology will continue to advance incrementally over a broad field. Making coke from low-grade coking coal by the addition of glue-containing components, such as waste plastic or ash-free coals to produce strong coke, will continue to develop. There is also a need for new technology to convert thermal coals into coking coals.

The use of coal as an alternative raw material to crude oil for the production of fuels and chemicals has had limited and only localized industrial success to date. This is mainly due to the process economics, high capital cost, and technical complexity of coal-to-liquid processes. The availability of cheap crude oil erodes the incentive and the justification to invest in coal liquefaction technology. As the crude oil price is inherently unpredictable and susceptible to market manipulation, it is difficult to predict whether coal liquefaction will become an industrially important technology in the short to medium term. Market fickleness and the political response to the changes in the oil market can both stimulate and decimate interest in the development of coal liquefaction technology (Crow et al. 1988). Viewed over a longer term, natural depletion of crude oil resources in relation to the increasing demand for liquid fuels and petrochemicals should in future provide fertile economic opportunities for coal liquefaction technology. However, a disruptive change that fundamentally alters the energy carrier used for transportation will render such a prediction moot.

As in the past, there may be strategic reasons for countries with abundant coal, but limited crude oil reserves, to invest in coal liquefaction technology. Some efforts in coal liquefaction are already seen in China. It may therefore be useful to point out the technical challenges that were identified in the past as key issues to be resolved in the future. These will become the future trends from a technical perspective, should coal liquefaction become a topic of interest.

23.6.1 Indirect Coal Liquefaction

Future developments in indirect coal liquefaction are linked to coal only through developments in coal gasification. The downstream processes in indirect liquefaction technology are synthesis gas based and largely independent of the raw material. For those interested in future trends in indirect liquefaction, independent of coal, discussions of the challenges and opportunities were recently presented for methanol-based (Bertau et al. 2014) and Fischer–Tropsch-based (Maitlis and De Klerk 2013) processes.

23.6.2 Direct Coal Liquefaction

Direct coal liquefaction, despite many decades of intense effort, has technical challenges remaining. A list with some of the more pressing technical issues is made. This list is not exhaustive or complete, but it captures important technical issues that will remain change drivers for future development of direct coal liquefaction technology.

There are two classes of technical issues that require future attention: lack of fundamental understanding and difficulty of practical implementation.

1. The direct liquefaction reaction is so complex that it defies accurate prediction, even based on exhaustive feed characterization. Many parameters have been identified that influence coal liquefaction performance, but as Van Krevelen (1993, p.773) puts it: "Process development in coal liquefaction is mainly an empirical art." These are not only strong words but, in the author's opinion, also truthful words. A similar opinion was expressed by Kabe et al. (2004, p.267): "the exact roles of catalyst and solvent in coal liquefaction remain ambiguous. Design in the coal liquefaction process is based mostly on experience."
2. Material science and mechanical engineering have key roles in the development of industrial coal liquefaction processes. Two of the most common problems that were identified during commercial development of direct coal liquefaction technology were erosion and abrasion (Lee 1979). If these issues cannot be solved, some redesign of the coal liquefaction process may be required to minimize these deleterious effects. Alpert and Wolk (1981) pointed out various mechanically challenging units and equipment designs, as well as inadequate knowledge about the characteristics of the solids at various stages of the liquefaction process.
3. Hydrogen is needed to produce coal liquids that can be used as transportation fuels (Chapter 15). The efficient and cheap production of hydrogen was highlighted as central issue that had to be addressed to make direct coal liquefaction a commercial reality (Alpert and Wolk 1981; Kabe et al. 2004).
4. Solids separation from coal liquids remains a serious challenge, even though many technologies were

developed for this separation (Khare and Dell'Amico 2013). The critical role of solids separation from coal liquids is not immediately apparent from conceptual designs or process flow diagrams. Yet, during development work and scale-up, it was raised as an important technical hurdle and source of inefficiency in a number of direct coal liquefaction processes (e.g., Johnson et al. 1973; Newman et al. 1977; Alpert and Wolk 1981). Inadequate knowledge of the solids in the process, which was mentioned earlier, is a contributing factor to the technical challenge of efficient solid–liquid separation.

It is anticipated that the fundamental knowledge important to direct coal liquefaction will gradually advance over time, but that many of the practical problems will remain. A snapshot of current activity was presented in the analysis of academic research in the coal literature (Mathews et al. 2013). Liquefaction was one of the main topics of research, but not supporting technologies such as the separation of coal liquefaction residues from coal liquids or the refining of coal liquids to final products. Limited progress is anticipated when dealing with practical problems that are more of an industrial nature and that require large-scale experimental investigations. Until the economics of coal-to-liquid conversion compared to conventional oil and gas production become favorable, advances made on issues related to process development will be limited.

REFERENCES

Alpert, S. B. and R. H. Wolk. 1981. Liquefaction processes. In *Chemistry of Coal Utilization, Second Supplemental Volume*. ed. Elliott, M. A., 1919–1990, Wiley-Interscience: New York.

Bertau, M., H. Offermanns, L. Plass, F. Schmidt, and H-J. Wernicke, eds. 2014. *Methanol: The Basic Chemical and Energy Feedstock of the Future. Asinger's Vision Today*, Springer: Heidelberg, Germany.

Bevan, R. R. 1981. Size reduction. In *Coal Handbook*. ed. Meyers, R. A., 173–208, Marcel Dekker: New York.

Crow, M., B. Bozeman, W. Meyer, and R. Shangraw Jr. 1988. *Synthetic Fuel Technology Development in the United States. A Retrospective Assessment*, Praeger: New York.

Enerdata. *Global Energy Statistical Yearbook 2014*, https://yearbook.enerdata.net, last accessed on December 27, 2014.

Higman, C. and M. Van der Burgt. 2008. *Gasification*, 2nd ed., Elsevier: Amsterdam, the Netherlands.

Hongve, D., S. Haaland, G. Riise, I. Blakar, and S. Norton. 2012. Decline of acid rain enhances mercury concentrations in fish. *Environ. Sci. Technol.* 46:2490–2491.

Johnson, C. A., M. C. Chervenak, E. S. Johanson, H. H. Stotler, O. Winter, and R. H. Wolk. 1973. Present status of the H-Coal® process. In *Symposium Papers. Clean Fuels from Coal*, 549–575, Institute of gas technology: Chicago, IL.

Kabe, T., A. Ishihara, E. W. Qian, I. P. Sutrisna, and Y. Kabe. 2004. *Coal and Coal-Related Compounds* (Studies in Surface Science and Catalysis 150), Elsevier: Amsterdam, the Netherlands.

Khare, S. and M. Dell'Amico. 2013. An overview of solids-liquid separation of residues from coal liquefaction processes. *Can. J. Chem. Eng.* 91:324–331.

Lee, E. S. 1979. Coal liquefaction. In *Coal Conversion Technology*. eds. Wen, C. Y. and E. S. Lee, 428–545, Addison-Wesley: Reading, MA.

Mahadevan, V. 1991. Investigations of the effect of chemical pretreatment on the comminution of coal. *Fuel Sci. Technol.* 10:49–55.

Maitlis, P. M. and A. de Klerk, eds. 2013. *Greener Fischer-Tropsch Processes for Fuels and Feedstocks*. Wiley-VCH: Weinheim, Germany.

Mathews, J. P., B. G. Miller, C. Song, H. H. Schobert, F. Botha, and R. B. Finkleman. 2013. The ebb and flow of US coal research 1970–2010 with a focus on academic institutions. *Fuel* 105:1–12.

Newman, J. O. H., S. Akhtar, and P. M. Yavorsky. 1977. Coagulation and filtration of solids from liquefied coal of Synthoil process. In *Liquid Fuels from Coal*, ed. Ellington, R., 183–200, Academic Press: New York.

Ross, B. and S. Amter. 2010. *The Polluters: The Making of Our Chemically Altered Environment*. Oxford University Press: Oxford.

Van Krevelen, D. W. 1993. *Coal. Typology–Physics–Chemistry–Constitution*, 3rd ed., Elsevier: Amsterdam, the Netherlands.

Weiss, B. M., R. R. Maddocks, and P. H. Howards. 1982. Chemical comminution for coal cleaning. In *Physical Cleaning of Coal. Present and Developing Methods*. ed. Liu, Y. A., 1–33, Marcel Dekker: New York.

Index

Note: Locator followed by '*f*' and '*t*' denotes figure and table in the text

D

E

F

G

H

N

O

P

T

T - #0551 - 071024 - C26 - 280/208/24 - PB - 9780367783303 - Gloss Lamination